Videos and Animations

Video 10.2

sites.sinauer.com/cooper7e/v10.2

Into the Nucleus Proteins pass through the nuclear pore complex by a selective transport process.

Throughout the book, you will see small boxes in the margins that refer to specific companion website videos and animations that are relevant to the topic being discussed. Below is the full list of videos and animations, along with the page on which each is referenced.

(Continued on next page)

Videos and Animations (continued)

The Cell
A Molecular Approach
Seventh Edition

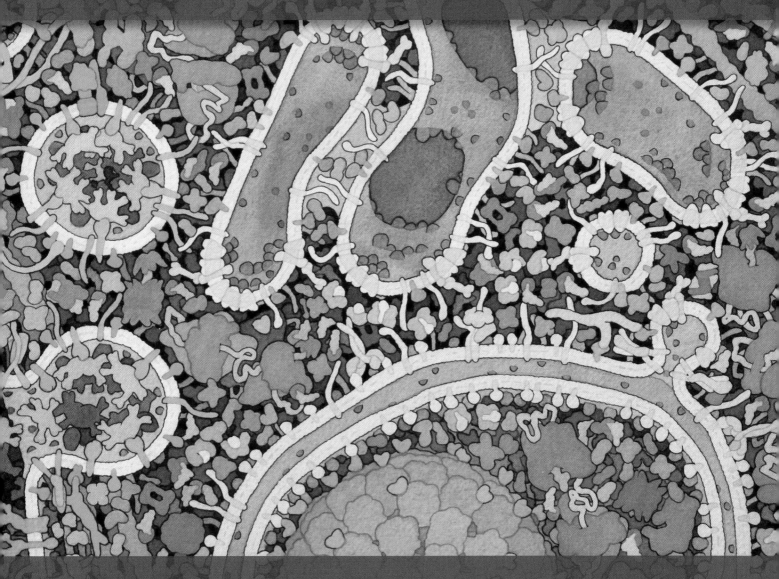

The Cell

A Molecular Approach

Seventh Edition

Geoffrey M. Cooper • Robert E. Hausman

BOSTON UNIVERSITY

 Sinauer Associates, Inc. • Publishers
Sunderland, Massachusetts U.S.A.

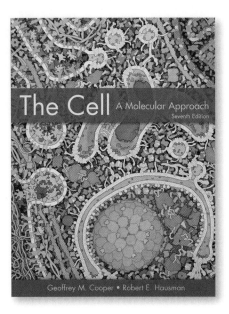

The Cover

The cover image shows the formation of an autophagosome, in which organelles and cytosol are engulfed in cytoplasmic membranes. Original painting by David S. Goodsell, based on the scientific design of Daniel J. Klionsky.

The Artist

David S. Goodsell is an Associate Professor of Molecular Biology at the Scripps Research Institute. His illustrated books, *The Machinery of Life* and *Our Molecular Nature*, explore biological molecules and their diverse roles within living cells, and his new book, *Bionanotechnology: Lessons from Nature*, presents the growing connections between biology and nanotechnology. More information may be found at: http://mgl.scripps.edu/people/goodsell

Address orders and requests for examination copies to: Sinauer Associates
P.O. Box 407, 23 Plumtree Road, Sunderland, MA 01375 U.S.A.
Phone: 413-549-4300
FAX: 413-549-1118
email: orders@sinauer.com
www.sinauer.com

Library of Congress Cataloging-in-Publication Data

Cooper, Geoffrey M., author.
 The cell : a molecular approach / Geoffrey M. Cooper, Robert E. Hausman.
-- Seventh edition.
 p. ; cm.
 Includes bibliographical references.
 ISBN 978-1-60535-290-9
 I. Hausman, Robert E., 1947-2015, author. II. Title.
 [DNLM: 1. Cells. 2. Cell Biology. 3. Cell Physiological Phenomena.
4. Molecular Biology. QU 300]
 QH581.2
 571.6--dc23
                                    2015032684
Printed in U.S.A
6 5 4 3 2

To my friend, colleague, and coauthor,
Robert E. Hausman
(1947–2015)

Geoffrey M. Cooper

Brief Table of Contents

Contents

Chapter 3 Bioenergetics and Metabolism 81

Chapter 4 Fundamentals of Molecular Biology 111

PART II
The Flow of Genetic Information   185

Chapter 8   RNA Synthesis and Processing   259

Chapter 9   Protein Synthesis, Processing, and Regulation   317

PART III
Cell Structure and Function   365

PART IV
Cell Regulation   599

Chapter 16   Cell Signaling   601

Preface

Learning cell biology can be a daunting task because the field is so vast and rapidly moving, characterized by a continual explosion of new information. The challenge is how to master the fundamental concepts without becoming bogged down in details. Students need to understand the principles of cell biology and be able to appreciate new advances, rather than just memorizing "the facts" as we see them today. At the same time, the material must be presented in sufficient depth to thoughtfully engage students and provide a sound basis for further studies. *The Cell* provides a balance of concepts and details designed to meet these needs of today's students and their teachers.

The Seventh Edition of *The Cell* continues the central goal of helping students understand the principles and concepts of contemporary cell biology while gaining an appreciation of the excitement and importance of research in this rapidly moving field. Our understanding of cell and molecular biology has progressed in many ways over the last three years, and these important advances have been incorporated into the current edition. Some of the most striking advances have continued to come from progress in genomics and proteomics. A new chapter in the current edition—Genomics, Proteomics, and Systems Biology—focuses on these rapidly advancing areas and includes discussion of the growing field of synthetic biology. Other notable advances covered in the current edition include super-resolution microscopy, genome editing by the CRISPR/Cas system, epigenetics, the organization of chromosome domains in the nucleus, and an updated discussion of prions and Alzheimer's disease.

These and other advances have been incorporated into the Seventh Edition of *The Cell* not only to provide current information but, even more importantly, to give students a sense of the excitement and challenges of research in our field. At the same time, I have sought to minimize details and focus on concepts in order to ensure that *The Cell* remains an accessible and readable text for undergraduates who are taking their first course in cell and molecular biology. Distinguishing features of *The Cell* include the Molecular Medicine and Key Experiment essays, which highlight clinical applications and describe seminal research papers, respectively. Together with additional experiments discussed throughout the text, these essays give the students a sense of how progress in our field is made and a feel for how hypotheses are framed and results interpreted. As in previous editions, each chapter also includes several short "FYIs" designed to highlight areas of interest or medical relevance, as well as a set of questions at the end of each chapter with answers at the back of the book. In keeping with *The Cell's* focus on experimental analysis, many of these questions ask the student to think of experiments or interpret results, as well as providing a review of the material. A new feature of this edition is the addition of Data Analysis Problems to the Instructor's Resource Library. These problems present data and figures from original research papers and engage students in the analysis of experimental methods and results. They also provide excellent material for discussions and opportunities for student participation in active learning.

My most important goal in writing has been to convey the excitement and challenges of research in contemporary cell and molecular biology. The opportunities in our field are greater than ever, and I hope *The Cell* stimulates today's students to contribute to the research on which future texts will be based.

Acknowledgments

I am particularly grateful to Alexandra Adams for carefully reviewing the entire text of the Seventh Edition of *The Cell*. The book has also benefited from the comments and suggestions of reviewers, colleagues, and instructors who used the previous edition. I am pleased to thank the following reviewers for their thoughtful comments and advice:

Sage Arbor*	*Marian University*
Rebecca Balish	*Miami University*
Esther Biswas-Fiss	*Thomas Jefferson University*
Edward Bonder	*Rutgers University*
Lucinda Carnell	*Central Washington University*
Xiao-Wen Cheng	*Miami University*
Heather Cook*	*Wagner College*
Gary Coombs*	*Waldorf College*
Paul Davis	*University of Nebraska at Omaha*
Jennifer Freytag*	*The Sage Colleges*
Michael Garcia	*University of Missouri*
David Gardner	*Marian University College of Osteopathic Medicine*
Brittany Gasper	*Florida Southern College*
Mark Grimes	*University of Montana*
Karen Guzman	*Campbell University*
Neil Haave*	*University of Alberta*
Joyce Hardy*	*Chadron State College*
Amber Heck	*Rocky Vista University College of Osteopathic Medicine*
Jose Hernandez	*Midwestern University*
Peter Hoffman	*Notre Dame of Maryland University*
Lisa Kadlec*	*Wilkes University*
Ondra Kielbasa	*Alvernia University*
Janet Kirkley	*Knox College*
Jon Lowrance	*Lipscomb University*
Kaushiki Menon	*Mt. San Antonio College*
Vida Mingo*	*Columbia College*
Craig Moyer	*Western Washington University*
Perpetua Muganda	*North Carolina A&T State University*
Laura Rhoads*	*State University of New York at Potsdam*
Robert Roberson	*Arizona State University*
German Rosas-Acosta	*The University of Texas at El Paso*
D. Scott Samuels	*University of Montana*
Ken Savage	*The University of British Columbia*
Alexander Schreiber	*St. Lawrence University*
Deborah Schulman	*Lake Erie College*

José Serrano-Moreno	*Western Washington University*
Vishal Shah	*Dowling College*
Mohammad Rafique Uddin	*LeMoyne–Owen College*
Shannon Vandaveer	*University of Kansas*
Thomas Wolkow	*University of Colorado, Colorado Springs*
Kathleen Wood*	*University of Mary Hardin-Baylor*
Robin Young	*The University of British Columbia*

*Reviewed chapter(s) and Data Analysis Problems.

It is also a pleasure to thank our publisher for continuing support. Chelsea Holabird did a superb editorial job. Jen Basil-Whitaker created a beautiful design and page layouts. Andy Sinauer, Dean Scudder, and Christopher Small were once again a real pleasure to work with.

Geoffrey M. Cooper
September, 2015

Organization and Features of
The Cell, Seventh Edition

The Cell has been designed to be an approachable and teachable text that can be covered in a single semester while allowing students to master the material in the entire book. It is assumed that most students will have had introductory biology and general chemistry courses, but will not have had previous courses in organic chemistry, biochemistry, or molecular biology. Several aspects of the organization and features of the book will help students to approach and understand its subject matter.

Organization

The Cell is divided into four parts, each of which is self-contained, so that the order and emphasis of topics can be easily varied according to the needs of individual courses.

Part I provides background chapters on the evolution of cells, methods for studying cells, the chemistry of cells, the fundamentals of modern molecular biology, and the fields of genomics and systems biology. For those students who have a strong background from either a comprehensive introductory biology course or a previous course in molecular biology, various parts of these chapters can be skipped or used for review.

Part II focuses on the molecular biology of cells and contains chapters dealing with genome organization and sequences; DNA replication, repair, and recombination; transcription and RNA processing; and the synthesis, processing, and regulation of proteins. The order of chapters follows the flow of genetic information (DNA→RNA→protein) and provides a concise but up-to-date overview of these topics.

Part III contains the core block of chapters on cell structure and function, including chapters on the nucleus, cytoplasmic organelles, the cytoskeleton, the plasma membrane, and the extracellular matrix. This part of the book starts with coverage of the nucleus, which puts the molecular biology of Part II within the context of the eukaryotic cell, and then works outward through cytoplasmic organelles and the cytoskeleton to the plasma membrane and the exterior of the cell. These chapters are relatively self-contained, however, and could be used in a different order should that be more appropriate for a particular course.

Finally, Part IV focuses on the exciting and fast-moving area of cell regulation, including coverage of topics such as cell signaling, the cell cycle, programmed cell death, and stem cells. This part of the book concludes with a chapter on cancer, which synthesizes the consequences of defects in basic cell regulatory mechanisms.

Features

Several pedagogical features have been incorporated into *The Cell* in order to help students master and integrate its contents. These features are reviewed below as a guide to students studying from this book.

CHAPTER ORGANIZATION Each chapter is divided into three to five major sections, which are further divided into a similar number of subsections. An outline listing the major sections at the beginning of each chapter provides a brief overview of its contents.

KEY TERMS AND GLOSSARY Key terms are identified as boldfaced words when they are introduced in each chapter. These key terms are reiterated in the chapter summary and defined in the glossary at the end of the book.

ILLUSTRATIONS AND MICROGRAPHS An illustration program of full-color art and micrographs has been carefully developed to complement and visually reinforce the text.

KEY EXPERIMENT AND MOLECULAR MEDICINE ESSAYS Each chapter contains either two Key Experiment essays or one Key Experiment and one Molecular Medicine essay. These features are designed to provide the student with a sense of both the experimental basis of cell and molecular biology and its applications to modern medicine. We have also found these essays to be a useful basis for student discussion sections, which can be accompanied with a review of the original paper upon which the Key Experiments are based.

FYIs Each chapter contains several sidebars that provide brief descriptive highlights of points of interest related to material covered in the text. The sidebars supplement the text and provide starting points for class discussion.

CHAPTER SUMMARIES Chapter summaries are organized in outline form corresponding to the major sections and subsections of each chapter. This section-by-section format is coupled with a list of the key terms introduced in each section, providing a succinct but comprehensive review of the material.

QUESTIONS AND ANSWERS Questions at the end of each chapter (with answers in the back of the book) are designed to further facilitate review of the material presented in the chapter and to encourage students to use this material to predict or interpret experimental results.

REFERENCES Comprehensive lists of references at the end of each chapter provide access to both reviews and selected papers from the primary literature. In order to help the student identify articles of interest, the references are organized according to chapter sections. Review articles and primary papers are distinguished by [R] and [P] designations, respectively. A particularly useful review article for each major chapter section is highlighted in bold.

ANIMATION AND VIDEO REFERENCES Boxes in the margin direct students to the website's animations and videos, with descriptions and direct Web links (URLs). The chapter summaries also include in-text references to the animations and videos.

Media and Supplements to Accompany
The Cell, Seventh Edition

For the Student

Companion Website (sites.sinauer.com/cooper7e)

The Companion Website to accompany *The Cell* provides students with a wide range of study and review materials and rich multimedia resources. The site is available free of charge (no access code required) and includes the following resources:

- *Animations*: Narrated animations of key concepts and processes

- *Videos*: A new collection of online videos (referenced throughout the book) helps students visualize complex cellular and molecular structures and processes.

- *Online Quizzes*: Two sets of questions for each chapter, both of which are assignable by the instructor. (Adopting instructors must register online in order for their students to access the quizzes.)

  - Multiple-choice quizzes test comprehension of the chapter's key material

  - Free-response questions ask students to apply what they have learned from the chapter.

- *Micrographs*: Interactive micrographs illustrating cellular structure

- *Flashcards*: A great way for students to learn and review the key terminology introduced in each chapter

- *Chapter Summaries*

- *Web Links*

- *Complete Glossary*

For the Instructor

(Available to qualified adopters)

Instructor's Resource Library

The Seventh Edition Instructor's Resource Library includes a wide range of digital resources to aid in planning your course, presenting your lectures, and assessing your students. The IRL includes the following:

- *Textbook Figures & Tables*: All available as both high- and low-resolution JPEGs
- *PowerPoint Resources*:
    - Figures and tables
    - Complete lecture presentations
    - Supplemental photos
- *Animations*: The entire collection of animations from the Companion Website, for use in lecture
- *Supplemental Photos*: Over 100 additional micrographs
- *Online Quiz Questions*: Multiple choice and free-response questions from the Companion Website, with answers and feedback
- *Test File*, in Microsoft Word format (see below)
- *Computerized Test Bank* (see below)
- *Data Analysis Problems*: New for the Seventh Edition, these problems engage students in analyzing data and figures from research papers.
- *Chapter Outlines and Key Terms*

Test File

(Included in the Instructor's Resource Library)
Revised and updated for the Seventh Edition, the Test File includes a collection of over 1,300 multiple-choice, fill-in-the-blank, true/false, and short-answer questions covering the full range of content in every chapter. New for the Seventh Edition, all questions are referenced to Bloom's Taxonomy, making it easier for instructors to select the specific types of questions they want when building an assessment.

Computerized Test Bank

(Included in the Instructor's Resource Library)
The entire test file plus all of the online quiz questions are provided in Blackboard's Diploma software. Diploma makes it easy to assemble quizzes and exams from any combination of publisher-provided questions and instructor-created questions. In addition, quizzes and exams can be exported to many different course management systems, such as Blackboard and Moodle.

Online Quizzing

The Cell's Companion Website features pre-built chapter quizzes (see above) that report into an online gradebook. Adopting instructors have access to these quizzes and can choose to either assign them or let students use them for review. (Instructors must register in order for their students to be able to take the quizzes.) Instructors also have the ability to add their own questions and create their own quizzes.

Course Management System Support

Using the Computerized Test Bank provided in the Instructor's Resource Library, instructors can easily create and export quizzes and exams (or the entire test bank) for import into many common course management system, including Blackboard, Moodle, and Desire2Learn/Brightspace.

Additional Media Resources

The following videos are available to qualified adopters of the text:

- Fink, *CELLebration* (DVD, ISBN 978-0-87893-330-3)
- Pickett-Heaps and Pickett-Heaps, *Diatoms: Life in Glass Houses* (DVD, ISBN 0-9586081-7-2)
- Pickett-Heaps and Pickett-Heaps, *The Dynamics and Mechanics of Mitosis* (DVD, ISBN 978-0-9775222-3-1)
- Pickett-Heaps and Pickett-Heaps, *Living Cells: Structure, Function, Diversity* (DVD, ISBN 0-9775222-2-9)
- Sardet, Larsonneur, and Koch, *Voyage Inside the Cell* (DVD, ISBN 978-0-87893-755-4)

Value Options

eBook

The Cell is available for purchase as an eBook in several different formats, including VitalSource, Yuzu, RedShelf, and BryteWave. The eBook can be purchased as either a 180-day rental or a permanent (non-expiring) subscription. All major mobile devices are supported. For details on the eBook platforms offered, please visit www.sinauer.com/ebooks.

Looseleaf Textbook

(ISBN 978-1-60535-540-5)
The Cell is also available in a three-hole punched, looseleaf format. Students can take just the sections they need to class and can easily integrate instructor material with the text.

An Overview of Cells and Cell Research

Understanding the molecular biology of cells is one of the most active and fundamental areas of research in the biological sciences. This is true not only from the standpoint of basic science, but also with respect to the many applications of cell and molecular biology to medicine, biomedical engineering, biotechnology, and agriculture. Especially with the ability to obtain rapid sequences of complete genomes of individuals, progress in cell and molecular biology is opening new horizons in the practice of medicine. Striking examples include the identification of genes that contribute to individual susceptibility to a variety of common diseases, such as heart disease, rheumatoid arthritis, and diabetes; the development of new drugs specifically targeted to interfere with the growth of cancer cells; and the potential use of stem cells to replace damaged tissues and treat patients suffering from conditions like diabetes, Parkinson's disease, Alzheimer's disease, and spinal cord injuries.

Because cell and molecular biology is such a rapidly growing field of research, it is important to understand its experimental basis as well as the current state of our knowledge. This chapter will therefore focus on how cells are studied, as well as review some of their basic properties. Appreciating the similarities and differences between cells is particularly important to understanding cell biology. The first section of this chapter discusses both the unity and the diversity of present-day cells in terms of their evolution from a common ancestor. On the one hand, all cells share common fundamental properties that have been conserved throughout evolution. For example, all cells employ DNA as their genetic material, are surrounded by plasma membranes, and use the same basic mechanisms for energy metabolism. On the other hand, present-day cells have evolved a variety of different lifestyles. Many organisms, such as bacteria, amoebas, and yeasts, consist of single cells that are capable of independent self-replication. More complex organisms are composed of collections of cells that function in a coordinated manner, with different cells specialized to perform particular tasks. The human body, for example, is composed of more than 200 different kinds of cells, each specialized for such distinctive functions as memory, sight, movement, and digestion. The diversity exhibited by the many different kinds of cells is striking; for example, consider the differences between bacteria and the cells of the human brain.

The fundamental similarities between different types of cells provide a unifying theme to cell biology, allowing the basic principles learned from experiments with one kind of cell to be extrapolated and generalized to other cell types. Several kinds of cells and organisms are widely used to study different aspects of cell and molecular biology; the second section of this chapter discusses some of the properties of these cells that make them particularly valuable as experimental models. Finally, it is important to recognize that progress in cell biology depends heavily on the availability of experimental tools that allow scientists to make new observations or conduct novel kinds of experiments. This introductory chapter therefore concludes with a discussion of some of the experimental approaches used to study cells, as well as a review of some of the major historical developments that have led to our current understanding of cell structure and function.

The Origin and Evolution of Cells

Cells are divided into two main classes, initially defined by whether they contain a nucleus. **Prokaryotic cells** lack a nuclear envelope; **eukaryotic cells** have a nucleus in which the genetic material is separated from the cytoplasm. Prokaryotic cells are generally smaller and simpler than eukaryotic cells; in addition to the absence of a nucleus, their genomes are less complex and they do not contain cytoplasmic organelles (Table 1.1). In spite of these differences, the same basic molecular mechanisms govern the lives of both prokaryotes and eukaryotes, indicating that all present-day cells are descended from a single primordial ancestor. How did this first cell develop? And how did the complexity and diversity exhibited by present-day cells evolve?

The first cell

It appears that life first emerged at least 3.8 billion years ago, approximately 750 million years after Earth was formed. How life originated and how the first cell came into being are matters of speculation, since these events cannot be reproduced in the laboratory. Nonetheless, several types of experiments provide important evidence bearing on some steps of the process.

It was first suggested in the 1920s that simple organic molecules could form and spontaneously polymerize into macromolecules under the conditions thought to exist in primitive Earth's atmosphere. At the time life arose, the atmosphere of Earth is thought to have contained little or no free oxygen, instead consisting principally of CO_2 and N_2 in addition to smaller amounts of gases such as H_2, H_2S, and CO. Such an atmosphere provides reducing conditions in which organic molecules, given a source of energy such as sunlight or electrical discharge, can form spontaneously. The spontaneous formation of organic molecules was first demonstrated experimentally in the 1950s when Stanley Miller (then a graduate student) showed that the discharge of electric sparks into a mixture of H_2, CH_4, and NH_3, in the presence of water, leads to the formation of a variety of organic molecules, including several

Table 1.1   Prokaryotic and Eukaryotic Cells

Characteristic	Prokaryote	Eukaryote
Nucleus	Absent	Present
Diameter of a typical cell	≈1 μm	10–100 μm
Cytoplasmic organelles	Absent	Present
DNA content (base pairs)	1×10^6 to 5×10^6	1.5×10^7 to 5×10^9
Chromosomes	Single circular DNA molecule	Multiple linear DNA molecules

Figure 1.1 Spontaneous formation of organic molecules   Water vapor was refluxed through an atmosphere consisting of H_2, CH_4, and NH_3, into which electric sparks were discharged. Analysis of the reaction products revealed the formation of a variety of organic molecules, including the amino acids alanine, aspartic acid, glutamic acid, and glycine.

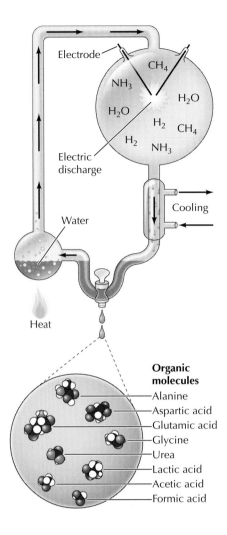

amino acids (**Figure 1.1**). Although Miller's experiments did not precisely reproduce the conditions of primitive Earth, they clearly demonstrated the plausibility of the spontaneous synthesis of organic molecules, providing the basic materials from which the first living organisms arose.

The next step in evolution was the formation of macromolecules. The monomeric building blocks of macromolecules have been demonstrated to polymerize spontaneously under plausible prebiotic conditions. Heating dry mixtures of amino acids, for example, results in their polymerization to form polypeptides. But the critical characteristic of the macromolecule from which life evolved must have been the ability to replicate itself. Only a macromolecule capable of directing the synthesis of new copies of itself would have been capable of reproduction and further evolution.

Of the two major classes of informational macromolecules in present-day cells (nucleic acids and proteins), only the nucleic acids are capable of directing their own self-replication. Nucleic acids can serve as templates for their own synthesis as a result of specific base pairing between complementary nucleotides (**Figure 1.2**). A critical step in understanding molecular evolution was thus reached in the early 1980s, when it was discovered in the laboratories of Sid Altman and Tom Cech that RNA is capable of catalyzing a number of chemical reactions, including the polymerization of nucleotides. Further studies have extended the known catalytic activities of RNA, including the description of RNA molecules that direct the synthesis of a new RNA strand from an RNA template. RNA is thus uniquely able to both serve as a template for and to catalyze its own replication. Consequently, RNA is generally believed to have been the initial genetic system, and an early stage of chemical evolution is thought to have been based on self-replicating RNA molecules—a

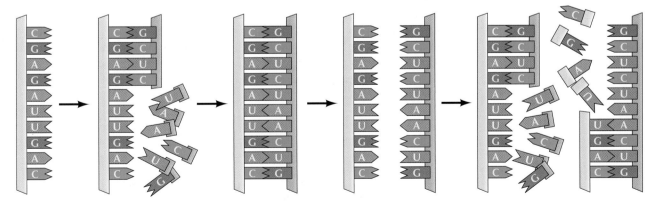

Figure 1.2 Self-replication of RNA   Complementary pairing between nucleotides (adenine [A] with uracil [U] and guanine [G] with cytosine [C]) allows one strand of RNA to serve as a template for the synthesis of a new strand with the complementary sequence.

period of evolution known as the **RNA world**. Ordered interactions between RNA and amino acids then evolved into the present-day genetic code, and DNA eventually replaced RNA as the genetic material.

As discussed further in Chapter 4, all present-day cells use DNA as the genetic material and employ the same basic mechanisms for DNA replication and expression of the genetic information. **Genes** are the functional units of inheritance, corresponding to segments of DNA that encode proteins or RNA molecules. The nucleotide sequence of a gene is copied into RNA by a process called **transcription**. For RNAs that encode proteins, their nucleotide sequence is then used to specify the order of amino acids in a protein by a process called **translation**.

The first cell is presumed to have arisen by the enclosure of self-replicating RNA in a membrane composed of **phospholipids** (**Figure 1.3**). As discussed in detail in the next chapter, phospholipids are the basic components of all present-day biological membranes, including the plasma membranes of both prokaryotic and eukaryotic cells. The key characteristic of the phospholipids that form membranes is that they are **amphipathic** molecules, meaning that one portion of the molecule is soluble in water and another portion is not. Phospholipids have long, water-insoluble (**hydrophobic**) hydrocarbon chains joined to water-soluble (**hydrophilic**) head groups that contain phosphate. When placed in water, phospholipids spontaneously aggregate into a bilayer with their phosphate-containing head groups on the outside in contact with water and their hydrocarbon tails in the interior in contact with each other. Such a phospholipid bilayer forms a stable barrier between two aqueous compartments—for example, separating the interior of the cell from its external environment.

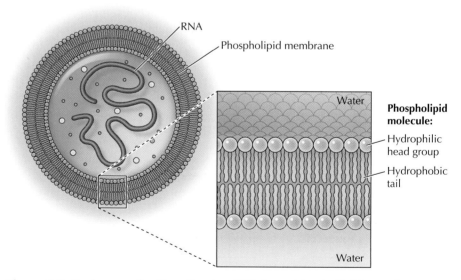

Figure 1.3 Enclosure of self-replicating RNA in a phospholipid membrane The first cell is thought to have arisen by the enclosure of self-replicating RNA and associated molecules in a membrane composed of phospholipids. Each phospholipid molecule has two long hydrophobic tails attached to a hydrophilic head group. The hydrophobic tails are buried in the lipid bilayer; the hydrophilic heads are exposed to water on both sides of the membrane.

The enclosure of self-replicating RNA and associated molecules in a phospholipid membrane would thus have maintained them as a unit, capable of self-reproduction and further evolution. RNA-directed protein synthesis may already have evolved by this time, in which case the first cell would have consisted of self-replicating RNA and its encoded proteins.

The evolution of metabolism

Because cells originated in a sea of organic molecules, they were able to obtain food and energy directly from their environment. But such a situation is self-limiting, so cells needed to evolve their own mechanisms for generating energy and synthesizing the molecules necessary for their replication. The generation and controlled utilization of metabolic energy is central to all cell activities, and the principal pathways of energy metabolism (discussed in detail in Chapter 3) are highly conserved in present-day cells. All cells use **adenosine 5′-triphosphate (ATP)** as their source of metabolic energy to drive the synthesis of cell constituents and carry out other energy-requiring activities, such as movement (e.g., muscle contraction). The mechanisms used by cells for the generation of ATP are thought to have evolved in three stages, corresponding to the evolution of glycolysis, photosynthesis, and oxidative metabolism (**Figure 1.4**). The development of these metabolic pathways changed Earth's atmosphere, thereby altering the course of further evolution.

In the initially anaerobic atmosphere of Earth, the first energy-generating reactions presumably involved the breakdown of organic molecules in the absence of oxygen. These reactions are likely to have been a form of present-day **glycolysis**—the anaerobic breakdown of glucose to lactic acid, with the net energy gain of two molecules of ATP. In addition to using ATP as their source of intracellular chemical energy, all present-day cells carry out glycolysis, consistent with the notion that these reactions arose very early in evolution.

Glycolysis provided a mechanism by which the energy in preformed organic molecules (e.g., glucose) could be converted to ATP, which could then be used

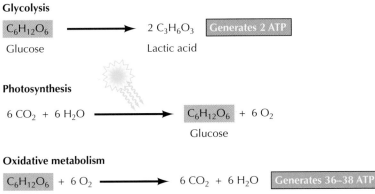

Glycolysis

$$C_6H_{12}O_6 \longrightarrow 2\ C_3H_6O_3 \quad \boxed{\text{Generates 2 ATP}}$$
Glucose   Lactic acid

Photosynthesis

$$6\ CO_2\ +\ 6\ H_2O \longrightarrow C_6H_{12}O_6\ +\ 6\ O_2$$
                              Glucose

Oxidative metabolism

$$C_6H_{12}O_6\ +\ 6\ O_2 \longrightarrow 6\ CO_2\ +\ 6\ H_2O \quad \boxed{\text{Generates 36–38 ATP}}$$
Glucose

Figure 1.4 Generation of metabolic energy   Glycolysis is the anaerobic breakdown of glucose to lactic acid. Photosynthesis utilizes energy from sunlight to drive the synthesis of glucose from CO_2 and H_2O, with the release of O_2 as a by-product. The O_2 released by photosynthesis is used in oxidative metabolism, in which glucose is broken down to CO_2 and H_2O, releasing much more energy than can be obtained from glycolysis.

as a source of energy to drive other metabolic reactions. The development of **photosynthesis** is generally thought to have been the next major evolutionary step, which allowed the cell to harness energy from sunlight and provided independence from the utilization of preformed organic molecules. The first photosynthetic bacteria probably utilized H_2S to convert CO_2 to organic molecules—a pathway of photosynthesis still used by some bacteria. The use of H_2O as a donor of electrons and hydrogen for the conversion of CO_2 to organic compounds evolved later and had the important consequence of changing Earth's atmosphere. The use of H_2O in photosynthetic reactions produces the by-product free O_2; this mechanism is thought to have been responsible for making O_2 abundant in Earth's atmosphere, which occurred about 2.4 billion years ago.

The release of O_2 as a consequence of photosynthesis changed the environment in which cells evolved and is commonly thought to have led to the development of **oxidative metabolism**. Alternatively, oxidative metabolism may have evolved before photosynthesis, with the increase in atmospheric O_2 then providing a strong selective advantage for organisms capable of using O_2 in energy-producing reactions. In either case, O_2 is a highly reactive molecule, and oxidative metabolism, utilizing this reactivity, has provided a mechanism for generating energy from organic molecules that is much more efficient than anaerobic glycolysis. For example, the complete oxidative breakdown of glucose to CO_2 and H_2O yields energy equivalent to that of 36 to 38 molecules of ATP, in contrast to the 2 ATP molecules formed by anaerobic glycolysis (see Figure 1.4). With few exceptions, present-day cells use oxidative reactions as their principal source of energy.

Present-day prokaryotes

Present-day prokaryotes are divided into two groups—**archaebacteria and bacteria**—which diverged early in evolution. Some archaebacteria live in extreme environments that are unusual today but may have been prevalent in primitive Earth. For example, thermoacidophiles live in hot sulfur springs with temperatures as high as 80°C and pH values as low as 2. The bacteria include the common forms of present-day prokaryotes—a large group of organisms that live in a wide range of environments, including soil, water, and other organisms (e.g., human pathogens).

Most prokaryotic cells are spherical, rod-shaped, or spiral, with diameters of 1 to 10 μm. Their DNA contents range from about 0.6 million to 5 million base pairs, an amount sufficient to encode about 5000 different proteins (Table 1.2). The largest and most complex prokaryotes are the **cyanobacteria**—bacteria in which photosynthesis evolved.

The structure of a typical prokaryotic cell is illustrated by *Escherichia coli* (*E. coli*), a common inhabitant of the human intestinal tract (Figure 1.5). The cell is rod-shaped, about 1 μm in diameter and about 2 μm long. Like most other prokaryotes, *E. coli* is surrounded by a rigid **cell wall** composed of polysaccharides and peptides. Beneath the cell wall is the **plasma membrane**, which is a bilayer of phospholipids and associated proteins. Whereas the cell wall is porous and readily penetrated by a variety of molecules, the plasma

Existence of organisms in extreme conditions has led to the hypothesis that life could exist in similar environments elsewhere in the solar system. The field of astrobiology (or exobiology) seeks to find signs of this extraterrestrial life.

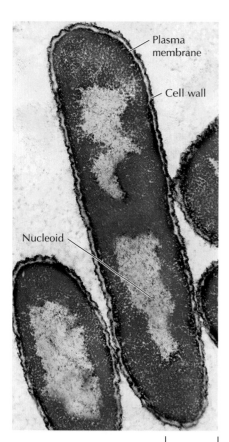

Figure 1.5 Electron micrograph of *E. coli*   The cell is surrounded by a cell wall, beneath which is the plasma membrane. DNA is located in the nucleoid. Artificial color has been added.

Table 1.2   Cell Genomes

Organism	Haploid DNA content (millions of base pairs)	Protein-coding genes
Archaebacteria		
Methanococcus jannaschii	1.7	1700
Bacteria		
Mycoplasma	0.6	470
E. coli	4.6	4200
Cyanobacterium	3.6	3200
Unicellular eukaryotes		
Saccharomyces cerevisiae (yeast)	12	6000
Dictyostelium discoideum	34	12,000
Paramecium	72	39,500
Chlamydomonas	118	14,500
Volvox	138	14,500
Plants		
Arabidopsis thaliana	125	26,000
Corn	2200	33,000
Apple	740	57,000
Animals		
Caenorhabditis elegans (nematode)	97	19,000
Drosophila melanogaster (fruit fly)	180	14,000
Chicken	1200	20–23,000
Zebrafish	1700	26,000
Xenopus tropicalis	1700	20–21,000
Mouse	3000	20–22,000
Human	3000	21,000

membrane provides the functional separation between the inside of the cell and its external environment. The DNA of *E. coli* is a single circular molecule in the nucleoid, which, in contrast to the nucleus of eukaryotes, is not surrounded by a membrane separating it from the cytoplasm. The cytoplasm contains approximately 30,000 **ribosomes** (the sites of protein synthesis), which account for its granular appearance.

Eukaryotic cells

Like prokaryotic cells, all eukaryotic cells are surrounded by a plasma membrane and contain ribosomes. However, eukaryotic cells are much more complex and contain a nucleus and a variety of cytoplasmic organelles (**Figure 1.6**). The largest and most prominent organelle of eukaryotic cells is the **nucleus**, with a diameter of approximately 5 μm. The nucleus contains the genetic information of the cell, which in eukaryotes is organized as linear rather than circular DNA molecules. The nucleus is the site of DNA replication and of RNA synthesis; the translation of RNA into proteins takes place on ribosomes in the cytoplasm.

In addition to a nucleus, eukaryotic cells contain a variety of membrane-enclosed organelles within their cytoplasm. These organelles provide compartments in which different metabolic activities are localized. Eukaryotic

Animal cell

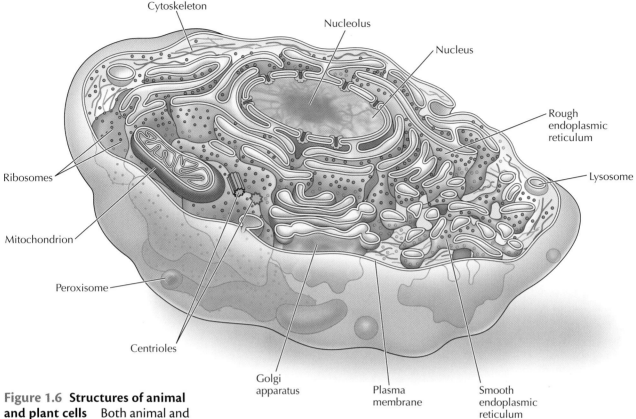

Figure 1.6 **Structures of animal and plant cells**   Both animal and plant cells are surrounded by a plasma membrane and contain a nucleus, a cytoskeleton, and many cytoplasmic organelles in common. Plant cells are also surrounded by a cell wall and contain chloroplasts and large vacuoles.

cells are generally much larger than prokaryotic cells, frequently having a cell volume at least a thousandfold greater. The compartmentalization provided by cytoplasmic organelles is what allows eukaryotic cells to function efficiently. Two of these organelles, **mitochondria** and **chloroplasts**, play critical roles in energy metabolism. Mitochondria, which are found in almost all eukaryotic cells, are the sites of oxidative metabolism and are thus responsible for generating most of the ATP derived from the breakdown of organic molecules. Chloroplasts are the sites of photosynthesis and are found only in the cells of plants and green algae. **Lysosomes** and **peroxisomes** also provide specialized metabolic compartments for the digestion of macromolecules and for various oxidative reactions, respectively. In addition, most plant cells contain large **vacuoles** that perform a variety of functions, including the digestion of macromolecules and the storage of both waste products and nutrients.

Because of the size and complexity of eukaryotic cells, the transport of proteins to their correct destinations within the cell is a formidable task. Two cytoplasmic organelles, the **endoplasmic reticulum** and the **Golgi apparatus**, are specifically devoted to the sorting and transport of proteins destined for secretion, incorporation into the plasma membrane, and incorporation into lysosomes. The endoplasmic reticulum is an extensive network of intracellular membranes, extending from the nuclear membrane throughout the cytoplasm. It functions not only in the processing and transport of proteins, but also in the synthesis of lipids. From the endoplasmic reticulum, proteins

Plant cell

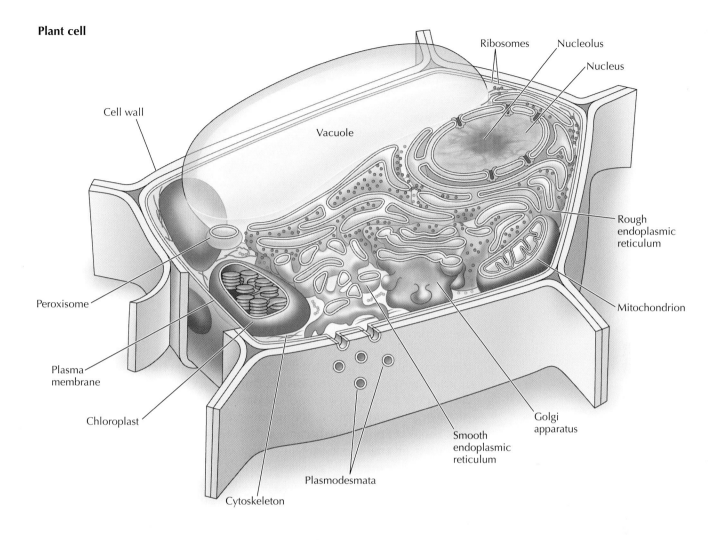

are transported within small membrane vesicles to the Golgi apparatus, where they are further processed and sorted for transport to their final destinations. In addition to this role in protein transport, the Golgi apparatus serves as a site of lipid synthesis and (in plant cells) as the site of synthesis of some of the polysaccharides that compose the cell wall.

The internal organization of eukaryotic cells is maintained by the **cytoskeleton**, a network of protein filaments extending throughout the cytoplasm. The cytoskeleton provides the structural framework of the cell, determining cell shape and the general organization of the cytoplasm. In addition, the cytoskeleton is responsible for the movements of entire cells (e.g., the contraction of muscle cells) and for the intracellular transport and positioning of organelles and other structures, including the movements of chromosomes during cell division.

The origin of eukaryotes

A critical step in the evolution of eukaryotic cells was the acquisition of membrane-enclosed subcellular organelles, allowing the development of the complexity characteristic of these cells. The organelles of eukaryotes are thought to have arisen by **endosymbiosis**—one cell living inside another. In

Certain present-day marine protists engulf algae to serve as endosymbionts that carry out photosynthesis for their hosts.

particular, eukaryotic organelles are thought to have evolved from prokaryotic cells living inside the ancestors of eukaryotes.

The hypothesis that eukaryotic cells evolved by endosymbiosis is particularly well supported by studies of mitochondria and chloroplasts, which are thought to have evolved from bacteria living in larger cells. Both mitochondria and chloroplasts are similar to bacteria in size and, like bacteria, they reproduce by dividing in two. Most important, both mitochondria and chloroplasts contain their own DNA, which encodes some of their components. The mitochondrial and chloroplast DNAs are replicated each time the organelle divides, and the genes they encode are transcribed within the organelle and translated on organelle ribosomes. Mitochondria and chloroplasts thus contain their own genetic systems, which are distinct from the nuclear genome of the cell. Furthermore, the ribosomes and ribosomal RNAs of these organelles are more closely related to those of bacteria than to those encoded by the nuclear genomes of eukaryotes.

An endosymbiotic origin for these organelles is now generally accepted, with mitochondria thought to have evolved from aerobic bacteria and chloroplasts from photosynthetic bacteria, such as the cyanobacteria. The acquisition of aerobic bacteria would have provided an anaerobic cell with the ability to carry out oxidative metabolism. The acquisition of photosynthetic bacteria would have provided the ability to perform photosynthesis, thereby affording nutritional independence. Thus, these endosymbiotic associations were highly advantageous to their partners and were selected for in the course of evolution. Through time, most of the genes originally present in these bacteria apparently became incorporated into the nuclear genome of the cell, so only a few components of mitochondria and chloroplasts are still encoded by the organelle genomes.

The precise origin of eukaryotic cells remains an unsettled and controversial issue in our understanding of early evolution. Studies of their DNA sequences indicate that the archaebacteria and bacteria are as different from each other as either is from present-day eukaryotes. Therefore, a very early event in evolution appears to have been the divergence of two lines of descent from a common prokaryotic ancestor, giving rise to present-day archaebacteria and bacteria. However, it has proven difficult to determine whether eukaryotes evolved from bacteria or from archaebacteria. Surprisingly, some eukaryotic genes are more similar to bacterial genes, whereas others are more similar to archaebacterial genes. The genome of eukaryotes thus appears to consist of some genes derived from bacteria and others from archaebacteria, rather than reflecting the genome of either ancestor. Curiously, most eukaryotic genes related to informational processes (such as DNA replication, transcription, and protein synthesis) were derived from archaebacteria, whereas most eukaryotic genes related to basic cell operational processes (such as glycolysis and amino acid biosynthesis) were derived from bacteria.

One hypothesis to explain the mosaic nature of eukaryotic genomes proposes that the genome of eukaryotes arose from a fusion of archaebacterial and bacterial genomes (**Figure 1.7**). According to this proposal, an endosymbiotic association between a bacterium and an archaebacterium was followed by fusion of the two prokaryotic genomes, giving rise to an ancestral eukaryotic genome with contributions from both bacteria and archaebacteria. The simplest version of this hypothesis is that an initial endosymbiotic relationship of a bacterium living inside an archaebacterium gave rise not only to

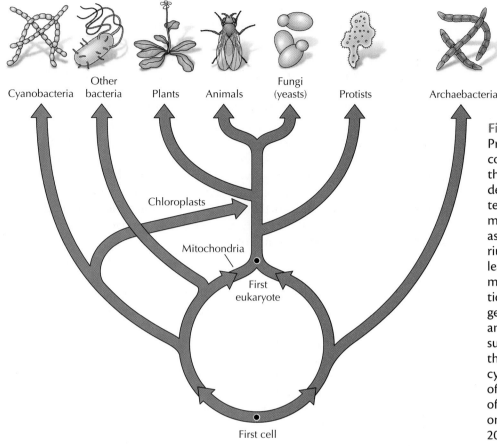

Cyanobacteria   **Other bacteria**   **Plants**   **Animals**   **Fungi (yeasts)**   **Protists**   **Archaebacteria**

Chloroplasts

Mitochondria

First eukaryote

First cell

Figure 1.7 Evolution of cells Present-day cells evolved from a common prokaryotic ancestor that diverged along two lines of descent, giving rise to archaebacteria and bacteria. Eukaryotic cells may have arisen by endosymbiotic association of an aerobic bacterium with an archaebacterium, leading to the development of mitochondria as well as the formation of a eukaryotic genome with genes derived from both bacteria and archaebacteria. Chloroplasts subsequently evolved as a result of the endosymbiotic association of a cyanobacterium with the ancestor of plants. The model for formation of the first eukaryotic cell is based on M. C. Rivera and J. A. Lake, 2004. *Nature* 431: 152.

mitochondria but also to the genome of eukaryotic cells, containing genes derived from both prokaryotic ancestors.

The development of multicellular organisms

Many eukaryotes are unicellular organisms that, like bacteria, consist of only single cells capable of self-replication. The simplest eukaryotes are the **yeasts**, which contain only slightly more genes than many bacteria (see Table 1.2). Although yeasts are more complex than bacteria, they are much smaller and simpler than the cells of animals or plants. For example, the commonly studied yeast ***Saccharomyces cerevisiae*** is about 6 μm in diameter and contains 12 million base pairs of DNA (**Figure 1.8**). Other unicellular eukaryotes, however, are far more complex cells, with substantially larger and more complex genomes. They include organisms specialized to perform a variety of tasks, including photosynthesis, movement, and the capture and ingestion of other organisms as food. The ciliated protozoan *Paramecium*, for example, is a large, complex cell that can be up to 350 μm in length and is specialized for movement and feeding on bacteria and

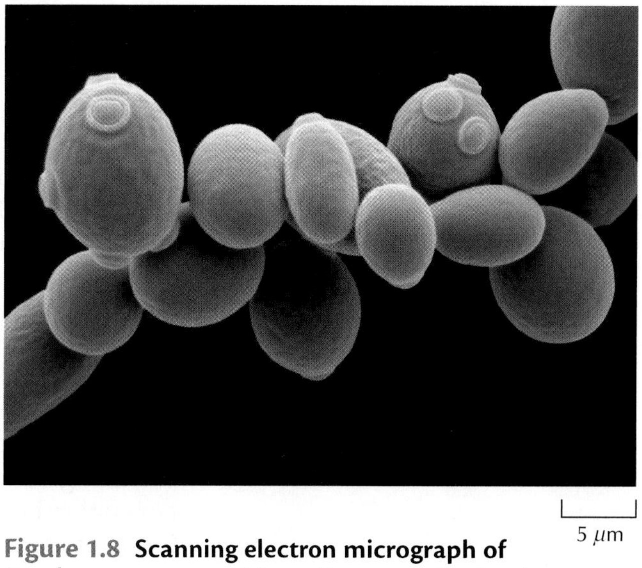

5 μm

Figure 1.8 Scanning electron micrograph of Saccharomyces cerevisiae Yeasts are the simplest eukaryotes. Artificial color has been added to the micrograph.

(A)

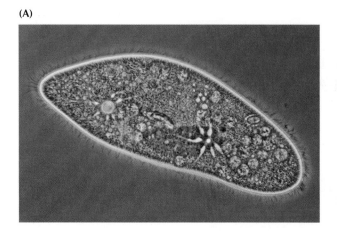

(B)

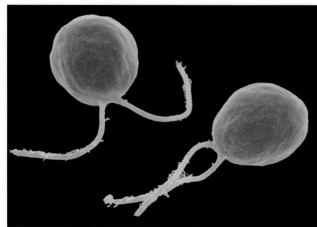

Figure 1.9 Light micrograph of *Paramecium* and scanning electron micrograph of *Chlamydomonas*   (A) *Paramecium* and (B) *Chlamydomas* are examples of unicellular eukaryotes that are more complex than yeast.

Video 1.1

sites.sinauer.com/cooper7e/v1.1

Paramecium Feeding *Paramecium caudatum* feeds on bacteria and other small cells via a deep oral groove that contains compound oral cilia used to draw in food.

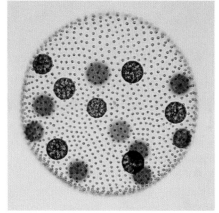

Figure 1.10 Multicellular green algae   *Volvox* consists of approximately 16 germ cells and 2000 somatic cells embedded in a gelatinous matrix. (Courtesy of David Kirk.)

yeast (**Figure 1.9**). Surprisingly, the *Paramecium* genome contains almost twice as many genes as humans (see Table 1.2), illustrating the fact that neither genome size nor gene number is directly related to the complexity of an organism—an unexpected result of genome sequencing projects that will be discussed further in Chapters 5 and 6. Other unicellular eukaryotes, such as the green alga *Chlamydomonas* (see Figure 1.9), contain chloroplasts and are able to carry out photosynthesis.

The evolution of multicellular organisms from unicellular eukaryotes is thought to have occurred 1 to 2 billion years ago. The algae, for example, contain both unicellular and multicellular species. The multicellular green alga *Volvox* contains cells of two different types: approximately 16 large germ cells and 2000 somatic cells that resemble the unicellular *Chlamydomonas* (**Figure 1.10**). Both *Chlamydomonas* and *Volvox* have genomes of similar size and complexity, about 10 times larger than that of yeast and containing 14,000–15,000 genes (see Table 1.2). Another example of the transition to multicellularity is provided by the amoeba *Dictyostelium discoideum*, which is able to alternate between unicellular and multicellular forms depending on the availability of food (**Figure 1.11**).

Increasing cell specialization and division of labor among the cells of simple multicellular organisms then led to the complexity and diversity observed in the many types of cells that make up present-day plants and animals, including human beings. Plants are composed of fewer cell types than are animals, but each different kind of plant cell is specialized to perform specific tasks required by the organism as a whole (**Figure 1.12**). The cells of plants are organized into three main tissue systems: ground tissue, dermal tissue, and vascular tissue. The ground tissue contains parenchyma cells, which carry out most of the metabolic reactions of the plant, including photosynthesis. Ground tissue also contains two specialized cell types

(A)

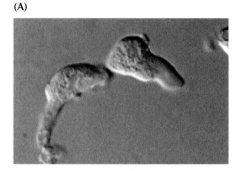

(B)

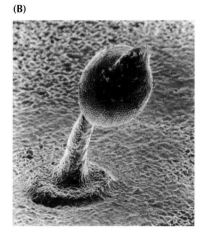

Figure 1.11 *Dictyostelium discoideum*
If food is unavailable, the unicellular amoebae (A) aggregate to form a multicellular fruiting body, specialized for the dispersal of spores (B). (A, courtesy of David Knecht, University of Connecticut.)

(collenchyma cells and sclerenchyma cells) that are characterized by thick cell walls and provide structural support to the plant. Dermal tissue covers the surface of the plant and is composed of epidermal cells, which form a protective coat and allow the absorption of nutrients. Finally, several types of elongated cells form the vascular system (the xylem and phloem), which is responsible for the transport of water and nutrients throughout the plant.

(A) Parenchyma cells

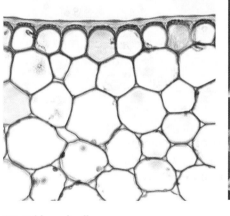

(B) Collenchyma cells

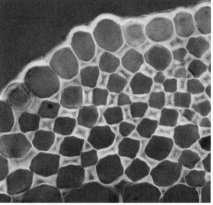

(C) Epidermal cells

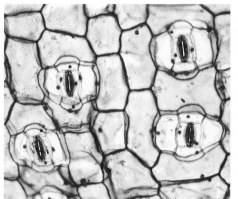

(D) Xylem vessel elements and tracheids

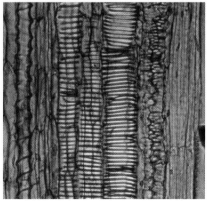

Figure 1.12 Light micrographs of representative plant cells
(A) Parenchyma cells (from corn stem) are responsible for photosynthesis and other metabolic reactions. (B) Collenchyma cells (from spinach leaf vein) are specialized for support and have thickened cell walls. (C) Epidermal cells on the surface of a dayflower leaf. Tiny pores (stomata) are flanked by specialized cells called guard cells. (D) Vessel elements and tracheids of a squash stem are elongated cells that are arranged end to end to form vessels of the xylem.

The cells found in animals are considerably more diverse than those of plants. The human body, for example, is composed of more than 200 different kinds of cells, which are generally considered to be components of five main types of tissues: epithelial tissue, connective tissue, blood, nervous tissue, and muscle (**Figure 1.13**). **Epithelial cells** form sheets that cover the

(A) i Mouth

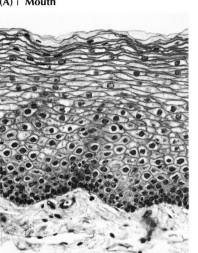

(A) ii Bile duct

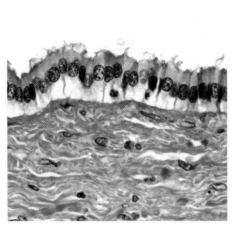

(A) iii Intestine

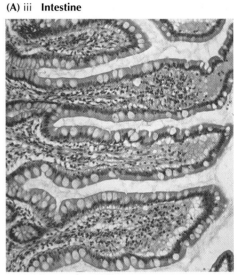

Figure 1.13 Light micrographs of representative animal cells
(A) Epithelial cells of the mouth (a thick, multilayered sheet), bile duct, and intestine. (B) Fibroblasts are connective tissue cells characterized by their elongated spindle shape. (C) Erythrocytes, granulocytes, lymphocytes, and monocytes in human blood.

(B)

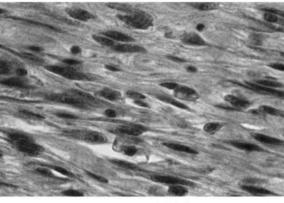

(C) Erythrocyte

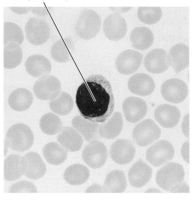

Granulocyte

Lymphocyte

Monocyte

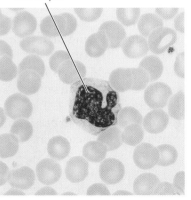

surface of the body and line the internal organs. There are many different types of epithelial cells, each specialized for a specific function, including protection (the skin), absorption (e.g., cells lining the small intestine), and secretion (e.g., cells of the salivary gland). Connective tissues include bone, cartilage, and adipose tissue, each of which is formed by different types of cells (osteoblasts, chondrocytes, and adipocytes, respectively). The loose connective tissue that underlies epithelial layers and fills the spaces between organs and tissues in the body is formed by another cell type, the **fibroblast**. Blood contains several different types of cells: red blood cells (**erythrocytes**) function in oxygen transport, and white blood cells (**granulocytes**, **monocytes**, **macrophages**, and **lymphocytes**) function in inflammatory reactions and the immune response. Nervous tissue is composed of supporting cells and nerve cells, or **neurons**, which are highly specialized to transmit signals throughout the body. Various types of sensory cells, such as cells of the eye and ear, are further specialized to receive external signals from the environment. Finally, several different types of muscle cells are responsible for the production of force and movement.

The evolution of animals clearly involved the development of considerable diversity and specialization at the cellular level. Understanding the mechanisms that control the growth and differentiation of such a complex array of specialized cells, starting from a single fertilized egg, is one of the major challenges facing contemporary cell and molecular biology.

Cells as Experimental Models

The evolution of present-day cells from a common ancestor has important implications for cell and molecular biology as an experimental science. Because the fundamental properties of all cells have been conserved during evolution, the basic principles learned from experiments performed with one type of cell are generally applicable to other cells. On the other hand, because of the diversity of present-day cells, many kinds of experiments can be more readily undertaken with one type of cell than with another. Several different kinds of cells and organisms are commonly used as experimental models to study various aspects of cell and molecular biology. The features of some of these cells that make them particularly advantageous as experimental models are discussed in the sections that follow. The availability of complete genome sequences further enhances the value of these organisms as model systems in understanding the molecular biology of cells.

E. coli

Because of their comparative simplicity, prokaryotic cells (bacteria) are ideal models for studying many fundamental aspects of biochemistry and molecular biology. The most thoroughly studied species of bacteria is *Escherichia coli* (*E. coli*), which has long been the favored organism for investigation of the basic mechanisms of molecular genetics. Most of our present concepts of molecular biology—including our understanding of DNA replication, the genetic code, gene expression, and protein synthesis—derive from studies of this humble bacterium.

E. coli has been especially useful to molecular biologists because of both its relative simplicity and the ease with which it can be propagated and studied in the laboratory. The genome of *E. coli*, for example, consists of approximately 4.6 million base pairs and contains about 4000 genes. The human genome is nearly a thousand times larger (approximately 3 billion base pairs) and is thought to contain about 21,000 protein-coding genes (see Table 1.2). The small size of the *E. coli* genome provides obvious advantages for genetic analysis.

Molecular genetic experiments are further facilitated by the rapid growth of *E. coli* under well-defined laboratory conditions. Under optimal culture conditions, *E. coli* divide every 20 minutes. Moreover, a clonal population of *E. coli*, in which all cells are derived by division of a single cell of origin, can be readily isolated as a colony grown on semisolid agar-containing medium (**Figure 1.14**). Because bacterial colonies containing as many as 10^8 cells can develop overnight, selecting genetic variants of an *E. coli* strain—for example, mutants that are resistant to an antibiotic such as penicillin—is easy and rapid. The ease with which such mutants can be selected and analyzed was critical to the success of experiments that defined the basic principles of molecular genetics, discussed in Chapter 4.

The nutrient mixtures in which *E. coli* divide most rapidly include glucose, salts, and various organic compounds, such as amino acids, vitamins, and nucleic acid precursors. However, *E. coli* can also grow in much simpler media consisting only of salts, a source of nitrogen (such as ammonia), and a source of carbon and energy (such as glucose). In such a medium, the bacteria grow a little more slowly (with a division time of about 40 minutes) because they must synthesize all their own amino acids, nucleotides, and other organic compounds. The ability of *E. coli* to carry out these biosynthetic reactions in simple defined media has made them extremely useful in elucidating the biochemical pathways involved. Thus, the rapid growth and simple nutritional requirements of *E. coli* have greatly facilitated fundamental experiments in both molecular biology and biochemistry.

Yeasts

Although bacteria have been an invaluable model for studies of many conserved properties of cells, they obviously cannot be used to study aspects of cell structure and function that are unique to eukaryotes. Yeasts, the simplest eukaryotes, have a number of experimental advantages similar to those of *E. coli*. Consequently, yeasts have provided a crucial model for studies of many fundamental aspects of eukaryotic cell biology.

The genome of the most frequently studied yeast, *Saccharomyces cerevisiae*, consists of 12 million base pairs of DNA and contains about 6000 genes. Although the yeast genome is approximately three times larger than that of *E. coli*, it is far more manageable than the genomes of more complex eukaryotes, such as humans. Yet, even in its simplicity, the yeast cell exhibits the typical features of eukaryotic cells (**Figure 1.15**): It contains a distinct nucleus surrounded by a nuclear membrane, its genomic DNA is organized as 16 linear chromosomes, and its cytoplasm contains subcellular organelles.

Yeasts can be readily grown in the laboratory and can be studied by many of the same molecular genetic approaches that have proved so successful with *E. coli*. Although yeasts do not replicate as rapidly as bacteria, they still divide as frequently as every 2 hours and they can easily be grown as

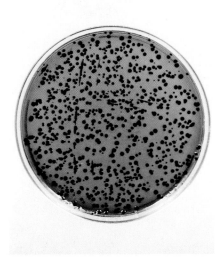

Figure 1.14 Bacterial colonies Photograph of colonies of *E. coli* growing on the surface of an agar-containing medium.

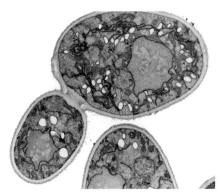

Figure 1.15 Transmission electron micrograph of *Saccharomyces cerevisiae* Yeast are the simplest model for studying eukaryotic cells.

colonies from a single cell. Consequently, yeasts can be used for a variety of genetic manipulations similar to those that can be performed using bacteria.

These features have made yeast cells the most approachable eukaryotic cells from the standpoint of molecular biology. Yeast mutants have been important in understanding many fundamental processes in eukaryotes, including DNA replication, transcription, RNA processing, protein sorting, and the regulation of cell division, as will be discussed in subsequent chapters. The unity of molecular cell biology is made abundantly clear by the fact that the general principles of cell structure and function revealed by studies of yeasts apply to all eukaryotic cells.

Caenorhabditis elegans

The unicellular yeasts are important models for studies of eukaryotic cells, but understanding the development of multicellular organisms requires the experimental analysis of plants and animals—organisms that are more complex. The nematode *Caenorhabditis elegans* (Figure 1.16) possesses several notable features that make it one of the most widely used models for studies of animal development and cell differentiation.

Although the genome of *C. elegans* (close to 100 million base pairs) is larger than those of unicellular eukaryotes, it is smaller and more manageable than the genomes of most animals. Despite its relatively small size, however, the genome of *C. elegans* contains approximately 19,000 genes—more than three times the number of genes in yeast, and nearly the same number of protein-coding genes in humans. Biologically, *C. elegans* is a relatively simple multicellular organism: Adult worms consist of only 959 somatic cells, plus 1000 to 2000 germ cells. In addition, *C. elegans* can be easily grown and subjected to genetic manipulations in the laboratory.

The simplicity of *C. elegans* has enabled the course of its development to be studied in detail by microscopic observation. Such analyses have successfully traced the embryonic origin and lineage of all the cells in the adult worm. Genetic studies have also identified many of the mutations responsible for developmental abnormalities, leading to the isolation and characterization of critical genes that control nematode development and differentiation. Importantly, similar genes have also been found to function in complex animals (including humans), making *C. elegans* an important model for studies of animal development.

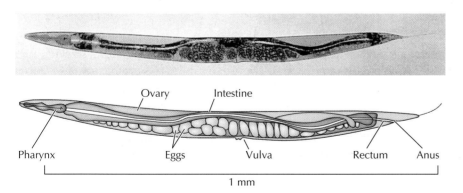

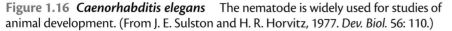

Figure 1.16 *Caenorhabditis elegans* The nematode is widely used for studies of animal development. (From J. E. Sulston and H. R. Horvitz, 1977. *Dev. Biol.* 56: 110.)

Figure 1.17 *Drosophila melanogaster* The fruit fly is a key model for genetics and developmental biology.

Drosophila melanogaster

Like *C. elegans*, the fruit fly **Drosophila melanogaster** (Figure 1.17) has been a crucial model organism in developmental biology. The genome of *Drosophila* is 180 million base pairs, larger than that of *C. elegans*, but the *Drosophila* genome only contains about 14,000 genes. Furthermore, *Drosophila* can be easily maintained and bred in the laboratory, and its short reproductive cycle (about 2 weeks) makes it a very useful organism for genetic experiments. Many fundamental concepts of genetics—such as the relationship between genes and chromosomes—were derived from studies of *Drosophila* early in the twentieth century (see Chapter 4).

Extensive genetic analysis of *Drosophila* has uncovered many genes that control development and differentiation, and current methods of molecular biology have allowed the functions of these genes to be analyzed in detail. Consequently, studies of *Drosophila* have led to striking advances in understanding the molecular mechanisms that govern animal development, particularly with respect to formation of the body plan of complex multicellular organisms. As with *C. elegans*, similar genes and mechanisms exist in vertebrates, validating the use of *Drosophila* as a major experimental model in contemporary developmental biology.

Arabidopsis thaliana

The study of plant molecular biology and development is an active and expanding field of considerable economic importance as well as intellectual interest. Since the genomes of plants cover a range of complexity comparable to that of animal genomes (see Table 1.2), an optimal model for studies of plant development would be a relatively simple organism with some of the advantageous properties of *C. elegans* and *Drosophila*. The small flowering plant **Arabidopsis thaliana** (mouse-ear cress) (Figure 1.18) meets these criteria and is therefore widely used as a model to study the molecular biology of plants.

Arabidopsis is notable for its genome of only about 125 million base pairs. Although *Arabidopsis* contains a total of about 26,000 genes, many of these are repeated, so its number of unique genes is approximately 15,000—a complexity similar to that of *C. elegans* and *Drosophila*. In addition, *Arabidopsis* is relatively easy to grow in the laboratory, and methods for molecular genetic manipulations of this plant have been developed. These studies have led to the identification of genes involved in various aspects of plant development, such as the development of flowers. Analysis of these genes points to many similarities, but also to striking differences, between the mechanisms that control the development of plants and animals.

Vertebrates

The most complex animals are the vertebrates, including humans and other mammals. The human genome is approximately 3 billion base pairs—about 20 to 30 times larger than the genomes of *C. elegans*, *Drosophila*, or *Arabidopsis*—and contains about 21,000 protein-coding genes. Moreover, the human body is composed of more than 200 different kinds of specialized cell types. This complexity makes the vertebrates difficult to study from the standpoint of cell and molecular biology, but much of the interest in biological sciences nonetheless stems from the desire to understand the human organism. Moreover, an understanding of many questions of immediate practical

Figure 1.18 *Arabidopsis thaliana* *Arabidopsis* is a model for studying the molecular biology of plants.

importance (e.g., in medicine) must be based directly on studies of human (or closely related) cell types.

One important approach to studying human and other mammalian cells is to grow isolated cells in culture, where they can be manipulated under controlled laboratory conditions. The use of cultured cells has allowed studies of many aspects of mammalian cell biology, including experiments that have elucidated the mechanisms of DNA replication, gene expression, protein synthesis and processing, and cell division. Moreover, the ability to culture cells in chemically defined media has allowed studies of the signaling mechanisms that normally control cell growth and differentiation within the intact organism.

The specialized properties of some highly differentiated cell types have made them important models for studies of particular aspects of cell biology. Muscle cells, for example, are highly specialized to undergo contraction, producing force and movement. Because of this specialization, muscle cells are crucial models for studying cell movement at the molecular level. Another example is provided by nerve cells (neurons), which are specialized to conduct electrochemical signals over long distances. In humans, nerve cell axons may be more than a meter long, and some invertebrates, such as the squid, have giant neurons with axons as large as 1 mm in diameter. Because of their highly specialized structure and function, these giant neurons have provided important models for studies of ion transport across the plasma membrane, and of the role of the cytoskeleton in the transport of cytoplasmic organelles.

The frog *Xenopus laevis* is an important model for studies of early vertebrate development. *Xenopus* eggs are unusually large cells, with a diameter of approximately 1 mm (**Figure 1.19**). Because those eggs develop outside of the mother, all stages of development from egg to tadpole can be readily studied in the laboratory. In addition, *Xenopus* eggs can be obtained in large numbers, facilitating biochemical analysis. Because of these technical advantages, *Xenopus* has been widely used in studies of developmental biology and has provided important insights into the molecular mechanisms that control development, differentiation, and embryonic cell division.

1 mm

Figure 1.19 Eggs of the frog *Xenopus laevis*   The large eggs of *Xenopus* are useful for studies of early vertebrate development. (Courtesy of Michael Danilchik and Kimberly Ray.)

(A)

(B)

Figure 1.20 Zebrafish   (A) A 24-hour-old embryo. (B) An adult fish. (A, courtesy of Charles Kimmel, University of Oregon.)

The **zebrafish** (Figure 1.20) possesses a number of advantages for genetic studies of vertebrate development. These small fish are easy to maintain in the laboratory and they reproduce rapidly, with a generation time of 3–4 months. In addition, the embryos develop outside of the mother and are transparent, so that early stages of development can be easily observed. Powerful methods have been developed to facilitate the isolation of mutations affecting zebrafish development, and several thousand such mutations have now been identified. Because the zebrafish is an easily studied vertebrate, it promises to bridge the gap between humans and the simpler invertebrate systems, such as *C. elegans* and *Drosophila*.

Among mammals, the mouse is the most suitable for genetic analysis. Although the technical difficulties in studying mouse genetics (compared, for example, with the genetics of yeasts or *Drosophila*) are formidable, many mutations affecting mouse development have been identified. Most important, recent advances in molecular biology have enabled the production of genetically engineered mice in which specific mutant genes have been introduced into the mouse germ line, allowing the functions of these genes to be studied in the context of the whole animal. The suitability of the mouse as a model for human development is indicated not only by the similarity of the mouse and human genomes but also by the fact that mutations in homologous genes result in similar developmental defects in both species—piebaldism (a defect in pigmentation) offering a striking example (Figure 1.21).

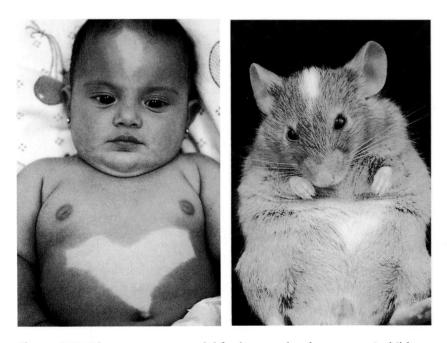

Figure 1.21 The mouse as a model for human development   A child and a mouse show similar defects in pigmentation (piebaldism) as a result of mutations in a gene required for normal migration of melanocytes (the cells responsible for skin pigmentation) during embryonic development. (Courtesy of R. A. Fleischman, Markey Cancer Center, University of Kentucky.)

Tools of Cell Biology

As in all experimental sciences, research in cell biology depends on the laboratory methods that can be used to study cell structure and function. Many important advances in understanding cells have directly followed the development of new methods that have opened novel avenues of investigation. An appreciation of the experimental tools available to the cell biologist is thus critical to understanding both the current status and future directions of this rapidly moving area of science. Some of the important general methods of cell biology are described in the sections that follow. Other experimental approaches, including the methods of molecular biology, genomics and proteomics, will be discussed in later chapters.

Light microscopy

Because most cells are too small to be seen by the naked eye, the study of cells has depended heavily on the use of microscopes. Indeed, the very discovery of cells arose from the development of the microscope: Robert Hooke first coined the term "cell" following his observations of a piece of cork with a simple light microscope in 1665 (Figure 1.22). Using a microscope that magnified objects up to about 300 times their actual size, Antony van Leeuwenhoek, in the 1670s, was able to observe a variety of different types of cells, including sperm, red blood cells, and bacteria. The proposal of the cell theory by Matthias Schleiden and Theodor Schwann in 1838 may be seen as the birth of contemporary cell biology. Microscopic studies of plant tissues by Schleiden and of animal tissues by Schwann led to the same conclusion: All organisms are composed of cells. Shortly thereafter, it was recognized that cells are not formed *de novo* but arise only from division of preexisting cells. Thus, the cell achieved its current recognition as the fundamental unit of all living organisms because of observations made with the light microscope.

The light microscope remains a basic tool of cell biologists, with technical improvements allowing the visualization of ever-increasing details of cell structure. Contemporary light microscopes are able to magnify objects up to about a thousand times. Since most cells are between 1 and 100 μm in diameter, they can be observed by light microscopy, as can some of the larger subcellular organelles, such as nuclei, chloroplasts, and mitochondria. However, the light microscope is not powerful enough to reveal fine details of cell structure, for which **resolution**—the ability of a microscope to distinguish objects separated by small distances—is even more important than magnification. Images can be magnified as much as desired (for example, by projection onto a large screen), but such magnification does not increase the level of detail that can be observed.

The diffraction of light limits the resolution of the light microscope to approximately 0.2 μm; two objects separated by less than this distance appear as a single image, rather than being distinguished from one another. This theoretical limitation of light microscopy is determined by two factors—the wavelength (λ) of visible light and the light-gathering power of the microscope lens (numerical aperture, *NA*)—according to the following equation:

$$\text{Resolution} = \frac{0.61\lambda}{NA}$$

The wavelength of visible light is 0.4 to 0.7 μm, so the value of λ is fixed at approximately 0.5 μm for the light microscope. The numerical aperture can

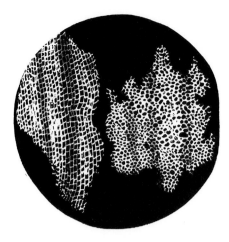

Figure 1.22 The cellular structure of cork A reproduction of Robert Hooke's drawing of a thin slice of cork examined with a light microscope. The "cells" that Hooke observed were actually only the cell walls remaining from cells that had long since died.

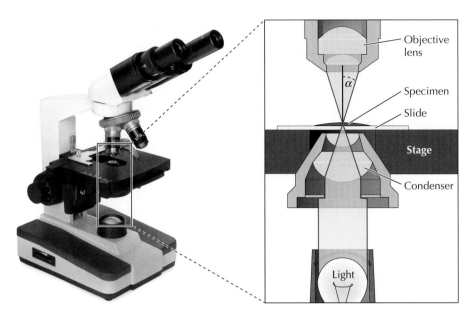

Figure 1.23 Numerical aperture Light is focused on the specimen by the condenser lens and then collected by the objective lens of the microscope. The numerical aperture is determined by the angle of the cone of light entering the objective lens (α) and by the refractive index of the medium (usually air or oil) between the lens and the specimen.

be envisioned as the size of the cone of light that enters the microscope lens after passing through the specimen (**Figure 1.23**). It is given by the equation

$$NA = \eta \sin \alpha$$

where η is the refractive index of the medium through which light travels between the specimen and the lens. The value of η for air is 1.0, but it can be increased to a maximum of approximately 1.4 by using an oil-immersion lens to view the specimen through a drop of oil. The angle α corresponds to half the width of the cone of light collected by the lens. The maximum value of α is 90°, at which $\sin \alpha = 1$, so the highest possible value for the numerical aperture is 1.4.

The theoretical limit of resolution of the light microscope can therefore be calculated as follows:

$$\text{Resolution} = \frac{0.61 \times 0.5}{1.4} = 0.22 \ \mu\text{m}$$

Microscopes achieving this level of resolution had already been made by the end of the nineteenth century, reaching the theoretical limit of resolution with the light microscope. However, as discussed below, new approaches have led to the development of novel methods (super-resolution microscopy) that have substantially increased the resolving power of fluorescence microscopy to reach beyond this limit.

Several different types of light microscopy are routinely used to study various aspects of cell structure. The simplest is **bright-field microscopy**, in which light passes directly through the cell and the ability to distinguish different parts of the cell depends on contrast resulting from the absorption of

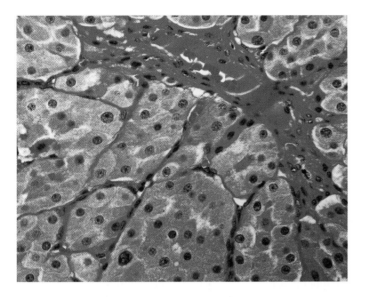

Figure 1.24 Bright-field micrograph of stained tissue Section of a benign kidney tumor.

visible light by cell components. In many cases, cells are stained with dyes that react with proteins or nucleic acids in order to enhance the contrast between different parts of the cell. Prior to staining, specimens are usually treated with fixatives (such as alcohol, acetic acid, or formaldehyde) to stabilize and preserve their structures. The examination of fixed and stained tissues by bright-field microscopy is the standard approach for the analysis of tissue specimens in histology laboratories (Figure 1.24). Such staining procedures kill the cells, however, and therefore are not suitable for many experiments in which the observation of living cells is desired.

Without staining, the direct passage of light does not provide sufficient contrast to distinguish many parts of the cell, limiting the usefulness of bright-field microscopy. However, optical variations of the light microscope can be used to enhance the contrast between light waves passing through regions of the cell with different densities. The two most common methods for visualizing living cells are **phase-contrast microscopy** and **differential interference-contrast microscopy** (Figure 1.25). Both kinds of microscopy use optical systems that convert variations in density or thickness between different parts of the cell to differences in contrast that can be seen in the final image. In bright-field microscopy, transparent structures (such as the nucleus) have little contrast because they absorb light poorly. However, light is slowed down as it passes through these structures so that its phase is altered compared with light that has passed through the surrounding cytoplasm. Phase-contrast and differential interference-contrast microscopy convert these differences in phase to differences in contrast, thereby yielding improved images of live, unstained cells.

The power of the light microscope has been considerably expanded by the use of video cameras and computers for image analysis and processing. Such

(A)

(B)

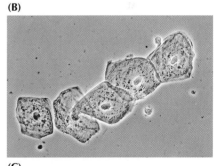

(C)

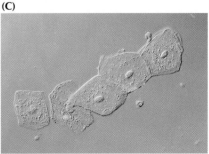

50 μm

Figure 1.25 Microscopic observation of living cells Photomicrographs of human cheek cells obtained with (A) bright-field, (B) phase-contrast, and (C) differential interference-contrast microscopy. (Courtesy of Mort Abramowitz, Olympus America, Inc.)

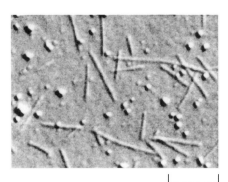

2.5 μm

Figure 1.26 Video-enhanced differential interference-contrast microscopy Electronic image processing allows the visualization of single microtubules. (Courtesy of E. D. Salmon, University of North Carolina, Chapel Hill.)

electronic image-processing systems can substantially enhance the contrast of images obtained with the light microscope, allowing the visualization of small objects that otherwise could not be detected. For example, video-enhanced differential interference-contrast microscopy has allowed visualization of the movement of organelles along microtubules, which are cytoskeletal protein filaments with a diameter of only 0.025 μm (**Figure 1.26**). However, this enhancement does not overcome the theoretical limit of resolution of the light microscope, approximately 0.2 μm. Thus, although video enhancement allows the visualization of microtubules, the microtubules appear as blurred images at least 0.2 μm in diameter and an individual microtubule cannot be distinguished from a bundle of adjacent structures.

Light microscopy has been brought to the level of molecular analysis by methods for labeling specific molecules so that they can be visualized within cells. Specific genes or RNA transcripts can be detected by hybridization with nucleic acid probes of complementary sequence, and proteins can be detected using appropriate antibodies (see Chapter 4). Both nucleic acid probes and antibodies can be labeled with a variety of tags that allow their visualization in the light microscope, making it possible to determine the location of specific molecules within individual cells.

Fluorescence microscopy is widely used to study the intracellular distribution of molecules (**Figure 1.27**). A fluorescent dye is used to label the molecule of interest within either fixed or living cells. The fluorescent dye is

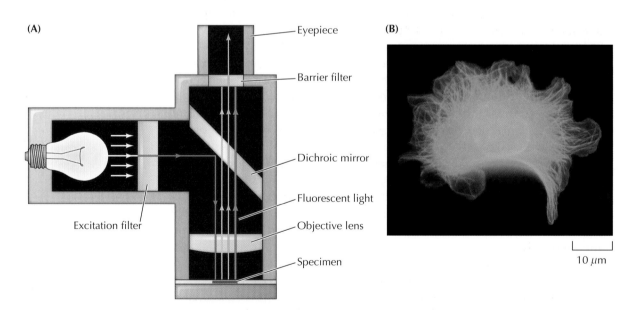

(A)

Eyepiece

Barrier filter

Dichroic mirror

Fluorescent light

Objective lens

Specimen

Excitation filter

(B)

10 μm

Figure 1.27 Fluorescence microscopy (A) Light passes through an excitation filter to select light of the wavelength (e.g., blue) that excites the fluorescent dye. A dichroic mirror then deflects the excitation light down to the specimen. The fluorescent light emitted by the specimen (e.g., green) then passes through the dichroic mirror and a second filter (the barrier filter) to select light of the wavelength emitted by the dye. (B) Fluorescence micrograph of a newt lung cell in which the DNA is stained blue and microtubules in the cytoplasm are stained green.

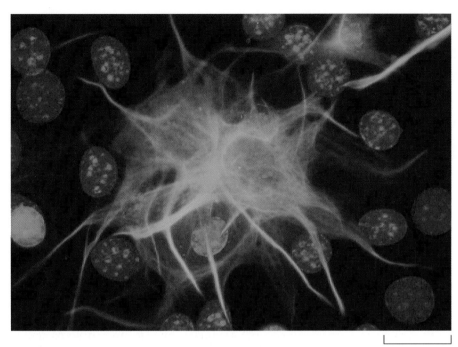

Figure 1.28 Fluorescence microscopy of a protein labeled with GFP A microtubule-associated protein fused to GFP was introduced into mouse neurons in culture and visualized by fluorescence microscopy. Nuclei are stained blue. (From A. Cariboni, 2004. *Nature Cell Biol.* 6: 929.)

5 μm

a molecule that absorbs light at one wavelength and emits light at a second wavelength. This fluorescence is detected by illuminating the specimen with a wavelength of light that excites the fluorescent dye and then using appropriate filters to detect the specific wavelength of light that the dye emits. Fluorescence microscopy can be used to study a variety of molecules within cells. One frequent application is to label antibodies directed against a specific protein with fluorescent dyes, so that the intracellular distribution of the protein can be determined.

A revolutionary advance in fluorescence microscopy came with the use of the **green fluorescent protein (GFP)** of jellyfish to visualize proteins within living cells. GFP can be fused to any protein of interest using standard methods of recombinant DNA, and the GFP-tagged protein can then be expressed in cells and detected by fluorescence microscopy, without the need to fix and stain the cell as would be required for the detection of proteins with antibodies. Because of its versatility, the use of GFP has become widespread in cell biology, and has been used to study the localization of a wide variety of proteins within living cells (**Figure 1.28**). Several related fluorescent proteins with blue, yellow, or red emissions are also available, further expanding the utility of this technique by allowing different proteins to be visualized simultaneously.

A variety of methods have been developed to follow the movement and interactions of GFP-labeled proteins within living cells. One widely used method for studying the movements of GFP-labeled proteins is **fluorescence recovery after photobleaching (FRAP)** (**Figure 1.29**). In this technique, a region

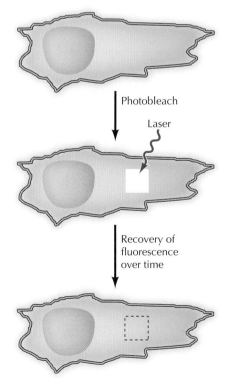

Figure 1.29 Fluorescence recovery after photobleaching (FRAP) A region of a cell expressing a GFP-labeled protein is bleached with a laser. Fluorescence recovers over time as unbleached GFP-labeled molecules diffuse into the bleached region. The rate of recovery of fluorescence therefore provides a measurement of the rate of protein movement within the cell.

No interaction

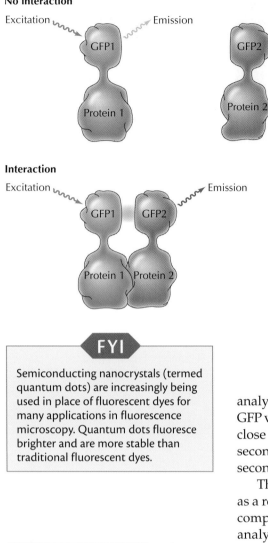

Interaction

FYI

Semiconducting nanocrystals (termed quantum dots) are increasingly being used in place of fluorescent dyes for many applications in fluorescence microscopy. Quantum dots fluoresce brighter and are more stable than traditional fluorescent dyes.

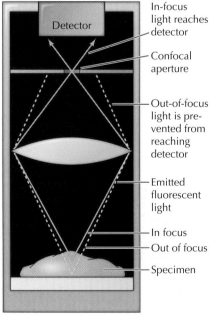

Figure 1.30 Fluorescence resonance energy transfer (FRET) Two proteins are fused to different variants of GFP (GFP1 and GFP2) with distinct wavelengths for excitation and emission, chosen such that the light emitted by GFP1 excites GFP2. Cells are then illuminated with a wavelength of light that excites GFP1. If the proteins do not interact, light emitted by GFP1 will be detected. However, if the proteins do interact, GFP1 will excite GFP2, and light emitted by GFP2 will be detected.

of interest in a cell expressing a GFP-labeled protein is bleached by exposure to high-intensity light. Fluorescence recovers over time due to the movement of unbleached GFP-labeled molecules into the bleached region, allowing the rate at which the protein moves within the cell to be determined.

The interactions of two proteins with one another within a cell can be analyzed by a technique called **fluorescence resonance energy transfer (FRET)** (Figure 1.30). In FRET experiments, the two proteins of interest are coupled to different fluorescent dyes, such as two variants of GFP. The GFP variants are chosen to absorb and emit distinct wavelengths of light, such that the light emitted by one GFP variant excites the second. Interaction between the two proteins can then be detected by illuminating the cell with a wavelength of light that excites the first GFP variant and analyzing the wavelength of emitted light. If the proteins coupled to these GFP variants interact within the cell, the fluorescent molecules will be brought close together and the light emitted by the first GFP variant will excite the second, resulting in emission of light at the wavelength characteristic of the second variant of GFP.

The images obtained by conventional fluorescence microscopy are blurred as a result of out-of-focus fluorescence. These images can be improved by a computational approach called image deconvolution, in which a computer analyzes images obtained from different depths of focus and generates a sharper image than would have been expected from a single focal point. Alternatively, **confocal microscopy** allows images of increased contrast and detail to be obtained by analyzing fluorescence from only a single point in the specimen. A small point of light, usually supplied by a laser, is focused on the specimen at a particular depth. The emitted fluorescent light is then collected using a detector, such as a video camera. Before the emitted light reaches the detector, however, it must pass through a pinhole aperture (called a confocal aperture) placed at precisely the point where light emitted from the chosen depth of the specimen comes to a focus (Figure 1.31). Consequently, only light emitted from the plane of focus is able to reach the detector. Scanning across the specimen generates a two-dimensional image of the plane of focus, a much sharper image than that obtained with

Figure 1.31 Confocal microscopy A pinpoint of light is focused on the specimen at a particular depth, and emitted fluorescent light is collected by a detector. Before reaching the detector, the fluorescent light emitted by the specimen must pass through a confocal aperture placed at the point where light emitted from the chosen depth of the specimen comes into focus. As a result, only in-focus light is detected.

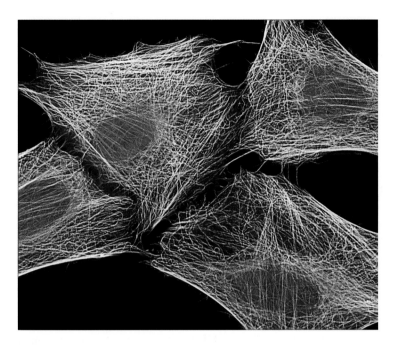

Figure 1.32 Confocal micrograph of human cells Microtubules are stained with a yellow fluorescent dye, actin filaments are stained blue, and nuclei are stained red.

standard fluorescence microscopy (Figure 1.32). Moreover, a series of images obtained at different depths can be used to reconstruct a three-dimensional image of the sample.

Multi-photon excitation microscopy is an alternative to confocal microscopy that can be applied to living cells. The specimen is illuminated with a wavelength of light such that excitation of the fluorescent dye requires the simultaneous absorption of two or more photons (Figure 1.33). The probability of two photons simultaneously exciting the fluorescent dye is only significant at the point in the specimen upon which the input laser beam is focused, so fluorescence is only emitted from the plane of focus of the input light. This highly localized excitation automatically provides three-dimensional resolution, without the need for passing the emitted light through a pinhole aperture, as in confocal microscopy. Moreover, the localization of excitation minimizes damage to the specimen, allowing three-dimensional imaging of living cells.

Figure 1.33 Two-photon excitation microscopy Simultaneous absorption of two photons is required to excite the fluorescent dye. This only occurs at the point in the specimen upon which the input light is focused, so fluorescent light is only emitted from the chosen depth.

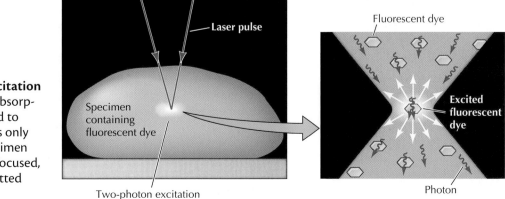

Electron microscopy

Because of the limited resolution of the light microscope, analysis of the details of cell structure has required the use of more powerful microscopic techniques—namely electron microscopy, which was developed in the 1930s and first applied to biological specimens by Albert Claude, Keith Porter, and George Palade in the 1940s and 1950s. The electron microscope can achieve a much greater resolution than that obtained with the light microscope because the wavelength of electrons is shorter than that of light. The wavelength of electrons in an electron microscope can be as short as 0.004 nm—about 100,000 times shorter than the wavelength of visible light. Theoretically, this wavelength could yield a resolution of 0.002 nm, but such a resolution cannot be obtained in practice, because resolution is determined not only by wavelength, but also by the numerical aperture of the microscope lens. Numerical aperture is a limiting factor for electron microscopy because inherent properties of electromagnetic lenses limit their aperture angles to about 0.5 °, corresponding to numerical apertures of only about 0.01. Thus, under optimal conditions, the resolving power of the electron microscope is approximately 0.2 nm. Moreover, the resolution that can be obtained with biological specimens is further limited by their lack of inherent contrast. Consequently, for biological samples the practical limit of resolution of the electron microscope is 1 to 2 nm. Although this resolution is much less than that predicted simply from the wavelength of electrons, it represents more than a hundredfold improvement over the resolving power of the light microscope.

Two types of electron microscopy—transmission and scanning—are widely used to study cells. In principle, **transmission electron microscopy** is similar to the observation of stained cells with the bright-field light microscope. Specimens are fixed and stained with salts of heavy metals, which provide contrast by scattering electrons. A beam of electrons is then passed through the specimen and focused to form an image on a fluorescent screen. Electrons that encounter a heavy metal ion as they pass through the sample are deflected and do not contribute to the final image, so stained areas of the specimen appear dark.

Specimens to be examined by transmission electron microscopy can be prepared by either positive or negative staining. In positive staining, tissue specimens are cut into thin sections and stained with heavy metal salts (such as osmium tetroxide, uranyl acetate, and lead citrate) that react with lipids, proteins, and nucleic acids. These heavy metal ions bind to a variety of cell structures, which consequently appear dark in the final image (**Figure 1.34**). Alternative positive-staining procedures can also be used to identify specific macromolecules within cells. For example, antibodies labeled with electron-dense heavy metals (such as gold particles) are frequently used to determine the subcellular location of specific proteins in the electron microscope. This method is similar to the use of antibodies labeled with fluorescent dyes in fluorescence microscopy. Three-dimensional views of structures with resolutions of 2–10 nm can also be obtained using the technique of **electron tomography**, which generates three-dimensional images by computer analysis of multiple two-dimensional images obtained over a range of viewing directions.

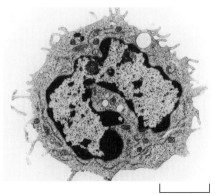

5 μm

Figure 1.34 Positive staining Transmission electron micrograph of a white blood cell positively stained by heavy metal salts.

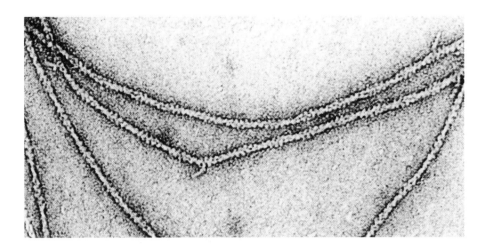

Figure 1.35 Negative staining
Transmission electron micrograph
of negatively stained actin filaments.
(Courtesy of Roger Craig, University
of Massachusetts Medical Center.)

Negative staining is useful for the visualization of intact biological structures such as bacteria, isolated subcellular organelles, and macromolecules (Figure 1.35). In this method, the biological specimen is deposited on a supporting film, and a heavy metal stain is allowed to dry around its surface. The unstained specimen is then surrounded by a film of electron-dense stain, producing an image in which the specimen appears light against a stained dark background.

Metal shadowing is another technique used in transmission electron microscopy to visualize the surface of isolated subcellular structures or macromolecules (Figure 1.36). The specimen is coated with a thin layer of evaporated metal, such as platinum. The metal is sprayed onto the specimen from an angle so that surfaces of the specimen that face the source of evaporated metal molecules are coated more heavily than others. This differential

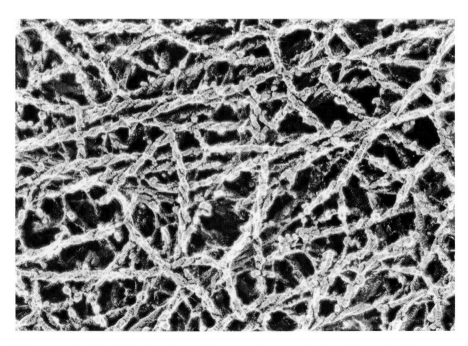

Figure 1.36 Metal shadowing   Electron micrograph of actin/myosin filaments of the cytoskeleton prepared by metal shadowing.

(A)

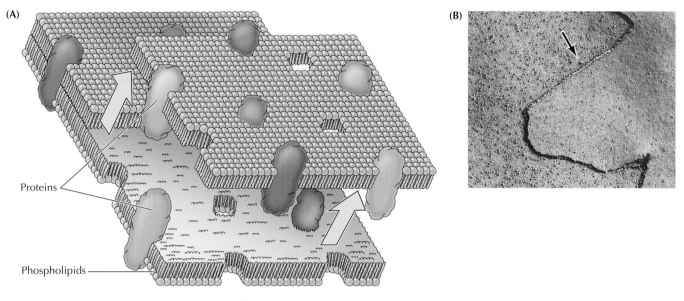

Proteins

Phospholipids

(B)

Figure 1.37 Freeze fracture   (A) Freeze fracture splits the lipid bilayer, leaving proteins embedded in the membrane associated with one of the two membrane halves. (B) Micrograph of freeze-fractured plasma membranes of two adjacent cells. Proteins that span the bilayer appear as intramembranous particles (arrow).

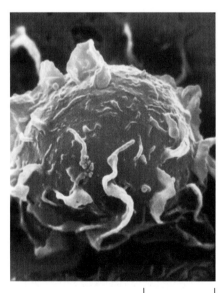

5 μm

Figure 1.38 Scanning electron microscopy   Scanning electron micrograph of a macrophage, where the electron beam scans across, not through, the specimen.

coating creates a shadow effect, giving the specimen a three-dimensional appearance in electron micrographs.

The preparation of samples by **freeze fracture**, in combination with metal shadowing, has been particularly important in studies of membrane structure. Specimens are frozen in liquid nitrogen (at –196°C) and then fractured with a knife blade. This process frequently splits the lipid bilayer, revealing the interior faces of a cell membrane (Figure 1.37). The specimen is then shadowed with platinum, and the biological material is dissolved with acid, producing a metal replica of the surface of the sample. Examination of such replicas in the electron microscope reveals many surface bumps, corresponding to proteins that span the lipid bilayer. A variation of freeze fracture called freeze etching allows visualization of the external surfaces of cell membranes in addition to their interior faces.

The second type of electron microscopy, **scanning electron microscopy**, is used to provide a three-dimensional image of cells (Figure 1.38). In scanning electron microscopy the electron beam does not pass through the specimen. Instead, the surface of the cell is coated with a heavy metal, and a beam of electrons is used to scan across the specimen. Electrons that are scattered or emitted from the sample surface are collected to generate a three-dimensional image as the electron beam moves across the cell.

Super-resolution light microscopy

Although electron microscopy provides higher resolution than light microscopy, it suffers from several drawbacks. Electron microscopes are expensive and technically difficult to operate, the fixation and staining procedures needed in electron microscopy can introduce artifacts, and the electron microscope cannot be used for examination of living specimens.

An exciting advance in recent years has thus been the development of **super-resolution light microscopy** techniques that break the diffraction barrier and increase the resolution of fluorescence microscopy to the range of 10-100 nm: about tenfold less than the theoretical limit of resolution of the light microscope.

Several methods of super-resolution light microscopy have been developed based on the use of fluorescent probes that shift the limit of resolution from the wavelength of visible light to the molecular level. A good example is provided by the method known as STORM (stochastic optical reconstruction microscopy), which was developed in 2006 and has a resolution of approximately 20 nm. The principle underlying STORM and related methods is to produce high resolution by compiling individual images from thousands to millions of individual fluorescent molecules (**Figure 1.39**). In standard fluorescence microscopy, all of the fluorescent probes in a sample fluoresce at the same time. The fluorescent images of the individual molecules overlap to yield a blurred image with resolution limited by the diffraction of light. In

The 2014 Nobel prize in Chemistry was awarded to Eric Betzig, Stefan Hell, and William Moerner for the development of super-resolution light microscopy.

Video 1.2

sites.sinauer.com/cooper7e/v1.2
Super-Resolution Light Microscopy
Super-resolution microscopy techniques allow the capture of images with a higher resolution than the diffraction limit of visible light.

(A)

Conventional fluorescence

STORM

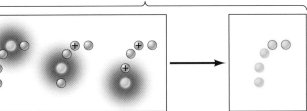

Individual molecules are all fluorescent— too close to resolve.

Random activation of individual fluorescent molecules…

…results in a composite super-resolution image.

(B)

Conventional

3 μm   500 nm   500 nm

STORM

3 μm   500 nm   500 nm

Figure 1.39 Super-resolution light microscopy (A) In conventional fluorescence microscopy, all of the fluorescent molecules in a sample fluoresce at the same time, yielding a blurred image. In STORM, only a small random fraction of the probes are fluorescent at any one time. Multiple images are obtained over time such that the individual fluorescent molecules can be resolved from one another and a super-resolution composite image is constructed. (B) Comparison of conventional and STORM microscopy of microtubules. (From M. Bates et al., 2007. *Science* 317: 1749; courtesy of B. Huang, Harvard University.)

contrast, STORM utilizes fluorescent probes that can be switched between the dark and fluorescent states. At any one time, only a small random fraction of the probes are fluorescent, such that the individual fluorescent molecules can be resolved from one another. Capturing many such images over time yields a series of snapshots, with different individual molecules emitting light in each picture. These multiple images can then be used to construct a composite image whose resolution is limited by the precision with which each fluorescent molecule is located, rather than by the diffraction of light.

The methods of super-resolution microscopy have been expanded to include three-dimensional imaging, simultaneous imaging of multiple different molecules using distinctly colored fluorescent probes, and super-resolution imaging of live cells. In addition to cytoskeletal structures, super-resolution microscopy has revealed fine-detailed structures of the nucleus, chromatin, sites of cell attachment, and the plasma membrane. It is likely that future research will result in the development of methods with even higher resolution than those presently available, allowing visualization of molecular events within living cells.

Subcellular fractionation

Although the electron microscope has allowed detailed visualization of cell structure, microscopy alone is not sufficient to define the functions of the various components of eukaryotic cells. To address many of the questions concerning the function of subcellular organelles, it has proven necessary to isolate the organelles of eukaryotic cells in a form that can be used for biochemical studies. This is usually accomplished by **differential centrifugation**—a method developed largely by Albert Claude, Christian de Duve, and their colleagues in the 1940s and 1950s to separate the components of cells on the basis of their size and density.

The first step in subcellular fractionation is the disruption of the plasma membrane under conditions that do not destroy the internal components of the cell. Several methods are used, including sonication (exposure to high-frequency sound), grinding in a mechanical homogenizer, or treatment with a high-speed blender. All these procedures break the plasma membrane and the endoplasmic reticulum into small fragments, while leaving other components of the cell (such as nuclei, lysosomes, peroxisomes, mitochondria, and chloroplasts) intact.

The suspension of broken cells (called a lysate or homogenate) is then fractionated into its components by a series of centrifugations in an **ultracentrifuge**, which rotates samples at very high speeds (over 100,000 rpm) to produce forces up to 500,000 times greater than gravity. This force causes cell components to move toward the bottom of the centrifuge tube and form a pellet (a process called sedimentation) at a rate that depends on their size and density, with the largest and heaviest structures sedimenting most rapidly (**Figure 1.40**). Usually the cell homogenate is first centrifuged at a low speed, which sediments only unbroken cells and the largest subcellular structures—the nuclei. Thus, an enriched fraction of nuclei can be recovered

Animation 1.1

sites.sinauer.com/cooper7e/a1.1
Subcellular Fractionation After cells are broken apart, the subcellular constituents are fractionated using different types of centrifugation.

Figure 1.40 Subcellular fractionation Cells are disrupted (lysed) and subcellular components are separated by a series of centrifugations at increasing speeds. Following each centrifugation, the organelles that have sedimented to the bottom of the tube are recovered in the pellet. The supernatant (remaining solution) is then recentrifuged at a higher speed to sediment the next-largest organelles. ▶

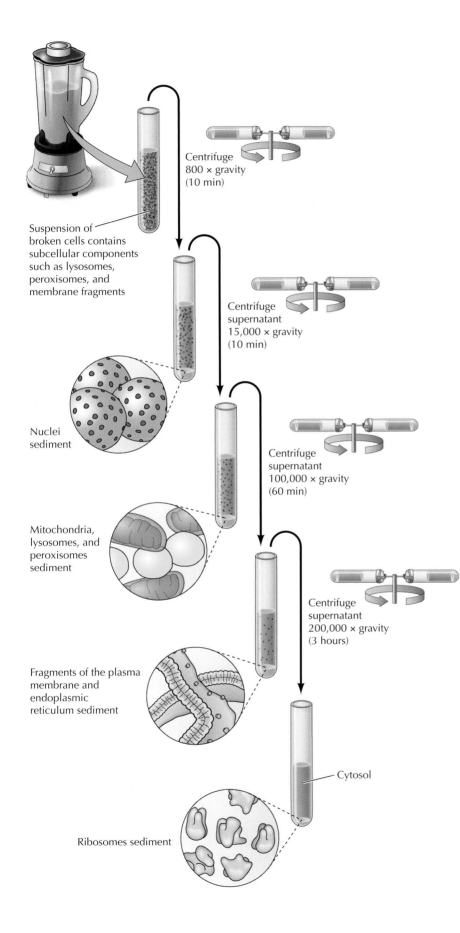

Suspension of broken cells contains subcellular components such as lysosomes, peroxisomes, and membrane fragments

Centrifuge
800 × gravity
(10 min)

Centrifuge
supernatant
15,000 × gravity
(10 min)

Nuclei sediment

Centrifuge
supernatant
100,000 × gravity
(60 min)

Mitochondria, lysosomes, and peroxisomes sediment

Centrifuge
supernatant
200,000 × gravity
(3 hours)

Fragments of the plasma membrane and endoplasmic reticulum sediment

Cytosol

Ribosomes sediment

from the pellet of such a low-speed centrifugation while the other cell components remain suspended in the supernatant (the remaining solution). The supernatant is then centrifuged at a higher speed to sediment mitochondria, chloroplasts, lysosomes, and peroxisomes. Recentrifugation of the supernatant at an even higher speed sediments fragments of the plasma membrane and the endoplasmic reticulum. A fourth centrifugation at a still higher speed sediments ribosomes, leaving only the soluble portion of the cytoplasm (the cytosol) in the supernatant.

The fractions obtained from differential centrifugation correspond to enriched, but still not pure, organelle preparations. A greater degree of purification can be achieved by **density-gradient centrifugation**, in which organelles are separated by sedimentation through a gradient of a dense substance, such as sucrose. In **velocity centrifugation**, the starting material is layered on top of the sucrose gradient (**Figure 1.41**). Particles of different sizes sediment through the gradient at different rates, moving as discrete bands. Following centrifugation, the collection of individual fractions of the

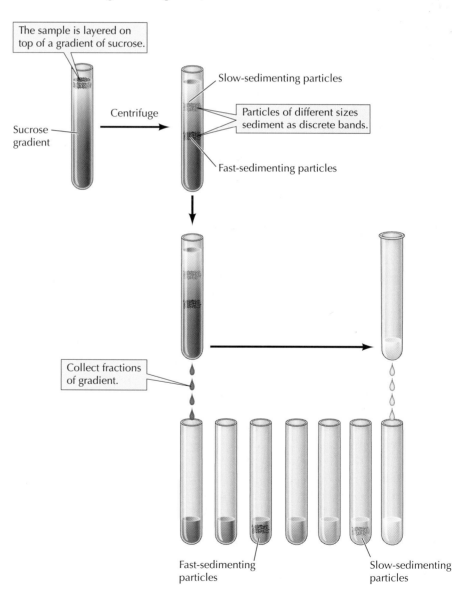

Figure 1.41 Velocity centrifugation in a density gradient   The sample is layered on top of a gradient of sucrose, and particles of different sizes sediment through the gradient as discrete bands. The separated particles can then be collected in individual fractions of the gradient, which can be obtained simply by puncturing the bottom of the centrifuge tube and collecting drops.

gradient provides sufficient resolution to separate organelles of similar size, such as mitochondria, lysosomes, and peroxisomes.

Equilibrium centrifugation in density gradients can be used to separate subcellular components on the basis of their buoyant density, independent of their size and shape. In this procedure, the sample is centrifuged in a gradient containing a high concentration of sucrose or cesium chloride. Rather than being separated on the basis of their sedimentation velocity, the sample particles are centrifuged until they reach an equilibrium position at which their buoyant density is equal to that of the surrounding sucrose or cesium chloride solution. Such equilibrium centrifugations are useful in separating different types of membranes from one another and are sufficiently sensitive to separate macromolecules that are labeled with different isotopes. A classic example, discussed in Chapter 4, is the analysis of DNA replication by separating DNA molecules containing heavy and light isotopes of nitrogen (^{15}N and ^{14}N) using equilibrium centrifugation in cesium chloride gradients.

Growth of animal cells in culture

The ability to study cells depends largely on how readily they can be grown and manipulated in the laboratory. Although the process is technically far more difficult than the culture of bacteria or yeasts, a wide variety of animal and plant cells can be grown and manipulated in culture. Such *in vitro* cell culture systems have enabled scientists to study cell growth and differentiation, as well as to perform genetic manipulations required to understand gene structure and function.

Animal cell cultures are initiated by the dispersion of a piece of tissue into a suspension of its component cells, which is then added to a culture dish containing nutrient media. Most animal cell types, such as fibroblasts and epithelial cells, attach and grow on the plastic surface of dishes used for cell culture (Figure 1.42). Because they contain rapidly growing cells, embryos or tumors are frequently used as starting material. Embryo fibroblasts grow particularly well in culture and consequently are one of the most widely studied types of animal cells. Under appropriate conditions, however, many specialized cell types can also be grown in culture, allowing their differentiated properties to be studied in a controlled experimental environment. **Embryonic stem (ES) cells** are a particularly notable example. These cells are established in culture from early embryos and maintain their ability to differentiate into all of the cell types present in adult organisms. Consequently, embryonic stem cells have played an important role in studying the functions of a variety of genes in mouse development, as well as offering the possibility of contributing to the treatment of human diseases by providing a source of tissue for transplantation therapies.

The culture media required for the propagation of animal cells are much more complex than the minimal media sufficient to support the growth of bacteria and yeasts. In addition to salts and glucose, the media used for animal cell cultures contain various amino acids and vitamins, which the cells cannot make for themselves. The growth media for most animal cells in culture also include serum, which serves as a source of polypeptide growth factors, which are required to stimulate cell division. Several such growth factors have been identified. They serve as critical regulators of cell growth and differentiation in multicellular organisms, providing signals by which

10 μm

Figure 1.42 Animal cells in culture
Scanning electron micrograph of human fibroblasts attached to the surface of a culture dish (artificial color has been added).

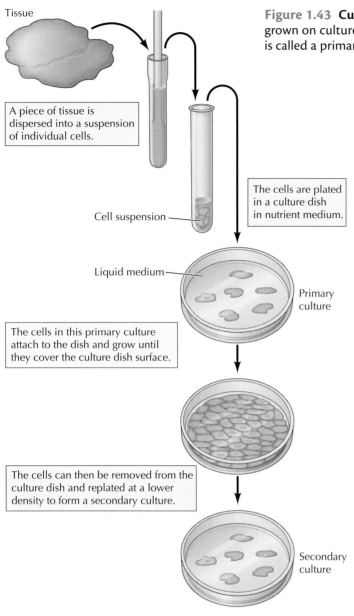

Tissue

A piece of tissue is dispersed into a suspension of individual cells.

Cell suspension

The cells are plated in a culture dish in nutrient medium.

Liquid medium

Primary culture

The cells in this primary culture attach to the dish and grow until they cover the culture dish surface.

The cells can then be removed from the culture dish and replated at a lower density to form a secondary culture.

Secondary culture

Figure 1.43 Culture of animal cells   Cells obtained from a tissue are grown on culture dishes in nutrient medium. The first culture established is called a primary culture.

different cells communicate with each other. For example, an important function of skin fibroblasts in the intact animal is to proliferate when needed to repair damage resulting from a cut or wound. Their division is triggered by a growth factor released from platelets during blood clotting, thereby stimulating proliferation of fibroblasts in the neighborhood of the damaged tissue. The identification of individual growth factors has made possible the culture of a variety of cells in serum-free media (media in which serum has been replaced by the specific growth factors required for proliferation of the cells in question).

The initial cell cultures established from a tissue are called **primary cultures** (Figure 1.43). The cells in a primary culture usually grow until they cover the culture dish surface. They can then be removed from the dish and replated at a lower density to form secondary cultures. This process can be repeated many times, but most normal cells cannot be grown in culture indefinitely. For example, normal human fibroblasts can usually be cultured for 50 to 100 population doublings, after which they stop growing and die. In contrast, embryonic stem cells and cells derived from tumors frequently proliferate indefinitely in culture and are referred to as permanent or immortal **cell lines**. In addition, a number of immortalized rodent cell lines have been isolated from cultures of normal fibroblasts. Instead of dying as most of their counterparts do, a few cells in these cultures continue proliferating indefinitely, forming cell lines like those derived from tumors. Such permanent cell lines have been particularly useful for many types of experiments because they provide a continuous and uniform source of cells that can be manipulated, cloned, and indefinitely propagated in the laboratory. The first human cell line to be established were HeLa cells, which were isolated from a cervical cancer in 1951 and have been used in thousands of laboratories studying many aspects of human cell biology.

Even under optimal conditions, the division time of most actively growing animal cells is on the order of 20 hours—ten times longer than the division time of yeasts. Consequently, experiments with cultured animal cells are more difficult and take much longer than those with bacteria or yeasts. For example, the growth of a visible colony of animal cells from a single cell takes a week or more, whereas colonies of *E. coli* or yeast develop from single

Key Experiment

HeLa Cells

Tissue Culture Studies of the Proliferative Capacity of Cervical Carcinoma and Normal Epithelium

George O. Gey, Ward D. Coffman and Mary T. Kubicek

Johns Hopkins Hospital and University, Baltimore, MD

Cancer Research, Volume 12, 1952, pages 264–265

The Context

The earliest cell cultures involved the growth of cells from fragments of tissue that were embedded in clots of plasma— a culture system that was far from suitable for experimental analysis. In the 1940s, a major advance was made by the establishment of cell lines that grew from isolated cells attached to the surface of culture dishes. The first of these cell lines was a mouse line called L cells that was established in 1942. However, it proved considerably more difficult to establish cell lines of human origin, which were highly desired as a model for cancer research. Despite many attempts, scientists were unable to grow human cells in culture for more than a few weeks. The breakthrough came in 1951, when George Gey and his colleagues established the first human cell line, HeLa cells, from tissue of a cervical cancer.

The Experiments

HeLa cells were cultured from a biopsy of a cervical cancer taken

from a 30-year-old woman, Henrietta Lacks, in February 1951. Like many cancer specimens that Gey and his colleagues had previously attempted to establish in culture, the tissue sample from Henrietta Lacks was plated in medium containing chicken plasma, calf embryo extract and human umbilical blood. However, unlike many previous attempts, the cells from Henrietta Lacks' cancer grew rapidly and continued growing in culture, making HeLa cells the first established human cancer cell line. The story behind this cell line and much of the history of early cell culture is described in *The Immortal Life of Henrietta Lacks* by Rebecca Skloot, 2010.

The Impact

HeLa cells have become the most widely used cell line for cancer research and other studies of human cell biology. One of the first major uses of HeLa cells was in development of the polio vaccine, because it was found that the polio virus grew

George O. Gey

readily in these cells. They continue to be used today as a model system for virtually all aspects of human molecular and cellular biology, including studies of DNA replication, gene expression, cell division, virology, cancer, and cell signaling. As one indication of their importance, HeLa cells have been used in more than 70,000 published research papers. Moreover, many other human cell lines have now been established and these lines form the basis for contemporary studies of human cell biology.

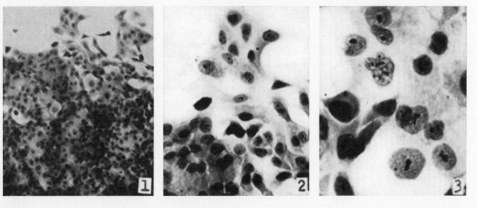

Photomicrographs of HeLa cells at magnifications of 65, 160, and 250×.
(From Scherer, W. F., J. T. Syverton and G. O. Gey. 1953. *J. Exp. Med.* 97: 695–710.)

cells overnight. Nonetheless, genetic manipulations of animal cells in culture have been indispensable to our understanding of cell structure and function.

Culture of plant cells

Plant cells can also be cultured in nutrient media containing appropriate growth regulatory molecules. In contrast to the polypeptide growth factors that regulate the proliferation of most animal cells, the growth regulators of plant cells are small molecules that can pass through the plant cell wall. When provided with appropriate mixtures of these growth regulatory molecules,

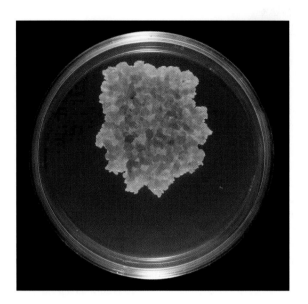

Figure 1.44 **Plant cells in culture** An undifferentiated mass of black Mexican sweet corn cells (a callus) growing on a solid medium.

many types of plant cells proliferate in culture, producing a mass of undifferentiated cells called a **callus** (Figure 1.44).

It is noteworthy that many plant cells are capable of forming all of the different cell types and tissues ultimately needed to regenerate an entire plant. Consequently, by appropriate manipulation of nutrients and growth regulatory molecules, undifferentiated plant cells in culture can be induced to form a variety of plant tissues, including roots, stems, and leaves. In many cases, even an entire plant can be regenerated from a single cultured cell. In addition to its theoretical interest, the ability to produce a new plant from a single cell that has been manipulated in culture makes it easy to introduce genetic alterations into plants, opening important possibilities for agricultural genetic engineering.

Viruses

Viruses are intracellular parasites that cannot replicate on their own. They reproduce by infecting host cells and usurping the cellular machinery to produce more virus particles. In their simplest forms, viruses consist only of genomic nucleic acid (either DNA or RNA) surrounded by a protein coat (Figure 1.45). Viruses are important in molecular and cellular biology because they provide simple systems that can be used to investigate the functions of cells. Because virus replication depends on the metabolism of the infected cells, studies of viruses have revealed many fundamental aspects of cell biology. Studies of bacterial viruses have contributed substantially to our understanding of the basic mechanisms of molecular genetics, and

(A)

(B)

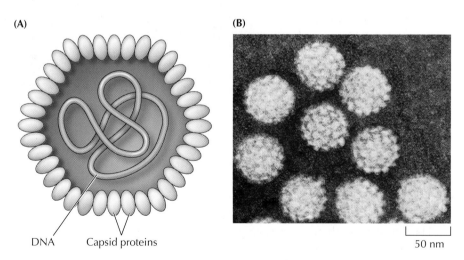

DNA   Capsid proteins

50 nm

Figure 1.45 **Structure of an animal virus** (A) Papillomavirus particles contain a small circular DNA molecule enclosed in a protein coat (the capsid). (B) Electron micrograph of human papillomavirus particles. Artificial color has been added.

Molecular Medicine

Viruses and Cancer

The Disease

Cancer is a family of diseases characterized by uncontrolled cell proliferation. The growth of normal animal cells is carefully regulated to meet the needs of the complete organism. In contrast, cancer cells grow in an unregulated manner, ultimately invading and interfering with the function of normal tissues and organs. Cancer is the second most common cause of death (next to heart disease) in the United States. Approximately one out of every three Americans will develop cancer at some point in life and, in spite of major advances in treatment, nearly one out of every four Americans ultimately die of this disease. Understanding the causes of cancer and developing more effective methods of cancer treatment therefore represent major goals of medical research.

Molecular and Cellular Basis

Cancer is now known to result from mutations in the genes that normally control cell proliferation. The major insights leading to identification of these genes came from studies of viruses that cause cancer in animals, the prototype of which was isolated by Peyton Rous in 1911. Rous found that sarcomas (cancers of connective tissues) in chickens could be transmitted by a virus, now known as Rous sarcoma virus, or RSV. Because RSV is a retrovirus with a genome of only 10,000 base pairs, it can be subjected to molecular analysis much more readily than the complex genomes of chickens or other animal cells. Such studies eventually led to identification of a specific cancer-causing gene (oncogene) carried by

the virus, and to the discovery of related genes in normal cells of all vertebrate species, including humans. Some cancers in humans are now known to be caused by viruses; others result from mutations in normal cell genes similar to the oncogene first identified in RSV.

Prevention and Treatment

The human cancers that are caused by viruses include cervical and other anogenital cancers (papillomaviruses), liver cancer (hepatitis B and C viruses), and some types of lymphomas (Epstein-Barr virus and human T-cell lymphotropic virus). Together, these virus-induced cancers account for about 20% of worldwide cancer incidence. In principle, these cancers could be prevented by vaccination against the responsible viruses, and considerable progress in this area has been made by the development of effective vaccines against hepatitis B virus and human papillomaviruses.

Other human cancers are caused by mutations in normal cell genes, most of which occur during the lifetime of the individual rather than from inheritance. Studies of cancer-causing viruses have led to the identification of many of the genes responsible for non-virus-induced cancers, and to an understanding of the molecular mechanisms responsible for cancer development. Major efforts are now under way to use these insights into the molecular and cellular biology of cancer to develop new approaches to cancer treatment. Indeed, the first designer drug effective in treating a human cancer (the drug imatinib or Gleevec, discussed in Chapter 19) was developed against a gene very similar to the oncogene of RSV.

Reference

Rous, P. 1911. A sarcoma of the fowl transmissible by an agent separable from the tumor cells. *J. Exp. Med.* 13: 397–411.

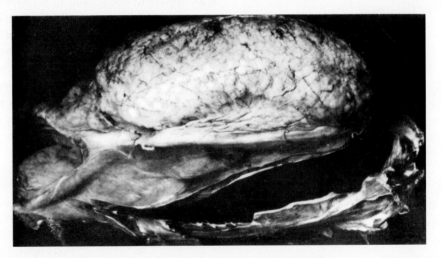

The transplantable tumor from which Rous sarcoma virus was isolated.

experiments with a plant virus (tobacco mosaic virus) first demonstrated the genetic potential of RNA. Animal viruses have provided particularly sensitive probes for investigations of various activities of eukaryotic cells.

The rapid growth and small genome size of bacteria make them excellent subjects for experiments in molecular biology, and bacterial viruses (**bacteriophages**) have simplified the study of bacterial genetics even further. One of the most important bacteriophages is T4, which infects and replicates in *E. coli*. Infection with a single particle of T4 leads to the formation of approximately 200 progeny virus particles in 20 to 30 minutes. The initially

Figure 1.46 Bacteriophage plaques T4 plaques are visible on a lawn of *E. coli*. Each plaque arises by the replication of a single virus particle.

Animal viruses are often used in gene therapy as carriers of genes to be introduced into cells (see Chapter 4).

infected cell then bursts (lyses), releasing progeny virus particles into the medium, where they can infect new cells. In a culture of bacteria growing on agar medium, the replication of T4 leads to the formation of a clear area of lysed cells (a plaque) in the lawn of bacteria (**Figure 1.46**). Just as infectious virus particles are easy to grow and assay, viral mutants—for example, viruses that will grow in one strain of *E. coli* but not another—are easy to isolate. Thus, T4 is manipulated even more readily than *E. coli* for studies of molecular genetics. Moreover, the genome of T4 is 23 times smaller than that of *E. coli*—approximately 0.2 million base pairs—further facilitating genetic analysis. Some other bacteriophages have even smaller genomes—the simplest consisting of RNA molecules of only about 3600 nucleotides. Bacterial viruses have thus provided extremely facile experimental systems for molecular genetics, and many fundamental principles of molecular biology are based on studies with these viruses.

Because of the increased complexity of the animal cell genome, viruses have been even more important in studies of animal cells than in studies of bacteria. Many animal viruses replicate and can be assayed by plaque formation in cell cultures, much as bacteriophages can. Moreover, the genomes of animal viruses are similar in complexity to those of bacterial viruses (ranging from approximately 3000 to 300,000 base pairs), so animal viruses are far more manageable than are their host cells.

There are many diverse animal viruses, each containing either DNA or RNA as their genetic material (**Table 1.3**). Most viruses with RNA genomes replicate by synthesizing new RNA copies of their genomes from RNA templates in infected cells. However, one family of animal viruses—the **retroviruses**—contain RNA genomes in their virus particles but synthesize a DNA copy of their genome in infected cells. These viruses provide a good example of the importance of viruses as models, because studies of the retroviruses are what first demonstrated the synthesis of DNA from RNA templates—a fundamental mode of genetic information transfer now known to occur in all eukaryotic cells. Other examples in which animal viruses have provided important models for investigations of their host cells include studies of DNA replication, transcription, RNA processing, and protein transport and secretion.

It is particularly noteworthy that infection by some animal viruses, rather than killing the host cell, converts a normal cell into a cancer cell. Studies of such cancer-causing viruses, first described by Peyton Rous in 1911, not only have provided the basis for our current understanding of cancer at the level of cell and molecular biology, but also have led to the elucidation of many of the molecular mechanisms that control animal cell growth and differentiation.

Table 1.3 Examples of Animal Viruses

Virus family	Representative member	Genome size (thousands of base pairs)
RNA genomes		
Picornaviruses	Poliovirus	7–8
Togaviruses	Rubella virus	12
Flaviviruses	Yellow fever virus	10
Paramyxoviruses	Measles virus	16–20
Orthomyxoviruses	Influenza virus	14
Retroviruses	Human immunodeficiency virus	9
DNA genomes		
Hepadnaviruses	Hepatitis B virus	3.2
Papovaviruses	Human papillomavirus	5–8
Adenoviruses	Adenovirus	36
Herpesviruses	Herpes simplex virus	120–200
Poxviruses	Vaccinia virus	130–280

SUMMARY	KEY TERMS

The Origin and Evolution of Cells

- *The first cell:* All present-day cells, both prokaryotes and eukaryotes, are descended from a single ancestor. The first cell is thought to have arisen at least 3.8 billion years ago as a result of enclosure of self-replicating RNA in a phospholipid membrane.

- *The evolution of metabolism:* The earliest reactions for the generation of metabolic energy were a form of anaerobic glycolysis. Photosynthesis then evolved, followed by oxidative metabolism.

- *Present-day prokaryotes:* Present-day prokaryotes are divided into two groups, the archaebacteria and the bacteria, which diverged early in evolution.

- *Eukaryotic cells:* Eukaryotic cells, which are larger and more complex than prokaryotic cells, contain a nucleus and cytoplasmic organelles.

- *The origin of eukaryotes:* Eukaryotic cells are thought to have evolved from symbiotic associations of prokaryotes. The genome of eukaryotes may have arisen from a fusion of bacterial and archaebacterial genomes.

- *The development of multicellular organisms:* The simplest eukaryotes are unicellular organisms, such as yeasts and amoebas. Multicellular organisms evolved from associations between such unicellular eukaryotes, and division of labor led to the development of the many kinds of specialized cells that make up present-day plants and animals. See Video 1.1.

Key terms (column)

prokaryotic cell, eukaryotic cell, RNA world, gene, transcription, translation, phospholipid, amphipathic, hydrophobic, hydrophilic

adenosine 5′-triphosphate (ATP), glycolysis, photosynthesis, oxidative metabolism

archaebacteria, bacteria, cyanobacteria, *Escherichia coli (E. coli)*, cell wall, plasma membrane, ribosome

nucleus, mitochondria, chloroplasts, lysosome, peroxisome, vacuole, endoplasmic reticulum, Golgi apparatus, cytoskeleton

endosymbiosis

yeast, *Saccharomyces cerevisiae*, epithelial cell, fibroblast, erythrocyte, granulocyte, monocyte, macrophage, lymphocyte, neuron

Cells as Experimental Models

- **E. coli:** Because of their genetic simplicity and ease of study, bacteria such as *E. coli* are particularly useful for investigation of fundamental aspects of biochemistry and molecular biology.

- **Yeasts:** As the simplest eukaryotic cells, yeasts are an important model for studying various aspects of eukaryotic cell biology.

- **Caenorhabditis elegans:** The nematode *C. elegans* is a simple multicellular organism that serves as an important model in developmental biology.

- **Drosophila melanogaster:** Because of extensive genetic analysis, studies of the fruit fly *Drosophila* have led to major advances in understanding animal development.

- **Arabidopsis thaliana:** The small flowering plant *Arabidopsis* is widely used as a model for studies of plant molecular biology and development.

- **Vertebrates:** Many kinds of vertebrate cells can be grown in culture, where they can be studied under controlled laboratory conditions. Specialized cell types, such as neurons and muscle cells, provide useful models for investigating particular aspects of cell biology. The frog *Xenopus laevis* and the zebra-fish are important models for studies of early vertebrate development, while the mouse is a mammalian species suitable for genetic analysis.

Caenorhabditis elegans

Drosophila melanogaster

Arabidopsis thaliana

Xenopus laevis, zebrafish

SUMMARY	KEY TERMS

Tools of Cell Biology

- *Light microscopy:* A variety of methods are used to visualize cells and sub-cellular structures and to determine the intracellular localization of specific molecules using the light microscope.

resolution, bright-field microscopy, phase-contrast microscopy, differential interference-contrast microscopy, fluorescence microscopy, green fluorescent protein (GFP), fluorescence recovery after photobleaching (FRAP), fluorescence resonance energy transfer (FRET), confocal microscopy, multi-photon excitation microscopy

- *Electron microscopy:* Electron microscopy, with a resolution that is approximately a hundredfold greater than that of light microscopy, is used to analyze details of cell structure.

transmission electron microscopy, electron tomography, metal shadowing, freeze fracture, scanning electron microscopy

- *Super-resolution light microscopy:* Several methods of super-resolution fluorescence microscopy break the diffraction barrier and increase the resolution of light microscopy to 10-100 nm. See Video 1.2.

super-resolution microscopy

- *Subcellular fractionation:* The organelles of eukaryotic cells can be isolated for biochemical analysis by differential centrifugation. See Animation 1.1.

differential centrifugation, ultracentrifuge, density-gradient centrifugation, velocity centrifugation, equilibrium centrifugation

- *Growth of animal cells in culture:* The propagation of animal cells in culture has allowed studies of the mechanisms that control cell growth and differentiation.

embryonic stem (ES) cell, primary culture, cell line

- *Culture of plant cells:* Cultured plant cells can differentiate to form specialized cell types and, in some cases, regenerate entire plants.

callus

- *Viruses:* Viruses provide simple models for studies of cell function.

bacteriophage, retrovirus

Refer To
The Cell
Companion Website
sites.sinauer.com/cooper7e
for quizzes, animations, videos, flashcards, and other study resources.

Questions

1. What did Stanley Miller's experiments show about the formation of organic molecules?

2. What kind of macromolecule is capable of directing its own replication?

3. Discuss the evidence that mitochondria and chloroplasts originated from bacteria that were engulfed by the precursor of eukaryotic cells.

4. Why is the evolution of photosynthesis thought to have favored the subsequent evolution of oxidative metabolism?

5. Given that the diameter of a *S. aureus* bacterial cell is 1 μm and the diameter of a human macrophage is 50 μm, how many *S. aureus* cells can fit inside a single human macrophage? Assume that the cells are spherical.

6. Which model organism provides the simplest system for studying eukaryotic DNA replication?

7. You are studying a gene involved in mammalian embryonic development. Which model organism would be best suited for your studies?

8. What resolution can be obtained with a light microscope if the specimen is viewed through air rather than through oil? Assume that the wavelength of visible light is 0.5 μm.

9. You are about to purchase a research microscope for your laboratory, and you have a choice between two different objective lenses. One has a magnification of 100 and a numerical aperture (*NA*) of 1.1. The other has a magnification of 60 and a numerical aperture of 1.3. Assuming that price is not a concern, which of these objectives would you choose?

10. What advantage does the use of green fluorescent protein (GFP) have over the use of fluorescent-labeled antibodies for studying the location and movement of a protein in cells?

11. Identify the different characteristics or properties of organelles that allow separation by velocity centrifugation as compared with equilibrium centrifugation in a sucrose gradient.

12. Why is serum usually required in the media used to grow animal cells in culture?

13. Distinguish between primary cell cultures and immortal cell lines.

14. Why is the ability to culture embryonic stem cells important?

References and Further Reading   (Key review articles for each major section are highlighted in **bold**.)

The Origin and Evolution of Cells

Andersson, S. G. E., A. Zomorodipour, J. O. Andersson, T. Sicheritz-Ponten, U. C. M. Alsmark, R. M. Podowski, A. K. Naslund, A.-S. Eriksson, H. H. Winkler and C. G. Kurland. 1998. The genome sequence of *Rickettsia prowazekii* and the origin of mitochondria. *Nature* 396: 133–140. [P]

Cech, T. R. 1986. A model for the RNA-catalyzed replication of RNA. *Proc. Natl. Acad. Sci. USA* 83: 4360–4363. [P]

Darnell, J. E. and W. F. Doolittle. 1986. Speculations on the early course of evolution. *Proc. Natl. Acad. Sci. USA* 83: 1271–1275. [P]

De Duve, C. 2007. The origin of eukaryotes: a reappraisal. *Nature Rev. Genet.* 8: 395–403. [R]

Dyall, S. D., M. T. Brown and P. J. Johnson. 2004. Ancient invasions: From endosymbionts to organelles. *Science* 304: 253–257. [R]

Joyce, G. F. 2007. A glimpse of biology's first enzyme. *Science* 315: 1507–1508. [R]

Kasting, J. F. and J. L. Siefert. 2002. Life and the evolution of Earth's atmosphere. *Science* 296: 1066–1068. [R]

Lincoln, T. A. and G. F. Joyce. 2009. Self-sustained replication of an RNA enzyme. *Science* 323:1229–1232. [P]

Margulis, L. 1992. *Symbiosis in Cell Evolution*. 2nd ed. New York: W. H. Freeman.

Martin, W. and T. M. Embley. 2004. Early evolution comes full circle. *Nature* 431: 134–137. [R]

Mast, F. D., L. D. Barlow, R. A. Rachubinski and J. B. Dacks. 2014. Evolutionary mechanisms for establishing eukaryotic cellular complexity. *Trends Cell Biol.* 24: 435–442. [R]

Miller, S. L. 1953. A production of amino acids under possible primitive Earth conditions. *Science* 117: 528–529. [P]

Prochnik, S. E. and 27 others. 2010. Genomic analysis of organismal complexity in the multicellular green alga *Volvox carteri*. *Science* 329: 223–226. [P]

Rivera, M. C. and J. A. Lake. 2004. The ring of life provides evidence for a genome fusion origin of eukaryotes. *Nature* 431: 152–155. [P]

Szostak, J. W., D. P. Bartel and P. L. Luisi. 2001. Synthesizing life. *Nature* 409: 387–390. [R]

Velasco, R. and 85 others. 2010. The genome of the domesticated apple (*Malus x domestica* Borkh.) *Nature Genet.* 42: 833–841. [P]

Williams, T. A., P. G. Foster, C. J. Cox and T. M. Embley. 2013. An archaeal origin of eukaryotes supports only two primary domains of life. *Nature* 504: 231–236. [R]

Wochner, A., J. Attwater, A. Coulson and P. Holliger. 2011. Ribozyme-catalyzed transcription of an active ribozyme. *Science* 332: 209–212. [P]

Cells as Experimental Models

Adams, M. D. and 194 others. 2000. The genome sequence of *Drosophila melanogaster*. *Science* 287: 2185–2195. [P]

Blattner, F. R. and 16 others. 1997. The complete genome sequence of *Escherichia coli* K–12. *Science* 277: 1453–1462. [P]

Goffeau, A. and 15 others. 1996. Life with 6000 genes. *Science* 274: 546–567. [P]

Hellsten, U. and 47 others. 2010. The genome of the western clawed frog *Xenopus tropicalis*. *Science* 328: 633–636. [P]

Howe, K. and 171 others. 2013. The zebrafish reference genome sequence and its relationship to the human genome. *Nature* 496: 498–503. [P]

International Human Genome Sequencing Consortium. 2004. Finishing the euchromatic sequence of the human genome. *Nature* 431: 931–945. [P]

Lieschke, G. J. and P. D. Currie. 2007. Animal models of human disease: zebrafish swim into view. *Nature Rev. Genet.* 8: 353–367. [R]

Meyerowitz, E. M. 2002. Plants compared to animals: The broadest comparative study of development. *Science* **295: 1482–1485. [R]**

Mouse Genome Sequencing Consortium. 2002. Initial sequence and comparative analysis of the mouse genome. *Nature* 420: 520–562. [P]

The *Arabidopsis* Genome Initiative. 2000. Analysis of the genome sequence of the flowering plant *Arabidopsis thaliana*. *Nature* 408: 796–815. [P]

The *C. elegans* Sequencing Consortium. 1998. Genome sequence of the nematode *C. elegans*: A platform for investigating biology. *Science* 282: 2012–2018. [P]

Tools of Cell Biology

Bowers, W. E. 1998. Christian de Duve and the discovery of lysosomes and peroxisomes. *Trends Cell Biol.* 8: 330–333 [R]

Cairns, J., G. S. Stent and J. D. Watson (eds.). 2007. *Phage and the Origins of Molecular Biology, The Centennial Edition.* Plainview, N.Y.: Cold Spring Harbor Laboratory Press.

Chalfie, M. 2009. GFP: lighting up life. *Proc. Natl. Acad. Sci. USA* 106: 10073–10080. [R]

Claude, A. 1975. The coming of age of the cell. *Science* 189: 433–435. [R]

De Duve, C. 1975. Exploring cells with a centrifuge. *Science* 189: 186–194. [R]

Flint, S. J., L. W. Enquist, V. R. Racaniello and A. M. Skalka. 2009. *Principles of Virology.* 3rd ed. Washington, DC: ASM Press.

Giepmans, B. N. G., S. R. Adams, M. H. Ellisman and R. Y. Tsien. 2006. The fluorescent toolbox for assessing protein location and function. *Science* 312: 217–224. [R]

Hosking, C. R. and J. Lippincott-Schwartz. 2009. The future's bright: imaging cell biology in the 21st century. *Trends Cell Biol.* **19: 553–554. [R]**

Huang, B., H. Babcock and X. Zhuang. 2010. Breaking the diffraction barrier: super-resolution imaging of cells. *Cell* 143: 1047–1058. [R]

Koster, A. J. and J. Klumperman. 2003. Electron microscopy in cell biology: Integrating structure and function. *Nature Rev. Molec. Cell Biol.* 4: SS6–SS10. [R]

Lippincott-Schwartz, J., N. Altan-Bonnet and G. H. Patterson. 2003. Photobleaching and photoactivation: Following protein dynamics in living cells. *Nature Cell Biol.* 5: S7–S14. [R]

Palade, G. 1975. Intracellular aspects of the process of protein synthesis. *Science* 189: 347–358. [R]

Porter, K. R., A. Claude and E. F. Fullam. 1945. A study of tissue culture cells by electron microscopy. *J. Exp. Med.* 81: 233–246. [P]

Rous, P. 1911. A sarcoma of the fowl transmissible by an agent separable from the tumor cells. *J. Exp. Med.* 13: 397–411. [P]

Salmon, E. D. 1995. VE-DIC light microscopy and the discovery of kinesin. *Trends Cell Biol.* 5: 154–158. [R]

Skloot, R. 2010. *The Immortal Life of Henrietta Lacks.* Penguin/Random House.

Molecules and Membranes

Cells are incredibly complex and diverse structures, capable not only of self-replication—the very essence of life—but also of performing a wide range of specialized tasks in multicellular organisms. Yet, cells obey the same laws of chemistry and physics that determine the behavior of nonliving systems. Consequently, modern cell biology seeks to understand cellular processes in terms of chemical and physical reactions.

This chapter considers the chemical composition of cells and the properties of the molecules that are ultimately responsible for all cellular activities. Proteins are given particular emphasis because of their diverse roles within the cell, including acting as enzymes that catalyze almost all biological reactions and serving as key components of cell membranes. Membranes are critical to cell structure and function because they serve as barriers that separate distinct aqueous compartments. For example, the plasma membrane separates the contents of a cell from the external environment, while the nuclear membrane separates the contents of the nucleus from the cytoplasm. Membranes consist of both lipids and proteins and are impermeable to most water-soluble molecules, with proteins carrying out the selective transport of molecules across the phospholipid bilayer.

The Molecules of Cells

Cells are composed of water, inorganic ions, and carbon-containing (organic) molecules. Water is the most abundant molecule in cells, accounting for 70% or more of total cell mass. Consequently, the interactions between water and the other constituents of cells are of central importance in biological chemistry. The critical property of water in this respect is that it is a polar molecule, in which the hydrogen atoms have a slight positive charge and the oxygen has a slight negative charge (**Figure 2.1**). Because of their polar nature, water molecules can form hydrogen bonds with each other or with other polar molecules, as well as interact with positively or negatively charged ions. As a result of these interactions, ions and polar molecules are readily soluble in water (**hydrophilic**). In contrast, nonpolar molecules, which cannot interact with water, are poorly soluble in an aqueous environment (**hydrophobic**). Consequently, nonpolar molecules tend to minimize their contact with water by associating closely with each other instead. As discussed later in this chapter, such interactions of polar and nonpolar molecules with water and with each other play crucial roles in the formation of biological structures, such as cell membranes. The inorganic ions of the cell, including sodium (Na^+), potassium (K^+), magnesium (Mg^{2+}), calcium (Ca^{2+}), phosphate (HPO_4^{2-}), chloride

(A)

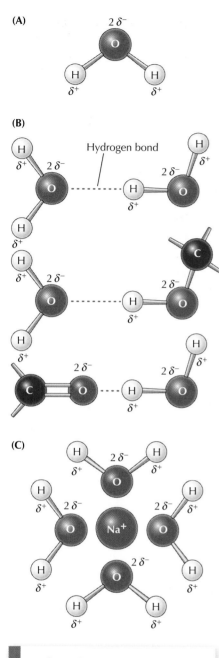

Figure 2.1 Characteristics of water (A) Water is a polar molecule, with a slight negative charge (δ^-) on the oxygen atom and a slight positive charge (δ^+) on the hydrogen atoms. Because of this polarity, water molecules can form hydrogen bonds (dashed lines) either with each other or with other polar molecules (B), in addition to interacting with charged ions (C).

(Cl^-), and bicarbonate (HCO_3^-), constitute 1% or less of the cell mass. These ions are involved in a number of aspects of cell metabolism, though, and thus play critical roles in cell function.

It is, however, the organic molecules that are the unique constituents of cells. Most of these organic compounds belong to one of four classes of molecules: carbohydrates, lipids, proteins, or nucleic acids. Proteins, nucleic acids, and most carbohydrates (the polysaccharides) are macromolecules formed by the joining (polymerization) of hundreds or thousands of low-molecular-weight precursors: amino acids, nucleotides, and simple sugars, respectively. Such macromolecules constitute 80 to 90% of the dry weight of most cells. Lipids are the other major constituent of cells. The remainder of the cell mass is composed of a variety of small organic molecules, including macromolecular precursors. The basic chemistry of cells can thus be understood in terms of the structures and functions of four major classes of organic molecules.

Carbohydrates

The **carbohydrates** include simple sugars as well as polysaccharides. These simple sugars, such as glucose, are the major nutrients of cells. As discussed in Chapter 3, their breakdown provides both a source of cellular energy and the starting material for the synthesis of other cell constituents. Polysaccharides are storage forms of sugars and form structural components of the cell. In addition, polysaccharides and shorter polymers of sugars act as markers for a variety of cell recognition processes, including the adhesion of cells to their neighbors and the transport of proteins to appropriate intracellular destinations.

The structures of representative simple sugars (**monosaccharides**) are illustrated in **Figure 2.2**. The basic formula for these molecules is $(CH_2O)_n$, from which the name carbohydrate is derived (C = "carbo" and H_2O = "hydrate"). The six-carbon ($n = 6$) sugar glucose ($C_6H_{12}O_6$) is especially important in cells, since it provides the principal source of cellular energy. Other simple sugars have between three and seven carbons, with three- and five-carbon sugars being the most common. Sugars containing five or more carbons can cyclize to form ring structures, which are the predominant forms of these molecules within cells. As illustrated in Figure 2.2, the cyclized sugars exist in two alternative forms (called α or β), depending on the configuration of carbon 1.

Monosaccharides can be joined together by dehydration reactions in which H_2O is removed and the sugars are linked by a **glycosidic bond** between two of their carbons (**Figure 2.3**). If only a few sugars are joined together, the resulting polymer is called an **oligosaccharide**. If a large number (hundreds or thousands) of sugars are involved, the resulting polymers are macromolecules called **polysaccharides**.

Two common polysaccharides—**glycogen** and **starch**—are the storage forms of carbohydrates for energy utilization in animal and plant cells, respectively. Both glycogen and starch are composed entirely of glucose

Animation 2.1

sites.sinauer.com/cooper7e/a2.1
Bond Formation The polymerization of sugars, amino acids, and nucleotides to form polysaccharides, polypeptides, and nucleic acids, respectively, occurs through covalent bonding.

FYI

The artificial sweetener sucralose (Splenda) is a synthetic derivative of common sugar in which some of the hydroxyl groups of the sugar are replaced with chlorine.

Triose sugars (C$_3$H$_6$O$_3$)

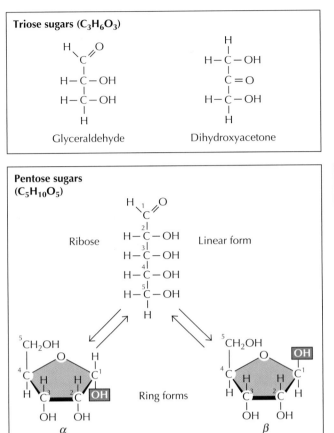

Glyceraldehyde   Dihydroxyacetone

Figure 2.2 Structure of simple sugars Representative sugars containing three, five, and six carbons (triose, pentose, and hexose sugars, respectively) are illustrated. Sugars with five or more carbons can cyclize to form rings, which exist in two alternative forms (α and β), depending on the configuration of carbon 1.

Pentose sugars (C$_5$H$_{10}$O$_5$)

Ribose   Linear form

Ring forms

α β

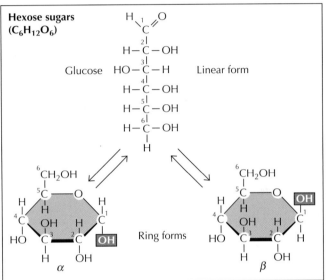

Hexose sugars (C$_6$H$_{12}$O$_6$)

Glucose   Linear form

Ring forms

α β

molecules in the α configuration (**Figure 2.4**). The principal linkage is between carbon 1 of one glucose and carbon 4 of a second. In addition, both glycogen and one form of starch (amylopectin) contain occasional $\alpha(1\rightarrow6)$ linkages, in which carbon 1 of one glucose is joined to carbon 6 of a second. As illustrated in Figure 2.4, these linkages lead to the formation of branches resulting from the joining of two separate $\alpha(1\rightarrow4)$ linked chains. Such branches are present in glycogen and amylopectin, although another form of starch (amylose) is an unbranched molecule.

The structures of glycogen and starch are thus basically similar, as is their function: to store glucose. **Cellulose**, in contrast, has a quite distinct function as the principal structural component of the plant cell wall. Perhaps surprisingly, then, cellulose is also composed entirely of glucose molecules. The glucose residues in cellulose, however, are in the β rather than the α configuration, and cellulose is an unbranched polysaccharide (see Figure 2.4). The linkage of glucose residues by $\beta(1\rightarrow4)$ rather than $\alpha(1\rightarrow4)$ bonds causes cellulose to form long extended chains that pack side by side to form fibers of great mechanical strength. The animal parallel of cellulose is **chitin**, a polysaccharide of modified glucose residues linked by β bonds, which forms the exoskeletons of crabs and insects.

In addition to their roles in energy storage and cell structure, oligosaccharides and polysaccharides are important in a variety

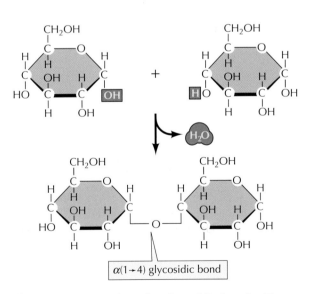

$\alpha(1\rightarrow4)$ glycosidic bond

Figure 2.3 Formation of a glycosidic bond Two simple sugars are joined by a dehydration reaction (a reaction in which water is removed). In the example shown, two glucose molecules in the α configuration are joined by a bond between carbons 1 and 4, which is therefore called an $\alpha(1\rightarrow4)$ glycosidic bond.

Amylopectin (starch)

Glycogen

Cellulose

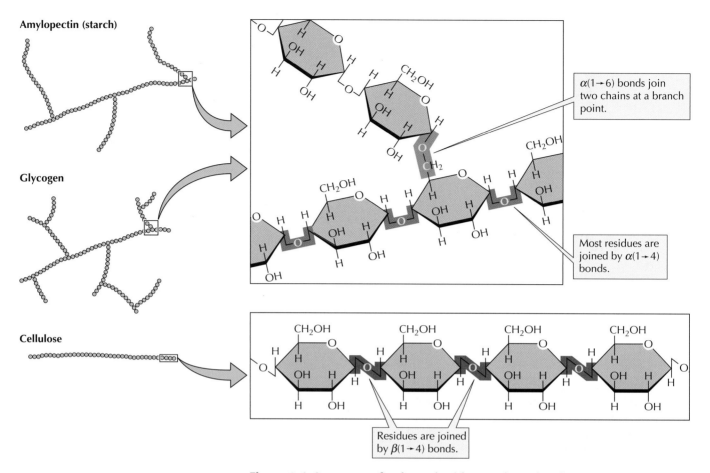

$\alpha(1\rightarrow6)$ bonds join two chains at a branch point.

Most residues are joined by $\alpha(1\rightarrow4)$ bonds.

Residues are joined by $\beta(1\rightarrow4)$ bonds.

Figure 2.4 Structure of polysaccharides   Polysaccharides are macromolecules consisting of hundreds or thousands of simple sugars. Starch, glycogen, and cellulose are all composed entirely of glucose residues, which are joined by $\alpha(1\rightarrow4)$ glycosidic bonds in starch and glycogen, but by $\beta(1\rightarrow4)$ bonds in cellulose. One form of starch (amylopectin) and glycogen also contain occasional $\alpha(1\rightarrow6)$ bonds, which serve as branch points by joining two separate $\alpha(1\rightarrow4)$ chains.

of informational processes. For example, oligosaccharides are frequently linked to proteins, where they play important roles in protein folding and serve as markers to target proteins for transport to the cell surface or incorporation into different subcellular organelles. Oligosaccharides and polysaccharides also serve as markers on the surface of cells, playing important roles in cell recognition and the interactions between cells in tissues of multicellular organisms.

Lipids

Lipids have three major roles in cells. First, they provide an important form of energy storage. Second, and of great importance in cell biology, lipids are the major components of cell membranes. Third, lipids play important roles in cell signaling, both as steroid hormones (e.g., estrogen and testosterone) and as messenger molecules that convey signals from cell surface receptors to targets within the cell.

The simplest lipids are **fatty acids**, which consist of long hydrocarbon chains, most frequently containing 16 or 18 carbon atoms, with a carboxyl

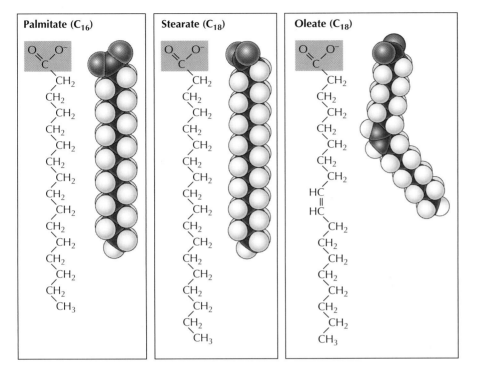

Figure 2.5 Structure of fatty acids Fatty acids consist of long hydrocarbon chains terminating in a carboxyl group (COO⁻). Palmitate and stearate are saturated fatty acids consisting of 16 and 18 carbons, respectively. Oleate is an unsaturated 18-carbon fatty acid containing a double bond between carbons 9 and 10. Note that the double bond introduces a kink in the hydrocarbon chain.

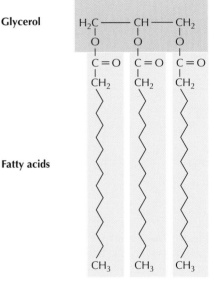

group (COO⁻) at one end (**Figure 2.5**). Unsaturated fatty acids contain one or more double bonds between carbon atoms; in saturated fatty acids all of the carbon atoms are bonded to the maximum number of hydrogen atoms. The long hydrocarbon chains of fatty acids contain only nonpolar C—H bonds, which are unable to interact with water. The hydrophobic nature of these fatty acid chains is responsible for much of the behavior of complex lipids, particularly in the formation of biological membranes.

Fatty acids are stored in the form of **triacylglycerols** (also called **triglycerides** or **fats**), which consist of three fatty acids linked to a glycerol molecule (**Figure 2.6**). Triacylglycerols are insoluble in water and therefore accumulate as fat droplets in the cytoplasm. When required, they can be broken down for use in energy-yielding reactions (see Chapter 3). It is noteworthy that fats are a more efficient form of energy storage than carbohydrates, yielding more than twice as much energy per weight of material broken down. Fats therefore allow energy to be stored in less than half the body weight than would be required to store the same amount of energy in carbohydrates—a particularly important consideration for animals because of their mobility.

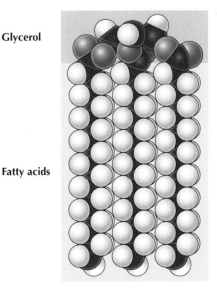

Figure 2.6 Structure of triacylglycerols Triacylglycerols (fats) contain three fatty acids joined to glycerol. In this example, all three fatty acids are palmitate, but triacylglycerols often contain a mixture of different fatty acids.

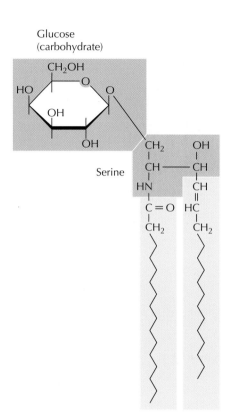

Glucose
(carbohydrate)

Serine

Figure 2.8 Structure of glycolipids
Two hydrocarbon chains are joined
to a polar head group formed from
serine and containing carbohydrates
(e.g., glucose).

Figure 2.7 Structure of phospholipids   Glycerol phospholipids contain two ▶
fatty acids joined to glycerol. The fatty acids may be different from each other.
The third carbon of glycerol is joined to a phosphate group (forming phosphatidic
acid), which in turn is frequently joined to another small polar molecule (forming
phosphatidylethanolamine, phosphatidylcholine, phosphatidylserine, or phosphatidylinositol). In sphingomyelin, two hydrocarbon chains are bound to a polar
head group formed from serine instead of glycerol.

Phospholipids, the principal components of cell membranes, consist of
two fatty acids joined to a polar head group (**Figure 2.7**). In the **glycerol
phospholipids**, the two fatty acids are bound to carbon atoms in glycerol,
as in triacylglycerols. The third carbon of glycerol, however, is bound to a
phosphate group, which is in turn frequently attached to another small polar
molecule, such as ethanolamine, choline, serine, or inositol. **Sphingomyelin**,
the only nonglycerol phospholipid in cell membranes, contains two hydrocarbon chains linked to a polar head group formed from serine rather than
from glycerol. All phospholipids have hydrophobic tails (consisting of the
two hydrocarbon chains), and hydrophilic head groups (consisting of the
phosphate group and its polar attachments). Consequently, phospholipids
are **amphipathic** molecules, part water-soluble and part water-insoluble.
This property of phospholipids is the basis for the formation of biological
membranes, as discussed later in this chapter.

In addition to phospholipids, many cell membranes contain **glycolipids**
and **cholesterol**. Glycolipids consist of two hydrocarbon chains linked to polar
head groups that contain carbohydrates (**Figure 2.8**). They are thus similar to
the phospholipids in their general organization as amphipathic molecules.
Cholesterol, in contrast, consists of four hydrocarbon rings rather than
linear hydrocarbon chains (**Figure 2.9**). The hydrocarbon rings are strongly

**Figure 2.9 Cholesterol and steroid
hormones**   Cholesterol, an important component of cell membranes,
is an amphipathic molecule because
of its polar hydroxyl group. Cholesterol is also a precursor to the steroid
hormones, such as testosterone and
estradiol (a form of estrogen). The
hydrogen atoms bonded to the ring
carbons are not shown in this figure.

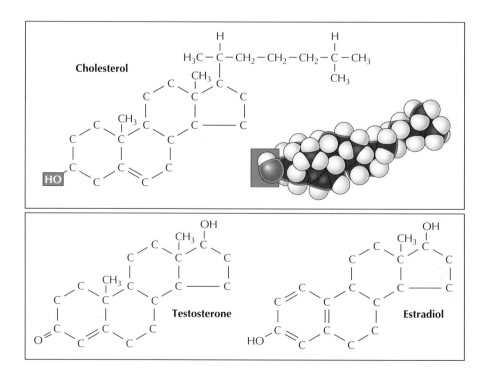

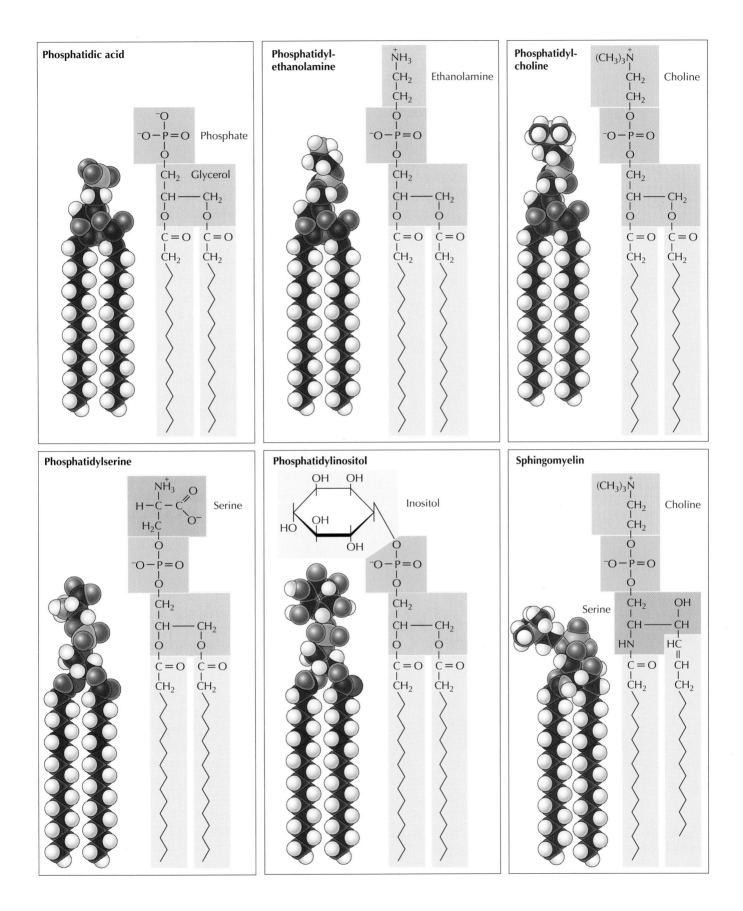

Phosphatidic acid

⁻O
|
⁻O−P=O   Phosphate
|
O
|
CH₂   Glycerol
|
CH ——— CH₂
| |
O   O
| |
C=O   C=O
| |
CH₂   CH₂

Phosphatidyl-ethanolamine

⁺NH₃
|
CH₂   Ethanolamine
|
CH₂
|
O
|
⁻O−P=O
|
O
|
CH₂
|
CH ——— CH₂
| |
O   O
| |
C=O   C=O
| |
CH₂   CH₂

Phosphatidyl-choline

(CH₃)₃N⁺
|
CH₂   Choline
|
CH₂
|
⁻O−P=O
|
O
|
CH₂
|
CH ——— CH₂
| |
O   O
| |
C=O   C=O
| |
CH₂   CH₂

Phosphatidylserine

NH₃⁺   O
| //
H−C−C
| \
H₂C   O⁻   Serine
|
O
|
⁻O−P=O
|
O
|
CH₂
|
CH ——— CH₂
| |
O   O
| |
C=O   C=O
| |
CH₂   CH₂

Phosphatidylinositol

OH   OH
| |

HO
OH
OH   Inositol

O
|
⁻O−P=O
|
O
|
CH₂
|
CH ——— CH₂
| |
O   O
| |
C=O   C=O
| |
CH₂   CH₂

Sphingomyelin

(CH₃)₃N⁺
|
CH₂   Choline
|
CH₂
|
O
|
⁻O−P=O
|
O
|
CH₂   OH
| |
Serine CH ——— CH
| |
HN   HC
| ‖
C=O   CH
| |
CH₂   CH₂

hydrophobic, but the hydroxyl (OH) group attached to one end of cholesterol is weakly hydrophilic, so cholesterol is also amphipathic.

In addition to their roles as components of cell membranes, lipids function as signaling molecules, both within and between cells. The **steroid hormones** (such as testosterone and estrogens) are derivatives of cholesterol (see Figure 2.9). These hormones are a diverse group of chemical messengers, all of which contain four hydrocarbon rings to which distinct functional groups are attached. Derivatives of phospholipids also serve as messenger molecules within cells, acting to convey signals from cell surface receptors to intracellular targets that regulate a wide range of cellular processes, including cell proliferation, movement, survival, and differentiation (see Chapter 16).

Nucleic acids

The nucleic acids—DNA and RNA—are the principal informational molecules of the cell. **Deoxyribonucleic acid (DNA)** has a unique role as the genetic material, which in eukaryotic cells is located in the nucleus. Different types of **ribonucleic acid (RNA)** participate in a number of cellular activities. **Messenger RNA (mRNA)** carries information from DNA to the ribosomes, where it serves as a template for protein synthesis. Two other types of RNA (**ribosomal RNA** and **transfer RNA**) are involved in protein synthesis. Still other kinds of RNAs are involved in the regulation of gene expression and in the processing and transport of both RNAs and proteins. In addition to acting as an informational molecule, RNA is also capable of catalyzing a number of chemical reactions. In present-day cells, these include reactions involved in both protein synthesis and RNA processing.

DNA and RNA are polymers of **nucleotides**, which consist of **purine** and **pyrimidine** bases linked to phosphorylated sugars (**Figure 2.10**). DNA contains two purines (**adenine** and **guanine**) and two pyrimidines (**cytosine** and **thymine**). Adenine, guanine, and cytosine are also present in RNA, but RNA contains **uracil** in place of thymine. The bases are linked to sugars (**2′-deoxyribose** in DNA, or **ribose** in RNA) to form

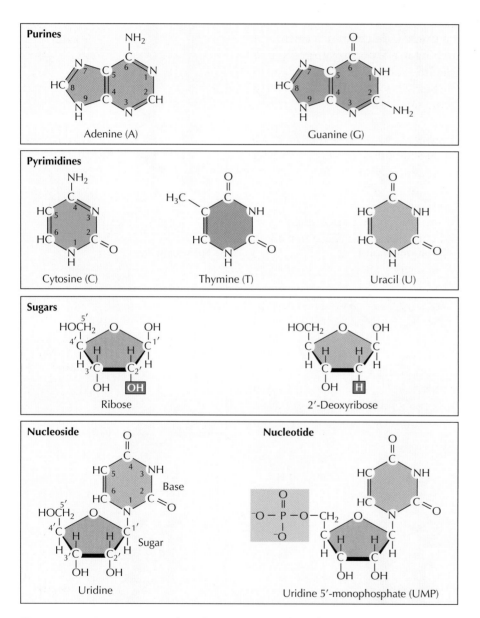

Figure 2.10 Components of nucleic acids Nucleic acids contain purine and pyrimidine bases linked to phosphorylated sugars. A nucleic acid base linked to a sugar alone is a nucleoside. Nucleo*tides* additionally contain one or more phosphate groups.

Figure 2.11 Polymerization of nucleotides Nucleotides are joined by phosphodiester bonds between 3' hydroxyl and 5' groups. The 5' end of a nucleotide chain terminates in a phosphate group and the 3' end terminates in a hydroxyl group.

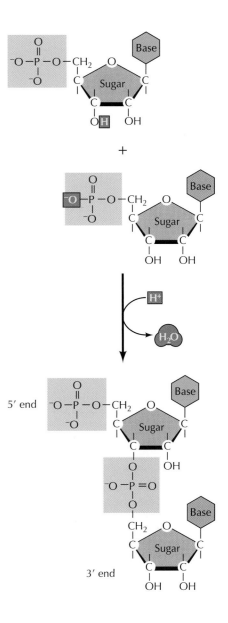

nucleosides. Nucleo*tides* additionally contain one or more phosphate groups linked to the 5' carbon of nucleoside sugars.

The polymerization of nucleotides to form nucleic acids involves the formation of **phosphodiester bonds** between the 5' phosphate of one nucleotide and the 3' hydroxyl of another (**Figure 2.11**). **Oligonucleotides** are small polymers containing only a few nucleotides; the large **polynucleotides** that make up cellular RNA and DNA may contain thousands or millions of nucleotides, respectively. It is important to note that a nucleotide chain has a sense of direction, with one end of the chain terminating in a 5' phosphate group and the other in a 3' hydroxyl group. Polynucleotides are always synthesized in the 5' to 3' direction, with a free nucleotide being added to the 3' OH group of a growing chain. By convention, the sequence of bases in DNA or RNA is also written in the 5' to 3' direction.

The information in DNA and RNA is conveyed by the order of the bases in the polynucleotide chain. DNA is a double-stranded molecule consisting of two polynucleotide chains running in opposite directions with respect to their 5' and 3' ends (see Chapter 4). The bases are on the inside of the molecule, and the two chains are joined by hydrogen bonds between complementary base pairs—guanine pairing with cytosine and adenine with thymine (**Figure 2.12**). The important consequence of such complementary base pairing is that one strand of DNA (or a strand of RNA) can act as a template to direct the synthesis of a complementary strand. Nucleic acids are thus uniquely capable of directing their own replication, allowing them to function as the fundamental informational molecules of the cell. The information carried by DNA and RNA directs the synthesis of specific proteins, which control most cellular activities.

Nucleotides are not only important as the building blocks of nucleic acids; they also play critical roles in other cell processes. Perhaps the most prominent example is adenosine 5'-triphosphate (ATP), which is the principal form of chemical energy within cells. Other nucleotides similarly function as carriers of either energy or reactive chemical groups in a wide variety

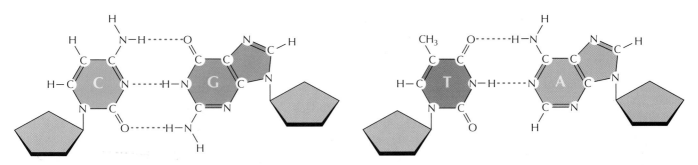

Figure 2.12 Complementary pairing between nucleic acid bases The formation of hydrogen bonds between bases on opposite strands of DNA leads to the specific pairing of cytosine (C) with guanine (G) and thymine (T) with adenine (A).

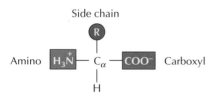

**Figure 2.13 Structure of amino
acids** Each amino acid consists of
a central carbon atom (the α carbon) bonded to a hydrogen atom, a
carboxyl group, an amino group, and
a specific side chain (designated R).
At physiological pH, both the carboxyl and amino groups are ionized,
as shown.

of metabolic reactions. In addition, some nucleotides (e.g., cyclic AMP) are
important signaling molecules within cells (see Chapter 16).

Proteins

While nucleic acids carry the genetic information of the cell, the primary re-
sponsibility of **proteins** is to execute the tasks directed by that information.
Proteins are the most diverse of all macromolecules, and each cell contains
thousands of different proteins, which perform a wide variety of functions.
The roles of proteins include serving as structural components of cells and
tissues, acting in the transport of small molecules (e.g., the transport of glucose and other nutrients into cells), transmitting information between cells
(e.g., protein hormones), and providing a defense against infection (e.g.,
antibodies). The most fundamental property of proteins, however, is their
ability to act as enzymes, which, as discussed in the next section, catalyze
nearly all the chemical reactions in biological systems. Thus, proteins direct
virtually all activities of the cell. The central importance of proteins in biological chemistry is indicated by their name, which is derived from the Greek
word *proteios*, meaning "of the first rank."

Proteins are polymers synthesized from 20 different **amino acids**. Each
amino acid consists of a carbon atom (called the α carbon) bonded to a
carboxyl group (COO^-), an amino group (NH_3^+), a hydrogen atom, and a
distinctive side chain (**Figure 2.13**). The specific chemical properties of the
different amino acid side chains determine the roles of each amino acid in
protein structure and function. The amino acids can be grouped into four broad
categories according to the properties of their side chains (**Figure 2.14**). Ten
amino acids have nonpolar side chains that do not interact with water. Glycine
is the simplest amino acid, with a side chain consisting of only a hydrogen
atom. Alanine, valine, leucine, and isoleucine have hydrocarbon side chains
consisting of up to four carbon atoms. The side chains of these amino acids
are hydrophobic and therefore tend to be located in the interior of proteins
or in membranes, where they are not in contact with water. Phenylalanine
and tryptophan similarly have very hydrophobic side chains containing
aromatic rings. Proline also has a hydrocarbon side chain, but it is unique
in that its side chain is bonded to the nitrogen of the amino group as well
as to the α carbon, forming a cyclic structure. The side chains of two amino
acids, cysteine and methionine, contain sulfur atoms. Methionine is quite
hydrophobic, but cysteine is less so because of its sulfhydryl (SH) group. As
discussed later, the sulfhydryl group of cysteine plays an important role in
protein structure because disulfide bonds can form between the side chains
of different cysteine residues.

Five amino acids have uncharged but polar side chains. These include
serine, threonine, and tyrosine (which have hydroxyl groups on their side
chains), as well as asparagine and glutamine (which have polar amide
[$O=C—NH_2$] groups). Because the polar side chains of these amino acids
can form hydrogen bonds with water, these amino acids are hydrophilic and
tend to be located on the outside of proteins.

The amino acids lysine, arginine, and histidine have side chains with
charged basic groups. Lysine and arginine are very basic amino acids, and
their side chains are positively charged in the cell. Consequently, they are very
hydrophilic and are found in contact with water on the surface of proteins.
Histidine can be either uncharged or positively charged at physiological

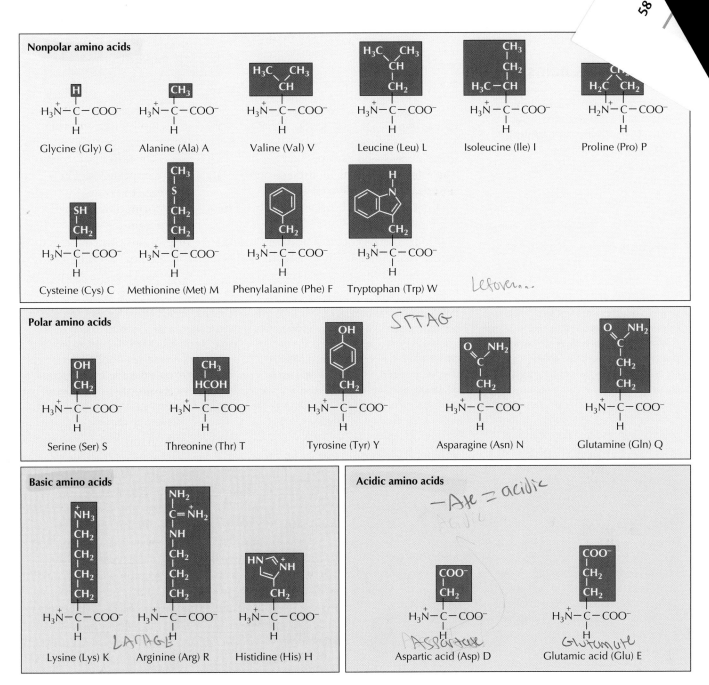

Figure 2.14 The amino acids The three-letter and one-letter abbreviations for each amino acid are indicated. The amino acids are grouped into four categories according to the properties of their side chains: nonpolar, polar, basic, and acidic.

pH, so it frequently plays an active role in enzymatic reactions involving the exchange of hydrogen ions, as discussed below.

Finally, two amino acids, aspartic acid and glutamic acid, have acidic side chains terminating in carboxyl groups. These amino acids are negatively charged within the cell and are therefore frequently referred to as aspartate and glutamate. Like the basic amino acids, these acidic amino acids are very hydrophilic and are usually located on the surface of proteins.

Video 2.1

sites.sinauer.com/cooper7e/v2.1

What Is a Protein? All proteins are made out of the same twenty amino acids, combined and folded in different ways—this 3D structure is critical for protein function.

The Folding of Polypeptide Chains

Reductive Cleavage of Disulfide Bridges in Ribonuclease

Michael Sela, Frederick H. White Jr., and Christian B. Anfinsen

National Institutes of Health, Bethesda, MD

Science, Volume 125, 1957, pages 691–692

The Context

Functional proteins are structurally far more complex than linear chains of amino acids. The formation of active enzymes or other proteins requires the folding of polypeptide chains into precise three-dimensional conformations. This difference between proteins and polypeptide chains raises critical questions with respect to understanding protein structure and function. How is the correct protein conformation chosen from the many possible conformations that could be adopted by a polypeptide, and what is the nature of the information that directs protein folding?

Classic experiments done by Christian Anfinsen and his colleagues provided the answers to these questions. Studying the enzyme ribonuclease, Anfinsen and his collaborators were able to demonstrate that denatured proteins can spontaneously re-fold into an active conformation. Therefore, all the information required to specify the correct three-dimensional conformation of a protein is contained in its primary amino acid sequence. A series of such experiments led Anfinsen to the conclusion that the native three-dimensional structure of a protein corresponds to its thermodynamically most stable conformation, as determined by interactions between its constituent amino acids. The original observations leading to the formulation of this critical principle were reported in this 1957 paper by Christian Anfinsen, Michael Sela, and Frederick White Jr.

The Experiments

The protein studied by Sela, White, and Anfinsen was bovine ribonuclease, a small protein of 124 amino acids that contains four disulfide (S—S) bonds between side chains of cysteine residues. The enzymatic activity of ribonuclease could be determined by its ability to degrade RNA to nucleotides, providing a ready assay for function of the native protein. This enzymatic activity was completely lost when ribonuclease was subjected to treatments that both disrupted the noncovalent bonds (e.g., hydrogen bonds) and cleaved the disulfide bonds by reducing them to sulfhydryl (SH) groups. The denatured protein thus appeared to be in a random inactive conformation.

Sela, White, and Anfinsen made the crucial observation that enzymatic activity reappears if the denatured protein is incubated under conditions that allow the polypeptide chain to refold and disulfide bonds to re-form. In these experiments the denaturing agents were removed, and the inactive enzyme was incubated at room temperature in a physiological buffer in the presence of O_2. This procedure led to oxidation of the sulfhydryl groups and the re-formation of disulfide bonds. During this process, the enzyme regained its catalytic activity, indicating that it had refolded to its native conformation (see figure). Because no other cellular components were present, all the information required for proper protein folding appeared to be contained in the primary amino acid sequence of the polypeptide chain.

The Impact

Further experiments defined conditions under which denatured ribonuclease completely regains its native structure and enzymatic activity, establishing the "thermodynamic hypothesis" of protein folding—that is, the native three-dimensional structure of a protein corresponds to the thermodynamically most stable state under physiological conditions. Thermodynamic stability is governed by interactions of the constituent amino acids, so the three-dimensional conformations of proteins are determined directly by their primary amino acid sequences. Since the order of nucleotides in DNA specifies the amino acid sequence of a polypeptide, it follows that the nucleotide sequence of a gene contains all the information needed to determine the three-dimensional structure of its protein product.

Christian B. Anfinsen

Although Anfinsen's work established the thermodynamic basis of protein folding, understanding the mechanism of this process remains an active area of research. Protein folding is extremely complex, and we are still unable to deduce the three-dimensional structure of a protein directly from its amino acid sequence. It is also important to note that the spontaneous folding of proteins *in vitro* is much slower than protein folding within the cell, which is assisted by enzymes (see Chapter 9). The protein-folding problem thus remains a central challenge in biological chemistry.

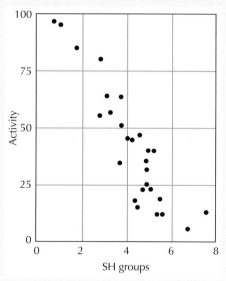

A summary of results of the renaturation experiments The enzymatic activity of ribonuclease is plotted as a function of the number of sulfhydryl (SH) groups present after various treatments. The formation of SH groups corresponds to the loss of disulfide bonds. Activity is expressed as percent activity of the native enzyme.

Figure 2.15 Formation of a peptide bond   The carboxyl group of one amino acid is linked to the amino group of a second.

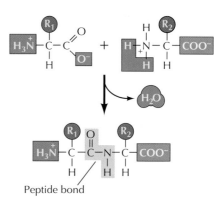

Peptide bond

Amino acids are joined together by **peptide bonds** between the α carboxyl group of one amino acid and the α amino group of a second (**Figure 2.15**). **Polypeptides** are linear chains of amino acids, usually hundreds or thousands of amino acids in length. Each polypeptide chain has two distinct ends, one terminating in an amino group (the amino, or N, terminus) and the other in a carboxyl group (the carboxy, or C, terminus). Polypeptides are synthesized from the amino to the carboxy terminus, and the sequence of amino acids in a polypeptide is written (by convention) in the same order.

The defining characteristic of proteins is that they are polypeptides with specific amino acid sequences. In 1953 Frederick Sanger was the first to determine the complete amino acid sequence of a protein, the hormone insulin. Insulin was found to consist of two polypeptide chains, joined by disulfide bonds between cysteine residues (**Figure 2.16**). Most important, Sanger's experiment revealed that each protein consists of a specific amino acid sequence. Protein sequences are now usually deduced from the sequences of mRNAs, and the complete amino acid sequences of over 100,000 proteins have been established. Each consists of a unique sequence of amino acids, determined by the order of nucleotides in a gene (see Chapter 4).

The amino acid sequence of a protein is only the first element of its structure. Rather than being extended chains of amino acids, proteins adopt distinct three-dimensional conformations that are critical to their function. These three-dimensional conformations of proteins are the result of interactions between their constituent amino acids, so the shapes of proteins are determined by their amino acid sequences. This was first demonstrated by experiments of Christian Anfinsen in which he disrupted the three-dimensional structures of proteins by treatments, such as heating, that break noncovalent bonds—a

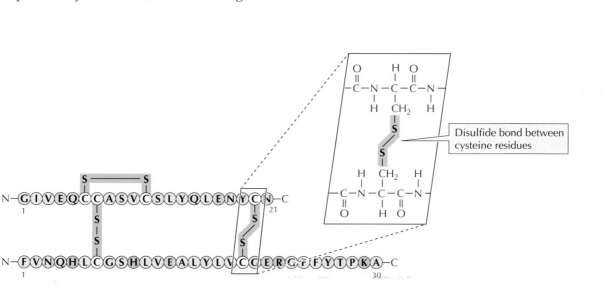

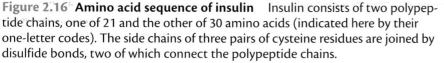

Figure 2.16 Amino acid sequence of insulin   Insulin consists of two polypeptide chains, one of 21 and the other of 30 amino acids (indicated here by their one-letter codes). The side chains of three pairs of cysteine residues are joined by disulfide bonds, two of which connect the polypeptide chains.

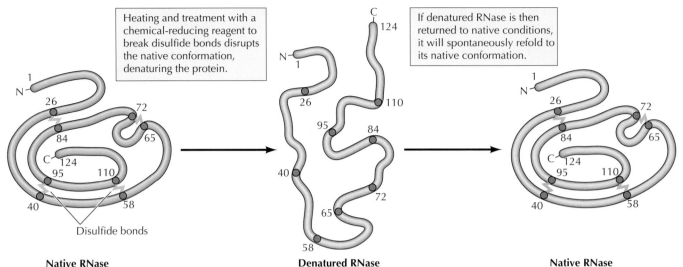

Heating and treatment with a chemical-reducing reagent to break disulfide bonds disrupts the native conformation, denaturing the protein.

If denatured RNase is then returned to native conditions, it will spontaneously refold to its native conformation.

Disulfide bonds

Native RNase

Denatured RNase

Native RNase

Figure 2.17 Protein denaturation and refolding Ribonuclease (RNase) is a protein of 124 amino acids (indicated by numbers). The protein is normally folded into its native conformation, which contains four disulfide bonds (indicated as paired circles representing the cysteine residues).

process called denaturation (**Figure 2.17**). Following incubation under milder conditions, such denatured proteins often spontaneously returned to their native conformations, indicating that these conformations were directly determined by the amino acid sequence.

The three-dimensional structure of proteins is most frequently analyzed by **X-ray crystallography**, a high-resolution technique that can determine the arrangement of individual atoms within a molecule. A beam of X-rays is directed at crystals of the protein to be analyzed, and the pattern of X-rays that pass through the protein crystal is detected on X-ray film. As the X-rays strike the crystal, they are scattered in characteristic patterns determined by the arrangement of atoms in the molecule. The structure of the molecule can therefore be deduced from the pattern of scattered X-rays (the diffraction pattern).

In 1958 John Kendrew was the first person to determine the three-dimensional structure of a protein, myoglobin—a relatively simple protein of 153 amino acids (**Figure 2.18**). Since then, more than 90,000 three-dimensional protein structures have been determined. Most proteins, like myoglobin, are globular molecules with polypeptide chains folded into compact structures, although some (such as the structural proteins of connective tissues) are long fibrous molecules. Analysis of the three-dimensional structures of these proteins has revealed several basic principles that govern protein folding, although protein structure is so complex that predicting the three-dimensional structure of a protein directly from its amino acid sequence is impossible.

Protein structure is generally described as having four levels. The **primary structure** of a protein is the sequence of amino acids in its polypeptide chain. The **secondary structure** is the regular arrangement of amino acids within localized regions of the polypeptide. Two types of secondary structure, which were first proposed by Linus Pauling and Robert Corey in 1951, are particularly common: the α **helix** and the β **sheet**. Both of these secondary structures are held together by hydrogen bonds between the CO and NH groups of peptide bonds. An α helix is formed when a region of a polypeptide chain coils around itself, with the CO group of one peptide bond forming a hydrogen bond with the NH group of a peptide bond located four residues downstream in the linear polypeptide chain (**Figure 2.19**). In contrast, a β sheet is formed when two parts of a polypeptide chain lie side by side with hydrogen bonds between them. Such β sheets can be formed between several

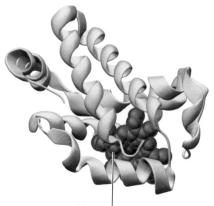

Heme group

Figure 2.18 Three-dimensional structure of myoglobin Myoglobin is a protein of 153 amino acids that is involved in oxygen transport. The polypeptide chain is folded around a heme group that serves as the oxygen-binding site.

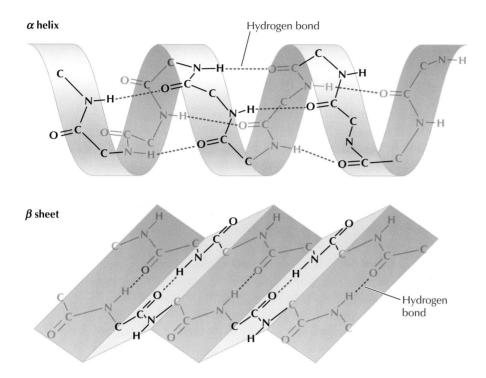

α helix

Hydrogen bond

β sheet

Hydrogen bond

Figure 2.19 Secondary structure of proteins   The most common types of secondary structure are the α helix and the β sheet. In an α helix, hydrogen bonds form between CO and NH groups of peptide bonds separated by four amino acid residues. In a β sheet, hydrogen bonds connect two parts of a polypeptide chain lying side by side. The amino acid side chains are not shown.

polypeptide strands, which can be oriented in either the same or opposite directions (parallel or antiparallel) with respect to each other.

Tertiary structure describes the overall folding of a single polypeptide chain as a result of interactions between the side chains of amino acids that lie in different regions of the primary sequence. The basic units of tertiary structure are called **domains**, which are folded three-dimensional structures usually containing between 50 and 200 amino acids (**Figure 2.20**). In most

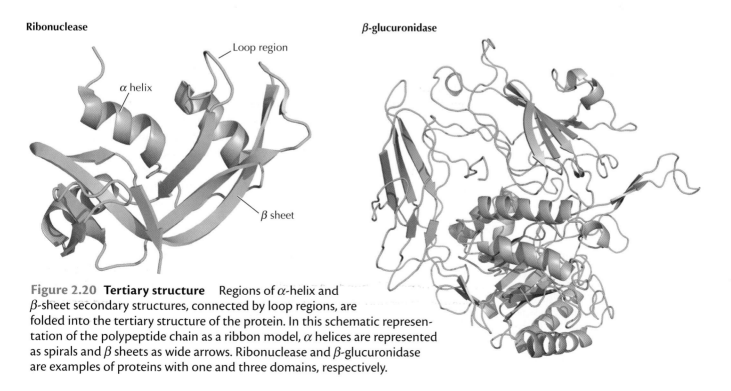

Ribonuclease

Loop region

α helix

β sheet

β-glucuronidase

Figure 2.20 Tertiary structure   Regions of α-helix and β-sheet secondary structures, connected by loop regions, are folded into the tertiary structure of the protein. In this schematic representation of the polypeptide chain as a ribbon model, α helices are represented as spirals and β sheets as wide arrows. Ribonuclease and β-glucuronidase are examples of proteins with one and three domains, respectively.

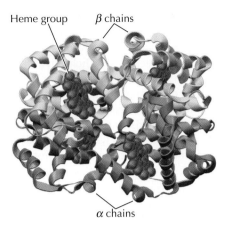

Heme group   β chains

α chains

Figure 2.21 Quaternary structure of hemoglobin Hemoglobin is composed of four polypeptide chains, each of which is bound to a heme group. The two α chains and the two β chains are identical.

proteins, combinations of α helices and β sheets, connected by loop regions of the polypeptide chain, fold into globular domains. Small proteins, such as ribonuclease or myoglobin, contain only a single domain; larger proteins, such as β-glucuronidase, may contain a number of different domains, and these are frequently associated with distinct functions.

A critical determinant of tertiary structure is the localization of hydrophobic amino acids in the interior of the protein and of hydrophilic amino acids on the surface, where they interact with water. The interiors of folded proteins thus consist mainly of hydrophobic amino acids arranged in α helices and β sheets; these secondary structures are found in the hydrophobic cores of proteins because hydrogen bonding neutralizes the polar character of the CO and NH groups of the polypeptide backbone. The loop regions connecting the elements of secondary structure are found on the surface of folded proteins, where the polar components of the peptide bonds form hydrogen bonds with water or with the polar side chains of hydrophilic amino acids. Interactions between polar amino acid side chains (hydrogen bonds and ionic bonds) on the protein surface are also important determinants of tertiary structure. In addition, the covalent disulfide bonds between the sulfhydryl groups of cysteine residues stabilize the folded structures of many cell surface or secreted proteins.

The fourth level of protein structure, **quaternary structure**, consists of the interactions between different polypeptide chains in proteins composed of more than one polypeptide. Hemoglobin, for example, is composed of four polypeptide chains held together by the same types of interactions that maintain tertiary structure (**Figure 2.21**).

The distinct chemical characteristics of the 20 different amino acids thus lead to considerable variation in the three-dimensional conformations of folded proteins. Consequently, proteins constitute an extremely complex and diverse group of macromolecules, suited to the wide variety of tasks they perform in cell biology.

Enzymes as Biological Catalysts

A fundamental task of proteins is to act as enzymes—catalysts that increase the rate of virtually all the chemical reactions within cells. Although RNAs are capable of catalyzing some reactions, most biological reactions are catalyzed by proteins. In the absence of enzymatic catalysis, most biochemical reactions are so slow that they would not occur under the mild conditions of temperature and pressure that are compatible with life. Enzymes accelerate the rates of such reactions by well over a millionfold, so reactions that would take years in the absence of catalysis can occur in fractions of seconds if catalyzed by the appropriate enzyme. Cells contain thousands of different enzymes, and their activities determine which of the many possible chemical reactions actually take place within the cell.

The catalytic activity of enzymes

Like all other catalysts, **enzymes** are characterized by two fundamental properties. First, they increase the rate of chemical reactions without themselves being consumed or permanently altered by the reaction. Second, they increase reaction rates without altering the chemical equilibrium between reactants and products.

FYI

Enzymes are remarkable catalysts. Some enzymes are capable of increasing the rate of a reaction by as much as 10^{18}-fold as compared with an uncatalyzed reaction, such that a process that would normally take billions of years can now be carried out in less than a second!

Animation 2.2

sites.sinauer.com/cooper7e/a2.2

Catalysts and Activation Energy In a chemical reaction, an enzyme lowers the amount of energy required to reach the transition state, allowing the reaction to proceed at an accelerated rate.

These principles of enzymatic catalysis are illustrated in the following example in which a molecule acted upon by an enzyme (referred to as a **substrate** [*S*]) is converted to a **product** (*P*) as a result of the reaction. In the absence of the enzyme, the reaction can be written as follows:

$$S \rightleftharpoons P$$

The chemical equilibrium between *S* and *P* is determined by the laws of thermodynamics (as discussed further in the next section of this chapter) and is represented by the ratio of the forward and reverse reaction rates (*S*→*P* and *P*→*S*, respectively). In the presence of the appropriate enzyme, the conversion of *S* to *P* is accelerated, but the equilibrium *between S* and *P* is unaltered. Therefore, the enzyme must accelerate both the forward and reverse reactions equally. The reaction can be written as follows:

$$S \overset{E}{\rightleftharpoons} P$$

Note that the enzyme (*E*) is not altered by the reaction, so the chemical equilibrium remains unchanged, determined solely by the thermodynamic properties of *S* and *P*.

The effect of the enzyme on such a reaction is best illustrated by the energy changes that must occur during the conversion of *S* to *P* (**Figure 2.22**). The equilibrium of the reaction is determined by the final energy states of *S* and *P*, which are unaffected by enzymatic catalysis. In order for the reaction to proceed, however, the substrate must first be converted to a higher energy state, called the **transition state**. The energy required to reach the transition state (the **activation energy**) constitutes a barrier to the progress of the reaction, limiting the rate of the reaction. Enzymes (and other catalysts) act by reducing the activation energy, thereby increasing the rate of reaction. The increased rate is the same in both the forward and reverse directions, since both must pass through the same transition state.

The catalytic activity of enzymes involves the binding of their substrates to form an enzyme-substrate complex (*ES*). The substrate binds to a specific region of the enzyme, called the **active site**. While bound to the active site, the substrate is converted into the product of the reaction, which is then released from the enzyme. The enzyme-catalyzed reaction can thus be written as follows:

$$S + E \rightleftharpoons ES \rightleftharpoons E + P$$

Note that *E* appears unaltered on both sides of the equation, so the equilibrium is unaffected. However, the enzyme provides a surface upon which the reactions converting *S* to *P* can occur more readily. This is a result of interactions between the enzyme and substrate that lower the energy of activation and favor formation of the transition state.

Mechanisms of enzymatic catalysis

The binding of a substrate to the active site of an enzyme is a very specific interaction. Active sites are clefts or grooves on the surface of an enzyme, usually composed of amino acids from different parts of the polypeptide chain that are brought together in the tertiary structure of the folded protein. Substrates initially bind to the active site by noncovalent interactions, including hydrogen bonds, ionic bonds, and hydrophobic interactions. Once a substrate is bound to the active site of an enzyme, multiple mechanisms can accelerate its conversion to the product of the reaction.

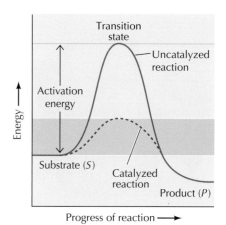

Figure 2.22 Catalyzed and uncatalyzed reactions The reaction illustrated is the simple conversion of a substrate *S* to a product *P*. Because the final energy state of *P* is lower than that of *S*, the reaction proceeds from left to right. For the reaction to occur, however, *S* must first pass through a higher energy transition state. The energy required to reach this transition state (the activation energy) represents a barrier to the progress of the reaction and thereby determines the rate at which the reaction proceeds. In the presence of a catalyst (e.g., an enzyme), the activation energy is lowered and the reaction proceeds at an accelerated rate.

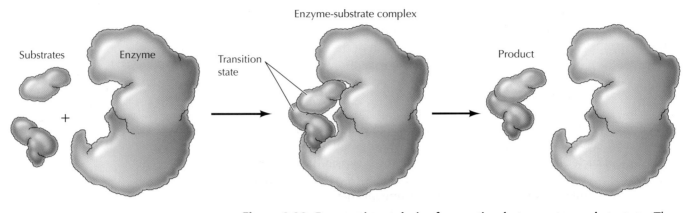

Figure 2.23 **Enzymatic catalysis of a reaction between two substrates** The enzyme provides a template upon which the two substrates are brought together in the proper position and orientation to react with each other.

Although the simple example discussed in the previous section involved only a single substrate molecule, most biochemical reactions involve interactions between two or more different substrates. For example, the formation of a peptide bond involves the joining of two amino acids. For such reactions, the binding of two or more substrates to the active site in the proper position and orientation accelerates the reaction (**Figure 2.23**). The enzyme provides a template upon which the reactants are brought together and properly oriented to favor the formation of the transition state in which they interact.

Enzymes also accelerate reactions by altering the conformation of their substrates to approach that of the transition state. The simplest model of enzyme-substrate interaction is the **lock-and-key model**, in which the substrate fits precisely into the active site (**Figure 2.24A**). In many cases, however, the configurations of both the enzyme and substrate are modified by substrate binding—a process called **induced fit** (**Figure 2.24B**). In such cases the conformation of the substrate is altered so that it more closely resembles that of the transition state. The stress produced by such distortion of the substrate can further facilitate its conversion to the transition state by weakening critical bonds. Moreover, the transition state is stabilized by its tight binding to the enzyme, thereby lowering the required energy of activation.

In addition to bringing multiple substrates together and distorting the conformation of substrates to approach the transition state, many enzymes participate directly in the catalytic process. In such cases, specific amino acid side chains in the active site may react with the substrate and form bonds with reaction intermediates. The acidic and basic amino acids are often involved in these catalytic mechanisms, as illustrated in the following discussion of chymotrypsin as an example of enzymatic catalysis.

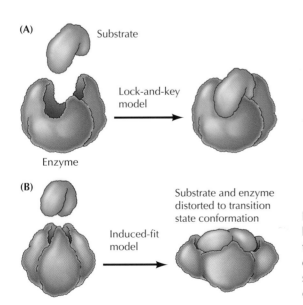

Figure 2.24 **Models of enzyme-substrate interaction** (A) In the lock-and-key model, the substrate fits precisely into the active site of the enzyme. (B) In the induced-fit model, substrate binding distorts the conformations of both substrate and enzyme. This distortion brings the substrate closer to the conformation of the transition state, thereby accelerating the reaction.

Chymotrypsin is a member of a family of enzymes (serine proteases) that digest proteins by catalyzing the hydrolysis of peptide bonds. The reaction can be written as follows:

$$\text{Protein} + \text{H}_2\text{O} \rightarrow \text{Peptide}_1 + \text{Peptide}_2$$

The different members of the serine protease family (including chymotrypsin, trypsin, elastase, and thrombin) have distinct substrate specificities; they preferentially cleave peptide bonds adjacent to different amino acids. For example, whereas chymotrypsin digests bonds adjacent to hydrophobic amino acids, such as tryptophan and phenylalanine, trypsin digests bonds next to basic amino acids, such as lysine and arginine. All the serine proteases, however, are similar in structure and use the same mechanism of catalysis. The active sites of these enzymes contain three critical amino acids—serine, histidine, and aspartate—that drive hydrolysis of the peptide bond. Indeed, these enzymes are called serine proteases because of the central role of the serine residue.

Substrates bind to the serine proteases by insertion of the amino acid adjacent to the cleavage site into a pocket at the active site of the enzyme (**Figure 2.25**). The nature of this pocket determines the substrate specificity of the different members of the serine protease family. For example, the binding pocket of chymotrypsin contains hydrophobic amino acids that interact with the hydrophobic side chains of its preferred substrates. In contrast, the binding pocket of trypsin contains a negatively charged acidic amino acid (aspartate), which is able to form an ionic bond with the lysine or arginine residues of its substrates.

Substrate binding positions the peptide bond to be cleaved adjacent to the active site serine (**Figure 2.26**). The proton of this serine is then transferred to the active site histidine. The conformation of the active site favors this proton transfer because the histidine interacts with the negatively charged aspartate residue. The serine reacts with the substrate, forming a tetrahedral transition state. The peptide bond is then cleaved, and the C-terminal portion of the substrate is released from the enzyme. However, the N-terminal

Proteases play a variety of roles in cells and viruses. Some HIV proteins are synthesized as precursors that need to be cleaved by a viral protease to make infectious viral particles. Inhibition of the protease blocks HIV replication, and drugs designed to inhibit the protease are currently used in the treatment of AIDS. (See Chapter 9.)

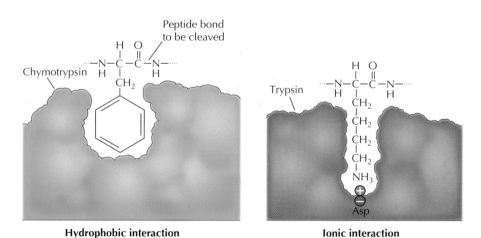

Hydrophobic interaction   **Ionic interaction**

Figure 2.25 Substrate binding by serine proteases   The amino acid adjacent to the peptide bond to be cleaved is inserted into a pocket at the active site of the enzyme. In chymotrypsin, the pocket binds hydrophobic amino acids; in trypsin, the binding pocket contains a negatively charged aspartate residue that binds basic amino acids via an ionic interaction.

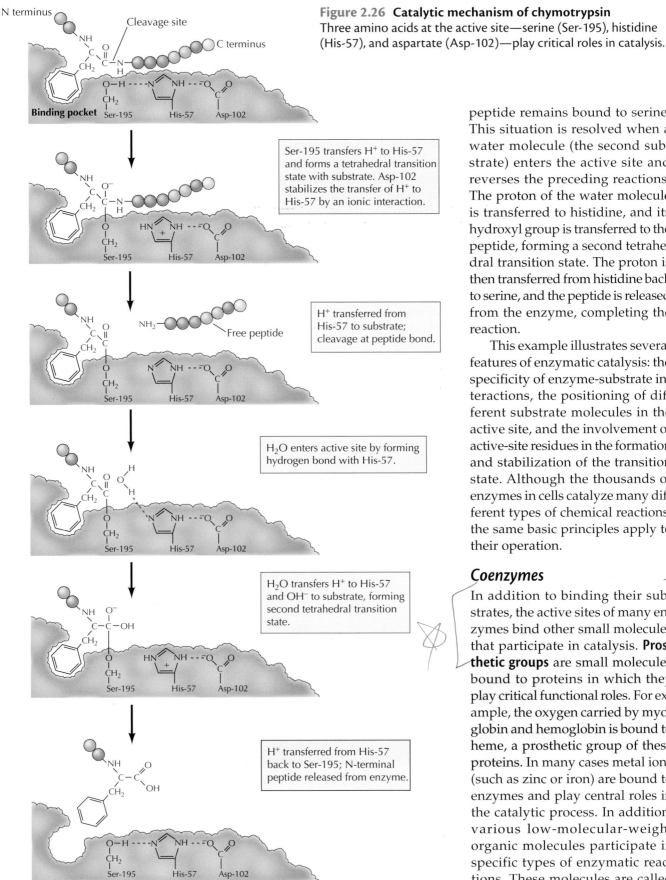

Figure 2.26 Catalytic mechanism of chymotrypsin
Three amino acids at the active site—serine (Ser-195), histidine (His-57), and aspartate (Asp-102)—play critical roles in catalysis.

Ser-195 transfers H⁺ to His-57 and forms a tetrahedral transition state with substrate. Asp-102 stabilizes the transfer of H⁺ to His-57 by an ionic interaction.

H⁺ transferred from His-57 to substrate; cleavage at peptide bond.

H₂O enters active site by forming hydrogen bond with His-57.

H₂O transfers H⁺ to His-57 and OH⁻ to substrate, forming second tetrahedral transition state.

H⁺ transferred from His-57 back to Ser-195; N-terminal peptide released from enzyme.

peptide remains bound to serine. This situation is resolved when a water molecule (the second substrate) enters the active site and reverses the preceding reactions. The proton of the water molecule is transferred to histidine, and its hydroxyl group is transferred to the peptide, forming a second tetrahedral transition state. The proton is then transferred from histidine back to serine, and the peptide is released from the enzyme, completing the reaction.

This example illustrates several features of enzymatic catalysis: the specificity of enzyme-substrate interactions, the positioning of different substrate molecules in the active site, and the involvement of active-site residues in the formation and stabilization of the transition state. Although the thousands of enzymes in cells catalyze many different types of chemical reactions, the same basic principles apply to their operation.

Coenzymes

In addition to binding their substrates, the active sites of many enzymes bind other small molecules that participate in catalysis. **Prosthetic groups** are small molecules bound to proteins in which they play critical functional roles. For example, the oxygen carried by myoglobin and hemoglobin is bound to heme, a prosthetic group of these proteins. In many cases metal ions (such as zinc or iron) are bound to enzymes and play central roles in the catalytic process. In addition, various low-molecular-weight organic molecules participate in specific types of enzymatic reactions. These molecules are called

coenzymes because they work together with enzymes to enhance reaction rates. In contrast to substrates, coenzymes are not irreversibly altered by the reactions in which they are involved. Rather, they are recycled and can participate in multiple enzymatic reactions.

Coenzymes serve as carriers of several types of chemical groups. A prominent example of a coenzyme is **nicotinamide adenine dinucleotide (NAD$^+$)**, which functions as a carrier of electrons in oxidation–reduction reactions (**Figure 2.27**). NAD$^+$ can accept a hydrogen ion (H$^+$) and two electrons (e$^-$) from one substrate, forming NADH. NADH can then donate these electrons to a second substrate, re-forming NAD$^+$. Thus, NAD$^+$ transfers electrons from the first substrate (which becomes oxidized) to the second substrate (which becomes reduced).

Several other coenzymes also act as electron carriers, and still others are involved in the transfer of a variety of additional chemical groups (e.g.,

(A)

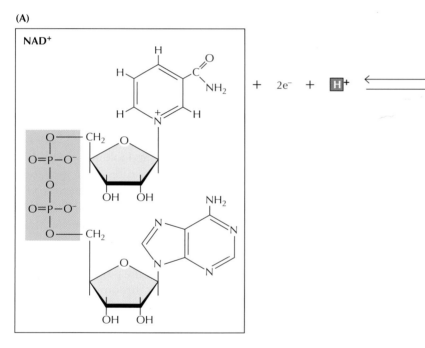

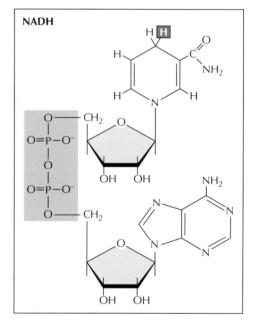

(B)

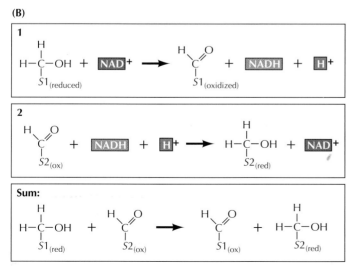

Figure 2.27 Role of NAD$^+$ in oxidation–reduction reactions (A) Nicotinamide adenine dinucleotide (NAD$^+$) acts as a carrier of electrons in oxidation–reduction reactions by accepting electrons (e$^-$) to form NADH. (B) For example, NAD$^+$ can accept electrons from one substrate (S1), yielding oxidized S1 plus NADH. The NADH formed in this reaction can then transfer its electrons to a second substrate (S2), yielding reduced S2 and regenerating NAD$^+$. The net effect is the transfer of electrons (carried by NADH) from S1 (which becomes oxidized) to S2 (which becomes reduced).

Table 2.1   Examples of Coenzymes and Vitamins

Coenzyme	Related vitamin	Chemical reaction
NAD⁺, NADP⁺	Niacin	Oxidation–reduction
FAD	Riboflavin (B2)	Oxidation–reduction
Thiamine pyrophosphate	Thiamine (B1)	Aldehyde group transfer
Coenzyme A	Pantothenate	Acyl group transfer
Tetrahydrofolate	Folate	Methyl group transfer
Pyridoxal phosphate	Pyridoxal (B6)	Transamination
Biotin	Biotin	Carboxylation and carboxyl group transfer

carboxyl groups and acyl groups; Table 2.1). The same coenzymes function together with a variety of different enzymes to catalyze the transfer of specific chemical groups between a wide range of substrates. Many coenzymes are closely related to vitamins, which contribute part or all of the structure of the coenzyme. Vitamins are not required by bacteria such as *E. coli* but are necessary components of the diets of humans and other higher animals, which have lost the ability to synthesize these compounds.

Regulation of enzyme activity

An important feature of most enzymes is that their activities are not constant but can be modulated. That is, the activities of enzymes can be regulated so that they function appropriately to meet the varied physiological needs that may arise during the life of the cell.

One common type of enzyme regulation is **feedback inhibition,** in which the product of a metabolic pathway inhibits the activity of an enzyme involved in its synthesis. For example, the amino acid isoleucine is synthesized by a series of reactions starting from the amino acid threonine (**Figure 2.28**). The first step in the pathway is catalyzed by the enzyme threonine deaminase, which is inhibited by isoleucine, the end product of the pathway. Thus, an adequate amount of isoleucine in the cell inhibits threonine deaminase, blocking further synthesis of isoleucine. If the concentration of isoleucine decreases, feedback inhibition is relieved, threonine deaminase is no longer inhibited, and additional isoleucine is synthesized. By regulating the activity of threonine deaminase, the cell synthesizes the necessary amount of isoleucine but avoids wasting energy on the synthesis of more isoleucine than is needed.

Feedback inhibition is one example of **allosteric regulation** in which enzyme activity is controlled by the binding of small molecules to regulatory sites on the enzyme (**Figure 2.29**). The term "allosteric regulation" derives from the fact that the regulatory molecules bind not to the catalytic site but to a distinct site on the protein (*allo* = "other" and *steric* = "site"). Binding of the regulatory molecule changes the conformation of the protein, which in turn alters the shape of the active site and the catalytic activity of the enzyme.

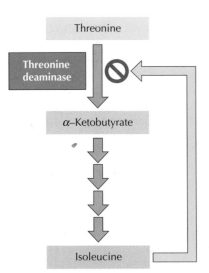

Figure 2.28 Feedback inhibition   The first step in the conversion of threonine to isoleucine is catalyzed by the enzyme threonine deaminase. The activity of this enzyme is inhibited by isoleucine, the end product of the pathway.

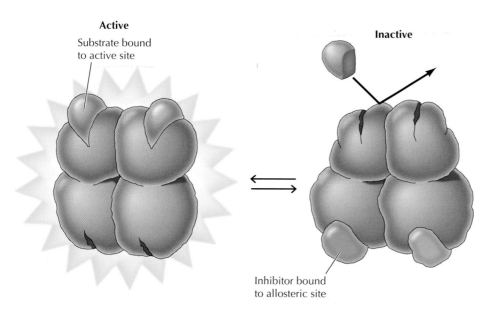

Active

Substrate bound
to active site

Inactive

Inhibitor bound
to allosteric site

Figure 2.29 Allosteric regulation In this example, enzyme activity is inhibited by the binding of a regulatory molecule to an allosteric site. In the absence of an inhibitor, the substrate binds to the active site of the enzyme and the reaction proceeds. The inhibitor induces a conformational change in the enzyme that prevents substrate binding. Most allosteric enzymes consist of multiple subunits.

In the case of threonine deaminase, binding of the regulatory molecule (isoleucine) inhibits enzymatic activity. In other cases, regulatory molecules serve as activators, stimulating rather than inhibiting their target enzymes.

The activities of enzymes can also be regulated by their interactions with other proteins and by covalent modifications, such as the addition of phosphate groups to serine, threonine, or tyrosine residues. **Phosphorylation** is a particularly common mechanism for regulating enzyme activity; the addition of phosphate groups either stimulates or inhibits the activities of many different enzymes (**Figure 2.30**). For example, muscle cells respond to epinephrine (adrenaline) by breaking down glycogen into glucose, thereby providing a source of energy for increased muscular activity. The breakdown of glycogen is catalyzed by the enzyme glycogen phosphorylase, which is

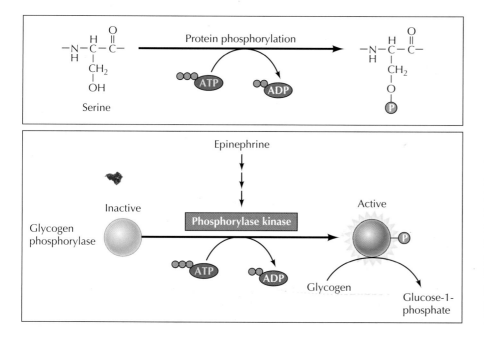

Protein phosphorylation

Serine

Epinephrine

Phosphorylase kinase

Glycogen
phosphorylase

Inactive

Active

Glycogen

Glucose-1-
phosphate

Figure 2.30 Protein phosphorylation Some enzymes are regulated by the addition of phosphate groups to the side-chain OH groups of serine (as shown here), threonine, or tyrosine residues. For example, the enzyme glycogen phosphorylase, which catalyzes the conversion of glycogen into glucose-1-phosphate, is activated by phosphorylation in response to the binding of epinephrine to muscle cells.

activated by phosphorylation in response to the binding of epinephrine to a receptor on the surface of the muscle cell. Protein phosphorylation plays a central role in controlling not only metabolic reactions but also many other cellular functions, including cell growth and differentiation.

Cell Membranes

The structure and function of cells are critically dependent on membranes, which not only separate the interior of the cell from its environment but also define the internal compartments of eukaryotic cells, including the nucleus and cytoplasmic organelles. The formation of biological membranes is based on the properties of lipids, and all cell membranes share a common structural organization: bilayers of phospholipids with associated proteins. These membrane proteins are responsible for many specialized functions; some act as receptors that allow the cell to respond to external signals, some are responsible for the selective transport of molecules across the membrane, and others participate in electron transport and oxidative phosphorylation. In addition, membrane proteins control the interactions between cells of multicellular organisms. The common structural organization of membranes thus underlies a variety of biological processes and specialized membrane functions, which will be discussed in detail in later chapters.

Membrane lipids

The fundamental building blocks of all cell membranes are phospholipids, which are amphipathic molecules, consisting of two hydrophobic fatty acid chains linked to a phosphate-containing hydrophilic head group (see Figure 2.7). Because their fatty acid tails are poorly soluble in water, phospholipids spontaneously form bilayers in aqueous solutions, with the hydrophobic tails buried in the interior of the membrane and the polar head groups exposed on both sides, in contact with water (**Figure 2.31**). Such **phospholipid bilayers** form a stable barrier between two aqueous compartments and represent the basic structure of all biological membranes.

Lipids constitute approximately 50% of the mass of most cell membranes, although this proportion varies depending on the type of membrane. Plasma membranes, for example, are approximately 50% lipid and 50% protein. The inner membrane of mitochondria, on the other hand, contains an unusually high fraction (about 75%) of protein, reflecting the abundance of protein complexes involved in electron transport and oxidative phosphorylation. The lipid composition of different cell membranes also varies. The plasma membrane of *E. coli* consists predominantly of the phospholipid phosphatidylethanolamine. Plasma membranes of animal cells are more complex, containing five major phospholipids—phosphatidylcholine, phosphatidylserine, phosphatidylethanolamine, phosphatidylinositol, and sphingomyelin. In addition to the phospholipids, glycolipids (which have carbohydrate head groups) and cholesterol generally correspond to about half of the lipid molecules of animal cell plasma membranes.

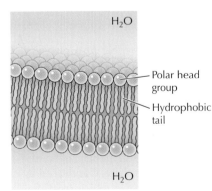

Figure 2.31 A phospholipid bilayer Phospholipids spontaneously form stable bilayers, with their polar head groups exposed to water and their hydrophobic tails buried in the interior of the membrane.

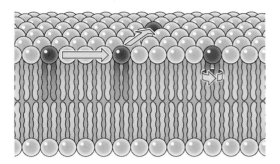

Figure 2.32 **Mobility of phospholipids in a membrane**   Individual phospholipids can rotate and move laterally within a bilayer.

An important property of lipid bilayers is that they behave as two-dimensional fluids in which individual molecules (both lipids and proteins) are free to rotate and move in lateral directions (**Figure 2.32**). Such fluidity is a critical property of membranes and is determined by both temperature and lipid composition. For example, the interactions between shorter fatty acid chains are weaker than those between longer chains, so membranes containing shorter fatty acid chains are less rigid and remain fluid at lower temperatures. Lipids containing unsaturated fatty acids similarly increase membrane fluidity because the presence of double bonds introduces kinks in the fatty acid chains, making them more difficult to pack together.

Because of its hydrocarbon ring structure (see Figure 2.9), cholesterol plays a distinct role in determining membrane fluidity. Cholesterol molecules insert into the bilayer with their polar hydroxyl groups close to the hydrophilic head groups of the phospholipids (**Figure 2.33**). The rigid hydrocarbon rings of cholesterol therefore interact with the regions of the fatty acid chains that are adjacent to the phospholipid head groups. This interaction decreases the mobility of the outer portions of the fatty acid chains, making this part of the membrane more rigid. On the other hand, insertion of cholesterol interferes with interactions between fatty acid chains, thereby maintaining membrane fluidity at lower temperatures.

Membrane proteins

Proteins are the other major constituents of cell membranes, constituting 25 to 75% of the mass of the various membranes of the cell. The basic model of membrane structure, proposed by Jonathan Singer and Garth Nicolson in 1972, views the membrane as a **fluid mosaic** in which proteins are inserted into a lipid bilayer (**Figure 2.34**). While phospholipids provide the fundamental structure of membranes, membrane proteins carry out the specific functions of the different membranes of the cell. These proteins are divided into two general classes, based on the nature of their association with the membrane. **Integral membrane proteins** are embedded directly within the lipid bilayer. **Peripheral membrane proteins** are not inserted into the lipid bilayer but are associated with the membrane indirectly, generally by interactions with integral membrane proteins.

Most integral membrane proteins (called **transmembrane proteins**) span the lipid bilayer, with portions exposed on both sides of the membrane. The

Figure 2.33 **Insertion of cholesterol in a membrane**   Cholesterol inserts into the membrane with its polar hydroxyl group close to the polar head groups of the phospholipids.

Video 2.2

sites.sinauer.com/cooper7e/v2.2
Ordering Effect of Cholesterol in a Lipid Bilayer Cholesterol molecules in a cell membrane influence the arrangement of the phospholipids in the bilayer.

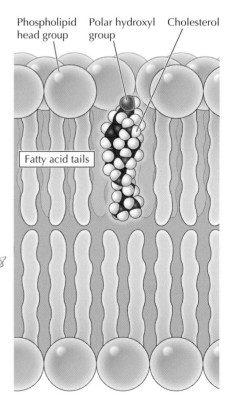

Phospholipid head group   Polar hydroxyl group   Cholesterol

Fatty acid tails

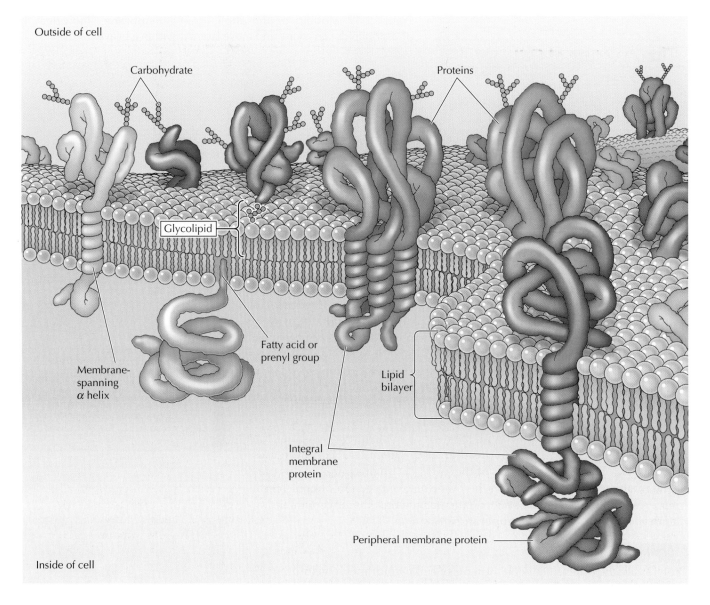

Outside of cell

Carbohydrate

Proteins

Glycolipid

Membrane-
spanning
α helix

Fatty acid or
prenyl group

Lipid
bilayer

Integral
membrane
protein

Peripheral membrane protein

Inside of cell

Figure 2.34 Fluid mosaic model of membrane structure Biological membranes consist of proteins inserted into a lipid bilayer. Integral membrane proteins are embedded in the membrane, usually via α-helical regions of 20–25 hydrophobic amino acids. Some transmembrane proteins span the membrane only once; others have multiple membrane-spanning regions. In addition, some proteins are anchored in the membrane by lipids that are covalently attached to the polypeptide chain. These proteins can be anchored to the extracellular face of the plasma membrane by glycolipids and to the cytosolic face by fatty acids or prenyl groups (see Chapter 9 for structures). Peripheral membrane proteins are not inserted in the membrane but are attached via interactions with integral membrane proteins or phospholipids.

membrane-spanning portions of these proteins are usually α-helical regions of 20 to 25 nonpolar amino acids. The hydrophobic side chains of these amino acids interact with the fatty acid chains of membrane lipids, and the formation of an α helix neutralizes the polar character of the peptide bonds,

The Structure of Cell Membranes

The Fluid Mosaic Model of the Structure of Cell Membranes

S. J. Singer and Garth L. Nicolson

University of California at San Diego and the Salk Institute for Biological Sciences, La Jolla, CA

Science, Volume 175, 1972, pages 720–731

The Context

Cell membranes play central roles in virtually all aspects of cell biology, serving as boundaries of the subcellular organelles of eukaryotic cells as well as defining the cell itself by separating its contents from the external environment. Understanding the structure and organization of cell membranes was thus a critical step in unraveling the molecular basis of cell behavior. In the 1960s, it was known that cell membranes were composed of proteins and lipids, but it was not clear how these molecules were organized. Moreover, there was a large amount of variation in the protein and lipid compositions of different cell membranes, leading some investigators to believe that different membranes might not share a common structural organization.

In contrast, Jonathan Singer and Garth Nicolson approached the problem of understanding membrane structure with the assumption that the same general principles would describe the organization of lipids and proteins in all cell membranes. They applied the principles of thermodynamics as well as integrating a variety of experimental results to develop the fluid mosaic model—a model that has stood the test of time and provided the structural basis for understanding the diverse functions of membranes in cell biology.

The Experiments

Considerations of thermodynamics were basic to Singer and Nicolson's development of the fluid mosaic model. In particular, they reasoned that the overall structure of membranes would be determined by a combination of hydrophobic and hydrophilic interactions. These interactions were recognized as determinants of the organization of phospholipid bilayers, and Singer and Nicolson assumed that similar interactions would govern the organization of membrane proteins. They therefore proposed that the nonpolar amino acid residues of membrane proteins would be sequestered within the membrane, like the fatty acid chains of the phospholipids, whereas polar groups of proteins would be exposed to the aqueous environment. These considerations argued strongly against an earlier model of membrane structure in which it was proposed that membranes had a three-layer structure consisting of an inner lipid bilayer sandwiched between two protein layers.

Singer and Nicolson instead proposed a mosaic structure, in which globular integral membrane proteins were inserted into a phospholipid bilayer (see figure). The proteins, like the phospholipids, were postulated to be amphipathic, with nonpolar regions inserted into the lipid bilayer and polar regions exposed to the aqueous environment. The lipid bilayer was thought to constitute the structural matrix of the membrane in which the proteins were embedded.

Several lines of experimental evidence were cited in support of the fluid mosaic model. First, electron microscopy of membranes using the freeze fracture technique provided direct evidence suggesting that proteins were embedded in the lipid bilayer (see Figure 1.37). Also as predicted by the fluid mosaic model, proteins were found to be randomly distributed in the plane of the membrane when detected by staining the surface of cells. And finally, as discussed in Chapter 14, experiments of Larry Frye and Michael Edidin demonstrated that proteins could move laterally through the membrane. Multiple experimental approaches thus indicated that membrane proteins were inserted into and

S. J. Singer

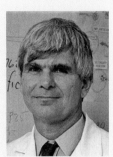
Garth L. Nicolson

able to move laterally through the lipid bilayer, providing strong support for the fluid mosaic model of membrane structure.

The Impact

Singer and Nicolson's fluid mosaic model has been abundantly supported by further experimentation, and continues to represent our basic understanding of the structure of cell membranes. Given the diverse roles of membranes in cell biology, the fluid mosaic model is thus fundamental to understanding virtually all aspects of cell behavior, including phenomena as diverse as energy metabolism, hormone action, and the transmission of information between nerve cells at a synapse. Initially based on thermodynamic considerations of the properties of lipids and proteins, Singer and Nicolson developed a model of membrane structure that has impacted all aspects of cell biology.

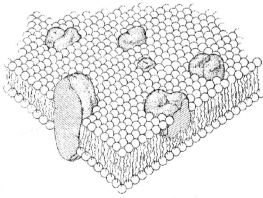

The lipid-globular protein mosaic model Globular proteins are distributed in a matrix of phospholipids.

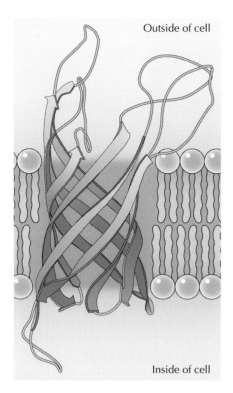

Outside of cell

Inside of cell

Figure 2.35 Structure of a β barrel Some transmembrane proteins span the phospholipid bilayer as β sheets folded into a barrel-like structure.

as discussed earlier in this chapter with respect to protein folding. The only other protein structure known to span lipid bilayers is the **β barrel**, formed by the folding of β sheets into a barrel-like structure, which is found in some transmembrane proteins of bacteria, chloroplasts, and mitochondria (**Figure 2.35**). Like the phospholipids, transmembrane proteins are amphipathic molecules, with their hydrophilic portions exposed to the aqueous environment on both sides of the membrane. Some transmembrane proteins span the membrane only once; others have multiple membrane-spanning regions. Most transmembrane proteins of eukaryotic plasma membranes have been modified by the addition of carbohydrates, which are exposed on the surface of the cell and may participate in cell-cell interactions.

Proteins can also be anchored in membranes by lipids that are covalently attached to the polypeptide chain (see Chapter 9). Distinct lipid modifications anchor proteins to the cytosolic and extracellular faces of the plasma membrane. Proteins can be anchored to the cytosolic face of the membrane either by the addition of a 14-carbon fatty acid (myristic acid) to their amino terminus or by the addition of either a 16-carbon fatty acid (palmitic acid) or 15- or 20-carbon prenyl groups to the side chains of cysteine residues. Alternatively, proteins are anchored to the extracellular face of the plasma membrane by the addition of glycolipids to their carboxy terminus.

Transport across cell membranes

The selective permeability of biological membranes to small molecules allows the cell to control and maintain its internal composition. Only small, uncharged molecules can diffuse freely through phospholipid bilayers (**Figure 2.36**). Small nonpolar molecules, such as O_2 and CO_2, are soluble in the lipid bilayer and therefore can readily cross cell membranes. Small, uncharged polar molecules, such as H_2O, also can diffuse through membranes, but larger

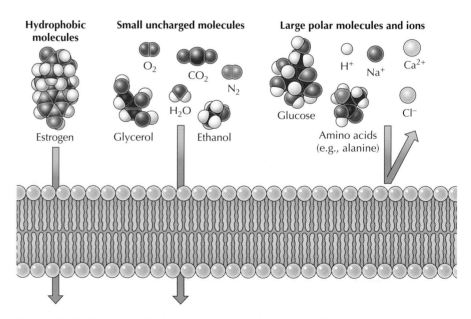

Figure 2.36 Permeability of phospholipid bilayers Gases, hydrophobic, and small uncharged molecules can diffuse freely through a phospholipid bilayer. However, the bilayer is impermeable to larger polar molecules (such as glucose and amino acids) and to ions.

uncharged polar molecules, such as glucose, cannot. Charged molecules, such as ions, are unable to diffuse through a phospholipid bilayer regardless of size; even H^+ ions cannot cross a lipid bilayer by free diffusion.

Although ions and most polar molecules cannot diffuse across a lipid bilayer, many such molecules (such as glucose) are able to cross cell membranes. These molecules pass across membranes via the action of specific transmembrane proteins, which act as transporters. Such transport proteins determine the selective permeability of cell membranes and thus play a critical role in membrane function. They contain multiple membrane-spanning regions that form a passage through the lipid bilayer, allowing polar or charged molecules to cross the membrane through a protein pore without interacting with the hydrophobic fatty acid chains of the membrane phospholipids.

There are two general classes of membrane transport proteins (**Figure 2.37**). **Channel proteins** form open pores through the membrane, allowing the free passage of any molecule of the appropriate size. Ion channels, for example, allow the passage of inorganic ions such as Na^+, K^+, Ca^{2+}, and Cl^- across the plasma membrane. The pores formed by these channel proteins are not permanently open; rather, they can be selectively opened and closed in response to extracellular signals, allowing the cell to control the movement of ions across the membrane. Such regulated ion channels have been particularly well studied in nerve and muscle cells, where they mediate the transmission of electrochemical signals.

In contrast to channel proteins, **carrier proteins** selectively bind and transport specific small molecules, such as glucose. Rather than forming open channels, carrier proteins act like enzymes to facilitate the passage of specific molecules across membranes. In particular, carrier proteins bind specific molecules and then undergo conformational changes that open channels through which the molecule to be transported can pass across the membrane and be released on the other side.

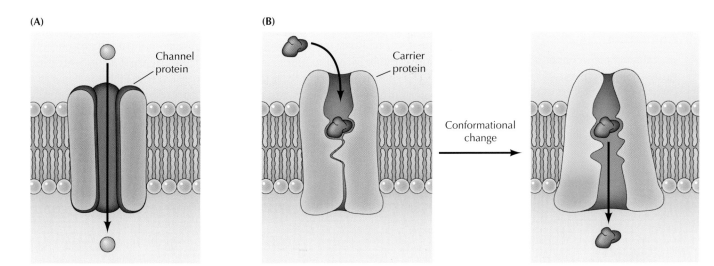

(A)   **(B)**

Figure 2.37 Channel and carrier proteins   (A) Channel proteins form open pores through which molecules of the appropriate size (e.g., ions) can cross the membrane. (B) Carrier proteins selectively bind the small molecule to be transported and then undergo a conformational change to release the molecule on the other side of the membrane.

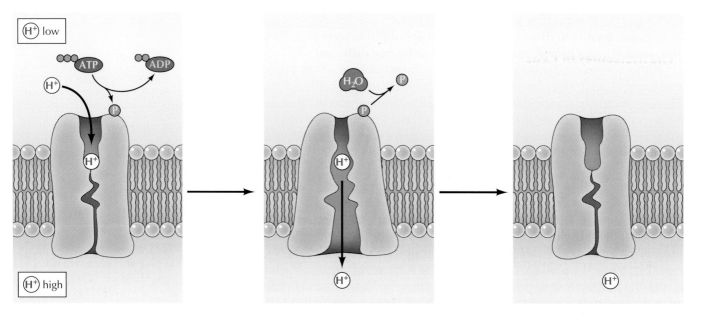

Figure 2.38 Model of active transport   Energy derived from the hydrolysis of ATP is used to transport H⁺ against the electrochemical gradient (from low to high H⁺ concentration). Binding of H⁺ is accompanied by phosphorylation of the carrier protein, which induces a conformational change that drives H⁺ transport against the electrochemical gradient. Release of H⁺ and hydrolysis of the bound phosphate group then restores the carrier to its original conformation.

Animation 2.4

sites.sinauer.com/cooper7e/a2.4

Passive Transport Molecules that cross a membrane in an energetically favorable direction do so by passive transport.

As described so far, molecules transported by either channel or carrier proteins cross membranes in the energetically favorable direction, as determined by concentration and electrochemical gradients—a process known as **passive transport**. However, carrier proteins also provide a mechanism through which the energy changes associated with transporting molecules across a membrane can be coupled to the use or production of other forms of metabolic energy, just as enzymatic reactions can be coupled to the hydrolysis or synthesis of ATP. For example, molecules can be transported in an energetically unfavorable direction across a membrane (e.g., against a concentration gradient) if their transport in that direction is coupled to ATP hydrolysis as a source of energy—a process called **active transport** (Figure 2.38). Membrane proteins can thus use the free energy stored as ATP to control the internal composition of the cell.

SUMMARY	KEY TERMS

The Molecules of Cells

- *Carbohydrates:* Carbohydrates include simple sugars and polysaccharides. Polysaccharides serve as storage forms of sugars, structural components of cells, and markers for cell recognition processes. See Animation 2.1.

hydrophilic, hydrophobic, carbohydrate, monosaccharide, glycosidic bond, oligosaccharide, polysaccharide, glycogen, starch, cellulose, chitin

- *Lipids:* Lipids are the principal components of cell membranes, and they serve as energy storage and signaling molecules. Phospholipids consist of two hydrophobic fatty acid chains linked to a hydrophilic phosphate-containing head group.

lipid, fatty acid, triacylglycerol, fats, phospholipid, glycerol phospholipid, sphingomyelin, amphipathic, glycolipid, cholesterol, steroid hormone

- *Nucleic acids:* Nucleic acids are the principal informational molecules of the cell. Both DNA and RNA are polymers of purine and pyrimidine nucleotides. Hydrogen bonding between complementary base pairs allows nucleic acids to direct their self-replication.

deoxyribonucleic acid (DNA), ribonucleic acid (RNA), messenger RNA (mRNA), ribosomal RNA(rRNA), transfer RNA(tRNA), nucleotide, purine, pyrimidine, adenine, guanine, cytosine, thymine, uracil, 2′-deoxyribose, ribose, nucleoside, phosphodiester bond, oligonucleotide, polynucleotide

- *Proteins:* Proteins are polymers of 20 different amino acids, each of which has a distinct side chain with specific chemical properties. Each protein has a unique amino acid sequence, which determines its three-dimensional structure. In most proteins, combinations of α helices and β sheets fold into globular domains with hydrophobic amino acids in the interior and hydrophilic amino acids on the surface. See Video 2.1.

protein, amino acid, peptide bond, polypeptide, X-ray crystallography, primary structure, secondary structure, α helix, β sheet, tertiary structure, domain, quaternary structure

Enzymes as Biological Catalysts

- *The catalytic activity of enzymes:* Almost all chemical reactions within cells are catalyzed by enzymes. See Animation 2.2.

enzyme, substrate, product, transition state, activation energy, active site

- *Mechanisms of enzymatic catalysis:* Enzymes increase reaction rates by binding substrates in the proper position, by altering the conformation of substrates to approach the transition state, and by participating directly in chemical reactions. See Animation 2.3.

lock-and-key model, induced fit

- *Coenzymes:* Coenzymes function in conjunction with enzymes to carry chemical groups between substrates.

prosthetic group, coenzyme, nicotinamide adenine dinucleotide (NAD$^+$)

- *Regulation of enzyme activity:* The activities of enzymes are regulated to meet the physiological needs of the cell. Enzyme activity can be controlled by the binding of small molecules, by interactions with other proteins, and by covalent modifications.

feedback inhibition, allosteric regulation, phosphorylation

Cell Membranes

- *Membrane lipids:* The basic structure of all cell membranes is a phospholipid bilayer. Membranes of animal cells also contain glycolipids and cholesterol. See Video 2.2.

phospholipid bilayer

- *Membrane proteins:* Proteins can either be inserted into the lipid bilayer or associated with the membrane indirectly, by protein-protein interactions. Some proteins span the lipid bilayer; others are anchored to one side of the membrane.

fluid mosaic model, integral membrane protein, peripheral membrane protein, transmembrane protein, β-barrel

- *Transport across cell membranes:* Lipid bilayers are permeable only to small, uncharged molecules. Ions and most polar molecules are transported across cell membranes by specific transport proteins, the action of which can be coupled to the hydrolysis or synthesis of ATP. See Animation 2.4.

channel protein, carrier protein, passive transport, active transport

Questions

1. What characteristics of water contribute to its function as the most abundant molecule in cells?

2. Although glycogen and cellulose are both composed of glucose monomers, they are chemically distinct. In what way?

3. What are the major functions of fats and phospholipids in cells?

4. In addition to serving as the building blocks for nucleic acids, what other important functions do nucleotides have in cells?

5. If you removed all the bases from an mRNA molecule, could it still function as an information-carrying molecule?

6. What was the experimental evidence that demonstrated that the information necessary for the folding of proteins is contained in the sequence of their amino acids?

7. How does the side chain of cysteine help in the folding of certain proteins?

8. What are the biological roles of cholesterol?

9. Why are the membrane-spanning regions of transmembrane proteins frequently α helical? What other protein structure is capable of spanning membranes?

10. Which of the following amino acids would most likely be found in a membrane-spanning α helix?

    a. Lysine

    b. Glutamine

    c. Aspartic acid

    d. Alanine

    e. Arginine

11. The binding pocket of trypsin contains an aspartate residue. How would changing this amino acid to lysine affect the enzyme's activity?

12. What characteristic of histidine allows it to play a role in enzymatic reactions involving the exchange of hydrogen ions?

13. How do carrier proteins differ from channel proteins?

References and Further Reading   (Key review articles for each major section are highlighted in **bold**.)

The Molecules of Cells

Anfinsen, C. B. 1973. Principles that govern the folding of protein chains. *Science* 181: 223–230. [P]

Berg, J. M., J. L. Tymoczko and L. Stryer. 2011. *Biochemistry*. 7th ed. New York: W. H. Freeman.

Branden, C. and J. Tooze. 1999. *Introduction to Protein Structure*. 2nd ed. New York: Garland.

Braselmann, E., J. L. Chaney and P. L. Clark. 2013. Folding the proteome. *Trends Biochem. Sci.* 38: 337–344. [R]

Dill, K. A. and J. L. MacCallum. 2012. The protein-folding problem, 50 years on. *Science* 338: 1042–1046. [R]

Kendrew, J. C. 1961. The three-dimensional structure of a protein molecule. *Sci. Am.* 205(6): 96–111. [R]

Nelson, D. L. and M. M. Cox. 2012. *Lehninger Principles of Biochemistry*. 6th ed. New York: W. H. Freeman.

Pauling, L., R. B. Corey, and H. R. Branson. 1951. The structure of proteins: Two hydrogen bonded configurations of the polypeptide chain. *Proc. Natl. Acad. Sci. USA* 37: 205–211. [P]

Petsko, G. A. and D. Ringe. 2003. *Protein Structure and Function*. Sunderland, Mass.: Sinauer.

Sanger, F. 1988. Sequences, sequences, and sequences. *Ann. Rev. Biochem.* 57: 1–28. [R]

Tanford, C. 1978. The hydrophobic effect and the organization of living matter. *Science* 200: 1012–1018. [R]

Enzymes as Biological Catalysts

Changeux, J.-P. 2013. 50 years of allosteric interactions: the twists and turns of the models. *Nature Rev. Mol. Cell Biol.* 14: 819–829. [R]

Fersht, A. 1999. *Structure and Mechanism in Protein Science: A Guide to Enzyme Catalysis and Protein Folding*. New York: W. H. Freeman.

Koshland, D. E. 1984. Control of enzyme activity and metabolic pathways. *Trends Biochem. Sci.* 9: 155–159. [R]

Laskowski, R. A., F. Gerick and J. M. Thornton. 2009. The structural basis of allosteric regulation in proteins. *FEBS Letters.* 583: 1692–1698. [R]

Lienhard, G. E. 1973. Enzymatic catalysis and transition-state theory. *Science* 180: 149–154. [R]

Lipscomb, W. N. 1983. Structure and catalysis of enzymes. *Ann. Rev. Biochem.* 52: 17–34. [R]

Monod, J., J.-P. Changeux and F. Jacob. 1963. Allosteric proteins and cellular control systems. *J. Mol. Biol.* 6: 306–329. [P]

Narlikar, G. J. and D. Herschlag. 1997. Mechanistic aspects of enzymatic catalysis: Lessons from comparison of RNA and protein enzymes. *Ann. Rev. Biochem.* 66: 19–59. [R]

Neurath, H. 1984. Evolution of proteolytic enzymes. *Science* 224: 350–357. [R]

Petsko, G. A. and D. Ringe. 2003. *Protein Structure and Function.* Sunderland, Mass: Sinauer.

Schramm, V. L. 1998. Enzymatic transition states and transition state analog design. *Ann. Rev. Biochem.* 67: 693–720. [R]

Cell Membranes

Bretscher, M. 1985. The molecules of the cell membrane. *Sci. Am.* 253(4): 100–108. [R]

Dowham, W. 1997. Molecular basis for membrane phospholipid diversity: Why are there so many lipids? *Ann. Rev. Biochem.* 66: 199–212. [R]

Englund, P. T. 1993. The structure and biosynthesis of glycosyl phosphatidylinositol protein anchors. *Ann. Rev. Biochem.* 62: 121–138. [R]

Farazi, T. A., G. Waksman and J. I. Gordon. 2001. The biology and enzymology of protein N-myristoylation. *J. Biol. Chem.* 276: 39501–39504. [R]

Lacapere, J.-J., E. Pebay-Peyroula, J.-M. Neumann and C. Etchebest. 2007. Determining membrane protein structures: still a challenge! *Trends Biochem. Sci.* 32: 259–270. [R],

Singer, S. J. 1990. The structure and insertion of integral proteins in membranes. *Ann. Rev. Cell Biol.* 6: 247–296. [R]

Singer, S. J. and G. L. Nicolson. 1972. The fluid mosaic model of the structure of cell membranes. *Science* 175: 720–731. [P]

Tamm, L. K., A. Arora and J. H. Kleinschmidt. 2001. Structure and assembly of β-barrel membrane proteins. *J. Biol. Chem.* 276: 32399–32402. [R]

Towler, D. A., J. I. Gordon, S. P. Adams and L. Glaser. 1988. The biology and enzymology of eukaryotic protein acylation. *Ann. Rev. Biochem.* 57: 69–99. [R]

White, S. H., A. S. Ladokhin, S. Jayasinghe and K. Hristova. 2001. How membranes shape protein structure. *J. Biol. Chem.* 276: 32395–32398. [R]

Zhang, F. L. and P. J. Casey. 1996. Protein prenylation: Molecular mechanisms and functional consequences. *Ann. Rev. Biochem.* 65: 241–269. [R]

Bioenergetics and Metabolism

Almost all cellular activities, including movement, membrane transport and the synthesis of cell constituents, require energy. Consequently, the generation and utilization of metabolic energy is fundamental to all of cell biology. All cells use ATP as their source of metabolic energy, and the mechanisms that cells use for the generation of ATP, either from the breakdown of organic molecules or from photosynthesis, are discussed in this chapter. In addition, this chapter presents an overview of the network of chemical reactions that constitute the metabolism of the cell and are responsible for the synthesis of major cell constituents, including carbohydrates, lipids, proteins, and nucleic acids.

Metabolic Energy and ATP

Energy is a basic requirement for many tasks that a cell must perform, so a large portion of the cell's activities is devoted to obtaining energy from the environment and using that energy to drive energy-requiring reactions. As discussed in the previous chapter, enzymes control the rates of virtually all chemical reactions within cells and therefore determine what chemical reactions can take place under the mild conditions of living cells. However, enzymes only control reaction rates; the equilibrium position of chemical reactions is not affected by enzymatic catalysis. Rather, the laws of thermodynamics govern chemical equilibria and determine the energetically favorable direction of all chemical reactions. Many of the reactions that must take place within cells are energetically unfavorable, and are therefore able to proceed only at the cost of additional energy input. Consequently, cells must constantly generate and expend energy derived from the environment.

Free energy and ATP

The energetics of biochemical reactions are best described in terms of the thermodynamic function called **Gibbs free energy (G)**, named for Josiah Willard Gibbs. The change in free energy (ΔG) of a reaction combines the effects of changes in enthalpy (the heat released or absorbed during a chemical reaction) and entropy (the degree of disorder resulting from a reaction) to predict whether or not a reaction is energetically favorable. All chemical reactions spontaneously proceed in the energetically favorable direction, accompanied by a decrease in free energy ($\Delta G < 0$). For example, consider a hypothetical reaction in which A is converted to B:

$$A \rightleftharpoons B$$

If $\Delta G < 0$, this reaction will proceed in the forward direction, as written. If $\Delta G > 0$, however, the reaction will proceed in the reverse direction, and *B* will be converted to *A*.

The ΔG of a reaction is determined not only by the intrinsic properties of reactants and products, but also by their concentrations and other reaction conditions (e.g., temperature). It is thus useful to define the free-energy change of a reaction under standard conditions. (Standard conditions are considered to be a 1-*M* concentration of all reactants and products, and 1 atm of pressure). The standard free-energy change ($\Delta G°$) of a reaction is directly related to its equilibrium position because the actual ΔG is a function of both $\Delta G°$ and the concentrations of reactants and products. For example, consider the reaction

$$A \rightleftharpoons B$$

The free-energy change can be written as follows:

$$\Delta G = \Delta G° + RT \ln [B]/[A]$$

where R is the gas constant and T is the absolute temperature.

At equilibrium, $\Delta G = 0$ and the reaction does not proceed in either direction. The equilibrium constant for the reaction ($K = [B]/[A]$ at equilibrium) is thus directly related to $\Delta G°$ by the above equation, which can be expressed as follows:

$$0 = \Delta G° + RT \ln K$$

or

$$\Delta G° = - RT \ln K$$

If the actual ratio $[B]/[A]$ is greater than the equilibrium ratio (K), $\Delta G > 0$ and the reaction proceeds in the reverse direction (conversion of *B* to *A*). On the other hand, if the ratio $[B]/[A]$ is less than the equilibrium ratio, $\Delta G < 0$ and *A* is converted to *B*.

The standard free-energy change ($\Delta G°$) of a reaction therefore determines its chemical equilibrium and predicts in which direction the reaction will proceed under any given set of conditions. For biochemical reactions, the standard free-energy change is usually expressed as $\Delta G°'$, which is the standard free-energy change of a reaction in aqueous solution at pH = 7—approximately the conditions within a cell.

Many biological reactions (such as the synthesis of macromolecules) are thermodynamically unfavorable ($\Delta G > 0$) under cellular conditions. In order for such reactions to proceed, an additional source of energy is required. For example, consider the reaction

$$A \rightleftharpoons B \quad \Delta G = +10 \text{ kcal/mol}$$

The conversion of *A* to *B* is energetically unfavorable, so the reaction proceeds in the reverse rather than the forward direction. However, the reaction can be driven in the forward direction by coupling the conversion of *A* to *B* with an energetically favorable reaction, such as:

$$C \rightleftharpoons D \quad \Delta G = -20 \text{ kcal/mol}$$

If these two reactions are combined, the coupled reaction can be written as follows:

$$A + C \rightleftharpoons B + D \quad \Delta G = -10 \text{ kcal/mol}$$

The ΔG of the combined reaction is the sum of the free-energy changes of its individual components, so the coupled reaction is energetically favorable and will proceed as written. Thus, the energetically unfavorable conversion of A to B is driven by coupling it to a second reaction associated with a large decrease in free energy. Enzymes are responsible for carrying out such coupled reactions in a coordinated manner.

The cell uses this basic mechanism to drive the many energetically unfavorable reactions that must take place in biological systems. **Adenosine 5′-triphosphate (ATP)** plays a central role in this process by acting as a store of free energy within the cell (**Figure 3.1**). The bonds between the phosphates in ATP are known as **high-energy bonds** because their hydrolysis is accompanied by a relatively large decrease in free energy. There is nothing special about the chemical bonds themselves; they are called high-energy bonds only because a large amount of free energy is released when they are hydrolyzed within the cell. In the hydrolysis of ATP to ADP plus phosphate (P_i), $\Delta G°' = -7.3$ kcal/mol. Recall, however, that $\Delta G°'$ refers to "standard conditions" in which the concentrations of all products and reactants are 1 M. Actual intracellular concentrations of P_i are approximately 10^{-2} M, and

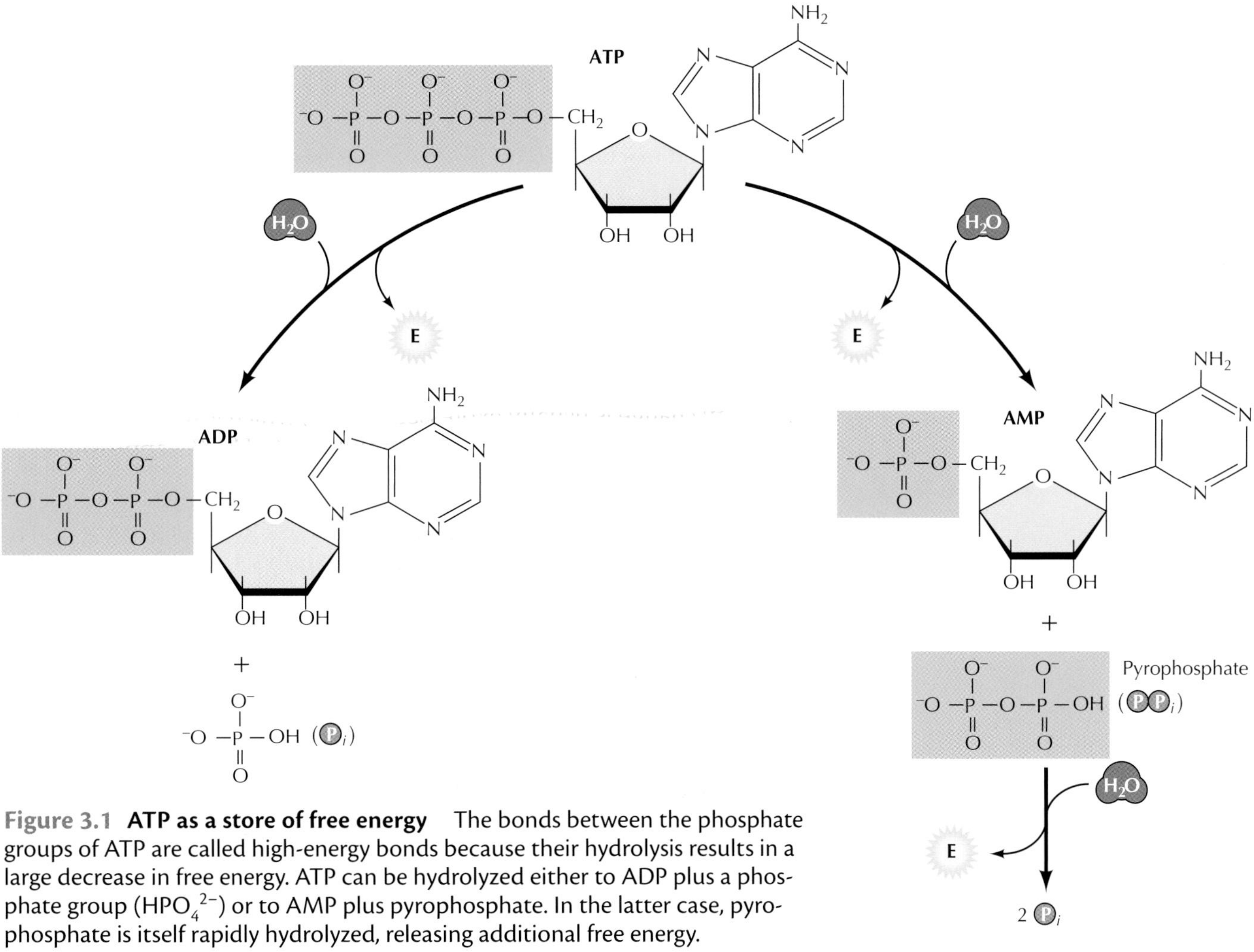

Figure 3.1 ATP as a store of free energy   The bonds between the phosphate groups of ATP are called high-energy bonds because their hydrolysis results in a large decrease in free energy. ATP can be hydrolyzed either to ADP plus a phosphate group (HPO_4^{2-}) or to AMP plus pyrophosphate. In the latter case, pyrophosphate is itself rapidly hydrolyzed, releasing additional free energy.

intracellular concentrations of ATP are higher than those of ADP. These differences between intracellular concentrations and those of the standard state favor ATP hydrolysis, so for ATP hydrolysis within a cell, ΔG is approximately –12 kcal/mol.

Alternatively, ATP can be hydrolyzed to AMP plus pyrophosphate (PP_i). This reaction yields about the same amount of free energy as the hydrolysis of ATP to ADP. However, the pyrophosphate produced by this reaction is then itself rapidly hydrolyzed, with a ΔG similar to that of ATP hydrolysis. Thus, the total free-energy change resulting from the hydrolysis of ATP to AMP is approximately twice that obtained by the hydrolysis of ATP to ADP. For comparison, the bond between the sugar and phosphate group of AMP, rather than having high energy, is typical of covalent bonds; for the hydrolysis of AMP, $\Delta G^{\circ\prime} = -3.3$ kcal/mol.

Because of the accompanying decrease in free energy, the hydrolysis of ATP can be used to drive other energy-requiring reactions within the cell. For example, the first reaction in glycolysis (discussed in the next section) is the conversion of glucose to glucose-6-phosphate. The reaction can be written as follows:

$$\text{Glucose + Phosphate } (HPO_4^{2-}) \rightarrow \text{Glucose-6-phosphate} + H_2O$$

Because this reaction is energetically unfavorable as written ($\Delta G^{\circ\prime} = +3.3$ kcal/mol), it must be driven in the forward direction by being coupled to ATP hydrolysis ($\Delta G^{\circ\prime} = -7.3$ kcal/mol):

$$ATP + H_2O \rightarrow ADP + HPO_4^{2-}$$

The combined reaction can be written as follows:

$$\text{Glucose + ATP} \rightarrow \text{Glucose-6-phosphate + ADP}$$

The free-energy change for this reaction is the sum of the free-energy changes for the individual reactions, so for the coupled reaction, $\Delta G^{\circ\prime} = -4.0$ kcal/mol, favoring glucose-6-phosphate formation.

Other molecules, including other nucleoside triphosphates (e.g., guanosine 5'-triphosphate, GTP), also have high-energy bonds and can be used as ATP is used to drive energy-requiring reactions. For most reactions, however, ATP provides the free energy. The energy-yielding reactions within the cell are therefore coupled to ATP synthesis, while the energy-requiring reactions are coupled to ATP hydrolysis. The high-energy bonds of ATP thus play a central role in cell metabolism by serving as a usable storage form of free energy to drive energy-requiring reactions. Hydrolysis of ATP can also drive the energy-requiring transport of molecules against a concentration gradient (active transport, see Figure 2.38) and cell movements, such as muscle contraction and the movements of chromosomes during mitosis (discussed in Chapter 13).

Glycolysis and the Krebs cycle

The breakdown of carbohydrates, particularly glucose, is a major source of cellular energy. The complete oxidative breakdown of glucose to CO_2 and H_2O can be written as follows:

$$C_6H_{12}O_6 + 6\,O_2 \rightarrow 6\,CO_2 + 6\,H_2O$$

The reaction yields a large amount of free energy: $\Delta G^{\circ\prime} = -686$ kcal/mol. To harness this free energy in usable form, glucose is oxidized within cells in a series of steps coupled to the synthesis of ATP.

Glycolysis, the initial stage in the breakdown of glucose, is common to virtually all cells. Glycolysis occurs in the absence of oxygen and can provide all the metabolic energy of anaerobic organisms. In aerobic cells, however, glycolysis is only the first stage in glucose degradation.

The reactions of glycolysis result in the breakdown of glucose into pyruvate, with the net gain of two molecules of ATP (**Figure 3.2**). The initial reactions in the pathway actually consume energy, using ATP to phosphorylate glucose to glucose-6-phosphate and then fructose-6-phosphate to fructose-1,6-bisphosphate. The enzymes that catalyze these two reactions—hexokinase and phosphofructokinase, respectively—are important regulatory points of the glycolytic pathway. The key control element is phosphofructokinase, which is inhibited by high levels of ATP. Inhibition of phosphofructokinase results in an accumulation of glucose-6-phosphate, which in turn inhibits hexokinase. Thus, when the cell has an adequate supply of metabolic energy available in the form of ATP, the breakdown of glucose is inhibited.

The reactions following the formation of fructose-1,6-bisphosphate constitute the energy-producing part of the glycolytic pathway. Cleavage of fructose-1,6-bisphosphate yields two molecules of the three-carbon sugar glyceraldehyde-3-phosphate, which is oxidized to 1,3-bisphosphoglycerate. The phosphate group of this compound has a very high free energy of hydrolysis ($\Delta G^{\circ\prime} = -11.5$ kcal/mol), so it is used in the next reaction of glycolysis to drive the synthesis of ATP from ADP. The product of this reaction, 3-phosphoglycerate, is then converted to phosphoenolpyruvate, the second high-energy intermediate in glycolysis. In the hydrolysis of the high-energy phosphate of phosphoenolpyruvate ($\Delta G^{\circ\prime} = -14.6$ kcal/mol), its conversion to pyruvate is coupled to the synthesis of ATP. Each molecule of glyceraldehyde-3-phosphate converted to pyruvate is thus coupled to the generation of two molecules of ATP; in total, four ATPs are synthesized from each starting molecule of glucose. Since two ATPs were required to prime the initial reactions, the net gain from glycolysis is two ATP molecules.

In addition to producing ATP, glycolysis converts two molecules of the coenzyme nicotinamide adenine dinucleotide (NAD^+) to NADH. In this reaction, NAD^+ acts as an oxidizing agent that accepts electrons from glyceraldehyde-3-phosphate. The NADH formed as a product must be recycled by serving as a donor of electrons for other oxidation–reduction reactions within the cell. In anaerobic conditions, the NADH formed during glycolysis is reoxidized to NAD^+ by the conversion of pyruvate to lactate or ethanol. In aerobic organisms, however, the NADH serves as an additional source of energy by donating its electrons to the electron transport chain, where those electrons are ultimately used to reduce O_2 to H_2O, coupled to the generation of additional ATP.

Glycolysis takes place in the cytosol. In eukaryotic cells, pyruvate is then transported into mitochondria, where its complete oxidation to CO_2 and H_2O yields most of the ATP derived from glucose breakdown. The next step in the metabolism of pyruvate is its oxidative decarboxylation in the presence of **coenzyme A (CoA-SH)**, which serves as a carrier of acyl groups in various

Animation 3.1

sites.sinauer.com/cooper7e/a3.1

Glycolysis Glycolysis is the initial stage in the breakdown of glucose, resulting in two molecules of pyruvate and a net gain of two molecules of ATP.

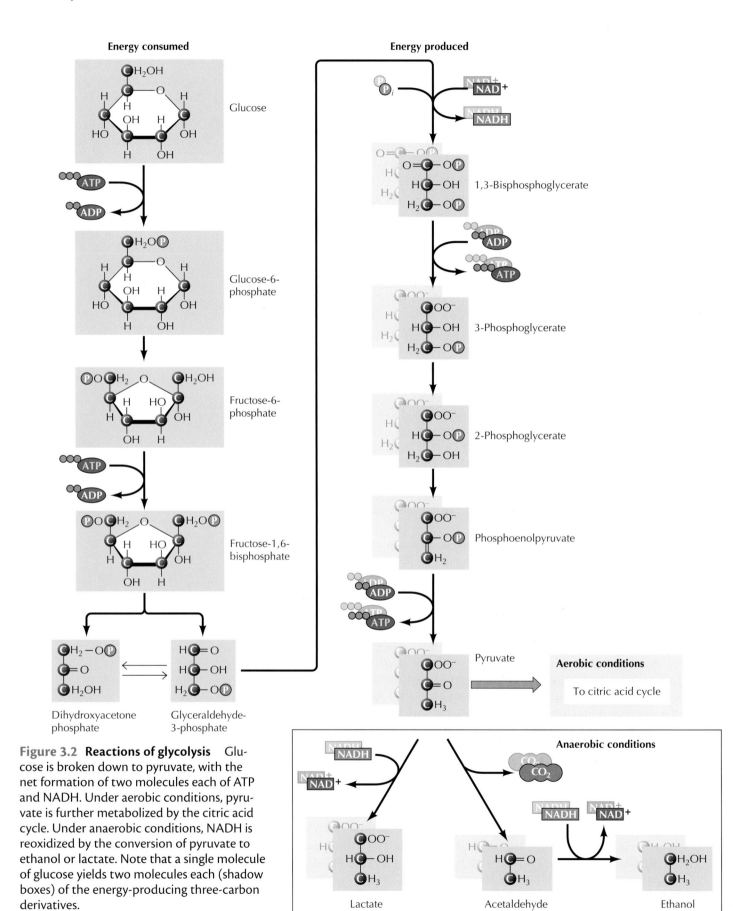

Figure 3.2 Reactions of glycolysis Glucose is broken down to pyruvate, with the net formation of two molecules each of ATP and NADH. Under aerobic conditions, pyruvate is further metabolized by the citric acid cycle. Under anaerobic conditions, NADH is reoxidized by the conversion of pyruvate to ethanol or lactate. Note that a single molecule of glucose yields two molecules each (shadow boxes) of the energy-producing three-carbon derivatives.

(A)

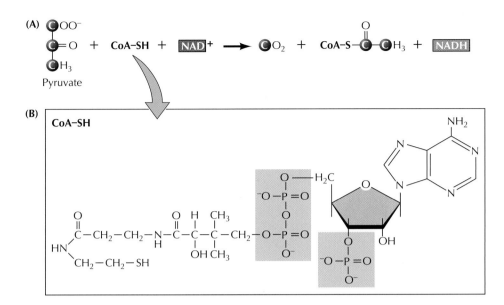

Pyruvate

(B)

CoA–SH

Figure 3.3 Oxidative decarboxylation of pyruvate (A) Pyruvate is converted to CO_2 and acetyl CoA, and one molecule of NADH is produced in the process. (B) Coenzyme A (CoA-SH) is a general carrier of activated acyl groups in a variety of reactions.

metabolic reactions (**Figure 3.3**). One carbon of pyruvate is released as CO_2, and the remaining two carbons are added to CoA-SH to form acetyl CoA. In the process, one molecule of NAD^+ is reduced to NADH.

The acetyl CoA formed by this reaction enters the **citric acid cycle** or **Krebs cycle** (**Figure 3.4**), which is the central pathway in oxidative metabolism. The two-carbon acetyl group combines with oxaloacetate (four carbons) to yield citrate (six carbons). Through eight further reactions, two carbons of citrate are completely oxidized to CO_2 and oxaloacetate is regenerated. During the cycle, one high-energy phosphate bond is formed in GTP, which is used directly to drive the synthesis of one ATP molecule. In addition, each turn of the cycle yields three molecules of NADH and one molecule of reduced **flavin adenine dinucleotide (FADH$_2$)**, which is another carrier of electrons in oxidation–reduction reactions.

The citric acid cycle completes the oxidation of glucose to six molecules of CO_2. Four molecules of ATP are obtained directly from each glucose molecule—two from glycolysis and two from the citric acid cycle (one for each molecule of pyruvate). In addition, ten molecules of NADH (two from glycolysis, two from the conversion of pyruvate to acetyl CoA, and six from the citric acid cycle) and two molecules of FADH$_2$ are formed. The remaining energy derived from the breakdown of glucose comes from the reoxidation of NADH and FADH$_2$, with their electrons being transferred through the electron transport chain to (eventually) reduce O_2 to H_2O. As discussed in the next section of this chapter, the oxidation of NADH yields sufficient energy to drive the synthesis of three molecules of ATP, while the oxidation of FADH$_2$ yields two ATP molecules. The complete oxidation of glucose thus yields a total of 38 molecules of ATP (4 directly from glycolysis and the citric acid cycle, 30 from oxidation of NADH and 4 from oxidation of FADH$_2$). In some cells, however, this yield is lower because the two molecules of NADH generated by glycolysis in the cytosol are unable to enter mitochondria directly. Instead,

Animation 3.2

sites.sinauer.com/cooper7e/a3.2
The Citric Acid Cycle The citric acid cycle is the central pathway in oxidative metabolism and completes the oxidation of glucose to six molecules of carbon dioxide.

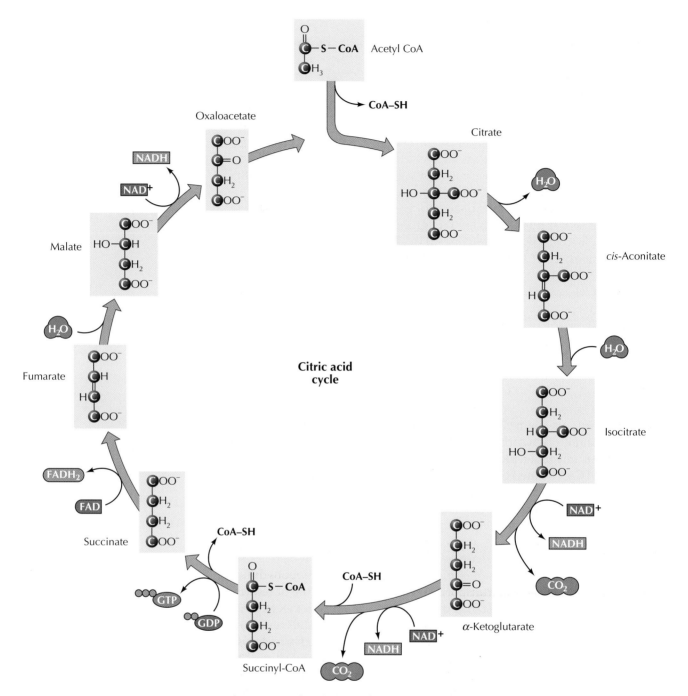

Figure 3.4 The citric acid cycle A two-carbon acetyl group is transferred from acetyl CoA to oxaloacetate, forming citrate. Two carbons of citrate are then oxidized to CO_2 and oxaloacetate is regenerated. Each turn of the cycle yields one molecule of GTP, three molecules of NADH, and one molecule of $FADH_2$.

their electrons must be transferred into the mitochondrion via a shuttle system that may result in these electrons entering the electron transport chain at the level of $FADH_2$. In such cases, the two molecules of NADH derived from glycolysis give rise to two rather than three molecules of ATP, reducing the total yield to 36 rather than 38 ATPs per molecule of glucose.

The derivation of energy from lipids

Energy in the form of ATP can be derived not only from the breakdown of glucose, but also from the degradation of other organic molecules, with glycolysis and the citric acid cycle again playing central roles. The two principal storage forms of energy within cells are polysaccharides and lipids, which can be broken down as needed by the cell to produce ATP. Polysaccharides are broken down into free sugars, which are then metabolized by glycolysis and the Krebs cycle. Lipids, however are an even more efficient energy storage molecule. Because lipids, which consist primarily of hydrocarbon chains, are more reduced than carbohydrates, the oxidation of lipids yields substantially more energy per weight of starting material.

Fats (triacylglycerols) are the major storage form of lipids. The first step in their utilization is their breakdown to glycerol and free fatty acids. Each fatty acid is joined to coenzyme A (CoA-SH), yielding a fatty acyl-CoA at the cost of one molecule of ATP (Figure 3.5). The fatty acids are then degraded in a stepwise oxidative process, two carbons at a time, yielding acetyl CoA plus a fatty acyl-CoA shorter by one two-carbon unit. Each round of oxidation also yields one molecule of NADH and one of $FADH_2$. The acetyl CoA then enters the citric acid cycle, and degradation of the remainder of the fatty acid continues in the same manner.

The breakdown of a 16-carbon fatty acid thus yields seven molecules of NADH, seven of $FADH_2$, and eight of acetyl CoA. In terms of ATP generation, this yield corresponds to 21 molecules of ATP derived from NADH (3×7), 14 ATPs from $FADH_2$ (2×7), and 96 from acetyl CoA (8×12). Since one ATP was used to start the process, the net gain is 130 ATPs per molecule of a 16-carbon fatty acid. Compare this yield with the net gain of 38 ATPs per molecule of glucose. Since the molecular weight of a saturated 16-carbon fatty acid is 256 and that of glucose is 180, the yield of ATP is approximately 2.5 times greater per gram of the fatty acid—hence, the advantage of lipids (fats) over polysaccharides (sugars) as energy storage molecules.

Electron Transport and Oxidative Phosphorylation

Most of the usable energy obtained from the breakdown of carbohydrates or fats is derived by electron transport and oxidative phosphorylation. For example, the breakdown of glucose by glycolysis and the citric acid cycle yields a total of four molecules of ATP, ten molecules of NADH, and two molecules of $FADH_2$. Electrons from NADH and $FADH_2$ are then transferred to molecular oxygen, which is coupled to the formation of an additional 34 ATP molecules by oxidative phosphorylation. The transfer of electrons from NADH to O_2 releases a large amount of free energy: $\Delta G°' = -52.5$ kcal/mol for each pair of electrons transferred. So that this energy can be harvested in usable form, the process takes place gradually by the passage of electrons through a series of carriers, which constitute the electron transport chain.

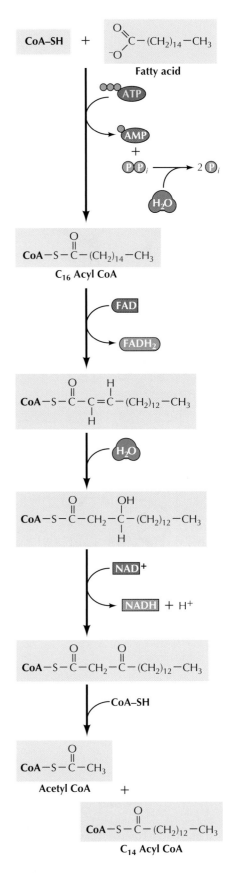

Figure 3.5 Oxidation of fatty acids The fatty acid (e.g., the 16-carbon saturated fatty acid palmitate) is initially joined to coenzyme A (CoA-SH) at the cost of one molecule of ATP. Degradation of the fatty acid then proceeds by stepwise oxidation (coupled to the formation of $FADH_2$ and NADH) and removal of a two-carbon unit as acetyl CoA.

The electron transport chain

The components of the **electron transport chain** are organized into four complexes in the plasma membrane of aerobic bacteria and in the inner mitochondrial membrane of eukaryotic cells (**Figure 3.6**). NADH is produced from the citric acid cycle in the mitochondrial matrix. Electrons from NADH then enter the electron transport chain in complex I, which consists of 44 polypeptide chains in eukaryotic cells. These electrons are initially transferred from NADH to flavin mononucleotide and then, through a series of iron-sulfur carriers, to coenzyme Q—an energy-yielding process with $\Delta G^{o\prime} = -16.6$ kcal/mol. **Coenzyme Q** (also called **ubiquinone**) is a small, lipid-soluble molecule that carries electrons from complex I through the membrane to complex III, which consists of about ten polypeptides. In complex III, electrons are transferred from cytochrome b to cytochrome c—an energy-yielding reaction with $\Delta G^{o\prime} = -10.1$

(A)

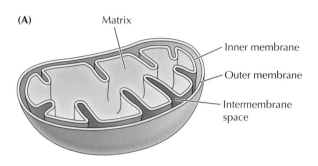

Matrix

Inner membrane

Outer membrane

Intermembrane space

(B)

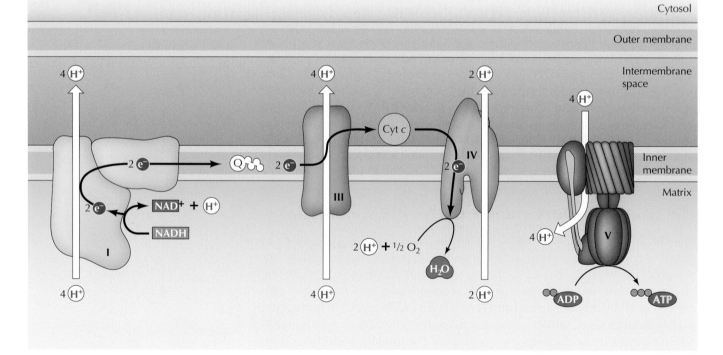

Figure 3.6 The electron transport chain in mitochondria (A) Mitochondria are bounded by a double-membrane system, consisting of inner and outer membranes. Folds of the inner membrane (cristae) extend into the matrix. NADH is produced from the Krebs cycle in the matrix. (B) Electrons from NADH enter the electron transport chain at complex I. The electrons are then transferred to complex III by coenzyme Q, and then to complex IV by cytochrome c, where they are transferred to O_2. The electron transfers in complexes I, III, and IV are associated with a decrease in free energy, which is used to pump protons from the matrix to the intermembrane space. This establishes a proton gradient across the inner membrane. The energy stored in the proton gradient is then used to drive ATP synthesis as the protons flow back to the matrix through complex V.

FYI

Rotenone, an inhibitor of electron transfer from complex I to coenzyme Q, is used as a broad spectrum insecticide.

kcal/mol. **Cytochrome *c*** then carries electrons to complex IV (**cytochrome oxidase**), where they are finally transferred to O_2 ($\Delta G^{o'}$ = –25.8 kcal/mol). The energy-yielding electron transfers at each of these complexes is coupled to the synthesis of ATP.

A distinct protein complex (complex II), which consists of four polypeptides, receives electrons from the citric acid cycle intermediate, succinate (**Figure 3.7**). These electrons are transferred to $FADH_2$, rather than to NADH, and then to coenzyme Q. From coenzyme Q, electrons are transferred to complex III and then to complex IV as already described. In contrast to the transfer of electrons from NADH to coenzyme Q at complex I, the transfer of electrons from $FADH_2$ to coenzyme Q is not associated with a significant decrease in free energy and, therefore, is not coupled to ATP synthesis. Consequently, the passage of electrons derived from $FADH_2$ through the electron transport chain yields free energy only at complexes III and IV.

The free energy derived from the passage of electrons through complexes I, III, and IV drives the synthesis of ATP by **oxidative phosphorylation**. Importantly, the mechanism by which the energy derived from these electron transport reactions is coupled to ATP synthesis is fundamentally different from the synthesis of ATP during glycolysis or the citric acid cycle. In the latter cases, a high-energy phosphate is transferred directly to ADP from the other substrate of an energy-yielding reaction. For example, in the final reaction of glycolysis, the high-energy phosphate of phosphoenolpyruvate is transferred to ADP, yielding pyruvate plus ATP (see Figure 3.2). Such direct

In addition to its role in electron transport, cytochrome *c* is a key regulator of programmed cell death in mammalian cells (see Chapter 18).

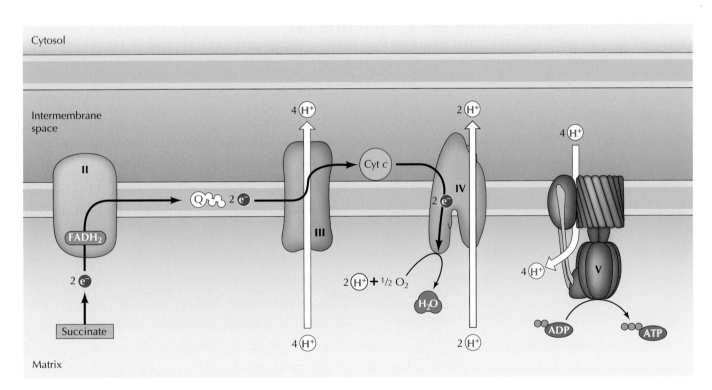

Figure 3.7 Transport of electrons from FADH₂   Electrons from succinate enter the electron transport chain via FADH₂ in complex II. They are then transferred to coenzyme Q and carried through the rest of the electron transport chain as described in Figure 3.6. The transfer of electrons from FADH₂ to coenzyme Q is not associated with a significant decrease in free energy, so protons are not pumped across the membrane at complex II, and free energy is produced only at complexes III and IV.

transfer of high-energy phosphate groups does not occur during oxidative phosphorylation. Instead, the energy derived from electron transport is coupled to the generation of a proton gradient across the inner mitochondrial membrane (or bacterial plasma membrane). The potential energy stored in this gradient is then harvested by a fifth protein complex, which couples the energetically favorable flow of protons back across the membrane to the synthesis of ATP.

Chemiosmotic coupling

The mechanism of coupling electron transport to ATP generation, **chemiosmotic coupling**, is a striking example of the relationship between structure and function in cell biology. The hypothesis of chemiosmotic coupling was first proposed in 1961 by Peter Mitchell, who suggested that ATP is generated by the use of energy stored in the form of proton gradients across biological membranes, rather than by direct chemical transfer of high-energy groups. Biochemists were initially highly skeptical of this proposal, and the chemiosmotic hypothesis took more than a decade to win general acceptance in the scientific community. Overwhelming evidence eventually accumulated in its favor, however, and chemiosmotic coupling is now recognized as the general mechanism of ATP generation, operating not only in mitochondria and bacteria, but also in chloroplasts during photosynthesis. The fundamental hypothesis of chemiosmotic coupling is that, just as ATP can be used to drive the active transport of small molecules against a concentration gradient (see Figure 2.38), the reverse reaction—the energetically favorable transport of small molecules (e.g., protons) down a concentration gradient—can be used to drive the synthesis of ATP.

Electron transport through complexes I, III, and IV is coupled to the transport of protons out of the interior of the mitochondrion (see Figure 3.6). Thus the energy-yielding reactions of electron transport establish a proton gradient across the inner mitochondrial membrane. Complexes I and IV act as proton pumps that transfer protons across the membrane as a result of conformational changes induced by electron transport. In complex III, protons are carried across the membrane by coenzyme Q, which accepts protons from the interior of mitochondria at complexes I or II and releases them on the other side of the membrane at complex III. Complexes I and III each transfer four protons across the membrane per pair of electrons. In complex IV, two protons per pair of electrons are pumped across the membrane and another two protons per pair of electrons are combined with O_2 to form H_2O within mitochondria. Thus the equivalent of four protons per pair of electrons is transported out of mitochondria at each of these three complexes. This transfer of protons across the mitochondrial membrane plays the critical role of converting the energy derived from the oxidation/reduction reactions of electron transport to the potential energy stored in a proton gradient.

Because protons are electrically charged particles, the potential energy stored in the proton gradient is electric as well as chemical in nature. The electric component corresponds to the voltage difference across the mitochondrial membrane, with the inside of the mitochondrion being negative and the outside being positive. The corresponding free energy is given by the equation

$$\Delta G = -F\Delta V$$

Key Experiment

The Chemiosmotic Theory

Coupling of Phosphorylation to Electron and Hydrogen Transfer by a Chemiosmotic Type of Mechanism

Peter Mitchell

University of Edinburgh, Edinburgh, Scotland

Nature, 1961, Volume 191, pages 144–148

The Context

By the 1950s it had been clearly established that oxidative phosphorylation involved the stepwise transfer of electrons through a series of carriers to molecular oxygen. But how the energy derived from these electron transfer reactions was converted to ATP remained a mystery. The natural assumption was that ADP was converted to ATP by direct transfer of high-energy phosphate groups from some other intermediate, as was known to occur during glycolysis. Thus it was postulated that high-energy intermediates were produced as a result of electron transfer reactions and that these intermediates drove ATP synthesis by phosphate group transfer.

The search for these postulated high-energy intermediates became a central goal of research during the 1950s and 1960s. But despite many false leads, no such intermediates were found. Moreover, several features of oxidative phosphorylation were difficult to reconcile with the orthodox hypothesis that ATP synthesis was driven by simple phosphate group transfer. In particular, phosphorylation was closely associated with membranes and was inhibited by a variety of compounds that disrupted membrane structure. These considerations led Peter Mitchell to propose a fundamentally different mechanism of energy coupling in which ATP synthesis was driven by an electrochemical gradient across a membrane rather than by the elusive high-energy intermediates sought by other investigators.

The Experiments

The fundamental proposal of the chemiosmotic hypothesis was that the "intermediate" that coupled electron transport to ATP synthesis was a proton electrochemical gradient across the membrane. Mitchell postulated that such a gradient was produced by electron transport and that the flow of protons back across the membrane in the energetically favorable direction was then coupled to ATP synthesis (see figure).

The hypothesis of chemiosmotic coupling clearly explained the lack of success in identifying a chemical high energy intermediate, as well as the fact that intact membranes were required to synthesize ATP. Yet it was a radical concept that went against the biochemical dogma of the time. In a concluding paragraph of this 1961 paper, Mitchell took a philosophical view of his revolutionary proposal:

> In the exact sciences, cause and effect are no more than events linked in sequence. Biochemists now generally accept the idea that metabolism is the cause of membrane transport. The underlying thesis of the hypothesis put forth here is that if the processes that we call metabolism and transport represent events in a sequence, not only can metabolism be the cause of transport, but also transport can be the cause of metabolism.

The Impact

Mitchell's hypothesis was greeted with skepticism and remained the subject of acrimonious debate for more than a decade. However, the wealth of supporting evidence obtained by Mitchell and his colleagues, as well as by other investigators, eventually led to general acceptance of the chemiosmotic hypothesis—which became known instead as the chemiosmotic theory. It is now accepted not only as the basis for the generation of ATP during oxidative phosphorylation and photosynthesis in bacteria, mitochondria, and chloroplasts, but also for the energy-requiring transport of a variety of molecules across cell membranes.

Mitchell's work was recognized with a Nobel Prize in 1978. The lecture he delivered on that occasion began as follows:

Peter Mitchell

Although I had hoped that the chemiosmotic rationale of vectorial metabolism and biological energy transfer might one day come to be generally accepted, it would have been presumptuous of me to expect it to happen. Was it not Max Planck who remarked that a new scientific idea does not triumph by convincing its opponents, but rather because its opponents eventually die? The fact that what began as the chemiosmotic hypothesis has now been acclaimed as the chemiosmotic theory ... has therefore both astonished and delighted me, particularly because those who were formerly my most capable opponents are still in the prime of their scientific lives.

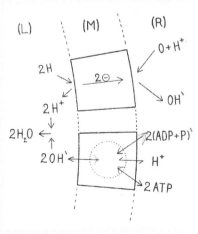

Mitchell's representation of chemiosmotic coupling between an electron transport system (top) and an ATP-generating system (bottom) in a membrane (M) enclosing aqueous phase L within aqueous phase R.

FYI

All newborn mammals (and certain adult mammals) contain a specialized tissue called brown fat. The mitochondria in cells of brown fat contain an uncoupling protein called thermogenin that utilizes the proton gradient to generate heat. Brown fat is very important for thermoregulation in neonates and in hibernating mammals.

where F is the Faraday constant and ΔV is the membrane potential. The additional free energy corresponding to the difference in proton concentration across the membrane is given by the equation

$$\Delta G = RT \ln \frac{[H^+]_i}{[H^+]_o}$$

where $[H^+]_i$ and $[H^+]_o$ refer, respectively, to the proton concentrations inside and outside the mitochondria.

In metabolically active cells, protons are typically pumped out of the matrix such that the proton gradient across the membrane corresponds to about one pH unit, or a tenfold lower concentration of protons within mitochondria (Figure 3.8). The pH of the mitochondrial matrix is therefore about 8, compared with the neutral pH (approximately 7) of the cytosol. This gradient also generates an electric potential of approximately 0.14 V across the membrane, with the inside of mitochondria being negative. Both the pH gradient and the electric potential drive protons back into the mitochondria from the cytosol, so they combine to form an **electrochemical gradient** across the mitochondrial membrane, corresponding to a ΔG of approximately –5 kcal/mol per proton.

Protons can pass freely through open channels in the mitochondrial outer membrane, but are only able to cross the inner membrane through a restricted

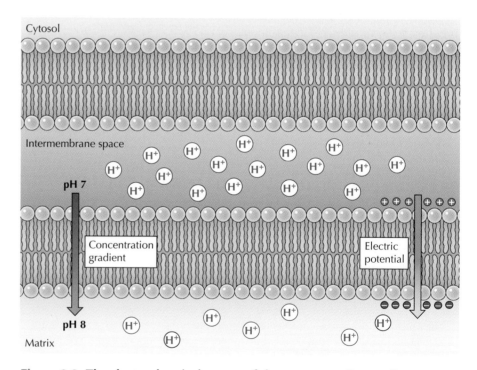

Figure 3.8 The electrochemical nature of the proton gradient Since protons are positively charged, the proton gradient established across the inner mitochondrial membrane has both chemical and electric components. The chemical component is the proton concentration (or pH) gradient, which corresponds to about a tenfold higher concentration of protons on the cytosolic side of the inner mitochondrial membrane (a difference of one pH unit). In addition, there is an electric potential across the membrane, resulting from the net increase in positive charge on the cytosolic side.

Figure 3.9 Structure of ATP synthase   The mitochondrial ATP synthase (complex V) consists of two multisubunit components, F_0 and F_1, which are linked by a slender stalk. F_0 spans the lipid bilayer, forming a spinning channel through which protons cross the membrane. One subunit of F_1 also spins and harvests the free energy derived from proton movement down the electrochemical gradient by catalyzing the synthesis of ATP.

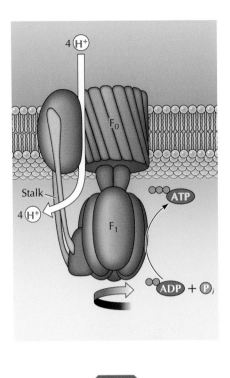

channel that couples proton flow to ATP synthesis. This allows the energy in the electrochemical gradient to be harnessed and converted to ATP as a result of the action of the fifth complex involved in oxidative phosphorylation, called complex V, or **ATP synthase** (see Figure 3.6). ATP synthase is organized into two structurally distinct components, F_0 and F_1, linked by a slender stalk (**Figure 3.9**). The F_0 portion is an electrically driven motor that spans the membrane and provides a channel through which protons are able to flow back into the mitochondrion. The energetically favorable return of protons to the mitochondrion is coupled to ATP synthesis by the spinning of a subunit of F_1, which catalyzes the synthesis of ATP from ADP and phosphate ions (P_i). Detailed structural studies have established the mechanism of ATP synthase action, which involves mechanical coupling between the F_0 and F_1 subunits. In particular, the flow of protons through F_0 drives the rotation of F_1, which acts as a rotary motor to drive ATP synthesis.

The flow of four protons back across the membrane through F_0 is required to drive the synthesis of one molecule of ATP by F_1, consistent with the proton transfers at complexes I, III, and IV each contributing sufficient free energy to the proton gradient to drive the synthesis of one ATP molecule. The oxidation of one molecule of NADH thus leads to the synthesis of three molecules of ATP, whereas the oxidation of $FADH_2$, which enters the electron transport chain at complex II, yields only two ATP molecules.

FYI

Another example of an electrochemical rotary motor is the bacterial flagellum.

Video 3.1

sites.sinauer.com/cooper7e/v3.1

ATP Synthase in Action ATP synthase generates ATP from ADP and phosphate using a rotary mechanism that turns about 150 times every second.

Photosynthesis

The generation of energy from oxidation of carbohydrates and lipids relies on the degradation of preformed organic compounds. The energy required for the synthesis of these compounds is ultimately derived from sunlight, which is harvested and used by plants and photosynthetic bacteria to drive the synthesis of carbohydrates. By converting the energy of sunlight to a usable form of chemical energy, photosynthesis is the ultimate source of metabolic energy in biological systems.

The overall equation of photosynthesis can be written as follows:

$$6\,CO_2 + 6\,H_2O \xrightarrow{\quad Light \quad} C_6H_{12}O_6 + 6\,O_2$$

The process is much more complex, however, and takes place in two distinct stages. In the first stage, called the **light reactions**, energy absorbed from sunlight drives the synthesis of ATP and NADPH (a coenzyme similar to NADH), coupled to the oxidation of H_2O to O_2. The ATP and NADPH generated by the light reactions drive the synthesis of carbohydrates from CO_2 and H_2O in the second set of reactions, called the **dark reactions**, because they do not require sunlight. In eukaryotic cells, both the light and dark reactions occur in chloroplasts.

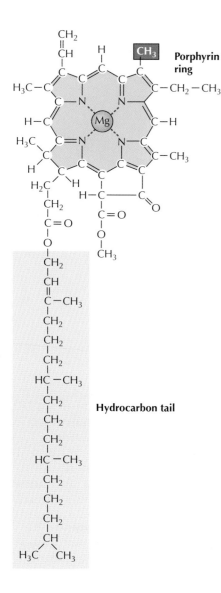

Porphyrin ring

Hydrocarbon tail

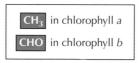

CH₃ in chlorophyll *a*

CHO in chlorophyll *b*

Figure 3.10 The structure of chlorophyll *a*
Chlorophylls consist of porphyrin ring structures linked to hydrocarbon tails. Chlorophyll *a* (shown here) differs by a single functional group in the porphyrin ring from chlorophyll *b*.

Electron transport

Sunlight is absorbed by photosynthetic pigments—in higher plants the most abundant of which are the **chlorophylls** (Figure 3.10). Absorption of light excites an electron to a higher energy state, thus converting the energy of sunlight to potential chemical energy. The photosynthetic pigments are organized into **photocenters** in the chloroplast membrane, each of which contains hundreds of pigment molecules (Figure 3.11). The many pigment molecules in each photocenter act as antennae to absorb light and transfer the energy of their excited electrons to a chlorophyll molecule that serves as a reaction center. The reaction center chlorophyll then transfers its high-energy electrons to an acceptor molecule in an electron transport chain. High-energy electrons are then transferred through a series of membrane carriers, coupled to the synthesis of ATP and NADPH (Figure 3.12).

The proteins involved in the light reactions of photosynthesis in plants are organized into four multiprotein complexes in a specialized internal membrane of chloroplasts (the thylakoid membrane) (Figure 3.13). Two of these complexes (**photosystems I** and **II**) absorb light and transfer energy

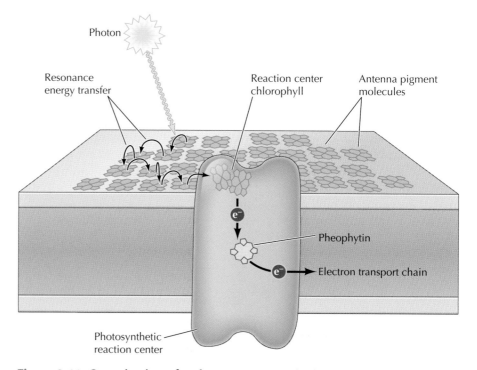

Figure 3.11 Organization of a photocenter   Each photocenter consists of hundreds of antenna pigment molecules, which absorb photons and transfer energy to a reaction center chlorophyll. The reaction center chlorophyll then transfers its excited electrons to an acceptor (pheophytin) in the electron transport chain.

Animation 3.3

sites.sinauer.com/cooper7e/a3.3

The Light Reactions During the light reactions of photosynthesis, energy absorbed from sunlight drives the synthesis of ATP and NADPH, coupled to the oxidation of H_2O to O_2.

Figure 3.12 The light reactions of photosynthesis Energy from sunlight is used to split H_2O to O_2. The high-energy electrons derived from this process are then transported through a series of carriers and used to convert $NADP^+$ to NADPH. Energy derived from the electron transport reactions also drives the synthesis of ATP.

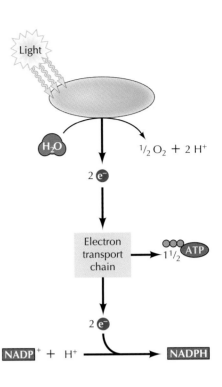

to reaction center chlorophylls. The high-energy electrons produced by absorption of light are then transferred through a series of carriers in both photosystems and in a third protein complex, the **cytochrome *bf* complex**. As in mitochondria, these electron transfers are coupled to the transfer of protons into the thylakoid lumen, establishing a proton gradient across the membrane. The energy stored in this proton gradient is then harvested by a fourth protein complex, ATP synthase, which (like the mitochondrial enzyme) couples proton flow back across the membrane to the synthesis of ATP.

(A)

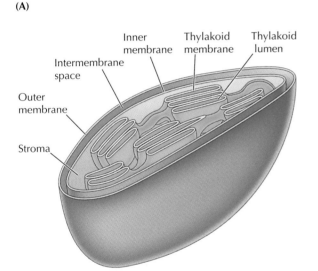

Figure 3.13 Electron transport and ATP synthesis in chloroplasts (A) In addition to the inner and outer membranes of the envelope, chloroplasts contain a third internal membrane system: the thylakoid membrane. These membranes divide chloroplasts into three internal compartments (the intermembrane space, stroma, and thylakoid lumen). (B) Four protein complexes in the thylakoid membrane function in electron transport and the synthesis of ATP and NADPH. Photons are absorbed by complexes of pigment molecules associated with photosystems I and II (PS I and PS II). At photosystem II, energy derived from photon absorption is used to split a water molecule within the thylakoid lumen. Electrons are then carried by plastoquinone (PQ) to the cytochrome *bf* complex, where they are transferred to a lower energy state and protons are pumped into the thylakoid lumen. Electrons are then transferred to photosystem I by plastocyanin (PC). At photosystem I, energy derived from light absorption again generates high energy electrons, which are transferred to ferredoxin (Fd) and used to reduce $NADP^+$ to NADPH in the stroma. ATP synthase then uses the energy stored in the proton gradient to convert ADP to ATP.

(B)

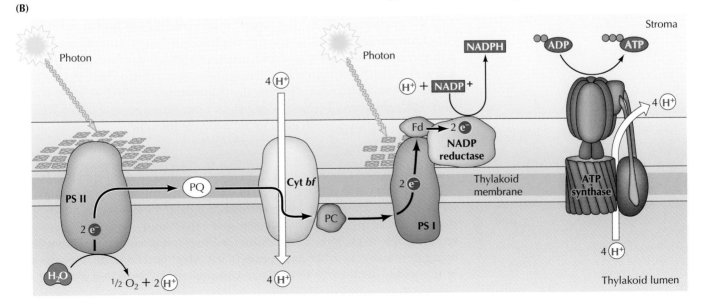

One important difference between electron transport in chloroplasts and that in mitochondria is that the energy derived from sunlight during photosynthesis not only is converted to ATP but also is used to generate the NADPH required for subsequent conversion of CO_2 to carbohydrates. This is accomplished by the use of two different photosystems in the light reactions of photosynthesis, one to generate ATP and the other to generate NADPH. Electrons are transferred sequentially between the two photosystems, with photosystem I acting to generate NADPH and photosystem II acting to generate a proton gradient which drives the synthesis of ATP.

The pathway of electron flow starts at photosystem II, where the energy derived from absorption of photons is used to split water molecules to molecular oxygen and protons, with the release of protons from H_2O establishing a proton gradient across the membrane (see Figure 3.13). The high-energy electrons derived from this process are transferred through a series of carriers to plastoquinone, a lipid-soluble carrier similar to coenzyme Q (ubiquinone) of mitochondria. Plastoquinone carries electrons from photosystem II to the cytochrome *bf* complex, within which electrons are transferred to plastocyanin (a peripheral membrane protein) and four additional protons are pumped into the thylakoid lumen. Electron transport through photosystem II is thus coupled to establishment of a proton gradient, which drives the chemiosmotic synthesis of ATP.

From the cytochrome *bf* complex, electrons are carried by plastocyanin to photosystem I, where the absorption of additional photons again generates high-energy electrons. Photosystem I, however, does not act as a proton pump; instead, it uses these high-energy electrons to reduce $NADP^+$ to NADPH. The reaction center chlorophyll of photosystem I transfers its excited electrons through a series of carriers to ferredoxin, a small protein on the outside of the thylakoid membrane. Ferredoxin can complex with the enzyme **NADP reductase**, which transfers electrons from ferrodoxin to $NADP^+$, generating NADPH. The passage of electrons through photosystems II and I thus generates both ATP and NADPH, which are used by the Calvin cycle enzymes to convert CO_2 to carbohydrates (discussed next).

A second electron transport pathway is called **cyclic electron flow**. In cyclic electron flow, light energy harvested at photosystem I is used for ATP synthesis rather than NADPH synthesis, thereby supplying additional ATP for other metabolic processes (Figure 3.14). Instead of being transferred to $NADP^+$, high-energy electrons from photosystem I are transferred to the cytochrome *bf* complex. Electron transfer through the cytochrome *bf* complex is then coupled, as in photosystem II, to the establishment of a proton gradient across the thylakoid membrane. Plastocyanin then returns these electrons to photosystem I in a lower energy state, completing a cycle of electron transport. Electron transfer from photosystem I can thus generate either ATP or NADPH, depending on the metabolic needs of the cell.

ATP synthesis

The chloroplast ATP synthase is similar to the mitochondrial enzyme. However, the energy stored in the proton gradient across the chloroplast thylakoid membrane, in contrast to the inner mitochondrial membrane, is almost entirely chemical in nature. This is because the thylakoid membrane, although impermeable to protons, differs from the inner mitochondrial membrane in being permeable to other ions, particularly Mg^{2+} and Cl^-. The free passage of these ions neutralizes the voltage component of the proton gradient, so

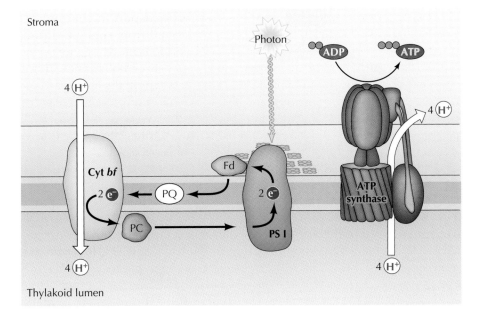

Figure 3.14 The pathway of cyclic electron flow Light energy absorbed at photosystem I (PS I) is used for ATP synthesis rather than NADPH synthesis. High-energy electrons generated by photon absorption are transferred via ferredoxin (Fd) and plastoquinone (PQ) to the cytochrome *bf* complex rather than to $NADP^+$. At the cytochrome *bf* complex, electrons are transferred to a lower energy state and protons are pumped into the thylakoid lumen. The electrons are then returned to photosystem I by plastocyanin (PC). Then, ATP synthase couples synthesis of ATP to the flow of protons from the thylakoid lumen to the stroma.

the energy derived from photosynthesis is conserved mainly as the difference in proton concentration (pH) across the membrane. However, because protons are transferred into a closed compartment (the thylakoid lumen), this difference in proton concentration can be quite large, corresponding to a differential of more than three pH units. Because of the magnitude of this pH differential, the total free energy stored across the chloroplast thylakoid membrane is similar to that stored across the mitochondrial membrane.

For each pair of electrons transported, two protons are transferred across the membrane at photosystem II and four at the cytochrome *bf* complex. Since four protons are needed to drive the synthesis of one molecule of ATP, passage of each pair of electrons through photosystems I and II by noncyclic electron flow yields 1.5 ATP molecules. Cyclic electron flow has a lower yield, corresponding to 1 ATP molecule per pair of electrons.

Synthesis of glucose

In the dark reactions, which take place in the chloroplast stroma, the ATP and NADPH produced from the light reactions drive the synthesis of carbohydrates from CO_2 and H_2O. One molecule of CO_2 at a time is added to a cycle of reactions—known as the **Calvin cycle** after its discoverer, Melvin Calvin—that leads to the formation of carbohydrates (**Figure 3.15**). Overall, the Calvin cycle consumes 18 molecules of ATP and 12 of NADPH for each molecule of glucose synthesized. Two electrons are needed to convert each molecule of $NADP^+$ to NADPH, so 24 electrons must pass through the electron transport chain to generate sufficient NADPH to synthesize one

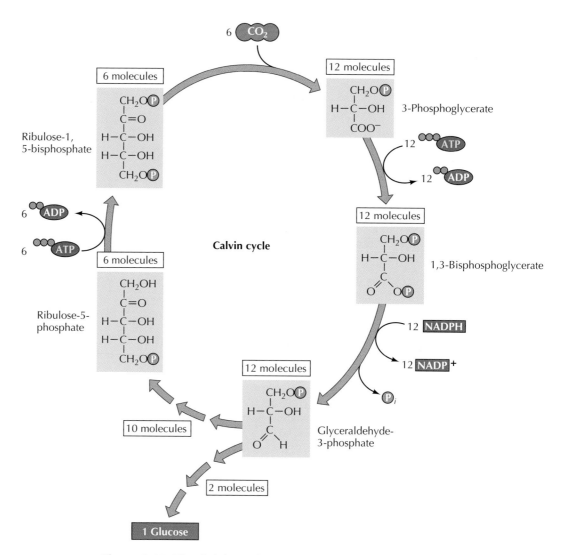

Figure 3.15 The Calvin cycle   Shown here is the synthesis of one molecule of glucose from six molecules of CO_2. Each molecule of CO_2 is added to ribulose-1,5-bisphosphate to yield two molecules each of 3-phosphoglycerate. These six molecules of CO_2 thus lead to the formation of 12 molecules of 3-phosphoglycerate, which are converted to 12 molecules of glyceraldehyde-3-phosphate at the cost of 12 molecules each of ATP and NADPH. Two molecules of glyceraldehyde-3-phosphate are then used for synthesis of glucose and ten molecules continue in the Calvin cycle to form six molecules of ribulose-5-phosphate. The cycle is then completed by the use of six additional ATP molecules for the synthesis of ribulose-1,5-bisphosphate.

molecule of glucose. These electrons are obtained by the conversion of 12 molecules of H_2O to six molecules of O_2, consistent with the formation of six molecules of O_2 for each molecule of glucose. Passage of the same 24 electrons through the electron transport chain can also generate the 18 ATPs that are required by the Calvin cycle. Under some conditions, however, the yield of ATP from electron transport may be less efficient and some of the ATP needed for the Calvin cycle may be generated by cyclic electron flow, which uses the energy derived from sunlight to synthesize ATP without the synthesis of NADPH (see Figure 3.14).

The Biosynthesis of Cell Constituents

The preceding sections reviewed the major metabolic reactions by which the cell obtains and stores energy in the form of ATP. This metabolic energy is then used to accomplish various tasks, including the synthesis of macromolecules and other cell constituents. Thus, energy derived from the breakdown of organic molecules (catabolism) is used to drive the synthesis of other required components of the cell. Most catabolic pathways involve the oxidation of organic molecules coupled to the generation of both energy (ATP) and reducing power (NADH). In contrast, biosynthetic (anabolic) pathways generally involve the use of both ATP and reducing power (often in the form of NADPH) for the production of new organic compounds. One major biosynthetic pathway, the synthesis of carbohydrates from CO_2 and H_2O during the dark reactions of photosynthesis, was discussed in the preceding section. Additional pathways leading to the biosynthesis of major cellular constituents (carbohydrates, lipids, proteins, and nucleic acids) are reviewed in the sections that follow.

Carbohydrates

In addition to being obtained directly from food or generated by photosynthesis, glucose can be synthesized from other organic molecules. In animal cells, glucose synthesis (**gluconeogenesis**) usually starts with lactate (produced by anaerobic glycolysis), amino acids (derived from the breakdown of proteins), or glycerol (produced by the breakdown of lipids). Plants (but not animals) are also able to synthesize glucose from fatty acids—a process that is particularly important during the germination of seeds, when energy stored as fats must be converted to carbohydrates to support growth of the plant. In both animal and plant cells, simple sugars are polymerized and stored as polysaccharides.

Gluconeogenesis (Figure 3.16) involves the conversion of pyruvate to glucose—essentially the reverse of glycolysis. However, as discussed earlier, the glycolytic conversion of glucose to pyruvate is an energy-yielding pathway, generating two molecules

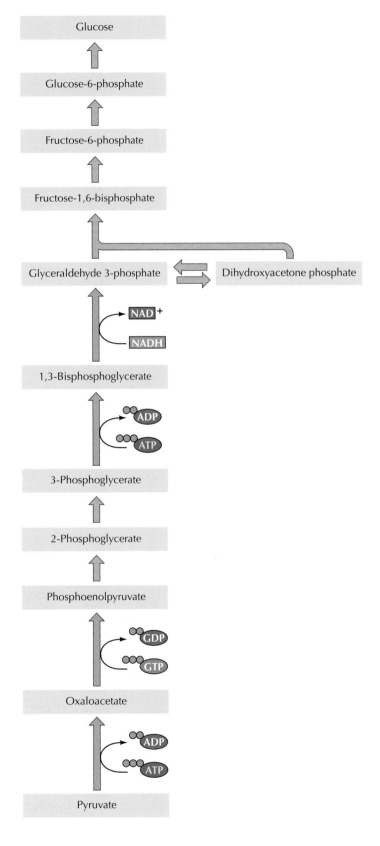

Figure 3.16 Gluconeogenesis Glucose is synthesized from two molecules of pyruvate, at the cost of four molecules of ATP, two molecules of GTP, and two molecules of NADH. The energy-requiring steps of gluconeogenesis are indicated by pink arrows.

each of ATP and NADH. Although some reactions of glycolysis are readily reversible, others will proceed only in the direction of glucose breakdown, because they are associated with a large decrease in free energy. These energetically favorable reactions of glycolysis are bypassed during gluconeogenesis by other reactions (catalyzed by different enzymes) that are coupled to the expenditure of ATP and NADH in order to drive them in the direction of glucose synthesis. Overall, the generation of glucose from two molecules of pyruvate requires four molecules of ATP, two molecules of GTP, and two molecules of NADH. This process is considerably more costly than the simple reversal of glycolysis (which would require two molecules of ATP and two of NADH), illustrating the additional energy required to drive the pathway in the direction of biosynthesis.

In both plant and animal cells, glucose is stored in the form of polysaccharides (starch and glycogen, respectively). The synthesis of polysaccharides, like that of all other macromolecules, is an energy-requiring reaction. As discussed in Chapter 2, the linkage of two sugars by a glycosidic bond can be written as a dehydration reaction, in which H_2O is removed (see Figure 2.3). Such a reaction, however, is energetically unfavorable and therefore unable to proceed in the forward direction. Consequently, the formation of a glycosidic bond must be coupled to an energy-yielding reaction, which is accomplished by the use of nucleotide sugars as intermediates in polysaccharide synthesis (**Figure 3.17**). Glucose is first phosphorylated in an ATP-driven reaction to glucose-6-phosphate, which is then converted to glucose-1-phosphate. Glucose-1-phosphate reacts with UTP (uridine triphosphate), yielding UDP-glucose plus pyrophosphate, which is hydrolyzed to phosphate with the release of additional free energy. UDP-glucose is an activated intermediate that then donates its glucose residue to a growing polysaccharide chain in an energetically favorable reaction. Thus, chemical energy in the form of ATP and UTP drives the synthesis of polysaccharides from simple sugars.

Lipids

Lipids are important energy storage molecules and the major constituent of biological membranes. They are synthesized from acetyl CoA, which is formed from the breakdown of carbohydrates, in a series of reactions that resemble the reverse of fatty acid oxidation. As with carbohydrate biosynthesis, however, the reactions leading to the synthesis of fatty acids differ from those involved in their degradation. They are driven in the biosynthetic direction by being coupled to the expenditure of both energy (in the form of ATP) and reducing power (in the form of

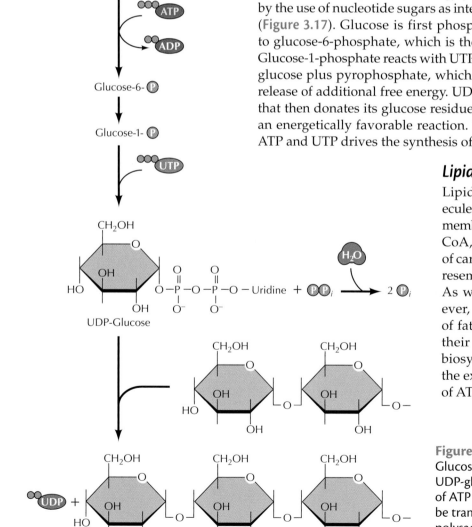

Figure 3.17 Synthesis of polysaccharides Glucose is first converted to an activated form, UDP-glucose, at the cost of one molecule each of ATP and UTP. The glucose residue can then be transferred from UDP-glucose to a growing polysaccharide chain in an energetically favorable reaction.

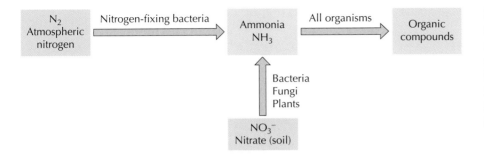

Figure 3.18 Assimilation of nitrogen into organic compounds Ammonia is incorporated into organic compounds by all organisms. Some bacteria are capable of converting atmospheric nitrogen to ammonia, and most bacteria, fungi, and plants can utilize nitrate from soil.

NADPH). Fatty acids are synthesized by the stepwise addition of two-carbon units derived from acetyl CoA to a growing chain. The addition of each of these two-carbon units requires the expenditure of one molecule of ATP and two molecules of NADPH.

The major product of fatty acid biosynthesis, which occurs in the cytosol of eukaryotic cells, is the 16-carbon fatty acid palmitate. The principal constituents of cell membranes (phospholipids, sphingomyelin, and glycolipids) are then synthesized from free fatty acids in the endoplasmic reticulum and Golgi apparatus (see Chapter 11).

Proteins

In contrast to carbohydrates and lipids, proteins (as well as nucleic acids) contain nitrogen in addition to carbon, hydrogen, and oxygen. Nitrogen is incorporated into organic compounds from different sources in different organisms (**Figure 3.18**). Some bacteria can use atmospheric N_2 by a process called **nitrogen fixation** in which N_2 is reduced to NH_3 at the expense of energy in the form of ATP. Although relatively few species of bacteria are capable of nitrogen fixation, most bacteria, fungi, and plants can use nitrate (NO_3^-), which is a common constituent of soil, by reducing it to NH_3 via electrons derived from NADH or NADPH. Finally, all organisms are able to incorporate ammonia (NH_3) into organic compounds.

NH_3 is incorporated into organic molecules primarily during the synthesis of the amino acids glutamate (Glu) and glutamine (Gln), which are derived from the citric acid cycle intermediate α-ketoglutarate. These amino acids then serve as donors of amino groups during the synthesis of the other amino acids, which are also derived from central metabolic pathways, such as glycolysis and the citric acid cycle (**Figure 3.19**). The raw material for amino acid synthesis is thus obtained from glucose, and the amino acids are synthesized at the cost of both energy

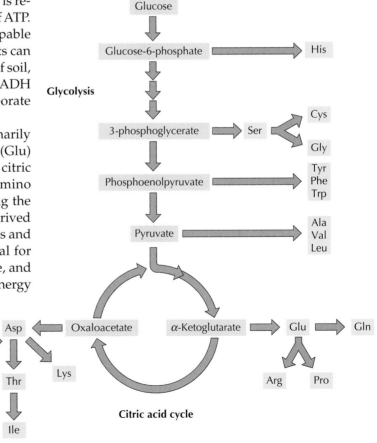

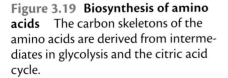

Figure 3.19 Biosynthesis of amino acids The carbon skeletons of the amino acids are derived from intermediates in glycolysis and the citric acid cycle.

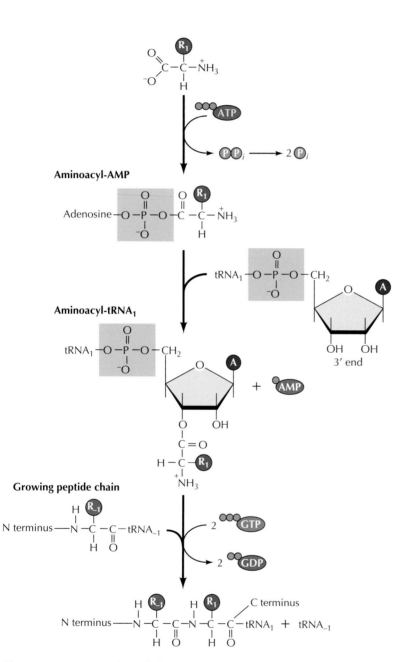

Aminoacyl-AMP

Aminoacyl-tRNA₁

Growing peptide chain

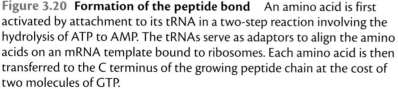

Figure 3.20 Formation of the peptide bond An amino acid is first activated by attachment to its tRNA in a two-step reaction involving the hydrolysis of ATP to AMP. The tRNAs serve as adaptors to align the amino acids on an mRNA template bound to ribosomes. Each amino acid is then transferred to the C terminus of the growing peptide chain at the cost of two molecules of GTP.

Table 3.1	Dietary Requirements for Amino Acids in Humans
Essential	**Nonessential**
Histidine	Alanine
Isoleucine	Arginine^a
Leucine	Asparagine
Lysine	Aspartate
Methionine	Cysteine
Phenylalanine	Glutamate
Threonine	Glutamine
Tryptophan	Glycine
Valine	Proline
	Serine
	Tyrosine

The essential amino acids must be obtained from dietary sources; the nonessential amino acids can be synthesized by human cells.

^{a}Although arginine is classified as a nonessential amino acid, growing children must obtain additional arginine from their diet.

(ATP) and reducing power (NADPH). Many bacteria and plants can synthesize all 20 amino acids. Humans and other mammals, however, can synthesize only about half of the required amino acids; the remainder must be obtained from dietary sources (Table 3.1).

The polymerization of amino acids to form proteins also requires energy. Like the synthesis of polysaccharides, the formation of the peptide bond can be considered a dehydration reaction, which must be driven in the direction of protein synthesis by being coupled to another source of metabolic energy. In the biosynthesis of polysaccharides, this coupling is accomplished through the conversion of sugars to activated intermediates, such as UDP-glucose. Amino acids are similarly activated before being used for protein synthesis.

A critical difference between the synthesis of proteins and that of polysaccharides is that the amino acids are incorporated into proteins in a unique order, specified by a gene. The order of nucleotides in a gene specifies the amino acid sequence of a protein via translation in which messenger RNA (mRNA) acts as a template for protein synthesis (see Chapter 4). Each amino acid is first attached to a specific transfer RNA (tRNA) molecule in a reaction coupled to ATP hydrolysis (Figure 3.20). The aminoacyl tRNAs then align on the mRNA template bound to ribosomes, and each

Key Experiment

Antimetabolites and Chemotherapy

Antagonists of Nucleic Acid Derivatives VI. Purines

Gertrude B. Elion, George H. Hitchings, and Henry Vanderwerff

Wellcome Research Laboratories, Tuckahoe, NY

Journal of Biological Chemistry, Volume 192, 1951, pages 505–518

The Context

Gertrude Elion and George Hitchings began a collaboration in 1944 that lasted over 20 years and led to the development of drugs that have proven effective for the treatment of cancer, gout, viruses, and parasitic infections. The principle of their approach to drug development was based on the antimetabolite theory, which initially proposed that some drugs active against bacteria functioned to prevent the utilization of essential nutrients (metabolites) by the bacterial cells. Elion and Hitchings suggested that the growth of rapidly dividing cells, such as cancer cells, might be inhibited by analogs of the nucleic acid bases that interfered with the normal synthesis of DNA. They proceeded to test this hypothesis by synthesizing a large number of compounds that were related to purines and testing their biological effects. These studies led to the finding that 6-mercaptopurine was a potent inhibitor of purine utilization in bacteria, which was rapidly followed by studies showing that it was an effective drug for treatment of acute leukemia in children.

The Experiments

Elion and Hitchings chose to test the potential activities of purine analogs on the growth of the bacterium *Lactobacillus casei*. In 1948, they found that the compound 2,6-diaminopurine inhibited the growth of

L. casei by interfering with normal purine metabolism. They extended these findings in their 1951 paper by testing the effects of 100 different purines, bearing substitutions of amino, hydroxyl, methyl, chloro, sulfhydryl, and other chemical groups at various positions of the purine ring. Two compounds, 6-mercaptopurine and 6-thioguanine, in which oxygen was substituted by sulfur at the 6-position of guanine and hypoxanthine, were found to be potent inhibitors of bacterial growth (see figure). These compounds were then found to be active inhibitors of the growth of a variety of rodent cancers, leading to the first trial of 6-mercaptopurine as a treatment of childhood leukemia. The results of this trial were a spectacular success, and 6-mercaptopurine was approved for treatment of childhood leukemia by the Food and Drug Administration in 1953, just over two years after the synthesis and initial demonstration of its activity as a purine antimetabolite in bacteria.

The Impact

The success of 6-mercaptopurine in treatment of childhood leukemia provided convincing evidence that inhibitors of nucleic acid metabolism could be effective drugs against cancer. This remains the case today, and 6-mercaptopurine is still one of the drugs being successfully used for leukemia treatment. In addition, a number of other antimetabolites that interfere with nucleic acid metabolism have been found to be useful in cancer therapy.

Gertrude Elion and George Hitchings continued their collaborative work on purine antimetabolites until Hitchings' retirement in 1967. In addition to 6-mercaptopurine, they developed drugs that are used for immunosuppression after tissue transplantation, for treatment of rheumatoid arthritis, for treatment of gout, and for treatment of parasitic infections. Following Hitchings'

Gertrude B. Elion

George H. Hitchings

retirement, Elion focused her research on treatment of viral infections and developed the first drug effective against human viral infections—the guanine analog acyclovir, which is highly active against herpes simplex and other herpes viruses. Subsequent success in development of nucleic acid antimetabolites as antiviral drugs includes the thymine analog AZT, which is widely used as an HIV inhibitor for treatment of AIDS. The pioneering work of Elion and Hitchings has thus opened many new areas of research and had a profound impact on the treatment of a wide range of diseases.

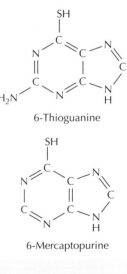

Structures of 6-thioguanine and 6-mercaptopurine.

amino acid is added to the C terminus of a growing peptide chain through a series of reactions that will be discussed in detail in Chapter 9. During the process, two molecules of GTP are also hydrolyzed, so the incorporation of each amino acid into a protein is coupled to the hydrolysis of one ATP and two GTP molecules.

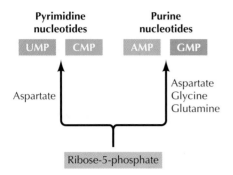

Figure 3.21 Biosynthesis of pyrimidine and purine nucleotides
Pyrimidine and purine nucleotides are synthesized from ribose-5-phosphate (a 5-carbon sugar) and several amino acids.

Nucleic acids

The precursors of nucleic acids, the nucleotides, are composed of phosphorylated five-carbon sugars joined to nucleic acid bases. Nucleotides can be synthesized from carbohydrates and amino acids; they can also be obtained from dietary sources or reused following nucleic acid breakdown. The starting point for nucleotide biosynthesis is the phosphorylated sugar ribose-5-phosphate, which is derived from glucose-6-phosphate. Divergent pathways then lead to the synthesis of purine and pyrimidine ribonucleotides, which are the immediate precursors for RNA synthesis (**Figure 3.21**). These ribonucleotides are converted to deoxyribonucleotides, which serve as the monomeric building blocks of DNA.

RNA and DNA are polymers of nucleoside monophosphates. As for other macromolecules, however, direct polymerization of nucleoside monophosphates is energetically unfavorable, and the synthesis of polynucleotides instead uses nucleoside triphosphates as activated precursors (**Figure 3.22**). A nucleoside 5'-triphosphate is added to the 3' hydroxyl group of a growing polynucleotide chain, with the release and subsequent hydrolysis of pyrophosphate serving to drive the reaction in the direction of polynucleotide synthesis.

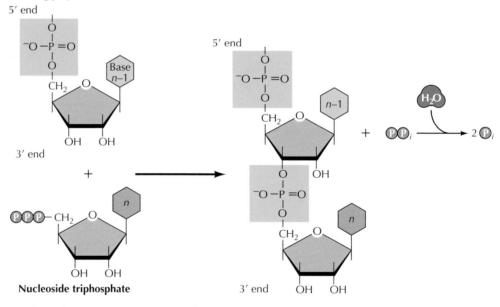

Figure 3.22 Synthesis of polynucleotides Nucleoside triphosphates are joined to the 3' end of a growing polynucleotide chain with the release of pyrophosphate.

SUMMARY	KEY TERMS

Metabolic Energy and ATP

- *Free energy and ATP:* ATP serves as a store of free energy, which is used to drive energy-requiring reactions within cells.

- *Glycolysis and the Krebs cycle:* The breakdown of glucose provides a major source of cellular energy. Glycolysis is the initial stage of glucose breakdown in all cells. In aerobic cells, the oxidation of glucose is then completed by the Krebs cycle, yielding 36 to 38 molecules of ATP. Most of this ATP is derived from electron transport reactions in which O_2 is reduced to H_2O. See Animations 3.1 and 3.2.

- *The derivation of energy from lipids:* ATP can also be derived from the breakdown of organic molecules other than glucose. Because fats are more reduced than carbohydrates, they provide a more efficient form of energy storage.

Gibbs free energy (*G*), adenosine 5′-triphosphate (ATP), high-energy bond

glycolysis, coenzyme A (CoA-SH), citric acid cycle, Krebs cycle, flavin adenine dinucleotide (FADH$_2$)

Electron Transport and Oxidative Phosphorylation

- *The electron transport chain:* Most of the energy derived from oxidative metabolism comes from the transfer of electrons from NADH and FADH$_2$ to O_2. In order to harvest this energy in usable form, electrons are transferred through a series of carriers organized into four protein complexes in the inner mitochondrial membrane.

- *Chemiosmotic coupling:* The energy-yielding reactions of electron transport are coupled to the generation of a proton gradient across the inner mitochondrial membrane in eukaryotic cells. The potential energy stored in this gradient is harvested by a fifth protein complex, ATP synthase, which couples ATP synthesis to the energetically favorable return of protons to the mitochondrion. See Video 3.1.

electron transport chain, coenzyme Q, ubiquinone, cytochrome *c*, cytochrome oxidase, oxidative phosphorylation

chemiosmotic coupling, electrochemical gradient, ATP synthase

Photosynthesis

- *Electron transport:* The energy required for the synthesis of organic molecules is ultimately derived from sunlight, which is harvested by plants and photosynthetic bacteria. In the first stage of photosynthesis, energy from sunlight is used to drive the synthesis of ATP and NADPH, coupled to the oxidation of H_2O to O_2. The ATP and NADPH produced by these reactions are then used to synthesize glucose from CO_2 and H_2O. Energy from sunlight is absorbed by chlorophylls, exciting electrons to a higher energy state. These high-energy electrons are then transferred through a series of membrane carriers, coupled to the synthesis of ATP and the reduction of NADP$^+$ to NADPH. See Animation 3.3.

- *ATP synthesis:* The chemiosmotic synthesis of ATP is driven by a proton gradient.

- *Synthesis of glucose:* Both ATP and NADPH are then used in the synthesis of glucose from CO_2, which takes place in chloroplasts. See Animation 3.4.

light reactions, dark reactions, chlorophyll, photocenter, photosystem I, photosystem II, cytochrome *bf* complex, NADP reductase, cyclic electron flow

Calvin cycle

The Biosynthesis of Cell Constituents

- *Carbohydrates:* Glucose can be synthesized from other organic molecules, using energy and reducing power in the forms of ATP and NADH, respectively. Additional ATP is then needed to drive the synthesis of polysaccharides from simple sugars.

gluconeogenesis

SUMMARY

- **Lipids:** Lipids are synthesized from acetyl CoA, which is formed from the breakdown of carbohydrates.

- **Proteins:** The amino acids are synthesized from intermediates in glycolysis and the citric acid cycle. Their polymerization to form proteins requires additional energy in the form of ATP and GTP.

- **Nucleic acids:** Pyrimidine and purine nucleotides are synthesized from carbohydrates and amino acids. Their polymerization to DNA and RNA is driven by the use of nucleoside triphosphates as activated precursors.

KEY TERMS

nitrogen fixation

Refer To

The Cell

Companion Website

sites.sinauer.com/cooper7e

for quizzes, animations, videos, flashcards, and other study resources.

Questions

1. Many biochemical reactions are energetically unfavorable under physiological conditions ($\Delta G^{\circ\prime} > 0$). How does the cell carry out these reactions?

2. Consider the reaction:
   Fructose-6-phosphate + HPO_4^{2-} ⇌ Fructose-1,6-bisphosphate + H_2O, $\Delta G^{\circ\prime} = +4$ kcal/mol.
   Knowing the standard free energy change for ATP hydrolysis (–7.3 kcal/mol), calculate the standard free energy change for the reaction catalyzed by phosphofructokinase.

3. Consider the reaction:
   $A \rightleftharpoons B + C$, in which $\Delta G^{\circ\prime} = +3.5$ kcal/mol. Calculate ΔG under intracellular conditions, given that the concentration of A is 10^{-2} M and the concentrations of B and C are each 10^{-3} M. In which direction will the reaction proceed in the cell? For your calculation, R = 1.98×10^{-3} kcal/mol/degree and T = 298 K (25°C). Note, $\ln(x) = 2.3 \log_{10}(x)$.

4. How would glycolysis be affected by an increase in the cellular concentration of ATP?

5. Yeast can grow under anaerobic as well as aerobic conditions. For every molecule of glucose consumed, how many molecules of ATP will be generated by yeast grown in anaerobic compared with aerobic conditions?

6. How do anaerobic organisms regenerate NAD^+ from NADH produced during glycolysis?

7. Why are lipids more efficient energy storage molecules than carbohydrates?

8. Assume that the electric potential across the inner mitochondrial membrane is dissipated, so the electrochemical gradient is composed solely of a proton concentration gradient corresponding to one pH unit. Calculate the free energy stored in this gradient. For your calculation, use R = 1.98×10^{-3} kcal/mol/deg, T = 298°K (25°C), and note that $\ln(x) = 2.3 \log_{10}(x)$.

9. What are the roles of coenzyme Q and cytochome c in the electron transport chain?

10. How many high-energy electrons are required to drive the synthesis of one molecule of glucose during photosynthesis, coupled to the formation of six molecules of O_2? How many molecules of ATP and NADPH are generated by passage of these electrons through photosystems I and II?

11. Differentiate between the light and dark reactions of photosynthesis.

12. Why is gluconeogenesis not a simple reversal of glycolysis?

References and Further Reading   (Key review articles for each major section are highlighted in **bold**.)

Metabolic Energy and ATP

Krebs, H. A. 1970. The history of the tricarboxylic cycle. *Perspect. Biol. Med.* 14: 154–170. [R]

Nicholls, D. G. and S. J. Ferguson. 2013. *Bioenergetics.* 4th ed. London: Academic Press.

Electron Transport and Oxidative Phosphorylation

Baradaran, R., J. M. Berrisford, G. S. Minhas and L. A. Sazanov. 2013. Crystal structure of the entire respiratory complex I. *Nature* 494: 443–448. [P]

Mitchell, P. 1979. Keilin's respiratory chain concept and its chemiosmotic consequences. *Science* 206: 1148–1159. [R]

Mitchell, P. 2011. Chemiosmotic coupling in oxidative and photosynthetic phosphorylation. 1966. *Biochim. Biophys. Acta* 1807: 1507–1538. [P]

Racker, E. 1980. From Pasteur to Mitchell: A hundred years of bioenergetics. *Fed. Proc.* 39: 210–215. [R]

Saraste, M. 1999. Oxidative phosphorylation at the fin de siècle. *Science* 283: 1488–1493. [R]

Sazanov, L. A. 2015. A giant molecular proton pump: structure and mechanism of respiratory complex I. *Nature Rev. Mol. Cell Biol.* 16: 375–388. [R]

Watanabe, R. and H. Noji. 2013. Chemomechanical coupling mechanism of F1-ATPase: catalysis and torque generation. *FEBS Letters* 587: 1030-1035. [R]

Photosynthesis

Arnon, D. I. 1984. The discovery of photosynthetic phosphorylation. *Trends Biochem. Sci.* 9: 258–262. [R]

Baniulis, D., E. Yamashita, H. Zhang, S.S. Hasan and W. A. Cramer. 2008. Structure-function of the cytochrome $b_6 f$ complex. *Phytochem. Phytobiol.* 84: 1349-1358. [R]

Calvin, M. 1962. The path of carbon in photosynthesis. *Science* 135: 879–889. [R]

Kargul, J., J. D. J. Olmos and T. Krupnik. 2012. Structure and function of photosystem I and its application in biomimetic solar-to-fuel systems. *J. Plant Physiol.* 169: 1639-1653. [R]

Schöttler, M. A., C. A. Albus and R. Bock. 2011. Photosystem I: its biogenesis and function in higher plants. *J. Plant Physiol.* 168: 1452–1461. [R]

Vinyard, D. J., G. M. Ananyev and G. C. Dismukes. 2013. Photosystem II: the reaction center of oxygenic photosynthesis. *Ann. Rev. Biochem.* 82: 577-606. [R]

The Biosynthesis of Cell Constituents

Berg, J. M., J. L. Tymoczko and L. Stryer. 2010. *Biochemistry.* 7th ed. New York: W. H. Freeman.

Canfield, D. E., A. N. Glazer and P. G. Falkowski. 2010. The evolution and future of Earth's nitrogen cycle. *Science* 330: 192–196. [R]

Nelson, D. L. and M. M. Cox. 2013. *Lehninger Principles of Biochemistry.* 6th ed. New York: W. H. Freeman.

Fundamentals of Molecular Biology

Contemporary molecular biology is concerned principally with understanding the mechanisms responsible for transmission and expression of the genetic information that governs cell structure and function. As reviewed in Chapter 1, all cells share a number of basic properties, and this underlying unity of cell biology is particularly apparent at the molecular level. Such unity has allowed scientists to choose simple organisms (such as bacteria) as models for many fundamental experiments, with the expectation that similar molecular mechanisms are operative in organisms as diverse as *E. coli* and humans. Numerous experiments have established the validity of this assumption, and it is now clear that the molecular biology of cells provides a unifying theme to understanding diverse aspects of cell behavior.

Initial advances in molecular biology were made by taking advantage of the rapid growth and readily manipulable genetics of simple bacteria, such as *E. coli*, and their viruses. The development of recombinant DNA then allowed both the fundamental principles and many of the experimental approaches first developed in prokaryotes to be extended to eukaryotic cells. The application of recombinant DNA technology has had a tremendous impact, initially allowing individual eukaryotic genes to be isolated and characterized in detail and more recently allowing the determination of the complete sequences of cellular genomes. The early development of molecular biology, recombinant DNA, and the experimental approaches used to investigate the function of eukaryotic genes are discussed in this chapter.

Heredity, Genes, and DNA

Perhaps the most fundamental property of all living things is the ability to reproduce. All organisms inherit the genetic information specifying their structure and function from their parents. Likewise, all cells arise from pre-existing cells, so the genetic material must be replicated and passed from parent to progeny cell at each cell division. How genetic information is replicated and transmitted from cell to cell and organism to organism thus represents a question that is central to all of biology. Consequently, elucidation of the mechanisms of genetic transmission and identification of the genetic material as DNA were discoveries that formed the foundation of our current understanding of biology at the molecular level.

Genes and chromosomes

The classical principles of genetics were deduced by Gregor Mendel in 1865, on the basis of the results of breeding experiments with peas. Mendel

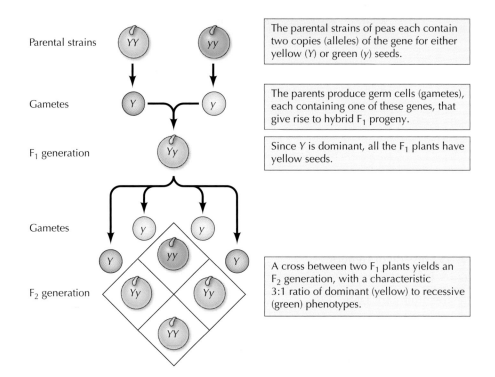

Parental strains — The parental strains of peas each contain two copies (alleles) of the gene for either yellow (Y) or green (y) seeds.

Gametes — The parents produce germ cells (gametes), each containing one of these genes, that give rise to hybrid F₁ progeny.

F₁ generation — Since Y is dominant, all the F₁ plants have yellow seeds.

Gametes

F₂ generation — A cross between two F₁ plants yields an F₂ generation, with a characteristic 3:1 ratio of dominant (yellow) to recessive (green) phenotypes.

Figure 4.1 Inheritance of dominant and recessive genes

studied the inheritance of a number of well-defined traits, such as seed color, and was able to deduce general rules for their transmission. In all cases, he could correctly interpret the observed patterns of inheritance by assuming that each trait is determined by a pair of inherited factors, which are now called **genes**. One gene copy (called an **allele**) specifying each trait is inherited from each parent. For example, breeding two strains of peas—one having yellow seeds, and the other green seeds—yields the following results (**Figure 4.1**). The parental strains each have two identical copies of the gene specifying yellow (Y) or green (y) seeds, respectively. The progeny plants are therefore hybrids, having inherited one gene for yellow seeds (Y) and one for green seeds (y). All these progeny plants (the first filial, or F₁, generation) have yellow seeds, so yellow (Y) is said to be **dominant** and green (y) **recessive**. The **genotype** (genetic composition) of the F₁ peas is thus Yy, and their **phenotype** (physical appearance) is yellow. If one F₁ offspring is bred with another, giving rise to F₂ progeny, the genes for yellow and green seeds segregate in a characteristic manner such that the ratio between F₂ plants with yellow seeds and those with green seeds is 3:1.

Mendel's findings, apparently ahead of their time, were largely ignored until 1900, when Mendel's laws were rediscovered and their importance recognized. Shortly thereafter, the role of **chromosomes** as the carriers of genes was proposed. It was realized that most cells of higher plants and animals are **diploid**—containing two copies of each chromosome. Formation of the germ cells (the sperm and egg), however, involves a unique type of cell division (**meiosis**) in which only one member of each chromosome pair is transmitted to each progeny cell (**Figure 4.2**). Consequently, the sperm and

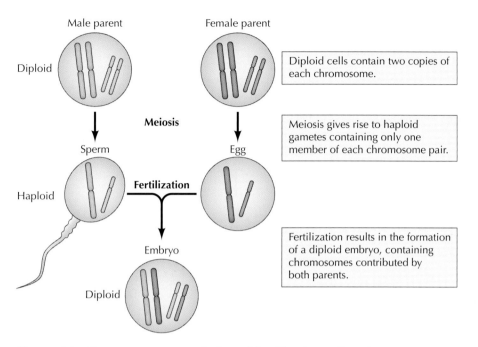

Male parent

Female parent

Diploid cells contain two copies of each chromosome.

Diploid

Meiosis

Sperm

Egg

Meiosis gives rise to haploid gametes containing only one member of each chromosome pair.

Haploid

Fertilization

Embryo

Diploid

Fertilization results in the formation of a diploid embryo, containing chromosomes contributed by both parents.

Figure 4.2 Chromosomes at meiosis and fertilization   Two chromosome pairs of a hypothetical organism are illustrated.

egg are **haploid**, containing only one copy of each chromosome. The union of these two haploid cells (gametes) at fertilization creates a new diploid organism, now containing one member of each chromosome pair derived from the male and one from the female parent. The behavior of chromosome pairs thus parallels that of genes, leading to the conclusion that genes are carried on chromosomes.

The fundamentals of mutation, genetic linkage, and the relationships between genes and chromosomes were largely established by experiments performed with the fruit fly, *Drosophila melanogaster*. *Drosophila* can be easily maintained in the laboratory, and they reproduce about every two weeks, which is a considerable advantage for genetic experiments. Indeed, these features continue to make *Drosophila* an organism of choice for genetic studies of animals, particularly the genetic analysis of development and differentiation.

In the early 1900s, a number of genetic alterations (**mutations**) were identified in *Drosophila*, usually affecting readily observable characteristics, such as eye color or wing shape. Breeding experiments indicated that some of the genes governing these traits are inherited independently of each other, suggesting that these genes are located on different chromosomes that segregate independently during meiosis (**Figure 4.3**). Other genes, however, are frequently inherited together as paired characteristics. Such genes are said to be linked to each other by virtue of being located on the same chromosome. The number of groups of linked genes is the same as the number of chromosomes (four in *Drosophila*), supporting the idea that chromosomes are carriers of the genes. By 1915, nearly 100 genes had been defined and mapped onto the four chromosomes of *Drosophila*, leading to general acceptance of the chromosomal basis of heredity.

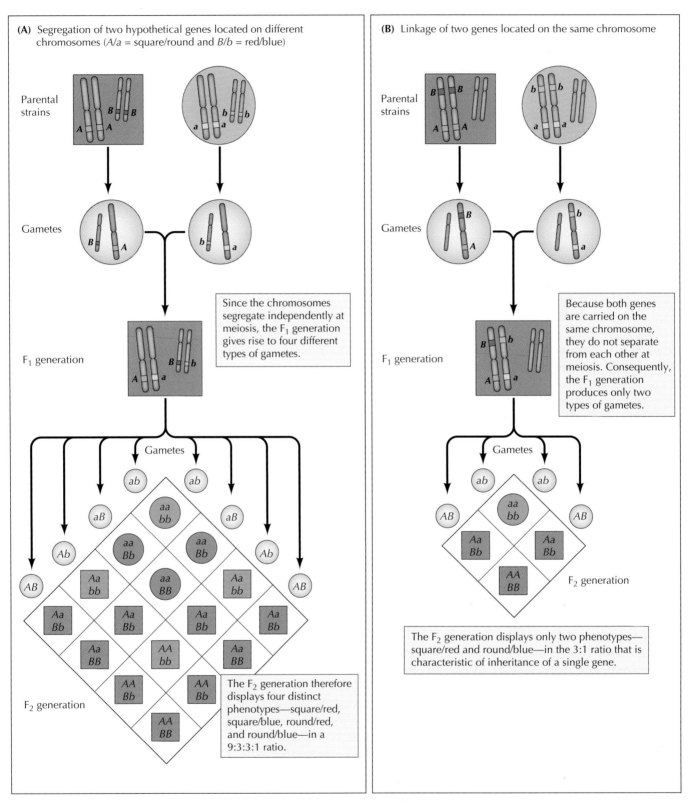

Figure 4.3 Gene segregation and linkage

Genes and enzymes

Early genetic studies focused on the identification and chromosomal localization of genes that control readily observable characteristics, such as the eye color of *Drosophila*. How these genes lead to the observed phenotypes, however, was unclear. The first insight into the relationship between genes and enzymes came in 1909, when it was realized that the inherited human disease phenylketonuria results from a genetic defect in metabolism of the amino acid phenylalanine. This defect was hypothesized to result from a deficiency in the enzyme needed to catalyze the relevant metabolic reaction, leading to the general suggestion that genes specify the synthesis of enzymes.

Clearer evidence linking genes with the synthesis of enzymes came from experiments of George Beadle and Edward Tatum, performed in 1941 with the fungus *Neurospora crassa*. In the laboratory, *Neurospora* can be grown on minimal or rich media similar to those discussed in Chapter 1 for the growth of *E. coli*. For *Neurospora*, minimal media consist only of salts, glucose, and biotin (a B-complex vitamin); rich media are supplemented with amino acids, vitamins, purines, and pyrimidines. Beadle and Tatum isolated mutants of *Neurospora* that grew normally on rich media but could not grow on minimal media. Each mutant was found to require a specific nutritional supplement, such as a particular amino acid, for growth. Furthermore, the requirement for a specific nutritional supplement correlated with the failure of the mutant to synthesize that particular compound. Thus each mutation resulted in a deficiency in a specific metabolic pathway. Since such metabolic pathways were known to be governed by enzymes, the critical conclusion from these experiments was that genes specified the structures of enzymes. In addition, some genes are now known to encode functional RNAs, such as transfer or ribosomal RNAs, rather than proteins.

Identification of DNA as the genetic material

Understanding the chromosomal basis of heredity and the relationship between genes and enzymes did not in itself provide a molecular explanation of the gene. Chromosomes contain proteins as well as DNA, and it was initially thought that genes were proteins. The first evidence leading to the identification of DNA as the genetic material came from studies in bacteria. These experiments represent a prototype for current approaches to defining the function of genes by introducing new DNA sequences into cells, as discussed later in this chapter.

The experiments that defined the role of DNA were derived from studies of the bacterium that causes pneumonia (*Pneumococcus*). Virulent strains of *Pneumococcus* are surrounded by a polysaccharide capsule that protects the bacteria from attack by the immune system of the host. Because the capsule gives bacterial colonies a smooth appearance in culture, encapsulated strains are denoted S. Mutant strains that have lost the ability to make a capsule form rough-edged colonies (denoted R) in culture and are no longer lethal when inoculated into mice. In 1928 it was observed that mice inoculated with nonencapsulated (R) bacteria plus heat-killed encapsulated (S) bacteria developed pneumonia and died. Importantly, the bacteria that were then isolated from these mice were of the S type. Subsequent experiments showed that a cell-free extract of S bacteria was similarly capable of converting (or transforming) R bacteria to the S state. Thus a substance in the S extract (called the transforming principle) was responsible for inducing the genetic **transformation** of R to S bacteria.

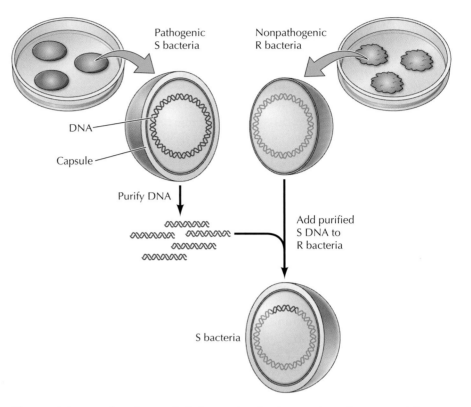

Figure 4.4 Transfer of genetic information by DNA DNA is extracted from a pathogenic strain of *Pneumococcus*, which is surrounded by a capsule and forms smooth colonies (S). Addition of the purified S DNA to a culture of nonpathogenic, non-encapsulated bacteria (R for "rough" colonies) results in the formation of S colonies. The purified DNA therefore contains the genetic information responsible for transformation of R to S bacteria.

In 1944 Oswald Avery, Colin MacLeod, and Maclyn McCarty established that the transforming principle was DNA, both by purifying it from bacterial extracts and by demonstrating that the activity of the transforming principle is abolished by enzymatic digestion of DNA—but not by digestion of proteins (**Figure 4.4**). Although these studies did not immediately lead to the acceptance of DNA as the genetic material, they were extended within a few years by experiments with bacterial viruses. In particular, it was shown that when a bacterial virus infects a cell, the viral DNA rather than the viral protein must enter the cell in order for the virus to replicate. Moreover, the parental viral DNA (but not the protein) is transmitted to progeny virus particles. The concurrence of these results with continuing studies of the activity of DNA in bacterial transformation led to acceptance of the idea that DNA is the genetic material.

The structure of DNA

Our understanding of the three-dimensional structure of DNA, deduced in 1953 by James Watson and Francis Crick, has been the basis for present-day molecular biology. At the time of Watson and Crick's work, DNA was known to be a polymer composed of four nucleic acid bases—two purines (adenine [A] and guanine [G]) and two pyrimidines (cytosine [C] and thymine [T])—linked to phosphorylated sugars. Given the central role of DNA as the genetic material, elucidation of its three-dimensional structure appeared critical to understanding its function. Watson and Crick's consideration of the problem was heavily influenced by Linus Pauling's description of hydrogen bonding and the α helix, a common element of the secondary structure of proteins (see Chapter 2). Moreover, experimental data on the structure of DNA were available from X-ray crystallography studies by Maurice Wilkins and Rosalind Franklin. Analysis of these data revealed that the DNA molecule is a helix that turns every 3.4 nm. In addition, the data showed that the distance between adjacent bases is 0.34 nm, so there are ten bases per turn of the helix. An important finding was that the diameter of the helix is approximately 2 nm, suggesting that it is composed of not one but two DNA chains.

From these data, Watson and Crick built their model of DNA (**Figure 4.5**). The central features of the model are that DNA is a double helix with the sugar–phosphate backbones on the outside of the molecule. The bases are on the inside, oriented such that hydrogen bonds are formed between

> **FYI**
>
> The structure of DNA described by Watson and Crick was a remarkable achievement, but not perfect. In their initial paper, they assumed that both A–T and G–C were paired by two hydrogen bonds, missing the third hydrogen bond in G–C base pairs.

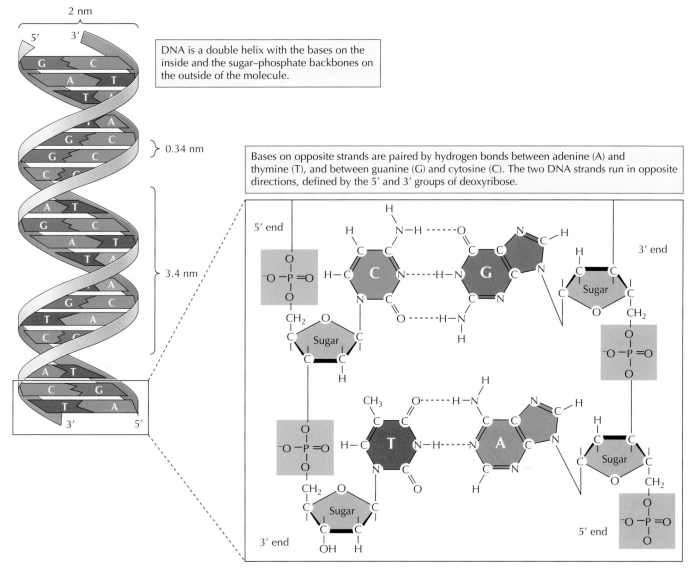

Figure 4.5 **The structure of DNA**

purines and pyrimidines on opposite chains. The base pairing is very specific: A always pairs with T and G with C. This specificity accounts for the earlier results of Erwin Chargaff, who had analyzed the base composition of various DNAs and found that the amount of adenine was always equal to that of thymine, and the amount of guanine to that of cytosine. Because of this specific base pairing, the two strands of a DNA molecule are complementary: Each strand contains all the information required to specify the sequences of bases on the other.

Replication of DNA

The discovery of complementary base pairing between DNA strands immediately suggested a molecular solution to the question of how the genetic material could direct its own replication—a process that is required each time a cell divides. It was proposed that the two strands of a DNA molecule could separate and serve as templates for synthesis of new complementary strands, the sequence of which would be dictated by the specificity of base pairing (Figure 4.6). The process is called **semiconservative replication** because one strand of parental DNA is conserved in each progeny DNA molecule.

Direct support for semiconservative DNA replication was obtained in 1958 as a result of elegant experiments performed by Matthew Meselson and Frank Stahl, in which DNA was labeled with isotopes that altered its density (Figure 4.7). *E. coli* were first grown in media containing the heavy isotope of nitrogen (^{15}N) in place of the normal light isotope (^{14}N). The DNA of these bacteria consequently contained ^{15}N and was heavier than that of bacteria grown in ^{14}N. Such heavy DNA could be separated from DNA containing ^{14}N by equilibrium centrifugation in a density gradient of cesium chloride (CsCl). This ability to separate heavy (^{15}N) DNA from light (^{14}N) DNA enabled the study of DNA synthesis. *E. coli* that had been grown in ^{15}N were transferred to media containing ^{14}N and allowed to replicate one more time. Their DNA was then extracted and analyzed by CsCl density gradient centrifugation. The results of this analysis indicated that all of the heavy DNA had been replaced by newly synthesized DNA with a density intermediate between that of heavy (^{15}N) and that of light (^{14}N) DNA molecules. The

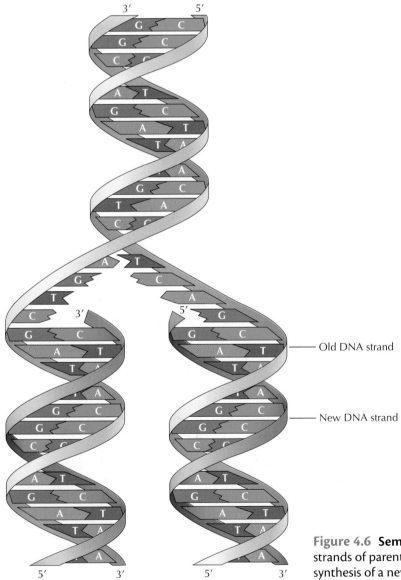

Old DNA strand

New DNA strand

Figure 4.6 Semiconservative replication of DNA The two strands of parental DNA separate, and each serves as a template for synthesis of a new daughter strand by complementary base pairing.

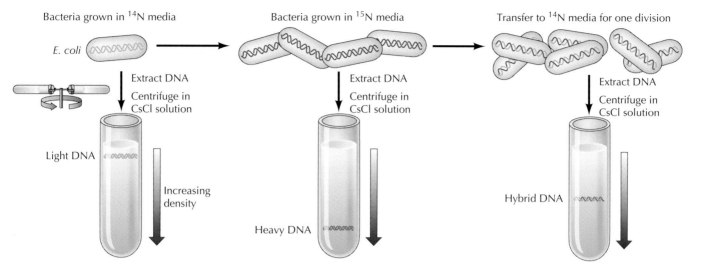

Figure 4.7 Experimental demonstration of semiconservative replication
Bacteria grown in medium containing the normal isotope of nitrogen (^{14}N) are transferred into medium containing the heavy isotope (^{15}N) and grown in this medium for several generations. They are then transferred back to medium containing ^{14}N and grown for one additional generation. DNA is extracted from these bacteria and analyzed by equilibrium ultracentrifugation in a CsCl solution. The CsCl sediments to form a density gradient, and the DNA molecules band at a position where their density is equal to that of the CsCl solution. DNA of the bacteria transferred from ^{15}N to ^{14}N medium for a single generation bands at a density intermediate between that of ^{15}N DNA and that of ^{14}N DNA, indicating that it represents a hybrid molecule with one heavy and one light strand.

implication was that during replication, the two parental strands of heavy DNA separated and served as templates for newly synthesized progeny strands of light DNA, yielding double-stranded molecules of intermediate density. This experiment thus provided direct evidence for semiconservative DNA replication, clearly underscoring the importance of complementary base pairing between strands of the double helix.

The ability of DNA to serve as a template for its own replication was further established with the demonstration that an enzyme purified from *E. coli* (**DNA polymerase**) could catalyze DNA replication *in vitro*. In the presence of DNA to act as a template, DNA polymerase was able to direct the incorporation of nucleotides into a complementary DNA molecule.

Expression of Genetic Information

Most genes act by determining the structure of proteins, which are responsible for directing cell metabolism through their activity as enzymes. The identification of DNA as the genetic material and the elucidation of its structure revealed that genetic information must be specified by the order of the four bases (A, C, G, and T) that make up the DNA molecule. Proteins, in turn, are polymers formed from 20 different amino acids, the sequence of which determines protein structure and function. The first direct link between a genetic mutation and an alteration in the amino acid sequence of a protein was made in 1957, when it was found that patients with the inherited disease sickle-cell anemia had hemoglobin molecules that differed from normal ones by a single amino acid substitution. Deeper understanding of the molecular relationship between DNA and proteins came, however, from a series of experiments that took advantage of *E. coli* and its viruses as genetic models.

Colinearity of genes and proteins

The simplest hypothesis to account for the relationship between genes and enzymes was that the order of nucleotides in DNA specified the order of amino acids in a protein. Mutations in a gene would correspond to alterations in the sequence of DNA, which might result from the substitution of one nucleotide for another or from the addition or deletion of nucleotides. These changes in the nucleotide sequence of DNA would then lead to corresponding changes in the amino acid sequence of the protein encoded by the gene in question. This hypothesis predicted that different mutations within a single gene could alter different amino acids in the encoded protein, and that the positions of mutations in a gene should reflect the positions of amino acid alterations in its protein product.

The rapid replication and the simplicity of the genetic system of *E. coli* were of major help in addressing these questions. A variety of mutants of *E. coli* could be isolated, including nutritional mutants that (like the *Neurospora* mutants discussed earlier) require particular amino acids for growth. Importantly, the rapid growth of *E. coli* made feasible the isolation and mapping of multiple mutants in a single gene, leading to the first demonstration of the linear relationship between genes and proteins. In these studies, Charles Yanofsky and his colleagues mapped a series of mutations in the gene that encodes an enzyme required for synthesis of the amino acid tryptophan. Analysis of the enzymes encoded by the mutant genes indicated that the relative positions of the amino acid alterations were the same as those of the corresponding mutations (**Figure 4.8**). Thus the sequence of amino acids in the protein was colinear with that of mutations in the gene, as expected if the order of nucleotides in DNA specifies the order of amino acids in proteins.

The role of messenger RNA

Although the sequence of nucleotides in DNA appeared to specify the order of amino acids in proteins, it did not necessarily follow that DNA

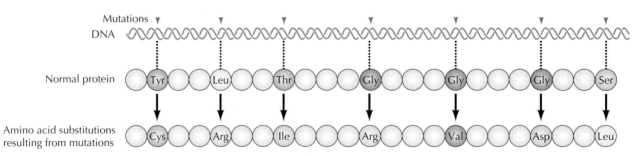

Figure 4.8 Colinearity of genes and proteins A series of mutations (red arrowheads) were mapped in the *E. coli* gene encoding tryptophan synthetase (top line). The amino acid substitutions resulting from each of the mutations were then determined by sequence analysis of the proteins of mutant bacteria (bottom line). These studies revealed that the order of mutations in DNA was the same as the order of amino acid substitutions in the encoded protein.

itself directs protein synthesis. Indeed, this appeared not to be the case, since DNA is located in the nucleus of eukaryotic cells, whereas protein synthesis takes place in the cytoplasm. Some other molecule was therefore needed to convey genetic information from DNA to the sites of protein synthesis (the ribosomes).

RNA appeared a likely candidate for such an intermediate because the similarity of its structure to that of DNA suggested that RNA could be synthesized from a DNA template (**Figure 4.9**). RNA differs from DNA in that it is single-stranded rather than double-stranded, its sugar component is ribose instead of deoxyribose, and it contains the pyrimidine base uracil (U) instead of thymine (T) (see Figure 2.10). However, neither the change in sugar nor the substitution of U for T alters base pairing, so the synthesis of RNA can be readily directed by a DNA template. Moreover, since RNA is located primarily in the cytoplasm, it appeared a logical intermediate to convey information from DNA to the ribosomes. These characteristics of RNA suggested a pathway for the flow of genetic information that is known as the **central dogma** of molecular biology:

$$DNA \rightarrow RNA \rightarrow Protein$$

According to this concept, RNA molecules are synthesized from DNA templates (a process called **transcription**), and proteins are synthesized from RNA templates (a process called **translation**).

Experimental evidence for the RNA intermediates postulated by the central dogma was obtained by Sidney Brenner, François Jacob, and Matthew Meselson in studies of *E. coli* infected with the bacteriophage T4. The synthesis of *E. coli* RNA stops following infection by T4, and the only new RNA synthesized in infected bacteria is transcribed from T4 DNA. This T4 RNA becomes associated with bacterial ribosomes, thus conveying the information from DNA to the site of protein synthesis. Because of their role as intermediates in the flow of genetic information, RNA molecules that serve as templates for protein synthesis are called **messenger RNAs (mRNAs)**. They are transcribed by an enzyme (**RNA polymerase**) that catalyzes the synthesis of RNA from a DNA template.

In addition to mRNA, two other types of RNA molecules are important in protein synthesis. **Ribosomal RNA (rRNA)** is a component of ribosomes, and **transfer RNAs (tRNAs)** serve as adaptor molecules that align amino acids along the mRNA template. The structures and functions of these molecules are discussed in the following section and in more detail in Chapters 8 and 9.

The genetic code

How is the nucleotide sequence of mRNA translated into the amino acid sequence of a protein? In this step of gene expression, genetic information is transferred between chemically unrelated types of macromolecules—nucleic acids and proteins—raising two new types of problems in understanding the action of genes.

First, since amino acids are structurally unrelated to the nucleic acid bases, direct complementary pairing between mRNA and amino acids during the incorporation of amino acids into proteins seemed impossible. How then could amino acids align on an mRNA template during protein synthesis? This question was solved by the discovery that tRNAs serve as adaptors

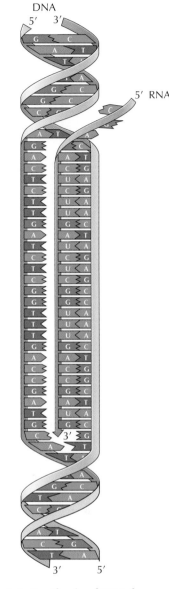

Figure 4.9 Synthesis of RNA from DNA The two strands of DNA unwind, and one is used as a template for synthesis of a complementary strand of RNA.

Animation 4.3

sites.sinauer.com/cooper7e/a4.3
The "Central Dogma" The central dogma of molecular biology, as first stated by Francis Crick, asserts that information flows from DNA to RNA, and then from RNA to protein.

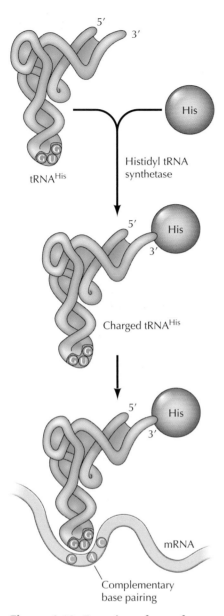

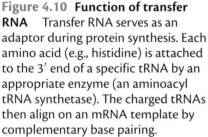

Figure 4.10 Function of transfer RNA Transfer RNA serves as an adaptor during protein synthesis. Each amino acid (e.g., histidine) is attached to the 3′ end of a specific tRNA by an appropriate enzyme (an aminoacyl tRNA synthetase). The charged tRNAs then align on an mRNA template by complementary base pairing.

between amino acids and mRNA during translation (**Figure 4.10**). Prior to its use in protein synthesis, each amino acid is attached by a specific enzyme to its appropriate tRNA. Base pairing between a recognition sequence on each tRNA and a complementary sequence on the mRNA then directs the attached amino acid to its correct position on the mRNA template.

The second problem in the translation of nucleotide sequence to amino acid sequence was determination of the **genetic code**. How could the information contained in the sequence of four different nucleotides be converted to the sequences of 20 different amino acids in proteins? Because 20 amino acids must be specified by only four nucleotides, at least three nucleotides must be used to encode each amino acid. Used singly, four nucleotides could encode only four amino acids and, used in pairs, four nucleotides could encode only sixteen (4^2) amino acids. Used as triplets, however, four nucleotides could encode 64 (4^3) different amino acids—more than enough to account for the 20 amino acids actually found in proteins.

Direct experimental evidence for the triplet code was obtained by studies of bacteriophage T4 bearing mutations in an extensively studied gene called *rII*. Phages with mutations in this gene form abnormally large plaques, which can be clearly distinguished from those formed by wild-type phages. Hence isolating and mapping a number of *rII* mutants was easy and led to the establishment of a detailed genetic map of this locus. Study of recombinants between *rII* mutants that had arisen by additions or deletions of nucleotides revealed that phages containing additions or deletions of one or two nucleotides always exhibited the mutant phenotype. Phages containing additions or deletions of three nucleotides, however, were frequently wild-type in function (**Figure 4.11**). These findings suggested that the gene is read in groups of three nucleotides, starting from a fixed point. Additions or deletions of one or two nucleotides would then alter the reading frame of the entire gene, leading to the coding of abnormal amino acids throughout

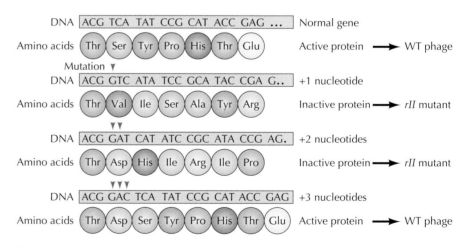

Figure 4.11 Genetic evidence for a triplet code A series of mutations consisting of additions of one, two, or three nucleotides were studied in the *rII* gene of bacteriophage T4. Additions of one or two nucleotides alter the reading frame of the remainder of the gene. Therefore all the subsequent amino acids are abnormal, and an inactive protein is produced, giving rise to mutant phage. Additions of three nucleotides, however, alter only a single amino acid. The reading frame of the remainder of the gene is normal, and an active protein giving rise to wild-type (WT) phage is produced.

the encoded protein. In contrast, additions or deletions of three nucleotides would lead to the addition or deletion of only a single amino acid; the rest of the amino acid sequence would remain unaltered, frequently yielding an active protein.

Deciphering the genetic code thus became a problem of assigning nucleotide triplets to their corresponding amino acids. This problem was approached using *in vitro* systems that could carry out protein synthesis (***in vitro* translation**). Cell extracts containing ribosomes, amino acids, tRNAs, and the enzymes responsible for attaching amino acids to the appropriate tRNAs (aminoacyl-tRNA synthetases) were known to catalyze the incorporation of amino acids into proteins. However, such protein synthesis depends on the presence of mRNA bound to the ribosomes, and purified mRNA can be added to direct protein synthesis in such systems. This allowed the genetic code to be deciphered by study of the translation of synthetic mRNAs of known base sequence.

The first such experiment, performed by Marshall Nirenberg and Heinrich Matthaei, involved the *in vitro* translation of a synthetic RNA polymer containing only uracil (Figure 4.12). This poly-U template was found to direct the incorporation of only a single amino acid—phenylalanine—into a polypeptide consisting of repeated phenylalanine residues. Therefore, the triplet UUU encodes the amino acid phenylalanine. Similar experiments with RNA polymers containing only single nucleotides established that AAA encodes lysine and CCC encodes proline. The remainder of the code was deciphered using RNA polymers containing mixtures of nucleotides, leading to the coding assignment of all 64 possible triplets (called **codons**) (Table 4.1). Of the 64 codons, 61 specify an amino acid; the remaining three (UAA, UAG, and UGA) are stop codons that signal the termination of protein synthesis. The code is degenerate; that is, many amino acids are specified by more than one codon. With few exceptions (discussed in Chapter 12), all organisms utilize

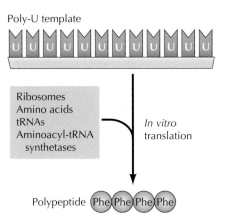

Figure 4.12 The triplet UUU encodes phenylalanine *In vitro* translation of a synthetic RNA consisting of repeated uracils (a poly-U template) results in the synthesis of a polypeptide containing only phenylalanine.

Table 4.1   The Genetic Code

First position	Second position				Third position
	U	C	A	G	
U	Phe	Ser	Tyr	Cys	U
	Phe	Ser	Tyr	Cys	C
	Leu	Ser	stop	stop	A
	Leu	Ser	stop	Trp	G
C	Leu	Pro	His	Arg	U
	Leu	Pro	His	Arg	C
	Leu	Pro	Gln	Arg	A
	Leu	Pro	Gln	Arg	G
A	Ile	Thr	Asn	Ser	U
	Ile	Thr	Asn	Ser	C
	Ile	Thr	Lys	Arg	A
	Met	Thr	Lys	Arg	G
G	Val	Ala	Asp	Gly	U
	Val	Ala	Asp	Gly	C
	Val	Ala	Glu	Gly	A
	Val	Ala	Glu	Gly	G

Animation 4.4

sites.sinauer.com/cooper7e/a4.4

DNA Mutations A single base-pair mutation in the coding region of a gene can result in a prematurely terminated polypeptide or in a polypeptide with an incorrect amino acid in its amino acid chain, depending on the particular mutation.

the same genetic code, providing strong support for the conclusion that all present-day cells evolved from a common ancestor.

RNA viruses and reverse transcription

With the elucidation of the genetic code, the fundamental principles of the molecular biology of cells appeared to have been established. According to the central dogma, the genetic material consists of DNA, which is capable of self-replication as well as being transcribed into mRNA, which serves in turn as the template for protein synthesis. However, as noted in Chapter 1, many viruses contain RNA rather than DNA as their genetic material, implying the use of other modes of information transfer.

RNA genomes were first discovered in plant viruses, many of which were found to be composed of only RNA and protein. Direct proof that RNA acts as the genetic material of these viruses was obtained in the 1950s by experiments demonstrating that RNA purified from tobacco mosaic virus could infect new host cells, giving rise to infectious progeny virus. The mode of replication of most viral RNA genomes was subsequently determined by studies of the RNA bacteriophages of *E. coli*. These viruses were found to encode a specific enzyme that could catalyze the synthesis of RNA from an RNA template (RNA-directed RNA synthesis), using the same mechanism of base pairing between complementary strands as is employed during DNA replication or transcription of RNA from DNA.

Although most animal viruses, such as poliovirus or influenza virus, were found to replicate by RNA-directed RNA synthesis, this mechanism did not appear to account for the replication of one family of animal viruses (the RNA tumor viruses), which were of particular interest because of their ability to cause cancer in infected animals. Although these viruses contain genomic RNA in their viral particles, experiments performed by Howard Temin in the early 1960s indicated that their replication requires DNA synthesis in infected cells, leading to the hypothesis that the RNA tumor viruses (now called **retroviruses**) replicate via synthesis of a DNA intermediate, called a DNA provirus (**Figure 4.13**). This hypothesis was initially met with widespread disbelief because it involves RNA-directed synthesis of DNA—a reversal of the central dogma. In 1970, however, Temin and David Baltimore independently discovered that retroviruses contain a novel enzyme that catalyzes the synthesis of DNA from an RNA template. In addition, clear-cut evidence for the existence of viral DNA sequences in infected cells was obtained. The synthesis of DNA from RNA, now called **reverse transcription**, was thus established as a mode of information transfer in biological systems.

Reverse transcription is important not only in the replication of retroviruses but also in at least two other broad aspects of molecular and cellular biology. First, reverse transcription is not restricted to retroviruses; it also occurs in cells, as discussed in Chapters 6 and 7. For example, reverse transcription is critical for replicating the ends of eukaryotic chromosomes. In addition, reverse transcription is frequently responsible for the transposition of DNA sequences from one chromosomal location to another. Indeed, the sequence of the human genome has revealed that approximately 40% of human genomic DNA is derived from reverse

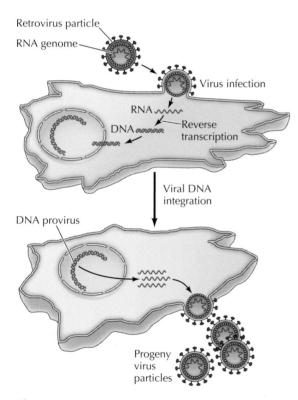

Figure 4.13 Reverse transcription and retrovirus replication   Retroviruses contain RNA genomes in their viral particles. When a retrovirus infects a host cell, however, a DNA copy of the viral RNA is synthesized via reverse transcription. This viral DNA is then integrated into chromosomal DNA of the host to form a DNA provirus, which is transcribed to yield progeny virus RNA.

Key Experiment

The DNA Provirus Hypothesis

Nature of the Provirus of Rous Sarcoma

Howard M. Temin

McArdle Laboratory, University of Wisconsin, Madison, WI

National Cancer Institute Monographs, Volume 17, 1964, pages 557–570

The Context

Rous sarcoma virus (RSV), the first cancer-causing virus to be described, was of considerable interest as an experimental system for studying the molecular biology of cancer. Howard Temin began his research in this area when, as a graduate student in 1958, he developed the first assay for the transformation of normal cells to cancer cells in culture following infection with RSV. The availability of such a quantitative *in vitro* assay provided the tool needed for further studies of both cell transformation and virus replication. As Temin proceeded with these studies, he made a series of unexpected observations indicating that the replication of RSV was fundamentally different from that of other RNA viruses. These experiments led to Temin's proposal of the DNA provirus hypothesis, which stated that the viral RNA was copied into DNA in infected cells—a proposal that ran directly counter to the universally accepted central dogma of molecular biology.

The Experiments

The DNA provirus hypothesis was based on several different types of experimental evidence. First, studies of cell transformation using mutants of RSV indicated that important characteristics of transformed cells were determined by genetic information of the virus. This information was regularly transmitted to daughter cells following cell division, even in the absence of virus replication. Temin therefore proposed that the viral genome was present in infected cells in a stably inherited form, which he called a provirus.

Evidence that the provirus was DNA was then derived from experiments with metabolic inhibitors. First, actinomycin D, which inhibits the synthesis of RNA from a DNA template, was found to interfere with virus production by RSV-infected cells (see figure). Second, inhibitors of DNA synthesis blocked early stages of cell infection by RSV. Thus, DNA synthesis appeared to be required early in infection, and DNA-directed RNA synthesis appeared to be needed subsequently for the production of progeny viruses, leading to the proposal that the provirus was a DNA copy of the viral RNA genome. Temin sought further evidence for this proposal by using nucleic acid hybridization to detect viral sequences in infected cell DNA, but the sensitivity of the available techniques was limited and the data were unconvincing.

The Impact

The DNA provirus hypothesis was thus proposed principally on the basis of genetic experiments and the effects of metabolic inhibitors. It was a radical proposal, which contradicted the accepted central dogma of molecular biology. In this setting, Temin's hypothesis that RSV replicated by the transfer of information from RNA to DNA not only failed to win the acceptance of the scientific community, but was met with general derision. Nonetheless, Temin persevered through the 1960s, continuing with experiments to test his hypothesis and providing increasingly convincing evidence in its support. These efforts culminated in 1970 with the discovery by Temin and Satoshi Mizutani, and at the same time by David Baltimore, of a viral enzyme, now known as reverse transcriptase, that synthesizes DNA from an RNA template—an unambiguous biochemical demonstration that the central dogma could be reversed.

Temin concluded his 1970 paper with the statement that the results "constitute strong evidence that the DNA provirus hypothesis is correct and that RNA tumour viruses have a DNA genome when they are in cells and an RNA genome when

Howard M. Temin

they are in virions. This result would have strong implications for theories of viral carcinogenesis and, possibly, for theories of information transfer in other biological systems." As Temin predicted, the discovery of RNA-directed DNA synthesis has led to major advances in our understanding of cancer, human retroviruses, and gene rearrangements. Reverse transcriptase has further provided a critical tool for cDNA cloning, thereby impacting virtually all areas of contemporary cell and molecular biology.

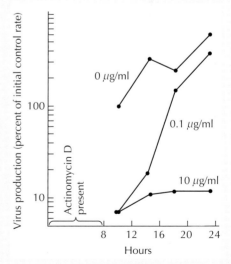

Effect of actinomycin D on RSV replication RSV-infected cells were cultured with the indicated concentrations of actinomycin D for 8 hours. Actinomycin D was then removed and the amount of virus produced was determined.

transcription. Second, enzymes that catalyze RNA-directed DNA synthesis (**reverse transcriptases**) can be used experimentally to generate DNA copies of any RNA molecule. The use of reverse transcriptase has thus allowed mRNAs of eukaryotic cells to be studied using the molecular approaches that are currently applied to the manipulation of DNA, as discussed in the following section.

Recombinant DNA

Classical experiments in molecular biology were strikingly successful in developing our fundamental concepts of the nature and expression of genes. Since these studies were based primarily on genetic analysis, their success depended largely on the choice of simple, rapidly replicating organisms (such as bacteria and viruses) as models. It was not clear, however, how these fundamental principles could be extended to provide a molecular understanding of the complexities of eukaryotic cells, since the genomes of most eukaryotes (e.g., the human genome) are up to a thousand times larger than that of *E. coli*. In the early 1970s, the possibility of studying such genomes at the molecular level seemed daunting. In particular, there appeared to be no way in which individual genes could be isolated and studied.

This obstacle to the progress of molecular biology was overcome by the development of recombinant DNA technology, which provided scientists with the ability to isolate, sequence, and manipulate individual genes derived from any type of cell. The application of recombinant DNA enabled detailed molecular studies of the structure and function of eukaryotic genes, thereby revolutionizing our understanding of cell biology.

Restriction endonucleases

The first step in the development of recombinant DNA technology was the characterization of **restriction endonucleases**—enzymes that cleave DNA at specific sequences. These enzymes were identified in bacteria, where they provide a defense against the entry of foreign DNA (e.g., from a virus) into the cell. Bacteria have a variety of restriction endonucleases that cleave DNA at more than 100 distinct recognition sites, each of which consists of a specific sequence of four to eight base pairs (examples are given in Table 4.2).

Table 4.2   Recognition Sites of Representative Restriction Endonucleases

Enzyme[a]	Source	Recognition site[b]
*Bam*HI	*Bacillus amyloliquefaciens* H	GGATCC
*Eco*RI	*Escherichia coli* RY13	GAATTC
*Hae*III	*Haemophilus aegyptius*	GGCC
*Hind*III	*Haemophilus influenzae* Rd	AAGCTT
*Hpa*I	*Haemophilus parainfluenzae*	GTTAAC
*Hpa*II	*Haemophilus parainfluenzae*	CCGG
*Mbo*I	*Moraxella bovis*	GATC
*Not*I	*Nocardia otitidis-caviarum*	GCGGCCGC
*Sfi*I	*Streptomyces fimbriatus*	GGCCNNNNNGGCC
*Taq*I	*Thermus aquaticus*	GATC

[a]Enzymes are named according to their species of isolation, followed by a number to distinguish different enzymes isolated from the same organism (e.g., *Hpa*I and *Hpa*II).

[b]Recognition sites show the sequence of only one strand of double-stranded DNA. "N" represents any base.

Animation 4.6

sites.sinauer.com/cooper7e/a4.6
Restriction Endonucleases   Restriction endonucleases cleave DNA at specific sequences, leaving staggered or blunt ends on the resulting DNA fragments.

Since restriction endonucleases digest DNA at specific sequences, they can be used to cleave a DNA molecule at unique sites. For example, the restriction endonuclease *Eco*RI recognizes the six-base-pair sequence GAATTC. This sequence is present at five sites in DNA of the bacteriophage λ, so *Eco*RI digests λ DNA into six fragments ranging from 3.6 to 21.2 kilobases long (1 kilobase, or kb = 1000 base pairs) (**Figure 4.14**). These fragments can be separated according to size by **gel electrophoresis**—a common method in which molecules are separated based on the rates of their migration in an electric field. A gel, usually formed from agarose or polyacrylamide, is placed between two buffer compartments containing electrodes. The sample (e.g., the mixture of DNA fragments to be analyzed) is then pipetted into preformed slots in the gel, and the electric field is turned on. Nucleic acids are negatively charged (because of their phosphate backbone), so they migrate toward the positive electrode. The gel acts like a sieve, selectively retarding the movement of larger molecules. Smaller molecules therefore move through

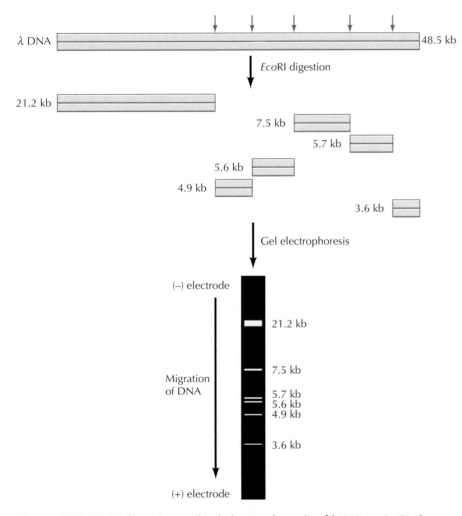

Figure 4.14 *Eco*RI digestion and gel electrophoresis of λ DNA *Eco*RI cleaves λ DNA at five sites (red arrows), yielding six DNA fragments. These fragments are then separated by electrophoresis in an agarose gel. The DNA fragments migrate toward the positive electrode, with smaller fragments moving more rapidly through the gel. Following electrophoresis, the DNA is stained with a fluorescent dye and photographed. The sizes of DNA fragments are indicated.

Figure 4.15 Restriction maps of λ and adenovirus DNAs   The locations of cleavage sites for *Bam*HI, *Eco*RI, and *Hind*III are shown in the DNAs of *E. coli* bacteriophage λ (48.5 kb) and human adenovirus-2 (35.9 kb).

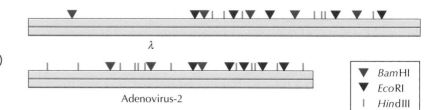

λ

Adenovirus-2

▼	*Bam*HI
▼	*Eco*RI
I	*Hind*III

Animation 4.7

sites.sinauer.com/cooper7e/a4.7
Recombinant DNA Molecules A DNA fragment is cloned by being inserted into a vector that replicates in a host cell.

the gel more rapidly, allowing a mixture of nucleic acids to be separated on the basis of size.

In addition to size, the order of restriction fragments can be determined by a variety of methods, yielding (for example) a map of the *Eco*RI sites in λ DNA. The locations of cleavage sites for multiple restriction endonucleases can be used to generate detailed **restriction maps** of DNA molecules, such as viral genomes (**Figure 4.15**). In addition, individual DNA fragments produced by restriction endonuclease digestion can be isolated following electrophoresis for further study—including determination of their DNA sequence. The DNAs of many viruses have been characterized by this approach.

Restriction endonuclease digestion alone, however, does not provide sufficient resolution for the analysis of larger DNA molecules, such as cellular genomes. A restriction endonuclease with a six-base-pair recognition site (such as *Eco*RI) cleaves DNA with a statistical frequency of once every 4096 base pairs ($1/4^6$). A molecule the size of λ DNA (48.5 kb) would therefore be expected to yield about ten *Eco*RI fragments, consistent with the results illustrated in Figure 4.14. However, restriction endonuclease digestion of larger genomes yields quite different results. For example, the human genome is approximately 3×10^6 kb long and is therefore expected to yield more than 500,000 *Eco*RI fragments. Such a large number of fragments cannot be separated from one another, so agarose gel electrophoresis of *Eco*RI-digested human DNA yields a continuous smear rather than a discrete pattern of DNA fragments. Because it is impossible to isolate single restriction fragments from such digests, restriction endonuclease digestion alone does not yield a source of homogeneous DNA suitable for further analysis. Quantities of such purified DNA fragments, however, can be obtained through molecular cloning.

Generation of recombinant DNA molecules

The basic strategy in **molecular cloning** is to insert a DNA fragment of interest (e.g., a segment of human DNA) into a DNA molecule (called a **vector**) that is capable of independent replication in a host cell. The result is a **recombinant molecule** or **molecular clone**, composed of the DNA insert linked to vector DNA sequences. Large quantities of the inserted DNA can be obtained if the recombinant molecule is allowed to replicate in an appropriate host. For example, fragments of human DNA can be cloned in plasmid vectors (**Figure 4.16**). **Plasmids** are small circular DNA molecules that can replicate independently—without being associated with chromosomal DNA—in bacteria.

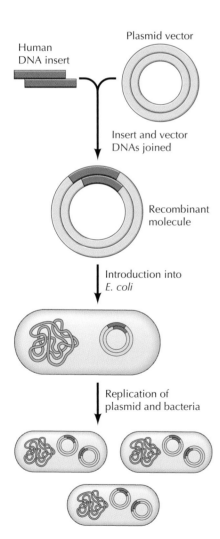

Figure 4.16 Generation of a recombinant DNA molecule   A fragment of human DNA is inserted into a plasmid DNA vector. The resulting recombinant molecule is then introduced into *E. coli*, where it replicates along with the bacteria to yield a population of bacteria carrying plasmids with the human DNA insert.

Recombinant plasmids carrying human DNA inserts can be introduced into *E. coli*, where they replicate along with the bacteria to yield millions of copies of plasmid DNA. The DNA of these plasmids can then be isolated, yielding large quantities of recombinant molecules containing a single fragment of human DNA. Whereas a typical DNA fragment 1 kb in length would represent less than 1 part in a million of human genomic DNA, it would represent approximately 1 part in 5 after being cloned in a plasmid vector. Moreover, the fragment can be easily isolated from the rest of the vector DNA by restriction endonuclease digestion and gel electrophoresis, allowing a pure fragment of human DNA to be analyzed and further manipulated.

The DNA fragments used to create recombinant molecules are usually generated by digestion with restriction endonucleases. Many of these enzymes cleave their recognition sequences at staggered sites, leaving overhanging or cohesive single-stranded tails that can associate with each other by complementary base pairing (**Figure 4.17**). The association between such paired complementary ends can be established permanently by treatment with

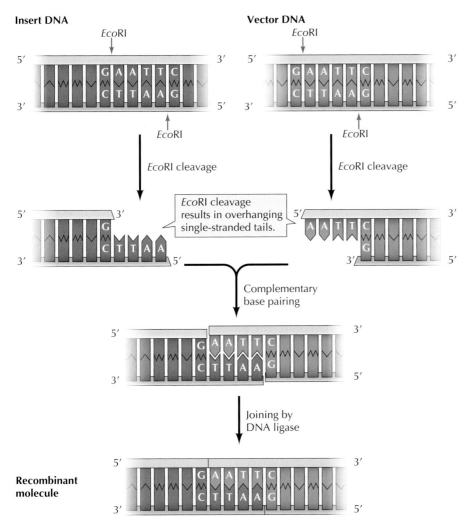

Figure 4.17 Joining of DNA molecules   Insert and vector DNAs are digested with a restriction endonuclease (such as *Eco*RI), which cleaves at staggered sites leaving overhanging single-stranded tails. Insert and vector DNAs can then associate by complementary base pairing, and covalent joining of the DNA strands by DNA ligase yields a recombinant molecule.

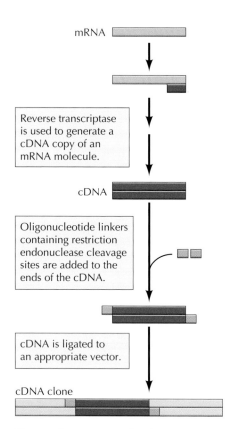

mRNA

Reverse transcriptase is used to generate a cDNA copy of an mRNA molecule.

cDNA

Oligonucleotide linkers containing restriction endonuclease cleavage sites are added to the ends of the cDNA.

cDNA is ligated to an appropriate vector.

cDNA clone

Figure 4.18 cDNA cloning

DNA ligase, an enzyme that seals breaks in DNA strands (see Chapter 7). Thus two different fragments of DNA (e.g., a human DNA insert and a plasmid DNA vector) prepared by digestion with the same restriction endonuclease can be readily joined to create a recombinant DNA molecule.

The fragments of DNA that can be cloned are not limited to those that terminate in restriction endonuclease cleavage sites. Synthetic DNA "linkers" containing desired restriction endonuclease sites can be added to the ends of any DNA fragment, allowing virtually any fragment of DNA to be ligated to a vector and isolated as a molecular clone.

Not only DNA, but also RNA sequences can be cloned (**Figure 4.18**). The first step is to synthesize a DNA copy of the RNA using the enzyme reverse transcriptase. The DNA product (called a **cDNA** because it is *c*omplementary to the RNA used as a template) can then be ligated to vector DNA as already described. Since eukaryotic genes are usually interrupted by noncoding sequences—or introns (see Chapter 6), which are removed from mRNA by splicing, the ability to clone cDNA as well as genomic DNA has been critical for understanding gene structure and function. Moreover, cDNA cloning allows the mRNA corresponding to a single gene to be isolated as a molecular clone.

Vectors for recombinant DNA

Depending on the size of the insert DNA and the purpose of the experiment, many different types of cloning vectors can be used for the generation of recombinant molecules. The basic vector systems used for the isolation and propagation of cloned DNAs are reviewed here. Other vectors developed for the expression of cloned DNAs and the introduction of recombinant molecules into eukaryotic cells are discussed in subsequent sections.

Plasmids are commonly used for cloning genomic or cDNA inserts of up to a few thousand base pairs. Plasmid vectors usually consist of 2 to 4 kb of DNA, including an **origin of replication**—the DNA sequence that signals the host cell DNA polymerase to replicate the DNA molecule. In addition, plasmid vectors carry genes that confer resistance to antibiotics (e.g., ampicillin resistance), so bacteria carrying the plasmids can be selected. For example, Figure 4.19 illustrates the isolation of human cDNA clones in a plasmid vector. A pool of cDNA fragments are ligated to restriction endonuclease-digested plasmid DNA. The resulting recombinant DNA molecules are then used to transform *E. coli.* Antibiotic-resistant colonies, which contain plasmid DNA, are selected. Since each recombinant plasmid yields a single antibiotic-resistant colony, the bacteria present in any given colony will contain a unique cDNA insert. Plasmid-containing bacteria can then be grown in large quantities and their DNA extracted. The small circular plasmid DNA molecules, of which there are often hundreds of copies per cell, can be separated from the bacterial chromosomal DNA; the result is purified plasmid DNA that is suitable for analysis of the cloned insert.

Bacteriophage λ vectors are also used for the isolation of either genomic or cDNA clones from eukaryotic cells, and will accommodate larger fragments of insert DNA than plasmids. In λ cloning vectors, sequences of the bacteriophage genome that are dispensable for virus replication have been removed and replaced with unique restriction sites for insertion of cloned DNA. These recombinant molecules can be introduced into *E. coli,* where they replicate to yield millions of progeny phages containing a single DNA insert. The DNA of these phages can then be isolated, yielding large quantities of recombinant molecules containing a single fragment of cloned DNA.

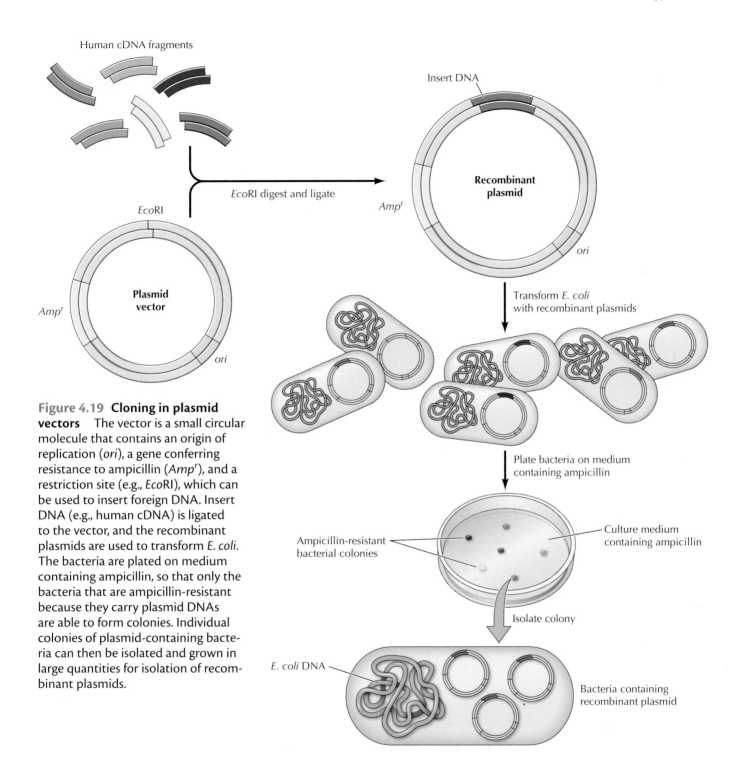

Figure 4.19 Cloning in plasmid vectors The vector is a small circular molecule that contains an origin of replication (*ori*), a gene conferring resistance to ampicillin (*Amp*ʳ), and a restriction site (e.g., *Eco*RI), which can be used to insert foreign DNA. Insert DNA (e.g., human cDNA) is ligated to the vector, and the recombinant plasmids are used to transform *E. coli*. The bacteria are plated on medium containing ampicillin, so that only the bacteria that are ampicillin-resistant because they carry plasmid DNAs are able to form colonies. Individual colonies of plasmid-containing bacteria can then be isolated and grown in large quantities for isolation of recombinant plasmids.

The DNA inserts can be as large as about 15 kb and still yield a recombinant genome that can be packaged into bacteriophage λ particles.

For many studies involving analysis of genomic DNA, it is desirable to clone larger fragments of DNA than are accommodated by plasmid or λ vectors. There are five major types of vectors that are used for this purpose (**Table 4.3**). **Cosmid** vectors accommodate inserts of 30–45 kb. These vectors contain bacteriophage λ sequences that allow efficient packaging of the

Table 4.3 Vectors for Cloning Large Fragments of DNA

Vector	DNA insert (kb)	Host cell
Cosmids	30–45	*E. coli*
Bacteriophage P1	70–100	*E. coli*
P1 artificial chromosomes (PACs)	130–150	*E. coli*
Bacterial artificial chromosomes (BACs)	120–300	*E. coli*
Yeast artificial chromosomes (YACs)	250–400	Yeast

cloned DNA into phage particles. In addition, cosmids contain origins of replication and the genes for antibiotic resistance that are characteristic of plasmids, so they are able to replicate as plasmids in bacterial cells. Two other types of vectors are derived from bacteriophage P1, rather than from bacteriophage λ. Bacteriophage P1 vectors, which will accommodate DNA fragments of 70–100 kb, contain sequences that allow recombinant molecules to be packaged *in vitro* into P1 phage particles and then to be replicated as plasmids in *E. coli*. **P1 artificial chromosome (PAC)** vectors also contain sequences of bacteriophage P1 but are introduced directly as plasmids into *E. coli* and will accommodate larger inserts of 130–150 kb. **Bacterial artificial chromosome (BAC)** vectors are derived from a naturally occurring plasmid of *E. coli* (called the F factor). The replication origin and other F factor sequences allow BACs to replicate as stable plasmids carrying inserts of 120–300 kb. Even larger fragments of DNA (250–400 kb) can be cloned in **yeast artificial chromosome (YAC)** vectors. These vectors contain yeast origins of replication as well as other sequences (centromeres and telomeres, discussed in Chapter 6) that allow them to replicate as linear chromosome-like molecules in yeast cells.

DNA sequencing

Molecular cloning allows the isolation of individual fragments of DNA in quantities suitable for detailed characterization, including the determination of nucleotide sequence. Indeed, determination of the nucleotide sequences of many genes has elucidated not only the structure of their protein products but also the properties of DNA sequences that regulate gene expression. Furthermore, the coding sequences of novel genes are frequently related to those of previously studied genes, and the functions of newly isolated genes can often be correctly deduced on the basis of such sequence similarities.

The basic method used for DNA sequencing, developed by Frederick Sanger in 1977, is based on premature termination of DNA synthesis resulting from the inclusion of chain-terminating **dideoxynucleotides** (which do not contain the deoxyribose 3′ hydroxyl group) in DNA polymerase reactions (**Figure 4.20**). DNA synthesis is initiated at a unique site on the cloned DNA from a synthetic primer. The DNA synthesis reaction includes each of the four dideoxynucleotides (A, C, G, and T), in addition to their normal counterparts. Each of the four dideoxynucleotides is labeled with a different fluorescent dye, so their incorporation into DNA can be monitored. Incorporation of a dideoxynucleotide stops further DNA synthesis because no 3′ hydroxyl

Animation 4.8

sites.sinauer.com/cooper7e/a4.8

Sequencing a DNA Strand One method of DNA sequencing involves a modified DNA synthesis reaction using chain-terminating fluorescent nucleotides, which can be identified by automated detection systems.

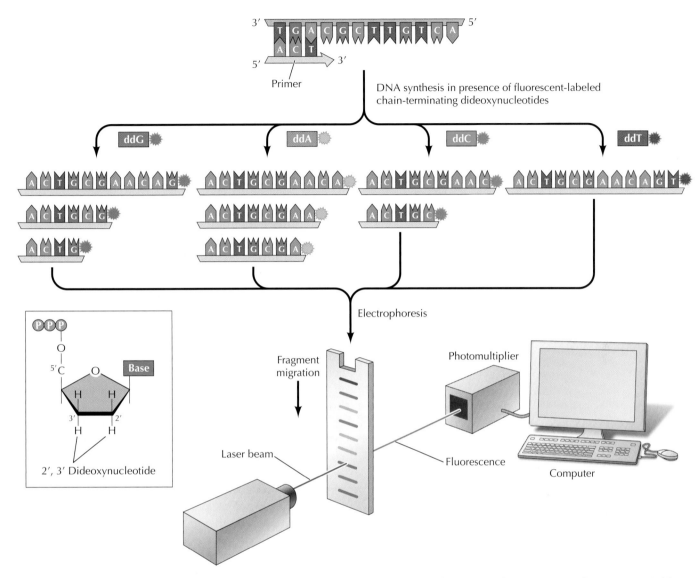

Figure 4.20 DNA sequencing Dideoxynucleotides, which lack OH groups at the 3′ as well as the 2′ position of deoxyribose, are used to terminate DNA synthesis at specific bases. These molecules are incorporated normally into growing DNA strands. Because they lack a 3′ OH, however, the next nucleotide cannot be added, so synthesis of that DNA strand terminates. DNA synthesis is initiated at a specific site with a primer. The reaction contains the four dideoxynucleotides, each labeled with a different fluorescent dye, as well as the four normal deoxynucleotides. When the dideoxynucleotide is incorporated, DNA synthesis stops, so the reaction yields a series of products extending from the primer to the base substituted by a fluorescent dideoxynucleotide. These products are then separated by gel electrophoresis. As the DNA strands migrate through the gel, they pass through a laser beam that excites the fluorescent labels on the dideoxynucleotides. The emitted light is detected by a photomultiplier, which is connected to a computer that collects and analyzes the data to determine the DNA sequence.

group is available for addition of the next nucleotide. Thus a series of labeled DNA molecules is generated, each terminating at the base represented by a specific fluorescent dideoxynucleotide. These fragments of DNA are then separated according to size by gel electrophoresis. As the newly synthesized DNA strands are electrophoresed through the gel, they pass through a laser beam that excites the fluorescent labels. The resulting emitted light is then detected by a photomultiplier, and a computer collects and analyzes the data. The size of each fragment is determined by its terminal dideoxynucleotide, marked by a specific color fluorescence, so the DNA sequence can be read from the order of fluorescent-labeled fragments as they migrate through the gel. High-throughput automated DNA sequencing using dideoxynucleotides has enabled the large-scale analysis required for determination of the sequences of complete genomes—including that of humans. However, recent years have seen the development of new methods that enable DNA to be sequenced far more rapidly and less expensively. As discussed in Chapter 5, these new approaches, generically referred to as **next-generation sequencing**, have allowed the sequencing of many individual genomes to be readily determined.

Expression of cloned genes

In addition to enabling determination of the nucleotide sequences of genes—and hence the amino acid sequences of their protein products—molecular cloning has provided new approaches to obtaining large amounts of proteins for structural and functional characterization. Many proteins of interest are present at only low levels in eukaryotic cells and therefore cannot be purified in significant amounts by conventional biochemical techniques. Given a cloned gene, however, this problem can be solved by the engineering of vectors that lead to high levels of gene expression in either bacteria or eukaryotic cells.

To express a eukaryotic gene in *E. coli*, the cDNA of interest is cloned into a plasmid or phage vector (called an **expression vector**) that contains sequences that drive transcription and translation of the inserted gene in bacterial cells (**Figure 4.21**). Inserted genes often can be expressed at levels high enough that the protein encoded by the cloned gene corresponds to as much as 10% of the total bacterial protein. Purifying the protein encoded by the cloned gene in quantities suitable for detailed biochemical or structural studies is then a straightforward matter.

It is frequently useful to express high levels of a cloned gene in eukaryotic cells rather than in bacteria. This mode of expression may be important, for example, to ensure that post-translational modifications of the protein (such as addition of carbohydrates or lipids) occur normally. Such protein expression in eukaryotic

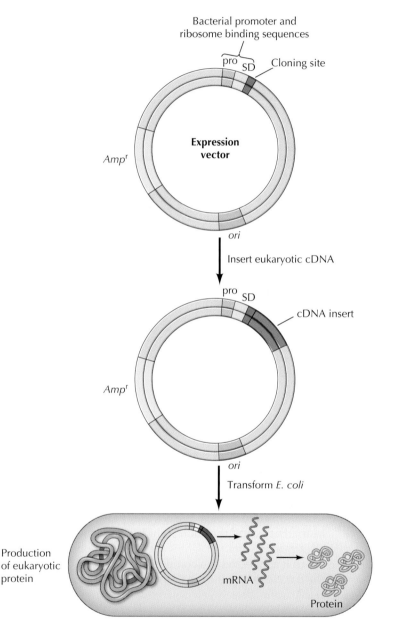

Figure 4.21 Expression of cloned genes in bacteria Expression vectors contain promoter sequences (pro) that direct transcription of inserted DNA in bacteria and sequences required for binding of mRNA to bacterial ribosomes (Shine-Dalgarno [SD] sequences, discussed in Chapter 9). A eukaryotic cDNA inserted adjacent to these sequences can be efficiently expressed in *E. coli*, resulting in production of eukaryotic proteins in transformed bacteria.

cells can be achieved, as in *E. coli*, by insertion of the cloned gene into a vector (usually derived from a virus) that directs high-level gene expression.

Detection of Nucleic Acids and Proteins

The advent of molecular cloning has enabled the isolation and characterization of individual genes from eukaryotic cells. Understanding the role of genes within cells, however, requires analysis of the intracellular organization and expression of individual genes and their encoded proteins. In this section, the basic procedures used for detection of specific nucleic acids and proteins are discussed. These approaches are important for a wide variety of studies, including analysis of gene expression and the localization of proteins to subcellular organelles.

Amplification of DNA by the polymerase chain reaction

Molecular cloning allows individual DNA fragments to be propagated in bacteria and isolated in large amounts. An alternative method to isolating large amounts of a single DNA molecule is the **polymerase chain reaction (PCR)**, which was developed by Kary Mullis in 1988. Provided that the DNA sequence is known, PCR can achieve a striking amplification of DNA via reactions carried out entirely *in vitro*. Essentially, DNA polymerase is used for repeated replication of a defined segment of DNA. The number of DNA molecules increases exponentially, doubling with each round of replication, so a substantial quantity of DNA can be obtained from a small number of initial template copies. For example, a single DNA molecule amplified through 30 cycles of replication would theoretically yield 2^{30} (approximately 1 billion) progeny molecules. Single DNA molecules can thus be amplified to yield readily detectable quantities of DNA that can be isolated by molecular cloning or further analyzed directly by nucleotide sequencing.

The general procedure for PCR amplification of DNA is illustrated in **Figure 4.22**. The starting material can be either a cloned DNA fragment or a mixture of DNA molecules—for example, total DNA from human cells. A specific region of DNA can be amplified from such a mixture, provided that the nucleotide sequence surrounding the region is known so that primers can be designed to initiate DNA synthesis at the desired point. Such primers are usually chemically synthesized oligonucleotides containing 15 to 20 bases of DNA. Two primers are used to initiate DNA synthesis in opposite directions from complementary DNA strands. The reaction is started by heating the template DNA to a high temperature (e.g., 95°C) so that the two strands separate. The temperature is then lowered to allow the primers to pair with their complementary sequences on the template strands. DNA polymerase then uses the primers to synthesize a new strand complementary to each template. Thus, in one cycle of amplification, two new DNA molecules are synthesized from one template molecule. The process can be repeated multiple times, with a twofold increase in DNA molecules resulting from each round of replication.

The multiple cycles of heating and cooling involved in PCR are performed by programmable heating blocks called thermocyclers. The DNA polymerases used in these reactions are heat-stable enzymes from bacteria such as *Thermus aquaticus*, which lives in hot springs at temperatures of about 75°C. These polymerases are stable even at the high temperatures used to separate the

FYI

PCR has become an extremely powerful forensic method. Amplification by PCR allows investigators to obtain a DNA profile from small samples of DNA at a crime scene.

Animation 4.9

sites.sinauer.com/cooper7e/a4.9
Polymerase Chain Reaction The polymerase chain reaction allows the production of millions of copies of a DNA fragment from just one starting DNA molecule.

Starting DNA

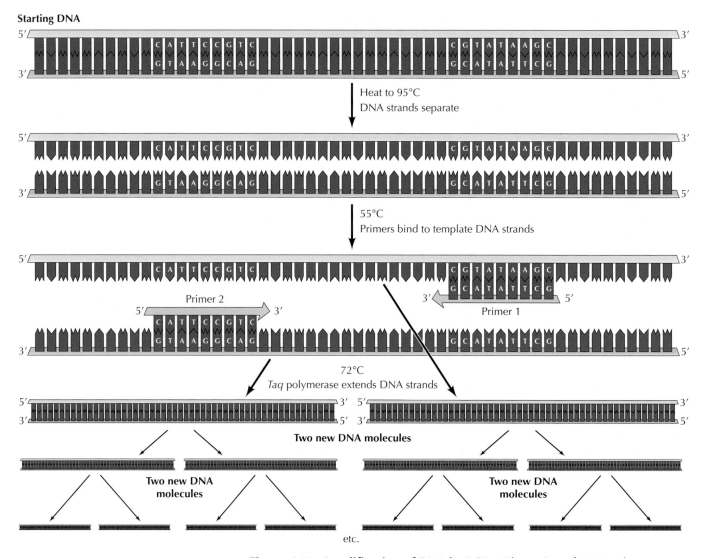

Figure 4.22 Amplification of DNA by PCR   The region of DNA to be amplified is flanked by two sequences used to prime DNA synthesis. The starting double-stranded DNA is heated to separate the strands and then cooled to allow primers (usually oligonucleotides of 15 to 20 bases) to bind to each strand of DNA. DNA polymerase from *Thermus aquaticus* (*Taq* polymerase) is used to synthesize new DNA strands starting from the primers, resulting in the formation of two new DNA molecules. The process can be repeated for multiple cycles, each resulting in a twofold amplification of DNA.

strands of double-stranded DNA, so PCR amplification can be performed rapidly and automatically. RNA sequences can also be amplified by this method if reverse transcriptase is used to synthesize a cDNA copy prior to PCR amplification.

If enough of the sequence of a gene is known so that primers can be specified, PCR amplification provides an extremely powerful method of detecting small amounts of specific DNA or RNA molecules in a complex mixture of other molecules. The only DNA molecules that will be amplified by PCR are those containing sequences complementary to the primers used in the reaction. Therefore PCR can selectively amplify a specific template from complex mixtures, such as total cell DNA or RNA.

Importantly, PCR can also be used quantitatively to determine the amounts of specific DNA or RNA molecules in a sample. These methods are based on measuring the formation of double-stranded DNA products during PCR amplification. A fluorescent dye that binds to double-stranded DNA, and only fluoresces when bound, is incorporated into the PCR reaction. Fluorescence is then monitored after each cycle of PCR, and the intensity of fluorescence is indicative of the amount of DNA that has been amplified. Since the amount of DNA amplified after a given number of cycles is determined by the amount of template that was initially present, this provides quantitation of the amount of target DNA or RNA that was present in the sample. Since fluorescence is measured continuously during the PCR reaction, this method is called **real-time PCR**. Its extraordinary sensitivity has made it important for a variety of applications, including the analysis of gene expression in cells and tissues.

Nucleic acid hybridization

The key to detection of specific nucleic acid sequences is base pairing between complementary strands of RNA or DNA. At high temperatures (e.g., 90 to 100°C), the complementary strands of DNA separate (denature), yielding single-stranded molecules. If such denatured DNA strands are then incubated under appropriate conditions (e.g., 65°C), they will renature to form double-stranded molecules as dictated by complementary base pairing—a process called **nucleic acid hybridization**. Nucleic acid hybrids can be formed between two strands of DNA, two strands of RNA, or one strand of DNA and one of RNA.

As discussed above, hybridization between the primers and the template DNA provides the specificity to PCR amplification. In addition, a variety of other methods use nucleic acid hybridization as a means for detecting DNA or RNA sequences that are complementary to any isolated nucleic acid—such as a cloned DNA sequence (**Figure 4.23**). The cloned DNA is labeled with either radioactive nucleotides or with modified nucleotides that can be detected by fluorescence or chemiluminescence. This labeled DNA is then used as a probe for hybridization to identify complementary DNA or RNA sequences, which are detected by virtue of the radioactivity, fluorescence, or luminescence of the resulting double-stranded hybrids.

Southern blotting (a technique developed by E. M. Southern) is widely used for detection of specific genes in cellular DNA (**Figure 4.24**). The DNA to be analyzed is digested with a restriction endonuclease, and the digested DNA fragments are separated by gel electrophoresis. The gel is then overlaid with a nitrocellulose filter or nylon membrane to which the DNA fragments

Figure 4.23 Detection of DNA by nucleic acid hybridization A specific sequence can be detected in total cell DNA by hybridization with a labeled DNA probe, containing radioactive nucleotides or modified nucleotides that can be detected by fluorescence or chemiluminescence. The DNA is denatured by heating to 95°C, yielding single-stranded molecules. The labeled probe is then added and the temperature is lowered to 65°C, allowing complementary DNA strands to renature by pairing with each other. The probe hybridizes to complementary sequences in cell DNA, which can then be detected by incorporation of the labeled probe into double-stranded molecules.

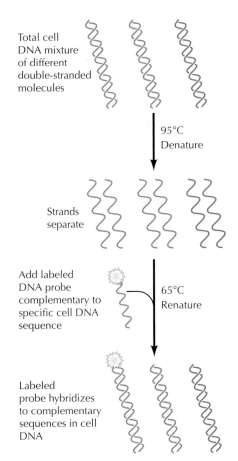

Total cell DNA mixture of different double-stranded molecules

95°C Denature

Strands separate

Add labeled DNA probe complementary to specific cell DNA sequence

65°C Renature

Labeled probe hybridizes to complementary sequences in cell DNA

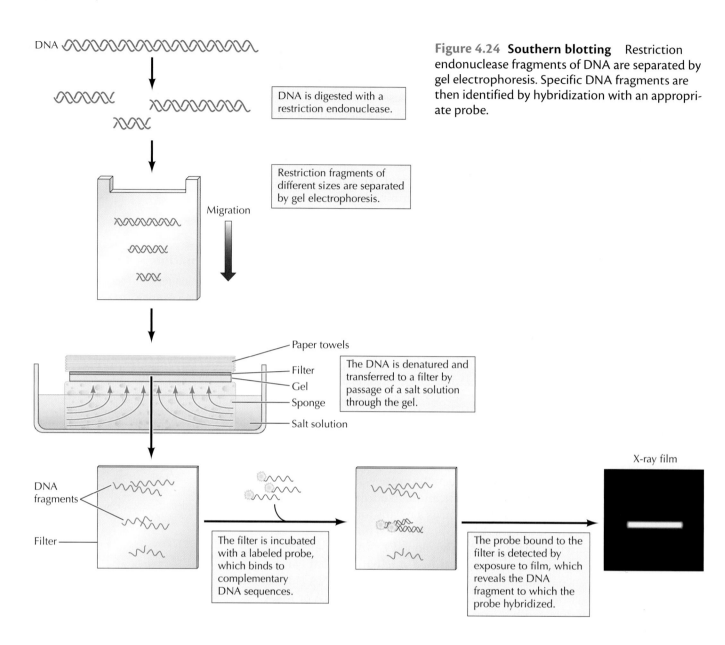

Figure 4.24 Southern blotting Restriction endonuclease fragments of DNA are separated by gel electrophoresis. Specific DNA fragments are then identified by hybridization with an appropriate probe.

DNA is digested with a restriction endonuclease.

Restriction fragments of different sizes are separated by gel electrophoresis.

Migration

The DNA is denatured and transferred to a filter by passage of a salt solution through the gel.

Paper towels
Filter
Gel
Sponge
Salt solution

DNA fragments

Filter

The filter is incubated with a labeled probe, which binds to complementary DNA sequences.

The probe bound to the filter is detected by exposure to film, which reveals the DNA fragment to which the probe hybridized.

X-ray film

are transferred (blotted) to yield a replica of the gel. The filter is then incubated with a labeled probe, which hybridizes to the DNA fragments that contain the complementary sequence, allowing visualization of these specific fragments of cell DNA.

Northern blotting is a variation of the Southern blotting technique (hence its name) that is used for detection of RNA instead of DNA. In this method, total cellular RNAs are extracted and fractionated according to size by gel electrophoresis. As in Southern blotting, the RNAs are transferred to a filter and detected by hybridization with a labeled probe. Northern blotting is frequently used in studies of gene expression—for example, to determine whether specific mRNAs are present in different types of cells.

Nucleic acid hybridization can also be used to identify molecular clones that contain specific cellular DNA inserts. The first step in isolation of either genomic or cDNA clones is frequently the preparation of **recombinant DNA libraries**—collections of clones that contain all the genomic or mRNA sequences

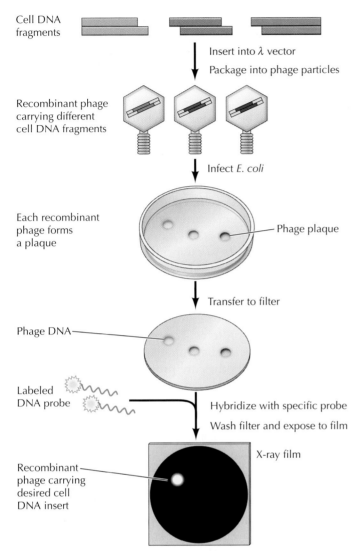

Cell DNA fragments

Insert into λ vector

Package into phage particles

Recombinant phage carrying different cell DNA fragments

Infect *E. coli*

Each recombinant phage forms a plaque

Phage plaque

Transfer to filter

Phage DNA

Labeled DNA probe

Hybridize with specific probe

Wash filter and expose to film

X-ray film

Recombinant phage carrying desired cell DNA insert

Figure 4.25 Screening a recombinant library by hybridization Fragments of cell DNA are cloned in a bacteriophage λ vector and packaged into phage particles, yielding a collection of recombinant phage carrying different cell inserts. The phages are used to infect bacteria, and the culture is overlaid with a filter. Some of the phages in each plaque are transferred to the filter, which is then hybridized with a labeled probe to identify the phage plaque containing the desired gene. The appropriate phage plaque can then be isolated from the original culture plate.

of a particular cell type (**Figure 4.25**). For example, a genomic library of human DNA might be prepared by cloning random DNA fragments of about 15 kb in a λ vector. Since the human genome is approximately 3 million kb, the complete human genome would be represented in a collection of approximately 500,000 such clones. Any gene for which a probe is available can then be isolated from such a recombinant library. The recombinant phages are plated on *E. coli*, and each phage replicates to produce a plaque on the lawn of bacteria. The plaques are then blotted onto filters in a process similar to the transfer of DNA from a gel to a filter during Southern blotting, and the filters are hybridized with a labeled probe to identify the phage plaques that contain the gene of interest. A variety of probes can be used for such experiments. For example, a cDNA clone can be used as a probe to isolate the corresponding genomic clone, or a gene cloned from one species (e.g., mouse) can be used to isolate a related gene from a different species (e.g., human). The appropriate plaque can then be isolated from the original plate in order to propagate the recombinant phage that carries the desired cell DNA insert. Similar procedures can be used to screen bacterial colonies carrying plasmid

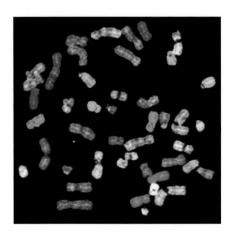

Figure 4.26 Fluorescence *in situ* hybridization   Hybridization of human chromosomes with chromosome-specific fluorescent probes that label each of the 24 chromosomes a different color. (Courtesy of Thomas Reid and Hesed Padilla-Nash, National Cancer Institute.)

Animation 4.12

sites.sinauer.com/cooper7e/a4.12

Colony Hybridization Bacteria can house recombinant DNA libraries, and specific recombinant DNA molecules within the library can be identified by the colony hybridization procedure—a procedure in which the DNA is incubated with a specific radioactive DNA probe.

Animation 4.13

sites.sinauer.com/cooper7e/a4.13

Monoclonal Antibodies Monoclonal antibodies can be made by inoculating an animal (such as a mouse) with an antigen, providing the animal time to mount an immune response, collecting the B cells from the animal, and then fusing the B cells to long-lived myeloma cells—thereby creating hybridomas that can be tested for the production of useful antibodies.

DNA clones, so specific clones can be isolated by hybridization from either phage or plasmid libraries.

Nucleic acid hybridization can be used to detect homologous DNA or RNA sequences not only in cell extracts but also in chromosomes or intact cells—a procedure called *in situ* **hybridization** (Figure 4.26). In this case, the hybridization of fluorescent probes to specific cells or subcellular structures is analyzed by microscopic examination. For example, labeled probes can be hybridized to intact chromosomes in order to identify the chromosomal regions that contain a gene of interest. *In situ* hybridization can also be used to detect specific mRNAs in different types of cells within a tissue.

Antibodies as probes for proteins

Studies of gene expression and function require the detection not only of DNA and RNA, but also of specific proteins. For these studies, **antibodies** take the place of nucleic acid probes as reagents that can selectively react with unique protein molecules. Antibodies are proteins produced by cells of the immune system (B lymphocytes) that react against molecules that the host organism recognizes as foreign substances (**antigens**)—for example, the protein coat of a virus. The immune systems of vertebrates are capable of producing millions of different antibodies, each of which specifically recognizes a unique antigen, which may be a protein, a carbohydrate, or a nonbiological molecule. An individual lymphocyte produces only a single type of antibody, but the antibody genes of different lymphocytes vary as a result of programmed gene rearrangements during development of the immune system (see Chapter 7). This variation gives rise to an array of lymphocytes with distinct antibody genes, programmed to respond to different antigens.

Antibodies can be generated by inoculation of an animal with any foreign protein. For example, antibodies against human proteins are frequently raised in rabbits. The sera of such immunized animals contain a mixture of antibodies (produced by different lymphocytes) that react against multiple sites on the immunizing antigen. However, single species of antibodies (**monoclonal antibodies**) can also be produced by the culturing of clonal lines of B lymphocytes from immunized animals (usually mice). Because each lymphocyte is programmed to produce only a single antibody, a clonal line of lymphocytes produces a monoclonal antibody that recognizes only a single antigenic determinant, thereby providing a highly specific immunological reagent.

Although antibodies can be raised against proteins purified from cells, other materials may also be used for immunization. For example, animals may be immunized with intact cells to raise antibodies against unknown proteins expressed by a specific cell type (e.g., a cancer cell). Such antibodies may then be used to identify proteins specifically expressed by the cell type used for immunization. In addition, antibodies are frequently raised against proteins expressed in bacteria as recombinant clones. In this way, molecular cloning allows the production of antibodies against proteins that may be difficult to isolate from eukaryotic cells. Moreover, antibodies can be raised against synthetic peptides that consist of only 10 to 15 amino acids,

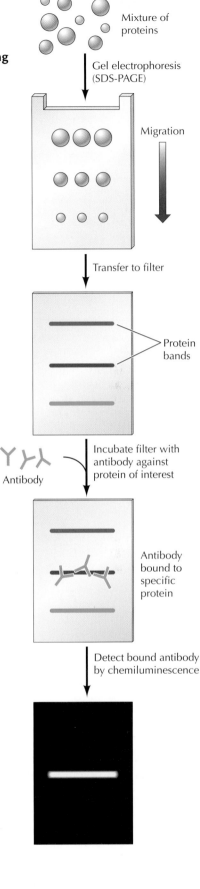

Figure 4.27 Immunoblotting

rather than against intact proteins. Therefore, once the sequence of a gene is known, antibodies against peptides synthesized from part of its predicted protein sequence can be produced. Because antibodies against these synthetic peptides frequently react with the intact protein as well, it is possible to produce antibodies against a protein starting with only the sequence of a gene.

Antibodies can be used in a variety of ways to detect proteins in cell extracts. A common method, **immunoblotting** (also called **Western blotting**) (Figure 4.27), is another variation of Southern blotting. Proteins in cell extracts are first separated according to size by gel electrophoresis. Because proteins have different shapes and charges, however, this process requires a modification of the methods used for electrophoresis of nucleic acids. Proteins are separated by a method known as **SDS-polyacrylamide gel electrophoresis (SDS-PAGE)** in which they are dissolved in a solution containing the negatively charged detergent sodium dodecyl sulfate (SDS). Each protein binds many detergent molecules, which denature the protein and give it an overall negative charge. Under these conditions, all proteins migrate toward the positive electrode—their rates of migration determined (like those of nucleic acids) only by size. Following electrophoresis, the proteins are transferred to a filter, which is then allowed to react with antibodies against the protein of interest. The antibody bound to the filter can be detected by a variety of methods, such as chemiluminescence, to identify the protein against which the antibody is targeted.

As discussed in Chapter 1, antibodies can be used to visualize proteins in intact cells, as well as in cell lysates. For example, cells can be stained with antibodies labeled with fluorescent dyes, and the subcellular localization of the antigenic proteins visualized by fluorescence microscopy (Figure 4.28).

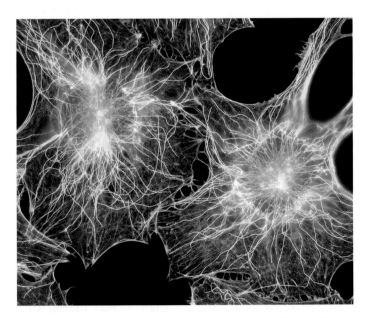

Figure 4.28 Immunofluorescence Human cells in culture were stained with immunofluorescent antibodies against actin (purple) and tubulin (yellow). Nuclei are stained with a green fluorescent dye.

Antibodies can also be labeled with tags that are visible in the electron microscope, such as heavy metals, allowing visualization of antigenic proteins at the ultrastructural level.

Gene Function in Eukaryotes

The recombinant DNA techniques discussed in the preceding sections provide powerful approaches to the isolation and detailed characterization of the genes of eukaryotic cells. Understanding the function of a gene, however, requires analysis of the gene within cells or intact organisms—not simply as a molecular clone in bacteria. In classical genetics, the function of genes has generally been revealed by the altered phenotypes of mutant organisms. The advent of recombinant DNA has added a new dimension to studies of gene function. Namely, it has become possible to investigate the function of a cloned gene directly by reintroducing the cloned DNA into cultured cells, as well as intact organisms, for functional analysis. These approaches can be coupled with the ability to introduce mutations in cloned DNA *in vitro*, extending the power of recombinant DNA to allow functional studies of the genes of complex eukaryotes.

Gene transfer in plants and animals

Although the cells of complex eukaryotes are not amenable to the simple genetic manipulations possible in bacteria and yeasts, gene function can still be assayed by the introduction of cloned DNA into plant and animal cells. Such experiments (generally called **gene transfer**) have proven critical to addressing a wide variety of questions, including studies of the mechanisms that regulate gene expression and protein processing. In addition, as discussed later in the book, gene transfer has enabled the identification and characterization of genes that control animal cell growth and differentiation, including a variety of genes responsible for the abnormal growth of human cancer cells.

The methodology for introduction of DNA into animal cells was initially developed for infectious viral DNAs and is therefore called **transfection** (a word derived from *trans*formation + in*fection*) (**Figure 4.29**). DNA can be introduced into animal cells in culture by a variety of methods, including direct microinjection into the cell nucleus, coprecipitation of DNA with calcium phosphate to form small particles that are taken up by the cells, incorporation of DNA into lipid vesicles (**liposomes**) that fuse with the plasma membrane, and exposure of cells to a brief electric pulse that transiently opens pores in the plasma membrane (**electroporation**). The DNA taken up by a high fraction of cells is transported to the nucleus, where it can be transcribed for several days—a phenomenon called **transient expression**. In a smaller fraction of cells (usually 1% or less), the foreign DNA becomes stably integrated into the cell genome and is transferred to progeny cells at cell division just as any other cell gene is. These stably transformed cells can be isolated if the transfected DNA contains a selectable marker, such as resistance to a drug that inhibits the growth of normal cells. Thus any cloned gene can be introduced into mammalian cells by being transferred together with a drug resistance marker that can be used to isolate stable transformants. The effects of such cloned genes on cell behavior—for example, cell growth or differentiation—can then be analyzed.

Animal viruses can also be used as vectors for more efficient introduction of cloned DNAs into cells. Retroviruses are particularly useful in this respect,

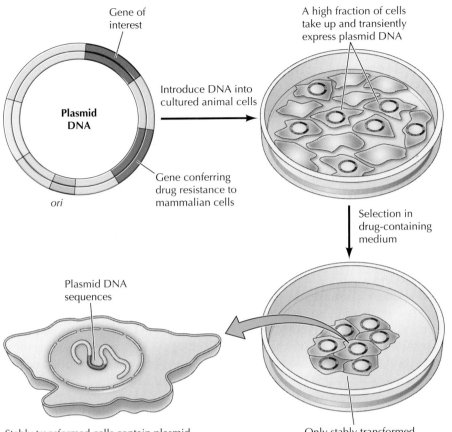

Figure 4.29 Introduction of DNA into animal cells A eukaryotic gene of interest is cloned in a plasmid containing a drug resistance marker that can be selected for in cultured animal cells. The plasmid DNA is taken up and expressed by a high fraction of the cells for a few days (transient expression). Stably transformed cells, in which the plasmid DNA becomes integrated into chromosomal DNA, can then be selected for their ability to grow in drug-containing medium.

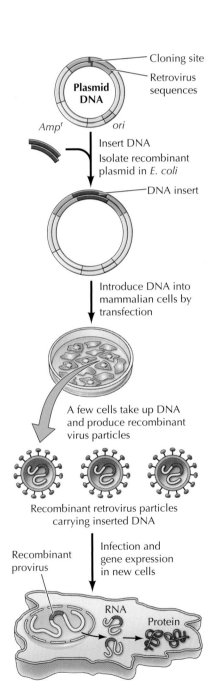

since their life cycle involves the stable integration of viral DNA into the genome of infected cells (Figure 4.30). Consequently, retroviral vectors can be used to efficiently introduce cloned genes into a wide variety of cell types, making them an important vehicle for a broad range of applications.

Cloned genes can also be introduced into the germ line of multicellular organisms, allowing them to be studied in the context of the intact animal rather than in cultured cells. One method used to produce mice that carry such foreign genes (**transgenic mice**) is the direct microinjection of cloned

Figure 4.30 Retroviral vectors The vector consists of retroviral sequences cloned in a plasmid that can be propagated in *E. coli*. Foreign DNA is inserted into the viral sequences, and recombinant plasmids are isolated in bacteria. Animal cells in culture are then transfected with the recombinant DNA. The DNA is taken up by a small fraction of the cells, which produce recombinant retroviruses carrying the inserted DNA. These recombinant retroviruses can be used to efficiently infect new cells, where the viral genome carrying the inserted gene integrates into chromosomal DNA as a provirus.

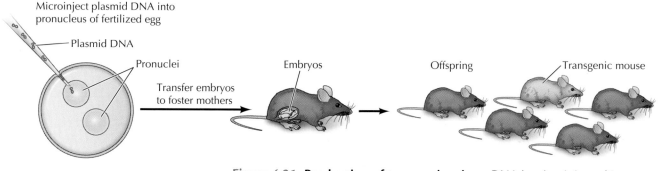

Microinject plasmid DNA into
pronucleus of fertilized egg

Plasmid DNA

Pronuclei

Transfer embryos
to foster mothers

Embryos

Offspring

Transgenic mouse

Figure 4.31 Production of transgenic mice   DNA is microinjected into a pronucleus of a fertilized mouse egg (fertilized eggs contain two pronuclei, one from
the egg and one from the sperm). The microinjected eggs are then transferred to
foster mothers and allowed to develop. Some of the offspring (transgenic) have
incorporated the injected DNA into their genome.

DNA into the pronucleus of a fertilized egg (Figure 4.31). The injected eggs
are then transferred to foster mothers and allowed to develop to term. In a
fraction of the progeny mice (often about 10%), the foreign DNA will have
integrated into the genome of the fertilized egg and is therefore present in
all cells of the animal. Since the foreign DNA is present in germ cells as well
as in somatic cells, it is transferred by breeding to new progeny mice just as
any other cell gene would be.

The properties of **embryonic stem (ES) cells** provide an alternative means
of introducing cloned genes into mice (Figure 4.32). ES cells can be established in culture from early mouse embryos. They can also be reintroduced
into early embryos, where they participate normally in development and
can give rise to cells in all tissues of the mouse—including germ cells. It
is thus possible to introduce cloned DNA into ES cells in culture, select
stably transformed cells, and then introduce these cells back into mouse
embryos. Such embryos give rise to offspring in which some cells are
derived from the normal embryo cells and some from the transfected ES
cells, referred to as chimeric—a mixture of two different cell types. In
some such mice, the transfected ES cells are incorporated into the germ
line. Breeding these mice therefore leads to the direct inheritance of the
transfected gene by their progeny.

Cloned DNAs can also be introduced into plant cells by a variety of
methods. Since many plants can be regenerated from single cultured cells
(see Chapter 1), transgenic plants can be established directly from cells into
which recombinant DNA has been introduced in culture—a much simpler
procedure than the production of transgenic animals. Indeed, many economically important types of plants, including corn, tomatoes, soybeans,
and potatoes, are transgenic varieties.

Mutagenesis of cloned DNAs

In classical genetic studies (e.g., in bacteria or yeasts), mutants are the key to
identifying genes and understanding their function by observing the altered
phenotype of mutant organisms. In such studies, mutant genes are detected
because they result in observable phenotypic changes—for example, temperature-sensitive growth or a specific nutritional requirement. The isolation
of genes by recombinant DNA, however, has opened a different approach
to mutagenesis. It is now possible to introduce any desired alteration into a

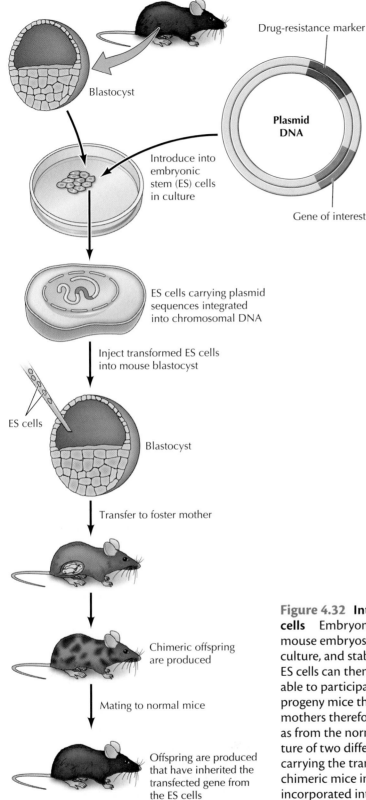

Drug-resistance marker

Plasmid
DNA

Gene of interest

Blastocyst

Introduce into
embryonic
stem (ES) cells
in culture

ES cells carrying plasmid
sequences integrated
into chromosomal DNA

Inject transformed ES cells
into mouse blastocyst

ES cells

Blastocyst

Transfer to foster mother

Chimeric offspring
are produced

Mating to normal mice

Offspring are produced
that have inherited the
transfected gene from
the ES cells

Figure 4.32 Introduction of genes into mice via embryonic stem cells Embryonic stem (ES) cells are cultured cells derived from early mouse embryos (blastocysts). DNA can be introduced into these cells in culture, and stably transformed ES cells can be isolated. These transformed ES cells can then be injected into a recipient blastocyst, where they are able to participate in normal development of the embryo. Some of the progeny mice that develop after transfer of injected embryos to foster mothers therefore contain cells derived from transformed ES cells, as well as from the normal cells of the blastocyst. Since these mice are a mixture of two different cell types, they are referred to as chimeric. Offspring carrying the transfected gene can then be produced by the breeding of chimeric mice in which descendants of the transformed ES cells have been incorporated into the germ line.

cloned gene and to determine the effect of the mutation on gene function. The ability to introduce specific mutations into cloned DNAs (*in vitro* **mutagenesis**) has proven to be a powerful tool in studying the expression and function of eukaryotic genes.

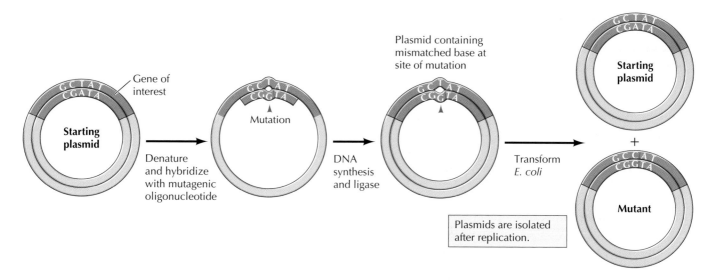

Figure 4.33 Mutagenesis with synthetic oligonucleotides An oligonucle-otide carrying the desired base alteration is used to prime DNA synthesis from plasmid DNA, and the DNA is circularized by incubation with DNA ligase. This process yields plasmids in which one strand contains the normal base and the other strand the mutant base. Replication of the plasmid DNAs after transforma-tion of *E. coli* therefore yields a mixture of both starting and mutant plasmids.

Cloned genes can be altered by many *in vitro* mutagenesis procedures, which can lead to the introduction of deletions, insertions, or single nucleotide alterations. The basic method of *in vitro* mutagenesis is the use of synthetic oligonucleotides to generate nucleotide changes in a DNA sequence (**Figure 4.33**). In this procedure a synthetic oligonucleotide bearing the desired mutation is used as a primer for DNA synthesis. Newly synthesized DNA molecules containing the mutation can then be isolated and characterized. For example, specific amino acids of a protein can be altered in order to characterize their role in protein function.

Variations of this approach, combined with the versatility of other methods for manipulating recombinant DNA molecules, can be used to introduce virtually any desired alteration in a cloned gene. The effects of such mutations on gene expression and function can then be determined by introduction of the gene into an appropriate cell type. *In vitro* mutagenesis has thus allowed detailed characterization of the functional roles of both the regulatory and protein-coding sequences of cloned genes.

Introducing mutations into cellular genes

Although the transfer of cloned genes into cells (particularly in combination with *in vitro* mutagenesis) provides a powerful approach to studying gene structure and function, such experiments fall short of defining the role of an unknown gene in a cell or intact organism. The cells used as recipients for transfer of cloned genes usually already have normal copies of the gene in their chromosomal DNAs, and these normal copies continue to perform their roles in the cell. Determining the biological role of a cloned gene therefore requires that the activity of the normal cellular gene copies be eliminated. Several approaches can be used to either inactivate the chromosomal copies

Figure 4.34 Gene inactivation by homologous recombination   A mutated copy of the cloned gene is introduced into cultured animal cells. The cloned gene may then replace the normal gene copy by homologous recombination, yielding a cell carrying the desired mutation in its chromosomal DNA.

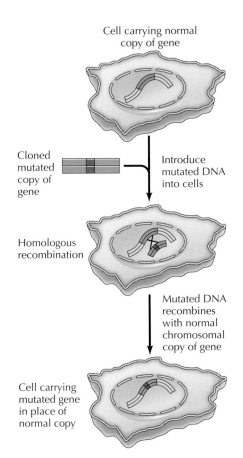

of a cloned gene or inhibit normal gene function, both in cultured cells and in transgenic mice.

Mutating chromosomal genes is based on the ability of a cloned gene introduced into a cell to undergo **homologous recombination** with its chromosomal copy (**Figure 4.34**). In homologous recombination, the cloned gene replaces the normal allele, so mutations introduced into the cloned gene *in vitro* become incorporated into the chromosomal copy of the gene. In the simplest case, mutations that inactivate (**knockout**) the cloned gene can be introduced in place of the normal gene copy in order to determine its role in cellular processes.

Recombination between transferred DNA and the homologous chromosomal gene is a rare event in mammalian cells, so inactivation of mammalian genes by this approach is technically difficult. However, various procedures have been developed to select and isolate the transformed cells in which homologous recombination has occurred, so it is feasible to inactivate any desired gene in mammalian cells. Importantly, genes can be readily inactivated in mouse (ES) cells, which can then be used to generate transgenic mice (**Figure 4.35**). These mice can be bred to yield progeny containing mutated

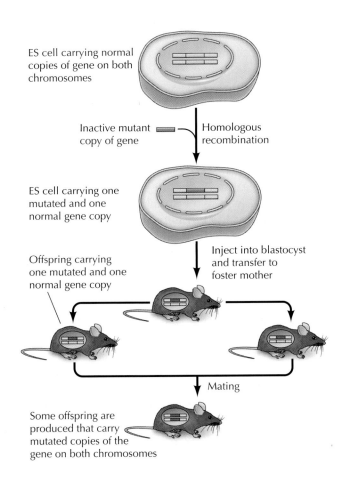

Figure 4.35 Production of mutant mice by homologous recombination in ES cells   Homologous recombination is used to replace a normal gene with an inactive mutant copy in embryonic stem (ES) cells, resulting in ES cells carrying one normal and one inactive copy of the gene on homologous chromosomes. These ES cells can be transferred to blastocysts and used to generate progeny mice as described in Figure 4.32. These mice, carrying one mutant and one normal gene copy, can then be mated to produce offspring carrying mutant copies of the gene on both chromosomes.

copies of the targeted gene on both homologous chromosomes, so the effects of inactivation of a gene can be investigated in the context of the intact animal. In addition, cells can be cultured from mouse embryos containing the mutated gene copies, so the functions of target genes can also be studied in cell culture. The biological activities of more than 7000 mouse genes have been investigated in this way, and such studies have been critically important in revealing the roles of many genes in mouse development. Methods have also been developed to conditionally knockout genes in specific mouse tissues, allowing the function of a gene to be studied in a defined cell type (for example, in nerve cells) rather than in all cells of the organism.

Genome engineering by the CRISPR/Cas system

The introduction of mutations into cellular genes by traditional methods of homologous recombination remains difficult and time-consuming. In order to facilitate the process, several approaches have been developed to increase the ease and efficiency of gene targeting by using site-specific nucleases to specifically cleave desired sequences in cellular genomes. The initial methods used recombinant proteins in which nucleases were linked to specially-designed DNA-binding domains that recognized DNA sequences of 20-30 bases, which were uniquely present within a desired target gene. Most recently, a method has been developed in which specific target sequences are recognized by an RNA molecule, rather than by a protein. Since RNAs homologous to any desired sequence can easily be synthesized, this approach (called the **CRISPR/Cas system**) is a versatile and effective method for introducing targeted gene mutations in mammalian cells.

The CRISPR (clustered regularly interspaced palindromic repeat)/Cas (CRISPR-associated systems) were discovered in bacteria, where they are used as a defense against foreign genetic elements, such as viruses or plasmids. They consist of CRISPR RNAs that recognize specific target sequences and Cas proteins that cleave the targeted DNA. The system was first applied to mammalian cells by two groups of researchers in 2013. Plasmids expressing the Cas9 nuclease and a guide RNA, containing sequences complementary to the targeted gene, are introduced into mammalian cells. The guide RNA forms a complex with Cas9, bringing Cas9 to the targeted gene (Figure 4.36). Depending on the variant of Cas9 used, it can either cut both strands of DNA, introducing a double-strand break, or only cut one strand, introducing a single-strand nick. If both strands are cleaved, the double-strand break is likely to be resealed with mutations introduced at the site of cleavage, frequently resulting in inactivation of the targeted gene. If only one strand is cleaved, the single-strand nick promotes efficient recombination if a homologous DNA sequence is also present in the cell. Thus, introducing a mutant copy of the targeted gene, together with a guide RNA and Cas9, leads to highly efficient introduction of specific mutations by homologous recombination. This approach to genetic engineering has been used successfully in a wide variety of cultured cells, including ES cells, allowing the rapid generation of mice carrying desired mutations.

Interfering with cellular gene expression

As an alternative to gene inactivation by homologous recombination, a variety of approaches can be used to specifically interfere with gene expression or function. One method that has been used to inhibit expression of a desired

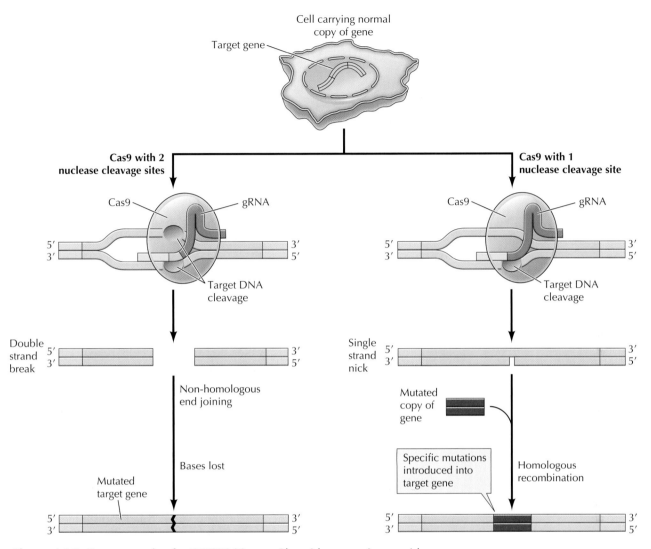

Figure 4.36 Gene targeting by CRISPR/Cas Plasmids expressing a guide RNA (gRNA) with sequences homologous to the target gene and the Cas9 nuclease are introduced into cells. The gRNA directs Cas9 to the target gene. Two variants of Cas9 are used. If Cas9 contains two nuclease cleavage sites, it introduces a double strand break in the targeted DNA. Resealing this break by non-homologous recombination introduces random mutations in the target gene. Alternatively, a variant of Cas9 with only one nuclease cleavage site plus a mutated copy of the target gene can be introduced into cells. In this case, Cas9 cleaves only one strand of the targeted DNA, leading to homologous recombination between the target gene and the mutated copy, resulting in introduction of a specific mutation in the target gene.

target gene is the introduction of **antisense nucleic acids** into cultured cells (**Figure 4.37**). RNA or single-stranded DNA complementary to the mRNA of the gene of interest (antisense) hybridizes with the mRNA and blocks its translation into protein. Moreover, the RNA-DNA hybrids resulting from the introduction of antisense DNA molecules are usually degraded within the cell. Antisense RNAs can be introduced directly into cells, or cells can be

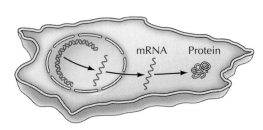

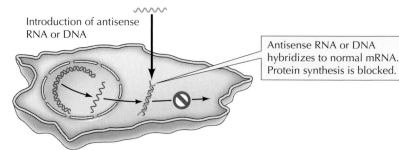

Introduction of antisense
RNA or DNA

mRNA   Protein

Antisense RNA or DNA
hybridizes to normal mRNA.
Protein synthesis is blocked.

Figure 4.37 Inhibition of gene expression by antisense RNA or DNA Antisense RNA or single-stranded DNA is complementary to the mRNA of a gene of interest. Antisense nucleic acids therefore form hybrids with their target mRNAs, blocking the translation of mRNA into protein and leading to mRNA degradation.

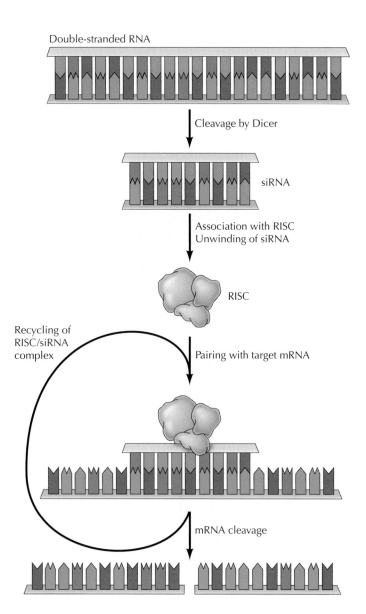

Double-stranded RNA

Cleavage by Dicer

siRNA

Association with RISC
Unwinding of siRNA

RISC

Recycling of
RISC/siRNA
complex

Pairing with target mRNA

mRNA cleavage

transfected with vectors that have been engineered to express antisense RNA. Antisense DNA is usually in the form of short oligonucleotides (about 20 bases long), which can either be transfected into cells or, in many cases, taken up by cells directly from the culture medium.

RNA interference (RNAi) has more recently emerged as an extremely effective and widely used method for interfering with gene expression at the mRNA level. RNA interference was first discovered in *C. elegans* in 1998, when Andrew Fire, Craig Mello and their colleagues found that injection of double-stranded RNA inhibited expression of a gene with a complementary mRNA sequence. This unexpected observation demonstrated an unanticipated role for double-stranded RNAs in gene regulation, which has subsequently been developed into a powerful experimental tool for inhibiting expression of target genes. When double-stranded RNAs are introduced into cells, they are cleaved into short double-stranded molecules (21–23 nucleotides) by an enzyme called Dicer (**Figure 4.38**). These short double-stranded molecules, called short interfering RNAs (siRNAs), then associate with a complex of proteins known as the RNA-induced silencing complex (RISC). Within this complex, the two strands of siRNA separate and the strand complementary to the mRNA (the antisense strand) guides the complex to the target

Figure 4.38 RNA interference Double-stranded RNA molecules are cleaved by the enzyme Dicer into short interfering RNAs (siRNAs). The siRNAs associate with the RNA-induced silencing complex (RISC) and are unwound. The antisense strand of siRNA then targets RISC to a homologous mRNA, which is cleaved by one of the RISC proteins. The RISC-siRNA complex is released and can participate in further cycles of mRNA degradation.

RNA Interference

Potent and Specific Genetic Interference by Double-Stranded RNA in *Caenorhabditis elegans*

Andrew Fire, SiQun Xu, Mary K. Montgomery, Steven A. Kostas, Samuel E. Driver, and Craig C. Mello

Carnegie Institute of Washington, Baltimore, MD, and Univ. of Mass. Medical School, Worcester, MA

Nature, Volume 391, 1998, pages 806–811

The Context

Antisense RNA was first introduced in 1984 and subsequently applied as an experimental method to inhibit gene expression in a variety of animal models, including *C. elegans*. It was thought that antisense RNA inhibited gene expression by hybridizing with a target (sense) mRNA and blocking its translation. However, the results of some control experiments were not consistent with this mechanism of action. In particular, Andrew Fire and his colleagues observed in 1991 that plasmids expressing sense as well as antisense RNAs interfered with gene expression in *C. elegans* muscle cells (*Development* 113:503). Similarly, in 1995, Su Guo and Kenneth Kemphues observed that injection of both sense and antisense RNAs interfered with the function of a *C. elegans* gene required for early embryonic development (*Cell* 81: 611). These results were surprising and unexplained, since sense RNA would not be expected to hybridize with a target mRNA. Fire, Craig Mello and their colleagues decided to investigate the reason for this unanticipated activity of sense RNA and, in this 1998 paper, discovered the phenomenon of RNA interference mediated by double-stranded RNA.

The Experiments

Plasmids engineered to produce either sense or antisense RNAs would also be expected to produce smaller amounts of the opposite strand, so Fire, Mello, and colleagues reasoned that small amounts of double-stranded RNA would be present in both sense and antisense RNA preparations. Therefore, they considered the possibility that double-stranded RNA might contribute to the inhibition of gene expression. To test this, they compared the activities of double-stranded RNA

hybrids with both sense and antisense single-stranded RNAs as inhibitors of several *C. elegans* genes. In all cases, double-stranded RNA was a much more potent inhibitor of gene function than antisense single-stranded RNA. In addition, injection of double-stranded RNA resulted in extensive degradation of the target mRNA, whereas single-stranded antisense RNA had only a minimal effect (see figure). By following up on surprising results from a control experiment, these investigators thus discovered that double-stranded RNA was a potent and specific tool for blocking gene expression.

The Impact

The serendipitous discovery of RNA interference not only led to the development of a new and highly effective experimental method for blocking gene function, but also to a fundamental change in our understanding of the mechanisms that regulate gene expression in eukaryotic cells. Double-stranded RNAs are indeed a potent experimental tool and are now used to interfere with gene expression in a wide range of experimental systems, including mammalian cells in culture. Even more striking, however, has been the growing recognition of the regulatory roles that double-stranded RNAs normally play in eukaryotic cells. In their 1998 paper, Fire and Mello predicted that "the mechanisms underlying RNA interference probably exist for a biological purpose. Genetic interference by dsRNA (double-stranded) could be used by the organism for physiological gene silencing." Further research has abundantly confirmed this prediction and double-stranded RNAs are now recognized as major regulators of gene expression in all eukaryotes. Approximately 1000 non-coding small double-stranded regulatory RNAs (called microRNAs) have been identified in mammals, and gene regulation by microRNAs has been shown to play an important role in a wide variety

of cellular processes, including cell signaling, cell survival, heart and brain development, and cancer. Each microRNA has multiple mRNA targets, and it is estimated that nearly half of human genes are regulated by microRNAs. The discovery of RNA interference, based on unexpected results of a control experiment, has thus revolutionized our understanding of gene regulation.

Andrew Fire   Craig Mello

Fire photo credit: Linda A. Cicero/Stanford News Service. Mello photo courtesy of University of Massachusetts Medical School.

(A) Uninjected control

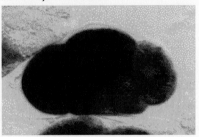

(B) Antisense RNA injected

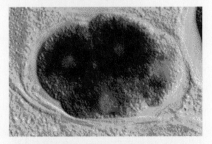

(C) Double-stranded RNA injected

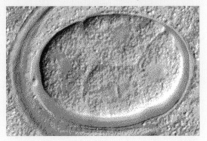

C. elegans embryos were injected with either antisense or double-stranded RNA homologous to *mex-3* mRNA. The level of *mex-3* mRNA in the embryos was then determined by *in situ* hybridization (indicated as a dark stain). Injection of double-stranded, but not antisense, RNA resulted in complete loss of the *mex-3* target mRNA.

Video 4.1

sites.sinauer.com/cooper7e/v4.1

RNA Interference RNA interference (RNAi) is an important process that regulates the activity of genes via small interfering RNAs (siRNAs) and microRNAs (miRNAs).

mRNA by complementary base pairing. The mRNA is then cleaved by one of the RISC proteins. The RISC-siRNA complex is released following degradation of the mRNA and can continue to participate in multiple rounds of mRNA cleavage, leading to effective destruction of the targeted mRNA.

RNAi has been established as a potent method for interfering with gene expression in *C. elegans, Drosophila, Arabidopsis,* and mammalian cells, and provides a relatively straightforward approach to investigating the function of any gene whose sequence is known. In addition, libraries of double-stranded RNAs or siRNAs that cover a large fraction of genes in the genome are being used to screen *C. elegans, Drosophila,* and human cells to identify novel genes involved in specific biological functions, such as cell growth or cell survival. Finally, RNAi is not only an experimental tool: as will be discussed in Chapters 8 and 9, RNA interference is also a major regulatory mechanism used by cells to control gene expression at both the transcriptional and translational levels.

In addition to inactivating a gene or inducing degradation of an mRNA, it is sometimes possible to interfere directly with the function of proteins within cells (**Figure 4.39**). One approach is to microinject antibodies that block the activity of the protein against which they are directed. Alternatively, some mutant proteins interfere with the function of their normal counterparts when they are expressed within the same cell—for example, by competing with the normal protein for binding to its target molecule. Cloned DNAs encoding such mutant proteins (called **dominant inhibitory** or **dominant negative mutants**) can be introduced into cells by gene transfer and used to study the effects of blocking normal gene function.

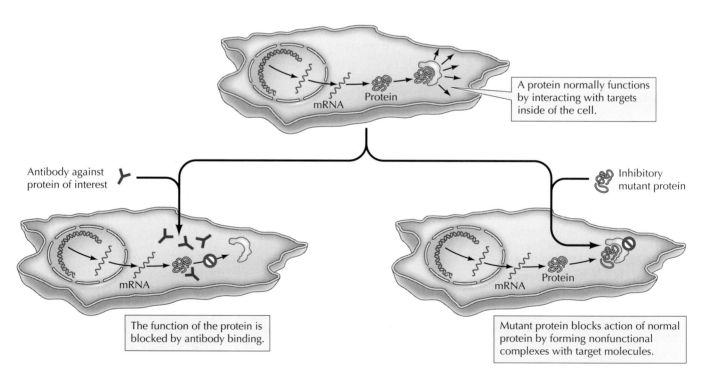

Figure 4.39 Direct inhibition of protein function Microinjected antibodies can bind to proteins within cells, thereby inhibiting their normal function. In addition, some mutant proteins interfere with the function of normal proteins—for example, by competing with the normal protein for interaction with target molecules.

SUMMARY	KEY TERMS

Heredity, Genes, and DNA

- *Genes and chromosomes:* Chromosomes are the carriers of genes.

  gene, allele, dominant, recessive, genotype, phenotype, chromosome, diploid, meiosis, haploid, mutation

- *Genes and enzymes:* Genes specify the amino acid sequence of proteins.

- *Identification of DNA as the genetic material:* DNA was identified as the genetic material by bacterial transformation experiments. See Animations 4.1 and 4.2.

  transformation

- *The structure of DNA:* DNA is a double helix in which hydrogen bonds form between purines and pyrimidines on opposite strands. Because of specific base pairing—A with T and G with C—the two strands of a DNA molecule are complementary in sequence.

- *Replication of DNA:* DNA replicates by semiconservative replication in which the two strands separate and each serves as a template for synthesis of a new progeny strand.

  semiconservative replication, DNA polymerase

Expression of Genetic Information

- *Colinearity of genes and proteins:* The order of nucleotides in DNA specifies the order of amino acids in proteins.

- *The role of messenger RNA:* Messenger RNA functions as an intermediate to convey information from DNA to the ribosomes, where it serves as a template for protein synthesis.

  central dogma, transcription, translation, messenger RNA (mRNA), RNA polymerase, ribosomal RNA (rRNA), transfer RNA (tRNA)

- *The genetic code:* Transfer RNAs serve as adaptors between amino acids and mRNA during translation. Each amino acid is specified by a codon consisting of three nucleotides. See Animations 4.3 and 4.4.

  genetic code, *in vitro* translation, codon

- *RNA viruses and reverse transcription:* DNA can be synthesized from RNA templates, as first discovered in retroviruses. See Animation 4.5.

  retrovirus, reverse transcription, reverse transcriptase

Recombinant DNA

- *Restriction endonucleases:* Restriction endonucleases cleave specific DNA sequences, yielding defined fragments of DNA molecules. See Animation 4.6.

  restriction endonuclease, gel electrophoresis, restriction map

- *Generation of recombinant DNA molecules:* Recombinant DNA molecules consist of a DNA fragment of interest ligated to a vector that is able to replicate independently in an appropriate host cell. See Animation 4.7.

  molecular cloning, vector, recombinant molecule, molecular clone, plasmid, DNA ligase, cDNA

- *Vectors for recombinant DNA:* A variety of vectors are used to clone different sizes of DNA fragments.

  origin of replication, cosmid, P1 artificial chromosome (PAC), bacterial artificial chromosome (BAC), yeast artificial chromosome (YAC)

- *DNA sequencing:* The nucleotide sequences of cloned DNA fragments can be readily determined. See Animation 4.8.

  dideoxynucleotide, next-generation sequencing

- *Expression of cloned genes:* The proteins encoded by cloned genes can be expressed at high levels in either bacteria or eukaryotic cells.

  expression vector

SUMMARY	KEY TERMS

Detection of Nucleic Acids and Proteins

- *Amplification of DNA by the polymerase chain reaction:* PCR allows the amplification and isolation of specific fragments of DNA *in vitro*, providing a sensitive method for detecting small amounts of specific DNA or RNA molecules. See Animation 4.9.

polymerase chain reaction (PCR), real-time PCR

- *Nucleic acid hybridization:* Nucleic acid hybridization allows the detection of specific DNA or RNA sequences by base pairing between complementary strands. See Animations 4.10–4.12.

nucleic acid hybridization, Southern blotting, Northern blotting, recombinant DNA library, *in situ* hybridization

- *Antibodies as probes for proteins:* Antibodies are used to detect specific proteins in cells or cell extracts. See Animation 4.13.

antibody, antigen, monoclonal antibody, immunoblotting, Western blotting, SDS-polyacrylamide gel electrophoresis (SDS-PAGE)

Gene Function in Eukaryotes

- *Gene transfer in plants and animals:* Cloned genes can be introduced into complex eukaryotic cells and multicellular organisms for functional analysis.

gene transfer, transfection, liposome, electroporation, transient expression, transgenic mice, embryonic stem (ES) cell

- *Mutagenesis of cloned DNAs:* In vitro mutagenesis of cloned DNAs is used to study the effect of engineered mutations on gene function.

in vitro mutagenesis

- *Introducing mutations into cellular genes:* Mutations can be introduced into chromosomal gene copies by homologous recombination with cloned DNA sequences.

homologous recombination, knockout

- *Genetic engineering by the CRISPR/Cas system:* The CRISPR/Cas system can be used to efficiently target desired genes for mutation or homologous recombination.

CRISPR/Cas system

- *Interfering with cellular gene expression:* The expression or function of specific genes can be blocked by antisense nucleic acids, RNA interference, or dominant inhibitory mutants. See Video 4.1.

antisense nucleic acid, RNA interference (RNAi), dominant inhibitory mutant, dominant negative mutant

---

Refer To

The Cell

Companion Website

sites.sinauer.com/cooper7e

for quizzes, animations, videos, flashcards, and other study resources.

Questions

1. How would you determine whether two different genes are linked in *Drosophila*?

2. You grow *E. coli* for several generations in media containing ^{15}N. The cells are then transferred to media containing ^{14}N and grown for two additional generations. What proportion of the DNA isolated from these cells will be heavy, light, or intermediate in density?

3. Addition or deletion of one or two nucleotides in the coding sequence of a gene produces a nonfunctional protein, whereas addition or deletion of three nucleotides often produces a protein with nearly normal function. Explain.

4. Describe the features that a yeast artificial chromosome must have in order to be used to clone a piece of human DNA cut with *Eco*RI.

5. You are studying an enzyme in which an active-site cysteine residue is encoded by the triplet UGU. How would mutating the third base to a C affect enzyme function? How about mutating it to an A?

6. Digestion of a 4-kb DNA molecule with *Eco*RI yields two fragments of 1 kb and 3 kb each. Digestion of the same molecule with *Hind*III yields fragments of 1.5 kb and 2.5 kb. Finally, digestion with *Eco*RI and *Hind*III in combination yields fragments of 0.5 kb, 1 kb, and 2.5 kb. Draw a restriction map indicating the positions of the *Eco*RI and *Hind*III cleavage sites.

7. Starting with DNA from a single sperm, how many copies of a specific gene sequence will be obtained after 10 cycles of PCR amplification? After 30 cycles?

8. You have cloned a fragment of human genomic DNA in a cosmid vector; approximately how many times would you expect the insert to be cut by the restriction enzyme *Bam*HI?

9. What is the minimum number of bacterial artificial chromosome (BAC) clones required to make a genomic library of human DNA?

10. How would you expect actinomycin D to affect the replication of influenza virus?

11. What is the critical feature of a cloning vector that would allow you to isolate stably transfected mammalian cells?

12. Nucleic acids have a net negative charge and can be separated by gel electrophoresis on the basis of their size. In contrast, different proteins have different charges. How then can proteins be separated on the basis of size by electrophoresis?

13. How does the CRISPR/Cas system increase the efficiency of homologous recombination in targeted genes?

References and Further Reading   (Key review articles for each major section are highlighted in **bold**.)

Heredity, Genes, and DNA

Avery, O. T., C. M. MacLeod and M. McCarty. 1944. Studies on the chemical nature of the substance inducing transformation of pneumococcal types. *J. Exp. Med.* 79: 137–158. [P]

Franklin, R. E. and R. G. Gosling. 1953. Molecular configuration in sodium thymonucleate. *Nature* 171: 740–741. [P]

Kornberg, A. 1960. Biologic synthesis of deoxyribonucleic acid. *Science* 131: 1503–1508. [P]

Kresge, N., R. D. Simoni and R. L. Hill. 2005. Launching the age of biochemical genetics with *Neurospora*: The work of George Wells Beadle. *J. Biol. Chem.* 280: e9–e11. [R]

Lehman, I. R. 2003. Discovery of DNA polymerase. *J. Biol. Chem.* 278: 34733–34738. [R]

Meselson, M. and F. W. Stahl. 1958. The replication of DNA in *Escherichia coli*. *Proc. Natl. Acad. Sci. USA* 44: 671–682. [P]

Watson, J. D. and F. H. C. Crick. 1953. Genetical implications of the structure of deoxyribonucleic acid. *Nature* 171: 964–967. [P]

Watson, J. D. and F. H. C. Crick. 1953. Molecular structure of nucleic acids: A structure for deoxyribose nucleic acid. *Nature* 171: 737–738. [P]

Wilkins, M. H. F., A. R. Stokes and H. R. Wilson. 1953. Molecular structure of deoxypentose nucleic acids. *Nature* 171: 738–740. [P]

Expression of Genetic Information

Baltimore, D. 1970. RNA-dependent DNA polymerase in virions of RNA tumour viruses. *Nature* 226: 1209–1211. [P]

Brenner, S., F. Jacob and M. Meselson. 1961. An unstable intermediate carrying information from genes to ribosomes for protein synthesis. *Nature* 190: 576–581. [P]

Crick, F. H. C., L. Barnett, S. Brenner and R. J. Watts-Tobin. 1961. General nature of the genetic code for proteins. *Nature* 192: 1227–1232. [P]

Ingram, V. M. 1957. Gene mutations in human hemoglobin: The chemical difference between normal and sickle cell hemoglobin. *Nature* 180: 326–328. [P]

Nirenberg, M. 2004. Historical review: Deciphering the genetic code—a personal account. *Trends Biochem. Sci.* 29: 46–54. [R]

Nirenberg, M. and P. Leder. 1964. RNA code-words and protein synthesis. *Science* 145: 1399–1407. [P]

Nirenberg, M. W. and J. H. Matthaei. 1961. The dependence of cell-free protein synthesis in *E. coli* upon naturally occurring or synthetic polyribonucleotides. *Proc. Natl. Acad. Sci. USA* 47: 1588–1602. [P]

Temin, H. M. and S. Mizutani. 1970. RNA-dependent DNA polymerase in virions of Rous sarcoma virus. *Nature* 226: 1211–1213. [P]

Yanofsky, C. 2007. Establishing the triplet nature of the genetic code. *Cell* 128: 815–818. [R]

Recombinant DNA

Burke, D. T., G. F. Carle and M. V. Olson. 1987. Cloning of large segments of exogenous DNA into yeast by means of artificial chromosome vectors. *Science* 236: 806–812. [P]

Cohen, S. N., A. C. Y. Chang, H. W. Boyer and R. B. Helling. 1973. Construction of biologically functional bacterial plasmids *in vitro*. *Proc. Natl. Acad. Sci. USA* 70: 3240–3244. [P]

Green, M. R. and J. Sambrook. 2012. *Molecular Cloning: A Laboratory Manual*. 4th ed. Plainview, NY: Cold Spring Harbor Laboratory Press.

Nathans, D. and H. O. Smith. 1975. Restriction endonucleases in the analysis and restructuring of DNA molecules. *Ann. Rev. Biochem.* 44: 273–293. [R]

Sanger, F., S. Nicklen and A. R. Coulson. 1977. DNA sequencing with chain-terminating inhibitors. *Proc. Natl. Acad. Sci. USA* 74: 5463–5467. [P]

Detection of Nucleic Acids and Proteins

Ausubel, F. M. ed. 1995. *Current Protocols in Molecular Biology*. New York: J. Wiley and Sons.

Caruthers, M. H. 1985. Gene synthesis machines: DNA chemistry and its uses. *Science* 230: 281–285. [R]

Grunstein, M. and D. S. Hogness. 1975. Colony hybridization: A method for the isolation of cloned DNAs that contain a specific gene. *Proc. Natl. Acad. Sci. USA* 72: 3961–3965. [P]

Harlow, E. and D. Lane. 1999. *Using Antibodies: A Laboratory Manual*. Cold Spring Harbor, NY: Cold Spring Harbor Laboratory Press.

Kohler, G. and C. Milstein. 1975. Continuous cultures of fused cells secreting antibody of predefined specificity. *Nature* 256: 495–497. [P]

Saiki, R. K., D. H. Gelfand, S. Stoffel, S. J. Scharf, R. Higuchi, G. T. Horn, K. B. Mullis and H. A. Erlich. 1988. Primer-directed enzymatic amplification of DNA with a thermostable DNA polymerase. *Science* 239: 487–491. [P]

Southern, E. M. 1975. Detection of specific sequences among DNA fragments separated by gel electrophoresis. *J. Mol. Biol.* 98: 503–517. [P]

Gene Function in Eukaryotes

Branda, C. S. and S. M. Dymecki. 2004. Talking about a revolution: The impact of site-specific recombinases on genetic analyses in mice. *Devel. Cell* 6: 7–28. [R]

Bronson, S. K. and O. Smithies. 1994. Altering mice by homologous recombination using embryonic stem cells. *J. Biol. Chem.* 269: 27155–27158. [R]

Capecchi, M. R. 2005. Gene targeting in mice: functional analysis of the mammalian genome for the twenty-first century. *Nature Rev. Genet.* 6: 507–512. [R]

Cong, L., F. A. Ran, D. Cox, S. Lin, R.Barretto, N. Habib, P. D. Hsu, X. Wu, W. Jiang, L. A. Marraffini and F. Zhang. 2013. Multiplex genome engineering using CRISPR/Cas systems. *Science* 339: 819–823. [P]

Doudna, J. A. and E. Charpentier. 2014. The new frontier of genome engineering with CRISPR-Cas9. *Science* 346: 1258096. [R]

Fire, A., S. Xu, M. K. Montgomery, S. A. Kostas, S. E. Driver and C. C. Mello. 1998. Potent and specific genetic interference by double-stranded RNA in *Caenorhabditis elegans*. *Nature* 391: 806–811. [P]

Herskowitz, I. 1987. Functional inactivation of genes by dominant negative mutations. *Nature* 329: 219–222. [R]

Hsu, P. D., E. S. Lander and F. Zhang. 2014. Development and applications of CRISPR-Cas9 for genome engineering. *Cell* 157: 1262–1278. [R]

Izant, J. G. and H. Weintraub. 1984. Inhibition of thymidine kinase gene expression by antisense RNA: A molecular approach to genetic analysis. *Cell* 36: 1007–1015. [P]

Kim, H. and J.-S. Kim. 2014. A guide to genome engineering with programmable nucleases. *Nature Rev. Genet.* 15: 321–334. [R]

Kuhn, R., F. Schwenk, M. Aguet and K. Rajewsky. 1995. Inducible gene targeting in mice. *Science* 269: 1427–1429. [P]

Mali, P., L. Yang, K. M. Esvelt, J. Aach, M. Guell, J. E. DiCarlo, J. E. Norville and G. M. Church. 2013. RNA-guided human genome engineering via Cas9. *Science* 339: 823–826. [P]

Mello, C. C. and D. Conte Jr. 2004. Revealing the world of RNA interference. *Nature* 431: 338–342. [R]

Nilsen, T. W. 2007. Mechanisms of microRNA-mediated gene regulation in animal cells. *Trends Genet.* 23:243–249. [R]

Novina, C. D. and P. A. Sharp. 2004. The RNAi revolution. *Nature* 430: 161–164. [R]

Palmiter, R. D. and R. L. Brinster. 1986. Germ-line transformation of mice. *Ann. Rev. Genet.* 20: 465–499. [R]

Sampson, T. R. and D. S. Weiss. 2013. Exploiting CRISPR/Cas systems for biotechnology. *Bioessays* 36: 34–38. [R]

Smith, M. 1985. *In vitro* mutagenesis. *Ann. Rev. Genet.* 19: 423–462. [R]

Genomics, Proteomics, and Systems Biology

Recent years have seen major changes in the way scientists approach cell and molecular biology, with large-scale experimental and computational approaches being applied to understand the complexities of biological systems. Traditionally, cell and molecular biologists studied one or a few genes or proteins at a time. This was changed by genome sequencing projects, which introduced large-scale experimental approaches that generated vast amounts of data to the study of biological systems. The complete genome sequences of a wide variety of organisms, including many individual humans, provide a wealth of information that forms a new framework for studies of cell and molecular biology and opens new possibilities in medical practice. Not only can the sequences of complete genomes be obtained and analyzed, but it is also now possible to undertake large-scale analyses of all of the RNAs and proteins expressed in a cell. These global experimental approaches form the basis of the new field of systems biology, which seeks a quantitative understanding of the integrated behavior of complex biological systems. This chapter considers the development of these new technologies and their impact on understanding the molecular biology of cells.

Genomes and Transcriptomes

Obtaining the complete sequence of the human genome was the first large-scale experimental project undertaken in the life sciences. When it was initiated in the 1980s, it appeared a daunting task to completely sequence all three billion base pairs of the human genome. After all, at that the time the largest DNA molecule to have been sequenced was a viral genome of less than 200,000 bases. The Human Genome Project became the largest collaborative undertaking in biology and yielded an initial draft sequence in 2001, with a more refined complete sequence of the human genome published in 2004. Along the way, the complete genome sequences of many other species were obtained, providing important insights into genome evolution. Moreover, tremendous advances in the technology of DNA sequencing have been made, and new sequencing methodologies allow rapid and economical sequencing of individual genomes or transcribed RNAs. These advances have changed the way scientists think about the structure and function of our genomes, as well as allowing new approaches to disease diagnosis and treatment based on personal genome sequencing.

The genomes of bacteria and yeast

The first complete sequence of a cellular genome, reported in 1995 by a team of researchers led by Craig Venter, was that of the bacterium *Haemophilus influenzae*, a common inhabitant of the human respiratory tract. The genome of *H. influenzae* is a circular molecule containing approximately 1.8×10^6 base pairs, more than 1000 times smaller than the human genome. Once the complete DNA sequence was obtained, it was analyzed to identify the genes encoding rRNAs, tRNAs, and proteins. Potential protein-coding regions were identified by computer analysis of the DNA sequence to detect **open-reading frames**—long stretches of nucleotide sequence that can encode polypeptides because they contain none of the three chain-terminating codons (UAA, UAG, and UGA). Since these chain-terminating codons occur randomly once in every 21 codons (three chain-terminating codons out of 64 total), open-reading frames that extend for more than 100 codons usually represent functional genes. This analysis identified 1743 potential protein-coding regions in the *H. influenzae* genome as well as six copies of rRNA genes and 54 different tRNA genes (**Figure 5.1**). The predicted coding sequences have an average size of approximately 900 base pairs, so they cover about 1.6 Mb of DNA, corresponding to nearly 90% of the genome of *H. influenzae*. The genome of *E. coli* is approximately twice the size of *H. influenzae*, 4.6×10^6 base pairs long and containing about 4200 genes, again with nearly 90% of the DNA used as protein-coding sequence. The use of almost all the DNA to encode proteins is typical of bacterial genomes, more than 2000 of which have now been sequenced.

A model for a simple eukaryotic genome, which was sequenced in 1996, is found in the yeast *Saccharomyces cerevisiae*. As discussed in Chapter 1, yeast are simple unicellular organisms, but they have all of the characteristics of eukaryotic cells. The genome of *S. cerevisiae* consists of 12×10^6 base pairs, which is about 2.5 times the size of the genome of *E. coli*. It contains about 6000

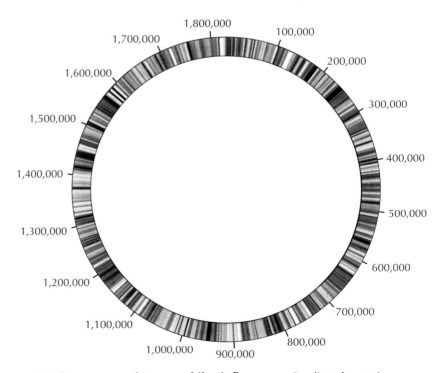

Figure 5.1 The genome of *Haemophilus influenzae*   Predicted protein-coding regions are designated by colored bars. Numbers indicate base pairs of DNA. (From R. D. Fleischmann et al., 1995. *Science* 269: 496.)

protein-coding genes. Thus, despite the greater complexity of a eukaryotic cell, yeast contain only about 50% more genes than *E. coli*. Protein-coding sequences account for approximately 70% of total yeast DNA, so yeast, like bacteria, have a high density of protein-coding sequence.

The genomes of Caenorhabditis elegans, Drosophila melanogaster, *and* Arabidopsis thaliana

The next major advance in genomics was the sequencing of the genomes of the relatively simple multicellular organisms, *C. elegans, Drosophila,* and *Arabidopsis*. Distinctive features of each of these organisms make them important models for genome analysis: *C. elegans* and *Drosophila* are widely used for studies of animal development, and *Drosophila* has been especially well analyzed genetically. Likewise, *Arabidopsis* is a model for studies of plant molecular biology and development. The genomes of all three of these organisms are approximately 100×10^6 base pairs, about ten times larger than the yeast genome but 30 times less than the genome of humans. The determinations of their complete sequences in 1998 (*C. elegans*) and 2000 (*Drosophila* and *Arabidopsis*) were major steps forward, which extended genome sequencing from unicellular bacteria and yeasts to multicellular organisms.

An unanticipated result of sequencing these genomes was that they contained fewer protein-coding genes than expected relative to bacterial or yeast genomes (Table 5.1). The *C. elegans* genome is 97×10^6 base pairs and contains about 19,000 predicted protein-coding sequences—approximately eight times the amount of DNA but only three times the number of genes in yeast. When the *Drosophila* genome was sequenced in 2000, it was surprising to find that it contains only about 14,000 protein-coding genes—substantially fewer than the number of genes in *C. elegans*, even though *Drosophila* is a more complex organism. Moreover, it is striking that a complex animal like *Drosophila* has only a little more than twice the number of genes found in yeast. Protein-coding sequences correspond to only about 10% of the *Drosophila* genome and 25% of the *C. elegans* genome, compared with 70% of the

> **Video 5.1**
>
> sites.sinauer.com/cooper7e/v5.1
> **A Brief Introduction to *C. elegans***
> *C. elegans* is an important model organism used in the study of molecular and developmental biology.

Table 5.1   Representative Genomes

Organism	Genome size (Mb)[a]	Number of protein-coding genes	Protein-coding sequence
Bacteria			
H. influenza	1.8	1700	90%
E. coli	4.6	4200	88%
Yeasts			
S. cerevisiae	12	6000	70%
Invertebrates			
C. elegans	97	19,000	25%
Drosophila	180	14,000	10%
Plants			
Arabidopsis thaliana	125	26,000	25%
Mammals			
Human	3000	21,000	1.2%

[a]Mb = millions of base pairs

yeast genome, so the larger sizes of the *Drosophila* and *C. elegans* genomes are substantially due to increased amounts of non-protein-coding sequences rather than protein-coding genes.

The completion of the genome sequence of *Arabidopsis thaliana* in 2000 extended genome sequencing from animals to plants. The *Arabidopsis* genome, approximately 125×10^6 base pairs of DNA, contains approximately 26,000 protein-coding genes—significantly more genes than were found in either *C. elegans* or *Drosophila*. Even more genes have since been found in other plant genomes. For example, the genome of the apple contains about 57,000 protein-coding genes, further emphasizing the lack of relationship between gene number and complexity of an organism.

The human genome

For many scientists, the ultimate goal of the Human Genome Project was determination of the complete nucleotide sequence of the human genome: approximately 3×10^9 base pairs of DNA. Because of its large size, determination of the human genome sequence was a phenomenal undertaking, and its publication in draft form in 2001 was heralded as a scientific achievement of historic magnitude.

The draft sequences of the human genome published in 2001 were produced by two independent teams of researchers, each using different approaches. Both of these sequences were initially incomplete drafts in which approximately 90% of the genome had been sequenced and assembled. Continuing efforts then closed the gaps and improved the accuracy of the draft sequences, leading to publication of a high-quality human genome sequence in 2004 by the International Human Genome Sequencing Consortium.

A major surprise from the genome sequence was the unexpectedly low number of human genes (see Table 5.1). The human genome consists of only about 21,000 protein-coding genes, which is not much larger than the number of genes in simpler animals like *C. elegans* and *Drosophila* and fewer than in *Arabidopsis* or other plants. Whereas protein-coding sequences correspond to the majority of the genomes of bacteria and yeast and 10–25% of the genomes of *Drosophila* and *C. elegans*, they represent only about 1% of the human genome. The nature of the additional sequences in the human genome and their roles in gene regulation, which may contribute more to biological complexity than simply the number of genes, are discussed in the next chapter.

Over 40% of the predicted human proteins are related to proteins in simpler sequenced eukaryotes, including yeast, *Drosophila* and *C. elegans*. Many of these conserved proteins function in basic cellular processes, such as metabolism, DNA replication and repair, transcription, translation, and protein trafficking. Most of the proteins that are unique to humans are made up of protein domains that are also found in other organisms, but these domains are arranged in novel combinations to yield distinct proteins in humans. Compared with *Drosophila* and *C. elegans*, the human genome contains expanded numbers of genes involved in functions related to the greater complexity of vertebrates, such as the immune response, the nervous system, and blood clotting, as well as increased numbers of genes involved in development, cell signaling, and the regulation of gene expression.

The genomes of other vertebrates

In addition to the human genome, a large and growing number of vertebrate genomes have been sequenced, including the genomes of fish, frogs,

Key Experiment

The Human Genome

Initial Sequencing and Analysis of the Human Genome

International Human Genome Sequencing Consortium

Nature, Volume 409, 2001, pages 860–921

The Sequence of the Human Genome

J. Craig Venter and 273 others

Science, Volume 291, 2001, pages 1304–1351

The Context

The idea of sequencing the entire human genome was first conceived in the mid-1980s. It was initially met with broad skepticism among biologists, most of whom felt it was simply not a feasible undertaking. At the time, the largest genome that had been completely sequenced was that of Epstein-Barr virus, which totaled approximately 180,000 base pairs of DNA. From this perspective, sequencing the human genome, which is almost 20,000 times larger, seemed inconceivable to many. However, the idea of such a massive project in biology captivated the imagination of others, including Charles DeLisi, who was then head of the Office of Health and Environmental Research at the U.S. Department of Energy. In 1986 DeLisi succeeded in launching the Human Genome Initiative as a project within the Department of Energy.

The project gained broader support in 1988 when it was endorsed by a committee of the U.S. National Research Council. This committee recommended a broader effort, including sequencing the genomes of several model organisms and the parallel development of detailed genetic and physical maps of the human chromosomes. This effort was centered at the U.S. National Institutes of Health, initially under the direction of James Watson (codiscoverer of the structure of DNA), and then under the leadership of Frances Collins.

The first complete genome to be sequenced was that of the bacterium *Haemophilus influenzae*, reported by Craig Venter and colleagues in 1995. Venter had

been part of the genome sequencing effort at the National Institutes of Health but had left to head a nonprofit company, The Institute for Genomic Research, in 1991. In the meantime, considerable progress had been made in mapping the human genome, and the initial sequence of *H. influenzae* was followed by the sequences of other bacteria, yeast, and *C. elegans* in 1998.

In 1998 Venter formed a new company, Celera Genomics, and announced plans to use advanced sequencing technologies to obtain the entire human genome sequence in three years. Collins and other leaders of the publicly funded genome project responded by accelerating their efforts, resulting in a race that eventually led to the publication of two draft sequences of the human genome in February 2001.

The Experiments

The two groups of scientists used different approaches to obtain the human genome sequence. The publicly funded team, The International Human Genome Sequencing Consortium, headed by Eric Lander,

Craig Venter

Eric Lander

sequenced DNA fragments derived from bacterial artificial chromosome (BAC) clones that had been previously mapped to human chromosomes, similar to the approach used to determine the sequence of the yeast and *C. elegans* genomes (see figure). In contrast, the Celera Genomics team used a whole-genome shotgun sequencing approach that Venter and colleagues had first used to sequence the genome of *H. influenzae*. In this approach, DNA fragments were sequenced at random, and overlaps between fragments were then used to reassemble a complete genome sequence. Both sequences covered only the euchromatin

(Continued on next page)

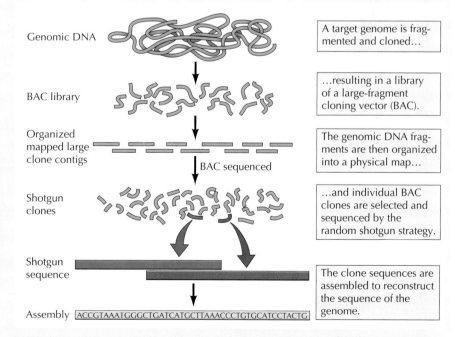

Genomic DNA

A target genome is fragmented and cloned...

BAC library

...resulting in a library of a large-fragment cloning vector (BAC).

Organized mapped large clone contigs

The genomic DNA fragments are then organized into a physical map...

BAC sequenced

Shotgun clones

...and individual BAC clones are selected and sequenced by the random shotgun strategy.

Shotgun sequence

The clone sequences are assembled to reconstruct the sequence of the genome.

Assembly   ACCGTAAATGGGCTGATCATGCTTAAACCCTGTGCATCCTACTG

Strategy for genome sequencing using bacterial artificial chromosome (BAC) clones that had been organized into overlapping clusters (contigs) and mapped to human chromosomes.

portion of the human genome—approximately 2900 million base pairs (Mb) of DNA—with the heterochromatin repeat-rich portion of the genome (approximately 300 Mb) remaining unsequenced.

Both of these initially published versions were draft, rather than completed, sequences. Subsequent efforts completed the sequence, leading to publication of a highly accurate sequence of the human genome in 2004.

The Impact

Several important conclusions immediately emerged from the human genome

sequences. Most strikingly, the number of human genes was surprisingly small and appeared to be between 20,000 and 25,000 in the completed sequence. This unexpected result has led to the recognition of the roles of non-protein coding sequences in our genome, particularly with respect to the multiple mechanisms by which they regulate gene expression.

Beyond the immediate conclusions drawn in 2004, the sequence of the human genome, together with the genome sequences of other organisms, has provided a new basis for biology and medicine. The impact of the genome sequence has been

and continues to be felt in discovering new genes and their functions, understanding gene regulation, elucidating the basis of human diseases, and developing new strategies for prevention and treatment based on the genetic makeup of individuals. Knowledge of the human genome may ultimately contribute to meeting what Venter and colleagues refer to as "The real challenge of human biology... to explain how our minds have come to organize thoughts sufficiently well to investigate our own existence."

chickens, dogs, rodents, and primates (**Figure 5.2**). The genomes of these other vertebrates are similar in size to the human genome and contain a similar number of genes. Their sequences provide interesting comparisons to that of the human genome and are proving useful in facilitating studies

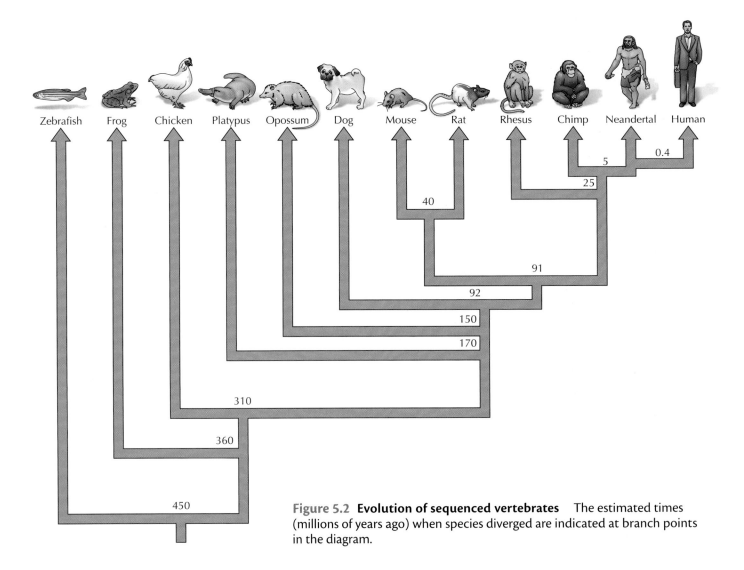

Figure 5.2 Evolution of sequenced vertebrates The estimated times (millions of years ago) when species diverged are indicated at branch points in the diagram.

of different model organisms and in identifying a variety of different types of functional sequences, including regulatory elements that control gene expression. For example, a comparison of the human, mouse, chicken, and zebrafish genomes indicates that about half of the protein-coding genes are common to all vertebrates, whereas approximately 3000 genes are unique to each of these four species (**Figure 5.3**).

The mammalian genomes that have been sequenced, in addition to the human genome, include the genomes of the platypus, opossum, mouse, rat, dog, rhesus macaque, and chimpanzee. As discussed in earlier chapters, the mouse is the key model system for experimental studies of mammalian genetics and development, while the rat is an important model for human physiology and medicine. Mice, rats, and humans have 90% of their genes in common, so the availability of the mouse and rat genome sequences provide essential databases for research in these areas. The many distinct breeds of pet dogs make the sequence of the dog genome particularly important in understanding the genetic basis of morphology, behavior, and a variety of complex diseases that afflict both dogs and humans. There are approximately 300 breeds of dogs, which differ in their physical and behavioral characteristics as well as in their susceptibility to a variety of diseases, including several types of cancer, blindness, deafness, and metabolic disorders. These characteristics are highly specific properties of different breeds, greatly facilitating identification of the responsible genes. For example, as discussed in Chapter 6, studies of breed-specific differences in dog legs led to the identification of a specific type of gene rearrangement responsible for this characteristic. Recent analyses of dog genomes have also identified genes responsible for coat color and for the body size of small breeds. Similar types of analysis are under way to understand the genetic basis of multiple diseases, including several types of cancer, that are common in some breeds of dogs. Since many of these diseases afflict both dogs and humans, the results of these studies can be expected to impact human health as well as veterinary medicine. In the future, we can also expect genetic analysis of behavior in dogs. Since many canine behaviors, such as separation anxiety, are also common in humans, psychologists may have much to learn from the species that has been our closest companion for thousands of years.

The sequences of the genomes of other primates, including the chimpanzee, bonobo, orangutan, and rhesus macaque, may help pinpoint the unique features of our genome that distinguish humans from other primates. Interestingly, however, comparison of these sequences does not suggest an easy answer to the question of what makes us human. The genome sequences of humans and chimpanzees are about 99% identical. Perhaps surprisingly, the sequence differences between humans and chimpanzees frequently alter the coding sequences of genes, leading to changes in the amino acid sequences of most proteins. Although many of these amino acid changes may not affect protein function, it appears that there are changes in the structure as well as in the expression of thousands of genes between chimpanzees and humans, so identifying those differences that are key to the origin of humans is not a simple task.

The genome of Neandertals, our closest evolutionary relatives, has also been recently sequenced. It is estimated that Neandertals and modern humans diverged about 300,000 to 400,000 years ago. The genomes of Neandertals and modern humans are more than 99.9% identical, significantly more closely related to each other than either is to chimpanzees. Interestingly, these differences alter the coding sequence of only about 90 genes that are conserved in

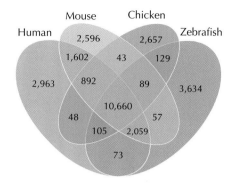

Figure 5.3 Comparison of vertebrate genomes The number of genes shared between human, mouse, chicken, and zebrafish genomes is indicated. (From K. Howe et al., 2013. *Nature* 496: 498.)

modern humans. These include genes that are involved in the skin, skeletal development, metabolism, and cognition. Further studies of these genes may elucidate their potential roles in the evolution of modern humans.

Next-generation sequencing and personal genomes

The human genome and most of the genomes of model organisms discussed above were sequenced using the dideoxynucleotide technique discussed in Chapter 4, first described by Fred Sanger in 1977 (see Figure 4.20). Automation of this basic method increased its speed and capacity to allow whole genome sequencing, culminating in the successful completion of a high-quality human genome sequence in 2004. However, even with robust automation, genome sequencing by this approach was slow and expensive, so that the sequencing of a complete genome was a major undertaking. For example, the initial sequencing of the human genome took 15 years at a cost of approximately $3 billion. Starting around 2005, a number of new sequencing methods, collectively called **next-generation sequencing**, were developed that have substantially increased the speed and lowered the cost of genome sequencing (Figure 5.4). Since 2001, the cost of sequencing a human genome has decreased more than 10,000 times—from approximately $100 million to several thousand. The speed of sequencing has increased even more, so it is now possible to sequence a complete human genome in a few days. These dramatic changes in sequencing technology have opened the door to sequencing the complete genomes of large numbers of different individuals, allowing new approaches to understanding the genetic basis for many of the diseases that afflict mankind, including cancer, heart disease, and degenerative diseases of the nervous system such as Parkinson's and Alzheimer's disease. In addition, understanding our unique genetic makeup as individuals is expected to lead to the development of new tailor-made strategies for disease prevention and treatment.

Next-generation DNA sequencing (also called massively parallel sequencing) refers to several different methods in which millions of templates are

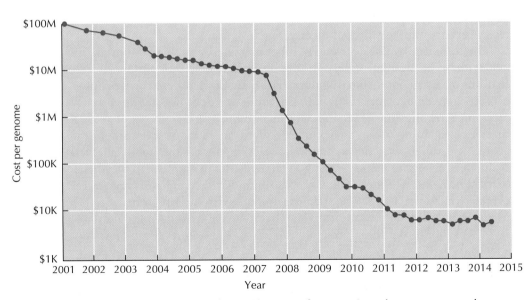

Figure 5.4 Progress in DNA sequencing The cost of sequencing a human genome has dropped from approximately $100 million in 2001 to several thousand dollars in 2015. (Data from the National Human Genome Research Institute.)

sequenced simultaneously in a single reaction. The basic general strategy of these methods is illustrated in Figure 5.5. First, the DNA is fragmented and adapter sequences, which serve as primers for amplification and sequencing reactions, are added to the ends of each fragment. Single DNA molecules are then attached to a solid surface and amplified by PCR to produce clusters of spatially separated templates. Millions of templates can then be sequenced in parallel by using lasers to monitor the incorporation of fluorescent nucleotides. The sequences derived from this large collection of overlapping fragments can then be assembled to yield a continuous genome sequence. Alternatively, if a known genome sequence (for example, the human genome) is already available, it can be used as a reference genome on which fragment sequences from a particular individual can be aligned.

The first individual human genomes to be sequenced included those of Craig Venter and James Watson, reported in 2007 and 2008. Since then, the genome sequences of thousands of individuals have been determined. With the cost of sequencing an individual genome now in the range of several thousand dollars, it can be anticipated that personal genome sequencing will become part of medical practice. This will allow therapies to be specifically tailored to the needs of individual patients, both with respect to disease prevention and treatment. Perhaps the best current example, discussed in Chapter 19, is the development of new drugs for cancer treatment, which are specifically targeted against mutations that can be identified by sequencing the cancer genomes of individual patients. In the future, we may expect genome sequencing of healthy people to play an important role in disease prevention by identifying genes that confer susceptibility to disease, followed by taking appropriate measures to intervene. For example, genome sequencing could identify women carrying mutations in genes that confer a high risk

Figure 5.5 Next-generation sequencing Cellular DNA is fragmented and adapters are ligated to the ends of each fragment. Single molecules are then anchored to a solid surface and amplified by PCR, forming millions of clusters of molecules. Four color-labeled reversible chain terminating nucleotides are added together with DNA polymerase and a primer that recognizes the adapter sequence. Incorporation of a labeled nucleotide into each cluster of DNA molecules is detected by a laser. Unincorporated nucleotides are removed, chain termination is reversed, and the cycle is repeated to obtain the sequences of millions of clusters simultaneously.

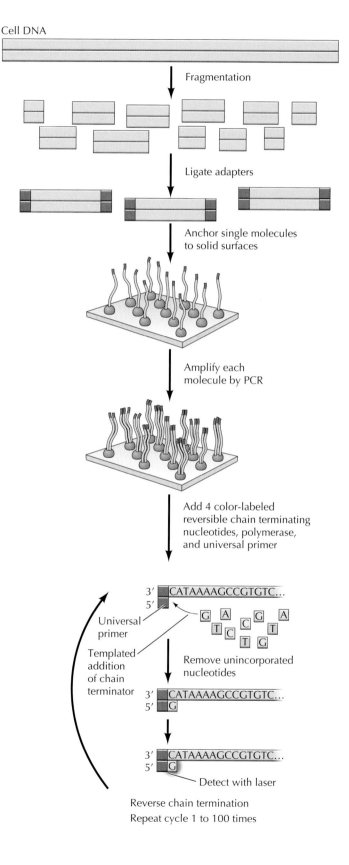

for development of breast cancer, which might be prevented by mastectomy. There is also little doubt that continuing progress in genomics will not only lead to increasing applications in medicine, but will also help us to elucidate the contribution of our genes to other unique characteristics, such as athletic ability or intelligence, and to better understand the interactions between genes and environment that lead to complex human behaviors.

Global analysis of gene expression

The availability of complete genome sequences has enabled researchers to study gene expression on a genome-wide global level. It is thus now possible to analyze all of the RNAs that are transcribed in a cell (the **transcriptome**), rather than analyzing the expression of one gene at a time. One commonly used method for global expression analysis is hybridization to **DNA microarrays**, which allows expression of tens of thousands of genes to be analyzed simultaneously. A DNA microarray consists of a glass or silicon chip onto which oligonucleotides are printed by a robotic system in small spots at a high density (**Figure 5.6**). Each spot on the array consists of a single oligonucleotide. Tens of thousands of unique probes can be printed onto a typical chip, so it is readily possible to produce DNA microarrays containing sequences representing all of the genes in cellular genomes. As illustrated in Figure 5.6, one widespread application of DNA microarrays is in studies of gene expression; for example, a comparison of the genes expressed by two different types of cells. In an experiment of this type, cDNA probes are

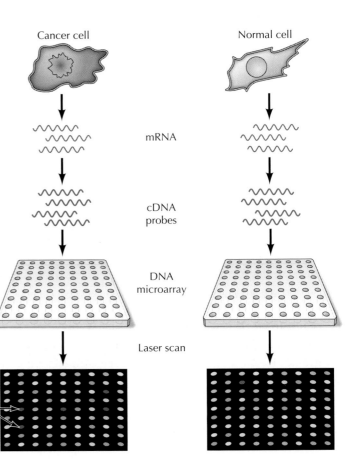

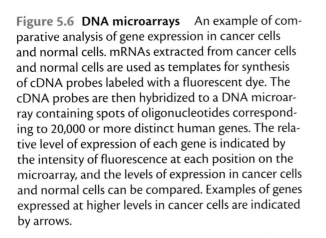

Figure 5.6 DNA microarrays An example of comparative analysis of gene expression in cancer cells and normal cells. mRNAs extracted from cancer cells and normal cells are used as templates for synthesis of cDNA probes labeled with a fluorescent dye. The cDNA probes are then hybridized to a DNA microarray containing spots of oligonucleotides corresponding to 20,000 or more distinct human genes. The relative level of expression of each gene is indicated by the intensity of fluorescence at each position on the microarray, and the levels of expression in cancer cells and normal cells can be compared. Examples of genes expressed at higher levels in cancer cells are indicated by arrows.

synthesized from the mRNAs expressed in each of the two cell types (e.g., cancer cells and normal cells) by reverse transcription. The cDNAs are labeled with fluorescent dyes and hybridized to DNA microarrays in which 20,000 or more human genes are represented by oligonucleotide spots. The arrays are then analyzed using a high-resolution laser scanner, and the relative extent of transcription of each gene is indicated by the intensity of fluorescence at the appropriate spot on the array. It is then possible to analyze the relative levels of gene expression between the cancer cells and normal cells by comparing the intensity of hybridization of their cDNAs to oligonucleotides representing each of the genes in the cell.

The continuing development of next-generation sequencing has also made it feasible to use DNA sequencing to determine and quantify all of the RNAs expressed in a cell. In this approach, called **RNA-seq**, cellular mRNAs are isolated, converted to cDNAs by reverse transcription, and subjected to next-generation sequencing (**Figure 5.7**). In contrast to microarray analysis, RNA-seq reveals the complete extent of transcribed sequences in a cell, rather than just detecting those that hybridize to a probe on a microarray. The frequency with which individual sequences are detected in RNA-seq is proportional to the quantity of RNA in the cell, so this analysis determines the abundance as well as the identity of all transcribed sequences. The sensitivity of RNA-seq is high enough to allow analysis at the single cell level, so the transcriptomes of individual cells can be determined.

One of the surprises revealed by RNA-seq analysis has been that many more RNAs are transcribed than are accounted for by the protein-coding genes of human cells. As discussed in the next chapter, these studies have led to the identification of new classes of RNAs that play critical roles in gene regulation.

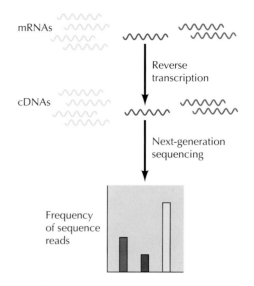

Figure 5.7 RNA-seq Cellular mRNAs are reverse transcribed to cDNAs, which are subjected to next-generation sequencing. The results yield the sequences of all mRNAs in a cell. The relative amount of each mRNA is indicated by the frequency at which its sequence is represented in the total number of sequences read.

Proteomics

The analysis of cell genomes and transcriptomes are only the first steps in understanding the workings of a cell. Since proteins are directly responsible for carrying out almost all cell activities, it is necessary to understand not only what proteins can be encoded by a cell's genome but also what proteins are actually expressed and how they function within the cell. A complete understanding of cell function therefore requires not only the analysis of the sequence and transcription of its genome, but also a systematic analysis of its protein complement. This large-scale analysis of cell proteins (**proteomics**) has the goal of identifying and quantifying all of the proteins expressed in a given cell (the **proteome**), as well as establishing the localization of these proteins to different subcellular organelles and elucidating the networks of interactions between proteins that govern cell activities.

Identification of cell proteins

The number of distinct species of proteins in eukaryotic cells is typically far greater than the number of genes. This arises because many genes can be expressed to yield several distinct mRNAs, which encode different polypeptides as a result of alternative splicing (discussed in Chapter 6). In addition, proteins can be modified in a variety of different ways, including the addition of phosphate groups, carbohydrates, and lipid molecules. The human genome, for example, contains approximately 20,000 different

FYI

Mass spectrometry has many uses, including testing the blood samples of athletes for the presence of certain performance-enhancing drugs.

protein-coding genes, and the number of these genes expressed in any given cell is thought to be around 10,000. However, because of alternative splicing and protein modifications, it is estimated that these genes can give rise to more than 100,000 different proteins. In addition, these proteins can be expressed at a wide range of levels. Characterization of the complete protein complement of a cell, the goal of proteomics, thus represents a considerable challenge. Although major progress has been made in the last several years, substantial technological hurdles remain to be overcome before a complete characterization of cell proteomes can be achieved.

The first technology developed for the large-scale separation of cell proteins was **two-dimensional gel electrophoresis** (**Figure 5.8**). A mixture of cell proteins is first subjected to electrophoresis in a tube with a pH gradient running from end to end. Each protein migrates according to charge until it reaches a pH at which the charge of the protein is neutralized, as determined by the protein's content of acidic and basic amino acids. After denaturation and SDS-binding, the proteins are then subjected to electrophoresis in a second dimension under conditions where they separate according to size, with lower molecular weight proteins moving more rapidly through a gel. This approach is capable of resolving several thousand protein species from a cell extract. However, this is much less than the total number of proteins expressed in mammalian cells, and it is important to note that two-dimensional gel electrophoresis is biased towards the detection of the most abundant cell proteins.

The major tool currently used in proteomics is **mass spectrometry**, which was developed in the 1990s as a powerful method of protein identification

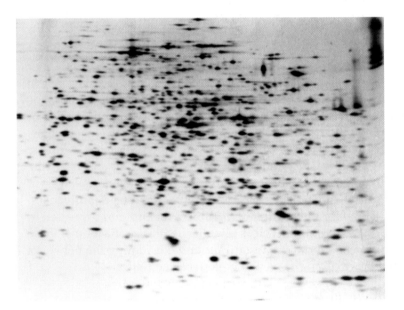

Figure 5.8 Two-dimensional gel electrophoresis   A mixture of cell proteins is first subjected to electrophoresis in a pH gradient so that they separate according to charge (horizontal axis). The proteins then undergo another electrophoresis in a second dimension under conditions where they separate by size (vertical axis). The gel is stained to reveal spots corresponding to distinct protein species. The example shown is a gel of proteins from *E. coli*. (From P. H. O'Farrell. 1975. *J. Biol. Chem.* 250: 4007; courtesy of Patrick H. O'Farrell.)

Figure 5.9 Identification of proteins by mass spectrometry
A protein is digested with a protease that cleaves it into small peptides. The peptides are then ionized and analyzed in a mass spectrometer, which determines the mass-to-charge ratio of each peptide. The results are displayed as a mass spectrum, which is compared to a database of theoretical mass spectra of all known proteins for protein identification.

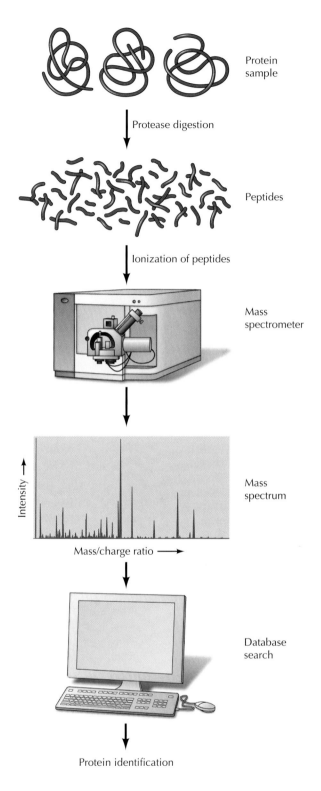

(**Figure 5.9**). Proteins to be analyzed, which could be excised from two-dimensional gels, are digested with a protease to cleave the proteins into small fragments (peptides) in the range of approximately 20 amino acid residues long. A commonly used protease is trypsin, which cleaves proteins at the carboxy-terminal side of lysine and arginine residues. The peptides are then ionized by irradiation with a laser or by passage through a field of high electrical potential and introduced into a mass spectrometer, which measures the mass-to-charge ratio of each peptide. This generates a mass spectrum in which individual tryptic peptides are indicated by a peak corresponding to their mass-to-charge ratio. Computer algorithms can then be used to compare the experimentally determined mass spectrum with a database of theoretical mass spectra representing tryptic peptides of all known proteins, allowing identification of the unknown protein.

More detailed sequence information than just the mass of the peptides can be obtained by **tandem mass spectrometry** (**Figure 5.10**). In this technique, individual peptides from the initial mass spectrum are automatically selected to enter a "collision cell" in which they are partially degraded by random breakage of peptide bonds. A second mass spectrum of the partial degradation products of each peptide is then determined. Because each amino acid has a unique molecular weight, the amino acid sequence of the peptide can be deduced from these data. Protein modifications, such as phosphorylation, can also be identified because they alter the mass of the modified amino acid.

Although powerful, the two-dimensional gel/mass spectrometry approach is limited. As noted above, two-dimensional gels favor detection of the most abundant cell proteins, and membrane proteins are characteristically difficult to resolve by this approach. Because of these limitations, it appears that two-dimensional gels are capable of resolving proteins corresponding to only several hundred genes, representing a small fraction of all cell proteins. An alternative approach is the use of mass spectrometry to analyze mixtures of proteins, thereby eliminating the initial separation of proteins by two-dimensional gel electrophoresis. In this approach, called "shotgun mass spectrometry," a mixture of cell proteins is digested with a protease (e.g., trypsin), and the complex mixture of peptides is subjected to sequencing by tandem mass spectrometry. The sequences of individual peptides

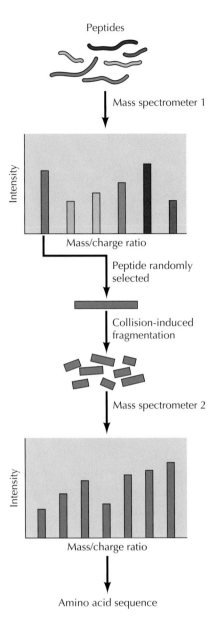

Peptides

Mass spectrometer 1

Intensity

Mass/charge ratio

Peptide randomly selected

Collision-induced fragmentation

Mass spectrometer 2

Intensity

Mass/charge ratio

Amino acid sequence

Figure 5.10 Tandem mass spectrometry A mixture of peptides are separated in a mass spectrometer 1. A randomly selected peptide is then fragmented by collision-induced breakage of peptide bonds. The fragments, which differ by single amino acids, are then separated in a second mass spectrometer 2. Since the fragments differ by single amino acids, the amino acid sequence of the peptide can be deduced.

are then used for database searching to identify the proteins present in the starting mixture. Additional methods have been developed to compare the amounts of proteins in two different samples, allowing a quantitative analysis of protein levels in different types of cells or in cells that have been subjected to different treatments. Although several problems with the sensitivity and accuracy of these methods remain to be solved, the analysis of complex mixtures of proteins by "shotgun" mass spectrometry provides a powerful approach to the systematic analysis of cell proteins.

Global analysis of protein localization

Understanding the function of eukaryotic cells requires not only the identification of the proteins expressed in a given cell type, but also characterization of the locations of those proteins within the cell. As reviewed in Chapter 1, eukaryotic cells contain a nucleus and a variety of subcellular organelles. Systematic analysis of the proteins present in these organelles is an important goal of proteomic approaches to cell biology.

The protein composition of a variety of organelles has been determined by combining classical cell biology methods with mass spectrometry. Organelles of interest are isolated from cells by subcellular fractionation techniques, as discussed in Chapter 1 (see Figures 1.39 and 1.40). The proteins present in isolated organelles can then be determined by mass spectrometry. The proteome of a variety of organelles and large subcellular structures, such as nucleoli, have been characterized by this approach—some examples are illustrated in Table 5.2. For example, more than 700 different proteins have been identified by mass spectrometry of isolated mitochondria and approximately 200 different proteins identified in lysosomes. By performing such studies with organelles isolated from cells of different tissues or grown under different conditions, it is also possible to determine the changes in protein composition that are associated with different cell types or physiological states.

Table 5.2 Protein composition of cellular structures

Structure	Number of proteins identified[a]
Plasma membrane	860–2000
Endoplasmic reticulum	140
Mitochondria	400–750
Lysosome	200
Centrosome	114
Peroxisome	70
Golgi apparatus	400

[a]Ranges for numbers of identified proteins represent the results obtained in different cell types.

Source: Data from J. R. Yates III et al., 2005. *Nature Rev. Mol. Cell Biol.* 6: 702.

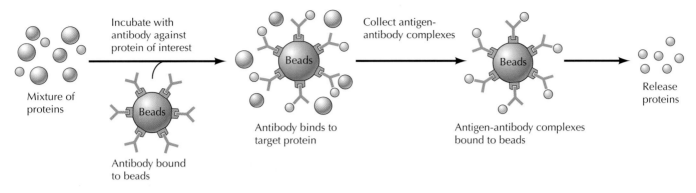

Figure 5.11 Immunoprecipitation A mixture of cell proteins is incubated with an antibody bound to beads. The antibody forms complexes with the protein (green) against which it is directed (the antigen). These antigen-antibody complexes are collected on the beads and the target protein is isolated.

Protein interactions

Proteins almost never act alone within the cell. Instead, they generally function by interacting with other proteins in protein complexes and networks. Elucidating the interactions between proteins therefore provides important clues as to the function of novel proteins, as well as helps to understand the complex networks of protein interactions that govern cell behavior. Along with global studies of subcellular localization, the systematic analysis of protein complexes and interactions has therefore become an important goal of proteomics.

One approach to the analysis of protein complexes is to isolate a protein from cells under gentle conditions, so that it remains associated with the proteins it normally interacts with inside the cell. Typically, an antibody against a protein of interest would be used to isolate that protein from a cell extract by **immunoprecipitation** (**Figure 5.11**). A cell extract is incubated with an antibody, which binds to its antigenic target protein. The resulting antigen-antibody complexes can then be isolated. If the cell extract has been prepared under gentle conditions so that the target protein remains associated with the proteins it normally interacts with inside the cell, then these interacting proteins will also be present in the immunoprecipitates. Such immunoprecipitated protein complexes can then be analyzed, for example by mass spectrometry, to identify not only the protein against which the antibody was directed, but also other proteins with which it was associated in the cell extract (**Figure 5.12**). This approach to analysis of protein interactions has been used to characterize

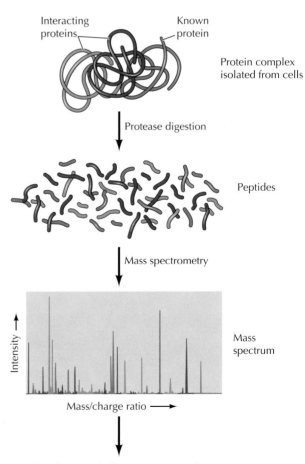

Figure 5.12 Analysis of protein complexes A known protein (blue) is isolated from cells as a complex with other interacting proteins (yellow and red). The entire complex can be analyzed by mass spectrometry to identify the interacting proteins.

a variety of protein-protein interactions in different types of cells and under different physiological conditions, leading to the identification of numerous interactions between proteins involved in processes such as cell signaling or gene expression.

Alternative approaches to systematic analyses of protein complexes include screens for protein interactions *in vitro* as well as genetic screens that detect interactions between pairs of proteins that are introduced into yeast cells. Expression of cloned genes in yeast is particularly useful because simple methods of yeast genetics can be employed to identify proteins that interact with one another. In this type of analysis, called the **yeast two-hybrid system,** two different cDNAs (for example, from human cells) are joined to two distinct domains of a protein that stimulates expression of a target gene in yeast (**Figure 5.13**). Yeast are then transformed with the hybrid cDNA

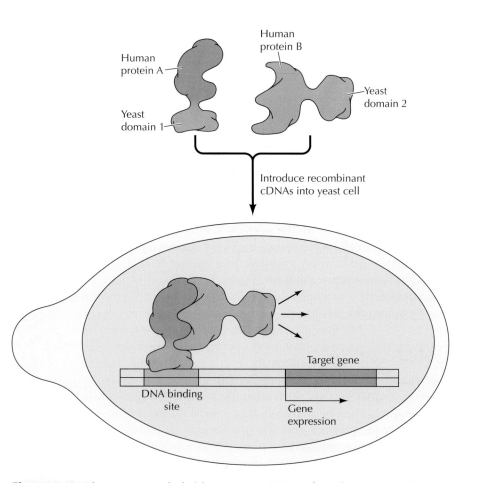

Figure 5.13 The yeast two-hybrid system cDNAs of two human proteins are cloned as fusions with two domains (designated 1 and 2) of a yeast protein that stimulates transcription of a target gene. The two recombinant cDNAs are introduced into a yeast cell. If the two human proteins interact with each other, they bring the two domains of the yeast protein together. Domain 1 binds DNA sequences at a site upstream of the target gene, and domain 2 stimulates target gene transcription. The interaction between the two human proteins can thus be detected by expression of the target gene in transformed yeast.

Figure 5.14 A protein interaction map of ***Drosophila*** Interactions among 2346 proteins are depicted, with each protein represented as a circle placed according to its subcellular localization. (From L. Giot et al., 2003. *Science* 302: 1727.)

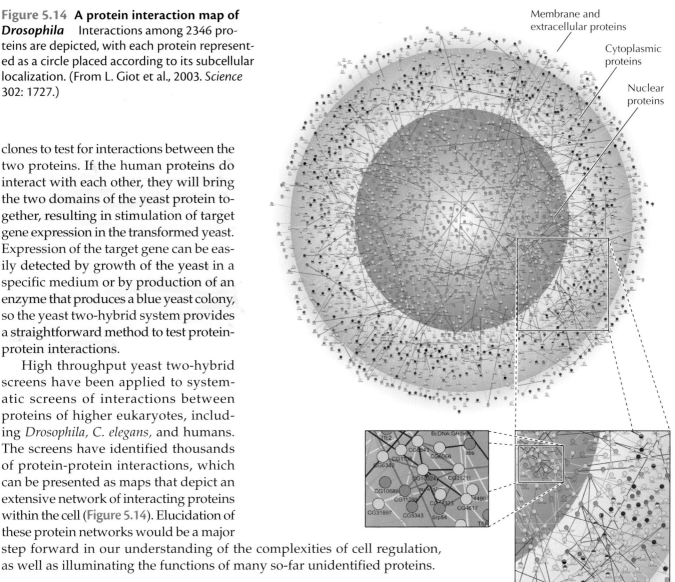

Membrane and
extracellular proteins

Cytoplasmic
proteins

Nuclear
proteins

clones to test for interactions between the two proteins. If the human proteins do interact with each other, they will bring the two domains of the yeast protein together, resulting in stimulation of target gene expression in the transformed yeast. Expression of the target gene can be easily detected by growth of the yeast in a specific medium or by production of an enzyme that produces a blue yeast colony, so the yeast two-hybrid system provides a straightforward method to test protein-protein interactions.

High throughput yeast two-hybrid screens have been applied to systematic screens of interactions between proteins of higher eukaryotes, including *Drosophila, C. elegans,* and humans. The screens have identified thousands of protein-protein interactions, which can be presented as maps that depict an extensive network of interacting proteins within the cell (**Figure 5.14**). Elucidation of these protein networks would be a major step forward in our understanding of the complexities of cell regulation, as well as illuminating the functions of many so-far unidentified proteins.

Systems Biology

The genome sequencing projects have led to a fundamental change in the way in which many problems in biology are being approached, with large-scale experimental approaches that generate vast amounts of data now in common use. Handling the enormous amounts of data generated by whole genome sequencing required sophisticated computational analysis and spawned the new field of **bioinformatics**. This field lies at the interface between biology and computer science and is focused on developing the computational methods needed to analyze and extract useful biological information from the sequence of billions of bases of DNA. Other types of large-scale biological experimentation, including global analysis of gene expression and proteomics, similarly yield vast amounts of data, far beyond the scope of traditional biological experimentation. These large-scale experimental approaches form the basis of the new field of **systems biology**, which seeks a quantitative understanding of the integrated dynamic behavior of complex

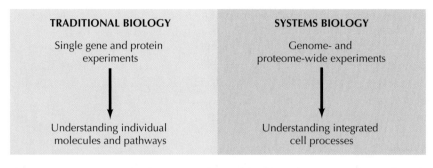

TRADITIONAL BIOLOGY	SYSTEMS BIOLOGY
Single gene and protein experiments	Genome- and proteome-wide experiments
↓	↓
Understanding individual molecules and pathways	Understanding integrated cell processes

Figure 5.15 Systems biology   Traditional biological experiments study individual molecules and pathways. Systems biology uses global experimental data for quantitative modeling of integrated systems and processes.

biological systems and processes. In contrast to traditional approaches, systems biology is characterized by the use of large scale datasets for quantitative experimental analysis and modeling (**Figure 5.15**). Some of the research areas that are amenable to large-scale experimentation, bioinformatics, and systems biology are discussed below.

Systematic screens of gene function

The identification of all of the genes in an organism opens the possibility for a large-scale systematic analysis of gene function. One approach is to systematically inactivate (or knockout) each gene in the genome by homologous recombination with an inactive mutant allele (see Figure 4.35). Complete collections of strains with mutations in all known genes are available for several model organisms, including *E. coli*, yeast, *Drosophila*, *C. elegans,* and *Arabidopsis thaliana*. These collections of mutant strains can be analyzed to determine which genes are involved in any biological property of interest. A large-scale international project to systematically knockout all genes in the mouse is also under way, and targeted mutagenesis has now indicated functions of more than 7000 mouse genes.

Alternatively, large-scale screens based on RNA interference (RNAi) are being used to systematically dissect gene function in a variety of organisms, including *Drosophila, C. elegans,* and mammalian cells in culture. In RNAi screens, double-stranded RNAs are used to induce degradation of the homologous mRNAs in cells (see Figure 4.38). With the availability of complete genome sequences, libraries of double-stranded RNAs can be designed and used in genome-wide screens to identify all of the genes involved in any biological process that can be assayed in a high-throughput manner. For example, genome-wide RNAi analysis can be used to identify genes required for the growth and viability of cells in culture (**Figure 5.16**). Individual double-stranded RNAs from the genome-wide library are tested in microwells in a high-throughput format to identify those that interfere with the growth of cultured cells, thereby characterizing the entire set of genes that are required for cell growth or survival

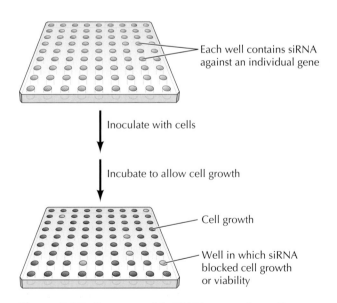

Each well contains siRNA against an individual gene

Inoculate with cells

Incubate to allow cell growth

Cell growth

Well in which siRNA blocked cell growth or viability

Figure 5.16 Genome-wide RNAi screen for cell growth and viability   Each microwell contains siRNA corresponding to an individual gene. Tissue culture cells are added to each well and incubated to allow cell growth. Those wells in which cells fail to grow identify genes required for cell growth or viability.

under particular sets of conditions. Similar RNAi screens have been used to identify genes involved in a variety of biological processes, including cell signaling pathways, protein degradation, and transmission at synapses in the nervous system. More recently, genome-wide screens using the CRISPR/Cas system (see Figure 4.36) have similarly been applied to systematically identify sets of genes in human cells that are responsible for properties such as survival or resistance to anticancer drugs.

Regulation of gene expression

Genome sequences can, in principle, reveal not only the protein-coding sequences of genes, but also the regulatory elements that control gene expression. As discussed in subsequent chapters, regulation of gene expression is critical to many aspects of cell function, including the development of complex multicellular organisms. Understanding the mechanisms that control gene expression, including transcription and alternative splicing, is therefore a central undertaking in contemporary cell and molecular biology, and the availability of genome sequences contributes substantially to this task. Unfortunately, it is far more difficult to identify gene regulatory sequences than it is to identify protein-coding sequences. Most regulatory elements are short sequences of DNA, typically spanning only about 10 base pairs. Consequently, sequences resembling regulatory elements occur frequently by chance in genomic DNA, so physiologically significant elements cannot be identified from DNA sequence alone. The identification of functional regulatory elements and elucidation of the signaling networks that control gene expression therefore represent major challenges in bioinformatics and systems biology.

The availability of genomic sequences has enabled scientists to undertake global studies of gene expression, using microarrays or RNA-seq (see Figures 5.6 and 5.7), in which the expression levels of all genes in a cell can be assayed simultaneously. Such studies of global patterns of transcription have become extremely valuable in revealing the overall changes in gene regulation associated with discrete cell behaviors, such as cell differentiation or the response of cells to a particular hormone or growth factor. Since genes that are coordinately regulated within a cell may be controlled by similar mechanisms, analyzing changes in the expression of multiple genes can help pinpoint shared regulatory elements. Moreover, changes in gene expression that occur over time can reveal networks of gene expression.

A variety of computational approaches are also being used to characterize functional regulatory elements. One approach is comparative analysis of the genome sequences of related organisms. This is based on the assumption that functionally important sequences are conserved in evolution, whereas nonfunctional segments of DNA diverge more rapidly. Approximately 5% of the genome sequence is conserved among mammals: since ~1.2% of the genome corresponds to protein-coding sequence, the remaining 4% may represent functionally important regulatory sequences. For example, computational analysis to identify noncoding sequences that are conserved between the human, mouse, rat, and dog genomes has proven useful in delineating sequences that control gene transcription (**Figure 5.17**). In addition, functional regulatory elements often occur in clusters,

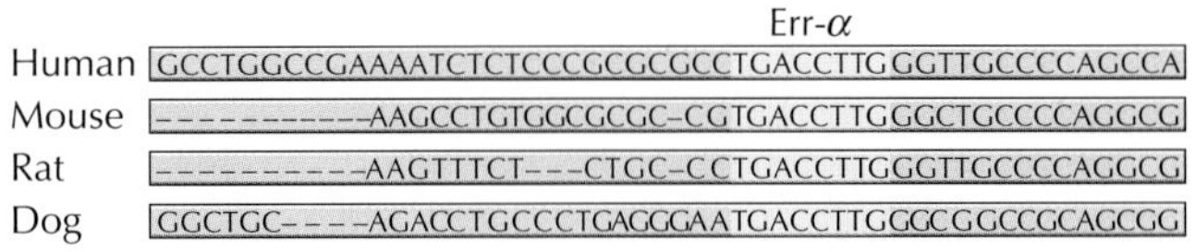

Figure 5.17 Conservation of functional gene regulatory elements Human, mouse, rat, and dog sequences near the transcription start site of a gene contain a functional regulatory element that binds the transcriptional regulatory protein Err-α. These sequences (highlighted in yellow) are conserved in all four genomes, whereas the surrounding sequences are not. (After X. Xie et al., 2005. *Nature* 434: 338.)

reflecting the fact that genes are generally regulated by the interactions of multiple transcription factors (see Chapter 8). Computer algorithms designed to detect clusters of transcription factor binding sites in genomic DNA have also proven useful in identifying sequences that regulate gene expression.

As discussed further in Chapter 8, large-scale experimental approaches for genome-wide analysis of the binding sites of regulatory proteins have also been developed. In addition, specific modifications of histones that are associated with transcriptionally-active genes or gene regulatory regions have been characterized, and genome-wide analysis of the sites of these histone modifications can be used to provide a global experimental identification of gene regulatory sequences. The application of these global approaches provides a new window to understanding the control of gene expression in complex eukaryotic cells. A prominent example is a large scale project called ENCODE (*Enc*yclopedia *of D*NA *E*lements), which utilized RNA-seq to characterize all transcribed RNAs as well as global methods for determining gene regulatory sequences in 147 different types of human cells. The results of this global analysis have changed our views of the structure and function of mammalian genomes, as will be discussed in subsequent chapters. Most notably, even though protein coding-sequences account for only about 1% of the human genome, more than 50% of our genome is transcribed, with many of these transcribed noncoding sequences playing important roles in gene regulation.

Networks

Classical experimental biology is focused on single genes or proteins, which often act sequentially to catalyze a series of reactions that constitute a pathway. Several metabolic pathways consisting of such series of reactions were discussed in Chapter 3. For example, glycolysis (the breakdown of glucose to pyruvate) is mediated by the sequential action of nine different enzymes (see Figure 3.2). Signaling pathways act similarly to transmit information from the environment, such as the presence of a hormone, to targets within the cell. For example, the hormone epinephrine signals the breakdown of glycogen to glucose in muscle cells (see Figure 2.30). This response to epinephrine is mediated by a signaling pathway consisting of six different proteins (**Figure 5.18**).

The activities of individual pathways have been elucidated by classical methods of analysis of small numbers of genes or proteins. However, signaling within the cell is far more complicated than the activities of individual pathways. Metabolic or signaling pathways do not operate in isolation; rather, there is extensive crosstalk between different pathways, so that multiple pathways interact with one another to form **networks** within the cell. Computational modeling of such networks is currently a major challenge in systems biology, which will be necessary to understand the dynamic response of cells to their environment.

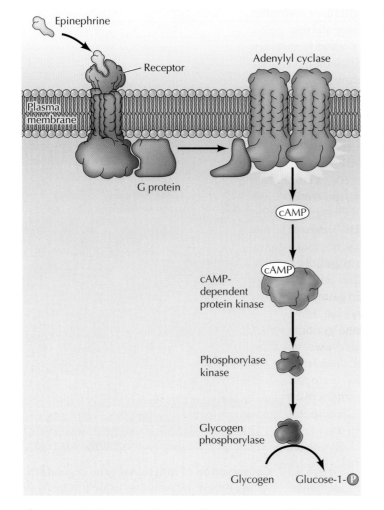

Figure 5.18 Example of a signaling pathway The binding of epinephrine (adrenaline) to its cell surface receptor triggers a signaling pathway that leads to the breakdown of glycogen to glucose-1-phosphate.

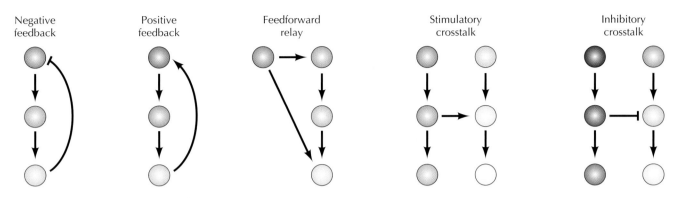

Figure 5.19 **Elements of signaling networks**   In feedback loops, a downstream element of a pathway either inhibits (negative feedback) or stimulates (positive feedback) an upstream element. In feedforward relays, an upstream element of a pathway stimulates both its immediate target and another element further downstream. Crosstalk occurs when an element of one pathway either stimulates or inhibits an element of a second pathway.

Some of the ways in which pathways can connect within a network are illustrated in **Figure 5.19**. The activity of many pathways is controlled by **feedback loops**. An example of a negative feedback loop is provided by feedback inhibition of metabolic pathways (see Figure 2.28). In addition to negative feedback loops, signaling networks can contain positive feedback loops and **feedforward relays** in which the activity of one component of a pathway stimulates a distant downstream component. Individual pathways communicate to form networks by **crosstalk**, which refers to the interaction of one pathway with another. The crosstalk between different pathways can be either positive (where one pathway stimulates the other) or negative (where one pathway inhibits the other).

A full understanding of cell signaling will require the development of network models that predict the dynamic behavior of the interconnected pathways that ultimately result in a biological response. In this view of the cell as an integrated system, mathematical models and computer simulations will clearly be needed to deal with the complexity of the problem. For example, many signaling pathways (discussed in Chapter 16) ultimately affect gene expression. The human genome encodes more than 4000 proteins that function in these pathways, so the potential for cross-regulation between pathways formed from combinations of these elements is enormous. A recently developed model of a gene regulatory network responsible for development of an embryonic cell lineage in sea urchins provides a graphical representation of this complexity (**Figure 5.20**). Although clearly a daunting task, understanding cell regulatory networks in quantitative terms using mathematical and computational approaches that view the cell as an

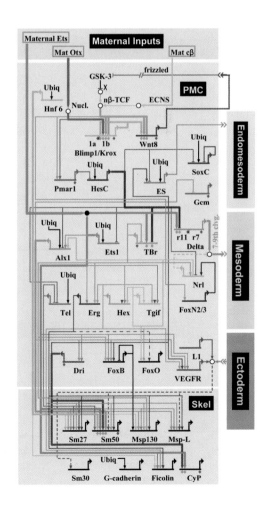

Figure 5.20 **A gene regulatory network**   The network includes all regulatory genes required for development of the embryonic cells that differentiate into skeletal cells of the sea urchin. (From P. Oliveri, Q. Tu and E. H. Davidson, 2008. *Proc. Natl. Acad. Sci. USA* 105: 5955.)

integrated biological system is a critically important problem at the cutting edge of research in cell biology.

Synthetic biology

Synthetic biology is an engineering approach to understanding and manipulating biological systems. Rather than studying existing (natural) biological systems, the goal of synthetic biology is to design and create new (unnatural or synthetic) systems. By designing such new systems, synthetic biologists hope to not only create useful products but also to better understand how the behavior of existing cells is controlled.

One approach taken by synthetic biologists is to synthesize new molecules with biological properties. An example is provided by the synthesis of RNA molecules that are capable of self-replication, which provided a critical demonstration of the ability of RNA to serve as the first self-replicating molecule in prebiotic evolution (see Chapter 1). Alternatively, many synthetic biologists engineer new systems using components of existing cells, which can be reassembled in different ways that result in distinct functional properties. The engineering approach of synthetic biology is not only important in designing biological systems with useful functions but is also complementary to the analytical approach of systems biology in understanding naturally occurring cells. The ability to engineer a novel biological system tests and expands our understanding of the principles that govern the function of natural systems.

The first systems designed by synthetic biologists were genetic circuits engineered in *E. coli* by the laboratories of James Collins and Stanislas Leibler in 2000. One of these systems, called a genetic toggle switch, was designed to confer stability and memory on a network regulating gene expression. The system consists of three genes: a reporter gene encoding an easily detectable indicator of gene expression and two regulatory genes encoding repressors that can be inactivated by the addition of small molecule inducers (**Figure 5.21**). The key feature responsible for the stability of this engineered circuit is that the two repressors control expression of each other as well as the

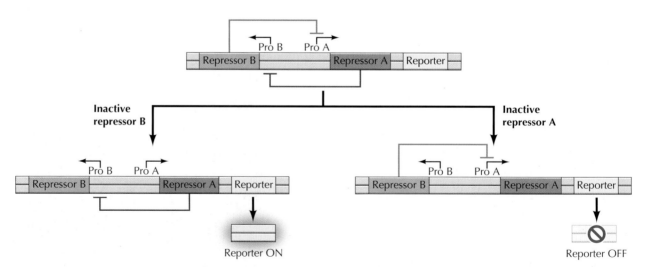

Figure 5.21 A genetic toggle switch The circuit includes genes encoding two repressors (A and B) that regulate each other and a reporter controlled by repressor B. Inactivation of repressor B leads to a stable state in which the reporter is expressed, whereas inactivation of repressor A leads to a stable state in which the reporter is repressed. (After T. S. Gardner, C. R. Cantor and J. J. Collins, 2000. *Nature* 403: 339.)

reporter. Repressor A blocks expression of repressor B, and repressor B blocks expression of both repressor A and the reporter gene. As an example of the system's stability, consider the effects of adding a small molecule inducer that inactivates repressor B. In the absence of repressor B, both the reporter gene and repressor A are expressed. Importantly, the system will remain in this state even if the inducer is withdrawn, because repressor A continues to prevent expression of repressor B. Conversely, if repressor A is inactivated, repressor B will be expressed and reporter gene expression as well as further expression of repressor A will be stably shut down. Thus, the system can switch between two stable states. Similar genetic circuits have since been engineered in eukaryotic models, including both yeast and mammalian cells. Engineering these systems has substantially advanced our understanding of how a regulatory circuit can alternate between two stable states—a common feature of the networks involved in many aspects of cell signaling and the regulation of cell proliferation, as will be discussed in later chapters.

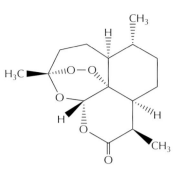

Figure 5.22 Structure of artemisinin The most effective treatment for malaria caused by the parasite *P. falciparum* is artemisinin in combination with other anti-malarial drugs.

The practical applications of synthetic biology include the engineering of metabolic pathways to efficiently produce therapeutic drugs. A good example is provided by the production of the antimalarial drug artemisinin (**Figure 5.22**). Malaria is a major global health problem, affecting more than 200 million people and responsible for more than 600,000 deaths per year. It is caused by infection with parasites belonging to the genus *Plasmodium*, most commonly *P. falciparum*. The most effective treatment for malaria caused by *P. falciparum* is artemisinin in combination with other anti-malarial drugs (called artemisinin-combination therapy or ACT). Unfortunately, artemisin is produced by a plant (sweet wormwood) that takes about eight months to grow to full size, and its supply has been unstable, leading to shortages and substantial price fluctuations. To address this problem and develop a non-botanical source of artemisin, Jay Keasling and his collaborators engineered strains of yeast that produced high-yields of artemisinic acid, which could then be efficiently converted to artemisinin by a chemical process. In 2013, the pharmaceutical company Sanofi announced the launch of a new facility to produce artemisinin by this method, and more than 16 million treatments have been produced since 2014.

The ultimate goal of synthetic biology might be viewed as the creation of a fully synthetic cell. A milestone was thus reached with the creation of the first cell with a completely synthetic genome in 2010. Starting from the known nucleotide sequence of the 1.08-Mb *Mycoplasma mycoides* genome, Craig Venter and his colleagues synthesized overlapping oligonucleotides corresponding to the complete genome sequence. These synthetic oligonucleotides were then assembled in several steps to yield a complete synthetic genome of 1,077,947 bases that also contained sequences required for propagation as a plasmid in yeast. The synthetic genome was propagated as a plasmid in yeast and then introduced into a different mycoplasma subspecies, *M. capricolum*, by gene transfer techniques. Cells containing the synthetic genome were selected by tetracycline resistance and propagated in culture. Sequencing their genomes indicated that they were entirely derived from the synthetic *M. mycoides* DNA, and cells with the synthetic genome were found to grow normally and show the morphology of normal *M. mycoides* (**Figure 5.23**). These mycoplasma thus represent the first cells with a purely synthetic genome. Since all of the

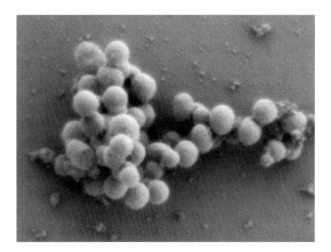

Figure 5.23 First cell with a synthetic genome Scanning electron micrograph of *M. mycoides* with a synthetic genome. (From D. G. Gibson et al., 2010. *Science* 329: 52.)

Molecular Medicine

Malaria and Synthetic Biology

The Disease

Malaria is caused by a parasite transmitted by infected mosquitoes and is the most significant human parasitic disease. Symptoms are severe chills, vomiting and fever, resembling flu. If the disease is left untreated, it may lead to severe complications and death. The World Health Organization estimates that there were 198 million cases of malaria and 584,000 deaths in 2013, with 90% of the deaths occurring in Africa.

Molecular and Cellular Basis

The parasites that cause malaria are protozoans belonging to the genus Plasmodium. Five species, *P. falciparum*, *P. malariae*, *P. ovale*, *P. vivax* and *P. knowlesi*, cause malaria in humans. Approximately 75% of cases and almost all deaths are caused by *P. falciparum* (see figure). The disease is transmitted by a bite of an Anopheles mosquito. The parasite then travels to the liver and replicates in liver cells, producing thousands of progeny. These progeny then infect and replicate in red blood cells, which lyse to release more parasites that infect additional red blood cells and undergo additional rounds of replication. The blood stage parasites cause the symptoms of the disease, and can also be picked up by a mosquito for further transmission.

Prevention and Treatment

Development of a vaccine against malaria is a major public health research priority, but no anti-malaria vaccine is currently available. Instead, efforts at disease prevention focus on measures to prevent mosquito bites, such as using insect repellants, sleeping under mosquito nets, or spraying insecticides. People traveling to areas where malaria is prevalent can take a variety of anti-malarial drugs, most of which are also used for treatment. However, prevention with these drugs is not practical for residents of areas where malaria exists because of their cost and the side effects of long term use.

The first effective treatment for malaria was quinine, which was replaced by chloroquine in the 1940s. Unfortunately, strains of *P. falciparum* that were resistant to chloroquine developed in the 1950s and have now become widespread. The most effective current therapy for *P. falciparum* malaria is artemisinin, which was discovered by the Chinese scientist Tu Youyou in the 1970s. It is administered as combination therapy with other antimalarial drugs, in order to prevent the development of strains of *P. falciparum* that are resistant to artemisinin.

Artemisinin was discovered in the leaves of *Artemisia annua* (sweet wormwood), which unfortunately provides a limited supply of the drug. The plants take about eight months to grow to full size and, because of instability in its availability, the market price for artemisinin has fluctuated over a tenfold range in recent years. The work of synthetic biologists has made an important contribution to this problem by engineering a strain of yeast that produces high levels of artemisinic acid, which can be efficiently converted chemically to artemisinin. Based on the efficient production of artemisinin by this engineered yeast, the pharmaceutical company Sanofi established a new production facility for artemisinin in 2013 and has produced more than 16 million treatments since 2014.

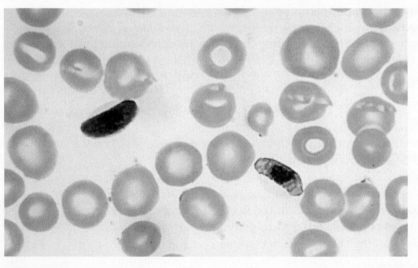

P. falciparum in a blood smear

FYI

The team that engineered the first yeast synthetic chromosome consisted largely of undergraduate researchers.

cellular proteins are specified by that genome, with the proteins present in the original recipient cell diluted out during replication, these mycoplasma also represent the first synthetic cells. Current efforts are under way to engineer a synthetic yeast genome as a model for design of a eukaryotic cell. Given the steadily decreasing costs of DNA synthesis, the possibility of engineering plant and animal genomes may also become feasible undertakings in the foreseeable future.

SUMMARY	KEY TERMS

Genomes and Transcriptomes

- ***The genomes of bacteria and yeast:*** Bacterial genomes are extremely compact, with protein-coding sequences accounting for nearly 90% of the DNA. The *E. coli* genome consists of 4.6×10^6 base pairs of DNA and contains about 4200 genes. The first eukaryotic genome to be sequenced was that of the yeast *S. cerevisiae*. The *S. cerevisiae* genome contains about 6000 genes, and protein-coding sequences account for approximately 70% of the genome.

open-reading frame

- ***The genomes of*** **Caenorhabditis elegans, Drosophila melanogaster,** *and* **Arabidopsis thaliana:** The genome of *C. elegans* contains about 19,000 protein-coding genes, which account for about 25% of the genome. The genome of *Drosophila* contains approximately 14,000 genes, with protein-coding sequences accounting for only about 10% of the genome. The numbers of genes in these species indicate that gene number is not simply related to the biological complexity of an organism. Likewise, the genome of the small flowering plant *Arabidopsis thaliana* contains approximately 26,000 genes—surprisingly more genes than in either *Drosophila* or *C. elegans*. Other plant genomes also contain large numbers of genes, including approximately 57,000 genes in the apple. The large number of genes in these plants further emphasizes the lack of relationship between gene number and complexity of an organism. See Video 5.1.

- ***The human genome:*** The human genome contains approximately 21,000 protein-coding genes—not much more than the number of genes found in simpler animals like *Drosophila* and *C. elegans*, and fewer than in *Arabidopsis* and other plants.

- ***The genomes of other vertebrates:*** The genomes of fish, chickens, mice, rats, dogs, rhesus macaques, chimpanzees, and Neandertals provide important comparisons to the human genome. All of these vertebrates contain similar numbers of genes.

- ***Next-generation sequencing and personal genomes:*** Enormous progress in the technology of DNA sequencing has now made it feasible to determine the complete sequence of individual genomes. See Animation 5.1.

next-generation sequencing

- ***Global analysis of gene expression:*** The RNAs expressed in a cell can be analyzed globally by hybridization to DNA microarrays or by next-generation sequencing. See Animation 5.2.

transcriptome, DNA microarray, RNA-seq

Proteomics

- ***Identification of cell proteins:*** Characterization of the complete protein complement of cells is a major goal of proteomics. Mass spectrometry provides a powerful tool for protein identification, which can be used to identify either isolated proteins or proteins present in mixtures.

proteomics, proteome, two-dimensional gel electrophoresis, mass spectrometry, tandem mass spectrometry

- ***Global analysis of protein localization:*** Isolated subcellular organelles can be analyzed by mass spectrometry to determine their protein constituents.

- ***Protein interactions:*** The purification of protein complexes from cells and analysis of interactions of proteins introduced into yeast can identify interacting proteins and may lead to elucidation of the complex networks of protein interactions that regulate cell behavior.

immunoprecipitation, yeast two-hybrid system

SUMMARY	KEY TERMS

Systems Biology

- **Systematic screens of gene function:** The genome sequencing projects have introduced large-scale experimental and computational approaches to research in cell and molecular biology. Genome-wide screens using RNA interference or the CRISPR/Cas system can systematically identify all of the genes in an organism that are involved in any biological process that can be assayed in a high-throughput format.

  bioinformatics, systems biology

- **Regulation of gene expression:** The identification of gene regulatory sequences and understanding the mechanisms that control gene expression are major challenges in bioinformatics and systems biology. These problems are being approached by genome-wide studies of gene expression, histone modification, and regulatory protein binding, combined with the development of computational approaches to identify functional regulatory elements.

- **Networks:** The activity of signaling pathways within the cell is regulated by feedback loops that control the extent and duration of signaling. Different signaling pathways also interact to regulate each other's activity. The extensive crosstalk between individual pathways leads to the formation of complex signaling networks that ultimately control cell behavior.

  network, feedback loop, feedforward relay, crosstalk

- **Synthetic biology:** Synthetic biology is an engineering approach to understanding biology, with the goal of designing new biological systems. The systems designed by synthetic biologists include genetic circuits, metabolic pathways, and synthetic genomes.

  synthetic biology

Refer To

The Cell

Companion Website

sites.sinauer.com/cooper7e

for quizzes, animations, videos, flashcards, and other study resources.

Questions

1. How can protein-coding sequences be identified in the DNA sequence of a genome?

2. How does the fraction of protein-coding sequence in the human genome compare to that in the yeast genome?

3. How do the number of genes in the human genome compare to the number of genes in apples?

4. Explain how mass spectrometry can be used to identify a protein.

5. How could you compare the protein composition of mitochondria from cancer cells and normal cells?

6. How can immunoprecipitation be used to study protein-protein interactions?

7. How could you use RNA interference to identify genes that might be good targets for the development of drugs for breast cancer treatment?

8. Explain how next-generation sequencing is used to study gene expression.

9. Why is it more difficult to identify regulatory sequences than protein-coding sequences?

10. How can synthetic biology contribute to understanding natural biological systems?

References and Further Reading (Key review articles for each major section are highlighted in **bold**.)

Genomes and Transcriptomes

Adams, M. D. and 194 others. 2000. The genome sequence of *Drosophila melanogaster*. *Science* 287: 2185–2195. [P]

Bentley, D. R. and 193 others. 2008. Accurate whole genome sequencing using reversible terminator chemistry. *Nature* 456: 53–59. [P]

Blattner, F. R. and 16 others. 1997. The complete genome sequence of *Escherichia coli* K–12. *Science* 277: 1453–1462. [P]

Collins, F. 2010. Has the revolution arrived? *Nature* 464: 674–675. [R]

Fleischmann, R. D., and 39 others. 1995. Whole-genome random sequencing and assembly of *Haemophilus influenzae* Rd. *Science* 269: 496–512. [P]

Hawkins, R. D., G. C. Hon and B. Ren. 2010. Next-generation genomics: an integrative approach. *Nature Rev. Genet.* 11: 476–486. [R]

Holt, R. A. and S. J. M. Jones. 2008. The new paradigm of flow cell sequencing. *Genome Res.* 18: 839–846. [R]

Howe, K. and 171 others. 2013. The zebrafish reference genome sequence and its relationship to the human genome. *Nature* 496: 498–503. [P]

International Human Genome Sequencing Consortium. 2001. Initial sequencing and analysis of the human genome. *Nature* 409: 860–921. [P]

International Human Genome Sequencing Consortium. 2004. Finishing the euchromatic sequence of the human genome. *Nature* 431: 931–945. [P]

Levy, S. and 30 others. 2007. The diploid genome sequence of an individual human. *PLoS Biology* 5: e254. [P]

Lindblad-Toh, K. and 46 others. 2005. Genome sequence, comparative analysis and haplotype structure of the domestic dog. *Nature* 438: 803–819. [P]

Mardis, E. R. 2011. A decade's perspective on DNA sequencing technology. *Nature* 470: 198–203. [R]

Metzker, M. L. 2010. Sequencing technologies—the next generation. *Nature Rev. Genet.* 11: 31–46. [R]

Mouse Genome Sequencing Consortium. 2002. Initial sequence and comparative analysis of the mouse genome. *Nature* 420: 520–562. [P]

Ozsolak, F. and P. M. Milos. 2011. RNA sequencing: advances, challenges and opportunities. *Nature Rev. Genet.* 12: 87–98. [R]

Pennisi, E. 2011. Green genomes. *Science* 332: 1373–1375. [R]

Prufer, K. and 40 others. 2012. The bonobo genome compared with the chimpanzee and human genomes. *Nature* 486: 527–531. [P]

Prufer, K. and 44 others. 2014. The complete genome sequence of a Neanderthal from the Altai Mountains. *Nature* 505: 43–49. [P]

Rat Genome Sequencing Project Consortium. 2004. Genome sequence of the Brown Norway rat yields insights into mammalian evolution. *Nature* 428: 493–521. [P]

Rhesus Macaque Genome Sequencing and Analysis Consortium. 2007. Evolutionary and biomedical insights from the rhesus macaque genome. *Science* 316: 222–234. [P]

Rogers, J. and R. A. Gibbs. 2014. Comparative primate genomics: emerging patterns of genome content and dynamics. *Nature Rev. Genet.* 5: 347–359. [R]

Stegle, O., S. A. Teichmann and J. C. Marioni. 2015. Computational and analytical challenges in single-cell transcriptomics. *Nature Rev. Genet.* 16: 133–145. [R]

The 1000 Genomes Project Consortium. 2010. A map of human genome variation from population-scale sequencing. *Nature* 467: 1061–1073. [P]

The *Arabidopsis* Genome Initiative. 2000. Analysis of the genome sequence of the flowering plant *Arabidopsis thaliana*. *Nature* 408: 796–815. [P]

The *C. elegans* Sequencing Consortium. 1998. Genome sequence of the nematode *C. elegans*: A platform for investigating biology. *Science* 282: 2012–2018. [P]

The Chimpanzee Sequencing and Analysis Consortium. 2005. Initial sequence of the chimpanzee genome and comparison with the human genome. *Nature* 437: 69–87. [P]

The ENCODE Project Consortium. 2012. An integrated encyclopedia of DNA elements in the human genome. *Nature* 489: 57–74. [P]

Venter, J. C. 2010. Multiple personal genomes await. *Nature* 464: 676–677. [R]

Venter, J. C. and 273 others. 2001. The sequence of the human genome. *Science* 291: 1304–1351. [P]

Wheeler, D. A. and 26 others. 2008. The complete genome of an individual by massively parallel DNA sequencing. *Nature* 452: 872–876. [P]

Proteomics

Ahmad, Y. and A. I. Lamond. 2014. A perspective on proteomics in cell biology. *Trends Cell Biol.* 24: 257–264. [R]

Ahrens, C. H., E. Brunner, E. Qeli, K. Basler, and R. Aebersold. 2010. Generating and navigating proteome maps using mass spectrometry. *Nature Rev. Mol. Cell Biol.* 11: 789–801. [R]

Breker, M. and M. Schuldiner. 2014. The emergence of proteome-wide technologies: systematic analysis of proteins comes of age. *Nature Rev. Mol. Cell Biol.* 15: 453–464. [R]

Domon, B. and R. Aebersold. 2006. Mass spectrometry and protein analysis. *Science* 312: 212–217. [R]

Gavin, A.-C. and 37 others. 2002. Functional organization of the yeast proteome by systematic analysis of protein complexes. *Nature* 415: 141–147. [P]

Gingras, A.-C., M. Gstaiger, B. Raught and R. Aebersold. 2007. Analysis of protein complexes using mass spectrometry. *Nature Rev. Mol. Cell Biol.* 8: 645–654. [R]

Giot, L. and 48 others. 2003. A protein interaction map of *Drosophila melanogaster*. *Science* 302: 1727–1736. [P]

Gstaiger, M. and R. Aebersold. 2009. Applying mass spectrometry-based proteomics to genetics, genomics and network biology. *Nature Rev. Genet.* 10: 617–627. [R]

Krogan, N. J. and 52 others. 2006. Global landscape of protein complexes in the yeast *Saccharomyces cerevisiae*. *Nature* 440: 637–643. [P]

Walther, T. C. and M. Mann. 2010. Mass spectrometry-based proteomics in cell biology. *J. Cell Biol.* 190: 491–500. [R]

Yates, J. R. III., A. Gilchrist, K. E. Howell and J. J. M. Bergeron. 2005. Proteomics of organelles and large cellular structures. *Nature Rev. Mol. Cell Biol.* 6: 702–714. [R]

Systems Biology

Annaluru, N. and 79 others. 2014. Total synthesis of a functional designer eukaryotic chromosome. *Science* 344: 55–58. [P]

Chuang, H.-Y., M. Hofree and T. Ideker. 2010. A decade of systems biology. *Ann. Rev. Cell Dev. Biol.* 26: 721–744. [R]

Elnitski, L., V. X. Jin, P. J. Farnham and S. J. M. Jones. 2006. Locating mammalian transcription factor binding sites: a survey of computational and experimental techniques. *Genome Res.* 16: 1455–1464. [R]

Elowitz, M. B. and S. A .Liebler. 2000. A synthetic oscillatory network of transcriptional regulators. *Nature* 403: 335–338. [P]

Eungdamrong, N. J. and R. Iyengar. 2004. Computational approaches for modeling regulatory cellular networks. *Trends Cell Biol.* 14: 661–669. [R]

Gardner, T. S., C. R. Cantor and J. J. Collins. 2000. Construction of a genetic toggle switch in *Escherichia coli*. *Nature* 403: 339–342. [P]

Ge, H., A. J. M. Walhout and M. Vidal. 2003. Integrating 'omic' information: A bridge between genomics and systems biology. *Trends Genet.* 19: 551–559. [R]

Gibson, D. G. and 23 others. 2010. Creation of a bacterial cell controlled by a chemically synthesized genome. *Science* 329: 52–56. [P]

Janes, K. A. and D. A. Lauffenburger. 2013. Models of signalling networks—what cell biologists can gain from them and give to them. *J. Cell Science* 126: 1913–1921. [R]

Kirschner, M. W. 2005. The meaning of systems biology. *Cell* 121: 503–504. [R]

Kitano, H. 2002. Systems biology: A brief overview. *Science* 295: 1662–1664. [R]

Komili, S. and P. Silver. 2008. Coupling and coordination in gene expression processes: a systems biology view. *Nature Rev. Genet.* 9: 38–48. [R]

Le Novère, N. 2015. Quantitative and logic modelling of molecular and gene networks. *Nature Rev. Genet.* 16: 146–158. [R]

Lienert, F., J. J. Lohmueller, A. Garg and P. A. Silver. 2014. Synthetic biology in mammalian cells: next generation research tools and therapeutics. *Nature Rev. Mol. Cell Biol.* 15: 95–107. [R]

Mohr, S. E., J. A. Smith, C. E. Shamu, R. A. Neumüller and N. Perrimon. 2014. RNAi screening comes of age: improved techniques and complementary approaches. *Nature Rev. Mol. Cell Biol.* 15: 591–600. [R]

Nurse, P. and J. Hayles. 2011. The cell in an era of systems biology. *Cell* 144: 850–854. [R]

Oliveri, P., Q. Tu and E. H. Davidson. 2008. Global regulatory logic for specification of an embryonic cell lineage. *Proc. Natl. Acad. Sci. USA* 105: 5955–5962. [P]

Paddon, C. J. and 49 others. 2013. High-level semi-synthetic production of the potent antimalarial artemisinin. *Nature* 496: 528–532. [P]

Papin, J. A., T. Hunter, B. O. Palsson and S. Subramaniam. 2005. Reconstruction of cellular signaling networks and analysis of their properties. *Nat. Rev. Mol. Cell Biol.* 6: 99–111. [R]

Shalem, O., N. E. Sanjana and F. Zhang. 2015. High-throughput functional genomics using CRISPR-Cas9. *Nature Rev. Genet.* 16: 299–311. [R]

Wang, T., J. J. Wei, D. M. Sabatini and E. S. Lander. 2014. Genetic screens in human cells using the CRISPR-Cas9 system. *Science* 343: 80–84. [P]

Wasserman, W. W. and A. Sandelin. 2004. Applied bioinformatics for the identification of regulatory elements. *Nature Rev. Genet.* 5: 276–287. [R]

White, J. K. and 30 others. 2013. Genome-wide generation and systematic phenotyping of knockout mice reveals new roles for many genes. *Cell* 154: 452–464. [R]

Xie, X., J. Lu, E. J. Kulbokas, T. R. Golub, V. Mootha, K. Lindblad-Toh, E. S. Lander and M. Kellis. 2005. Systematic discovery of regulatory motifs in human promoters and 3' UTRs by comparison of several mammals. *Nature* 434: 338–345. [P]

Zhu, X., M. Gerstein and M. Snyder. 2007. Getting connected: analysis and principles of biological networks. *Genes Dev.* 21: 1010–1024. [R]

Genes and Genomes

As the genetic material, DNA provides a blueprint that directs all cellular activities and specifies the developmental plan of multicellular organisms. An understanding of gene structure and function is therefore fundamental to an appreciation of the molecular biology of cells. The development of gene cloning represented a major step toward this goal, enabling scientists to dissect complex eukaryotic genomes and probe the functions of eukaryotic genes. As discussed in the preceding chapter, advances in DNA sequencing then brought us to the exciting point of knowing the complete genome sequences of hundreds of bacteria, of yeast, and of many species of plants and animals, including humans. This chapter will focus on the organization of eukaryotic genes and the types of sequences in the genomes of higher eukaryotes, many of which play important roles in gene regulation rather than encoding proteins.

The Structure of Eukaryotic Genes

The genomes of most eukaryotes are larger and more complex than those of prokaryotes (**Figure 6.1**). This larger size of eukaryotic genomes is not inherently surprising, since one would expect to find more genes in organisms that are more complex. However, the genome size of many eukaryotes does not appear to be related to genetic complexity. For example, the genomes of salamanders and lilies contain more than ten times the amount of DNA that is in the human genome, yet these organisms are clearly not ten times more complex than humans. Moreover, as discussed in Chapter 5, the number of genes in eukaryotic genomes is not simply related to either genome size or biological complexity. For example, the human genome is a thousandfold larger than that of *E. coli*, but it contains only about five times more genes. And although the genome of the small plant *Arabidopsis thaliana* is only about 5% the size of the human genome, it contains approximately 26,000 protein-coding genes, compared with 21,000 in the human genome.

These discrepancies have been resolved by the discovery that the genomes of most eukaryotic cells contain not only protein-coding sequences but also large amounts of DNA that does not code for proteins. The difference in the sizes of eukaryotic genomes primarily reflects differences in amounts of noncoding DNA, rather than differences in the numbers of protein-coding genes. Some of these non-protein coding sequences lie within eukaryotic genes, as discussed below. Others, which lie between protein-coding genes, are discussed in the next section of this chapter. Although these sequences do not encode proteins, many noncoding sequences play critical roles in cells, including regulating gene expression and significantly expanding the coding potential of our genomes by allowing genes to be expressed in alternate ways.

Figure 6.1 Genome size The range of sizes of the genomes of representative groups of organisms is shown on a logarithmic scale.

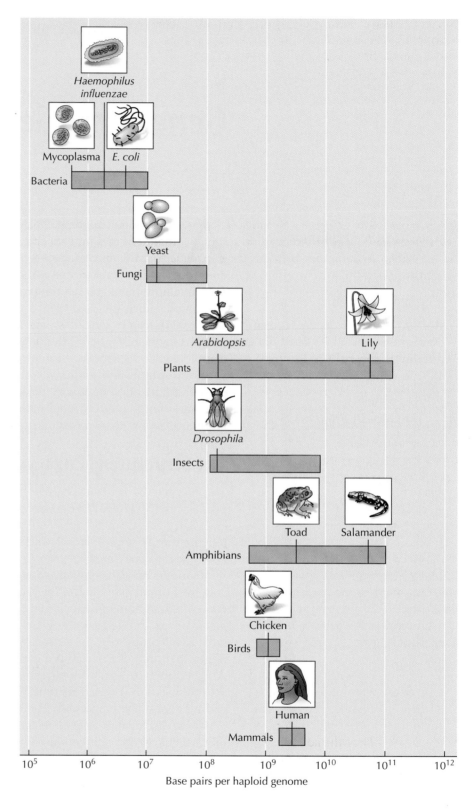

Base pairs per haploid genome

Introns and exons

In molecular terms, a **gene** can be defined as a segment of DNA that is expressed to yield a functional product, which may be either an RNA (e.g., ribosomal and transfer RNAs) or a polypeptide. While some of the noncoding DNA in eukaryotes is accounted for by DNA sequences that lie between

Figure 6.2 The structure of eukaryotic genes Most eukaryotic genes contain segments of coding sequences (exons) interrupted by noncoding sequences (introns). Both exons and introns are transcribed to yield a long primary RNA transcript. The introns are then removed by splicing to form the mature mRNA.

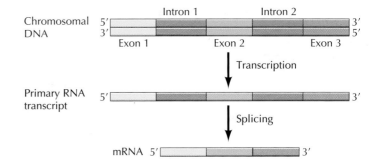

genes, most eukaryotic genes also include large amounts of noncoding DNA. Such genes have a split structure in which segments of coding sequence (called **exons**) are separated by noncoding sequences (intervening sequences, or **introns**) (**Figure 6.2**). The entire gene is transcribed to yield a long RNA molecule and the introns are then removed by splicing, so only exons are included in the mRNA.

Introns were first discovered in 1977, independently in the laboratories of Phillip Sharp and Richard Roberts, during studies of the replication of adenovirus in cultured human cells. Adenovirus is a useful model for studies of gene expression, both because the viral genome is only about 3.5×10^4 base pairs long and because adenovirus mRNAs are produced at high levels in infected cells. One approach used to characterize the adenovirus mRNAs was to determine the locations of the corresponding viral genes by examination of RNA-DNA hybrids in the electron microscope. Because RNA-DNA hybrids are distinguishable from single-stranded DNA, the positions of RNA transcripts on a DNA molecule can be determined. Surprisingly, such experiments revealed that adenovirus mRNAs do not hybridize to only a single region of viral DNA (**Figure 6.3**). Instead, a single mRNA molecule hybridizes to several separated regions of the viral genome. Thus the adenovirus mRNA does not correspond to an uninterrupted transcript of the template DNA; rather, the mRNA is assembled from several distinct blocks of sequences that originated from different parts of the viral DNA. This was subsequently shown to occur by **RNA splicing**, which will be discussed in detail in Chapter 8.

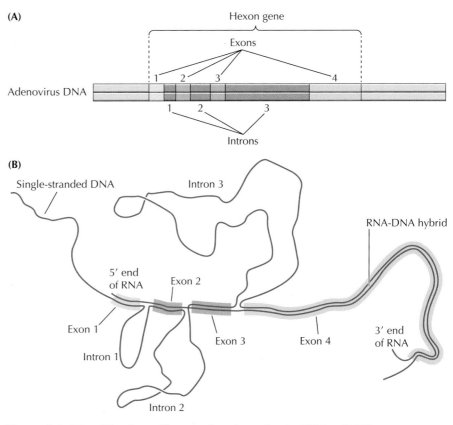

Figure 6.3 Identification of introns in adenovirus mRNA (A) The gene encoding the adenovirus hexon (a major structural protein of the viral particle) consists of four exons, interrupted by three introns. (B) This tracing illustrates an electron micrograph of a hypothetical hybrid between hexon mRNA and a portion of adenovirus DNA. The exons are seen as regions of RNA-DNA hybrid, which are separated by single-stranded DNA loops corresponding to the introns.

Key Experiment

The Discovery of Introns

Spliced Segments at the 5′ Terminus of Adenovirus 2 Late mRNA

Susan M. Berget, Claire Moore, and Phillip A. Sharp

Massachusetts Institute of Technology, Cambridge, MA

Proceedings of the National Academy of Sciences USA, Volume 74, 1977, pages 3171–3175

The Context

Prior to molecular cloning, little was known about mRNA synthesis in eukaryotic cells. However, it was clear that this process was more complex in eukaryotes than in bacteria. The synthesis of eukaryotic mRNAs appeared to require not only transcription but also processing reactions that modify the structure of primary transcripts. Most notably, eukaryotic mRNAs appeared to be synthesized as long primary transcripts, found in the nucleus, which were then cleaved to yield much shorter mRNA molecules that were exported to the cytoplasm.

These processing steps were generally assumed to involve the removal of sequences from the 5′ and 3′ ends of the primary transcripts. In this model, mRNAs embedded within long primary transcripts would be encoded by uninterrupted DNA sequences. This view of eukaryotic mRNA was changed radically by the discovery of splicing, made independently by Berget, Moore, and Sharp, and by Louise Chow, Richard Gelinas, Tom Broker, and Richard Roberts (An amazing sequence arrangement at the 5′ ends of adenovirus 2 messenger RNA, 1977. *Cell* 12: 1–8).

The Experiments

Both of the research groups that discovered splicing used adenovirus 2 to investigate mRNA synthesis in human cells. The major advantage of the virus is that it provides a model that is much simpler than the host cell. Viral DNA can be isolated directly from virus particles, and mRNAs encoding the viral structural proteins are present in such high amounts that they can be purified directly from infected cells. Berget, Moore, and Sharp focused their experiments on an abundant mRNA that encodes a viral structural polypeptide known as the hexon.

To map the hexon mRNA on the viral genome, purified mRNA was hybridized to adenovirus DNA and the hybrid molecules were examined by electron microscopy. As expected, the body of the hexon mRNA formed hybrids with restriction fragments of adenovirus DNA that had previously been shown to contain the hexon gene. Surprisingly, however, sequences at the 5′ end of hexon mRNA failed to hybridize to DNA sequences adjacent to those encoding the body of the message, suggesting that the 5′ end of the mRNA had arisen from sequences located elsewhere in the viral genome.

This possibility was tested by hybridization of hexon mRNA to a restriction fragment extending upstream of the hexon gene. The mRNA-DNA hybrids formed in this experiment displayed a complex loop structure (see figure). The body of the mRNA formed a long hybrid region with the previously identified hexon DNA sequences. Strikingly, the 5′ end of the hexon mRNA hybridized to three short upstream regions of DNA, which were separated from each other and from the body of the message by large single-stranded DNA loops. The sequences at the 5′ end of hexon mRNA thus appeared to be transcribed from three separate regions of the viral genome, which were spliced to the body of the mRNA during the processing of a long primary transcript.

The Impact

The discovery of splicing in adenovirus mRNA was quickly followed by similar

Phillip Sharp   Richard Roberts

experiments with cellular mRNAs, demonstrating that eukaryotic genes had a previously unexpected structure. Rather than being continuous, their coding sequences were interrupted by introns, which were removed from primary transcripts by splicing. Introns are now known to account for much of the DNA in eukaryotic genomes, and the roles of introns in the evolution and regulation of gene expression continue to be active areas of investigation. The discovery of splicing also stimulated intense interest in the mechanism of this unexpected RNA processing reaction. As discussed in Chapter 8, these studies have not only illuminated new mechanisms of regulating gene expression; they have also revealed novel catalytic activities of RNA and provided critical evidence supporting the hypothesis that early evolution was based on self-replicating RNA molecules. The unexpected structure of adenovirus mRNAs has thus had a major impact on diverse areas of cellular and molecular biology.

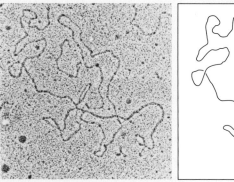

An electron micrograph and tracing of hexon mRNA hybridized to adenovirus DNA. The single-stranded loops designated A, B, and C correspond to introns.

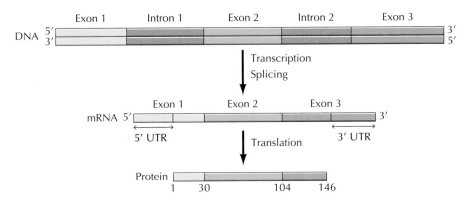

Figure 6.4 The mouse β-globin gene   This gene contains two introns that divide the coding region among three exons. Exon 1 encodes amino acids 1 to 30, exon 2 encodes amino acids 31 to 104, and exon 3 encodes amino acids 105 to 146. Exons 1 and 3 also contain untranslated regions (UTRs) at the 5′ and 3′ ends of the mRNA, respectively.

Soon after the discovery of introns in adenovirus, similar observations were made on cloned genes of eukaryotic cells. For example, electron microscopic analysis of RNA-DNA hybrids and subsequent nucleotide sequencing of cloned genomic DNAs and cDNAs indicated that the coding region of the mouse β-globin gene (which encodes the β subunit of hemoglobin) is interrupted by two introns that are removed from the mRNA by splicing (**Figure 6.4**).

The intron-exon structure of many eukaryotic genes is quite complicated, and the amount of DNA in the intron sequences is often greater than that in the exons. For example, an average human gene contains about 10 exons (totaling 4.3 kb), interrupted by introns (totaling 52 kb), resulting in approximately 56,000 base pairs (56 **kilobases**, or **kb**) of genomic DNA (Table 6.1). The exons include regions at both the 5′ and 3′ ends of the mRNA that are not translated into protein (**3′ and 5′ untranslated regions** or **UTRs**). These non-coding exons average about 2.6 kb and protein-coding sequence about 1.7 kb per gene. Thus, protein-coding sequences account for only about 3% and introns comprise approximately 93% of the average human gene.

Introns are present in most genes of complex eukaryotes, although they are not universal. Almost all histone genes, for example, lack introns, so introns are clearly not required for gene function in eukaryotic cells. In contrast, introns are found in only rare genes of prokaryotes, whose genomes consist almost entirely (~90%) of protein-coding sequences. It is noteworthy that many introns are conserved in genes of both plants and animals, indicating that they arose early in evolution, prior to the plant–animal divergence.

Table 6.1	Characteristics of the Average Human Gene[a]
Number of exons	10
Coding sequence	1700 base pairs
Total exon sequence	4300 base pairs
Intron sequence	52,000 base pairs

[a]Data are summarized from The ENCODE Project Consortium. 2012. *Nature* 489: 57.

Table 6.2 Introns in Representative Genomes

Organism	Genome size (Mb)[a]	Number of genes	Protein-coding sequence	Intron sequence	Introns per gene
Bacteria					
E. coli	4.6	4288	88%	NS[b]	NS
Yeasts					
S. cerevisiae	12	6000	70%	1%	0.04
Invertebrates					
C. elegans	97	19,000	25%	25%	5
Drosophila	180	13,600	13%	10%	4
Plants					
Arabidopsis thaliana	125	26,000	25%	25%	5
Mammals					
Human	3200	21,000	1.2%	35%	9

[a]Mb = millions of base pairs
[b]NS = not significant

In eukaryotes, the frequency of introns increases with genomic size and complexity (Table 6.2). For example, introns are present in only about 4% of the genes of the yeast S. cerevisiae, which consists of 12×10^6 base pairs and contains about 6000 protein-coding genes. Moreover, those S. cerevisiae genes that do contain introns usually have only a single small intron near the beginning of the gene. Thus yeast have compact genomes, similar to prokaryotes, with protein-coding sequences accounting for approximately 70% of total yeast DNA. In contrast, the genomes of C. elegans, Drosophila, and Arabidopsis, which are intermediate in size between the genomes of yeasts and humans (approximately 100×10^6 base pairs), also contain intermediate numbers of introns. Each of these organisms has an average of 4–5 introns per gene and the amount of intron sequence is similar to the amount of protein-coding sequence. The average gene in these organisms thus contains about 50% intron and 50% exon sequences, with introns accounting for approximately 15-25% of their genomes. As noted above, the average human gene contains about 90% intron and 10% exon sequences, with introns accounting for approximately 35% of the genome.

Roles of introns

Although introns are excised from completed mRNAs, they have many other important biological roles. First, many introns actually encode functional products, which may be either proteins or noncoding RNAs. In these cases (called nested genes) one gene is contained within an intron of a larger gene (Figure 6.5). Such nested genes, with proteins encoded by introns of a larger gene, were first identified in Drosophila, where they account for more than 5% of the total number of protein-coding genes. In the human genome, nested genes are less frequent: about 150 nested protein-coding genes have been identified, accounting for less than 1% of the total. Some nested genes in both the Drosophila and human genomes also encode pseudogenes or noncoding RNAs, including microRNAs and the small nucleolar RNAs that function in ribosomal RNA processing (discussed in the next section).

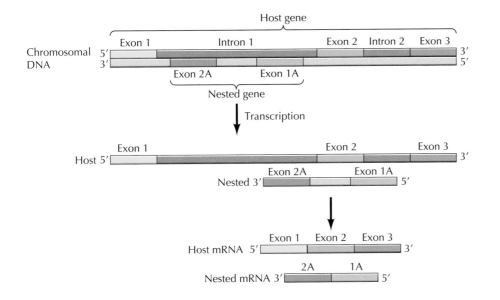

Figure 6.5 Nested genes   A nested gene is contained within an intron of a larger host gene. Transcription yields primary transcripts of both the host and nested genes, which are spliced to yield host gene and nested gene mRNAs.

Introns also contain regulatory sequences that control gene expression, which is critical to cell behavior. Since all cells in an organism contain the same genes, regulated differences in gene expression determine the difference between one type of cell and another: for example, between a human brain cell and a muscle cell. The first step in gene expression, transcription, is regulated by sequences that can reside in a variety of locations: upstream of genes, at distant locations in the genome, or within genes (**Figure 6.6**). Most of the transcriptional regulatory sequences that lie within genes are located either within the gene's first intron or within the 5′ untranslated region encoded by the first exon. Global studies indicate that approximately 10–20% of the human genome consists of sequences that act to regulate gene expression, as will be discussed in Chapter 8.

In addition to containing sequences that regulate transcription, sequences within introns regulate splicing, which is critical to the formation of functional mRNAs. Most notably, the presence of introns allows the exons of a gene to be joined in different combinations, resulting in the synthesis of different

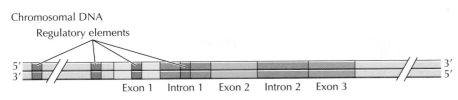

Figure 6.6 Transcriptional regulatory elements   Regulatory elements that control transcription of a gene may be located up to hundreds of kilobases away from the gene, immediately upstream of the gene, within noncoding exons or within introns.

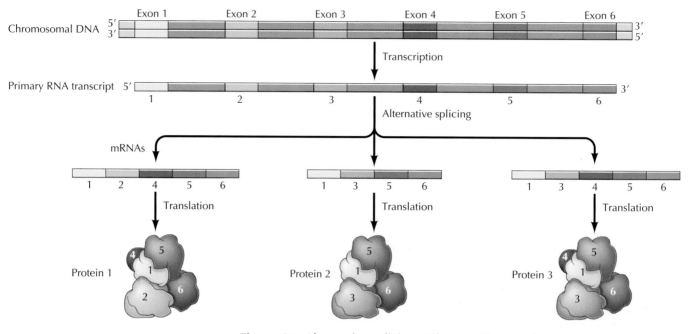

Figure 6.7 Alternative splicing The gene illustrated contains six exons, separated by five introns. Alternative splicing allows these exons to be joined in different combinations, resulting in the formation of three distinct mRNAs and proteins from the single primary transcript.

proteins from the same gene. This process, called **alternative splicing** (**Figure 6.7**), occurs frequently in the genes of complex eukaryotes. For example, most human genes can be alternatively spliced to yield between two and several thousand different mRNAs. On average, each human gene yields 6 alternatively spliced mRNAs, 4 of which encode different proteins. Alternative splicing thus allows the 21,000 human protein-coding genes to specify nearly 85,000 different proteins—substantially extending the functional repertoire of the human genome.

Noncoding Sequences

Protein-coding genes, including their introns, account for approximately 36% of the human genome. The rest of the human genome, as well as the genomes of other eukaryotes, are composed of several additional types of sequences. Although these sequences do not encode proteins, recent research has established that they are not nonfunctional. Although the roles and activities of these sequences are not yet fully understood, it appears that many play important roles in gene regulation. Others have been key to the evolution of eukaryotic genomes, and still others are involved in the structure and replication of eukaryotic chromosomes. The complexity of the genomes of higher eukaryotes is reflected more clearly in the extent and multiple activities of these noncoding sequences than it is in the number of protein-coding genes, so it is likely that unraveling the functions of noncoding sequences will be key to understanding the development and behavior of higher plants and animals.

Key Experiment

The ENCODE Project

An Integrated Encyclopedia of DNA Elements in the Human Genome

The ENCODE Project Consortium

Nature, Volume 489, 2012, pages 57–73

The Context

A surprising result from the sequence of the human genome published in 2004 (see Key Experiment Chapter 5) was the small number of protein-coding genes, estimated at that time as approximately 20,000–25,000. Protein coding sequences accounted for less than 1.5% of the genome—so what did the rest of the DNA do? Early views of our genome, in the 1980s, assumed that most of the DNA that did not encode proteins was non-functional. In this view, much of the genome was considered "junk DNA." This was questioned, however, on the grounds that non-protein coding sequences might play other important roles in the cell. The ENCODE project was launched in 2003 with the goal of defining all of the functional elements in the human genome. How much of our genome really was non-functional "junk DNA" and how much might have important functions other than encoding proteins, for example in gene regulation? The results of ENCODE have suggested that the vast majority of our genome is functional and have changed our views of mammalian genomes, particularly with respect to the prevalence of noncoding RNAs.

The Experiments

The first stage of ENCODE, initiated in 2003, was a pilot project to analyze 1% of the genome. The results of this initial phase, published in 2007, demonstrated the feasibility of the approach. Advances in DNA sequencing, in particular the development of next-generation sequencing technologies (discussed in Chapter 5), enabled the project to be scaled up to full genome analysis. This phase of ENCODE

has been a vast international collaboration, involving more than 400 scientists working at 32 different institutions worldwide. They used several different methods to analyze transcription and gene regulatory sequences on a genome-wide level in 147 different human cell lines. Transcription and RNA processing was characterized by RNA-seq, and gene regulatory sequences were analyzed both by determining the binding sites for gene regulatory proteins and identifying regions of chromatin characteristic of regulatory sequences. Since different genes are expressed in different cells, the results obtained with all 147 cell lines were integrated to determine the full extent of transcribed and regulatory sequences in the human genome.

Some of the results of the ENCODE project are summarized in the table. The most striking finding was that almost all of our genome, at least 80%, has a function that could be characterized by this genomic analysis. An unexpectedly large fraction of the genome, approximately 75%, was transcribed into RNA. This included transcripts derived from the 20,687 protein-coding genes identified by ENCODE, as well as 8801 short noncoding RNAs, 9640 long noncoding RNAs (lncRNAs), and transcripts of 863 pseudogenes. Regions of DNA that bound transcription factors encompassed 8.1% of the genome. However, only a limited number of transcription factors were included in these experiments and many differentiated cell types were not represented among the 147 cell lines studied, so this is expected to be an underestimate, corresponding to about half of the total amount of genome sequence that functions as binding sites for proteins that regulate gene expression.

The Impact

The most surprising result of EN-CODE is the extent to which our genome is functional, at least at the level of biochemical activity.

Especially unexpected is the extent of transcription of noncoding RNAs and the finding of a large number of genes encoding lncRNAs. These results have changed the way we think about our genome. Rather than viewing our genome as a small number of protein-coding sequences embedded in a vast excess of nonfunctional "junk DNA," it now appears that the majority of our genome consists of functional sequences, with abundant roles for noncoding RNAs and gene regulatory sequences.

Understanding the functions of the large number of noncoding RNAs identified by ENCODE remains a challenge for future research. Many have been shown to function as regulators of gene transcription, but functions for the majority of these transcripts have yet to be determined. Thus, the extent to which they truly represent biologically functional elements as well as a more complete understanding of their roles in cell behavior remain unanswered. Likewise, a full elucidation of gene regulatory sequences in our genome remains a goal of future research, which will require analysis of an even larger number of cell types representing the full extent of differentiated cells. Although much remains to be understood, ENCODE has laid the foundation for a new way of looking at mammalian genes and genomes.

Summary of ENCODE Results	
Protein-coding genes	20,687
Short noncoding RNAs	8801
Long noncoding RNAs	9640
Pseudogenes	11,224
Percentage of genome transcribed into RNA	74.7%
Percentage of genome-binding transcription factors	8.1%

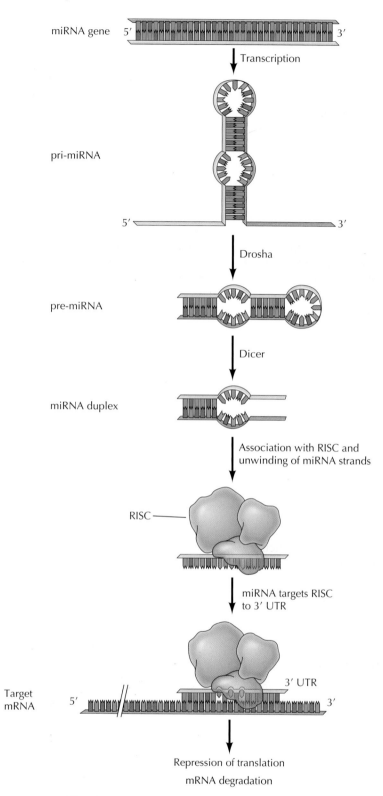

miRNA gene 5′ ... 3′

Transcription

pri-miRNA

5′ ... 3′

Drosha

pre-miRNA

Dicer

miRNA duplex

Association with RISC and unwinding of miRNA strands

RISC

miRNA targets RISC to 3′ UTR

3′ UTR

Target mRNA 5′ ... 3′

Repression of translation
mRNA degradation

Figure 6.8   miRNAs   miRNA genes are transcribed to yield primary transcripts (pri-miRNAs) that contain hairpin structures. Pri-miRNAs are sequentially cleaved by the nucleases Drosha and Dicer to yield double-stranded miRNAs of approximately 22 nucleotides. miRNAs associate with the RISC complex in which the two strands of the miRNA are unwound. The miRNA then targets RISC to the 3′ untranslated region (UTR) of a target mRNA, leading to repression of translation and mRNA degradation.

Noncoding RNAs

A major advance in our understanding of the functional elements of the human genome was made in 2012 by a large-scale project called EN-CODE (Encyclopedia of DNA Elements). The goal of the ENCODE project was to define the functions of the different types of sequences in the human genome. It involved the analysis of 147 different human cell lines using a variety of methods, including large-scale RNA sequencing (RNA-seq; see Figure 5.7) to characterize all of the sequences that were transcribed into RNA. An unexpected result of this extensive analysis was that about 75% of the human genome was transcribed. This is much more of the genome than can be accounted for by protein-coding genes, and these findings have led to the realization of the extensive roles that noncoding RNAs play in gene regulation. Some noncoding RNAs, such as transfer RNAs (tRNAs) and ribosomal RNAs (rRNAs) have been known for a long time and play key roles in protein synthesis (discussed in Chapter 4). Other noncoding RNAs function in splicing of pre-mRNAs and the processing of ribosomal RNAs, as will be discussed in subsequent chapters. Importantly, two types of noncoding RNAs have been more recently discovered and some of the most exciting advances in recent years have come from understanding their importance in gene regulation. The results of ENCODE suggest that the extent to which noncoding RNAs regulate gene expression may be much greater than previously appreciated.

As discussed in Chapter 4, **RNA interference (RNAi)** mediated by short double-stranded RNAs has become widely used as an experimental tool to block gene expression at the level of translation since its discovery in 1998 (see Figure 4.38). Regulation of translation by short double-stranded RNAs is not only an important experimental method, but is also normally used by cells to control mRNA translation and degradation. The endogenous noncoding RNAs that mediate RNA interference are **microRNAs (miRNAs)**, which are approximately 22 nucleotides in length. There may be as many as 1000 miRNAs encoded in the human genome, some of which lie within introns of protein-coding genes. The precursors of miRNAs are longer RNAs that fold into hairpin structures, which are then cleaved sequentially by the nucleases Drosha and Dicer to yield double-stranded RNAs of about 22 nucleotides (**Figure 6.8**). One strand of a miRNA is incorporated into the RNA-induced silencing complex (RISC), and the miRNA targets RISC to complementary mRNAs, where they inhibit translation and stimulate mRNA degradation. It is estimated that each miRNA can target up to 100 different mRNAs, so up to half of our protein-coding genes may be targets for regulation by miRNAs. Although their biological roles are not yet fully understood, miRNAs have been found to function in a variety of developmental processes, including development of early embryos, the nervous system, muscle, heart, lungs, and the immune system. miRNAs have also been shown to regulate cell proliferation and survival, and abnormal expression of miRNAs has been found to contribute to heart disease and several kinds of cancer.

In addition to miRNAs, **long noncoding RNAs (lncRNAs)**, which are noncoding RNAs greater than 200 nucleotides, have recently become recognized as major regulators of gene expression in eukaryotic cells. The phenomenon of **X chromosome inactivation** provides one of the first examples of the role of a lncRNA. In many animals, including humans, females have two X chromosomes, and males have one X and one Y chromosome. The X chromosome contains approximately 1000 genes that are not present on the much smaller Y chromosome. Thus females have twice as many copies of most X chromosome genes as males have. Despite this difference, female and male cells contain equal amounts of the proteins encoded by the majority of X chromosome genes. This results from a dosage compensation mechanism in which most of the genes on one of the two X chromosomes in female cells are silenced early in development. Consequently, only one copy of most genes located on the X chromosome is available for transcription in either female or male cells. The key element in X chromosome inactivation is a lncRNA called *Xist*, which is 17 kb in length. *Xist* RNA binds to the inactive X, blocking its transcription (**Figure 6.9**).

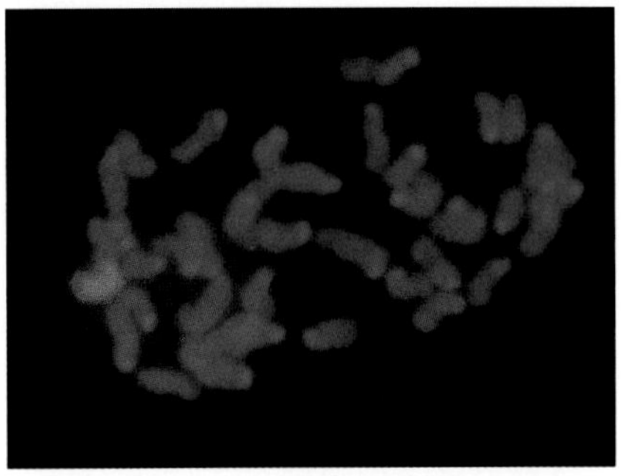

While a few examples like *Xist* were previously known, the extent to which lncRNAs are expressed in mammalian cells was recognized as a result of the large-scale ENCODE project that revealed the extensive transcription of mammalian genomes. The most recent high-throughput sequencing of human cellular RNAs has identified more than 50,000 lncRNAs, which are the products of distinct transcription units located outside of protein-coding genes. Recent studies have found that many lncRNAs (like *Xist*) may be functional regulators of gene expression, as will be discussed in Chapter 8. ENCODE also identified transcripts of almost 9000 small noncoding RNAs, the roles of most of which remain to be determined. It is particularly noteworthy that the number of noncoding RNAs that have now been identified in human cells is larger than the number of protein-coding genes, suggesting that these noncoding RNAs may contribute considerably to the biological complexity of higher eukaryotes.

Figure 6.9 X chromosome inactivation The inactive X chromosome is coated and inactivated by the lncRNA *Xist* (red). Chromosomes are stained with a dye that binds DNA (blue). (From J. T. Lee, 2009. *Genes Dev.* 23: 1831.)

Repetitive sequences

A large portion of complex eukaryotic genomes consists of highly repeated DNA sequences, which can be present in hundreds of thousands of copies per genome. There are several types of these highly repeated sequences (**Table 6.3**). One class (called **simple-sequence repeats**) consists of tandem arrays of up to thousands of copies of short sequences, ranging from 1 to 500 nucleotides. For example, one type of simple-sequence repeat in *Drosophila*

Table 6.3 Repetitive Sequences in the Human Genome

Type of sequence	Number of copies	Fraction of genome
Simple-sequence repeats[a]	>1,000,000	~10%
Retrotransposons		
LINEs	850,000	21%
SINEs	1,500,000	13%
Retrovirus-like elements	450,000	8%
DNA transposons	300,000	3%

[a]The content of simple-sequence repeats is estimated from the fraction of heterochromatin in the human genome.

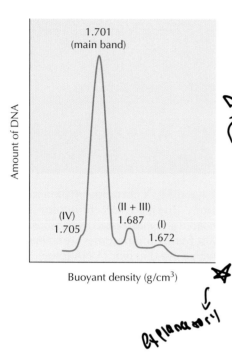

Figure 6.10 Satellite DNA   Equilibrium centrifugation of *Drosophila* DNA in a CsCl gradient separates satellite DNAs (designated I–IV) with buoyant densities (in g/cm³) of 1.672, 1.687, and 1.705 from the main band of genomic DNA (buoyant density 1.701).

consists of tandem repeats of the seven nucleotide unit ACAAACT. Because of their distinct base compositions, many simple-sequence DNAs can be separated from the rest of the genomic DNA by equilibrium centrifugation in CsCl density gradients. The density of DNA is determined by its base composition, with AT-rich sequences being less dense than GC-rich sequences (the molecular mass of a GC base pair is one dalton greater than an AT base pair). Therefore an AT-rich simple-sequence DNA bands in CsCl gradients at a lower density than the bulk of *Drosophila* genomic DNA (**Figure 6.10**). Since such repeat-sequence DNAs band as "satellites" separate from the main band of DNA, they are frequently referred to as **satellite DNAs**. These sequences are repeated millions of times per genome, accounting for about 40% of total genomic DNA in *Drosophila* and about 10% in humans. Simple-sequence DNAs are not transcribed and do not convey functional genetic information. Some, however, play important roles in chromosome structure, as discussed later in this chapter.

Other repetitive DNA sequences are scattered throughout the genome rather than being clustered as tandem repeats. These interspersed repetitive elements are a major contributor to genome size, accounting for approximately 45% of human genomic DNA (see Table 6.3). The two most prevalent classes of these sequences are called **short interspersed elements (SINEs)** and **long interspersed elements (LINEs)**. SINEs are 100–300 base pairs long. About 1.5 million such sequences are dispersed throughout the genome, accounting for approximately 13% of total human DNA. The major human LINEs are 4–6 kb long, although many repeated sequences derived from LINEs are shorter, with an average size of about 1 kb. There are approximately 850,000 repeats of LINE sequences in the genome, accounting for about 21% of human DNA.

Both SINEs and LINEs are examples of **transposable elements**, which are capable of moving to different sites in genomic DNA. SINEs and LINEs are **retrotransposons**, meaning that their transposition is mediated by reverse transcription (**Figure 6.11**). An RNA copy of a SINE or LINE is converted to DNA by reverse transcriptase within the cell, and the new DNA copy is integrated at a new site in the genome. A third class of interspersed repetitive sequences, which closely resemble retroviruses and are called **retrovirus-like elements**, also move within the genome by reverse transcription. Human retrovirus-like elements range from approximately 2–10 kb in length. There are approximately 450,000 retrovirus-like elements in the human genome, accounting for approximately 8% of human DNA. A fourth class of interspersed repetitive elements (**DNA transposons**) moves through the genome by being copied and reinserted as DNA sequences, rather than moving by reverse transcription. In the human genome, there are about 300,000 copies of DNA transposons, ranging from 80–3000 base pairs in length, and accounting for approximately 3% of human DNA.

Our understanding of retrotransposons, which account for approximately 42% of the human genome, has derived from studies of retroviruses. As

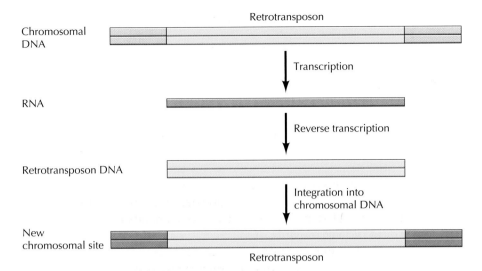

Retrotransposon

Chromosomal DNA

| Transcription

RNA

| Reverse transcription

Retrotransposon DNA

| Integration into chromosomal DNA

New chromosomal site

Retrotransposon

Figure 6.11 Movement of retrotransposons   A retrotransposon present at one site in chromosomal DNA is transcribed into RNA, and then converted back into DNA by reverse transcription. The retrotransposon DNA can then integrate into a new chromosomal site.

discussed in Chapter 4, retroviruses (for example, HIV) contain RNA genomes in their virus particles but replicate via the synthesis of a DNA provirus in infected cells (see Figure 4.13). The provirus is synthesized by the viral enzyme reverse transcriptase and integrated into the chromosomal DNA of infected cells by a viral enzyme called integrase. The viral life cycle continues with transcription of the provirus, which yields viral genomic RNA as well as mRNAs that direct the synthesis of viral proteins (including reverse transcriptase and integrase). The genomic RNA is then packaged into viral particles, which are released from the host cell. These progeny viruses can infect a new cell, initiating another round of proviral DNA synthesis and integration. The net effect can be viewed as the movement of the provirus from one chromosomal site to another, via the synthesis and reverse transcription of an RNA intermediate.

Some retrotransposons, such as retrovirus-like elements, are structurally similar to retroviruses. They encode reverse transcriptase and integrase and can move to new chromosomal sites within the same cell via mechanisms similar to those involved in retrovirus replication. However, the retrovirus-like elements differ from retroviruses in that they are not packaged into infectious particles and therefore cannot spread from one cell to another. LINEs similarly encode reverse transcriptase and an enzyme involved in their integration into cellular DNA, although the detailed mechanism of their transposition differs from that of retroviruses and retrovirus-like elements. In contrast, SINEs, which are only 100–300 base pairs, do not encode reverse transcriptase or other proteins. SINEs arose by reverse transcription of small RNAs, including tRNAs and small cytoplasmic RNAs involved in protein transport. Since they do not include genes for reverse transcriptase or enzymes involved in DNA integration, their transposition presumably involves the action of reverse transcriptases and integration enzymes that are encoded elsewhere in the genome—probably by other retrotransposons, such as LINEs.

Although the highly repetitive SINEs and LINEs account for a significant fraction of genomic DNA, their transpositions to random sites in the genome are not likely to be useful for the cell in which they are located. These transposons induce mutations when they integrate at a new target site and, like mutations induced by other agents, most mutations resulting from transposon integration are expected to be harmful to the cell. Mutations resulting from the transposition of LINEs and SINEs have been associated with several inherited human diseases, including hemophilia, cystic fibrosis, muscular dystrophy, and some hereditary cancers. Recent genome-wide studies of human cancers have also found that transposition of LINEs and SINEs frequently occurs during the development of several types of non-inherited cancer. Many of the transpositions identified in these cancers affected genes that are involved in cancer development, suggesting that mutation of these genes by retrotransposons may have contributed to the disease.

On the other hand, some mutations resulting from the movement of transposable elements may be beneficial, contributing in a positive way to evolution of the species. For example, some retrotransposons in mammalian genomes have been found to contain regulatory sequences that control the expression of adjacent genes, so their movements have contributed to evolutionary changes in gene expression. Moreover, sequences of cellular DNA adjacent to LINEs are frequently carried along during the process of transposition. Consequently, the transposition of LINEs can result in the movement of cellular DNA sequences to new genomic sites. Since LINEs can integrate into active genes, the associated transposition of cellular DNA sequences can lead to the formation of new combinations of regulatory and/or coding sequences and contribute directly to the evolution of new genes.

In addition to the effects of their transposition, retrotransposons have played a major role in shaping the genome by stimulating DNA rearrangements resulting from recombination between repetitive sequences (**Figure 6.12**). For example, rearrangements of chromosomal DNA resulting from

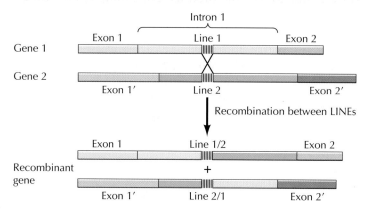

Figure 6.12 Recombination between repetitive sequences Recombination between repetitive sequences in different locations of the genome can lead to gene rearrangements. In the example shown, the repetitive sequences are located within introns, so recombination results in a rearrangement of the protein-coding sequences of two different genes.

recombination between LINEs integrated at different sites in the genome can lead to the formation of new genes. Transposable elements have played a major role in stimulating gene rearrangements that have contributed to the generation of genetic diversity, and thus have played key roles in shaping our genomes.

It is also notable that transposable elements are found within the sequences of the majority of lncRNAs. Moreover, SINE sequences within some lncRNAs have been found to directly contribute to their functions in gene regulation, suggesting that transposable elements within lncRNAs may play general roles in gene regulation.

Gene duplication and pseudogenes

Another factor contributing to the large size of eukaryotic genomes is that many genes are present in multiple copies, some of which are non-functional. In some cases, multiple copies of genes are needed to produce RNAs or proteins required in large quantities, such as ribosomal RNAs or histones. In other cases, distinct members of a group of related genes (called a **gene family**) may be transcribed in different tissues or at different stages of development. For example, the α and β subunits of hemoglobin are both encoded by gene families in the human genome, with different members of these families being expressed in embryonic, fetal, and adult tissues (**Figure 6.13**). Members of some gene families (e.g., the globin genes) are clustered within a region of DNA; members of other gene families are dispersed to different chromosomes.

Gene families are thought to have arisen by duplication of an original ancestral gene, with different members of the family then diverging as a consequence of mutations during evolution. Such divergence can lead to the evolution of related proteins that are optimized to function in different tissues or at different stages of development. For example, fetal globins have a higher affinity for O_2 than do adult globins—a difference that allows the fetus to obtain O_2 from the maternal circulation.

As might be expected, however, not all mutations enhance gene function. Some gene copies have instead sustained mutations that result in their loss of ability to produce a functional gene product. For example, the human

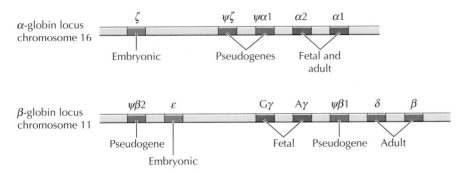

Figure 6.13 Globin gene families   Members of the human α- and β-globin gene families are clustered on chromosomes 16 and 11, respectively. Each family contains genes that are specifically expressed in embryonic, fetal, and adult tissues, in addition to nonfunctional gene copies (pseudogenes).

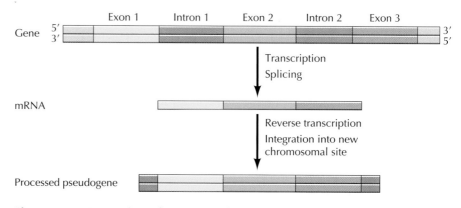

Figure 6.14 Formation of a processed pseudogene   A gene is transcribed and spliced to yield an mRNA from which the introns have been removed. The mRNA is copied by reverse transcription, yielding a cDNA copy lacking introns. Integration into chromosomal DNA results in formation of a processed pseudogene.

α- and β-globin gene families each contain genes that have been inactivated by mutations. Such nonfunctional gene copies (called **pseudogenes**) represent evolutionary relics that increase the size of eukaryotic genomes without making a functional genetic contribution. Recent studies have identified about 11,000 pseudogenes in the human genome.

Gene duplications can arise by two distinct mechanisms. The first is duplication of a segment of DNA, which can result in the transfer of a block of DNA sequence to a new location in the genome. Such duplications of DNA segments ranging from 1 kb to more than 50 kb are estimated to account for approximately 5% of the human genome. Duplications of the entire genome can also occur and are particularly common in plants. For example, the sequence of the *Arabidopsis* genome indicate that it has undergone at least two full-genome duplications, followed by the gradual loss of some of the duplicated genes. The large number of protein-coding genes in *Arabidopsis* (26,000) is the result of these duplications, as are the large sizes and high numbers of genes in other plant genomes.

Alternatively, genes can be duplicated by reverse transcription of an mRNA, followed by integration of the cDNA copy into a new chromosomal site (**retrotransposition**) (**Figure 6.14**). This mode of gene duplication, analogous to the transposition of repetitive elements that move via RNA intermediates, results in the formation of gene copies that lack introns and also lack the normal chromosomal sequences that direct transcription of the gene into mRNA. As a result, duplication of a gene by reverse transcription usually yields an inactive gene copy called a **processed pseudogene**. Processed pseudogenes account for the majority of the pseudogenes that have been identified in the human genome.

In contrast to processed pseudogenes, some genes that have been duplicated by reverse transcription remain functional after insertion at a new chromosomal location. An interesting example is the gene responsible for the short legs characteristic of some breeds of dogs, such as dachshunds and basset hounds (**Figure 6.15**). This is the result of the retrotransposition of a gene that inhibits bone growth. The transposed retrogene is active but abnormally expressed in its new location, probably as a result of regulation

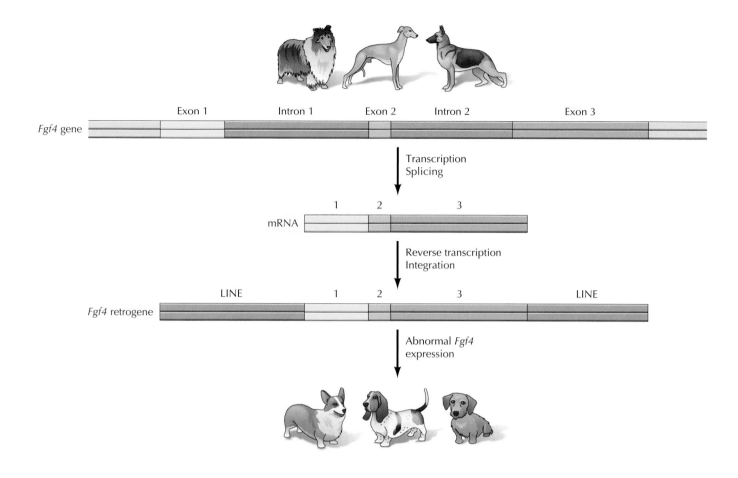

Figure 6.15 Transposition of a retrogene determines short legs in dog breeds   The *Fgf4* gene in dog breeds with long legs is expressed from its normal chromosomal location. In dogs with short legs, the gene has undergone a retrotransposition event and integrated in the middle of a LINE. This transposed retrogene is abnormally expressed, resulting in premature termination of bone growth and the short legs characteristic of some breeds.

by an adjacent LINE, which also illustrates the role of LINEs in regulating expression of adjacent genes. Because of the abnormal expression of the transposed gene, bone growth ceases prematurely and dogs develop short legs.

Chromosomes and Chromatin

Not only are the genomes of most eukaryotes much more complex than those of prokaryotes, but the DNA of eukaryotic cells is also organized differently from that of prokaryotic cells. The genomes of prokaryotes are contained in single chromosomes, which are usually circular DNA molecules. In contrast, the genomes of eukaryotes are composed of multiple chromosomes, each containing a linear molecule of DNA. Although the sizes and numbers of chromosomes vary considerably between different species (Table 6.4), their basic structure is the same in all eukaryotes. The DNA of eukaryotic cells is tightly bound to small basic proteins (histones) that package the DNA in

Animation 6.1

sites.sinauer.com/cooper7e/a6.1

Chromatin and Chromosomes In a eukaryotic cell, DNA is wrapped tightly around histone proteins (forming chromatin), and when a cell prepares for division, the chromatin coils upon itself multiple times to form compact chromosomes.

Table 6.4 Chromosome Numbers of Eukaryotic Cells

Organism	Genome size (Mb)[a]	Chromosome number[a]
Yeast (*Saccharomyces cerevisiae*)	12	16
Slime mold (*Dictyostelium*)	70	7
Arabidopsis thaliana	125	5
Corn	2200	10
Onion	15,000	8
Lily	50,000	12
Nematode (*Caenorhabditis elegans*)	97	6
Fruit fly (*Drosophila*)	180	4
Toad (*Xenopus laevis*)	3000	18
Lungfish	50,000	17
Chicken	1200	39
Mouse	3000	20
Cow	3000	30
Dog	3000	39
Human	3000	23

[a]Both genome size and chromosome number are for haploid cells. Mb = millions of base pairs.

an orderly way in the cell nucleus. This task is substantial, given the DNA content of most eukaryotes. For example, the total extended length of DNA in a human cell is nearly 2 meters, but this DNA must fit into a nucleus with a diameter of only 5 to 10 μm.

Chromatin

The complexes between eukaryotic DNA and proteins are called **chromatin**, which typically contains about twice as much protein as DNA. The major proteins of chromatin are the **histones**—small proteins containing a high proportion of basic amino acids (arginine and lysine) that facilitate binding to the negatively charged DNA molecule. There are five major types of histones—called H1, H2A, H2B, H3, and H4—which are very similar among different species of eukaryotes (Table 6.5). The histones are extremely abundant proteins in eukaryotic cells; together their mass is approximately equal to that of the cell's DNA. In addition, chromatin contains an approximately equal mass of a wide variety of nonhistone chromosomal proteins that are involved in a range of activities, including DNA replication and gene expression.

Table 6.5 The Major Histone Proteins

Histone	Molecular weight	Number of amino acids	Percentage lysine and arginine
H1	22,500	244	30.8
H2A	13,960	129	20.2
H2B	13,774	125	22.4
H3	15,273	135	22.9
H4	11,236	102	24.5

The basic structural unit of chromatin, the **nucleosome**, was described by Roger Kornberg in 1974 (**Figure 6.16**). Two types of experiments led to Kornberg's proposal of the nucleosome model. First, partial digestion of chromatin with micrococcal nuclease (an enzyme that degrades DNA) was found to yield DNA fragments approximately 200 base pairs long. In contrast, a similar digestion of naked DNA (not associated with proteins) yielded a continuous smear of randomly sized fragments. These results suggested that the binding of proteins to DNA in chromatin protects regions of the DNA from nuclease digestion, so that the enzyme can attack DNA only at sites separated by approximately 200 base pairs. Consistent with this notion, electron microscopy revealed that chromatin fibers have a beaded appearance, with the beads spaced at intervals of approximately 200 base pairs. Thus both the nuclease digestion and the electron microscopic studies suggested

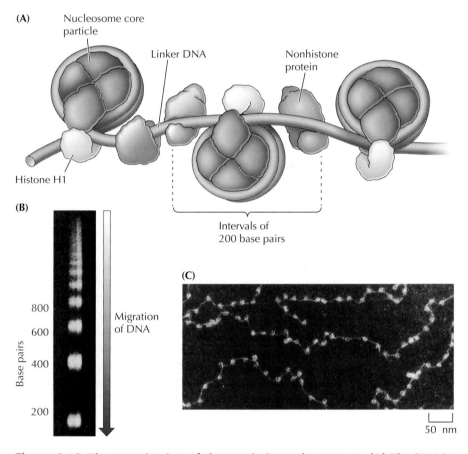

Figure 6.16 The organization of chromatin in nucleosomes (A) The DNA is wrapped around histones in nucleosome core particles and sealed by histone H1. Nonhistone proteins bind to the linker DNA between nucleosome core particles. (B) Gel electrophoresis of DNA fragments obtained by partial digestion of chromatin with micrococcal nuclease. The linker DNA between the nucleosome core particles is preferentially sensitive, so limited digestion of chromatin yields fragments corresponding to multiples of 200 base pairs. (C) An electron micrograph of an extended chromatin fiber, illustrating its beaded appearance. (B, courtesy of Roger Kornberg, Stanford University; C, courtesy of Ada L. Olins and Donald E. Olins, Oak Ridge National Laboratory.)

Video 6.1

sites.sinauer.com/cooper7e/v6.1
How DNA Is Packaged DNA is tightly coiled around proteins to form chromatin, which is then further looped and packaged, forming the chromosome.

(A)

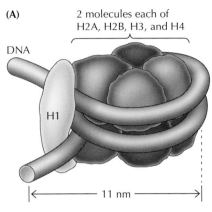

(B)

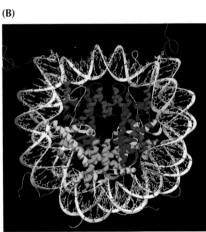

Figure 6.17 Structure of a chromatosome (A) A nucleosome core particle consists of 147 base pairs of DNA wrapped around a histone core consisting of two molecules each of H2A, H2B, H3, and H4. A chromatosome contains two full turns of DNA locked in place by one molecule of H1. (B) Model of the nucleosome core particle. The DNA backbones are shown in white. The histones are shown in red (H3), blue (H4), yellow (H2A), and green (H2B). (B, after C. A. Davey et al., 2002. *J. Mol. Biol.* 319: 1097.)

that chromatin is composed of repeating 200-base-pair units, which were called nucleosomes.

More extensive digestion of chromatin with micrococcal nuclease was found to yield particles (called **nucleosome core particles**) that correspond to the beads visible by electron microscopy. Detailed analysis of these particles has shown that they contain 147 base pairs of DNA wrapped 1.67 turns around a histone core consisting of two molecules each of H2A, H2B, H3, and H4 (the core histones) (**Figure 6.17**). One molecule of the fifth histone, H1, is bound to the DNA as it enters each nucleosome core particle. This forms a chromatin subunit known as a **chromatosome**, which consists of 166 base pairs of DNA wrapped around the histone core and held in place by H1 (a linker histone).

The packaging of DNA with histones yields a chromatin fiber approximately 10 nm in diameter that is composed of chromatosomes separated by linker DNA segments averaging about 50 base pairs in length (**Figure 6.18**). In the electron microscope, this 10-nm fiber has the beaded appearance

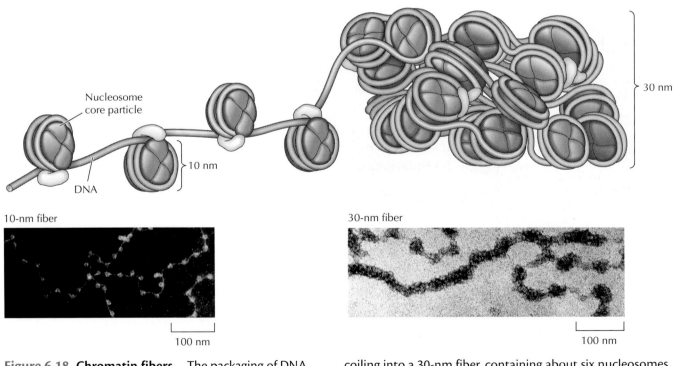

Figure 6.18 Chromatin fibers The packaging of DNA into nucleosomes yields a chromatin fiber approximately 10 nm in diameter. The chromatin is further condensed by coiling into a 30-nm fiber, containing about six nucleosomes per turn. (Photographs courtesy of Ada L. Olins and Donald E. Olins, Oak Ridge National Laboratory.)

that suggested the nucleosome model. Packaging of DNA into such a 10-nm chromatin fiber shortens its length approximately sixfold. The chromatin can then be further condensed by coiling into 30-nm fibers, resulting in a total condensation of about fiftyfold. Interactions between histone H1 molecules appear to play an important role in this stage of chromatin condensation, which is critical to determining the accessibility of chromosomal DNA for processes such as DNA replication and transcription. Folding of 30-nm fibers upon themselves can lead to further condensation of chromatin within the cell. Despite the importance of chromatin condensation, the structures of 30-nm fibers and more condensed chromatin states are not firmly established.

The extent of chromatin condensation varies during the life cycle of the cell and plays an important role in regulating gene expression, as will be discussed in Chapter 8. In interphase (nondividing) cells, most of the chromatin (called **euchromatin**) is relatively decondensed and distributed throughout the nucleus (**Figure 6.19**). During this period of the cell cycle, genes are transcribed and the DNA is replicated in preparation for cell division. Most of the euchromatin in interphase nuclei appears to be in the form of 30-nm, or somewhat more condensed 60- to 130-nm, chromatin fibers. Genes that are actively transcribed are in a more decondensed state that makes the DNA accessible to the transcription machinery. In contrast to euchromatin, about 10% of interphase chromatin (called **heterochromatin**) is in a highly condensed state that resembles the chromatin of cells undergoing mitosis. Heterochromatin is transcriptionally inactive and contains highly repeated DNA sequences, such as those present at centromeres and telomeres (described in the next

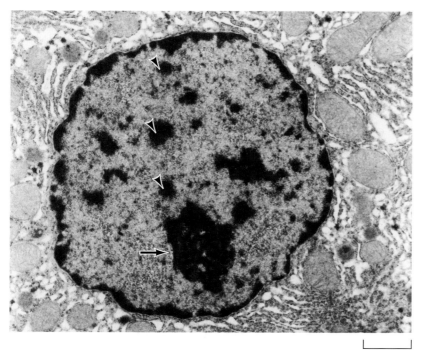

1 μm

Figure 6.19 Interphase chromatin   Electron micrograph of an interphase nucleus. The euchromatin is distributed throughout the nucleus. The heterochromatin is indicated by arrowheads and the nucleolus by an arrow. (Courtesy of Ada L. Olins and Donald E. Olins, Oak Ridge National Laboratory.)

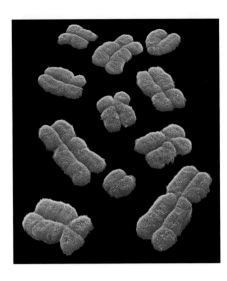

Figure 6.20 Chromatin condensation during mitosis Scanning electron micrograph of metaphase chromosomes. Artificial color has been added.

section) as well as genes that are not expressed in the particular type of cell examined (for example, inactive X chromosomes; see Figure 6.9).

As cells enter mitosis, their chromosomes become highly condensed so that they can be distributed to daughter cells. The chromatin in interphase nuclei is organized in loops, which are thought to fold upon themselves to form the compact metaphase chromosomes of mitotic cells in which the DNA has been condensed nearly ten-thousandfold (Figure 6.20). Such condensed chromatin can no longer be used as a template for RNA synthesis, so transcription ceases during mitosis. Electron micrographs indicate that the DNA in metaphase chromosomes is organized into large loops attached to a protein scaffold (Figure 6.21), but we currently understand neither the detailed structure of this highly condensed chromatin nor the mechanism of chromatin condensation.

Metaphase chromosomes are so highly condensed that their morphology can be studied using the light microscope (Figure 6.22). Several staining techniques yield characteristic patterns of alternating light and dark chromosome bands, which result from the preferential binding of stains or fluorescent dyes to AT-rich versus GC-rich DNA sequences. These bands are specific for each chromosome and appear to represent distinct chromosome regions. Genes can be localized to specific chromosome bands by *in situ* hybridization, indicating that the packaging of DNA into metaphase chromosomes is a highly ordered and reproducible process.

Centromeres

The **centromere** is a specialized region of the chromosome that plays a critical role in ensuring the correct distribution of duplicated chromosomes to daughter cells during mitosis (Figure 6.23). The cellular DNA is replicated

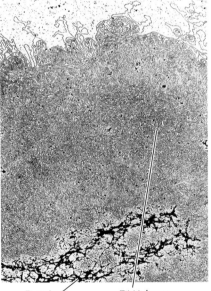

Protein scaffold   DNA loops

Figure 6.21 Structure of metaphase chromosomes An electron micrograph of DNA loops attached to the protein scaffold of metaphase chromosomes that have been depleted of histones. (From J. R. Paulson and U. K. Laemmli, 1977. *Cell* 12: 817.)

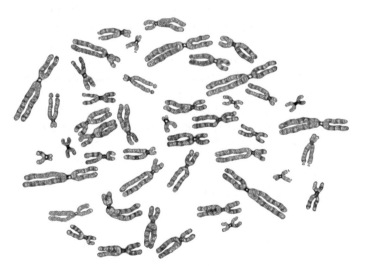

Figure 6.22 Human metaphase chromosomes A micrograph of human chromosomes spread from a metaphase cell.

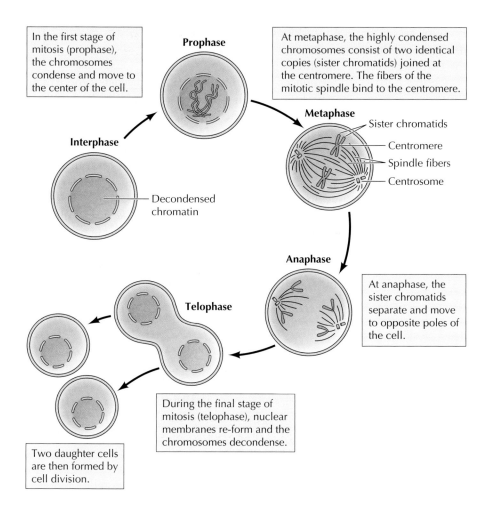

In the first stage of mitosis (prophase), the chromosomes condense and move to the center of the cell.

Prophase

Interphase

Decondensed chromatin

At metaphase, the highly condensed chromosomes consist of two identical copies (sister chromatids) joined at the centromere. The fibers of the mitotic spindle bind to the centromere.

Metaphase

Sister chromatids

Centromere

Spindle fibers

Centrosome

Anaphase

Telophase

At anaphase, the sister chromatids separate and move to opposite poles of the cell.

During the final stage of mitosis (telophase), nuclear membranes re-form and the chromosomes decondense.

Two daughter cells are then formed by cell division.

Figure 6.23 Chromosomes during mitosis   Since DNA replicates during interphase, the cell contains two identical duplicated copies of each chromosome prior to entering mitosis.

during interphase, resulting in the formation of two copies of each chromosome prior to the beginning of mitosis. As the cell enters mitosis, chromatin condensation leads to the formation of metaphase chromosomes consisting of two identical sister chromatids. These sister chromatids are held together at the centromere, which is seen as a constricted chromosomal region. As mitosis proceeds, microtubules of the mitotic spindle attach to the centromere, and the two sister chromatids separate and move to opposite poles of the spindle. At the end of mitosis, nuclear membranes re-form and the chromosomes decondense, resulting in the formation of daughter nuclei containing one copy of each parental chromosome.

The centromeres thus serve both as the sites where sister chromatids are joined and as the attachment sites for microtubules of the mitotic spindle. The proteins associated with centromeres form a specialized structure called the **kinetochore** (Figure 6.24). The binding of microtubules to kinetochore proteins mediates the attachment of chromosomes to the mitotic spindle.

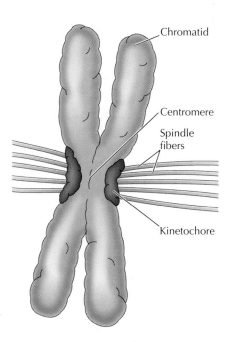

Chromatid

Centromere

Spindle fibers

Kinetochore

Figure 6.24 The centromere of a metaphase chromosome   The centromere is the region at which the two sister chromatids remain attached at metaphase. Specific proteins bind to centromeric DNA, forming the kinetochore, which is the site of spindle fiber attachment.

Proteins associated with the kinetochore then act as "molecular motors" that drive the movement of chromosomes along the spindle fibers, segregating the chromosomes to daughter nuclei.

Centromeric DNA sequences were initially defined in yeasts, where their function can be assayed by following the segregation of plasmids at mitosis (**Figure 6.25**). Plasmids that contain functional centromeres segregate like chromosomes and are equally distributed to daughter cells following mitosis. In the absence of a functional centromere, however, the plasmid does not segregate properly, and many daughter cells fail to inherit plasmid DNA. Assays of this type have enabled determination of the sequences required for centromere function. Such experiments first showed that the centromere sequences of the well-studied yeast *Saccharomyces cerevisiae* are contained in approximately 125 base pairs consisting of three sequence elements: two short sequences of 8 and 25 base pairs separated by 78 to 86 base pairs of very A/T-rich DNA (**Figure 6.26A**).

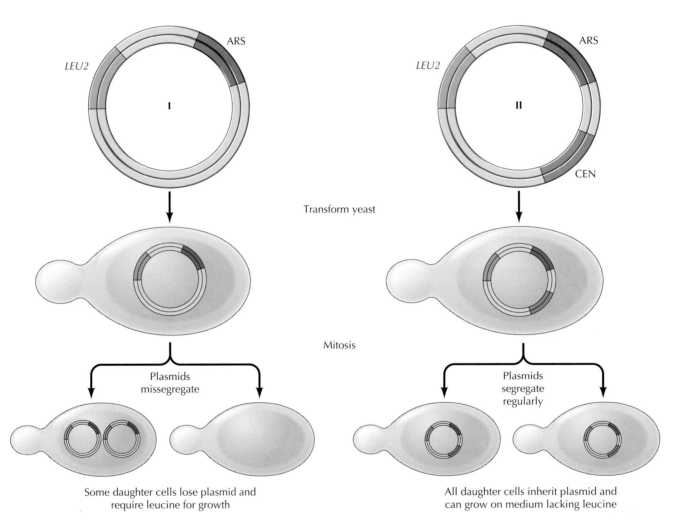

Figure 6.25 Assay of a centromere in yeast Both plasmids shown contain a selectable marker (*LEU2*) and DNA sequences that serve as origins of replication in yeast (ARS, which stands for autonomously replicating sequence). However, plasmid I lacks a centromere and is therefore frequently lost as a result of incorrect segregation during mitosis. In contrast, the presence of a centromere (CEN) in plasmid II ensures its regular transmission to daughter cells.

The short centromere sequences defined in *S. cerevisiae*, however, do not appear to reflect the situation in other eukaryotes, including the fission yeast *Schizosaccharomyces pombe*. Although *S. cerevisiae* and *S. pombe* are both yeasts, they appear to be as divergent from each other as either is from humans and are quite different in many aspects of their cell biology. These two yeast species thus provide complementary models for simple and easily studied eukaryotic cells. The centromeres of *S. pombe* span 40–100 kb of DNA; they are approximately a thousand times larger than those of *S. cerevisiae*. They consist of a central core of 4–7 kb of single-copy DNA flanked by repetitive sequences (**Figure 6.26B**). Not only the central core but also the flanking repeated sequences are required for centromere function, so the centromeres of *S. pombe* appear to be considerably more complex than those of *S. cerevisiae*.

Studies of a *Drosophila* chromosome provided the first characterization of a centromere in higher eukaryotes (**Figure 6.26C**). The *Drosophila* centromere spans 420 kb, most of which (more than 85%) consists of two highly repeated satellite DNAs with the sequences AATAT and AAGAG. The remainder of

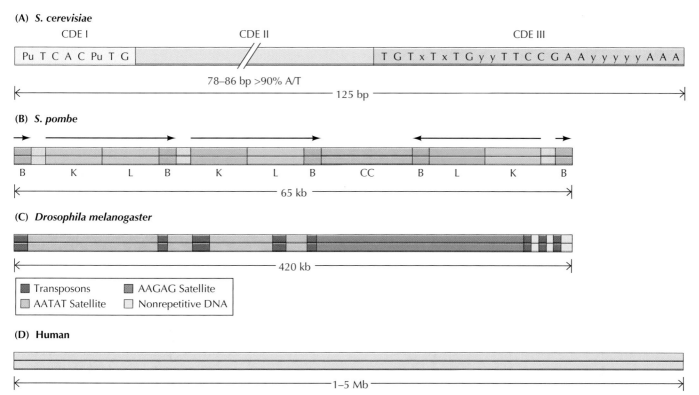

Figure 6.26 Centromeric DNA Sequences   (A) The *S. cerevisiae* centromere (CEN) sequences consist of two short conserved sequences (CDE I and CDE III) separated by 78 to 86 base pairs (bp) of A/T-rich DNA (CDE II). The sequences shown are consensus sequences derived from analysis of the centromere sequences of individual yeast chromosomes. Pu = A or G; x = A or T; y = any base. (B) The arrangement of sequences at the centromere of *S. pombe* chromosome II is illustrated. The centromere consists of a central core (CC) of unique-sequence DNA, flanked by tandem repeats of three repetitive sequence elements (B, K, and L). (C) A *Drosophila* centromere consists of two satellite sequences, transposable elements, and nonrepetitive DNA, spanning hundreds of kilobases (kb). (D) Human centromeres consist of tandem repeats of 171 base-pair A/T-rich α–satellite DNA, spanning 1 to 5 million base pairs (Mb).

the centromere consists of interspersed transposable elements (transposons), which are also found at other sites in the *Drosophila* genome, in addition to nonrepetitive regions of A/T-rich DNA.

Centromeres of other plants and animals are characterized by heterochromatin containing extensive arrays of highly repetitive sequences. In *Arabidopsis*, centromeres consist of 3 million base pairs of an A/T-rich 178-base pair satellite DNA. In humans and other primates the primary centromeric sequence is α-satellite DNA, which is a 171-base-pair A/T-rich sequence arranged in tandem repeats spanning 1–5 million base pairs (**Figure 6.26D**).

Despite extensive efforts, it has not been possible to identify specific DNA sequences that mediate centromere function in eukaryotes other than *S. cerevisiae*. On the other hand, it has been shown that the chromatin at centromeres has a unique structure. In particular, histone H3 is replaced in centromeric chromatin by an H3-like variant histone called **CENP-A**. CENP-A is uniformly present at the centromeres of all organisms that have been studied and CENP-A-containing nucleosomes are required for assembly of the other kinetochore proteins needed for centromere function. It thus appears that chromatin structure rather than a specific DNA sequence is the primary determinant of the identity and function of centromeres.

Importantly, the presence of a unique chromatin structure allows centromeres to be stably maintained at cell division, even in the absence of specific centromeric DNA sequences. This is an example of **epigenetic inheritance**—the transfer of information from parent to progeny that is *not* based on DNA sequence. In this and many other types of epigenetic inheritance (discussed in Chapter 8) the information is carried by histones (**Figure 6.27**). When chromosomal DNA replicates, the parental nucleosomes are distributed to the two progeny strands, so that CENP-A is

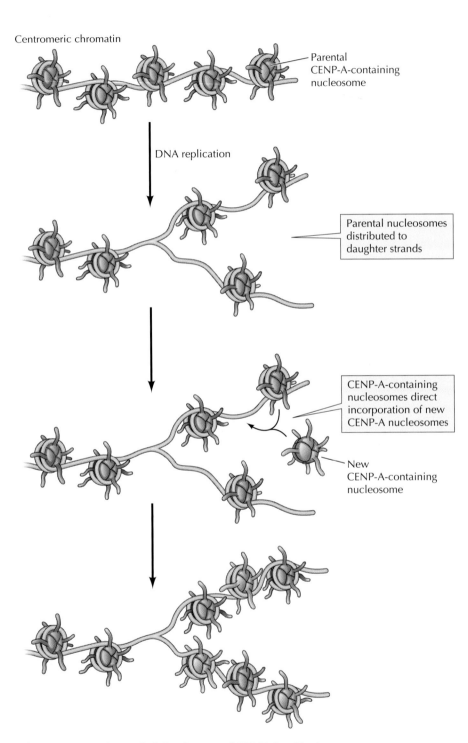

Centromeric chromatin

Parental CENP-A-containing nucleosome

DNA replication

Parental nucleosomes distributed to daughter strands

CENP-A-containing nucleosomes direct incorporation of new CENP-A nucleosomes

New CENP-A-containing nucleosome

Figure 6.27 Epigenetic inheritance of CENP-A Nucleosomes at centromeres contain the H3-like variant histone CENP-A. When DNA replicates, the parental CENP-A-containing nucleosomes are distributed to both progeny strands. These CENP-A-containing nucleosomes direct the incorporation of new CENP-A-containing nucleosomes, maintaining the chromatin structure of centromeres.

present in nucleosomes of the centromeres of both newly replicated chromosomes. These CENP-A-containing nucleosomes have been shown to direct the assembly of new CENP-A-containing nucleosomes into chromatin, so the chromatin structure of centromeres is maintained when cells divide.

Telomeres

The sequences at the ends of eukaryotic chromosomes, called **telomeres**, play critical roles in chromosome replication and maintenance. Telomeres were initially recognized as distinct structures because broken chromosomes were highly unstable in eukaryotic cells, implying that specific sequences are required at normal chromosomal termini. This was subsequently demonstrated by experiments in which telomeres from the protozoan *Tetrahymena* were added to the ends of linear molecules of yeast plasmid DNA. The addition of these telomeric DNA sequences allowed these plasmids to replicate as linear chromosome-like molecules in yeasts, demonstrating directly that telomeres are required for the replication of linear DNA molecules.

The telomere DNA sequences of a variety of eukaryotes are similar, consisting of repeats of a simple-sequence DNA containing clusters of G residues on one strand (Table 6.6). For example, the sequence of telomere repeats in humans and other mammals is TTAGGG, and the telomere repeat in *Tetrahymena* is TTGGGG. These sequences are repeated hundreds or thousands of times and terminate with a 3′ overhang of single-stranded DNA. The repeated sequences of telomere DNA of some organisms (including humans) form loops at the ends of chromosomes and bind a protein complex (shelterin) that protects the chromosome termini from degradation (Figure 6.28).

Telomeres play a critical role in replication of the ends of linear DNA molecules (see Chapter 7). DNA polymerase is able to extend a growing DNA chain but cannot initiate synthesis of a new chain at the terminus of a linear DNA molecule. Consequently, the ends of linear chromosomes cannot be replicated by the normal action of DNA polymerase. This problem has been solved by the evolution of a special enzyme, **telomerase**, which uses reverse transcriptase activity to replicate telomeric DNA sequences. Maintenance of telomeres appears to be an important factor in determining the lifespan and reproductive capacity of cells, so studies of telomeres and telomerase have the promise of providing new insights into conditions such as aging and cancer.

Table 6.6 Telomeric DNAs

Organism	Telomeric repeat sequence
Yeasts	
Saccharomyces cerevisiae	TG_{1-3}
Schizosaccharomyces pombe	$TTACG_{2-5}$
Protozoans	
Tetrahymena	TTGGGG
Dictyostelium	AG_{1-8}
Plant	
Arabidopsis	TTTAGGG
Mammal	
Human	TTAGGG

Cancer cells have high levels of telomerase, allowing them to maintain the ends of their chromosomes through indefinite divisions. Since normal somatic cells lack telomerase activity and do not divide indefinitely, drugs that inhibit telomerase are being developed as anti-cancer agents. (See Chapter 19.)

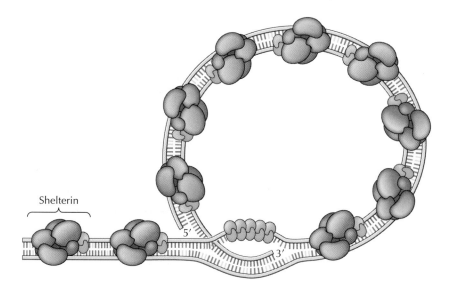

Shelterin

5′

3′

Figure 6.28 Structure of a telomere
Telomere DNA loops back on itself to form a circular structure and associates with a protein complex (shelterin) that protects the ends of chromosomes.

SUMMARY	**KEY TERMS**

The Structure of Eukaryotic Genes

- *Introns and exons:* Most eukaryotic genes have a split structure in which segments of coding sequence (exons) are interrupted by noncoding sequences (introns). In complex eukaryotes, introns account for more than ten times as much DNA as exons.

gene, exon, intron, RNA splicing, kilobase (kb), 5′ untranslated region (UTR), 3′ untranslated region (UTR)

- *Roles of introns:* Although they are removed from mRNAs, introns have multiple cellular functions. Some genes are encoded within introns of larger genes. In addition, introns contain regulatory sequences and allow alternative splicing.

nested gene, alternative splicing

Noncoding Sequences

- *Noncoding RNAs:* A variety of RNAs have functions other than encoding proteins. Some noncoding RNAs function in protein synthesis (tRNAs and rRNAs), mRNA splicing, and rRNA processing. In addition, miRNAs are important regulators of protein synthesis and a large number of lncRNAs have been identified, many of which function as regulators of gene expression.

RNA interference (RNAi), microRNA (miRNA), long noncoding RNA (lncRNA), X chromosome inactivation

- *Repetitive sequences:* Over 50% of mammalian DNA consists of highly repetitive DNA sequences, some of which are present in 10^5 to 10^6 copies per genome. These sequences include simple-sequence repeats as well as repetitive elements that have moved throughout the genome by either RNA or DNA intermediates.

simple-sequence repeat, satellite DNA, SINE, LINE, transposable element, retrotransposon, retrovirus-like element, DNA transposon

- *Gene duplication and pseudogenes:* Many eukaryotic genes are present in multiple copies, called gene families, which have arisen by duplication of ancestral genes. Some members of gene families function in different tissues or at different stages of development. Other members of gene families (pseudogenes) have been inactivated by mutations and no longer represent functional genes. Gene duplications can occur by duplication of DNA, ranging from a segment of DNA up to the entire genome. Approximately 5% of the human genome consists of duplicated DNA segments, whereas whole genome duplications are common in plants. Alternatively, gene duplication can occur by reverse transcription of an mRNA, giving rise to a processed pseudogene.

gene family, pseudogene, retrotransposition, processed pseudogene

Chromosomes and Chromatin

- *Chromatin:* The DNA of eukaryotic cells is wrapped around histones to form nucleosomes. Chromatin can be further compacted by the folding of nucleosomes into higher order structures, including the highly condensed metaphase chromosomes of cells undergoing mitosis. See Animation 6.1 and Video 6.1.

chromatin, histone, nucleosome, nucleosome core particle, chromatosome, euchromatin, heterochromatin

- *Centromeres:* Centromeres are specialized regions of eukaryotic chromosomes that serve as the sites where sister chromatids are joined and the sites of spindle fiber attachment during mitosis. Centromere function is determined by a variant H3-like histone, which is epigenetically maintained at cell division.

centromere, kinetochore, CENP-A, epigenetic inheritance

- *Telomeres:* Telomeres are specialized sequences required to maintain the ends of eukaryotic chromosomes.

telomere, telomerase

Questions

1. Many eukaryotic organisms have genome sizes that are much larger than their complexity would seem to require. Explain this paradox.

2. How were introns discovered during studies of adenovirus mRNAs?

3. How do intron sequences in the human genome increase the diversity of proteins expressed from the limited number of approximately 21,000 genes?

4. How can simple-sequence repetitive DNA be separated from the bulk of the nuclear DNA?

5. How do noncoding RNAs regulate gene expression?

6. How could you distinguish a processed pseudogene from a pseudogene formed by DNA duplication?

7. Yeast (*S. cerevisiae*) centromeres form a kinetochore that attaches to a single microtubule, whereas multiple microtubules are attached to the kinetochores of most animal cells. How does the structure of *S. cerevisiae* centromeres reflect this difference?

8. When circular plasmids are provided with a centromere sequence and inserted into yeast cells, they reproduce and segregate normally each cell division. However, if a linear chromosome is generated by cutting the plasmid at a single site with a restriction endonuclease, the plasmid genes are quickly lost from the yeast. Explain. What additional experiment could you perform to test your hypothesized explanation?

9. Approximately how many molecules of histone H1 are bound to yeast genomic DNA?

10. What is the average length of an intron in a human gene?

11. You have made a library in a plasmid vector containing complete human cDNAs. What is the expected average size of an insert?

12. Why do many species of plants have larger numbers of genes than humans?

Refer

The

Com
site
for
flashcards, a...
resources.

216   Chapter 6

Iyer, M. K. a
scape
hu
K

References and Further Reading   (Key review articles for each major section are highlighted in **bold**.)

The Structure of Eukaryotic Genes

Ben-Dov, C., B. Hartmann, J. Lundgren and J. Valcarel. 2008. Genome-wide analysis of alternative pre-mRNA splicing. *J. Biol. Chem.* 283: 1229–1233. [R]

Berget, S. M., C. Moore and P. A. Sharp. 1977. Spliced segments at the 5′ terminus of adenovirus 2 late mRNA. *Proc. Natl. Acad. Sci. USA* 74: 3171–3175. [P]

Chorev, M. and L. Carmel. 2012. The function of introns. *Front. Genet.* 3: 55. [R]

Chow, L. T., R. E. Gelinas, T. R. Broker and R. J. Roberts. 1977. An amazing sequence arrangement at the 5′ ends of adenovirus 2 messenger RNA. *Cell* 12: 1–8. [P]

International Human Genome Sequencing Consortium. 2004. Finishing the euchromatic sequence of the human genome. *Nature* 431: 931–945. [P]

Kumar, A. 2009. An overview of nested genes in eukaryotic genomes. *Eukaryotic Cell* 8: 1321–1329. [R]

Nilsen, T. W. and B. R. Graveley. 2010. Expansion of the eukaryotic proteome by alternative splicing. *Nature* 463: 457–463. [R]

The ENCODE Project Consortium. 2012. An integrated encyclopedia of DNA elements in the human genome. *Nature* 489: 57–74. [P]

Tilghman, S. M., P. J. Curtis, D. C. Tiemeier, P. Leder and C. Weissmann. 1978. The intervening sequence of a mouse β-globin gene is transcribed within the 15S β-globin mRNA precursor. *Proc. Natl. Acad. Sci. USA* 75: 1309–1313. [P]

Noncoding Sequences

Bartel, D. P. 2009. MicroRNAs: target recognition and regulatory functions. *Cell* 136: 215–233. [R]

Cordaux, R. and M. A. Batzer. 2009. The impact of retrotransposons on human genome evolution. *Nature Rev. Genet.* 10: 691–703. [R]

Fatica, A. and I. Bozzoni. 2014. Long noncoding RNAs: new players in cell differentiation and development. *Nature Rev. Genet.* 15: 7–21. [R]

Guttman, M, J. and 16 others. 2011. lincRNAs act in the circuitry controlling pluripotency and differentiation. *Nature* 477: 295–300. [P]

Inui, M., M. Graziano and S. Piccolo. 2010. MicroRNA control of signal transduction. *Nature Rev. Mol. Cell. Biol.* 11: 252–263. [R]

...d 18 others. 2015. The land-...f long noncoding RNAs in the ...man transcriptome. *Nature Genet.* 47: ...9–208. [P]

...zazian, H. H., Jr. 2004. Mobile elements: Drivers of genome evolution. *Science* 303: 1626–1632. [R]

Kugel, J. F. and J. A. Goodrich. 2011. Noncoding RNAs: key regulators of mammalian transcription. *Trends Biochem. Sci.* 37: 144–151. [R]

Lee, E. and 16 others. 2012. Landscape of somatic retrotransposition in human cancers. *Science* 337: 967–971. [P]

Nagano, T. and P. Fraser. 2011. No-nonsense functions for long noncoding RNAs. *Cell* 145: 178–181. [R]

Parker, H. G. and 16 others. 2009. An expressed *Fgf4* retrogene is associated with breed-defining chrondrodysplasia in domestic dogs. *Science* 325: 995–998. [P]

Pennisi, E. 2011. Green genomes. *Science* 332: 1373–1375. [R]

Shukla, R. and 22 others. 2013. Endogenous retrotransposition activates oncogenic pathways in hepatocellular carcinoma. *Cell* 153: 101–111. [P]

Sutter, N. B. and 20 others. 2007. A single *IGF1* allele is a major determinant of small size in dogs. *Science* 316: 112–115. [P]

The ENCODE Project Consortium. 2012. An integrated encyclopedia of DNA elements in the human genome. *Nature* 489: 57–74. [P]

Ulitsky, I. and D. P. Bartel. 2013. lincRNAs: genomics, evolution, and mechanisms. *Cell* 154: 26–46. [R]

Wang, K. C. and H. Y. Chang. 2011. Molecular mechanisms of long noncoding RNAs. *Mol. Cell* 43: 904–914. [R]

Yang, L., J. E. Froberg and J. T. Lee. 2014. Long noncoding RNAs: fresh perspectives into the RNA world. *Trends Biochem. Sci.* 39: 35-43. [R]

Zhang, Z. and M. Gerstein. 2004. Large-scale analysis of pseudogenes in the human genome. *Curr. Opin. Genet. Dev.* 14: 328–335. [R]

Chromosomes and Chromatin

Allshire, R. C. and G. H. Karpen. 2008. Epigenetic regulation of centromeric chromatin: old dogs, new tricks? *Nature Rev. Genet.* 9: 923–937. [R]

Black, B. E. and E. A. Bassett. 2008. The histone variant CENP-A and centromere specification. *Curr. Opin. Cell Biol.* 20: 91–100 [R]

Blackburn, E. H. 2005. Telomeres and telomerase: Their mechanisms of action and the effects of altering their functions. *FEBS Letters* 579: 859–862. [R]

Blasco, M. A. 2007. Telomere length, stem cells and aging. *Nature Chem. Biol.* 3: 640–649. [R]

Espinoza, C. A. and B. Ren. 2011. Mapping higher order structure of chromatin domains. *Nature Genet.* 43: 615–616. [R]

Gilson, E. and V. Geli. 2007. How telomeres are replicated. *Nature Rev. Mol. Cell Biol.* 8: 825–838. [R]

Jain, D. and J. P. Cooper. 2010. Telomeric strategies: means to an end. *Ann. Rev. Genet.* 44: 243–269. [R]

Kornberg, R. D. 1974. Chromatin structure: A repeating unit of histones and DNA. *Science* 184: 868–871. [P]

Lampert, F. and S. Westermann. 2011. A blueprint for kinetochores—new insights into the molecular mechanics of cell division. *Nature Rev. Mol. Cell Biol.* 12: 407–412. [R]

Luger, K., A. W. Mader, R. K. Richmond, D. F. Sargent and T. J. Richmond. 1997. Crystal structure of the nucleosome core particle at 2.8 Å resolution. *Nature* 389: 251–260. [P]

Morris, C. A. and D. Moazed. 2007. Centromere assembly and propagation. *Cell* 128: 647–650. [R]

Ohta, S., L. Wood, J.-C. Bukowski-Wills, J. Rappsilber and W. C. Earnshaw. 2011. Building mitotic chromosomes. *Curr. Opin. Cell Biol.* 23: 114–121. [R]

O'Sullivan, R. J. and J. Karlseder. 2010. Telomeres: protecting chromosomes against genomic instability. *Nature Rev. Mol. Cell Biol.* 11: 171–181. [R]

Paulson, J. R. and U. K. Laemmli. 1977. The structure of histone-depleted metaphase chromosomes. *Cell* 12: 817–828. [P]

Tremethick, D. J. 2007. Higher-order structures of chromatin: the elusive 30 nm fiber. *Cell* 128: 651–654. [R]

Van Holde, K. and J. Zlatanova. 2007. Chromatin fiber structure: where is the problem now? *Sem. Cell Dev. Biol.* 18: 651–658. [R]

Verdaasdonk, J. S. and K. Bloom. 2011. Centromeres: unique chromatin structures that drive chromosome segregation. *Nature Rev. Mol. Cell Biol.* 12: 320–332. [R]

CHAPTER 7

Replication, Maintenance, and Rearrangements of Genomic DNA

The fundamental biological process of reproduction requires the faithful transmission of genetic information from parent to offspring. Thus, the accurate replication of genomic DNA is essential to the lives of all cells and organisms. Each time a cell divides, its entire genome must be duplicated, and complex enzymatic machinery is required to copy the large DNA molecules that make up both prokaryotic and eukaryotic chromosomes. In addition, cells have evolved mechanisms to correct mistakes that sometimes occur during DNA replication and to repair DNA damage that can result from the action of environmental agents, such as radiation. Abnormalities in these processes result in a failure of accurate replication and maintenance of genomic DNA—a failure that can have disastrous consequences, such as the development of cancer.

Despite the importance of accurate DNA replication and maintenance, cell genomes are far from static. In order for species to evolve, mutations and gene rearrangements are needed to maintain genetic variation between individuals. Rearrangements of DNA sequences within the genome are also thought to contribute to evolution by creating novel combinations of genetic information. In addition, some DNA rearrangements are programmed to regulate gene expression during the differentiation and development of individual cells and organisms. In humans, a prominent example is the rearrangement of antibody genes during development of the immune system. A careful balance between maintenance and variation of genetic information is thus critical to the development of individual organisms as well as the evolution of a species.

DNA Replication

As discussed in Chapter 4, DNA replication is a semiconservative process in which each parental strand serves as a template for the synthesis of a new complementary daughter strand. The central enzyme involved is DNA polymerase, which catalyzes the joining of deoxyribonucleotide triphosphates (dNTPs) to form the growing DNA chain. However, DNA replication is much more complex than a single enzymatic reaction. Other proteins are involved, and proofreading mechanisms are required to ensure that the accuracy of replication is compatible with the low frequency of errors that is needed for cell reproduction. Additional proteins and specific DNA sequences are also needed, both to initiate replication and to copy the ends of eukaryotic chromosomes.

DNA polymerases

DNA polymerase was first identified in lysates of *E. coli* by Arthur Kornberg in 1956. The ability of this enzyme to accurately copy a DNA template

Video 7.1

sites.sinauer.com/cooper7e/v7.1

Mechanism of DNA Replication
In DNA replication, both strands of the double helix act as templates for the formation of new DNA molecules, using two different mechanisms.

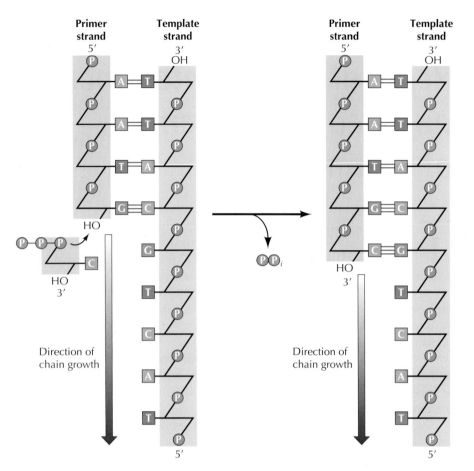

Figure 7.1 The reaction catalyzed by DNA polymerase All known DNA polymerases add a deoxyribonucleotide triphosphate to the 3' hydroxyl group of a growing DNA chain (the primer strand).

provided a biochemical basis for the mode of DNA replication that was initially proposed by Watson and Crick, so its isolation represented a landmark discovery in molecular biology. Ironically, the first DNA polymerase to be identified (now called DNA polymerase I) is not the major enzyme responsible for *E. coli* DNA replication. Instead, polymerase I is principally involved in repair of damaged DNA, and it is now clear that both prokaryotic and eukaryotic cells contain multiple different DNA polymerases that play distinct roles in DNA replication and repair. In bacteria, DNA polymerase III is the major polymerase responsible for DNA replication. Eukaryotic cells contain three DNA polymerases (α, δ, and ε) that function in replication of nuclear DNA. A distinct DNA polymerase (γ) is localized to mitochondria and is responsible for replication of mitochondrial DNA.

All known DNA polymerases share two fundamental properties that have critical implications for DNA replication (**Figure 7.1**). First, all polymerases synthesize DNA only in the 5' to 3' direction, adding a dNTP to the 3' hydroxyl group of a growing chain. Second, DNA polymerases can add a new deoxyribonucleotide only to a preformed primer strand that is hydrogen-bonded to the template; they are not able to initiate DNA synthesis *de novo* by catalyzing the polymerization of free dNTPs. In this respect, DNA polymerases differ from RNA polymerases, which can initiate the synthesis of a new strand of RNA in the absence of a primer. As discussed later in this chapter, these properties of DNA polymerases appear critical for maintaining the high fidelity of DNA replication that is required for cell reproduction.

The replication fork

DNA molecules in the process of replication were first analyzed by John Cairns in experiments which *E. coli* were grown in the presence of radioactive thymidine, which allowed subsequent visualization of newly replicated DNA by autoradiography (**Figure 7.2**). In some cases, complete circular molecules undergoing replication could be observed. These DNA molecules contained two **replication forks**, representing the regions of active DNA synthesis. At each fork the parental strands of DNA separated and were used as templates for the synthesis of two new daughter strands.

The synthesis of new DNA strands complementary to both strands of the parental molecule posed an important problem to understanding the biochemistry of DNA replication. Since the two strands of double-helical DNA run in opposite (antiparallel) directions, continuous synthesis of two new strands at the replication fork would require that one strand be synthesized in the 5′ to 3′ direction while the other was synthesized in the opposite (3′ to 5′) direction. But DNA polymerase catalyzes the polymerization of dNTPs only in the 5′ to 3′ direction. How, then, can the other progeny strand of DNA be synthesized?

This enigma was resolved by experiments showing that only *one* strand of DNA is synthesized in a continuous manner in the direction of overall DNA replication; the other is formed from short (1–3 kb), discontinuous pieces of DNA that are synthesized backward with respect to the direction of movement of the replication fork (**Figure 7.3**). These small pieces of newly synthesized DNA (called **Okazaki fragments** after

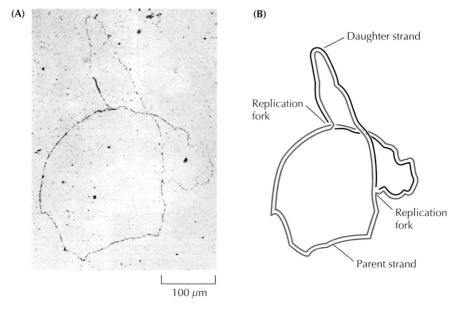

(A)

(B)

100 μm

Figure 7.2 Replication of *E. coli* DNA   (A) An autoradiograph showing bacteria that were grown in [³H]thymidine for two generations to label the DNA, which was then extracted and visualized by exposure to photographic film. (B) This schematic illustrates the two replication forks shown in (A). (From J. Cairns, 1963. *Cold Spring Harbor Symp. Quant. Biol.* 28: 43.)

their discoverer, the Japanese biochemist Reiji Okazaki) are joined by the action of **DNA ligase**, forming a new intact DNA strand. The continuously synthesized strand is called the **leading strand**, since its elongation in the

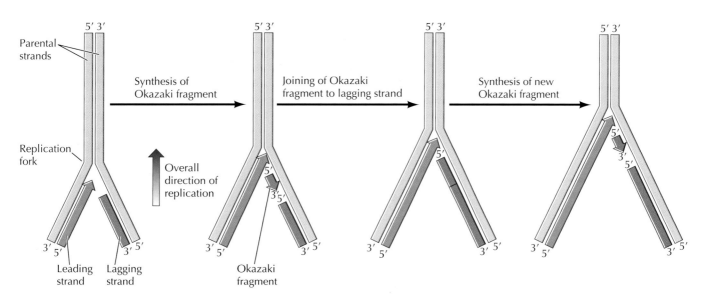

Figure 7.3 Synthesis of leading and lagging strands of DNA   The leading strand is synthesized continuously in the direction of replication fork movement. The lagging strand is synthesized in small pieces (Okazaki fragments) backward from the overall direction of replication. The Okazaki fragments are then joined by the action of DNA ligase.

direction of replication fork movement exposes the template used for the synthesis of Okazaki fragments (the **lagging strand**).

Although the discovery of discontinuous synthesis of the lagging strand provided a mechanism for the elongation of both strands of DNA at the replication fork, it raised another question: Since DNA polymerase requires a primer and cannot initiate synthesis *de novo*, how is the synthesis of Okazaki fragments initiated? The answer is that short fragments of RNA serve as primers for DNA replication (**Figure 7.4**). In contrast to DNA synthesis, the synthesis of RNA can initiate *de novo*, and an enzyme called **primase** synthesizes short fragments of RNA (e.g., three to ten nucleotides long) complementary to the lagging strand template at the replication fork. Okazaki fragments are then synthesized via extension of these RNA primers by DNA polymerase. An important consequence of such RNA priming is that newly synthesized Okazaki fragments contain an RNA-DNA joint, the discovery of which provided critical evidence for the role of RNA primers in DNA replication.

To form a continuous lagging strand of DNA, the RNA primers must eventually be removed from the Okazaki fragments and replaced with DNA (**Figure 7.5**). In prokaryotes, RNA primers are removed by the action of polymerase I. In addition to its DNA polymerase activity, DNA polymerase I acts as an **exonuclease** that can hydrolyze DNA (or RNA) in either the 3' to 5' or 5' to 3' direction. The action of DNA polymerase I as a 5' to 3' exonuclease removes ribonucleotides from the 5' ends of Okazaki fragments, allowing them to be replaced with deoxyribonucleotides to yield fragments consisting entirely of DNA. In eukaryotic cells, RNA primers are removed by the combined action of **RNase H**, an enzyme that degrades the RNA strand of RNA-DNA hybrids, and 5' to 3' exonucleases. The resulting gaps are then

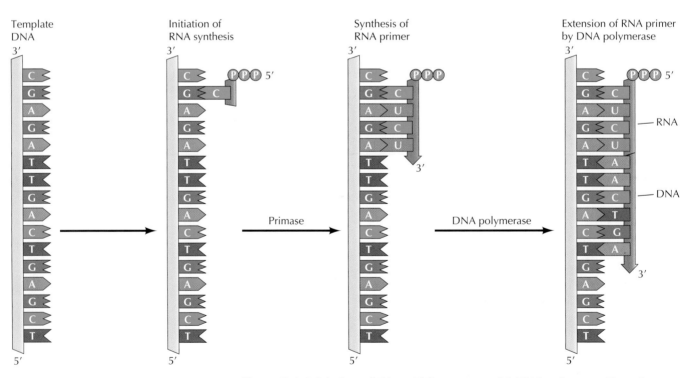

Figure 7.4 Initiation of Okazaki fragments with RNA primers   Short fragments of RNA serve as primers that can be extended by DNA polymerase.

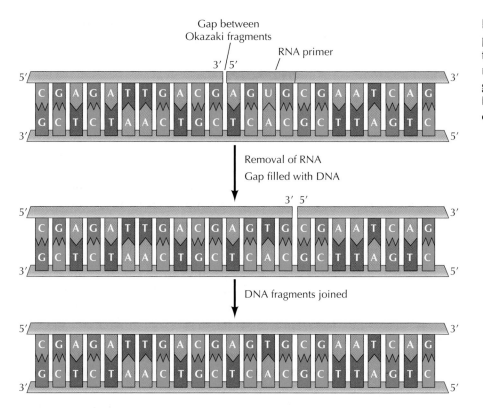

Figure 7.5 Removal of RNA primers and joining of Okazaki fragments   RNA primers are removed and DNA polymerase fills the gaps between Okazaki fragments with DNA. The resultant DNA fragments can then be joined by DNA ligase.

filled by DNA polymerase δ and the DNA fragments are joined by DNA ligase, yielding an intact lagging strand.

As noted earlier, different DNA polymerases play distinct roles at the replication fork in both prokaryotic and eukaryotic cells (**Figure 7.6**). In *E. coli*, polymerase III is the major replicative polymerase, functioning in the synthesis both of the leading strand of DNA and of Okazaki fragments

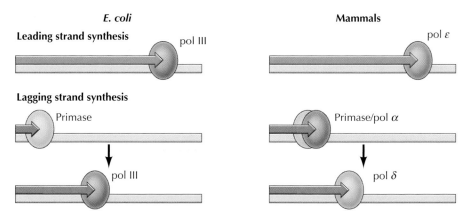

Figure 7.6 Roles of DNA polymerases in *E. coli* and mammalian cells   The leading strand is synthesized by polymerase III (pol III) in *E. coli* and by polymerase ε (pol ε) in mammalian cells. In *E. coli*, lagging strand synthesis is initiated by primase, and RNA primers (red) are extended by polymerase III. In mammalian cells, lagging strand synthesis is initiated by a complex of primase with polymerase α (pol α). The short RNA-DNA fragments synthesized by this complex are then extended by polymerase δ.

by the extension of RNA primers. In eukaryotic cells, three distinct DNA polymerases (α, δ, and ε) are involved in the replication of nuclear DNA. The roles of these DNA polymerases have been studied in two types of experiments. First, replication of the DNA of some animal viruses, such as SV40, can be studied in cell-free extracts, allowing direct biochemical analysis of the activities of different DNA polymerases as well as other proteins involved in DNA replication. Second, polymerases α, δ, and ε are found in yeasts as well as in mammalian cells, enabling the use of the powerful approaches of yeast genetics (see Chapter 4) to test their biological roles. Biochemical analysis has established that polymerase α is found in a complex with primase, and functions in conjunction with primase to synthesize short RNA-DNA fragments during lagging strand synthesis and at origins of replication. Polymerases δ and ε then act as the major replicative polymerases and are thought to be responsible for synthesis of the lagging and leading strands, respectively.

Not only polymerases and primase but also a number of other proteins act at the replication fork. These additional proteins have been identified both by the analysis of *E. coli* and yeast mutants defective in DNA replication and by the purification of the mammalian proteins required for *in vitro* replication of SV40 DNA. One class of proteins required for replication binds to DNA polymerases, increasing the activity of the polymerases and causing them to remain bound to the template DNA so that they continue synthesis of a new DNA strand. Both *E. coli* polymerase III and eukaryotic polymerases δ and ε are associated with sliding-clamp proteins (proliferating cell nuclear antigen [PCNA] in eukaryotes) that load the polymerase onto the primer and maintain its stable association with the template (**Figure 7.7**). The sliding-clamp proteins are themselves loaded onto DNA at the junction between the primer and template by clamp-loading proteins (replication factor C [RFC] in eukaryotes). The clamp-loading proteins use energy derived from ATP hydrolysis to open the sliding clamps. The clamp

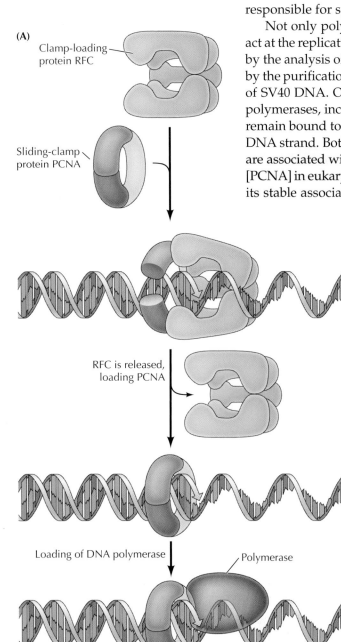

(A)

Clamp-loading protein RFC

Sliding-clamp protein PCNA

RFC is released, loading PCNA

Loading of DNA polymerase

Polymerase

Figure 7.7 Polymerase accessory proteins
(A) A complex of the clamp-loading and sliding-clamp proteins (RFC and PCNA in mammalian cells, respectively) binds DNA at the junction between primer and template. RFC is then released, loading PCNA onto the DNA. DNA polymerase then binds to PCNA. (B) Model of PCNA bound to DNA. (B, from T. S. Krishna et al., 1994. *Cell* 79: 1233.)

(B)

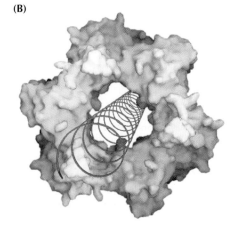

Figure 7.8 Action of helicases and single-stranded DNA-binding proteins Helicases unwind the two strands of parental DNA ahead of the replication fork. The unwound DNA strands are then stabilized by single-stranded DNA-binding proteins so that they can serve as templates for new DNA synthesis.

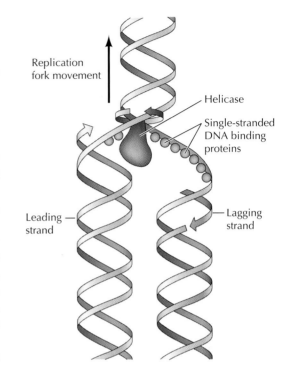

Replication fork movement

Helicase

Single-stranded DNA binding proteins

Leading strand

Lagging strand

loader then releases the sliding clamp protein, which forms a ring around the template DNA. The sliding-clamp protein then loads the DNA polymerase onto DNA at the primer-template junction. The ring formed by the sliding clamp maintains the association of the polymerase with its template as replication proceeds, allowing the uninterrupted synthesis of many thousands of nucleotides of DNA.

Other proteins unwind the template DNA and stabilize single-stranded regions (Figure 7.8). **Helicases** are enzymes that catalyze the unwinding of parental DNA, coupled to the hydrolysis of ATP, ahead of the replication fork. **Single-stranded DNA-binding proteins** (e.g., eukaryotic replication protein A [RPA]) then stabilize the unwound template DNA, keeping it in an extended single-stranded state so that it can be copied by the DNA polymerase.

As the strands of parental DNA unwind, the DNA ahead of the replication fork is forced to rotate. Unchecked, this rotation would cause circular DNA molecules (such as SV40 DNA or the *E. coli* chromosome) to become twisted around themselves, eventually blocking replication (Figure 7.9). This problem is solved by **topoisomerases**, enzymes that catalyze the reversible breakage and rejoining of DNA

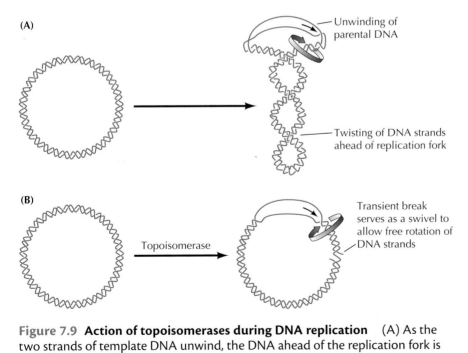

(A)

Unwinding of parental DNA

Twisting of DNA strands ahead of replication fork

(B)

Topoisomerase

Transient break serves as a swivel to allow free rotation of DNA strands

Figure 7.9 Action of topoisomerases during DNA replication (A) As the two strands of template DNA unwind, the DNA ahead of the replication fork is forced to rotate in the opposite direction, causing circular molecules to become twisted around themselves. (B) This problem is solved by topoisomerases, which catalyze the reversible breakage and rejoining of DNA strands. The transient breaks introduced by these enzymes serve as swivels that allow the two strands of DNA to rotate freely around each other.

FYI

An inhibitor of type II topoisomerase, a drug called etoposide, is used as a chemotherapeutic agent for the treatment of several types of cancer (see Chapter 19).

strands. There are two types of these enzymes: Type I topoisomerases break just one strand of DNA; type II topoisomerases introduce simultaneous breaks in both strands. The breaks introduced by type I and type II topoisomerases serve as "swivels" that allow the two strands of template DNA to rotate freely around each other so that replication can proceed without twisting the DNA ahead of the fork (see Figure 7.9). Although eukaryotic chromosomes are composed of linear rather than circular DNA molecules, their replication also requires topoisomerases; otherwise, the complete chromosomes would have to rotate continually during DNA synthesis.

Type II topoisomerase is needed not only to unwind DNA but also to unravel newly replicated DNA molecules that become intertwined with each other. In eukaryotic cells, topoisomerase II appears to be involved in mitotic chromosome condensation. In addition, studies of yeast mutants, as well as experiments in *Drosophila* and mammalian cells, indicate that topoisomerase II is required for the separation of daughter chromatids at mitosis, suggesting that it functions to untangle newly replicated loops of DNA in the chromosomes of eukaryotes.

The enzymes involved in DNA replication act in a coordinated manner to synthesize both leading and lagging strands of DNA simultaneously at the replication fork (Figure 7.10). This task is accomplished by the association of pairs of the replicative DNA polymerases with subunits of the clamp-loading protein, which is also associated with helicase at the fork. One molecule of DNA polymerase then acts in synthesis of the leading strand while the other acts in synthesis of the lagging strand. The lagging-strand template is folded at the replication fork so that the polymerase subunit engaged in lagging-strand synthesis moves in the same overall direction as the other subunit, which is synthesizing the leading strand. When the lagging-strand polymerase reaches the end of an Okazaki fragment, it dissociates from DNA. Because it is bound to the clamp loader, it can then be efficiently reloaded onto a new sliding clamp at the primer-template junction to initiate synthesis of another Okazaki fragment.

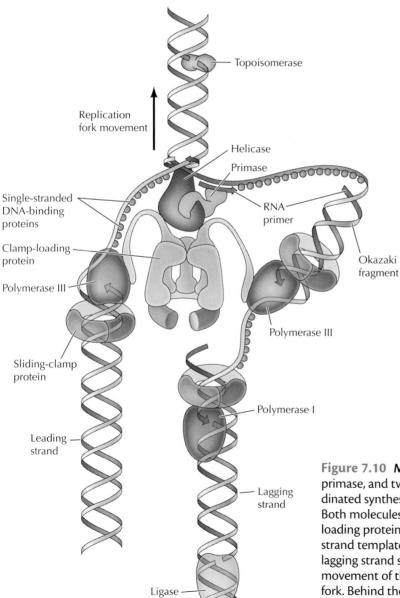

Figure 7.10 Model of the *E. coli* replication fork Helicase, primase, and two molecules of DNA polymerase III carry out coordinated synthesis of both the leading and lagging strands of DNA. Both molecules of DNA polymerase are associated with the clamp-loading protein, which is bound to helicase at the fork. The lagging strand template is folded so that the polymerase responsible for lagging strand synthesis moves in the same direction as overall movement of the fork. Topoisomerase acts as a swivel ahead of the fork. Behind the fork, RNA primers are removed by DNA polymerase I and Okazaki fragments are joined by DNA ligase.

In eukaryotic cells, nucleosomes are also disrupted during DNA replication. The nucleosomes of the parental chromatin are divided between the two daughter strands of DNA, and new histones are then added to reassemble nucleosomes by additional proteins (chromatin assembly factors) that travel along with the replication fork.

The fidelity of replication

The accuracy of DNA replication is critical to cell reproduction, and estimates of mutation rates for a variety of genes indicate that the frequency of errors during replication corresponds to less than one incorrect base per 10^9 nucleotides incorporated. This error frequency is much lower than would be predicted simply on the basis of complementary base pairing. In particular, the free energy differences resulting from the changes in hydrogen bonding between correctly matched and mismatched bases are only large enough to favor the formation of correctly matched base pairs by about a hundredfold. Consequently, base selection determined simply by hydrogen bonding between complementary bases would result in an error frequency corresponding to the incorporation of one incorrect base in approximately every 100 nucleotides of newly synthesized DNA. The much higher degree of fidelity actually achieved results largely from the activities of DNA polymerase.

One mechanism by which DNA polymerase increases the fidelity of replication is by helping to select the correct base for insertion into newly synthesized DNA. The polymerase does not simply catalyze incorporation of whatever nucleotide is hydrogen bonded to the template strand. Instead, it actively discriminates against incorporation of a mismatched base by adapting to the conformation of a correct base pair. In particular, recent structural studies of several DNA polymerases indicate that the binding of correctly matched dNTPs induces conformational changes in DNA polymerase that lead to the incorporation of the nucleotide into DNA. This ability of DNA polymerase to select matched nucleotides for incorporation appears to increase the accuracy of replication about a thousandfold, reducing the expected error frequency to approximately 10^{-5}.

The other major mechanism responsible for the accuracy of DNA replication is the **proofreading** activity of DNA polymerase. The replicative DNA polymerases (eukaryotic polymerases δ and ε and *E. coli* polymerase III) have an exonuclease activity that can hydrolyze DNA in the 3' to 5' direction. This 3' to 5' exonuclease operates in the reverse direction of DNA synthesis, and participates in proofreading newly synthesized DNA (**Figure 7.11**). Proofreading is effective because DNA polymerase requires a primer and is not able to initiate synthesis *de novo*. Primers that are hydrogen bonded to the template are preferentially used, so when an incorrect base is incorporated, it is likely to be removed by the 3' to 5' exonuclease activity rather than being used to continue synthesis. The 3' to 5' exonucleases of the replicative DNA

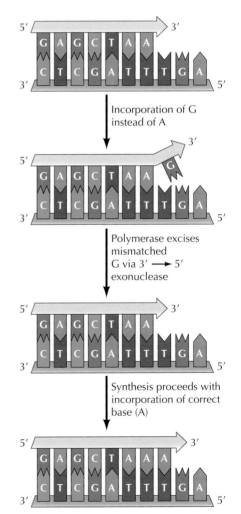

Figure 7.11 Proofreading by DNA polymerase   G is incorporated in place of A as a result of mispairing with T on the template strand. Because it is mispaired, the 3' terminal G is not hydrogen bonded to the template strand. This mismatch at the 3' terminus of the growing chain is recognized and excised by the 3' to 5' exonuclease activity of DNA polymerase, which requires a primer hydrogen bonded to the template strand in order to continue synthesis. Following excision of the mismatched G, DNA synthesis can proceed with incorporation of the correct base (A).

polymerases selectively excise mismatched bases that have been incorporated at the end of a growing DNA chain, thereby increasing the accuracy of replication by a hundred- to a thousandfold.

The importance of proofreading may explain the fact that DNA polymerases require primers and catalyze the growth of DNA strands only in the 5′ to 3′ direction. When DNA is synthesized in the 5′ to 3′ direction, the energy required for polymerization is derived from hydrolysis of the 5′ triphosphate group of a free dNTP as it is added to the 3′ hydroxyl group of the growing chain (see Figure 7.1). If DNA were to be extended in the 3′ to 5′ direction, the energy of polymerization would instead have to be derived from hydrolysis of the 5′ triphosphate group of the terminal nucleotide already incorporated into DNA. This arrangement would eliminate the possibility of proofreading, because removal of a mismatched terminal nucleotide would also remove the 5′ triphosphate group needed as an energy source for further chain elongation. Thus, although the ability of DNA polymerase to extend a primer only in the 5′ to 3′ direction appears to make replication a complicated process, it is necessary for ensuring accurate duplication of the genetic material.

Combined with the ability to discriminate against the insertion of mismatched bases, the proofreading activity of DNA polymerases is sufficient to reduce the error frequency of replication to about one mismatched base per 10^7 to 10^8 nucleotides. Additional mechanisms (discussed in the section "DNA Repair") act to remove mismatched bases that have been incorporated into newly synthesized DNA, further reducing the error rate to less than 10^{-9}.

Origins and the initiation of replication

The replication of both prokaryotic and eukaryotic DNAs starts at sites called **origins of replication**, which serve as binding sites for proteins that initiate the replication process. The first origin to be defined was that of *E. coli*, in which genetic analysis indicated that replication always begins at a unique site on the bacterial chromosome. The *E. coli* origin has since been studied in detail and found to consist of 245 base pairs of DNA, elements of which serve as binding sites for proteins required to initiate DNA replication (**Figure 7.12**). The key step is the binding of an initiator protein to specific DNA sequences

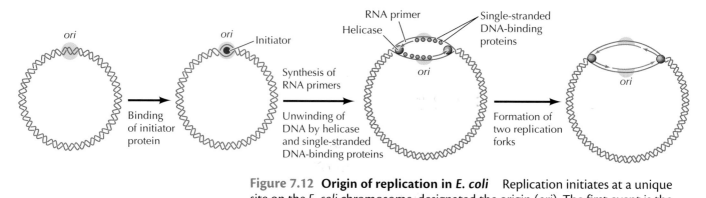

Figure 7.12 Origin of replication in *E. coli* Replication initiates at a unique site on the *E. coli* chromosome, designated the origin (*ori*). The first event is the binding of an initiator protein to *ori* DNA, which leads to partial unwinding of the template. The DNA continues to unwind by the actions of helicase and single-stranded DNA-binding proteins, and RNA primers are synthesized by primase. The two replication forks formed at the origin then move in opposite directions along the circular DNA molecule.

within the origin. The initiator protein begins to unwind the origin DNA and recruits the other proteins involved in DNA synthesis. Helicase and single-stranded DNA-binding proteins then act to continue unwinding and exposing the template DNA, and primase initiates the synthesis of leading strands. Two replication forks are formed and move in opposite directions along the circular *E. coli* chromosome.

The origins of replication of several animal viruses, such as SV40, have been studied as models for the initiation of DNA synthesis in eukaryotes. SV40 has a single origin of replication (consisting of 64 base pairs) that functions both in infected cells and in cell-free systems. Replication is initiated by a virus-encoded protein (called T antigen) that binds to the origin and also acts as a helicase. A single-stranded DNA-binding protein is required to stabilize the unwound template, and the DNA polymerase α-primase complex then initiates DNA synthesis.

Although single origins are sufficient to direct the replication of bacterial and viral genomes, multiple origins are needed to replicate the much larger genomes of eukaryotic cells within a reasonable period of time. For example, the entire genome of *E. coli* (4×10^6 base pairs) is replicated from a single origin in approximately 30 minutes. If mammalian genomes (3×10^9 base pairs) were replicated from a single origin at the same rate, DNA replication would require about 3 weeks (30,000 minutes). The problem is further exacerbated by the fact that the rate of DNA replication in mammalian cells is actually about tenfold lower than in *E. coli*, probably as a result of the packaging of eukaryotic DNA in chromatin. Nonetheless, the genomes of mammalian cells are typically replicated within a few hours, necessitating the use of thousands of replication origins.

The presence of multiple replication origins in eukaryotic cells was first demonstrated by the exposure of cultured mammalian cells to radioactive thymidine for different time intervals, followed by autoradiography to detect newly synthesized DNA. The results of such studies indicated that DNA synthesis is initiated at multiple sites, from which it then proceeds in both directions along the chromosome (**Figure 7.13**). The replication origins in mammalian cells are spaced at intervals of approximately 50–300 kb; thus, the human genome has about 30,000 origins of replication. The genomes of simpler eukaryotes also have multiple origins; for example, replication in yeasts initiates at origins separated by intervals of approximately 40 kb.

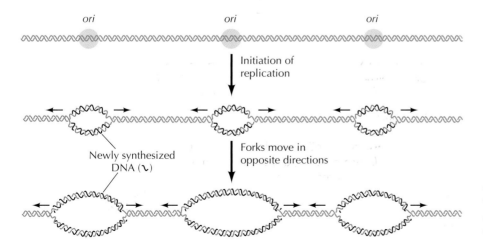

ori   ori   ori

Initiation of replication

Forks move in opposite directions

Newly synthesized DNA (⌁)

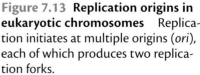

Figure 7.13 Replication origins in eukaryotic chromosomes Replication initiates at multiple origins (*ori*), each of which produces two replication forks.

The origins of replication of eukaryotic chromosomes were first studied in the yeast *S. cerevisiae*, in which they were identified as sequences that can support the replication of plasmids in transformed cells (**Figure 7.14**). This has provided a functional assay for these sequences, and several such elements (called **autonomously replicating sequences**, or **ARSs**) have been isolated. Their

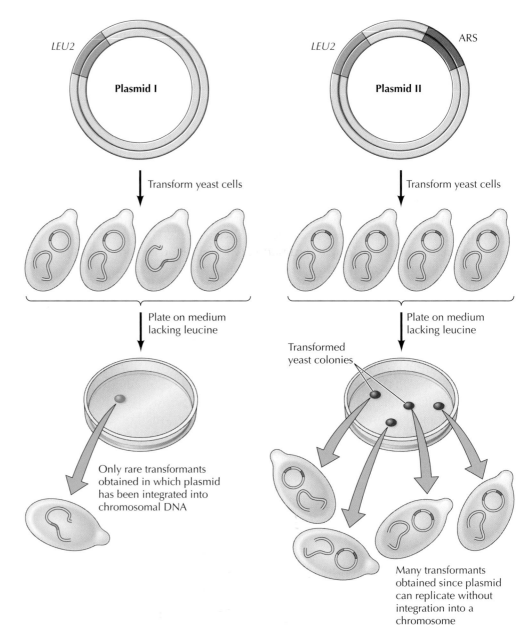

Figure 7.14 Identification of origins of replication in yeast Both plasmids I and II contain a selectable marker gene (*LEU2*) that allows transformed cells to grow on medium lacking leucine. Only plasmid II, however, contains an origin of replication (ARS). The transformation of yeasts with plasmid I yields only rare transformants in which the plasmid has integrated into chromosomal DNA. Plasmid II, however, is able to replicate without integration into a yeast chromosome (autonomous replication), so many more transformants result from its introduction into yeast cells.

role as origins of replication has been verified by direct biochemical analysis, not only in plasmids but also in yeast chromosomal DNA.

Functional ARS elements span about 100 base pairs, including an 11-base-pair core sequence common to many different ARSs (**Figure 7.15**). This core sequence is essential for ARS function and has been found to be the binding site of a six-subunit protein complex (called the **origin recognition complex**, or **ORC**) that is required for initiation of DNA replication at *S. cerevisiae* origins. The ORC complex functions to recruit other proteins (including the MCM DNA helicase) to the origin, leading to the initiation of replication. The mechanism of initiation of DNA replication in *S. cerevisiae* thus appears similar to that in bacteria and eukaryotic viruses; that is, an initiator protein binds to specific DNA sequences that define an origin of replication.

Subsequent studies have shown that the role of ORC proteins as initiators of replication is conserved in all eukaryotes, from yeasts to mammals. However, replication origins in other eukaryotes are much less well defined than the ARS elements of *S. cerevisiae*. In the fission yeast *S. pombe*, origin sequences are spread over 0.5–1 kb of DNA. The *S. pombe* origins lack the clearly defined ORC binding site of the *S. cerevisiae* ARS elements, but they contain repeats of A/T-rich sequences that appear to serve as binding sites for the *S. pombe* ORC complex. In *Drosophila* and mammalian cells, however, the ORC proteins do not appear to recognize specific DNA sequences. Instead, the sites of ORC binding and initiation of DNA replication in higher

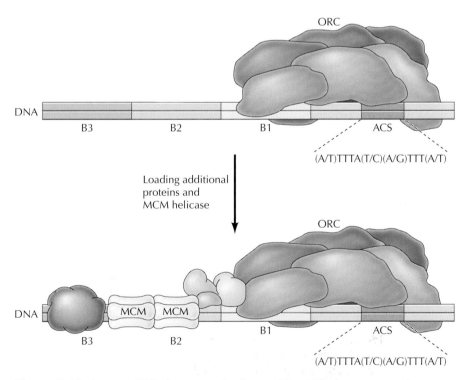

Figure 7.15 A yeast ARS element   An *S. cerevisiae* ARS element contains an 11-base-pair ARS consensus sequence (ACS) and three additional elements (B1, B2, and B3) that contribute to origin function. The origin recognition complex (ORC) binds to the ACS and B1. ORC then recruits additional proteins, including the MCM DNA helicase, to the origin.

eukaryotes may be determined by aspects of chromatin structure that remain to be elucidated by future research.

Telomeres and telomerase: Maintaining the ends of chromosomes

Because DNA polymerases extend primers only in the 5′ to 3′ direction, they are unable to copy the extreme 5′ ends of linear DNA molecules. Consequently, special mechanisms are required to replicate the terminal sequences of the linear chromosomes of eukaryotic cells. These sequences (**telomeres**) consist of tandem repeats of simple-sequence DNA (see Chapter 6). They are maintained by the action of a unique enzyme called **telomerase**, which is able to catalyze the synthesis of telomeres in the absence of a DNA template.

Telomerase is a **reverse transcriptase**, one of a class of DNA polymerases, first discovered in retroviruses (see Chapter 4) that synthesize DNA from an RNA template. Importantly, telomerase carries its own template RNA, which is complementary to the telomere repeat sequences, as part of the enzyme complex. The use of this RNA as a template allows telomerase to generate multiple copies of the telomeric repeat sequences, thereby maintaining telomeres in the absence of a conventional DNA template to direct their synthesis.

The mechanism of telomerase action was initially elucidated in 1985 by Carol Greider and Elizabeth Blackburn during studies of the protozoan *Tetrahymena* (**Figure 7.16**). The *Tetrahymena* telomerase is complexed to a 159-nucleotide-long RNA that includes the sequence 3′–AACCCCAAC–5′. This sequence is complementary to the *Tetrahymena* telomeric repeat (5′–TTGGGG–3′) and serves as the template for the synthesis of telomeric DNA. The use of this RNA as a template allows telomerase to extend the 3′ end of chromosomal DNA by one repeat unit beyond its original length. The complementary strand can then be synthesized by the polymerase α-primase complex using conventional RNA priming. Removal of the RNA primer leaves an overhanging 3′ end of chromosomal DNA, which can form loops at the ends of eukaryotic chromosomes (see Figure 6.28).

Telomerase has been identified in a variety of eukaryotes, and genes encoding telomerase RNAs have been cloned from *Tetrahymena*, yeasts, mice, and humans. In each case, the telomerase RNA contains sequences complementary to the telomeric repeat sequence of that organism (see Table 6.6). Moreover, the introduction of mutant telomerase RNA genes into yeasts has been shown to result in corresponding alterations of the chromosomal telomeric repeat sequences, directly demonstrating the function of telomerase in maintaining the termini of eukaryotic chromosomes.

Defects in telomerase and the normal maintenance of telomeres are associated with several human diseases. The activity of telomerase is regulated in dividing cells to maintain telomeres at their normal length. In human cells, for example, telomeres span approximately 10 kb with a 3′ single-strand overhang of 150–200 bases. Although this length is maintained in germ cells, most somatic cells do not have sufficiently high levels of telomerase to maintain the length of their telomeres for an indefinite number of cell divisions. Consequently, telomeres gradually shorten as cells age, and this shortening eventually leads to cell death or senescence. Interestingly, several premature aging syndromes are characterized by an abnormally high rate of telomere loss, and some of these diseases have been found to be caused directly by mutations in telomerase. Conversely, cancer cells express abnormally high levels of telomerase, allowing them to continue dividing indefinitely. Mutations in regulatory sequences that result in increased expression of telomerase are

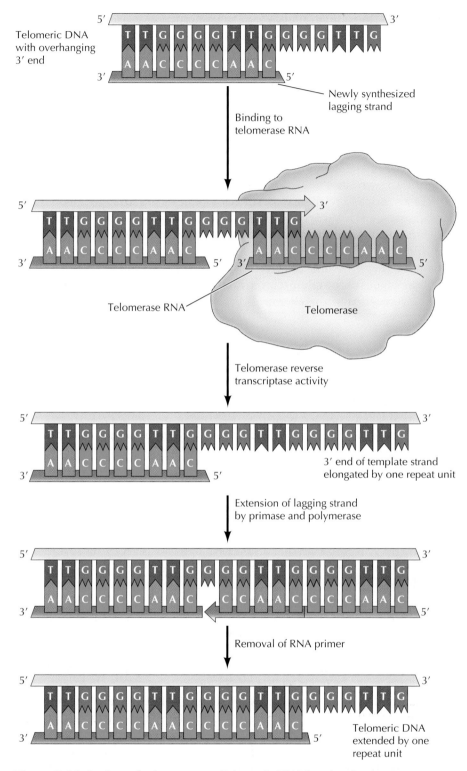

Figure 7.16 Action of telomerase   Telomeric DNA is a simple repeat sequence with an overhanging 3′ end. Telomerase carries its own RNA molecule, which is complementary to telomeric DNA, as part of the enzyme complex. The overhanging end of telomeric DNA binds to the telomerase RNA, which then serves as a template for extension of the template strand by one repeat unit. The lagging strand of telomeric DNA can then be elongated by conventional RNA priming and DNA polymerase activity.

common in some human skin cancers (melanomas), suggesting that elevated telomerase levels can contribute directly to cancer development.

DNA Repair

DNA, like any other molecule, can undergo a variety of chemical reactions. Because DNA uniquely serves as a permanent copy of the cell genome, however, changes in its structure are of much greater consequence than are alterations in other cell components, such as RNAs or proteins. Mutations can result from the incorporation of incorrect bases during DNA replication. In addition, various chemical changes occur in DNA either spontaneously (Figure 7.17) or as a result of exposure to chemicals or radiation (Figure 7.18). Such damage to DNA can block replication or transcription, and can result in a high frequency of mutations—consequences that are unacceptable from the standpoint of cell reproduction. To maintain the integrity of their genomes, cells have therefore had to evolve mechanisms to repair damaged DNA. These mechanisms of DNA repair can be divided into two general classes: (1) direct reversal of the chemical reaction responsible for DNA damage,

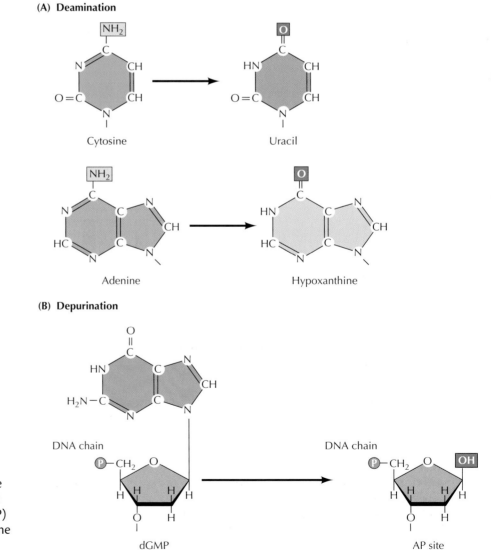

(A) Deamination

Cytosine → Uracil

Adenine → Hypoxanthine

(B) Depurination

DNA chain

dGMP → AP site

DNA chain

Figure 7.17 Spontaneous damage to DNA There are two major forms of spontaneous DNA damage: (A) deamination (loss of amino groups) of adenine and cytosine, and (B) depurination (loss of purine bases) resulting from cleavage of the bond between the purine bases and deoxyribose, leaving an apurinic (AP) site in DNA. dGMP = deoxyguanosine monophosphate.

and (2) removal of the damaged bases followed by their replacement with newly synthesized DNA. Where DNA repair fails, additional mechanisms have evolved to enable cells to cope with the damage.

Direct reversal of DNA damage

Most damage to DNA is repaired by removal of the damaged bases followed by resynthesis of the excised region. Some lesions in DNA, however, can be

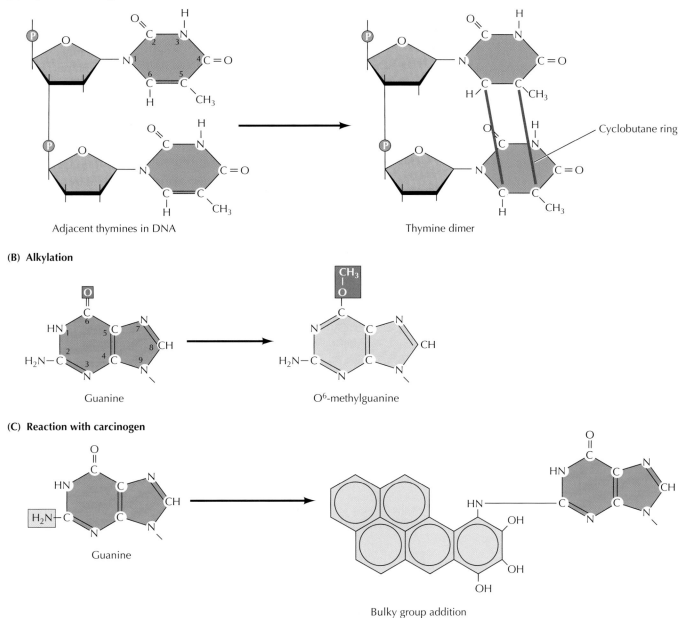

(A) Exposure to UV light

Adjacent thymines in DNA

Thymine dimer

Cyclobutane ring

(B) Alkylation

Guanine

O^6-methylguanine

(C) Reaction with carcinogen

Guanine

Bulky group addition

Figure 7.18 Examples of DNA damage induced by radiation and chemicals (A) UV light induces the formation of pyrimidine dimers in which two adjacent pyrimidines (e.g., thymines) are joined by a cyclobutane ring structure. (B) Alkylation is the addition of methyl or ethyl groups to various positions on the DNA bases. In this example, alkylation of the O^6 position of guanine results in formation of O^6-methylguanine. (C) Many carcinogens (e.g., benzo[α]pyrene) react with DNA bases, resulting in the addition of large bulky chemical groups to the DNA molecule.

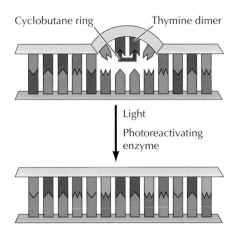

Cyclobutane ring   Thymine dimer

Light

Photoreactivating enzyme

Figure 7.19 Direct repair by photoreactivation UV-induced thymine dimers can be repaired by photoreactivation, in which energy from visible light is used to split the bonds forming the cyclobutane ring.

repaired by direct reversal of the damage, which may be a more efficient way of dealing with specific types of DNA damage that occur frequently. Only a few types of DNA damage are repaired in this way, particularly pyrimidine dimers resulting from exposure to ultraviolet (UV) light and alkylated guanine residues that have been modified by the addition of methyl or ethyl groups at the O^6 position of the purine ring.

UV light is one of the major sources of damage to DNA and is also the most thoroughly studied form of DNA damage in terms of repair mechanisms. Its importance is illustrated by the fact that exposure to solar UV irradiation is the cause of almost all skin cancer in humans. The major type of damage induced by UV light is the formation of **pyrimidine dimers**, in which adjacent pyrimidines on the same strand of DNA are joined by the formation of a cyclobutane ring resulting from saturation of the double bonds between carbons 5 and 6 (see Figure 7.18A). The formation of such dimers distorts the structure of the DNA chain and blocks transcription or replication past the site of damage, so their repair is closely correlated with the ability of cells to survive UV irradiation. One mechanism of repairing UV-induced pyrimidine dimers is direct reversal of the dimerization reaction. The process is called **photoreactivation** because energy derived from visible light is utilized to break the cyclobutane ring structure (**Figure 7.19**). The original pyrimidine bases remain in DNA, now restored to their normal state. As might be expected from the fact that solar UV irradiation is a major source of DNA damage for diverse cell types, the repair of pyrimidine dimers by photoreactivation is common to a variety of prokaryotic and eukaryotic cells, including *E. coli*, yeasts, and some species of plants and animals. Curiously, however, photoreactivation is not universal; placental mammals (including humans) lack this mechanism of DNA repair.

Another form of direct repair deals with damage resulting from the reaction between alkylating agents and DNA. Alkylating agents are reactive compounds that can transfer methyl or ethyl groups to a DNA base, thereby chemically modifying the base (see Figure 7.18B). A particularly important type of damage is methylation of the O^6 position of guanine, because the product, O^6-methylguanine, forms complementary base pairs with thymine instead of cytosine. This lesion can be repaired by an enzyme (called O^6-methylguanine methyltransferase) that transfers the methyl group from O^6-methylguanine to a cysteine residue in its active site (**Figure 7.20**). The potentially mutagenic chemical modification is thus removed, and the original guanine is restored.

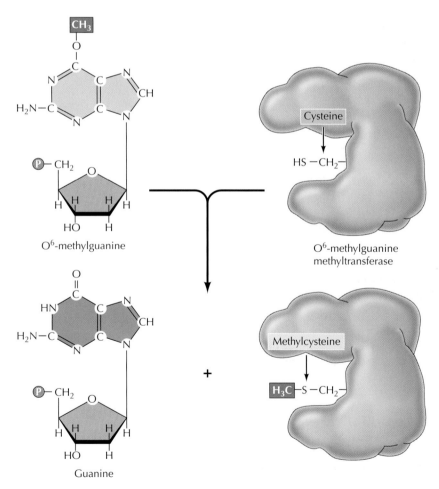

O^6-methylguanine

O^6-methylguanine methyltransferase

Cysteine

Methylcysteine

Guanine

Figure 7.20 Repair of O^6-methylguanine O^6-methylguanine methyltransferase transfers the methyl group from O^6-methylguanine to a cysteine residue in the enzyme's active site.

Enzymes that catalyze this direct repair reaction are widespread in both prokaryotes and eukaryotes, including humans.

Excision repair

Although direct repair is an efficient way of dealing with particular types of DNA damage, excision repair is a more general means of repairing a wide variety of chemical alterations to DNA. Consequently, the various types of excision repair are the most important DNA repair mechanisms in both prokaryotic and eukaryotic cells. In excision repair, the damaged DNA is recognized and removed, either as free bases or as nucleotides. The resulting gap is then filled in by synthesis of a new DNA strand, using the undamaged complementary strand as a template. Three types of excision repair—base-excision, nucleotide-excision, and mismatch repair—enable cells to cope with a variety of different kinds of DNA damage.

BASE-EXCISION REPAIR The repair of uracil-containing DNA is a good example of **base-excision repair**, in which single damaged bases are recognized and removed from the DNA molecule (**Figure 7.21**). Uracil can arise in DNA by two mechanisms: (1) Uracil (as dUTP [deoxyuridine triphosphate]) is occasionally incorporated in place of thymine during DNA synthesis, and (2) uracil can be formed in DNA by the deamination of cytosine (see Figure 7.17A). The second mechanism is of much greater biological significance because it alters the normal pattern of complementary base pairing and thus represents a mutagenic event. The excision of uracil in DNA is catalyzed by **DNA glycosylase**, an enzyme that cleaves the bond linking the base (uracil) to the deoxyribose of the DNA backbone. This reaction yields free uracil and an apyrimidinic site—a sugar with no base attached. DNA glycosylases also recognize and remove other abnormal bases, including hypoxanthine formed by the deamination of adenine, alkylated purines other than O^6-alkylguanine, and bases damaged by oxidation or ionizing radiation.

The result of DNA glycosylase action is the formation of an apyrimidinic or apurinic site (generally called an AP site) in DNA. Similar AP sites are formed as a result of the spontaneous loss of purine bases (see Figure 7.17B), which occurs at a significant rate under normal cellular conditions. For example, each cell in the human body is estimated to lose several thousand purine bases daily. These sites are repaired by **AP endonuclease**, which cleaves adjacent to the AP site (see Figure 7.21). The remaining deoxyribose moiety is then removed, and the resulting single-base gap is filled by DNA polymerase and ligase.

NUCLEOTIDE-EXCISION REPAIR Whereas DNA glycosylases recognize only specific forms of damaged bases, other excision repair systems recognize

Figure 7.21 Base-excision repair In this example, uracil (U) has been formed by deamination of cytosine (C) and is therefore opposite a guanine (G) in the complementary strand of DNA. The bond between uracil and the deoxyribose is cleaved by a DNA glycosylase, leaving a sugar with no base attached in the DNA (an AP site). This site is recognized by AP endonuclease, which cleaves the DNA chain. The remaining deoxyribose is removed by deoxyribosephosphodiesterase. The resulting gap is then filled by DNA polymerase and sealed by ligase, leading to incorporation of the correct base (C) opposite the G.

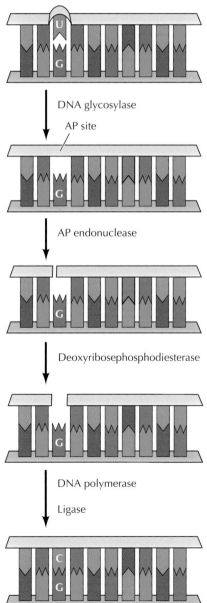

DNA containing U formed by deamination of C

DNA glycosylase

AP site

AP endonuclease

Deoxyribosephosphodiesterase

DNA polymerase

Ligase

a wide variety of damaged bases that distort the DNA molecule, including UV-induced pyrimidine dimers and bulky groups added to DNA bases as a result of the reaction of many carcinogens with DNA (see Figure 7.18C). This widespread form of DNA repair is known as **nucleotide-excision repair**, because the damaged bases (e.g., a thymine dimer) are removed as part of an oligonucleotide containing the lesion (**Figure 7.22**).

In *E. coli*, nucleotide-excision repair is catalyzed by the products of three genes (*uvrA*, *uvrB*, and *uvrC*) that were identified because mutations at these loci result in extreme sensitivity to UV light. The protein UvrA recognizes damaged DNA and recruits UvrB and UvrC to the site of the lesion. UvrB and UvrC then cleave on the 3' and 5' sides of the damaged site, respectively,

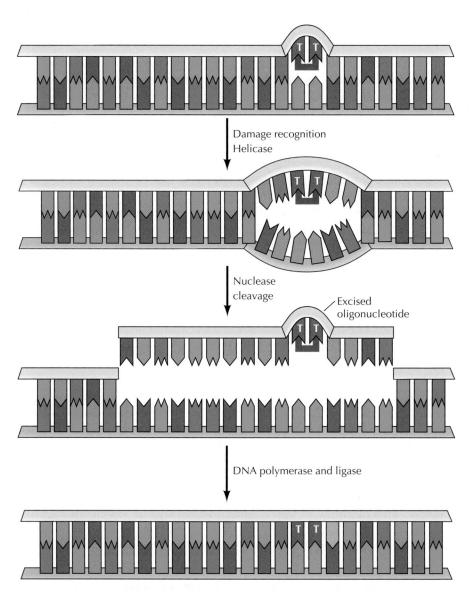

Figure 7.22 Nucleotide-excision repair of thymine dimers   Damaged DNA is recognized and then unwound around the site of damage by a helicase. The DNA is then cleaved on both sides of a thymine dimer by 3' and 5' nucleases, resulting in excision of an oligonucleotide containing the damaged bases. The gap is then filled by DNA polymerase and sealed by ligase.

thus excising an oligonucleotide consisting of 12 or 13 bases. The UvrABC complex is frequently called an **excinuclease**, a name that reflects its ability to directly *exci*se an oligonucleotide. The action of a helicase is then required to remove the damage-containing oligonucleotide from the double-stranded DNA molecule, and the resulting gap is filled by DNA polymerase and sealed by ligase.

Nucleotide-excision repair systems have also been studied extensively in eukaryotes, including yeasts, rodents, and humans. In yeasts, as in *E. coli*, several genes involved in DNA repair (called *RAD* genes for *rad*iation sensitivity) have been identified by the isolation of mutants with increased sensitivity to UV light. In humans, DNA repair genes have been identified largely by studies of individuals suffering from inherited diseases resulting from deficiencies in the ability to repair DNA damage. The most extensively studied of these diseases is xeroderma pigmentosum (XP), a rare genetic disorder that affects approximately one in 250,000 people. Individuals with this disease are extremely sensitive to UV light and develop multiple skin cancers on the regions of their bodies that are exposed to sunlight. In 1968, James Cleaver made the key discovery that cultured cells from XP patients were deficient in the ability to carry out nucleotide-excision repair. This observation not only provided the first link between DNA repair and cancer, but also suggested the use of XP cells as an experimental system to identify human DNA repair genes. The identification of human DNA repair genes has been accomplished by studies not only of XP cells, but also of two other human diseases resulting from DNA repair defects (Cockayne's syndrome and trichothiodystrophy) and of UV-sensitive mutants of rodent cell lines. The availability of mammalian cells with defects in DNA repair has allowed the cloning of repair genes based on the ability of wild-type alleles to re-store normal UV sensitivity to mutant cells in gene transfer assays, thereby opening the door to experimental analysis of nucleotide-excision repair in mammalian systems.

Molecular cloning has identified seven different repair genes (designated *XPA* through *XPG*) that are mutated in cases of xeroderma pigmentosum, as well as genes that are mutated in Cockayne's syndrome, trichothiodys-trophy, and UV-sensitive mutants of rodent cells. The proteins encoded by these mammalian DNA repair genes are closely related to proteins encoded by yeast *RAD* genes, indicating that nucleotide-excision repair is highly conserved throughout eukaryotes. With cloned yeast and mammalian repair genes available, it has been possible to purify their encoded proteins and develop *in vitro* systems to study their roles in the repair process (**Figure 7.23**). The initial step in excision repair in mammalian cells involves recognition of disrupted base pairing by XPC. This is followed by the cooperative bind-ing of XPA, the single-stranded DNA binding protein replication protein A (RPA) discussed earlier in this chapter (see Figure 7.8), and a multisubunit transcription factor called TFIIH, which is required to initiate transcription of eukaryotic genes (see Chapter 8). In some cases, XPE may also play a role

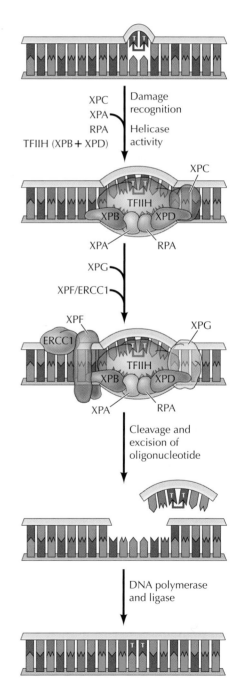

Figure 7.23 Nucleotide-excision repair in mammalian cells   DNA damage (e.g., a thymine dimer) is recognized by XPC, followed by binding of XPA, RPA, and TFIIH, which contains the XPB and XPD helicases. Following unwinding of the DNA by XPB and XPD, the XPG and XPF/ERCC1 endonucleases are recruited and the DNA is cleaved, excising the damaged oligonucleotide. The resulting gap is filled by DNA polymerase and sealed by ligase.

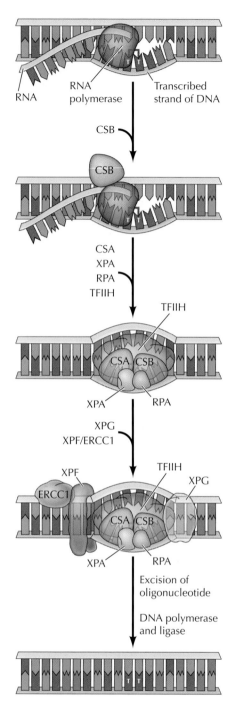

RNA
RNA
polymerase
Transcribed
strand of DNA

CSB

CSB

CSA
XPA
RPA
TFIIH

TFIIH

CSA CSB

XPA   RPA

XPG
XPF/ERCC1

XPF   TFIIH
                        XPG
ERCC1

CSA CSB

XPA   RPA

Excision of
oligonucleotide

DNA polymerase
and ligase

Figure 7.24 Transcription-coupled repair in mammalian cells   RNA polymerase is stalled by DNA damage (e.g., a thymine dimer) in the transcribed strand of DNA. The stalled polymerase is recognized by CSB at the site of damage. CSB then recruits CSA, XPA, RPA and TFIIH, and excision repair proceeds as described in Figure 7.23.

in damage recognition. Two of the subunits of TFIIH are the XPB and XPD proteins, which act as helicases to unwind approximately 25 base pairs of DNA around the site of damage. The XPG protein is then recruited to the complex followed by recruitment of XPF as a heterodimer with ERCC1 (a repair protein identified in UV-sensitive rodent cells). XPF/ERCC1 and XPG are endonucleases, which cleave DNA on the 5' and 3' sides of the damaged site, respectively. This cleavage excises an oligonucleotide consisting of approximately 30 bases. The resulting gap is then filled by DNA polymerase δ and sealed by ligase.

TRANSCRIPTION-COUPLED REPAIR   Although damaged DNA can be recognized throughout the genome, an alternative form of nucleotide-excision repair, called **transcription-coupled repair**, is specifically dedicated to repairing damage within actively transcribed genes (**Figure 7.24**). A connection between transcription and repair was first suggested by experiments showing that transcribed strands of DNA are repaired more rapidly than nontranscribed strands in both *E. coli* and mammalian cells. Since DNA damage blocks transcription, this transcription-repair coupling is thought to be advantageous by allowing the cell to preferentially repair damage to genes that are actively expressed. In both *E. coli* and mammalian cells, the mechanism of transcription-repair coupling involves recognition of RNA polymerase stalled at a lesion in the DNA strand being transcribed. In *E. coli*, the stalled RNA polymerase is recognized by a protein called transcription-repair coupling factor, which displaces RNA polymerase and recruits the UvrABC excinuclease to the site of damage. In mammalian cells, transcription-coupled repair involves recognition of stalled RNA polymerase by a protein called CSB, which is encoded by a gene responsible for Cockayne's syndrome. In contrast to patients with xeroderma pigmentosum, patients with Cockayne's syndrome are specifically defective in transcription-coupled repair, consistent with the role of CSB as a transcription-repair coupling factor. Following recognition of stalled RNA polymerase, CSB recruits another protein involved in Cockayne's syndrome (CSA) as well as XPA, RPA, and TFIIH to the site of DNA damage, and nucleotide excision repair proceeds as outlined in Figure 7.23.

MISMATCH REPAIR   A distinct excision repair system recognizes mismatched bases that are incorporated during DNA replication. Many such mismatched bases are removed by the proofreading activity of DNA polymerase. The ones that are missed are subject to later correction by the **mismatch repair** system, which scans newly replicated DNA. If a mismatch is found, the enzymes of this repair system are able to identify and excise the mismatched base specifically from the newly replicated DNA strand, allowing the error to be corrected and the original sequence restored.

In *E. coli*, the ability of the mismatch repair system to distinguish between parental DNA and newly synthesized DNA is based on the fact that DNA of this bacterium is modified by the methylation of adenine residues within

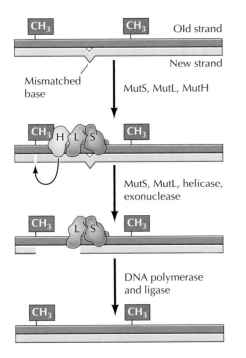

Figure 7.25 Mismatch repair in *E. coli* The mismatch repair system detects and excises mismatched bases in newly replicated DNA, which is distinguished from the parental strand because it has not yet been methylated. MutS binds to the mismatched base, followed by MutL. The binding of MutL activates MutH, which cleaves the unmodified strand opposite a site of methylation. MutS and MutL, together with a helicase and an exonuclease, then excise the portion of the unmodified strand that contains the mismatch. The gap is then filled by DNA polymerase and sealed by ligase.

> **Video 7.2**
>
> **sites.sinauer.com/cooper7e/v7.2**
> **Methyl-Directed Mismatch Repair**
> In *E. coli*, mismatch repair is facilitated by the lack of methylation on the newly-synthesized DNA strands.

the sequence GATC to form 6-methyladenine (**Figure 7.25**). Since methylation occurs after replication, newly synthesized DNA strands are not methylated and thus can be specifically recognized by the mismatch repair enzymes. Mismatch repair is initiated by the protein MutS, which recognizes the mismatch and forms a complex with two other proteins called MutL and MutH. The MutH endonuclease then cleaves the unmethylated DNA strand at a GATC sequence. MutL and MutS then act together with an exonuclease and a helicase to excise the DNA between the strand break and the mismatch, with the resulting gap being filled by DNA polymerase and ligase.

Eukaryotes have a similar mismatch repair system, although the mechanism by which eukaryotic cells identify newly replicated DNA differs from that used by *E. coli*. Eukaryotic cells do not have a homolog of MutH and the strand specificity of mismatch repair is not determined by DNA methylation. Instead, the presence of single-strand breaks in newly replicated DNA appears to specify the strand to be repaired. Eukaryotic homologs of MutS and MutL (MSH and MLH, respectively) bind to the mismatched base, as in *E. coli*, and direct excision of the DNA between a strand break and the mismatch (**Figure 7.26**). The newly synthesized lagging strand could be identified by nicks at either end of Okazaki fragments, whereas the leading strand might be identified by its growing 3' end.

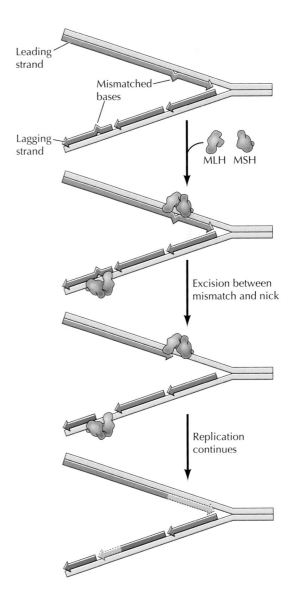

Figure 7.26 Mismatch repair in mammalian cells Eukaryotic homologs of MutS and MutL (MSH and MLH, respectively) bind to mismatched bases and direct excision of the DNA between the mismatch and a break in the DNA strand. Breaks in the newly-synthesized lagging strand are present at either end of Okazaki fragments, and a break in the leading strand is present at its growing 3' end.

The importance of this repair system is dramatically illustrated by the fact that mutations in *MSH* and *MLH* are responsible for a common type of inherited colon cancer (hereditary nonpolyposis colorectal cancer, or HNPCC). HNPCC is one of the most common inherited diseases; it affects as many as one in 200 people and is responsible for about 15% of all colorectal cancers in this country. The relationship between HNPCC and defects in mismatch repair was discovered in 1993, when two groups of researchers cloned the human homolog of *MutS* (*MSH*) and found that mutations in this gene were responsible for about half of all HNPCC cases. Subsequent studies have shown that most of the remaining cases of HNPCC are caused by mutations in one of three human genes that are homologs of *MutL* (*MLH*). Defects in these genes appear to result in a high frequency of mutations in other cell genes, with a correspondingly high likelihood that some of these mutations will eventually lead to the development of cancer.

Translesion DNA synthesis

The direct reversal and excision repair systems act to correct DNA damage before replication, so that replicative DNA synthesis can proceed using an undamaged DNA strand as a template. Should these systems fail, however, the cell has alternative mechanisms for dealing with damaged DNA at the replication fork. Pyrimidine dimers and many other types of lesions cannot be copied by the normal action of DNA polymerases, so replication is blocked at the sites of such damage. However, cells also possess several specialized DNA polymerases that are capable of replicating across a site of DNA damage. The replication of damaged DNA by these specialized polymerases, called **translesion DNA synthesis**, provides a mechanism by which the cell can bypass DNA damage at the replication fork, which can then be corrected after replication is complete.

The first specialized DNA polymerase responsible for translesion DNA synthesis was discovered in *E. coli* in 1999. This enzyme, called DNA polymerase V, is induced in response to extensive UV irradiation and can synthesize a new DNA strand across from a thymine dimer (**Figure 7.27**). Two other specialized *E. coli* DNA polymerases, polymerases II and IV, are similarly induced by DNA damage and function in translesion synthesis. Eukaryotic cells also contain multiple specialized DNA polymerases, with at least five such enzymes having been shown to function in translesion synthesis in mammalian cells. These specialized DNA polymerases replace a normal polymerase that is stalled at a site of DNA damage. They can then synthesize across the site of damage, after which they are themselves replaced by the normal replicative polymerase.

All of the specialized DNA polymerases exhibit low fidelity when copying undamaged DNA and lack 3′ to 5′ proofreading activity (see Figure 7.11), so their error rates range from 100 to 10,000 times higher than the error rates of the normal replicative DNA polymerases (e.g., polymerase III in *E. coli* or polymerases δ and ε in eukaryotes). However, they do display some selectivity in inserting the correct base opposite certain lesions, such

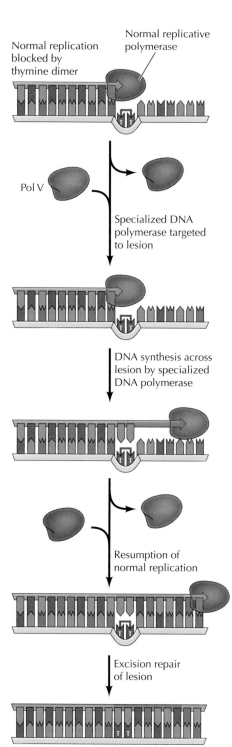

Normal replication blocked by thymine dimer

Normal replicative polymerase

Pol V

Specialized DNA polymerase targeted to lesion

DNA synthesis across lesion by specialized DNA polymerase

Resumption of normal replication

Excision repair of lesion

Figure 7.27 Translesion DNA synthesis Normal replication is blocked by a thymine dimer, but specialized DNA polymerases such as DNA polymerase V (pol V) recognize and continue DNA synthesis across the lesion. Replication can then be resumed by the normal replicative DNA polymerase, and the thymine dimer subsequently removed by nucleotide-excision repair (see Figure 7.22).

Molecular Medicine

Colon Cancer and DNA Repair

The Disease

Cancers of the colon and rectum (colorectal cancers) are some of the most common types of cancer in Western countries, accounting for about 150,000 cancer cases per year in the United States (approximately 10% of the total cancer incidence). Most colon cancers (like other types of cancer) are not inherited diseases; that is, they are not transmitted directly from parent to offspring. However, two inherited forms of colon cancer have been described. In both of these syndromes, the inheritance of a cancer susceptibility gene results in a very high likelihood of cancer development. One inherited form of colon cancer (familial adenomatous polyposis) is extremely rare, accounting for less than 1% of total colon cancer incidence. The second inherited form of colon cancer (hereditary nonpolyposis colorectal cancer, or HNPCC) is much more common and accounts for up to 15% of all colon cancer cases. Indeed, HNPCC is one of the most common inherited diseases, affecting as many as one in 200 people. Although colon cancers are the most common manifestation of this disease, affected individuals also suffer an increased incidence of other types of cancer, including cancers of the ovaries and endometrium.

Molecular and Cellular Basis

Like other cancers, colorectal cancer results from mutations in genes that regulate cell proliferation, leading to the uncontrolled growth of cancer cells. In most cases these mutations occur sporadically in somatic cells. In hereditary cancers, however, inherited germ-line mutations predispose the individual to cancer development.

A striking advance was made in 1993 with the discovery that a gene responsible for approximately 50% of HNPCC cases encodes an enzyme involved in mismatch repair of DNA; this gene is a human homolog of the *E. coli MutS* gene. Subsequent studies have shown that three other genes, responsible for most remaining cases of HNPCC, are homologs of *MutL* and thus are also involved in the mismatch repair pathway. Defects in these genes appear to result in a high frequency of mutations in other cell genes, with a correspondingly high likelihood that some of these mutations will eventually lead to the development of cancer by affecting genes that regulate cell proliferation.

Prevention and Treatment

As with other inherited diseases, identification of the genes responsible for HNPCC allows individuals at risk for this inherited cancer to be identified by genetic testing. Moreover, prenatal genetic diagnosis may be of great importance to carriers of HNPCC mutations who are planning a family. However, the potential benefits of detecting these mutations are not limited to preventing the transmission of mutant genes to the next generation; their detection may also help prevent the development of cancer in affected individuals.

In terms of disease prevention, a key characteristic of colon cancer is that it develops gradually over several years. Early diagnosis of the disease substantially improves the chances for patient survival. The initial stage of colon cancer development is the outgrowth of small benign polyps, which eventually become malignant and invade the surrounding connective tissue. Prior to the development of malignancy, however, polyps can be easily removed surgically, effectively preventing the outgrowth of a malignant tumor. Polyps and early stages of colon cancer can be detected by examination of the colon with a thin lighted tube (colonoscopy), so frequent colonoscopy of HNPCC patients may allow polyps to be removed before cancer develops. In addition, several drugs are being tested as potential inhibitors of colon cancer development, and these drugs may be of significant benefit to HNPCC patients. By allowing the timely application of such preventive measures, the identification of mutations responsible for HNPCC may make a significant contribution to disease prevention.

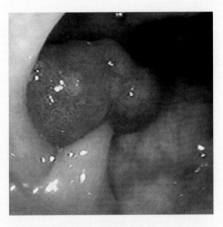

A colon polyp visualized by colonoscopy.

---

as thymine dimers, in damaged DNA. Thus, the specialized polymerases are able to insert correct bases opposite some forms of DNA damage, although they are "error-prone" in inserting bases opposite other forms of damaged DNA or in the synthesis of DNA from a normal undamaged template.

Repair of double-strand breaks

Double-strand breaks are a particularly dangerous form of DNA damage, because the continuity of the DNA molecule is disrupted by breaks in both strands. Such double-strand breaks occur naturally during DNA replication, for example, when DNA polymerase at the replication fork encounters a nick in the template strand of DNA. In addition, double-strand breaks are induced

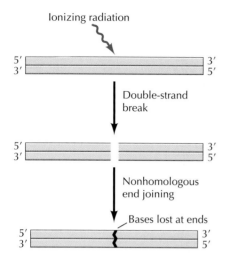

Figure 7.28 Repair of double-strand breaks by nonhomologous end joining Ionizing radiation and some chemicals induce double-strand breaks in DNA. These breaks can be repaired by rejoining the ends of the broken molecule by nonhomologous end joining. This frequently leads to the loss of bases, though, around the site of damage.

by ionizing radiation (such as X-rays) and some chemicals that damage DNA by introducing breaks in opposite strands.

Because double-strand breaks affect both strands of DNA, they cannot be repaired by mechanisms that require DNA synthesis across a site of damage. Instead, double-strand breaks are repaired by a distinct mechanism, **recombinational repair**, which rejoins the broken strands. Cells utilize two major pathways of recombinational repair. In the simplest case, double-strand breaks can be repaired simply by rejoining the broken ends of a single DNA molecule (**Figure 7.28**). However, this leads to a high frequency of errors resulting from deletion of bases around the site of damage.

Alternatively, double-strand breaks can be repaired by **homologous recombination** with DNA sequences on an undamaged chromosome, which provides a mechanism for repairing such damage and restoring the normal DNA sequence (**Figure 7.29**). In eukaryotic cells, this mechanism of repair is only operative following DNA replication, when the newly replicated sister

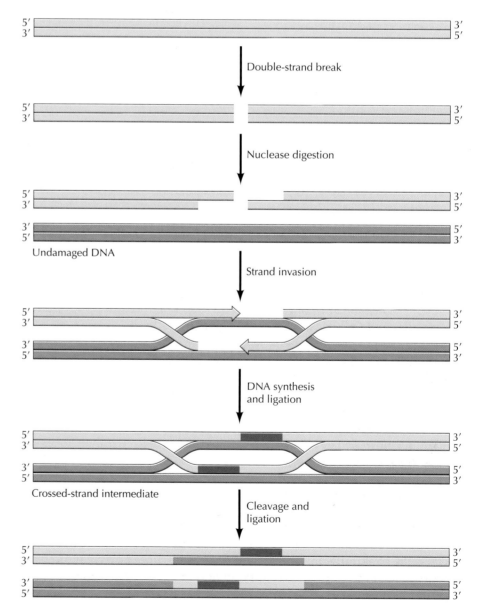

Figure 7.29 Repair of double-strand breaks by homologous recombination Double-strand breaks can be repaired by homologous recombination with an undamaged chromosome, leading to restoration of the normal DNA sequence. Both strands of DNA at a double-strand break are digested by nucleases in the 5′ to 3′ direction. The overhanging 3′ ends then invade the other parental molecule by homologous base pairing. The gaps are then filled by repair synthesis and sealed by ligation, yielding a crossed-strand intermediate. Cleavage and ligation of the crossed strands then yields recombinant molecules.

(A) RecA   **(B) Rad51**

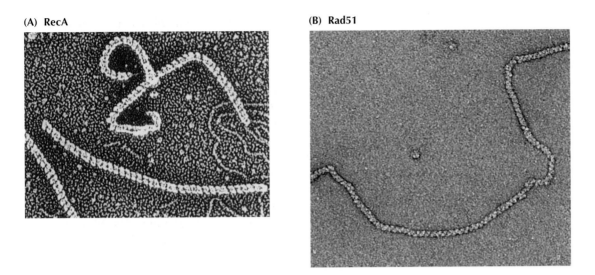

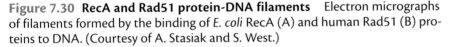

Figure 7.30 RecA and Rad51 protein-DNA filaments   Electron micrographs of filaments formed by the binding of E. coli RecA (A) and human Rad51 (B) proteins to DNA. (Courtesy of A. Stasiak and S. West.)

chromatids remain associated with each other (see Figure 6.23). Homologous recombination results from the breakage and rejoining of two parental DNA molecules, leading to reassortment of the genetic information of the two parental chromosomes. The process is initiated at double-strand breaks, both during DNA repair and during recombination between homologous chromosomes during meiosis (discussed in Chapter 17). Both strands of DNA at the double-strand break are first resected by nucleases that digest DNA in the 5' to 3' direction, yielding overhanging 3' single-stranded ends. These single strands then invade the other parental molecule by homologous base pairing. The gaps are then filled by repair synthesis and the strands are joined by ligation to yield a recombinant molecule.

The key protein involved in homologous recombination is **RecA** in E. coli or **Rad51** in eukaryotic cells. These proteins bind to single-stranded DNA, forming protein-DNA filaments (**Figure 7.30**). They then bind a second, double-stranded DNA molecule, forming a complex between the two DNAs, and catalyze the exchange of strands between homologous sequences. It is noteworthy that one of the genes responsible for inherited breast cancer (BRCA2) encodes a protein that acts to recruit Rad51 to the single-stranded DNA generated at double-strand breaks, suggesting that defects in this type of DNA repair can lead to the development of one of the most common cancers in women.

DNA Rearrangements

Recombination between homologous chromosomes, either during meiosis or DNA repair, results in the exchange of DNA between chromosomes without altering the arrangement of genes within the genome. In contrast, other types of recombinational events lead to rearrangements of genomic DNA. Some of these DNA rearrangements are important in controlling gene expression in specific cell types; others may play an evolutionary role by contributing to genetic diversity.

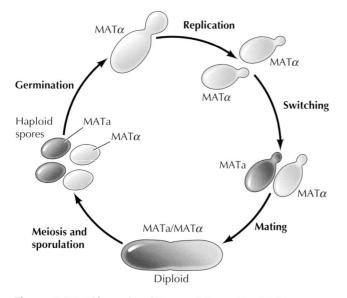

Figure 7.31 Life cycle of *S. cerevisiae* Haploid MATα yeast replicate to generate a MATα population. Switching at the mating type locus results in a mixture of MATa and MATα haploids, which can mate to form a MATa/MATα diploid. Under conditions of nutrient starvation, diploid yeast undergo meiosis to produce four haploid spores. Under favorable conditions, the spores germinate and replicate as haploids.

The discovery that genes can move to different chromosomal locations came from Barbara McClintock's studies of corn in the 1940s. Purely on the basis of genetic analysis, McClintock described novel genetic elements that could move to different locations in the genome and alter the expression of adjacent genes. Nearly three decades elapsed, however, before the physical basis of McClintock's work was elucidated by the discovery of transposable elements in bacteria and the notion of movable genetic elements became widely accepted by scientists. Several types of DNA rearrangements, including the transposition of elements initially described by McClintock, are now recognized in both prokaryotic and eukaryotic cells. Moreover, we now know that transposable elements constitute a large fraction of the genomes of plants and animals, including nearly half of the human genome (see Table 6.3).

As discussed in Chapter 6, transposable elements can move to random new sites in chromosomal DNA and have played a major role in the evolution of our genome. However, they do not generally play a useful role for the cells in which they transpose. In contrast, programmed DNA rearrangements, discussed below, can play important roles in development and regulation of gene expression.

Yeast mating types

Control of yeast mating type provides a well-studied example of a programmed DNA rearrangement mediated by homologous recombination between specific sequences (**site-specific recombination**). *S. cerevisiae* can replicate by mitosis as either haploid or diploid cells (**Figure 7.31**). The diploid state confers advantages under conditions of stress, such as the unavailability of nutrients, because the diploid cells can undergo meiosis and form four haploid spores, which can then germinate when conditions improve. Diploids can be produced by the mating of two haploid cells of different mating types, designated either MATa or MATα, depending on which allele of the mating-type locus (*MAT*) is expressed. Importantly, haploid yeast are able to switch between the MATa and MATα mating types, which allows cells derived from a single colony to generate a mixed population of MATa and MATα yeast that are able to form diploids.

The molecular basis of mating type switching involves site-specific recombination between three mating type loci on *S. cerevisiae* chromosome 3 (**Figure 7.32**). One of these loci (*MAT*) is transcriptionally active and the identity of the allele (*MAT*a or *MAT*α) in the *MAT* locus determines the phenotype of the haploid cell. The other two loci (*HMR*a and *HML*α) are transcriptionally inactive and contain non-expressed alleles of *MAT*a and *MAT*α, respectively. Mating type switching is mediated by recombination between the active allele in the *MAT* locus and the alternate inactive allele at *HMR*a or *HML*α. The process is initiated by an endonuclease called HO, which specifically cleaves DNA at the active *MAT* locus to introduce a double-strand break. The single strands are then resected by 5′ to 3′ exonucleases and the gap is filled by synthesis of a new copy using the inactive alternate *HMR*a or *HML*α allele as the donor. The result is the replacement of the original *MAT* allele (e.g.,

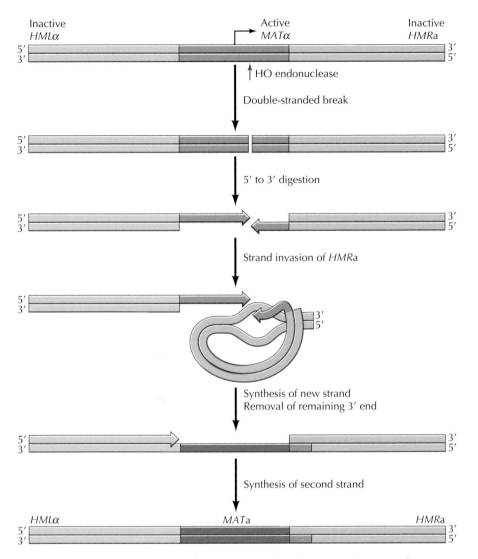

Figure 7.32 Mating type switching   *S. cerevisiae* have an active MAT locus (*MATα* in the example shown) and two inactive loci, *HMLα* and *HMRa*. Switching is initiated by the HO endonuclease, which introduces a double-strand break in the active *MAT* locus. Both DNA strands at the break are digested by nucleases in the 5′ to 3′ direction, leaving overhanging 3′ ends. One of these 3′ ends invades the alternate inactive allele, in this case *HMRa*, and primes synthesis of a new strand using *HMRa* as a template. The remaining 3′ end is removed and a second strand of DNA is synthesized. The result is the synthesis of a new copy in the *MAT* locus, resulting in a switch to *MATa*.

MATα) with its alternate allele (e.g., *MATa*), while both *HMRa* and *HMLα* remain in their original state.

Antibody genes

Development of the vertebrate immune system, which recognizes foreign substances (**antigens**) and provides protection against infectious agents, is a prime example of programmed DNA rearrangement in higher eukaryotes. There are two major classes of immune responses, which are mediated by

Rearrangement of Immunoglobulin Genes

Evidence for Somatic Rearrangement of Immunoglobulin Genes Coding for Variable and Constant Regions

Nobumichi Hozumi and Susumu Tonegawa

Basel Institute for Immunology, Basel, Switzerland

Proceedings of the National Academy of Sciences, USA, Volume 73, 1976, pages 3628–3632

The Context

The ability of the vertebrate immune system to recognize a seemingly infinite variety of foreign molecules implies that lymphocytes can produce a correspondingly vast array of antibodies. Since this antibody diversity is key to immune recognition, understanding the mechanism by which an apparently unlimited number of distinct immunoglobulins are encoded in genomic DNA is a central issue in immunology.

Prior to the experiments of Hozumi and Tonegawa, protein sequencing of multiple immunoglobulins had demonstrated that both heavy and light chains consist of distinct variable and constant regions. Genetic studies further indicated that mice inherit only single copies of the constant-region genes. These observations first led to the proposal that immunoglobulins are encoded by multiple variable-region genes that can associate with a single constant-region gene. The discovery of immunoglobulin gene rearrangements by Hozumi and Tonegawa provided the first direct experimental support for this hypothesis and laid the groundwork for understanding the molecular basis of antibody diversity.

The Experiments

Hozumi and Tonegawa tested the possibility that the genes encoding immunoglobulin variable and constant regions were joined at the DNA level during lymphocyte development. Their experimental approach was to use restriction endonuclease digestion to compare the organization of variable-region and constant-region sequences

in DNAs extracted from mouse embryos and from cells of a mouse plasmacytoma (a B lymphocyte tumor that produces a single species of immunoglobulin). Embryo and plasmacytoma DNAs were digested with the restriction endonuclease *Bam*HI, and DNA fragments of different sizes were separated by electrophoresis in an agarose gel. The gel was then cut into slices, and DNA extracted from each slice was hybridized with radiolabeled probes that had been prepared from immunoglobulin mRNA isolated from the plasmacytoma cells. Two probes were used, corresponding either to the complete immunoglobulin mRNA or to the 3' half of the mRNA, consisting only of constant-region sequences.

The critical result was that completely different patterns of variable-region and constant-region sequences were detected in embryo versus plasmacytoma DNAs (see figure). In embryo DNA, the complete probe hybridized to two *Bam*HI

Susumu Tonegawa

fragments of approximately 8.6 and 5.6 kb, respectively. Only the 8.6-kb fragment hybridized to the 3' probe, suggesting that the 8.6-kb fragment contained constant-region sequences and the 5.6-kb fragment contained variable-region sequences. In striking contrast, both probes hybridized to only a single 3.4-kb fragment in plasmacytoma DNA. The interpretation of these results was that the variable and constant-region sequences were separated in embryo

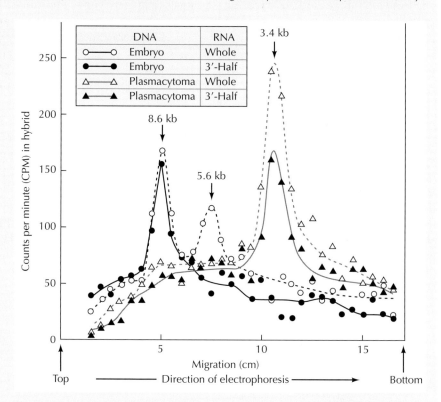

Gel electrophoresis of embryo and plasmacytoma DNAs digested with BamHI and hybridized to probes corresponding to either the whole or the 3' half of the plasmacytoma mRNA. Data are presented as the radioactivity detected in hybrid molecules with DNA from each gel slice.

B and T lymphocytes. B lymphocytes secrete antibodies (**immunoglobulins**) that react with soluble antigens; T lymphocytes express cell surface proteins (called **T cell receptors**) that react with antigens expressed on the surfaces of other cells. The key feature of both immunoglobulins and T cell receptors is their enormous diversity, which enables different antibody or T cell receptor molecules to recognize a vast array of foreign antigens. For example, each individual is capable of producing more than 10^{11} different antibody molecules, which is far in excess of the total number of genes in mammalian genomes (~21,000). Rather than being encoded in germ-line DNA, these diverse antibodies (and T cell receptors) are encoded by unique lymphocyte genes that are formed during development of the immune system as a result of site-specific recombination between distinct segments of immunoglobulin and T cell receptor genes.

The role of site-specific recombination in the formation of immunoglobulin genes was first demonstrated by Susumu Tonegawa in 1976. Immunoglobulins consist of pairs of identical heavy and light polypeptide chains (**Figure 7.33**). Both the heavy and light chains are composed of C-terminal constant regions and N-terminal variable regions. The variable regions, which have different amino acid sequences in different immunoglobulin molecules, are responsible for antigen binding, and it is the diversity of variable-region amino acid sequences that allows different

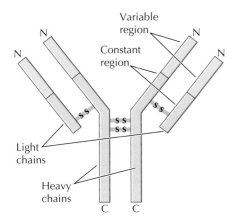

Figure 7.33 Structure of an immunoglobulin   Immunoglobulins are composed of two heavy chains and two light chains, joined by disulfide bonds. Both the heavy and the light chains consist of N-terminal variable and C-terminal constant regions.

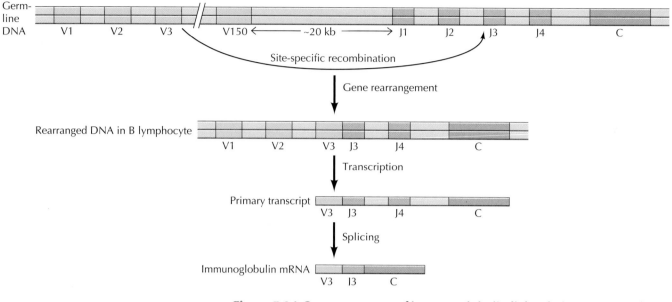

Figure 7.34 Rearrangement of immunoglobulin light-chain genes Each light-chain gene (mouse *k* light chains are illustrated) consists of a variable region (V), a joining region (J), and a constant region (C). There are approximately 150 different V regions, which are separated from J and C regions by about 20 kb in germ-line DNA. During the development of B lymphocytes, site-specific recombination joins one of the V regions to one of the four J regions. This rearrangement activates transcription, resulting in the formation of a primary transcript containing the rearranged VJ region together with the remaining J regions and C region. The remaining unused J regions and the introns between J and C regions are then removed by splicing, yielding a functional mRNA.

Animation 7.2

sites.sinauer.com/cooper7e/a7.2
Light-Chain Genes During B cell development, site-specific recombination joins regions of the immunoglobulin light-chain genes, leading to the production of unique immunoglobulin light chains.

Animation 7.3

sites.sinauer.com/cooper7e/a7.3
Heavy-Chain Genes Site-specific recombination between three regions leads to the formation of immunoglobulin heavy-chain genes.

individual antibodies to recognize unique antigens. Although every individual is capable of producing a vast spectrum of different antibodies, each B lymphocyte produces only a single type of antibody. Tonegawa's key discovery was that each antibody is encoded by unique genes formed by site-specific recombination during B lymphocyte development. These gene rearrangements create different immunoglobulin genes in different individual B lymphocytes, so the population of approximately 10^{12} B lymphocytes in the human body includes cells capable of producing antibodies against a diverse array of foreign antigens.

The genes that encode immunoglobulin light chains consist of three regions: a V region that encodes the 95 to 96 N-terminal amino acids of the polypeptide variable region; a joining (J) region that encodes the 12 to 14 C-terminal amino acids of the polypeptide variable region; and a C region that encodes the polypeptide constant region (**Figure 7.34**). The major class of light-chain genes in the mouse is formed from combinations of approximately 150 V regions and 4 J regions with a single C region. Site-specific recombination during lymphocyte development leads to a gene rearrangement in which a single V region recombines with a single J region to generate a functional light-chain gene. Different V and J regions are rearranged in different B lymphocytes, so the possible combinations of 150 V regions with 4 J regions can generate approximately 600 (4 × 150) unique light chains.

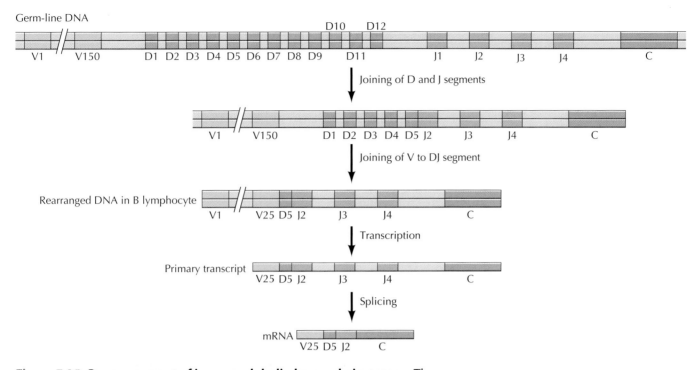

Germ-line DNA

Rearranged DNA in B lymphocyte

Primary transcript

mRNA

Figure 7.35 Rearrangement of immunoglobulin heavy-chain genes   The heavy-chain genes contain D regions in addition to V, J, and C regions. First the D and J segments join. Then a V segment is joined to the rearranged DJ region. The introns between J and C regions are removed by splicing to yield heavy-chain mRNA.

The heavy-chain genes include a fourth region (known as the diversity, or D, region), which encodes amino acids lying between V and J (**Figure 7.35**). Assembly of a functional heavy-chain gene requires two recombination events: A D region first recombines with a J region, and a V region then recombines with the rearranged DJ segment. In the mouse, there are about 150 heavy-chain V regions, 12 D regions, and 4 J regions, so the total number of heavy chains that can be generated by the recombination events is about 7,200 ($150 \times 12 \times 4$).

Combinations between the 600 different light chains and 7,200 different heavy chains formed by site-specific recombination can generate approximately 4×10^6 different immunoglobulin molecules. This diversity is further increased because the joining of immunoglobulin gene segments often involves the loss or gain of one to several nucleotides, as discussed below. The mutations resulting from these deletions and insertions increase the diversity of immunoglobulin variable regions more than a hundredfold, corresponding to the formation of about 10^5 different light chains and 10^6 heavy chains, which can then combine to form approximately 10^{11} distinct antibodies.

T cell receptors similarly consist of two chains (called α and β), each of which contains variable and constant regions (**Figure 7.36**). The genes encoding these polypeptides are generated by recombination between V and J segments (the α chain) or between V, D, and J segments (the β chain), analogous to the formation of immunoglobulin genes. Site-specific recombination between these distinct segments of DNA, in combination with mutations introduced

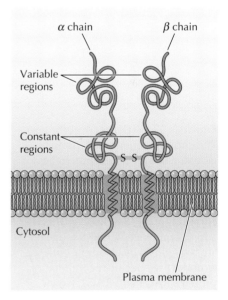

Figure 7.36 Structure of a T cell receptor   T cell receptors consist of two polypeptide chains (α and β) that span the plasma membrane and are joined by disulfide bonds. Both the α and β chains are composed of variable and constant regions.

during recombination, generates a degree of diversity in T cell receptors that is similar to that in immunoglobulins.

V(D)J recombination is mediated by a complex of two proteins, called RAG1 and RAG2, which are specifically expressed in lymphocytes. The RAG proteins recognize recombination signal (RS) sequences adjacent to the coding sequences of each gene segment, and initiate DNA rearrangements by introducing a double-strand break between the RS sequences and the coding sequences (**Figure 7.37**). The coding ends of the gene segments are then joined to yield a rearranged immunoglobulin or T cell receptor gene. Since these double-strand breaks are joined by a nonhomologous end-joining process (see Figure 7.28), the joining reaction is frequently accompanied by a loss of nucleotides. In addition, lymphocytes contain a specialized enzyme

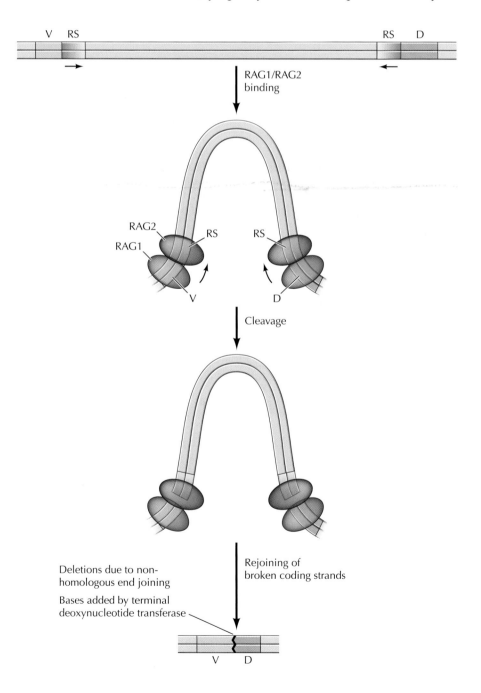

Figure 7.37 V(D)J recombination The coding sequences of immunoglobulin and T cell receptor genes (e.g., a V and D segment) are flanked by short recombination signal (RS) sequences, which are in opposite orientations at the 5′ and 3′ ends of the coding sequences. The RS sequences are recognized by a complex of the lymphocyte-specific recombination proteins RAG1 and RAG2, which cleave the DNA between the coding sequence and the RS sequence. The broken coding strands are then rejoined to yield a rearranged gene segment. Mutations result from the loss of bases at the ends during nonhomologous end joining, as well as from the addition of bases by terminal deoxynucleotidyl transferase.

(terminal deoxynucleotidyl transferase) that adds nucleotides at random to the ends of DNA molecules, so mutations corresponding to both the loss and the gain of nucleotides are introduced during the joining reaction. As noted above, these mutations contribute substantially to the diversity of immunoglobulins and T cell receptors.

Still further antibody diversity is generated after the formation of rearranged immunoglobulin genes by two processes that occur only in B lymphocytes: class switch recombination and somatic hypermutation. **Class switch recombination** results in the association of rearranged V(D)J regions with different heavy-chain constant regions, leading to the production of antibodies with distinct functional roles in the immune response. Mammals produce four different classes of immunoglobulins—IgM, IgG, IgE, and IgA—with heavy chains encoded by a variable region joined to the $C\mu$, $C\gamma$, $C\varepsilon$, and $C\alpha$ constant regions, respectively. In addition, there are several subclasses of IgG that are encoded by different $C\gamma$ regions. The different classes of immunoglobulins are specialized to remove antigens in different ways. IgM activates complement (a group of serum proteins that destroy invading cells or viruses), so IgM antibodies are an effective first line of defense against bacterial or viral infections. IgG antibodies, the most abundant immunoglobulins in serum, not only activate complement but also bind receptors on phagocytic cells. In addition, IgG antibodies can cross the placenta from the maternal circulation, providing immune protection to the fetus. IgA antibodies are secreted into a variety of bodily fluids, such as nasal mucus and saliva, where they can bind and eliminate invading bacteria or viruses to prevent infection. IgA antibodies are also secreted into the milk of nursing mothers, so they provide immune protection to newborns. IgE antibodies are effective in protecting against parasitic infections, and are also the class of antibodies responsible for allergies.

The initial V(D)J rearrangement produces a variable region joined to $C\mu$ resulting in the production of IgM antibodies. Class switch recombination then transfers a rearranged variable region to a new downstream constant region, with deletion of the intervening DNA (**Figure 7.38**). Recombination occurs between highly repetitive sequences in switch (S) regions that are located immediately upstream of each C region. The switch regions are 2–10 kb in length and recombination can take place anywhere within these regions, so class switching is more properly referred to as a region-specific rather than a site-specific recombination event. Because the switch regions are within introns,

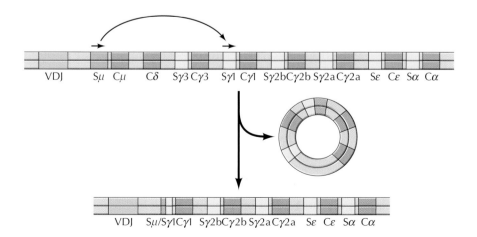

Figure 7.38 Class switch recombination   Class switching takes place by recombination between repetitive switch (S) regions upstream of a series of constant (C) regions in the heavy-chain locus (the mouse locus is shown). In the example shown, a V(D)J region is transferred from $C\mu$ to $C\gamma1$ by recombination between the $S\mu$ and $S\gamma1$ switch regions. The intervening DNA is excised as a circular molecule.

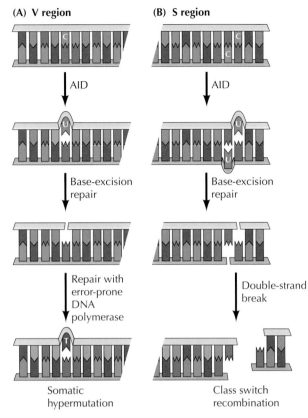

(A) V region

AID

Base-excision repair

Repair with error-prone DNA polymerase

Somatic hypermutation

(B) S region

AID

Base-excision repair

Double-strand break

Class switch recombination

Figure 7.39 Role of AID in somatic hypermutation and class switch recombination The activation-induced deaminase (AID) enzyme deaminates C to U in DNA. Removal of U by base-excision repair (see Figure 7.21) leaves a single-strand gap in the DNA. (A) In the variable (V) regions, errors during repair may lead to somatic hypermutation, possibly resulting from the action of a specialized error-prone DNA polymerase (see Figure 7.27). (B) In the switch (S) regions, the excision of bases on opposite strands may result in double-strand breaks that stimulate class switch recombination.

the precise site at which class switch recombination takes place does not affect the immunoglobulin coding sequence.

Somatic hypermutation increases the diversity of immunoglobulins by producing multiple mutations within rearranged variable regions of both heavy and light chains. These mutations, principally single-base substitutions, occur with frequencies as high as 10^{-3}, approximately a million times higher than normal rates of spontaneous mutation. They lead to the production of immunoglobulins with a substantially increased affinity for antigen, and therefore are an important contributor to an effective immune response.

Class switch recombination and somatic hypermutation are novel types of programmed genetic alterations, and the molecular mechanisms involved are very active areas of investigation. A key player in both processes is an enzyme called **activation-induced deaminase (AID)**, which was discovered by Tasuku Honjo and his collaborators in 1999. AID is expressed only in B lymphocytes, and it is required for both class switch recombination and somatic hypermutation. AID catalyzes the deamination of cytosine in DNA to form uracil (**Figure 7.39**). Its action results in the conversion of C→U in the variable regions and switch regions of immunoglobulin genes, leading to class switch recombination and somatic hypermutation, respectively. The mechanisms by which C→U mutations stimulate these processes are not completely understood, but an important step appears to be the removal of U by base-excision repair (see Figure 7.21). In the switch regions (which have a high content of GC base pairs) double-strand breaks are formed by the excision of uracil residues on opposite strands. Class switch recombination then results from the rejoining of these breaks in different switch regions by nonhomologous end joining (see Figure 7.28). In the variable regions, somatic hypermutation is thought to result from a high frequency of errors during repair of the C→U mutations, possibly resulting from repair by specialized error-prone DNA polymerases (see Figure 7.27). Although the details of these processes remain to be elucidated, it is clear that AID is an extremely interesting enzyme with the novel role of introducing mutations into DNA at a specific stage of development.

Gene amplification

The DNA rearrangements that have been discussed so far alter the position of a DNA sequence within the genome. **Gene amplification** may be viewed as a different type of alteration in genome structure; it increases the number of copies of a gene within a cell. Gene amplification results from repeated rounds of DNA replication, yielding multiple copies of a particular region (**Figure 7.40**). The amplified DNA sequences can be found either as free extrachromosomal molecules or as tandem arrays of sequences within a chromosome. In either case, the result is increased expression of the amplified gene, simply because more copies of the gene are available to be transcribed.

In some cases, gene amplification is responsible for developmentally programmed increases in gene expression. The prototypical example is

DNA

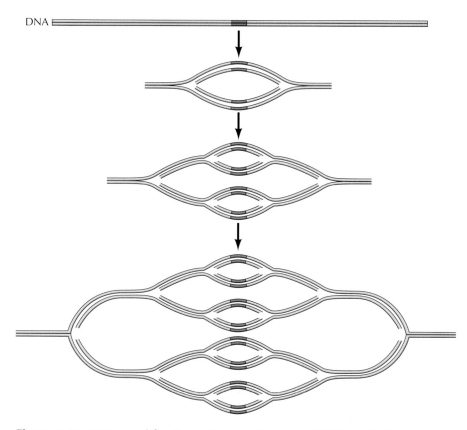

Figure 7.40 DNA amplification   Repeated rounds of DNA replication yield multiple copies of a particular chromosomal region.

amplification of the ribosomal RNA genes in amphibian oocytes (unfertilized eggs). Eggs are extremely large cells, with correspondingly high requirements for protein synthesis. Amphibian oocytes in particular are about a million times larger in volume than typical somatic cells and must support large amounts of protein synthesis during early development. This requires increased synthesis of ribosomal RNAs, which is accomplished in part by amplification of the ribosomal RNA genes. The genes for ribosomal RNA are already present in several hundred copies per genome, so that enough ribosomal RNA can be produced to meet the needs of somatic cells. In amphibian eggs, these genes are amplified an additional two-thousandfold, to approximately 1 million copies per oocyte. Another example of programmed gene amplification occurs in *Drosophila*, where the genes that encode eggshell proteins (chorion genes) are amplified in ovarian cells to support the requirement for large amounts of these proteins. Like other programmed gene rearrangements, however, gene amplification is a relatively infrequent event that occurs in highly specialized cell types; it is not a common mechanism of gene regulation.

Gene amplification also occurs as an abnormal event in cancer cells, where it frequently results in the increased expression of genes that drive cell proliferation (oncogenes), leading to uncontrolled cell growth and tumor development (see Chapter 19). Thus, as with other types of DNA rearrangements, gene amplification can have either beneficial or deleterious consequences for the cell or organism in which it occurs.

SUMMARY	KEY TERMS

DNA Replication

- *DNA polymerases:* Different DNA polymerases play distinct roles in DNA replication and repair in both prokaryotic and eukaryotic cells. All known DNA polymerases synthesize DNA only in the 5′ to 3′ direction by the addition of deoxyribonucleotide triphosphates (dNTPs) to a preformed primer strand of DNA. See Video 7.1.

  DNA polymerase

- *The replication fork:* Parental strands of DNA separate and serve as templates for the synthesis of two new strands at the replication fork. One new DNA strand (the leading strand) is synthesized in a continuous manner; the other strand (the lagging strand) is formed by the joining of small fragments of DNA that are synthesized backward with respect to the overall direction of replication. DNA polymerases and various other proteins act in a coordinated manner to synthesize both leading and lagging strands of DNA. See Animation 7.1.

  replication fork, Okazaki fragment, DNA ligase, leading strand, lagging strand, primase, exonuclease, RNase H, helicase, single-stranded DNA-binding protein, topoisomerase

- *The fidelity of replication:* DNA polymerases increase the accuracy of replication both by selecting the correct base for insertion and by proofreading newly synthesized DNA to eliminate mismatched bases.

  proofreading

- *Origins and the initiation of replication:* DNA replication starts at specific origins of replication, which contain binding sites for proteins that initiate the process. In higher eukaryotes, origins may be defined by chromatin structure rather than DNA sequence.

  origin of replication, autonomously replicating sequence (ARS), origin recognition complex (ORC)

- *Telomeres and telomerase: maintaining the ends of chromosomes:* Telomeric repeat sequences at the ends of chromosomes are maintained by the action of a reverse transcriptase (telomerase) that carries its own template RNA.

  telomere, telomerase, reverse transcriptase

DNA Repair

- *Direct reversal of DNA damage:* A few types of common DNA lesions, such as pyrimidine dimers and alkylated guanine residues, are repaired by direct reversal of the damage.

  pyrimidine dimer, photoreactivation

- *Excision repair:* Most types of DNA damage are repaired by excision of the damaged DNA. The resulting gap is filled by newly synthesized DNA, using the undamaged complementary strand as a template. In base-excision repair, specific types of single damaged bases are removed from the DNA molecule. In contrast, nucleotide excision repair systems recognize a wide variety of lesions that distort the structure of DNA and remove the damaged bases as part of an oligonucleotide. A third excision repair system specifically removes mismatched bases from newly synthesized DNA strands. See Video 7.2.

  base-excision repair, DNA glycosylase, AP endonuclease, nucleotide-excision repair, excinuclease, transcription-coupled repair, mismatch repair

- *Translesion DNA synthesis:* Specialized DNA polymerases are capable of replicating DNA across from a site of DNA damage, although the action of these polymerases may result in a high frequency of incorporation of incorrect bases.

  translesion DNA synthesis

- *Repair of double-strand breaks:* Double-strand breaks are repaired by recombination to rejoin the damaged strands. Recombinational repair can occur either by homologous recombination with an undamaged chromosome or by nonhomologous rejoining of the broken ends of a single DNA molecule.

  double-strand break, recombinational repair, homologous recombination, RecA, Rad51

SUMMARY

KEY TERMS

DNA Rearrangements

- *Yeast mating types:* Switching between yeast mating types is mediated by site-specific recombination between active and inactive alleles of the mating type locus.

- *Antibody genes:* In vertebrates, site-specific recombination plays a critical role in generating immunoglobulin and T cell receptor genes during development of the immune system. Additional diversity is provided to immunoglobulin genes by somatic hypermutation and class switch recombination. See Animations 7.2 and 7.3.

- *Gene amplification:* Gene amplification results from repeated replication of a chromosomal region. In some cases, gene amplification provides a mechanism for increasing gene expression during development. Gene amplification also frequently occurs in cancer cells, where it can result in the elevated expression of genes that contribute to uncontrolled cell proliferation.

site-specific recombination

antigen, immunoglobulin, T cell receptor, class switch recombination, somatic hypermutation, activation-induced deaminase (AID)

gene amplification

Questions

1. You have isolated a temperature-sensitive strain of *E. coli* with a mutation in DNA polymerase I. What defects, if any, would you observe in bacteria carrying this mutation at high temperature?

2. Compare and contrast the actions of topoisomerases I and II.

3. What is the approximate number of Okazaki fragments synthesized during replication of the yeast genome?

4. Why is the synthesis of Okazaki fragments initiated by primase instead of by a DNA polymerase?

5. What is the function of 3′ to 5′ exonuclease activity in DNA polymerases? What would be the consequence of mutating this activity of DNA polymerase III on the fidelity of replication of *E. coli* DNA?

6. How would you test a sequence of DNA in a yeast cell to determine whether it contains an origin of replication?

7. Yeast cells require telomerase to completely replicate their genome, whereas *E. coli* does not need this special enzyme. Why?

8. What mechanism does a cell use to repair double-strand breaks in its DNA? How does this differ from the repair of single-strand breaks?

9. Patients with xeroderma pigmentosum suffer an extremely high incidence of skin cancer but have not been found to have correspondingly high incidences of cancers of internal organs (e.g., colon cancer). What might this suggest about the kinds of DNA damage responsible for most internal cancers?

10. Many of the drugs in clinical use for the treatment of AIDS are inhibitors of the HIV reverse transcriptase. What other cellular processes might be affected by these inhibitors?

11. The mismatch repair system in humans includes homologs of *E. coli* *MutS* and *MutL* but not of *MutH*. Why?

12. What phenotype would you predict for a mutant mouse lacking one of the genes required for site-specific recombination in lymphocytes?

Refer To

The Cell

Companion Website

sites.sinauer.com/cooper7e

for quizzes, animations, videos, flashcards, and other study resources.

References and Further Reading   (Key review articles for each major section are highlighted in **bold**.)

DNA Replication

Annunziato, A. T. 2005. Split decision: What happens to nucleosomes during DNA replication? *J. Biol. Chem.* 280: 12065–12068. [R]

Armanios, M. and E. H. Blackburn. 2012. The telomere syndromes. *Nature Rev. Genet.* 13: 693–704. [R]

Bell, S. P. and A. Dutta. 2002. DNA replication in eukaryotic cells. *Ann. Rev. Biochem.* 71: 333–374. [R]

Bernandes de Jesus, B. and M. A. Blasco. 2013. Telomerase at the intersection of cancer and aging. *Trends Genet.* 29: 513–520.[R]

Cozzarelli, N. R., G. J. Cost, M. Nollmann, T. Uiard and J. E. Stray. 2006. Giant proteins that move DNA: bullies of the genomic playground. *Nature Mol. Cell Biol.* 7: 580–588. [R]

Frick, D. N. and C. C. Richardson. 2001. DNA primases. *Ann. Rev. Biochem.* 70: 39–80. [R]

Gilbert, D. M. 2004. In search of the holy replicator. *Nature Rev. Mol. Cell Biol.* 5: 1–8. [R]

Greider, C. W. and E. H. Blackburn. 1985. Identification of a specific telomere terminal transferase activity in *Tetrahymena* extracts. *Cell* 43: 405–413. [P]

Horn, S., A. Figl, P. S. Rachakonda, C. Fischer, A. Sucker, A. Gast, S. Kadel, I. Moll, E. Nagore, K. Hemminki, D. Schadendorf and R. Kumar. 2013. *TERT* promoter mutations in familial and sporadic melanoma. *Science* 339: 959-961. [P]

Huang, F. W., E. Hodis, M. J. Xu, G. V. Kryukov, L. Chin and L. A. Garraway. 2013. Highly recurrent *TERT* promoter mutations in human melanoma. *Science* 339: 957-959. [P]

Huberman, J. A. and A. D. Riggs. 1968. On the mechanism of DNA replication in mammalian chromosomes. *J. Mol. Biol.* 32: 327–341. [P]

Johnson, A. and M. O'Donnell. 2005. Cellular DNA replicases: components and dynamics at the replication fork. *Ann. Rev. Biochem.* 74: 283–315. [R]

Kornberg, A. 2000. Ten commandments: lessons from the enzymology of DNA replication. *J. Bacteriol.* 182: 3613–3618 [R]

Machida, Y. J., J. L. Hamlin and A. Dutta. 2005. Right place, right time, and only once: replication initiation in metazoans. *Cell* 123: 13–24. [R]

McCulloch, S. D. and T. A. Kunkel. 2008. The fidelity of DNA synthesis by eukaryotic replicative and translesion synthesis polymerases. *Cell Res.* 18: 148–161. [R]

Mechali, M. 2010. Eukaryotic DNA replication origins: many choices for appropriate answers. *Nature Rev. Mol. Cell Biol.* 11: 728–738. [R]

Moldovan, G.-L., B. Pfander and S. Jentsch. 2007. PCNA, the maestro of the replication fork. *Cell* 129: 665–679. [R]

Nitiss, J. L. 2009. DNA topoisomerase II and its growing repertoire of biological functions. *Nature Rev. Cancer* 9: 327–337. [R]

Robinson, N. P. and S. D. Bell. 2005. Origins of DNA replication in the three domains of life. *FEBS J.* 272: 3757–3766. [R]

Stillman, B. 2008. DNA polymerases at the replication fork in eukaryotes. *Mol. Cell* 30: 259–260. [R]

Stinthcomb, D. T., K. Struhl and R. W. Davis. 1979. Isolation and characterization of a yeast chromosomal replicator. *Nature* 282: 39–43. [P]

DNA Repair

Cleaver, J. E. 1968. Defective repair replication of DNA in xeroderma pigmentosum. *Nature* 218: 652–656. [P]

David, S. S., V. L. O'Shea and S. Kundu. 2007. Base-excision repair of oxidative DNA damage. *Nature* 447: 941–950. [R]

Essen, L. O. and T. Klar. 2006. Light-driven DNA repair by photolyases. *Cell. Mol. Life Sci.* 63: 1266–1277. [R]

Fishel, R., M. K. Lescoe, M. R. S. Rao, N. G. Copeland, N. A. Jenkins, J. Garber, M. Kane and R. Kolodner. 1993. The human mutator gene homolog *MSH2* and its association with hereditary nonpolyposis colon cancer. *Cell* 75: 1027–1038. [P]

Friedberg, E. C., A. R. Lehmann and R. P. P. Fuchs. 2005. Trading places: How do DNA polymerases switch during translesion DNA synthesis? *Mol. Cell* 18: 499–505. [R]

Friedberg, E. C., G. C. Walker, W. Siede, R. D. Wood, R. A. Schultz and T. Ellenberger. 2005. *DNA Repair and Mutagenesis.* Washington, D.C.: ASM Press.

Hanawalt, P. C. and G. Spivak. 2008. Transcription-coupled DNA repair: two decades of progress and surprises. *Nature Rev. Mol. Cell Biol.* 9: 958–970. [R]

Hegde, M. L., T. K. Hazra and S. Mitra. 2008. Early steps in the DNA base excision/single-strand interruption repair pathway in mammalian cells. *Cell Res.* 18: 27–47. [R]

Jensen, R.B., A. Carreira and S. C. Kowalczykowski. 2010. Purified human BRCA2 stimulates RAD51-mediated recombination. *Nature* 467: 678–683. [P]

Jiricny, J. 2006. The multifaceted mismatch repair system. *Nature Rev. Mol. Cell Biol.* 7: 335–346. [R]

Leach, F. S. and 34 others. 1993. Mutations of a *mutS* homolog in hereditary nonpolyposis colorectal cancer. *Cell* 75: 1215–1225. [P]

Lehmann, A. R., A. Niimi, T. Ogi, S. Brown, S. Sabbioneda, J. F. Wing, P. L. Kannouche and C. M. Green. 2007. Translesion synthesis: Y-family polymerases and the polymerase switch. *DNA Repair* 6: 891–899. [R]

Lieber, M. R. 2008. The mechanism of human nonhomologous DNA end joining. *J. Biol. Chem.* 283: 1–5. [R]

Marteijn, J. A., H. Lans, W. Vermeulen and J. H. Hoeijmakers. 2014. Understanding nucleotide excision repair and its roles in cancer and ageing. *Nature Rev. Mol. Cell Biol.* 15: 465–481. [R]

Mimitou, E. P. and L. S. Symington. 2009. Nucleases and helicases take center stage in homologous recombination. *Trends Biochem. Sci.* 34: 264–272. [R]

Pena-Diaz, J. and J. Jiricny. 2012. Mammalian mismatch repair: error-free or error-prone? *Trends Biochem. Sci.* 37: 206–214. [R]

San Filippo, J., P. Sung and H. Klein. 2008. Mechanism of eukaryotic homologous recombination. *Ann. Rev. Biochem.* 77: 9.1–9.29. [R]

Savery, N. J. 2007. The molecular mechanism of transcription-coupled DNA repair. *Trends Microbiol.* 15: 326–333. [R]

Sedgwick, B., P. A. Bates, J. Paik, S. C. Jacobs and T. Lindahl. 2007. Repair of alkylated DNA: recent advances. *DNA Repair* 6: 429–442. [R]

Thorlund, T. and S. C. West. 2007. BRCA2: a universal recombinase regulator. *Oncogene* 26: 7720–7730. [R]

Yang, W. and R. Woodgate. 2007. What a difference a decade makes: insights into translesion DNA synthesis. *Proc. Natl. Acad. Sci. USA* 104: 15591–15598. [R]

DNA Rearrangements

Chaudhuri, J. and F. W. Alt. 2004. Class-switch recombination: Interplay of transcription, DNA deamination and DNA repair. *Nature Rev. Immunol.* 4: 541–552. [R]

Claycomb, J. M. and T. L. Orr-Weaver. 2005. Developmental gene amplification: Insights into DNA replication and gene expression. *Trends Genet.* 21: 149–162. [R]

Di Noia, J. M. and M. S. Neuberger. 2007. Molecular mechanisms of antibody somatic hypermutation. *Ann. Rev. Biochem.* 76: 1–22. [R]

Haber, J. E. 2012. Mating-type genes and *MAT* switching in *Saccharomyces cerevisiae. Genetics* 191: 33-64. [R]

Honjo, T., H. Nagaoka, R. Shinkura and M. Muramatsu. 2005. AID to overcome the limitations of genomic information. *Nature Immunol.* 6: 655–661. [R]

Hozumi, N. and S. Tonegawa. 1976. Evidence for somatic rearrangement of immunoglobulin genes coding for variable and constant regions. *Proc. Natl. Acad. Sci. USA* 73: 3628–3632. [P]

Jung, D. and F. W. Alt. 2006. Mechanism and control of V(D)J recombination at the immunoglobulin heavy chain locus. *Ann. Rev. Immunol.* 24: 541–570. [R]

Maizels, N. 2005. Immunoglobulin gene diversification. *Ann. Rev. Genet.* 39: 23–46. [R]

Maul, R. W. and P. J. Gearhart. 2010. AID and somatic hypermutation. *Adv. Immunol.* 105: 159–191. [R]

Schatz, D. G. and Y. Ji. 2011. Recombination centres and the orchestration of V[D]J recombination. *Nature Rev. Immunol.* 11: 251–263. [R]

Teng, G. and F. N. Papavasiliou. 2007. Immunoglobulin somatic hypermutation. *Ann. Rev. Genet.* 41: 107–120. [R]

Tower, J. 2004. Developmental gene amplification and origin regulation. *Ann. Rev. Genet.* 38: 273–304. [R]

RNA Synthesis and Processing

Chapters 6 and 7 discussed the organization and maintenance of genomic DNA, which can be viewed as the set of genetic instructions governing all cellular activities. These instructions are implemented via the synthesis of RNAs and proteins. Importantly, the behavior of a cell is determined not only by what genes it inherits but also by which of those genes are expressed at any given time. Regulation of gene expression allows cells to adapt to changes in their environments and is responsible for the distinct activities of the multiple differentiated cell types that make up complex plants and animals. Almost all cells in multicellular organisms have the same genomic DNA. Muscle cells and liver cells, for example, contain the same genes; the functions of these cells are determined not by differences in their genomes, but by regulated patterns of gene expression that govern development and differentiation.

The first step in expression of a gene, the transcription of DNA into RNA, is the initial level at which gene expression is regulated in both prokaryotic and eukaryotic cells. RNAs in eukaryotic cells are then modified in various ways—for example, introns are removed by splicing—to convert the primary transcript into its functional form. Different types of RNA play distinct roles in cells: Messenger RNAs (mRNAs) serve as templates for protein synthesis; ribosomal RNAs (rRNAs) and transfer RNAs (tRNAs) function in mRNA translation. Still other noncoding RNAs function in gene regulation, mRNA splicing, rRNA processing, and protein sorting in eukaryotes. In fact, some of the most exciting advances in recent years have pertained to the roles of noncoding RNAs as regulators of gene expression in eukaryotic cells. Transcription and RNA processing are discussed in this chapter. The final step in gene expression, the translation of mRNA to protein, is the subject of Chapter 9.

Transcription in Bacteria

As in most areas of molecular biology, studies of *E. coli* have provided the model for subsequent investigations of transcription in eukaryotic cells. As reviewed in Chapter 4, mRNA was discovered first in *E. coli*. *E. coli* was also the first organism from which RNA polymerase was purified and studied. The basic mechanisms by which transcription is regulated were likewise elucidated by pioneering experiments in *E. coli* in which regulated gene expression allows the cell to respond to variations in the environment, such as changes in the availability of nutrients. An understanding of transcription in *E. coli* has thus provided the foundation for studies of the far more complex mechanisms that regulate gene expression in eukaryotic cells.

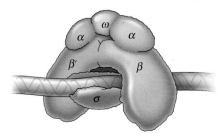

Figure 8.1 *E. coli* **RNA polymerase** The complete enzyme consists of six subunits: two α, one β, one β', one ω, and one σ. The σ subunit is relatively weakly bound and can be dissociated from the other five subunits, which constitute the core polymerase.

Animation 8.1

sites.sinauer.com/cooper7e/a8.1

Transcription Transcription is DNA-directed RNA synthesis, catalyzed by the enzyme RNA polymerase.

RNA *polymerase and transcription*

The principal enzyme responsible for RNA synthesis is **RNA polymerase**, which catalyzes the polymerization of ribonucleotide triphosphates (NTPs) as directed by a DNA template. The synthesis of RNA is similar to that of DNA, and like DNA polymerase, RNA polymerase catalyzes the growth of RNA chains always in the 5' to 3' direction. Unlike DNA polymerase, however, RNA polymerase does not require a preformed primer to initiate the synthesis of RNA. Instead, transcription initiates *de novo* at specific sites at the beginning of genes. The initiation process is particularly important because this is a major step at which transcription is regulated.

RNA polymerase, like DNA polymerase, is a complex enzyme made up of multiple polypeptide chains. The intact bacterial enzyme consists of five different types of subunits, called α, β, β', ω, and σ (**Figure 8.1**). The σ subunit is relatively weakly bound and can be separated from the other subunits, yielding a core polymerase consisting of two α, one β, one β', and one ω subunits. The core polymerase is fully capable of catalyzing the polymerization of NTPs into RNA, indicating that σ is not required for the basic catalytic activity of the enzyme. However, the core polymerase does not bind specifically to the DNA sequences that signal the normal initiation of transcription; therefore the σ subunit is required to identify the correct sites for transcription initiation. The selection of these sites is a critical element of transcription because synthesis of a functional RNA must start at the beginning of a gene. Most bacteria, including *E. coli*, have several different σ's that direct RNA polymerase to different classes of transcription start sites under different conditions—for example, under conditions of starvation as opposed to nutrient availability.

The DNA sequence to which RNA polymerase binds to initiate transcription of a gene is called the **promoter**. The DNA sequences involved in promoter function were first identified by comparisons of the nucleotide sequences of a series of different genes isolated from *E. coli*. These comparisons revealed that the region upstream (5') of the transcription initiation site contains two sets of sequences that are similar in a variety of genes. These common sequences encompass six nucleotides each and are located approximately 10 and 35 base pairs upstream of the transcription start site (**Figure 8.2**). They are called the –10 and –35 elements, denoting their position relative to the transcription initiation site, which is defined as the +1 position. The sequences at the –10 and –35 positions in different promoters are not identical, but they are all similar enough to establish consensus sequences—the bases most frequently found at each position.

Several types of experimental evidence support the functional importance of the –10 and –35 promoter elements. First, genes with promoters that differ from the consensus sequences are transcribed less efficiently than genes whose promoters match the consensus sequences more closely. In addition, direct analysis of the sites at which RNA polymerase binds to promoters have shown that the polymerase generally binds to an approximately 60-base-pair region, extending from –40 to +20 (i.e., from 40 nucleotides upstream to 20 nucleotides downstream of the transcription start site). The σ subunit binds specifically to sequences in both the –35 and –10 promoter regions, substantiating the importance of these sequences in promoter function. In addition, some *E. coli*

Figure 8.2 Sequences of *E. coli* promoters *E. coli* promoters are characterized by two sets of sequences located 10 and 35 base pairs upstream of the transcription start site (+1). The consensus sequences shown correspond to the bases most frequently found in different promoters.

promoters have a third sequence, located upstream of the –35 region, that serves as a specific binding site for the RNA polymerase α subunit.

In the absence of σ, RNA polymerase binds nonspecifically to DNA with low affinity. The role of σ is to direct the polymerase to promoters by binding specifically to both the –35 and –10 sequences, leading to the initiation of transcription at the beginning of a gene (Figure 8.3). The initial binding

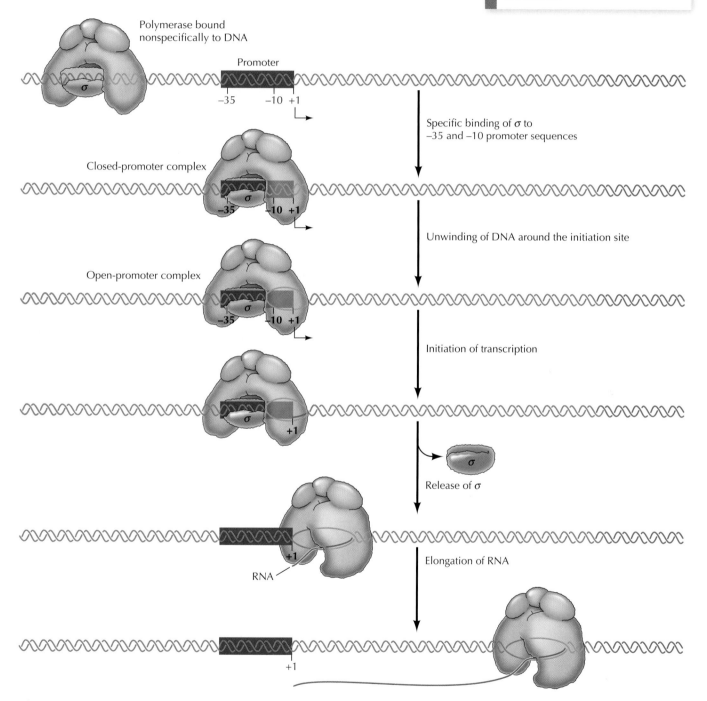

Figure 8.3 Transcription by *E. coli* RNA polymerase
The polymerase initially binds nonspecifically to DNA and migrates along the molecule until the σ subunit binds to the –35 and –10 promoter elements, forming a closed-promoter complex. The polymerase then unwinds DNA around the initiation site, forming an open-promoter complex, and transcription is initiated by the polymerization of free NTPs. The σ subunit then dissociates from the core polymerase, which migrates along the DNA and elongates the growing RNA chain.

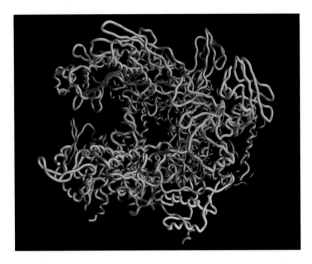

Figure 8.4 Structure of bacterial RNA polymerase The α subunit of the polymerase is colored green, with β blue, β' pink, and ω yellow. (Courtesy of Seth Darst, Rockefeller University.)

between the polymerase and a promoter is referred to as a closed-promoter complex because the DNA is not unwound. The polymerase then unwinds 12–14 bases of DNA, from about –12 to +2, to form an open-promoter complex in which single-stranded DNA is available as a template for transcription. Transcription is initiated by the joining of two free NTPs. After addition of about the first ten nucleotides, σ is released from the polymerase, which then leaves the promoter and moves along the template DNA to continue elongation of the growing RNA chain.

During elongation, the polymerase remains associated with its template while it continues synthesis of mRNAs. As it travels, the polymerase unwinds the template DNA ahead of it and rewinds the DNA behind it, maintaining an unwound region of about 15 base pairs in the region of transcription. Within this unwound portion of DNA, 8–9 bases of the growing RNA chain are bound to the complementary template DNA strand. High-resolution structural analysis of bacterial RNA polymerase indicates that the β and β' subunits form a crab-claw-like structure that grips the DNA template (**Figure 8.4**). An internal channel between the β and β' subunits accommodates approximately 20 base pairs of DNA and contains the polymerase active site, at which RNA synthesis occurs.

RNA synthesis continues until the polymerase encounters a termination signal, at which point transcription stops, the RNA is released from the polymerase, and the enzyme dissociates from its DNA template. There are two alternative mechanisms for termination of transcription in *E. coli*. The simplest and most common type of termination signal consists of a symmetrical inverted repeat of a GC-rich sequence followed by approximately seven A residues (**Figure 8.5**). Transcription of the

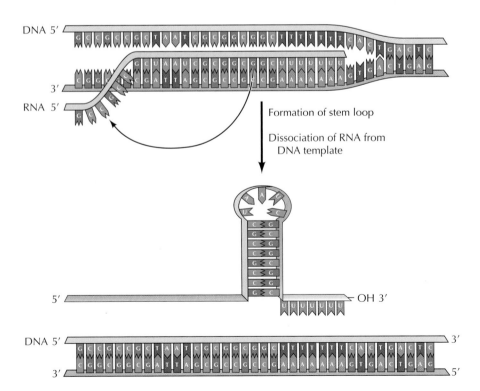

Formation of stem loop

Dissociation of RNA from DNA template

Figure 8.5 Transcription termination The termination of transcription is signaled by a GC-rich inverted repeat followed by seven A residues. The inverted repeat forms a stable stem-loop structure in the RNA, terminating transcription and leading to dissociation of the RNA from the DNA template.

GC-rich inverted repeat results in the formation of a segment of RNA that can form a stable stem-loop structure by complementary base pairing. The formation of such a self-complementary structure in the RNA disrupts its association with the DNA template and terminates transcription. Because hydrogen bonding between A and U is weaker than that between G and C, the presence of A residues downstream of the inverted repeat sequences is thought to facilitate the dissociation of the RNA from its template. Alternatively, the transcription of some genes is terminated by a specific termination protein (called Rho), which binds extended segments (greater than 60 nucleotides) of single-stranded RNA. Since mRNAs in bacteria become associated with ribosomes and are translated while they are being transcribed, such extended regions of single-stranded RNA are exposed only at the end of an mRNA.

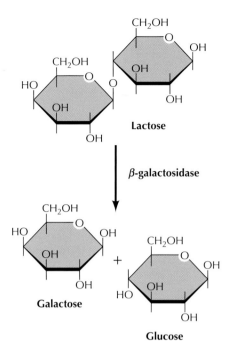

Figure 8.6 Metabolism of lactose β-galactosidase catalyzes the hydrolysis of lactose to galactose and glucose.

Repressors and negative control of transcription

Transcription can be regulated at the stages of both initiation and elongation, but most transcriptional regulation in bacteria operates at the level of initiation. The pioneering studies of gene regulation in *E. coli* were carried out by François Jacob and Jacques Monod in the 1950s. These investigators and their colleagues analyzed the expression of enzymes involved in the metabolism of lactose, which can be used as a source of carbon and energy via cleavage to galactose and glucose (**Figure 8.6**). The enzyme that catalyzes the cleavage of lactose (β-galactosidase) and other enzymes involved in lactose metabolism are expressed only when lactose is available for use by the bacteria. Otherwise, the cell is able to economize by not investing energy in the synthesis of unnecessary RNAs and proteins. Thus lactose induces the synthesis of enzymes involved in its own metabolism. In addition to requiring β-galactosidase, lactose metabolism involves the products of two other closely linked genes: lactose permease, which transports lactose into the cell, and a transacetylase, which is thought to inactivate toxic thiogalactosides that are transported into the cell along with lactose by the permease.

On the basis of purely genetic experiments, Jacob and Monod deduced the mechanism by which the expression of these genes was regulated, thereby formulating a model that remains fundamental to our understanding of transcriptional regulation. The genes encoding β-galactosidase, permease, and transacetylase are expressed as a single unit, called an **operon** (**Figure 8.7**). Studies of mutants that were

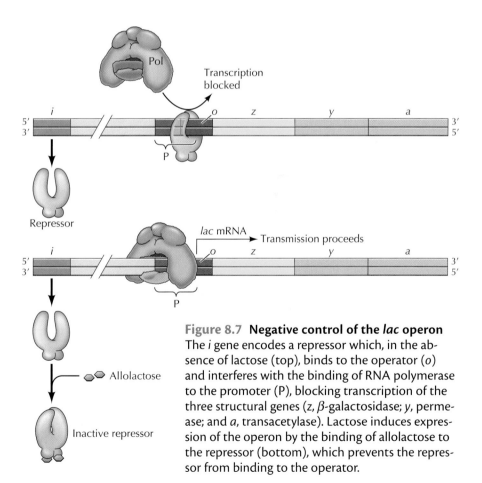

Figure 8.7 Negative control of the *lac* operon The *i* gene encodes a repressor which, in the absence of lactose (top), binds to the operator (*o*) and interferes with the binding of RNA polymerase to the promoter (P), blocking transcription of the three structural genes (*z*, β-galactosidase; *y*, permease; and *a*, transacetylase). Lactose induces expression of the operon by the binding of allolactose to the repressor (bottom), which prevents the repressor from binding to the operator.

defective in regulation of these genes identified two distinct loci, called *o* and *i*, that controlled expression of the operon. Transcription of the operon is controlled by *o* (the **operator**), which is adjacent to the transcription initiation site. The *i* gene, which is not physically linked to the operon, encodes a protein that regulates transcription by binding to the operator DNA. Importantly, mutants that fail to make a functional *i* gene product result in constitutive expression of the operon even when lactose is not available. This result implies that the normal *i* gene product is a **repressor**, which blocks transcription when bound to *o*. The addition of lactose leads to induction of the operon because allolactose (a metabolite of lactose) binds to the repressor, thereby preventing it from binding to the operator DNA.

Confirmation of this basic model has since come from a variety of experiments, including Walter Gilbert's isolation, in the 1960s, of the *lac* repressor and analysis of its binding to operator DNA. Molecular analysis has defined the operator as approximately 20 base pairs of DNA, starting a few bases before the transcription initiation site. The repressor binds to this region, blocking transcription by interfering with the binding of RNA polymerase to the promoter.

The central principle of gene regulation exemplified by the lactose operon is that control of transcription is mediated by the interaction of regulatory proteins with specific DNA sequences. This general mode of regulation is broadly applicable to both prokaryotic and eukaryotic cells. Regulatory sequences like the operator are called **cis-acting control elements**, because they affect the expression of linked genes on the same DNA molecule. On the other hand, proteins like the repressor can affect the expression of genes located on other chromosomes within the cell. The *lac* operon is an example of negative control because binding of the repressor blocks transcription. This, however, is not always the case; there are also many cases of positive control in which regulatory proteins are activators rather than inhibitors of transcription.

Positive control of transcription

The best-studied example of positive control in *E. coli* is the effect of glucose on the expression of genes that encode enzymes involved in the breakdown (catabolism) of other sugars (including lactose) that provide alternative sources of carbon and energy. Glucose is preferentially utilized; so as long as glucose is available, enzymes involved in catabolism of alternative energy sources are not expressed. For example, if *E. coli* are grown in medium containing both glucose and lactose, the *lac* operon is not induced and only glucose is used by the bacteria. Thus glucose represses the *lac* operon even in the presence of the normal inducer (lactose).

Glucose repression (generally called catabolite repression) is now known to be mediated by a positive control system, which is coupled to levels of cyclic AMP (cAMP) (**Figure 8.8**). In bacteria, the enzyme adenylyl cyclase,

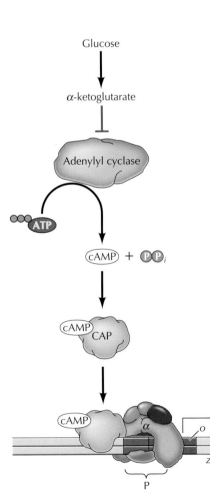

Figure 8.8 Control of the *lac* operon by glucose Glucose is metabolized to α-ketoglutarate, which inhibits adenylyl cyclase, the enzyme that converts ATP to cyclic AMP (cAMP). But if glucose is low, cyclic AMP binds to the catabolite activator protein (CAP) and stimulates its binding to regulatory sequences of various operons concerned with the metabolism of alternative sugars, such as lactose. CAP interacts with the α subunit of RNA polymerase to facilitate the binding of polymerase to the promoter (P).

which converts ATP to cAMP, is inhibited by α-ketoglutarate, an intermediate in the breakdown of glucose via the citric acid cycle (see Figure 3.4). When glucose is available, α-ketoglutarate is produced and adenylyl cyclase is inhibited. However, in the absence of glucose or other efficiently used sources of carbon and energy, levels of α-ketoglutarate decrease, leading to activation of adenylyl cyclase and synthesis of cAMP. cAMP then binds to a transcriptional regulatory protein called catabolite activator protein (CAP). The binding of cAMP stimulates the binding of CAP to its target DNA sequences, which in the *lac* operon are located approximately 60 bases upstream of the transcription start site. CAP then interacts with the α subunit of RNA polymerase, facilitating the binding of polymerase to the promoter and activating transcription.

Eukaryotic RNA Polymerases and General Transcription Factors

Although transcription proceeds by the same fundamental mechanisms in all cells, it is considerably more complex in eukaryotic cells than in bacteria. This is reflected in three distinct differences between the bacterial and eukaryotic systems. First, whereas all genes are transcribed by a single core RNA polymerase in bacteria, eukaryotic cells contain three nuclear RNA polymerases that transcribe distinct classes of genes. Second, eukaryotic RNA polymerases need to interact with a variety of additional proteins to specifically initiate and regulate transcription. Finally, transcription in eukaryotes takes place on chromatin, and regulation of chromatin structure is an important factor in regulating the transcriptional activity of eukaryotic genes. This increased complexity of eukaryotic transcription presumably facilitates the sophisticated regulation of gene expression needed to direct the activities of the many different cell types of multicellular organisms.

Eukaryotic RNA polymerases

Eukaryotic cells contain three distinct nuclear RNA polymerases that transcribe different classes of genes (Table 8.1). Protein-coding genes are transcribed by RNA polymerase II to yield mRNAs. RNA polymerase II also transcribes microRNAs (miRNAs) and long noncoding RNAs (lncRNAs), which are critical regulators of gene expression in eukaryotic cells. Ribosomal RNAs (rRNAs) and transfer RNAs (tRNAs) are transcribed by RNA polymerases I and III. RNA polymerase I is specifically devoted to transcription of the three largest species of rRNAs, which are designated 28S, 18S, and 5.8S according to their rates of sedimentation during velocity centrifugation. RNA polymerase III transcribes the genes for tRNAs and for the smallest species of ribosomal RNA (5S rRNA). Some of the small RNAs involved in splicing and protein transport (snRNAs and scRNAs) are also transcribed by RNA polymerase III, while others are polymerase II transcripts. In addition, separate RNA polymerases (which are similar to bacterial RNA polymerases) are found in chloroplasts and mitochondria, where they specifically transcribe the DNAs of those organelles.

   All three of the nuclear RNA polymerases are complex enzymes, consisting of 12 to 17 different subunits each. Although they recognize different promoters and transcribe distinct

Table 8.1 Classes of Genes Transcribed by Eukaryotic RNA Polymerases

Type of RNA synthesized	RNA polymerase
Nuclear genes	
mRNA	II
miRNA	II
lncRNA	II
tRNA	III
rRNA	
5.8S, 18S, 28S	I
5S	III
snRNA and scRNA	II and III[a]
Mitochondrial genes	Mitochondrial[b]
Chloroplast genes	Chloroplast[b]

[a]Some small nuclear (sn) and small cytoplasmic (sc) RNAs are transcribed by polymerase II and others by polymerase III.

[b]The mitochondrial and chloroplast RNA polymerases are similar to bacterial enzymes.

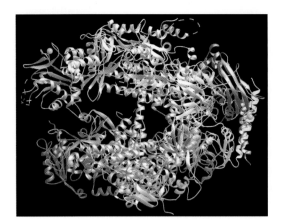

Figure 8.9 Structure of yeast RNA polymerase II Individual subunits are distinguished by colors. (From P. D. Kramer et al., 2001. *Science* 292: 1863.)

classes of genes, they share several features in common with each other as well as with bacterial RNA polymerase. In particular, all three eukaryotic RNA polymerases contain nine conserved subunits, five of which are related to the α, β, β', and ω subunits of bacterial RNA polymerase. The structure of yeast RNA polymerase II is strikingly similar to that of the bacterial enzyme (Figure 8.9), suggesting that all RNA polymerases utilize fundamentally conserved mechanisms to transcribe DNA.

General transcription factors and initiation of transcription by RNA polymerase II

Because RNA polymerase II is responsible for the synthesis of mRNA from protein-coding genes, it has been the focus of most studies of transcription in eukaryotes. Early attempts at studying this enzyme indicated that its activity is different from that of bacterial RNA polymerase. The accurate transcription of bacterial genes that can be accomplished *in vitro* simply by the addition of purified RNA polymerase to DNA containing a promoter is not possible in eukaryotic systems. The basis of this difference was elucidated in 1979, when Robert Roeder and his colleagues discovered that RNA polymerase II is able to initiate transcription only if additional proteins are added to the reaction. Thus transcription in the eukaryotic system appeared to require distinct initiation factors that (in contrast to bacterial σ factors) were not associated with the polymerase.

Biochemical fractionation of nuclear extracts subsequently led to the identification of specific proteins (called **transcription factors**) that are required for RNA polymerase II to initiate transcription. **General transcription factors** are involved in transcription from most polymerase II promoters and therefore constitute part of the basic transcription machinery. Additional gene-specific transcription factors (discussed later in the chapter) bind to DNA sequences that control the expression of individual genes and are thus responsible for regulating gene expression. It is estimated that about 10% of the genes in the human genome encode transcription factors, emphasizing the importance of these proteins.

The promoters of genes transcribed by polymerase II contain several different sequence elements surrounding their transcription sites (Figure 8.10). The first

Figure 8.10 Formation of a polymerase II preinitiation complex *in vitro* Sequence elements at polymerase II promoters include the TATA box (consensus sequence TATAA) 25 to 30 nucleotides upstream of the transcription start site, the TFIIB recognition element (BRE) approximately 35 nucleotides upstream of the transcription start site, the initiator (Inr) element, which spans the transcription start site, and several elements downstream of the transcription start site (the DCE, MTE, and DPE). Formation of a transcription complex is initiated by the binding of transcription factor TFIID. One subunit of TFIID, the TATA-binding protein or TBP, binds to the TATA box; other subunits (TBP-associated factors or TAFs) bind to the Inr and downstream promoter elements. TFIIB(B) then binds to TBP as well as to BRE sequences, followed by binding of the polymerase in association with TFIIF(F). Finally, TFIIE(E) and TFIIH(H) associate with the complex. CTD is the C-terminal domain of the largest subunit of RNA polymerase.

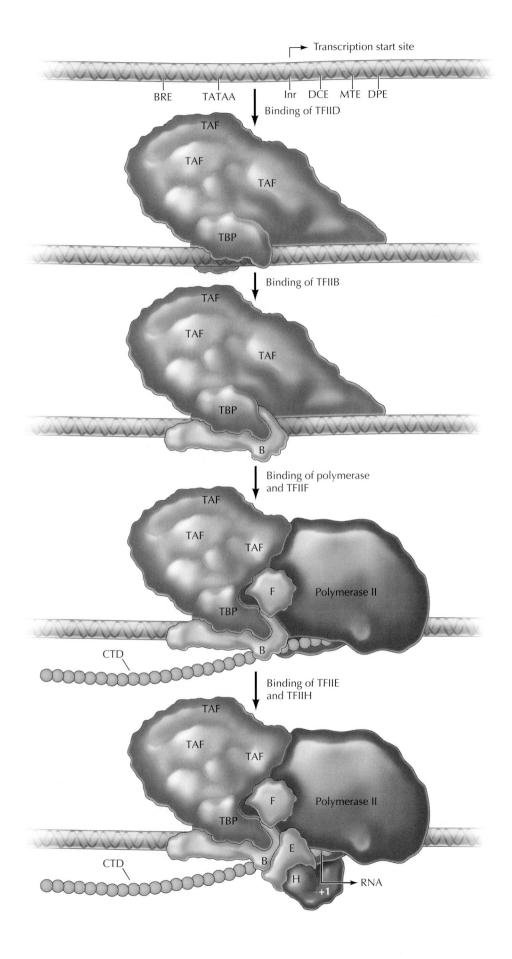

of these elements to be identified was a sequence similar to TATAA located 25 to 30 nucleotides upstream of the transcription start site. This sequence (called the **TATA box**) resembles the –10 sequence element of *E. coli* promoters, and was initially thought to be a general characteristic of the promoters of genes transcribed by RNA polymerase II. However, more recent genome-wide studies have shown that TATA boxes are present in the promoters of only 10 to 20% of RNA polymerase II promoters. Additional sequence elements in the promoters of genes transcribed by RNA polymerase II include the initiator (Inr) elements, which span transcription start sites; TFIIB recognition elements (BRE), which are located upstream of transcription start sites; and several promoter elements located downstream of transcription start sites (the DCE, MTE, and DPE). The promoters of different genes contain different combinations of these core promoter elements, which appear to function together to bind general transcription factors.

Five general transcription factors are minimally required for initiation of transcription by RNA polymerase II in reconstituted *in vitro* systems (see Figure 8.10). The first step in formation of a transcription complex is the binding of a general transcription factor called TFIID to the promoter (TF indicates transcription factor; II indicates polymerase II). TFIID is itself composed of multiple subunits, including the **TATA-binding protein (TBP)** and 13 or 14 other polypeptides, called **TBP-associated factors (TAFs)**. TBP binds specifically to the TATA box while other subunits of TFIID (TAFs) bind to the Inr, DCE, MTE, and DPE sequences. The binding of TFIID is followed by recruitment of a second general transcription factor (TFIIB), which binds to TBP as well as to BRE sequences. TFIIB in turn serves as a bridge to RNA polymerase II, which binds to the TBP-TFIIB complex in association with a third factor, TFIIF.

Following recruitment of RNA polymerase II to the promoter, the binding of two additional factors (TFIIE and TFIIH) completes formation of the preinitiation complex. TFIIH is a multisubunit factor that appears to play at least two important roles. First, two subunits of TFIIH are helicases, which unwind DNA around the initiation site. (These subunits of TFIIH are the XPB and XPD proteins, which are also required for nucleotide excision repair, as discussed in Chapter 7). Another subunit of TFIIH is a protein kinase that phosphorylates repeated sequences present in the C-terminal domain of the largest subunit of RNA polymerase II. The polymerase II C-terminal domain (or CTD) consists of tandem repeats (26 repeats in yeast and 52 in humans) of seven amino acids with the consensus sequence Tyr-Ser-Pro-Thr-Ser-Pro-Ser. Phosphorylation of serine-5 (the fifth amino acid) in these CTD repeats by the TFIIH protein kinase releases the polymerase from its association with the preinitiation complex and leads to the initiation of transcription.

Although the recruitment of five general transcription factors and RNA polymerase II described above represents the minimal system required for transcription *in vitro*, additional factors are needed to stimulate transcription within the cell. These factors include a large protein complex, called **Mediator,** which consists of more than 20 distinct subunits and interacts both with general transcription factors and with RNA polymerase (**Figure 8.11**). The Mediator complex not only stimulates basal transcription, but also plays a key role in linking the general transcription factors to the gene-specific transcription factors that regulate gene expression. The Mediator proteins are released from the polymerase following assembly of the preinitiation complex and phosphorylation of the polymerase C-terminal domain. The phosphorylated

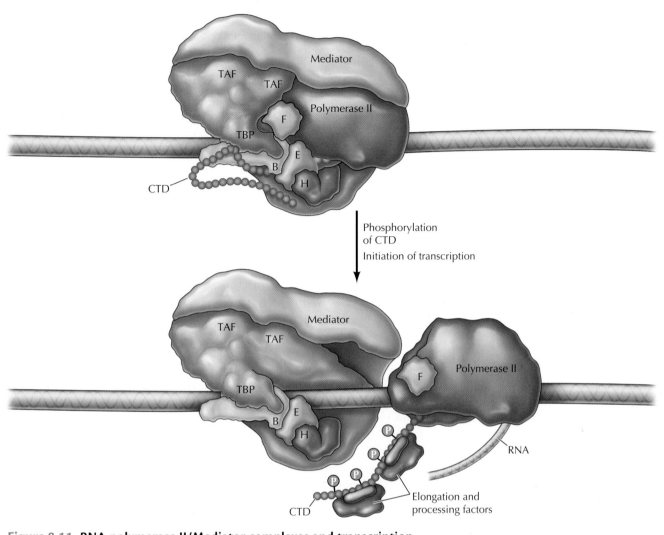

Figure 8.11 RNA polymerase II/Mediator complexes and transcription initiation   RNA polymerase II is associated with Mediator proteins, as well as with the general transcription factors, at the promoter (see Figure 8.10). The Mediator complex binds to the nonphosphorylated C-terminal domain (CTD) of polymerase II and is released following phosphorylation of the CTD when transcription initiates. The phosphorylated CTD then binds elongation and processing factors that facilitate mRNA synthesis and processing.

CTD then binds other proteins that facilitate transcriptional elongation and function in mRNA processing, as discussed later in this chapter.

Transcription by RNA polymerases I and III

As previously discussed, distinct RNA polymerases are responsible for the transcription of genes encoding ribosomal RNAs, transfer RNAs, and some small noncoding RNAs in eukaryotic cells. Like RNA polymerase II, these other eukaryotic RNA polymerases also require additional transcription factors to associate with appropriate promoter sequences.

RNA polymerase I is devoted solely to the transcription of ribosomal RNA genes, which are present in tandem repeats. Transcription of these genes yields a large 45S pre-rRNA, which is then processed to yield the 28S, 18S, and 5.8S

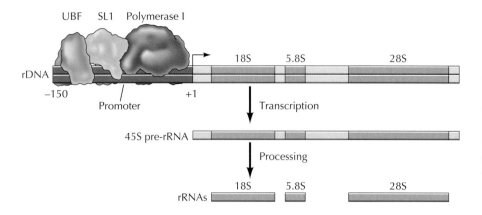

Figure 8.12 Transcription of the ribosomal RNA gene   Ribosomal DNA (rDNA) is transcribed to yield a large RNA molecule (45S pre-rRNA), which is then cleaved into 28S, 18S, and 5.8S rRNAs. Two transcription factors, UBF (upstream binding factor), and SL1 (selectivity factor 1), bind cooperatively to the rDNA promoter and recruit RNA polymerase I to form an initiation complex.

rRNAs (**Figure 8.12**). The promoters of ribosomal RNA genes span about 150 base pairs just upstream of the transcription initiation site. These promoter sequences are recognized by two transcription factors, UBF (upstream binding factor), and SL1 (selectivity factor 1), which bind cooperatively to the promoter and recruit polymerase I to form an initiation complex.

The genes for 5S rRNA, tRNAs, and some of the small nuclear RNAs (snRNAs) involved in splicing and protein transport are transcribed by polymerase III. These genes are transcribed from three distinct classes of promoters, two of which lie within, rather than upstream of, the transcribed sequence (**Figure 8.13**). TFIIIA (which is the first transcription factor to have been purified) initiates assembly of a transcription complex by binding to specific DNA sequences in the 5S rRNA promoter. This binding is followed by the binding of TFIIIC, TFIIIB, and the polymerase. The

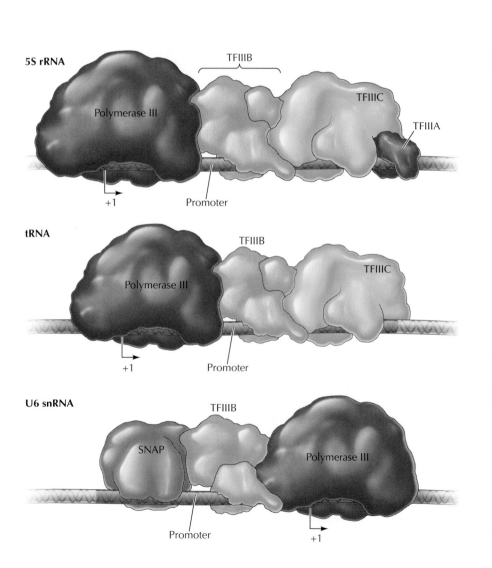

Figure 8.13 Transcription of RNA polymerase III genes   Genes transcribed by polymerase III are expressed from three types of promoters. The promoters of 5S rRNA and tRNA genes are downstream of the transcription initiation site. Transcription of the 5S rRNA gene is initiated by the binding of TFIIIA, followed by the binding of TFIIIC, TFIIIB, and the polymerase. Promoters of tRNA and snRNA genes do not contain binding sites for TFIIIA and are recognized by other factors (TFIIIC and SNAP, respectively) that recruit TFIIIB and RNA polymerase.

promoters for the tRNA genes and other genes (such as snRNAs) transcribed by RNA polymerase III are recognized by other factors (TFIIIC and SNAP, respectively) that recruit TFIIIB and the polymerase to the transcription complex.

Regulation of Transcription in Eukaryotes

The control of gene expression is far more complex in eukaryotes than in bacteria, although some of the same basic principles apply. As in bacteria, transcription in eukaryotic cells is controlled by proteins that bind to specific regulatory sequences and modulate the activity of RNA polymerase. An important difference between transcriptional regulation in bacteria and eukaryotes, however, results from the packaging of eukaryotic DNA into chromatin, which limits its availability as a template for transcription. As a result, modifications of chromatin structure play key roles in the control of transcription in eukaryotic cells. A particularly exciting area of current research is based on the discovery that noncoding RNAs, as well as proteins, regulate transcription in eukaryotic cells via modifications in chromatin structure.

cis-*acting regulatory sequences: promoters and enhancers*

As already discussed, transcription in bacteria is regulated by the binding of proteins to *cis*-acting sequences (e.g., the *lac* operator) that control the transcription of adjacent genes. Similar *cis*-acting sequences regulate the expression of eukaryotic genes. These sequences have been identified in mammalian cells largely by the use of gene transfer assays to study the activity of suspected regulatory regions of cloned genes (**Figure 8.14**). The eukaryotic regulatory sequences are usually added upstream of a reporter gene that encodes an easily detectable enzyme, such as firefly luciferase (the enzyme responsible for bioluminescence). The expression of the reporter gene following its transfer into cultured cells then provides a sensitive assay for the ability of the cloned regulatory sequences to direct transcription. Biologically active regulatory regions can thus be identified, and *in vitro* mutagenesis can be used to determine the roles of specific sequences within the region.

Genes transcribed by RNA polymerase II have core promoter elements, including the TATA box and the Inr sequence, that serve as specific binding sites for general transcription factors. Other *cis*-acting sequences serve as binding sites for a wide variety of other gene-specific regulatory factors that control the expression of individual genes. These *cis*-acting regulatory sequences are sometimes, though not always, located upstream of the transcription start site. For example, two regulatory sequences that are found in many eukaryotic genes were identified by studies of the promoter of the herpes simplex virus gene that encodes thymidine kinase (**Figure 8.15**). Both of these sequences are located within 100 base pairs upstream of the transcription start site: Their consensus sequences are CCAAT and GGGCGG (called a GC box). Specific proteins that bind to these sequences and stimulate transcription have since been identified.

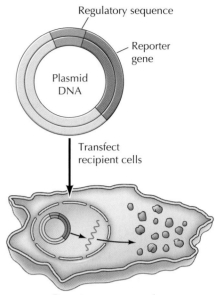

Figure 8.14 Identification of eukaryotic regulatory sequences The regulatory sequence of a cloned eukaryotic gene is ligated to a reporter gene that encodes an easily detectable enzyme. The resulting plasmid is then introduced into cultured recipient cells by transfection. An active regulatory sequence directs transcription of the reporter gene, expression of which is then detected in the transfected cells.

Figure 8.15 A eukaryotic promoter The promoter of the thymidine kinase gene of herpes simplex virus contains three sequence elements upstream of the TATA box that are required for efficient transcription: two GC boxes (consensus sequence GGGCGG) and a CCAAT box.

Figure 8.16 The SV40 enhancer The SV40 promoter for early gene expression contains a TATA box and six GC boxes arranged in three sets of repeated sequences. In addition, efficient transcription requires an upstream enhancer consisting of two 72-base-pair (bp) repeats.

In contrast to the relatively simple organization of CCAAT and GC boxes in the herpes thymidine kinase promoter, many genes in mammalian cells are controlled by regulatory sequences located farther away (sometimes hundreds of kilobases) from the transcription start site. These sequences, called **enhancers**, were first identified during studies of the promoter of another virus, SV40 (**Figure 8.16**). In addition to a TATA box and a set of six GC boxes, two 72-base-pair repeats located farther upstream are required for efficient transcription from this promoter. These sequences were found to stimulate transcription from other promoters as well as from that of SV40, and, surprisingly, their activity depended on neither their distance nor their orientation with respect to the transcription initiation site (**Figure 8.17**). They could stimulate transcription when placed either upstream or downstream of the promoter, in either a forward or backward orientation.

The ability of enhancers to function even when separated by long distances from transcription initiation sites at first suggested that they work by mechanisms different from those of promoters. However, this has turned out not to be the case: Enhancers, like promoters, function by binding transcription factors that

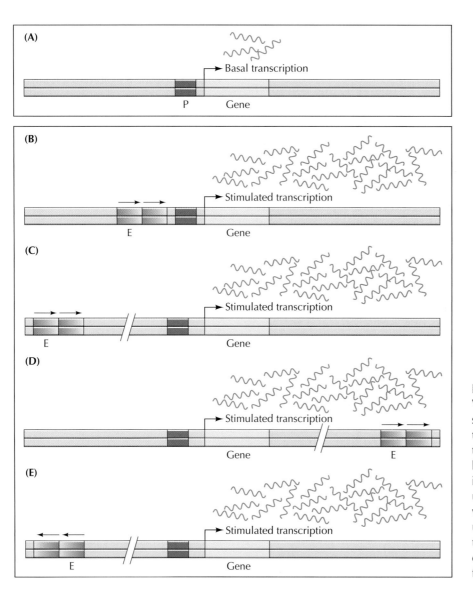

Figure 8.17 Action of enhancers Without an enhancer, the gene is transcribed at a low basal level (A). Addition of an enhancer, E—for example, the SV40 72-base-pair repeats—stimulates transcription. The enhancer is active not only when placed just upstream of the promoter (B), but also when inserted up to several kilobases upstream or downstream from the transcription start site (C and D). In addition, enhancers are active in either the forward or backward orientation (E).

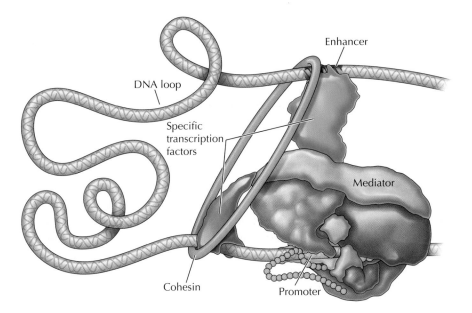

Figure 8.18 DNA looping   Transcription factors bound at a distant enhancer are able to interact with Mediator or general transcription factors at the promoter because the intervening DNA can form loops, stabilized by cohesin.

then regulate RNA polymerase. This is possible because of DNA looping, which allows a transcription factor bound to a distant enhancer to interact with proteins associated with Mediator or general transcription factors (e.g., TFIID) at the promoter (**Figure 8.18**). The formation of these DNA loops is facilitated by the protein cohesin, which acts to maintain the association between sister chromatids following DNA replication (discussed in Chapter 17). Transcription factors bound to distant enhancers can thus work by the same mechanisms as those bound adjacent to promoters, so there is no fundamental difference between the actions of enhancers and those of *cis*-acting regulatory sequences adjacent to transcription start sites. Interestingly, although enhancers were first identified in mammalian cells, they have subsequently been found in bacteria—an unusual instance in which studies of eukaryotes served as a model for the simpler bacterial systems.

The binding of specific transcriptional regulatory proteins to enhancers is responsible for the control of gene expression during development and differentiation, as well as during the response of cells to hormones and growth factors. A well-studied example is the enhancer that controls the transcription of immunoglobulin genes in B lymphocytes. Gene transfer experiments have established that the immunoglobulin enhancer is active in lymphocytes, but not in other types of cells. Thus this regulatory sequence is at least partly responsible for tissue-specific expression of the immunoglobulin genes in the appropriate differentiated cell type.

An important aspect of enhancers is that they usually contain multiple sequence elements that bind different transcriptional regulatory proteins. These proteins work together to regulate gene expression. The immunoglobulin heavy-chain enhancer, for example, spans approximately 200 base pairs and contains at least nine distinct sequence elements that serve as protein-binding sites (**Figure 8.19**). The overall activity of the enhancer reflects the combined action of the proteins associated with each of its individual sequence elements.

Figure 8.19 The immunoglobulin enhancer The immunoglobulin heavy-chain enhancer spans about 200 bases and contains nine functional sequence elements (E, μE1–5, π, μB, and OCT), which together stimulate transcription in B lymphocytes.

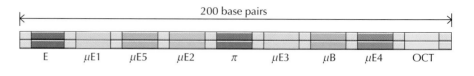

Enhancers can be identified not only as binding sites for transcription factors, but also by characteristic features of chromatin, which will be discussed later in this chapter. Global analyses based on these characteristics have identified 500,000 to over 1 million enhancers in the human genome, accounting for 10% or more of total genomic DNA. There are clearly many more enhancers than genes, emphasizing the importance of these regulatory elements in determining biological behavior. This point is further made by the observation that many mutations linked to human diseases affect enhancers rather than protein-coding sequences. In any given cell type, it is estimated that on the order of 50,000 enhancers may be active, with multiple enhancers working together to regulate each individual gene. The program of gene regulation in higher eukaryotes is thus far more complex than might be suggested by the limited number of protein-coding genes in our genome.

Although DNA looping allows enhancers to act at a considerable distance from promoters, the activity of any given enhancer is specific for the promoter of its appropriate target gene. This specificity is maintained by **insulators**, which were initially characterized as sequences that divide chromosomes into independent domains and prevent enhancers from acting on promoters located in a different domain. Considerable progress in understanding the action of insulators and the nature of chromatin domains has been made recently using new methods (called chromosome conformation capture or 3C) that determine the sites of interaction between different chromosome regions within a cell. Analyses of these interactions indicate that genomes are divided into a series of discrete chromosomal domains, called **topologically associating domains** or **TADs** (Figure 8.20). Enhancers and promoters within a domain interact frequently with each other, but only rarely with elements in other domains. In humans, TADs range from approximately one hundred to several thousand kb.

The main protein that binds insulators in vertebrates is **CTCF**, which has been found to play a key role as an architectural protein in regulating the organization of the genome (see Figure 8.20). The boundaries of TADs contain multiple binding sites for CTCF, which appears to be involved in establishing these domains and restricting the actions of enhancers. Conversely, CTCF can act together with cohesin to promote the formation of loops and facilitate enhancer/promoter interactions within a TAD. These distinct actions of CTCF may depend on its interactions with other proteins. For example, the boundaries between TADs are also enriched for binding sites for TFIIIC, which may act together with CTCF and cohesin to establish these boundaries. Given the importance of enhancer/promoter interactions in mammalian cells, understanding the regulation of CTCF and how it functions to coordinate long-range chromosome interactions remains an exciting avenue of future research.

Transcription factor binding sites

The binding sites of transcriptional regulatory proteins in promoter or enhancer sequences have been identified by several types of experiments.

A major hurdle for gene therapy is that introduced genes are often aberrantly regulated or inactivated because of the nearby chromatin structure. The addition of insulator elements is a potential solution to this problem.

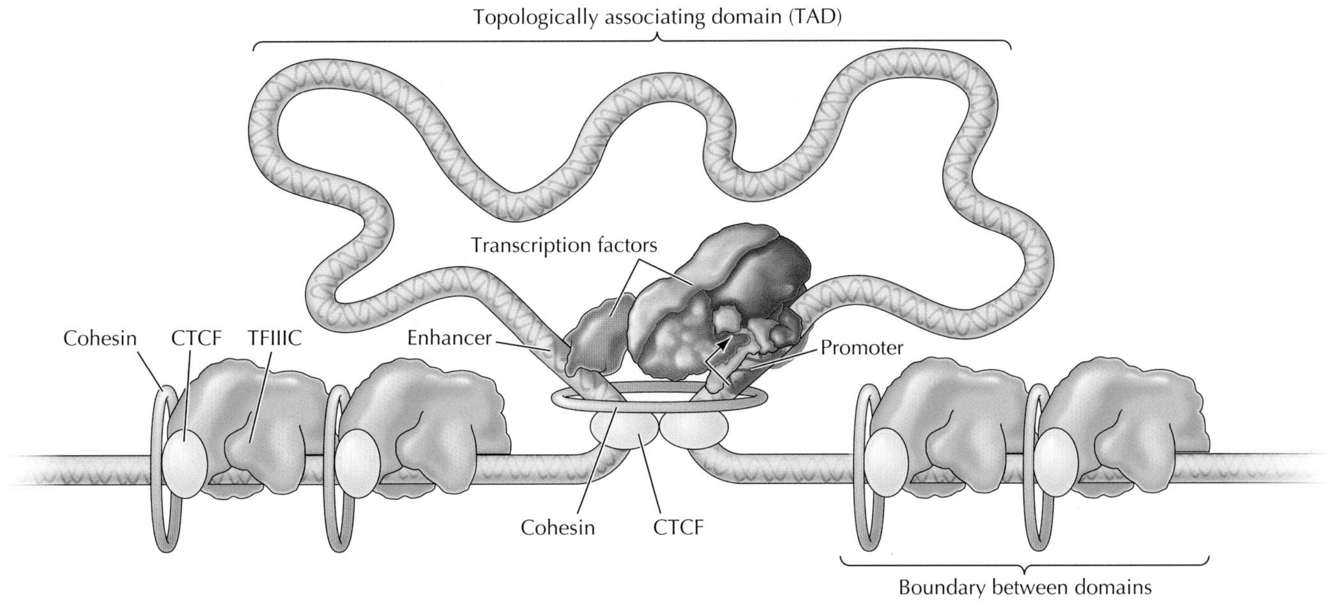

Topologically associating domain (TAD)

Transcription factors

Cohesin   CTCF   TFIIIC   Enhancer

Promoter

Cohesin   CTCF

Boundary between domains

Figure 8.20 Chromosomal domains and CTCF   Enhancers and promoters within a topologically associating domain (TAD) interact frequently with each other, but only rarely with elements in other domains. The boundaries of TADs contain multiple binding sites for CTCF and for TFIIIC, which may act to establish these boundaries and restrict enhancers. Within a domain, CTCF can act together with cohesin to promote the formation of loops and facilitate enhancer/promoter interactions.

One common approach is the **electrophoretic-mobility shift assay,** in which a radiolabeled DNA fragment is incubated with a protein preparation and then subjected to electrophoresis through a nondenaturing gel (**Figure 8.21**). Protein binding is detected as a decrease in the electrophoretic mobility of

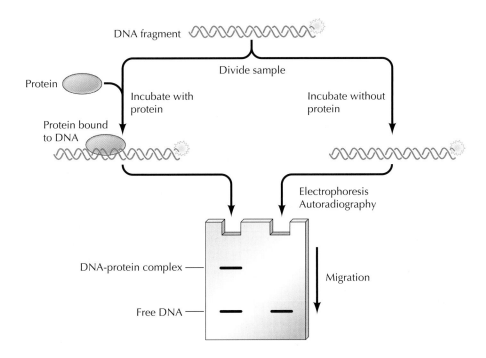

DNA fragment

Protein

Divide sample

Incubate with protein

Incubate without protein

Protein bound to DNA

Electrophoresis
Autoradiography

DNA-protein complex

Free DNA

Migration

Figure 8.21 Electrophoretic-mobility shift assay   A sample containing radiolabeled fragments of DNA is divided in two, and one half of the sample is incubated with a protein that binds to a specific DNA sequence. Samples are then analyzed by electrophoresis in a nondenaturing gel so that the protein remains bound to DNA. Protein binding is detected by the slower migration of DNA-protein complexes compared to that of free DNA. Only a fraction of the DNA in the sample is actually bound to protein, so both DNA-protein complexes and free DNA are detected following incubation of the DNA with protein.

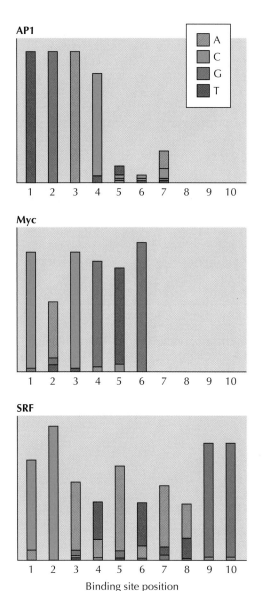

Binding site position

Figure 8.22 Representative transcription factor binding sites The binding sites of three mammalian transcription factors (AP1, Myc, and SRF) are shown as pictograms in which the frequency of each nucleotide is represented by the height of the corresponding letter at each position of the binding site.

the DNA fragment, since its migration through the gel is slowed by the bound protein. The combined use of electrophoretic-mobility shift assays and detailed mapping of protein-DNA interactions has led to the characterization of transcription factor-binding sites within enhancers and promoters of eukaryotic cells.

The binding sites of most transcription factors consist of short DNA sequences, typically spanning 6–10 base pairs. In most cases, these binding sites are degenerate, meaning that the transcription factor will bind not only to the consensus sequence but also to sequences that differ from the consensus at one or more positions. It is thus common to represent transcription factor binding sites as pictograms, representing the frequency of each base at all positions of known binding sites for a given factor (**Figure 8.22**). Because of their short degenerate nature, sequences matching transcription factor binding sites occur frequently in genomic DNA, so physiologically significant regulatory sequences cannot be identified from DNA sequence alone.

An important experimental approach for determining the regions of DNA that bind a transcription factor within the cell has been provided by **chromatin immunoprecipitation** (**Figure 8.23**). Cells are first treated with formaldehyde, which cross-links proteins to DNA. As a result, transcription factors are covalently linked to the DNA sequences to which they were bound within the living cell. Chromatin is then extracted and sheared to fragments of about 500 base pairs. Fragments of DNA linked to a transcription factor of interest can then be isolated by immunoprecipitation with an antibody against the transcription factor (see Figure 5.11). The formaldehyde cross-links are then reversed, and the immunoprecipitated DNA isolated and analyzed to determine the sites to which the specific transcription factor was bound within the cell. The binding of a transcription factor to a specific regulatory element can be tested by using PCR to detect the sequence of interest in chromatin immunoprecipitates (see Figure 4.22). Alternatively, a genome-wide analysis of the immunoprecipitated DNA fragments, either by large-scale DNA sequencing or by hybridization to microarrays (see Figures 5.5 and 5.6) can be used to identify all of the sites to which a transcription factor is bound within a cell.

Transcriptional regulatory proteins

A variety of transcriptional regulatory proteins have been isolated based on their binding to specific DNA sequences. One of the prototypes of eukaryotic transcription factors was initially identified by Robert Tjian and his colleagues during studies of the transcription of SV40 DNA. This factor (called Sp1, for specificity protein 1) was found to stimulate transcription via binding to GC boxes in the SV40 promoter (see Figure 8.16). Importantly, the specific binding of Sp1 to the GC box not only established the action of Sp1 as a sequence-specific transcription factor,

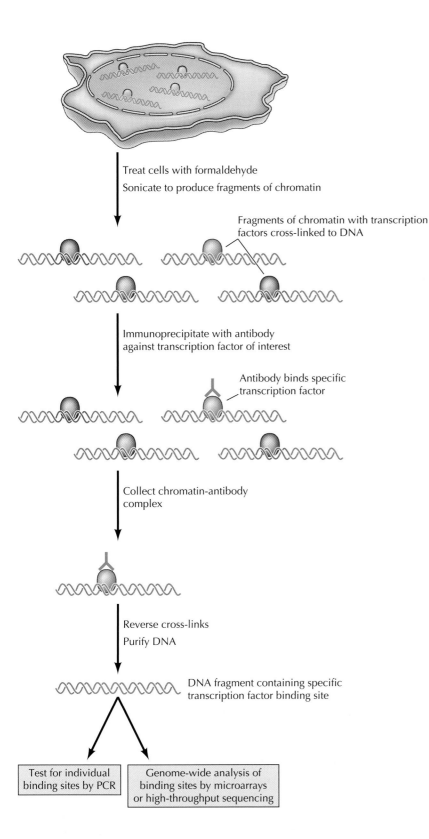

Figure 8.23 Chromatin immuno-precipitation Cells are treated with formaldehyde to cross-link DNA and proteins and then sonicated to produce fragments of chromatin. The chromatin fragments are incubated with an antibody against a specific transcription factor, and chromatin fragments bound to the antibody are collected as described in Figure 5.11. The cross-links are then reversed, and DNA is purified to yield DNA fragments containing binding sites for the specific transcription factor of interest. The immunoprecipitated fragments can be analyzed either by PCR to test for the presence of a specific DNA sequence or by global methods, such as hybridization to microarrays or high-throughput DNA sequencing, to identify all of the binding sites for the transcription factor within the genome.

but also suggested a general approach to the purification of transcription factors. The isolation of these proteins initially presented a formidable challenge because they are present in very small quantities (e.g., only 0.001% of total cell protein) that are difficult to purify by conventional biochemical

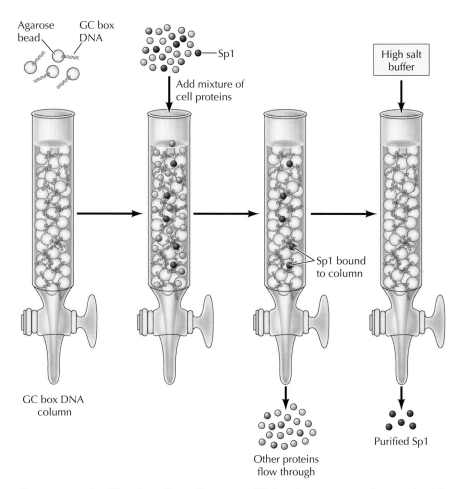

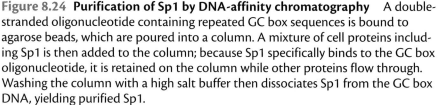

Figure 8.24 Purification of Sp1 by DNA-affinity chromatography   A double-stranded oligonucleotide containing repeated GC box sequences is bound to agarose beads, which are poured into a column. A mixture of cell proteins including Sp1 is then added to the column; because Sp1 specifically binds to the GC box oligonucleotide, it is retained on the column while other proteins flow through. Washing the column with a high salt buffer then dissociates Sp1 from the GC box DNA, yielding purified Sp1.

techniques. This problem was overcome in the purification of Sp1 by **DNA-affinity chromatography** (Figure 8.24). Multiple copies of oligonucleotides corresponding to the GC box sequence were bound to a solid support, and cell extracts were passed through the oligonucleotide column. Because Sp1 bound to the GC box with high affinity, it was specifically retained on the column while other proteins were not. Highly purified Sp1 could thus be obtained and used for further studies.

The general method of DNA-affinity chromatography, first optimized for the purification of Sp1, has been used successfully to isolate a wide variety of sequence-specific DNA-binding proteins from eukaryotic cells. Genes encoding other transcription factors have been isolated by screening cDNA expression libraries to identify recombinant proteins that bind to specific DNA sequences. The cloning and sequencing of transcription factor cDNAs has led to our current understanding of the structure and function of these critical regulatory proteins.

Key Experiment

Isolation of a Eukaryotic Transcription Factor

Affinity Purification of Sequence-Specific DNA-Binding Proteins

James T. Kadonaga and Robert Tjian

University of California at Berkeley, Berkeley, CA

Proceedings of the National Academy of Sciences, USA, 1986, Volume 83, pages 5889–5893

The Context

Starting with studies of the *lac* operon by François Jacob and Jacques Monod in the 1950s, it became clear that transcription is regulated by proteins that bind to specific DNA sequences. One of the prototype systems for studies of gene expression in eukaryotic cells was the monkey virus SV40 in which several regulatory DNA sequences were identified in the early 1980s. In 1983 William Dynan and Robert Tjian first demonstrated that one of these sequence elements (the GC box) is the specific binding site of a protein detectable in nuclear extracts of human cells. This protein (called Sp1, for specificity protein 1) not only binds to the GC box sequence; it also stimulates transcription *in vitro*, demonstrating that it is a sequence-specific transcriptional activator.

To study the mechanism of Sp1 action, it then became necessary to obtain the transcription factor in pure form and eventually to clone the *Sp1* gene. The isolation of pure Sp1 thus became a high priority, but it also posed a daunting technical challenge. Sp1 and other transcription factors appeared to represent only about 0.001% of total cell protein, so they could not readily be purified by conventional biochemical techniques. James Kadonaga and Robert Tjian solved this problem by developing a method of DNA-affinity chromatography that led to the purification not only of Sp1 but also of many other eukaryotic transcription factors, thereby opening the door to molecular analysis of transcriptional regulation in eukaryotic cells.

The Experiments

The DNA-affinity chromatography method developed by Kadonaga and Tjian exploited the specific high-affinity binding of Sp1 to the GC box sequence—GGGCGG. Synthetic oligonucleotides containing multiple copies of this sequence were coupled to solid beads, and a crude nuclear extract was passed through a column consisting of beads linked to GC box DNA. The beads were then washed to remove proteins that had failed to bind specifically to the oligonucleotides. Finally, the beads were washed with a high salt buffer (0.5 M KCl), which disrupted the binding of Sp1 to DNA, thereby releasing Sp1 from the column.

Gel electrophoresis demonstrated that the crude nuclear extract initially applied to the column was a complex mixture of proteins (see figure). In contrast, approximately 90% of the protein that was recovered after two cycles of DNA-affinity chromatography corresponded to only two polypeptides, which were identified as Sp1 by DNA binding and by their activity in *in vitro* transcription assays. Thus Sp1 had been successfully purified by DNA-affinity chromatography.

The Impact

In their 1986 paper, Kadonaga and Tjian stated that the DNA-affinity chromatography technique "should be generally applicable for the purification of other sequence-specific DNA binding proteins." This prediction has been amply verified; many eukaryotic transcription factors have been purified by this method. The genes that encode still other transcription factors have been isolated by an alternative approach (developed independently in 1988 in the laboratories of Phillip Sharp and Steven McKnight) in which cDNA expression libraries are screened with oligonucleotide probes to detect recombinant proteins that bind specifically to the desired DNA sequences. The ability to isolate sequence-specific DNA-binding proteins by these methods has led to detailed characterization of the structure and function of a wide variety of transcriptional

Robert Tjian

regulatory proteins, providing the basis for our current understanding of gene expression in eukaryotic cells.

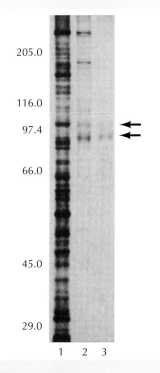

Purification of Sp1 Gel electrophoresis of proteins initially present in the crude nuclear extract (lane 1) and of proteins obtained after either one or two sequential cycles of DNA-affinity chromatography (lanes 2 and 3, respectively). The sizes of marker proteins (in kilodaltons) are indicated to the left of the gel, and the Sp1 polypeptides are indicated by arrows.

YI

...tion factors with
...ns designed to bind
...es within the genome
have bee... ...oped. By ligating different effector domains to the DNA binding domain, the target gene can be either activated or repressed. This technology can be applied to gene therapy and the development of transgenic plants and animals.

Structure and function of transcriptional activators

Because transcription factors are central to the regulation of gene expression, understanding the mechanisms of their action has been a major area of research in cell and molecular biology. The most thoroughly studied of these proteins are **transcriptional activators**, which, like Sp1, bind to regulatory DNA sequences and stimulate transcription. In general, these factors consist of two independent domains: One region of the protein specifically binds DNA; the other region stimulates transcription by interacting with other proteins, including Mediator or other components of the transcriptional machinery (**Figure 8.25**). The basic function of the DNA-binding domain is to anchor the transcription factor to the proper site on DNA; the activation domain then independently stimulates transcription through protein-protein interactions.

Many different transcription factors have now been identified in eukaryotic cells, with about 2000 encoded in the human genome. They contain many distinct types of DNA-binding domains, some of which are illustrated in **Figure 8.26**. The most common is the **zinc finger domain**, which contains repeats of cysteine and histidine residues that bind zinc ions and fold into looped structures ("fingers") that bind DNA. These domains were initially identified in the polymerase III transcription factor TFIIIA, but are also common among transcription factors that regulate polymerase II promoters, including Sp1. Other examples of transcription factors that contain zinc finger domains are the **steroid hormone receptors**, which regulate gene transcription in response to hormones such as estrogen and testosterone.

The **helix-turn-helix** motif was first recognized in bacterial DNA-binding proteins, including the *E. coli* catabolite activator protein (CAP). In these proteins, one helix makes most of the contacts with DNA, while the other helices lie across the complex to stabilize the interaction. In eukaryotic cells,

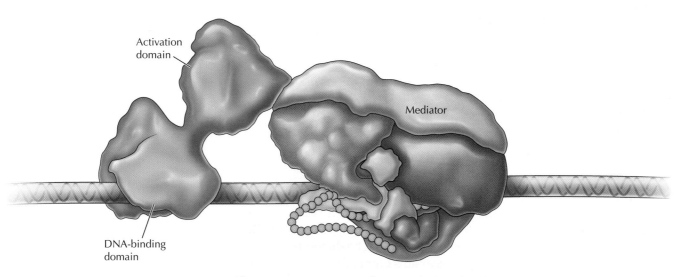

Figure 8.25 Structure of transcriptional activators Transcriptional activators consist of two independent domains. The DNA-binding domain recognizes a specific DNA sequence, and the activation domain interacts with Mediator or other components of the transcriptional machinery.

(A) Zinc fingers

Zinc ion

β sheet

α helix

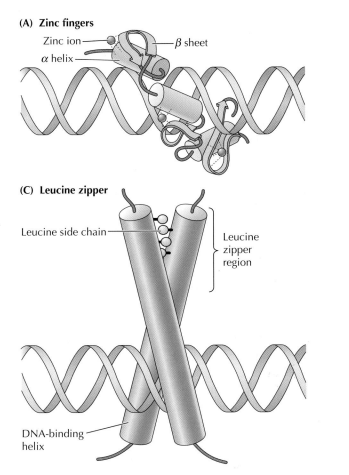

(C) Leucine zipper

Leucine side chain

Leucine zipper region

DNA-binding helix

(B) Helix-turn-helix

(D) Helix-loop-helix

Loop

DNA-binding helix

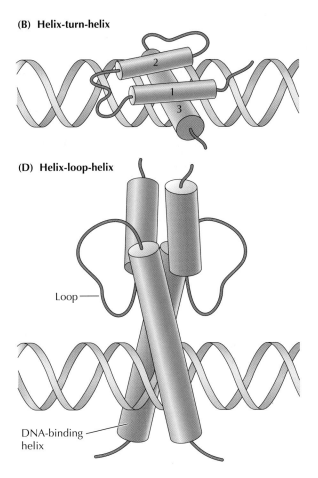

Figure 8.26 Examples of DNA-binding domains   (A) Zinc finger domains consist of loops in which an α helix and a β sheet coordinately bind a zinc ion. (B) Helix-turn-helix domains consist of three (or in some cases four) helical regions. One helix (helix 3) makes most of the contacts with DNA, while helices 1 and 2 lie on top and stabilize the interaction. (C) The DNA-binding domains of leucine zipper proteins are formed from two distinct polypeptide chains. Interactions between the hydrophobic side chains of leucine residues exposed on one side of a helical region (the leucine zipper) are responsible for dimerization. Immediately following the leucine zipper is a DNA-binding helix, which is rich in basic amino acids. (D) Helix-loop-helix domains are similar to leucine zippers, except that the dimerization domains of these proteins each consist of two helical regions separated by a loop.

helix-turn-helix proteins include the **homeodomain** proteins, which play critical roles in the regulation of gene expression during embryonic development.

Two other families of DNA-binding proteins, **leucine zipper** and **helix-loop-helix** proteins, contain DNA-binding domains formed by dimerization of two polypeptide chains. The leucine zipper contains four or five leucine residues spaced at intervals of seven amino acids, resulting in their hydrophobic side chains being exposed at one side of a helical region. This region serves as the dimerization domain for the two protein subunits, which are held together by hydrophobic interactions between the leucine side chains. Immediately

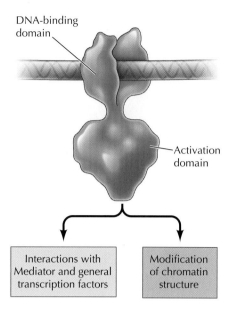

DNA-binding domain

Activation domain

Interactions with Mediator and general transcription factors

Modification of chromatin structure

Figure 8.27 Action of transcriptional activators Eukaryotic activators stimulate transcription by two mechanisms: (1) they interact with Mediator proteins and general transcription factors to facilitate the assembly of a transcription complex and stimulate transcription, and (2) they interact with coactivators that facilitate transcription by modifying chromatin structure.

following the leucine zipper is a region rich in positively charged amino acids (lysine and arginine) that binds DNA. The helix-loop-helix proteins are similar in structure, except that their dimerization domains are each formed by two helical regions separated by a loop. An important feature of both leucine zipper and helix-loop-helix transcription factors is that different members of each family can dimerize with one another. Thus the combination of distinct protein subunits can form an expanded array of factors that can differ both in DNA sequence recognition and in transcription-stimulating activities. Both leucine zipper and helix-loop-helix proteins play important roles in regulating tissue-specific and inducible gene expression, and the formation of dimers between different members of these families is a critical aspect of the control of their function.

The activation domains of transcription factors are not as well characterized as their DNA-binding domains. Some, called acidic activation domains, are rich in negatively charged residues (aspartate and glutamate); others are rich in proline or glutamine residues. The activation domains of eukaryotic transcription factors stimulate transcription by two distinct mechanisms (Figure 8.27). First, they interact with Mediator proteins and general transcription factors, such as TFIIB or TFIID, to recruit RNA polymerase and facilitate the assembly of a transcription complex on the promoter, similar to transcriptional activators in bacteria (see Figure 8.8). In addition, eukaryotic transcription factors interact with a variety of **coactivators** that stimulate transcription by modifying chromatin structure, as discussed later in this chapter.

Eukaryotic repressors

Transcription in eukaryotic cells is regulated by repressors as well as by activators. Like their bacterial counterparts, eukaryotic repressors bind to specific DNA sequences and inhibit transcription. In some cases, eukaryotic repressors simply interfere with the binding of other transcription factors to DNA (Figure 8.28A). For example, the binding of a repressor near the transcription start site can block the interaction of RNA polymerase or general transcription factors with the promoter, which is similar to the action of repressors in bacteria. Other repressors compete with activators for binding to specific regulatory sequences. Some such repressors contain the same DNA-binding domain as the activator but lack its activation domain. As a result, their binding to a promoter or enhancer blocks the binding of the activator, thereby inhibiting transcription.

In contrast to repressors that simply interfere with activator binding, many repressors (called active repressors) contain specific functional domains that inhibit transcription via protein-protein interactions (Figure 8.28B). Such active repressors play key roles in the regulation of transcription in animal cells, in many cases serving as critical regulators of cell growth and differentiation. As with transcriptional activators, several distinct types of repression domains

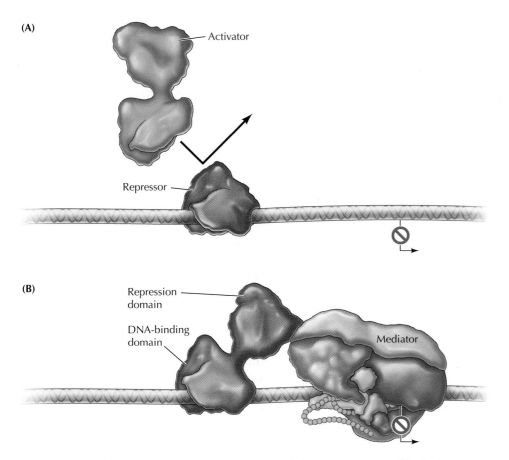

(A)

Activator

Repressor

(B)

Repression domain

DNA-binding domain

Mediator

Figure 8.28 Action of eukaryotic repressors   (A) Some repressors block the binding of activators to regulatory sequences. (B) Other repressors have active repression domains that inhibit transcription by interactions with Mediator proteins or general transcription factors, as well as with corepressors that act to modify chromatin structure.

have been identified. The functional targets of repressors are also diverse: Repressors can inhibit transcription by interacting with specific activator proteins, with Mediator proteins or general transcription factors, and with **corepressors** that act by modifying chromatin structure.

Regulation of elongation

The transcription of many genes is regulated at the level of formation of a preinitiation complex and initiation of transcription (see Figures 8.10 and 8.11, respectively). However, transcription can also be regulated at the level of elongation both by direct modulation of the activity of RNA polymerase and by effects on chromatin structure. The importance of control of elongation has been highlighted by recent studies indicating that a large number of genes in both *Drosophila* and human cells are characterized by molecules of RNA polymerase II that have initiated transcription but are stalled immediately downstream of promoters. These RNA polymerases are "poised" to continue transcription in response to appropriate signals. Interestingly, many of the genes on which such poised polymerases have been found are regulated by extracellular signals or function during development, suggesting an important role for the control of transcriptional elongation during development and differentiation.

Transcription by RNA polymerase II initiates following phosphorylation of CTD serine-5 by the TFIIH protein kinase (**Figure 8.29**). Following initiation, the polymerase synthesizes a short region of RNA and then pauses near the beginning of the gene, typically within about 50 nucleotides of the transcription start site. The arrest of the polymerase at this point results from the association of negative regulatory factors, including NELF (negative elongation factor) and DSIF that prevent further transcription. Continuation of transcription is dependent on the action of another factor called P-TEFb (positive transcription-elongation factor-b). P-TEFb contains a protein kinase that phosphorylates NELF and DSIF as well as serine-2 of the RNA polymerase CTD. This leads to the initiation of productive elongation and the association of additional elongation and processing factors with the CTD.

It appears that recruitment of P-TEFb is the major regulatory step in the control of elongation, and that at least some transcriptional activators associate with P-TEFb and recruit it to promoters. For example, paused polymerases have been identified at approximately 30% of the genes that are transcribed in human embryonic stem cells. These genes are activated by the transcription factor c-Myc, which binds and recruits P-TEFb to sites near their promoters. Notably, c-Myc is one of the transcription factors that plays a major role in human cancers (discussed in Chapter 19), further highlighting the importance of regulation of transcriptional elongation.

Chromatin and Epigenetics

As noted in the preceding discussion, both activators and repressors regulate transcription in eukaryotes not only by interacting with Mediator and other components of the transcriptional machinery but also by inducing changes in the structure of chromatin. Rather than being present within the nucleus as naked DNA, the DNA of all eukaryotic cells is tightly bound to histones. This packaging of eukaryotic DNA in chromatin has important consequences in terms of its availability as a template for transcription, so chromatin structure is a critical aspect of gene expression in eukaryotic cells. Importantly, histones can be modified in a variety of ways that affect the transcriptional activity of chromatin, so histone modification is a key mechanism for regulating the expression of eukaryotic genes. Moreover, many modifications of histones are stably inherited when cells divide, so they provide a mechanism through which patterns of gene expression can be transmitted to progeny cells following mitosis.

Histone modifications

The basic structural unit of chromatin is the nucleosome, which consists of 147 base pairs of DNA wrapped around two molecules each of histones H2A, H2B, H3, and H4, with one molecule of histone H1 bound to the DNA as it enters the nucleosome core particle (see Figure 6.17). The chromatin is then further condensed by being coiled into higher order structures organized into large loops of DNA. The packaging of eukaryotic DNA in chromatin limits its availability for transcription, affecting both the ability of transcription factors to bind DNA and the ability of RNA polymerase to transcribe through a chromatin template. Actively transcribed genes are found in

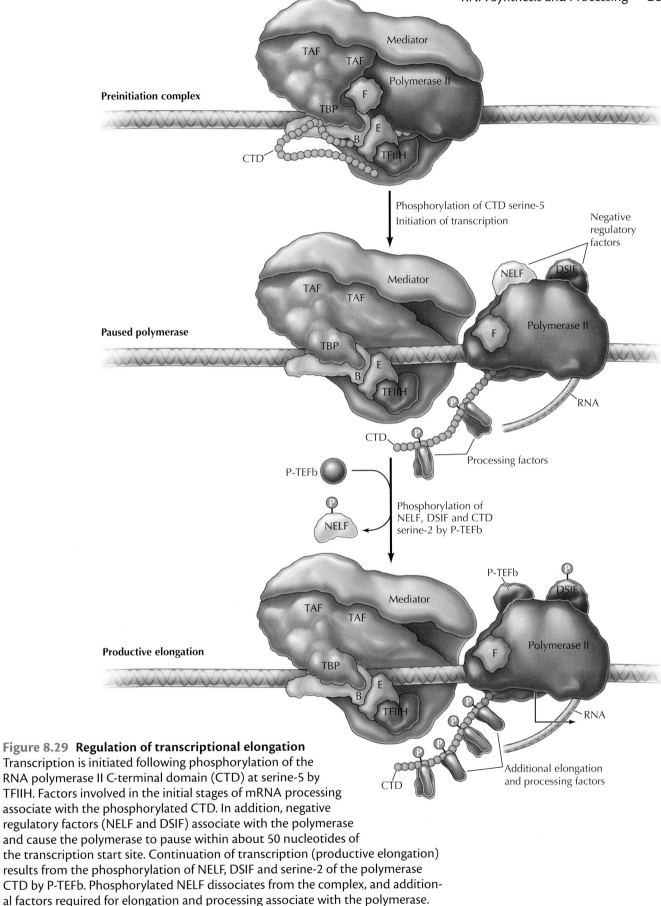

Figure 8.29 Regulation of transcriptional elongation
Transcription is initiated following phosphorylation of the
RNA polymerase II C-terminal domain (CTD) at serine-5 by
TFIIH. Factors involved in the initial stages of mRNA processing
associate with the phosphorylated CTD. In addition, negative
regulatory factors (NELF and DSIF) associate with the polymerase
and cause the polymerase to pause within about 50 nucleotides of
the transcription start site. Continuation of transcription (productive elongation)
results from the phosphorylation of NELF, DSIF and serine-2 of the polymerase
CTD by P-TEFb. Phosphorylated NELF dissociates from the complex, and additional factors required for elongation and processing associate with the polymerase.

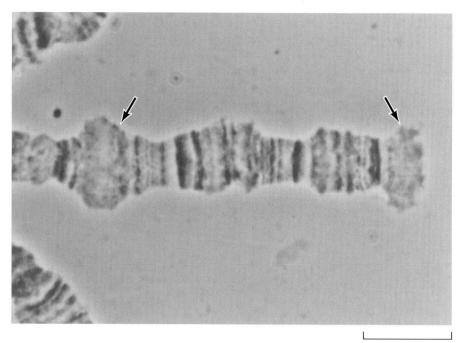

10 μm

Figure 8.30 Decondensed chromosome regions in *Drosophila* A light micrograph showing decondensed regions of polytene chromosomes (arrows), which are active in RNA synthesis. (Courtesy of Joseph Gall, Carnegie Institution.)

relatively decondensed regions of chromatin, which facilitates their transcription (Figure 8.30).

The structure of chromatin can be altered both by modifications of histones and by rearrangements of nucleosomes. The first type of histone modification to be described, **histone acetylation**, has been correlated with transcriptionally active chromatin in a wide variety of cell types (Figure 8.31). The core histones (H2A, H2B, H3, and H4) have two domains: an amino-terminal tail, which extends outside of the nucleosome, and a histone-fold, which is involved in interactions with other histones and in wrapping DNA around the nucleosome core particle. The amino-terminal tail domains are rich in lysine and can be modified by acetylation at specific lysine residues. This neutralizes the positive charge of lysine residues and appears to contribute to transcriptional activation by relaxing chromatin structure and increasing the availability of the DNA template to transcription factors and RNA polymerase.

Studies from two groups of researchers in 1996 provided direct links between histone acetylation and transcriptional regulation by demonstrating that transcriptional activators and repressors are associated with histone acetyltransferases and deacetylases, respectively. This association was first revealed by cloning a gene encoding a histone acetyltransferase from *Tetrahymena*. Unexpectedly, the sequence of this histone acetyltransferase was closely related to a previously known yeast transcriptional coactivator called Gcn5p. Further experiments revealed that Gcn5p has histone acetyltransferase activity, suggesting that transcriptional activation results directly from histone acetylation. These results have been extended by demonstrations that histone acetyltransferases are also associated with a number of mammalian transcriptional coactivators, as well as with the general transcription factor

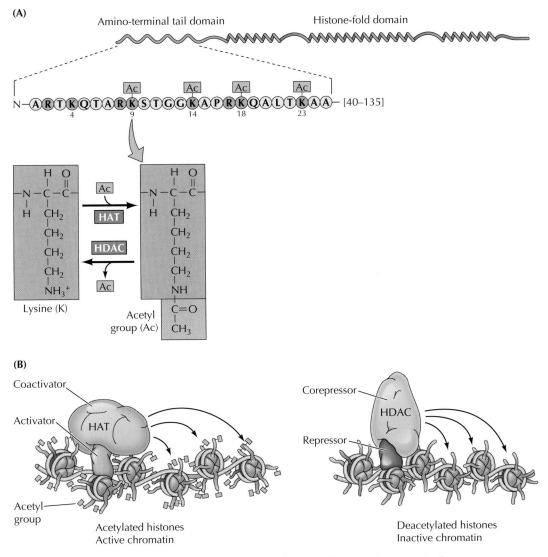

Figure 8.31 Histone acetylation (A) The core histones have amino-terminal tails, which extend outside of the nucleosome, and histone-fold domains, which interact with other histones and with DNA in the nucleosome. The amino-terminal tails of the core histones (e.g., H3) are modified by the addition of acetyl groups (Ac) to the side chains of specific lysine residues. (B) Transcriptional activators and repressors are associated with coactivators and corepressors, which have histone acetyltransferase (HAT) and histone deacetylase (HDAC) activities, respectively. Histone acetylation is characteristic of actively transcribed chromatin.

TFIID. Conversely, many transcriptional corepressors in both yeast and mammalian cells function as histone deacetylases, which remove the acetyl groups from histone tails. Histone acetylation is thus targeted directly by both transcriptional activators and repressors, indicating that it plays a key role in regulation of eukaryotic gene expression.

Histones are modified not only by acetylation, but also by several other modifications, including methylation of lysine and arginine residues, phosphorylation of serine residues, and the addition of small peptides (ubiquitin and SUMO, discussed in Chapter 9) to lysine residues. Like acetylation, these

Figure 8.32 Patterns of histone modification Transcriptional activity of chromatin is affected by methylation and phosphorylation of specific amino acid residues in histone tails, as well as by their acetylation. Distinct patterns of histone modification are characteristic of transcriptionally active and inactive chromatin.

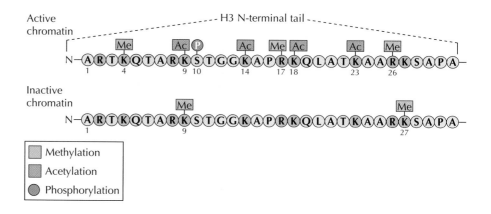

modifications occur at specific amino acid residues in the histone tails and are associated with changes in transcriptional activity (**Figure 8.32**). For example, transcriptionally active chromatin is associated with several specific modifications of histone H3, including methylation of lysine-4, phosphorylation of serine-10, acetylation of lysines 9, 14, 18, and 23, and methylation of arginines 17 and 26. It appears that histone modifications can affect gene expression both by altering the properties of chromatin and by providing binding sites for other proteins that act to either activate or repress transcription. For example, acetylation of lysine residues neutralizes their positive charge and facilitates transcription by relaxing chromatin structure. In addition, acetylated lysines, as well as the other modifications of histones that are associated with transcriptional activation, serve as binding sites for a variety of proteins that stimulate transcription.

In contrast to these modifications of active chromatin, methylation of H3 lysines 9 and 27 is associated with repression and chromatin condensation. Enzymes that catalyze the methylation of these lysine residues are recruited to target genes by corepressors. The methylated H3 lysine-9 and -27 residues then serve as binding sites for proteins that induce chromatin condensation, directly linking this histone modification to transcriptional repression and the formation of heterochromatin.

Promoters and enhancers are marked by distinct chromatin features that have been used for the global identification of these regions. Both promoters and enhancers are free of nucleosomes, so that their DNA is accessible for the binding of transcription factors (**Figure 8.33**). These nucleosome-free regions are

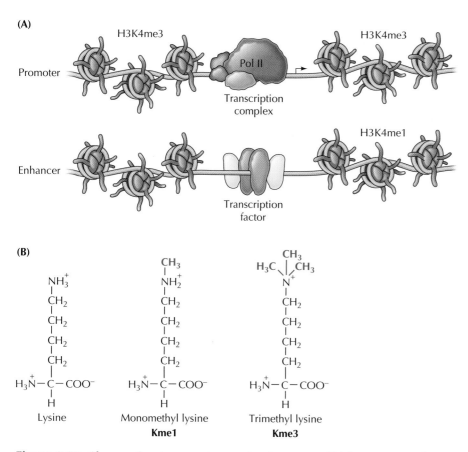

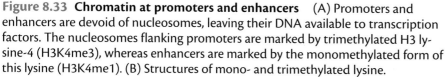

Figure 8.33 Chromatin at promoters and enhancers (A) Promoters and enhancers are devoid of nucleosomes, leaving their DNA available to transcription factors. The nucleosomes flanking promoters are marked by trimethylated H3 lysine-4 (H3K4me3), whereas enhancers are marked by the monomethylated form of this lysine (H3K4me1). (B) Structures of mono- and trimethylated lysine.

sensitive to digestion with DNase and can thus be identified as **DNase hypersensitive sites** in the genome. The nucleosomes flanking these regions have different histone modifications at promoters compared with enhancers. For example, promoters are marked by the trimethylated form of H3 lysine-4 (H3K4me3), whereas enhancers are characterized by the monomethylated form of this lysine (H3K4me1).

Chromatin remodeling factors

In contrast to the enzymes that regulate chromatin structure by modifying histones, **chromatin remodeling factors** are protein complexes that use energy derived from the hydrolysis of ATP to alter the contacts between DNA and histones (**Figure 8.34**). One mechanism by which chromatin remodeling factors act is to catalyze the sliding of histone octamers along the DNA molecule, thereby repositioning nucleosomes to change the accessibility of specific DNA sequences to transcription factors. Alternatively, chromatin remodeling factors may act by inducing changes in the conformation of nucleosomes, again affecting the ability of specific DNA sequences to interact with transcriptional regulatory proteins. Finally, chromatin remodeling factors can eject histones from the DNA, leaving nucleosome-free regions, which are commonly found at enhancers and promoters (see Figure 8.33). Like histone

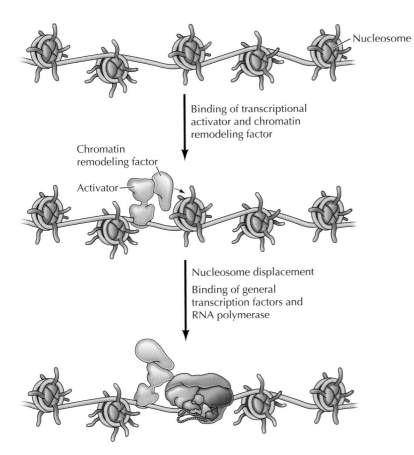

Figure 8.34 Chromatin remodeling factors Chromatin remodeling factors alter the arrangement or structure of nucleosomes. For example, a chromatin remodeling factor recruited in association with a transcriptional activator could facilitate the binding of general transcription factors and RNA polymerase to chromatin by displacing nucleosomes from the promoter.

modifying enzymes, chromatin remodeling factors can be recruited to DNA in association with either transcriptional activators or repressors, and can alter the arrangement of nucleosomes to either stimulate or inhibit transcription.

The recruitment of histone modifying enzymes and chromatin remodeling factors by transcriptional activators stimulates the initiation of transcription by altering the chromatin structure of enhancer and promoter regions. However, following the initiation of transcription, RNA polymerase is still faced with the problem of transcriptional elongation through a chromatin template. This is facilitated by **elongation factors** that become associated with the phosphorylated C-terminal domain of RNA polymerase II when productive elongation is initiated (see Figure 8.29). These elongation factors include both histone modifying enzymes (e.g., histone acetylases) and chromatin remodeling factors that transiently displace nucleosomes during transcription.

Histones and epigenetic inheritance

An important feature of the modifications of histone tails is that the modified histones serve as binding sites for proteins that themselves catalyze histone modifications. Histone modifications can thus regulate one another, leading to the establishment of stable patterns of modified chromatin. Consequently, histone modification provides a mechanism for **epigenetic inheritance**—the transmission of information that is not contained within the sequence of DNA to daughter cells at cell division. This is particularly critical for the maintenance of differentiated cell types during the development of multicellular organisms, since it provides a mechanism by which the distinct programs of gene expression characteristic of each cell type can be transmitted to progeny cells during mitosis. As discussed in Chapter 6, parental nucleosomes are distributed to the two progeny strands when chromosomal DNA replicates. Consequently, modified histones will be transferred from the parental to both progeny chromosomes (Figure 8.35). These modified histones can then direct similar modifications of newly synthesized histones, maintaining a pattern of histone modification characteristic of either active or inactive chromatin.

A good example is provided by the repression of key regulatory genes that determine cell fate during *Drosophila* development. Once these genes become repressed in specific cells of the developing embryo, that repression is passed on to daughter cells through many subsequent cell divisions. Repression is maintained through these embryonic cell divisions as a result of methylation of H3 lysine 27 by the **Polycomb proteins** (Figure 8.36). The Polycomb proteins consist of two complexes, designated Polycomb Repressive Complex (PRC) 1 and 2. One of the proteins in PRC1 binds to methylated H3 lysine 27, whereas one of the proteins in PRC2 is an enzyme that methylates H3 lysine 27. As a result, methylation of H3 lysine 27 can spread to adjacent nucleosomes and is maintained when cells divide. Genes that are repressed via this chromatin mark in a parental cell thus remain repressed in daughter cells.

The boundaries between regions of chromatin characterized by some histone modifications, including methylation of H3 lysine 27, correlate with the topologically associating domains (TADs) which, as discussed earlier, restrict the actions of enhancers and promoters (see Figure 8.20). For example, repressed regions of chromatin characterized by methylation of H3 lysine 27 are often separated from adjacent transcriptionally-active chromatin regions by CTCF binding sites. The architectural protein CTCF thus may play a role in establishing the boundaries of domains of chromatin modification, as well as the boundaries of TADs.

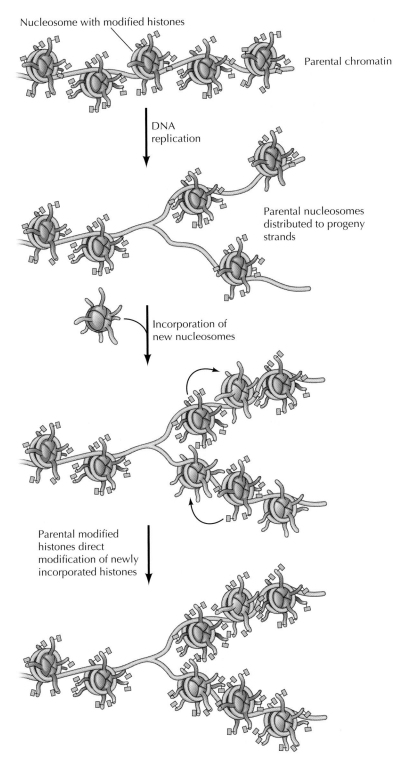

Nucleosome with modified histones

Parental chromatin

DNA replication

Parental nucleosomes distributed to progeny strands

Incorporation of new nucleosomes

Parental modified histones direct modification of newly incorporated histones

Figure 8.35 Epigenetic inheritance of histone modifications Parental nucleosomes, containing modified histones, are distributed to both strands at DNA replication. The parental modified histones can then direct similar modifications of newly synthesized histones incorporated into the progeny chromosomes.

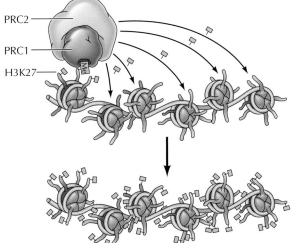

PRC2

PRC1

H3K27

Figure 8.36 Polychrome proteins Polycomb Repressive Complex (PRC) 1 binds to methylated H3 lysine 27 (H3K27) and PRC2 methylates lysine 27 residues of adjacent nucleosomes. The complex of PRC1 and PRC2 can thus spread methylation of H3 lysine 27 to neighboring nucleosomes.

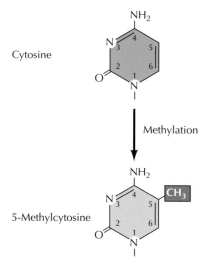

Cytosine

Methylation

5-Methylcytosine

Figure 8.37 DNA methylation A methyl group is added to the 5-carbon position of cytosine residues in DNA.

DNA *methylation*

The methylation of DNA is another general mechanism for epigenetic control of transcription in eukaryotes. Cytosine residues in DNA of fungi, plants, and animals can be modified by the addition of methyl groups at the 5-carbon position (Figure 8.37). DNA is methylated specifically at the cytosines (C) that precede guanines (G) in the DNA chain (CpG dinucleotides), and this methylation is correlated with transcriptional repression. Methylation commonly occurs within transposable elements, and it appears that methylation plays a key role in suppressing the movement of transposons throughout the genome. In addition, DNA methylation is associated with transcriptional repression of mammalian genes involved in development and differentiation, in concert with alterations in chromatin structure.

DNA methylation, like the modification of histones, is an important mechanism of epigenetic inheritance. Patterns of DNA methylation are stably maintained following DNA replication by an enzyme that specifically methylates CpG sequences of a daughter strand that is hydrogen-bonded to a methylated parental strand (Figure 8.38). Thus, a gene that has been methylated and repressed in a parental cell remains methylated and repressed in progeny cells. Methylation can also be reversed by the TET family enzymes, which catalyze the oxidation of 5-methylcytosine to 5-formylcytosine and 5-carboxylcytosine, which are excised and replaced by cytosine via DNA repair (Figure 8.39). Methylation can thus be actively reversed, as well as epigenetically transmitted.

One important regulatory role of DNA methylation has been established in the phenomenon known as **genomic imprinting**, which controls the expression of some genes involved in the development of mammalian embryos. In most cases, both the paternal and maternal alleles of a gene are expressed in diploid cells. However, there are some imprinted genes (about 75 have

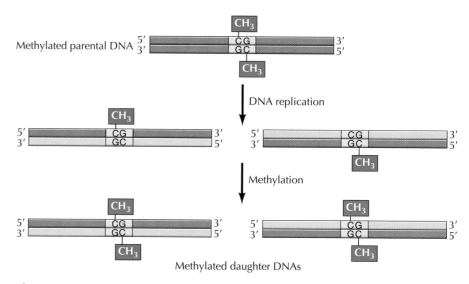

Figure 8.38 Maintenance of methylation patterns In parental DNA, both strands are methylated at complementary CpG sequences. Following replication, only the parental strand of each daughter molecule is methylated. The newly synthesized daughter strands are then methylated by an enzyme that specifically recognizes CpG sequences opposite a methylation site.

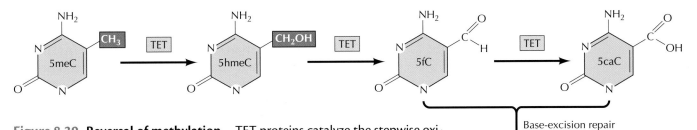

Figure 8.39 Reversal of methylation TET proteins catalyze the stepwise oxidation of 5-methylcytosine (5meC) to 5-hydroxymethylcytosine (5hmeC), 5-formylcytosine (5fC), and 5-carboxylcytosine (5caC). 5fC and 5caC are removed and replaced by cytosine (C) via base-excision repair (see Figure 7.21).

been described in mice and humans) whose expression depends on whether they are inherited from the mother or from the father. In some cases, only the paternal allele of an imprinted gene is expressed, and the maternal allele is transcriptionally inactive. For other imprinted genes, the maternal allele is expressed and the paternal allele is inactive.

DNA methylation appears to play a key role in distinguishing between the paternal and maternal alleles of imprinted genes. A good example is the gene *H19*, which is transcribed only from the maternal copy (**Figure 8.40**). The *H19* gene is specifically methylated during the development of male, but not female, germ cells. The union of sperm and egg at fertilization therefore yields an embryo containing a methylated paternal allele and an unmethylated maternal allele of the gene. The paternal *H19* allele remains methylated, and transcriptionally inactive, in embryonic cells and somatic tissues. However, the paternal *H19* allele becomes demethylated in the germ line, allowing a new pattern of methylation to be established for transmittal to the next generation. Interestingly, *H19* encodes a long noncoding RNA (lncRNA) that regulates expression of other imprinted genes.

Noncoding RNAs

A series of recent advances have highlighted the roles of noncoding RNAs, including microRNAs (miRNAs) and **long noncoding RNAs (lncRNAs)** in gene regulation. miRNAs (20–30 nucleotides) generally act by the RNA interference pathway (see Figure 4.38) to inhibit translation or induce degradation of homologous mRNAs, and gene regulation at this post-transcriptional level will be discussed later in this chapter and in Chapter 9. As discussed in Chapter 6, global analyses

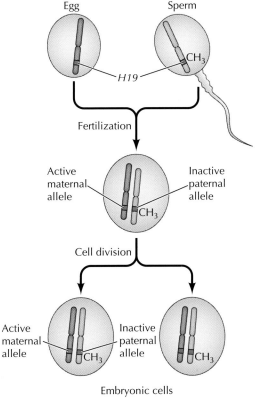

Figure 8.40 Genomic imprinting The *H19* gene is specifically methylated during development of male germ cells. Therefore sperm contain a methylated *H19* allele and eggs contain an unmethylated allele. Following fertilization and subsequent cell divisions, the paternal allele remains methylated and transcriptionally inactive, so only the unmethylated maternal allele is expressed in the embryo.

of transcription in human cells has also focused attention on the roles of lncRNAs (>200 nucleotides) in gene regulation. More than 50,000 lncRNAs are encoded in the human genome. Although the functions of the majority of this unexpectedly large number of lncRNAs remain to be determined, it appears that at least one important function of lncRNAs is to regulate gene expression at the transcriptional level.

Many lncRNAs form complexes with proteins that modify chromatin and recruit these chromatin-modifying complexes to their sites of transcription, thereby regulating the expression of neighboring genes. In many cases, including the *Xist* lncRNA, which mediates X chromosome inactivation in mammals (see Chapter 6), and several lncRNAs involved in imprinting (see Figure 8.40), lncRNAs act to repress their target genes by forming complexes with Polycomb Repressive Complex 2 (PRC2). This leads to methylation of histone H3 lysine 27 and transcriptional silencing (**Figure 8.41A**). Other lncRNAs repress their target genes by forming complexes with a histone H3 lysine 9 methylase, which also induces transcriptional silencing (see Figure 8.32), or with DNA methyltransferases, which establish repressive DNA methylation. Importantly however, not all lncRNAs repress transcription of their target genes—some lncRNAs are associated with chromatin modifiers that activate transcription, such as histone acetylases and enzymes that methylate histone H3 lysine 4 (see Figure 8.32). Thus, lncRNAs can associate with a number of different chromatin-modifying enzymes and can function as either repressors or activators of their target genes.

Although many lncRNAs act in *cis* to regulate the expression of neighboring genes by recruiting chromatin modifiers to their own sites of transcription, some lncRNAs act in *trans* to affect the expression of distant target genes (**Figure 8.41B**). For example, the lncRNA *HOTAIR* is an important developmental regulator that targets hundreds of genes throughout the human genome. It recruits both PRC2 and a complex that demethylates histone H3 lysine 4, thus inducing two repressive chromatin modifications on its target genes. The mechanism by which *trans*-acting lncRNAs, such as *HOTAIR*, find their target genes remains to be determined. One possibility is that the lncRNAs recognize their target DNA sequences by homologous base pairing. Alternatively, lncRNAs may form complexes with proteins, such as transcription factors, that target the lncRNAs to specific DNA sequences.

Although recruitment of chromatin-modifying complexes is one common mechanism of lncRNA action,

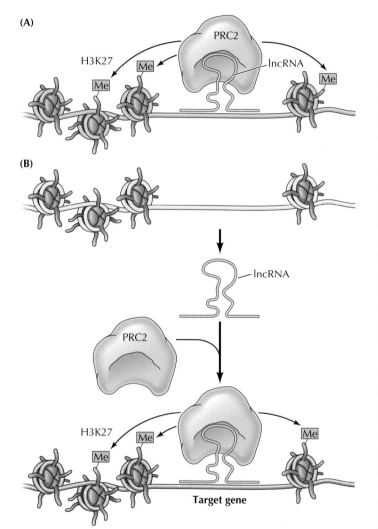

Figure 8.41 Action of lncRNAs (A) Many lncRNAs recruit Polycomb Repressive Complex 2 (PRC2) to their sites of transcription, leading to methylation of H3 lysine 27 (H3K27) and repression of neighboring genes. (B) Some lncRNAs act in *trans* by recruiting chromatin-modifying complexes (e.g., PRC2) to distant target genes, either by binding to homologous DNA sequences or to DNA binding proteins (e.g., transcription factors).

it is not the only way in which lncRNAs can regulate gene expression. Some lncRNAs regulate transcription by interactions with transcription factors, Mediator, or components of the basal transcription machinery (e.g., TFIIB). Still other lncRNAs interact with proteins that regulate post-transcriptional levels of gene expression, such as the processing or translation of mRNAs. It is also noteworthy that active enhancers are transcribed to yield noncoding RNAs termed enhancer RNAs (eRNAs). Although eRNAs may represent nonfunctional byproducts of normal transcription, recent evidence suggests that at least some eRNAs may play a role in transcriptional regulation of adjacent genes. The mechanisms and extent of gene regulation by multiple different types of noncoding RNA remains an active and largely unexplored area of investigation.

RNA Processing and Turnover

Although transcription is the first and most highly regulated step in gene expression, it is usually only the beginning of the series of events required to produce a functional RNA. Most newly synthesized RNAs must be modified in various ways to be converted to their functional forms. Bacterial mRNAs are an exception; they are used immediately as templates for protein synthesis while still being transcribed. However, the primary transcripts of both rRNAs and tRNAs undergo a series of processing steps in bacteria as well as in eukaryotic cells. Primary transcripts of eukaryotic mRNAs similarly undergo extensive modifications, including the removal of introns by splicing, before they are transported from the nucleus to the cytoplasm to serve as templates for protein synthesis. Regulation of these processing steps provides an additional level of control of gene expression, as does regulation of the rates at which different mRNAs are subsequently degraded within the cell.

Processing of ribosomal and transfer RNAs

The basic processing of ribosomal and transfer RNAs in bacteria and eukaryotic cells is similar, as might be expected given the fundamental roles of these RNAs in protein synthesis. As discussed previously, eukaryotes have four species of ribosomal RNAs (see Table 8.1), three of which (the 5.8S, 18S, and 28S rRNAs) are derived by cleavage of a single long precursor transcript, called a **pre-rRNA** (Figure 8.42). Bacteria have three ribosomal RNAs (23S, 16S, and 5S), which are equivalent

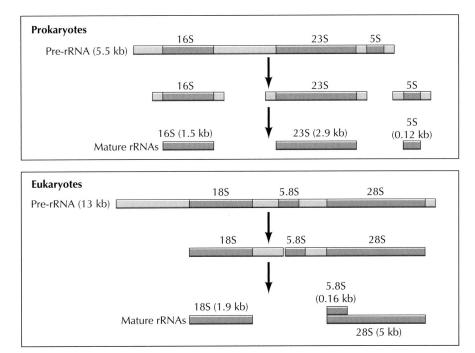

Figure 8.42 Processing of ribosomal RNAs Bacteria contain three rRNAs (16S, 23S, and 5S), which are formed by cleavage of a pre-rRNA transcript. Eukaryotic cells (e.g., human cells) contain four rRNAs. One of these (5S rRNA) is transcribed from a separate gene; the other three (18S, 5.8S, and 28S) are derived from a common pre-rRNA. Following cleavage, the 5.8S rRNA (which is unique to eukaryotes) becomes hydrogen-bonded to 28S rRNA.

(A)

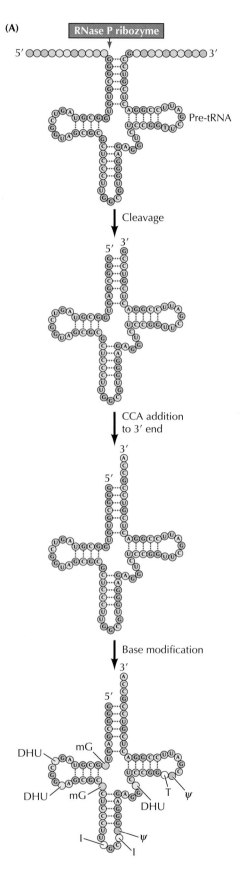

to the 28S, 18S, and 5S rRNAs of eukaryotic cells and are also formed by the processing of a single pre-rRNA transcript. The only rRNA that is not processed extensively is the 5S rRNA in eukaryotes, which is transcribed from a separate gene.

Bacterial and eukaryotic pre-rRNAs are processed in several steps. Initial cleavages of bacterial pre-rRNA yield separate precursors for the three individual rRNAs; these are then further processed by secondary cleavages to the final products. In eukaryotic cells, pre-rRNA is first cleaved at a site adjacent to the 5.8S rRNA on its 5′ side, yielding two separate precursors that contain the 18S and the 28S plus 5.8S rRNAs, respectively. Further cleavages then convert these to their final products, with the 5.8S rRNA becoming hydrogen-bonded to the 28S molecule. In addition to these cleavages, rRNA processing involves the addition of methyl groups to the bases and sugar moieties of specific nucleotides and the conversion of some uridines to pseudouridines. Processing of rRNA takes place within the nucleolus of eukaryotic cells, and will be discussed in detail in Chapter 10.

Like rRNAs, tRNAs in both bacteria and eukaryotes are synthesized as longer precursor molecules (**pre-tRNAs**), some of which contain several individual tRNA sequences (**Figure 8.43**). In bacteria, some tRNAs are

Figure 8.43 Processing of transfer RNAs (A) Transfer RNAs are derived from pre-tRNAs, some of which contain several individual tRNA molecules. Cleavage at the 5′ end of the tRNA is catalyzed by the RNase P ribozyme; cleavage at the 3′ end is catalyzed by a conventional protein RNase. A CCA terminus is then added to the 3′ end of many tRNAs in a post-transcriptional processing step. Finally, some bases are modified at characteristic positions in the tRNA molecule. In this example, these modified nucleosides include dihydrouridine (DHU), methylguanosine (mG), inosine (I), ribothymidine (T), and pseudouridine (ψ). (B) Structure of modified bases. Dihydrouridine, ribothymidine, , and pseudouridine are formed by modification of uridines in tRNA. Inosine and methylguanosine are formed by the modification of guanosines.

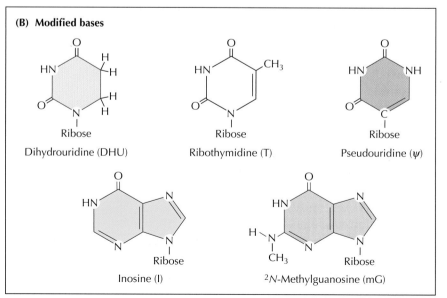

(B) Modified bases

Dihydrouridine (DHU)

Ribothymidine (T)

Pseudouridine (ψ)

Inosine (I)

^2N-Methylguanosine (mG)

included in the pre-rRNA transcripts. The processing of the 5' end of pre-tRNAs involves cleavage by an enzyme called **RNase P**, which is of special interest because it is a prototypical model of a reaction catalyzed by an RNA enzyme. RNase P consists of RNA and protein molecules, both of which are required for maximal activity. In 1983 Sidney Altman and his colleagues demonstrated that the isolated RNA component of RNase P is itself capable of catalyzing pre-tRNA cleavage. These experiments established that RNase P is a **ribozyme**—an enzyme in which RNA rather than protein is responsible for catalytic activity.

The 3' end of tRNAs is generated by the action of a conventional protein RNase, but the processing of this end of the tRNA molecule also involves an unusual activity: the addition of a CCA terminus. All tRNAs have the sequence CCA at their 3' ends. This sequence is the site of amino acid attachment, so it is required for tRNA function during protein synthesis. The CCA terminus is encoded in the DNA of some tRNA genes, but in others it is not, instead being added as an RNA processing step by an enzyme that recognizes and adds CCA to the 3' end of all tRNAs that lack this sequence.

Another unusual aspect of tRNA processing is the extensive modification of bases in tRNA molecules. Approximately 10% of the bases in tRNAs are altered to yield a variety of modified nucleotides at specific positions in tRNA molecules (see Figure 8.43). The functions of most of these modified bases are unknown, but some play important roles in protein synthesis by altering the base-pairing properties of the tRNA molecule (see Chapter 9).

Some pre-tRNAs, as well as pre-rRNAs in a few organisms, contain introns that are removed by splicing. In contrast to other splicing reactions, which involve the activities of catalytic RNAs (discussed in the section on mRNA processing), tRNA splicing is mediated by conventional protein enzymes. An endonuclease cleaves the pre-tRNA at the splice sites to excise the intron, followed by joining of the exons to form a mature tRNA molecule.

Processing of mRNA in eukaryotes

In contrast to the processing of ribosomal and transfer RNAs, the processing of messenger RNAs represents a major difference between bacteria and eukaryotic cells. In bacteria, ribosomes have immediate access to mRNA and translation begins on the nascent mRNA chain while transcription is still in progress. In eukaryotes, mRNA synthesized in the nucleus must first be transported to the cytoplasm before it can be used as a template for protein synthesis. Moreover, the initial products of transcription in eukaryotic cells (**pre-mRNAs**) are extensively modified before export from the nucleus. Throughout processing in the nucleus, transport to the cytoplasm, translation, and eventual degradation, mRNA molecules are associated with proteins to form messenger ribonucleoprotein particles (**mRNPs**). These RNA-binding proteins are responsible for mRNA processing, transport, and the regulation of mRNA function.

The processing of mRNA includes modification of both ends of the initial transcript, as well as the removal of introns from its middle (**Figure 8.44**). Rather than occurring as independent events following synthesis of a pre-mRNA, these processing reactions are coupled to transcription so that mRNA synthesis and processing are closely coordinated steps in gene expression. The C-terminal domain (CTD) of RNA polymerase II plays a key role

Animation 8.2

sites.sinauer.com/cooper7e/a8.2

RNA Processing Eukaryotic cells must process the RNA product of transcription to form a mature, translatable messenger RNA.

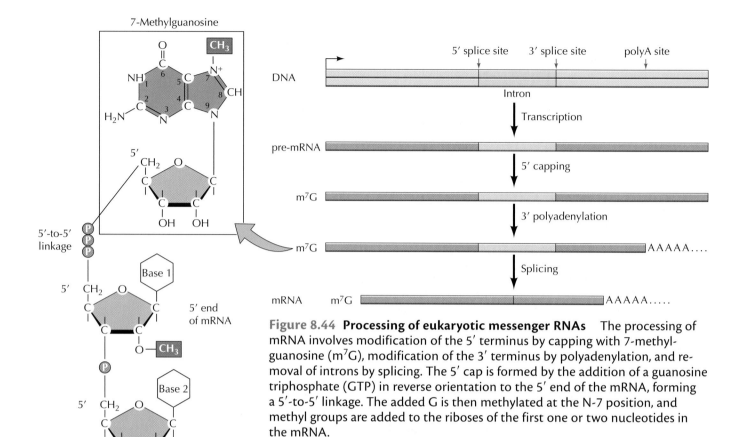

Figure 8.44 Processing of eukaryotic messenger RNAs The processing of mRNA involves modification of the 5′ terminus by capping with 7-methylguanosine (m⁷G), modification of the 3′ terminus by polyadenylation, and removal of introns by splicing. The 5′ cap is formed by the addition of a guanosine triphosphate (GTP) in reverse orientation to the 5′ end of the mRNA, forming a 5′-to-5′ linkage. The added G is then methylated at the N-7 position, and methyl groups are added to the riboses of the first one or two nucleotides in the mRNA.

in coordinating these processes by serving as a binding site for the enzyme complexes involved in mRNA processing. The association of these processing enzymes with the CTD of polymerase II accounts for their specificity in processing mRNAs; polymerases I and III lack a CTD, so their transcripts are not processed by the same enzyme complexes.

The first step in mRNA processing is the modification of the 5′ end of the transcript by the addition of a structure called a **7-methylguanosine cap**. The enzymes responsible for capping are recruited to the phosphorylated CTD following initiation of transcription, and the cap is added after transcription of the first 20 to 30 nucleotides of the RNA. Capping is initiated by the addition of a guanosine triphosphate (GTP) in reverse orientation to the 5′ terminal nucleotide of the RNA. Then methyl groups are added to this G residue and to the ribose moieties of one or two 5′ nucleotides of the RNA chain. The 5′ cap stabilizes the RNA, as well as aligning eukaryotic mRNAs on the ribosome during translation (see Chapter 9).

The 3′ end of most eukaryotic mRNAs is defined not by termination of transcription but by cleavage of the primary transcript and addition of a **poly-A tail**—a processing reaction called **polyadenylation** (Figure 8.45). The signals for polyadenylation include a highly conserved hexanucleotide (AAUAAA in mammalian cells), which is located 10 to 30 nucleotides upstream of the site of polyadenylation, and a U- or GU- rich downstream sequence element. These sequences are recognized by a complex of proteins,

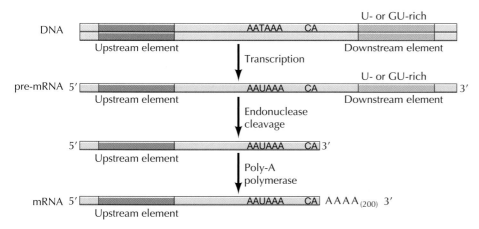

Figure 8.45 Formation of the 3′ ends of eukaryotic mRNAs   Polyadenylation signals in mammalian cells consist of the hexanucleotide AAUAAA in addition to upstream and downstream (U- or GU-rich) elements. An endonuclease cleaves the pre-mRNA 10 to 30 nucleotides downstream of the AAUAAA, usually at a CA sequence. Poly-A polymerase then adds a poly-A tail consisting of about 200 adenines (A) to the 3′ end of the RNA.

including an endonuclease that cleaves the RNA chain and a separate poly-A polymerase that adds a poly-A tail of about 200 nucleotides to the transcript. These processing enzymes are associated with the phosphorylated CTD of RNA polymerase II, and travel with the polymerase all the way from the transcription initiation site. Recognition of the polyadenylation signal leads to termination of transcription and cleavage and polyadenylation of the mRNA, followed by degradation of the RNA that has been synthesized downstream of the site of poly-A addition.

Almost all mRNAs in eukaryotes are polyadenylated, and poly-A tails have been shown to regulate both translation and mRNA stability. In addition, polyadenylation plays an important regulatory role in early development, where changes in the length of poly-A tails control mRNA translation. For example, many mRNAs are stored in unfertilized eggs in an untranslated form with short poly-A tails (usually 30 to 50 nucleotides long). Fertilization stimulates the lengthening of the poly-A tails of these stored mRNAs, which in turn activates their translation and the synthesis of proteins required for early embryonic development.

The most striking modification of pre-mRNAs is the removal of introns by splicing. As discussed in previous chapters, the coding sequences of most eukaryotic genes are interrupted by noncoding sequences (introns) that are precisely excised from the mature mRNA. In mammals, most genes contain multiple introns, which typically account for about ten times more pre-mRNA sequences than do the exons. The unexpected discovery of introns in 1977 generated an active research effort directed toward understanding the mechanism of splicing, which had to be highly specific to yield functional mRNAs. Further studies of splicing have not only illuminated new mechanisms of gene regulation; they have also revealed novel catalytic activities of RNA molecules and led to an understanding of how alternative splicing expands the repertoire of mammalian genomes.

Splicing mechanisms

The key to understanding pre-mRNA splicing was the development of *in vitro* systems that efficiently carried out the splicing reaction (**Figure 8.46**). Pre-mRNAs were synthesized *in vitro* by the cloning of structural genes (with their introns) adjacent to promoters for bacteriophage RNA polymerases, which could readily be isolated in large quantities. Transcription of these plasmids could then be used to prepare large amounts of pre-mRNAs that, when added to nuclear extracts of mammalian cells, were found to be correctly spliced. As with transcription, the use of such *in vitro* systems has allowed splicing to be analyzed in much greater detail than would have been possible in intact cells.

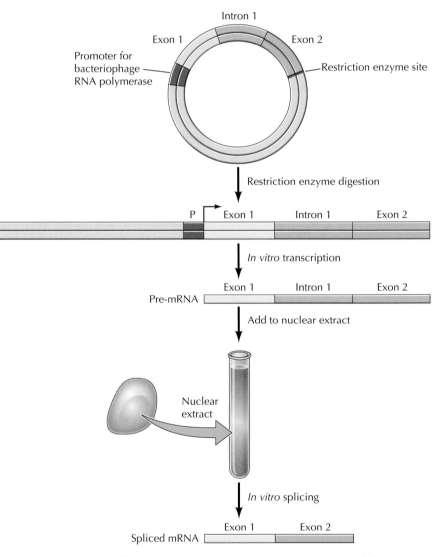

Figure 8.46 *In vitro* splicing   A gene containing an intron is cloned downstream of a promoter (P) recognized by a bacteriophage RNA polymerase. The plasmid is digested with a restriction enzyme that cleaves at the 3′ end of the inserted gene to yield a linear DNA molecule. This DNA is then transcribed *in vitro* with the bacteriophage polymerase to produce pre-mRNA. Splicing reactions can then be studied *in vitro* by addition of this pre-mRNA to nuclear extracts of mammalian cells.

Analysis of the reaction products and intermediates formed *in vitro* revealed that pre-mRNA splicing proceeds in two steps (Figure 8.47). First, the pre-mRNA is cleaved at the 5′ splice site, and the 5′ end of the intron is joined to an adenine nucleotide within the intron (near its 3′ end). In this step, an unusual bond forms between the 5′ end of the intron and the 2′ hydroxyl group of the adenine nucleotide. The resulting intermediate is a lariat-like structure in which the intron forms a loop. The second step in splicing then proceeds with simultaneous cleavage at the 3′ splice site and ligation of the two exons. The intron is thus excised as a lariat-like structure, which is then linearized and degraded within the nucleus of intact cells.

These reactions define three critical sequence elements of pre-mRNAs: sequences at the 5′ splice site, sequences at the 3′ splice site, and sequences within the intron at the branch point (the point at which the 5′ end of the intron becomes ligated to form the lariat-like structure). Pre-mRNAs contain similar consensus sequences at each of these positions, allowing the splicing apparatus to recognize pre-mRNAs and carry out the cleavage and ligation reactions involved in the splicing process.

Biochemical analysis of nuclear extracts has revealed that splicing takes place in large complexes, called **spliceosomes**, composed of proteins and RNAs. The RNA components of the spliceosome are five types of **small nuclear RNAs (snRNAs)** called U1, U2, U4, U5, and U6. These snRNAs, which range in size from approximately 50 to nearly 200 nucleotides, are complexed with six to ten protein molecules to form **small nuclear ribonucleoprotein particles (snRNPs)**, which play central roles in the splicing process. The U1, U2, and U5 snRNPs each contain a single snRNA molecule, whereas U4 and U6 snRNAs are complexed to each other in a single snRNP.

Video 8.3

sites.sinauer.com/cooper7e/v8.3
RNA Splicing The splicing process removes non-coding regions (introns) from the RNA transcript, leaving only protein-coding regions in the final mRNA.

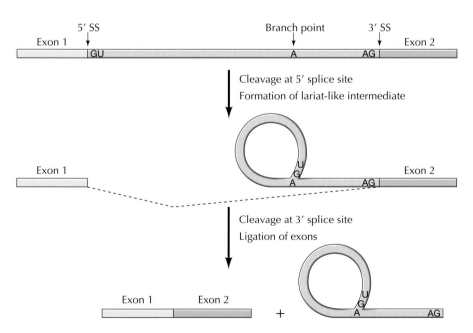

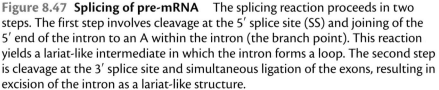

Figure 8.47 Splicing of pre-mRNA   The splicing reaction proceeds in two steps. The first step involves cleavage at the 5′ splice site (SS) and joining of the 5′ end of the intron to an A within the intron (the branch point). This reaction yields a lariat-like intermediate in which the intron forms a loop. The second step is cleavage at the 3′ splice site and simultaneous ligation of the exons, resulting in excision of the intron as a lariat-like structure.

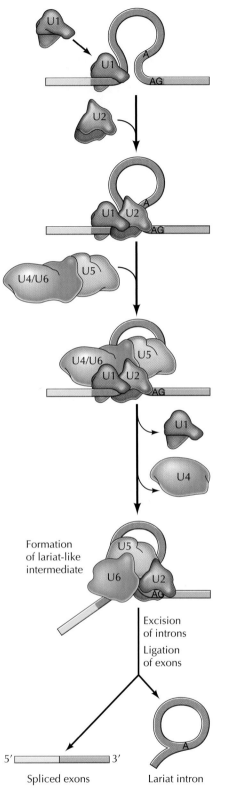

Figure 8.48 Assembly of the spliceosome The first step in spliceosome assembly is the binding of U1 snRNP to the 5′ splice site (SS), followed by the binding of U2 snRNP to the branch point. A preformed complex consisting of U4/U6 and U5 snRNPs then enters the spliceosome. U4 and U1 dissociate from the spliceosome and U6 catalyzes both formation of the lariat-like intermediate and ligation of the exons.

The first step in spliceosome assembly is the binding of U1 snRNP to the 5′ splice site of pre-mRNA (**Figure 8.48**). This recognition of 5′ splice sites involves base pairing between the 5′ splice site consensus sequence and a complementary sequence at the 5′ end of U1 snRNA (**Figure 8.49**). U2 snRNP then binds to the branch point, by similar complementary base pairing between U2 snRNA and branch point sequences. A preformed complex consisting of U4/U6 and U5 snRNPs is then incorporated into the spliceosome. The splicing reaction is then accompanied by rearrangements of the snRNAs, with the splicing reactions catalyzed by U6.

Not only do the snRNAs recognize consensus sequences at the branch points and splice sites of pre-mRNAs, but they also catalyze the splicing reaction directly. The catalytic role of RNAs in splicing was demonstrated by the discovery that some RNAs are capable of **self-splicing**; that is, they can catalyze the removal of their own introns in the absence of other protein or RNA factors. Self-splicing was first described by Tom Cech and his colleagues during studies of the 28S rRNA of the protozoan *Tetrahymena*. This RNA contains an intron of approximately 400 bases that is precisely

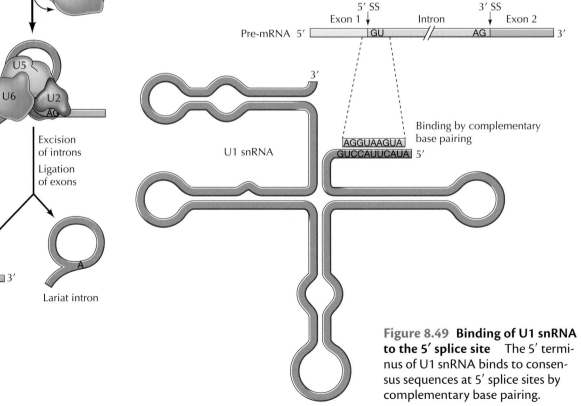

Figure 8.49 Binding of U1 snRNA to the 5′ splice site The 5′ terminus of U1 snRNA binds to consensus sequences at 5′ splice sites by complementary base pairing.

The Discovery of snRNPs

Antibodies to small nuclear RNAs complexed with proteins are produced by patients with systemic lupus erythematosus

Michael R. Lerner and Joan A. Steitz
Yale University, New Haven, CT
Proceedings of the National Academy of Sciences, USA, 1979, Volume 76, pages 5495–5499

The Context

The discovery of introns in 1977 implied that a totally unanticipated processing reaction was required to produce mRNA in eukaryotic cells. Introns had to be precisely excised from pre-mRNA, followed by the joining of exons to yield a mature mRNA molecule. Given the unexpected nature of pre-mRNA splicing, understanding the mechanism of the splicing reaction captivated the attention of many molecular biologists. One of the major steps in elucidating this mechanism was the discovery of snRNPs and their involvement in pre-mRNA splicing.

Small nuclear RNAs were first identified in eukaryotic cells in the late 1960s. However, the function of snRNAs remained unknown. In this 1979 paper, Michael Lerner and Joan Steitz demonstrated that the most abundant snRNAs were present as RNA-protein complexes called snRNPs. In addition, they provided the first suggestion that these RNA-protein complexes might function in pre-mRNA splicing. This identification of snRNPs led to a variety of experiments that confirmed their roles and elucidated the mechanism by which pre-mRNA splicing takes place.

The Experiments

The identification of snRNPs was based on the use of antisera from patients with systemic lupus erythematosus, an autoimmune disease in which patients produce antibodies against their own normal cell constituents. Many of the antibodies produced by systemic lupus erythematosus patients are directed against components of the nucleus, including DNA, RNA, and histones. The discovery of snRNPs arose

from studies in which Lerner and Steitz sought to characterize two antigens, called ribonucleoprotein (RNP) and Sm, which were recognized by antibodies from systemic lupus erythematosus patients. Indirect data suggested that RNP consisted of both protein and RNA, as its name implies, but neither RNP nor Sm had been characterized at the molecular level.

To identify possible RNA components of the RNP and Sm antigens, nuclear RNAs of mouse cells were radiolabeled with ^{32}P and immunoprecipitated with antisera from different systemic lupus erythematosus patients (see Figure 5.11). Six specific species of snRNAs were found to be selectively immunoprecipitated by antisera from different patients but not by serum from a normal control patient (see figure). Anti-Sm serum immunoprecipitated all six of these snRNAs, which were designated U1a, U1b, U2, U4, U5, and U6. Anti-RNP serum immunoprecipitated only U1a and U1b, and serum from a third patient (which had been characterized as mostly anti-RNP) immunoprecipitated U1a, U1b, and U6. The immunoprecipitated snRNAs were further characterized by sequence analysis, which demonstrated that U1a, U1b, and U2 were identical to the most abundant snRNAs previously reported in mammalian nuclei, with U1a and U1b representing sequence variants of a single species of U1 snRNA present in human cells. In contrast, the U4, U5, and U6 snRNAs were newly identified by Lerner and Steitz in these experiments.

Importantly, the immunoprecipitation of these snRNAs demonstrated that they were components of RNA-protein complexes. The anti-Sm serum, which immunoprecipitated all six of the snRNAs, had previously been shown to be directed against a protein antigen. Similarly, protein was known to be required for antigen recognition by anti-RNP serum. Moreover, Lerner and Steitz showed that none of the snRNAs could be immunoprecipitated if protein was first removed by extraction of the RNAs with phenol. Further analysis of cells in which proteins had been radiolabeled with ^{35}S-methionine identified seven prominent nuclear proteins that were immunoprecipitated along with the snRNAs by anti-Sm and anti-RNP sera. These data

Joan Steitz

therefore indicated that each of the six snRNAs was present in an snRNP complex with specific nuclear proteins.

The Impact

The finding that snRNAs were components of snRNPs that were recognized by specific antisera opened a new approach to studying snRNA function. Lerner and Steitz noted that a "most intriguing" possible role for snRNAs might be in pre-mRNA splicing,

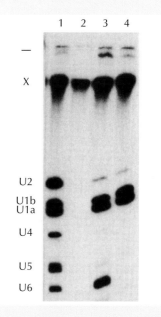

Immunoprecipitation of snRNAs with antisera from systemic lupus erythematosus patients Lane 1, anti-Sm; lane 2, normal control serum; lane 3, antiserum recognizing primarily the RNP antigen; lane 4, anti-RNP. Note that a nonspecific RNA designated X is present in all immunoprecipitates, including the control.

(Continued on next page)

and pointed out that sequences near the 5' terminus of U1 snRNA were complementary to splice sites.

Steitz and her colleagues then proceeded with a series of experiments that established the critical involvement of snRNPs in splicing. These studies included more extensive sequence analysis that demonstrated the complementarity of

conserved 5' sequences of U1 snRNA to the consensus sequences of 5' splice sites, suggesting that U1 functioned in 5' splice site recognition. In addition, antisera against snRNPs were used to demonstrate that U1 was required for pre-mRNA splicing, both in isolated nuclei and in *in vitro* splicing extracts. Further studies have gone on to show that the snRNAs themselves play

critical roles not only in the identification of splice sites, but also as catalysts of the splicing reaction. The initial discovery that snRNAs were components of snRNPs that could be recognized by specific antisera thus opened the door to understanding the mechanism of pre-mRNA splicing.

removed following incubation of the pre-rRNA in the absence of added proteins. Further studies have revealed that splicing is catalyzed by the intron, which acts as a ribozyme to direct its own excision from the pre-rRNA molecule. The discovery of self-splicing of *Tetrahymena* rRNA, together with the studies of RNase P already discussed, provided the first demonstrations of the catalytic activity of RNA.

Additional studies have revealed self-splicing RNAs in mitochondria, chloroplasts, and bacteria. These self-splicing RNAs are divided into two classes on the basis of their reaction mechanisms (**Figure 8.50**). The first step

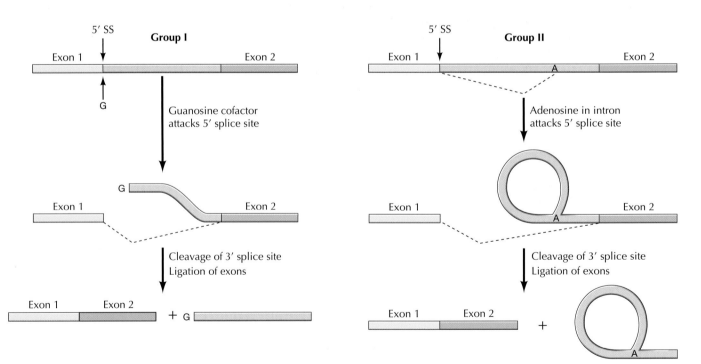

Figure 8.50 Self-splicing introns Group I and group II self-splicing introns are distinguished by their reaction mechanisms. In group I introns, the first step in splicing is cleavage of the 5' splice site by reaction with a guanosine (G) cofactor. The result is a linear intermediate with a G added to the 5' end of the intron. In group II introns (as in pre-mRNA splicing), the first step is cleavage of the 5' splice site by reaction with adenosine (A) within the intron, forming a lariat-like intermediate. In both cases, the second step is simultaneous cleavage of the 3' splice site and ligation of the exons.

in splicing for group I introns (e.g., *Tetrahymena* pre-rRNA) is cleavage at the 5′ splice site mediated by a guanosine (G) cofactor. The 3′ end of the free exon then reacts with the 3′ splice site to excise the intron as a linear RNA. In contrast, the self-splicing reactions of group II introns (which are found within some mRNA, rRNA, and tRNA genes of bacteria, mitochondria, and chloroplasts) closely resemble those characteristic of nuclear pre-mRNA splicing in which cleavage of the 5′ splice site results from attack by an adenosine (A) nucleotide in the intron. As with pre-mRNA splicing, the result is a lariat-like intermediate, which is then excised.

The similarity between spliceosome-mediated pre-mRNA splicing and self-splicing of group II introns strongly suggested that the active catalytic components of the spliceosome were RNAs rather than proteins. In particular, these similarities suggested that pre-mRNA splicing was catalyzed by the snRNAs of the spliceosome. Continuing studies of pre-mRNA splicing have provided clear support for this view, and it has now been established that U6 snRNA catalyzes both steps in pre-mRNA splicing. Pre-mRNA splicing is thus considered to be an RNA-based reaction, catalyzed by U6 snRNA acting analogously to group II self-splicing introns. Within the cell, protein components of the snRNPs are also required, however, and participate in both assembly of the spliceosome and the splicing reaction.

A number of protein splicing factors that are not snRNP components also play critical roles in spliceosome assembly, particularly in identification of the correct splice sites in pre-mRNAs. Mammalian pre-mRNAs typically contain multiple short exons (an average of 150 nucleotides in humans) separated by much larger introns (average of 3500 nucleotides). Introns frequently contain many sequences that resemble splice sites, so the splicing machinery must be able to identify the appropriate 5′ and 3′ splice sites at intron/exon boundaries to produce a functional mRNA. Splicing factors serve to direct spliceosomes to the correct splice sites by binding to specific RNA sequences and then recruiting U1 and U2 snRNPs to the appropriate sites on pre-mRNA by protein-protein interactions. For example, the SR splicing factors bind to specific sequences within exons and act to recruit U1 snRNP to the 5′ splice site (**Figure 8.51**). SR proteins also interact with another splicing factor (U2AF), which binds to pyrimidine-rich sequences at 3′ splice sites and recruits U2 snRNP to the branch point. In addition to recruiting the components of the spliceosome to pre-mRNA, splicing factors couple splicing to transcription

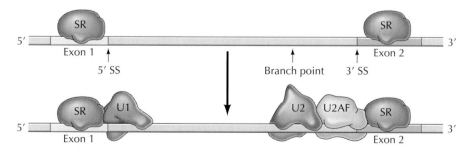

Figure 8.51 Role of splicing factors in spliceosome assembly   SR splicing factors bind to specific sequences within exons. The SR proteins recruit U1 snRNP to the 5′ splice site and an additional splicing factor (U2AF) to the 3′ splice site. U2AF then recruits U2 snRNP to the branch point.

by associating with the phosphorylated CTD of RNA polymerase II. This anchoring of the splicing machinery to RNA polymerase is thought to be important in ensuring that exons are joined in the correct order as the pre-mRNA is synthesized.

Alternative splicing

The central role of splicing in the processing of pre-mRNA opens the possibility of regulation of gene expression by control of the splicing machinery. Since most pre-mRNAs contain multiple introns, different mRNAs can be produced from the same gene by different combinations of 5′ and 3′ splice sites. The possibility of joining exons in varied combinations provides a novel means of controlling gene expression by generating multiple mRNAs (and therefore multiple proteins) from the same pre-mRNA. This process, called **alternative splicing**, occurs frequently in genes of complex eukaryotes. For example, it is estimated that nearly all human genes produce multiple transcripts that are alternatively spliced, with up to several thousand mRNAs arising from some genes. Each human gene produces an average of six alternatively spliced mRNAs, four of which encode different proteins. Alternative splicing thus results in a considerable increase in the diversity of proteins that can be encoded by the approximately 21,000 genes in mammalian genomes. Because patterns of alternative splicing can vary in different tissues and in response to extracellular signals, alternative splicing provides an important mechanism for tissue-specific and developmental regulation of gene expression.

One well-studied example of tissue-specific alternative splicing is provided by sex determination in *Drosophila*, where alternative splicing of the same pre-mRNA determines whether a fly is male or female (**Figure 8.52**). Alternative splicing of the pre-mRNA of a gene called *transformer* (*tra*) is controlled by a protein (SXL) that is only expressed in female flies. The *transformer* pre-mRNA

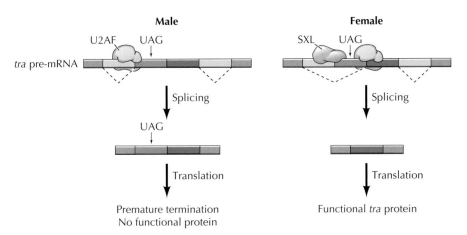

Figure 8.52 Alternative splicing in *Drosophila* sex determination   Alternative splicing of *transformer* (*tra*) pre-mRNA is regulated by the SXL protein, which is only expressed in female flies. In males, the first exon of *tra* pre-mRNA is joined to a 3′ splice site that yields a second exon containing a translation termination codon (UAG), so no *tra* protein is expressed. In females, the binding of SXL protein blocks the binding of U2AF to this 3′ splice site, resulting in the use of an alternative site further downstream in exon 2. This alternative 3′ splice site is downstream of the translation termination codon, so the mRNA expressed in females directs the synthesis of functional *tra* protein.

has three exons, but a different second exon is incorporated into the mRNA as a result of using alternate 3′ splice sites in the two different sexes. In males, exon 1 is joined to the most upstream of these 3′ splice sites, which is selected by the binding of the U2AF splicing factor. In females, the SXL protein binds to this 3′ splice site, blocking the binding of U2AF. Consequently, the upstream 3′ splice site is skipped in females, and exon 1 is instead joined to an alternate 3′ splice site that is further downstream. The exon 2 sequences included in the male *transformer* mRNA contain a translation termination codon (UAG), so no protein is produced. This termination codon is not included in the female mRNA, so female flies express a functional *transformer* protein, which acts as a key regulator of sex determination.

The alternative splicing of *transformer* illustrates the action of a repressor (the SXL protein) that functions by blocking the binding of a splicing factor (U2AF), and a large group of proteins similarly regulate alternative splicing by binding to silencer sequences in pre-mRNAs. In other cases, alternative splicing is controlled by activators that recruit splicing factors to splice sites that would otherwise not be recognized. The best-studied splicing activators are members of the SR protein family (see Figure 8.51), which bind to specific splicing enhancer sequences.

Multiple mechanisms can thus regulate alternative splicing, and variations in alternative splicing make a major contribution to the diversity of proteins expressed during development and differentiation. One of the most striking examples is the alternative splicing of a *Drosophila* cell surface protein (called Dscam) that is involved in specifying connections between neurons. The *Dscam* gene contains four sets of alternative exons, with a single exon from each set being incorporated into the spliced mRNA (**Figure 8.53**). These exons can be joined in any combination, so alternative splicing can potentially yield 38,016 different mRNAs and proteins from this single gene—more than twice the total number of genes in the *Drosophila* genome. These alternatively spliced forms of Dscam provide individual neurons with an identity code that is essential in establishing the connections between neurons needed for development of the fly brain.

RNA editing

RNA editing refers to RNA processing events (other than splicing) that alter the protein-coding sequences of some mRNAs. This unexpected form of RNA processing was first discovered in mitochondrial mRNAs of trypanosomes

> **FYI**
>
> The mammalian ear contains hair cells that are tuned to respond to sounds of different frequency. The tuning of hair cells is thought to be mediated in part by the alternative splicing of a gene encoding a channel protein.

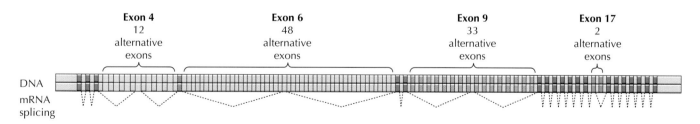

Figure 8.53 Alternative splicing of *Dscam* The *Dscam* gene (which encodes the *Drosophila* cell surface protein Dscam) contains four sets of alternative exons: 12 for exon 4, 48 for exon 6, 33 for exon 9, and 2 for exon 17. Any single exon from each of these sets can be incorporated into the mature mRNA, so alternative splicing can produce a total of 38,016 different mRNAs (12 × 48 × 33 × 2 = 38,016). One pattern of alternative splicing is illustrated by the dashed lines.

in which U residues are added and deleted at multiple sites along the pre-mRNA in order to generate the mRNA. More recently, editing has also been described in mitochondrial mRNAs of other organisms, chloroplast mRNAs of higher plants, and nuclear mRNAs of some mammalian genes.

Editing in mammalian nuclear mRNAs, as well as in mitochondrial and chloroplast RNAs of higher plants, involves single base changes as a result of base modification reactions, similar to those involved in tRNA processing. In mammalian cells, RNA editing reactions include the deamination of cytosine to uridine and of adenosine to inosine. One of the best-studied examples is editing of the mRNA for apolipoprotein B, which transports lipids in the blood. In this case, tissue-specific RNA editing results in two different forms of apolipoprotein B (Figure 8.54). In humans, Apo-B100 (4536 amino acids) is synthesized in the liver by translation of the unedited mRNA. However, a shorter protein Apo-B48 (2152 amino acids) is synthesized in the intestine as a result of translation of an edited mRNA in which a C has been changed to a U by deamination. This alteration changes the codon for glutamine (CAA) in the unedited mRNA to a translation termination codon (UAA) in the edited mRNA, resulting in synthesis of the shorter Apo-B protein. Tissue-specific editing of Apo-B mRNA thus results in the expression of structurally and functionally different proteins in the liver and intestine. The full-length Apo-B100 produced by the liver transports lipids in the circulation; Apo-B48 functions in the absorption of dietary lipids by the intestine.

RNA editing by the deamination of adenosine to inosine is the most common form of nuclear RNA editing in mammals. This form of editing plays an important role in the nervous system, where A-to-I editing results in single amino acid changes in ion channels and receptors on the surface of neurons. For example, the mRNAs encoding receptors for the neurotransmitter serotonin can be edited at up to five sites, yielding 24 different versions of the receptor with different signaling activities. The importance of A-to-I editing in the nervous system is further demonstrated by the finding that *C. elegans*, *Drosophila*, and mouse mutants lacking the editing enzyme suffer from a variety of neurological defects. Recent genome-wide sequencing studies have

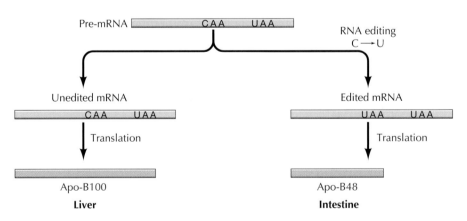

Figure 8.54 Editing of apolipoprotein B mRNA In human liver, unedited mRNA is translated to yield a 4536-amino-acid protein called Apo-B100. In human intestine, however, the mRNA is edited by a base modification that changes a specific C to a U. This modification changes the codon for glutamine (CAA) to a termination codon (UAA), resulting in synthesis of a shorter protein (Apo-B48, consisting of only 2152 amino acids).

further indicated that RNA editing may occur much more frequently than currently recognized, usually affecting noncoding sequences within introns or untranslated regions rather than coding sequences in mammalian mRNAs.

RNA degradation

The processing steps discussed in the previous section result in the formation of mature mRNAs, which are then transported to the cytoplasm and function to direct protein synthesis. However, most of the sequences transcribed into pre-mRNA are instead degraded within the nucleus. Over 90% of pre-mRNA sequences are introns, which are degraded within the nucleus following their excision by splicing. This is carried out by an enzyme that recognizes the unique 2′-5′ bond formed at the branch point, as well as by enzymes that recognize either the 5′ or 3′ ends of RNA molecules and catalyze degradation of the RNA in either direction. The 5′ and 3′ ends of processed mRNAs are protected from this degradation machinery by capping and polyadenylation, respectively, while the unprotected ends of introns are recognized and degraded.

In addition to degrading introns, cells possess quality-control systems that detect and degrade aberrant mRNAs produced as a result of errors during transcription. The best-studied of these quality control mechanisms is **nonsense-mediated mRNA decay**, which leads to the degradation of mRNAs that lack complete open-reading frames. This eliminates defective mRNA molecules and prevents the synthesis of abnormal truncated proteins. Nonsense-mediated mRNA decay is triggered when ribosomes encounter premature termination codons, which are identified as termination codons that occur more than about 50 nucleotides upstream of an exon-exon junction in spliced mRNAs. The presence of such premature termination codons halts translation and leads to degradation of the defective mRNA.

What may be considered the final aspect of the processing of an RNA molecule is its eventual degradation in the cytoplasm. Since the intracellular level of any RNA is determined by a balance between synthesis and degradation, the rate at which individual RNAs are degraded is another level at which gene expression can be controlled. Both ribosomal and transfer RNAs are very stable, and this stability largely accounts for the high levels of these RNAs (greater than 90% of all RNA) in both prokaryotic and eukaryotic cells. In contrast, bacterial mRNAs are rapidly degraded, usually having half-lives of only 2 to 3 minutes. This rapid turnover of bacterial mRNAs allows the cell to respond quickly to alterations in its environment, such as changes in the availability of nutrients required for growth.

In eukaryotic cells, however, different mRNAs are degraded at different rates, providing an additional parameter to the regulation of eukaryotic gene expression. The half-lives of mRNAs in mammalian cells vary from less than 30 minutes to approximately 20 hours. The unstable mRNAs frequently code for regulatory proteins, such as transcription factors, whose levels within the cell vary rapidly in response to environmental stimuli. In contrast, mRNAs encoding structural proteins or central metabolic enzymes generally have long half-lives.

The cytoplasmic degradation of most eukaryotic mRNAs is initiated by shortening of their poly-A tails (**Figure 8.55**). The deadenylated mRNAs are then degraded either by nucleases that act from the 3′ end or by removal of the 5′ cap and degradation of the RNA by nucleases acting from the 5′ end. Rapidly degraded mRNAs often contain specific AU-rich sequences near their

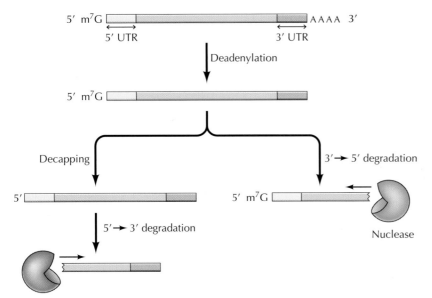

Figure 8.55 mRNA degradation
Degradation of mRNAs is usually initiated by shortening the poly-A tail (deadenylation). This is followed either by removal of the 5′ cap and degradation from the 5′ end (5′ to 3′ degradation) or by degradation from the 3′ end (3′ to 5′ degradation).

3′ ends that serve as binding sites for proteins that can either stabilize these mRNAs or target them for degradation. The activities of these RNA-binding proteins are regulated by extracellular signals, such as growth factors and hormones, so cells can control the rates of degradation of specific mRNAs in response to the environment. In addition, the degradation of many mRNAs is regulated by miRNAs, which stimulate mRNA degradation as well as inhibit translation (discussed in Chapter 9). Thus, although transcription is the first level at which gene expression is regulated, variations in the rate of mRNA degradation also play an important role in controlling steady-state levels of mRNAs within the cell.

SUMMARY

KEY TERMS

Transcription in Bacteria

- *RNA polymerase and transcription:* E. coli RNA polymerase consists of α, β, β', ω, and σ subunits. Transcription is initiated by the binding of σ to promoter sequences. After synthesis of about the first ten nucleotides of RNA, the core polymerase dissociates from σ and travels along the template DNA as it elongates the RNA chain. Transcription then continues until the polymerase encounters a termination signal. See Animation 8.1 and Video 8.1.

RNA polymerase, promoter

- *Repressors and negative control of transcription:* The prototype model for gene regulation in bacteria is the *lac* operon, which is regulated by the binding of a repressor to specific DNA sequences overlapping the promoter.

operon, operator, repressor, *cis*-acting control element

- *Positive control of transcription:* Some bacterial genes are regulated by transcriptional activators rather than repressors.

Eukaryotic RNA Polymerases and General Transcription Factors

- *Eukaryotic RNA polymerases:* Eukaryotic cells contain three distinct nuclear RNA polymerases that transcribe genes encoding mRNAs, miRNAs and lncRNAs (polymerase II), rRNAs (polymerases I and III), and tRNAs (polymerase III).

SUMMARY	KEY TERMS

- *General transcription factors and initiation of transcription by RNA polymerase II:* Eukaryotic RNA polymerases do not bind directly to promoter sequences; they require additional proteins (general transcription factors) to initiate transcription. The promoter sequences of genes transcribed by RNA polymerase II are recognized by the general transcription factor TFIID, which recruits additional transcription factors and RNA polymerase to the promoter.

transcription factor, general transcription factor, TATA box, TATA-binding protein (TBP), TBP-associated factor (TAF), Mediator

- *Transcription by RNA polymerases I and III:* RNA polymerases I and III also require additional transcription factors to bind to the promoters of rRNA, tRNA, and some snRNA genes.

Regulation of Transcription in Eukaryotes

- cis-*acting regulatory sequences: promoters and enhancers:* Transcription of eukaryotic genes is controlled by proteins that bind to regulatory sequences, which can be located hundreds of kilobases away from the transcription start site. Enhancers typically contain binding sites for multiple proteins that work together to regulate gene expression.

enhancer, insulator, topologically associating domain (TAD), CTCF

- *Transcription factor binding sites:* Eukaryotic transcription factors bind to short DNA sequences, usually 6–10 base pairs, in promoters or enhancers.

electrophoretic-mobility shift assay, chromatin immunoprecipitation

- *Transcriptional regulatory proteins:* Many eukaryotic transcription factors have been isolated on the basis of their binding to specific DNA sequences.

DNA-affinity chromatography

- *Structure and function of transcriptional activators:* Transcriptional activators are modular proteins, consisting of distinct DNA-binding and activation domains. DNA-binding domains mediate association with specific regulatory sequences; activation domains stimulate transcription by interacting with Mediator proteins and general transcription factors, as well as with coactivators that modify chromatin structure.

transcriptional activator, zinc finger domain, steroid hormone receptor, helix-turn-helix, homeodomain, leucine zipper, helix-loop-helix, coactivator

- *Eukaryotic repressors:* Gene expression in eukaryotic cells is regulated by repressors as well as by activators. Some repressors interfere with the binding of activators or general transcription factors to DNA. Other repressors contain discrete repression domains that inhibit transcription by interacting with Mediator proteins, general transcription factors, transcriptional activators, or corepressors that affect chromatin structure.

corepressor

- *Regulation of elongation:* Transcription is regulated at the level of elongation as well as initiation. Many genes have molecules of polymerase II that have initiated transcription but then paused immediately downstream of the promoter. These paused polymerases are poised to resume transcription in response to appropriate extracellular signals.

Chromatin and Epigenetics

- *Histone modifications:* Modification of histones is tightly linked to transcriptional regulation. Enzymes that catalyze histone acetylation are associated with transcriptional activators, whereas histone deacetylases are associated with repressors. Histones are also modified by phosphorylation and methylation, and specific modifications of histones affect gene expression by serving as binding sites for other regulatory proteins. Distinct patterns of histone modication mark promoters and enhancers. See Video 8.2.

histone acetylation, DNase hypersensitive site

- *Chromatin remodeling factors:* Chromatin remodeling factors facilitate the binding of transcription factors to DNA by altering the arrangement or structures of nucleosomes.

chromatin remodeling factor

SUMMARY	KEY TERMS

- *Histones and epigenetic inheritance:* Histone modification is a major mechanism for epigenetic inheritance.

epigenetic inheritance, Polycomb proteins

- *DNA methylation:* Methylation of cytosine residues can inhibit the transcription of eukaryotic genes and is important in silencing transposable elements. Regulation of gene expression by methylation also plays an important role in genomic imprinting, which controls the transcription of some genes involved in mammalian development.

genomic imprinting

- *Noncoding RNAs:* A large number of long noncoding RNAs (lncRNAs) have been identified, many of which may function as transcriptional regulators. Many lncRNAs act by forming complexes with proteins that regulate chromatin modification and recruiting these complexes to their target genes.

long noncoding RNA (lncRNA)

RNA Processing and Turnover

- *Processing of ribosomal and transfer RNAs:* Ribosomal and transfer RNAs are derived by cleavage of long primary transcripts in both bacteria and eukaryotic cells. Methyl groups are added to rRNAs, and various bases are modified in tRNAs.

pre-rRNA, pre-tRNA, RNase P, ribozyme

- *Processing of mRNA in eukaryotes:* Eukaryotic pre-mRNAs are modified by the addition of 7-methylguanosine caps and 3′ poly-A tails, in addition to the removal of introns by splicing. See Animation 8.2.

pre-mRNA, mRNP, 7-methylguanosine cap, poly-A tail, polyadenylation

- *Splicing mechanisms:* Splicing of nuclear pre-mRNAs takes place in large complexes, called spliceosomes, composed of proteins and small nuclear RNAs (snRNAs). The snRNAs recognize sequences at the splice sites of pre-mRNAs and catalyze the splicing reaction. Some mitochondrial, chloroplast, and bacterial RNAs undergo self-splicing in which the splicing reaction is catalyzed by intron sequences. See Video 8.3.

spliceosome, small nuclear RNA (snRNA), small nuclear ribonucleoprotein particle (snRNP), self-splicing

- *Alternative splicing:* Exons can be joined in various combinations as a result of alternative splicing, which provides an important mechanism for tissue-specific control of gene expression in complex eukaryotes.

alternative splicing

- *RNA editing:* Some mRNAs are modified by processing events that alter their protein-coding sequences. Editing of mitochondrial mRNAs in some protozoans involves the addition and deletion of U residues at multiple sites in the molecule. RNA editing in mammalian cells involves the deamination of cytosine to uridine and of adenosine to inosine.

RNA editing

- *RNA degradation:* Introns are degraded within the nucleus, and abnormal mRNAs lacking complete open-reading frames are eliminated by nonsense-mediated mRNA decay. Functional mRNAs in eukaryotic cells are degraded at different rates, providing an additional mechanism for control of gene expression. In some cases, rates of mRNA degradation are regulated by extracellular signals.

nonsense-mediated mRNA decay

Questions

1. What is the role of sigma (σ) factors in bacterial RNA synthesis?

2. What is the major mechanism for termination of *E. coli* mRNAs?

3. How does the *lac* repressor affect transcription of the *lac* operon?

4. You are comparing the requirements for *in vitro* transcription of two RNA polymerase II genes, one containing a TATA box and the other containing only an Inr sequence. Does transcription from these promoters require TBP or TFIID?

5. How do enhancers differ from promoters as *cis*-acting regulatory sequences in eukaryotes?

6. You are studying the enhancer of a gene that normally is expressed only in neurons. Constructs in which this enhancer is linked to a reporter gene are expressed in neuronal cells but not in fibroblasts. However, if you mutate a specific sequence element within the enhancer, you find expression in both fibroblasts and neuronal cells. What type of regulatory protein do you expect binds the sequence element?

7. What are the functions of insulators?

8. How do lncRNAs regulate transcription?

9. What property of Sp1 was exploited for its purification by DNA affinity chromatography? How would you determine that the purified protein is indeed Sp1?

10. You have developed an *in vitro* splicing reaction in which unspliced pre-mRNA is processed to the mature mRNA. What results would you expect if you added anti-Sm antiserum to the reaction?

11. How could you distinguish enhancers from promoters in global experimental analysis?

12. What is the function of splicing factors that are not components of snRNPs?

13. How are two structurally and functionally different forms of apolipoprotein B synthesized in human liver and intestine?

14. What is nonsense-mediated mRNA decay? What is its significance in the cell?

> **Refer To**
>
> **The Cell**
> Companion Website
> **sites.sinauer.com/cooper7e**
> for quizzes, animations, videos, flashcards, and other study resources.

References and Further Reading   (Key review articles for each major section are highlighted in **bold**.)

Transcription in Bacteria

Borukhov, S., J. Lee and O. Laptenko. 2005. Bacterial transcription elongation factors: New insights into molecular mechanism of action. *Molec. Microbiol.* 55: 1315–1324. [R]

Gilbert, W. and B. Muller-Hill. 1966. Isolation of the *lac* repressor. *Proc. Natl. Acad. Sci. USA* 56: 1891–1899. [P]

Jacob, F. 2011. The birth of the operon. *Science* 332:767. [R]

Jacob, F. and J. Monod. 1961. Genetic and regulatory mechanisms in the synthesis of proteins. *J. Mol. Biol.* 3: 318–356. [P]

Lawson, C. L., D. Swigon, K. S. Murakami, S. A. Darst, H. M. Berman and R. H. Ebright. 2004. Catabolite activator protein: DNA binding and transcription activation. *Curr. Opin. Struc. Biol.* 14: 10–20. [R]

Mathew, R. and D. Chetterji. 2006. The evolving story of the ω subunit of bacterial RNA polymerase. *Trends Microbiol.* 14: 450–455. [R]

Mooney, R. A., S. A. Darst and R. Landick. 2005. Sigma and RNA polymerase: an on-again, off-again relationship? *Mol. Cell* 20: 335–345. [R]

Murakami, K. S. and S. A. Darst. 2003. Bacterial RNA polymerases: The wholo story. *Curr. Opin. Struc. Biol.* 13: 31–39. [R]

Ptashne, M. and A. Gann. 2002. Genes and Signals. Cold Spring Harbor, N.Y.: Cold Spring Harbor Laboratory Press.

Rabinowitz, J. D. and T. J. Silhavy. 2013. Systems biology: Metabolite turns master regulator. *Nature* 500: 283–284. [R]

Richardson, J. P. 2003. Loading Rho to terminate transcription. *Cell* 114: 157–159. [R]

Wilson, C. J., H. Zhan, L. Swint-Kruse and K. S. Matthews. 2007. The lactose repressor system: paradigms for regulation, allosteric behavior and protein folding. *Cell. Mol. Life. Sci.* 64: 3–16. [R]

Zhang, G., E. A. Campbell, E. A., L. Minakhin, C. Richter, K. Severinov and S. A. Darst. 1999. Crystal structure of *Thermus aquaticus* core RNA polymerase at 3.3 Å resolution. *Cell* 98: 811–824. [P]

Eukaryotic RNA Polymerases and General Transcription Factors

Allen, B. L. and D. J. Taatjes. 2015. The Mediator complex: a central integrator of transcription. *Nature Rev. Mol. Cell Biol.* 16: 155–166. [R]

Buratowski, S. 2009. Progression through the RNA polymerase II CTD cycle. *Mol. Cell* 36: 541–546. [R]

Ebright, R. H. 2000. RNA polymerase: Structural similarities between bacterial RNA polymerase and eukaryotic RNA polymerase II. *J. Mol. Biol.* 304: 687–698. [R]

Goodrich, J. A. and R. Tjian. 2010. Unexpected roles for core promoter recognition factors in cell-type-specific transcription and gene regulation. *Nature Rev. Genet.* 11: 549–558. [R]

Kim, Y.-J. and J. T. Lis. 2005. Interactions between subunits of *Drosophila* Mediator and activator proteins. *Trends Biochem. Sci.* 30: 245–249. [R]

Kornberg, R. D. 2007. The molecular basis of eukaryotic transcription. *Proc. Natl. Acad. Sci. USA* 104: 12955–12961. [R]

Kuehner, J. N., E. L. Pearson and C. Moore. 2011. Unravelling the means to an end: RNA polymerase II transcription termination. *Nature Rev. Mol. Cell Biol.* 12: 1–12. [R]

Malik, S. and R. G. Roeder. 2010. The metazoan Mediator co-activator complex as an integrative hub for transcriptional regulation. *Nature Rev. Genet.* 11: 761–772. [R]

Matsui, T., J. Segall, P. A. Weil and R. G. Roeder. 1980. Multiple factors are required for accurate initiation of transcription by purified RNA polymerase II. *J. Biol. Chem.* 255: 11992–11996. [P]

Richard, P. and J. L. Manley. 2009. Transcription termination by nuclear RNA polymerases. *Genes Dev.* 23: 1247–1269. [R]

Roy, A. L. and D. S. Singer. 2015. Core promoters in transcription: old problem, new insights. *Trends Biochem. Sci.* 40: 165–171. [R]

Russell, J. and J. C. B. M. Zomerdijk. 2005. RNA polymerase I-directed rDNA transcription, life and works. *Trends Biochem. Sci.* 30: 87–96. [R]

Schramm, L. and N. Hernandez. 2002. Recruitment of RNA polymerase III to its target promoters. *Genes Dev.* 16: 2593–2620. [R]

Weil, P. A., D. S. Luse, J. Segall and R. G. Roeder. 1979. Selective and accurate initiation of transcription at the Ad2 major late promoter in a soluble system dependent on purified RNA polymerase II and DNA. *Cell* 18: 469–484. [P]

Regulation of Transcription in Eukaryotes

Buecker, C. and J. Wysocka. 2012. Enhancers as information integration hubs in development: lessons from genomics. *Trends Genet.* 28: 276–284. [R]

De Laat, W. and D. Duboule. 2013. Topology of mammalian developmental enhancers and their regulatory landscapes. *Nature* 502: 499–506. [R]

Espinoza, C. A. and B. Ren. 2011. Mapping higher order structure of chromatin domains. *Nature Genet.* 43: 615–616. [R]

Fuda, N. J., M. B. Ardehali and J. T. Lis. 2009. Defining mechanisms that regulate RNA polymerase II transcription *in vivo*. *Nature* 461: 186–192. [R]

Kadonaga, J. T. 2004. Regulation of RNA polymerase II transcription by sequence-specific DNA binding factors. *Cell* 116: 247–257. [R]

Lonard, D. M. and B. W. O'Malley. 2005. Expanding functional diversity of the coactivators. *Trends Biochem. Sci.* 30: 126–132. [R]

Maksimenko, O. and P. Georgiev. 2014. Mechanisms and proteins involved in long-distance interactions. *Front. Genet.* 5:28 [R]

Ong, C.-T. and V. G. Corces. 2014. CTCF: an architectural protein bridging genome topology and function. *Nature Rev. Genet.* 15: 234–246. [R]

Price, D. H. 2010. Regulation of RNA polymerase II elongation by c-Myc. *Cell* 141: 399–400. [R]

Shlyueva, D., G. Stampfel and A. Stark. 2014. Transcriptional enhancers: from properties to genome-wide predictions. *Nature Rev. Genet.* 15: 272–286. [R]

Chromatin and Epigenetics

Badeaux, A. I. and Y. Shi. 2013. Emerging roles for chromatin as a signal integration and storage platform. *Nature Rev. Mol. Cell Biol.* 14:211–224. [R]

Brownell, J. E., J. Zhou, T. Ranalli, R. Kobayashi, D. G. Edmondson, S. Y. Roth and C. D. Allis. 1996. *Tetrahymena* histone acetyltransferase A: A homolog to yeast Gcn5p linking histone acetylation to gene activation. *Cell* 84: 843–851. [P]

Calo, E. and J. Wysocka. 2013. Modification of enhancer chromatin: what, how, and why? *Mol. Cell* 49: 825–837. [R]

D'Urso, A. and J. H. Brickner. 2014. Mechanisms of epigenetic memory. *Trends Genet.* 30: 230–236. [R]

Fatica, A. and I. Bozzoni. 2014. Long noncoding RNAs: new players in cell differentiation and development. *Nature Rev. Genet.* 15: 7–21. [R]

Guttman, M, J. and 16 others. 2011. lincRNAs act in the circuitry controlling pluripotency and differentiation. *Nature* 477: 295–300. [P]

Ho, L. and G. R. Crabtree. 2010. Chromatin remodeling during development. *Nature* 463: 474–484. [R]

Holoch, D. and D. Moazed. 2015. RNA-mediated epigenetic regulation of gene expression. *Nature Rev. Genet.* 16: 71–84. [R]

Iyer, M. K. and 18 others. 2015. The landscape of long noncoding RNAs in the human transcriptome. *Nature Genet.* 47: 199–208. [P]

Jones, P. A. 2012. Functions of DNA methylation: islands, start sites, gene bodies and beyond. *Nature Rev. Genet.* 13: 484–492. [R]

Kugel, J. F. and J. A. Goodrich. 2011. Noncoding RNAs: key regulators of mammalian transcription. *Trends Biochem. Sci.* 37: 144–151. [R]

Moazed, D. 2011. Mechanisms for the inheritance of chromatin states. *Cell* 146: 510–518. [R]

Piccolo, F. M. and A. G. Fisher. 2014. Getting rid of DNA methylation. *Trends Cell Biol.* 24: 136–143. [R]

Rinn, J. L. and H. Y. Chang. 2012. Genome regulation by long noncoding RNAs. *Ann. Rev. Biochem.* 81: 145–166. [R]

Saha, A., J. Wittmeyer and B. R. Cairns. 2006. Chromatin remodeling: the industrial revolution of DNA around histones. *Nature Rev. Mol. Cell Biol.* 7: 437–447. [R]

Schwartz, Y. B. and V. Pirrotta. 2013. A new world of Polycombs: unexpected partnerships and emerging functions. *Nature Rev. Genet.* 14: 853–864. [R]

Taunton, J., C. A. Hassig and S. L. Schreiber. 1996. A mammalian histone deacetylase related to the yeast transcriptional regulator Rpd3p. *Science* 272: 408–411. [P]

Wang, K. C. and H. Y. Chang. 2011. Molecular mechanisms of long noncoding RNAs. *Mol. Cell* 43: 904–914. [R]

Wu, H. and Y. Zhang. 2014. Reversing DNA methylation: mechanisms, genomics, and biological functions. *Cell* 156: 45–68. [R]

Yang, L., J. E. Froberg and J. T. Lee. 2014. Long noncoding RNAs: fresh perspectives into the RNA world. *Trends Biochem. Sci.* 39: 35–43. [R]

RNA Processing and Turnover

Blanc, V. and N. O. Davidson. 2003. C-to-U RNA editing: Mechanisms leading to genetic diversity. *J. Biol. Chem.* 278: 1395–1398. [R]

Chen, M. and J. L. Manley. 2009. Mechanisms of alternative splicing regulation: insights from molecular and genomics approaches. *Nature Rev. Mol. Cell Biol.* 10: 741–754. [R]

Evans, D., S. M. Marquez and N. R. Pace. 2006. RNase P: interface of the RNA and protein worlds. *Trends Biochem. Sci.* 31: 333–341. [R]

Fica, S. M., N. Tuttle, T. Novak, N.-S. Li, J. Lu, P. Koodathingal, Q. Dai, J. P. Staley and J. A. Piccirilli. 2013. RNA catalyses nuclear pre-mRNA splicing. *Nature* 503: 229–234. [P]

Garneau, N. L., J. Wilusz and C. J. Wilusz. 2007. The highways and byways of mRNA decay. *Nature Rev. Mol. Cell Biol.* 8: 113–126. [R]

Guo, H., N. T. Ingolia, J. S. Weissman and D. P. Bartel. 2010. Mammalian microRNAs predominantly act to decrease target mRNA levels. *Nature* 466: 835–841. [P]

Guerrier-Takada, C., K. Gardiner, T. Marsh, N. Pace and S. Altman. 1983. The RNA moiety of ribonuclease P is the catalytic subunit of the enzyme. *Cell* 35: 849–857. [P]

Haugen, P., D. M. Simon and D. Bhattacharya. 2005. The natural history of group I introns. *Trends Genet.* 21: 111–119. [R]

Hundley, H. A. and B. L. Bass. 2010. ADAR editing in double-stranded UTRs and other noncoding RNA sequences. *Trends Biochem. Sci.* 35: 377–383 [R]

Kornblihtt, A. R., I. E. Schor, M. Alló, G. Dujardin, E. Petrillo and M. J. Muñoz. 2013. Alternative splicing: a pivotal step between eukaryotic transcription and translation. *Nature Rev. Mol. Cell Biol.* 14: 153–165. [R]

Kruger, K., P. J. Grabowski, A. Zaug, A. J. Sands, D. E. Gottschling and T. R. Cech. 1982. Self-splicing RNA: Autoexcision and autocyclization of the ribosomal RNA intervening sequence of *Tetrahymena*. *Cell* 31: 147–157. [P]

Kuehner, J. N., E. L. Pearson and C. Moore. 2011. Unravelling the means to an end: RNA polymerase II transcription termination. *Nature Rev. Mol. Cell Biol.* 12: 1–12. [R]

Li, J. B. and G. M. Church. 2013. Deciphering the functions and regulation of brain-enriched A-to-I RNA editing. *Nature Neurosci.* 16: 1518–1522. [R]

Mitchell, S. F. and R. Parker. 2014. Principles and properties of eukaryotic mRNPs. *Mol. Cell* 54: 547–558. [R]

Nilsen, T. W. and B. R. Graveley. 2010. Expansion of the eukaryotic proteome by alternative splicing. *Nature* 463: 457–463. [R]

Padgett, R. A., S. M. Mount, J. A. Steitz and P. A. Sharp. 1983. Splicing of messenger RNA precursors is inhibited by antisera to small nuclear ribonucleoprotein. *Cell* 35: 101–107. [P]

Proudfoot, N. J. 2011. Ending the message: poly(A) signals then and now. *Genes Dev.* 25:1770–1782.

Richard, P. and J. L. Manley. 2009. Transcription termination by nuclear RNA polymerases. *Genes Dev.* 23: 1247–1269. [R]

Schmucker, D. 2007. Molecular diversity of Dscam: recognition of molecular identity in neuronal wiring. *Nature Rev. Neurosci.* 8: 915–920. [R]

Schoenberg, D. R. and L. E. Maquat. 2012. Regulation of cytoplasmic mRNA decay. *Nature Rev. Genet.* 13: 246–259. [R]

Wahl, M. C., C. L. Will and R. Luhrmann. 2009. The spliceosome: design principles of a dynamic RNP machine. *Cell* 136: 701–718. [R]

Protein Synthesis, Processing, and Regulation

Transcription and RNA processing are followed by translation, the synthesis of proteins as directed by mRNA templates. Proteins are the active players in most cell processes, implementing the myriad tasks that are directed by the information encoded in genomic DNA. Protein synthesis is thus the final stage of gene expression. However, the translation of mRNA is only the first step in the formation of a functional protein. The polypeptide chain must then fold into the appropriate three-dimensional conformation and, frequently, undergo various processing steps before being converted to its active form. These processing steps, particularly in eukaryotes, are intimately related to the sorting and transport of different proteins to their appropriate destinations within the cell.

Gene expression is controlled not only at the level of transcription (see Chapter 8), but also at the level of translation, and this control is an important element of gene regulation. Of even broader significance, however, are the mechanisms that control the activities of proteins within cells. Once synthesized, most proteins can be regulated in response to extracellular signals by either covalent modifications or by association with other molecules. In addition, the levels of proteins within cells can be controlled by differential rates of protein degradation. These multiple controls of both the amounts and activities of intracellular proteins ultimately regulate all aspects of cell behavior.

Translation of mRNA

Proteins are synthesized from mRNA templates by a process that has been highly conserved throughout evolution (reviewed in Chapter 4). All mRNAs are read in the 5′ to 3′ direction, and polypeptide chains are synthesized from the amino to the carboxy terminus. Each amino acid is specified by three bases (a codon) in the mRNA, according to a nearly universal genetic code. The basic mechanics of protein synthesis are also the same in all cells: Translation is carried out on ribosomes, with tRNAs serving as adaptors between the mRNA template and the amino acids being incorporated into protein. Protein synthesis thus involves interactions between three types of RNA molecules (mRNA templates, tRNAs, and rRNAs), as well as various proteins that are required for translation.

Transfer RNAs

During translation, each of the 20 amino acids must be aligned with its corresponding codon on the mRNA template. All cells contain a variety of

tRNAs that serve as adaptors for this process. As might be expected, given their common function in protein synthesis, different tRNAs share similar overall structures. However, they also possess unique identifying sequences that allow the correct amino acid to be attached and aligned with the appropriate codon in mRNA.

Transfer RNAs are approximately 70 to 80 nucleotides long and have characteristic cloverleaf structures that result from complementary base pairing between different regions of the molecule (**Figure 9.1**). X-ray crystallography studies have further shown that all tRNAs fold into similar compact L shapes, which are required for the tRNAs to fit onto ribosomes during the translation process. The adaptor function of the tRNAs involves two separate regions of the molecule. All tRNAs have the sequence CCA at their 3′ terminus, and amino acids are covalently attached to the ribose of the terminal adenosine. The mRNA template is then recognized by the **anticodon** loop, located at the other end of the folded tRNA, which binds to the appropriate codon by complementary base pairing.

The incorporation of the correctly encoded amino acids into proteins depends on the attachment of each amino acid to an appropriate tRNA, as well as on the specificity of codon-anticodon base pairing. The attachment of amino acids to specific tRNAs is mediated by a group of enzymes called **aminoacyl tRNA synthetases**, which were discovered by Paul Zamecnik and Mahlon Hoagland in 1957. Each of these 20 enzymes recognizes a single amino acid, as well as the correct tRNA (or tRNAs) to which that amino acid should be attached. The reaction proceeds in two steps (**Figure 9.2**).

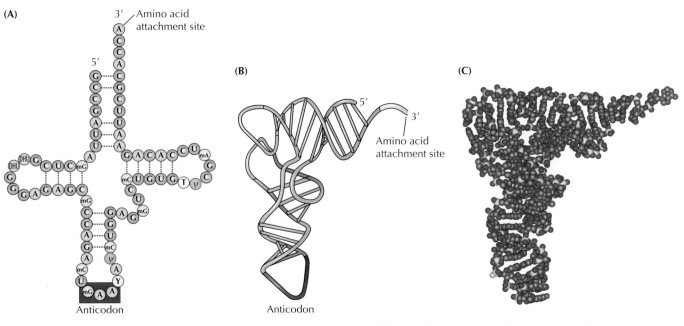

Figure 9.1 Structure of tRNAs The structure of yeast phenylalanyl tRNA is illustrated in open "cloverleaf" form (A) to show complementary base pairing. Modified bases are indicated as mG, methylguanosine; mC, methylcytosine; DHU, dihydrouridine; T, ribothymidine; Y, a modified purine (usually adenosine); and ψ, pseudouridine. The folded form of the molecule is shown in (B) and a space-filling model in (C).

Figure 9.2 Attachment of amino acids to tRNAs In the first reaction step, the amino acid is joined to AMP, forming an aminoacyl AMP intermediate. In the second step, the amino acid is transferred to the 3′ CCA terminus of the acceptor tRNA and AMP is released. Both steps of the reaction are catalyzed by aminoacyl tRNA synthetases.

First, the amino acid is activated by reaction with ATP to form an aminoacyl AMP synthetase intermediate. The activated amino acid is then joined to the 3′ terminus of the tRNA. The aminoacyl tRNA synthetases must be highly selective enzymes that recognize both individual amino acids and specific base sequences that identify the correct acceptor tRNAs. In some cases, the high fidelity of amino acid recognition results in part from a proofreading function by which incorrect aminoacyl AMPs are hydrolyzed rather than being joined to tRNA during the second step of the reaction. Recognition of the correct tRNA by the aminoacyl tRNA synthetase is also highly selective; the synthetase recognizes specific nucleotide sequences (in most cases including the anticodon) that uniquely identify each species of tRNA.

After being attached to tRNA, an amino acid is aligned on the mRNA template by complementary base pairing between the mRNA codon and the anticodon of the tRNA. Codon-anticodon base pairing is somewhat less stringent than the standard A-U and G-C base pairing discussed in preceding chapters. The significance of this unusual base pairing in codon-anticodon recognition relates to the redundancy of the genetic code. Of the 64 possible codons, three are stop codons that signal the termination of translation; the other 61 encode amino acids (see Table 4.1). Thus most of the amino acids are specified by more than one codon. In part, this redundancy results from the attachment of many amino acids to more than one species of tRNA. Cells contain about 40 different tRNAs that serve as acceptors for the 20 different amino acids. In addition, some tRNAs are able to recognize more than one codon in mRNA, as a result of nonstandard base pairing (called wobble) between the tRNA anticodon and the third position of some complementary codons (**Figure 9.3**). Relaxed base pairing at this position results partly from the formation of G-U base pairs and partly from the modification of guanosine to inosine in the anticodons of several tRNAs during processing (see Figure 8.43). Inosine can base-pair with either U, C, or A in the third position, so its inclusion in the anticodon allows a single tRNA to recognize three different codons in mRNA templates.

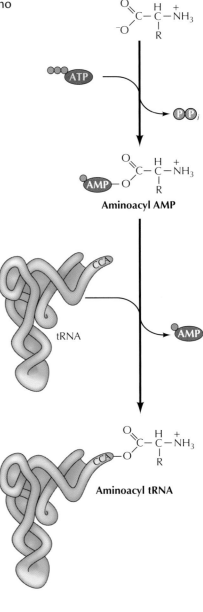

Aminoacyl AMP

tRNA

Aminoacyl tRNA

The ribosome

Ribosomes are the sites of protein synthesis in both prokaryotic and eukaryotic cells. First characterized as particles detected by ultracentrifugation of cell lysates, ribosomes are usually designated according to their rates of sedimentation: 70S for bacterial ribosomes and 80S for the somewhat larger ribosomes of eukaryotic cells. Both prokaryotic and eukaryotic ribosomes are composed of two distinct subunits, each containing characteristic proteins and **rRNAs**. The fact that cells typically contain many ribosomes reflects the central importance of protein synthesis in cell metabolism. *E. coli*, for example, contain about 20,000 ribosomes, which account for approximately 25% of the dry weight of the cell, and rapidly growing mammalian cells contain about 10 million ribosomes.

FYI

In some cases, selenocysteine, which has been termed the 21st amino acid, is inserted at stop codons. Selenocysteine is a selenium-containing amino acid that is found in proteins from diverse organisms, including humans. Another modified amino acid, pyrrolysine, can also be inserted at stop codons. Pyrrolysine is a modified lysine found in proteins from some archaeabacteria and is important in the production of methane.

Figure 9.3 Nonstandard codon–anticodon base pairing Base pairing at the third codon position is relaxed, allowing G to pair with U, and inosine (I) in the anticodon to pair with U, C, or A. Two examples of abnormal base pairing, allowing phenylalanyl (Phe) tRNA to recognize either UUC or UUU codons and alanyl (Ala) tRNA to recognize GCU, GCC, or GCA, are illustrated.

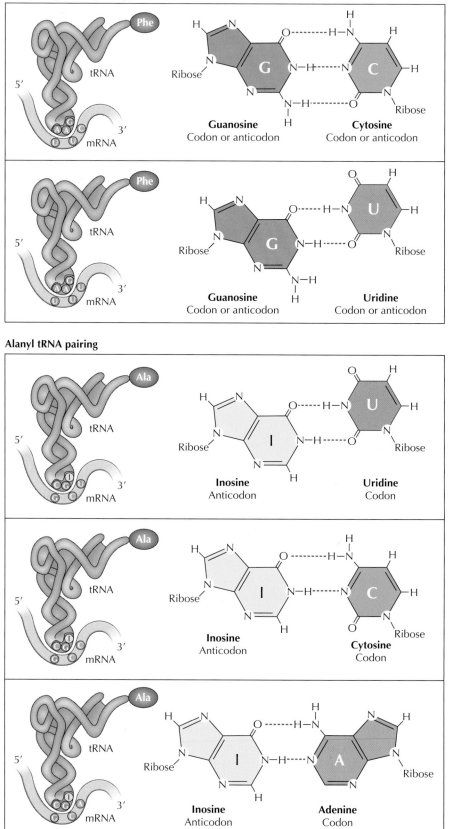

The general structures of prokaryotic and eukaryotic ribosomes are similar, although they differ in some details (**Figure 9.4**). The large subunit (designated 50S) of *E. coli* ribosomes consists of the 23S and 5S rRNAs and 34 proteins; the small subunit (30S) is composed of the 16S rRNA and 21 proteins. Each ribosome contains one copy of the rRNAs and one copy of each of the ribosomal proteins, with one exception: One protein of the 50S subunit is present in four copies. The subunits of eukaryotic ribosomes are larger and contain more proteins than their prokaryotic counterparts. The large subunit (60S) of eukaryotic ribosomes is composed of the 28S, 5.8S, and 5S rRNAs and 46 proteins; the small subunit (40S) contains the 18S rRNA and 33 proteins. Because of their large size and complexity, high-resolution structural analysis of prokaryotic ribosomes by X-ray crystallography was not accomplished until 2000, when structures of both the 50S and 30S subunits were first reported. The crystal structure of eukaryotic ribosomes followed in 2010. As discussed

(A)

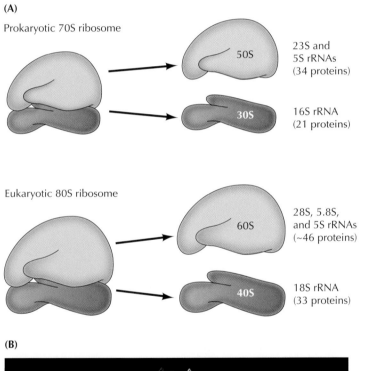

(B)

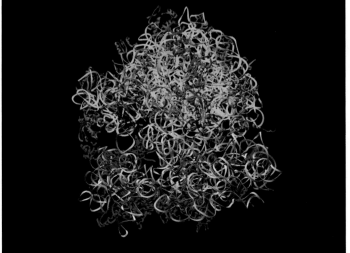

Figure 9.4 Ribosome structure   (A) Components of prokaryotic and eukaryotic ribosomes. Intact prokaryotic and eukaryotic ribosomes are designated 70S and 80S, respectively, on the basis of their sedimentation rates in ultracentrifugation. They consist of large and small subunits, which contain both ribosomal proteins and rRNAs. (B) High-resolution X-ray crystal structure of a prokaryotic 70S ribosome. The 30S subunit is blue (rRNA is light blue; ribosomal proteins are dark blue), and the 50S subunit is green (rRNA is light green; ribosomal proteins are dark green). (B, after R. M. Voorhees et al., 2009. *Nat. Struct. Mol. Biol.* 16: 528.)

below, understanding the structure of ribosomes at the atomic level has had a major impact on our understanding of ribosome function.

A noteworthy feature of ribosomes is that they can be formed *in vitro* by self-assembly of their RNA and protein constituents. As first described in 1968 by Masayasu Nomura, purified ribosomal proteins and rRNAs can be mixed together and, under appropriate conditions, will re-form a functional ribosome. Although ribosome assembly *in vivo* (particularly in eukaryotic cells) is considerably more complicated, the ability of ribosomes to self-assemble *in vitro* has provided an important experimental tool, allowing analysis of the roles of individual proteins and rRNAs.

Like tRNAs, rRNAs form characteristic secondary structures by complementary base pairing (**Figure 9.5**). In association with ribosomal proteins,

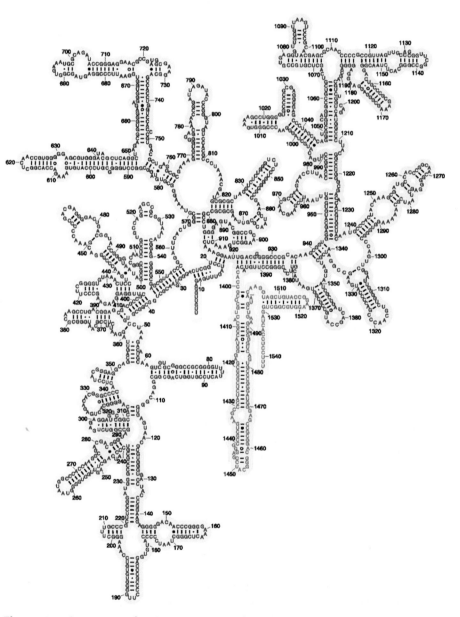

Figure 9.5 Structure of 16S rRNA   Complementary base pairing results in the formation of a distinct secondary structure. (From M. M. Yusupov et al., 2001. *Science* 292: 883.)

the rRNAs fold further, into distinct three-dimensional structures. Initially, rRNAs were thought to play a structural role, providing a scaffold upon which ribosomal proteins assemble. However, with the discovery of the catalytic activity of other RNA molecules (e.g., RNase P and the self-splicing introns discussed in Chapter 8), the possible catalytic role of rRNA became widely considered. Consistent with this hypothesis, rRNAs were found to be absolutely required for the *in vitro* assembly of functional ribosomes. On the other hand, the omission of many ribosomal proteins resulted in a decrease, but not a complete loss, of ribosomal activity.

Direct evidence for the catalytic activity of rRNA first came from experiments of Harry Noller and his colleagues in 1992. These investigators demonstrated that the large ribosomal subunit is able to catalyze the formation of peptide bonds (the peptidyl transferase reaction) even after approximately 90% of the ribosomal proteins have been removed by standard protein extraction procedures. In contrast, treatment with RNase completely abolishes peptide bond formation, providing strong support for the hypothesis that the formation of a peptide bond is an RNA-catalyzed reaction. However, some of the ribosomal proteins could not be removed under conditions that left the ribosomal RNA intact, so the role of ribosomal proteins as catalysts of peptide bond formation could not be definitively ruled out.

Unambiguous evidence that protein synthesis is catalyzed by rRNA came from the first high-resolution structural analysis of the 50S ribosomal subunit, which was reported by Peter Moore, Thomas Steitz, and their colleagues in 2000 (**Figure 9.6**). This atomic-level view of ribosome

FYI

Although the sequence of ribosomal RNAs is highly conserved, substitutions have occurred throughout the course of evolution. The comparison of rRNA sequences is a powerful tool in determining evolutionary relationships between different species.

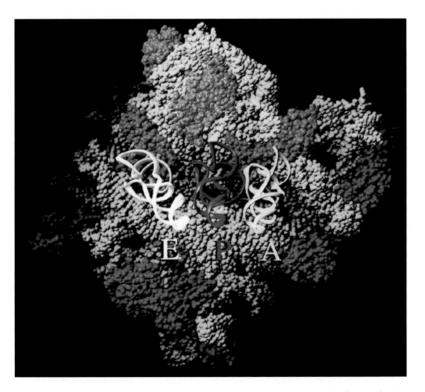

Figure 9.6 Structure of the 50S ribosomal subunit   A high-resolution model of the 50S ribosomal subunit with three tRNA molecules bound to the E (exit), P (peptidyl), and A (aminoacyl) sites of the ribosome (see Figure 9.13). Ribosomal proteins are shown in purple and the rRNA in aqua. (From P. Nissen et al., 2000. *Science* 289: 920.)

structure revealed that ribosomal proteins were strikingly absent from the site at which the peptidyl transferase reaction occurred, making it evident that rRNA was responsible for catalyzing peptide bond formation. Further studies have illuminated the mechanism by which rRNA catalyzes the peptidyl transferase reaction and shown that the large ribosomal subunit functions as a ribozyme, with the fundamental reaction of protein synthesis being catalyzed by RNA. Rather than being the primary catalytic constituents of ribosomes, ribosomal proteins are now thought to play a largely structural role.

The direct involvement of rRNA in the peptidyl transferase reaction has important evolutionary implications. RNAs are thought to have been the first self-replicating macromolecules (see Chapter 1). This notion is strongly supported by the fact that ribozymes, such as RNase P and self-splicing introns, can catalyze reactions that involve RNA substrates. The role of rRNA in the formation of peptide bonds extends the catalytic activities of RNA beyond self-replication to direct involvement in protein synthesis. Additional studies indicate that the *Tetrahymena* rRNA ribozyme can catalyze the attachment of amino acids to RNA, lending credence to the possibility that the original aminoacyl tRNA synthetases were RNAs rather than proteins. The ability of RNA molecules to catalyze the reactions required for protein synthesis as well as for self-replication may provide an important link for understanding the early evolution of cells.

The organization of mRNAs and the initiation of translation

Although the mechanisms of protein synthesis in prokaryotic and eukaryotic cells are similar, there are also differences, particularly in the signals that determine the positions at which synthesis of a polypeptide chain is initiated on an mRNA template (**Figure 9.7**). Translation does not simply begin at the 5′ end of the mRNA; it starts at specific initiation sites. The 5′

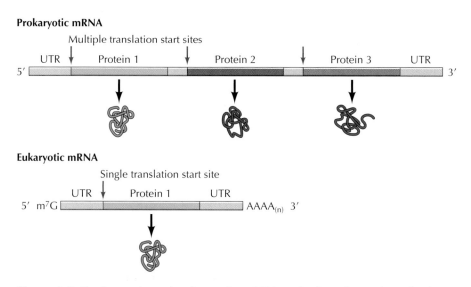

Figure 9.7 Prokaryotic and eukaryotic mRNAs Both prokaryotic and eukaryotic mRNAs contain untranslated regions (UTRs) at their 5′ and 3′ ends. Eukaryotic mRNAs also contain 5′ 7-methylguanosine (m^7G) caps and 3′ poly-A tails. Prokaryotic mRNAs are frequently polycistronic: They encode multiple proteins, each of which is translated from an independent start site. Eukaryotic mRNAs are usually monocistronic, encoding only a single protein.

terminal portions of both prokaryotic and eukaryotic mRNAs are therefore noncoding sequences, referred to as **5′ untranslated regions (UTR)**. Eukaryotic mRNAs usually encode only a single polypeptide chain, but many prokaryotic mRNAs encode multiple polypeptides that are synthesized independently from distinct initiation sites. For example, the *E. coli lac* operon consists of three genes that are translated from the same mRNA (see Figure 8.7). Messenger RNAs that encode multiple polypeptides are called **polycistronic**, whereas **monocistronic** mRNAs encode a single polypeptide chain. Finally, both prokaryotic and eukaryotic mRNAs end in noncoding 3′ untranslated regions.

In both prokaryotic and eukaryotic cells, translation always initiates with the amino acid methionine, usually encoded by AUG. Alternative initiation codons, such as GUG, are used occasionally, but when they occur at the beginning of a polypeptide chain, these codons direct the incorporation of methionine rather than the amino acid they normally encode (GUG normally encodes valine). In most bacteria, protein synthesis is initiated with a modified methionine residue (*N*-formylmethionine), whereas unmodified methionines initiate protein synthesis in eukaryotes (except in mitochondria and chloroplasts, whose ribosomes resemble those of bacteria).

The signals that identify initiation codons are different in prokaryotic and eukaryotic cells, consistent with the distinct functions of polycistronic and monocistronic mRNAs (**Figure 9.8**). Initiation codons in bacterial mRNAs are preceded by a specific sequence (called a **Shine-Dalgarno sequence**, after its discoverers, John Shine and Lynn Dalgarno) that aligns the mRNA on the ribosome for translation by base-pairing with a complementary sequence near the 3′ terminus of 16S rRNA. This base-pairing interaction enables bacterial ribosomes to initiate translation not only at the 5′ end of an mRNA

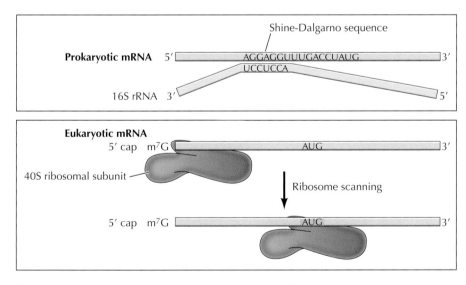

Figure 9.8 Signals for translation initiation   Initiation sites in prokaryotic mRNAs are characterized by a Shine-Dalgarno sequence that precedes the AUG initiation codon. Base pairing between the Shine-Dalgarno sequence and a complementary sequence near the 3′ terminus of 16S rRNA aligns the mRNA on the ribosome. In contrast, eukaryotic mRNAs are bound to the 40S ribosomal subunit by their 5′ 7-methylguanosine caps. The ribosome then scans along the mRNA until it encounters an AUG initiation codon.

but also at the internal initiation sites of polycistronic messages. In contrast, ribosomes recognize most eukaryotic mRNAs by binding to the 7-methyl-guanosine cap at their 5′ terminus (see Figure 8.44). The ribosomes then scan downstream of the 5′ cap until they encounter the initiation codon (usually AUG). Sequences that surround AUGs affect the efficiency of initiation, so in some cases the first AUG in the mRNA is bypassed and translation initiates at an AUG farther downstream.

It should also be noted that several viral and cellular mRNAs have internal ribosome entry sites at which translation can initiate by ribosome binding to an internal position on the mRNA. It appears that this alternative mechanism of initiation functions under conditions of cell stress to allow the selective translation of some mRNAs when the normal mode of initiation at the 5′ cap is inhibited, as discussed later in this chapter. There appear to be multiple molecular mechanisms by which translation initiates at these internal ribosome entry sites, and understanding these mechanisms remains an active area of investigation.

The process of translation

Translation is generally divided into three stages: initiation, elongation, and termination (**Figure 9.9**). In both prokaryotes and eukaryotes the first step of the initiation stage is the binding of a specific initiator methionyl tRNA and the mRNA to the small ribosomal subunit. The large ribosomal subunit then joins the complex, forming a functional ribosome on which elongation of the polypeptide chain proceeds. A number of specific non-ribosomal proteins are also required for the various stages of the transla-tion process (**Table 9.1**).

The initiation of translation in bacteria starts with a 30S ribosomal subunit bound to two **initiation factors**, IF1 and IF3 (**Figure 9.10**). The third initiation factor (IF2, which is bound to GTP), mRNA, and initiator *N*-formylmethionyl tRNA then join the complex, with IF2 specifically interacting with the initia-tor tRNA. The tRNA then binds to the initiation codon of the mRNA, and IF1 and IF3 are released. A 50S ribosomal subunit then associates with the complex, triggering the hydrolysis of the GTP bound to IF2, and the release of IF2 (bound to GDP) from the complex. The result is the formation of a 70S

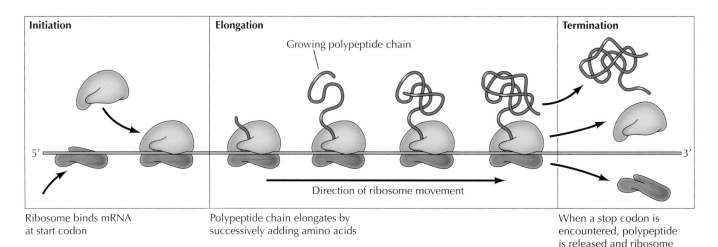

Initiation	Elongation	Termination

Growing polypeptide chain

5′   3′

Direction of ribosome movement

Ribosome binds mRNA at start codon

Polypeptide chain elongates by successively adding amino acids

When a stop codon is encountered, polypeptide is released and ribosome dissociates

Figure 9.9 Overview of translation

Table 9.1   Translation Factors

Role	Prokaryotes	Eukaryotes
Initiation	IF1, IF2, IF3	eIF1, eIF1A, eIF2, eIF2B, eIF3, eIF4A, eIF4B, eIF4E, eIF4G, eIF4H, eIF5, eIF5B
Elongation	EF-Tu, EF-Ts, EF-G	eEF1α, eEF1$\beta\gamma$, eEF2
Termination	RF1, RF2, RF3	eRF1, eRF3

initiation complex (with mRNA and initiator tRNA bound to the ribosome) that is ready to begin peptide bond formation during the elongation stage of translation.

Initiation in eukaryotes is more complicated and requires at least 12 proteins (each consisting of multiple polypeptide chains), which are designated eIFs (eukaryotic initiation factors; see Table 9.1). The factors eIF1, eIF1A, and eIF3 bind to the 40S ribosomal subunit, and eIF2 (in a complex with

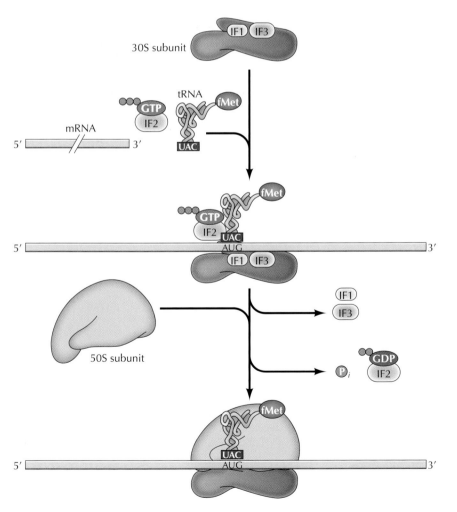

Figure 9.10 Initiation of translation in bacteria   The initiation factors IF1 and IF3 are initially bound to the 30S ribosomal subunit. The mRNA, the initiator *N*-formylmethionyl (fMet) tRNA, and IF2 (bound to GTP) then join the complex. IF1 and IF3 are then released, and a 50S subunit binds to the complex, triggering the hydrolysis of bound GTP and the release of IF2 bound to GDP.

Figure 9.11 Initiation of translation in eukaryotic cells Initiation factors eIF1, eIF1A, and eIF3 bind to the 40S ribosomal subunit. The initiator methionyl tRNA is bound by eIF2 (complexed to GTP), and forms a complex with the 40S subunit and eIF5. The mRNA is brought to the 40S subunit by eIF4E (which binds to the 5′ cap), eIF4G (which binds to both eIF4E at the 5′ cap and PABP at the 3′ poly-A tail), eIF4A, and eIF4B. The ribosome then scans down the mRNA to identify the first AUG initiation codon. Scanning requires energy and is accompanied by ATP hydrolysis. When the initiating AUG is identified, eIF5 triggers the hydrolysis of GTP bound to eIF2, followed by the release of eIF2 (complexed to GDP) and other initiation factors. The 60S ribosomal subunit then joins the 40S complex, facilitated by eIF5B.

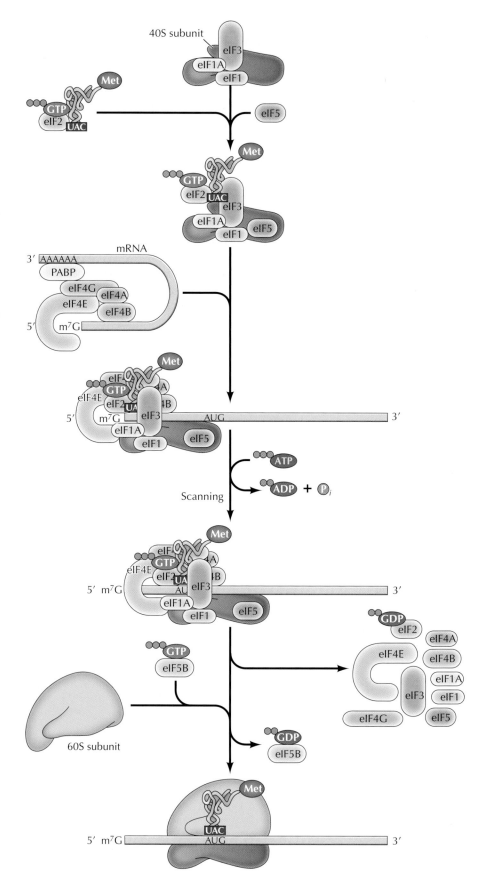

GTP) binds to the initiator methionyl tRNA (**Figure 9.11**). A preinitiation complex is then formed by association of the 40S subunit with eIF5 and the initiator tRNA. The mRNA is recognized and brought to the ribosome by the eIF4 group of factors. The 5′ cap of the mRNA is recognized by eIF4E, which forms a complex with eIF4A and eIF4G. eIF4G also binds to poly-A binding protein (PABP), which is associated with the poly-A tail at the 3′ end of the mRNA. Eukaryotic initiation factors thus recognize both the 5′ and 3′ ends of mRNAs, accounting for the stimulatory effect of polyadenylation on translation. The initiation factors eIF4E and eIF4G, in association with eIF4A and eIF4B, then bring the mRNA to the 40S ribosomal subunit, with eIF4G interacting with eIF3. The 40S ribosomal subunit, in association with the bound methionyl tRNA and eIFs, then scans the mRNA to identify the AUG initiation codon. When the AUG codon is reached, eIF5 triggers the hydrolysis of GTP bound to eIF2. Initiation factors (including eIF2 bound to GDP) are then released, and eIF5B (initially bound to GTP) facilitates the binding of a 60S subunit (associated with GTP hydrolysis) to form the 80S initiation complex of eukaryotic cells.

As noted earlier in this chapter, some viral and cellular eukaryotic mRNAs have internal ribosome entry sites (IRESs) at which translation can initiate independently of the 5′ cap. For viral mRNAs, this can be mediated either by the binding of IRES sequences directly to eIF4G, complexed to eIF4A, or to 40S ribosomal subunits (**Figure 9.12**). The mechanism of action of IRESs in cellular mRNAs is not yet understood, but is thought to involve binding of an eIF4G-eIF4A complex.

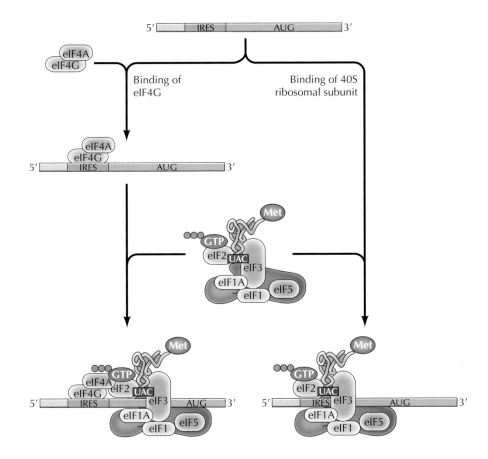

Figure 9.12 Initiation of translation at internal ribosome entry sites Internal ribosome entry sites (IRESs) in some mRNAs can be recognized by eIF4G in complex with eIF4A, followed by recruitment of the 40S ribosomal subunit complexed to the initiator methionyl tRNA bound by eIF2. Alternatively, IRESs in other mRNAs are recognized directly by the 40S ribosomal subunit.

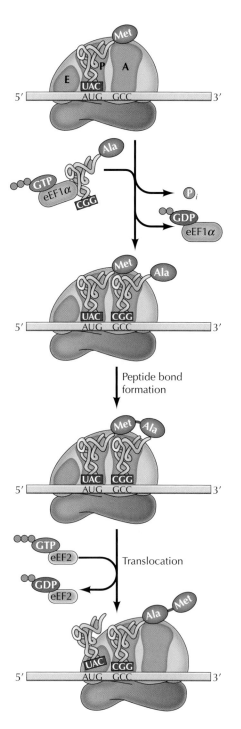

Figure 9.13 Elongation stage of translation The ribosome has three tRNA-binding sites, designated P (peptidyl), A (aminoacyl), and E (exit). The initiating methionyl tRNA is positioned at the P site, leaving an empty A site. The second aminoacyl tRNA (e.g., alanyl tRNA) is then brought to the A site by eEF1α (complexed with GTP). Following GTP hydrolysis, eEF1α (complexed with GDP) leaves the ribosome, with alanyl tRNA inserted into the A site. A peptide bond is then formed, resulting in the transfer of methionine to the aminoacyl tRNA at the A site. The ribosome then moves three nucleotides along the mRNA. This movement translocates the peptidyl (Met-Ala) tRNA to the P site and the uncharged tRNA to the E site, leaving an empty A site ready for addition of the next amino acid. Translocation is mediated by eEF2, coupled to GTP hydrolysis. The process, illustrated here for eukaryotes, is very similar in prokaryotes. (Table 9.1 gives the names of the prokaryotic elongation factors.)

After the initiation complex has formed, translation proceeds by elongation of the polypeptide chain. The mechanism of elongation in prokaryotic and eukaryotic cells is very similar (**Figure 9.13**). The ribosome has three sites for tRNA binding, designated the P (peptidyl), A (aminoacyl), and E (exit) sites. The initiator methionyl tRNA is bound at the P site. The first step in elongation is the binding of the next aminoacyl tRNA to the A site by pairing with the second codon of the mRNA. The aminoacyl tRNA is escorted to the ribosome by an **elongation factor** (EF-Tu in prokaryotes, eEF1α in eukaryotes), which is complexed to GTP. The selection of the correct aminoacyl tRNA for incorporation into the growing polypeptide chain is the critical step that determines the accuracy of protein synthesis. Although this selection is based on base pairing between the codon on mRNA and the anticodon on tRNA, base pairing alone is not sufficient to account for the accuracy of protein synthesis, which has an error rate of less than 10^{-3}. This accuracy is provided by a "decoding center" in the small ribosomal subunit, which recognizes correct codon-anticodon base pairs and discriminates against mismatches. Insertion of a correct aminoacyl tRNA into the A site triggers a conformational change that induces the hydrolysis of GTP bound to eEF1α and release of the elongation factor bound to GDP. A notable result of the recent structural studies of the ribosome is that recognition of correct codon-anticodon pairing in the decoding center, like peptidyl transferase activity, is principally based on the activity of ribosomal RNA rather than proteins.

Once eEF1α has left the ribosome, a peptide bond can be formed between the initiator methionyl tRNA at the P site and the second aminoacyl tRNA at the A site. This reaction is catalyzed by the large ribosomal subunit, with the rRNA playing a critical role (as already discussed). The result is the transfer of methionine to the aminoacyl tRNA at the A site of the ribosome, forming a peptidyl tRNA at this position and leaving the uncharged initiator tRNA at the P site. The next step in elongation is translocation, which requires another elongation factor (EF-G in prokaryotes, eEF2 in eukaryotes) and is again coupled to GTP hydrolysis. During translocation, the ribosome moves three nucleotides along the mRNA, positioning the next codon in an empty A site. This step translocates the peptidyl tRNA from the A site to the P site, and the uncharged tRNA from the P site to the E site. The ribosome is then left with a peptidyl tRNA bound at the P site, and an empty A site. The binding of a new aminoacyl tRNA to the A site then induces the release of the uncharged tRNA from the E site, leaving

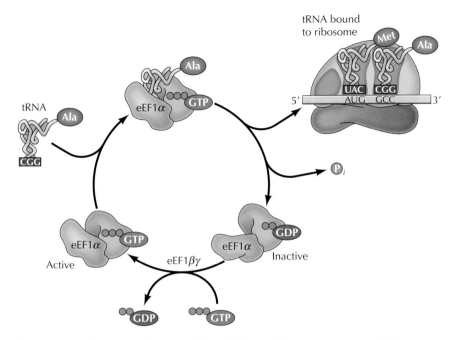

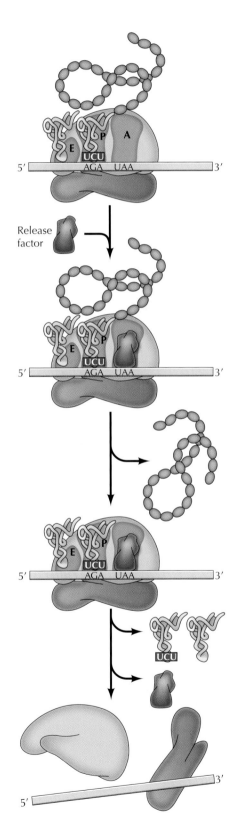

Figure 9.14 Regeneration of eEF1α/GTP eEF1α complexed to GTP escorts the aminoacyl tRNA to the ribosome. The bound GTP is hydrolyzed as the correct tRNA is inserted, so eEF1α complexed to GDP is released. The eEF1α/GDP complex is inactive and unable to bind another tRNA. In order for translation to continue, the active eEF1α/GTP complex must be regenerated by another factor, eEF1βγ, which stimulates the exchange of the bound GDP for free GTP.

the ribosome ready for insertion of the next amino acid in the growing polypeptide chain.

As elongation continues, the eEF1α (or EF-Tu) that is released from the ribosome bound to GDP must be reconverted to its GTP form (**Figure 9.14**). This conversion requires a third elongation factor, eEF1βγ (EF-Ts in prokaryotes), which binds to the eEF1α/GDP complex and promotes the exchange of bound GDP for GTP. This exchange results in the regeneration of eEF1α/GTP, which is now ready to escort a new aminoacyl tRNA to the A site of the ribosome, beginning a new cycle of elongation. The regulation of eEF1α by GTP binding and hydrolysis illustrates a common means of the regulation of protein activities. As will be discussed in later chapters, similar mechanisms control the activities of a wide variety of proteins involved in the regulation of cell growth and differentiation, as well as in protein transport and secretion.

Elongation of the polypeptide chain continues until a stop codon (UAA, UAG, or UGA) is translocated into the A site of the ribosome. Cells do not contain tRNAs with anticodons complementary to these termination signals; instead, they have **release factors** that recognize the signals and terminate protein synthesis (**Figure 9.15**). Prokaryotic cells contain two release factors that recognize termination codons (see Table 9.1): RF1 recognizes UAA or

Figure 9.15 Termination of translation A termination codon (e.g., UAA) at the A site is recognized by a release factor rather than by a tRNA. The result is the release of the completed polypeptide chain, followed by the dissociation of tRNA and mRNA from the ribosome.

(A)

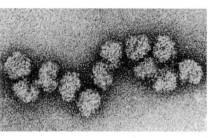

50 nm

(B)

Growing polypeptide chain

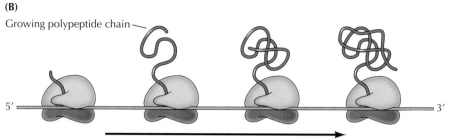

5′   3′

Direction of ribosome movement

Figure 9.16 Polysomes   Messenger RNAs are translated by a series of multiple ribosomes (a polysome). (A) Electron micrograph of a eukaryotic polysome. (B) Schematic of a generalized polysome. Note that the ribosomes closer to the 3′ end of the mRNA have longer polypeptide chains. (A, from M. Boublik et al., 1990. *The Ribosome*, p. 117. Courtesy of American Society for Microbiology.)

UAG, and RF2 recognizes UAA or UGA. In eukaryotic cells a single release factor (eRF1) recognizes all three termination codons. Both prokaryotic and eukaryotic cells also contain release factors (RF3 and eRF3, respectively) that do not recognize specific termination codons but act together with RF1 (or eRF1) and RF2. The release factors bind to a termination codon at the A site and stimulate hydrolysis of the bond between the tRNA and the polypeptide chain at the P site, resulting in release of the completed polypeptide from the ribosome. The tRNA is then released, and the ribosomal subunits and the mRNA template dissociate.

A single mRNA molecule can be translated simultaneously by several ribosomes in both prokaryotic and eukaryotic cells. Once one ribosome has moved away from the initiation site, another can bind to the mRNA and begin synthesis of a new polypeptide chain. Thus mRNAs are usually translated by a series of ribosomes, spaced at intervals of about 100 to 200 nucleotides (**Figure 9.16**). The group of ribosomes bound to an mRNA molecule is called a polyribosome, or **polysome**. Each ribosome within the group functions independently to synthesize a separate polypeptide chain.

Regulation of translation

Although transcription is the initial level at which gene expression is controlled, regulation of the translation of mRNAs also plays a key role in modulating gene expression. The translation of particular mRNAs can be regulated by both translational repressor proteins and noncoding microRNAs, which have become recognized as central regulators of gene expression in eukaryotic cells. In addition, the global translational activity of cells is modulated in response to cell stress, nutrient availability, and growth factor stimulation.

One mechanism of translational regulation is the binding of repressor proteins (which block translation) to specific mRNA sequences. A well understood example of this mechanism in eukaryotic cells is regulation of the synthesis of ferritin, a protein that stores iron within the cell (**Figure 9.17**). The translation of ferritin mRNA is regulated by the supply of iron: More ferritin is synthesized if iron is abundant. This regulation is mediated by iron regulatory proteins (IRPs), which, in the absence of iron, bind to a sequence

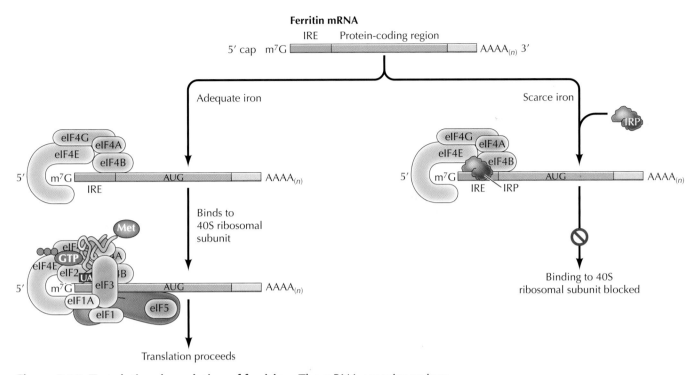

Figure 9.17 Translational regulation of ferritin The mRNA contains an iron response element (IRE) near its 5′ cap. In the presence of adequate supplies of iron, translation of the mRNA proceeds normally. If iron is scarce, however, a protein (called the iron regulatory protein, or IRP) binds to the IRE, blocking translation by interfering with binding of the mRNA to the 40S ribosomal subunit.

(the iron response element, or IRE) in the 5′ untranslated region of ferritin mRNA, blocking its translation. In the presence of iron, the iron regulatory proteins no longer bind to the IRE and ferritin translation is able to proceed. The IRE is located within 40 nucleotides of the 5′ cap, and it appears that the iron regulatory proteins repress translation by interfering with the binding of the 40S ribosomal subunit to the mRNA.

Translation can also be regulated by proteins that bind to specific sequences in the 3′ untranslated regions of some mRNAs. In most cases, these translational repressors function by binding to the initiation factor eIF4E, interfering with the interaction of eIF4E and eIF4G and inhibiting the initiation of translation (**Figure 9.18**). Proteins that bind to the 3′ untranslated regions of mRNAs are also responsible for localizing mRNAs to specific regions of cells, allowing proteins to be produced in specific subcellular locations. Localization of mRNAs is an important part of translational regulation in a variety of cell types, including eggs, embryos, nerve cells, and moving fibroblasts.

> **FYI**
>
> Specific mRNAs are localized at synapses between neurons. The regulated translation of these localized mRNAs appears to be required for the formation and maintenance of synapses and may play a role in learning and memory.

Figure 9.18 Translational repressor binding to 3′ untranslated sequences Translational repressors can bind to regulatory sequences in the 3′ untranslated region (UTR) and inhibit translation by binding to the initiation factor eIF4E, bound to the 5′ cap. This interferes with translation by blocking the formation of a normal initiation complex (see Figure 9.11).

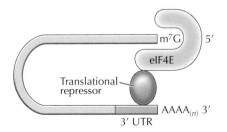

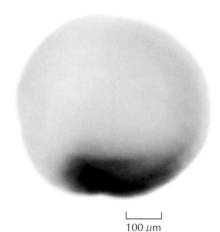

L_____I
100 μm

Figure 9.19 Localization of mRNA in *Xenopus* oocytes   *In situ* hybridization illustrating the localization of *Xlerk* mRNA to a specific region (the vegetal cortex) of *Xenopus* oocytes. (Courtesy of James Deshler, Boston University.)

For example, localization of mRNAs to specific regions of eggs or embryos plays an important role in development by allowing the encoded proteins to be synthesized at the appropriate sites in the developing embryo (**Figure 9.19**). The localization of mRNAs is coupled to regulation of their translation, so that their encoded proteins are synthesized when the mRNA becomes properly localized at the appropriate developmental stage.

Translational regulation is particularly important during early development. As discussed in Chapter 8, a variety of mRNAs are stored in oocytes (unfertilized eggs) in an untranslated form; the translation of these stored mRNAs is activated at fertilization or later stages of development. One mechanism of such translational regulation is the controlled polyadenylation of oocyte mRNAs. Many untranslated mRNAs are stored in oocytes with short poly-A tails (approximately 30 to 50 nucleotides). These stored mRNAs are initially synthesized with long poly-A tails of approximately 200 nucleotides, like other mRNAs (see Figure 8.45). However, they are subsequently recognized by a translational repressor protein that binds to specific sequences in their 3' untranslated regions and leads to removal of most of the poly-A tail. These stored mRNA are subsequently recruited for translation at the appropriate stage of development by the lengthening of their poly-A tails to several hundred nucleotides. This allows the binding of poly-A binding protein (PABP), which stimulates translation by interacting with eIF4G (see Figure 9.11).

As discussed in earlier chapters, one of the most exciting advances in recent years has been the elucidation of the role of noncoding RNAs in regulating gene expression. Since its discovery in 1998, **RNA interference (RNAi)** mediated by short double-stranded RNAs has been widely used as an experimental tool to block gene expression at the level of translation (see Figure 4.38). In addition, it is normally used by cells as an important mechanism of translational regulation. The two major types of short RNAs that mediate RNA interference are **short interfering RNAs (siRNAs)** and **microRNAs (miRNAs)**, which are both approximately 22 nucleotides in length but differ in their origins. siRNAs are produced by cleavage of longer double-stranded RNAs by the nuclease Dicer. The longer double-stranded precursors of siRNAs may be introduced into cells experimentally or originate within cells from overlapping transcripts. In contrast, the precursors of miRNAs are transcribed by RNA polymerase II as primary transcripts of approximately 70 nucleotides, which are then cleaved sequentially by the nucleases Drosha and Dicer to yield double-stranded RNAs of about 22 nucleotides (see Figure 6.8). One strand of both miRNAs and siRNAs is incorporated into the RNA-induced silencing complex (RISC), and the siRNAs or miRNAs target RISC to complementary mRNAs. Whereas siRNAs generally pair perfectly with their targets and induce cleavage of the targeted mRNA by a component of RISC (see Figure 4.38), most miRNAs form mismatched duplexes with sequences within the 3' untranslated regions of their target mRNAs (**Figure 9.20**). In these cases, the miRNA/RISC complex does not cleave the mRNA, but instead represses translation and targets the mRNA for degradation by stimulating deadenylation. The mechanism of translational repression by miRNAs is not yet fully understood but appears to involve inhibition of translation initiation. Recent data further indicate that miRNAs may interfere with the activity of eIF4A, which functions to unwind secondary structures in the mRNA and allow ribosome scanning to the initiation codon. However, further work is needed to fully elucidate the

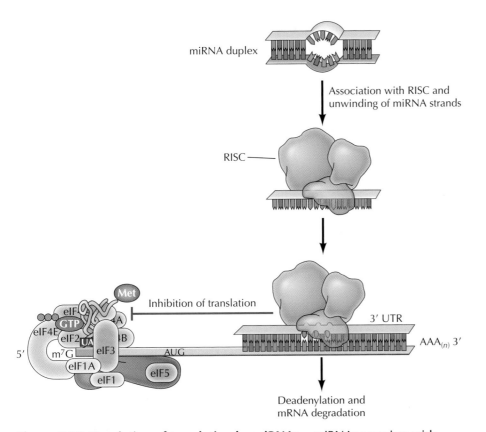

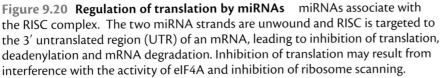

Figure 9.20 Regulation of translation by miRNAs   miRNAs associate with the RISC complex. The two miRNA strands are unwound and RISC is targeted to the 3′ untranslated region (UTR) of an mRNA, leading to inhibition of translation, deadenylation and mRNA degradation. Inhibition of translation may result from interference with the activity of eIF4A and inhibition of ribosome scanning.

mechanism for translational inhibition by miRNAs and to determine how this is coupled to stimulation of mRNA degradation.

The importance of miRNAs in gene regulation is enormous. It is estimated that as many as 1000 miRNAs are encoded in mammalian genomes. It further appears that each miRNA can target up to 100 different mRNAs, and it is estimated that half of our protein-coding genes may be targets for regulation by miRNAs. Although their biological roles are still being explored, miRNAs have been found to function in a variety of developmental processes and to play important roles in the regulation of cell proliferation and survival. Moreover, abnormal expression of miRNAs has been found to contribute to cancer and other diseases. Thus, understanding both the mechanism of miRNA action and the full extent of their biological roles are highly active areas of investigation.

Another mechanism of translational regulation in eukaryotic cells, resulting in global effects on overall translational activity rather than effects on the translation of specific mRNAs, involves modulation of the activity of initiation factors, particularly eIF2 and eIF4E. As already discussed, eIF2 (complexed with GTP) binds to the initiator methionyl tRNA, bringing it to the ribosome. The subsequent release of eIF2 is accompanied by GTP hydrolysis, leaving eIF2 as an inactive GDP complex. To participate in another cycle of initiation, the eIF2/GTP complex must be regenerated by the exchange of bound GDP for GTP (**Figure 9.21**). This exchange is mediated by another factor, eIF2B. The control of eIF2 activity by GTP binding and hydrolysis is thus similar to that of eEF1α (see Figure 9.14). However, the regulation of eIF2 provides a critical control point in a variety of eukaryotic cells. In particular, both eIF2 and eIF2B can be phosphorylated by regulatory protein kinases. These phosphorylations inhibit the exchange of bound GDP for GTP, thereby inhibiting initiation of translation. For example, if mammalian cells are subjected to

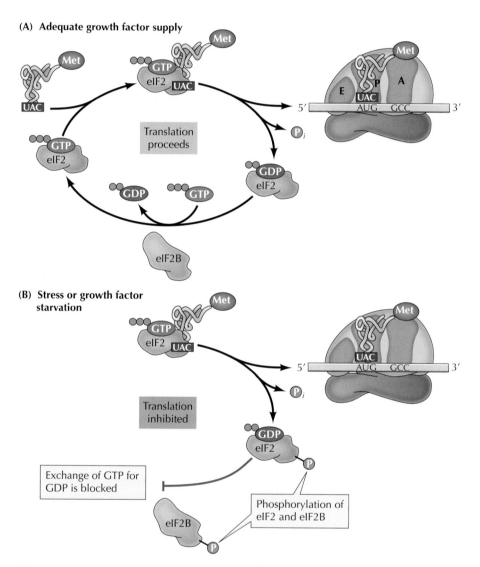

Figure 9.21 Regulation of translation by phosphorylation of eIF2 and eIF2B
(A) The active form of eIF2 (complexed with GTP) escorts initiator methionyl tRNA to the ribosome (see Figure 9.11). The eIF2 is then released from the ribosome in an inactive GDP-bound form. In order to continue translation, eIF2 must be reactivated by eIF2B, which stimulates the exchange of GTP for the bound GDP. (B) Translation can be inhibited (for example, if cells are stressed or starved of growth factors) by regulatory protein kinases that phosphorylate either eIF2 or eIF2B. These phosphorylations block the exchange of GTP for GDP, so eIF2/GTP cannot be regenerated.

stress or starved of growth factors, protein kinases that phosphorylate eIF2 and eIF2B become activated, inhibiting further protein synthesis.

Regulation of the activity of eIF4E, which binds to the 5′ cap of mRNAs, is another critical point at which growth factors act to control protein synthesis (**Figure 9.22**). For example, growth factors that stimulate protein synthesis in mammalian cells activate protein kinases that phosphorylate regulatory proteins (called eIF4E binding proteins, or 4E-BPs) that bind to eIF4E. In the absence of the appropriate growth factors, the nonphosphorylated 4E-BPs bind to eIF4E and inhibit translation by interfering with the interaction between eIF4E and eIF4G. When growth factors are present in adequate supply, phosphorylation of the 4E-BPs prevents their interaction with eIF4E, leading to increased rates of translation initiation.

It is noteworthy that global inhibition of translation can actually promote the selective translation of some mRNAs. For example, mRNAs containing internal ribosome entry sites may be selectively translated under conditions that inhibit eIF4E and thereby lead to inhibition of cap-dependent initiation (see Figure 9.12). Such selective translation of a subset of mRNAs may help cells cope with conditions of cellular stress—such as starvation, growth-factor deprivation, or DNA damage.

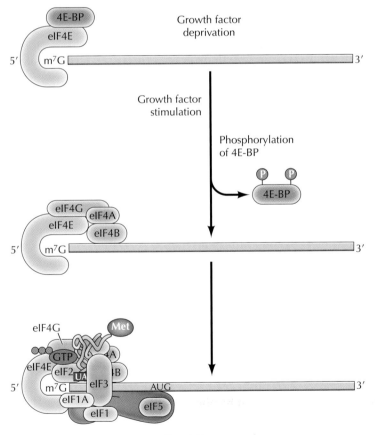

Figure 9.22 Regulation of eIF4E   In the absence of growth factors, translation is inhibited by eIF4E binding proteins (4E-BPs), which bind to eIF4E and block its interaction with eIF4G. Growth factor stimulation, however, leads to the phosphorylation of 4E-BPs, which then dissociate from eIF4E, allowing translation to proceed.

Protein Folding and Processing

Translation completes the flow of genetic information within the cell. The sequence of nucleotides in DNA has now been converted to the sequence of amino acids in a polypeptide chain. The synthesis of a polypeptide, however, is not equivalent to the production of a functional protein. To be useful, polypeptides must fold into distinct three-dimensional conformations, and in many cases multiple polypeptide chains must assemble into a functional complex. In addition, many proteins undergo further modifications, including cleavage and the covalent attachment of carbohydrates and lipids, that are critical for the function and correct localization of proteins within the cell.

Chaperones and protein folding

The three-dimensional conformations of proteins result from interactions between the side chains of their constituent amino acids, as reviewed in Chapter 2. The classic principle of protein folding is that all the information required for a protein to adopt the correct three-dimensional conformation is provided by its amino acid sequence. This was initially established by Christian Anfinsen's experiments demonstrating that denatured RNase can spontaneously refold *in vitro* to its active conformation (see Figure 2.17). Protein folding thus appeared to be a self-assembly process that did not require additional cellular factors. More recent studies, however, have shown that this is not an adequate description of protein folding within the cell. The proper folding of proteins within cells is mediated by the activities of other proteins.

Proteins that facilitate the folding of other proteins are called molecular **chaperones**. The term "chaperone" was first used by Ron Laskey and his colleagues to describe a protein (nucleoplasmin) that is required for the assembly of nucleosomes from histones and DNA. Nucleoplasmin binds to histones and mediates their assembly into nucleosomes, but nucleoplasmin itself is not incorporated into the final nucleosome structure. Chaperones thus act as catalysts that facilitate assembly without being part of the assembled complex. Subsequent studies have extended the concept to include proteins that mediate a variety of other assembly processes, particularly protein folding.

It is important to note that chaperones do not convey additional information required for the folding of polypeptides into their correct three-dimensional conformations; the folded conformation of a protein is determined solely by its amino acid sequence. Rather, chaperones catalyze protein folding by assisting the self-assembly process. They appear to function by binding to and stabilizing unfolded or partially folded polypeptides that are intermediates along the pathway leading to the final correctly folded state. In the absence of chaperones, unfolded or partially folded polypeptide chains would be unstable within the cell. The binding of chaperones stabilizes these unfolded polypeptides, thereby preventing incorrect folding or aggregation and allowing the polypeptide chain to fold into its correct conformation.

A good example is provided by chaperones that bind to nascent polypeptide chains that are still being translated on ribosomes, thereby preventing incorrect folding or aggregation of the amino-terminal portion of the polypeptide before synthesis of the chain is finished (**Figure 9.23**). Proteins fold into domains of approximately 100 to 300 amino acids, so it is necessary to protect the nascent chain from aberrant folding or aggregation with other proteins until synthesis of the entire domain is complete and the protein can fold into its correct conformation. Chaperone binding stabilizes the amino-terminal

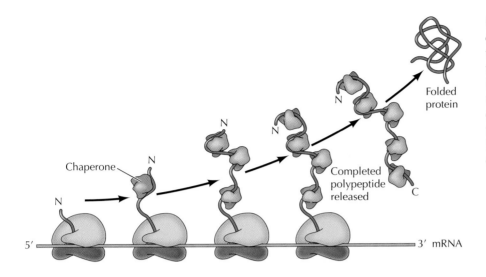

Figure 9.23 Action of chaperones during translation   Chaperones bind to the amino (N) terminal portion of the nascent polypeptide chain, stabilizing it in an unfolded configuration until synthesis of the polypeptide is completed. The completed protein is then released from the ribosome and is able to fold into its correct three-dimensional conformation.

portion in an unfolded conformation until the rest of the polypeptide chain is synthesized and the completed protein can fold correctly. Chaperones also stabilize unfolded polypeptide chains during their transport into subcellular organelles—for example, during the transfer of proteins into mitochondria from the cytosol (**Figure 9.24**). Proteins are transported across the mitochondrial membrane in partially unfolded conformations that are stabilized by chaperones in the cytosol. Chaperones within the mitochondrion then facilitate transfer of the polypeptide chain across the membrane and its subsequent folding within the organelle. In addition, chaperones are involved in the assembly of proteins that consist of multiple polypeptide chains and in the assembly of macromolecular structures (e.g., nucleoplasmin).

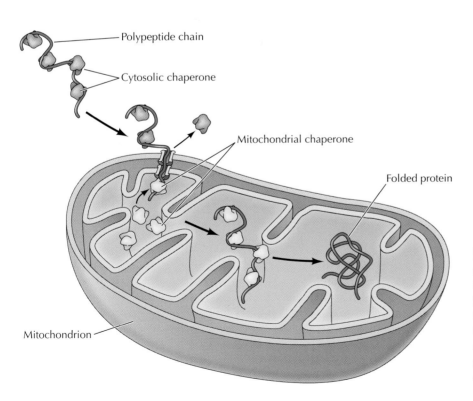

Figure 9.24 Action of chaperones during protein transport   A partially unfolded polypeptide is transported from the cytosol to a mitochondrion. Cytosolic chaperones stabilize the unfolded configuration. Mitochondrial chaperones facilitate transport and subsequent folding of the polypeptide chain within the organelle.

Many of the proteins now known to function as chaperones were initially identified as heat-shock proteins, a group of proteins expressed in cells that have been subjected to elevated temperatures. The heat-shock proteins (abbreviated Hsp) are thought to stabilize and facilitate the refolding of proteins that have been partially denatured as a result of exposure to elevated temperatures. Two families of chaperone proteins, the Hsp70 chaperones and the chaperonins, act in a general pathway of protein folding in both bacteria and eukaryotic cells (**Figure 9.25**). Members of the Hsp70 and chaperonin families are found in the cytosol and in subcellular organelles (endoplasmic reticulum, mitochondria, and chloroplasts) of eukaryotic cells, as well as in bacteria. Members of the Hsp70 family stabilize unfolded polypeptide chains during translation (see, for example, Figure 9.23) as well as during the transport of polypeptides into a variety of subcellular compartments, such as mitochondria and the endoplasmic reticulum. These proteins bind to short hydrophobic segments (approximately seven amino acid residues) of unfolded polypeptides, maintaining the polypeptide chain in an unfolded configuration and preventing aggregation.

The unfolded polypeptide chain is then transferred from Hsp70 chaperones to a chaperonin, within which protein folding takes place, yielding a protein correctly folded into its functional three-dimensional conformation. The chaperonins consist of multiple protein subunits arranged in two stacked rings to form a double-chambered structure. Unfolded polypeptide chains are shielded from the cytosol within the chamber of the chaperonin. In this isolated environment, protein folding can proceed while aggregation of unfolded segments of the polypeptide chain with other unfolded polypeptides is prevented.

Both bacteria and eukaryotic cells also contain additional families of chaperones, and the number of chaperones is considerably larger in eukaryotes. For example, an alternative pathway for the folding of some proteins in the cytosol and endoplasmic reticulum of eukaryotic cells involves the sequential actions of Hsp70 and Hsp90 family members. The majority of substrates for folding by Hsp90 are proteins that are involved in cell signaling, including receptors for steroid hormones and a variety of protein kinases.

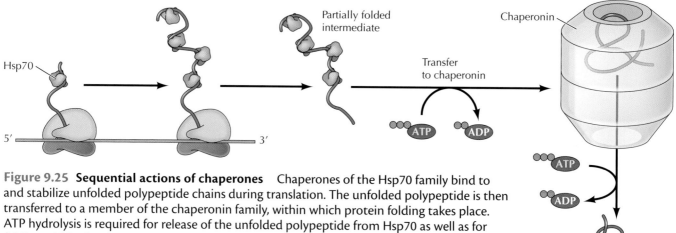

Figure 9.25 Sequential actions of chaperones Chaperones of the Hsp70 family bind to and stabilize unfolded polypeptide chains during translation. The unfolded polypeptide is then transferred to a member of the chaperonin family, within which protein folding takes place. ATP hydrolysis is required for release of the unfolded polypeptide from Hsp70 as well as for folding within the chaperonin.

Protein misfolding diseases

Defects in protein folding are responsible for a large number of common diseases, collectively called **protein misfolding diseases**. In some cases, the misfolding of a protein leads to disease as a result of reduced levels of functional protein in a cell. For example, cystic fibrosis is caused by mutations in a protein called CFTR (cystic fibrosis transmembrane conductance regulator), which is responsible for the transport of chloride ions across the plasma membranes of several types of epithelial cells, including those lining the respiratory tract (see Molecular Medicine, Chapter 14). Defective chloride transport as a result of these mutations leads to obstruction of the respiratory tract by thick plugs of mucus, leading to recurrent infections and the death of most patients from lung disease. The majority of cystic fibrosis cases are due to deletion of a critical amino acid (phenylalanine at position 508) of CFTR. Deletion of this amino acid disrupts the normal interactions of CFTR with chaperones in the endoplasmic reticulum, leading to defective protein folding and reduced levels of functional CFTR in the plasma membranes of affected cells.

In contrast to cystic fibrosis, a large number of protein misfolding diseases, including Alzheimer's disease, Parkinson's disease, and type 2 diabetes, are associated with the aggregation of misfolded proteins (Table 9.2). In each of these diseases, specific proteins misfold to form fibrous aggregates called **amyloids**. In contrast to their normal globular conformations, misfolded proteins aggregate to form insoluble amyloid fibrils characterized by β-sheet structures (Figure 9.26A). Alzheimer's disease, for example, is characterized by amyloid plaques in the brains of patients (Figure 9.26B). These plaques form as a result of aggregation of amyloid-β protein (Aβ), which is a cleavage fragment of a transmembrane protein called amyloid precursor protein. As indicated in Table 9.2, the formation of amyloid aggregates of other proteins is associated with a variety of neurodegenerative diseases as well as with many other disorders, including type 2 diabetes and systemic amyloidoses.

Table 9.2 Representative Diseases Associated with Protein Aggregation

Disease	Aggregating Protein
Neurodegenerative diseases	
Alzheimer's disease	Amyloid-β
Parkinson's disease	α-Synuclein
Huntington's disease	Huntingtin
Amyotrophic lateral sclerosis	Superoxide dismutase
Spongiform encephalopathies	Prion protein
Non-neurodegenerative localized diseases	
Type 2 diabetes	Amylin
Cataracts	Crystallins
Injection-localized amyloidosis	Insulin
Systemic diseases	
Amyloid light-chain amyloidosis	Immunoglobulin light chain
Amyloid A amyloidosis	Serum amyloid A protein
Senile systemic amyloidosis	Transthyretin

Video 9.3

sites.sinauer.com/cooper7e/v9.3
Mechanisms of Alzheimer's Disease
Alzheimer's disease is characterized by the formation of beta amyloid plaques and neurofibrillary tangles, which impede neuronal communication.

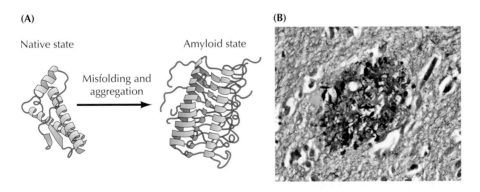

(A) Native state → Amyloid state

Misfolding and aggregation

Figure 9.26 Protein aggregation and amyloid formation (A) A native globular protein misfolds and aggregates to form an amyloid fibril, with a β-sheet structure. (B) An amyloid plaque from the brain of a patient with Alzheimer's disease.

Molecular Medicine

Alzheimer's Disease

The Disease

Alzheimer's disease was named for the Bavarian psychiatrist Dr. Alois Alzheimer who, in 1906, recognized distinct lesions in the brain tissue of a woman who died of a mental illness with symptoms including memory loss. It is a progressive form of dementia that gradually destroys memory, reasoning and the ability to perform everyday activities. The disease is ultimately fatal and people in the final stages of Alzheimer's disease are bedridden, unable to communicate, and completely dependent on others for their care. Alzheimer's disease affects more than 5 million Americans and is the sixth leading cause of death in the United States, resulting in more than 500,000 deaths per year. The disease is usually diagnosed in people over age 65 and it is estimated that approximately 11% of people in this age group have Alzheimer's disease.

Molecular and Cellular Basis

The two lesions identified by Dr. Alois Alzheimer and now recognized as the hallmark of Alzheimer's disease are neurofibrillary tangles and amyloid plaques (see figure). Neurofibrillary tangles are misfolded aggregates of a protein called tau, which is normally associated with microtubules. Amyloid plaques are aggregates of misfolded amyloid-β protein (Aβ), which is a cleavage fragment of a transmembrane protein called amyloid precursor protein (APP). The role of Aβ in the disease is indicated by studies of rare inherited cases of Alzheimer's. Three

genes responsible for inherited Alzheimer's disease result in increased production of Aβ, either as a result of structural mutations in APP or in the proteases responsible for APP cleavage, directly linking Aβ to disease development. In addition, the role of Aβ is supported by transgenic mouse models, in which overproduction of Aβ leads to neurodegeneration and pathologies similar to the human disease. However, the mechanism by which Aβ causes neurodegeneration remains to be established. Current research indicates that soluble aggregates of misfolded Aβ, which are precursors to amyloid plaques, are toxic to neurons and it is thought that these aggregates, rather than the amyloid plaques themselves, may be the primary agent responsible for disease development.

Prevention and Treatment

There is currently no cure or means to prevent Alzheimer's disease. Treatments are limited to helping patients maintain mental functions and delaying progression of the disease. Four drugs that regulate the activity of neurotransmitters are currently used for Alzheimer's treatment. These drugs can help to maintain brain function, but they are effective for only about half of Alzheimer's patients and only delay progression of the disease for six months to a year. Research is focused on the development of drugs that would interfere directly with the pathogenesis of Alzheimer's disease, for example by inhibiting production of Aβ or interfering with Aβ aggregation—but the realization of such drugs remains a goal for the future.

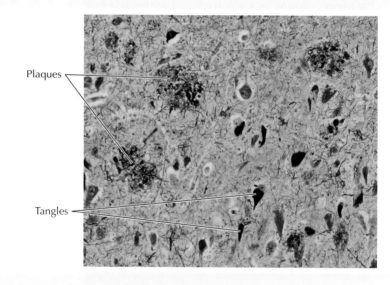

Plaques

Tangles

Amyloid plaques and neurofibrillary tangles in the brain of an Alzheimer's disease patient.

It is likely that protein aggregation plays different roles in the development of different diseases. The systemic amyloidoses, for example, are a group of diseases in which different proteins (such as immunoglobulin light chains) form large amounts of amyloid aggregates that are distributed throughout the body and become deposited in a variety of organs. In this case, disruption of the structure and function of vital organs (such as the heart and kidneys) by these amyloid aggregates results in the disease. In contrast, the role of amyloid formation in the pathogenesis of neurodegenerative diseases is less clear. These diseases may result from cellular damage induced by misfolded proteins, rather than from the bulk formation of amyloid aggregates. For example, current evidence suggests that Alzheimer's disease may result from the toxicity of soluble forms of Aβ oligomers, rather than being caused by the accumulation of insoluble amyloid plaques.

Studies of a group of rare transmissible neurodegenerative diseases, called transmissible spongiform encephalopathies, led to the characterization of **prions** as misfolded proteins that were capable of self-replication. These diseases include scrapie in sheep, mad cow disease, Creutzfeldt-Jakob disease, and kuru. In contrast to bacteria or viruses, the infectious agents responsible for these diseases were found to be resistant to treatment with agents that inactivate nucleic acids, such as nucleases or UV-irradiation, but were sensitive to treatments that degraded proteins. In 1982, Stanley Prusiner isolated an infectious particle, composed only of protein, that caused scrapie. He termed the particle a prion, designating a proteinaceous infectious particle that lacks nucleic acids. The infectivity of prions is based on amyloid formation by the prion protein, PrP. PrP is expressed in mammalian cells in its normal α-helical form, termed PrP^C. In its infectious form, PrP forms a misfolded amyloid structure, designated PrP^{Sc}. Importantly, PrP^{Sc} can propagate by inducing the misfolding of PrP^C proteins to the amyloid PrP^{Sc} state (**Figure 9.27**). Thus, PrP^{Sc} can infect a cell and "replicate" by inducing autocatalytic amyloid formation of endogenous PrP^C—a novel form of propagation that does not require any nucleic acid in the infectious particle. Importantly, the proteins that form amyloids in the neurodegenerative diseases discussed above (e.g., Alzheimer's disease) are similar to prions in that the misfolded proteins also stimulate further misfolding and aggregation. Prion diseases may therefore also encompass these common forms of neurodegenerative disease.

Enzymes that catalyze protein folding

In addition to chaperones that facilitate protein folding by binding to and stabilizing partially folded intermediates, cells contain at least two types of enzymes that act as chaperones by catalyzing protein folding. The formation of disulfide bonds between cysteine residues is important in stabilizing the folded structures of many proteins (see Figure 2.16). **Protein disulfide isomerase (PDI)**, which was discovered by Christian Anfinsen in 1963, catalyzes disulfide bond formation (**Figure 9.28**). For proteins that contain multiple

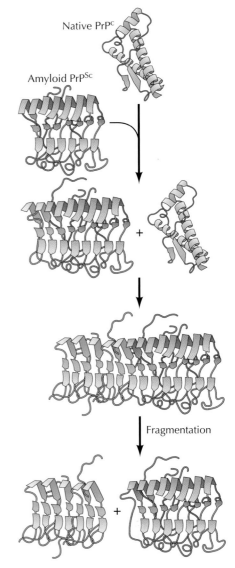

Figure 9.27 Prion propagation The prion protein PrP^{Sc} propagates by inducing the misfolding of native PrP^C to the amyloid PrP^{Sc} state. The growing amyloid fibers fragment to yield new PrP^{Sc} particles.

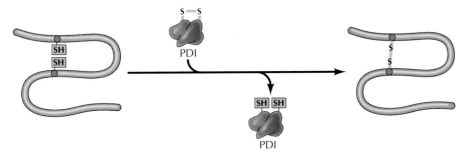
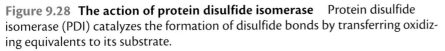

Figure 9.28 The action of protein disulfide isomerase   Protein disulfide isomerase (PDI) catalyzes the formation of disulfide bonds by transferring oxidizing equivalents to its substrate.

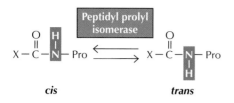

Figure 9.29 The action of peptidyl prolyl isomerase Peptidyl prolyl isomerase catalyzes the isomerization of peptide bonds that involve proline between the *cis* and *trans* conformations.

cysteine residues, PDI also plays an important role by promoting rapid exchanges between paired disulfides, thereby allowing the protein to attain the pattern of disulfide bonds that is compatible with its stably folded conformation. Disulfide bonds are generally restricted to secreted proteins and some membrane proteins because the cytosol contains reducing agents that maintain cysteine residues in their reduced (–SH) form, thereby preventing the formation of disulfide (S–S) linkages. In eukaryotic cells, disulfide bonds generally form in the endoplasmic reticulum in which an oxidizing environment is maintained. PDI is a critical chaperone and catalyst of protein folding in the endoplasmic reticulum and is one of the most abundant proteins in that organelle.

The second enzyme that plays a role in protein folding catalyzes the isomerization of peptide bonds that involve proline residues (**Figure 9.29**). Proline is an unusual amino acid in that the equilibrium between the *cis* and *trans* conformations of peptide bonds that precede proline residues is only slightly in favor of the *trans* form. In contrast, peptide bonds between other amino acids are almost always in the *trans* form. Isomerization between the *cis* and *trans* configurations of prolyl-peptide bonds, which could otherwise represent a rate-limiting step in protein folding, is catalyzed by the enzyme **peptidyl prolyl isomerase**. This enzyme is widely distributed in both prokaryotic and eukaryotic cells and plays an important role in the folding of some proteins.

Protein cleavage

Cleavage of the polypeptide chain (**proteolysis**) is an important step in the maturation of many proteins. A simple example is removal of the initiator methionine from the amino terminus of many polypeptides, which occurs soon after the amino terminus of the growing polypeptide chain emerges from the ribosome. Additional chemical groups, such as acetyl groups or fatty acid chains (discussed shortly), are then frequently added to the amino-terminal residues.

Proteolytic modifications of the amino terminus also play a part in the translocation of many proteins across membranes, including secreted proteins in both bacteria and eukaryotes as well as proteins destined for incorporation into the plasma membrane, lysosomes, mitochondria, and chloroplasts of eukaryotic cells. These proteins are targeted for transport to their destinations by amino-terminal sequences that are removed by proteolytic cleavage as the protein crosses the membrane. For example, amino-terminal **signal sequences**, usually about 20 amino acids long, target many secreted proteins to the plasma membrane of bacteria or to the endoplasmic reticulum of eukaryotic cells while translation is still in progress (**Figure 9.30**). The signal sequence, which consists predominantly of hydrophobic amino acids, is inserted into a membrane channel as it emerges from the ribosome. The remainder of the polypeptide chain passes through the channel as translation proceeds. The signal sequence is then cleaved by a specific membrane protease (**signal peptidase**), and the mature protein is released. In eukaryotic cells, the translocation of growing polypeptide chains into the endoplasmic reticulum is the first step in targeting proteins for secretion, incorporation into the plasma membrane, or incorporation into lysosomes. The mechanisms that direct the transport of proteins to these destinations, as well as the role of other targeting sequences in directing the import of proteins into mitochondria and chloroplasts, will be discussed in detail in Chapters 11 and 12.

In other important instances of proteolytic processing, active enzymes or hormones form via cleavage of larger precursors. Insulin, which is synthesized

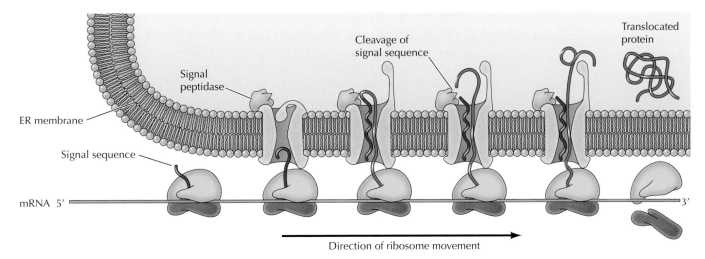

Figure 9.30 The role of signal sequences in membrane translocation   Signal sequences target the translocation of polypeptide chains across the plasma membrane of bacteria or into the endoplasmic reticulum of eukaryotic cells (shown here). The signal sequence, a stretch of hydrophobic amino acids at the amino terminus of the polypeptide chain, inserts into a membrane channel as it emerges from the ribosome. The rest of the polypeptide is then translocated through the channel and the signal sequence is cleaved by the action of signal peptidase, releasing the mature translocated protein.

as a longer precursor polypeptide, is a good example. Insulin forms by two cleavages. The initial precursor (preproinsulin) contains an amino-terminal signal sequence that targets the polypeptide chain to the endoplasmic reticulum (**Figure 9.31**). Removal of the signal sequence during transfer to the endoplasmic reticulum yields a second precursor, called proinsulin. This precursor is then converted to insulin (which consists of two chains held together by disulfide bonds) by proteolytic removal of an internal peptide.

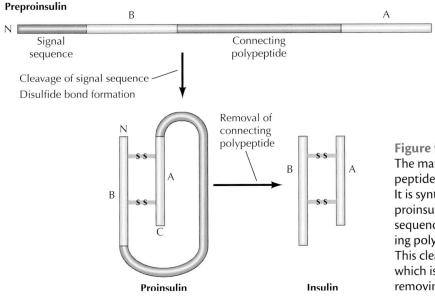

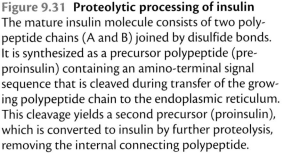

Figure 9.31 Proteolytic processing of insulin The mature insulin molecule consists of two polypeptide chains (A and B) joined by disulfide bonds. It is synthesized as a precursor polypeptide (preproinsulin) containing an amino-terminal signal sequence that is cleaved during transfer of the growing polypeptide chain to the endoplasmic reticulum. This cleavage yields a second precursor (proinsulin), which is converted to insulin by further proteolysis, removing the internal connecting polypeptide.

Other proteins activated by similar cleavage processes include digestive enzymes, proteins involved in blood clotting, and a cascade of proteases that regulate programmed cell death in animals.

It is interesting to note that the proteins of many animal viruses are derived from the cleavage of larger precursors. One particularly important example of the role of proteolysis in virus replication is provided by HIV. In the replication of HIV, a virus-encoded protease cleaves precursor polypeptides to form the viral structural proteins. Because of its central role in virus replication, the HIV protease (in addition to reverse transcriptase) is an important target for the development of drugs used for treating AIDS. Indeed, such protease inhibitors are now among the most effective agents available for combating this disease.

Glycosylation

Many proteins, particularly in eukaryotic cells, are modified by the addition of carbohydrates, a process called **glycosylation**. The proteins to which carbohydrate chains have been added (called **glycoproteins**) are usually secreted or localized to the cell surface, although many nuclear and cytosolic proteins are also glycosylated. The carbohydrate moieties of glycoproteins play important roles in protein folding in the endoplasmic reticulum, in the targeting of proteins for delivery to the appropriate intracellular compartments, and as recognition sites in cell-cell interactions.

Most glycoproteins are classified as either N-linked or O-linked, depending on the site of attachment of the carbohydrate side chain (**Figure 9.32**). In N-linked glycoproteins, the carbohydrate is attached to the nitrogen atom in the side chain of asparagine. In O-linked glycoproteins, the oxygen atom in the side chain of serine or threonine is the site of carbohydrate attachment. The sugars directly attached to these positions are usually either N-acetylglucosamine or N-acetylgalactosamine, respectively. In addition, mannose residues can be attached to tryptophan residues in some proteins by carbon–carbon bonds.

Most glycoproteins in eukaryotic cells are destined either for secretion or for incorporation into the plasma membrane. These proteins are usually transferred into the endoplasmic reticulum while their translation is still in progress. Glycosylation is also initiated in the endoplasmic reticulum before translation is complete. The first step is the transfer of a common oligosaccharide consisting of 14 sugar residues (two N-acetylglucosamine, nine mannose, and three glucose) to an asparagine residue of the growing polypeptide chain (**Figure 9.33**). The oligosaccharide is assembled within the endoplasmic reticulum on a lipid carrier (**dolichol phosphate**). It is then transferred as an intact unit to an acceptor asparagine (Asn) residue within the sequence Asn-X-Ser or Asn-X-Thr (where X is any amino acid other than proline). As discussed in Chapter 11, the common N-linked oligosaccharide is then modified by both the removal and addition of carbohydrate residues, so that proteins can ultimately have a variety of different N-linked oligosaccharides.

N-linkage

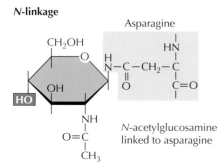

O-linkage

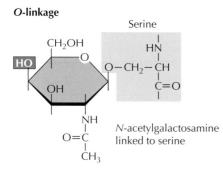

Figure 9.32 Linkage of carbohydrate side chains to glycoproteins The carbohydrate chains of N-linked glycoproteins are attached to asparagine; those of O-linked glycoproteins are attached to either serine (shown) or threonine. The sugars joined to the amino acids are usually either N-acetylglucosamine (N-linked) or N-acetylgalactosamine (O-linked).

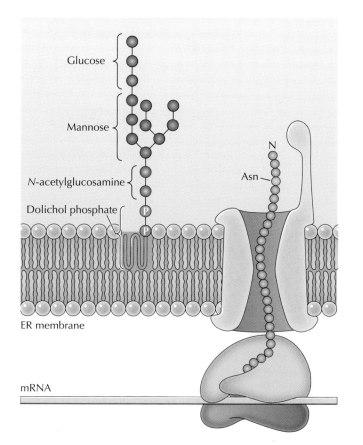

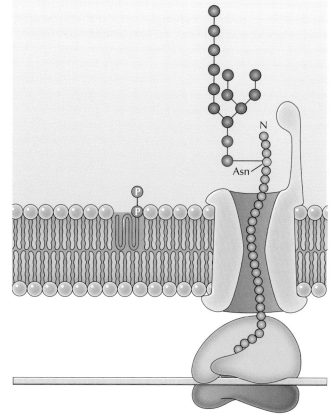

Figure 9.33 Synthesis of *N*-linked glycoproteins The first step in glycosylation is the addition of an oligosaccharide consisting of 14 sugar residues to a growing polypeptide chain in the endoplasmic reticulum (ER). The oligosaccharide (consisting of two *N*-acetylglucosamine, nine mannose, and three glucose residues) is assembled on a lipid carrier (dolichol phosphate) in the ER membrane. It is then transferred as a unit to an acceptor asparagine (Asn) residue of the polypeptide.

Most *O*-linked oligosaccharides are added within the Golgi apparatus. In contrast to the *N*-linked oligosaccharides, *O*-linked oligosaccharides are formed by the addition of one sugar at a time and usually consist of only a few residues (**Figure 9.34**). Some *O*-linked oligosaccharides, however, are long chains containing many sugars, such as the secreted proteoglycans of the extracellular matrix (discussed in Chapter 15). Many cytoplasmic and nuclear proteins, including a variety of transcription factors, are also modified by the addition of single *O*-linked *N*-acetylglucosamine residues, catalyzed by a different enzyme system in the cytosol. Glycosylation of these cytoplasmic and nuclear proteins is thought to play a role in regulating their activities, although the consequences of *O*-glycosylation remain to be fully understood.

Attachment of lipids

Some proteins in eukaryotic cells are modified by the attachment of lipids to the polypeptide chain. Such modifications frequently target and anchor these proteins to the plasma membrane, with which the hydrophobic lipid is able to interact (see Figure 2.34). Three general types of lipid

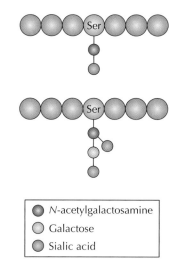

Figure 9.34 Examples of *O*-linked oligosaccharides *O*-linked oligosaccharides usually consist of only a few carbohydrate residues, which are added one sugar at a time.

Figure 9.35 Addition of a fatty acid by N-myristoylation The initiating methionine is removed, leaving glycine at the N terminus of the polypeptide chain. Myristic acid (a 14-carbon fatty acid) is then added.

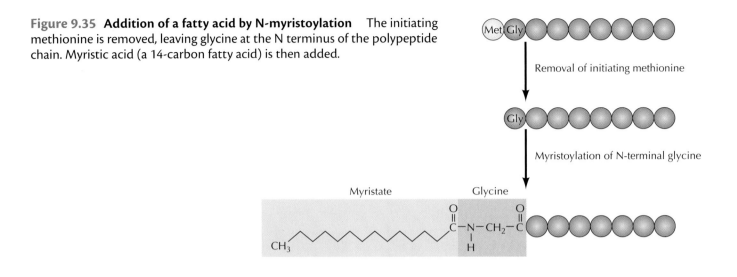

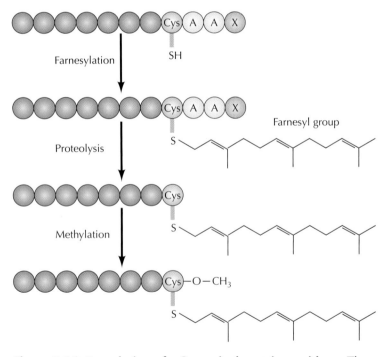

Figure 9.36 Prenylation of a C-terminal cysteine residue The type of prenylation shown affects Ras proteins and proteins of the nuclear envelope (nuclear lamins). These proteins terminate with a cysteine residue (Cys) followed by two aliphatic amino acids (A) and any other amino acid (X) at the C terminus. The first step in their modification is addition of the 15-carbon farnesyl group to the side chain of cysteine (farnesylation). This step is followed by proteolytic removal of the three C-terminal amino acids and methylation of the cysteine, which is now at the C terminus.

additions—N-myristoylation, prenylation, and palmitoylation—are common in eukaryotic proteins associated with the cytosolic face of the plasma membrane. A fourth type of modification, the addition of glycolipids, plays an important role in anchoring some cell surface proteins to the extracellular face of the plasma membrane.

In some proteins, a fatty acid is attached to the amino terminus of the growing polypeptide chain during translation. In this process, called **N-myristoylation**, myristic acid (a 14-carbon fatty acid) is attached to an N-terminal glycine residue (**Figure 9.35**). The glycine is usually the second amino acid incorporated into the polypeptide chain; the initiator methionine is removed by proteolysis before fatty acid addition. Many proteins that are modified by N-myristoylation are associated with the inner face of the plasma membrane, and the role of the fatty acid in this association has been clearly demonstrated by analysis of mutant proteins in which the N-terminal glycine is changed to an alanine. This substitution prevents myristoylation and blocks the function of the mutant proteins by inhibiting their membrane association.

Lipids can also be attached to the side chains of cysteine, serine, and threonine residues. One important example of this type of modification is **prenylation**, in which specific types of lipids (prenyl groups) are attached to the sulfur atoms in the side chains of cysteine residues located near the C terminus of the polypeptide chain (**Figure 9.36**). Many plasma membrane-associated proteins involved in the control of cell growth and differentiation are modified in this way, including

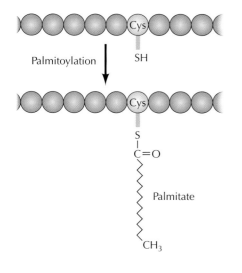

Figure 9.37 Palmitoylation Palmitic acid (a 16-carbon fatty acid) is added to the side chain of an internal cysteine residue.

the Ras oncogene proteins, which are responsible for the uncontrolled growth of many human cancers (see Chapter 19). Prenylation of these proteins proceeds by three steps. First, the prenyl group is added to a cysteine located three amino acids from the carboxy terminus of the polypeptide chain. The prenyl groups added in this reaction are either farnesyl (15 carbons, as shown in Figure 9.36) or geranylgeranyl (20 carbons). The amino acids following the cysteine residue are then removed, leaving cysteine at the carboxy terminus. Finally, a methyl group is added to the carboxyl group of the C-terminal cysteine residue.

In the third type of fatty acid modification, **palmitoylation**, palmitic acid (a 16-carbon fatty acid) is added to sulfur atoms of the side chains of internal cysteine residues (**Figure 9.37**). Like N-myristoylation and prenylation, palmitoylation plays an important role in the association of some proteins with the cytosolic face of the plasma membrane.

Finally, lipids linked to oligosaccharides (**glycolipids**) are added to the C-terminal carboxyl groups of some proteins, where they serve as anchors that attach the proteins to the external face of the plasma membrane. Because the glycolipids attached to these proteins contain phosphatidylinositol, they are usually called **glycosylphosphatidylinositol** (or **GPI**) **anchors** (**Figure 9.38**). The oligosaccharide portions of GPI anchors are attached to the terminal carboxyl group of polypeptide chains. The inositol head group of phosphatidylinositol is in turn attached to the oligosaccharide, so the carbohydrate serves as a bridge between the protein and the fatty acid chains of the phospholipid. The GPI anchors are synthesized and added to proteins as a preassembled unit within the endoplasmic reticulum. Their addition is accompanied by cleavage of a peptide consisting of about 20 amino acids from the C terminus of the polypeptide chain. The modified protein is then transported to the cell

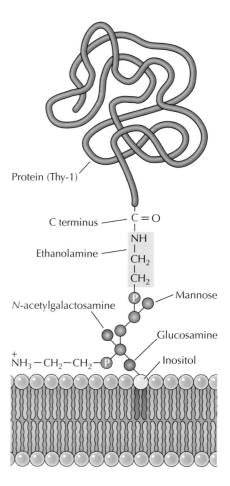

Figure 9.38 Structure of a GPI anchor The GPI anchor, attached to the C terminus, anchors the protein in the plasma membrane. The anchor is joined to the C-terminal amino acid by an ethanolamine, which is linked to an oligosaccharide that consists of mannose, N-acetylgalactosamine, and glucosamine residues. The oligosaccharide is in turn joined to the inositol head group of phosphatidylinositol. The two fatty acid chains of the lipid are embedded in the plasma membrane. The GPI anchor shown here is that of a rat protein, Thy-1.

surface, where the fatty acid chains of the GPI anchor mediate its attachment to the plasma membrane.

Regulation of Protein Function

A critical function of proteins is their activity as enzymes, which are needed to catalyze almost all biological reactions. Regulation of enzyme activity thus plays a key role in governing cell behavior. This is accomplished in part at the level of gene expression, which determines the amount of any enzyme (protein) synthesized by the cell. A further level of control is then obtained by regulation of protein function, which allows the cell to regulate not only the amounts but also the activities of its protein constituents. Regulation of the activities of some of the proteins involved in transcription and translation has already been discussed in this and the preceding chapter, and many further examples of regulated protein function in the control of cell behavior will be evident throughout the remainder of this book. This section discusses three general mechanisms by which the activities of cellular proteins are controlled.

Regulation by small molecules

Most enzymes are controlled by changes in their conformation, which in turn alter catalytic activity. In many cases such conformational changes result from the binding of small molecules, such as amino acids or nucleotides that regulate enzyme activity. This type of regulation commonly is responsible for controlling metabolic pathways by feedback inhibition. For example, the end products of many biosynthetic pathways (e.g., amino acids) inhibit the enzymes that catalyze the first step in their synthesis, thus ensuring an adequate supply of the product while preventing the synthesis of excess amounts (**Figure 9.39**). Feedback inhibition is an example of **allosteric regulation**, in which a regulatory molecule binds to a site on an enzyme that is distinct from the catalytic site (*allo* = "other"; *steric* = "site"). The binding of such a regulatory molecule alters the conformation of the protein, thereby changing the shape of the catalytic site and affecting catalytic activity (see Figure 2.29). Many transcription factors (discussed in Chapter 8) are also regulated by the binding of small molecules. For example, the binding of lactose to the *E. coli lac* repressor induces a conformational change that prevents the repressor from binding DNA (see Figure 8.7). In eukaryotic cells, steroid hormones similarly control gene expression by binding to transcriptional regulatory proteins.

The regulation of translation factors such as eEF1α by GTP binding (see Figure 9.14) illustrates another common mechanism by which the activities of intracellular proteins are controlled. In this case, the GTP-bound form of the protein is its active conformation, while the GDP-bound form is inactive. Many cellular proteins are similarly regulated by GTP or GDP binding. These proteins include the Ras oncogene proteins, which have been studied intensively because of their roles in the control of cell proliferation and in human cancers. X-ray crystallography of these proteins has

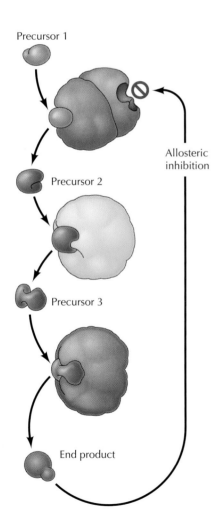

Precursor 1

Allosteric inhibition

Precursor 2

Precursor 3

End product

Figure 9.39 Feedback inhibition The end product of a biochemical pathway acts as an allosteric inhibitor of the enzyme that catalyzes the first step in its synthesis.

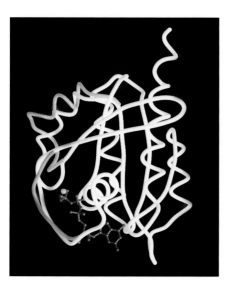

Figure 9.40 Conformational differences between active and inactive Ras proteins The Ras proteins alternate between active GTP-bound and inactive GDP-bound forms. The major effect of GTP binding versus GDP binding is alteration of the conformation of two regions of the molecule, designated the switch I and switch II regions. The backbone of the GTP complex is shown here in white; the backbone of the GDP complex of switch I and switch II are in blue and yellow, respectively. The guanine nucleotide is in red and Mg^{2+} is the yellow sphere. (Courtesy of Sung-Hou Kim, University of California, Berkeley.)

been particularly interesting, revealing subtle but functionally important conformational differences between the inactive GDP-bound and active GTP-bound forms (Figure 9.40). This small difference in protein conformation determines whether Ras (in the active GTP-bound form) can interact with its target molecule, which signals the cell to divide. The importance of such subtle differences in protein conformation is dramatically illustrated by the fact that mutations in *ras* genes contribute to the development of about 25% of human cancers. Such mutations alter the structure of the Ras proteins so that they are locked in the active GTP-bound conformation and continually signal cell division, thereby driving the uncontrolled growth of cancer cells. In contrast, normal Ras proteins alternate between the GTP- and GDP-bound conformations, such that they are active only following stimulation by the hormones and growth factors that normally control cell proliferation in multicellular organisms.

Protein phosphorylation and other modifications

The examples discussed in the previous section involve noncovalent associations of proteins with small-molecule inhibitors or activators. Since no covalent bonds form, the binding of these regulatory molecules to the protein is readily reversible, allowing the cell to respond rapidly to environmental changes. The activity of many proteins, however, is also regulated by covalent modifications. One example of this type of regulation is the activation of some enzymes by proteolytic cleavage of inactive precursors. As noted previously in this chapter, digestive enzymes and proteins involved in blood clotting and programmed cell death are regulated by this mechanism. Since proteolysis is irreversible, however, it provides a means of controlling enzyme activation rather than of turning proteins on and off in response to changes in the environment. In contrast, other covalent modifications—particularly phosphorylation—are readily reversible within the cell and function, as allosteric regulation does, to reversibly activate or inhibit a wide variety of cellular proteins in response to environmental signals.

Protein phosphorylation is catalyzed by **protein kinases**, most of which transfer phosphate groups from ATP to the hydroxyl groups of the side chains of serine, threonine, or tyrosine residues (**Figure 9.41**). The protein kinases are one of the largest protein families in eukaryotes, accounting for approximately 2% of eukaryotic genes. Most protein kinases phosphorylate either serine and threonine or tyrosine residues: These enzymes are called **serine/threonine kinases** or **tyrosine kinases**, respectively. Protein phosphorylation is reversed by **protein phosphatases**, which catalyze the hydrolysis of phosphorylated amino acid residues. Like protein kinases, most protein phosphatases are specific either for serine and threonine or for tyrosine residues, although some protein phosphatases recognize all three phosphoamino acids.

The combined action of protein kinases and protein phosphatases mediates the reversible phosphorylation of many cellular proteins. Frequently, protein kinases function as components of signal transduction pathways in which one kinase activates a second kinase, which may act on yet another kinase. The sequential action of a series of protein kinases can transmit a signal received at the cell surface to target proteins within the cell, resulting in changes in cell behavior in response to environmental stimuli.

The prototype of the action of protein kinases came from studies of glycogen metabolism by Ed Fischer and Ed Krebs in 1955. In muscle cells the hormone epinephrine (adrenaline) signals the breakdown of glycogen to glucose-1-phosphate, providing an available source of energy for increased muscular activity. Glycogen breakdown is catalyzed by the enzyme glycogen

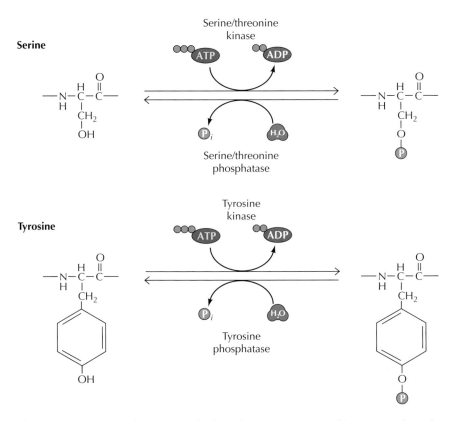

Figure 9.41 Protein kinases and phosphatases Protein kinases catalyze the transfer of a phosphate group from ATP to the side chains of serine and threonine (serine/threonine kinases) or tyrosine (tyrosine kinases) residues. Protein phosphatases catalyze the removal of phosphate groups from the same amino acids by hydrolysis.

The Discovery of Tyrosine Kinases

Transforming Gene Product of Rous Sarcoma Virus Phosphorylates Tyrosine

Tony Hunter and Bartholomew M. Sefton

The Salk Institute, San Diego, CA

Proceedings of the National Academy of Sciences, USA, 1980, Volume 77, pages 1311–1315

Tony Hunter   Bartholomew Sefton

The Context

Following its isolation in 1911, Rous sarcoma virus (RSV) became the first virus that was generally accepted to cause tumors in animals (see Molecular Medicine box, Chapter 1). Several features of RSV then made it an attractive model for studying the development of cancer. In particular, the small size of the RSV genome offered the hope of identifying specific viral genes responsible for inducing the abnormal proliferation that is characteristic of cancer cells. This goal was reached in the 1970s when it was established that a single RSV gene (called *src* for sarcoma) is required for tumor induction. Importantly, a closely related *src* gene was also found to be part of the normal genetic complement of a variety of vertebrates, including humans. Since the viral Src protein is responsible for driving the uncontrolled proliferation of cancer cells, it appeared that understanding Src function would yield crucial insights into the molecular bases of both cancer induction and the regulation of normal cell proliferation.

In 1977 Ray Erikson and his colleagues identified the Src protein by immunoprecipitation (see Figure 5.11) with antisera from animals bearing RSV-induced tumors. Shortly thereafter, it was found that incubation of Src immunoprecipitates with radioactive ATP resulted in phosphorylation of the immunoglobulin molecules. Src therefore appeared to be a protein kinase, clearly implicating protein phosphorylation in the control of cell proliferation.

All previously studied protein kinases phosphorylated serine or threonine residues, which were also the only phospho-amino acids to have been detected in animal cells. However, Walter Eckhardt and Tony Hunter observed in 1979 that the oncogenic protein of another animal tumor virus (polyomavirus) was phosphorylated on a tyrosine residue. Hunter and Bartholomew Sefton therefore tested the possibility that Src might phosphorylate tyrosine, rather than serine/threonine, residues in its substrate proteins. Their experiments demonstrated that Src does indeed function as a tyrosine kinase—an activity now recognized as playing a central role in cell signaling pathways.

The Experiments

Hunter and Sefton identified the amino acid phosphorylated by Src by incubating Src immunoprecipitates with [32P]-labeled ATP. The amino acid that was phosphorylated by Src in the substrate protein (in this case, immunoglobulin) therefore became radioactively labeled. The immunoglobulin was then isolated and hydrolyzed to yield individual amino acids, which were analyzed by electrophoresis and chromatography methods that separated phosphotyrosine, phosphoserine, and phosphothreonine (see figure). The radioactive amino acid detected in these experiments was phosphotyrosine, indicating that Src specifically phosphorylates tyrosine residues.

Further experiments showed that the normal cell Src protein, as well as viral Src, functioned as a tyrosine kinase in immunoprecipitation assays. In addition, Hunter and Sefton extended these *in vitro* experiments by demonstrating the presence of phosphotyrosine in proteins extracted from whole cells. In normal cells, phosphotyrosine accounted for only about 0.03% of total phosphoamino acids (the rest being phosphoserine and phosphothreonine), explaining why it had previously escaped detection. However, phosphotyrosine was about ten times more abundant in cells that were infected with RSV, suggesting that increased tyrosine kinase activity of the viral Src protein was responsible for its ability to induce abnormal cell proliferation.

The Impact

The discovery that Src was a tyrosine kinase both identified a new protein kinase activity and established this activity as being related to the control of cell proliferation. The results of Hunter and Sefton's work were followed by demonstrations that many other tumor virus proteins also function as tyrosine kinases, generalizing the link between tyrosine phosphorylation and the abnormal proliferation of cancer cells. In addition, continuing studies have identified numerous tyrosine kinases that function in a variety of signaling pathways in normal cells. Studies of the mechanism by which a virus causes cancer in chickens thus revealed a previously unknown enzymatic activity that plays a central role in the signaling pathways that regulate animal cell growth, survival, and differentiation. Moreover, as discussed in Chapter 19, tyrosine kinases encoded by oncogenes have provided the most promising targets to date for development of specific drugs against cancer cells.

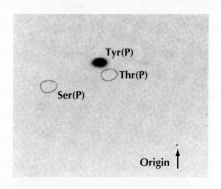

Identification of phosphotyrosine in immunoglobulin phosphorylated by Src   An immunoprecipitate containing RSV Src was incubated with [32P]-ATP. The immunoglobulin was then isolated and hydrolyzed. Amino acids in the hydrolysate were separated by electrophoresis and chromatography on a cellulose thin-layer plate. The positions of 32P-labeled amino acids were determined by exposing the plate to X-ray film. Circles indicate the positions of unlabeled phospho-amino acids that were included as markers. Note that the principal 32P-labeled amino acid is phosphotyrosine [Tyr (P)].

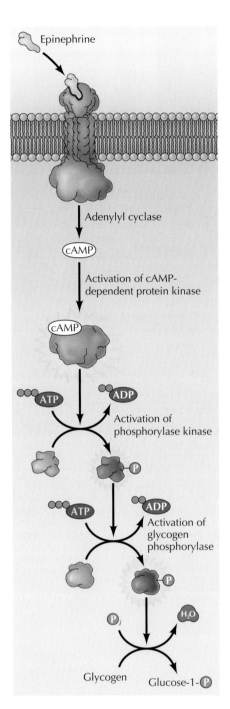

Epinephrine

Adenylyl cyclase

cAMP

Activation of cAMP-
dependent protein kinase

cAMP

ATP   ADP

Activation of
phosphorylase kinase

ATP   ADP

Activation of
glycogen
phosphorylase

P$_i$ H$_2$O

Glycogen   Glucose-1-P

Figure 9.42 Regulation of glycogen breakdown by protein phosphorylation
The binding of epinephrine (adrenaline) to its cell surface receptor triggers the production of cyclic AMP (cAMP), which activates cAMP-dependent protein kinase. cAMP-dependent protein kinase phosphorylates and activates phosphorylase kinase, which in turn phosphorylates and activates glycogen phosphorylase. Glycogen phosphorylase then catalyzes the breakdown of glycogen to glucose-1-phosphate.

phosphorylase, which is regulated by a protein kinase (Figure 9.42). Epinephrine binds to a cell surface receptor that triggers the conversion of ATP to cyclic AMP (cAMP), which then binds to and activates a protein kinase, called cAMP-dependent protein kinase. This kinase phosphorylates and activates a second protein kinase, called phosphorylase kinase. Phosphorylase kinase in turn phosphorylates and activates glycogen phosphorylase, leading to glucose production. The activating phosphorylations of both phosphorylase kinase and glycogen phosphorylase can be reversed by specific phosphatases, so removal of the initial stimulus (epinephrine) inhibits further glycogen breakdown.

The signaling pathway that leads to activation of glycogen phosphorylase is initiated by allosteric regulation due to the binding of small molecules to their targets—epinephrine binding to its cell surface receptor and cAMP binding to cAMP-dependent protein kinase. The signal is then transmitted to its intracellular target by the sequential action of protein kinases. Similar signaling pathways, in which protein kinases and phosphatases play central roles, are involved in regulating almost all aspects of the behavior of eukaryotic cells (see Chapters 16 and 17). Aberrations in these pathways, frequently involving abnormalities of tyrosine kinases, are also responsible for many diseases associated with improper regulation of cell growth and differentiation, including the development of cancer. Indeed, the first tyrosine kinase was discovered in 1980 during studies of the oncogenic proteins of animal tumor viruses—in particular, Rous sarcoma virus—by Tony Hunter and Bartholomew Sefton. Subsequent studies have implicated abnormalities of tyrosine kinases in the development of many kinds of human cancer, and small molecule inhibitors of these enzymes are currently among the most promising drugs being developed in new approaches to cancer treatment.

Although phosphorylation is the most common and best studied type of covalent modification that regulates protein activity, several other types of protein modifications by small molecules also play important roles (Figure 9.43). These include acetylation of lysine residues, methylation of lysine and arginine residues (discussed in Chapter 8 with respect to histones), nitrosylation (addition of NO groups) to cysteine residues, and glycosylation of serine and threonine residues. Recent proteomic studies have shown that lysine acetylation is a particularly common modification that affects thousands of proteins in both prokaryotic and eukaryotic cells, including many metabolic enzymes as well as transcriptional regulatory proteins.

In addition to modification by small molecules, some proteins are regulated by the covalent attachment of polypeptides. The first polypeptide found to act in this way was **ubiquitin**, a 76-amino-acid polypeptide that is highly conserved in all eukaryotes. Ubiquitin was initially found to target proteins for degradation, as discussed in the next section of this chapter. However, it is now recognized that the modification of proteins by addition of ubiquitin and other ubiquitin-like proteins, such as **SUMO** (small ubiquitin-related modifier), serves a variety of functions.

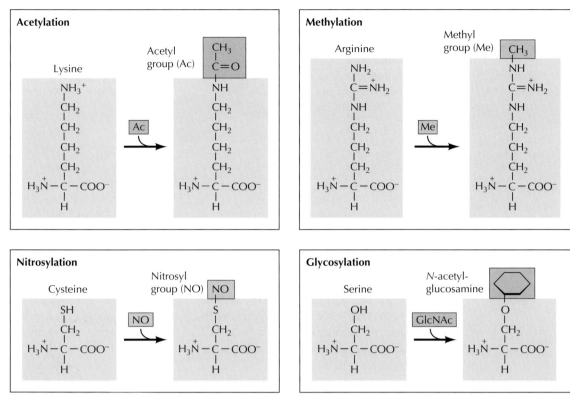

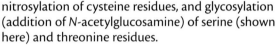

Figure 9.43 Modification of proteins by small molecules   In addition to phosphorylation, proteins can be regulated by acetylation of lysine residues, methylation of lysine and arginine residues, nitrosylation of cysteine residues, and glycosylation (addition of N-acetylglucosamine) of serine (shown here) and threonine residues.

Ubiquitin is attached to the amino groups of the side chains of lysine residues. The selective addition of ubiquitin to target proteins (ubiquitylation) is a multistep process (**Figure 9.44**). First, ubiquitin is activated by being attached to a ubiquitin-activating enzyme, E1. The ubiquitin is then transferred to a second enzyme, called ubiquitin-conjugating enzyme (E2). The ubiquitin is then transferred to the target protein by E2 complexed with a third protein, called ubiquitin ligase or E3. E3s mediate the selective recognition of target proteins by binding to both a substrate and an E2. Mammalian cells contain only two ubiquitin E1s, but have approximately 40 E2s and about 600 E3s. Importantly, ubiquitin can also be removed from target proteins by deubiquitylation enzymes, so ubiquitylation is a reversible protein modification. SUMO and other ubiquitin-like proteins are added to and removed from their targets by similar mechanisms.

The modification of proteins by ubiquitin and SUMO affects a variety of functions. Histone modification by these polypeptides is one mechanism for regulating the transcriptional activity of chromatin, as discussed in Chapter 8. Ubiquitylation is also important in the regulation of protein kinases, proteins involved in DNA repair, and in the control of endocytosis and vesicle trafficking (see Chapters 11 and 14). Many of the proteins modified by SUMO are transcription factors and other nuclear proteins, whose localization is affected by sumoylation. An important example of a protein modified by SUMO is Ran GTPase-activating protein (Ran GAP). As discussed in Chapter 10, Ran GAP is associated with nuclear pore complexes and is

Figure 9.44 Modification of proteins by ubiquitin Ubiquitin is first activated by the enzyme E1. Activated ubiquitin is then transferred to one of several different ubiquitin-conjugating enzymes (E2). A ubiquitin ligase (E3) then associates with both E2 and a substrate protein to direct the transfer of ubiquitin to a specific target. Ubiquitin can be removed by a deubiquitylation enzyme.

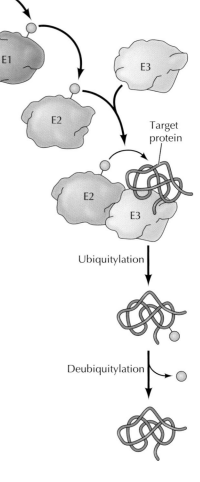

required for the import of proteins, such as transcription factors, from the cytosol to the nucleus. The addition of SUMO to Ran GAP is responsible for its association with the nuclear pore complex, and thus for all protein trafficking between the cytoplasm and nucleus of eukaryotic cells.

Protein–protein interactions

Many proteins consist of multiple subunits, each of which is an independent polypeptide chain. In some proteins the subunits are identical; other proteins are composed of two or more distinct polypeptides. In either case, interactions between the polypeptide chains are important in regulation of protein activity. The importance of these interactions is evident in many allosteric enzymes in which the binding of a regulatory molecule alters protein conformation by changing the interactions between subunits.

Many other enzymes are similarly regulated by protein-protein interactions. A good example is cAMP-dependent protein kinase, which is composed of two regulatory and two catalytic subunits (**Figure 9.45**). In this state, the enzyme is inactive; the regulatory subunits inhibit the enzymatic activity of the catalytic subunits. The enzyme is activated by cAMP, which binds to the regulatory subunits and induces a conformational change leading to dissociation of the complex; the free catalytic subunits are then enzymatically active protein kinases. Cyclic AMP thus acts as an allosteric regulator by altering protein–protein interactions. As discussed in later chapters, other protein–protein interactions, which can themselves be regulated by the binding of small molecules and protein modifications (including phosphorylation

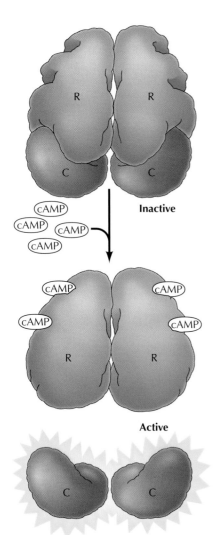

Figure 9.45 Regulation of cAMP-dependent protein kinase In the inactive state, the enzyme consists of two regulatory (R) and two catalytic (C) subunits. Cyclic AMP (cAMP) binds to the regulatory subunits, inducing a conformational change that leads to their dissociation from the catalytic subunits, leaving the catalytic subunits enzymatically active

and the addition of ubiquitin and SUMO), play critical roles in the control of many different aspects of cell behavior.

Protein Degradation

The levels of proteins within cells are determined not only by rates of synthesis but also by rates of degradation. The half-lives of proteins within cells vary widely, from minutes to several days, and differential rates of protein degradation are an important aspect of cell regulation. Many rapidly degraded proteins function as regulatory molecules, such as transcription factors. The rapid turnover of these proteins is necessary to allow their levels to change quickly in response to external stimuli. Other proteins are rapidly degraded in response to specific signals, providing another mechanism for the regulation of intracellular enzyme activity. In addition, faulty or damaged proteins are recognized and rapidly degraded within cells, thereby eliminating the consequences of mistakes made during protein synthesis. In eukaryotic cells, two major pathways—the ubiquitin-proteasome pathway and lysosomal proteolysis—mediate protein degradation.

The ubiquitin-proteasome pathway

The major pathway of selective protein degradation in eukaryotic cells uses ubiquitin as a marker that targets cytosolic and nuclear proteins for rapid proteolysis. As discussed earlier, ubiquitin is a 76-amino-acid polypeptide that can be attached to amino groups of lysine residues and regulates the activities of a variety of target proteins. Proteins can also be targeted for degradation by the addition of multiple ubiquitins to form a polyubiquitin chain, which is catalyzed by some E3s (**Figure 9.46**). Such polyubiquinated proteins are recognized and degraded by a large, multisubunit protease complex, called the **proteasome**.

A number of proteins that control fundamental cellular processes, such as gene expression and cell proliferation, are targets for regulated ubiquitylation and proteolysis. An interesting example of such controlled degradation is provided by proteins (known as cyclins) that regulate progression through the division cycle of eukaryotic cells (**Figure 9.47**). The entry of all eukaryotic cells into mitosis is controlled in part by cyclin B, which is a regulatory subunit of a protein kinase called Cdk1 (see Chapter 17). The association of cyclin B with Cdk1 is required for activation of the Cdk1 kinase, which initiates the events of mitosis (including chromosome condensation and nuclear envelope breakdown) by phosphorylating various cellular proteins. Cdk1 also activates a ubiquitin ligase that targets cyclin B for degradation toward the end of mitosis. This degradation of cyclin B inactivates Cdk1, allowing the cell to exit mitosis and progress to interphase of the next cell cycle. The ubiquitylation of cyclin B is a highly selective reaction, targeted by a nine-amino-acid cyclin B sequence called the destruction box. Mutations of this sequence prevent cyclin B proteolysis and lead to the arrest of dividing cells in mitosis, demonstrating the importance of regulated protein degradation in controlling the fundamental process of cell division.

Figure 9.46 The ubiquitin-proteasome pathway   Proteins are marked for rapid degradation by the covalent attachment of multiple molecules of ubiquitin to form a polyubiquitin chain, mediated by some ubiquitin ligases (E3s). The polyubiquitylated proteins are degraded by a protease complex (the proteasome).

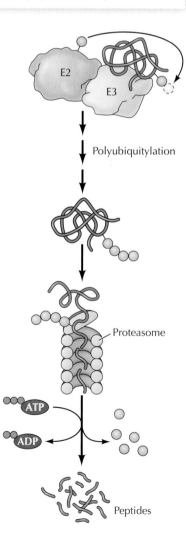

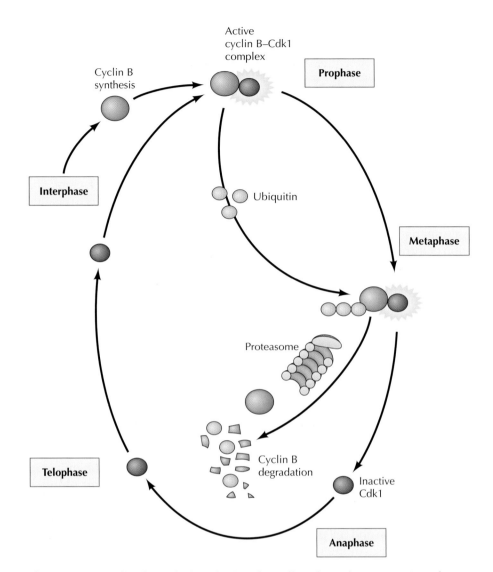

Figure 9.47 Cyclin degradation during the cell cycle   The progression of eukaryotic cells through the division cycle is controlled in part by the synthesis and degradation of cyclin B, which is a regulatory subunit of the Cdk1 protein kinase. Synthesis of cyclin B during interphase leads to the formation of an active cyclin B–Cdk1 complex, which induces entry into mitosis. Rapid degradation of cyclin B by the proteasome then leads to inactivation of the Cdk1 kinase, allowing the cell to exit mitosis and return to interphase of the next cell cycle.

Lysosomal proteolysis

The other major pathway of protein degradation in eukaryotic cells involves the uptake of proteins by **lysosomes**. Lysosomes are membrane-enclosed organelles that contain an array of digestive enzymes, including several proteases (see Chapter 11). They have several roles in cell metabolism, including the digestion of extracellular proteins taken up by endocytosis as well as the turnover of cytoplasmic organelles and cytosolic proteins.

The containment of proteases and other digestive enzymes within lysosomes prevents uncontrolled degradation of the contents of the cell.

Therefore, in order to be degraded by lysosomal proteolysis, cellular proteins must first be taken up by lysosomes. The principal pathway for this uptake of cellular proteins, **autophagy**, involves the formation of vesicles (autophagosomes) in which small areas of cytoplasm or cytoplasmic organelles are enclosed in membranes, most likely derived from the endoplasmic reticulum (**Figure 9.48**). These vesicles then fuse with lysosomes, and the degradative lysosomal enzymes digest their contents. The uptake of most proteins into autophagosomes appears to be nonselective, so it results in the eventual slow degradation of long-lived cytoplasmic proteins. Approximately 1% of cell proteins per hour are degraded by this mechanism. In addition, some organelles, such as damaged mitochondria, can be selectively targeted for autophagic degradation—often as a result of being marked by ubiquitylation.

Autophagy is regulated, both in response to the availability of nutrients and during development of multicellular organisms. Autophagy is generally activated under conditions of nutrient starvation, allowing cells to degrade nonessential proteins and organelles so that their components can be reutilized. In addition, autophagy plays an important role in many developmental processes, such as insect metamorphosis, which involve extensive tissue remodeling and degradation of cellular components. As discussed in Chapter 18, autophagy also plays an important role in programmed cell death, and defects in autophagy have been linked to several human diseases, including neurodegenerative diseases and cancer.

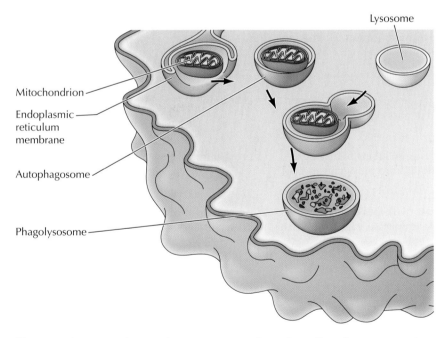

Figure 9.48 Autophagy   Lysosomes contain various digestive enzymes, including proteases. Lysosomes take up cellular proteins by fusion with autophagosomes, which are formed by the enclosure of areas of cytoplasm or organelles (e.g., a mitochondrion) in membranes derived from the endoplasmic reticulum. This fusion yields a phagolysosome, which digests the contents of the autophagosome.

SUMMARY	KEY TERMS

Translation of mRNA

- *Transfer RNAs:* Transfer RNAs serve as adaptors that align amino acids on the mRNA template. Aminoacyl tRNA synthetases attach amino acids to the appropriate tRNAs, which then bind to mRNA codons by complementary base pairing.

tRNA, anticodon, aminoacyl tRNA synthetase

- *The ribosome:* Ribosomes consist of two subunits, which are composed of proteins and ribosomal RNAs. The 23S rRNA is the catalyst of peptide bond formation.

rRNA

- *The organization of mRNAs and the initiation of translation:* Translation of both prokaryotic and eukaryotic mRNAs initiates with a methionine residue. In bacteria, initiation codons are preceded by a sequence that aligns the mRNA on the ribosome by base pairing with 16S rRNA. In eukaryotes, most initiation codons are identified by scanning from the 5′ end of the mRNA, which is recognized by its 7-methylguanosine cap.

untranslated region (UTR), polycistronic, monocistronic, Shine-Dalgarno sequence

- *The process of translation:* Translation is initiated by the binding of methionyl tRNA and mRNA to the small ribosomal subunit. The large ribosomal subunit then joins the complex, and the polypeptide chain elongates until the ribosome reaches a termination codon in the mRNA. A variety of nonribosomal factors are required for initiation, elongation, and termination of translation in both prokaryotic and eukaryotic cells. See Animation 9.1 and Video 9.1.

initiation factor, elongation factor, release factor, polysome

- *Regulation of translation:* Translation of specific mRNAs can be regulated by the binding of repressor proteins and by noncoding microRNAs. Controlled polyadenylation of mRNA is also an important mechanism for the regulation of translation during early development. In addition, the general translational activity of cells can be regulated by modification of initiation factors.

RNA interference (RNAi), short interfering RNA (siRNA), microRNA (miRNA)

Protein Folding and Processing

- *Chaperones and protein folding:* Molecular chaperones facilitate protein folding by binding to and stabilizing unfolded or partially folded polypeptide chains. See Video 9.2.

chaperone

- *Protein misfolding diseases:* A variety of diseases, including neurodegenerative diseases such as Alzheimer's disease, result from the aggregation of misfolded proteins. See Video 9.3.

protein misfolding disease, amyloid, prion

- *Enzymes that catalyze protein folding:* At least two types of enzymes, protein disulfide isomerase and peptidyl prolyl isomerase, catalyze protein folding.

protein disulfide isomerase (PDI), peptidyl prolyl isomerase

- *Protein cleavage:* Proteolysis is an important step in the processing of many proteins. For example, secreted proteins and proteins incorporated into most eukaryotic organelles are targeted to their destinations by amino-terminal sequences that are removed by proteolytic cleavage as the polypeptide chain crosses the membrane.

proteolysis, signal sequence, signal peptidase

- *Glycosylation:* Many eukaryotic proteins, particularly secreted and plasma membrane proteins, are modified by the addition of carbohydrates in the endoplasmic reticulum and Golgi apparatus.

glycosylation, glycoprotein, dolichol phosphate

- *Attachment of lipids:* Covalently attached lipids frequently target and anchor proteins to the plasma membrane.

N-myristoylation, prenylation, palmitoylation, glycolipid, glycosylphosphatidylinositol (GPI) anchor

SUMMARY	KEY TERMS

Regulation of Protein Function

- *Regulation by small molecules:* Many proteins are regulated by the binding of small molecules, such as amino acids and nucleotides, which induce changes in protein conformation and activity.

allosteric regulation

- *Protein phosphorylation and other modifications:* Reversible phosphorylation, which controls the activities of a wide variety of cellular proteins, results from the action of protein kinases and phosphatases. Proteins are also regulated by attachment of other small molecules and polypeptides.

protein kinase, serine/threonine kinase, tyrosine kinase, protein phosphatase, ubiquitin, SUMO

- *Protein–protein interactions:* Interactions between polypeptide chains are important in the regulation of allosteric enzymes and other cellular proteins.

Protein Degradation

- *The ubiquitin-proteasome pathway:* The major pathway of selective protein degradation in eukaryotic cells uses polyubiquitin chains as a marker that targets proteins for rapid proteolysis by the proteasome. See Animation 9.2.

proteasome

- *Lysosomal proteolysis:* Lysosomal proteases degrade extracellular proteins taken up by endocytosis and are responsible for the degradation of cytoplasmic organelles and long-lived cytosolic proteins by autophagy. Autophagy is activated as a response to cell starvation and plays an important role in development and programmed cell death.

lysosome, autophagy

Questions

1. *E. coli* contain 64 different codons in their mRNAs, 61 of which code for amino acids. How can they synthesize proteins when they have only about 40 different tRNAs?

2. You wish to express a cloned eukaryotic cDNA in bacteria. What type of sequence must you add for the mRNA to be translated on prokaryotic ribosomes?

3. Discuss the evidence that ribosomal RNA is the major catalytic component of the ribosome.

4. What effect would an inhibitor of polyadenylation have on protein synthesis in fertilized eggs?

5. What are chaperones? Why is it beneficial for the synthesis of heat-shock proteins to be induced by exposure of cells to elevated temperatures?

6. You are interested in studying a protein expressed on the surface of liver cells. How could treatment of these cells with a phospholipase (an enzyme that cleaves phospholipids) enable you to determine whether your protein is a transmembrane protein or one that is attached to the cell surface by a GPI anchor?

7. What was the first evidence that ubiquitylation and degradation of specific proteins by proteasomes requires a specific target sequence on the protein?

8. Does ubiquitylation of a protein always signal its destruction by the proteasome?

Refer To

The Cell
Companion Website
sites.sinauer.com/cooper7e
for quizzes, animations, videos, flashcards, and other study resources.

9. How do miRNAs regulate the translation of specific mRNAs?

10. What is the function of 3′ untranslated regions in mRNAs?

11. Why is regulated proteolytic cleavage important for the activity of certain proteins?

12. How does the ribosome ensure that the correct aminoacyl tRNA is inserted opposite a codon?

13. You are studying the pathway responsible for the secretion of ribonuclease (RNase) in cultured pancreatic cells, by assaying the activity of secreted RNase in the culture medium. How would the expression of siRNA targeted against protein disulfide isomerase (PDI) affect the amount of active RNase you detect in your experiments?

References and Further Reading   (Key review articles for each major section are highlighted in **bold**.)

Translation of mRNA

Aitken, C. E. and J. R. Lorsch. 2012. A mechanistic overview of translation initiation in eukaryotes. *Nature Struc. Mol. Biol.* 19: 568–576. [R]

Ban, N., P. Nissen, J. Hansen, P. B. Moore and T. A. Steitz. 2000. The complete atomic structure of the large ribosomal subunit at 2.4 Å resolution. *Science* 289: 905–920. [P]

Bartel, D. P. 2009. MicroRNAs: target recognition and regulatory functions. *Cell* 136: 215–233. [R]

Ben-Shem, A., L. Jenner, G. Yusupova and M. Yusupov. 2010. Crystal structure of the eukaryotic ribosome. *Science* 330: 1203–1209. [P]

Beringer, M. and M. V. Rodnina. 2007. The ribosomal peptidyl transferase. *Mol. Cell* 26: 311–321. [R]

Buxbaum, A. R., G. Haimovich and R. H. Singer. 2015. In the right place at the right time: visualizing and understanding mRNA localization. *Nature Rev. Mol. Cell Biol.* 16: 95–109. [R]

Crick, F. H. C. 1966. Codon-anticodon pairing: The wobble hypothesis. *J. Mol. Biol.* 19: 548–555. [P]

Fabian, M. R. and N. Sonenberg. 2012. The mechanics of miRNA-mediated gene silencing: a look under the hood of miRISC. *Nature Struc. Mol. Biol.* 19: 586–593. [R]

Gebauer, F. and M. W. Hentze. 2004. Molecular mechanisms of translational control. *Nature Rev. Mol. Cell Biol.* 5: 827–835. [R]

Holcik, M. and N. Sonenberg. 2005. Translational control in stress and apoptosis. *Nature Rev. Cell Mol. Biol.* 6: 318–327. [R]

Holt, C. E. and S. L. Bullock. 2009. Subcellular mRNA localization in animal cells and why it matters. *Science* 326: 1212–1216. [R]

Ibba, M. and D. Soll. 2004. Aminoacyl-tRNAs: Setting the limits of the genetic code. *Genes Dev.* 18: 731–738. [R]

Jackson, R. J., C. U. T. Hellen and T. V. Pestova. 2010. The mechanism of eukaryotic translation initiation and principles of its regulation. *Nature Rev. Mol. Cell Biol.* 10: 113–127. [R]

Laursen, B. S., H. P. Sorensen, K. K. Mortensen and H. U. Sperling-Petersen. 2005. Initiation of protein synthesis in bacteria. *Microbiol. Mol. Biol. Rev.* 69: 101–123. [R]

Martin, K. C. and A. Ephrussi. 2009. mRNA localization: gene expression in the spatial dimension. *Cell* 136: 719–730. [R]

Nilsen, T. W. 2007. Mechanisms of microRNA-mediated gene regulation in animal cells. *Trends Genet.* 23: 243–249. [R]

Nissen, P., J. Hansen, N. Ban, P. B. Moore and T. A. Steitz. 2000. The structural basis of ribosome activity in peptide bond synthesis. *Science* 289: 920–930. [P]

Noller, H. F., V. Hoffarth and L. Zimniak. 1992. Unusual resistance of peptidyl transferase to protein extraction procedures. *Science* 256: 1416–1419. [P]

Nomura, M. 1997. Reflections on the days of ribosome reconstitution research. *Trends Biochem. Sci.* 22: 275–279. [R]

Richter, J. D. 2007. CPEB: a life in translation. *Trends Biochem. Sci.* 32: 279–285. [R]

Richter, J. D. and N. Sonenberg. 2005. Regulation of cap-dependent translation by eIF4E inhibitory proteins. *Nature* 433: 477–480. [R]

Sonenberg, N., and A. G. Hinnesbusch. 2009. Regulation of translation initiation in eukaryotes: mechanisms and biological targets. *Cell* 136: 731–745. [R]

Spriggs, K. A., M. Bushell and A. E. Willis. 2010. Translational regulation of gene expression during conditions of cell stress. *Mol. Cell* 40: 228–237. [R]

Steitz, T. A. 2008. A structural understanding of the dynamic ribosome machine. *Nature Rev. Mol. Cell Biol.* 9: 242 – 253. [R]

Wu, L. and J. G. Belasco. 2008. Let me count the ways: mechanisms of gene regulation by miRNAs and siRNAs. *Mol. Cell* 29: 1–7. [R]

Protein Folding and Processing

Bulleid, N. J. and L. Ellgaard. 2011. Multiple ways to make disulfides. *Trends Biochem. Sci.* 36: 485–492. [R]

Ellis, R. J. 2006. Molecular chaperones: assisting assembly in addition to folding. *Trends Biochem. Sci.* 31: 395–401. [R]

Farazi, T. A., G. Waksman and J. I. Gordon. 2001. The biology and enzymology of protein N-myristoylation. *Ann. Rev. Biochem.* 276: 39501–39504. [R]

Hebert, D. N., S. C. Garman and M. Molinari. 2005. The glycan code of the endoplasmic reticulum: Asparagine-linked carbohydrates as protein maturation and quality-control tags. *Trends Cell Biol.* 15: 364–370. [R]

Helenius, A. and M. Aebi. 2004. Roles of N-linked glycans in the endoplasmic reticulum. *Ann. Rev. Biochem.* 73: 1019–1049. [R]

Horwich, A. L., W. A. Fenton, E. Chapman and G. W. Farr. 2007. Two families of chaperonin: physiology and mechanism. *Ann. Rev. Cell Dev. Biol.* 23: 115–145. [R]

Iyer, S. P. N. and G. W. Hart. 2003. Dynamic nuclear and cytoplasmic glycosylation: Enzymes of O-GlcNAc cycling. *Biochem.* 42: 2493–2499. [R]

Jonas, S. and E. Izaurralde. 2015. Towards a molecular understanding of microRNA-mediated gene silencing. *Nature Rev. Genet.* 16: 421–433. [R]

Knowles, T. P. J., M. Vendruscolo and C. M. Dobson. 2014. The amyloid state and its association with protein misfolding diseases. *Nature Rev. Mol. Cell Biol.* 15: 384–396. [R]

Mayer, M. P. 2010. Gymnastics of molecular chaperones. *Mol. Cell* 39: 321–331. [R]

Ohtsubo, K. and J. D. Marth. 2006. Glycosylation in cellular mechanisms of health and disease. *Cell* 126: 855–867. [R]

Paetzel, M., A. Karla, N. C. Strynadka and R. E. Dalbey. 2002. Signal peptidases. *Chem. Rev.* 102: 4549–4580. [R]

Prusiner, S. B. 1998. Prions. *Proc. Natl. Acad. Sci. USA* 95: 13363–13383. [R]

Prusiner, S. B. 2013. Biology and genetics of prions causing neurodegeneration. *Ann. Rev. Genet.* 47: 601–623. [R]

Selkoe, D. J. 2011. Alzheimer's disease. *Cold Spring Harbor Perspect. Biol.* 3:a004457. [R]

Soto, C. 2012. Transmissible proteins: expanding the prion heresy. *Cell* 149: 968–977. [R]

Taipale, M., D. F. Jarosz and S. Lindquist. 2010. HSP90 at the hub of protein homeostasis: emerging mechanistic insights. *Nature Rev. Mol. Cell Biol.* 11: 515–528. [R]

Udenfriend, S. and K. Kodukula. 1995. How glycosylphosphatidylinositol-anchored membrane proteins are made. *Ann. Rev. Biochem.* 64: 563–591. [R]

Wilkinson, B. and H. F. Gilbert. 2004. Protein disulfide isomerase. *Biochim. Biophys. Acta* 1699: 35–44. [R]

Wright, L. P. and M. R. Philips. 2006. CAAX modification and membrane targeting of Ras. *J. Lipid Res.* 47: 883–891. [R]

Yan, A. and W. J. Lennarz. 2005. Unraveling the mechanism of protein N-glycosylation. *J. Biol. Chem.* 280: 3121–3124. [R]

Yebenes, H., P. Mesa, I. G. Munoz, G. Montoya and J. M. Valpuesta. 2011. Chaperonins: two rings for folding. *Trends Biochem. Sci.* 36: 424–432. [R]

Young, J. C., V. R. Agashe, K. Siegers and F. U. Hartl. 2004. Pathways of chaperone-mediated protein folding in the cytosol. *Nature Rev. Mol. Cell Biol.* 5: 781–791. [R]

Regulation of Protein Function

Alonso, A., J. Sasin, N. Bottini, I. Friedberg, I. Friedberg, A. Osterman, A. Godzik, T. Hunter, J. Dixon and T. Mustelin. 2004. Protein tyrosine phosphatases in the human genome. *Cell* 117: 699–711. [R]

Barford, D. 1996. Molecular mechanisms of the protein serine/threonine phosphatases. *Trends Biochem. Sci.* 21: 407–412. [R]

Chen, Z. J. and L. J. Sun. 2009. Nonproteolytic functions of ubiquitin in cell signaling. *Mol. Cell* 33: 275–286. [R]

Choudhary, C., B. T. Weinert, Y. Nishida, E. Verdin and M. Mann. 2014. The growing landscape of lysine acetylation links metabolism and cell signalling. *Nature Rev. Mol. Cell Biol.* 15: 536–550. [R]

Fauman, E. B. and M. A. Saper. 1996. Structure and function of the protein tyrosine phosphatases. *Trends Biochem. Sci.* 21: 413–417. [R]

Fischer, E. H. and E. G. Krebs. 1989. Commentary on "The phosphorylase β to α converting enzyme of rabbit skeletal muscle." *Biochim. Biophys. Acta* 1000: 297–301. [R]

Gareau, J. R. and C. D. Lima. 2010. The SUMO pathway: emerging mechanisms that shape specificity, conjugation and recognition. *Nature Rev. Mol. Cell Biol.* 11: 861–871. [R]

Hanks, S. K., A. M. Quinn and T. Hunter. 1988. The protein kinase family: Conserved features and deduced phylogeny of the catalytic domains. *Science* 241: 42–52. [R]

Hess, D. T., A. Matsumoto, S.-O. Kim, H. E. Marshall and J. S. Stamler. 2005. Protein S-nitrosylation: Purview and parameters. *Nature Rev. Mol. Cell Biol.* 6: 150–166. [R]

Hochstrasser, M. 2009. Origin and function of ubiquitin-like proteins. *Nature* 458: 422–429. [R]

Hunter, T. 1995. Protein kinases and phosphatases: The yin and yang of protein phosphorylation and signaling. *Cell* 80: 225–236. [R]

Huse, M. and J. Kuriyan. 2002. The conformational plasticity of protein kinases. *Cell* 109: 275–282. [R]

Manning, G., G. D. Plowman, T. Hunter and S. Sudarsanam. 2002. Evolution of protein kinase signaling from yeast to man. *Trends Biochem. Sci.* 27: 514–520. [R]

Marianayagam, N. J., M. Sunde and J. M. Matthews. 2004. The power of two: Protein dimerization in biology. *Trends Biochem. Sci.* 29: 618–625. [R]

Milburn, M. V., L. Tong, A. M. DeVos, A. Brunger, Z. Yamaizumi, S. Nishimura and S.-H. Kim. 1990. Molecular switch for signal transduction: Structural differences between active and inactive forms of protooncogenic ras proteins. *Science* 247: 939–945. [P]

Monod, J., J.-P. Changeux and F. Jacob. 1963. Allosteric proteins and cellular control systems. *J. Mol. Biol.* 6: 306–329. [P]

Ulrich, H. D. and H. Walden. 2010. Ubiquitin signaling in DNA replication and repair. *Nature Rev. Mol. Cell Biol.* 11: 479–489. [R]

Vetter, I. R. and A. Wittinghofer. 2001. The guanine nucleotide-binding switch in three dimensions. *Science* 294: 1299–1304. [R]

Protein Degradation

Bedford, L., S. Paine, P. W. Sheppard, R. J. Mayer and J. Roelofs. 2010. Assembly, structure, and function of the 26S proteasome. *Trends Cell Biol.* 20: 391–401. [R]

Boya, P., F. Reggiori and P. Codogno. 2013. Emerging regulation and functions of autophagy. *Nature Cell Biol.* 15: 713–720. [R]

Glotzer, M., A. W. Murray and M. W. Kirschner. 1991. Cyclin is degraded by the ubiquitin pathway. *Nature* 349: 132–138. [P]

Kleiger, G. and T. Mayor. 2014. Perilous journey: a tour of the ubiquitin-proteasome system. *Trends Cell Biol.* 24: 352–359. [R]

Mizushima, N. and B. Levine. 2010. Autophagy in mammalian development and differentiation. *Nature Cell Biol.* 12: 823–830. [R]

Mizushima, N. and M. Komatsu. 2011. Autophagy: renovation of cells and tissues. *Cell* 147: 728–741. [R]

Pickart, C. M. and R. E. Cohen. 2004. Proteasomes and their kin: Proteases in the machine age. *Nature Rev. Mol. Cell Biol.* 5: 177–187. [R]

Reed, S. I. 2003. Ratchets and clocks: The cell cycle, ubiquitylation and protein turnover. *Nature Rev. Mol. Cell Biol.* 4: 855–864. [R]

PART III

Cell Structure and Function

The Nucleus

The presence of a nucleus is the principal feature that distinguishes eukaryotic from prokaryotic cells. By housing the cell's genome, the nucleus serves both as the repository of genetic information and as the cell's control center. DNA replication, transcription, and RNA processing all take place within the nucleus, with only the final stage of gene expression (translation) localized to the cytoplasm.

By separating the genome from the cytoplasm, the nuclear envelope allows gene expression to be regulated by mechanisms that are unique to eukaryotes. Whereas prokaryotic mRNAs are translated while their transcription is still in process, eukaryotic mRNAs undergo several forms of posttranscriptional processing before being transported from the nucleus to the cytoplasm. The presence of a nucleus thus allows gene expression to be regulated by posttranscriptional mechanisms, such as alternative splicing. By limiting the access of selected proteins to the genetic material, the nuclear envelope also provides novel opportunities for the control of gene expression at the level of transcription. For example, the expression of some eukaryotic genes is controlled by the regulated transport of transcription factors from the cytoplasm to the nucleus—a form of transcriptional regulation unavailable to prokaryotes. Separation of the genome from the site of mRNA translation thus plays a central role in eukaryotic gene expression.

The Nuclear Envelope and Traffic between the Nucleus and the Cytoplasm

The nuclear envelope separates the contents of the nucleus from the cytoplasm and provides the structural framework of the nucleus. The two envelope membranes, acting as barriers that prevent the free passage of molecules between the nucleus and the cytoplasm, maintain the nucleus as a distinct biochemical compartment. The only channels through the nuclear envelope are provided by the nuclear pore complexes, which allow the regulated exchange of molecules between the nucleus and the cytoplasm. The selective traffic of proteins and RNAs through the nuclear pore complexes not only establishes the internal composition of the nucleus, but also plays a critical role in regulating eukaryotic gene expression.

Structure of the nuclear envelope

The **nuclear envelope** has a complex structure consisting of two nuclear membranes, an underlying nuclear lamina, and nuclear pore complexes (**Figure 10.1**). The nucleus is surrounded by a system of two concentric membranes, called the inner and outer **nuclear membranes**. The outer nuclear membrane

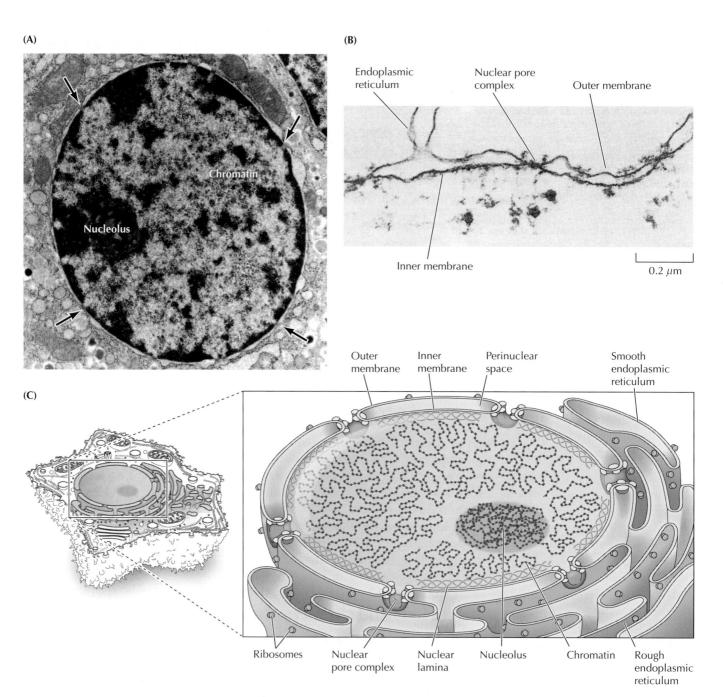

Figure 10.1 The nuclear envelope (A) An electron micrograph of a nucleus. The inner and outer nuclear membranes are joined at nuclear pore complexes (arrows). (B) An electron micrograph illustrating the continuity of the outer nuclear membrane with the endoplasmic reticulum. (C) Schematic of the nuclear envelope. The inner nuclear membrane is lined by the nuclear lamina, which serves as an attachment site for chromatin. (B, courtesy of Dr. Werner W. Franke, German Cancer Research Center, Heidelberg.)

is continuous with the endoplasmic reticulum, so the space between the inner and outer nuclear membranes is directly connected with the lumen of the endoplasmic reticulum. The outer nuclear membrane is functionally similar to the membranes of the endoplasmic reticulum (see Chapter 11) and has ribosomes bound to its cytoplasmic surface, but it differs slightly in

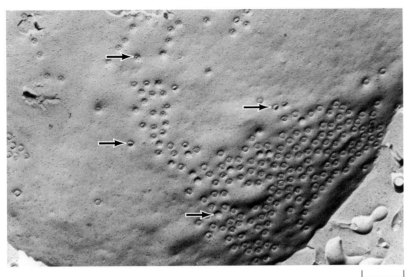

Figure 10.2 Electron micrograph showing nuclear pores Many nuclear pores (arrows) are visible in this freeze-fracture preparation of the nuclear envelope.

0.5 µm

protein composition, being enriched in membrane proteins that bind the cytoskeleton and lacking the proteins that give the endoplasmic reticulum its tubular organization. In contrast, the inner nuclear membrane contains about 60 specific integral membrane proteins, such as those that bind the nuclear lamina (discussed below).

The critical function of the nuclear membranes is to act as a barrier that separates the contents of the nucleus from the cytoplasm. Like other cell membranes, each nuclear membrane is a phospholipid bilayer permeable only to small nonpolar molecules (see Figure 2.36). Other molecules are unable to diffuse through the bilayer. The inner and outer nuclear membranes are joined at nuclear pore complexes—the sole channels through which small polar molecules and macromolecules pass through the nuclear envelope (**Figure 10.2**). As discussed in the next section, the nuclear pore complex is a complicated structure that is responsible for the selective traffic of proteins and RNAs between the nucleus and the cytoplasm.

Underlying the inner nuclear membrane is the **nuclear lamina**, a fibrous meshwork that provides structural support to the nucleus. The nuclear lamina of *Xenopus* oocytes appears as a distinct meshwork of perpendicular fibers (**Figure 10.3**), although it is unclear whether the lamina of somatic cells has a similar structure. The nuclear lamina is composed of 60- to 80-kilodalton (kd) fibrous proteins called **lamins**, along with associated proteins. Plant cells have a similar fibrous network comprised of unrelated proteins. Lamins are a class of intermediate filament proteins; the

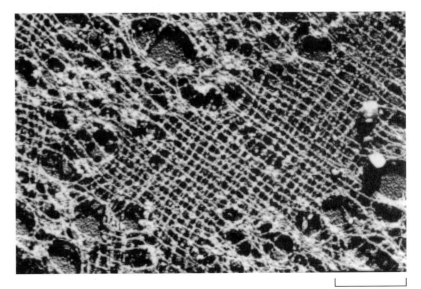

Figure 10.3 Electron micrograph of the nuclear lamina The lamina of *Xenopus* oocytes is a meshwork of filaments underlying the inner nuclear membrane. (From U. Aebi et al., 1986. *Nature* 323: 560.)

0.5 µm

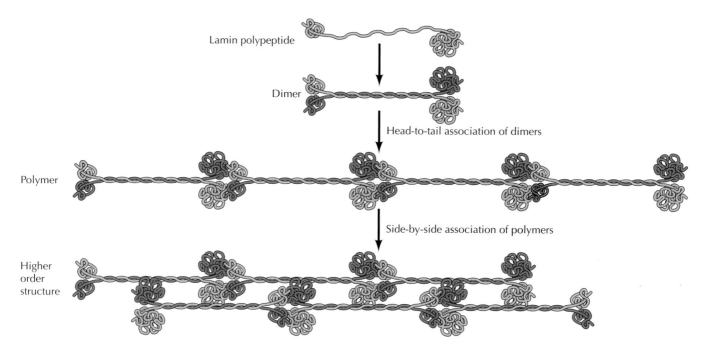

Lamin polypeptide

Dimer

Head-to-tail association of dimers

Polymer

Side-by-side association of polymers

Higher
order
structure

Figure 10.4 Lamin assembly   The lamin polypeptides form dimers in which the central α-helical regions of two polypeptide chains are wound around each other. Further assembly involves the head-to-tail association of dimers to form linear polymers and the side-by-side association of polymers to form higher order structures.

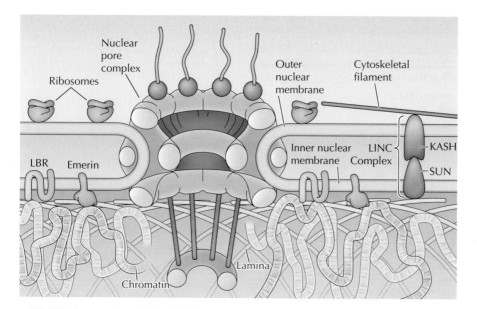

Figure 10.5 The nuclear lamina   The inner nuclear membrane contains several integral membrane proteins, such as emerin, the lamin B receptor (LBR), and the SUN proteins, that interact with nuclear lamins. The SUN proteins bind the KASH proteins in the outer nuclear membrane, making up the LINC complex that connects the nuclear lamina to the cytoskeleton. The lamins and inner nuclear membrane proteins are also associated with chromatin through interactions of emerin and the lamin B receptor with chromatin-associated proteins.

other classes are found in the cytoskeleton (see Chapter 13). Like other intermediate filament proteins, the lamins associate with each other to form higher order structures (Figure 10.4). The first stage of this association is the interaction of two lamins to form a dimer in which the α-helical regions of two polypeptide chains are wound around each other in a structure called a coiled coil. These lamin dimers then associate with each other to form the nuclear lamina.

The association of lamins with the inner nuclear membrane is facilitated by the posttranslational addition of lipid—in particular, prenylation of C-terminal cysteine residues (see Figure 9.36). In addition, the lamins bind to specific inner nuclear membrane proteins, such as emerin and the lamin B receptor, and are directly connected to the cytoskeleton by protein complexes (called LINC complexes) that span the inner and outer nuclear membranes (Figure 10.5). The nuclear

Nuclear Lamina Diseases

The Diseases

In 1966 Alan Emery and Fritz E. Dreifuss described a new muscular dystrophy linked to the X-chromosome. Early in the disease the elbows, neck, and heels of affected individuals become stiff, and often there is a conduction block in the heart. These symptoms are seen by age ten and include "toewalking" because of stiff Achilles' tendons in the heels and difficulty bending the elbows. Heart problems develop by age 20 and may require a pacemaker. There is a gradual wasting and weakness of the shoulder and upper arm muscles and the calf muscles of the legs, but this occurs slowly and is often not a problem until late in life.

Nearly 30 years after their discovery, researchers showed that mutations in a novel transmembrane protein were responsible for this X-linked Emery–Dreifuss muscular dystrophy. They named the protein emerin after Alan Emery. Soon several groups found that emerin was a protein localized to the inner nuclear membrane and absent in patients with the X-linked Emery–Dreifuss muscular dystrophy. This was unexpected; mutations in a nuclear envelope protein expressed in all cells apparently caused a tissue-specific disease. While all cells in the body were missing the protein, the pathology occurred only in muscle. Subsequent investigators found that the same dystrophy could also be inherited in a non-sex-linked manner. Families with this non-sex-linked Emery–Dreifuss muscular dystrophy had mutations in the single gene (LMNA) encoding nuclear lamins A and C. So mutations in one of two genes—one coding for an inner nuclear membrane protein and one coding for a major nuclear lamin—caused clinically identical muscular dystrophy.

More surprising was that parallel investigations on different diseases—Dunnigan-type partial lipodystrophy, Charcot–Marie–Tooth disorder type 2B1, and a disease that causes premature aging (Hutchinson-Gilford progeria syndrome)—traced them to different mutations in the LMNA gene. Previously, physicians classified these as distinct diseases based on their clinical features and modes of inheritance. Recent work shows that mutations in another protein in the inner nuclear membrane, the lamin B receptor, are the basis for Pelger–Huët anomaly.

Molecular and Cellular Basis

Most biologists thought that mutations in lamins would cause generalized defects in nuclear architecture and serious problems in rapidly dividing cells. However, only minor aberrations of nuclear structure occur in the patients. Thus the puzzle is how mutations in nuclear lamins or lamin-binding proteins cause different tissue-specific diseases. The answer is not yet known but there are two major hypotheses. The first is the "gene expression" hypothesis. This posits that the correct interaction of the two lamin proteins, A and C, with the nuclear envelope is essential for normal tissue-specific expression of certain genes. Transcriptionally inactive genes are located preferentially at the nuclear periphery, whereas expressed genes are concentrated in the center of the nucleus with cell-type specificity. Thus the basis of these diseases would be a change in gene expression caused by defective protein interactions.

In the "mechanical stress" hypothesis, the mutations in the nuclear lamin-emerin complex are thought to weaken the structural integrity of an integrated cytoskeletal network. In all cells the lamina, inner nuclear membrane, and nuclear pore complexes are tightly connected. This hypothesis, which works best for the muscular dystrophies, suggests that through filaments attached to the nuclear pore complex, the lamina could be connected indirectly with the muscle cell cytoskeleton.

Prevention and Treatment

The discovery that mutations in commonly expressed proteins of the nuclear lamina complex cause different inherited tissue-specific diseases came as a surprise and changed the way scientists look at nuclear lamins. Further research is needed to learn whether the bases of the pathologies in each of these diseases is mechanical stability of the nuclear envelope or misregulation of gene expression. However, the known molecular nature of the diseases greatly simplifies their diagnosis and makes eventual treatment more likely. The development of a mouse model where the gene is knocked out represents a first step. As the embryos develop they show symptoms of Emery–Dreifuss muscular dystrophy. Finally, researchers are now aware that several slow-developing congenital diseases may be new members of the nuclear "laminopathies."

Reference

Davidson, P. M. and J. Lammerding. 2014. Broken nuclei—lamins, nuclear mechanics and disease. *Trends Cell Biol.* 24: 247–256.

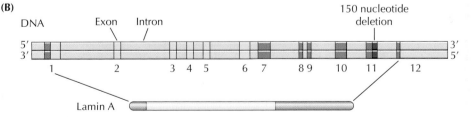

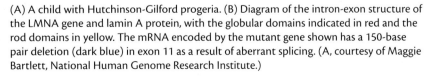

(A) A child with Hutchinson-Gilford progeria. (B) Diagram of the intron-exon structure of the LMNA gene and lamin A protein, with the globular domains indicated in red and the rod domains in yellow. The mRNA encoded by the mutant gene shown has a 150-base pair deletion (dark blue) in exon 11 as a result of aberrant splicing. (A, courtesy of Maggie Bartlett, National Human Genome Research Institute.)

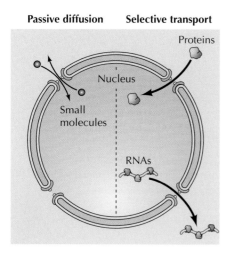

Passive diffusion   Selective transport

Proteins

Nucleus

Small
molecules

RNAs

Figure 10.6 Molecular traffic through nuclear pore complexes Small molecules are able to pass freely through the nuclear pore complex by passive diffusion. In contrast, macromolecules (proteins and RNAs) are recognized by specific signals and selectively transported either from nucleus to cytoplasm or from cytoplasm to nucleus.

lamina is also associated with chromatin through interactions of chromatin-associated proteins with emerin and the lamin B receptor. As discussed later in this chapter, the interaction of chromatin with the nuclear envelope plays an important role in localizing non-transcribed heterochromatin to the periphery of the nucleus.

The nuclear pore complex

The **nuclear pore complexes** are the only channels through which small polar molecules, ions, and macromolecules (proteins and RNAs) can travel between the nucleus and the cytoplasm. The nuclear pore complex is an extremely large structure with a diameter of about 120 nm and an estimated molecular mass of approximately 125 million daltons—about 30 times the size of a ribosome. In vertebrates, the nuclear pore complex is composed of multiple copies of about 30 different pore proteins (called nucleoporins). By controlling the traffic of molecules between the nucleus and the cytoplasm, the nuclear pore complex plays a fundamental role in the physiology of all eukaryotic cells. RNAs synthesized in the nucleus must be efficiently exported to the cytoplasm where they function in protein synthesis. Conversely, proteins required for nuclear functions (e.g., transcription factors) must be transported to the nucleus from their sites of synthesis in the cytoplasm. In addition, many proteins shuttle continuously between the nucleus and the cytoplasm.

Depending on their size and structure, molecules can travel through the nuclear pore complex by one of two mechanisms (**Figure 10.6**). Small molecules and some proteins with molecular mass less than approximately 40 kd diffuse freely through the pore in either direction: cytoplasm to nucleus or nucleus to cytoplasm. Most proteins and RNAs, however, pass through the nuclear pore complex by a selective transport process. These proteins and RNAs are recognized by specific signals that direct their transport either from nucleus to cytoplasm or from cytoplasm to nucleus.

Visualization of nuclear pore complexes by electron microscopy reveals a structure with eightfold symmetry organized around a large central channel (**Figure 10.7**), which is the route through which proteins and RNAs cross the nuclear envelope. Detailed structural studies, including computer-based image

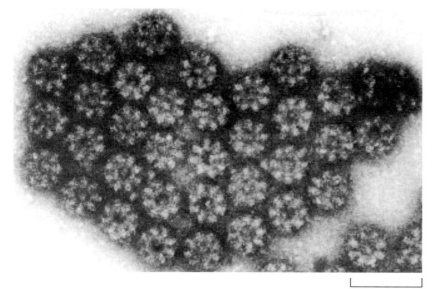

0.2 μm

Figure 10.7 Electron micrograph of nuclear pore complexes In this face-on view, isolated nuclear pore complexes appear to consist of eight structural subunits surrounding a central channel. (Courtesy of Dr. Ron Milligan, The Scripps Research Institute.)

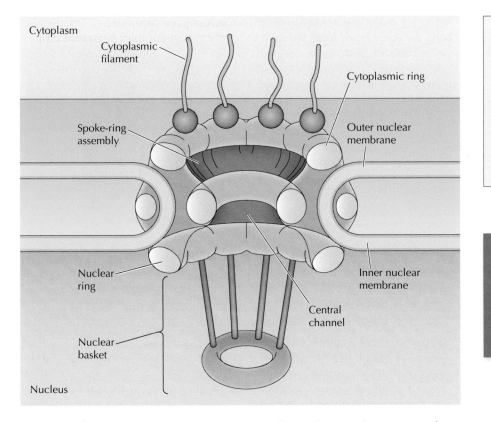

Figure 10.8 **Model of the nuclear pore complex**   The complex consists of an assembly of eight spokes attached to rings on the cytoplasmic and nuclear sides of the nuclear envelope. The spoke-ring assembly surrounds a central channel. Cytoplasmic filaments extend from the cytoplasmic ring, and filaments forming the nuclear basket extend from the nuclear ring.

analysis, have led to the development of three-dimensional models of the nuclear pore complex (**Figure 10.8**). These studies show that the nuclear pore complex consists of an assembly of eight spokes arranged around a central channel. The spokes are connected to rings at the nuclear and cytoplasmic surfaces, and the spoke-ring assembly is anchored within the nuclear envelope at sites of fusion between the inner and outer nuclear membranes. Protein filaments extend from both the cytoplasmic and nuclear rings, forming a distinct basketlike structure on the nuclear side.

Selective transport of proteins to and from the nucleus

Several million macromolecules selectively pass between the nucleus and the cytoplasm every minute. The basis for selective traffic across the nuclear envelope was first understood for proteins imported from the cytoplasm to the nucleus. Such proteins are responsible for all aspects of genome structure and function; they include histones, DNA polymerases, RNA polymerases, transcription factors, splicing factors, and many others. These proteins are targeted to the nucleus by specific amino acid sequences called **nuclear localization signals**, which are recognized by **nuclear transport receptors** that direct protein transport through the nuclear pore complex.

The first nuclear localization signal to be mapped in detail was characterized by Alan Smith and colleagues in 1984. These investigators studied simian virus 40 (SV40) T antigen, a virus-encoded protein that initiates viral

Key Experiment

Identification of Nuclear Localization Signals

A Short Amino Acid Sequence Able to Specify Nuclear Location

Daniel Kalderon, Bruce L. Roberts, William D. Richardson, and Alan E. Smith
National Institute for Medical Research, Mill Hill, London
Cell, Volume 39, 1984, pages 499–509

The Context

Maintaining the nucleus as a distinct biochemical compartment requires a mechanism by which proteins are segregated between the nucleus and the cytoplasm. Studies in the 1970s established that small molecules diffuse rapidly across the nuclear envelope but that most proteins are unable to do so. It therefore appeared likely that nuclear proteins are specifically recognized and selectively imported to the nucleus from their sites of synthesis on cytoplasmic ribosomes.

Earlier experiments of Günter Blobel and his colleagues had established that proteins are targeted to the endoplasmic reticulum by signal sequences consisting of short stretches of amino acids (see Chapter 11). In this 1984 paper, Alan Smith and his colleagues extended this principle to the targeting of nuclear proteins by identifying a short amino acid sequence that serves as a nuclear localization signal.

The Experiments

The viral protein SV40 T antigen was used as a model for studies of nuclear localization in animal cells. T antigen is a 94-kd protein that is required for SV40 DNA replication and is normally localized to the nucleus of SV40-infected cells. Previous studies in both Alan Smith's laboratory and in the laboratory of Janet Butel (Lanford and Butel, 1984, *Cell* 37: 801–813) had shown that mutation of Lys-128 to either Thr or Asn prevented the normal nuclear accumulation of T antigen in both rodent and monkey cells. Rather than being transported to the nucleus, these mutant T antigens remained in the cytoplasm, suggesting

that Lys-128 was part of a nuclear localization signal. Smith and colleagues tested this hypothesis using two distinct experimental approaches.

First, they determined the effects of different deletions on the subcellular localization of T antigen. Mutant T antigens bearing deletions that eliminated amino acids either between residues 1 and 126 or between residue 136 and the C terminus were found to accumulate normally in the nucleus. In contrast, a mutant with a deletion of amino acids 127 to 132 remained in the cytoplasm. Thus the amino acid sequence extending from residue 127 to 132 appeared to be responsible for nuclear localization of T antigen.

To determine whether this amino acid sequence was able to target other proteins to the nucleus the investigators constructed chimeras in which the T antigen amino acid sequence was fused to proteins that were normally cytoplasmic. These experiments established that the addition of T antigen amino acids 126 to 132 to either β-galactosidase or pyruvate kinase is sufficient to specify the nuclear accumulation of these otherwise cytoplasmic proteins (see figure). This short amino acid sequence of SV40 T antigen thus functions as a nuclear localization signal, which is both necessary and sufficient to target proteins for nuclear import.

Alan Smith

The Impact

As Smith and colleagues suggested in their 1984 paper, the nuclear localization signal of SV40 T antigen has proved to "represent a prototype of similar sequences in other nuclear proteins." By targeting proteins for nuclear import, these signals are key to establishing the biochemical identity of the nucleus and maintaining the fundamental division of eukaryotic cells into nuclear and cytoplasmic compartments. Nuclear localization signals are now known to be recognized by cytoplasmic receptors that transport their substrate proteins through the nuclear pore complex. The identification of nuclear localization signals was thus a key advance in understanding nuclear protein import.

(A) 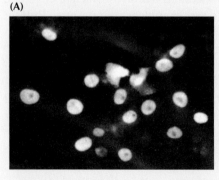   **(B)**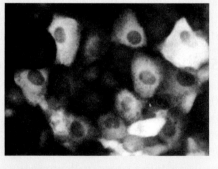

Cells were microinjected with plasmid DNAs encoding chimeric proteins in which SV40 amino acids were fused to pyruvate kinase. Cellular localization of the fusion proteins was then determined by immunofluorescence microscopy. (A) The fusion protein contains an intact SV40 nuclear localization signal (amino acids 126 to 132). (B) The nuclear localization signal has been inactivated by deletion of amino acids 131 and 132.

- example
- Definition

DNA replication in infected cells (see Chapter 7). As expected for a replication protein, T antigen is normally localized to the nucleus. The signal responsible for its nuclear localization was first identified by the finding that mutation of a single lysine residue prevents nuclear import, resulting instead in the accumulation of T antigen in the cytoplasm. Subsequent studies defined the T antigen nuclear localization signal as the seven-amino-acid sequence Pro-Lys-Lys-Lys-Arg-Lys-Val. Not only was this sequence necessary for the nuclear transport of T antigen but its addition to other, normally cytoplasmic, proteins was sufficient to direct their accumulation in the nucleus.

Nuclear localization signals have since been identified in many other proteins. Many of these sequences, like that of T antigen, are short stretches rich in basic amino acid residues (lysine and arginine). Often, however, the amino acids that form the nuclear localization signal are close together but not immediately adjacent to each other. For example, the nuclear localization signal of nucleoplasmin (a protein involved in chromatin assembly) consists of two parts: a Lys-Arg pair followed by four lysines located ten amino acids farther downstream (**Figure 10.9**). Both the Lys-Arg and Lys-Lys-Lys-Lys sequences are required for nuclear targeting, but the ten amino acids between these sequences can be mutated without affecting nuclear localization. Because this nuclear localization sequence is composed of two separated elements, it is called **bipartite**. Many nuclear localization signals, like those of T antigen and nucleoplasmin, consist of basic amino acid residues—often termed the "classical" nuclear localization signal. In addition, some proteins contain other nuclear localization signals, which are recognized by distinct transport receptors.

Proteins with a classical nuclear localization signal are recognized by nuclear transport receptors called **importins**, because they carry proteins through the nuclear pore complex into the nucleus. The importins work in conjunction with a GTP-binding protein called **Ran**, which controls the directionality of movement through the nuclear pore. Ran is one of several types of small GTP-binding proteins whose conformation and activity are regulated by GTP binding and hydrolysis. Other examples of GTP-binding

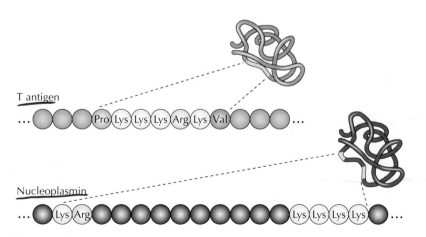

Figure 10.9 Nuclear localization signals The T antigen nuclear localization signal is a single stretch of amino acids. In contrast, the nuclear localization signal of nucleoplasmin is bipartite, consisting of a Lys-Arg sequence, followed by a Lys-Lys-Lys-Lys sequence located ten amino acids farther downstream.

proteins were discussed in Chapter 9, including several of the translation factors involved in protein synthesis (see Figures 9.14 and 9.21) and Ras (see Figure 9.40). The conformations and activities of these proteins are regulated by whether they are bound to GTP or GDP. In the case of Ran, enzymes that stimulate the hydrolysis of GTP to GDP are associated with the cytoplasmic filaments of the nuclear pore complex, whereas enzymes that stimulate the exchange of GDP for GTP are associated with chromatin within the nucleus. Consequently, there is a high concentration of Ran/GTP in the nucleus, which determines the directionality of nuclear transport.

The classical cycle of nuclear import is summarized in **Figure 10.10**. Protein import begins when an importin binds to the nuclear localization signal of

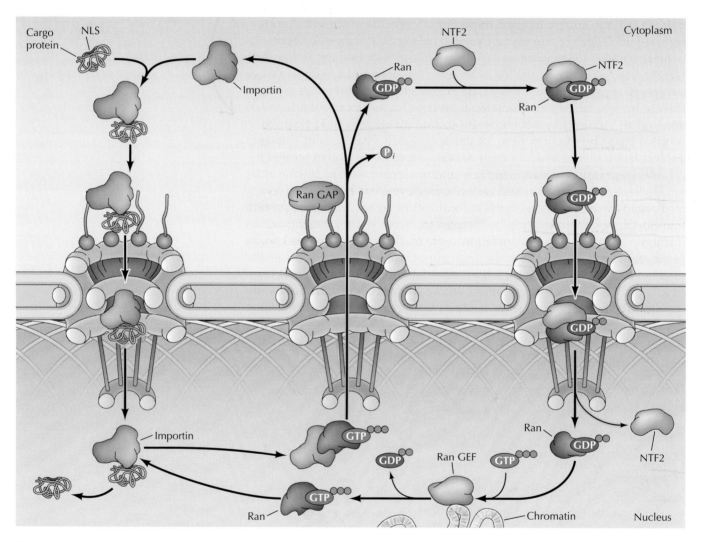

Figure 10.10 Protein import through the nuclear pore complex   Transport begins when the nuclear localization sequence (NLS) of a cargo protein is recognized by an importin. The cargo/importin complex binds to nuclear pore proteins in the cytoplasmic filaments and is transported through the pore. At the nuclear side of the envelope, Ran/GTP binds to the importin, disrupting the cargo/importin complex and releasing the cargo protein into the nucleus. The importin-Ran/GTP complex is re-exported through the nuclear pore and the GTPase-activating protein (Ran GAP) associated with cytoplasmic filaments hydrolyzes the GTP on Ran to GDP, releasing the importin. Ran/GDP is then transported back to the nucleus in association with its own import receptor, NTF2. In the nucleus, Ran GEF (bound to chromatin) stimulates the exchange of GDP bound to Ran for GTP, leading to the conversion of Ran/GDP to Ran/GTP and maintaining a high concentration of Ran/GTP within the nucleus.

a cargo protein in the cytoplasm. This cargo/importin complex then binds to proteins in the cytoplasmic filaments of the nuclear pore complex, and transport proceeds via the interaction of importins with proteins that line the central channel of the pore. Once the cargo/importin complex reaches the nuclear side of the envelope, Ran/GTP binds to importin. This causes a change in the conformation of the importin, which disrupts the cargo/importin complex, displacing the cargo protein and releasing it into the nucleus.

The cycle is maintained by the export of the importin-Ran/GTP complex back through the nuclear pore complex. In the cytoplasm the GTP is hydrolyzed to GDP. This releases the importin so that it can bind to a new cargo protein in the cytoplasm and participate in another round of transport. The Ran/GDP formed in the cytoplasm is then transported back to the nucleus by its own import receptor (a protein called NTF2), where Ran/GTP is regenerated by the action of Ran guanine-nucleotide exchange factor.

Some proteins remain within the nucleus following their import from the cytoplasm, but many others shuttle back and forth between the nucleus and the cytoplasm. Some of these proteins act as carriers in the transport of other molecules, such as RNAs; others coordinate nuclear and cytoplasmic functions (e.g., by regulating the activities of transcription factors). Proteins are targeted for export from the nucleus by specific amino acid sequences, called **nuclear export signals**, which are often rich in the hydrophobic amino acid leucine. Like nuclear localization signals, nuclear export signals are recognized by receptors within the nucleus—**exportins**, which direct protein transport through the nuclear pore complex to the cytoplasm. Like importins, many exportins are members of a family of nuclear transport receptors known as **karyopherins** (Table 10.1).

Exportins bind to Ran, which is required for nuclear export as well as for nuclear import (**Figure 10.11**). However, Ran/GTP promotes the formation of stable complexes between exportins and their cargo proteins, whereas it dissociates the complexes between importins and their cargos. This effect of Ran/GTP binding on exportins dictates the movement of proteins containing nuclear export signals (NES) from the nucleus to the cytoplasm. Thus exportins form stable complexes with their cargo proteins in association with Ran/GTP within the nucleus. Following transport to

Table 10.1   Examples of Karyopherins

Karyopherin	Substrates
Import	
Importin (Kapα/Kapβ1 dimer)	Proteins with a basic amino acid nuclear localization signal (e.g., nucleoplasmin)
Snurportin/Kapβ1	snRNPs (U1, U2, U4, U5)
Transportin/Kapβ2	mRNA binding proteins, ribosomal proteins
Importin7/Kapβ1 dimer	Histone H1, ribosomal proteins
Export	
Crm1	Proteins with a leucine-rich nuclear export signal, snurportin, snRNAs, ribosomal subunits
CAS	Kapα
Exportin-t	tRNAs
Exportin5	miRNAs

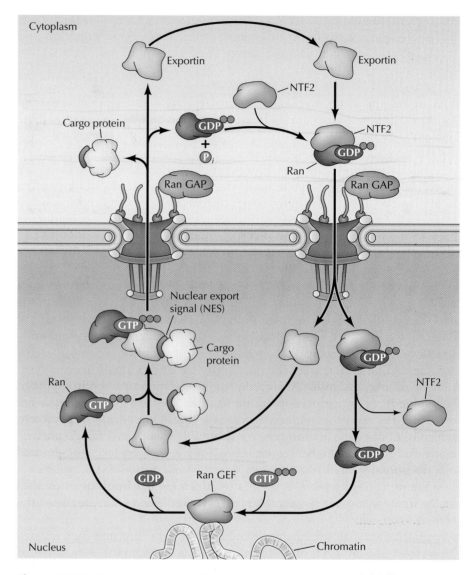

Figure 10.11 Nuclear export Complexes between cargo proteins bearing nuclear export signals (NES), exportins, and Ran/GTP form in the nucleus. Following transport through the nuclear pore complex, Ran GAP stimulates the hydrolysis of bound GTP, leading to formation of Ran/GDP and release of the cargo protein and exportin into the cytoplasm. Ran/GDP and the exportin are then returned to the nucleus. Ran/GTP is regenerated by Ran GEF bound to chromatin.

the cytosolic side of the nuclear envelope, GTP hydrolysis and release of Ran/GDP leads to dissociation of the cargo protein, which is released into the cytoplasm. Exportins, as well as Ran/GDP, are then recycled through the nuclear pore complex for reuse.

Transport of RNAs

Whereas many proteins are selectively transported from the cytoplasm to the nucleus, the RNAs involved in protein synthesis are exported from the nucleus to the cytoplasm. RNAs are transported through the nuclear pore

(A) **(B)** **(C)** **(D)**

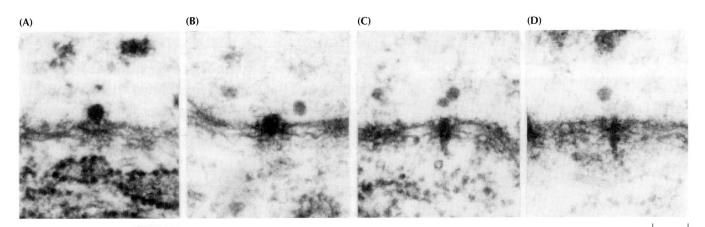

0.1 µm

Figure 10.12 Transport of a ribonucleoprotein complex   Insect salivary gland cells produce large ribonucleoprotein complexes (RNPs), which contain 35–40 kb of RNA and have a total mass of approximately 30 million daltons. This series of electron micrographs shows (A) the attachment of such an RNP to a nuclear pore complex and (B–D) the unfolding of the RNA during its transloca-tion to the cytoplasm. (From H. Mehlin et al., 1992. *Cell* 69: 605.)

complex as ribonucleoprotein complexes (RNPs) (Figure 10.12). The export of tRNAs, rRNAs, and miRNAs is mediated by specific karyopherin exportins (see Table 10.1), as described in Figure 10.11. tRNAs and miRNA precursors are exported from the nucleus by exportin-t and exportin5, respectively, which bind directly to the RNAs (see Table 10.1). Ribosomal RNAs are first associated with ribosomal proteins in the nucleolus, and nascent 40S and 60S ribosomal subunits are then separately transported to the cytoplasm (see Figure 10.28) by the exportin Crm1. Their export from the nucleus is mediated by nuclear export signals present on proteins within the ribosomal subunit complex.

mRNAs are exported by a distinct mechanism that does not involve karyopherins and is independent of Ran (Figure 10.13). Pre-mRNAs are as-sociated with a set of at least 20 proteins throughout their processing in the nucleus and eventual transport to the cytoplasm. These proteins include a distinct mRNA exporter complex, which is recruited to pre-mRNAs in the nucleus in concert with the completion of splicing and polyadenylation. The exporter complex then transports the mRNAs through the nuclear pores. Directionality of the process is established by a RNA helicase localized to the cytoplasmic face of the nuclear pore complex. The helicase remodels the mRNA and removes the exporter complex on the cytoplasmic side of the pore. This releases the mRNA into the cytoplasm and prevents its transport back into the nucleus.

In contrast to mRNAs, miRNAs, tRNAs, and rRNAs, which function in the cytoplasm, many noncoding RNAs, including the snRNAs involved in

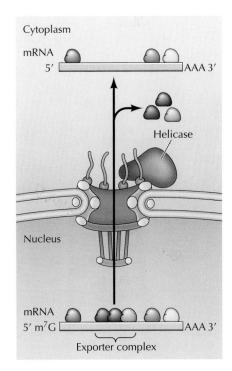

Cytoplasm

mRNA
5′ ⬛⬛⬛⬛⬛⬛ AAA 3′

Helicase

Nucleus

mRNA
5′ m⁷G ⬛⬛⬛⬛⬛⬛ AAA 3′
Exporter complex

Figure 10.13 mRNA export   Following processing, mRNAs are bound by an exporter complex, which mediates their transport through the nuclear pore com-plex. A helicase associated with the cytoplasmic face of the nuclear pore complex releases the mRNA into the cytoplasm.

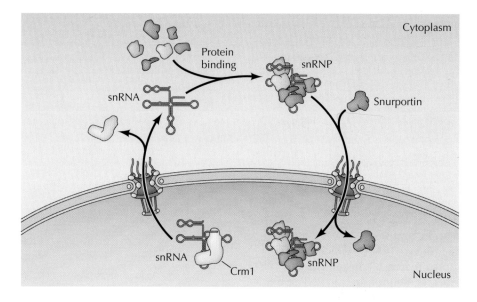

Figure 10.14 Transport of snRNAs between nucleus and cytoplasm Small nuclear RNAs (snRNAs) are initially exported from the nucleus to the cytoplasm by an exportin (Crm1). In the cytoplasm, the snRNAs associate with proteins to form snRNPs, which are recognized by an importin (snurportin) and transported back to the nucleus.

CRM1 [handwritten]

Types of RNA active in nucleus [handwritten]

pre-mRNA splicing and the snoRNAs involved in rRNA processing (discussed later in this chapter), function within the nucleus. snoRNAs remain in the nucleus, but snRNAs are initially transported from the nucleus to the cytoplasm by the exportin Crm1 (**Figure 10.14**). In the cytoplasm, the snRNAs associate with proteins to form snRNPs. Sequences present on the snRNP proteins are then recognized by an importin called snurportin, which mediates their transport back to the nucleus.

Regulation of nuclear protein import

The transport of proteins to the nucleus is an important level at which the activities of nuclear proteins can be controlled. Transcription factors, for example, are functional only when they are present in the nucleus, so regulation of their import to, and export from, the nucleus is a novel means of controlling gene expression. As will be discussed in Chapter 16, the regulated nuclear import of both transcription factors and protein kinases plays an important role in controlling cell behavior by providing a mechanism through which signals received at the cell surface can be transmitted to the nucleus.

In one mechanism of regulation, transcription factors (or other proteins) associate with cytoplasmic proteins that mask their nuclear localization signals; because their signals are no longer recognizable, these proteins remain in the cytoplasm. A good example is provided by the transcription factor NF-κB, which is activated in response to a variety of extracellular signals in mammalian cells (**Figure 10.15**). In unstimulated cells, NF-κB is found as an inactive complex with an inhibitory protein (IκB) in the cytoplasm. Binding to IκB masks the NF-κB nuclear localization signal, thus preventing NF-κB from being transported to the nucleus. In stimulated cells, IκB is phosphorylated and degraded by ubiquitin-mediated proteolysis (see Figure 9.46), allowing NF-κB to enter the nucleus and activate transcription of its target genes.

The nuclear import of other transcription factors is regulated directly by their phosphorylation rather than by association with inhibitory proteins (see Figure 10.15). For example, the yeast transcription factor Pho4 is phosphorylated at a serine residue adjacent to its nuclear localization signal. Phosphorylation at this site inhibits Pho4 by interfering with its nuclear

Mentioned in Lecture? D [handwritten]

Q. [handwritten]

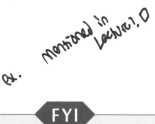

FYI

An Ebola virus protein blocks the antiviral response by selectively inhibiting nuclear import of a transcription factor.

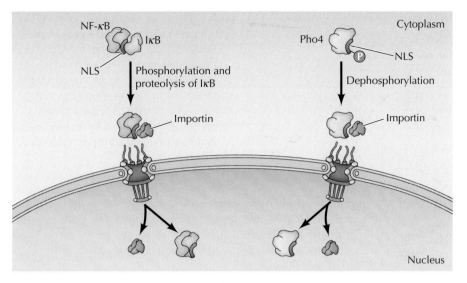

Figure 10.15 **Regulation of nuclear import of transcription factors** The transcription factor NF-κB is maintained as an inactive complex with IκB, which masks its nuclear localization sequence (NLS) in the cytoplasm. In response to appropriate extracellular signals, IκB is phosphorylated and degraded by proteolysis, allowing the import of NF-κB to the nucleus. The yeast transcription factor Pho4 is maintained in the cytoplasm by phosphorylation in the vicinity of its nuclear localization sequence. Regulated dephosphorylation exposes the NLS and allows Pho4 to be transported to the nucleus.

import. Under appropriate conditions, though, regulated dephosphorylation of this site activates Pho4 by permitting its translocation to the nucleus.

The Organization of Chromosomes

Chromatin becomes highly condensed during mitosis to form the compact metaphase chromosomes that are distributed to daughter nuclei. During interphase, most of the chromatin decondenses and is distributed throughout the nucleus. Although interphase chromatin appears to be uniformly distributed, the chromosomes actually occupy distinct regions of the nucleus and are organized such that the transcriptional activity of a gene is correlated with its position. In addition, the key activities of DNA replication and transcription take place in clustered regions within the nucleus.

Chromosome territories

The nonrandom distribution of chromatin within the interphase nucleus was first suggested in 1885 by Carl Rabl, who proposed that each chromosome occupies a distinct territory, with centromeres and telomeres attached to opposite sides of the nuclear envelope (**Figure 10.16**). This basic model of chromosome organization was confirmed nearly a hundred years later (in

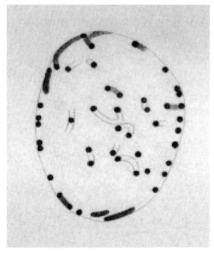

Figure 10.16 **Chromosome organization** Reproduction of hand-drawn sketches by Carl Rabl of chromosomes in salamander cells. (A) Complete chromosomes. (B) Telomeres only (located at the nuclear membrane). (From C. Rabl, 1885. *Morphologisches Jahrbuch* 10: 214.)

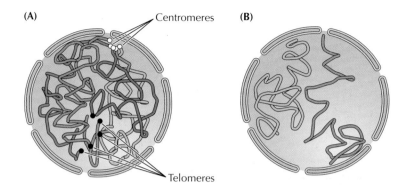

Figure 10.17 Organization of *Drosophila* chromosomes (A) A model of the nucleus, showing the five chromosome arms in different colors. The positions of telomeres and centromeres are indicated. (B) The two arms of chromosome 3 are shown to illustrate the topological separation between chromosomes. (From D. Mathog et al., 1984. *Nature* 308: 414.)

1984) by detailed studies of polytene chromosomes in *Drosophila* salivary glands. Rather than randomly winding around one another, each chromosome was found to occupy a discrete region of the nucleus, called a chromosome territory (Figure 10.17).

More recent studies have probed the organization of chromosomes within the nucleus using two complementary approaches. First, *in situ* hybridization with fluorescent probes that are specific for repeated sequences on individual chromosomes has been used to visualize the location of chromosomes within a nucleus, confirming that individual chromosomes occupy distinct territories within the nuclei of mammalian cells (Figure 10.18). Second, the associations between chromosomes in living cells have been analyzed by chromosome conformation capture (3C) techniques. As discussed in Chapter 8, these methods identify the sites of interactions between chromosomal regions by cross-linking interacting DNA sequences, which can then be amplified and identified by high-throughput sequencing. The global views of chromosome organization provided by 3C methods at the molecular level have confirmed and extended the results obtained by fluorescence *in situ* hybridization (FISH). Most interactions identified by 3C techniques occur between regions located on the same chromosome, rather than between different chromosomes, consistent with the conclusion that chromosomes occupy distinct territories within the nucleus. It further appears that the centromere divides each chromosome into distinct domains, so that the two arms of each chromosome do not interact with one another.

Chromatin localization and transcriptional activity

Most of the chromatin in interphase cells (**euchromatin**) is decondensed and transcriptionally-active, but some of the chromatin (**heterochromatin**) is highly condensed and not transcribed. Heterochromatin includes DNA

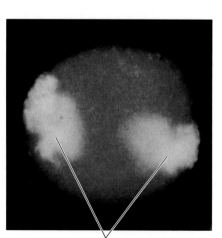

Copies of chromosome 4

Figure 10.18 Organization of chromosomes in the mammalian nucleus Fluorescent probes to repeated sequences on chromosome 4 were hybridized to a human cell. The two copies of chromosome 4, identified by yellow fluorescence, occupy distinct territories in the nucleus. (From A. I. Lamond and W. C. Earnshaw, 1999. *Science* 280: 547.)

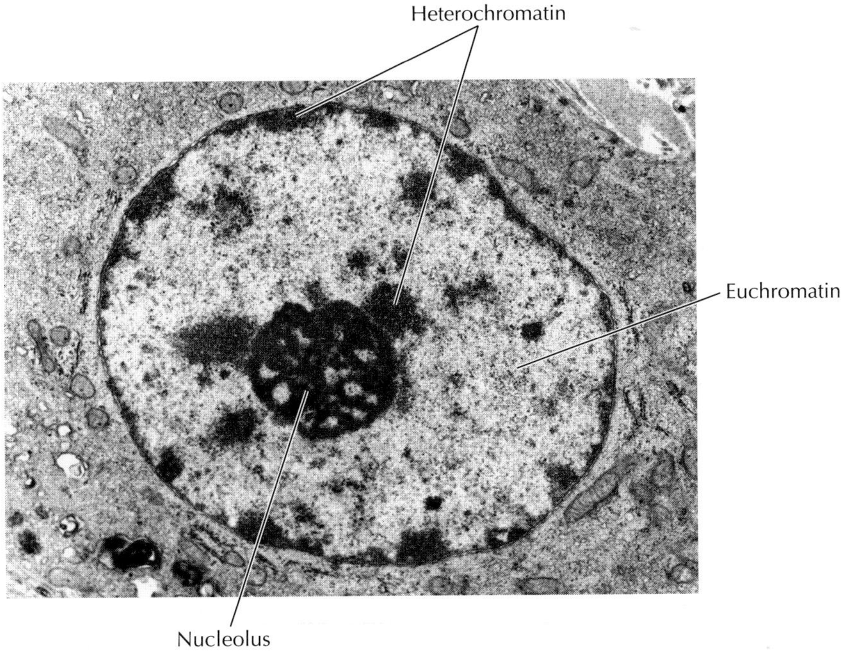

Heterochromatin

Euchromatin

Nucleolus

Figure 10.19 Heterochromatin in interphase nuclei   The euchromatin is distributed throughout the nucleus. The heterochromatin is associated with the nuclear envelope and the nucleolus.

sequences that are generally not transcribed, such as the highly repetitive sequences present at centromeres and telomeres, as well as genes that are not currently transcribed in the cell being examined but are transcribed in other differentiated cell types, at different times in the cell cycle, or in response to different extracellular signals, such as hormones. The transcriptional activity of chromosomal domains correlates with their location in the nucleus. Transcriptionally-inactive heterochromatin is frequently associated with the nuclear envelope or periphery of the nucleolus, whereas transcriptionally-active chromatin is preferentially localized to the interior of the nucleus (Figure 10.19).

These microscopic observations of chromatin localization have been extended by more refined studies of the localization of transcriptionally-active versus -inactive chromatin domains. For example, some human chromosomes are rich in transcribed genes whereas others are gene-poor, containing relatively few genes. A comparison of the territories occupied by such chromosomes indicates that gene-rich chromosomes are located in the center of the nucleus, whereas gene-poor chromosomes are found at the nuclear periphery (Figure 10.20A). Moreover, FISH probes that are specific for active genes indicate that transcribed regions of a chromosome extend out of the bulk territory of that chromosome, towards the interior of the nucleus

FYI

Ciliated protozoa contain two types of nuclei: a polyploid macronucleus that contains transcriptionally active genes, and one or more diploid transcriptionally inactive micronuclei that participate in sexual reproduction.

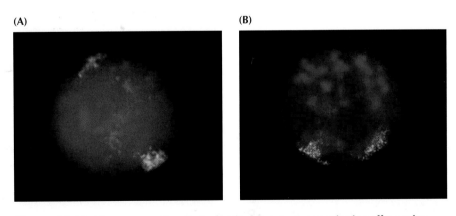

(A)

(B)

Figure 10.20 Fluorescent *in situ* hybridization to transcriptionally-active and-inactive chromosomes   (A) Fluorescent *in situ* hybridization with probes for the gene-rich chromosome 19 (red) and the gene-poor chromosome 18 (green). (B) Hybridization with probes spanning the entire mouse chromosome 2 (green) or only the exon sequences of that chromosome (red). (From W. A. Bickmore and B. van Steensel, 2013. *Cell* 152: 1270.)

(Figure 10.20B). This suggests that actively transcribed genes can loop out from their chromosomal territories, perhaps facilitating transcription or RNA processing.

Chromosome conformation capture (3C) analysis has indicated that genomes are divided into a series of topologically associating domains (TADs), such that regions within a domain interact frequently with one another, but only rarely interact with regions within other domains (see Figure 8.20). In humans, these chromosome domains range from approximately one hundred to several thousand kb and are thought to correspond to regions within which promoters and enhancers can interact, as well as to chromosomal regions characterized by histone modifications associated with repressed versus actively-transcribed chromatin. The boundaries of these chromosomal domains contain multiple binding sites for cohesin and CTCF, which is thought to act as an architectural protein that regulates the organization of the genome.

Chromatin domains that are localized to the periphery of the nucleus have been identified by several methods, including the use of chromatin immunoprecipitation (see Figure 8.23) to isolate regions of DNA that are associated with nuclear lamins. In human cells, between 1000 and 1500 chromosomal domains that interact with the nuclear lamina have been identified. These domains (called **lamina-associated domains** or **LADs**) range from approximately 10 to 10,000 kb and cover a total of about 40% of the human genome. Consistent with the localization of transcriptionally-inactive heterochromatin at the nuclear periphery, the genes found in LADs are generally transcriptionally repressed. Notably, LADs align well with chromatin domains defined by histone modification and with the topologically associating domains (TADs) that restrict enhancer/promoter interactions. It thus appears that LADs correspond to transcriptionally-inactive heterochromatin domains that are associated with the nuclear lamina. The mechanisms responsible for the interaction of chromatin with the nuclear lamina are not yet fully understood and may include the binding of chromatin to both lamins and proteins of the inner nuclear membrane, such as emerin and the lamin B receptor (see Figure 10.5). Interactions that mediate association of heterochromatin with the nuclear lamina include the binding of lamin B receptor to heterochromatin protein 1 (HP1). HP1 binds to methylated histone H3 lysine-9 residues, which are characteristic of transcriptionally-inactive chromatin (see Figure 8.32). Likewise, emerin interacts with a histone deacetylase, which is also associated with transcriptional repression and chromatin condensation.

In addition to the localization of heterochromatin at the nuclear lamina, the nucleolus is also surrounded by heterochromatin (see Figure 10.19). The regions of chromatin associated with the nucleolus (called **nucleolus-associated domains** or **NADs**) have been identified by sequencing the DNA attached to isolated nucleoli. Interestingly, the DNA sequences found in NADs substantially overlap with those found in LADs. This suggests that transcriptionally-inactive genes can be localized either to the nuclear lamina or to the periphery of nucleoli in different cells.

Taken together, these findings suggest that chromosomes are divided into domains corresponding to transcriptional units and regions of chromatin modification. Transcriptionally-active domains are generally found in the interior of the nucleus, whereas domains of transcriptionally-inactive chromatin are generally localized to either the nuclear lamina or the nucleolus (Figure 10.21). An interesting question that remains to be answered is to what

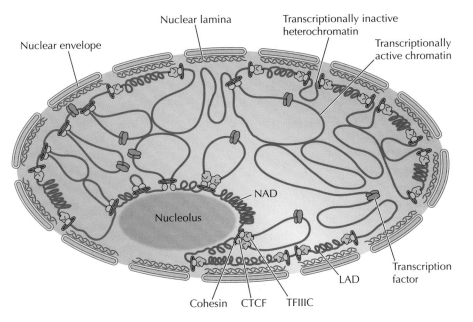

Figure 10.21 **Distribution of transcriptionally-active and -inactive chromatin** Domains of transcriptionally-inactive heterochromatin are generally associated with either the lamina (LADs) or the nucleolus (NADs). Transcriptionally-active domains are localized to the interior of the nucleus. Domain boundaries contain multiple binding sites for CTCF and cohesin.

extent the localization of a gene in the nucleus is a cause or an effect of its transcriptional activity.

Replication and transcription factories

The functional organization of chromatin is further indicated by the localization of key processes to clustered regions of the nucleus, which represent large complexes of the proteins involved in a particular process. DNA replication, for example, takes place within discrete regions called **replication factories**. Replication factories contain clustered sites of DNA replication within which the replication of multiple DNA molecules takes place. These discrete sites of DNA replication were initially defined by experiments in which newly synthesized DNA was visualized within cell nuclei by labeling cells with bromodeoxyuridine, an analog of thymidine that is incorporated into DNA and then detected by staining with fluorescent antibodies (**Figure 10.22A**). These clustered sites of newly synthesized DNA also represent concentrated sites of the proteins involved in DNA replication (**Figure 10.22B**), indicating that they are large complexes of proteins engaged in DNA synthesis. There are several

Figure 10.22 **Replication factories**   (A) Newly replicated DNA was labeled by a brief exposure of cells to bromodeoxyuridine (BrdU), which is incorporated into DNA in place of thymidine. This substitution allows detection of newly synthesized DNA by immunofluorescence following staining with an antibody against BrdU. Note that the newly replicated DNA (red) is present in discrete clusters distributed throughout the nucleus. (B) The same clusters represent concentrated sites of the GFP-labeled (green) DNA replication protein PCNA (see Figure 7.7). (From H. Leonhart et al., 2000. *J. Cell Biol.* 149: 271.)

(A) BrdU   **(B) PCNA**

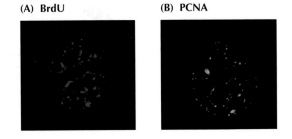

Figure 10.23 Transcription factories Transcription factories are clustered sites of RNA polymerases and transcription factors in which multiple domains of active chromatin are transcribed.

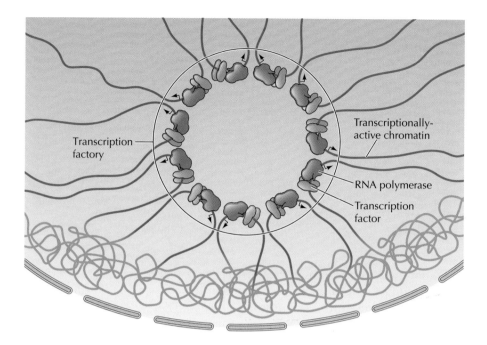

hundred such sites in the nuclei of mammalian cells. Since approximately 4000 origins of replication can be active in a diploid mammalian cell at any given time, each of these clustered sites of DNA replication must contain multiple replication forks, probably on the order of 5 to 50 replication forks per factory.

Transcription also occurs at clustered sites (**transcription factories**) that contain newly synthesized RNA and are highly enriched in active RNA polymerases and transcription factors. It is estimated that a transcription factory contains approximately eight molecules of RNA polymerase. Several chromosomal loci can share the same transcription factory and it is thought that active genes loop out and are pulled through a factory as transcription occurs (Figure 10.23). In some cases, coregulated genes may be transcribed in the same factory. For example, immunoglobulin genes from different chromosomes colocalize in the same transcription factory in lymphocytes that are producing large amounts of antibodies. Thus, transcription factories may not only increase the efficiency of gene expression, but may also facilitate the coordinated regulation of related genes.

Nuclear Bodies

The nucleus contains a variety of discrete organelles, referred to as **nuclear bodies**. Like cytoplasmic organelles (e.g., mitochondria), nuclear bodies compartmentalize the nucleus and serve to concentrate proteins and RNAs that function in specific nuclear processes. However, in contrast to cytoplasmic organelles, nuclear bodies are not enclosed by membranes but are instead maintained by protein-protein and protein-RNA interactions. As a result, nuclear bodies are dynamic structures whose contents are able to exchange with the rest of the nucleus.

The most prominent nuclear body is the nucleolus (see Figure 10.1), which functions in rRNA synthesis and ribosome production. A variety of other nuclear bodies are involved in the regulation of transcription, RNA

Table 10.2 Examples of Nuclear Bodies

Nuclear body	Number per nucleus	Function
Nucleolus	1–4	rRNA transcription, processing and ribosome assembly
Cajal body	0–10	snRNP assembly
Clastosome	0–3	Proteasomal proteolysis
Histone locus body	2–4	Transcription and processing of histone pre-mRNAs
Speckle	20–50	Storage of pre-mRNA splicing factors
PML body	10–30	Transcriptional regulation, DNA repair
Polycomb body	10–20	Gene silencing

processing, and assembly of ribonucleoprotein complexes (Table 10.2). The nucleolus and examples of three other nuclear bodies that function in regulation of gene expression and in the assembly, maturation, and storage of components of the pre-mRNA splicing machinery are discussed below. However, the roles of these and many other nuclear bodies are not yet fully understood, so elucidating this aspect of the structure and function of the nucleus remains an active area of research.

The nucleolus and rRNA

The **nucleolus** is the site of rRNA transcription and processing as well as ribosome assembly. As discussed in the preceding chapter, cells require large numbers of ribosomes to meet their needs for protein synthesis. Actively growing mammalian cells, for example, contain 5 million to 10 million ribosomes that must be synthesized each time the cell divides. The nucleolus is a ribosome production factory, designed to fulfill the need for regulated and efficient production of rRNAs and assembly of the ribosomal subunits. Eukaryotic ribosomes contain four types of RNA designated the 5S, 5.8S, 18S, and 28S rRNAs (see Figure 9.4). The 5.8S, 18S, and 28S rRNAs are transcribed as a single unit within the nucleolus by RNA polymerase I, yielding a 45S ribosomal precursor RNA (see Figure 8.12). The 45S pre-rRNA is processed to the 18S rRNA of the 40S (small) ribosomal subunit and to the 5.8S and 28S rRNAs of the 60S (large) ribosomal subunit. Transcription of the 5S rRNA, which is also found in the 60S ribosomal subunit, takes place outside the nucleolus in higher eukaryotes and is catalyzed by RNA polymerase III.

To meet the need for transcription of large numbers of rRNA molecules, all cells contain multiple copies of the rRNA genes. The human genome, for example, contains about 200 copies of the gene that encodes the 5.8S, 18S, and 28S rRNAs and approximately 2000 copies of the gene that encodes 5S rRNA. The genes for 5.8S, 18S, and 28S rRNAs are clustered in tandem arrays on five different human chromosomes (chromosomes 13, 14, 15, 21, and 22); the 5S rRNA genes are present in a single tandem array on chromosome 1.

The nucleolus is organized around the chromosomal regions that contain the genes for the 5.8S, 18S, and 28S rRNAs, which are therefore called **nucleolar organizing regions**. The formation of nucleoli after cell division requires the transcription of 45S pre-rRNA, which appears to lead to the fusion of small prenucleolar bodies that contain processing factors and other components of

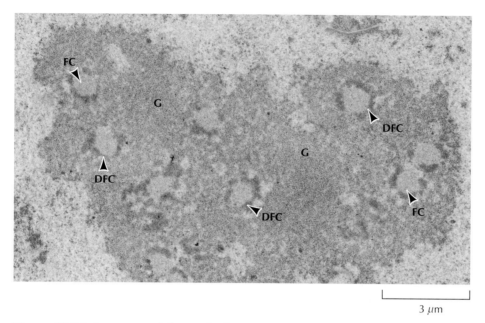

the nucleolus. In most cells, the initially separate nucleoli then fuse to form a single nucleolus. Morphologically, nucleoli consist of three regions: the fibrillar center, dense fibrillar component, and granular component (**Figure 10.24**). These different regions represent the sites of progressive stages of rRNA transcription, processing, and ribosome assembly. The genes encoding rRNA are localized in the fibrillar center and transcribed at the interface with the dense fibrillar component, where pre-rRNA is processed. The assembly of ribosomal subunits then takes place in the granular component. The size of the nucleolus depends on the metabolic activity of the cell, with large nucleoli found in cells that are actively engaged in

Figure 10.24 Structure of the nucleolus An electron micrograph illustrating the fibrillar center (FC), dense fibrillar component (DFC), and granular component (G) of a nucleolus. (Courtesy of David L. Spector, Cold Spring Harbor Laboratory.)

protein synthesis. This variation is due primarily to differences in the size of the granular component, reflecting the levels of ribosome assembly.

Each nucleolar organizing region contains a cluster of tandemly repeated rRNA genes separated from each other by nontranscribed DNA (**Figure 10.25A**).

(A)

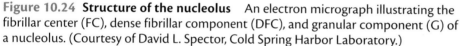

(B)

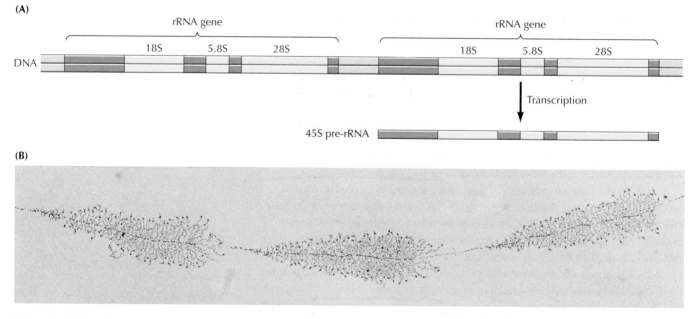

Figure 10.25 Ribosomal RNA genes (A) Each rRNA gene is a single transcription unit containing the 18S, 5.8S, and 28S rRNAs. The rRNA genes are organized in tandem arrays, separated by nontranscribed DNA. (B) An electron micrograph showing the transcription of three rRNA genes. Each gene is surrounded by an array of growing RNA chains, resulting in a Christmas tree-like appearance. (Courtesy of O. L. Miller Jr.)

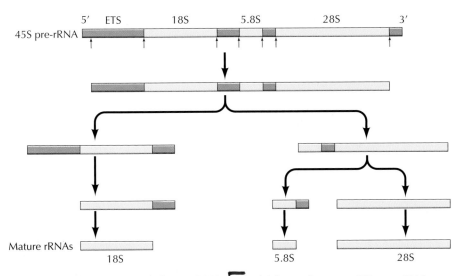

Figure 10.26 Processing of pre-rRNA The higher eukaryote 45S pre-rRNA transcript is processed via a series of cleavages to yield the mature 18S, 5.8S, and 28S rRNAs species.

These genes are very actively transcribed by RNA polymerase I, allowing their transcription to be readily visualized by electron microscopy. In such electron micrographs, each of the tandemly arrayed rRNA genes is surrounded by densely packed growing RNA chains forming a structure that looks like a Christmas tree (Figure 10.25B). The high density of growing RNA chains reflects that of RNA polymerase molecules, which are present at a maximal density of approximately one polymerase per hundred base pairs of template DNA.

In higher eukaryotes the primary transcript of the rRNA genes is the large 45S pre-rRNA, which is processed via a series of cleavages to produce the 18S, 5.8S, and 28S rRNAs (Figure 10.26). In addition to cleavage, the processing of pre-rRNA involves a substantial amount of modification resulting both from the addition of methyl groups to specific ribose residues and from the conversion of uridine to pseudouridine (see Figure 8.43). In animal cells, pre-rRNA processing involves the methylation of approximately 100 ribose residues and the formation of about 100 pseudouridines. Most of these modifications occur during or shortly after synthesis of the pre-rRNA, although a few take place at later stages of pre-rRNA processing.

The processing of pre-rRNA requires the action of both proteins and RNAs that are localized to the nucleolus. Nucleoli contain more than 300 proteins and about 200 **small nucleolar RNAs (snoRNAs)** that function in pre-rRNA processing. Like the spliceosomal snRNAs involved in pre-mRNA processing (discussed in Chapter 8), the snoRNAs are complexed with proteins, forming snoRNPs. Individual snoRNPs consist of single snoRNAs associated with eight to ten proteins. The snoRNPs then assemble on the pre-rRNA to form processing complexes in a manner analogous to the formation of spliceosomes on pre-mRNA.

The action of snoRNAs differs from that of the snRNAs that catalyze pre-mRNA splicing (discussed in Chapter 8). In contrast, snoRNAs function as guide RNAs to direct protein enzymes to the sites of specific modifications of pre-rRNA molecules. There are two families of snoRNAs that associate with different proteins, which catalyze either ribose methylation or pseudouridine formation (Figure 10.27). The snoRNAs contain short sequences of

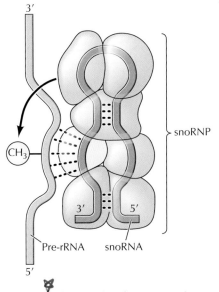

Figure 10.27 Role of snoRNAs in base modification of pre-rRNA The snoRNAs contain short sequences complementary to rRNA. Base pairing between snoRNAs and pre-rRNA targets the enzymes that catalyze pre-rRNA modification (e.g., methylation) to the appropriate sites.

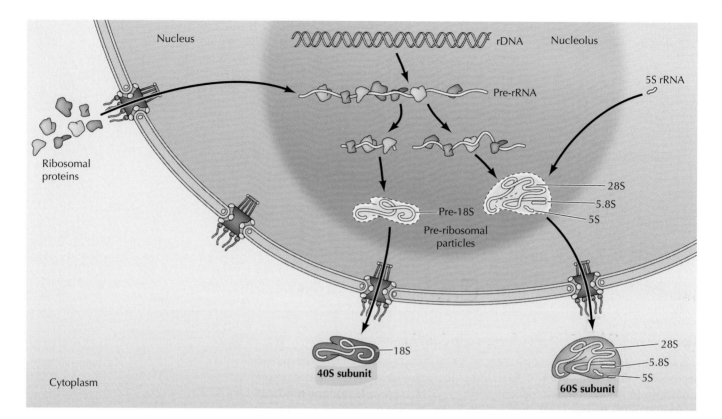

Figure 10.28 Ribosome assembly   Ribosomal proteins are imported to the nucleolus from the cytoplasm and begin to assemble on pre-rRNA prior to its cleavage. As the pre-rRNA is processed, additional ribosomal proteins and the 5S rRNA (which is synthesized elsewhere in the nucleus) assemble to form pre-ribosomal particles. The final steps of maturation follow the export of pre-ribosomal particles to the cytoplasm, yielding the 40S and 60S ribosomal subunits.

approximately 15 nucleotides that are complementary to 18S or 28S rRNA. Importantly, these regions of complementarity include the sites of base modification in the rRNA. By base pairing with specific regions of the pre-rRNA, the snoRNAs act to target the proteins responsible for methylation or pseudouridylation to the correct site on the pre-rRNA molecule.

The formation of ribosomes involves the assembly of the ribosomal precursor RNA with both ribosomal proteins and 5S rRNA (Figure 10.28). The genes that encode ribosomal proteins are transcribed outside the nucleolus by RNA polymerase II, yielding mRNAs that are translated on cytoplasmic ribosomes. The ribosomal proteins are then transported from the cytoplasm to the nucleolus where they are assembled with rRNAs to form pre-ribosomal particles. Although the genes for 5S rRNA are also transcribed outside the nucleolus—in this case by RNA polymerase III—5S rRNAs similarly are assembled into pre-ribosomal particles within the nucleolus. The final stages of ribosomal subunit maturation follow the export of pre-ribosomal particles to the cytoplasm, forming the active 40S and 60S subunits of eukaryotic ribosomes.

(A)

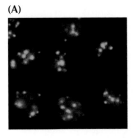

Figure 10.29 Polycomb bodies (A) Nuclei of human cells expressing a GFP-labeled Polycomb protein. (B) Polycomb bodies contain clusters of repressed genes modified by methylation of histone H3 lysine 27 (H3K27me). (A, from V. Pirrotta and H.-B. Li, 2012. *Curr. Opin. Genet. Dev.* 22:101.)

(B)

Polychrome proteins   Polycomb body

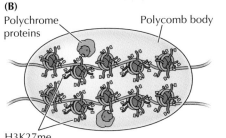

H3K27me

Polycomb bodies: *Centers of transcriptional repression*

As discussed in Chapter 8, Polycomb proteins are responsible for transcriptional repression of a variety of genes via the methylation of histone H3 lysine 27 residues. The Polycomb proteins consist of two complexes: one binds to methylated H3 lysine 27 residues and the other is an enzyme that methylates adjacent H3 lysine 27 residues. As a result, methylation of H3 lysine 27 can spread to adjacent nucleosomes to generate domains of repressed chromatin (see Figure 8.36). Analysis of the nuclear localization of Polycomb proteins by immunofluorescence indicates that they are concentrated in a relatively small number of domains, called **Polycomb bodies**, rather than being distributed throughout the nucleus (**Figure 10.29A**).

Polycomb bodies are frequently associated with heterochromatin, consistent with their role in gene repression. In some cases, Polycomb bodies may be formed as a result of Polycomb repression of a series of adjacent genes. However, at least some Polycomb bodies contain clusters of repressed domains from different chromosomal regions (**Figure 10.29B**). Thus, analogous to transcription factories (see Figure 10.23), Polycomb bodies are clustered sites of coordinately repressed genes.

Cajal bodies and speckles: *Processing and storage of snRNPs*

Both Cajal bodies and nuclear speckles play important roles in the assembly and storage of the snRNPs that are responsible for pre-mRNA splicing. Cajal bodies, first described by Ramón y Cajal in 1906, are small nuclear bodies that are involved in assembly of snRNPs and other RNA-protein complexes. As discussed earlier in this chapter, snRNAs are first exported to the cytoplasm, where they associate with proteins to form snRNPs that return to the nucleus (see Figure 10.14). The snRNPs are then concentrated in Cajal bodies, where the final stages of snRNP maturation occur. This includes modification of the snRNAs by ribose methylation and pseudouridylation, similar to the modifications of pre-rRNAs that take place in the nucleolus. Consistent with the processing of snRNAs in Cajal bodies, the enzyme responsible for RNA methylation (fibrillarin) is concentrated in Cajal bodies as well as in nucleoli

(A)   **(B)**

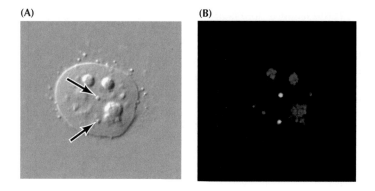

Figure 10.30 **Cajal bodies in the nucleus**   (A) Differential interference contrast microscope image of the nucleus of a human cell. The arrows indicate the two Cajal bodies. (B) Immunofluorescent staining of the same nucleus with antibodies to the proteins coilin (green) and fibrillarin (red). Fibrillarin (the protein responsible for methylation of RNAs) is present in both the nucleoli and in the Cajal bodies. Coilin (a major Cajal body protein) is detectable only in the Cajal bodies. (From J. G. Gall, 2000. *Ann. Rev. Cell Dev. Biol.* 16: 273.)

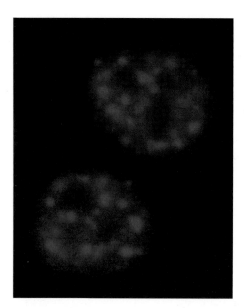

(Figure 10.30). In addition, Cajal bodies contain specific small Cajal body-specific RNAs (scaRNAs). The scaRNAs are related to the snoRNAs in nucleoli and similarly serve as guide RNAs to direct the ribose methylation and pseudouridylation of snRNAs in Cajal bodies.

In addition to their role in snRNP assembly, Cajal bodies appear to play a role in assembly of telomerase, which is essential for replication of the ends of chromosomal DNA. As discussed in Chapter 7, telomerase is a complex between an RNA that recognizes the sequences at the ends of chromosomes (telomeres) and a reverse transcriptase that uses the telomerase RNA as a template to copy those terminal sequences (see Figure 7.16). Cajal bodies may promote assembly of the RNA-protein telomerase complex as well as facilitating its delivery to telomeres.

Following assembly and maturation in Cajal bodies, snRNPs are transferred to **speckles**, which are discrete nuclear bodies containing components of the mRNA splicing machinery (Figure 10.31). Splicing factors as well as snRNPs are concentrated in speckles rather than being distributed uniformly throughout the nucleus. It is thought that the speckles are storage sites of splicing components, which are then recruited to actively transcribed genes where pre-mRNA processing occurs.

Figure 10.31 **Nuclear speckles**   Staining with immunofluorescent antibodies indicates that splicing factors are concentrated in discrete bodies within the nucleus—termed nuclear speckles. (Courtesy of David L. Spector, Cold Spring Harbor Laboratory.)

SUMMARY

KEY TERMS

The Nuclear Envelope and Traffic between the Nucleus and the Cytoplasm

- *Structure of the nuclear envelope:* The nuclear envelope separates the contents of the nucleus from the cytoplasm, maintaining the nucleus as a distinct biochemical compartment that houses the genetic material and serves as the site of transcription and RNA processing in eukaryotic cells. The nuclear envelope consists of the inner and outer nuclear membranes (which are joined at nuclear pore complexes) and an underlying nuclear lamina.

nuclear envelope, nuclear membrane, nuclear lamina, lamin

SUMMARY	KEY TERMS

- **The nuclear pore complex:** Nuclear pore complexes are large structures that provide the only routes through which molecules can travel between the nucleus and the cytoplasm. Small molecules diffuse freely through the nuclear pore complex. Macromolecules are selectively transported either from nucleus to cytoplasm or from cytoplasm to nucleus. See Video 10.1.

nuclear pore complex

- **Selective transport of proteins to and from the nucleus:** Proteins destined for import to the nucleus contain nuclear localization signals that are recognized by receptors that direct transport through the nuclear pore complex. Proteins that shuttle back and forth between the nucleus and the cytoplasm contain nuclear export signals that target them for transport from the nucleus to the cytoplasm. In most cases, the small GTP-binding protein Ran is required for translocation through the nuclear pore complex and determines the directionality of transport. See Animation 10.1 and Video 10.2.

nuclear localization signal, nuclear transport receptor, importin, Ran, nuclear export signal, exportin, karyopherin

- **Transport of RNAs:** RNAs are transported through the nuclear pore complex as ribonucleoprotein complexes. Messenger RNAs, ribosomal RNAs, and transfer RNAs are exported from the nucleus to function in protein synthesis. Small nuclear RNAs are initially transported from the nucleus to the cytoplasm where they associate with proteins to form RNPs; they then return to the nucleus.

- **Regulation of nuclear protein import:** The activity of some proteins, such as transcription factors, is controlled by regulation of both their import to, and export from, the nucleus.

The Organization of Chromosomes

- **Chromosome territories:** Individual chromosomes occupy distinct territories within the nucleus and are divided into large looped domains that function as independent units.

- **Chromatin localization and transcriptional activity:** Transcriptionally-inactive heterochromatin is frequently associated with the nuclear envelope or nucleolus, whereas transcriptionally-active chromatin is localized to the interior of the nucleus.

euchromatin, heterochromatin, lamin-associated domain (LAD), nucleolus-associated domain (NAD)

- **Replication and transcription factories:** DNA replication takes place within large complexes containing multiple replication forks. Transcription also occurs at clustered sites that are enriched in RNA polymerases and transcription factors.

replication factory, transcription factory

Nuclear Bodies

- **The nucleolus and rRNA:** Several types of nuclear bodies compartmentalize the nucleus and serve to concentrate proteins and RNAs involved in a variety of aspects of gene expression. The nucleolus is associated with the genes for ribosomal RNAs and is the site of rRNA transcription, rRNA processing, and ribosome assembly. Processing of pre-rRNA is mediated by small nucleolar RNAs (snoRNAs).

nuclear body, nucleolus, nucleolar organizing region, small nucleolar RNA (snoRNA)

- **Polycomb bodies: Centers of transcriptional repression:** The Polycomb proteins, which repress a variety of genes via histone methylation, are concentrated in clusters that repress multiple chromatin domains.

Polycomb body

- **Cajal bodies and speckles: Processing and storage of snRNPs:** Cajal bodies are involved in snRNA modification and the assembly of snRNPs, as well as assembly of telomerase. Nuclear speckles are storage sites of snRNPs and other components of the pre-mRNA splicing machinery.

Cajal body, speckle

Questions

1. By separating transcription from translation, the nuclear envelope allows eukaryotes to regulate gene expression by processes that are not found in prokaryotes. What are these regulatory mechanisms that are unique to eukaryotes?

2. What roles do lamins play in nuclear structure and function?

3. You inject a frog egg with two globular proteins, one 15 kd and the other 100 kd—both of which lack nuclear localization signals. Will either protein enter the nucleus?

4. What determines the directionality of nuclear import?

5. Describe how the activity of a transcription factor can be regulated by nuclear import.

6. You are studying a transcription factor that is regulated by phosphorylation of serine residues that inactivates its nuclear localization signal. How would mutating these serines to alanines affect subcellular localization of the transcription factor and expression of its target gene?

7. How would mutational inactivation of the nuclear export signal of a protein that normally shuttles back and forth between the nucleus and cytoplasm affect its subcellular distribution?

8. DNA replication appears to occur in specific sites or replication factories. How would you locate these sites in mammalian cells in culture?

9. How did Alan Smith and colleagues demonstrate that the T antigen amino acid sequence 126 to 132 was sufficient for nuclear accumulation of the protein?

10. What is the significance of nuclear speckles?

11. What is the function of snoRNAs?

12. How would RNAi against human exportin-t affect human fibroblasts in culture?

References and Further Reading   (Key review articles for each major section are highlighted in **bold**.)

The Nuclear Envelope and Traffic between the Nucleus and the Cytoplasm

Bank, E. M. and Y. Gruenbaum. 2011. The nuclear lamina and heterochromatin: a complex relationship. *Biochem. Soc. Trans.* 39: 1705–1709. [R]

Burke, B. and C. L. Stewart. 2013. The nuclear lamins: flexibility in function. *Nature Rev. Mol. Cell Biol.* 14: 13–24. [R]

Chook, Y. M. and K. E. Suel. 2011. Nuclear import by karyopherin-betas: recognition and inhibition. *Biochim. Biophys. Acta* 1813: 1593–1606. [R]

Gerace, L. and M. D. Huber. 2012. Nuclear lamina at the crossroads of the cytoplasm and nucleus. *J. Struc. Biol.* 177: 24–31. [R]

Grossman, E., O. Medalia and M. Zwerger. 2012. Functional architecture of the nuclear pore complex. *Ann. Rev. Biophys.* 41: 557–584. [R]

Grunwald, D., R. H. Singer, and M. Rout. 2011. Nuclear export dynamics of RNA-protein complexes. *Nature* 475: 333–341. [R]

Guttler, T. and D. Gorlich. 2011. Ran-dependent nuclear export mediators: a structural perspective. *EMBO J.* 30: 3457–3474. [R]

Kohler, A. and E. Hurt. 2007. Exporting RNA from the nucleus to the cytoplasm. *Nature Rev. Mol. Cell Biol.* 8: 761–773. [R]

Matera, A. G., R. M. Terns and M. P. Terns. 2007. Non-coding RNAs: lessons from the small nuclear and small nucleolar RNAs. *Nature Rev. Mol. Cell Biol.* 8: 209–220. [R]

Simon, D. N. and K. L. Wilson. 2011. The nucleoskeleton as a genome-associated dynamic 'network of networks.' *Nature Rev. Mol. Cell Biol.* 12: 695–708. [R]

Starr, D. A. and H. N. Fridolfsson. 2010. Interactions between nuclei and the cytoskeleton are mediated by SUN-KASH nuclear-envelope bridges. *Ann. Rev. Cell Dev. Biol.* 26: 421–444. [R]

Stewart, M. 2007. Molecular mechanism of the nuclear protein import cycle. *Nature Rev. Mol. Cell Biol.* 8: 195–208. [R]

Stewart, M. 2010. Nuclear export of mRNA. *Trends Biochem. Sci.* 35: 609–617. [R]

Strambio-De-Castillia, C., M. Niepel and M. P. Rout. 2010. The nuclear pore complex: bridging nuclear transport and gene regulation. *Nature Rev. Mol. Cell Biol.* 11: 490–500. [R]

Walde, S. and R. H. Kehlenback. 2010. The Part and the Whole: functions of nucleoporins in nucleocytoplasmic transport. *Trends Cell Biol.* 20: 461–469. [R]

The Organization of Chromosomes

Amendola, M. and B. van Steensel. 2014. Mechanisms and dynamics of nuclear lamina-genome interactions. *Curr. Opin. Cell Biol.* 28: 61–68. [R]

Bickmore, W. A. 2013. The spatial organization of the human genome. *Ann. Rev. Genomics Hum. Genet.* 14: 67–84. [R]

Bickmore, W. A. and B. van Steensel. 2013. Genome architecture: domain organization of interphase chromosomes. *Cell* 152: 1270–1284. [R]

Buckley, M. S. and J. T. Lis. 2014. Imaging RNA polymerase II transcription sites in living cells. *Curr. Opin. Genet. Dev.* 25: 126–130. [R]

Cremer, T. and M. Cremer. 2010. Chromosome territories. *Cold Spring Harb. Perspect. Biol.* 2: a003889. [R]

Edelman, L. B. and P. Fraser. 2012. Transcription factories: genetic programming in three dimensions. *Curr. Opin. Genet. Dev.* 22: 110–114. [R]

Mekhail, K. and D. Moazed. 2010. The nuclear envelope in genome organization, expression and stability. *Nature Rev. Mol. Cell Biol.* 11: 317–328. [R].

Padeken, J. and P. Heun. 2014. Nucleolus and nuclear periphery: velcro for heterochromatin. *Curr. Opin. Cell Biol.* 28: 54–60. [R]

Misteli, T. 2013. The cell biology of genomes: bringing the double helix to life. *Cell* 152: 1209–1212. [R]

Park, S.-K., Y. Xiang, X. Feng and W. T. Garrard. 2014. Pronounced cohabitation of active immunoglobulin genes from three different chromosomes in transcription factories during maximal antibody synthesis. *Genes Dev.* 28: 1159–164. [P]

Sexton, T. and G. Cavalli. 2015. The role of chromosome domains in shaping the functional genome. *Cell* 160: 1049–1059. [R]

Van Bortle, K. and V. G. Corces. 2012. Nuclear organization and genome function. *Ann. Rev. Cell Dev. Biol.* 28: 163–187. [R]

Nuclear Bodies

Bernardi, R. and P. P. Pandolfi. 2007. Structure, dynamics and functions of promyelocytic leukaemia nuclear bodies. *Nature Rev. Mol. Cell Biol.* 8: 1006–1016. [R]

Bratkovic, T. and B. Rogelj. 2014. The many faces of small nucleolar RNAs. *Biochim. Biophys. Acta* 1839: 438–443. [R]

Dundr, M. 2012. Nuclear bodies: multifunctional companions of the genome. *Curr. Opin. Cell Biol.* 24: 415–422. [R]

Mao, Y. S., B. Zhang and D. L. Spector. 2011. Biogenesis and function of nuclear bodies. *Trends Genet*. 27: 295–306. [R]

Nizami, Z., S. Deryusheva and J. G. Gall. 2010. The Cajal body and histone locus body. *Cold Spring Harbor Perspect. Biol.* 2:a000653. [R]

Pederson, T. 2011. The nucleolus. *Cold Spring Harbor Perspect. Biol.* 3:a000638. [R]

Pirrotta, V. and H.-B. Li. 2012. A view of nuclear Polycomb bodies. *Curr. Opin. Genet. Dev.* 22: 101–109. [R]

Sleeman, J. E. and L. Trinkle-Mulcahy. 2014. Nuclear bodies: new insights into assembly/dynamics and disease relevance. *Curr. Opin. Cell Biol.* 28: 76–83. [R]

Spector, D. L. and A. I. Lamond. 2011. Nuclear speckles. *Cold Spring Harbor Perspect. Biol.* 3:a000646. [R]

CHAPTER 11

Protein Sortin[g] [and] Transport

The Endoplasmic R[eticulum, Golgi] Apparatus, and Lysosomes

In addition to the presence of a nucleus, eukaryotic cells have a variety of membrane-enclosed organelles within their cytoplasm. These organelles provide discrete compartments in which specific cellular activities take place, and the resulting subdivision of the cytoplasm allows eukaryotic cells to function efficiently in spite of their large size—at least a thousand times the volume of bacteria.

Because of the complex internal organization of eukaryotic cells, the sorting and targeting of proteins to their appropriate destinations are considerable tasks. The first step of protein sorting takes place while translation is still in progress. Proteins destined for the endoplasmic reticulum, the Golgi apparatus, lysosomes, the plasma membrane, and secretion from the cell are synthesized on ribosomes that are bound to the membrane of the endoplasmic reticulum. As translation proceeds, the polypeptide chains are transported into the endoplasmic reticulum where protein folding and processing take place. From the endoplasmic reticulum, proteins are transported in vesicles to the Golgi apparatus where they are further processed and sorted for transport to endosomes, lysosomes, the plasma membrane, or secretion from the cell. Some of these organelles also participate in the sorting and transport of proteins being taken up from outside the cell (see Chapter 14). The endoplasmic reticulum, Golgi apparatus, endosomes, and lysosomes are thus distinguished from other cytoplasmic organelles by their common involvement in protein processing and connection by vesicular transport. About one-third of cellular proteins are processed in the endoplasmic reticulum, highlighting the importance of this pathway in cell physiology.

The Endoplasmic Reticulum

The **endoplasmic reticulum (ER)** is a network of membrane-enclosed tubules and sacs (cisternae) that extends from the nuclear membrane throughout the cytoplasm (Figure 11.1). The entire endoplasmic reticulum is enclosed by a continuous membrane and is the largest organelle of most eukaryotic cells. Its membrane may account for about half of all cell membranes, and the space enclosed by the ER (the lumen, or cisternal space) may represent about 10% of the total cell volume. As discussed below, there are two contiguous membrane domains within the ER that perform different functions within the cell. The **rough ER**, which is covered by ribosomes on its outer (cytosolic) surface, functions in protein processing. The **smooth ER** is not associated with ribosomes and is involved in lipid, rather than protein, metabolism.

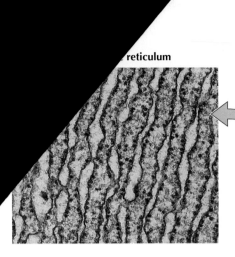 reticulum

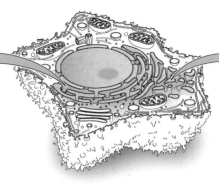

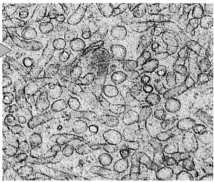 **(B) Smooth endoplasmic reticulum**

Figure 11.1 The endoplasmic reticulum (ER) (A) Electron micrograph of rough ER in rat liver cells. Ribosomes are attached to the cytosolic face of the ER membrane. (B) Electron micrograph of smooth ER in Leydig cells of the testis, which are active in steroid hormone synthesis.

The endoplasmic reticulum and protein secretion

The role of the endoplasmic reticulum in protein processing and sorting was first demonstrated by George Palade and his colleagues in the 1960s (**Figure 11.2**). These investigators studied the fate of newly synthesized proteins in specialized cells of the pancreas (pancreatic acinar cells) that secrete digestive enzymes into the small intestine. Because most of the

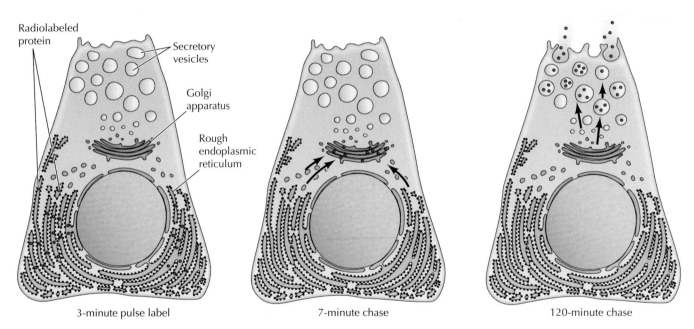

Figure 11.2 The secretory pathway Pancreatic acinar cells, which secrete most of their newly synthesized proteins into the digestive tract, were labeled with radioactive amino acids to study the intracellular pathway taken by secreted proteins. After a 3-minute incubation with radioactive amino acids (a "pulse"), autoradiography revealed that newly synthesized proteins were localized to the rough ER. Following further incubation with nonradioactive amino acids (a "chase"), proteins were found to move from the ER to the Golgi apparatus and then, within secretory vesicles, to the plasma membrane and cell exterior.

protein synthesized by these cells is secreted, Palade and coworkers were able to study the pathway taken by secreted proteins simply by labeling newly synthesized proteins with radioactive amino acids in a procedure known as a "pulse-chase" experiment. The location of the radiolabeled proteins within the cell was then determined by autoradiography and electron microscopy, revealing the cellular sites involved in the events leading to protein secretion. After a brief exposure of pancreatic acinar cells to radioactive amino acids (a "pulse"), newly synthesized proteins were detected in the rough ER, which was therefore identified as the site of synthesis of proteins destined for secretion. If the cells were then incubated for a short time in media containing nonradioactive amino acids (a "chase"), the radiolabeled proteins were detected in the Golgi apparatus. Following longer chase periods, the radiolabeled proteins traveled from the Golgi apparatus to the cell surface in **secretory vesicles**, which then fused with the plasma membrane to release their contents outside of the cell.

These experiments defined a pathway taken by secreted proteins—the **secretory pathway**: rough ER → Golgi → secretory vesicles → cell exterior. Further studies extended these results and demonstrated that this pathway is not restricted to proteins destined for secretion from the cell. Portions of it are shared by proteins destined for other compartments. Plasma membrane and lysosomal proteins also travel from the rough ER to the Golgi and then to their final destinations. Still other proteins travel through the initial steps of the secretory pathway but are then retained and function within either the ER or the Golgi apparatus.

The entrance of proteins into the ER thus represents a major branch point for the traffic of proteins within eukaryotic cells (**Figure 11.3**). Proteins destined for secretion or incorporation into the ER, Golgi apparatus, lysosomes, or

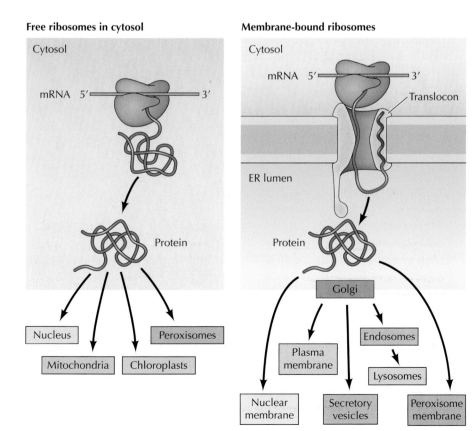

Figure 11.3 Overview of protein sorting   In higher eukaryotic cells, the initial sorting of proteins to the ER takes place while translation is in progress. Proteins synthesized on free ribosomes either remain in the cytosol or are transported to the nucleus, mitochondria, chloroplasts, or peroxisomes. In contrast, proteins synthesized on membrane-bound ribosomes are translocated directly into the ER through the translocon. These proteins contain signal sequences (indicated in red) that are cleaved during translocation. Proteins that are translocated into the ER may be either retained within the ER or transported to nuclear membranes, peroxisomal membranes or the Golgi apparatus and, from there, to endosomes, lysosomes, the plasma membrane, or the cell exterior via secretory vesicles.

plasma membrane are initially targeted to the ER, as are nuclear envelope and peroxisomal membrane proteins. In mammalian cells most proteins are transferred into the ER while they are being translated on membrane-bound ribosomes. In contrast, proteins destined to remain in the cytosol or to be incorporated into mitochondria, chloroplasts, or the interior of the nucleus or peroxisomes are synthesized on free ribosomes and released into the cytosol when their translation is complete.

Targeting proteins to the endoplasmic reticulum

Proteins can be translocated into the ER either during their synthesis on membrane-bound ribosomes (cotranslational translocation) or after their translation has been completed on free ribosomes in the cytosol (post-translational translocation). In mammalian cells, most proteins enter the ER cotranslationally, whereas both cotranslational and posttranslational pathways are used in yeast. The first step in the cotranslational pathway is the association of the ribosome-mRNA complex with the ER. Ribosomes are targeted for binding to the ER membrane by the amino acid sequence of the polypeptide chain being synthesized, rather than by intrinsic properties of the ribosome itself. Free and membrane-bound ribosomes are functionally indistinguishable, and protein synthesis generally initiates on ribosomes that are free in the cytosol. Ribosomes engaged in the synthesis of proteins that are destined for secretion are then targeted to the endoplasmic reticulum by a **signal sequence** at the amino terminus of the growing polypeptide chain. These signal sequences are short stretches of hydrophobic amino acids that are cleaved from the polypeptide chain during its transfer into the ER lumen.

The general role of signal sequences in targeting proteins to their appropriate locations within the cell was first elucidated by studies of the import of secretory proteins into the ER. These experiments used *in vitro* preparations of rough ER, which were isolated from cell extracts by density-gradient centrifugation (**Figure 11.4**). When cells are disrupted, the ER breaks up into

Animation 11.1

sites.sinauer.com/cooper7e/a11.1

Cotranslational Targeting of Secretory Proteins to the ER In mammals, proteins enter the ER primarily by a cotranslational pathway, a process that requires a signal sequence on the newly forming protein.

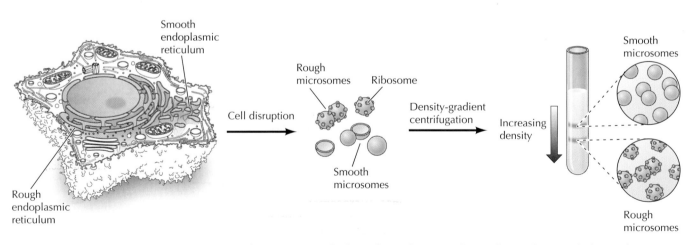

Figure 11.4 Isolation of rough ER When cells are disrupted, the ER fragments into small vesicles called microsomes. The microsomes derived from the rough ER (rough microsomes) are covered with ribosomes on their outer surface. Because ribosomes contain large amounts of RNA, the rough microsomes are denser than smooth microsomes and can be isolated by equilibrium density-gradient centrifugation (discussed in Chapter 1).

small vesicles called **microsomes**. Because the vesicles derived from the rough ER are covered with ribosomes, they can be separated from similar vesicles derived from the smooth ER or from other membranes (e.g., the plasma membrane). In particular, the large amount of RNA within ribosomes increases the density of the membrane vesicles to which they are attached, allowing purification of vesicles derived from the rough ER (rough microsomes) by equilibrium centrifugation in density gradients.

Günter Blobel and David Sabatini first proposed in 1971 that the signal for ribosome attachment to the ER might be an amino acid sequence near the amino terminus of the growing polypeptide chain. This hypothesis was supported by the results of *in vitro* translation of mRNAs encoding secreted proteins (such as immunoglobulins) (**Figure 11.5**). If an mRNA encoding a secreted protein was translated on free ribosomes *in vitro*, it was found that the protein produced was slightly larger than the normal secreted protein. If microsomes were added to the system, however, the *in vitro* translated protein was incorporated into the microsomes and cleaved to the correct size. These experiments led to a more detailed formulation of the signal hypothesis, which proposed that an amino terminal signal sequence targets the polypeptide chain to the microsomes and is then cleaved by a microsomal protease. Many subsequent findings have substantiated this model, including recombinant DNA experiments demonstrating that addition of a

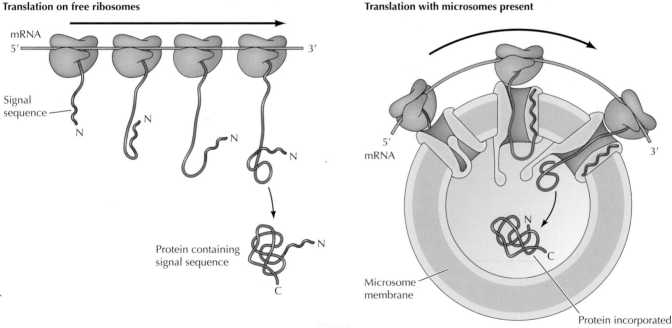

Translation on free ribosomes

Translation with microsomes present

Figure 11.5 Incorporation of secretory proteins into microsomes Secretory proteins are targeted to the ER by a signal sequence at their amino (N) terminus, which is removed during incorporation of the growing polypeptide chain into the ER. This was demonstrated by experiments showing that translation of secretory protein mRNAs on free ribosomes yielded proteins that retained their signal sequences and were therefore slightly larger than the normal secreted proteins. However, when microsomes were added to the system, the growing polypeptide chains were incorporated into the microsomes and the signal sequences were removed by proteolytic cleavage.

The Signal Hypothesis

Transfer of Proteins across Membranes. I. Presence of Proteolytically Processed and Unprocessed Nascent Immunoglobulin Light Chains on Membrane-Bound Ribosomes of Murine Myeloma

Günter Blobel and Bernhard Dobberstein

Rockefeller University, New York

Journal of Cell Biology, 1975, Volume 67, pages 835–851

The Context

How are specific polypeptide chains transferred across the appropriate membranes? Studies in the 1950s and 1960s indicated that secreted proteins are synthesized on membrane-bound ribosomes and transferred across the membrane during their synthesis. However, this did not explain why ribosomes—engaged in the synthesis of secreted proteins—attach to membranes, while ribosomes synthesizing cytosolic proteins do not. A hypothesis to explain this difference was first suggested by Günter Blobel and David Sabatini in 1971. At that time, they proposed that (1) mRNAs to be translated on membrane-bound ribosomes contain a unique set of codons just 3′ of the translation initiation site, (2) translation of these codons yields a unique sequence at the amino terminus of the growing polypeptide chain (the signal sequence), and (3) the signal sequence triggers attachment of the ribosome to the membrane. In 1975 Blobel and Dobberstein reported a series of experiments that provided critical support for this notion. In addition, they proposed "a somewhat more detailed version of this hypothesis, henceforth referred to as the signal hypothesis."

The Experiments

Myelomas are cancers of B lymphocytes that actively secrete immunoglobulins, so they provide a good model for studies of secreted proteins. Previous studies in César Milstein's laboratory had shown that the proteins produced by *in vitro* translation of immunoglobulin light-chain mRNA

contain about 20 amino acids at their amino terminus that are not present in the secreted light chains. This result led to the suggestion that these amino acids direct binding of the ribosome to the membrane. To test this idea, Blobel and Dobberstein investigated the synthesis of light chains by membrane-bound ribosomes from myeloma cells.

As expected from earlier work, *in vitro* translation of light-chain mRNA on free ribosomes yielded a protein that was larger than the secreted light chain (see figure). In contrast, *in vitro* translation of mRNA associated with membrane-bound ribosomes from myeloma cells yielded a protein that was the same size as the normally secreted light chain. Moreover, the light chains synthesized by ribosomes that remained bound to microsomes were resistant to digestion by added proteases, indicating that the light chains had been transferred into the microsomes.

These results indicated that an amino-terminal signal sequence is removed by a microsomal protease as growing polypeptide chains are transferred across the membrane. The results were interpreted in terms of a more detailed version of the signal hypothesis. As stated by Blobel and Dobberstein, "the essential feature of the signal hypothesis is the occurrence of a unique sequence of codons, located immediately to the right of the initiation codon, which is present only in those mRNAs whose translation products are to be transferred across a membrane."

The Impact

The selective transfer of proteins across membranes is critical to the maintenance of the membrane-enclosed organelles of eukaryotic cells. To maintain the identity of these organelles, proteins must be translocated specifically across the appropriate membranes. The signal hypothesis provided the conceptual basis for understanding this phenomenon. Not only has this basic model been firmly substantiated for the transfer of secreted proteins into the endoplasmic reticulum, but it also has provided the framework for understanding the targeting of proteins to all the

Günter Blobel

other membrane-enclosed compartments of the cell, thereby impacting virtually all areas of cell biology.

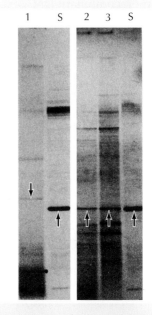

In vitro translation of immunoglobulin light-chain mRNA on free ribosomes (lane 1) yields a product that migrates slower than secreted light chains (lane S) in gel electrophoresis. In contrast, light chains synthesized by *in vitro* translation on membrane-bound ribosomes (lane 2) are the same size as secreted light chains. In addition, the products of *in vitro* translation on membrane-bound ribosomes were unaffected by subsequent digestion with proteases (lane 3), indicating that they were protected from the proteases by insertion into microsomes.

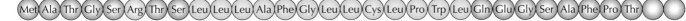

Cleavage site of
signal peptidase

Met Ala Thr Gly Ser Arg Thr Ser Leu Leu Leu Ala Phe Gly Leu Leu Cys Leu Pro Trp Leu Gln Glu Gly Ser Ala Phe Pro Thr

Figure 11.6 The signal sequence of growth hormone Most signal sequences (e.g., for growth hormone) contain a stretch of hydrophobic amino acids (yellow) preceded by basic residues (e.g., arginine, green).

signal sequence to a normally nonsecreted protein is sufficient to direct the incorporation of the recombinant protein into the rough ER.

The mechanism by which secretory proteins are targeted to the ER during their translation (the cotranslational pathway) is now well understood. The signal sequences span about 15–40 amino acids, including a stretch of 7–12 hydrophobic residues, usually located at the amino terminus of the polypeptide chain (Figure 11.6). As they emerge from the ribosome, signal sequences are recognized and bound by the **signal recognition particle (SRP)** consisting of six polypeptides and a small cytoplasmic RNA (**SRP RNA**). The SRP binds the ribosome as well as the signal sequence, inhibiting further translation and targeting the entire complex (the SRP, ribosome, mRNA, and growing polypeptide chain) to the rough ER by binding to the **SRP receptor** on the ER membrane (Figure 11.7). Binding to the receptor releases

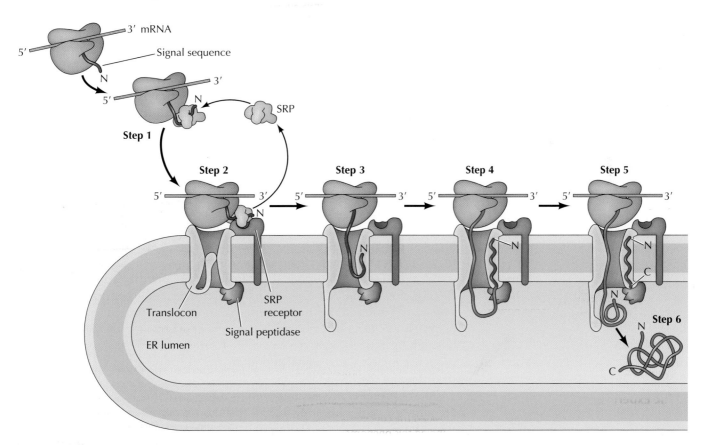

Figure 11.7 Cotranslational targeting of secretory proteins to the ER Step 1: As the signal sequence emerges from the ribosome, it is recognized and bound by the signal recognition particle (SRP). Step 2: The SRP escorts the complex to the ER membrane where it binds to the SRP receptor. Step 3: The SRP is released, the ribosome binds to the translocon, and insertion of the signal sequence opens the translocon. Step 4: Translation resumes and the signal sequence is cleaved by signal peptidase. Step 5: Continued translation drives translocation of the growing polypeptide chain across the membrane. Step 6: The completed polypeptide chain is released within the ER lumen.

(A) Top view

(B) Lateral view

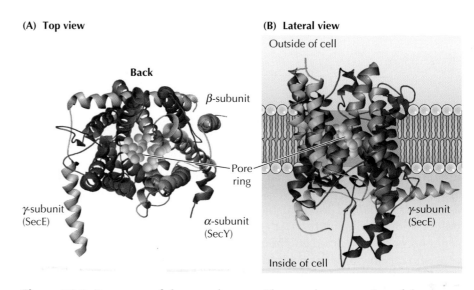

Figure 11.8 Structure of the translocon The translocon consists of three transmembrane subunits, shown in green, purple, and yellow. (A) Top view from the cytosol, showing the plug in the translocon channel. (B) Lateral view of the translocon inserted into the ER membrane. (After E. Park and T. A. Rapoport, 2012. *Ann. Rev. Biophys.* 41:21.)

the SRP from both the ribosome and the signal sequence of the growing polypeptide chain. The ribosome then binds to a protein translocation complex or **translocon** in the ER membrane, and the signal sequence is inserted into a membrane channel.

Key insights into the process of translocation through the ER membrane came from determination of the translocon structure by Tom Rapoport and his colleagues in 2004. In both yeast and mammalian cells, the translocons through the ER membrane are complexes of three transmembrane proteins called the Sec61 proteins (**Figure 11.8**). The yeast and mammalian translocon proteins are closely related to the plasma membrane proteins that translocate secreted polypeptides in bacteria, demonstrating a striking conservation of the protein secretion machinery in prokaryotic and eukaryotic cells. Insertion of the signal sequence opens the translocon by moving a plug away from the translocon channel. This allows the growing polypeptide chain to be transferred through the translocon as translation proceeds. Thus the process of protein synthesis directly drives the transfer of growing polypeptide chains through the translocon and into the ER. As translocation proceeds, the signal sequence is cleaved by **signal peptidase** and the polypeptide is released into the lumen of the ER.

Many proteins in yeast, as well as a few proteins in mammalian cells, are targeted to the ER after their translation is complete (posttranslational translocation) rather than being transferred into the ER during synthesis on membrane-bound ribosomes. These proteins are synthesized on free cytosolic ribosomes, and their posttranslational incorporation into the ER does not require the SRP. Instead, their signal sequences are recognized by distinct receptor proteins (the Sec62/63 complex) associated with the translocon in the ER membrane (**Figure 11.9**). Cytosolic Hsp70 and Hsp40 chaperones are

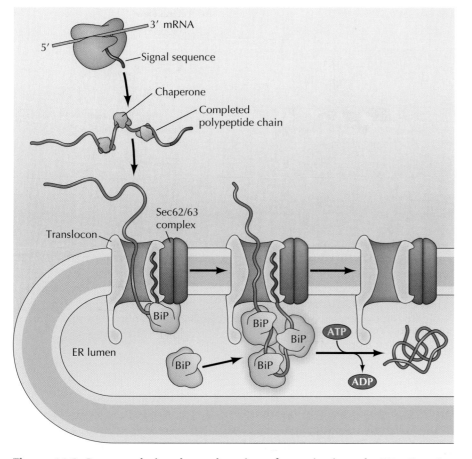

Figure 11.9 Posttranslational translocation of proteins into the ER   Proteins destined for posttranslational import to the ER are synthesized on free ribosomes and maintained in an unfolded conformation by cytosolic chaperones. They are recognized by the Sec62/63 complex and their signal sequences are inserted into the translocon in the ER membrane. The Sec63 protein is also associated with a chaperone protein called BiP. The binding of multiple molecules of BiP and their release (coupled to ATP hydrolysis) drives translocation of the polypeptide chain into the ER.

required to maintain the polypeptide chains in an unfolded conformation so they can enter the translocon, and another Hsp70 chaperone within the ER (called BiP) is required to pull the polypeptide chain through the channel and into the ER. Multiple molecules of BiP bind to the polypeptide chain as it is translocated into the ER, preventing it from sliding backwards and effectively pulling it through the channel. The posttranslational translocation of proteins into the ER is thus driven by BiP, whereas the cotranslational translocation of growing polypeptide chains is driven directly by the process of protein synthesis.

Insertion of proteins into the ER membrane

Proteins destined for secretion from the cell or residence within the lumen of the ER, Golgi apparatus, or lysosomes are translocated across the ER membrane and released into the lumen of the ER as already described. However,

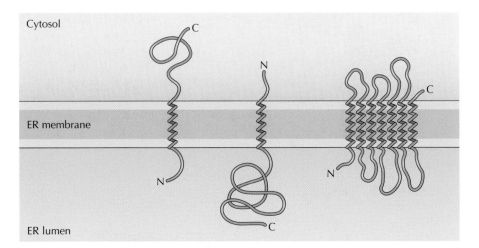

Cytosol

ER membrane

ER lumen

Figure 11.10 Orientations of membrane proteins Integral membrane proteins span the membrane via α-helical regions of 20 to 25 hydrophobic amino acids, which can be inserted in a variety of orientations. The proteins at left and center each span the membrane once, but they differ in whether the carboxy (C) or amino (N) terminus is on the cytosolic side. On the right is an example of a protein that has multiple membrane-spanning regions.

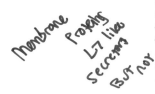

proteins destined for incorporation into the plasma membrane or the membranes of these compartments are initially inserted into the ER membrane instead of being released into the lumen. From the ER membrane, they proceed to their final destination along the same pathway as that of secretory proteins: ER → Golgi → plasma membrane or lysosomes, transported as membrane components rather than as soluble proteins.

Integral membrane proteins are embedded in the membrane by hydrophobic sequences that span the phospholipid bilayer (see Figure 2.34). The membrane-spanning portions of these proteins are usually α helical regions consisting of 20 to 25 hydrophobic amino acids. The formation of an α helix maximizes hydrogen bonding between the peptide bonds, and the hydrophobic amino acid side chains interact with the fatty acid tails of the phospholipids in the bilayer. However, different integral membrane proteins vary in how they are inserted (Figure 11.10). For example, whereas some integral membrane proteins span the membrane only once, others have multiple membrane-spanning regions. In addition, some proteins are oriented in the membrane with their carboxy terminus on the cytosolic side; others have their amino terminus exposed to the cytosol. These orientations of membrane proteins are established as the growing polypeptide chains are translocated into the ER. The lumen of the ER is topologically equivalent to the exterior of the cell, so the domains of plasma membrane proteins that are exposed on the cell surface correspond to the regions of polypeptide chains that are translocated into the ER lumen (Figure 11.11).

Most proteins are inserted into the ER membrane by the SRP/Sec61 cotranslational pathway described above. Many of the proteins are inserted directly into the ER membrane by internal transmembrane sequences that are recognized and brought to the translocon by SRP, but not cleaved by

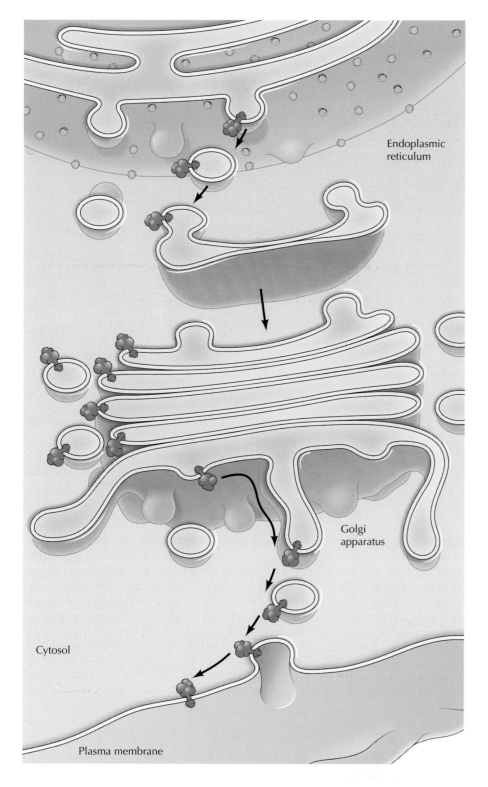

Endoplasmic
reticulum

Golgi
apparatus

Cytosol

Plasma membrane

Figure 11.11 Topology of the secretory pathway   The lumens of the endoplasmic reticulum and Golgi apparatus are topologically equivalent to the exterior of the cell. Consequently, those portions of polypeptide chains that are translocated into the ER are exposed on the cell surface following transport to the plasma membrane.

(A)

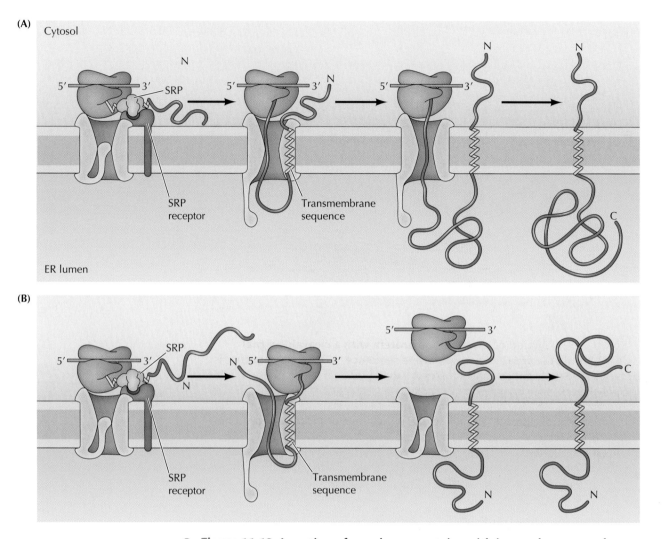

(B)

○ **Figure 11.12 Insertion of membrane proteins with internal transmembrane sequences** Internal transmembrane sequences can lead to the insertion of polypeptide chains in either orientation in the ER membrane. (A) The transmembrane sequence directs insertion of the polypeptide such that its amino (N) terminus is exposed on the cytosolic side. The transmembrane sequence exits the translocon to anchor the protein in the lipid bilayer and the remainder of the polypeptide chain is translocated into the ER as translation proceeds. (B) Other internal transmembrane sequences are oriented to direct the transfer of the amino-terminal portion of the polypeptide across the membrane. Continued translation results in a protein that spans the ER membrane with its amino terminus in the lumen and its carboxy (C) terminus in the cytosol.

signal peptidase (**Figure 11.12**). Instead, these transmembrane α helices exit the translocon laterally and anchor proteins in the ER membrane. The hydrophobic transmembrane sequence signals a change in the translocon, causing the membrane-spanning helices of the translocon to open and allowing the hydrophobic transmembrane domain of the protein to exit the translocon into the lipid bilayer. Importantly, transmembrane sequences can be oriented so as to direct the translocation of either the amino or carboxy

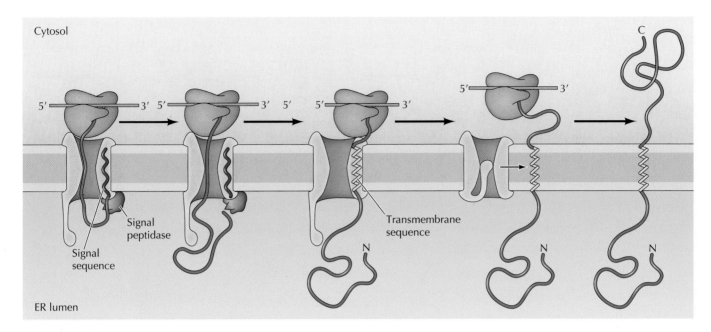

Cytosol

Signal
peptidase

Signal
sequence

Transmembrane
sequence

ER lumen

Figure 11.13 Insertion of a membrane protein with a cleavable signal sequence and an internal transmembrane sequence   The signal sequence is cleaved as the polypeptide chain crosses the membrane, so the amino terminus of the polypeptide chain is exposed in the ER lumen. However, translocation of the polypeptide chain across the membrane is halted when the translocon recognizes a transmembrane sequence. This allows the protein to exit the translocon laterally and become anchored in the ER membrane. Continued translation results in a membrane-spanning protein with its carboxy terminus on the cytosolic side.

terminus of the polypeptide chain across the membrane. Therefore, depending on the orientation of the transmembrane sequence, proteins inserted into the membrane by this mechanism can have either their amino or carboxy terminus exposed to the cytosol.

Some transmembrane proteins oriented with their carboxy termini exposed to the cytosol are inserted into the ER membrane by an alternative mechanism (Figure 11.13). These proteins have a normal amino terminal signal sequence, which is cleaved by signal peptidase during translocation of the polypeptide chain through the translocon. The growing amino terminal portion of the polypeptide is then translocated into the ER as translation proceeds. Translocation of the polypeptide is halted by a membrane-spanning α helix in the middle of the protein, which anchors the polypeptide in the membrane. The exit of this membrane-spanning α helix from the translocon blocks further translocation of the polypeptide, so the carboxy terminal portion of the growing polypeptide chain remains in the cytosol. This mechanism is preferentially utilized by proteins with large amino terminal domains inserted into the ER, since the translocation of these amino terminal domains across the membrane is driven cotranslationally. as trans. goes on

Proteins that span the membrane multiple times are inserted as a result of a series of transmembrane sequences with alternating orientations. For example, an internal transmembrane sequence can result in membrane insertion

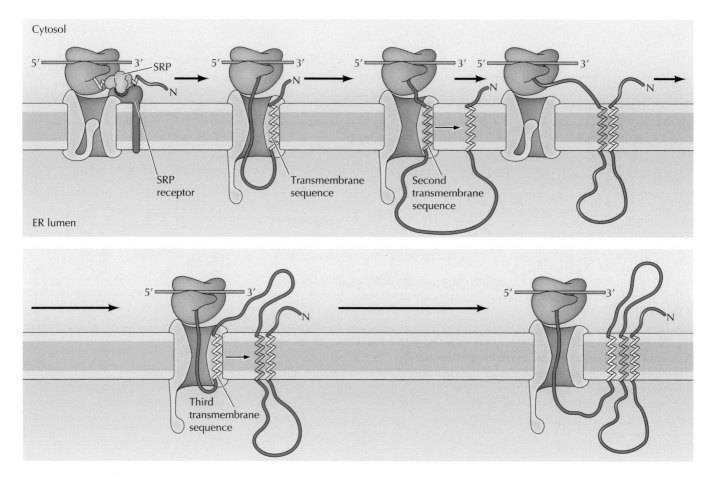

Figure 11.14 Insertion of a protein that spans the membrane multiple times In this example, an internal transmembrane sequence results in insertion of the polypeptide chain with its amino terminus on the cytosolic side of the membrane. Translation proceeds until a second transmembrane sequence is encountered. This causes the polypeptide chain to form a loop within the lumen of the ER; translation continues in the cytosol. A third transmembrane sequence reopens the channel, triggering reinsertion of the polypeptide chain into the translocon and forming a loop in the cytosol. The process can be repeated many times, resulting in the insertion of proteins with multiple membrane-spanning regions.

of a polypeptide chain with its amino terminus on the cytosolic side (**Figure 11.14**). Continuing translation then drives the synthesis of the polypeptide chain inside the ER lumen. If a second membrane-spanning domain is then encountered, the polypeptide chain will exit the translocon, forming a loop in the ER lumen, and protein synthesis will continue on the cytosolic side of the membrane. If a third transmembrane sequence is encountered, the growing polypeptide chain will again be inserted into the ER, forming another looped domain on the cytosolic side of the membrane. This can be followed by yet more transmembrane domains, resulting in the insertion of proteins that span the membrane multiple times, with looped domains exposed on both the lumenal and cytosolic sides.

Some proteins with a transmembrane sequence at their carboxy terminus are inserted into the ER membrane by an alternative posttranslational

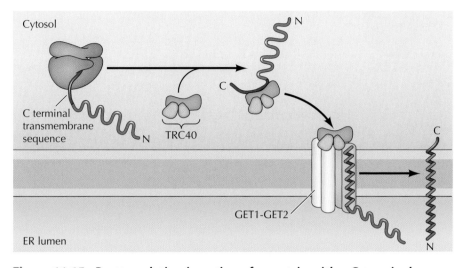

Figure 11.15 Posttranslation insertion of a protein with a C-terminal transmembrane sequence Proteins with a transmembrane sequence at their carboxy (C) terminus are not recognized by SRP because the C-terminal sequence does not exit the ribosome until translation is complete. Instead, these proteins are recognized posttranslationally by the targeting factor TRC40, which brings them to the GET1-GET2 receptor. They are inserted into the ER membrane with their short C-terminal domains on the cytosolic side.

pathway. These proteins cannot be recognized by the SRP because their carboxy-terminal transmembrane domain does not emerge from the ribosome until translation is terminated and the completed polypeptide chain is released from the ribosome. Instead, the transmembrane domains of these proteins are recognized by a distinct targeting factor (called TRC40 or GET3) once translation is complete (**Figure 11.15**). TRC40 then escorts the transmembrane protein to the ER membrane, where it is inserted via the GET1-GET2 receptor.

As discussed below, most transmembrane proteins destined for other compartments in the secretory pathway are delivered to them in transport vesicles. However, proteins destined for the nuclear envelope (which is continuous with the ER) move laterally in the plane of the membrane rather than being transported by vesicles. Recent studies suggest that inner nuclear membrane proteins (such as emerin or the lamin B receptor; see Chapter 10) contain specific transmembrane sequences that signal their transport to the inner nuclear membrane, where they are retained by interactions with nuclear components such as lamins or chromatin.

Protein folding and processing in the ER

The folding of polypeptide chains into their correct three-dimensional conformations, the assembly of polypeptides into multisubunit proteins, and the covalent modifications involved in protein processing were discussed in Chapter 9. For proteins that enter the secretory pathway, many of these events occur either during translocation across the ER membrane or within the ER lumen. One such processing event is the proteolytic cleavage of the signal sequence as the polypeptide chain is translocated across the ER membrane. The ER is also the site of protein folding, assembly of multisubunit

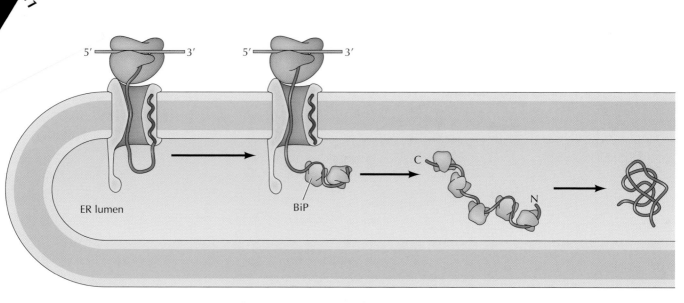

ER lumen

BiP

C

N

Figure 11.16 Protein folding in the ER The molecular chaperone BiP binds to polypeptide chains as they cross the ER membrane and facilitates protein folding and assembly within the ER.

proteins, disulfide bond formation, *N*-linked glycosylation, and the addition of glycolipid anchors to some plasma membrane proteins. In fact, the primary role of lumenal ER proteins is to assist the folding and assembly of newly translocated polypeptides.

As already discussed, proteins are translocated across the ER membrane as unfolded polypeptide chains while their translation is still in progress. These polypeptides, therefore, fold into their three-dimensional conformations within the ER, assisted by molecular chaperones that facilitate the folding of polypeptide chains (see Chapter 9). The Hsp70 chaperone, BiP, is thought to bind to the unfolded polypeptide chain as it crosses the membrane and then mediate protein folding and the assembly of multisubunit proteins within the ER (**Figure 11.16**). Correctly assembled proteins are released from BiP (and other chaperones) and are available for transport to the Golgi apparatus. Abnormally folded or improperly assembled proteins are targets for degradation, as will be discussed later.

The formation of disulfide bonds between the side chains of cysteine residues is an important aspect of protein folding and assembly within the ER. These bonds generally do not form in the cytosol, which is characterized by a reducing environment that maintains most cysteine residues in their reduced (—SH) state. In the ER, however, an oxidizing environment promotes disulfide (S—S) bond formation, and disulfide bonds formed in the ER play important roles in the structure of secreted and cell surface proteins. Disulfide bond formation is facilitated by the enzyme **protein disulfide isomerase (PDI)** (see Figure 9.28), which is located in the ER lumen.

Proteins are also glycosylated on specific asparagine residues (*N*-linked glycosylation) within the ER while their translation is still in process (**Figure 11.17**). As discussed in Chapter 9 (see Figure 9.33), oligosaccharide units consisting of 14 sugar residues are added to acceptor asparagine residues of growing polypeptide chains as they are translocated into the ER. The

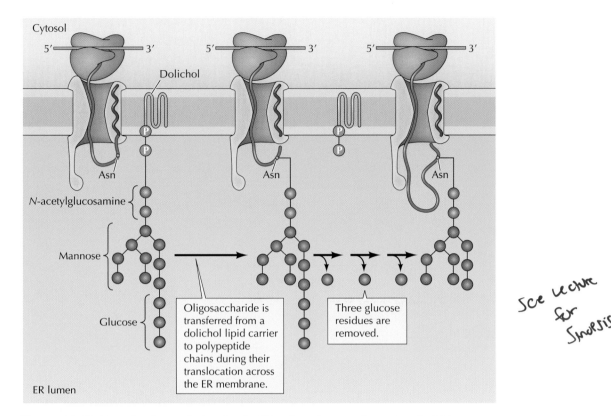

See lecture for Snopsis

Figure 11.17 Protein glycosylation in the ER   Proteins are glycosylated within the ER by the addition of a 14-sugar oligosaccharide to an acceptor asparargine (Asn) residue. Three glucose residues are removed while the protein is still within the ER.

oligosaccharide is synthesized on a lipid (dolichol) carrier anchored in the ER membrane. It is then transferred as a unit to acceptor asparagine residues in the consensus sequence Asn-X-Ser/Thr by a membrane-bound enzyme called oligosaccharyl transferase. Three glucose residues are removed while the protein is still within the ER, and the protein is modified further after being transported to the Golgi apparatus (discussed later in this chapter). Glycosylation helps to prevent protein aggregation in the ER and provides signals that promote protein folding and subsequent sorting in the secretory pathway.

Some proteins are attached to the plasma membrane by glycolipids rather than by membrane-spanning regions of the polypeptide chain. Because these membrane-anchoring glycolipids contain phosphatidylinositol, they are 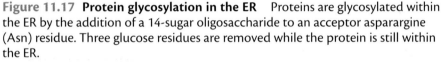 called **glycosylphosphatidylinositol (GPI) anchors**, the structure of which was illustrated in Figure 9.38. The GPI anchors are assembled in the ER membrane. They are then added immediately after completion of protein synthesis to the carboxy terminus of some proteins that are retained in the membrane by a C-terminal hydrophobic sequence (**Figure 11.18**). The C-terminal sequence of the protein is cleaved and exchanged for the GPI anchor, so these proteins remain attached to the membrane only by their associated glycolipid. Like transmembrane proteins, they are transported to the cell surface as membrane components via the secretory pathway. Their orientation within the

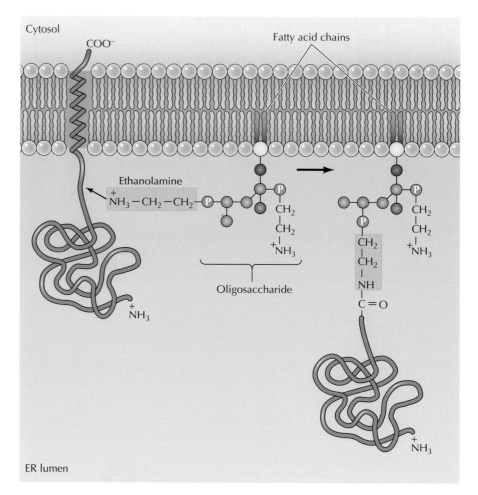

Figure 11.18 Addition of GPI anchors   Glycosylphosphatidylinositol (GPI) anchors contain two fatty acid chains, an oligosaccharide portion consisting of inositol and other sugars, and ethanolamine (see Figure 9.38 for a more detailed structure). The GPI anchors are assembled in the ER and added to polypeptides anchored in the membrane by a carboxy-terminal membrane-spanning region. The membrane-spanning region is then cleaved, and the new carboxy terminus is joined to the NH_2 group of ethanolamine immediately after translation is completed, leaving the protein attached to the membrane by the GPI anchor.

ER dictates that GPI-anchored proteins are exposed on the outside of the cell, with the GPI anchor mediating their attachment to the plasma membrane.

Quality control in the ER

Many proteins synthesized in the ER are rapidly degraded, primarily because they fail to fold correctly. Protein folding in the ER is slow and inefficient, and many proteins never reach their correctly folded conformations. Such misfolded proteins are removed from the ER by a process referred to as **ER-associated degradation (ERAD)**, in which misfolded proteins are identified, returned from the ER to the cytosol, and degraded by the ubiquitin-proteasome system.

Because they assist proteins in correct folding, chaperones and protein processing enzymes in the ER lumen often act as sensors of misfolded proteins. One well-characterized pathway of glycoprotein folding involves the related chaperones **calnexin** and **calreticulin**, which are located in the ER lumen and the ER membrane, respectively. Both calnexin and calreticulin recognize the partially-processed oligosaccharides from which two terminal glucose residues have been removed (**Figure 11.19**). Calnexin or calreticulin, in association with a protein disulfide isomerase and a peptidyl prolyl

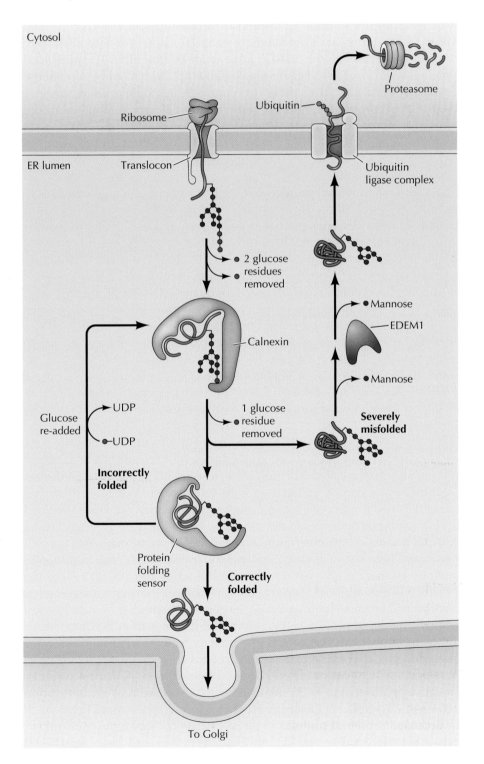

Figure 11.19 Glycoprotein folding by calnexin   As the glycoprotein exits the translocon, two glucose residues are removed, allowing calnexin to bind and assist in folding. Removal of the remaining glucose residue terminates the interaction with calnexin, releasing the glycoprotein. A protein-folding sensor then assesses the extent of folding of the glycoprotein by monitoring exposed hydrophobic regions. If none are found, the glycoprotein is correctly folded and proceeds to exit the ER. If the glycoprotein is incorrectly folded, the folding sensor, which is a glucosyltransferase, will add back a glucose residue, allowing the glycoprotein to again bind calnexin for another attempt at correct folding. However, if too many hydrophobic regions are exposed and the protein cannot be properly folded, the protein is recognized by EDEM1, which removes mannose residues. This prevents the protein from being bound by calnexin and the protein is targeted back to the cytosol through a ubiquitin ligase complex in the ER membrane. The protein is ubiquitylated at the cytosolic side of this complex and degraded in the proteasome.

isomerase (see Figures 9.28 and 9.29), then assist the glycoprotein in folding into its correct conformation. The glycoprotein is then released by removal of the third terminal glucose residue from the oligosaccharide. This allows the glycoprotein to be recognized by a protein-folding sensor, which monitors whether the glycoprotein has achieved a fully folded state. If so, the glycoprotein moves on to exit the ER and travel to the Golgi. However, if the glycoprotein is not correctly folded, the folding sensor will add back a glucose residue to the oligosaccharide, allowing it to cycle back to calnexin or calreticulin for another attempt at correct folding. If the glycoprotein fails to fold correctly after repeated cycles and/or is irreversibly misfolded, it is instead targeted to the ERAD pathway for degradation. The misfolded protein is recognized by an enzyme called EDEM1, which removes mannose residues from the oligosaccharide. The removal of mannose residues prevents the misfolded proteins from being sent back to calnexin or calreticulin, and the misfolded proteins are transferred to a transmembrane complex with ubiquitin ligase activity. They are then translocated through this complex back to the cytosol, where they are ubiquitylated and degraded in the proteasome (see Figure 9.46).

The level of unfolded proteins in the ER is monitored in order to coordinate the protein-folding capacity of the ER with the physiological needs of the cell. This regulation is mediated by a signaling pathway known as the **unfolded protein response (UPR)**, which is activated if an excess of unfolded proteins accumulates in the ER (**Figure 11.20**). Activation of the unfolded protein response pathway leads to expansion of the ER and production of additional chaperones to meet the need for increased protein folding, as well as a transient reduction in the amount of newly synthesized proteins entering the ER. If these changes are insufficient to adjust protein folding in the ER to a normal level, sustained activity of the unfolded protein response leads to programmed cell death (see Chapter 18), thereby eliminating cells that are unable to properly fold proteins from the body.

In mammalian cells, the unfolded protein response results from the activity of three signaling molecules or stress sensors in the ER membrane (IRE1, ATF6, and PERK), which are activated by unfolded proteins in the ER lumen (see Figure 11.20). Activation of IRE1 (which is the only sensor in yeast) results in cleavage and activation of the mRNA encoding a transcription factor called XBP1 on the cytosolic side of the ER membrane. The XBP1 transcription factor then induces expression of genes involved in the unfolded protein response, including chaperones, enzymes involved in lipid synthesis, and ERAD proteins. The second sensor, ATF6, is itself a transcription factor, which is sequestered in the ER membrane in unstressed cells. Unfolded proteins in the ER signal the transport of ATF6 to the Golgi apparatus, where it is cleaved to release the active form of ATF6 into the cytosol. ATF6 then translocates to the nucleus and induces the expression of additional unfolded protein response genes. The third sensor, PERK, is a protein kinase that phosphorylates and inhibits translation initiation factor eIF2 (see Figure 9.21). This results in a general decrease in protein synthesis and a reduction in the load of unfolded proteins entering the ER. In addition, inhibition of eIF2 leads to the preferential translation of some mRNAs. As a result, activation of PERK leads to increased expression of a transcription factor ATF4, which further contributes to the induction of genes involved in the unfolded protein response.

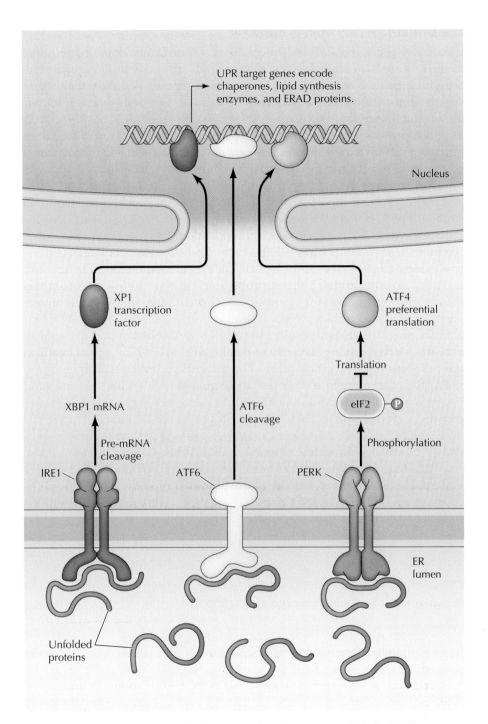

Figure 11.20 Unfolded protein response (UPR) Unfolded proteins activate three receptors in the ER membrane. The first, IRE1, cleaves pre-mRNA of a transcription factor (XBP1). This leads to synthesis of XBP1, which translocates to the nucleus and stimulates transcription of UPR target genes. The second receptor, ATF6, is cleaved to release the active ATF6 transcription factor. The third receptor, PERK, is a protein kinase that phosphorylates the translation factor eIF2. This inhibits general translation, reducing the amount of protein entering the ER. It also results in preferential translation of the transcription factor ATF4, which further contributes to the induction of UPR target genes encoding chaperones, enzymes involved in lipid synthesis and ERAD proteins.

The smooth ER and lipid synthesis

In addition to its activities in the processing of secreted and membrane proteins, the ER is the major site at which membrane lipids are synthesized in eukaryotic cells. Because they are extremely hydrophobic, membrane lipids are synthesized in association with already existing cellular membranes rather than in the aqueous environment of the cytosol. Although some lipids are synthesized in association with other membranes, most are synthesized in the ER and then transported from the ER to their ultimate destinations in other membranes.

As discussed in Chapter 2, the membranes of eukaryotic cells are composed of three main types of lipids: phospholipids, glycolipids, and cholesterol (see Figures 2.7, 2.8 and 2.9). Most of the phospholipids, which are the basic structural components of the membrane, are derived from glycerol. They are synthesized on the cytosolic side of the ER membrane from water-soluble cytosolic precursors (**Figure 11.21**). Fatty acids are first transferred from coenzyme A carriers to glycerol-3-phosphate by membrane-bound enzymes, and the resulting phospholipid (phosphatidic acid) is inserted into the membrane. Enzymes on the cytosolic face of the ER membrane then convert phosphatidic acid to diacylglycerol and catalyze the addition of different polar head groups, resulting in formation of phosphatidylcholine, phosphatidylserine, phosphatidylethanolamine, and phosphatidylinositol. The synthesis of these phospholipids on the cytosolic side of the ER membrane allows the hydrophobic fatty acid chains to remain buried in the membrane

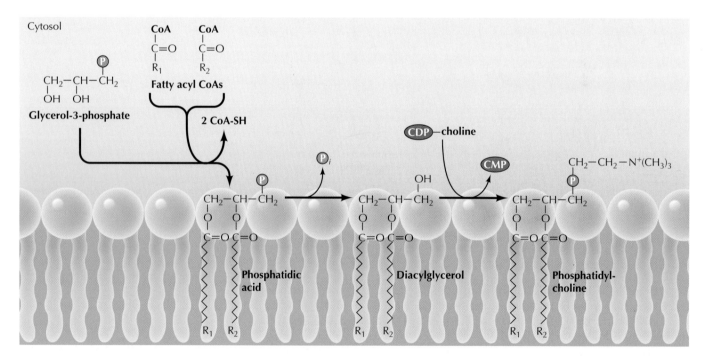

Figure 11.21 Synthesis of a phospholipid Glycerol phospholipids are synthesized in the ER membrane from cytosolic precursors. Two fatty acids linked to coenzyme A (CoA) carriers are first joined to glycerol-3-phosphate, yielding phosphatidic acid, which is simultaneously inserted into the membrane. A phosphatase then converts phosphatidic acid to diacylglycerol, which is converted to phosphatidylcholine by addition of a polar phosphocholine head group.

while membrane-bound enzymes catalyze their reactions with water-soluble precursors (e.g., CDP-choline) in the cytosol. Because of this topography, however, new phospholipids are added only to the cytosolic half of the ER membrane (**Figure 11.22**). To maintain a stable membrane, some of these newly synthesized phospholipids must therefore be transferred to the other (lumenal) half of the ER bilayer. This transfer, which requires the passage of a polar head group through the membrane, is facilitated by membrane proteins called **flippases**. By catalyzing the rapid translocation of phospholipids across the ER membrane, the flippases ensure even growth of both halves of the bilayer. Different families of these enzymes, some of which are specific for particular phospholipids, are found in virtually all cell membranes.

In addition to its role in synthesis of the glycerol phospholipids, the ER also serves as the major site of synthesis of cholesterol and ceramide, which is converted to either glycolipids or sphingomyelin (the only membrane phospholipid not derived from glycerol) in the Golgi apparatus. The ER is thus responsible for synthesis of either the final products or the precursors of all the major lipids of eukaryotic membranes.

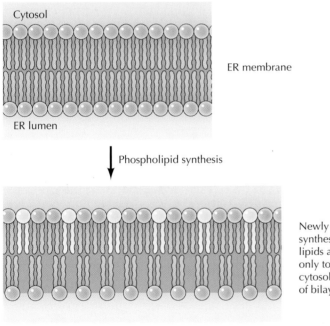

Cytosol

ER membrane

ER lumen

Phospholipid synthesis

Newly synthesized lipids added only to cytosolic half of bilayer

Lipid transfer facilitated by flippases

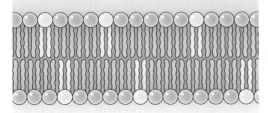

Growth of both halves of phospholipid bilayer

Figure 11.22 Translocation of phospholipids across the ER membrane Because phospholipids are synthesized on the cytosolic side of the ER membrane, they are added only to the cytosolic half of the bilayer. They are then translocated across the membrane by phospholipid flippases, resulting in even growth of both halves of the phospholipid bilayer.

Smooth ER is abundant in cell types that are particularly active in lipid metabolism. For example, steroid hormones are synthesized (from cholesterol) in the ER, so large amounts of smooth ER are found in steroid-producing cells, such as those in the testis and ovary. In addition, smooth ER is abundant in the liver where it contains enzymes that metabolize various lipid-soluble compounds. These detoxifying enzymes inactivate a number of potentially harmful drugs (e.g., phenobarbital) by converting them to water-soluble compounds that can be eliminated from the body in the urine. The smooth ER is thus involved in multiple aspects of the metabolism of lipids and lipid-soluble compounds.

Export of proteins and lipids from the ER

Both proteins and phospholipids travel along the secretory pathway in transport vesicles, which bud from the membrane of one organelle and then fuse with the membrane of another. Thus molecules are exported from the ER in vesicles that bud from a specialized region of the ER, called the ER exit site or ERES (**Figure 11.23**). The vesicles that bud from the ER fuse to form the ER–Golgi intermediate compartment (ERGIC), from which cargo is transported

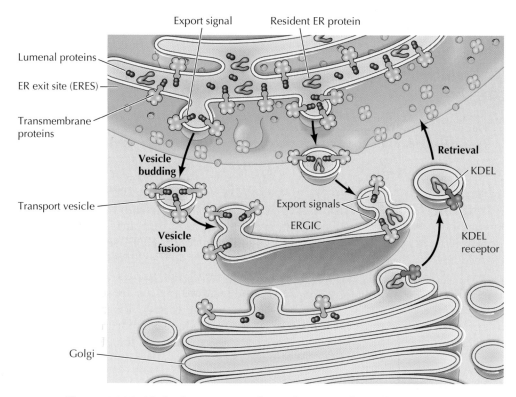

Figure 11.23 Vesicular transport from the ER to the Golgi Proteins and lipids are carried from the ER to the Golgi in transport vesicles that bud from ER exit sites (ERES), fuse to form the vesicles and tubules of the ER–Golgi intermediate compartment (ERGIC), and are then carried to the Golgi. Lumenal ER proteins targeted for the Golgi are bound by transmembrane proteins that are selectively packaged into vesicles. Resident ER proteins destined to remain in the lumen of the ER are marked by KDEL retrieval sequences at their carboxy terminus. If these proteins are exported from the ER to the Golgi, they are recognized by a recycling receptor in the ERGIC or the Golgi and selectively returned to the ER.

to the Golgi apparatus. Subsequent steps in the secretory pathway involve transport between different compartments of the Golgi and from the Golgi to endosomes, lysosomes, or the plasma membrane. In most cases, proteins within the lumen of one organelle are packaged into budding transport vesicles and then released into the lumen of the recipient organelle following vesicle fusion. Membrane proteins and lipids are transported similarly, and it is significant that their topological orientation is maintained as they travel from one membrane-enclosed organelle to another. For example, the domains of a protein exposed on the cytosolic side of the ER membrane will also be exposed on the cytosolic side of the Golgi and plasma membranes, whereas protein domains exposed on the lumenal side of the ER membrane will be exposed on the lumenal side of the Golgi and on the exterior of the cell (see Figure 11.11).

Transmembrane proteins that are targeted for transport from the ER to the Golgi have a variety of peptide and carbohydrate signals that direct their efficient packaging into transport vesicles at ER exit sites. Likewise, most lumenal proteins that are exported from the ER are bound to transmembrane proteins that are selectively packaged into vesicles (see Figure 11.23). The bound lumenal proteins are then released in the ER–Golgi intermediate compartment or Golgi.

Unmarked proteins in the ER can also be packaged into vesicles and transported to the Golgi by a default pathway, which does not require specific signals for protein packaging. Thus, not only proteins destined for export, but also proteins that function within the ER (including lumenal proteins such as BiP, protein disulfide isomerase, and other enzymes discussed earlier) can be packaged into transport vesicles and exported to the Golgi apparatus. If these proteins were allowed to proceed along the secretory pathway, they would be lost to the cell. To prevent this, these resident ER proteins are recognized in the ER–Golgi intermediate compartment or Golgi and transported back to the ER by a retrieval pathway (see Figure 11.23). For example, many proteins that function in the ER (e.g., BiP) have a targeting sequence Lys-Asp-Glu-Leu (KDEL, in the single-letter code) at their carboxy terminus that directs their retrieval back to the ER. If this sequence is deleted from a protein that normally functions in the ER, the mutated protein is instead transported to the Golgi and secreted from the cell. Conversely, addition of the KDEL sequence to the carboxy terminus of proteins that are normally secreted blocks their secretion. Some ER transmembrane proteins are similarly marked by short C-terminal sequences that contain two lysine residues (KKXX sequences). Proteins bearing the KDEL and KKXX sequences bind to specific recycling receptors in the membranes of the ER–Golgi intermediate compartment or the Golgi complex, are packaged into vesicles, and returned to the ER.

The Golgi Apparatus

The **Golgi apparatus**, or **Golgi complex**, functions as a factory in which proteins received from the ER are further processed and sorted for transport to their eventual destinations: endosomes, lysosomes, the plasma membrane, or secretion. In addition, as noted earlier, most glycolipids and sphingomyelin are synthesized within the Golgi. In plant cells, the Golgi apparatus further serves as the site at which the complex polysaccharides of the cell wall are synthesized. The Golgi apparatus is thus involved in processing the broad range of cellular constituents that travel along the secretory pathway.

Animation 11.2

sites.sinauer.com/cooper7e/a11.2
Organization of the Golgi The Golgi apparatus is composed of flattened membrane-enclosed sacs that receive proteins from the ER, process them, and sort them to their eventual destinations.

Organization of the Golgi

In most cells, the Golgi is composed of flattened membrane-enclosed sacs (cisternae) and associated vesicles (**Figure 11.24**). A striking feature of the Golgi apparatus is its distinct polarity in both structure and function. Proteins from the ER enter at its *cis* (entry) face , which is usually oriented toward the nucleus. They are then transported through the Golgi and exit from its *trans* (exit) face. As they pass through the Golgi, proteins are modified and sorted for transport to their eventual destinations within the cell.

The Golgi consists of multiple discrete compartments, which are commonly viewed as corresponding to four functionally distinct regions: the **cis**, **medial**, and **trans compartments** and the **trans-Golgi network** (see Figure 11.24). Distinct processing and sorting events occur in an ordered sequence within different Golgi compartments. Proteins from the ER–Golgi intermediate compartment enter the *cis* compartment of the Golgi apparatus where modification of proteins, lipids, and polysaccharides begins. After progress through the medial and *trans* compartments, where further modification takes place, they move to the *trans*-Golgi network, which acts as a sorting and distribution center, directing molecular traffic to endosomes, lysosomes, the plasma membrane, or the cell exterior.

The mechanism by which proteins move through the Golgi apparatus has been a long-standing area of controversy. Two general models have been proposed, but current experimental evidence is inconclusive. In the *stable*

(A)

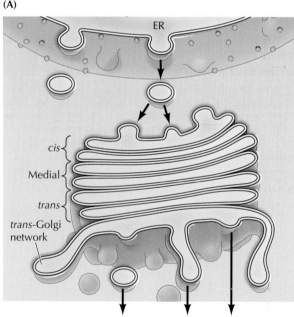

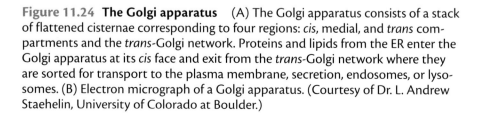

(B)

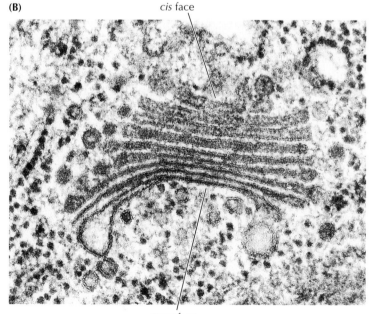

Figure 11.24 The Golgi apparatus (A) The Golgi apparatus consists of a stack of flattened cisternae corresponding to four regions: *cis*, medial, and *trans* compartments and the *trans*-Golgi network. Proteins and lipids from the ER enter the Golgi apparatus at its *cis* face and exit from the *trans*-Golgi network where they are sorted for transport to the plasma membrane, secretion, endosomes, or lysosomes. (B) Electron micrograph of a Golgi apparatus. (Courtesy of Dr. L. Andrew Staehelin, University of Colorado at Boulder.)

(A) Stable cisternae model

(B) Cisternal maturation model

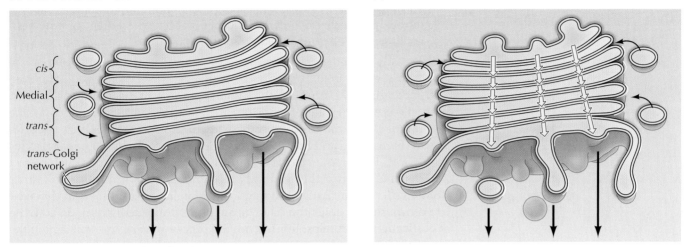

Figure 11.25 Transport through the Golgi   In the stable cisternae model (A), proteins are carried in the *cis* to *trans* direction in transport vesicles. In the cisternal maturation model (B), proteins are carried through the Golgi in the *cis* to *trans* direction within the Golgi cisternae, which progressively mature and move through the Golgi apparatus. In both models, transport vesicles also carry Golgi resident proteins back to earlier compartments for reuse.

cisternae model (Figure 11.25A), proteins are carried between Golgi cisternae in transport vesicles. The cistenae are stable structures, with the enzymes active in each Golgi cisterna being retained in that cisterna or recycled by a retrieval pathway if they are transported out of the cisterna in which they function. In the alternative *cisternal maturation* model (Figure 11.25B), proteins are carried through compartments of the Golgi within the Golgi cisternae, which gradually mature and progressively move through the Golgi in the *cis* to *trans* direction, rather than in transport vesicles. Instead of carrying proteins through the Golgi in the *cis* to *trans* direction, the transport vesicles associated with the Golgi apparatus function to return Golgi resident proteins back to earlier Golgi compartments. Both of these models explain some but not all characteristics of Golgi transport and further studies are needed to resolve the controversial nature of Golgi trafficking.

Protein glycosylation within the Golgi

Protein processing within the Golgi involves extensive modification of the carbohydrate portions of glycoproteins. In mammals, the Golgi apparatus contains more than 250 enzymes that catalyze the addition of different sugars to glycoproteins. These enzymes are located in different compartments of the Golgi, so carbohydrate processing takes place in an ordered fashion as a glycoprotein travels through the Golgi complex.

One aspect of this processing is the modification of the *N*-linked oligosaccharides that were added to proteins in the ER. As discussed earlier in this chapter, proteins are modified by the addition of an oligosaccharide consisting of 14 sugar residues within the ER (see Figure 11.17). Three glucose residues are removed within the ER and the *N*-linked oligosaccharides of these glycoproteins are then subject to extensive further modifications as the

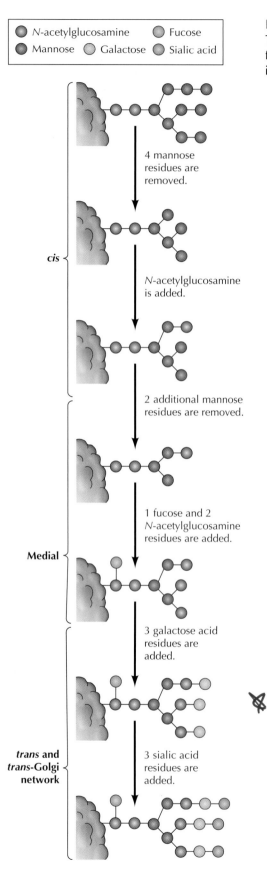

Legend:
- ● N-acetylglucosamine
- ● Mannose
- ● Fucose
- ○ Galactose
- ● Sialic acid

cis

4 mannose residues are removed.

N-acetylglucosamine is added.

2 additional mannose residues are removed.

Medial

1 fucose and 2 N-acetylglucosamine residues are added.

3 galactose acid residues are added.

trans and trans-Golgi network

3 sialic acid residues are added.

Figure 11.26 Processing of N-linked oligosaccharides in the Golgi
The N-linked oligosaccharides of glycoproteins transported from the ER are further modified by an ordered sequence of reactions catalyzed by enzymes in different compartments of the Golgi.

protein travels through the Golgi apparatus. In the example shown in Figure 11.26, mannose residues are removed and N-acetylglucosamine is added in the *cis* Golgi compartment. Additional mannose residues are removed and fucose plus additional N-acetylglucosamine residues are added in the medial Golgi, followed by addition of galactose and sialic acid residues in the *trans* compartment and *trans*-Golgi network. Importantly, different glycoproteins are modified in different ways during their passage through the Golgi, depending on both the structure of the protein and on the processing enzymes present in the Golgi complexes of each cell type. Consequently, proteins can emerge from the Golgi with a variety of different N-linked oligosaccharides.

The processing of the N-linked oligosaccharide of lysosomal proteins differs from that of secreted and plasma membrane proteins. Rather than the initial removal of mannose residues, proteins destined for incorporation into lysosomes are modified by mannose phosphorylation. In the first step of this reaction, N-acetylglucosamine phosphates are added to specific mannose residues while the protein is in the *cis* Golgi compartment (Figure 11.27). This is followed by removal of the N-acetylglucosamine group, leaving **mannose-6-phosphate** residues on the N-linked oligosaccharide. Because of this modification these residues are not removed during further processing. Instead, the phosphorylated mannose residues are specifically recognized by a mannose-6-phosphate receptor in the *trans*-Golgi network, which directs the transport of these proteins to endosomes and on to lysosomes. The phosphorylation of mannose residues is thus a critical step in sorting lysosomal proteins to their correct intracellular destinations. The specificity of this process resides in the enzyme that catalyzes the first step in the reaction sequence—the selective addition of N-acetylglucosamine phosphates to lysosomal proteins. This enzyme recognizes a structural determinant that is present on lysosomal proteins but not on proteins destined for the plasma membrane or secretion. This recognition determinant is not a simple sequence of amino acids; rather, it is formed in the folded protein by the juxtaposition of amino acid sequences from different regions of the polypeptide chain. In contrast to the signal sequences that direct protein translocation to the ER, the recognition determinant that leads to mannose phosphorylation—and thus ultimately targets proteins to lysosomes—depends on the three-dimensional conformation of the folded protein. Such determinants are called **signal patches**, in contrast to the linear targeting signals discussed earlier in this chapter.

Proteins can also be modified by the addition of carbohydrates to the side chains of acceptor serine and threonine residues within specific sequences of amino acids (O-linked glycosylation) (see Figure 9.32). These modifications take place in the Golgi apparatus by the sequential addition of single sugar residues. Some O-linked oligosaccharides consist of only a few sugar residues, whereas others are long chains

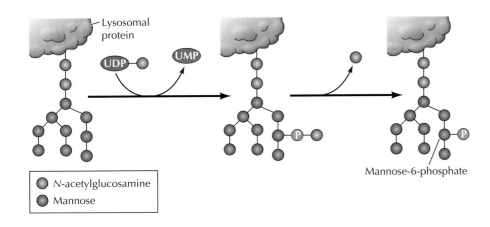

Figure 11.27 Targeting of lysosomal proteins by phosphorylation of mannose residues Proteins destined for incorporation into lysosomes are specifically recognized and modified by the addition of phosphate groups to the number 6 position of mannose residues. In the first step of the reaction, *N*-acetylglucosamine phosphates are transferred to mannose residues from UDP-*N*-acetylglucosamine. The *N*-acetylglucosamine group is then removed, leaving a mannose-6-phosphate.

of many sugars. Proteoglycans, secreted proteins that are important components of the extracellular matrix (discussed in Chapter 15), are an example of extensive *O*-glycosylation. Their processing in the Golgi apparatus involves the addition of 100 or more carbohydrate chains to a polypeptide, with each chain consisting of up to 100 sugar residues that are further modified by the addition of sulfate groups (**Figure 11.28**).

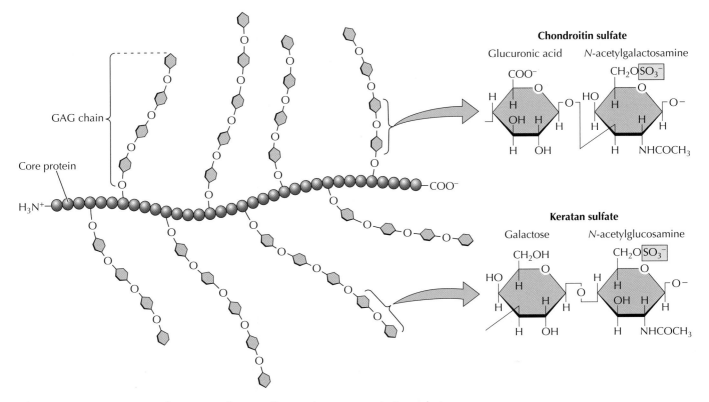

Figure 11.28 Structure of a proteoglycan Proteoglycans contain long chains of glycosaminoglycans (GAGs), which are repeating disaccharides (e.g., chondroitin sulfate and keratan sulfate), joined to a core protein. Each GAG many contain up to 100 sugar residues and as many as 100 GAGs may be linked to a core protein.

Lipid and polysaccharide metabolism in the Golgi

In addition to its activities in processing and sorting glycoproteins, the Golgi apparatus functions in lipid metabolism—in particular, in the synthesis of glycolipids and sphingomyelin. As discussed earlier, the glycerol phospholipids, cholesterol, and ceramide are synthesized in the ER. Sphingomyelin and glycolipids are then synthesized from ceramide in the Golgi apparatus (**Figure 11.29**). Sphingomyelin (the only nonglycerol phospholipid in cell membranes) is synthesized by the transfer of a phosphorylcholine group from phosphatidylcholine to ceramide. Alternatively, the addition of carbohydrates to ceramide can yield a variety of different glycolipids.

Sphingomyelin is synthesized on the lumenal surface of the Golgi, but glucose is added to ceramide on the cytosolic side. Glucosylceramide then apparently flips, however, and additional carbohydrates are added on the lumenal side of the membrane. Glycolipids are not able to translocate across the Golgi membrane, so they are found only in the lumenal half of the Golgi bilayer as is most sphingomyelin. Following vesicular transport they are correspondingly localized to the exterior half of the plasma membrane, with their polar head groups exposed on the cell surface. As will be discussed in

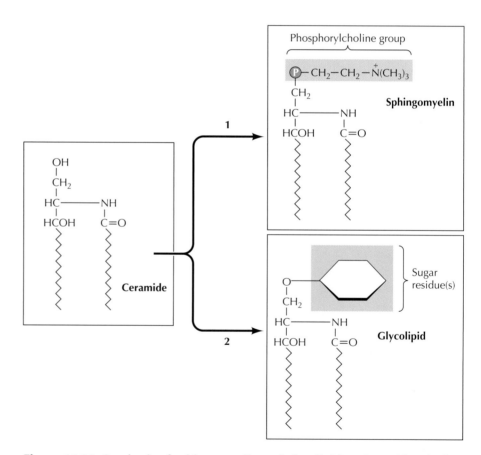

Figure 11.29 Synthesis of sphingomyelin and glycolipids Ceramide, which is synthesized in the ER, is converted either to sphingomyelin (a phospholipid) or to glycolipids in the Golgi apparatus. In the first reaction, a phosphorylcholine group is transferred from phosphatidylcholine to ceramide. Alternatively, a variety of different glycolipids can be synthesized by the addition of one or more sugar residues (e.g., glucose).

Chapter 14, the oligosaccharide portions of glycolipids are important surface markers in cell–cell recognition.

In plant cells, the Golgi apparatus has the additional task of serving as the site where complex polysaccharides of the cell wall are synthesized. As discussed further in Chapter 15, the plant cell wall is composed of three major types of polysaccharides. Cellulose, the predominant constituent, is a simple linear polymer of glucose residues. It is synthesized at the cell surface by enzymes in the plasma membrane. The other cell wall polysaccharides (hemicelluloses and pectins), however, are complex branched chain molecules that are synthesized in the Golgi apparatus and then transported in vesicles to the cell surface. The synthesis of these cell wall polysaccharides is a major cellular function, and as much as 80% of the metabolic activity of the Golgi apparatus in plant cells may be devoted to polysaccharide synthesis.

Protein sorting and export from the Golgi apparatus

Proteins as well as lipids and polysaccharides are transported from the Golgi apparatus to their final destinations through the secretory pathway. This involves the sorting of proteins into different kinds of transport vesicles, which bud from the *trans*-Golgi network and deliver their contents to the appropriate cellular locations (**Figure 11.30**). Some proteins are carried from the Golgi to the plasma membrane, either directly or via endosomes as an intermediate compartment. Other proteins are secreted from the cell and still others are specifically targeted to other intracellular destinations, such as lysosomes in animal cells or vacuoles in yeast and plants.

Proteins that function within the Golgi apparatus must be retained within the appropriate compartments of that organelle rather than being transported along the secretory pathway. Most resident Golgi proteins are transmembrane proteins that function in glycosylation and are retained within the Golgi by signals

Video 11.2

sites.sinauer.com/cooper7e/v11.2
Budding from the Golgi Apparatus
Transport vesicles bud from the *trans*-Golgi network and deliver their secretory cargo to their destinations.

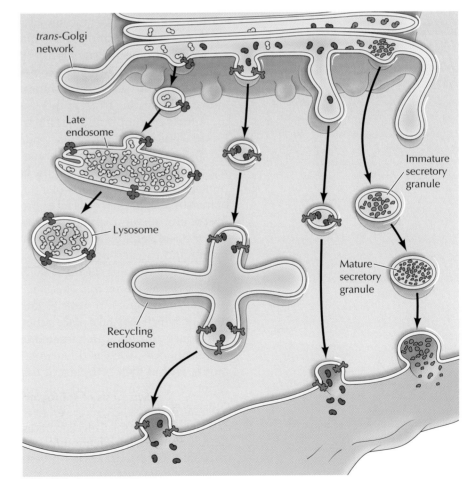

Figure 11.30 Transport from the Golgi apparatus Proteins are sorted in the *trans*-Golgi network and transported in vesicles to their final destinations. Proteins can be transported to the plasma membrane either directly or via recycling endosomes. In addition, proteins can be sorted into distinct secretory granules for regulated secretion. Alternatively, proteins can be targeted to late endosomes, which develop into lysosomes.

within their membrane spanning and cytoplasmic domains. In addition, like resident ER proteins, signals in some Golgi proteins mediate their retrieval from subsequent compartments along the secretory pathway, including transport from endosomes back to the *trans*-Golgi network.

Transport from the Golgi apparatus to the cell surface can occur by at least three routes (see Figure 11.30). Simplest is direct transport from the *trans*-Golgi network to the plasma membrane, which leads to the continuous secretion of proteins from the cell as well as the incorporation of proteins and lipids into the plasma membrane. In addition, proteins can be transported from the Golgi to the plasma membrane via an intermediate of recycling endosomes, which are one of three types of endosomes in animal cells (discussed later in this chapter).

Some cells also possess a distinct regulated secretory pathway in which specific proteins are secreted in response to environmental signals (see Figure 11.30). Examples of regulated secretion include the release of hormones from endocrine cells, the release of neurotransmitters from neurons, and the release of digestive enzymes from the pancreatic acinar cells discussed at the beginning of this chapter (see Figure 11.2). Proteins are sorted into the regulated secretory pathway in the *trans*-Golgi network where they are packaged into immature secretory granules. Many of the proteins that are sorted into these granules selectively aggregate within the *trans*-Golgi network as a result of structural characteristics that drive aggregation at the slightly acidic pH (~6.4) characteristic of the *trans*-Golgi network. These proteins further condense within the immature secretory granules, which develop into mature secretory granules. The mature secretory granules then store their contents until specific signals direct their fusion with the plasma membrane. For example, the digestive enzymes produced by pancreatic acinar cells are stored in mature secretory granules until the presence of food in the stomach and small intestine triggers their secretion.

A further complication in the transport of proteins to the plasma membrane arises in epithelial cells, which are polarized when they are organized into tissues. The plasma membrane of such cells is divided into two separate regions, the **apical domain** and the **basolateral domain**, which contain specific proteins related to their particular functions. For example, the apical membrane of intestinal epithelial cells faces the lumen of the intestine and is specialized for the efficient absorption of nutrients; the remainder of the cell is covered by the basolateral membrane (**Figure 11.31**). Distinct domains of the plasma membrane are present not only in epithelial cells but also in other cell types. Thus proteins leaving the *trans*-Golgi network must be selectively transported to these distinct domains. This is accomplished by the selective packaging of proteins into transport vesicles targeted for either the apical or basolateral domain, which may take place either in the *trans*-Golgi network or in recycling endosomes. Proteins are targeted to the basolateral domain by short amino acid sequences, such as dileucine (LL) or tyrosine-containing hydrophobic motifs, within their cytoplasmic domains. Protein sorting to apical domains is not yet fully understood, but appears to be directed by targeting signals consisting of GPI anchors (see Figure 11.18) and carbohydrate modifications rather than amino acid sequences.

The best-characterized pathway of protein sorting in the Golgi is the selective transport of proteins to lysosomes. As already discussed, lumenal lysosomal proteins are marked by mannose-6-phosphates that are formed

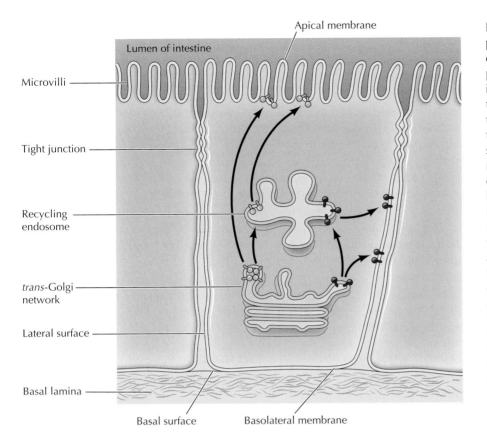

Lumen of intestine
Apical membrane
Microvilli
Tight junction
Recycling endosome
trans-Golgi network
Lateral surface
Basal lamina
Basal surface
Basolateral membrane

Figure 11.31 Transport to the plasma membrane of polarized cells The plasma membranes of polarized epithelial cells are divided into apical and basolateral domains. In this example (intestinal epithelium), the apical surface of the cell faces the lumen of the intestine, the lateral surfaces are in contact with neighboring cells, and the basal surface rests on a sheet of extracellular matrix (the basal lamina). The apical membrane is characterized by the presence of microvilli, which facilitate the absorption of nutrients by increasing surface area. Specific proteins are targeted to the apical or basolateral membranes either in the *trans*-Golgi network or in a recycling endosome. Tight junctions between neighboring cells maintain the identity of the apical and basolateral membranes by preventing the diffusion of proteins between these domains.

by modification of their *N*-linked oligosaccharides shortly after entry into the Golgi apparatus (see Figure 11.27). Transmembrane receptors in the *trans*-Golgi network then recognize these mannose-6-phosphate residues. The receptors contain sequences in their cytosolic domains that direct the packaging of the receptor/lysosomal enzyme complexes into transport vesicles destined for late endosomes, which subsequently mature into lysosomes.

In yeasts and plant cells, both of which lack lysosomes, proteins are transported from the Golgi apparatus to an additional destination: the **vacuole** (Figure 11.32). Vacuoles assume the functions of lysosomes in these cells as well as perform a variety of other tasks, such as the storage of nutrients and the maintenance of turgor pressure and osmotic balance. In contrast to lysosomal targeting, proteins are directed to vacuoles by short peptide sequences instead of by carbohydrate markers.

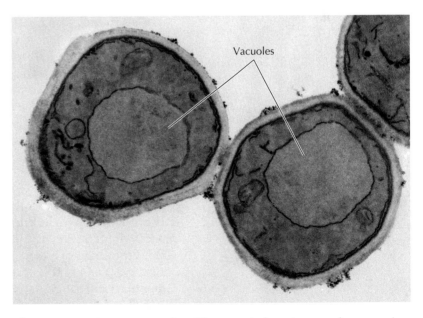

Vacuoles

Figure 11.32 A yeast vacuole The vacuole functions as a lysosome in yeast.

The Mechanism of Vesicular Transport

Transport vesicles play a central role in the traffic of molecules between different membrane-enclosed compartments of the secretory pathway. As discussed later in this chapter and in Chapter 14, vesicles are similarly involved in the transport of materials taken up at the cell surface. Vesicular transport is thus a major cellular activity, responsible for molecular traffic between a variety of specific membrane-enclosed compartments. The selectivity of such transport is therefore key to maintaining the functional organization of the cell. For example, lysosomal enzymes must be transported specifically from the Golgi apparatus to lysosomes—not to the plasma membrane or to the ER. Some of the signals that target proteins to specific organelles, such as lysosomes, were discussed earlier in this chapter. These proteins are transported within vesicles, so the specificity of transport is based on the selective packaging of the intended cargo into vesicles that recognize and fuse only with the appropriate target membrane. Because of the central importance of vesicular transport to the organization of eukaryotic cells, understanding the molecular mechanisms that control vesicle packaging, budding, targeting, and fusion is a major area of research in cell biology.

Experimental approaches to understanding vesicular transport

Our basic understanding of the molecular mechanisms of vesicular transport came from the convergence of three distinct experimental approaches in different biological systems: (1) isolation of yeast mutants that are defective in protein transport and sorting; (2) reconstitution of vesicular transport in cell-free systems; and (3) biochemical analysis of synaptic vesicles, which are responsible for the regulated secretion of neurotransmitters by neurons. Each of these experimental systems has distinct advantages for understanding particular aspects of the transport process and their exploitation revealed similar molecular mechanisms that regulate vesicular transport in cells as different as yeasts and mammalian neurons.

Yeasts are advantageous in studying the secretory pathway because they are readily amenable to genetic analysis. In particular, Randy Schekman and his colleagues, in 1980, pioneered the isolation of yeast mutants defective in vesicular transport. These include mutants that are defective at various stages of protein secretion (*sec* mutants), mutants that are unable to transport proteins to the vacuole, and mutants that are unable to retain resident ER proteins. The isolation of such mutants in yeasts led directly to the molecular cloning and analysis of the corresponding genes, thereby identifying a number of proteins involved in various steps of the secretory pathway.

Biochemical studies of vesicular transport using reconstituted systems complemented these genetic studies and enabled the direct isolation of transport proteins from mammalian cells. The first cell-free transport system was developed in 1980 by James Rothman and colleagues, who analyzed protein transport between compartments of the Golgi apparatus. Similar reconstituted systems have been developed to analyze transport between other compartments, including transport from the ER to the Golgi and transport from the Golgi to secretory vesicles, vacuoles, and the plasma membrane. The development of these *in vitro* systems has enabled biochemical studies of the transport process and functional analysis of proteins identified by

FYI

Randy Schekman, James Rothman, and Thomas Südhof shared the 2013 Nobel Prize in Physiology or Medicine "for their discoveries of machinery regulating vesicle traffic, a major transport system in our cells."

mutations in yeasts, as well as direct isolation of some of the proteins involved in vesicle budding and fusion.

Insights into the molecular mechanisms of vesicular transport have also come from studies of synaptic transmission in neurons, a specialized form of regulated secretion. A synapse is the junction of a neuron with another cell, which may be either another neuron or an effector, such as a muscle cell. Information is transmitted across the synapse by chemical neurotransmitters, such as acetylcholine, which are stored within the neuron in **synaptic vesicles**. Stimulation of the transmitting neuron triggers the fusion of these synaptic vesicles with the plasma membrane, causing neurotransmitters to be released into the synapse and stimulating the postsynaptic neuron or effector cell. Synaptic vesicles are extremely abundant in the brain, allowing them to be purified in large amounts for biochemical analysis. Studies of the proteins isolated from synaptic vesicles in Thomas Sudhof's laboratory, many of which are closely related to proteins that had been shown to play critical roles in vesicular transport by yeast genetics and reconstitution experiments, provided critical insights into the molecular mechanisms of vesicle fusion.

Our understanding of vesicular transport has been furthered by studies using GFP fusion proteins, which have allowed transport vesicles carrying specific proteins to be visualized by immunofluorescence as they move through the secretory pathway (Figure 11.33). In these experiments, cells are transfected with cDNA constructs encoding secretory proteins tagged with green fluorescent protein (GFP) (see Figure 1.28). The progress of the GFP-labeled proteins through the secretory pathway can then be followed in living cells, allowing characterization of many aspects of the dynamics and molecular interactions involved in vesicular transport.

Cargo selection, coat proteins, and vesicle budding

Most of the transport vesicles that carry proteins from the ER to the Golgi and subsequent compartments are coated with cytosolic coat proteins and thus

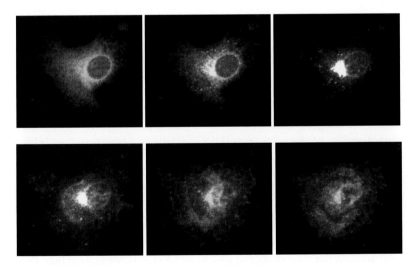

Figure 11.33 Visualization of transport vesicles in a living cell   A time-lapse sequence showing movement of GFP-labeled vesicles from the ER to the Golgi and from the Golgi to the plasma membrane. (Courtesy of Jennifer Lippincott-Schwartz.)

Video 11.4

sites.sinauer.com/cooper7e/v11.4

Vesicle Trafficking Transport vesicles move between the ER, Golgi, and cell membrane along cytoskeletal filaments.

are called coated vesicles. Assembly of the coat proteins drives the budding of vesicles containing selected cargo proteins from the donor membrane (**Figure 11.34**). The vesicles then travel along cytoskeletal filaments to their targets as described in Chapter 13. The coats are removed at the target membrane, allowing the membranes to fuse, and the vesicles empty their lumenal cargo and insert their membrane proteins into the target membrane.

Three families of vesicle coat proteins have been characterized: **clathrin**, **COPI**, and **COPII** (COP indicates coat protein). Vesicles coated with these different proteins are involved in transport between different compartments

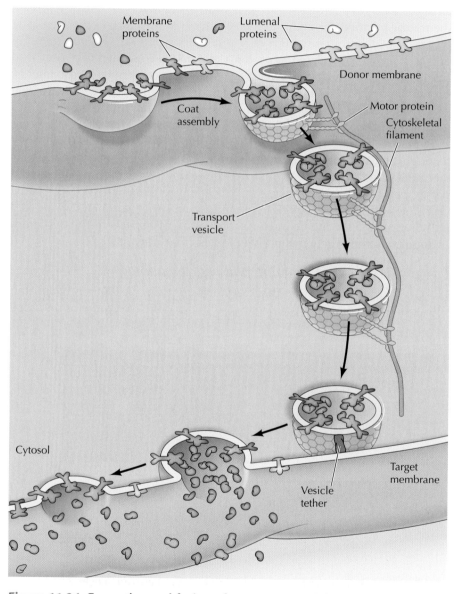

Figure 11.34 Formation and fusion of a transport vesicle   Membrane proteins and lumenal secretory proteins with their receptors are collected into selected regions of a donor membrane where the formation of a cytosolic coat results in the budding of a transport vesicle. The vesicle is transported by motor proteins along cytoskeletal filaments to its target. The transport vesicle then docks at its target membrane, the coat is removed, and the vesicle fuses with its target.

of the secretory pathway (**Figure 11.35**). **COPII-coated vesicles** carry proteins from the ER to the ER–Golgi intermediate compartment (ERGIC) and on to the Golgi apparatus, budding from the transitional ER and carrying their cargo forward along the secretory pathway. In contrast, **COPI-coated vesicles** bud from the ERGIC or the Golgi apparatus and carry their cargo backwards, returning resident proteins to earlier compartments of the secretory pathway. Thus, COPI is the coat protein on the retrieval vesicles that return resident ER proteins back to the ER from the ERGIC or the Golgi, and on the vesicles that retrieve Golgi processing enzymes back from the later Golgi compartments. Finally, **clathrin-coated vesicles** are responsible for transport in both

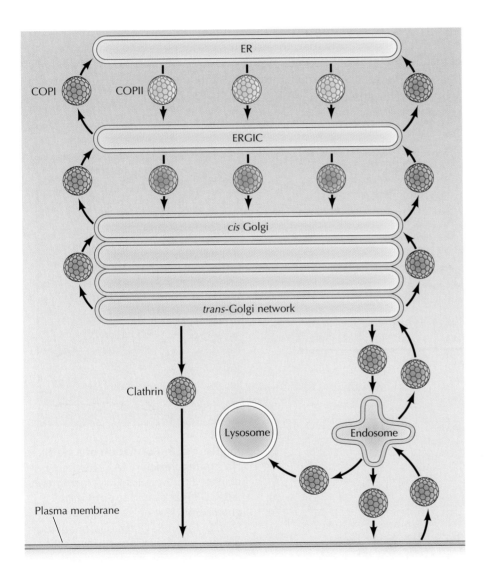

Figure 11.35 Transport by coated vesicles   COPII-coated vesicles carry cargo from the ER to the Golgi, and clathrin-coated vesicles carry cargo outward from the *trans*-Golgi network. Clathrin-coated vesicles also carry cargo back from the plasma membrane to endosomes and other organelles such as the *trans*-Golgi network and lysosomes. COPI-coated vesicles retrieve ER-resident proteins from the ERGIC and *cis* Golgi and carry Golgi-resident enzymes back from the *trans* Golgi to earlier Golgi cisternae.

directions between the *trans*-Golgi network, endosomes, lysosomes, and the plasma membrane.

The formation of coated vesicles is regulated by small GTP-binding proteins (Arf and Sar), which are related to Ras and Ran. Arf functions in the formation of COPI- and clathrin-coated vesicles budding from the Golgi apparatus and Sar functions in the formation of COPII-coated vesicles budding from the ER. These GTP-binding proteins recruit adaptor proteins that mediate vesicle assembly by interacting both with cargo proteins and with vesicle coat proteins. For example, **Figure 11.36** illustrates the role of Arf in the assembly of clathrin-coated vesicles at the *trans*-Golgi network. First, Arf/GDP is converted to the active GTP-bound form by a guanine nucleotide exchange factor localized to the *trans*-Golgi network membrane. Arf/GTP then recruits an adaptor protein that nucleates both cargo selection and coat assembly. In particular, the adaptor protein binds to sequences in the cytosolic domains of transmembrane proteins that signal their export from the Golgi. As discussed earlier, the transmembrane proteins include receptors

(A)

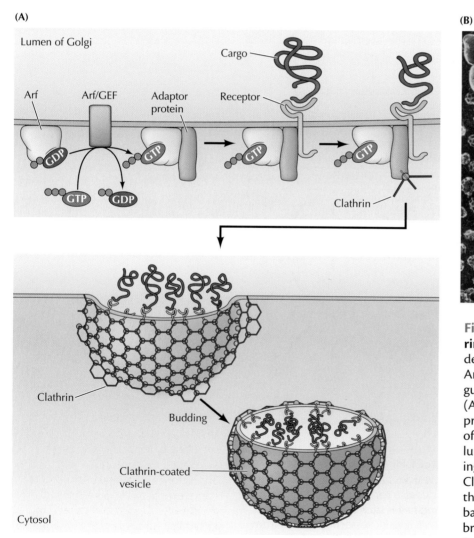

(B)

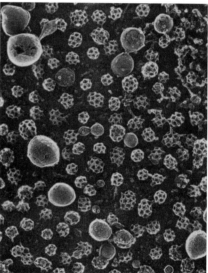

Figure 11.36 Formation of a clathrin-coated vesicle (A) After being delivered to the *trans*-Golgi membrane, Arf/GDP is activated to Arf/GTP by a guanine nucleotide exchange factor (Arf/GEF). Arf/GTP recruits an adaptor protein, which binds to the cytosolic tail of a transmembrane receptor with its lumenal cargo and also serves as a binding site for assembly of a clathrin coat. Clathrin consists of three protein chains that associate with each other to form a basketlike lattice that distorts the membrane and drives vesicle budding. (B) Scanning electron micrograph of clathrin-coated vesicles. (Courtesy of Tomas Kirchhausen, Harvard University.)

for lumenal cargo proteins, which are thereby selected for incorporation into vesicles. In addition, the adaptor protein recruits clathrin and initiates the assembly of the vesicle coat. Clathrin plays a structural role in vesicle budding by assembling into a basketlike lattice structure that distorts the membrane and initiates the bud.

Vesicle fusion

The fusion of a transport vesicle with its target involves two types of events. First, the transport vesicle must recognize the correct target membrane; for example, a vesicle carrying lysosomal enzymes has to deliver its cargo only to lysosomes. Second, the vesicle and target membranes must fuse, delivering the contents of the vesicle to the target organelle. Research over the last several years has elucidated a process of vesicle fusion in which specific recognition between a vesicle and its target (tethering) is mediated by interactions between proteins on the vesicle and the target membranes, followed by additional protein-protein interactions that drive fusion of the phospholipid bilayers.

The initial interaction between transport vesicles and specific target membranes is mediated by **tethering factors** and small-GTP binding proteins (**Rab** proteins). More than 60 different Rab proteins have been identified and shown to function in specific vesicle transport processes (Table 11.1). Different Rab proteins or combinations of Rab proteins mark different organelles and transport vesicles. Vesicle Rab proteins in the active GTP-bound state bind membrane tethering factors, providing the initial bridge between the target and vesicle membranes (Figure 11.37). Tethering factors also bind coat proteins, which contributes to their interaction with vesicles. In addition, the tethering factors may stimulate the formation of complexes between transmembrane proteins called **SNAREs** on the vesicle and target membranes, leading to membrane fusion.

SNAREs were initially identified in James Rothman's laboratory by biochemical analysis of reconstituted vesicular transport systems from mammalian cells. Analysis of the proteins involved in vesicle fusion in these systems led Rothman and his colleagues to hypothesize that vesicle fusion is mediated by interactions between specific pairs of SNAREs on the vesicle

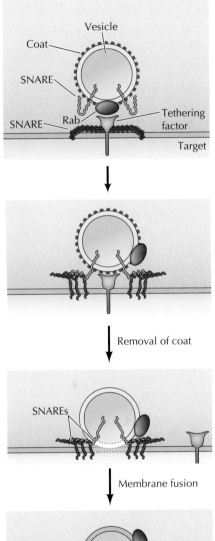

Figure 11.37 Vesicle docking and fusion A Rab protein on the vesicle membrane binds to a tethering factor associated with the target membrane. This is followed by the formation of complexes between SNAREs on the vesicle and target membranes. The coiled-coil domains of the SNAREs zip together, bringing the vesicle and target membranes into close proximity and the membranes fuse.

Table 11.1	Representative Rab Proteins
Rab protein	**Transport step**
Rab1A and B	ER to Golgi, intra-Golgi
Rab2A and B	ER to Golgi
Rab3A-D	Exocytosis
Rab4A and B	Endosome to plasma membrane
Rab5A-C	Early endosome fusion
Rab6A-C	Endosome to Golgi, intra-Golgi, Golgi to ER
Rab7A and B	Late endosome to lysosome
Rab8A and B	*Trans*-Golgi network and recycling endosome to plasma membrane
Rab10	*Trans*-Golgi network and recycling endosome to basolateral plasma membrane
Rab14	*Trans*-Golgi network and recycling endosome to apical plasma membrane
Rab22A	Recycling endosome to plasma membrane

and target membranes. Subsequent research has confirmed that formation of complexes between vesicle and target SNAREs is required for fusion of vesicle and target membranes and further shown that SNARE-SNARE pairing provides the energy to drive fusion of the phospholipid bilayers. All SNARE proteins have a long central coiled-coil domain like that found in nuclear lamins (see Figure 10.4). As in the lamins, this domain binds strongly to other coiled-coil domains and, in effect, zips the SNAREs on vesicle and target membranes together, bringing the two membranes into direct contact and leading to fusion of the lipid bilayers.

Lysosomes

Lysosomes are membrane-enclosed organelles that contain an array of enzymes capable of breaking down all types of biological polymers—proteins, nucleic acids, carbohydrates, and lipids. Lysosomes function as the digestive system of the cell, serving both to degrade material taken up from outside the cell and to digest obsolete components of the cell itself. In their simplest form, lysosomes are visualized as dense spherical vacuoles, but they can display considerable variation in size and shape as a result of differences in the materials that have been taken up for digestion (**Figure 11.38**). Lysosomes thus represent morphologically diverse organelles defined by the common function of degrading intracellular material.

Lysosomal acid hydrolases

Lysosomes contain about 60 different degradative enzymes that can hydrolyze proteins, DNA, RNA, polysaccharides, and lipids. Mutations in the genes that encode these enzymes are responsible for more than 30 different human genetic diseases, which are called **lysosomal storage diseases** because undegraded material accumulates within the lysosomes of affected individuals.

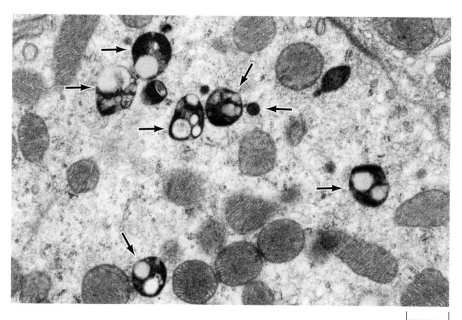

0.5 µm

Figure 11.38 Electron micrograph of lysosomes and mitochondria in a mammalian cell   Lysosomes are indicated by arrows. Notice the variation in size and shape, determined by differences in the materials taken up for digestion.

Molecular Medicine

Gaucher Disease

The Disease

Gaucher disease is the most common of the lysosomal storage diseases, which are caused by a failure of lysosomes to degrade substances that they normally break down. The resulting accumulation of nondegraded compounds leads to an increase in the size and number of lysosomes within the cell, eventually resulting in cellular malfunction and pathological consequences to affected organs. There are three types of Gaucher disease, which differ in severity and nervous system involvement. In the most common form of the disease (type 1), the nervous system is not involved; the disease is manifest as spleen and liver enlargement and development of bone lesions. Many patients with this form of the disease have no serious symptoms, and their life span is unaffected. Type 1 disease is very common in the Ashkenazi Jewish population, where it has a frequency of about 1 in 1000 people. The more severe forms of the disease (types 2 and 3) are much rarer and found in both Jewish and non-Jewish populations. The most devastating is type 2 disease, in which extensive neurological involvement is evident in infancy, and patients die early in life. Type 3 disease, intermediate in severity between types 1 and 2, is characterized by the onset of neurological symptoms (including dementia and spasticity) by about age ten.

Molecular and Cellular Basis

Gaucher disease is caused by a deficiency of the lysosomal enzyme glucocerebrosidase, which catalyzes the hydrolysis of glucosylceramide to glucose and ceramide (see figure). This enzyme deficiency was demonstrated in 1965, and the responsible gene was cloned in 1985. Since then more than 100 different mutations responsible for Gaucher disease have been identified. Interestingly, the severity of the disease

can be largely predicted from the nature of these mutations. For example, patients with a mutation leading to the relatively conservative amino acid substitution of serine for asparagine have type 1 disease; patients with a mutation leading to substitution of proline for leucine have more severe enzyme deficiencies and develop either type 2 or 3 disease.

Except for the very rare type 2 and 3 forms of the disease, the only cells affected in Gaucher disease are macrophages. Because their function is to eliminate aged and damaged cells by phagocytosis, macrophages continually ingest large amounts of lipids, which are normally degraded in lysosomes. Deficiencies of glucocerebrosidase are therefore particularly evident in macrophages of both the spleen and the liver, consistent with these organs being the primary sites affected in most cases of Gaucher disease.

Prevention and Treatment

Gaucher disease is a prime example of a disease that can be treated by enzyme replacement therapy in which exogenous administration of an enzyme is used to correct an enzyme defect. This approach to treatment of lysosomal storage diseases was suggested by Christian de Duve in the

1960s, based on the idea that exogenously administered enzymes might be taken up by endocytosis and transported to lysosomes. In type 1 Gaucher disease, this approach is particularly attractive because the single target cell is the macrophage. In the 1970s it was discovered that macrophages express cell surface receptors that bind mannose residues on extracellular glycoproteins and then internalize these proteins by endocytosis. This finding suggested that exogenously administered glucocerebrosidase could be specifically targeted to macrophages by modifications that would expose mannose residues. The first therapy to be developed used enzyme prepared from human placenta; clinical studies demonstrated its effectiveness and it was approved for the treatment of Gaucher disease in 1991. Subsequently, enzyme produced by recombinant DNA technology was developed for therapy and is now used instead of the human enzyme to provide greater availability and reduce the possibility of contamination with infectious agents, such as hepatitis viruses.

Reference

Deegan, P. B. and T. M. Cox. 2012. Imiglucerase in the treatment of Gaucher disease: a history and perspective. *Drug Des. Dev. Ther.* 6: 81–106.

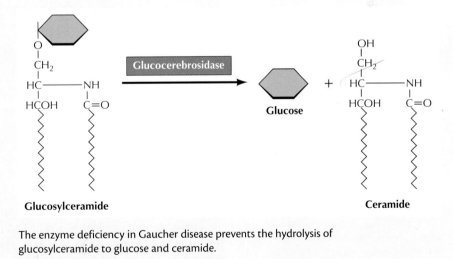

The enzyme deficiency in Gaucher disease prevents the hydrolysis of glucosylceramide to glucose and ceramide.

Most of these diseases result from deficiencies in single lysosomal enzymes. For example, Gaucher disease (the most common of these disorders) results from a mutation in the gene that encodes a lysosomal enzyme required for the breakdown of glycolipids. An intriguing exception is I-cell disease,

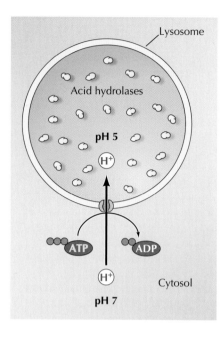

Lysosome

Acid hydrolases

pH 5

H+

ATP

ADP

H+

Cytosol

pH 7

Figure 11.39 Organization of the lysosome Lysosomes contain a variety of acid hydrolases that are active at the acidic pH maintained within the lysosome but not at the neutral pH of the cytosol. The acidic internal pH of lysosomes results from the action of a proton pump in the lysosomal membrane, which imports protons from the cytosol coupled to ATP hydrolysis.

which is caused by a deficiency in the enzyme that catalyzes the first step in the tagging of lysosomal enzymes with mannose-6-phosphate in the Golgi apparatus (see Figure 11.28). The result is a general failure of lysosomal enzymes to be incorporated into lysosomes.

Most lysosomal enzymes are acid hydrolases, which are active at the acidic pH (about 5) that is maintained within lysosomes but not at the neutral pH (about 7.2) characteristic of the rest of the cytoplasm (**Figure 11.39**). The requirement of these lysosomal hydrolases for acidic pH provides double protection against uncontrolled digestion of the contents of the cytosol; even if the lysosomal membrane were to break down, the released acid hydrolases would be inactive at the neutral pH of the cytosol. To maintain their acidic internal pH, lysosomes must actively concentrate H+ ions (protons). This is accomplished by a proton pump in the lysosomal membrane, which actively transports protons into the lysosome from the cytosol. This pumping requires expenditure of energy in the form of ATP hydrolysis in order to maintain approximately a hundredfold higher H+ concentration inside the lysosome than in the cytosol.

Endocytosis and lysosome formation

One of the major functions of lysosomes is the digestion of material taken up from outside the cell by **endocytosis**, which is discussed in detail in Chapter 14. This role of lysosomes relates not only to their function but also to their formation. In particular, lysosomes are formed when transport vesicles from the *trans*-Golgi network fuse with a late endosome, which contains molecules taken up by endocytosis at the plasma membrane.

Endosomes represent an intersection between the secretory pathway, through which lysosomal proteins are processed, and the endocytic pathway, through which extracellular molecules are taken up at the cell surface (**Figure 11.40**). As noted earlier in this chapter, animal cells have three types of endosomes: early endosomes, recycling endosomes, and late endosomes. Early endosomes receive vesicles formed by endocytosis at the plasma membrane. They separate molecules targeted for recycling back to the plasma membrane from those destined for degradation in lysosomes. The molecules to be recycled (for example, cell surface receptors, as discussed in Chapter 14) are then passed to recycling endosomes and back to the plasma membrane. In contrast, molecules destined for degradation are transported to multivesicular bodies and then to late endosomes.

Lysosomal acid hydrolases are also transported to late endosomes from the *trans*-Golgi network. An important change during the early to late endosome transition is the lowering of the internal pH to about 5.5, which plays a key role in the delivery of lysosomal acid hydrolases. As discussed, lysosomal proteins are targeted to late endosomes by mannose-6-phosphate residues, which are recognized by mannose-6-phosphate receptors in the *trans*-Golgi network and packaged into clathrin-coated vesicles. Following fusion of these transport vesicles with late endosomes, the acidic internal pH causes the hydrolases to dissociate from the mannose-6-phosphate receptor.

FYI

The antimalarial drug chloroquine is uncharged and able to cross membranes at physiological pH. At acidic pH, chloroquine is positively charged and accumulates within the mosquito's digestive vacuoles (analogous to lysosomes). The high concentration of chloroquine within the digestive vacuoles is thought to be responsible for its activity against the parasite.

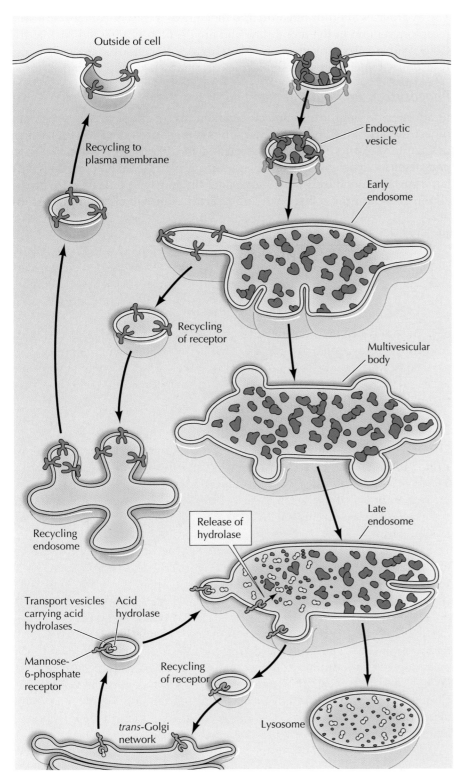

Figure 11.40 Endocytosis and lysosome formation
Molecules are taken up from outside the cell in endocytic vesicles, which fuse with early endosomes. Membrane receptors are recycled to the plasma membrane through recycling endosomes. Early endosomes, though, can mature to multivesicular bodies and late endosomes. Transport vesicles carrying acid hydrolases from the *trans*-Golgi network then fuse with late endosomes, which mature into lysosomes as they acquire a full complement of lysosomal enzymes. The acid hydrolases dissociate from the mannose-6-phosphate receptors when the transport vesicles fuse with late endosomes, and the mannose-6-phosphate receptors are recycled back to the *trans*-Golgi network.

The hydrolases are thus released into the lumen of the endosome, while the receptors remain in the membrane and are eventually recycled back to the Golgi. With a full complement of acid hydrolases, late endosomes develop into lysosomes, which digest the molecules originally taken up by endocytosis.

Phagocytosis and autophagy

In addition to degrading molecules taken up by endocytosis, lysosomes digest material derived from two other routes: phagocytosis (discussed in Chapter 14) and autophagy (discussed in Chapter 9) (**Figure 11.41**). In **phagocytosis**, specialized cells, such as macrophages, take up and degrade large particles, including bacteria, cell debris, and aged cells that need to be eliminated from the body. Such large particles are taken up in phagocytic

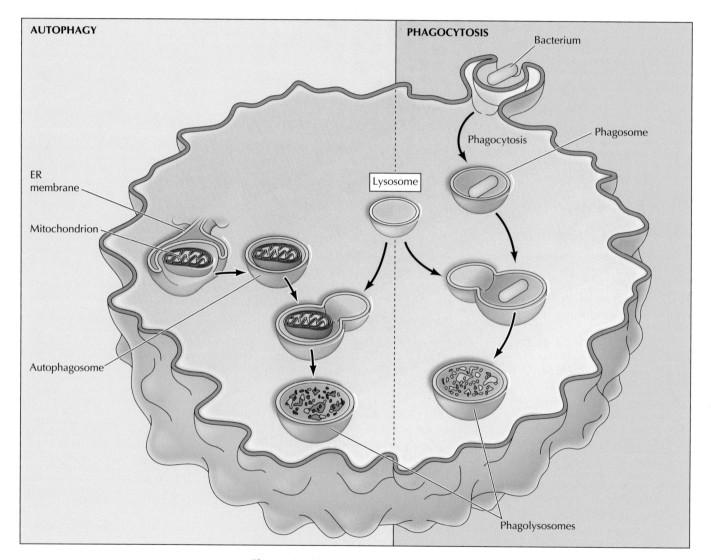

Figure 11.41 Lysosomes in phagocytosis and autophagy In phagocytosis, large particles (such as bacteria) are taken up into phagocytic vacuoles or phagosomes. In autophagy, regions of the cytoplasm or internal organelles (such as mitochondria) are enclosed by membranes derived from the endoplasmic reticulum, forming autophagosomes. Both phagosomes and autophagosomes fuse with lysosomes to form large phagolysosomes in which their contents are digested.

vacuoles (**phagosomes**), which then fuse with lysosomes, resulting in digestion of their contents. The lysosomes formed in this way (**phagolysosomes**) can be quite large and heterogeneous since their size and shape is determined by the content of material that is being digested.

Lysosomes are also responsible for **autophagy**, the turnover of the cell's own components (see Figure 9.48). In contrast to phagocytosis, autophagy is a function of all cells and results in the gradual degradation of long-lived proteins. The first step in autophagy appears to be the enclosure of a small area of cytoplasm or a cytoplasmic organelle (e.g., a mitochondrion) in a membrane derived from the endoplasmic reticulum. The resulting vesicle (an **autophagosome**) then fuses with a lysosome, and its contents are digested (see Figure 11.41).

Autophagy not only leads to the continuous turnover of cellular constituents, but it can also be regulated during development and in response to stress. For example, autophagy plays an important role in many developmental processes, such as insect metamorphosis, which involve extensive tissue remodeling and degradation of cellular components. In addition, autophagy is activated when cells are deprived of nutrients. Under conditions of nutrient starvation, autophagy allows cells to degrade nonessential macromolecules so that their components can be reutilized. In other circumstances, however, autophagy plays an important role in programmed cell death, as discussed in Chapter 18.

SUMMARY

KEY TERMS

The Endoplasmic Reticulum

- ***The endoplasmic reticulum and protein secretion:*** The endoplasmic reticulum is the first branch point in protein sorting. In mammalian cells, proteins destined for secretion, lysosomes, or the plasma membrane are translated on membrane-bound ribosomes and transferred into the rough ER as their translation proceeds. See Video 11.1.

endoplasmic reticulum (ER), rough ER, smooth ER, secretory vesicle, secretory pathway

- ***Targeting proteins to the endoplasmic reticulum:*** Proteins can be targeted to the ER either while their translation is still in progress or following completion of translation in the cytosol. In mammalian cells, most proteins are translocated into the ER while they are being translated on membrane-bound ribosomes. Ribosomes engaged in the synthesis of secreted proteins are targeted to the endoplasmic reticulum by signal sequences at the amino terminus of the polypeptide chain. Growing polypeptide chains are then translocated into the ER through protein channels and released into the ER lumen by cleavage of the signal sequence. See Animation 11.1.

signal sequence, microsome, signal recognition particle (SRP), SRP RNA, SRP receptor, translocon, signal peptidase

- ***Insertion of proteins into the ER membrane:*** Integral membrane proteins of the plasma membrane or the membranes of the ER, Golgi apparatus, and lysosomes are initially inserted into the membrane of the ER. Rather than being translocated into the ER lumen, these proteins are anchored by membrane-spanning α helices that stop the transfer of the growing polypeptide chain across the membrane.

- ***Protein folding and processing in the ER:*** Polypeptide chains are folded into their correct three-dimensional conformations within the ER. The ER is also the site of *N*-linked glycosylation and addition of GPI anchors.

protein disulfide isomerase (PDI), glycosylphosphatidylinositol (GPI) anchor

SUMMARY	KEY TERMS
• **Quality control in the ER:** Many secretory proteins are not folded correctly the first time. Chaperones detect incorrectly folded proteins and recycle them through the folding pathway. Those proteins that cannot be correctly folded are diverted from the secretory pathway and marked for degradation. The unfolded protein response pathway monitors the amount of unfolded protein in the ER and regulates cellular activities to maintain appropriate protein folding capacity.	ER-associated degradation (ERAD), calnexin, calreticulin, unfolded protein response (UPR)
• **The smooth ER and lipid synthesis:** The ER is the major site of lipid synthesis in eukaryotic cells, and the smooth ER is abundant in cells that are active in lipid metabolism and detoxification of lipid-soluble drugs.	flippase
• **Export of proteins and lipids from the ER:** Proteins and lipids are transported in vesicles from the ER to the Golgi apparatus. Targeting sequences mediate the selective packaging of exported proteins into vesicles that transport them to the Golgi. Resident ER proteins are marked by other targeting sequences that signal their return from the Golgi to the ER by a recycling pathway.	

The Golgi Apparatus

• **Organization of the Golgi:** The Golgi apparatus functions in protein processing and sorting as well as in the synthesis of lipids and polysaccharides. Proteins are transported from the endoplasmic reticulum to the *cis* compartment of the Golgi. Distinct processing events take place as proteins move through the medial and *trans* compartments to the *trans*-Golgi network, where proteins are sorted and packaged into vesicles for transport to endosomes, the plasma membrane, or the exterior of the cell. See Animation 11.2.	Golgi apparatus, Golgi complex, *trans*-Golgi network
• **Protein glycosylation within the Golgi:** The *N*-linked oligosaccharides added to proteins in the ER are modified within the Golgi. Those proteins destined for lysosomes are specifically phosphorylated on mannose residues, and mannose-6-phosphate serves as a targeting signal that directs their transport to lysosomes from the *trans*-Golgi network. *O*-linked glycosylation also takes place within the Golgi.	mannose-6-phosphate, signal patch
• **Lipid and polysaccharide metabolism in the Golgi:** The Golgi apparatus is the site of synthesis of glycolipids, sphingomyelin, and the complex polysaccharides of plant cell walls.	apical domain, basolateral domain, vacuole
• **Protein sorting and export from the Golgi apparatus:** Proteins are sorted in the *trans*-Golgi network for packaging into transport vesicles targeted for secretion, the plasma membrane, endosomes, lysosomes, or yeast and plant vacuoles. In polarized cells, proteins are specifically targeted to the apical and basolateral domains of the plasma membrane. See Videos 11.2 and 11.3.	

The Mechanism of Vesicular Transport

• **Experimental approaches to understanding vesicular transport:** The mechanism of vesicular transport has been elucidated through studies of yeast mutants, reconstituted cell-free systems, synaptic vesicles, and visualization of protein traffic in living cells using GFP-labeled proteins.	synaptic vesicle
• **Cargo selection, coat proteins, and vesicle budding:** The cytoplasmic surfaces of most vesicles are coated with proteins that drive vesicle budding. The specific molecules to be transported are selected by complexes of small GTP-binding proteins and adaptor proteins that associate with the coat proteins. See Video 11.4.	clathrin, COPI, COPII, COPI-coated vesicle, COPII-coated vesicle, clathrin-coated vesicle

SUMMARY	KEY TERMS

- **Vesicle fusion:** The initial interaction between vesicles and their target membranes is mediated by the binding of tethering factors to Rab proteins. Subsequent interactions between transmembrane proteins on vesicle and target membranes lead to membrane fusion.

tethering factor, Rab, SNARE

Lysosomes

- **Lysosomal acid hydrolases:** Lysosomes contain an array of acid hydrolases that degrade proteins, nucleic acids, polysaccharides, and lipids. These enzymes function specifically at the acidic pH maintained within lysosomes.

lysosome, lysosomal storage disease

- **Endocytosis and lysosome formation:** Extracellular molecules taken up by endocytosis are transported to early endosomes. Molecules destined for degradation and then transported to late endosomes, which mature to lysosomes as lysosomal acid hydrolases, are delivered from the Golgi.

endocytosis, endosome

- **Phagocytosis and autophagy:** Lysosomes are responsible for the degradation of large particles taken up by phagocytosis and for the digestion of the cell's own components by autophagy.

phagocytosis, phagosome, phagolysosome, autophagy, autophagosome

Questions

1. What was the original experimental evidence for the secretory pathway from rough ER → Golgi apparatus → secretory vesicles → secreted protein?

2. How did *in vitro* translation of mRNAs provide evidence for the existence of a signal sequence that targets secretory proteins to the rough endoplasmic reticulum?

3. Compare and contrast cotranslational and posttranslational translocation of polypeptide chains into the endoplasmic reticulum.

4. Sec61 is a critical component of the protein channel through the ER membrane. In Sec61 mutant yeast, what is the fate of proteins that are normally localized to the Golgi apparatus?

5. Why are the carbohydrate groups of glycoproteins always exposed on the surface of the cell?

6. What would be the effect of mutating the KDEL sequence of a resident ER protein like BiP? Would this effect be similar or different from that of mutating the KDEL receptor protein?

7. How is a lysosomal protein targeted to a lysosome? What effect would the addition of a lysosome-targeting signal patch have on the subcellular localization of a protein that is normally cytosolic? How would it affect localization of a protein that is normally secreted?

8. What is the predicted fate of lysosomal acid hydrolases in I-cell disease in which cells are deficient in the enzyme required for formation of mannose-6-phosphate residues?

9. What processes result in glycolipids and sphingomyelin being found in the outer—but not the inner—half of the plasma membrane bilayer?

Refer To

The Cell

Companion Website

sites.sinauer.com/cooper7e

for quizzes, animations, videos, flashcards, and other study resources.

10. A patient comes to your clinic with an accumulation of glucocerebrosides in macrophage lysosomes. What is your diagnosis, and what therapy would you suggest?

11. Lysosomes contain powerful hydrolytic enzymes, which are transported there from the site of their synthesis in the ER via the Golgi apparatus. Why don't these enzymes damage the constituents of these organelles?

12. What is the source of energy for fusion between target and vesicle membranes?

References and Further Reading
(Key review articles for each major section are highlighted in **bold**.)

The Endoplasmic Reticulum

Aebi, M., R. Bernasconi, S. Clerc and M. Molinari. 2010. N-glycan structures: recognition and processing in the ER. *Trends Biochem. Sci.* 35: 74–82. [R]

Blobel, G. and B. Dobberstein. 1975. Transfer of proteins across membranes. I. Presence of proteolytically processed and unprocessed nascent immunoglobulin light chains on membrane-bound ribosomes of murine myeloma. *J. Cell Biol.* 67: 835–851. [P]

Braakman, I. and N. J. Bulleid. 2011. Protein folding and modification in the mammalian endoplasmic reticulum. *Ann. Rev. Biochem.* 80: 71–99. [R]

Chen, S., P. Novick and S. Ferro-Novick. 2013. ER structure and function. *Curr. Opin. Cell Biol.* 25: 428–433. [R]

Dancourt, J. and C. Barlowe. 2010. Protein sorting receptors in the early secretory pathway. *Ann. Rev. Biochem.* 79: 777–802. [R]

Ferris, S. P., V. K. Kodali and R. J. Kaufman. 2014. Glycoprotein folding and quality-control mechanisms in protein-folding diseases. *Dis. Model Mech.* 7: 331–341. [R]

Goyal, U. and C. Blackstone. 2013. Untangling the web: mechanisms underlying ER network formation. *Biochim. Biophys. Acta* 1833: 2492–2498. [R

Hegde, R. S. and R. J. Keenan. 2011. Tail-anchored membrane protein insertion into the endoplasmic reticulum. *Nature Rev. Mol. Cell Biol.* 12: 787–798. [R]

Hetz, C. 2012. The unfolded protein response: controlling cell fate decisions under ER stress and beyond. *Nature Rev. Mol. Cell Biol.* 13: 89–102. [R]

Park, E. and T. A. Rapoport. 2012. Mechanisms of Sec61/SecY-mediated protein translocation across membranes. *Ann. Rev. Biophys.* 41: 21–40. [R]

Presley, J. F., N. B. Cole, T. A. Schroer, K. Hirschberg, K. J. Zaal and J. Lippincott-Schwartz. 1997. ER-to-Golgi transport visualized in living cells. *Nature* 389: 81–85. [P]

Ron, D. and P. Walter. 2007. Signal integration in the endoplasmic reticulum unfolded protein response. *Nature Rev. Mol. Cell Biol.* 8: 519–529. [R]

Sebastian, T. T., R. D. Baldridge, P. Xu and T. R. Graham. 2012. Phospholipid flippases: building asymmetric membranes and transport vesicles. *Biochim. Biophys. Acta* 1821: 1068–1077. [R]

Shao, S. and R. S. Hegde. 2011. Membrane protein insertion at the endoplasmic reticulum. *Ann. Rev. Cell Dev. Biol.* 27: 25–56. [R]

Smith, M. H., H. L. Ploegh and J. S. Weissman. 2011. Road to ruin: targeting proteins for degradation in the endoplasmic reticulum. *Science* 334: 1086–1090. [R]

van den Berg, B., W. M. Clemens Jr., I. Collinson, Y. Modis, E. Hartmann, S. C. Harrison and T. A. Rapoport. 2004. X-ray structure of a protein-conducting channel. *Nature* 427: 36–44. [P]

Walter, P. and D. Ron. 2011. The unfolded protein response: from stress pathway to homeostatic regulation. *Science* 334: 1081–1085. [R]

The Golgi Apparatus

Armstrong, J. 2010. Yeast vacuoles: more than a model lysosome. *Trends Cell Biol.* 20: 580–585. [R]

Banfield, D. K. 2011. Mechanisms of protein retention in the Golgi. *Cold Spring Harb. Perspect. Biol.* 3: a005264. [R]

Baranski, T. J., P. L. Faust and S. Kornfeld. 1990. Generation of a lysosomal enzyme targeting signal in the secretory protein pepsinogen. *Cell* 63: 281–291. [P]

De Matteis, M.A. and A. Luini. 2008. Exiting the Golgi complex. *Nature Rev. Mol. Cell Biol.* 9: 273–284. [R]

Glick, B. S. and A. Luini. 2011. Models for Golgi traffic: a critical assessment. *Cold Spring Harbor Perspect. Biol.* 3: a005215. [R]

Guo, Y., D. W. Sirkis and R. Schekman. 2014. Protein sorting at the *trans*-Golgi network. *Ann. Rev. Cell Dev. Biol.* 30: 169–206. [R]

Kienzle, C. and J. von Blume. 2014. Secretory cargo sorting at the *trans*-Golgi network. *Trends Cell Biol.* 24: 584–593. [R]

Lowe, M. 2011. Structural organization of the Golgi apparatus. *Curr. Opin. Cell Biol.* 23: 85–93. [R]

Matsuura-Tokita, K., M. Takeuchi, A. Ichihara, K. Mikuriya and A. Nakano. 2006. Live imaging of yeast Golgi cisternal maturation. *Nature* 441: 1007–1010. [P]

Morriswood, B. and G. Warren. 2013. Stalemate in the Golgi battle. *Science* 341: 1465–1466. [R]

Pfeffer, S. R. 2009. Multiple routes of protein transport from endosomes the *trans* Golgi network. *FEBS Letters* 583: 3811–3816. [R]

Stanley, P. 2011. Golgi glycosylation. *Cold Spring Harbor Perspect. Biol.* 3: a005199. [R]

Xiang, L., E. Etxeberria and W. Van den Ende. 2013. Vacuolar protein sorting mechanisms in plants. *FEBS J.* 280: 979–993. [R]

The Mechanism of Vesicular Transport

Bröcker, C., S. Engelbrecht-Vandré and C. Ungermann. 2010. Multisubunit tethering complexes and their role in membrane fusion. *Current Biol.* 20: R943–R952. [R]

Cai, H., K. Reinisch and S. Ferro-Novick. 2007. Coats, tethers, Rabs, and SNAREs work together to mediate the intracellular destination of a transport vesicle. *Dev. Cell* 12: 671–682. [R]

Carmosino, M., G. Valenti, M. Caplan and M. Svelto. 2010. Polarized traffic towards the cell surface: how to find the route. *Biol. Cell* 102: 75–91. [R]

Faini, M., R. Beck, F. T. Wieland and J. A. G. Briggs. 2013. Vesicle coats: structure, function, and general principles of assembly. *Trends Cell Biol.* 23: 279–288. [R]

Fries, E., and J. E. Rothman. 1980. Transport of vesicular stomatitis virus glycoprotein in a cell-free extract. *Proc. Natl. Acad. Sci. U.S.A.* 77: 3870–3874. [P]

Gillingham, A. K. and S. Munro. 2007. The small G proteins of the Arf family and their regulators. *Ann. Rev. Cell Dev. Biol.* 23: 579–611. [R]

Hong, W. J. and S. Lev. 2014. Tethering the assembly of SNARE complexes. *Trends Cell Biol.* 24: 35–43. [R]

Hutagalung, A. H. and P. J. Novick. 2011. Role of Rab GTPases in membrane traffic and cell physiology. *Physiol. Rev.* 91: 119–149. [R]

Mellman, I. and S. D. Emr. 2013. A Nobel Prize for membrane traffic: vesicles find their journey's end. *J. Cell Biol.* 203: 559–561. [R]

Novick, P., C. Field and R. Schekman. 1980. Identification of 23 complementation groups required for post-translational events in the yeast secretory pathway. *Cell* 21: 205–215. [P]

Pucadyil, T. J. and S. L. Schmid. 2009. Conserved functions of membrane active GTPases in coated vesicle formation. *Science* 325: 1217–1220. [R]

Söllner, T., S. W. Whiteheart, M. Brunner, H. Erdjument-Bromage, S. Geromanos, P. Tempst and J. E. Rothman. 1993. SNAP receptors implicated in vesicle targeting and fusion. *Nature* 362: 318–324. [P]

Zanetti, G., K. B. Pahuja, S. Studer, S. Shim and R. Schekman. 2012. COPII and the regulation of protein sorting in mammals. *Nature Cell Biol.* 14: 20–28. [R]

Lysosomes

Bissig, C. and J. Gruenberg. 2013. Lipid sorting and multivesicular endosome biogenesis. *Cold Spring Harbor Perspect. Biol.* 5:a016816 [R]

Coutinho, M. F., M. J. Prata and S. Alves. 2012. Mannose-6-phosphate pathway: A review on its role in lysosomal function and dysfunction. *Mol. Genet. Metab.* 105: 542–550. [R]

Forgac, M. 2007. Vacuolar ATPases: rotary proton pumps in physiology and pathophysiology. *Nature Rev. Mol. Cell Biol.* 8: 917–929. [R]

Luzio, J. P., P. R. Pryor and N. A. Bright. 2007. Lysosomes: fusion and function. *Nature Rev. Mol. Cell Biol.* 8: 622–632. [R]

Neufeld, E. F. 1991. Lysosomal storage diseases. *Ann. Rev. Biochem.* 60: 257–280. [R]

Rubinsztein, D. C., T. Shpilka and Z. Elazar. 2012. Mechanisms of autophagosome biogenesis. *Current Biol.* 22: R29–R34. [R]

Saftig, P. and J. Klumperman. 2009. Lysosome biogenesis and lysosomal membrane proteins: trafficking meets function. *Nature Rev. Mol. Cell Biol.* 10: 623–635. [R]

Settembre, C., A. Fraldi, D. L. Medina and A. Ballabio. 2013. Signals from the lysosome: a control centre for cellular clearance and energy metabolism. *Nature Rev. Mol. Cell Biol.* 14: 283–296. [R]

Shen, H.-M. and N. Mizushima. 2014. At the end of the autophagic road: an emerging understanding of lysosomal functions in autophagy. *Trends Biochem. Sci.* 39: 61–71. [R]

Weidberg, H., E. Shvets and Z. Elazar. 2011. Biogenesis and cargo selectivity of autophagosomes. *Ann. Rev. Biochem.* 80: 125–156. [R]

Mitochondria, Chloroplasts, and Peroxisomes

In addition to being involved in protein sorting and transport, cytoplasmic organelles provide specialized compartments in which a variety of metabolic activities take place. The generation of metabolic energy is a major activity of all cells, and two cytoplasmic organelles are specifically devoted to energy metabolism and the production of ATP. Mitochondria are responsible for generating most of the useful energy derived from the breakdown of lipids and carbohydrates, and chloroplasts use energy captured from sunlight to generate both ATP and the reducing power needed to synthesize carbohydrates from CO_2 and H_2O. The third organelle discussed in this chapter, the peroxisome, contains enzymes involved in a variety of different metabolic pathways, including the breakdown of fatty acids and photorespiration.

Mitochondria, chloroplasts, and peroxisomes differ from the organelles discussed in the preceding chapter not only in their functions but also in their mechanism of assembly. Rather than being synthesized on membrane-bound ribosomes and translocated into the endoplasmic reticulum, most proteins destined for mitochondria, chloroplasts, and peroxisomes are synthesized on free ribosomes in the cytosol and imported into their target organelles as completed polypeptide chains. Mitochondria and chloroplasts also contain their own genomes, which include some genes that are transcribed and translated within the organelle. Protein sorting to the cytoplasmic organelles discussed in this chapter is thus distinct from the pathways of vesicular transport that connect the endoplasmic reticulum, Golgi apparatus, lysosomes, and plasma membrane.

Mitochondria

Mitochondria play a critical role in the generation of metabolic energy in eukaryotic cells. As reviewed in Chapter 3, they are responsible for most of the useful energy derived from the breakdown of carbohydrates and fatty acids, which is converted to ATP by the process of oxidative phosphorylation. Most mitochondrial proteins are translated on free cytosolic ribosomes and imported into the organelle by specific targeting signals. In addition, mitochondria are unique among the cytoplasmic organelles already discussed in that they contain their own DNA, which encodes tRNAs, rRNAs, and some mitochondrial proteins. The assembly of mitochondria thus involves proteins encoded by their own genomes and translated within the organelle, as well as proteins encoded by the nuclear genome and imported from the cytosol.

Video 12.1

sites.sinauer.com/cooper7e/v12.1
Mitochondrial Networks
In most cells, mitochondria exist as part of a large interconnected network.

Video 12.2

sites.sinauer.com/cooper7e/v12.2
Mitochondrial Dynamics
Continual fusion and fission events re-model the mitochondrial network and modify morphology and function.

Organization and function of mitochondria

Mitochondria are surrounded by a double-membrane system, consisting of inner and outer mitochondrial membranes separated by an intermembrane space (**Figure 12.1**) The inner membrane forms numerous folds (**cristae**), which extend into the interior (or **matrix**) of the organelle. Each of these components plays distinct functional roles, with the matrix and inner membrane representing the major working compartments of mitochondria.

The matrix contains the mitochondrial genetic system as well as the enzymes responsible for the central reactions of oxidative metabolism (**Figure 12.2**). As discussed in Chapter 3, the oxidative breakdown of glucose and fatty acids is the principal source of metabolic energy in animal cells. The initial stages of glucose metabolism (glycolysis) occur in the cytosol, where glucose is converted to pyruvate (see Figure 3.2). Pyruvate is then transported into the mitochondria, where its complete oxidation to CO_2 yields the bulk of usable energy (ATP) obtained from glucose metabolism. This involves the initial oxidation of pyruvate to acetyl CoA, which is then broken down to CO_2 via the citric acid cycle (see Figures 3.3 and 3.4). The oxidation of fatty acids also yields acetyl CoA (see Figure 3.5), which is similarly metabolized by the citric acid cycle in mitochondria. The enzymes of the citric acid cycle (located in the matrix of mitochondria) thus are central players in the oxidative breakdown of both carbohydrates and fatty acids.

The oxidation of acetyl CoA to CO_2 is coupled to the reduction of NAD^+ and FAD to NADH and $FADH_2$, respectively. Most of the energy derived from oxidative metabolism is then produced by the process of oxidative phosphorylation (discussed in detail in Chapter 3), which takes place in the inner mitochondrial membrane. The high-energy electrons from NADH and $FADH_2$ are transferred through a series of carriers in the membrane to molecular oxygen (see Figures 3.6 and 3.7). The energy derived from these electron transfer reactions is converted to potential energy stored in a proton gradient across the membrane, which is then used to drive ATP synthesis. The inner mitochondrial membrane thus represents the principal site of ATP generation, and this critical role is reflected in its structure. First, its surface area is substantially increased by its folding into cristae. In addition, the inner

Figure 12.1 Structure of a mitochondrion Mitochondria are bounded by a double-membrane system, consisting of inner and outer membranes. Folds of the inner membrane (cristae) extend into the matrix.

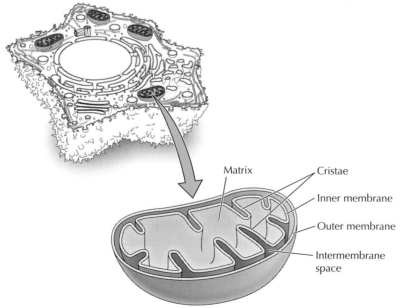

Matrix
Cristae
Inner membrane
Outer membrane
Intermembrane space

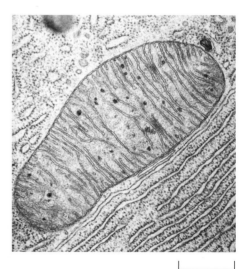

0.5 μm

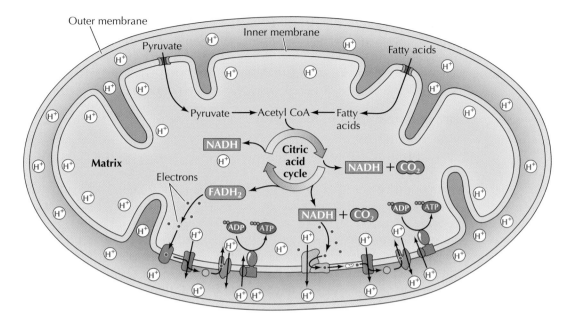

Figure 12.2 Oxidative metabolism in mitochondria Pyruvate and fatty acids are imported from the cytosol and converted to acetyl CoA in the mitochondrial matrix. Acetyl CoA is then oxidized to CO_2 via the citric acid cycle, coupled to the reduction of NAD^+ and FAD to NADH and $FADH_2$, respectively. The high-energy electrons from NADH and $FADH_2$ are transferred through a series of carriers in the inner membrane to molecular oxygen, coupled to the generation of a proton gradient across the membrane. Energy stored in the proton gradient is then used to drive ATP synthesis.

FYI

The integrity of the mitochondrial outer membrane plays a critical role in regulating programmed cell death (see Chapter 18).

mitochondrial membrane contains an unusually high percentage (greater than 70%) of proteins, which are involved in oxidative phosphorylation as well as in the transport of metabolites (e.g., pyruvate and fatty acids) between the cytosol and mitochondria. Otherwise, the inner membrane is impermeable to most ions and small molecules—a property critical to maintaining the proton gradient that drives oxidative phosphorylation.

In contrast to the inner membrane, the outer mitochondrial membrane is highly permeable to small molecules. This is because it contains proteins called **porins**, which form channels that allow the free diffusion of molecules smaller than about 1000 daltons. The composition of the intermembrane space is therefore similar to the cytosol with respect to ions and small molecules. Consequently, the inner mitochondrial membrane is the functional barrier to the passage of small molecules between the cytosol and the matrix, and it maintains the proton gradient that drives oxidative phosphorylation.

As the major sources of cellular energy, the effective functioning of mitochondria is critical for cell viability. In many cells this requires selectively positioning mitochondria to locations of high-energy use, such as synapses in nerve cells. In addition, mitochondria are not static organelles but are constantly fusing with one another and dividing. In most cells mitochondria exist as part of a large interconnected network (**Figure 12.3**). Continual fusion and fission events remodel this network of mitochondria within the cell, and

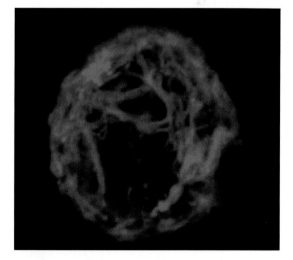

Figure 12.3 A mitochondrial network Fluorescence micrograph of a cell expressing a mitochondrial protein labeled with red fluorescent protein (RFP). The mitochondria form an interconnected network. (Courtesy of Jennifer Lippincott-Schwartz and Kasturi Mitra.)

modify mitochondrial morphology and function. Mitochondrial fusion allows the exchange of genetic material among mitochondria within the cell, whereas fission is important in the distribution of mitochondria between daughter cells at cell division and in facilitating the transport of mitochondria to areas of the cell where production of energy is in high demand.

The genetic system of mitochondria

Mitochondria contain their own genetic system, which is separate and distinct from the nuclear genome of the cell. As reviewed in Chapter 1, mitochondria are thought to have evolved from bacteria that developed a symbiotic relationship in which they lived within larger cells (**endosymbiosis**). The genomes of living organisms that are most similar to the mitochondrial genome are those of free-living α-proteobacteria, which code for approximately 7000 proteins. An intracellular parasite like the α-proteobacterium *Rickettsia prowazekii* has a smaller genome, about 830 protein-coding genes. Like mitochondria, *Rickettsia prowazekii* is able to reproduce only within eukaryotic cells, but unlike mitochondria it still transcribes and translates most of its own genes.

Mitochondrial genomes are usually circular DNA molecules like those of bacteria, which are present in multiple copies per organelle. They vary considerably in size between different species. Most present-day mitochondrial genomes encode only a small number of proteins that are essential components of the oxidative phosphorylation system. In addition, mitochondrial genomes encode all of the ribosomal RNAs and most of the transfer RNAs needed for translation of these protein-coding sequences within mitochondria. Other mitochondrial proteins are encoded by nuclear genes, which have been transferred to the nucleus from the ancestral mitochondrial genome.

The genomes of human and most other animal mitochondria are only about 16 kb, but substantially larger mitochondrial genomes are found in yeasts (approximately 80 kb) and plants (more than 200 kb). However, these larger mitochondrial genomes are composed predominantly of noncoding sequences and do not appear to contain significantly more genetic information. For example, the mitochondrial genome of the plant *Arabidopsis thaliana* is approximately 370 kb but it encodes only 31 proteins: just more than twice the number encoded by human mitochondrial DNA.

The human mitochondrial genome encodes 13 proteins involved in electron transport and oxidative phosphorylation (**Figure 12.4**). The mitochondrial genes encoding these proteins are transcribed and translated within mitochondria, which contain their own ribosomes and tRNAs. Human mitochondrial DNA encodes 16S and 12S rRNAs and 22 tRNAs, which are required for translation of the proteins encoded by the organelle genome. The two rRNAs are the only RNA components of animal and yeast mitochondrial ribosomes, in contrast with the three rRNAs of bacterial ribosomes (23S,

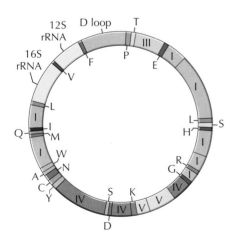

Figure 12.4 The human mitochondrial genome The genome contains 13 protein coding sequences, which are designated as components of respiratory complexes I, III, IV, or V. In addition, the genome contains genes for 16S and 12S rRNAs and for 22 tRNAs, which are designated by the one letter code for the corresponding amino acid. The region of the genome designated "D loop" contains an origin of DNA replication and transcriptional promoter sequences.

Table 12.1 Differences between the Universal and Mitochondrial Genetic Codes

Codon	Universal code	Human mitochondrial code
UGA	Stop	Trp
AGA	Arg	Stop
AGG	Arg	Stop
AUA	Ile	Met

Note: Other codons vary from the universal code in yeast and plant mitochondria.

16S, and 5S). Plant mitochondrial DNAs, however, also encode a third rRNA: 5S. The mitochondria of plants and protozoans also differ in importing and utilizing tRNAs encoded by the nuclear as well as the mitochondrial genome, whereas in animal mitochondria, all the tRNAs are encoded by the organelle genome. In contrast to the rRNAs and tRNAs encoded by the mitochondrial genome, nuclear genes encode all of the proteins needed for transcription and translation, including mitochondrial RNA polymerase, ribosomal proteins, and translation factors.

The small number of tRNAs encoded by the mitochondrial genome highlights an important feature of the mitochondrial genetic system—the use of a slightly different genetic code, which is distinct from the "universal" genetic code used by both prokaryotic and eukaryotic cells (Table 12.1). As discussed in Chapter 4, there are 64 possible triplet codons, of which 61 encode the 20 different amino acids incorporated into proteins (see Table 4.1). Many tRNAs in both prokaryotic and eukaryotic cells are able to recognize more than a single codon in mRNA because of "wobble," which allows some mispairing between the tRNA anticodon and the third position of certain complementary codons (see Figure 9.3). However, at least 30 different tRNAs are required to translate the universal code according to the wobble rules. Yet human mitochondrial DNA encodes only 22 tRNA species, and these are the only tRNAs used for translation of mitochondrial mRNAs. This is accomplished by an extreme form of wobble in which U in the anticodon of the tRNA can pair with any of the four bases in the third codon position of mRNA, allowing four codons to be recognized by a single tRNA. In addition, some codons specify different amino acids in mitochondria than in the universal code.

Like the DNA of nuclear genomes, mitochondrial DNA can be altered by mutations, which are frequently deleterious to the organelle. Since almost all the mitochondria of fertilized eggs are contributed by the oocyte rather than by the sperm, germline mutations in mitochondrial DNA are transmitted to the next generation by the mother. Such mutations have been associated with a number of diseases. For example, Leber's hereditary optic neuropathy, a disease that leads to blindness, can be caused by mutations in mitochondrial genes that encode components of the electron transport chain. In addition, the progressive accumulation of mutations in mitochondrial DNA during the lifetime of individuals has been suggested to contribute to the process of aging.

Molecular Medicine

Diseases of Mitochondria: Leber's Hereditary Optic Neuropathy

The Disease

Leber's hereditary optic neuropathy (LHON) is a rare inherited disease that results in blindness because of degeneration of the optic nerve. Vision loss usually occurs between the ages of 15 and 35, and is generally the only manifestation of the disease. Not all individuals who inherit the genetic defects responsible for LHON develop the disease, and females are affected less frequently than males. This propensity to affect males might suggest that LHON is an X-linked disease. This is not the case, however, because males never transmit LHON to their offspring. Instead, the inheritance of LHON is entirely by maternal transmission. This characteristic is consistent with cytoplasmic rather than nuclear inheritance of LHON, since the cytoplasm of fertilized eggs is derived almost entirely from the oocyte.

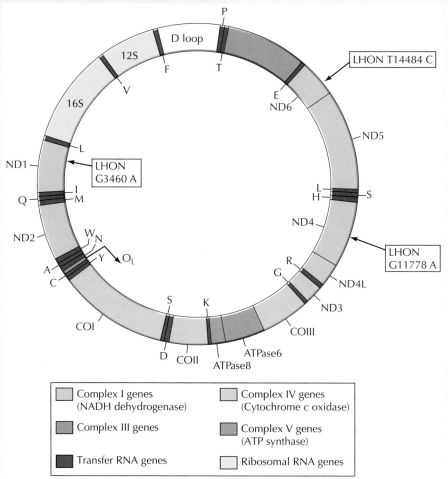

Complex I genes (NADH dehydrogenase)

Complex III genes

Transfer RNA genes

Complex IV genes (Cytochrome c oxidase)

Complex V genes (ATP synthase)

Ribosomal RNA genes

Molecular and Cellular Basis

In 1988 Douglas Wallace and his colleagues identified a mutation in the mitochondrial DNA of LHON patients. This mutation (at base pair 11778) affects one of the subunits of complex I of the electron transport chain (NADH dehydrogenase). Subsequent studies have revealed that mutations in three of the subunits of mitochondrial-encoded NADH dehydrogenase (see Figure) account for most cases of LHON.

The mutations causing LHON reduce the capacity of mitochondria to carry out oxidative phosphorylation and generate ATP. This has the greatest effect on those tissues that are most dependent on oxidative phosphorylation, so defects in components of mitochondria can lead to clinical manifestations in specific organs, rather than to systemic disease. The central nervous system (including the brain and optic nerve) is most highly dependent on oxidative metabolism, consistent with blindness being the primary clinical manifestation of LHON. As already noted, inheritance of LHON mutations does not

always lead to development of the disease; only about 10% of females and 50% of males possessing a mutation suffer vision loss. Since mitochondria normally undergo fission and fusion, mixtures of normal and defective genomes will be distributed to mitochondrial networks and individual mitochondria within a cell. Many individuals who bear predominantly mutant mitochondrial DNAs still fail to develop the disease, suggesting that additional genetic or environmental factors, which have yet to be identified, play a significant role in the development of LHON.

Prevention and Treatment

The identification of mitochondrial DNA mutations responsible for LHON allows molecular diagnosis of the disease, which can be important in establishing a definitive diagnosis of patients without a family history. However, the detection of mutations

in mitochondrial DNA is of little value for screening members of affected families or for family planning. This contrasts to the utility of detecting inherited mutations of nuclear genes, where molecular analysis can determine whether a family member or embryo has inherited a mutant or wildtype allele. In LHON, however, mutant mitochondria are present in large numbers and are maternally transmitted to all offspring. As noted above, not all such offspring develop the disease, but this cannot be predicted by genetic analysis.

The finding that LHON is caused by mutations of mitochondrial DNA suggests the potential of new therapies. One approach is metabolic therapy intended to enhance oxidative phosphorylation by administration of substrates or cofactors in the electron transport pathway, such as succinate or coenzyme Q. Alternatively, antioxidants and factors that promote

Molecular Medicine

survival of neurons are being considered as therapies for LHON. Finally, gene therapy strategies are being pursued, including the possibility of relocating a normal gene allele to the nucleus. An appropriate targeting signal would be added to direct the gene product to mitochondria, where it could

substitute for the defective mitochondrial-encoded protein.

References

Brown, M. D., D. S. Voljavec, M. T. Lott, I. MacDonald and D. C. Wallace. 1992. Leber's hereditary optic neuropathy: A model for mitochondrial neurode-generative diseases. *FASEB J.* 6: 2791–2799.

Farrar, G. J., N. Chadderton, P. F. Kenna and S. Millington-Ward. 2013. Mitochondrial disorders: aetiologies, model systems, and candidate thera-pies. *Trends Genet.* 29: 488–497.

Protein import and mitochondrial assembly

Although their genomes encode only 13 proteins, mammalian mitochondria contain about 1500 different proteins that are encoded by the nuclear genome of the cell. In contrast to the RNA components of the mitochondrial transla-tion apparatus (rRNAs and tRNAs), most mitochondrial genomes do not encode the proteins required for DNA replication, transcription, or transla-tion. Instead, the genes that encode proteins required for the replication and expression of mitochondrial DNA are contained in the nucleus. In addition, the nucleus contains the genes that encode most of the mitochondrial proteins required for oxidative phosphorylation and all of the enzymes involved in mitochondrial metabolism (e.g., enzymes of the citric acid cycle). Some of these genes were transferred to the nucleus from the original prokaryotic ancestor of mitochondria.

The proteins encoded by nuclear genes (~99% of mitochondrial proteins) are synthesized on free cytosolic ribosomes and imported into mitochondria as completed polypeptide chains. Because of the double-membrane structure of mitochondria, the import of proteins is considerably more complicated than the transfer of a polypeptide across a single phospholipid bilayer. Proteins targeted to the matrix have to cross both the outer and inner mitochondrial membranes, while other proteins need to be sorted to distinct compartments within the organelle, including the inner membrane, the outer membrane, and the intermembrane space. Several classes of targeting signals direct proteins to these different mitochondrial compartments.

The most thoroughly studied targeting sequences are amino-terminal **presequences** of 15 to 55 amino acids that direct the import of proteins to the mitochondrial matrix as well as target some proteins to the inner membrane (**Figure 12.5**). The presequences of mitochondrial proteins, first characterized by Gottfried Schatz, contain multiple positively charged amino acid residues, usually in an amphipathic α helix, and are removed by proteolytic cleavage following their import into the organelle. The first step in protein import is the binding of these presequences to a protein complex on the surface of mitochondria that directs translocation across the outer membrane (the *t*ranslocase of the *o*uter *m*embrane, or **Tom complex**). Following transloca-tion through the Tom complex, these proteins are transferred to a second protein complex, called Tim23, in the inner membrane (one of two different *t*ranslocases of the *i*nner *m*embrane, or **Tim complexes**). Mitochondrial matrix proteins are then translocated across the inner membrane through Tim23. Alternatively, proteins that contain a transmembrane sequence exit the Tim23 channel laterally and insert into the inner membrane.

Translocation of proteins containing presequences through Tim23 requires the electrochemical potential established across the inner mitochondrial membrane during electron transport. As discussed in Chapter 3, the transfer

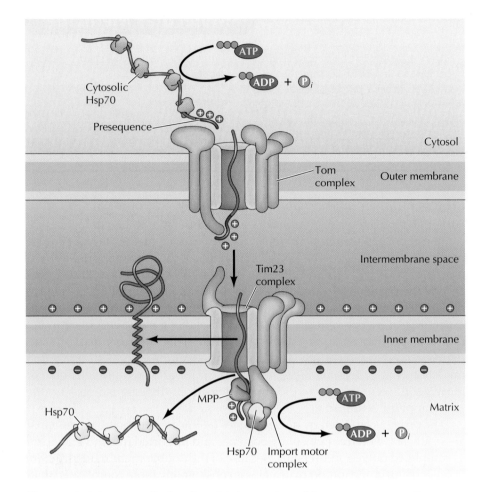

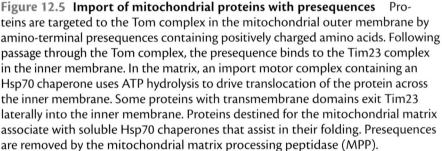

Figure 12.5 Import of mitochondrial proteins with presequences Proteins are targeted to the Tom complex in the mitochondrial outer membrane by amino-terminal presequences containing positively charged amino acids. Following passage through the Tom complex, the presequence binds to the Tim23 complex in the inner membrane. In the matrix, an import motor complex containing an Hsp70 chaperone uses ATP hydrolysis to drive translocation of the protein across the inner membrane. Some proteins with transmembrane domains exit Tim23 laterally into the inner membrane. Proteins destined for the mitochondrial matrix associate with soluble Hsp70 chaperones that assist in their folding. Presequences are removed by the mitochondrial matrix processing peptidase (MPP).

of high-energy electrons from NADH and $FADH_2$ to molecular oxygen is coupled to the transfer of protons from the mitochondrial matrix to the intermembrane space. Since protons are charged particles, this transfer establishes an electric potential across the inner membrane, with the matrix being negative (see Figure 3.8). During protein import, this electric potential drives translocation of the positively charged presequence.

To be translocated across the mitochondrial membrane, proteins must be at least partially unfolded. Consequently, protein import into mitochondria requires molecular chaperones in addition to the membrane proteins involved in translocation (see Figure 12.5). On the cytosolic side, members of the Hsp70 family of chaperones both maintain proteins in a partially unfolded state and present them to the Tom complex. As they cross the inner membrane, the unfolded polypeptide chains are bound by another Hsp70 chaperone, which

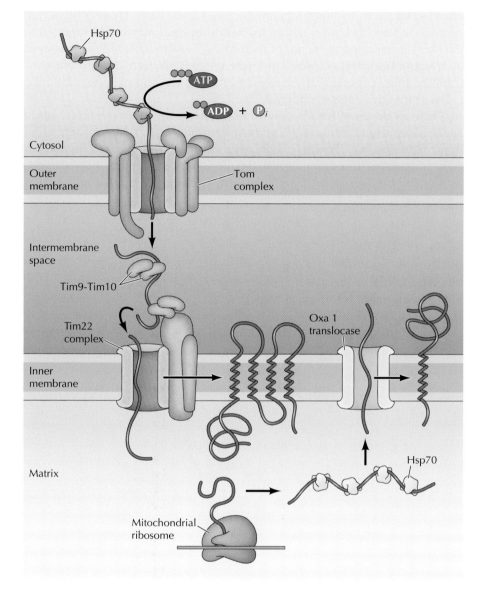

Figure 12.6 Protein targeting to the mitochondrial inner membrane
Inner membrane proteins with multiple transmembrane domains have internal signal sequences, rather than N-terminal presequences. Following translocation across the outer membrane, these proteins are bound by mobile Tim9-Tim10 chaperones in the intermembrane space, which transfer the protein to the Tim22 complex in the inner membrane. Internal transmembrane sequences halt translocation, and the protein is transferred laterally into the inner membrane. Some inner membrane proteins are encoded by the mitochondrial genome. These proteins are translated on mitochondrial ribosomes in the matrix and targeted to the inner membrane via the Oxa1 translocase.

is part of a motor complex that drives protein import. The presequence is then cleaved by the **matrix processing peptidase (MPP)** and the polypeptide chain is bound by other matrix chaperones that facilitate its folding.

Proteins can be targeted to the inner membrane not only by presequences, but also by other signals (**Figure 12.6**). Many proteins in the inner membrane are multiple-pass transmembrane proteins that serve as transporters to exchange nucleotides and ions between the mitochondria and the cytosol. These proteins do not contain presequences but instead have multiple internal mitochondrial import signals. Proteins marked by these sequences cross the outer membrane through the Tom complex, but instead of being transferred to Tim23, they are recognized by mobile chaperones (Tim9-Tim10) in the intermembrane space. The Tim9-Tim10 chaperones then escort these proteins to the second translocase of the inner membrane (Tim22). The proteins are then partially translocated through Tim22 before internal transmembrane sequences cause them to exit the Tim22 pore laterally and insert into the inner membrane. In addition to the inner membrane proteins that are imported into mitochondria, some inner membrane proteins are encoded by the mitochondrial genome (see Figure 12.4). These proteins are synthesized on ribosomes within the mitochondrial matrix and targeted to the Oxa1 translocase in the inner membrane, which is related to a bacterial membrane insertion system. The proteins are then translocated outward from the matrix towards the intermembrane space, and exit Oxa1 laterally to insert into the inner membrane.

Proteins destined for the outer membrane or the intermembrane space are also imported through the Tom complex (**Figure 12.7**). Many outer membrane proteins (e.g., porins) are β-barrel proteins (see Figure 2.35), which pass through the Tom complex into the intermembrane space. They are then recognized by the mobile Tim9-Tim10 chaperones and carried to a second translocon complex, called the SAM (*s*orting and *a*ssembly *m*achinery) complex, which mediates their insertion into the outer membrane. In addition, some outer membrane proteins with α-helical transmembrane domains are inserted into the membrane by a distinct pathway mediated by the outer membrane protein Mim1. Proteins are targeted to the intermembrane space by cysteine-rich sequences that are recognized by specific chaperones after the proteins exit the Tom complex.

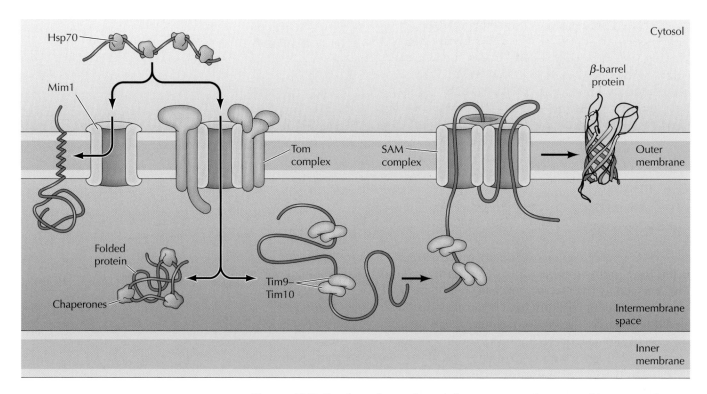

Figure 12.7 Sorting of proteins to the outer membrane and intermembrane space Single transmembrane domain proteins destined for the outer membrane are inserted via the outer membrane protein Mim1. Precursors of β-barrel proteins of the outer membrane are transferred through the Tom complex and bound by Tim9-Tim10 chaperone proteins in the intermembrane space. They are then directed to a second translocon complex, called the SAM (*s*orting and *a*ssembly *m*achinery) complex, and inserted into the outer membrane. Proteins destined for the intermembrane space are translocated through the Tom complex and recognized by specific intermembrane space chaperones.

Mitochondrial lipids

Not only the proteins but also most of the lipids of mitochondrial membranes are imported from the cytosol. In animal cells, spingolipids, cholesterol, phosphatidylcholine, phosphatidylinositol, and phosphatidylserine are synthesized in the ER and transported to mitochondria. The mitochondria then synthesize phosphatidylethanolamine from phosphatidylserine. Mitochondria also catalyze the synthesis of the unusual phospholipid **cardiolipin**, which contains four fatty acid chains (**Figure 12.8**). Cardiolipin is localized to the inner membrane of mitochondria, where it acts to improve the efficiency of oxidative phosphorylation, in part by restricting proton flow across the membrane.

The transfer of lipids between the ER and mitochondria takes place at sites of close contact between the ER and mitochondrial membranes and is thought to be mediated by **phospholipid transfer proteins**, which extract single phospholipid molecules from the membrane of the ER (**Figure 12.9**). The lipid can then be transported through the aqueous environment of the cytosol, buried in a hydrophobic binding site of the protein, and released when the complex reaches a new membrane, such as that of mitochondria. Lipids are exchanged between the mitochondrial outer and inner membranes at sites of contact between them.

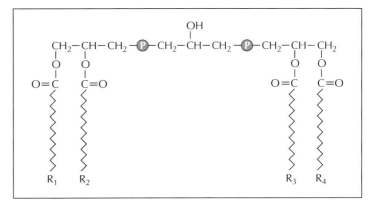

Figure 12.8 Structure of cardiolipin Cardiolipin, an unusual "double" phospholipid containing four fatty acid chains, is found primarily in the inner mitochondrial membrane.

Transport of metabolites across the inner membrane

The function of mitochondria in the generation of metabolic energy requires the efficient transport of small molecules into and out of mitochondria. For example, the ATP synthesized within mitochondria has to be exported to the cytosol, while ADP and P_i need to be imported from the cytosol for ATP synthesis to continue. The outer membrane of mitochondria is freely permeable to small molecules, but many proteins of the inner membrane are small molecule transporters (see Figure 12.6). The energy required to drive the transport of small molecules into and out of mitochondria is provided by the electrochemical gradient generated by proton pumping across the inner mitochondrial membrane during the process of oxidative phosphorylation (see Figures 3.6 and 3.7).

Figure 12.9 Phospholipid transfer proteins Phospholipid transfer proteins extract phospholipid molecules from the endoplasmic reticulum (ER) membrane and transport them to the mitochondrial outer membrane.

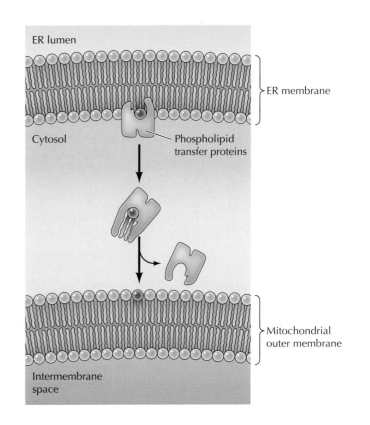

Figure 12.10 Transport of metabolites across the mitochondrial inner membrane The transport of small molecules across the inner membrane of mitochondria is mediated by membrane-spanning proteins and driven by the electrochemical gradient. For example, ATP is exported from mitochondria to the cytosol by a transporter that exchanges it for ADP. The voltage component of the electrochemical gradient drives this exchange: ATP carries a greater negative charge (−4) than ADP (−3), so ATP is exported from the mitochondrial matrix to the cytosol, while ADP is imported into mitochondria. In contrast, the transport of phosphate (P_i) and pyruvate is coupled to an exchange for hydroxyl ions (OH^-); in this case, the pH component of the electrochemical gradient drives the export of hydroxyl ions, coupled to the transport of P_i and pyruvate into mitochondria.

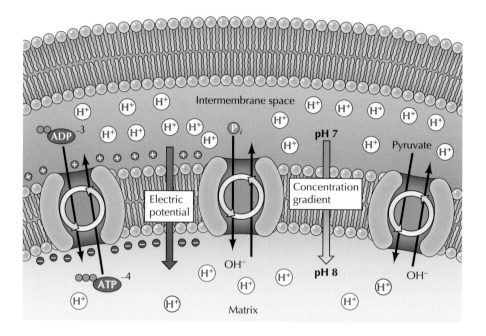

As discussed in Chapter 3, the proton gradient across the inner mitochondrial membrane has both chemical and electrical components, and both are used to drive the transport of metabolites (Figure 12.10). For example, the transport of ATP and ADP across the inner membrane is mediated by an integral membrane protein, the adenine nucleotide translocator, which transports one molecule of ADP into the mitochondrion in exchange for one molecule of ATP transferred from the mitochondrion to the cytosol. Because ATP carries more negative charge than ADP (−4 compared with −3), this exchange is driven by the voltage component of the electrochemical gradient. Since the proton gradient establishes a positive charge on the cytosolic side of the membrane, the export of ATP in exchange for ADP is energetically favorable.

The synthesis of ATP within the mitochondrion requires phosphate ions (P_i) as well as ADP, so P_i must also be imported from the cytosol. This is mediated by another membrane transport protein, which imports phosphate ($H_2PO_4^-$) and exports hydroxyl ions (OH^-). This exchange is electrically neutral because both phosphate and hydroxyl ions have a charge of −1. However, the exchange is driven by the proton concentration gradient; the higher pH within mitochondria corresponds to a higher concentration of hydroxyl ions, favoring their translocation to the cytosolic side of the membrane.

Energy from the electrochemical gradient is similarly used to drive the transport of other metabolites into mitochondria. For example, the import of pyruvate from the cytosol (where it is produced by glycolysis) is mediated by a transport protein that exchanges pyruvate for hydroxyl ions. Other intermediates of the citric acid cycle are able to shuttle between mitochondria and the cytosol by similar exchange mechanisms.

Chloroplasts and Other Plastids

Chloroplasts, the organelles responsible for photosynthesis, are in many respects similar to mitochondria. Both chloroplasts and mitochondria function to generate metabolic energy, have evolved by endosymbiosis, contain their own genetic systems, and replicate by division. However, chloroplasts are larger and more complex than mitochondria, and they perform several critical tasks in addition to the generation of ATP. Most importantly, chloroplasts are responsible for the photosynthetic conversion of CO_2 to carbohydrates. In addition, chloroplasts synthesize amino acids, fatty acids, and the lipid components of their own membranes. The reduction of nitrite (NO_2^-) to ammonia (NH_3), an essential step in the incorporation of nitrogen into organic compounds, also occurs in chloroplasts. Moreover, chloroplasts are only one of several types of related organelles (plastids) that play a variety of roles in plant cells.

The structure and function of chloroplasts

Plant chloroplasts are large organelles (5 to 10 μm long) that, like mitochondria, are bounded by a double membrane called the chloroplast envelope (**Figure 12.11**). In addition to the inner and outer membranes of the envelope, chloroplasts have a third internal membrane system, called the thylakoid membrane. The **thylakoid membrane** forms a network of flattened discs called thylakoids, which are frequently arranged in stacks called grana. Because

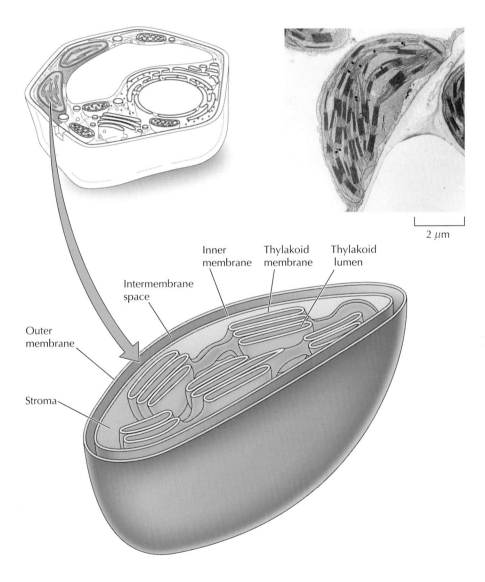

2 μm

Figure 12.11 Structure of a chloroplast In addition to the inner and outer membranes of the envelope, chloroplasts contain a third internal membrane system: the thylakoid membrane. These membranes divide chloroplasts into three internal compartments (the intermembrane space, stroma, and thylakoid lumen).

of this three-membrane structure, the internal organization of chloroplasts is more complex than that of mitochondria. In particular, their three membranes divide chloroplasts into three distinct internal compartments: (1) the intermembrane space between the two membranes of the chloroplast envelope; (2) the **stroma**, which lies inside the envelope but outside the thylakoid membrane; and (3) the thylakoid lumen.

Despite this greater complexity, the membranes of chloroplasts have clear functional similarities with those of mitochondria—as expected, given the role of both organelles in the generation of ATP. The outer membrane of the chloroplast envelope, like that of mitochondria, contains porins and is therefore freely permeable to small molecules. In contrast, the inner membrane is impermeable to ions and metabolites, which are therefore able to enter chloroplasts only via specific membrane transporters. These properties of the inner and outer membranes of the chloroplast envelope are similar to the inner and outer membranes of mitochondria: In both cases the inner membrane restricts the passage of molecules between the cytosol and the interior of the organelle. The chloroplast stroma is also equivalent in function to the mitochondrial matrix: It contains the chloroplast genetic system and a variety of metabolic enzymes, including those responsible for the critical conversion of CO_2 to carbohydrates during photosynthesis.

The major difference between chloroplasts and mitochondria, in terms of both structure and function, is the thylakoid membrane. This membrane is of central importance in chloroplasts, where it fills the role of the mitochondrial inner membrane in electron transport and the generation of ATP (**Figure 12.12**). The inner membrane of the chloroplast envelope (which is not folded

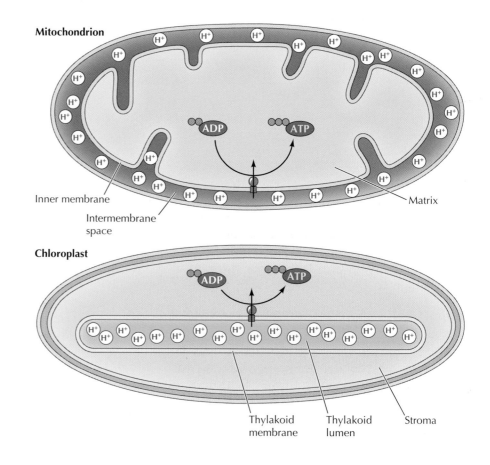

Figure 12.12 Chemiosmotic generation of ATP in chloroplasts and mitochondria In mitochondria, electron transport generates a proton gradient across the inner membrane, which is then used to drive ATP synthesis in the matrix. In chloroplasts, the proton gradient is generated across the thylakoid membrane and used to drive ATP synthesis in the stroma.

Mitochondrion

Inner membrane

Intermembrane space

Matrix

Chloroplast

Thylakoid membrane

Thylakoid lumen

Stroma

into cristae) does not function in either electron transport or photosynthesis. Instead, the chloroplast electron transport system is located in the thylakoid membrane, and protons are pumped across this membrane from the stroma to the thylakoid lumen. The resulting electrochemical gradient then drives ATP synthesis as protons cross back into the stroma. In terms of its role in generation of metabolic energy, the thylakoid membrane of chloroplasts is thus equivalent to the inner membrane of mitochondria.

The chloroplast genome

Like mitochondria, chloroplasts contain their own genetic system, reflecting their evolutionary origins from photosynthetic bacteria. The genomes of chloroplasts are similar to those of mitochondria in that they consist of circular DNA molecules present in multiple copies per organelle. However, chloroplast genomes are larger and more complex than those of mitochondria, ranging from 100–200 kb and containing approximately 150 genes.

Although there is considerable variability between the chloroplast genomes of different species, they generally include a common set of genes, including both RNAs and proteins involved in gene expression, as well as a variety of proteins that function in photosynthesis and metabolism (Table 12.2). Both the ribosomal and transfer RNAs used for translation of chloroplast mRNAs are encoded by the organelle genome. These include three rRNAs (23S, 16S, and 5S), similar to bacterial rRNAs, and 27-31 tRNAs. In contrast to the smaller number of tRNAs encoded by the mitochondrial genome, the chloroplast tRNAs are sufficient to translate all the mRNA codons according to the universal genetic code. In addition to these RNA components of the translation system, the chloroplast genome encodes about 40 ribosomal proteins, which represent most of the proteins of chloroplast ribosomes. At least three of the four subunits of bacterial RNA polymerase are also encoded by chloroplast genomes, although additional factors needed for chloroplast gene expression are encoded in the nucleus.

Chloroplast genomes also encode approximately 50 proteins that are involved in photosynthesis, including components of photosystems I and II, of the cytochrome bf complex, of ATP synthase, and of NADP reductase dehydrogenase (see Figure 3.13). Additional genes in chloroplast genomes encode proteins involved in metabolism, protein folding and membrane insertion. It is noteworthy that one of the subunits of $ribulose$ bisphosphate $c$arboxylase/$o$xygenase (rubisco) is encoded by chloroplast DNA. Rubisco is the critical enzyme that catalyzes the addition of CO_2 to ribulose-1,5-bisphosphate during the Calvin cycle (see Figure 3.15). Not only is it the major protein component of the chloroplast stroma, but it is also thought to be the single most abundant protein on Earth.

Import and sorting of chloroplast proteins

Although chloroplasts encode more of their own proteins than mitochondria, about 3000 or 95% of chloroplast proteins are still encoded by nuclear genes. As with mitochondria, these proteins are synthesized on cytosolic ribosomes and then imported into chloroplasts as completed polypeptide chains. They must then be sorted to their appropriate locations within chloroplasts—an even more complicated task than protein

Table 12.2   Genes Encoded by Chloroplast DNA

Function	Number of genes
Gene expression	
rRNAs (23S, 16S, 5S)	3
tRNAs	30
Ribosomal proteins	40
RNA polymerase subunits	4
Photosynthesis	
Photosystem I	8
Photosystem II	18
Cytochrome bf complex	8
ATP synthase	8
NADP reductase dehydrogenase	11
Metabolism	11
Protein folding and membrane insertion	14

sorting in mitochondria, since chloroplasts contain three separate membranes that divide them into three distinct internal compartments.

Most proteins are targeted for import into chloroplasts by N-terminal sequences, usually from 30 to 100 amino acids, called **transit peptides**, which direct protein translocation across the two membranes of the chloroplast envelope and are then removed by proteolytic cleavage (**Figure 12.13**). Transit peptides direct proteins to the translocase of the chloroplast outer member (the **Toc complex**). In contrast to the presequences for mitochondrial import, transit peptides are not positively charged and the chloroplast inner membrane does not have a strong electric potential. Protein import into chloroplasts requires Hsp70 molecules to keep the protein in an unfolded state. In addition, some Toc proteins bind and hydrolyze GTP, providing an additional source of energy for translocation. After proteins are transported through the Toc complex, they are transferred to the translocase of the chloroplast inner membrane (the **Tic complex**) and transported into the stroma. A chloroplast Hsp93 chaperone, associated with the stromal side of the Tic complex, acts

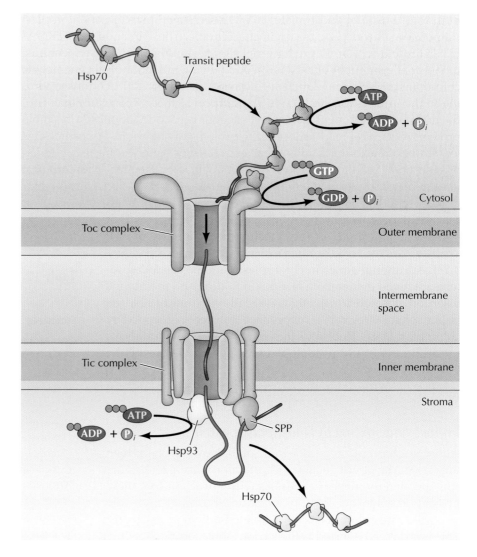

Figure 12.13 Import of proteins into the chloroplast stroma Proteins with N-terminal transit peptides are targeted to the Toc complex in the chloroplast outer membrane. Passage through the outer membrane requires ATP hydrolysis by Hsp70 and hydrolysis of GTP by Toc proteins. Once through the chloroplast outer membrane, the transit peptide is passed to the Tic complex in the inner membrane. The protein is drawn through the Tic complex by the action of an Hsp93 chaperone and the transit peptide is removed by the chloroplast stromal processing peptidase (SPP).

to draw the protein through the inner membrane. In the stroma, the transit peptide is cleaved by a **stromal processing peptidase (SPP)**, and the protein associates with stromal Hsp70 chaperones.

The pathways for incorporation of proteins into the membranes of the chloroplast envelope are not fully understood. Outer membrane proteins do not have transit peptides and are targeted directly to the membrane, where the Toc channel appears to mediate their integration. Some inner membrane proteins are imported by the Toc complex and then targeted to the inner membrane by transmembrane domains that promote their lateral exit from the Tic channel. Other inner membrane proteins are first imported to the stroma and then redirected to the inner membrane, similar to some inner membrane proteins of mitochondria (see Figure 12.6). However, the chloroplast machinery that mediates this pathway is not yet characterized.

Proteins targeted to the thylakoid lumen are transported in two steps. They are first imported into the stroma, as described above, and then targeted for translocation across the thylakoid membrane by a second signal sequence, which is exposed following cleavage of the transit peptide. Two different pathways then act to import proteins from the stroma to the thylakoid lumen (**Figure 12.14**). The Sec pathway evolved from the pathway used for protein secretion by bacteria and is thus also related to the pathway used to translocate proteins into the endoplasmic reticulum (discussed in Chapter 11). In this pathway, the thylakoid signal sequence is recognized by the SecA protein and translocated as an unfolded protein through the Sec translocon in an ATP-dependent manner. The second pathway used by chloroplasts, the twin-arginine translocation (Tat) pathway, is also related to a bacterial

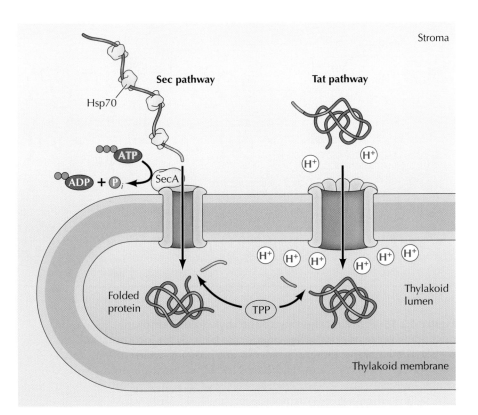

Figure 12.14 Import of proteins into the thylakoid lumen   Two pathways transfer proteins from the stroma to the thylakoid lumen. In the Sec pathway, the SecA protein recognizes a thylakoid signal sequence and targets the protein to the Sec translocon, using energy derived from ATP hydrolysis to transfer the protein to the lumen. In the twin-arginine translocation (Tat) pathway, a thylakoid signal sequence containing two arginines near the amino terminus targets fully folded proteins to the Tat translocon, which transfers them into the thylakoid lumen using energy derived from the proton gradient across the thylakoid membrane. In the lumen, the thylakoid signal sequences are cleaved by a thylakoid processing protease (TPP).

membrane translocation system. In this pathway, proteins are recognized by a twin-arginine signal sequence and transported across the thylakoid membrane in a fully-folded state. The energy for transport in the Tat pathway is provided by the proton gradient across the thylakoid membrane. In both pathways, the thylakoid signal sequences are cleaved by the thylakoid processing protease (TPP) in the thylakoid lumen.

Proteins can be targeted to the thylakoid membrane by at least three pathways (Figure 12.15). First, some proteins with a transmembrane sequence can exit the Sec translocon laterally. Alternatively, some proteins are incorporated into the thylakoid membrane via the Alb3 insertion machinery, which is related to Oxa1 of the mitochondrial inner membrane (see Figure 12.6). These proteins are targeted to Alb3 by a chloroplast signal recognition particle (cpSRP), which can also target thylakoid membrane proteins that are encoded by the chloroplast genome and translated by

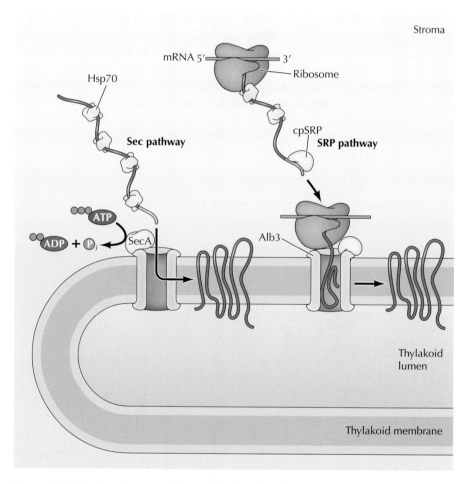

Figure 12.15 Protein targeting to the thylakoid membrane Some transmembrane proteins exit the Sec translocase laterally for incorporation into the thylakoid membrane. Other transmembrane proteins, which may be either imported into the stroma or synthesized by chloroplast ribosomes within the stroma, are recognized by the chloroplast signal recognition particle (cpSRP) and inserted into the thylakoid membrane by the Alb3 translocase.

ribosomes in the stroma. Finally, the insertion of many thylakoid membrane proteins is independent of either the Sec or SRP pathways and may occur by a "spontaneous" mechanism that does not involve any known transport machinery.

Other plastids

Chloroplasts are only one, albeit the most prominent, member of a larger family of plant organelles called **plastids**. All plastids contain the same genome as chloroplasts, but they differ in both structure and function. Chloroplasts are specialized for photosynthesis and are unique in that they contain the internal thylakoid membrane system. Other plastids, which are involved in different aspects of plant cell metabolism (such as synthesis of amino acids, fatty acids and lipids, plant hormones, nucleotides, vitamins, and secondary metabolites), are bound by the two membranes of the plastid envelope but lack both the thylakoid membranes and other components of the photosynthetic apparatus.

The different types of plastids are frequently classified according to the kinds of pigments they contain. Chloroplasts are so named because they contain chlorophyll. **Chromoplasts** (Figure 12.16A) lack chlorophyll but contain carotenoids and are responsible for the yellow, orange, and red colors of some flowers and fruits. **Leucoplasts** are nonpigmented plastids, which store a variety of energy sources in nonphotosynthetic tissues. **Amyloplasts** (Figure 12.16B) and **elaioplasts** are examples of leucoplasts that store starch and lipids, respectively.

All plastids, including chloroplasts, develop from **proplastids**, small (0.5–1 μm in diameter) undifferentiated organelles present in the rapidly dividing cells of plant roots and shoots. Proplastids then develop into the various types of mature plastids according to the needs of differentiated cells. In addition, several types of mature plastids are able to change from one type

Animation 12.1

sites.sinauer.com/cooper7e/a12.1

From Proplastid to Chloroplast A proplastid with inner and outer membranes develops into a mature chloroplast with a third membrane system—the thylakoid membranes.

(A)

(B)

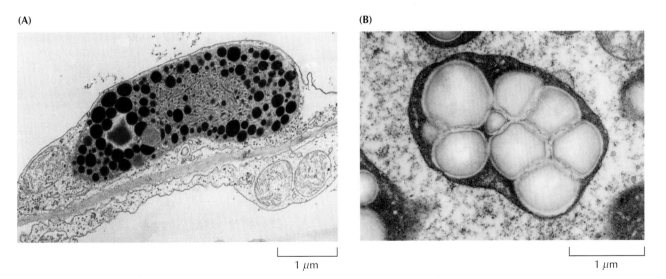

1 μm

1 μm

Figure 12.16 Electron micrographs of chromoplasts and amyloplasts
(A) Chromoplasts contain lipid droplets in which carotenoids are stored.
(B) Amyloplasts contain large starch granules.

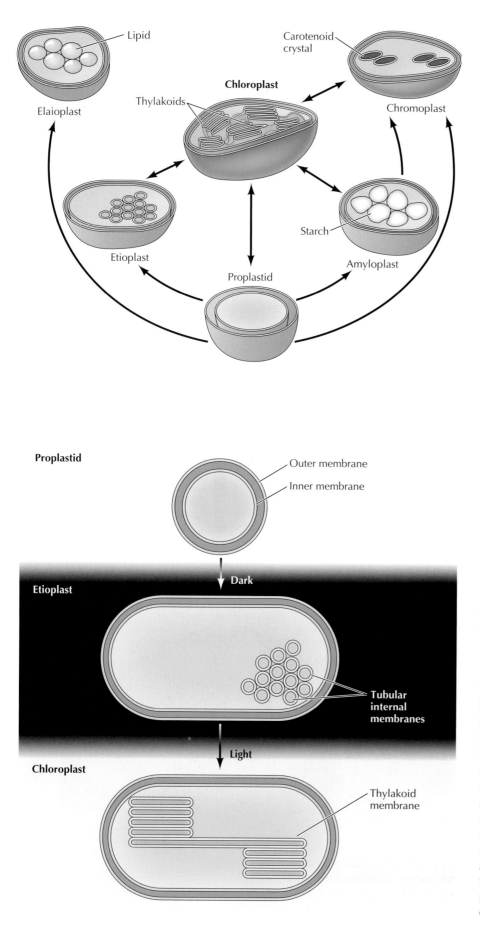

Figure 12.17 Interconversions of plastids Proplastids can develop into chloroplasts, chromoplasts, etioplasts, elaioplasts and amyloplasts. Different plastids can also convert to other types.

to another (**Figure 12.17**). Chromoplasts develop from chloroplasts, for example, during the ripening of fruit (e.g., tomatoes). During this process, chlorophyll and the thylakoid membranes break down, while new types of carotenoids are synthesized.

An interesting feature of plastids is that their development is controlled both by environmental signals and by intrinsic programs of cell differentiation. In the photosynthetic cells of leaves, for example, proplastids develop into chloroplasts (**Figure 12.18**). During this process, the thylakoid membrane is formed by vesicles budding from the inner membrane of the plastid envelope and the various components of the photosynthetic apparatus are synthesized and assembled. However, chloroplasts develop only in the presence of light. If plants are kept in the dark, the development of proplastids in leaves is arrested at an intermediate stage (called **etioplasts**) in which a semicrystalline array of

Figure 12.18 Development of chloroplasts Chloroplasts develop from proplastids in the photosynthetic cells of leaves. Proplastids contain only the inner and outer envelope membranes; the thylakoid membrane is formed by vesicle budding from the inner membrane during chloroplast development. If the plant is kept in the dark, chloroplast development is arrested at an intermediate stage (etioplasts). Etioplasts lack chlorophyll and contain semicrystalline arrays of tubular internal membranes. In the presence of light, they continue their development to chloroplasts.

tubular internal membranes has formed (**Figure 12.19**). These membranes contain precursors to chlorophyll, but chlorophyll has not been synthesized. If dark-grown plants are then exposed to light, the etioplasts continue their development to chloroplasts.

Since most of the proteins present in chloroplasts and other plastids are encoded by nuclear genes, the control of plastid development requires the coordination of gene expression between the plastid and nuclear genomes. Chloroplasts and other plastids therefore send signals to the nucleus that regulate the transcription of nuclear genes encoding plastid proteins, such as proteins involved in photosynthesis. Understanding the mechanisms responsible for plastid signaling to the nucleus is a challenging problem in plant molecular biology.

Peroxisomes

Peroxisomes are small (0.1–1 μm), single-membrane-enclosed organelles (**Figure 12.20**) that contain enzymes involved in a variety of metabolic reactions, including several aspects of energy metabolism. Most human cells contain 100 to 1000 peroxisomes, depending on the metabolic activity of the cell. Peroxisomes do not have their own genomes and all their proteins are synthesized from the nuclear genome. Similar to mitochondria and chloroplasts, most peroxisomal proteins are synthesized on free ribosomes and then imported into peroxisomes as completed polypeptide chains. However, some membrane proteins are transferred to peroxisomes from the endoplasmic reticulum. Also like mitochondria and chloroplasts, peroxisomes can replicate by division. However, unlike those organelles, peroxisomes can be rapidly regenerated even if entirely lost to the cell. While many mitochondrial and plastid proteins resemble those of prokaryotes, reflecting their endosymbiotic origin, the proteins of peroxisomes are typical eukaryotic proteins.

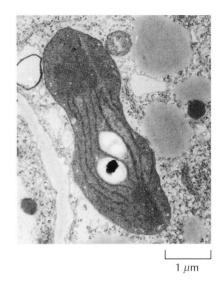

Figure 12.19 Electron micrograph of an etioplast

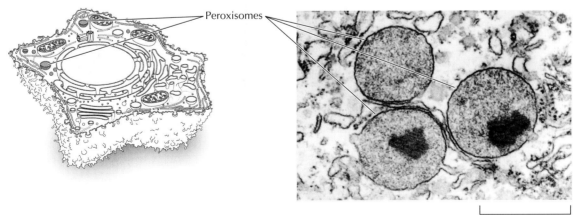

Figure 12.20 Electron micrograph of peroxisomes   Three peroxisomes from rat liver are shown. Two contain dense regions, which are paracrystalline arrays of the enzyme urate oxidase.

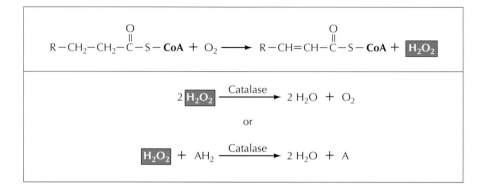

Figure 12.21 Fatty acid oxidation in peroxisomes The oxidation of a fatty acid is accompanied by the production of hydrogen peroxide (H_2O_2) from oxygen. The hydrogen peroxide is decomposed by catalase, either by conversion to water and oxygen or by oxidation of another organic compound (represented as AH_2).

Functions of peroxisomes

Human peroxisomes contain more than 50 different enzymes that play critical roles in several metabolic pathways, some of which are specific to particular tissues. Peroxisomes originally were defined as organelles that carry out oxidation reactions leading to the production of hydrogen peroxide. Because hydrogen peroxide is harmful to the cell, peroxisomes also contain the enzyme **catalase**, which decomposes hydrogen peroxide either by converting it to water or by using it to oxidize another organic compound. A variety of substrates are broken down by such oxidative reactions in peroxisomes, including uric acid, amino acids, purines, methanol, and fatty acids. The oxidation of fatty acids (**Figure 12.21**) is a particularly important example, since it provides a major source of metabolic energy. In animal cells, fatty acids are oxidized in both peroxisomes and mitochondria, but in yeasts and plants, fatty acid oxidation is restricted to peroxisomes. Mammalian peroxisomes are involved particularly in the catabolism of very long chain and branched fatty acids.

In addition to providing a compartment for oxidation reactions, peroxisomes are involved in biosynthesis of lipids. In animal cells, cholesterol and dolichol are synthesized in peroxisomes as well as in the ER. In the liver, peroxisomes are also involved in the synthesis of bile acids, which are derived from cholesterol. In addition, peroxisomes contain enzymes required for the synthesis of **plasmalogens**—a family of phospholipids in which one of the hydrocarbon chains is joined to glycerol by an ether bond rather than by an ester bond (**Figure 12.22**). Plasmalogens are important membrane components in some tissues, particularly heart and brain, although they are absent in others.

Peroxisomes play two particularly important roles in plants. First, peroxisomes in seeds are responsible for the conversion of stored fatty acids to carbohydrates, which is critical to providing energy and raw materials for

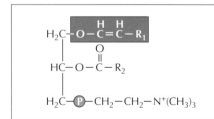

Figure 12.22 Structure of a plasmalogen The plasmalogen shown is analogous to phosphatidylcholine. However, one of the fatty acid chains is joined to glycerol by an ether, rather than by an ester, bond.

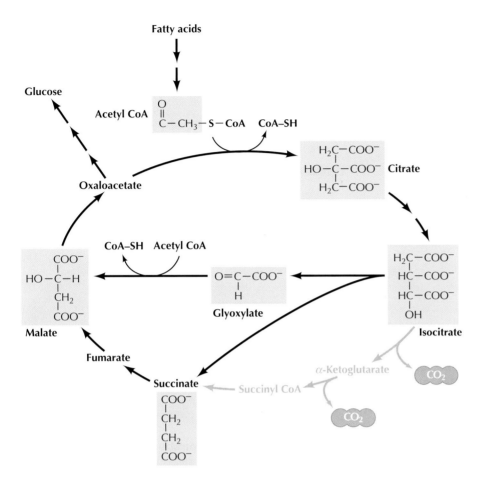

Figure 12.23 The glyoxylate cycle Plants are capable of synthesizing carbohydrates from fatty acids via the glyoxylate cycle, which is a variant of the citric acid cycle (see Figure 3.4). As in the citric acid cycle, acetyl CoA combines with oxaloacetate to form citrate, which is converted to isocitrate. However, instead of being degraded to CO_2 and α-ketoglutarate (gray arrows), isocitrate is converted to succinate and glyoxylate. Glyoxylate then reacts with another molecule of acetyl CoA to yield malate, which is converted to oxaloacetate and used for glucose synthesis.

growth of the germinating plant. This occurs via a series of reactions termed the **glyoxylate cycle**, which is a variant of the citric acid cycle (**Figure 12.23**). The peroxisomes in which this takes place are sometimes called **glyoxysomes**. Second, peroxisomes in leaves are involved in **photorespiration**, which serves to metabolize a side product formed during photosynthesis (**Figure 12.24**). CO_2 is converted to carbohydrates during photosynthesis via a series of reactions called the Calvin cycle (see Figure 3.15). The first step is the addition of CO_2 to the five-carbon sugar ribulose-1,5-bisphosphate, yielding two molecules of 3-phosphoglycerate (three carbons each). However, the enzyme involved (ribulose bisphosphate carboxylase/oxygenase, or rubisco) sometimes catalyzes the addition of O_2 instead of CO_2, producing one molecule of 3-phosphoglycerate and one molecule of phosphoglycolate (two carbons) (see Figure 12.24). When this occurs, the phosphoglycolate is first converted to glycolate and then transferred to peroxisomes, where it is oxidized and converted to glycine. Glycine is then transferred to mitochondria, where two molecules of glycine are converted to one molecule of serine, with the

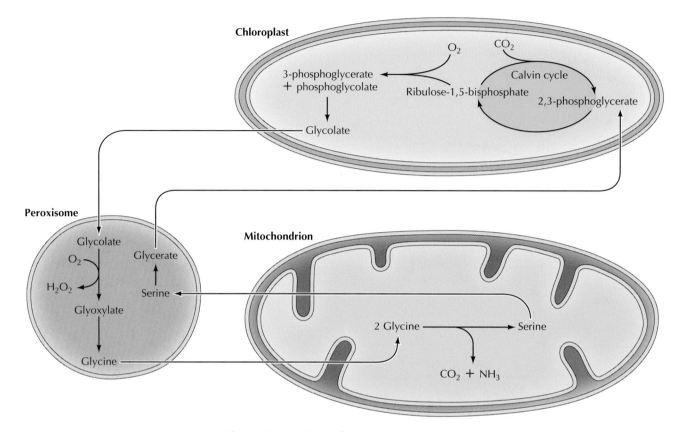

Figure 12.24 Role of peroxisomes in photorespiration During photosynthesis CO_2 is converted to carbohydrates by the Calvin cycle, which initiates with the addition of CO_2 to the five-carbon sugar ribulose-1,5-bisphosphate. However, the enzyme involved sometimes catalyzes the addition of O_2 instead, resulting in production of the two-carbon compound phosphoglycolate. Phosphoglycolate is converted to glycolate, which is then transferred to peroxisomes, where it is oxidized and converted to glycine. Glycine is then transferred to mitochondria and converted to serine. The serine is returned to peroxisomes and converted to glycerate, which is transferred back to chloroplasts.

loss of CO_2 and NH_3. The serine is then returned to peroxisomes, where it is converted to glycerate. Finally, the glycerate is transferred back to chloroplasts, where it reenters the Calvin cycle. Since the occasional utilization of O_2 in place of CO_2 is an inherent property of rubisco, photorespiration is a general accompaniment of photosynthesis. Peroxisomes thus play an important role by allowing most of the carbon in glycolate to be recovered and utilized.

Peroxisome assembly

The assembly of peroxisomes is mediated by an intersection of proteins synthesized on free cytosolic ribosomes and proteins transported to peroxisomes from the ER. Internal peroxisomal proteins are synthesized on free cytosolic ribosomes and then imported into peroxisomes, but many peroxisomal transmembrane proteins are transported to peroxisomes from the ER. These transmembrane proteins include proteins involved in the transport of metabolites as well as proteins called **peroxins** or **Pex proteins**, which are involved in peroxisome assembly. Mutations in peroxins have been identified

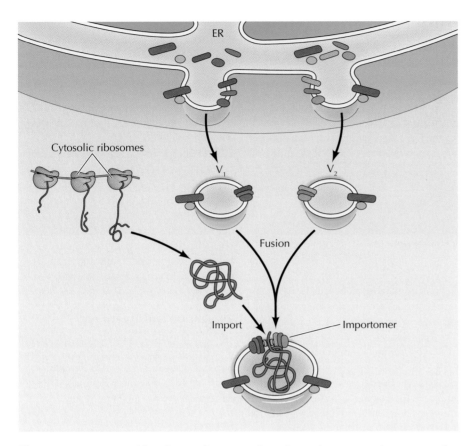

Figure 12.25 Assembly of peroxisomes Peroxisomal transmembrane proteins are derived from the ER. Two different types of vesicles (designated V_1 and V_2) carrying distinct transmembrane proteins fuse to form a functional peroxisome. The combination of these different proteins forms the machinery (importomer) through which internal peroxisome proteins synthesized on cytosolic ribosomes can be imported.

in yeast as well as in association with a variety of diseases in humans (**peroxisome biogenesis disorders**) that result from defective peroxisome assembly. As discussed below, some of the peroxins form membrane channels through which the internal (matrix) proteins of peroxisomes are imported.

Most peroxisome transmembrane proteins are first translocated into the ER and inserted into the ER membrane as discussed in Chapter 11. These proteins then bud in vesicles from a specialized region of the ER called the peroxisomal ER (**Figure 12.25**). Unlike other vesicles budding from the ER, vesicles budding from the peroxisomal ER do not have a COPII coat and their formation may be driven by peroxins. Rather than being targeted to the Golgi apparatus, vesicles from the peroxisomal ER fuse to form peroxisomes into which matrix proteins can be imported. Formation of a functional peroxisome requires the fusion of two different types of vesicles containing different classes of peroxins, both of which are needed for the import of matrix proteins.

Peroxisome matrix proteins are translated on free cytosolic ribosomes and then imported into peroxisomes as completed and folded polypeptides. They are targeted to peroxisomes by two pathways, which are conserved from yeasts to humans. Most proteins are targeted to peroxisomes by the

Molecular Medicine

Peroxisome Biogenesis Disorders

The Disease

A variety of diseases result from mutations that impair the metabolic activities that take place in peroxisomes. Some of these diseases result from mutations in the genes encoding single peroxisomal enzymes. In contrast, peroxisome biogenesis disorders result from mutations in the Pex proteins responsible for peroxisome assembly and thus have the potential of leading to dysfunction of many or all of the enzymes localized to peroxisomes. Peroxisome biogenesis disorders are recessive genetic diseases with an incidence of about 1 in 50,000. There are two distinct types of these disorders: the Zellweger spectrum disorders and rhizomelic chondrodysplasia punctata (RCDP) type 1. The Zellweger spectrum disorders vary in severity from Zellweger syndrome (the most severe) to the intermediate and milder forms, neonatal leukodystrophy (NALD) and infantile Refsum disease (IRD), respectively. Patients with Zellweger syndrome usually die in infancy, whereas patients with NALD may survive to their teens and those with IRD may reach adulthood. Symptoms include delayed development, characteristic abnormalities of facial features, loss of hearing and vision, and liver dysfunction. RCDP type 1 is distinct from the Zellweger syndrome disorders and is characterized by shortening of the limbs, small stature, cataracts and psychomotor retardation. Some patients with RCDP die in infancy while others survive to adulthood.

Molecular and Cellular Basis

Mutations in 14 different *Pex* genes, which are required for the assembly of peroxisomes, have been found to give rise to peroxisome biogenesis disorders (see Table). RCDP type 1 results specifically from mutations of *Pex7*, which encodes the cytosolic receptor for the small number of matrix proteins targeted to peroxisomes by the PTS2 targeting signal. Notably, other types of RCDP (types 2 and 3) are caused by defects in single peroxisomal enzymes involved in the synthesis of plasmalogens, indicating that the major defect in RCDP type 1 is a failure in the import of PTS2-targeted enzymes required for plasmalogen synthesis. Mutations in 13 other *Pex* genes are responsible for the Zellweger spectrum disorders. The proteins affected in Zellweger spectrum disorders include Pex5, which is the cytosolic receptor for the majority of peroxisomal enzymes targeted by the PTS1 signal. Zellweger spectrum disorders can also result from mutations in several other *Pex* genes encoding proteins that serve to dock Pex5 at the peroxisomal membrane or that function in recycling Pex5 after its cargo has been delivered to peroxisomes (see Figure 12.26). In addition, Zellweger spectrum disorders can be caused by defects in several Pex proteins involved in the import of peroxisome membrane proteins (see Figure 12.27). Finally, mutations in a *Pex* gene encoding a protein required for division of peroxisomes can lead to Zellweger spectrum disorders. Different mutations in these *Pex* genes can result in severe Zellweger syndrome or the less severe NALD or IRD presentations. The severity of the disorder is not directly related to which gene is mutated, but to the nature of the mutation and the corresponding effect on peroxisome assembly. Mutations that completely destroy Pex protein function result in severe disease, whereas less severe disease results from amino acid substitutions that reduce but do not eliminate function of the mutated Pex protein.

Prevention and Treatment

There is currently neither a cure nor standard treatment for peroxisome biogenesis disorders. Patients are offered supportive care aimed at alleviating symptoms of the disease.

Reference

Waterham, H. R. and M. S. Ebberink. 2012. Genetics and molecular basis of human peroxisome biogenesis disorders. *Biochim. Biophys. Acta* 1822: 1430–1441.

Peroxin Mutations in Peroxisome Biogenesis Disorders

Pex gene	Protein function	Disease
Pex7	Cytosolic receptor for PTS2-targeted matrix proteins	RCDP type 1
Pex5	Cytosolic receptor for PTS1-targeted matrix proteins	Zellweger spectrum disorders
Pex1, 2, 6, 10, 12, 26	Recycling of Pex5	
Pex13, 14	Pex5 docking	
Pex19	Cytosolic receptor for peroxisome membrane proteins	
Pex3, 16	Pex19 docking and insertion of membrane proteins	
Pex11β	Division of peroxisomes	

simple amino acid sequence Ser-Lys-Leu at their carboxy terminus (peroxisome targeting signal 1, or PTS1). A small number of proteins are targeted by a sequence of nine amino acids near their amino terminus (peroxisome targeting signal 2, or PTS2). PTS1 and PTS2 are recognized by distinct cytosolic receptors, the peroxins Pex5 and Pex7, respectively. The process

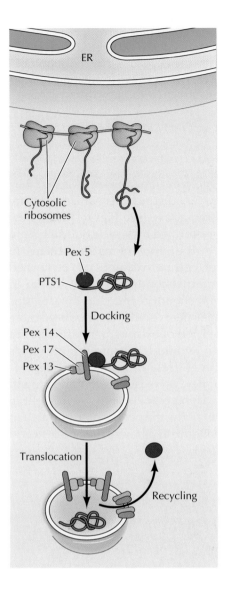

Figure 12.26 Import of peroxisomal matrix proteins   Peroxisomal matrix proteins containing the targeting signal PTS1 are recognized by the cytosolic receptor Pex5. The Pex5/cargo complex then binds to a docking complex (Pex13, Pex14, and Pex17) on the peroxisome membrane. Pex5 and Pex14 form a pore in the membrane and the cargo protein is translocated into the peroxisome. Pex5 is then recycled to the cytosol.

of protein translocation into peroxisomes, which is noteworthy in that it mediates the passage of large proteins and protein complexes across the membrane, has been studied in detail for proteins with the PTS1 targeting signal (**Figure 12.26**). The Pex5/cargo complex first binds to a docking complex, which consists of three other peroxins (Pex13, Pex14, and Pex17) on the peroxisome membrane. Pex5 and Pex14 then form a pore in the membrane through which the cargo is translocated into the peroxisome. Pex5 is then recycled by other Pex proteins in the membrane and used for further rounds of cargo import.

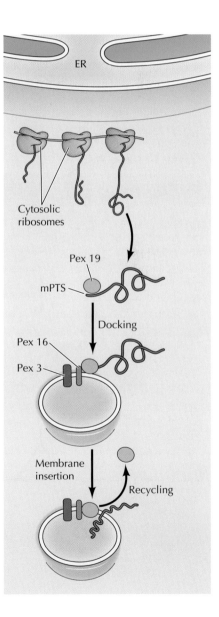

Not only internal matrix proteins, but also some peroxisome membrane proteins are synthesized on cytosolic ribosomes rather than being transported from the ER (**Figure 12.27**). These proteins have a membrane peroxisome targeting signal (mPTS) that is recognized by a distinct cytosolic receptor, Pex19. The Pex19/cargo complexes are recognized by Pex3 and Pex16 in the peroxisome membrane and Pex19 mediates the insertion of its cargo proteins into the lipid bilayer as well as their assembly into complexes.

Figure 12.27 Import of peroxisomal membrane proteins from the cytosol Peroxisome membrane proteins synthesized on cytosolic ribosomes have a targeting signal (mPTS) that is recognized by the Pex19 cytosolic receptor. The Pex19/cargo complexes then bind to Pex3 and Pex16 in the peroxisome membrane and Pex19 mediates the insertion of its cargo and the assembly of protein complexes in the lipid bilayer. Pex19 is then recycled to the cytosol.

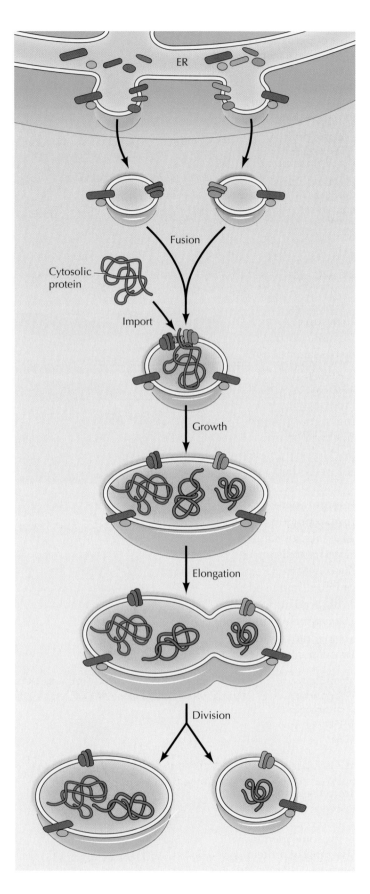

Figure 12.28 Formation of new peroxisomes   New peroxisomes can be formed *de novo* by the fusion of vesicles budding from the ER followed by the import of cytosolic proteins (see Figure 12.25). In addition, new peroxisomes can be formed by the growth and division of old ones.

An interesting feature of peroxisomes is that they can be formed by two distinct mechanisms (**Figure 12.28**). As discussed above (see Figure 12.25), new peroxisomes can be formed by vesicle budding from the ER, followed by the import of matrix proteins. In addition, however, new peroxisomes can be formed by the growth and division of old ones, similar to the division of mitochondria and chloroplasts. It is noteworthy that new peroxisomes are produced more rapidly by the division of existing peroxisomes, but *de novo* formation by budding from the ER yields peroxisomes with entirely new contents. Both *de novo* formation of new peroxisomes and the growth and division of existing peroxisomes are thought to contribute to maintaining the peroxisome population of the cell by responding to different metabolic needs.

SUMMARY	KEY TERMS

Mitochondria

- *Organization and function of mitochondria:* Mitochondria, which play a critical role in the generation of metabolic energy, are surrounded by a double-membrane system. The matrix contains the enzymes of the citric acid cycle; the inner membrane contains protein complexes involved in electron transport and oxidative phosphorylation. In contrast to the inner membrane, the outer membrane is freely permeable to small molecules. See Videos 12.1 and 12.2.

  mitochondria, cristae, matrix, porin

- *The genetic system of mitochondria:* Mitochondria contain their own genomes, which encode rRNAs, tRNAs, and some of the proteins that are involved in oxidative phosphorylation.

  endosymbiosis

- *Protein import and mitochondrial assembly:* Most mitochondrial proteins are encoded by the nuclear genome. These proteins are translated on free ribosomes and imported into mitochondria as completed polypeptide chains. Positively charged presequences target proteins for import to the mitochondrial matrix.

  presequence, Tom complex, Tim complex, matrix processing peptidase (MPP)

- *Mitochondrial lipids:* Most phospholipids are carried to mitochondria from the endoplasmic reticulum by phospholipid transfer proteins. Mitochondria also synthesize cardiolipin, which is an unusual phospholipid containing four fatty acids.

  phospholipid transfer protein, cardiolipin

- *Transport of metabolites across the inner membrane:* Many proteins of the inner membrane transport small molecules into and out of mitochondria. Potential energy stored in the proton gradient drives the transport of ATP, ADP, and other metabolites into and out of mitochondria.

Chloroplasts and Other Plastids

- *The structure and function of chloroplasts:* Chloroplasts are large organelles that function in photosynthesis and a variety of other metabolic activities. Like mitochondria, chloroplasts are bounded by a double-membrane envelope. In addition, chloroplasts have an internal thylakoid membrane, which is the site of electron transport and the chemiosmotic generation of ATP. See Video 12.3.

  chloroplast, thylakoid membrane, stroma

- *The chloroplast genome:* Chloroplast genomes contain approximately 150 genes, including genes encoding rRNAs, tRNAs, ribosomal proteins, RNA polymerase, and proteins involved in photosynthesis and metabolism.

- *Import and sorting of chloroplast proteins:* Most chloroplast proteins are synthesized on free ribosomes in the cytosol and targeted for import to chloroplasts by amino-terminal transit peptides. Most proteins incorporated into the thylakoid lumen are first imported into the chloroplast stroma and then targeted for transport across the thylakoid membrane.

  transit peptide, Toc complex, Tic complex, stromal processing peptidase (SPP)

- *Other plastids:* The chloroplast is only one member of a family of related organelles, all of which contain the same genome. Other plastids serve to store energy sources, such as starch and lipids, and function in other aspects of plant metabolism. See Animation 12.1.

  plastid, chromoplast, leucoplast, amyloplast, elaioplast, proplastid, etioplast

Peroxisomes

- *Functions of peroxisomes:* Peroxisomes are small organelles, bounded by a single membrane, that contain enzymes involved in a variety of metabolic reactions, including fatty acid oxidation, lipid biosynthesis, the glyoxylate cycle, and photorespiration.

  peroxisome, catalase, plasmalogen, glyoxylate cycle, glyoxysome, photorespiration

SUMMARY

- *Peroxisome assembly:* Most transmembrane proteins are transported to peroxisomes from the ER, whereas internal peroxisomal proteins are synthesized on free ribosomes in the cytosol and imported into peroxisomes as completed and folded polypeptide chains. Peroxins can be formed both *de novo* from the ER and by growth and division of existing peroxisomes.

KEY TERMS

peroxin, Pex protein, peroxisome biogenesis disorder

Refer To

The Cell

Companion Website

sites.sinauer.com/cooper7e

for quizzes, animations, videos, flashcards, and other study resources.

Questions

1. What two properties of the mitochondrial inner membrane allow it to have unusually high metabolic activity?

2. According to standard "wobble" rules, protein synthesis requires a minimum of 30 different tRNAs. How do human mitochondria manage to translate mRNAs into proteins with only 22 tRNA species?

3. What roles do molecular chaperones play in mitochondrial protein import?

4. What provides the energy needed to transport metabolites into and out of mitochondria?

5. How is the thylakoid membrane of chloroplasts related to the inner membrane of mitochondria?

6. Why don't the transit peptides of chloroplast proteins, in contrast to the presequences of mitochondrial proteins, need to be positively charged?

7. You are studying a protein that is found in the thylakoid lumen and have identified a signal sequence that targets the protein for translocation across the thylakoid membrane. If you mutate the gene by deleting that signal sequence, where will the protein be located?

8. How do chloroplasts differ from chromoplasts?

9. How are membrane proteins targeted to peroxisomes?

10. What effect would siRNA against Pex5 have on peroxisome assembly?

References and Further Reading   (Key review articles for each major section are highlighted in **bold**.)

Mitochondria

Andreux, P. A., R. H. Houtkooper and J. Auwerx. 2013. Pharmacological approaches to restore mitochondrial function. *Nature Rev. Drug Discovery* 12: 465–483. [R]

Blackstone, C. and C. R. Chang. 2011. Mitochondria unite to survive. *Nature Cell Biol.* 13: 521–522. [R]

Chacinska, A., C. M. Koehler, D. Milenkovic, T. Lithgow and N. Pfanner. 2009. Importing mitochondrial proteins: machineries and mechanisms. *Cell* 138: 628–644. [R]

Chen, X. J. and R. A. Butow. 2005. The organization and inheritance of the mitochondrial genome. *Nature Rev. Genet.* 6: 815–825. [R]

Dudek, J., P. Rehling and M. van der Laan. 2013. Mitochondrial protein import: common principles and physiological networks. *Biochim. Biophys. Acta* 1833: 274–285. [R]

Farrar, G. J., N. Chadderton, P. F. Kenna and S. Millington-Ward. 2013. Mitochondrial disorders: aetiologies, model systems, and candidate therapies. *Trends Genet.* 29: 488–497.

Friedman, J. R. and J. Nunnari. 2014. Mitochondrial form and function. *Nature* 505: 335–343. [R]

Lev, S. 2010. Non-vesicular lipid transport by lipid-transport proteins and beyond. *Nature Rev. Mol. Cell Biol.* 11: 739–750. [R]

Mishra, P. and D. C. Chan. 2014. Mitochondrial dynamics and inheritance during cell division, development and disease. *Nature Rev. Mol. Cell Biol.* 15: 634–646. [R]

Neupert, W. and J. M. Herrmann. 2007. Translocation of proteins into mitochondria. *Ann. Rev. Biochem.* 76: 723–749. [R]

Park, C. B. and N.G. Larsson. 2011. Mitochondrial DNA mutations in disease and aging. *J. Cell Biol.* 193: 809–818. [R]

Schmidt, O., N. Pfanner and C. Meisinger. 2010. Mitochondrial protein import: from proteomics to functional mechanisms. *Nature Rev. Mol. Cell Biol.* 11: 655–667. [R]

Schulz, C., A. Schendzielorz and P. Rehling. 2015. Unlocking the presequence import pathway. *Trends Cell Biol.* 25: 265–275. [R]

Tatsuta, T., M. Scharwey and T. Langer. 2014. Mitochondrial lipid trafficking. *Trends Cell Biol.* 24: 44–52. [R]

Chloroplasts and Other Plastids

Albiniak, A. M., J. Baglieri and C. Robinson. 2012. Targeting of lumenal proteins across the thylakoid membrane. *J. Exptl. Botany* 63: 1689–1698. [R]

Andres, C., B. Agne and F. Kessler. 2010. The TOC complex: preprotein gateway to the chloroplast. *Biochim. Biophys. Acta* 1803: 715–723. [R]

Flores-Pérez, U. and P. Jarvis. 2013. Molecular chaperone involvement in chloroplast protein import. *Biochim. Biophys. Acta* 1833: 332–340. [R]

Green, B. R. 2011. Chloroplast genomes of photosynthetic eukaryotes. *Plant J.* 66: 34–44. [R]

Inoue, K. 2011. Emerging roles of the chloroplast outer envelope membrane. *Trends Plant Sci.* 16: 550–557. [R]

Jarvis, P. and E. López-Juez. 2013. Biogenesis and homeostasis of chloroplasts and other plastids. *Nature Rev. Mol. Cell Biol.* 14: 787–802. [R]

Kovacs-Bogdan, E., J. Soll and B. Bolter. 2010. Protein import into chloroplasts: the Tic complex and its regulation. *Biochim. Biophys. Acta* 1803: 740–747. [R]

Lee, D. W., C. Jung and I. Hwang. 2013. Cytosolic events involved in chloroplast protein targeting. *Biochim. Biophys. Acta* 1833: 245–252. [R]

Li, H. M. and C.-C. Chiu. 2010. Protein transport into chloroplasts. *Ann. Rev. Plant Biol.* 61: 157–180. [R]

Shi, L.-X. and S. M. Theg. 2013. The chloroplast protein import system: from algae to trees. *Biochim. Biophys. Acta* 1833: 314–331. [R]

Wang, P. and R. E. Dalbey. 2011. Inserting membrane proteins: The YidC/Oxa1/Alb3 machinery in bacteria, mitochondria, and chloroplasts. *Biochim. Biophys. Acta* 1808: 866–875. [R]

Peroxisomes

Agrawal, G. and S. Subramani. 2013. Emerging role of the endoplasmic reticulum in peroxisome biogenesis. *Front. Physiol.* 4: 286. [R]

Hasan, S., H. W. Platta and R. Erdmann. 2013. Import of proteins into the peroxisomal matrix. *Front. Physiol.* 4: 261. [R]

Hoepfner, D., D. Schildknegt, L. Braakman, P. Philippsen and H. F. Tabak. 2005. Contribution of the endoplasmic reticulum to peroxisome formation. *Cell* 122: 85–95. [P]

Smith, J. J. and J. D. Aitchison. 2013. Peroxisomes take shape. *Nature Rev. Mol. Cell Biol.* 14: 803–817. [R]

Tabak, H. F., I. Braakman and A. van der Zand. 2013. Peroxisome formation and maintenance are dependent on the endoplasmic reticulum. *Ann. Rev. Biochem.* 82: 723–744. [R]

Wanders, R. J. A. 2014. Metabolic functions of peroxisomes in health and disease. *Biochimie* 98: 36–44. [R]

Waterham, H. R. and M. S. Ebberink. 2012. Genetics and molecular basis of human peroxisome biogenesis disorders. *Biochim. Biophys. Acta* 1822: 1430–1441. [R]

CHAPTER 13

The Cytoskeleton and Cell Movement

The membrane-enclosed organelles discussed in the preceding chapters constitute one level of the organizational substructure of eukaryotic cells. A further level of organization is provided by the cytoskeleton, which consists of a network of protein filaments extending throughout the cytoplasm. The cytoskeleton provides a structural framework for the cell, serving as a scaffold that determines cell shape, the positions of organelles, and the general organization of the cytoplasm. In addition to playing this structural role, the cytoskeleton is responsible for cell movements. These include not only the movement of entire cells, but also the internal transport of organelles and other structures (such as mitotic chromosomes) through the cytoplasm. Importantly, the cytoskeleton is much less rigid and permanent than its name implies. Rather, it is a dynamic structure that is continually reorganized as cells move and change shape—for example, during cell division.

The cytoskeleton is composed of three principal types of protein filaments: actin filaments, microtubules, and intermediate filaments, which are held together and linked to subcellular organelles and the plasma membrane by a variety of accessory proteins. This chapter discusses the structure and organization of each of these three major components of the cytoskeleton as well as the roles of actin filaments and microtubules in cell motility, organelle transport, cell division, and other types of cell movements.

Structure and Organization of Actin Filaments

The most abundant cytoskeletal protein of most cells is **actin**, which polymerizes to form actin filaments—thin, flexible fibers approximately 7 nm in diameter and up to several micrometers in length (**Figure 13.1**). Within the cell, actin filaments (also called **microfilaments**) are organized into higher order structures, forming bundles or three-dimensional networks with the properties of semisolid gels. The assembly and disassembly of actin filaments, their cross-linking into bundles and networks, and their association with other cell structures (such as the plasma membrane) are regulated by a variety of **actin-binding proteins**, which are critical components of the actin cytoskeleton (**Table 13.1**). Actin filaments are particularly abundant beneath the plasma membrane where they form a network that provides mechanical support, determines cell shape, and allows movement of the cell surface, thereby enabling cells to migrate, engulf particles, and divide.

Assembly and disassembly of actin filaments

Actin was first isolated from muscle cells, in which it constitutes approximately 20% of total cell protein, in 1942. Although actin was initially thought to be

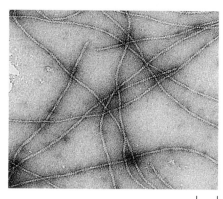

Figure 13.1 Actin filaments Electron micrograph of actin filaments. (Courtesy of Roger Craig, University of Massachusetts Medical Center.)

50 nm

uniquely involved in muscle contraction, it is now known to be an extremely abundant protein (typically 5 to 10% of total protein) in all types of eukaryotic cells. Yeasts have only a single actin gene, but higher eukaryotes have several distinct types of actin, which are encoded by different members of the actin gene family. Mammals, for example, have six distinct actin genes: Four are expressed in different types of muscle (skeletal, cardiac, and smooth muscle) and two are expressed in nonmuscle cells. All of the actins, however, are very similar in amino acid sequence and have been highly conserved throughout the evolution of eukaryotes. Yeast actin, for example, is 90% identical in amino acid sequence to the actins of mammalian cells. Actin-related proteins also play important roles in bacteria, including participating in cell division and determining cell shape.

The three-dimensional structures of both individual actin molecules and actin filaments were determined in 1990 by Kenneth Holmes, Wolfgang Kabsch, and their colleagues. Individual actin molecules are globular proteins of 375 amino acids (43 kd). Each actin monomer (**globular [G] actin**) has tight binding sites that mediate head-to-tail interactions with two other actin monomers, so actin monomers polymerize to form filaments (**filamentous [F] actin**) (**Figure 13.2**). Each monomer is rotated by 166° in the filaments, which therefore have the appearance of a double-stranded helix. Because all the actin monomers are oriented in the same direction, actin filaments have a distinct polarity and their ends (called barbed [or plus] ends and pointed [or minus] ends) are distinguishable from one another. This polarity of actin filaments is important both in their assembly and in establishing a specific direction of myosin movement relative to actin, as discussed later in the chapter.

The behavior of actin monomers and filaments is regulated within cells to take advantage of the intrinsic properties of actin. Under physiological conditions, actin monomers polymerize to form filaments. The first step in actin polymerization (called nucleation) is the formation of a small aggregate consisting of three actin monomers. Actin filaments are then able to grow by the reversible addition of monomers to both ends, but the barbed end elongates five to ten times faster than the pointed end. The actin monomers bind ATP (see Figure 13.2), which is hydrolyzed to ADP following filament assembly. Although ATP is not required to provide energy for

Animation 13.1

sites.sinauer.com/cooper7e/a13.1
Assembly of an Actin Filament
Actin monomers polymerize to form actin filaments, a process that is reversible and is regulated within the cell by actin-binding proteins.

Table 13.1   Examples of Actin-Binding Proteins

Cellular role	Representative proteins
Monomer binding	Profilin, twinfilin
Filament initiation and polymerization	Arp2/3, formin
End capping	CapZ, tropomodulin
Filament stabilization	Nebulin, tropomyosin
Filament cross-linking	α-actinin, filamin, fimbrin, villin
Actin filament linkage to other proteins	Dystrophin, spectrin, talin, vinculin
Filament severing	Cofilin, gelsolin

(A)

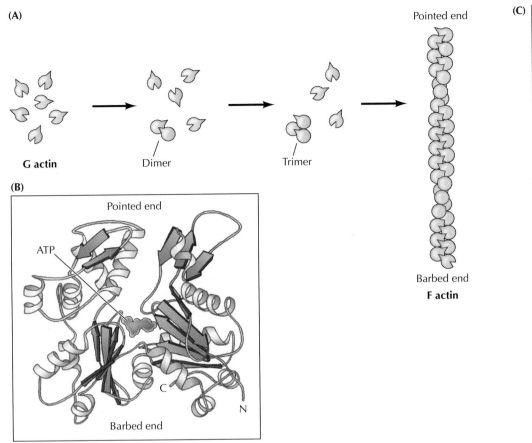

(C)

Figure 13.2 Assembly and structure of actin filaments (A) Actin monomers (G actin) polymerize to form actin filaments (F actin). The first step is the formation of dimers and trimers, which then grow by the addition of monomers to both ends. (B) Structure of an actin monomer with ATP bound. (C) Space-filling model of F actin. Fourteen actin monomers are represented in different colors. (C, after data from Chen et al., 2002. *J. Struct. Biol.* 138: 92.)

polymerization, actin monomers to which ATP is bound polymerize more readily than those to which ADP is bound. As a result, ATP-actin monomers are added rapidly to the barbed ends of filaments, and the ATP is then hydrolyzed to ADP after polymerization. The ADP-actin is less tightly bound and can dissociate from the pointed end. This can result in the phenomenon known as **treadmilling**, in which ATP-actin is added to the barbed end while ADP-actin dissociates from the pointed end of a filament (Figure 13.3). Treadmilling illustrates the dynamic behavior of actin filaments, which, as discussed below, is critical in regulating the structure and function of actin filaments within the cell.

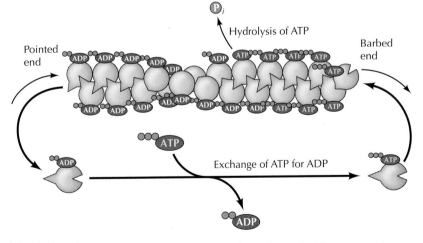

Figure 13.3 Treadmilling and the role of ATP in actin filament polymerization Actin bound to ATP associates with the rapidly growing barbed ends, and the ATP is then hydrolyzed to ADP within the filament. Because ADP-actin dissociates from filaments more readily than ATP-actin, actin monomers bound to ADP dissociate from the pointed end while monomers bound to ATP are added to the barbed end.

FYI

Phalloidin is derived from the death cap mushroom (*Amanita phalloides*). This mushroom also produces α-amanitin, a specific inhibitor of RNA polymerase II. The toxins produced by the death cap and related mushrooms primarily attack the liver and kidney and can cause death if left untreated. (See Chapter 8.)

It is noteworthy that several drugs useful in cell biology act by binding to actin and affecting its polymerization. For example, the **cytochalasins** bind to the barbed ends of actin filaments and block their elongation. This results in changes in cell shape as well as inhibition of some types of cell movements (e.g., cell division following mitosis), indicating that actin polymerization is required for these processes. Another drug, **phalloidin**, binds tightly to actin filaments and prevents their dissociation into individual actin molecules. Phalloidin labeled with a fluorescent dye is frequently used to visualize actin filaments by fluorescence microscopy.

Within the cell, the assembly and disassembly of actin filaments is regulated by actin-binding proteins. Some of these proteins initiate the formation of actin filaments or stimulate their growth. Others stabilize actin filaments by binding along their length or by capping their ends and preventing the dissociation of actin monomers. Conversely, some proteins act to disassemble actin filaments by severing them and stimulating their depolymerization.

The principal proteins that stimulate the initiation and elongation of actin filaments are **formin** and the **Arp2/3 complex** (*a*ctin-*r*elated *p*rotein). The activities of these proteins are regulated in response to a variety of signals to determine where filaments are formed within the cell. Formins are a family of proteins that bind ATP-actin and nucleate the initial polymerization of actin monomers (Figure 13.4). They then move along the growing filament, adding new monomers to the barbed end. Formins nucleate long unbranched actin filaments that make up stress fibers, the contractile ring, filopodia, and the thin filaments of muscle cells (discussed later in this chapter). Formins are associated with another actin-binding protein called **profilin**, which binds actin monomers and stimulates the exchange of bound ADP for ATP. This promotes actin polymerization by increasing the local concentration of ATP-actin.

The Arp2/3 proteins initiate growth of branched actin filaments, which play a key role in driving cell movement at the plasma membrane (Figure 13.5). The Arp2/3 proteins bind ATP-actin near the barbed ends of filaments and initiate the formation of a new branch. The activity of the Arp2/3 proteins is

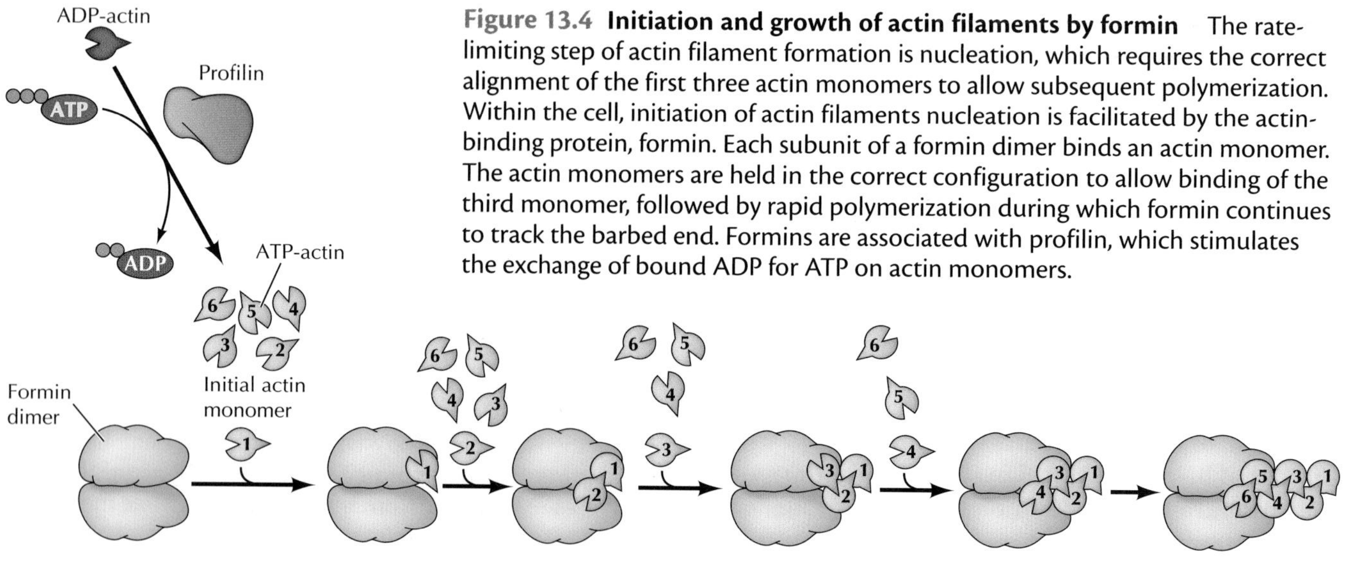

Figure 13.4 Initiation and growth of actin filaments by formin The rate-limiting step of actin filament formation is nucleation, which requires the correct alignment of the first three actin monomers to allow subsequent polymerization. Within the cell, initiation of actin filaments nucleation is facilitated by the actin-binding protein, formin. Each subunit of a formin dimer binds an actin monomer. The actin monomers are held in the correct configuration to allow binding of the third monomer, followed by rapid polymerization during which formin continues to track the barbed end. Formins are associated with profilin, which stimulates the exchange of bound ADP for ATP on actin monomers.

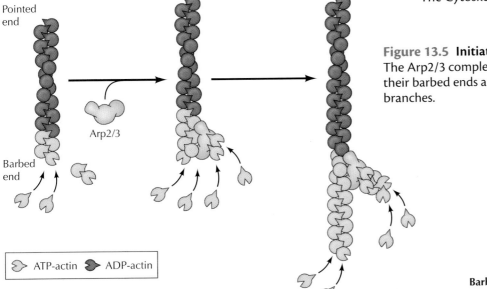

Figure 13.5 Initiation of actin filament branches The Arp2/3 complex binds to actin filaments near their barbed ends and initiates the formation of branches.

controlled by several other cellular proteins, so that the formation of branched actin filaments is regulated by the physiological needs of the cell.

Many actin filaments are relatively stable within cells due to capping proteins that bind to their ends and to filament-stabilizing proteins, such as members of the **tropomyosin** family (**Figure 13.6**). Tropomyosins are 30–36 kd fibrous proteins that bind lengthwise along the groove of actin filaments. Because of alternative splicing, four tropomyosin genes encode more than 40 different tropomyosin proteins, which are major determinants of the location of actin filaments within the cell. Other proteins that bind along the length of actin filaments serve to cross-link actin filaments into bundles or networks or to mediate the association of actin filaments with other cellular proteins. Several examples of these actin-binding proteins are discussed later in this chapter in terms of their specific roles in cell physiology.

Other actin-binding proteins remodel or modify, rather than stabilize, existing filaments. An example is the actin-binding protein **cofilin**, which severs actin filaments (**Figure 13.7**). This generates new ends of the filaments,

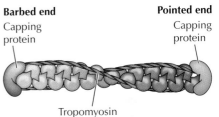

Figure 13.6 Stabilization of actin filaments   Actin filaments can be stabilized by capping proteins that bind to their barbed or pointed ends and by filament-stabilizing proteins (e.g., tropomyosin) that bind along their length.

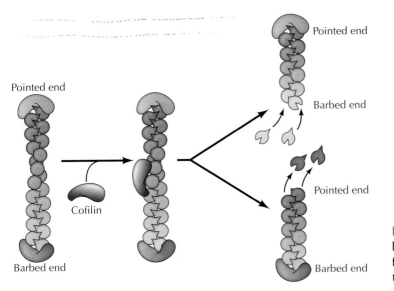

Figure 13.7 Filament severing by cofilin   Cofilin binds to and severs actin filaments. The newly formed filament ends are then available for the polymerization or depolymerization of actin monomers.

which are then accessible either for depolymerization from their pointed ends or growth by addition of new actin monomers to their barbed ends.

As might be expected, the activities of these proteins are controlled by a variety of cell signaling mechanisms, allowing actin polymerization to be appropriately regulated in response to environmental signals. Cofilin, profilin, formin, and the Arp2/3 complex (as well as other actin-binding proteins) can thus act together to promote the rapid turnover of actin filaments and remodeling of the actin cytoskeleton that is required for a variety of cell movements and changes in cell shape, as discussed later in this chapter. This is a major undertaking, and in some cell types, actin filament assembly and disassembly are responsible for half the hydrolysis and turnover of ATP in the cell.

Organization of actin filaments

Individual actin filaments are assembled into two general types of structures called **actin bundles** and **actin networks**, which play different roles in the cell (**Figure 13.8**). In bundles, the actin filaments are cross-linked into closely packed parallel arrays. In networks, the actin filaments are cross-linked in orthogonal arrays that form three-dimensional meshworks with the properties of semisolid gels. As mentioned previously, the formation of these structures is governed by a variety of actin-binding proteins that cross-link actin filaments in distinct patterns (see Table 13.1).

All of the actin-binding proteins involved in cross-linking contain at least two domains that bind actin, allowing them to bind and cross-link two different actin filaments. The nature of the association between these filaments is then determined by the size and shape of the cross-linking proteins (see Figure 13.8). The proteins that cross-link actin filaments into bundles (called **actin-bundling proteins**) usually are small rigid proteins that force the filaments to align closely with one another. In contrast, the proteins that

Figure 13.8 Actin bundles and networks (A) Electron micrograph of actin bundles (arrowheads) projecting from the actin network (arrows) underlying the plasma membrane of a macrophage. The bundles support cell surface projections called filopodia (see Figure 13.19). (B) Schematic organization of a bundle and a network. Actin filaments in bundles are cross-linked into parallel arrays by small proteins that align the filaments closely with one another. In contrast, networks are formed by large flexible proteins that cross-link orthogonal filaments. (A, courtesy of John H. Hartwig, Brigham & Women's Hospital.)

(A)

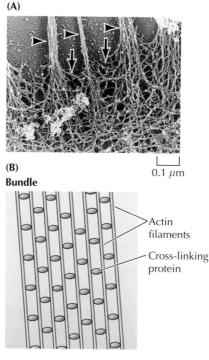

0.1 µm

(B)
Bundle

Actin filaments

Cross-linking protein

Network

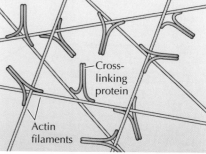

Cross-linking protein

Actin filaments

Figure 13.9 Actin-bundling proteins   Actin filaments are associated into two types of bundles by different actin-bundling proteins. Fimbrin has two adjacent actin-binding domains (ABD) and cross-links actin filaments into closely packed parallel bundles in which the filaments are approximately 14 nm apart. In contrast, the two separated actin-binding domains of α-actinin dimers cross-link filaments into more loosely spaced contractile bundles in which the filaments are separated by 40 nm. Both fimbrin and α-actinin contain two related Ca²⁺-binding domains, and α-actinin contains four repeated α-helical spacer domains.

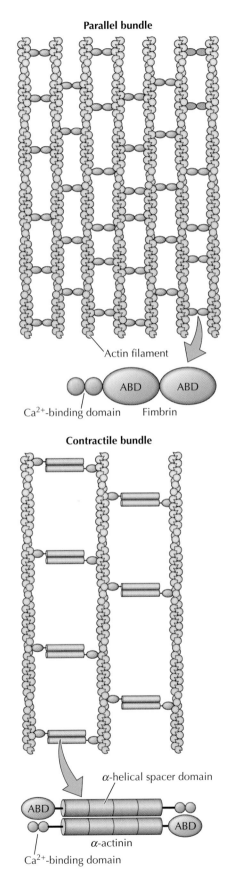

Parallel bundle

Actin filament

ABD   ABD

Ca²⁺-binding domain   Fimbrin

Contractile bundle

α-helical spacer domain

ABD

ABD

Ca²⁺-binding domain   α-actinin

organize actin filaments into networks tend to be large flexible proteins that can cross-link perpendicular filaments.

There are at least two structurally and functionally distinct types of actin bundles involving different actin-bundling proteins (**Figure 13.9**). The first type of bundle, containing closely spaced actin filaments aligned in parallel (called **parallel bundles**), supports projections of the plasma membrane, such as microvilli (see Figures 13.17 and 13.18). In these bundles, all the filaments have the same polarity, with their barbed ends adjacent to the plasma membrane. An example of a bundling protein involved in the formation of these structures is **fimbrin**, which was first isolated from intestinal microvilli and later found in surface projections of a wide variety of cell types. Fimbrin is a 68-kd protein containing two adjacent actin-binding domains. It binds to actin filaments as a monomer, holding two parallel filaments close together.

The second type of actin bundle is composed of filaments that are more widely spaced, allowing the bundle to contract. The more open structure of these bundles (which are called **contractile bundles**) reflects the properties of the cross-linking protein, **α-actinin**. In contrast to fimbrin, α-actinin binds to actin as a dimer, each subunit of which is a 102-kd protein containing a single actin-binding site. Filaments cross-linked by α-actinin are consequently separated by a greater distance than those cross-linked by fimbrin (40 nm apart instead of 14 nm). The increased spacing between filaments allows the motor protein myosin to interact with the actin filaments in these bundles, which (as discussed later) enables the bundles to contract.

The actin filaments in networks are held together by large actin-binding proteins, such as **filamin** (**Figure 13.10**). Filamin binds actin as a dimer of two 280-kd subunits. The actin-binding domains and dimerization domains are

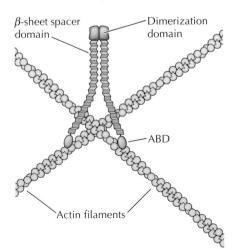

β-sheet spacer domain

Dimerization domain

ABD

Actin filaments

Figure 13.10 Actin networks and filamin   Filamin is a dimer of two large (280 kd) subunits, forming a flexible V-shaped molecule (blue) that cross-links actin filaments into orthogonal networks. The carboxy-terminal dimerization domain is separated from the amino-terminal actin-binding domain (ABD) by repeated β-sheet spacer domains.

at opposite ends of each subunit, so the filamin dimer is a flexible V-shaped molecule with actin-binding domains at the ends of each arm. As a result, filamin forms cross-links between orthogonal actin filaments, creating a loose three-dimensional meshwork. As discussed in the next section, such networks of actin filaments underlie the plasma membrane and support the surface of the cell.

Association of actin filaments with the plasma membrane

Actin filaments are highly concentrated at the periphery of the cell where they form a three-dimensional network beneath the plasma membrane (see Figure 13.8). This network of actin filaments and associated actin-binding proteins (called the **cell cortex**) determines cell shape and is involved in a variety of cell surface activities, including movement. The association of the actin cytoskeleton with the plasma membrane is thus central to cell structure and function.

Human red blood cells (erythrocytes) have proven particularly useful for studies of both the plasma membrane (discussed in the next chapter) and the cortical cytoskeleton. The principal advantage of these cells is that they contain no nucleus or internal organelles, so their plasma membrane and associated proteins can be easily isolated without contamination by the various internal membranes that are abundant in other cell types. In addition, human erythrocytes lack other cytoskeletal components (microtubules and intermediate filaments), so the cortical actin cytoskeleton is the principal determinant of their distinctive shape as biconcave discs (**Figure 13.11**).

The major protein that provides the structural basis for the cortical cytoskeleton in erythrocytes is the actin-binding protein **spectrin** (**Figure 13.12**). Spectrins are members of the large calponin family of actin-binding proteins, which includes α-actinin, filamin, and fimbrin. All spectrins are tetramers consisting of two distinct polypeptide chains called α and β,

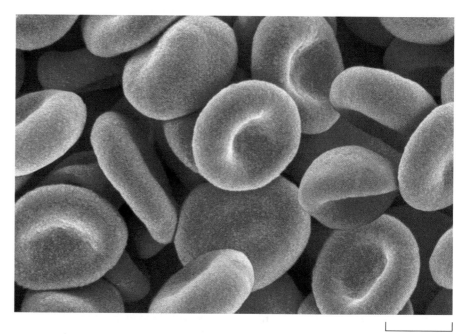

5 μm

Figure 13.11 Morphology of red blood cells Scanning electron micrograph of human red blood cells illustrating their biconcave shape. Artificial color has been added.

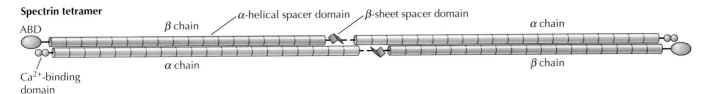

Spectrin tetramer

ABD

β chain

α-helical spacer domain

β-sheet spacer domain

α chain

α chain

β chain

Ca^{2+}-binding domain

Figure 13.12 Structure of spectrin   Spectrin is a tetramer consisting of two α and two β chains. Each β chain has a single actin-binding domain (ABD) at its amino terminus. Both α and β chains contain multiple repeats of α-helical spacer domains, which separate the two actin-binding domains of the tetramer. Each α chain has two Ca^{2+}-binding domains at its carboxy terminus.

with molecular weights of 240 and 220 kd, respectively. The α and β chains associate to form tetramers with two actin-binding domains separated by approximately 200 nm. The ends of the spectrin tetramers then associate with short actin filaments, resulting in the spectrin-actin network that forms the cortical cytoskeleton of red blood cells (**Figure 13.13**). Although spectrin itself binds phospholipids in the plasma membrane, the major link between the spectrin-actin network and the plasma membrane is provided by a protein called **ankyrin**, which binds both to spectrin and to the cytoplasmic domain of an abundant transmembrane protein called band 3. An additional link between the spectrin-actin network and the plasma membrane is provided by protein 4.1, which binds to spectrin-actin junctions as well as to the cytoplasmic domain of glycophorin (another abundant transmembrane protein).

Other types of cells contain linkages between the cortical cytoskeleton and the plasma membrane that are similar to those observed in red blood

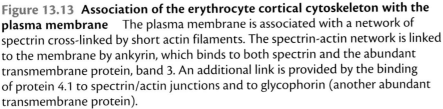
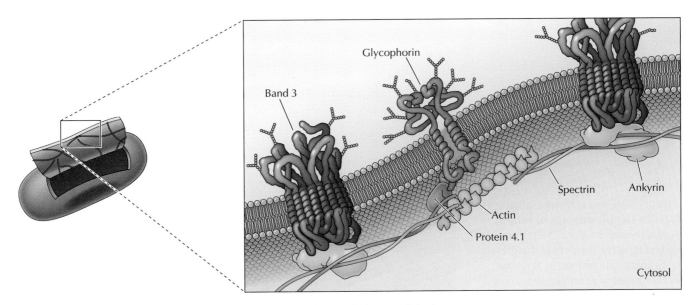

Glycophorin

Band 3

Spectrin

Ankyrin

Actin

Protein 4.1

Cytosol

Figure 13.13 Association of the erythrocyte cortical cytoskeleton with the plasma membrane   The plasma membrane is associated with a network of spectrin cross-linked by short actin filaments. The spectrin-actin network is linked to the membrane by ankyrin, which binds to both spectrin and the abundant transmembrane protein, band 3. An additional link is provided by the binding of protein 4.1 to spectrin/actin junctions and to glycophorin (another abundant transmembrane protein).

Figure 13.14 Stress fibers and focal adhesions Fluorescence microscopy of fibroblasts in which actin stress fibers are stained magenta and focal adhesions are stained bright green with antibody against vinculin.

cells. Proteins related to spectrin, ankyrin, and protein 4.1 are expressed in a wide range of cell types where they fulfill functions analogous to those described for erythrocytes. For example, another member of the calponin family, **dystrophin**, is of particular interest because it is the product of the gene responsible for two types of muscular dystrophy (Duchenne and Becker). These X-linked inherited diseases result in progressive degeneration of skeletal muscle, and patients with the more severe form of the disease (Duchenne muscular dystrophy) usually die in their teens or early twenties. Molecular cloning of the gene responsible for this disorder revealed that it encodes a large protein (427 kd) that is either absent or abnormal in patients with Duchenne or Becker muscular dystrophy, respectively. The sequence of dystrophin identified it as an actin-binding protein. Like spectrin, dystrophin links actin filaments to transmembrane proteins of the muscle cell plasma membrane. These transmembrane proteins in turn link the cytoskeleton to the extracellular matrix, which plays an important role in maintaining cell stability during muscle contraction.

In contrast to the uniform surface of red blood cells, most cells have specialized regions of the plasma membrane that form contacts with adjacent cells or the extracellular matrix (discussed in Chapter 15). These regions can also serve as attachment sites for bundles of actin filaments that anchor the cytoskeleton to areas of cell contact. These attachments of actin filaments are particularly evident in fibroblasts maintained in tissue culture (**Figure 13.14**). Such cultured fibroblasts secrete extracellular matrix proteins that stick to the surface of the culture dish. The fibroblasts then attach to this extracellular matrix on the culture dish via the binding of transmembrane proteins (called **integrins**). The sites of attachment are discrete regions (called **focal adhesions**) that also serve as attachment sites for large bundles of actin filaments, called **stress fibers**.

Stress fibers are contractile bundles of actin filaments, cross-linked by α-actinin and stabilized by tropomysosin, which anchor the cell and exert tension against the substratum. They are attached to the plasma membrane at focal adhesions via interactions with integrin. These complex associations are mediated by several other proteins, including **talin** and **vinculin** (**Figure 13.15**). For example, both talin and α-actinin bind to the cytoplasmic domains of integrins. Talin also binds to vinculin and both proteins also bind actin.

Figure 13.15 Attachment of stress fibers to the plasma membrane at focal adhesions Focal adhesions are mediated by the binding of integrins to the extracellular matrix. Stress fibers (bundles of actin filaments cross-linked by α-actinin) are bound to the cytoplasmic domain of integrins by complex associations involving a number of proteins. Two possible associations are illustrated: (1) talin binds to both integrin and vinculin, and both talin and vinculin bind to actin, and (2) integrin binds to α-actinin.

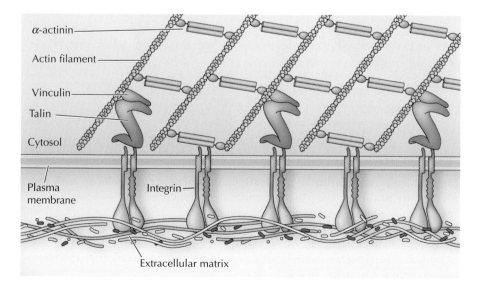

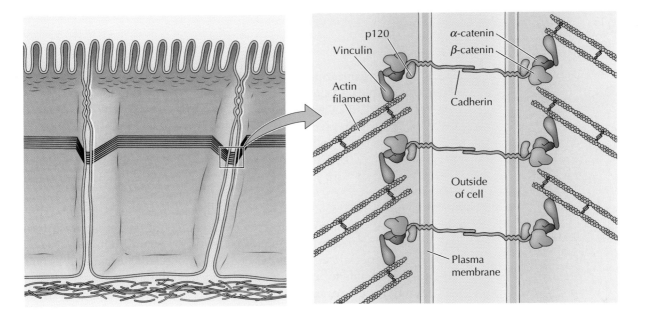

Figure 13.16 Attachment of actin filaments to adherens junctions Cell-cell contacts at adherens junctions are mediated by cadherins, which serve as sites for attachment of actin filaments. In sheets of epithelial cells, these junctions form a continuous belt of actin filaments around each cell. The transmembrane cadherins bind β-catenin and p120, which regulates the stability of the junction. β-catenin also binds α-catenin, which mediates the association of actin filaments with adherens junctions by interacting with vinculin.

> **FYI**
>
> In addition to its structural role at adherens junctions, β-catenin is a critical component of the Wnt signaling pathway, where it functions as a transcriptional activator (see Chapter 16).

Other proteins found at focal adhesions also participate in the attachment of actin filaments, and a combination of these interactions is responsible for the linkage of actin filaments to the plasma membrane.

The actin cytoskeleton is similarly anchored to regions of cell–cell contact called **adherens junctions** (Figure 13.16). In sheets of epithelial cells, these junctions form a continuous beltlike structure (called a **circumferential belt**) around each cell in which an underlying bundle of actin filaments is linked to the plasma membrane. Contact between cells at adherens junctions is mediated by transmembrane proteins called **cadherins**, which are discussed further in Chapter 15. The cadherins form a complex with cytoplasmic proteins called **catenins**, which interact with vinculin and other actin-binding proteins to anchor actin filaments to the plasma membrane.

Microvilli

The surfaces of most cells have a variety of protrusions or extensions that are involved in cell movement, phagocytosis, or specialized functions, such as absorption of nutrients. Most of these cell surface extensions are based on actin filaments, which are organized into either relatively permanent or rapidly rearranging bundles or networks.

The best-characterized of these actin-based cell surface protrusions are **microvilli**, fingerlike extensions of the plasma membrane that are particularly abundant on the surfaces of cells involved in absorption, such as the epithelial cells lining the intestine (Figure 13.17). The microvilli of these cells

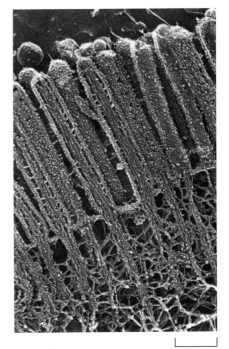

0.25 μm

Figure 13.17 Electron micrograph of microvilli The microvilli of intestinal epithelial cells are fingerlike projections of the plasma membrane. They are supported by actin bundles anchored in a dense region of the cortex called the terminal web. (Courtesy of Nobutaka Hirokawa.)

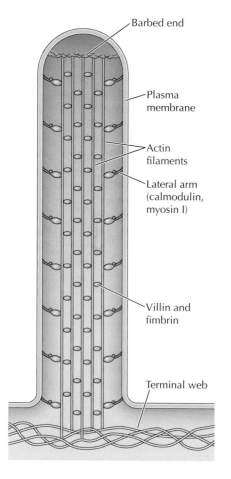

- Barbed end
- Plasma membrane
- Actin filaments
- Lateral arm (calmodulin, myosin I)
- Villin and fimbrin
- Terminal web

Figure 13.18 Organization of microvilli The core actin filaments of microvilli are cross-linked into closely packed bundles by villin and fimbrin. They are attached to the plasma membrane along their length by lateral arms, consisting of calmodulin in association with myosin I. The barbed ends of the actin filaments are embedded in a cap of proteins at the tip of the microvillus.

form a layer on the apical surface (called a **brush border**) that consists of approximately a thousand microvilli per cell and increases the exposed surface area available for absorption by ten to twentyfold. In addition to their role in absorption, specialized forms of microvilli, the **stereocilia** of auditory hair cells, are responsible for hearing by detecting sound vibrations.

Their abundance and ease of isolation have facilitated detailed structural analysis of intestinal microvilli, which contain closely packed parallel bundles of 20 to 30 actin filaments (**Figure 13.18**). The filaments in these bundles are cross-linked in part by fimbrin, an actin-bundling protein (discussed earlier, see Figure 13.9) that is present in surface projections of a variety of cell types. However, the major actin-bundling protein in intestinal microvilli is **villin**, a 95-kd protein present in microvilli of specialized epithelial cells, such as those lining the intestine and kidney tubules and in cell junctions. Along their length, the actin bundles of microvilli are attached to the plasma membrane by lateral arms consisting of the calcium-binding protein calmodulin in association with myosin I, which may be involved in movement of the plasma membrane along the actin bundle of the microvillus. At their base, the actin bundles are anchored in a spectrin-rich region of the actin cortex called the terminal web, which cross-links and stabilizes the microvilli.

Cell surface protrusions and cell movement

In contrast to microvilli, many surface protrusions are transient structures that form in response to environmental stimuli and are involved in cell movement. The movement of cells across a surface represents a basic form of cell locomotion employed by a wide variety of different kinds of cells. Examples include the crawling of amoebas, the migration of embryonic cells during development, the invasion of tissues by white blood cells to fight infection, the migration of cells involved in wound healing, and the spread of cancer cells during metastasis. Similar types of movement are also responsible for phagocytosis and for the extension of nerve cell processes during development of the nervous system.

All of these movements are based on local extensions of the plasma membrane that extend from the leading edge of a moving cell (**Figure 13.19**). **Pseudopodia** are extensions of moderate width, based on actin filaments cross-linked into a three-dimensional network, that are responsible for phagocytosis and for the movement of amoebas across a surface. **Lamellipodia** are broad, sheetlike extensions at the leading edge of fibroblasts, which contain a network of actin filaments. **Filopodia** are very thin projections of the plasma membrane, supported by actin bundles, that extend from lamellipodia. The formation and retraction of these structures during cell movement is based on the regulated assembly and disassembly of actin filaments.

Video 13.1

sites.sinauer.com/cooper7e/v13.1
Lamellipodia Lamellipodia are sheet-like extensions at the leading edge of fibroblasts.

FYI

Certain pathogenic bacteria (e.g., *Listeria*) subvert the normal dynamics of the actin cytoskeleton to move through the host cell cytoplasm and infect neighboring cells

(A)

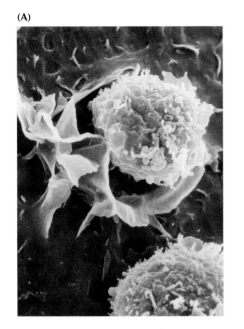

(B)

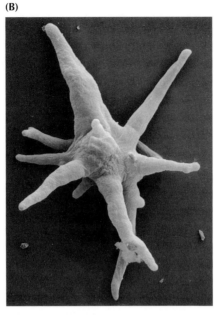

(C)

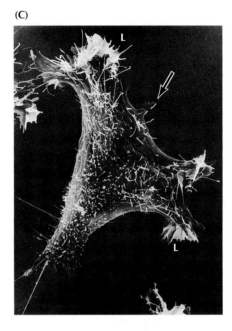

Figure 13.19 Examples of cell surface projections involved in phagocytosis and movement   (A) Scanning electron micrograph showing pseudopodia of a macrophage engulfing a tumor cell during phagocytosis. (B) An amoeba with several extended pseudopodia. (C) A tissue culture cell illustrating lamellipodia (L) at the leading edge of a fibroblast and filopodia (arrow).

> **Video 13.2**
>
> **sites.sinauer.com/cooper7e/v13.2**
>
> **Cell Movement**   Cell movement across a surface involves extension of the leading edge, attachment, and retraction.

Cell movement or the extension of long cellular processes involves a coordinated series of movements, which can be viewed in several stages (**Figure 13.20**). First, protrusions such as pseudopodia, lamellipodia, or filopodia must be extended to establish a leading edge of the cell. The initial movements of most cells are mediated by the extension of filopodia, which are subsequently incorporated into lamellipodia. These extensions must then attach to the substratum across which the cell is moving. Finally, the trailing edge of the cell must dissociate from the substratum and retract into the cell body. A variety of experiments indicate that extension of the leading edge involves the branching and polymerization of actin filaments. For example, inhibition of actin polymerization (e.g., by treatment with cytochalasin) blocks the formation of cell surface protrusions. As described below, the process that underlies the extension of cell protrusions is the force of actin filament polymerization against the plasma membrane at the leading edge.

In most cases, cells move in response to signals from other cells or the environment. For example, in wound healing, cells at the edge of a cut move across the extracellular matrix to cover the wound. The formation of cell surface protrusions in response to extracellular stimuli is regulated by small GTP-binding proteins of the **Rho** family (Cdc42, Rho, and Rac), which are

Figure 13.20 Cell migration   The movement of cells across a surface can be viewed as three stages of coordinated movements: (1) extension of the leading edge, (2) attachment of the leading edge to the substratum, and (3) retraction of the rear of the cell into the cell body.

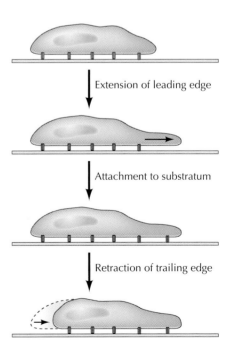

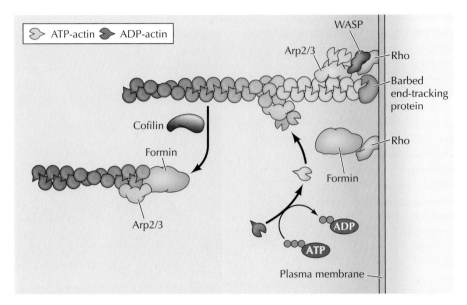

Figure 13.21 Actin filament remodeling at the leading edge Rho family members activate WASP and Arp2/3 as well as formin, leading to the initiation and elongation of branched actin filaments. Profilin activates ADP-actin monomers through ADP/ATP exchange to promote filament extension. Cleavage of existing filaments by cofilin provides new barbed ends for the initiation and growth of new filaments.

related to the Ras proteins and discussed further in Chapter 16. Signals that stimulate cell movement activate receptors in the plasma membrane, leading to activation of Rho family members, which function to promote actin polymerization (**Figure 13.21**). Rho family members activate proteins of the **WASP** family, which stimulate the Arp2/3 complex and initiate the growth of branched actin filaments. The Rho family members also activate formins, which extend the branched filaments initiated by the Arp2/3 complex as well as initiating linear filaments. Cofilin also plays an important role in remodeling the actin network by cleaving existing filaments and providing new barbed ends to support filament branching and growth. The growing actin filaments push against the plasma membranes and drive the formation of cell surface protrusions. At high local concentrations of ATP-actin, generated by profilin, the growth of the barbed ends of actin filaments is energetically favorable and can generate considerable force. While a single filament would not exert enough force to extend the cell membrane, many filaments can easily do so.

As the new actin filaments extend into the growing cell process, they provide pathways (together with reorganized microtubules) for the delivery of vesicles containing lipids and proteins needed for continued extension. Cell attachment to a surface then requires rebuilding cell–substratum adhesions. For slow-moving cells, such as epithelial cells or fibroblasts, attachment involves the formation of focal adhesions (see Figure 13.15). Cells moving more rapidly, such as amoebas or white blood cells, form more diffuse contacts with the substratum, the molecular composition of which is not known.

Among the cargo proteins delivered to the growing cell protrusion by actin filaments and microtubules are actin-bundling proteins, needed for the formation of actin bundles and stress fibers, along with focal adhesion proteins—such as talin and vinculin. Reconstruction of focal adhesions occurs in two steps: the appearance of small focal complexes containing a few actin filaments attached to integrins, and the growth of those focal complexes into mature focal adhesions (illustrated in Figure 13.15). Vinculin and talin are activated by contact with plasma membrane lipids. They activate integrins to bind to the extracellular matrix as well as connect the integrins to actin filaments. The development of focal adhesions also requires the formation of stress fibers (see Figure 13.14), which is stimulated by Rho as a result of activation of both formin and myosin II (discussed in the next section of this chapter).

The final stage of cell migration—retraction of the trailing edge—involves the action of small GTP-binding proteins of the Arf (see Chapter 11) and Rho families, which regulate the breakdown of existing focal adhesions and stimulate endocytosis of the plasma membrane at the trailing edge of the cell. The stress fibers connected to the new focal adhesions at the leading edge then pull the trailing edge of the cell forward.

Myosin Motors

Actin filaments, often in association with myosin, are responsible for many types of cell movements. **Myosin** is the prototype of a **molecular motor**—a protein that converts chemical energy in the form of ATP to mechanical energy, thus generating force and movement. The most striking variety of such movement is muscle contraction, which has provided the model for understanding actin–myosin interactions and the motor activity of myosin molecules. However, interactions of actin and myosin are responsible not only for muscle contraction but also for a variety of movements of nonmuscle cells, including cell division and the intracellular transport of vesicles and organelles, so these interactions play a central role in cell biology.

Muscle contraction

Muscle cells are highly specialized for a single task—contraction—and it is this specialization in structure and function that has made muscle the prototype for studying movement at the cellular and molecular levels. There are three distinct types of muscle cells in vertebrates: skeletal muscle, which is responsible for all voluntary movements; cardiac muscle, which pumps blood from the heart; and smooth muscle, which is responsible for involuntary movements of organs such as the stomach, intestine, uterus, and blood vessels. In both skeletal and cardiac muscle, the contractile elements of the cytoskeleton are present in highly organized arrays that give rise to characteristic patterns of cross-striations. It is the characterization of these structures in skeletal muscle that has led to our current understanding of muscle contraction and other myosin-based cell movements at the molecular level.

Skeletal muscles are bundles of **muscle fibers**, which are single large cells (approximately 50 μm in diameter and up to several centimeters in length) formed by the fusion of many individual cells during development (**Figure 13.22**). Most of the cytoplasm consists of **myofibrils**, which are cylindrical

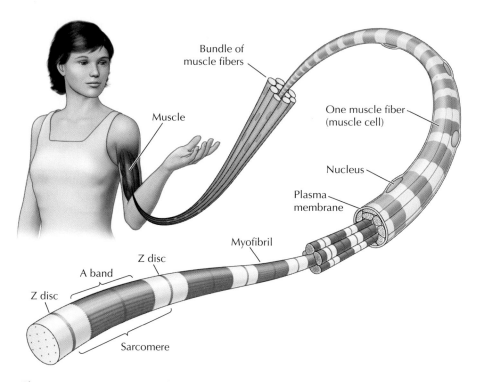

Figure 13.22 Structure of muscle cells   Muscles are composed of bundles of single large cells (called muscle fibers) that form by cell fusion and contain multiple nuclei. Each muscle fiber contains many myofibrils, which are bundles of actin and myosin filaments organized into a chain of repeating units called sarcomeres, bounded by Z discs.

bundles of two types of filaments: thick filaments of myosin (about 15 nm in diameter) and thin filaments of actin (about 7 nm in diameter). Each myofibril is organized as a chain of contractile units called **sarcomeres**, which are responsible for the striated appearance of skeletal and cardiac muscle.

The sarcomeres (which are approximately 2.3 μm long) consist of several distinct regions discernible by electron microscopy, which provided critical insights into the mechanism of muscle contraction (**Figure 13.23**). The ends of each sarcomere are defined by the Z disc. Within each sarcomere, dark bands (called A bands because they are anisotropic when viewed with polarized light) alternate with light bands (called I bands for isotropic). These bands correspond to the presence or absence of myosin filaments. The I bands contain only thin actin filaments, whereas the A bands contain thick filaments consisting of myosin. The actin and myosin filaments overlap in peripheral regions of the A band, whereas a middle region (called the H zone) contains only myosin. The actin filaments are attached at their barbed ends to the Z disc, which includes the cross-linking protein α-actinin. The myosin filaments are anchored at the M line in the middle of the sarcomere.

The basis for understanding muscle contraction is the **sliding filament model**, first proposed in 1954 both by Andrew Huxley and Ralph Niedergerke and by Hugh Huxley and Jean Hanson (**Figure 13.24**). During muscle contraction each sarcomere shortens, bringing the Z discs closer together. There is no change in the width of the A band, but both the I bands and the H zone almost completely disappear. These changes are explained by the actin and

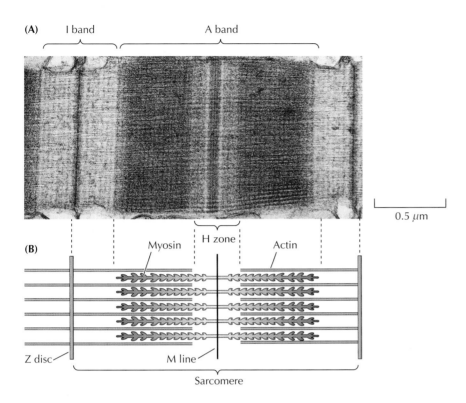

Figure 13.23 Structure of the sarcomere (A) Electron micrograph of a sarcomere. (B) Diagram showing the organization of actin (thin) and myosin (thick) filaments in the indicated regions.

myosin filaments sliding past one another so that the actin filaments move into the A band and H zone. Muscle contraction thus results from an interaction between the actin and myosin filaments that generates their movement relative to one another. The molecular basis for this interaction is the binding of myosin to actin filaments, allowing myosin to function as a motor that drives filament sliding.

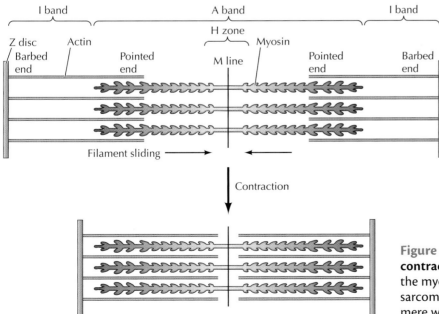

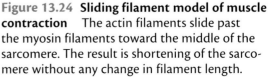

Figure 13.24 Sliding filament model of muscle contraction The actin filaments slide past the myosin filaments toward the middle of the sarcomere. The result is shortening of the sarcomere without any change in filament length.

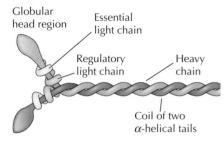

Figure 13.25 Myosin II The myosin II molecule consists of two heavy chains and two pairs of light chains (called the essential and regulatory light chains). The heavy chains have globular head regions and long α-helical tails, which coil around each other to form dimers.

The type of myosin present in muscle (**myosin II**) is a very large protein (about 500 kd) consisting of two identical heavy chains (about 200 kd each) and two pairs of light chains (about 20 kd each) (**Figure 13.25**). Each heavy chain consists of a globular head region and a long α-helical tail. The α-helical tails of two heavy chains twist around each other in a coiled-coil structure to form a dimer, and two light chains associate with the neck of each head region to form the complete myosin II molecule.

The thick filaments of muscle consist of several hundred myosin molecules associated in a parallel staggered array by interactions between their tails. (**Figure 13.26**). The globular heads of myosin bind actin, forming cross-bridges between the thick and thin filaments. It is important to note that the orientation of myosin molecules in the thick filaments reverses at the M line of the sarcomere. The polarity of actin filaments (which are attached to Z discs at their barbed ends) similarly reverses at the M line, so the relative orientation of myosin and actin filaments is the same on both halves of the sarcomere. Contraction results from movement of the myosin head groups along the actin filament in the direction of the barbed end. This movement slides the actin filaments from both sides of the sarcomere toward the M line, shortening the sarcomere and resulting in muscle contraction.

The activity of myosin as a molecular motor is powered by the myosin head groups, which bind and hydrolyze ATP to provide the energy that drives movement. This translation of chemical energy to movement is mediated by changes in the shape of myosin resulting from ATP binding. ATP hydrolysis drives repeated cycles of interaction between myosin heads and actin. During each cycle, conformational changes in myosin result in the movement of myosin heads along actin filaments.

A molecular model for myosin function has been derived both from the three-dimensional structure of myosin and from *in vitro* studies of myosin movement along actin filaments (**Figure 13.27**). The cycle starts with myosin (in the absence of ATP) tightly bound to actin. ATP binding dissociates the myosin-actin complex and the hydrolysis of ATP then induces a conformational change in myosin. This change affects the neck region of myosin that binds the light chains (see Figure 13.25), which acts as a lever arm to displace the

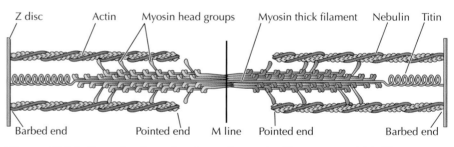

Figure 13.26 Organization of actin and myosin filaments Thick filaments are formed by the association of several hundred myosin II molecules in a staggered array. The globular heads of myosin bind actin, forming cross-bridges between the myosin and actin filaments. The orientation of both actin and myosin filaments reverses at the M line, so their relative polarity is the same on both sides of the sarcomere. Molecules of a large protein called titin extend from the Z disc to the M line and act as springs to keep the myosin filaments centered in the sarcomere. Molecules of another protein (nebulin) extend from the Z disc in association with actin and serve to organize and stabilize the actin filaments.

Animation 13.2

sites.sinauer.com/cooper7e/a13.2

A Thin Filament The thin filaments of skeletal muscle consist of actin filaments decorated with tropomyosin-troponin complexes covering the myosin binding sites.

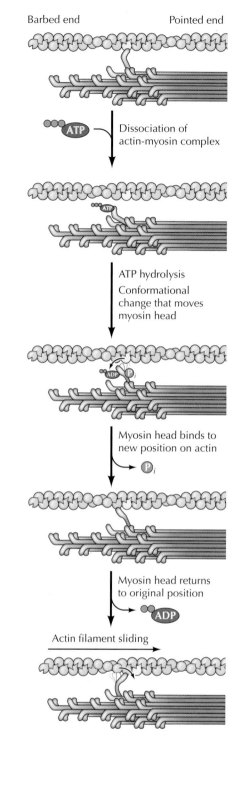

Figure 13.27 Model for myosin action The binding of ATP dissociates myosin from actin. ATP hydrolysis then induces a conformational change that alters the position of the myosin head group. This is followed by binding of the myosin head to a new position on the actin filament and release of P_i. The return of the myosin head to its original conformation, coupled to the release of ADP, drives actin filament sliding.

Dissociation of actin-myosin complex

ATP hydrolysis Conformational change that moves myosin head

Myosin head binds to new position on actin

Myosin head returns to original position

Actin filament sliding

myosin head by about 5 nm. The products of hydrolysis (ADP and P_i) remain bound to the myosin head, which is said to be in the "cocked" position. The myosin head then rebinds at a new position on the actin filament, resulting in the release of P_i. This triggers the "power stroke" in which ADP is released and the myosin head returns to its initial conformation, thereby sliding the actin filament toward the M line of the sarcomere.

The contraction of skeletal muscle is triggered by nerve impulses, which stimulate the release of Ca^{2+} from the **sarcoplasmic reticulum**—a specialized network of internal membranes (similar to the endoplasmic reticulum) that stores high concentrations of Ca^{2+} ions. The release of Ca^{2+} from the sarcoplasmic reticulum increases the concentration of Ca^{2+} in the cytosol from approximately 10^{-7} to 10^{-5} M. The increased Ca^{2+} concentration signals muscle contraction via the action of two actin filament binding proteins: tropomyosin and **troponin** (Figure 13.28). In striated muscle, each tropomyosin molecule

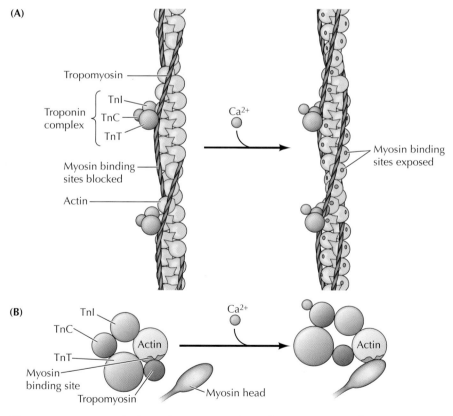

(A)

Tropomyosin

Troponin complex { TnI, TnC, TnT }

Myosin binding sites blocked

Actin

Ca^{2+}

Myosin binding sites exposed

(B)

TnI

TnC

TnT

Myosin binding site

Tropomyosin

Actin

Ca^{2+}

Actin

Myosin head

Figure 13.28 Association of tropomyosin and troponins with actin filaments (A) Tropomyosin binds lengthwise along actin filaments, and in striated muscle, is associated with a complex of three troponins: troponin I (TnI), troponin C (TnC), and troponin T (TnT). In the absence of Ca^{2+}, the tropomyosin-troponin complex blocks the binding of myosin to actin. Binding of Ca^{2+} to TnC shifts the complex, relieving this inhibition and allowing contraction to proceed. (B) Cross-sectional view.

is bound to troponin, which is a complex of three polypeptides: troponin I (inhibitory), troponin C (Ca^{2+}-binding), and troponin T (tropomyosin-binding). When the concentration of Ca^{2+} is low, the complex of the troponins with tropomyosin blocks the interactions of almost all actins with the myosin head groups, so the muscle does not contract. At high concentrations, Ca^{2+} binding to troponin C shifts the position of the complex, allowing access of the myosin head groups to an increasing number of actins and allowing contraction to proceed.

Contractile assemblies of actin and myosin in nonmuscle cells

Contractile assemblies of actin and myosin resembling small-scale versions of muscle fibers are also present in nonmuscle cells. As in muscle, the actin filaments in these contractile assemblies are interdigitated with bipolar filaments of myosin II, consisting of 15 to 20 myosin II molecules, which produce contraction by sliding the actin filaments relative to one another (**Figure 13.29**). In contrast to actin filaments in striated muscle, the actin/myosin filaments in nonmuscle cells are not organized into sarcomeres.

Two examples of contractile assemblies in nonmuscle cells—stress fibers and circumferential belts—were discussed earlier with respect to attachment of the actin cytoskeleton to regions of cell–substratum and cell–cell contacts (see Figures 13.15 and 13.16). The contraction of stress fibers produces tension across the cell, allowing the cell to pull on a substratum (e.g., the extracellular matrix) to which it is anchored. The contraction of adhesion belts alters the shape of epithelial cell sheets: a process that is particularly important during embryonic development when sheets of epithelial cells fold into structures such as tubes.

The most dramatic example of actin-myosin contraction in nonmuscle cells, however, is provided by **cytokinesis**—the division of a cell into two cells following mitosis (**Figure 13.30**). Toward the end of mitosis in yeast and animal cells, a **contractile ring** consisting of actin filaments and myosin II is assembled by a membrane-bound myosin just underneath the plasma membrane. Its contraction pulls the plasma membrane progressively inward, constricting the center of the cell and pinching it in two. The thickness of the contractile ring remains constant as it contracts, implying that actin filaments disassemble as contraction proceeds. The ring then disperses completely following cell division. Interestingly, even though plant cells divide by an

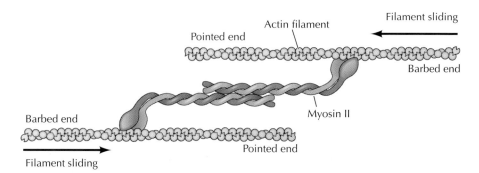

Figure 13.29 Contractile assemblies in nonmuscle cells Bipolar filaments of myosin II produce contraction by sliding actin filaments in opposite directions.

Figure 13.30 Cytokinesis Following completion of mitosis (nuclear division), a contractile ring consisting of actin filaments and myosin II divides the cell in two.

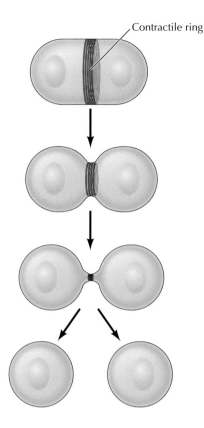

entirely different mechanism (see Chapter 15), actin assembles at the position of cell division. This role of actin appears to be evolutionarily ancient, as a bacterial actin-like protein also functions in cell division. However, bacteria are not known to have myosins or other motor proteins.

The regulation of actin-myosin contraction in striated muscle, discussed earlier, is mediated by the binding of Ca^{2+} to troponin. In nonmuscle cells and in smooth muscle, however, contraction is regulated primarily by phosphorylation of one of the myosin II light chains—called the regulatory light chain (Figure 13.31). Phosphorylation of the regulatory light chain in these cells has at least two effects: It promotes the assembly of myosin into filaments, and it increases myosin catalytic activity enabling contraction to proceed. The enzyme that catalyzes this phosphorylation, called **myosin light-chain kinase**, is itself regulated by association with the Ca^{2+}-binding protein **calmodulin**. Increases in cytosolic Ca^{2+} promote the binding of calmodulin to the kinase, resulting in phosphorylation of the myosin regulatory light chain. Increases in cytosolic Ca^{2+} are thus responsible, albeit indirectly, for activating myosin in smooth muscle and nonmuscle cells, as well as in striated muscle.

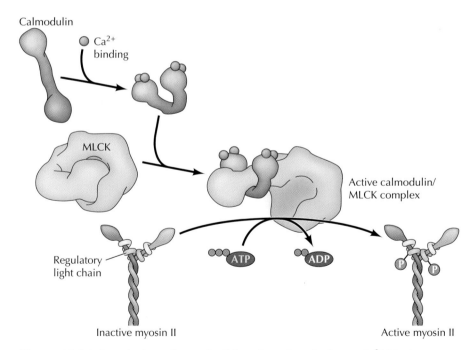

Figure 13.31 Regulation of myosin II by phosphorylation Ca^{2+} binds to calmodulin, which in turn binds to myosin light-chain kinase (MLCK). The active calmodulin-MLCK complex then phosphorylates the myosin II regulatory light chain, converting myosin from an inactive to an active state.

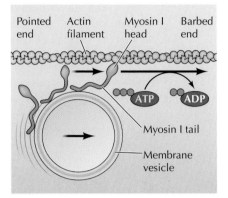

Pointed end | Actin filament | Myosin I head | Barbed end

ATP ADP

Myosin I tail

Membrane vesicle

Figure 13.32 Myosin I Myosin I contains a head group that acts as an ATP-driven motor, similar to myosin II, but it has a comparatively short tail and does not form dimers or filaments. Although it cannot induce contraction, myosin I can move along actin filaments (toward the barbed end), carrying a variety of cargoes (such as membrane vesicles) attached to its tail.

Unconventional myosins

In addition to myosin II (the two-headed myosins found primarily in muscle cells), more than 20 other types of myosin are found in nonmuscle cells. These **unconventional myosins** do not form filaments and are not involved in contraction. However, they are important in a variety of cell movements, including the transport of membrane vesicles, organelles, and other cargo along actin filaments.

One example of unconventional myosins is provided by **myosin I** (**Figure 13.32**). The myosin I proteins contain a globular head group that acts as a molecular motor, like that of myosin II, but myosin I proteins are much smaller molecules (about 110 kd in mammalian cells) that lack the long tail of myosin II. Rather than forming dimers, their tails bind to other structures, such as membrane vesicles or organelles. The movement of myosin I along an actin filament can then transport its attached cargo. One function of myosin I, discussed earlier, is to help link actin bundles to the plasma membrane of intestinal microvilli (see Figure 13.18). In these structures, the motor activity of myosin I moves the plasma membrane along the actin bundles toward the tip of the microvillus. Additional functions of myosin I include the transport of vesicles and organelles along actin filaments and the movement of the plasma membrane during phagocytosis and pseudopod extension.

Other unconventional myosins form dimers, like myosin II. For example, myosin V functions as a two-headed dimer to transport vesicles and other cargo along actin filaments (**Figure 13.33**). It plays a particularly important role in the transport of organelles and macromolecules within neurons, where myosin V is required to provide the new membrane components for extension of cell processes. All of the myosins move toward the barbed end of actin filaments, with the exception of myosin VI, which moves toward the pointed ends. In addition to transporting cargo, some unconventional myosins participate in actin filament reorganization or anchor actin filaments to the plasma membrane.

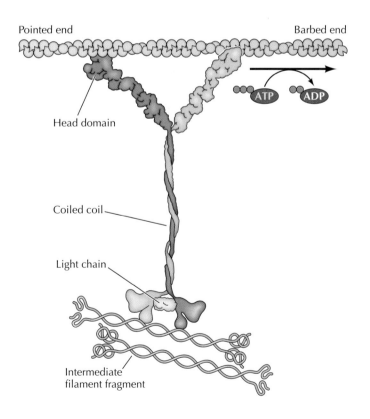

Pointed end

Barbed end

ATP ADP

Head domain

Coiled coil

Light chain

Intermediate filament fragment

Figure 13.33 Myosin V Myosin V is a two-headed myosin like myosin II. It transports organelles and other cargo (for example, intermediate filaments) toward the barbed ends of actin filaments, driven by ATP hydrolysis. The model shown is based on X-ray crystallography data (R. D. Vale, 2003. *Cell* 112: 467).

Microtubules

Microtubules, the second principal component of the cytoskeleton, are rigid hollow rods approximately 25 nm in diameter. Like actin filaments, microtubules are dynamic structures that undergo continual assembly and disassembly within the cell. They function to determine cell shape and participate in a variety of cell movements, including some forms of cell locomotion, the intracellular transport of organelles, and the separation of chromosomes during mitosis.

Structure and dynamic organization of microtubules

Microtubules are composed of a single type of globular protein called **tubulin**. The building blocks of microtubules are tubulin dimers consisting of two closely related 55-kd polypeptides: α-tubulin and β-tubulin. Like actin, both α- and β-tubulin are encoded by small families of related genes (six and seven genes, respectively). In addition, a third type of tubulin (γ-tubulin) is concentrated in the centrosome where it plays a critical role in initiating microtubule assembly (discussed shortly). The evolutionary ancestor of all plant and animal tubulins appears to be a protein similar to the bacterial protein FtsZ, which forms a ring structure beneath the plasma membrane and directs synthesis of the bacterial cell wall.

Tubulin dimers polymerize to form microtubules, which generally consist of 13 linear protofilaments assembled around a hollow core (**Figure 13.34**). The protofilaments, which are composed of head-to-tail arrays of tubulin dimers, are arranged in parallel. Consequently, microtubules (like actin filaments) are polar structures with two distinct ends, referred to as the plus and minus ends. This polarity determines the direction of molecular motor movement along microtubules, just as the polarity of actin filaments defines the direction of myosin movement.

Microtubules are highly dynamic structures that can undergo rapid cycles of assembly and disassembly. Both α- and β-tubulin bind GTP (see Figure 13.34), which functions analogously to the ATP bound to actin to regulate polymerization. In particular, the GTP bound to β-tubulin (but not to

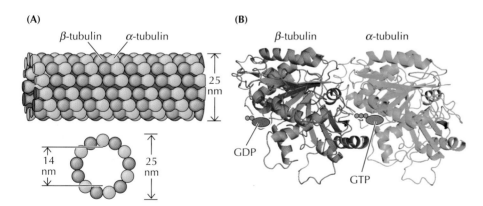

(A)

β-tubulin α-tubulin

25 nm

14 nm   25 nm

(B)

β-tubulin α-tubulin

GDP

GTP

Figure 13.34 Structure of microtubules   (A) Dimers of α- and β-tubulin polymerize to form microtubules, which are composed of 13 protofilaments arranged around a hollow core. (B) Structure of a tubulin dimer, with GTP bound to α-tubulin and GDP bound to β-tubulin.

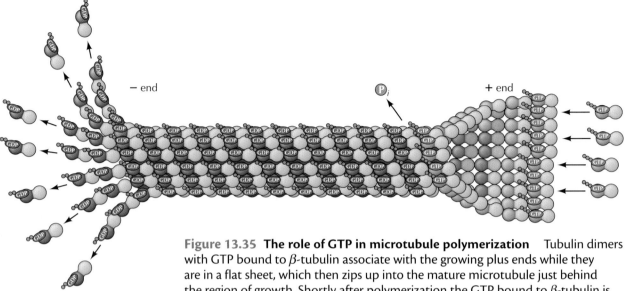

Figure 13.35 The role of GTP in microtubule polymerization Tubulin dimers with GTP bound to β-tubulin associate with the growing plus ends while they are in a flat sheet, which then zips up into the mature microtubule just behind the region of growth. Shortly after polymerization the GTP bound to β-tubulin is hydrolyzed to GDP. Since GDP-bound tubulin is less stable in the microtubule, the dimers at the minus end rapidly dissociate.

α-tubulin) is hydrolyzed to GDP shortly after polymerization (**Figure 13.35**). This GTP hydrolysis weakens the binding affinity of tubulin dimers for each other, which can lead to the rapid dissociation of GDP-bound tubulin from the ends of microtubules. Consequently, the minus ends of microtubules within cells must be protected to prevent rapid depolymerization, usually by anchoring microtubule minus ends in the microtubule organizing center or centrosome, as discussed below.

In microtubules that have been stabilized at the minus end, rapid GTP hydrolysis at the plus end results in a behavior known as **dynamic instability** in which individual microtubules alternate between cycles of growth and shrinkage (called rescue and catastrophe) (**Figure 13.36**). Whether a microtubule grows or shrinks is determined in part by the rate of tubulin addition relative to the rate of GTP hydrolysis. As long as new GTP-bound tubulin dimers are added more rapidly than GTP is hydrolyzed, the microtubule retains a GTP cap at its plus end and microtubule growth continues. However, if the rate of polymerization slows, the GTP bound to tubulin at the plus end of the microtubule will be hydrolyzed to GDP. If this occurs, the GDP-bound tubulin will dissociate, resulting in rapid depolymerization and shrinkage of the microtubule (catastrophe).

Dynamic instability, described by Tim Mitchison and Marc Kirschner in 1984, allows for the continual and rapid turnover of many microtubules within the cell, some of which have half-lives of only several minutes. As discussed later, this rapid turnover of microtubules is particularly critical for the remodeling of the cytoskeleton that occurs during mitosis. Because of the central role of microtubules in mitosis, drugs that affect microtubule assembly are useful not only as experimental tools in cell biology but also in the treatment of cancer. **Colchicine** and **colcemid** are examples of commonly used experimental drugs that bind tubulin and inhibit microtubule polymerization, which in turn blocks mitosis. Two related drugs (**vincristine** and **vinblastine**) are used in cancer chemotherapy because they selectively inhibit rapidly dividing cells. Another useful drug, **taxol**, stabilizes microtubules

Growth   **Shrinkage**

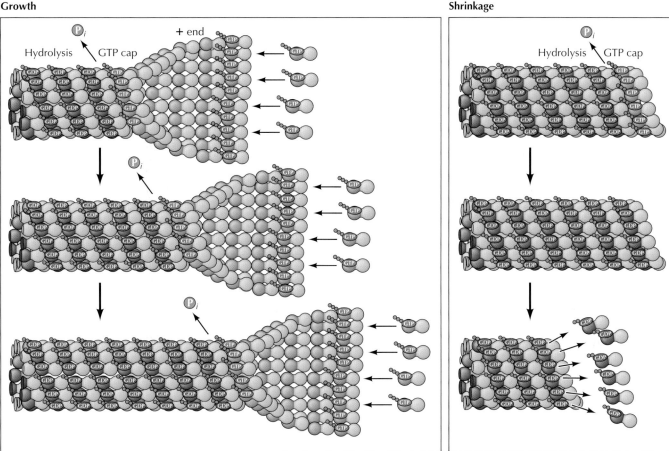

Figure 13.36 Dynamic instability of microtubules Dynamic instability results from the hydrolysis of GTP bound to β-tubulin during or shortly after polymerization, which reduces its binding affinity for adjacent molecules. Growth of microtubules continues as long as new GTP-bound tubulin molecules are added more rapidly than GTP is hydrolyzed, so a GTP cap is retained at the growing end. However, if GTP is hydrolyzed more rapidly than new subunits are added, the presence of GDP-bound tubulin at the plus end of the microtubule leads to disassembly and shrinkage.

rather than inhibiting their assembly. Such stabilization also blocks cell division, and taxol is used as an anticancer agent as well as an experimental tool.

Microtubule-associated proteins (MAPs) regulate the dynamic behavior of microtubules within cells, just as actin-binding proteins regulate the behavior of actin filaments (Table 13.2). As already discussed, the minus ends

FYI

Taxol was originally isolated from the bark of Pacific yew trees. The Pacific yew is one of the slowest growing trees, and several trees were needed to extract enough taxol for the treatment of a single patient, limiting the availability of the drug. Fortunately, taxol and its derivatives are now produced economically via a semisynthetic process wherein a compound similar to taxol is extracted from related trees and modified chemically.

Table 13.2   Microtubule-Associated Proteins (MAPs)

Cellular role	Representative proteins
Microtubule initiation	γ-tubulin ring complex
Plus-end polymerization	XMAP215
Plus-end depolymerization	Kinesins 8 and 13
Severing	Katanin
Lengthwise stabilization	Tau, MAP1, MAP2, MAP4,
Rescue	CLASP

Animation 13.3

sites.sinauer.com/cooper7e/a13.3
Microtubule Assembly Tubulin dimers can polymerize as well as depolymerize to assemble or break down microtubules.

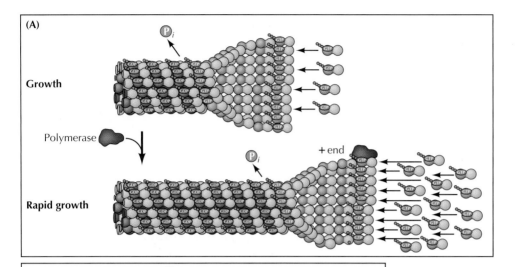

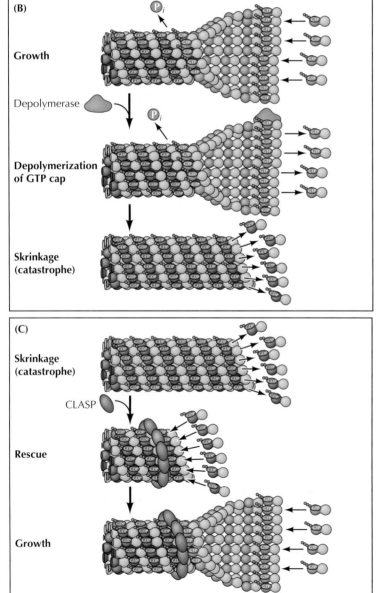

Figure 13.37 Roles of microtubule-associated proteins in dynamic instability The dynamic behavior of microtubules is regulated by microtubule-associated proteins (MAPs) that bind to their plus ends. (A) Polymerases accelerate growth by increasing incorporation of GTP-bound tubulin. (B) In contrast, depolymerases dissociate GTP-tubulin from the plus end, leading to microtubule shrinkage (catastrophe). (C) CLASP proteins rescue microtubules from catastrophe by stopping disassembly and restarting growth.

of microtubules are stabilized by proteins that prevent depolymerization. Growth or shrinkage of the plus ends of microtubules is regulated by MAPs that favor either polymerization or depolymerization of tubulin dimers (**Figure 13.37**). One class of MAPs are polymerases, which bind to the plus ends of microtubules and accelerate their growth by about tenfold. In contrast, depolymerases stimulate shrinkage by accelerating the dissociation of GTP-tubulin from the plus ends of microtubules. Another class of MAPs, the CLASP proteins, act to suppress microtubule catastrophe and promote rescue, apparently by stopping the disassembly of microtubules and restarting their assembly. The opposing actions of these MAPs can thus control the transition between microtubule growth and shrinkage in response to the needs of the cell. Several MAPs are plus-end tracking proteins, which bind to microtubule plus-ends and can mediate the attachment of microtubules to other cellular structures (such as the plasma membrane or the endoplasmic reticulum) as well as regulating microtubule dynamics.

Assembly of microtubules

In animal cells, most microtubules extend outward from the **centrosome** (first described by Theodor Boveri in 1888), which is located adjacent to the nucleus near the center of interphase (nondividing) cells (**Figure 13.38**). During mitosis, microtubules similarly extend outward from duplicated centrosomes to form the mitotic spindle, which is responsible for the separation and distribution of chromosomes to daughter cells. The centrosome thus plays a key role as a **microtubule-organizing center** in determining the intracellular distribution of microtubules in animal cells. In fungi, spindle pole bodies function similarly to centrosomes. However, plant cells do not have an organized centrosome or other form of microtubule-organizing center. Instead, microtubules in most plant cells are dispersed and form an array underlying the plasma membrane, where they play a key role in guiding synthesis of plant cell walls (discussed in Chapter 15).

The centrosome serves as the initiation site for the assembly of microtubules in animal cells, which then grow outward toward the periphery of the cell with their minus ends anchored in the centrosome. This can be clearly visualized in cells that have been treated with colcemid to disassemble their microtubules (**Figure 13.39**). When the drug is removed, the cells recover and new microtubules can be seen growing outward from the centrosome. Thus the initiation of microtubule growth at the centrosome establishes the polarity of microtubules within the cell. Note that microtubules grow by the addition of tubulin to their plus ends, which extend outward from the centrosome toward the cell periphery, so the role of the centrosome is to *initiate* microtubule growth. The key protein in the centrosome is γ-tubulin, a minor species of tubulin first identified in fungi, which nucleates assembly of microtubules. γ-tubulin is associated with eight or more other proteins in a ring-shaped structure

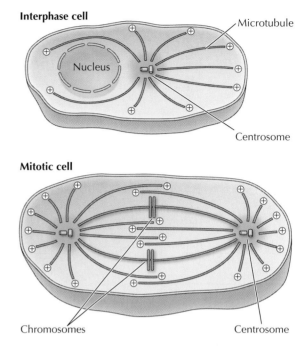

Figure 13.38 Intracellular organization of microtubules The minus ends of microtubules are anchored in the centrosome. In interphase cells, the centrosome is located near the nucleus and microtubules extend outward to the cell periphery. During mitosis, duplicated centrosomes separate and microtubules reorganize to form the mitotic spindle.

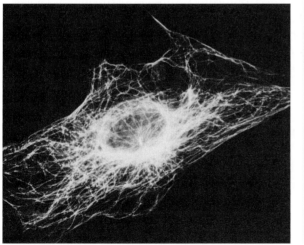

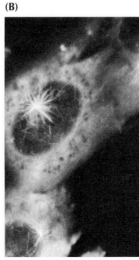

10 μm

Figure 13.39 Growth of microtubules from the centrosome Microtubules in mouse fibroblasts are visualized by immunofluorescence microscopy using an antibody against tubulin. (A) The distribution of microtubules in a normal interphase cell. (B) This cell was treated with colcemid for one hour to disassemble microtubules. The drug was then removed and the cell allowed to recover for 30 minutes, allowing the visualization of new microtubules growing out of the centrosome. (From M. Osborn and K. Weber, 1976. *Proc. Natl. Acad. Sci. USA* 73: 867.)

(A)

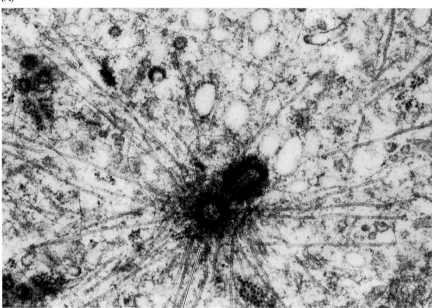

(B)

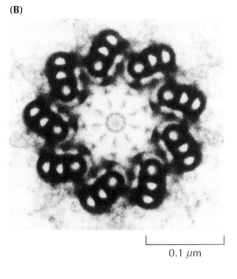

0.1 μm

Pericentriolar material

Figure 13.40 Structure of centrosomes   (A) Electron micrograph of a centrosome showing microtubules radiating from the pericentriolar material that surrounds a pair of centrioles. (B) Transverse section of a centriole illustrating its nine triplets of microtubules.

called the **γ-tubulin ring complex**. This is thought to act as a seed for rapid microtubule growth, bypassing the rate-limiting nucleation step.

The centrosomes of most animal cells contain a pair of **centrioles**, oriented perpendicular to each other and surrounded by **pericentriolar material** (Figure 13.40). The centrioles are cylindrical structures containing nine triplets of microtubules organized around a central cartwheel-like structure. They also function at the cell surface as the basal bodies of cilia and flagella (discussed later in the chapter). However, while centrioles are necessary to form cilia and flagella, they are not needed for the microtubule-organizing functions of the centrosome and are not found in many unicellular eukaryotes and most meiotic animal cells (such as mouse eggs). It is the pericentriolar material, not the centrioles, that contains the γ-tubulin ring complexes and functions to initiate microtubule assembly and anchor microtubule minus ends.

Organization of microtubules within cells

As discussed above, the behavior of microtubules within the cell is regulated by MAPs (see Table 13.2). Many MAPs control dynamic instability by binding to microtubule plus ends, but others act to either sever microtubules or to stabilize microtubules by binding along their length. Microtubule stability is also regulated by extensive post-translational modification of tubulin. Both α- and β-tubulin can be modified by phosphorylation, acetylation, palmitoylation, removal and subsequent addition of carboxy-terminal tyrosines, and addition of multiple glutamines or glycines. These post-translational modifications affect microtubule behavior by providing sites for the binding of specific MAPs.

Interactions with MAPs allow the cell to stabilize microtubules in particular locations and provide an important mechanism for determining cell shape and polarity. Many MAPs are cell-type specific. Some of the earliest identified were MAP1, MAP2, and tau (isolated from neuronal cells), and MAP4, which is present in all non-neuronal vertebrate cell types. The tau protein has been extensively studied because it is the main component of one of the characteristic lesions found in the brains of Alzheimer's patients (see Molecular Medicine Chapter 9).

A good example of the association of stable microtubules with cell polarity is provided by nerve cells, which consist of two distinct types of processes (axons and dendrites) extending from a cell body (**Figure 13.41**). Both axons and dendrites are supported by stable microtubules, together with the neurofilaments discussed in the next section of this chapter. However, the microtubules in nerve cells are not anchored in the centrosome. Rather, the microtubules in nerve cells are released from the centrosome and both their plus and minus ends terminate in the cytoplasm, stabilized by capping

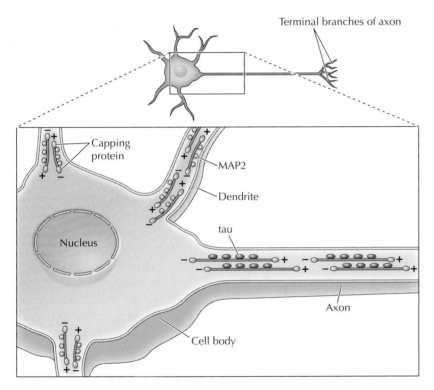

Figure 13.41 Organization of microtubules in nerve cells   Two distinct types of processes extend from the cell body of nerve cells (neurons). Dendrites are short processes that receive stimuli from other nerve cells. The single long axon then carries impulses from the cell body to other cells, which may be either other neurons or an effector cell, such as a muscle. Stable microtubules in both axons and dendrites terminate in the cytoplasm rather than being anchored in the centrosome. In dendrites, microtubules are oriented in both directions, with their plus ends pointing both toward and away from the cell body. In contrast, all of the axon microtubules are oriented with their plus ends pointing toward the tip of the axon. Microtubule-associated proteins (MAPs) cap both the plus and minus ends, as well as stabilize microtubules by binding along their length (MAP2 in dendrites and tau in axons).

proteins. Moreover, microtubules are organized by distinct MAPs in axons and dendrites. In axons, the microtubules are all oriented with their plus ends away from the cell body, similar to the general orientation of microtubules in other cell types. In dendrites, the microtubules are oriented in both directions; some plus ends point toward the cell body and some point toward the cell periphery. These distinct microtubule arrangements are paralleled by differences in MAPs: Axons contain tau proteins, but no MAP2, whereas dendrites contain MAP2, but no tau proteins, and it appears that these differences in MAP distribution play a role in the distinct organization of stable microtubules in axons and dendrites.

Microtubule Motors and Movement

Microtubules are responsible for a variety of cell movements, including the intracellular transport and positioning of membrane vesicles and organelles, the separation of chromosomes at mitosis, and the beating of cilia and flagella. As discussed with actin filaments earlier in the chapter, this occurs in two ways: polymerization and depolymerization of the microtubules and the action of motor proteins that utilize energy derived from ATP hydrolysis to produce force and movement. Members of two large families of motor proteins—the **kinesins** and the **dyneins**—are responsible for powering the variety of movements in which microtubules participate.

Microtubule motor proteins

Dynein and kinesin, the two types of microtubule motor proteins, move along microtubules in opposite directions—most dyneins toward the minus end and kinesins toward the plus end (**Figure 13.42**). The first of these microtubule motor proteins to be identified was dynein, which was isolated by Ian

> **Animation 13.4**
>
> sites.sinauer.com/cooper7e/a13.4
>
> **Kinesin** Kinesin is a motor protein that moves vesicles and organelles toward the plus ends of microtubules.

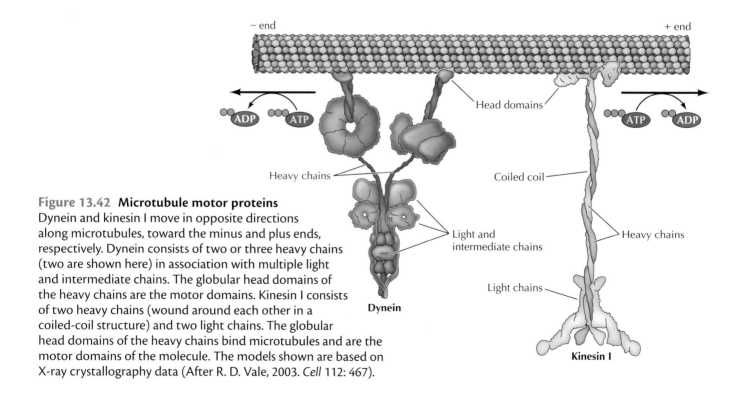

Figure 13.42 Microtubule motor proteins
Dynein and kinesin I move in opposite directions along microtubules, toward the minus and plus ends, respectively. Dynein consists of two or three heavy chains (two are shown here) in association with multiple light and intermediate chains. The globular head domains of the heavy chains are the motor domains. Kinesin I consists of two heavy chains (wound around each other in a coiled-coil structure) and two light chains. The globular head domains of the heavy chains bind microtubules and are the motor domains of the molecule. The models shown are based on X-ray crystallography data (After R. D. Vale, 2003. *Cell* 112: 467).

The Isolation of Kinesin

Identification of a Novel Force-Generating Protein, Kinesin, Involved in Microtubule-Based Motility

Ronald D. Vale, Thomas S. Reese, and Michael P. Sheetz

National Institute of Neurological and Communicative Disorders and Stroke, Marine Biological Laboratory, Woods Hole, MA; University of Connecticut Health Center, Farmington, CT; Stanford University School of Medicine, Stanford, CA

Cell, Volume 42, 1985, pages 39–50

The Context

The transport and positioning of cytoplasmic organelles is key to the organization of eukaryotic cells, so understanding the mechanisms responsible for vesicle and organelle transport is a fundamental question in cell biology. In 1982 Robert Allen, Scott Brady, Ray Lasek, and their colleagues used video-enhanced microscopy to visualize the movement of organelles along cytoplasmic filaments in squid giant axons, both *in vivo* and in a cell-free system. These filaments were then identified as microtubules by electron microscopy, but the motor proteins responsible for organelle movement were unknown. The only microtubule motor identified at the time was axonemal dynein, which was present only in cilia and flagella. In 1985 Ronald Vale, Thomas Reese, and Michael Sheetz described the isolation of a novel motor protein, kinesin, which was responsible for the movement of organelles along microtubules. Similar experiments were reported at the same time by Scott Brady (*Nature*, 1985, 317: 73–75).

The Experiments

Two experimental strategies were key to the isolation of kinesin. The first, based on the work of Allen and his colleagues, was the use of an *in vitro* system in which motor protein activity could be detected. Vale, Reese, and Sheetz used a system in which proteins in the cytoplasm of axons were found to power the movement of microtubules across the surface of glass coverslips. This cell-free system provided a

sensitive and rapid functional assay for the activity of kinesin as a molecular motor.

The second important approach used in the isolation of kinesin was to take advantage of its binding to microtubules. Lasek and Brady (*Nature*, 1985, 316: 645–647) had made the key observation that *in vitro* movements of

T. S. Reese   M. P. Sheetz

R. D. Vale

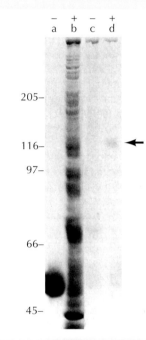

Binding of a motor protein to microtubules in the presence of AMP-PNP Protein samples were analyzed by electrophoresis through a polyacrylamide gel, stained, and photographed. Molecular weights of marker proteins are indicated in kilodaltons at the left. Lane a represents purified microtubules in which only tubulin (55 kd) is detected. Lane b is a soluble extract of squid axon cytoplasm containing many different polypeptides. This soluble extract was incubated with microtubules either without AMP-PNP (lane c) or with AMP-PNP (lane d). Microtubules were then recovered and incubated with ATP to release bound proteins, which were subjected to electrophoresis. Note that a 110-kd polypeptide (arrow) was specifically bound to microtubules in the presence of AMP-PNP and then released by ATP. The proteins bound in the presence of AMP-PNP and released by ATP were also found to induce microtubule movement (indicated by "+" above the gel).

microtubules require ATP and are inhibited by an ATP analog (adenylyl imidodiphosphate, AMP-PNP) that cannot be hydrolyzed and therefore does not provide a usable source of energy. In the presence of AMP-PNP, organelles remained attached to microtubules, suggesting that the motor protein responsible for organelle movement might also remain bound to microtubules under these conditions.

On the basis of these findings, Vale and colleagues incubated microtubules with cytoplasmic proteins from squid axon in the presence of AMP-PNP. The microtubules were then recovered and incubated with ATP to release proteins that were specifically bound in the presence of the nonhydrolyzable ATP analog. This experiment identified a 110-kd polypeptide that was bound to microtubules in the presence of AMP-PNP and released by subsequent incubation with ATP (see figure). Furthermore, the proteins bound to microtubules and then released by ATP were also shown to support the movement of microtubules *in vitro*. Binding to microtubules in the presence of AMP-PNP thus provided

(Continued on next page)

an efficient approach to isolation of the motor protein, which was shown by further biochemical studies to contain the 110-kd polypeptide complexed to polypeptides of 60 to 70 kd. By a similar approach, a related protein was also purified from bovine brain. The authors concluded that these proteins "represent a novel class of motility proteins that are structurally as well as enzymatically distinct from dynein, and we propose to call these translocators kinesin (from the Greek *kinein*, to move)."

The Impact

The movement of vesicles and organelles along microtubules is fundamental to the organization of eukaryotic cells, so the motor proteins responsible for these movements play critical roles in cell biology. Subsequent experiments using *in vitro* assays similar to those described here established that kinesin moves along microtubules in the plus-end direction, whereas cytoplasmic dynein is responsible for the transport of vesicles and organelles in the

minus-end direction. Moreover, a large family of kinesin-related proteins has since been identified. In addition to their roles in vesicle and organelle transport, members of these families of motor proteins are responsible for the separation and distribution of chromosomes during mitosis. The identification of kinesin thus opened the door to understanding a variety of microtubule-based movements that are critical to the structure and function of eukaryotic cells.

Gibbons in 1965. The purification of this form of dynein (called **axonemal dynein**) was facilitated because it is a highly abundant protein in cilia, just as the abundance of myosin facilitated its isolation from muscle cells. The identification of other microtubule-based motors, however, was more problematic because the proteins responsible for processes such as chromosome movement and organelle transport are present at comparatively low concentrations in the cytoplasm. Isolation of these proteins therefore depended on the development of new experimental methods to detect the activity of molecular motors in cell-free systems.

The development of *in vitro* assays for cytoplasmic motor proteins was based on the use of **video-enhanced microscopy** (developed by Robert Allen and Shinya Inoué in the early 1980s) to study the movement of membrane vesicles and organelles along microtubules in squid axons. In this method, a video camera is used to increase the contrast of images obtained with the light microscope, substantially improving the detection of small objects and allowing the movement of organelles to be followed in living cells. Using this approach, Allen, Scott Brady, and Ray Lasek demonstrated in 1982 that organelle movements also took place in a cell-free system in which the plasma membrane had been removed and a cytoplasmic extract had been spread on a glass slide. These observations led to the development of an *in vitro* reconstructed system, which provided an assay capable of detecting cellular proteins responsible for organelle movement. In 1985, Brady, as well as Ronald Vale, Thomas Reese, and Michael Sheetz, capitalized on these developments to identify kinesin (now known as "conventional" kinesin, or kinesin I) as a novel microtubule motor protein present in both squid axons and bovine brains.

Further studies demonstrated that kinesin I translocates along microtubules in only a single direction—toward the plus end. Because the plus ends of microtubules in axons are all oriented away from the cell body (see Figure 13.41), the movement of kinesin I in this direction transports vesicles and organelles away from the cell body toward the tip of the axon. Within intact axons, however, vesicles and organelles also had been observed to move back toward the cell body, implying that a different motor protein might be responsible for movement along microtubules in the opposite direction—toward the minus end. Consistent with this prediction, further experiments showed that a previously identified microtubule-associated protein related to the dynein isolated from cilia (axonemal dynein) was in fact the motor protein that moved along microtubules towards the minus end. It is now known as **cytoplasmic dynein**.

Kinesin I is a molecule of approximately 380 kd consisting of two heavy chains (120 kd each) and two light chains (64 kd each) (see Figure 13.42). The heavy chains have long α-helical regions that wind around each other in a coiled-coil structure. The amino-terminal globular head domains of the heavy chains are the motor domains of the molecule. They bind to both microtubules and ATP, the hydrolysis of which provides the energy required for movement. Although the motor domain of kinesin (approximately 340 amino acids) is much smaller than that of myosin (about 850 amino acids), X-ray crystallography indicates that kinesin and myosin are structurally related and share a similar molecular mechanism for generating force and movement. The tail portion of the kinesin molecule consists of the light chains in association with the carboxy-terminal domains of the heavy chains. This portion of kinesin is responsible for binding to other cell components (such as membrane vesicles and organelles) that are transported along microtubules by the action of kinesin motors.

Like the myosins, the kinesins are a family of related motor proteins. Eighteen different kinesins are encoded in the genome of *C. elegans*, and there are 45 kinesins known in humans. Most of these, like kinesin I, move along microtubules in the plus-end direction. However, some kinesins move in the opposite direction toward the minus end, and some kinesins do not move along microtubules at all but act as microtubule depolymerizing enzymes. There is a direct correlation between where in the molecule the motor domain is located and the direction of movement along the microtubule; plus-end-directed kinesins have N-terminal motor domains, minus-end-directed kinesins have C-terminal motor domains, and those that depolymerize rather than moving along microtubules have motor domains in the middle of the heavy chain (middle motor kinesins). Different members of the kinesin family vary in the sequences of their carboxy-terminal tails and are responsible for the movements of different types of cargo along microtubules.

There are several types of axonemal dyneins, which power the beating of cilia, and at least two types of cytoplasmic dyneins, which are responsible for the movement of most cargo towards the minus ends of microtubules. Cytoplasmic dynein is an extremely large molecule (up to 2000 kd), which consists of two or three heavy chains (each about 500 kd) complexed with a variable number of light and intermediate chains, which range from 14 to 120 kd (see Figure 13.42). As in kinesin, the heavy chains of dynein form globular ATP-binding motor domains that are responsible for movement along microtubules. However, the motor domains of dynein are unrelated to those of myosin and kinesin, and the molecular mechanism of dynein action as a molecular motor is less well understood. The basal portion of cytoplasmic dynein, including the light and intermediate chains, binds to other subcellular structures, such as organelles and vesicles. There are several types of light and intermediate chains, which transport distinct cargoes. Specific dyneins and kinesins are also thought to recognize different microtubules by the post-translational modifications of tubulin.

Video 13.4

sites.sinauer.com/cooper7e/v13.4
Vesicle Movement along Microtubules Vesicles are transported through the cytoplasm along microtubules by kinesin and dynein.

Cargo transport and intracellular organization

One of the major roles of microtubules, like actin filaments, is to transport macromolecules, membrane vesicles, and organelles through the cytoplasm of eukaryotic cells. As already discussed, such cytoplasmic organelle transport is particularly evident in nerve cell axons, which may extend more than a meter in length. Organelles and membrane vesicles containing proteins and RNAs must be transported from the cell body to the axon. Using video-enhanced microscopy, the transport of membrane vesicles and organelles in both directions can be visualized along axon microtubules where kinesin and dynein carry their cargoes to and from the tips of the axons, respectively. For example, secretory vesicles containing neurotransmitters are carried from the Golgi apparatus to the terminal branches of the axon by kinesin. In the reverse direction, cytoplasmic dynein transports endocytic vesicles from the axon back to the cell body.

Microtubules similarly transport macromolecules, membrane vesicles, and organelles in other types of cells. Because microtubules are usually oriented with their minus end anchored in the centrosome and their plus end extending toward the cell periphery, different members of the kinesin family and cytoplasmic dynein transport cargo in opposite directions through the cytoplasm (**Figure 13.43**). Kinesin I and other plus-end-directed members of the kinesin family carry their cargo toward the cell periphery, whereas cytoplasmic dyneins and minus-end-directed members of the kinesin family transport materials toward the center of the cell. Cargo selection can be very specific. Kinesin II, a plus-end-directed motor, transports selected mRNAs toward the cell cortex in *Xenopus* oocytes, a process that appears to be general to all vertebrates. Kinesin I similarly transports actin mRNA in fibroblasts, and dynein moves specific mRNAs toward one side of the *Drosophila* embryo. Specific dyneins and kinesins are also thought to recognize different microtubules by the post-translational modifications of tubulin. Moreover, several types of molecular motors associate with a given cargo at the same time, allowing precise positioning of the cargo. For example, a plus-end directed kinesin, a minus-end directed kinesin, and perhaps a dynein or a myosin might bind the same vesicle, allowing transport in alternating directions along either microtubules or actin filaments. A future challenge is to determine how transport is controlled by switching among the different motors.

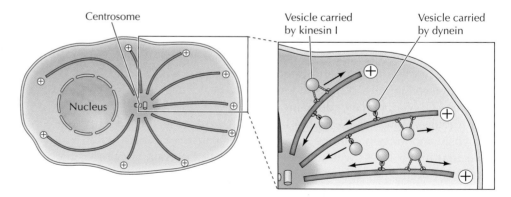

Figure 13.43 Transport of vesicles along microtubules Kinesin I and other plus-end-directed kinesins transport vesicles and organelles in the direction of microtubule plus ends, which extend toward the cell periphery. In contrast, dynein and minus-end-directed members of the kinesin family carry their cargo in the direction of microtubule minus ends, which are anchored in the center of the cell.

In addition to transporting membrane vesicles in the endocytic and secretory pathways, microtubules and associated motor proteins position membrane-enclosed organelles (such as the endoplasmic reticulum, Golgi apparatus, lysosomes, peroxisomes, and mitochondria) within the cell. For example, the endoplasmic reticulum extends to the periphery of the cell in association with microtubules (**Figure 13.44**). Drugs that depolymerize microtubules cause the endoplasmic reticulum to retract toward the cell center, indicating that association with microtubules is required to maintain the endoplasmic reticulum in its extended state. This positioning of the endoplasmic reticulum appears to involve the action of kinesin I (or possibly multiple members of the kinesin family), which pulls the endoplasmic reticulum along microtubules in the plus-end direction, toward the cell periphery. Similarly, kinesin appears to play a key role in the positioning of lysosomes away from the center of the cell, and three different members of the kinesin family have been implicated in the movements of mitochondria.

Conversely, cytoplasmic dynein is thought to play a role in positioning the Golgi apparatus. The Golgi apparatus is located in the center of the cell near the centrosome. If microtubules are disrupted, either by a drug or when

Video 13.5

sites.sinauer.com/cooper7e/v13.5
Mitochondrial Movement Microtubules and associated motor proteins move mitochondria within the cell.

(A)   **(B)**

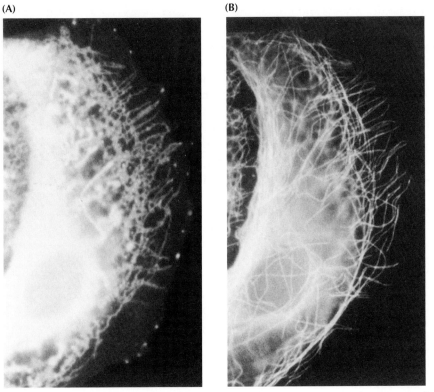

10 µm

Figure 13.44 Association of the endoplasmic reticulum with microtubules
Fluorescence microscopy of the endoplasmic reticulum (A) and microtubules (B) in an epithelial cell. The endoplasmic reticulum is stained with a fluorescent dye, and microtubules are stained with an antibody against tubulin. Note the close correlation between the endoplasmic reticulum and microtubules at the periphery of the cell. (From M. Terasaki, L. B. Chen, and K. Fujiwara, 1986. *J. Cell Biol.* 103: 1557.)

the cell enters mitosis, the Golgi breaks up into small vesicles that disperse throughout the cytoplasm. When the microtubules re-form, the Golgi apparatus also reassembles, with the Golgi vesicles apparently being transported to the center of the cell (toward the minus end of the microtubules) by cytoplasmic dynein. Movement along microtubules is thus responsible not only for vesicle transport but also for establishing the positions of membrane-enclosed organelles within the cytoplasm of eukaryotic cells.

Cilia and flagella

Cilia and **flagella** are microtubule-based projections of the plasma membrane that are responsible for movement of a variety of eukaryotic cells. Cilia are more widespread, being found on almost all animal cells. Many bacteria also have flagella, but these prokaryotic flagella are quite different from those of eukaryotes. Bacterial flagella (which are not discussed further here) are protein filaments projecting from the cell surface, rather than projections of the plasma membrane supported by microtubules.

Eukaryotic cilia and flagella are very similar structures, each with a diameter of approximately 0.25 μm (**Figure 13.45**). Many cells are covered by

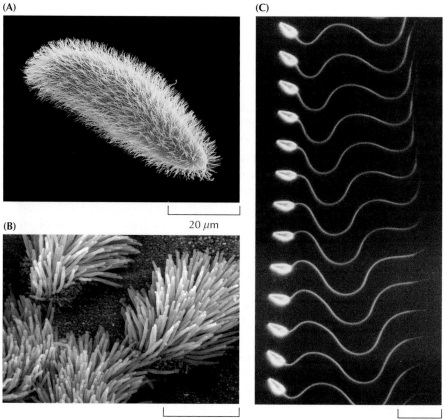

(A)

(B)

(C)

 20 μm

5 μm

 10 μm

Figure 13.45 Examples of cilia and flagella (A) Scanning electron micrograph showing numerous cilia covering the surface of *Paramecium*. (B) Scanning electron micrograph of ciliated epithelial cells lining the surface of a trachea. (C) Multiple-flash photograph (500 flashes per second) showing the wavelike movement of a sea urchin sperm flagellum. (C, courtesy of C. J. Brokaw, California Institute of Technology.)

(A)

(B)

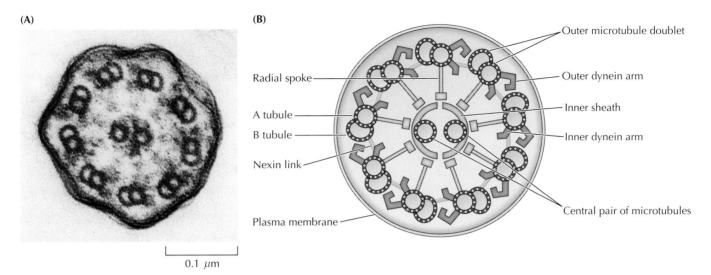

Radial spoke

A tubule

B tubule

Nexin link

Plasma membrane

Outer microtubule doublet

Outer dynein arm

Inner sheath

Inner dynein arm

Central pair of microtubules

0.1 μm

Figure 13.46 Structure of the axoneme of cilia and flagella   (A) Computer-enhanced electron micrograph of a cross-section of the axoneme of a rat sperm flagellum. (B) Schematic cross-section of an axoneme. The nine outer microtubule doublets consist of a complete A tubule, containing 13 protofilaments, and an incomplete B tubule, containing 10 or 11 protofilaments. The outer doublets are joined to each other by nexin links and to the central pair of microtubules by radial spokes. Each outer microtubule doublet is associated with inner and outer dynein arms.

numerous cilia, which are about 10 μm in length. Cilia beat in a coordinated back-and-forth motion, which either moves the cell through fluid or moves fluid over the surface of the cell. For example, the cilia of some protozoans (such as *Paramecium*) are responsible both for cell motility and for sweeping food organisms over the cell surface and into the oral cavity. In animals, an important function of cilia is to move fluid or mucus over the surface of epithelial cell sheets. A good example is provided by the ciliated cells lining the respiratory tract, which clear mucus and dust from the respiratory passages. Flagella differ from cilia in their length (they can be as long as 200 μm) and in their wavelike pattern of beating. Cells usually have only one or two flagella, which are responsible for the locomotion of a variety of protozoans and of sperm.

The fundamental structure of both cilia and flagella is the **axoneme**, which is composed of microtubules and their associated proteins (**Figure 13.46**). The microtubules are arranged in a characteristic "9 + 2" pattern in which a central pair of microtubules is surrounded by nine outer microtubule doublets. The two fused microtubules of each outer doublet are distinct: One (called the A tubule) is a complete microtubule consisting of 13 protofilaments; the other (the B tubule) is incomplete, containing only 10 or 11 protofilaments fused to the A tubule. The outer microtubule doublets are connected to the central pair by radial spokes and to each other by links of a protein called **nexin**. In addition, two arms of dynein are attached to each A tubule, and it is the motor activity of these axonemal dyneins that drives the beating of cilia and flagella.

The minus ends of the microtubules of cilia and flagella are anchored in a centriole called a **basal body**, which contains nine triplets of microtubules

Video 13.8

sites.sinauer.com/cooper7e/v13.8
Coordinated Motion of Cilia In animals, the coordinated motion of cilia can move fluid or mucus over a surface.

(A)

(B)

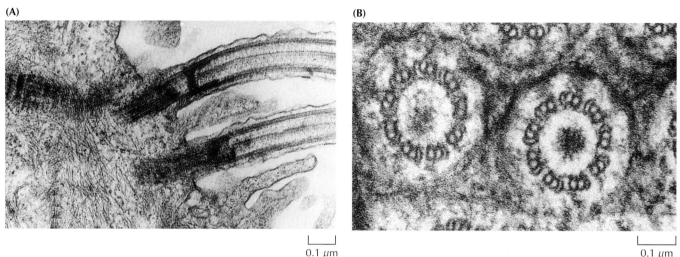

0.1 μm

0.1 μm

Figure 13.47 Electron micrographs of basal bodies (A) A longitudinal view of cilia anchored in basal bodies. (B) A cross-section of basal bodies. Each basal body consists of nine triplets of microtubules.

(**Figure 13.47**). Centrioles were discussed earlier as components of the centrosome, in which their function is complex and poorly understood. Basal bodies, however, play a clear role in organization of the axoneme microtubules. Basal bodies are derived from centrioles that have been transported to the plasma membrane. The outer microtubule doublets of the axoneme are then formed by extension of two of the microtubules present in each triplet of the basal body. Basal bodies thus serve to initiate the growth of axonemal microtubules as well as anchor cilia and flagella to the surface of the cell.

The movements of cilia and flagella result from the sliding of outer microtubule doublets relative to one another, powered by the motor activity of the axonemal dyneins (**Figure 13.48**). The dynein bases bind to the A tubules, while the dynein head groups bind to the B tubules of adjacent doublets. Movement of the dynein head groups in the minus-end direction then causes the A tubule of one doublet to slide toward the basal end of the adjacent B tubule. Because the microtubule doublets in an axoneme are connected by nexin links, the sliding of one doublet along another causes them to bend, forming the basis of the beating movements of cilia and flagella. It is apparent, however, that the activities of dynein molecules in different regions of the axoneme must be carefully regulated to

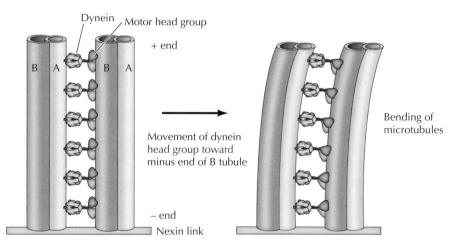

Figure 13.48 Movement of microtubules in cilia and flagella The bases of dynein arms are attached to A tubules, and the motor head groups interact with the B tubules of adjacent doublets. Movement of the dynein head groups in the minus-end direction (toward the base of the cilium) then causes the A tubule of one doublet to slide toward the base of the adjacent B tubule. Because both microtubule doublets are connected by nexin links, this sliding movement forces them to bend.

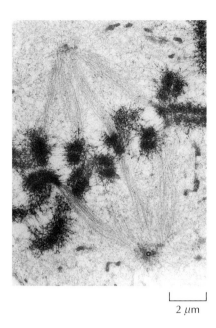

Figure 13.49 Electron micrograph of the mitotic spindle   The spindle microtubules are attached to condensed chromosomes at metaphase.

2 μm

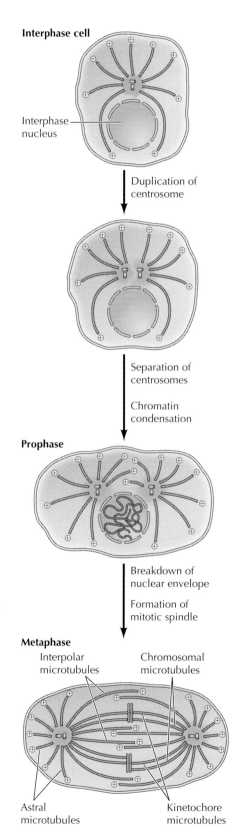

produce the coordinated beating of cilia and the wavelike oscillations of flagella—a process about which little is understood.

Reorganization of microtubules during mitosis

As noted earlier, microtubules completely reorganize during mitosis, providing a dramatic example of the importance of their dynamic instability. During mitosis, dynamic instability accelerates as a result of increased rates of both microtubule catastrophe and growth. The microtubule array present in interphase cells thus disassembles and the free tubulin subunits are reassembled to form the **mitotic spindle**, which is responsible for the separation of daughter chromosomes (**Figure 13.49**). This restructuring of the microtubule cytoskeleton is directed by duplication of the centrosome to form two separate microtubule-organizing centers at opposite poles of the mitotic spindle.

The centrioles and other components of the centrosome are duplicated in interphase cells, but they remain together on one side of the nucleus until the beginning of mitosis (**Figure 13.50**). The two centrosomes then separate and move to opposite sides of the nucleus, forming the two poles of the mitotic spindle. As the cell enters mitosis, the dynamics of microtubule assembly and disassembly also change dramatically. First,

Figure 13.50 Formation of the mitotic spindle   The centrioles and centrosomes duplicate during interphase. During prophase of mitosis, the duplicated centrosomes separate and move to opposite sides of the nucleus. The nuclear envelope then disassembles, and microtubules reorganize to form the mitotic spindle. Kinetochore microtubules are attached to the kinetochores of condensed chromosomes and chromosomal microtubules are attached to their ends. Interpolar microtubules overlap with each other in the center of the cell, and astral microtubules extend outward to the cell periphery. At metaphase, the condensed chromosomes are aligned at the center of the spindle.

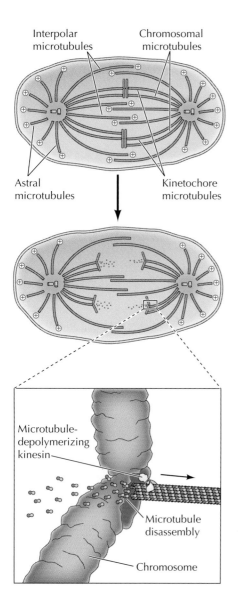

Figure 13.51 **Anaphase A chromosome movement** Chromosomes move toward the spindle poles along the kinetochore and chromosomal microtubules. Chromosome movement is driven by kinesins associated with the chromosomes, which act to depolymerize and shorten both the kinetochore and chromosomal microtubules.

the rate of microtubule disassembly increases about tenfold, resulting in overall depolymerization and shrinkage of microtubules. At the same time, the number of microtubules emanating from the centrosome increases by five- to tenfold. In combination, these changes result in disassembly of the interphase microtubules and the outgrowth of large numbers of short microtubules from the centrosomes.

As first proposed by Marc Kirschner and Tim Mitchison in 1986, formation of the mitotic spindle involves the selective stabilization of some of the microtubules radiating from the centrosomes. These microtubules are of four types, three of which make up the mitotic spindle. **Kinetochore microtubules** attach to the condensed chromosomes of mitotic cells at their centromeres, which are associated with specific proteins to form the kinetochore (see Figure 6.24). Attachment to the kinetochore stabilizes the plus ends of these microtubules, which as discussed below play a critical role in separation of the mitotic chromosomes. Also emanating from the centrosomes are the **chromosomal microtubules**, which connect to the ends of the chromosomes via chromokinesin. The third type of microtubules found in the mitotic spindle (**interpolar microtubules**) are not attached to chromosomes. Instead, the interpolar microtubules are stabilized by overlapping with each other in the center of the cell. **Astral microtubules** extend outward from the centrosomes to the cell periphery, with their plus ends anchored in the cell cortex. As discussed later, both the interpolar and astral microtubules contribute to chromosome movement by pushing the spindle poles apart.

As mitosis proceeds, the condensed chromosomes first align on the metaphase plate and then separate, with the two chromatids of each chromosome being pulled to opposite poles of the spindle. Chromosome movement is mediated by motor proteins associated with the kinetochore and chromosomal microtubules, as will be discussed shortly. In the final stage of mitosis, nuclear envelopes re-form, the chromosomes decondense, and cytokinesis takes place. Each daughter cell then contains one centrosome, which nucleates the formation of a new network of interphase microtubules.

Chromosome movement

After the two centrosomes move to opposite sides of the cell at the beginning of mitosis, the duplicated chromosomes attach to kinetochore and chromosomal microtubules and align on the metaphase plate, equidistant from the two spindle poles. This alignment of chromosomes is mediated by rapid growth of the kinetochore microtubules and capture of the kinetochores by plus-end tracking proteins. In addition, the chromosome ends are pushed toward the metaphase plate by chromokinesin moving along the chromosomal microtubules. Once all of the chromosomes have aligned on the metaphase plate, the links between the sister chromatids are severed and anaphase begins, with the sister chromatids separating and moving to opposite poles of the spindle. Chromosome movement proceeds by two distinct mechanisms, referred to as anaphase A and anaphase B, which involve different types of spindle microtubules.

Anaphase A consists of the movement of chromosomes toward the spindle poles along the kinetochore and chromosomal microtubules, which shorten as chromosome movement proceeds (**Figure 13.51**). Movement of chromosomes along the spindle microtubules in the minus-end direction (toward the centrosomes) is driven by chromosome-associated kinesins

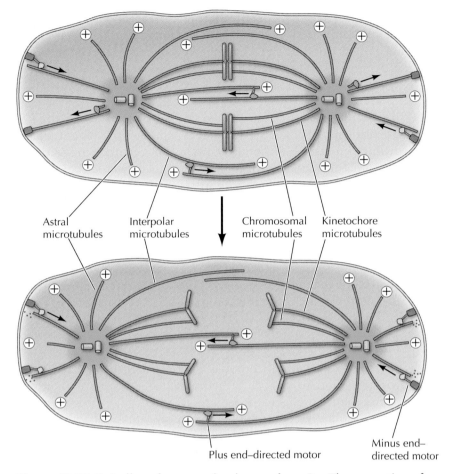

Astral
microtubules

Interpolar
microtubules

Chromosomal
microtubules

Kinetochore
microtubules

Plus end–directed motor

Minus end–
directed motor

Figure 13.52 Spindle pole separation in anaphase B   The separation of spindle poles results from two types of movement. First, overlapping interpolar microtubules slide past each other to push the spindle poles apart, driven by the action of plus-end-directed motor proteins. Second, the spindle poles are pulled apart by the astral microtubules, driven by minus-end-directed motors anchored to the cell cortex.

bound to the plus ends of microtubules. These kinesins act as microtubule-depolymerizing enzymes that disassemble and shorten both the kinetochore and chromosomal microtubules as the chromosomes move to opposite poles of the spindle.

Anaphase B refers to the separation of the spindle poles themselves (**Figure 13.52**). Spindle-pole separation is accompanied by elongation of the interpolar microtubules and is similar to the initial separation of duplicated centrosomes to form the spindle poles at the beginning of mitosis (see Figure 13.50). During anaphase B the overlapping interpolar microtubules slide against one another, pushing the spindle poles apart. This type of movement results from the action of plus-end–directed kinesins, which cross-link interpolar microtubules and move them toward the plus end of their overlapping microtubule, away from the opposite spindle pole. In addition, the spindle poles are pulled apart by the astral microtubules. This type of movement involves the action of cytoplasmic dynein anchored to the cell cortex. The movement of this anchored dynein along astral microtubules in the minus-end direction pulls the spindle poles apart toward the periphery of the cell. There is a simultaneous shrinkage of the astral microtubules by depolymerases, leading to separation of the spindle poles and their movement to the periphery of the cell, prior to the formation of two daughter cells at the end of mitosis.

Intermediate Filaments

Intermediate filaments have diameters of 10–12 nm, which is intermediate between the diameters of the two other principal elements of the cytoskeleton—actin filaments (about 7 nm) and microtubules (about 25 nm). In contrast to actin filaments and microtubules, the intermediate filaments are not directly involved in cell movements. Instead, they appear to both play a structural role by providing mechanical strength to cells and tissues and provide a scaffold for the localization of cellular processes, including intracellular signaling. Some bacteria contain proteins related to eukaryote intermediate filaments, which assemble into filaments underlying the bacterial plasma membrane. However, while cytoplasmic intermediate filaments are characteristic of most multicellular animals, they are not found in yeast, plants, and some insects.

Intermediate filament proteins

Whereas actin filaments and microtubules are polymers of a single type of protein (actin or tubulin, respectively), intermediate filaments are composed of several families of proteins with a common structural organization that are expressed in different types of cells. More than 70 different intermediate filament proteins have been identified and classified into five groups based on similarities between their amino acid sequences (Table 13.3). Types I and II correspond to two groups of **keratins**, each consisting of about 15 different proteins, which are expressed in epithelial cells. Each type of epithelial cell synthesizes at least one type I (acidic) and one type II (neutral/basic) keratin, which copolymerize to form filaments. Some type I and type II keratins (called hard keratins) are used for production of structures such as hair, nails, and horns. The other type I and type II keratins (soft keratins) are abundant

Table 13.3 Intermediate Filament Proteins

Type	Protein	Size (kd)	Site of expression
I	Acidic keratins	40–60	Epithelial cells
II	Neutral or basic keratins	50–70	Epithelial cells
III	Vimentin	54	Fibroblasts, white blood cells, and other cell types
	Desmin	53	Muscle cells
	Glial fibrillary acidic protein	51	Glial cells
	Peripherin	57	Peripheral neurons
IV	Neurofilament proteins		
	NF-L	67	Neurons
	NF-M	150	Neurons
	NF-H	200	Neurons
	α-internexin	66	Neurons
	Nestin	200	Stem cells
V	Nuclear lamins	60–75	Nuclear lamina of all cell types

in the cytoplasm of epithelial cells, with different keratins being expressed in various differentiated cell types.

The type III intermediate filament proteins include **vimentin**, which is found in several different kinds of cells, including fibroblasts, smooth muscle cells, and white blood cells. Unlike actin filaments, vimentin forms a network extending out from the nucleus toward the cell periphery. Another type III protein, **desmin**, is specifically expressed in muscle cells where it connects the Z discs of individual contractile elements. A third type III intermediate filament protein is specifically expressed in glial cells (which support neurons), and a fourth is expressed in neurons of the peripheral nervous system.

The type IV intermediate filament proteins include the three **neurofilament (NF) proteins** (designated NF-L, NF-M, and NF-H for light, medium, and heavy, respectively). These proteins, in association with α-internexin, form the major intermediate filaments of many types of mature neurons. They are particularly abundant in the axons of motor neurons and are thought to play a critical role in supporting these long, thin processes, which can extend more than a meter in length. Another type IV protein, nestin, is expressed during embryonic development in several types of stem cells. Nestins differ from other intermediate filaments in that they only polymerize if other intermediate filaments are present in the cell.

The type V intermediate filament proteins are the nuclear lamins. Rather than being part of the cytoskeleton, the lamins are components of the nucleus, where they form an orthogonal meshwork underlying the nuclear membrane (see Figure 10.5).

Assembly of intermediate filaments

Despite considerable diversity in size and amino acid sequence, intermediate filament proteins share a common structural organization (**Figure 13.53**). All have a central α-helical rod domain of approximately 310 amino acids (350 amino acids in the nuclear lamins). This central rod domain is flanked by amino- and carboxy-terminal domains, which vary among the different intermediate filament proteins in size, sequence, and secondary structure. The α-helical rod domain plays a central role in filament assembly, while the variable head and tail domains presumably determine the specific functions of the different intermediate filament proteins.

The first stage of filament assembly is the formation of dimers in which the central rod domains of

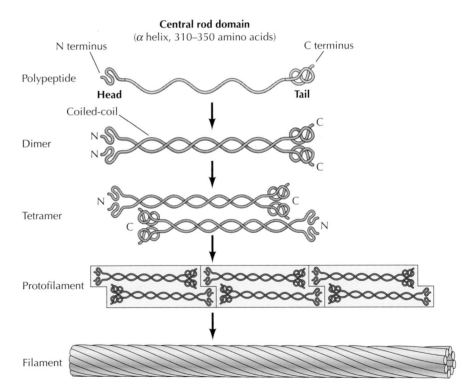

Figure 13.53 Structure and assembly of intermediate filaments   Intermediate filament proteins contain central α-helical rod domains and N-terminal head and C-terminal tail domains. The central rod domains of two polypeptides wind around each other in a coiled-coil structure to form dimers. Dimers then associate in a staggered antiparallel fashion to form tetramers. Tetramers associate end-to-end to form protofilaments and laterally to form filaments. Each filament contains approximately eight protofilaments wound around each other in a ropelike structure.

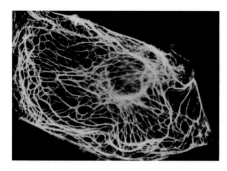

Figure 13.54 Intracellular organization of keratin filaments Micrograph of an epithelial cell stained with fluorescent antibodies to keratin (green). The keratin filaments extend from a ring surrounding the nucleus to the plasma membrane.

two polypeptide chains are wound around each other in a coiled-coil structure similar to that formed by myosin II heavy chains. The dimers of cytoskeletal intermediate filaments then associate in a staggered antiparallel fashion to form tetramers, which can assemble end-to-end to form protofilaments. A common step is the interaction of approximately eight protofilaments wound around each other in a ropelike structure. Because they are assembled from antiparallel tetramers, both ends of intermediate filaments are equivalent. Consequently, in contrast to actin filaments and microtubules, intermediate filaments are apolar—they do not have distinct ends, such as the barbed and pointed ends of actin filaments or the plus and minus ends of microtubules.

Filament assembly requires interactions between specific types of intermediate filament proteins. For example, keratin filaments are always assembled from heterodimers containing one type I and one type II polypeptide. In contrast, the type III proteins can assemble into filaments containing only a single polypeptide (e.g., vimentin) or consisting of two different type III proteins (e.g., vimentin plus desmin). The type III proteins do not, however, form copolymers with the keratins. Intermediate filaments are generally more stable than actin filaments or microtubules and do not exhibit the dynamic behaviors controlled by nucleotide binding that are associated with these other elements of the cytoskeleton. However, intermediate filaments are dynamic within the cell. Most intermediate filament proteins can be modified by phosphorylation, which regulates their assembly and disassembly. One example is phosphorylation of the nuclear lamins (see Chapter 17), which results in disassembly of the nuclear lamina and breakdown of the nuclear envelope during mitosis. Cytoplasmic intermediate filaments, such as vimentin, are also phosphorylated, which can lead to their disassembly and reorganization in dividing or migrating cells.

Intracellular organization of intermediate filaments

Intermediate filaments form an elaborate network in the cytoplasm of most cells, extending from a ring surrounding the nucleus to the plasma membrane (**Figure 13.54**). Both keratin and vimentin filaments attach to the nuclear envelope, apparently serving to position and anchor the nucleus within the cell. In addition, intermediate filaments can associate not only with the plasma membrane but also with the other elements of the cytoskeleton, actin filaments and microtubules. Intermediate filaments thus provide a scaffold that integrates the components of the cytoskeleton and organizes the internal structure of the cell.

The keratin filaments of epithelial cells are tightly anchored to the plasma membrane at two areas of specialized cell contacts: **desmosomes** and **hemidesmosomes** (**Figure 13.55**). Desmosomes are junctions between adjacent cells at which cell-cell contacts are mediated by transmembrane proteins related to the cadherins. On their cytoplasmic side, desmosomes are associated with a characteristic dense plaque of intracellular proteins to which keratin filaments are attached. These attachments are mediated by desmoplakin, a member of a family of proteins called **plakins** that bind intermediate filaments and link them to other cellular structures. Hemidesmosomes are morphologically similar junctions between epithelial cells and underlying connective tissue at which keratin filaments are linked by different members of the plakin family (e.g., plectin) to integrins. Desmosomes and hemidesmosomes thus anchor intermediate filaments to regions of cell-cell and cell-substratum contact, respectively, similar to the attachment of the actin cytoskeleton to the plasma membrane at adherens junctions and focal adhesions. It is important to note that the keratin filaments anchored to both sides of desmosomes serve as a

(A)

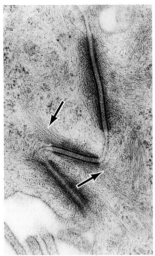

Figure 13.55 Attachment of intermediate filaments to desmosomes and hemidesmosomes (A) Electron micrograph illustrating keratin filaments (arrows) attached to the dense plaques of intracellular protein on both sides of a desmosome. (B) Schematic of a desmosome. The desmosomal cadherins (desmoglein and desmocollin) link adjoining cells to intermediate filaments through plakoglobin, plakophilin, and desmoplakin. (C) Schematic of a hemidesmosome. The integrin $\alpha_6\beta_4$ links the extracellular matrix to intermediate filaments through plectin. BP180 and BP230 regulate hemidesmosome assembly and stability.

(B) Desmosome

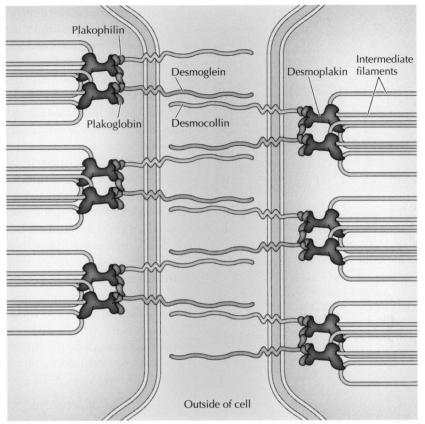

(C) Hemidesmosome

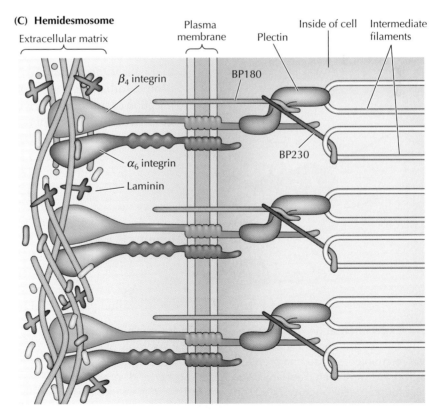

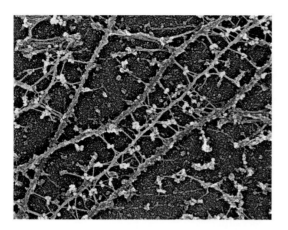

Figure 13.56 Electron micrograph of plectin bridges between intermediate filaments and microtubules Micrograph of a fibroblast stained with antibody against plectin. The micrograph has been artificially colored to show plectin (green), antibodies against plectin (yellow), intermediate filaments (blue), and microtubules (red). (Courtesy of Tatyana Svitkina and Gary Borisy, University of Wisconsin, Madison.)

mechanical link between adjacent cells in an epithelial layer, thereby providing mechanical stability to the entire tissue.

In addition to linking intermediate filaments to cell junctions, some plakins link intermediate filaments to other elements of the cytoskeleton. Plectin, for example, binds actin filaments and microtubules in addition to intermediate filaments, so it can provide bridges between these cytoskeletal components (**Figure 13.56**). These bridges to intermediate filaments are thought to brace and stabilize actin filaments and microtubules, thereby increasing the mechanical stability of the cell.

Two types of intermediate filaments—desmin and the neurofilaments—play specialized roles in muscle and nerve cells, respectively. Desmin connects the individual actin-myosin assemblies of muscle cells both to one another and to the plasma membrane, thereby linking the actions of individual contractile elements. Several mutations in the desmin gene result in multiple muscle defects, including early onset cardiomyopathy. Neurofilaments are the major intermediate filaments in most mature neurons. They are particularly abundant in the long axons of motor neurons where they appear to be anchored to actin filaments and microtubules by neuronal members of the plakin family. Neurofilaments are thought to play an important role in providing mechanical support and stabilizing other elements of the cytoskeleton in these long, thin extensions of nerve cells.

Functions of intermediate filaments: Keratins and diseases of the skin

Although intermediate filaments have long been thought to provide structural support to the cell, direct evidence for their function has only recently been obtained. Some cells in culture make no intermediate filament proteins, indicating that these proteins are not required for the growth of cells *in vitro*. Similarly, injection of cultured cells with antibody against vimentin disrupts intermediate filament networks without affecting cell growth or movement. Therefore, it is thought that the primary role of intermediate filaments is to strengthen the cytoskeleton of cells in the tissues of multicellular organisms, where they are subjected to a variety of mechanical stresses that do not affect cells in the isolated environment of a culture dish.

Experimental evidence for such an *in vivo* role of intermediate filaments was first provided in 1991 by studies in the laboratory of Elaine Fuchs. These investigators used transgenic mice to investigate the *in vivo* effects of expressing a keratin deletion mutant encoding a truncated polypeptide that disrupted the formation of normal keratin filaments (**Figure 13.57**). This mutant keratin gene was introduced into transgenic

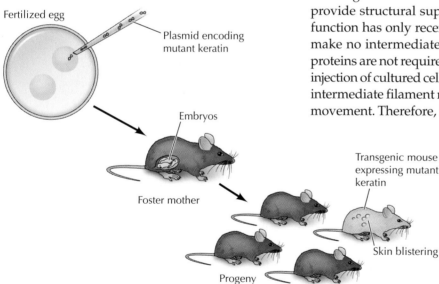

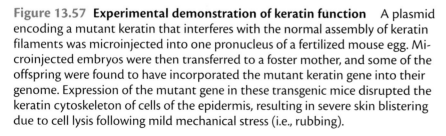

Figure 13.57 Experimental demonstration of keratin function A plasmid encoding a mutant keratin that interferes with the normal assembly of keratin filaments was microinjected into one pronucleus of a fertilized mouse egg. Microinjected embryos were then transferred to a foster mother, and some of the offspring were found to have incorporated the mutant keratin gene into their genome. Expression of the mutant gene in these transgenic mice disrupted the keratin cytoskeleton of cells of the epidermis, resulting in severe skin blistering due to cell lysis following mild mechanical stress (i.e., rubbing).

Key Experiment

Expression of Mutant Keratin Causes Abnormal Skin Development

Mutant Keratin Expression in Transgenic Mice Causes Marked Abnormalities Resembling a Human Genetic Skin Disease

Robert Vassar, Pierre A. Coulombe, Linda Degenstein, Kathryn Albers, and Elaine Fuchs

University of Chicago

Cell, 1991, Volume 64, pages 365–380

The Context

By 1991 intermediate filaments in epithelial cells were well-known and the developmental appearance of different forms of type I and type II keratins in the skin was being studied. What remained a mystery was the function of intermediate filaments. While all vertebrate cells contain intermediate filaments, cell lines that lack them survive in culture and continue to proliferate. Thus whatever function intermediate filaments had, they were important not for cells in culture but perhaps for cells within the tissues of multicellular organisms.

Elaine Fuchs and coworkers knew that during early development the epithelium of the skin expresses keratin 5 (type I) and keratin 14 (type II). Because keratins polymerize as heterodimers, it was thought that expression of an abnormal protein might interfere with the formation of normal intermediate filaments. Fuchs and her colleagues initially tested this possibility in cultured skin cells and demonstrated that expression of a truncated keratin 14 interfered with keratin filament formation. This suggested that similar expression of the mutant keratin in transgenic mice might cause a defect in the intermediate filament network in the skin cells of an embryo. If this occurred, it would provide a test of the role of intermediate filaments in an intact tissue.

In the experiments described here, Fuchs and her colleagues demonstrated that expression of a mutant keratin in transgenic mice not only disrupted the intermediate filament network of skin cells but also led to severe defects in the organization and tissue stability of the skin. These experiments thus provided the first demonstration of a physiological role for intermediate filaments.

The Experiments

For their earlier experiments in cultured cells, Fuchs and colleagues had constructed a mutant keratin 14 gene in which a truncated protein missing 30% of the central α-helical domain and all of the carboxy-terminal tail was expressed from the normal keratin promoter. To investigate the role of keratin 14 in early mouse development, they introduced the plasmid encoding this mutant keratin 14 into fertilized mouse eggs, which were transferred to foster mothers and allowed to develop into offspring. All of the offspring were analyzed for keratin 14, and some were found to be transgenic and express the mutant keratin 14 protein.

Most of the transgenic animals died within 24 hours of birth. Those that survived longer showed severe skin abnormalities, including blisters due to epidermal cell lysis following mild mechanical trauma, such as rubbing of the skin. Analysis of stained tissue sections of the skin from transgenic animals demonstrated severe disorganization of the epidermis in the most affected animals (see figure) and patches of disorganized tissue in others. Patchy expression is characteristic of a mosaic animal where some tissue develops from normal embryo cells and some from cells carrying the transgene. By analyzing for the mutant protein, they found that the areas of disorganized tissue correlated with expression of the mutant keratin 14. In addition, there was a clear correlation between the amount of disorganized epidermis and susceptibility to skin damage and death during the trauma of birth.

Fuchs and her colleagues further noted that the pattern of tissue disorganization in the transgenic mice resembled that seen in a group of human skin diseases called epidermolysis bullosa simplex. Thus they compared sections of the transgenic mouse tissue with sections obtained from the skin of a human patient and found very

Elaine Fuchs

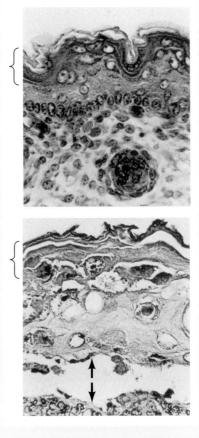

Skin from a normal and a transgenic mouse   (Top) Skin from a normal mouse showing the highly organized outer layers (brackets) with no intervening spaces between the intact underlying tissue. (Bottom) Skin from a transgenic mouse showing severe disruption of the outer layers, which contain spaces due to abrasion by mechanical trauma and are separated from the underlying tissue (arrows).

(Continued on next page)

Key Experiment

similar patterns of tissue disruption. From this, Fuchs and coworkers concluded that defects in keratins or intermediate filament-related proteins might be a cause of human genetic diseases of the skin.

The Impact

The skin abnormalities of these transgenic mice provided the first direct support for the presumed role of keratins in providing

mechanical strength to epithelial cells in tissues. It is now known that the intermediate filament cytoskeleton is critical for the structure of tissues such as skin, intestine, heart, and skeletal muscle that are subject to mechanical stress. In contrast, single-celled eukaryotes such as yeast and many small invertebrates survive perfectly well without intermediate filaments, often modifying the actin or tubulin cytoskeleton

to serve a structural role. The results of Fuchs and her colleagues also suggested a basis for several human diseases. In fact, there are now more than 17 different keratins known to be defective in human disease; these include keratin 5 and keratin 14, both of which were initially studied in the transgenic mice.

mice where it was expressed in basal cells of the epidermis and disrupted formation of a normal keratin cytoskeleton. This resulted in the development of severe skin abnormalities, including blisters due to epidermal cell lysis following mild mechanical trauma, such as rubbing of the skin. The skin abnormalities of these transgenic mice thus provided direct support for the presumed role of keratins in providing mechanical strength to epithelial cells in tissues. Subsequently, the same result has been shown in mice in which the gene for the same keratin was inactivated by homologous recombination.

These experiments also pointed to the molecular basis of a human genetic disease, epidermolysis bullosa simplex (EBS). Like the transgenic mice expressing mutant keratin genes, patients with this disease develop skin blisters resulting from cell lysis after minor trauma. This similarity prompted studies of the keratin genes in EBS patients, leading to the demonstration that EBS is caused by keratin gene mutations that interfere with the normal assembly of keratin filaments. Thus both experimental studies in transgenic mice and molecular analysis of a human genetic disease have demonstrated the role of keratins in allowing skin cells to withstand mechanical stress. Continuing studies have shown that mutations in other keratins are responsible for several other inherited skin diseases, which are similarly characterized by abnormal fragility of epidermal cells.

SUMMARY

KEY TERMS

Structure and Organization of Actin Filaments

- *Assembly and disassembly of actin filaments:* Actin filaments are formed by the head-to-tail polymerization of actin monomers into a helix. A variety of actin-binding proteins regulate the assembly and disassembly of actin filaments within the cell. See Animation 13.1.

actin, microfilament, actin-binding protein, globular [G] actin, filamentous [F] actin, treadmilling, cytochalasin, phalloidin, formin, Arp2/3 complex, profilin, tropomyosin, cofilin

- *Organization of actin filaments:* Actin filaments are cross-linked by actin-binding proteins to form bundles or three-dimensional networks.

actin bundle, actin network, actin-bundling protein, parallel bundle, fimbrin, contractile bundle, α-actinin, filamin

- *Association of actin filaments with the plasma membrane:* A network of actin filaments and other cytoskeletal proteins underlies the plasma membrane and determines cell shape. Actin bundles also attach to the plasma membrane and anchor the cell at regions of cell–cell and cell–substratum contact.

cell cortex, spectrin, ankyrin, dystrophin, integrin, focal adhesion, stress fiber, talin, vinculin, adherens junction, circumferential belt, cadherin, catenin

SUMMARY

- **Microvilli:** Actin filaments support permanent protrusions of the cell surface, such as microvilli and stereocilia.

- **Cell surface protrusions and cell movement:** Transient protrusions of the plasma membrane are responsible for phagocytosis and cell locomotion. Extension of cell protrusions is mediated by the growth of multiple actin filament branches at the leading edge of the cell. Cell locomotion is a complex process in which adhesions form at the ends of the new cell protrusions, the cell body is brought forward by the action of myosin II along stress fibers, and the trailing edge retracts into the cell body. See Videos 13.1–13.3.

Myosin Motors

- **Muscle contraction:** Studies of muscle established the role of myosin as a motor protein that uses the energy derived from ATP hydrolysis to generate force and movement. Muscle contraction results from the sliding of actin and myosin filaments past each other. ATP hydrolysis drives repeated cycles of interaction between myosin and actin during which conformational changes result in movement of the myosin head group along actin filaments. See Animation 13.2.

- **Contractile assemblies of actin and myosin in nonmuscle cells:** Assemblies of actin and myosin II are responsible for a variety of movements of nonmuscle cells, including cytokinesis.

- **Unconventional myosins:** Other types of myosin that do not function in contraction serve to transport membrane vesicles and organelles along actin filaments.

Microtubules

- **Structure and dynamic organization of microtubules:** Microtubules are formed by the reversible polymerization of tubulin. They display dynamic instability and undergo continual cycles of assembly and disassembly as a result of GTP hydrolysis following tubulin polymerization. See Animation 13.3.

- **Assembly of microtubules:** The microtubules in most animal cells extend outward from a microtubule-organizing center, or centrosome, located near the center of the cell. The centrosome usually contains a pair of centrioles surrounded by pericentriolar material. The growth of microtubules is initiated in the pericentriolar material, which then serves to anchor their minus ends.

- **Organization of microtubules within cells:** Selective stabilization of microtubules by post-translational modification of tubulin and binding of microtubule-associated proteins can determine cell shape and polarity, such as the extension of nerve cell axons and dendrites.

Microtubule Motors and Movement

- **Identification of microtubule motor proteins:** Two families of motor proteins, the kinesins and the dyneins, are responsible for movement along microtubules. Most kinesins move in the plus-end direction, whereas the dyneins and some members of the kinesin family move toward microtubule minus ends. See Animation 13.4.

KEY TERMS

microvillus, brush border, stereocilium, villin

pseudopodium, lamellipodium, filopodium, Rho, WASP

myosin, molecular motor, muscle fiber, myofibril, sarcomere, sliding filament model, myosin II, sarcoplasmic reticulum, troponin

cytokinesis, contractile ring, myosin light-chain kinase, calmodulin

unconventional myosin, myosin I

microtubule, tubulin, dynamic instability, colchicine, colcemid, vincristine, vinblastine, taxol, microtubule-associated protein (MAP)

centrosome, microtubule-organizing center, γ-tubulin ring complex, centriole, pericentriolar material

kinesin, dynein, axonemal dynein, video-enhanced microscopy, cytoplasmic dynein

SUMMARY	KEY TERMS

• **Cargo transport and intracellular organization:** Movement along microtubules transports macromolecules, membrane vesicles, and organelles through the cytoplasm, as well as positioning cytoplasmic organelles within the cell. See Videos 13.4 and 13.5.

• **Cilia and flagella:** Cilia and flagella are microtubule-based extensions of the plasma membrane. Their movements result from the sliding of microtubules driven by the action of dynein motors. See Videos 13.6–13.8.

cilium, flagellum, axoneme, nexin, basal body

• **Reorganization of microtubules during mitosis:** Microtubules reorganize at the beginning of mitosis to form the mitotic spindle, which is responsible for chromosome separation.

mitotic spindle, kinetochore microtubule, chromosomal microtubule, interpolar microtubule, astral microtubule

• **Chromosome movement:** The duplicated chromosomes align on the metaphase plate. During anaphase of mitosis, daughter chromosomes separate and move to opposite poles of the mitotic spindle.

anaphase A, anaphase B

Intermediate Filaments

• **Intermediate filament proteins:** Intermediate filaments are polymers of more than 70 different proteins that are expressed in various types of cells. They are not involved in cell movement but provide a scaffold for localization of cellular processes and provide mechanical support to cells and tissues.

intermediate filament, keratin, vimentin, desmin, neurofilament (NF) protein

• **Assembly of intermediate filaments:** Intermediate filaments are formed from dimers of two polypeptide chains wound around each other in a coiled-coil structure. The dimers then associate to form tetramers, which assemble into protofilaments. Intermediate filaments are formed from protofilaments wound around one another in a ropelike structure.

• **Intracellular organization of intermediate filaments:** Intermediate filaments form a network extending from a ring surrounding the nucleus to the plasma membrane of most cell types. In epithelial cells, intermediate filaments are anchored to the plasma membrane at regions of specialized cell contacts (desmosomes and hemidesmosomes). Intermediate filaments also play specialized roles in muscle and nerve cells.

desmosome, hemidesmosome, plakin

• **Functions of intermediate filaments: Keratins and diseases of the skin:** The importance of intermediate filaments in providing mechanical strength to cells in tissues has been demonstrated by the introduction of mutant keratin genes into transgenic mice. Similar keratin gene mutations are responsible for human skin diseases.

Refer To
The Cell
Companion Website
sites.sinauer.com/cooper7e
for quizzes, animations, videos, flashcards, and other study resources.

Questions

1. Why do actin filaments have a distinct polarity? Why is the polarity of actin filaments important in muscle contraction?

2. How would cytochalasin and phalloidin affect treadmilling of actin filaments?

3. How do cofilin, profilin, and the Arp2/3 complex regulate actin filament assembly and turnover?

4. Which bands or zones of a muscle sarcomere change length during contraction? Why doesn't the A band change length?

5. How does Ca^{2+} regulate the contraction of smooth muscle cells?

6. What key observation helped Vale and colleagues devise a strategy for the isolation of kinesin?

7. You are studying the transport of secretory vesicles containing insulin along microtubules in cultured pancreatic cells. How would treatment with colcemid affect the transport of these vesicles?

8. What is the cellular function of γ-tubulin?

9. How would expression of siRNAs targeted against vimentin affect the growth of fibroblasts in culture?

10. Why are intermediate filaments apolar, even though they are assembled from monomers that have distinct ends?

11. How would the removal of nexin affect the beating of cilia?

References and Further Reading   (Key review articles for each major section are highlighted in **bold**.)

Structure and Organization of Actin Filaments

Baines, A. J., H.-C. Lu and P. M. Bennett 2013. The protein 4.1 family: hub proteins in animals for organizing membrane proteins. *Biochim. Biophys. Acta* 1838: 605-619. [R]

Bravo-Cordero, J. J., M. A. O. Magalhaes, R. J. Eddy, L. Hodgson and J. Condeelis. 2013. Functions of cofilin in cell locomotion and invasion. *Nature Rev. Mol. Cell Biol.* 14: 405-415. [R]

Breitsprecher, D. and B. L. Goode. 2013. Formins at a glance. *J. Cell Sci.* 126: 1-7. [R]

Campellone, K. G. and M. D. Welch. 2010. A nucleator arms race: cellular control of actin assembly. *Nature Rev. Mol. Cell Biol.* 11: 237–251. [R]

Dominguez, R. and K. C. Holmes. 2011. Actin structure and function. *Ann. Rev. Biophys.* 40: 169–186. [R]

Fehon, R. G., A. I. McClatchey and A. Bretscher. 2010. Organizing the cell cortex: the role of ERM proteins. *Nature Rev. Mol. Cell Biol.* 11: 276–287. [R]

Kabsch, W., H. G. Mannherz, D. Suck, E. F. Pai and K. C. Holmes. 1990. Atomic structure of the actin: DNase I complex. *Nature* 347: 37–44. [P]

Ozyamak, E., J. M. Kollman and A. Komeili. 2013. Bacterial actins and their diversity. *Biochemistry* 52: 6928-6939. [R]

Pollard, T. D. and J. A. Cooper. 2009. Actin, a central player in cell shape and movement. *Science* 326: 1208–1212. [R]

Rahimov, F. and L. M. Kunkcl. 2013. Cellular and molecular mechanisms underlying muscular dystrophy. *J. Cell Biol.* 201: 499–510. [R]

Ridley, A. J. 2011. Life at the leading edge. *Cell* 145: 1012–1022. [R]

Rotty, J. D., C. Wu and J. E. Bear. 2013. New insights into the regulation and cellular functions of the ARP2/3 complex. *Nature Rev. Mol. Cell Biol.* 14: 7–12. [R]

Yonemura, S. 2011. Cadherin-actin interactions at adherens junctions. *Curr. Opin. Cell Biol.* 23:515–522. [R]

Zhou, A. X., J. H. Hartwig and L. M. Akyurek. 2010. Filamins in cell signaling, transcription and organ development. *Trends Cell Biol.* 20: 113–123. [R]

Myosin Motors

Finer, J. T., R. M. Simmons and J. A. Spudich. 1994. Single myosin molecule mechanics: Piconewton forces and nanometre steps. *Nature* 368: 113–119. [P]

Geeves, M. A. and K. C. Holmes. 1999. Structural mechanism of muscle contraction. *Ann. Rev. Biochem.* 68: 687–728. [R]

Greenberg, M. J. and E. M. Ostap. 2013. Regulation and control of myosin-I by the motor and light chain binding domains. *Trends Cell Biol.* 23: 81–89. [R]

Hammer, J. A. III and W. Wagner. 2013. Functions of class V myosins in neurons. *J. Biol. Chem.* 288: 28428–28434. [R]

Hartman, M. A., D. Finan, S. Sivaramakrishnan and J. A. Spudich. 2011. Principles of unconventional myosin function and targeting. *Ann. Rev. Cell Dev. Biol.* 27: 133–155. [R]

Huxley, A. F. and R. Niedergerke. 1954. Interference microscopy of living muscle fibres. *Nature* 173: 971–973. [P]

Huxley, H. E. 1969. The mechanism of muscle contraction. *Science* 164: 1356–1366. [R]

Huxley, H. E. and J. Hanson. 1954. Changes in the cross-striations of muscle contraction and their structural interpretation. *Nature* 173: 973–976. [P]

Kull, F. J. and S. A. Endow. 2013. Force generation by kinesin and myosin cytoskeletal motor proteins. *J. Cell Sci.* 126: 9–19. [R]

Rayment, I., H. M. Holden, M. Whittaker, C. B. Yohn, M. Lorenz, K. C. Kolmes and R. A. Milligan. 1993. Structure of the actin-myosin complex and its implications for muscle contraction. *Science* 261: 58–65. [P]

Rayment, I., W. R. Rypniewski, K. Schmidt-Base, R. Smith, D. R. Tomchick, M. M. Benning, D. A. Winkelmann, G. Wesenberg and H. M. Holden. 1993. Three-dimensional structure of myosin subfragment-1: A molecular motor. *Science* 261: 50–58. [P]

Sweeney, H. L. and A. Houdusse. 2010. Myosin VI rewrites the rules for myosin motors. *Cell* 141: 573–582. [R]

Microtubules

Akhmanova, A. and M. O. Steinmetz. 2008. Tracking the ends: a dynamic protein network controls the fate of microtubule tips. *Nature Rev. Mol. Cell Biol.* 9: 309–322. [R]

Al-Bassam, J. and F. Chang. 2011. Regulation of microtubule dynamics by TOG-domain proteins XMAP215/Dis1 and CLASP. *Trends Cell Biol.* 21: 604–614. [R]

Bornens, M. 2012. The centrosome in cells and organisms. *Science* 335: 422–426. [R]

Gonczy, P. 2012. Towards a molecular architecture of centriole assembly. *Nature Rev. Mol. Cell Biol.* 13: 425–435. [R]

Howard, J. and A. A. Hyman. 2009. Growth, fluctuation and switching at microtubule plus ends. *Nature Rev. Mol. Cell Biol.* 10: 569–574. [R]

Janke, C. and J. C. Bulinski. 2011. Post-translational regulation of the microtubule cytoskeleton: mechanisms and functions. *Nature Rev. Mol. Cell Biol.* 12: 773–785. [R]

Kueh, H. Y. and T. J. Mitchison. 2009. Structural plasticity in actin and tubulin polymer dynamics. *Science* 325: 960–963. [R]

Mennella, V., D. A. Agard, B. Huang and L. Pelletier. 2014. Amorphous no more: subdiffraction view of the pericentriolar material architecture. *Trends Cell Biol.* 24: 188–197. [R]

Mitchison, T. and M. Kirschner. 1984. Dynamic instability of microtubule growth. *Nature* 312: 237–242. [P]

Nogales, E., M. Whittaker, R. A. Milligan and K. H. Downing. 1999. High-resolution model of the microtubule. *Cell* 96: 79–88. [P]

Osborn, M. and K. Weber. 1976. Cytoplasmic microtubules in tissue culture cells appear to grow from an organizing structure towards the plasma membrane. *Proc. Natl. Acad. Sci. USA* 73: 867–871. [P]

Wasteneys, G. O. and J. C. Ambrose. 2009. Spatial organization of plant cortical microtubules: close encounters of the 2D kind. *Trends Cell Biol.* 19: 62–71. [R]

Microtubule Motors and Movement

Bettencourt-Dias, M., F. Hildebrandt, D. Pellman, G. Woods and S. A. Godinho. 2011. Centrosomes and cilia in human disease. *Trends Genet.* 27: 307–315. [R]

Brady, S. T. 1985. A novel brain ATPase with properties expected for the fast axonal motor. *Nature* 317: 73–75. [P]

Brady, S. T., R. J. Lasek and R. D. Allen. 1982. Fast axonal transport in extruded axoplasm from squid giant axon. *Science* 218: 1129–1131. [P]

Cross, R. A. and A. McAinsh. 2014. Prime movers: the mechanochemistry of mitotic kinesins. *Nature Rev. Mol. Cell Biol.* 15: 257–271. [R]

Gennerich, A. and R. D. Vale. 2009. Walking the walk: how kinesin and dynein coordinate their steps. *Curr. Opin. Cell Biol.* 21: 59–67. [R]

Gibbons, I. R. and A. Rowe. 1965. Dynein: A protein with adenosine triphosphatase activity from cilia. *Science* 149: 424–426. [P]

Hammond, J. W., D. Cai and K. J. Verhey. 2008. Tubulin modifications and their cellular functions. *Curr. Opin. Cell Biol.* 20: 71–76. [R]

Hirokawa, N., Y. Noda, Y. Tanaka and S. Niwa. 2009. Kinesin superfamily motor proteins and intracellular transport. *Nature Rev. Mol. Cell Biol.* 10: 682–696. [R]

Kardon, J. R. and R. D. Vale. 2009. Regulators of the cytoplasmic dynein motor. *Nature Rev. Mol. Cell Biol.* 10: 854–864. [R]

Lasek, R. J. and S. T. Brady. 1985. Attachment of transported vesicles to microtubules in axoplasm is facilitated by AMP-PNP. *Nature* 316: 645–647. [P]

Lüders, J. and T. Stearns. 2007. Microtubule-organizing centres: a re-evaluation. *Nature Rev. Mol. Cell Biol.* 8: 161–167. [R]

Marx, A., A. Hoenger and E. Mandelkow. 2009. Structures of kinesin motor proteins. *Cell Motil. Cytoskeleton* 66: 958–966. [R]

Nigg, E. A. and T. Stearns. 2011. The centrosome cycle: Centriole biogenesis, duplication and inherent asymmetries. *Nature Cell Biol.* 13: 1154–1160. [R]

Pedersen, L. B., I. R. Veland, J. M. Schroder and S. T. Christensen. 2008. Assembly of primary cilia. *Dev. Dyn.* 237: 1993–2006. [R]

Roberts, A. J., T. Kon, P. J. Knight, K. Sutoh and S. A. Burgess. 2013. Functions and mechanics of dynein motor proteins. *Nature Rev. Mol. Cell Biol.* 14: 713–726. [R]

Salmon, E. D. 1995. VE-DIC light microscopy and the discovery of kinesin. *Trends Cell Biol.* 5: 154–157. [R]

Scholey, J. M., I. Brust-Mascher and A. Mogilner. 2003. Cell division. *Nature* 422: 746–752. [R]

Vale, R. D. 2003. The molecular motor toolbox for intracellular transport. *Cell* 112: 467–480. [R]

Vale, R. D., T. S. Reese and M. P. Sheetz. 1985. Identification of a novel force-generating protein, kinesin, involved in microtubule-based motility. *Cell* 42: 39–50. [P]

Intermediate Filaments

Coulombe, P. A., M. E. Hutton, A. Letai, A. Hebert, A. S. Paller and E. Fuchs. 1991. Point mutations in human keratin 14 genes of epidermolysis bullosa simplex patients: Genetic and functional analyses. *Cell* 66: 1301–1311. [P]

Epstein, E. H., Jr., J. M. Bonifas, and A. L. Rothman. 1991. Epidermolysis bullosa simplex: Evidence in two families for keratin gene abnormalities. *Science* 254: 1202–1205. [P]

Fuchs, E. and D. W. Cleveland. 1998. A structural scaffolding of intermediate filaments in health and disease. *Science* 279: 514–519. [R]

Godsel, L. M., R. P. Hobbs and K. J. Green. 2008. Intermediate filament assembly: dynamics to disease. *Trends Cell Biol.* 18: 28–37. [R]

Goldfarb, L. G. and M. C. Dalakas. 2009. Tragedy in a heartbeat: malfunctioning desmin causes skeletal and cardiac muscle disease. *J. Clin. Invest* 119: 1806–1813. [R]

Graumann, P. L. 2007. Cytoskeletal elements in bacteria. *Ann. Rev. Microbiol.* 61: 589–618. [R]

Herrmann, H., H. Bar, L. Kreplak, S. V. Strelkov and U. Aebi. 2007. Intermediate filaments: from cell architecture to nanomechanics. *Nature Rev. Mol. Cell Biol.* 8: 562–573. [R]

Kim, S. and P. A. Coulombe. 2007. Intermediate filament scaffolds fulfill mechanical, organizational, and signaling functions in the cytoplasm. *Genes Dev.* 21: 1581–1597. [R]

Snider, N. T. and M. B. Omary. 2014. Post-translational modifications of intermediate filament proteins: mechanisms and functions. *Nature Rev. Mol. Cell Biol.* 15: 163–177. [R]

Suozzi, K. C., X. Wu and E. Fuchs. 2012. Spectraplakins: Master orchestrators of cytoskeletal dynamics. *J. Cell Biol.* 197: 465–475. [R]

Wiche, G. and L. Winter. 2011. Plectin isoforms as organizers of intermediate filament cytoarchitecture. *Bioarchitecture.* 1: 14–20. [R]

The Plasma Membrane

All cells—both prokaryotic and eukaryotic—are surrounded by a plasma membrane, which defines the boundary of the cell and separates its internal contents from the environment. By serving as a selective barrier to the passage of molecules, the plasma membrane determines the composition of the cytoplasm. This ultimately defines the very identity of the cell, so the plasma membrane is one of the most fundamental structures of cellular evolution. Indeed, as discussed in Chapter 1, the first cell is thought to have arisen by the enclosure of self-replicating RNA in a membrane of phospholipids.

The basic structure of the plasma membrane of present-day cells is the lipid bilayer, which is impermeable to most water-soluble molecules. The passage of ions and most organic molecules across the plasma membrane is therefore mediated by proteins, which are responsible for the selective traffic of molecules into and out of the cell. Other proteins of the plasma membrane control the interactions between cells of multicellular organisms and serve as sensors through which the cell receives signals from its environment. The plasma membrane thus plays a dual role: It both isolates the cytoplasm and mediates interactions between the cell and its environment.

Structure of the Plasma Membrane

Like all other cellular membranes, the plasma membrane consists of both lipids and proteins. The fundamental structure of the membrane is the lipid bilayer, which forms a stable barrier between two aqueous compartments. In the case of the plasma membrane, these compartments are the inside and the outside of the cell. Proteins embedded within the lipid bilayer carry out the specific functions of the plasma membrane, including selective transport of molecules and cell–cell recognition.

The lipid bilayer

The plasma membrane is the most thoroughly studied of all cell membranes, and it is largely through investigations of the plasma membrane that our current concepts of membrane structure have evolved. The plasma membranes of mammalian red blood cells (erythrocytes) have been particularly useful as a model for studies of membrane structure. Mammalian red blood cells do not contain nuclei or internal membranes, so they represent a source from which pure plasma membranes can be easily isolated for biochemical analysis. Indeed, studies of the red blood cell plasma membrane provided the first evidence that biological membranes consist of lipid bilayers. In 1925, two Dutch scientists (Edwin Gorter and F. Grendel) extracted the

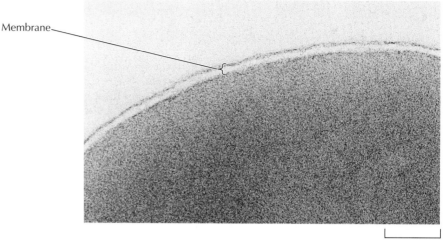

Membrane

20 nm

Figure 14.1 Bilayer structure of the plasma membrane Electron micrograph of a human red blood cell. Note the "railroad track" appearance of the plasma membrane. (Courtesy of J. David Robertson, Duke University Medical Center.)

membrane lipids from a known number of red blood cells corresponding to a known surface area of plasma membrane. They then determined the surface area occupied by a monolayer of the extracted lipid spread out at an air–water interface. The surface area of the lipid monolayer turned out to be twice that occupied by the erythrocyte plasma membranes, leading to the conclusion that the membranes consisted of lipid bilayers rather than monolayers.

The bilayer structure of the erythrocyte plasma membrane is clearly evident in high-magnification electron micrographs (**Figure 14.1**). The plasma membrane appears as two dense lines separated by an intervening space—a morphology frequently referred to as a "railroad track" appearance. This image results from the binding of the electron-dense heavy metals used as stains in transmission electron microscopy (see Chapter 1) to the polar head groups of the phospholipids, which therefore appear as dark lines. These dense lines are separated by the lightly stained interior portion of the membrane, which contains the hydrophobic fatty acid chains.

As discussed in Chapter 2, the membranes of animal cells contain five major phospholipids (**phosphatidylcholine**, **phosphatidylethanolamine**, **phosphatidylserine**, **phosphatidylinositol**, and **sphingomyelin**), which together account for about 50% of the lipids in plasma membranes (**Table 14.1**). These phospholipids are asymmetrically distributed between the two halves of the membrane bilayer (**Figure 14.2**). The outer leaflet of the plasma membrane consists mainly of phosphatidylcholine and sphingomyelin, whereas phosphatidylethanolamine and phosphatidylserine are the predominant phospholipids of the inner leaflet. As discussed in Chapter 11, sphingomyelin is synthesized on the lumenal surface of the Golgi membrane, which determines its location on the extracellular side (outer leaflet) of the plasma membrane. The other phospholipids, however, are synthesized in the endoplasmic reticulum where they are distributed symmetrically to the cytosolic and lumenal sides of the membrane (see Figure 11.22). The asymmetric orientations of these

Table 14.1	Lipid Composition of the Plasma Membrane
Lipid	**Mole percent**
Phosphatidylcholine	20
Phosphatidylethanolamine	11
Phosphatidylserine	4
Phosphatidylinositol	2
Cholesterol	49
Sphingomyelin	13
Glycolipids	1

Source: Data are for mammalian cells from G. van Meer, D. R. Voelker and G. W. Feigenson. 2008. *Nature Rev. Mol. Cell Biol.* 9: 112.

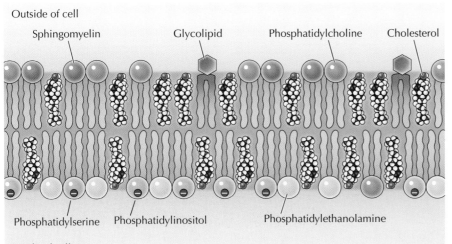

Figure 14.2 Lipid components of the plasma membrane The outer leaflet consists predominantly of phosphatidylcholine, sphingomyelin, and glycolipids, whereas the inner leaflet contains phosphatidylethanolamine, phosphatidylserine, and phosphatidylinositol. Cholesterol is distributed in both leaflets. The net negative charge of the head groups of phosphatidylserine and phosphatidylinositol is indicated. The structures of phospholipids, glycolipids, and cholesterol are shown in Figures 2.7, 2.8, and 2.9, respectively.

phospholipids in the plasma membrane are established by transporters that selectively translocate phosphatidylethanolamine and phosphatidylserine to the inner leaflet and phosphatidylcholine to the outer leaflet. The head group of phosphatidylserine is negatively charged, so its predominance in the inner leaflet results in a net negative charge on the cytosolic face of the plasma membrane. It is also noteworthy that the localization of phosphatidylserine to the inner leaflet plays an important role in programmed cell death, as will be discussed in Chapter 18.

In addition to the phospholipids, the plasma membranes of animal cells contain **glycolipids** and **cholesterol**. The glycolipids are found exclusively in the outer leaflet of the plasma membrane, with their carbohydrate portions exposed on the cell surface, as determined by their synthesis in the Golgi. They are relatively minor membrane components, constituting only a small fraction of the lipids of most plasma membranes. Cholesterol, on the other hand, is a major plasma membrane constituent of animal cells, being present in about the same molar amount as the phospholipids. Cholesterol has a rigid ring structure (see Figure 2.9) that inserts into a bilayer of phospholipids with its polar hydroxyl group close to the phospholipid head groups (see Figure 14.2). Because it has a higher affinity for sphingomyelin than for the other phospholipids, it is more concentrated in the outer leaflet. Bacteria and plant cells do not contain cholesterol, but instead contain related compounds (sterols or sterol-like lipids) that fulfill a similar function.

Two general features of lipid bilayers are critical to membrane function. First, the structure of phospholipids is responsible for the basic function of membranes as barriers between two aqueous compartments. Because the interior of the lipid bilayer is occupied by hydrophobic fatty acid chains, the membrane is impermeable to water-soluble molecules, including ions

FYI

During programmed cell death, phosphatidylserine translocates to the outer leaflet of the plasma membrane, where it signals the elimination of dying cells by phagocytosis (see Chapter 18).

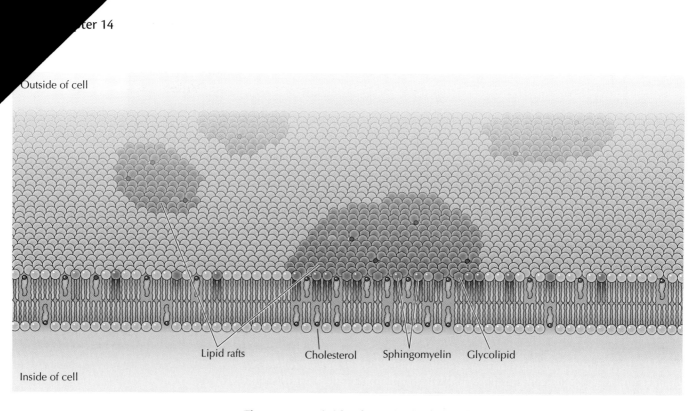

Outside of cell

Lipid rafts   Cholesterol   Sphingomyelin   Glycolipid

Inside of cell

Figure 14.3 Lipid rafts   Lipid rafts are formed by the interactions of sphingomyelin, glycolipids, and cholesterol.

and most biological molecules. Second, bilayers of the naturally occurring phospholipids are viscous fluids, not solids. The fatty acids of most natural phospholipids have one or more double bonds, which introduce kinks into the hydrocarbon chains and make them difficult to pack together. The long hydrocarbon chains of the fatty acids therefore move freely in the interior of the membrane, so the membrane itself is soft and flexible. In addition, both lipids and proteins are able to diffuse laterally within the membrane—a property that is critical for many membrane functions.

As discussed in Chapter 2, cholesterol has important effects on membrane fluidity. In addition, cholesterol is involved in the formation of membrane domains that mediate the clustering of a variety of proteins in the plasma membrane. Rather than diffusing freely in the membrane, cholesterol and the sphingolipids (sphingomyelin and glycolipids) tend to cluster in small semisolid patches known as **lipid rafts** (Figure 14.3). As discussed later in this chapter, lipid rafts play an important role in compartmentalizing the plasma membrane into discrete functional domains.

Plasma membrane proteins

While lipids are the fundamental structural elements of membranes, proteins are responsible for carrying out specific membrane functions. Most plasma membranes consist of approximately 50% lipid and 50% protein by weight, with the carbohydrate portions of glycolipids and glycoproteins constituting 5 to 10% of the membrane mass. Since proteins are much larger than lipids, this percentage corresponds to about one protein molecule per every 50 to 100 molecules of lipid.

In 1972, Jonathan Singer and Garth Nicolson proposed the **fluid mosaic model** of membrane structure, in which membranes are viewed as

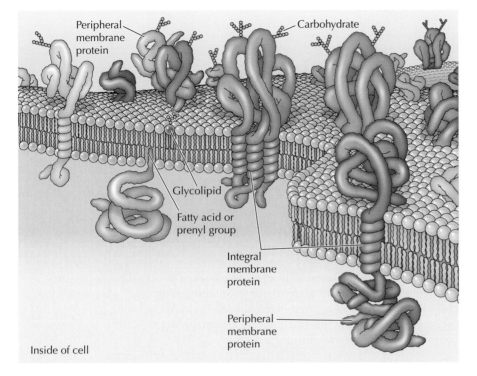

Figure 14.4 Fluid mosaic model of the plasma membrane Integral membrane proteins are inserted into the lipid bilayer, whereas peripheral proteins are bound to the membrane indirectly by protein–protein interactions. Most integral membrane proteins are transmembrane proteins with portions exposed on both sides of the lipid bilayer. The extracellular portions of these proteins are usually glycosylated, as are the peripheral membrane proteins bound to the external face of the membrane.

two-dimensional fluids with proteins inserted into lipid bilayers (**Figure 14.4**). One of the central features of the fluid mosaic model is that, because they are inserted into a fluid lipid bilayer, both proteins and lipids are able to diffuse laterally through the membrane. This lateral movement of membrane proteins was first shown directly by Larry Frye and Michael Edidin in 1970. Frye and Edidin fused human and mouse cells in culture to produce human–mouse cell hybrids (**Figure 14.5**). They then analyzed the distribution of proteins in the membranes of these hybrid cells using antibodies that specifically recognize proteins of human and mouse origin. These antibodies were labeled with different fluorescent dyes, so the human and mouse proteins could be distinguished by fluorescence microscopy. Within 40 minutes after fusion, the mouse and human proteins became intermixed over the surface of hybrid cells, indicating that they moved freely through the plasma membrane. The fluid mosaic model remains the basic paradigm for the organization of all biological membranes, although we now recognize

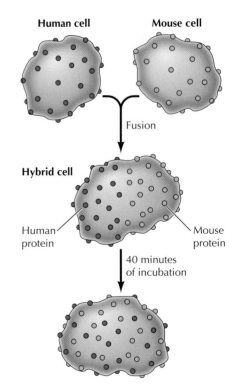

Figure 14.5 Mobility of membrane proteins Human and mouse cells were fused to produce hybrid cells. The distribution of cell surface proteins was then analyzed using anti-human and anti-mouse antibodies labeled with different fluorescent dyes (red and green, respectively). The human and mouse proteins had intermingled over the cell surface following 40 minutes of incubation.

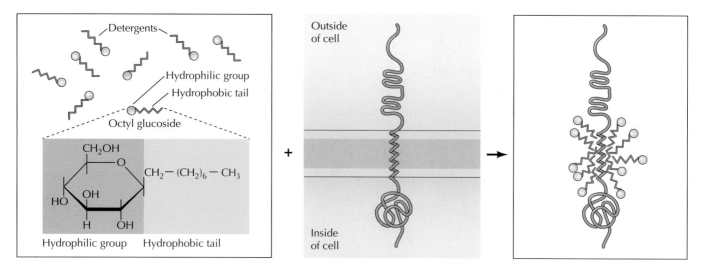

Figure 14.6 Solubilization of integral membrane proteins by detergents Detergents (e.g., octyl glucoside) are amphipathic molecules containing hydrophilic head groups and hydrophobic tails. The hydrophobic tails bind to the hydrophobic regions of integral membrane proteins, forming detergent-protein complexes that are soluble in aqueous solution.

that the plasma membrane is composed of several specialized domains, which are discussed in the next section.

Singer and Nicolson distinguished two classes of membrane-associated proteins, which they called **peripheral** and **integral membrane proteins**. Peripheral membrane proteins were operationally defined as proteins that dissociate from the membrane following treatments with polar reagents, such as solutions of extreme pH or high salt concentration that do not disrupt the phospholipid bilayer. Once dissociated from the membrane, peripheral membrane proteins are soluble in aqueous buffers. These proteins are not inserted into the hydrophobic interior of the lipid bilayer. Instead, they are indirectly associated with membranes through protein–protein interactions. These interactions frequently involve ionic bonds, which are disrupted by extreme pH or high salt.

In contrast to the peripheral membrane proteins, integral membrane proteins can be released only by treatments that disrupt the phospholipid bilayer. Portions of these integral membrane proteins are inserted into the lipid bilayer, so they can be dissociated only by reagents that disrupt hydrophobic interactions. The most commonly used reagents for solubilization of integral membrane proteins are detergents, which are small amphipathic molecules containing both hydrophobic and hydrophilic groups (**Figure 14.6**). The hydrophobic portions of detergents displace the membrane lipids and bind to the hydrophobic portions of integral membrane proteins. Because the other end of the detergent molecule is hydrophilic, the detergent–protein complexes are soluble in aqueous solutions.

Many integral proteins are **transmembrane proteins**, which span the lipid bilayer with portions exposed on both sides of the membrane. These proteins can be visualized in electron micrographs of plasma membranes prepared by the freeze-fracture technique (see Figure 1.37). In these specimens, the

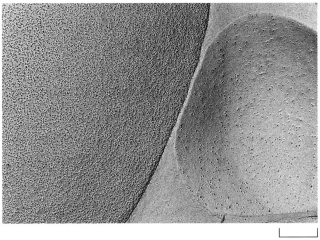

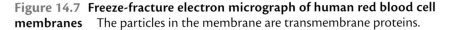

0.2 μm

Figure 14.7 Freeze-fracture electron micrograph of human red blood cell membranes   The particles in the membrane are transmembrane proteins.

membrane is split and separated into its two leaflets. Transmembrane proteins are then apparent as particles on the internal faces of the membrane (Figure 14.7).

The membrane-spanning portions of transmembrane proteins are usually α helices of 20 to 25 hydrophobic amino acids that are inserted into the membrane of the endoplasmic reticulum during synthesis of the polypeptide chain (see Figures 11.12–11.15). These proteins are then transported in membrane vesicles from the endoplasmic reticulum to the Golgi apparatus and from there to the plasma membrane. Carbohydrate groups are added to the polypeptide chains in both the endoplasmic reticulum and Golgi apparatus, so most transmembrane proteins of the plasma membrane are glycoproteins with their oligosaccharides exposed on the surface of the cell.

Studies of red blood cells have provided good examples of both peripheral and integral proteins associated with the plasma membrane. The membranes of human erythrocytes contain about a dozen major proteins, which were originally identified by gel electrophoresis of membrane preparations. Most of these are peripheral membrane proteins that have been identified as components of the cortical cytoskeleton, which underlies the plasma membrane and determines cell shape (see Chapter 13). For example, the most abundant peripheral membrane protein of red blood cells is spectrin, which is the major cytoskeletal protein of erythrocytes (see Figure 13.13). Other peripheral membrane proteins of red blood cells include actin, ankyrin, and band 4.1. Ankyrin serves as the principal link between the plasma membrane and the cytoskeleton by binding to both spectrin and the integral membrane protein band 3. An additional link between the membrane and the cytoskeleton is provided by band 4.1, which binds to the junctions of spectrin and actin, as well as to glycophorin (the other major integral membrane protein of erythrocytes).

The two major integral membrane proteins of red blood cells—glycophorin and band 3—provide well-studied examples of transmembrane protein

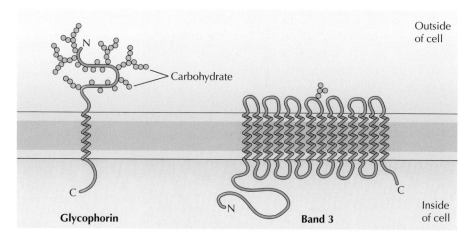

Figure 14.8 Integral membrane proteins of red blood cells Glycophorin is a dimer, with each polypeptide chain containing 131 amino acids and a single transmembrane α helix. It is heavily glycosylated, with oligosaccharides attached to 16 sites on the extracellular portion of the polypeptide chain. Band 3 (929 amino acids) has multiple transmembrane α helices and crosses the membrane 14 times.

structure (**Figure 14.8**). Glycophorin is a small glycoprotein of 131 amino acids, with a molecular weight of about 30,000, that crosses the membrane with a single membrane-spanning α helix of 23 amino acids. It is heavily glycosylated, containing about 50% carbohydrate, with the glycosylated amino-terminal portion of the polypeptide chain exposed on the cell surface. The other major transmembrane protein of red blood cells, band 3, is the anion transporter responsible for the passage of bicarbonate (HCO_3^-) and chloride (Cl^-) ions across the red blood cell membrane. The band 3 polypeptide chain is 929 amino acids with 14 membrane-spanning α-helical regions. Within the membrane, dimers of band 3 form globular structures containing internal channels through which ions are able to travel across the lipid bilayer.

Because of their amphipathic character and the need for detergents for their isolation, transmembrane proteins have been difficult to crystallize, as required for three-dimensional structural analysis by X-ray diffraction. However, progress has been made in this area and crystal structures of more than 400 membrane proteins have now been determined. The first transmembrane protein to be analyzed by X-ray crystallography was the photosynthetic reaction center of the bacterium *Rhodopseudomonas viridis*, whose structure was reported in 1985 (**Figure 14.9**). The reaction center contains three transmembrane proteins designated L, M, and H (light, medium, and heavy), according to their apparent sizes indicated by gel electrophoresis. The L and M subunits each have five membrane-spanning α helices. The H subunit has only a single transmembrane α helix, with the bulk of the polypeptide chain

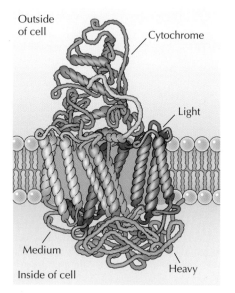

Figure 14.9 A bacterial photosynthetic reaction center The reaction center of *Rhodopseudomonas viridis* consists of three transmembrane proteins designated light (red), medium (yellow), and heavy (green). The L and M subunits each have five transmembrane α helices, whereas the H subunit has only one. The fourth subunit of the reaction center is a cytochrome (blue), which is a peripheral membrane protein.

on the cytosolic side of the membrane. The fourth subunit of the reaction center is a cytochrome, which is a peripheral membrane protein bound to the complex by protein–protein interactions.

In contrast to transmembrane proteins, a variety of proteins (some of which behave as integral membrane proteins) are anchored in the plasma membrane by covalently attached lipids or glycolipids (**Figure 14.10**). Members of one class of these proteins are inserted into the outer leaflet of the plasma

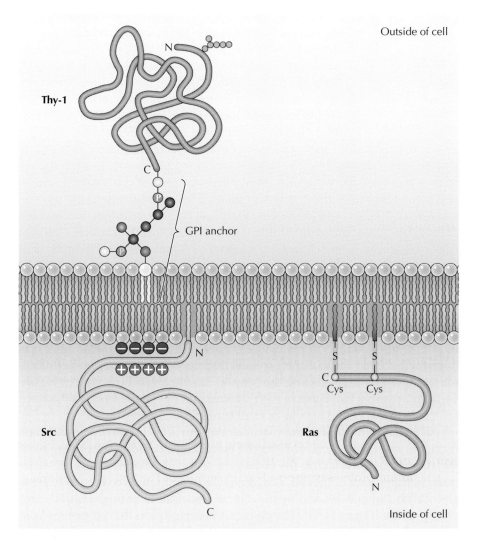

Figure 14.10 Examples of proteins anchored in the plasma membrane by lipids and glycolipids   Some proteins (e.g., the lymphocyte protein Thy-1) are anchored in the outer leaflet of the plasma membrane by GPI anchors added to their C terminus in the endoplasmic reticulum. These proteins are glycosylated and exposed on the cell surface. Other proteins are anchored in the inner leaflet of the plasma membrane following their translation on free cytosolic ribosomes. The Src protein is anchored by a myristoyl group attached to its N terminus. A positively charged region of Src also plays a role in membrane association, perhaps by interacting with the negatively charged head groups of phosphatidylserine. The Ras protein illustrated is anchored by a prenyl group attached to the side chain of a C-terminal cysteine and by a palmitoyl group attached to a cysteine located five amino acids upstream. The structures of these lipid and glycolipid groups are illustrated in Figures 9.34–9.37.

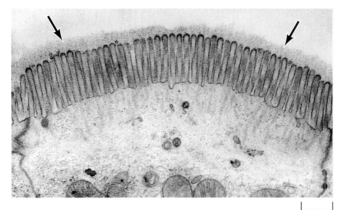

Figure 14.11 The glycocalyx An electron micrograph of intestinal epithelium illustrating the glycocalyx (arrows).

1 μm

membrane by **glycosylphosphatidylinositol (GPI) anchors**. GPI anchors are added to certain proteins that have been transferred into the endoplasmic reticulum and are anchored in the membrane by a C-terminal transmembrane region (see Figure 11.18). The transmembrane region is cleaved as the GPI anchor is added, so these proteins remain attached to the membrane only by the glycolipid. Since the polypeptide chains of GPI-anchored proteins are transferred into the endoplasmic reticulum, they are glycosylated and exposed on the surface of the cell following transport to the plasma membrane.

Other proteins are anchored in the inner leaflet of the plasma membrane by covalently attached lipids. Rather than being processed through the secretory pathway, these proteins are synthesized on free cytosolic ribosomes and then modified by the addition of lipids. These modifications include the addition of myristic acid (a 14-carbon fatty acid) to the amino terminus of the polypeptide chain, the addition of prenyl groups (15 or 20 carbons) to the side chains of carboxy-terminal cysteine residues, and the addition of palmitic acid (16 carbons) to the side chains of cysteine residues (see Figures 9.35–9.37). In some cases, these proteins (many of which behave as peripheral membrane proteins) are targeted to the plasma membrane by positively charged regions of the polypeptide chain as well as by the attached lipids. These positively charged protein domains interact with the negatively charged head groups of phosphatidylserine on the cytosolic face of the plasma membrane. It is noteworthy that many of the proteins anchored in the inner leaflet of the plasma membrane (including the Src and Ras proteins illustrated in Figure 14.10) play important roles in the transmission of signals from cell surface receptors to intracellular targets, as discussed in Chapter 16.

As already discussed, the extracellular portions of plasma membrane proteins can be heavily glycosylated. Likewise, the carbohydrate portions of glycolipids are exposed on the outer face of the plasma membrane. Consequently, the surface of the cell is covered by a carbohydrate coat known as the **glycocalyx**, which is formed by the oligosaccharides of both glycolipids and glycoproteins (**Figure 14.11**). The glycocalyx both protects the cell surface from ionic and mechanical stress and forms a barrier to invading microorganisms. In addition, the oligosaccharides of the glycocalyx participate in a variety of cell–cell interactions (discussed in Chapter 15).

Plasma membrane domains

The fluid mosaic model of the plasma membrane, proposed in 1972, has continued to provide the basis for our understanding of plasma membrane structure. However, a number of advances over the last 40 years have changed our understanding of the organization of the plasma membrane. In particular, the original fluid mosaic model viewed membranes as viscous fluids, with proteins free to move through the lipid bilayer. We now know that the mobility of many plasma membrane proteins is restricted and that, rather than being uniform, membranes are composed of distinct domains that play important structural and functional roles.

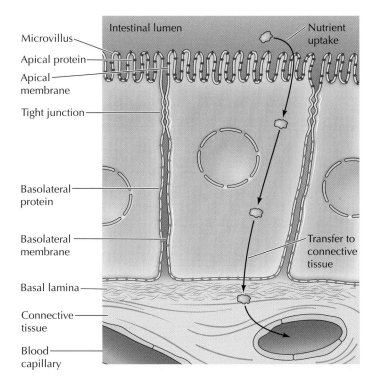

Microvillus

Apical protein

Apical membrane

Tight junction

Basolateral protein

Basolateral membrane

Basal lamina

Connective tissue

Blood capillary

Intestinal lumen

Nutrient uptake

Transfer to connective tissue

Figure 14.12 A polarized intestinal epithelial cell The apical surface of the cell contains microvilli and is specialized for absorption of nutrients from the intestinal lumen. The basolateral surface is specialized for the transfer of absorbed nutrients to the underlying connective tissue, which contains blood capillaries. Tight junctions separate the apical and basolateral domains of the plasma membrane and prevent membrane proteins from diffusing from one domain to the other.

Many cells are polarized, with different domains of their plasma membranes responsible for performing distinct activities. A prime example is provided by epithelial cells, which form sheets that cover the surface of the body and line the internal organs. Mammals have more than 100 different types of epithelial cells, each specialized for a specific function, such as protection (the skin), the uptake of nutrients (intestinal epithelial cells), and secretion (mammary epithelial cells). In order to carry out nutrient uptake and secretion in specific directions across the epithelial sheet, epithelial cells are polarized when they are organized into tissues, with different parts of the cell responsible for performing distinct functions. The plasma membranes of epithelial cells are thus divided into distinct **apical** and **basolateral domains** (Figure 14.12). For example, epithelial cells of the small intestine function to absorb nutrients from the digestive tract. The apical surface of these cells, which faces the intestinal lumen, is therefore covered by microvilli, which increase its surface area and facilitate nutrient absorption. The basolateral surface, which faces underlying connective tissue and the blood supply, is specialized to mediate the transfer of absorbed nutrients into the circulation. In order to maintain these distinct functions, plasma membrane proteins must be restricted to the appropriate domains of the cell surface. At least part of the mechanism by which this occurs involves the formation of tight junctions (discussed in Chapter 15) between adjacent cells of the epithelium. These junctions not only seal the space between cells but also serve as barriers to the movement of membrane lipids and proteins. As a result, proteins are able to diffuse within either the apical or basolateral domains of the plasma membrane but are not able to cross from one domain to the other.

The mobility of many plasma membrane proteins is restricted as a result of association with the cytoskeleton or with specialized lipid domains. These

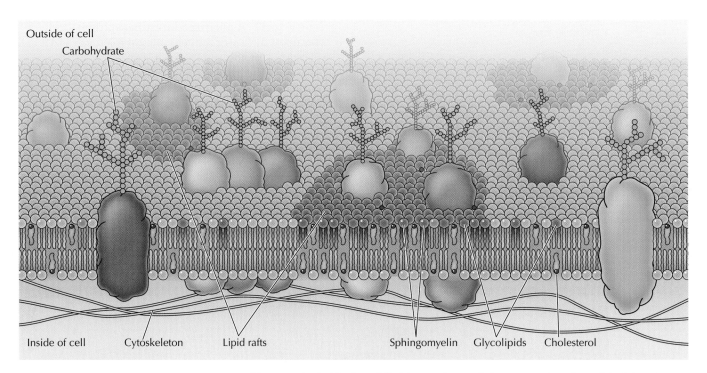

Outside of cell

Carbohydrate

Inside of cell   Cytoskeleton   Lipid rafts   Sphingomyelin   Glycolipids   Cholesterol

Figure 14.13 Updated fluid mosaic model An updated model of the plasma membrane, illustrating lipid rafts and the interactions of transmembrane proteins with the underlying actin cytoskeleton. After G. L. Nicolson, 2014. *Biochim. Biophys. Acta* 1838: 1451.

Video 14.2

sites.sinauer.com/cooper7e/v14.2
Lipid Raft Turnover Lipid rafts are small, transient, semisolid patches in the plasma membrane composed of cholesterol and sphingolipids.

---

limitations on the mobility of membrane proteins have led to the development of new models of plasma membrane structure, in which associations with the cytoskeleton and domains of specialized lipid-protein interactions are incorporated (**Figure 14.13**). For example, the major transmembrane proteins of red blood cells (glycophorin and band 3, see Figure 14.8) are both associated with the spectrin-actin cortical cytoskeleton that underlies the plasma membrane (see Figure 13.13). Consequently, they are not free to move in the lipid bilayer. Transmembrane proteins that are anchored to the cytoskeleton are not only restricted in their own mobility: They may also act as barriers that limit the mobility of other membrane proteins. Interactions with proteins on the surface of adjacent cells or with the extracellular matrix also restrict the mobility of some transmembrane proteins.

The free diffusion of some membrane proteins is restricted by their association with specialized lipid domains. As noted earlier, sphingomyelin, glycolipids, and cholesterol form clustered domains known as **lipid rafts** (see Figure 14.3). Lipid rafts are small (10–200 nm) transient structures that serve as platforms in which specific proteins can be concentrated to facilitate their interactions (see Figure 14.13). They are enriched in GPI-anchored proteins and in transmembrane proteins involved in a variety of functions, including cell signaling, cell movement, and endocytosis. Lipid rafts have been difficult to study because their small size is below the 200 nm limit of resolution of conventional light microscopy (discussed in Chapter1). However, recently developed methods of super-resolution microscopy, which provide resolution

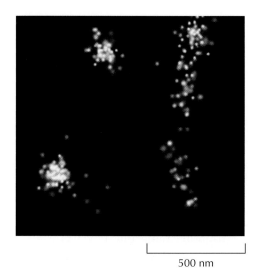

500 nm

Figure 14.14 Super-resolution microscopy of lipid rafts   Lipid rafts visualized with a fluorescent probe for cholesterol. (From H. Mizuno et al., 2011, *Chem. Sci.* 2: 1548.)

in the 10-100 nm range (see Figure 1.39), have enabled the visualization of lipid rafts in living cells (Figure 14.14). Lipid rafts also interact with the cytoskeleton, which can stabilize rafts and further divide the plasma membrane into specialized domains.

Caveolae (Latin for "little caves") are a subset of lipid rafts that were initially identified by electron microscopy as 60–80 nm invaginations of the plasma membrane (Figure 14.15). They are formed by oligomers of the membrane protein caveolin, which interacts with the cytoplasmic protein cavin. Caveolin also interacts with cholesterol and a high content of membrane cholesterol is required for the formation of caveolae. The number of

(A)   (B)

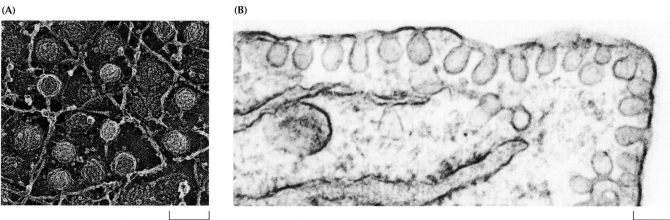

0.1 μm   0.2 μm

Figure 14.15 Caveolae   Electron micrographs of caveolae. (A, courtesy of John E. Heuser, Washington University School of Medicine; B, courtesy of R. G. W. Anderson, University of Texas Southwestern Medical School, Dallas.)

caveolae in the plasma membrane of different types of cells is highly variable: some cells lack detectable caveolae, whereas caveolae are abundant in other cell types, representing half of the plasma membrane in endothelial cells and adipocytes. They have been implicated in a variety of cellular activities including endocytosis, cell signaling, regulation of lipid transport, and protection of the plasma membrane against mechanical stress. These and other potential roles of caveolae are not yet fully understood and remain active areas of investigation.

Transport of Small Molecules

The internal composition of the cell is maintained because the plasma membrane is selectively permeable to small molecules. As discussed in Chapter 2, only gases (such as O_2 and CO_2), hydrophobic molecules (such as steroid hormones), and small polar but uncharged molecules (such as H_2O and ethanol) are able to passively diffuse across the plasma membrane (see Figure 2.36). Larger uncharged polar molecules such as glucose are unable to cross the plasma membrane by passive diffusion, as are charged molecules of any size (including amino acids and small ions such as H^+, Na^+, K^+, Ca^{2+} and Cl^-). Instead, the passage of these molecules across the membrane requires the activity of specific transport and channel proteins, which therefore control the traffic of most molecules into and out of the cell.

Facilitated diffusion and carrier proteins

Facilitated diffusion involves the movement of molecules in the direction determined by their relative concentrations inside and outside of the cell. No external source of energy is provided, so molecules travel across the membrane in the direction determined by their concentration gradients and in the case of charged molecules, by the electrical potential across the membrane. However, facilitated diffusion differs from passive diffusion in that the transported molecules do not dissolve in the phospholipid bilayer. Instead, their passage is mediated by proteins that enable the transported molecules to cross the membrane without directly interacting with its hydrophobic interior. Facilitated diffusion therefore allows polar and charged molecules, such as carbohydrates, amino acids, nucleosides, and ions, to cross the plasma membrane.

Two classes of proteins that mediate facilitated diffusion have generally been distinguished: carrier proteins and channel proteins. **Carrier proteins** bind specific molecules to be transported on one side of the membrane. They then undergo conformational changes that allow the molecule to pass through the membrane and be released on the other side. In contrast, **channel proteins** form open pores through the membrane, allowing the free diffusion of any molecule of the appropriate size and charge.

Carrier proteins are responsible for the facilitated diffusion of sugars, amino acids, and nucleosides across the plasma membranes of cells. The uptake of glucose, which serves as a primary source of metabolic energy, is one of the most important transport functions of the plasma membrane, and the glucose transporter provides a well-studied example of a carrier protein. Glucose transporters are a family of 14 proteins with overlapping specificities for glucose, fructose, and related molecules. They have 12 α-helical transmembrane segments—a structure typical of many carrier proteins. These transmembrane α helices contain predominantly hydrophobic amino acids, but several also contain polar amino acid residues that form the glucose-binding site in the interior of the protein. The glucose transporters function by alternating between two conformational states (**Figure 14.16**). In the first conformation, a glucose-binding site faces the outside of the cell. The binding of glucose to this exterior site induces a conformational change in the transporter such that the glucose-binding site now faces the interior of the cell. Glucose can then be released into the cytosol, followed by the return of the transporter to its original conformation.

Most cells, including erythrocytes, are exposed to extracellular glucose concentrations that are higher than those inside the cell, so facilitated diffusion results in the net inward transport of glucose. Once glucose is taken up by these cells it is rapidly metabolized so intracellular glucose concentrations remain low and glucose continues to be transported into the cell from the extracellular fluids. Because the conformational changes of glucose transporters

(A)

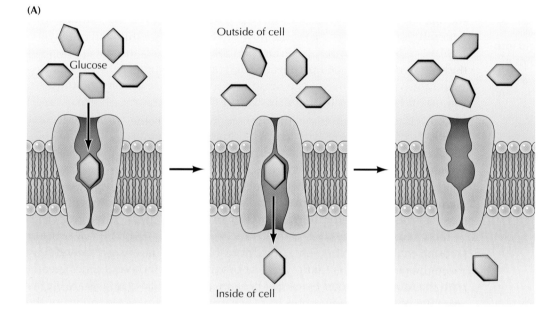

(B)

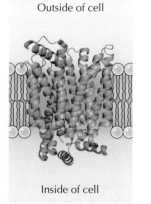

Figure 14.16 Facilitated diffusion of glucose   (A) The glucose transporter alternates between two conformations in which a glucose-binding site is alternately exposed on the outside and the inside of the cell. In the first conformation shown (left panel) glucose binds to a site exposed on the outside of the plasma membrane. The transporter then undergoes a conformational change such that the glucose-binding site faces the inside of the cell and glucose is released into the cytosol (middle panel). The transporter then returns to its original conformation (right panel). (B) Structure of a glucose transporter with the glucose-binding site facing the outside of the cell (left) and inside of the cell (right). (B, after D. Deng et al., 2015. *Nature.* doi: 10.1038/nature14655.)

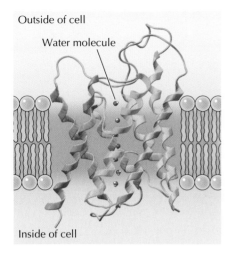

Outside of cell

Water molecule

Inside of cell

Figure 14.17 Structure of an aquaporin Aquaporins have six transmembrane α-helices (blue) that form a narrow aqueous channel through the membrane. Two loop domains form half helices (green) that fold into the channel. The channel contains a narrow pore that restricts passage to water molecules (red) in single file and excludes charged ions.

are reversible, however, glucose can be transported in the opposite direction simply by reversing the steps in Figure 14.16. Such reverse flow occurs, for example, in liver cells in which glucose is synthesized and released into the circulation. Some glucose transporters are regulated by interactions with other molecules. One of these, GLUT4, is the subject of intense study because it is the glucose transporter involved in metabolic syndrome and diabetes.

Ion channels

In contrast to carrier proteins, channel proteins form open pores in the membrane, allowing small molecules of the appropriate size and charge to pass freely through the lipid bilayer. One group of channel proteins discussed earlier is the porins, which permit the free passage of ions and small polar molecules through the outer membranes of bacteria, mitochondria, and chloroplasts (see Chapter 12). Gap junctions, which are discussed in Chapter 15, contain channel proteins that permit the passage of molecules between connected cells. The plasma membranes of a variety of different animal and plant cells also contain water channel proteins (**aquaporins**) through which water molecules are able to cross the membrane much more rapidly than they can diffuse through the phospholipid bilayer (**Figure 14.17**). Aquaporins generally function to increase the flow of water across epithelial cell layers. In plant cells, aquaporins play important roles in water transport up the stem and in the regulation of transpiration in leaves. In animal cells, aquaporins maintain water balance in the brain and extrude sweat from the skin. An important feature of all aquaporins is that they are impermeable to charged ions. Therefore, aquaporins allow the free passage of water without affecting the electrochemical gradient across the plasma membrane.

The best-characterized channel proteins, however, are the **ion channels**, which mediate the passage of ions across plasma membranes. Although ion channels are present in the membranes of all cells, they have been especially well studied in nerve and muscle cells, where their regulated opening and closing is responsible for the transmission of electric signals. Three properties of ion channels are central to their function (**Figure 14.18**). First, transport through channels is extremely rapid. More than a million ions per second flow through open channels—a flow rate approximately a thousand times greater than the rate of transport by carrier proteins. Second, ion channels are highly selective because narrow pores in the channel restrict passage to ions of the

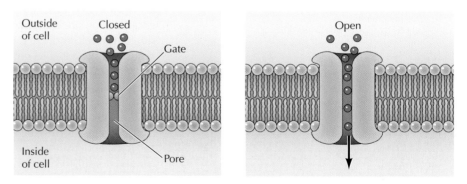

Figure 14.18 Model of an ion channel In the closed conformation, the flow of ions is blocked by a gate. Opening of the gate allows ions to flow rapidly through the channel. The channel contains a narrow pore that restricts passage to ions of the appropriate size and charge.

appropriate size and charge. Thus specific channel proteins allow the passage of Na^+, K^+, Ca^{2+}, and Cl^- across the membrane. Third, most ion channels are not permanently open. Instead, the opening of ion channels is regulated by "gates" that transiently open in response to specific stimuli. Some channels (called **ligand-gated channels**) open in response to the binding of neurotransmitters or other signaling molecules; others (**voltage-gated channels**) open in response to changes in electric potential across the plasma membrane.

The fundamental role of ion channels in the transmission of electric impulses was elucidated through a series of elegant experiments reported by Alan Hodgkin and Andrew Huxley in 1952. These investigators used the giant nerve cells of the squid as a model. The axons of these giant neurons have a diameter of about 1 mm, making it possible to insert electrodes and measure the changes in membrane potential that take place during the transmission of nerve impulses. Using this approach, Hodgkin and Huxley demonstrated that these changes in membrane potential result from the regulated opening and closing of Na^+ and K^+ channels in the plasma membrane.

The flow of ions through membrane channels is dependent on the establishment of ion gradients across the plasma membrane. All cells, including nerve and muscle, contain ion pumps (discussed in the next section) that use energy derived from ATP hydrolysis to actively transport ions across the plasma membrane. As a result, the ionic composition of the cytoplasm is substantially different from that of extracellular fluids (Table 14.2). For example, Na^+ is actively pumped out of cells while K^+ is pumped in. Therefore, in the squid axon the concentration of Na^+ is about 10 times higher in extracellular fluids than inside the cell, whereas the concentration of K^+ is approximately 20 times higher in the cytosol than in the surrounding medium.

Because ions are electrically charged, their transport results in the establishment of an electric gradient across the plasma membrane. In resting squid axons there is an electric potential of about 60 mV across the plasma membrane, with the inside of the cell negative with respect to the outside (Figure 14.19).

Table 14.2   Intracellular and Extracellular Ion Concentrations

Ion	Concentration (mM)	
	Intracellular	**Extracellular**
Squid axon		
K^+	400	20
Na^+	50	450
Cl^-	40–150	560
Ca^{2+}	0.0001	10
Mammalian cell		
K^+	140	5
Na^+	5–15	145
Cl^-	4	110
Ca^{2+}	0.0001	2.5–5

FYI

The squid giant axon is approximately 100 times larger in diameter than typical mammalian axons. Its large diameter allows high-speed transduction of nerve impulses, which is vital for the rapid escape response of squids from predators.

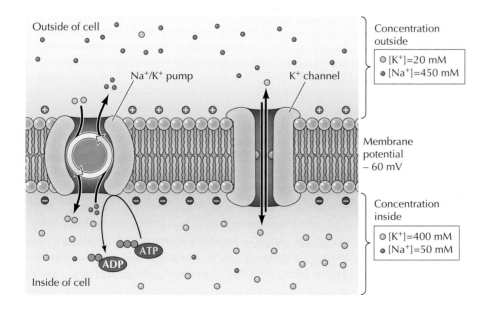

Figure 14.19 Ion gradients and resting membrane potential of the giant squid axon   Only the concentrations of Na^+ and K^+ are shown because these are the ions that function in the transmission of nerve impulses. Na^+ is pumped out of the cell while K^+ is pumped in, so the concentration of Na^+ is higher outside than inside of the axon, whereas the concentration of K^+ is higher inside than out. The resting membrane is more permeable to K^+ than to Na^+ or other ions because it contains open K^+ channels. The flow of K^+ through these channels makes the major contribution to the resting membrane potential of -60 mV, which is therefore close to the K^+ equilibrium potential.

This electric potential arises both from ion pumps and from the flow of ions through channels that are open in the resting cell plasma membrane. The plasma membrane of resting squid axons contains open K$^+$ channels, so it is more permeable to K$^+$ than to Na$^+$ or other ions. Consequently, the flow of K$^+$ makes the largest contribution to the resting membrane potential.

As discussed in Chapter 12, the flow of ions across a membrane is driven by both the concentration and voltage components of an electrochemical gradient. For example, the twentyfold higher concentration of K$^+$ inside the squid axon as compared with outside drives the flow of K$^+$ out of the cell. However, because K$^+$ is positively charged, this efflux of K$^+$ from the cell generates an electric potential across the membrane, with the inside of the cell becoming negatively charged. This membrane potential opposes the continuing flow of K$^+$ out of the cell, and the system approaches the equilibrium state in which the membrane potential balances the K$^+$ concentration gradient.

Quantitatively, the relationship between ion concentration and membrane potential is given by the **Nernst equation**:

$$V = \frac{RT}{zF} \ln \frac{C_o}{C_i}$$

where V is the equilibrium potential in volts, R is the gas constant, T is the absolute temperature, z is the charge of the ion, F is Faraday's constant, and C_o and C_i are the concentrations of the ion outside and inside of the cell, respectively. An equilibrium potential exists separately for each ion, and the membrane potential is determined by the flow of all the ions that cross the plasma membrane. However, because resting squid axons are more permeable to K$^+$ than to Na$^+$ or other ions (including Cl$^-$), the resting membrane potential (–60 mV) is close to the equilibrium potential determined by the intracellular and extracellular K$^+$ concentrations (–75 mV).

As nerve impulses (**action potentials**) travel along axons, the membrane depolarizes (**Figure 14.20**). The membrane potential changes from –60 mV to approximately +30 mV in less than a millisecond, after which it becomes negative again and returns to its resting value. These changes result from the rapid sequential opening and closing of voltage-gated Na$^+$ and K$^+$ channels. Relatively small initial changes in membrane potential (from –60 mV to about –40 mV) lead to the rapid but transient opening of Na$^+$ channels. This allows Na$^+$ to flow into the cell, driven by both its concentration gradient and the membrane potential. The sudden entry of Na$^+$ leads to a large change in membrane potential, which increases to nearly +30 mV, approaching the Na$^+$ equilibrium potential of approximately +50 mV. At this time, the Na$^+$ channels are inactivated and voltage-gated K$^+$ channels open, substantially increasing the permeability of the membrane to K$^+$. K$^+$ then flows rapidly out of the cell, driven by both the membrane potential and the K$^+$ concentration gradient, leading to a rapid decrease in membrane potential to about –75 mV (the K$^+$ equilibrium potential). This change in membrane potential inactivates the voltage-gated K$^+$ channels and the membrane potential returns to its resting level of –60 mV, determined by the flow of K$^+$ and other ions through the channels that remain open in unstimulated cells.

Depolarization of adjacent regions of the plasma membrane allows action potentials to travel down the length of nerve cell axons as electric signals, resulting in the rapid transmission of nerve impulses over long distances. For example, the axons of human motor neurons can be more than a meter

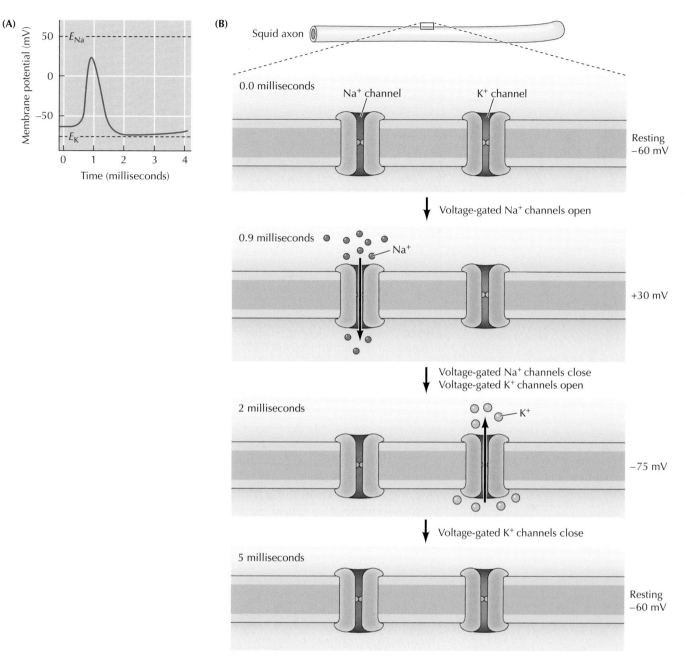

Figure 14.20 Membrane potential and ion channels during an action potential (A) Changes in membrane potential at one point on a squid giant axon following a stimulus. E_{Na} and E_K are the equilibrium potentials for Na$^+$ and K$^+$, respectively. (B) The membrane potential first increases as voltage-gated Na$^+$ channels open. The membrane potential then falls below its resting value as the Na$^+$ channels are inactivated and voltage-gated K$^+$ channels open. The voltage-gated K$^+$ channels are then inactivated, and the membrane potential returns to its resting value.

Animation 14.1

sites.sinauer.com/cooper7e/a14.1

A Chemical Synapse The arrival of a nerve impulse at the terminus of a neuron triggers synaptic vesicles to fuse with the plasma membrane, thereby releasing the neurotransmitter.

long. The arrival of action potentials at the terminus of most neurons then signals the release of neurotransmitters, such as acetylcholine, which carry signals between cells at a synapse (**Figure 14.21**).

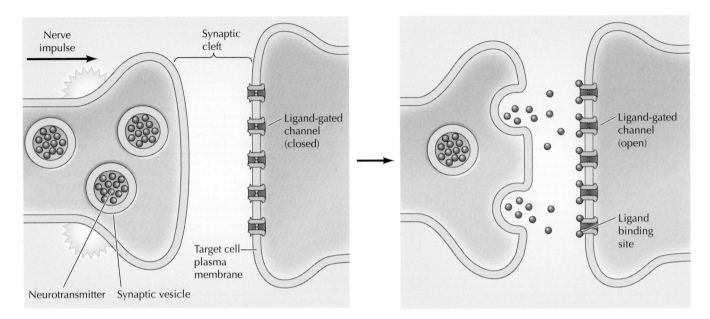

Figure 14.21 **Signaling by neurotransmitter release at a synapse** The arrival of a nerve impulse at the terminus of the neuron signals the fusion of synaptic vesicles with the plasma membrane, resulting in the release of neurotransmitters from the presynaptic cell into the synaptic cleft. The neurotransmitter binds to receptors and opens ligand-gated ion channels in the target cell plasma membrane.

FYI

Nicotinic acetylcholine receptors are so named because they respond to nicotine. Nicotine binds to the receptor and keeps the channel in its open state. Molecules that block the activity of nicotinic acetylcholine receptors are potent neurotoxins; these include several snake venom toxins and curare.

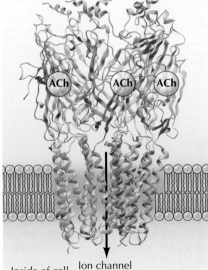

Figure 14.22 **Model of the nicotinic acetylcholine receptor** The mammalian receptor consists of five subunits (each with four transmembrane domains) arranged around a central pore. Three subunits are shown in the figure. The binding of acetylcholine to sites in the extracellular region of the receptor induces allosteric changes that open the ion channel gate. (After J.-P. Changeux and A. Taly, 2008. *Trends. Mol. Med.* 14: 93.)

Neurotransmitters released from presynaptic cells bind to receptors on the membranes of postsynaptic cells, where they act to open ligand-gated ion channels. One of the best-characterized of these channels is the nicotinic acetylcholine receptor of muscle cells. Binding of acetylcholine opens a channel that is permeable to both Na^+ and K^+. The combined flux of Na^+ and K^+ depolarizes the muscle cell membrane and triggers an action potential. The action potential then results in the opening of voltage-gated Ca^{2+} channels, leading to the increase in intracellular Ca^{2+} that signals contraction (see Figure 13.28).

This acetylcholine receptor, initially isolated from the electric organ of *Torpedo* rays in the 1970s, is the prototype of ligand-gated channels. All nicotinic acetylcholine receptors consist of five subunits arranged as a cylinder in the membrane (Figure 14.22). In its closed state, the channel pore is thought to be blocked by the side chains of hydrophobic amino acids. The binding of acetylcholine induces a conformational change in the receptor such that these hydrophobic side chains shift out of the channel, opening a pore that allows the passage of positively charged ions, including Na^+ and K^+. However, the channel remains impermeable to negatively charged ions, such as Cl^-, because it is lined by negatively charged amino acids.

A greater degree of ion selectivity is displayed by the voltage-gated Na^+ and K^+ channels. Na^+ channels are more than ten times more permeable

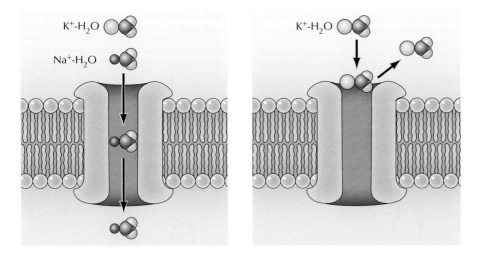

Figure 14.23 Ion selectivity of Na⁺ channels   A narrow pore permits the passage of Na⁺ bound to a single water molecule but interferes with the passage of K⁺ or larger ions.

to Na⁺ than to K⁺, whereas K⁺ channels are more than a thousand times more permeable to K⁺ than to Na⁺. The selectivity of the Na⁺ channel can be explained, at least in part, on the basis of a narrow pore that acts as a size filter. The ionic radius of Na⁺ (0.95 Å) is smaller than that of K⁺ (1.33 Å), and it is thought that the Na⁺ channel pore is narrow enough to interfere with the passage of K⁺ or larger ions (Figure 14.23).

K⁺ channels also have narrow pores, which prevent the passage of larger ions. However, since Na⁺ has a smaller ionic radius, this does not account for the selective permeability of these channels to K⁺. Selectivity of the K⁺ channel is based on a different mechanism, which was elucidated with the determination of the three-dimensional structure of a K⁺ channel by X-ray crystallography in 1998 (Figure 14.24). The channel pore contains a narrow

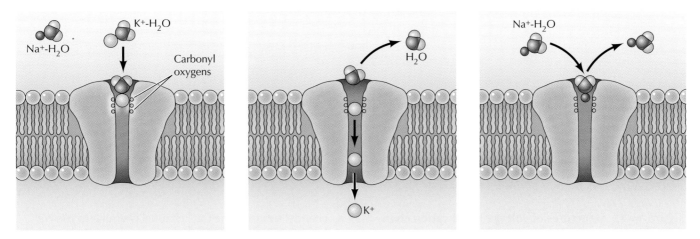

Figure 14.24 Selectivity of K⁺ channels   The K⁺ channel contains a narrow selectivity filter lined with carbonyl oxygen (C=O) atoms. The pore is just wide enough to allow the passage of dehydrated K⁺ from which all associated water molecules have been displaced as a result of interactions between K⁺ and these carbonyl oxygens. Na⁺ is too small to interact with the carbonyl oxygens of the selectivity filter, so it remains bound to water in a complex that is too large to pass through the channel pore.

selectivity filter that is lined with carbonyl oxygen (C=O) atoms from the polypeptide backbone. When a K⁺ ion enters the selectivity filter, interactions with these carbonyl oxygens displace the water molecules to which K⁺ is bound, allowing dehydrated K⁺ to pass through the pore. In contrast, a dehydrated Na⁺ is too small to interact with these carbonyl oxygens in the selectivity filter, which is held rigidly open. Consequently, Na⁺ remains bound to water molecules in a hydrated complex that is too large to pass through the channel.

Voltage-gated K⁺, Na⁺, and Ca²⁺ channels all belong to a large family of related proteins (**Figure 14.25**). For example, the genome sequence of *C. elegans*

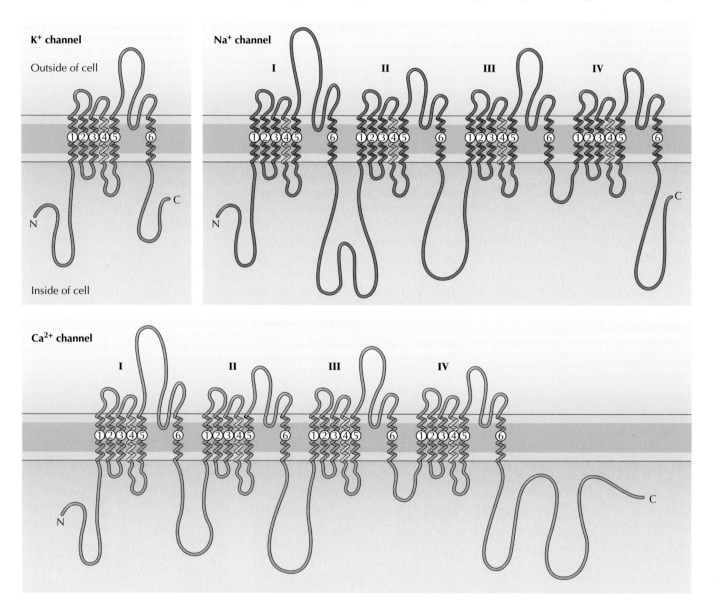

Figure 14.25 Structures of voltage-gated cation channels The K⁺, Na⁺, and Ca²⁺ channels belong to a family of related proteins. The K⁺ channel is formed from the association of four identical subunits, one of which is shown. The Na⁺ channel consists of a single polypeptide chain containing four repeated domains, each of which is similar to one K⁺ channel subunit. The Ca²⁺ channel is similar to the Na⁺ channel. Each subunit or domain contains six α helices. Helices 5 and 6 line the channel pore. Helix 4 contains multiple positively charged amino acids and acts as the voltage sensor that mediates channel opening in response to changes in membrane potential.

has revealed nearly 200 genes encoding ion channels, which presumably are needed to play diverse roles in cell signaling. K^+ channels consist of four identical subunits, each containing six transmembrane α helices. Na^+ and Ca^{2+} channels consist of a single polypeptide chain, but each polypeptide contains four repeated domains that correspond to the four K^+ channel subunits. The basic module consists of an ion pore domain between helices 5 and 6 and a voltage sensing module in helices 1-4. Helix 4 contains multiple positively charged amino acids and is thought to move in response to changes in the membrane potential, opening the ion pore.

A wide variety of ion channels (including Ca^{2+} and Cl^- channels) respond to different neurotransmitters or open and close with different kinetics following membrane depolarization. The concerted actions of these multiple channels are responsible for the complexities of signaling in the nervous system. Moreover, as discussed in Chapter 16, the roles of ion channels are not restricted to the electrically excitable cells of nerve and muscle; they also play critical roles in signaling in other cell types. The regulated opening and closing of ion channels thus provides cells with a sensitive and versatile mechanism for responding to a variety of environmental stimuli.

Active transport driven by ATP hydrolysis

The net flow of molecules by facilitated diffusion, through either carrier proteins or channel proteins, is always energetically downhill in the direction determined by electrochemical gradients across the membrane. In many cases, however, the cell must transport molecules against their concentration gradients. In **active transport**, energy provided by another coupled reaction (such as the hydrolysis of ATP) is used to drive the uphill transport of molecules in the energetically unfavorable direction.

The **ion pumps** responsible for maintaining gradients of ions across the plasma membrane provide important examples of active transport driven directly by ATP hydrolysis. As discussed earlier (see Table 14.2), the concentration of Na^+ is approximately ten times higher outside than inside of cells, whereas the concentration of K^+ is higher inside than out. These ion gradients are maintained by the **Na^+-K^+ pump** (also called the **Na^+-K^+ ATPase**) (**Figure 14.26**), which uses energy derived from ATP hydrolysis to transport Na^+ and K^+ against their electrochemical gradients. This process is a result of ATP-driven conformational changes in the pump (**Figure 14.27**). First, Na^+ ions bind to high-affinity sites inside the cell. This binding stimulates the hydrolysis of ATP and phosphorylation of the pump, inducing a conformational change that exposes the Na^+-binding sites to the outside of the cell and reduces their affinity for Na^+. Consequently, the bound Na^+ is released into the extracellular fluids. At the same time, high-affinity K^+ binding sites are exposed on the cell surface. The binding of extracellular K^+ to these sites then stimulates hydrolysis of the phosphate group bound to the pump, which induces a second conformational change, exposing the K^+ binding sites to the cytosol and lowering their binding affinity so that K^+ is released inside the cell. The pump has three binding sites for Na^+ and two for K^+, so each

Animation 14.2

sites.sinauer.com/cooper7e/a14.2
The Na^+-K^+ Pump The sodium and potassium ion gradients across the plasma membrane are maintained by the Na^+-K^+ pump, which hydrolyzes ATP to fuel the transport of these ions against their electrochemical gradients.

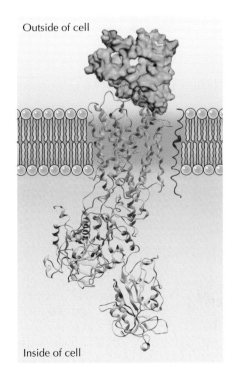

Outside of cell

Inside of cell

Figure 14.26 Structure of the Na^+-K^+ pump The Na^+-K^+ pump is a heterodimer with a ten-transmembrane domain α subunit (blue) and a single-transmembrane domain β subunit with an unstructured extracellular domain (green). The associated γ subunit (red) regulates the pump activity in a tissue-specific manner. (After J. P. Morth et al., 2007. *Nature* 450: 1043.)

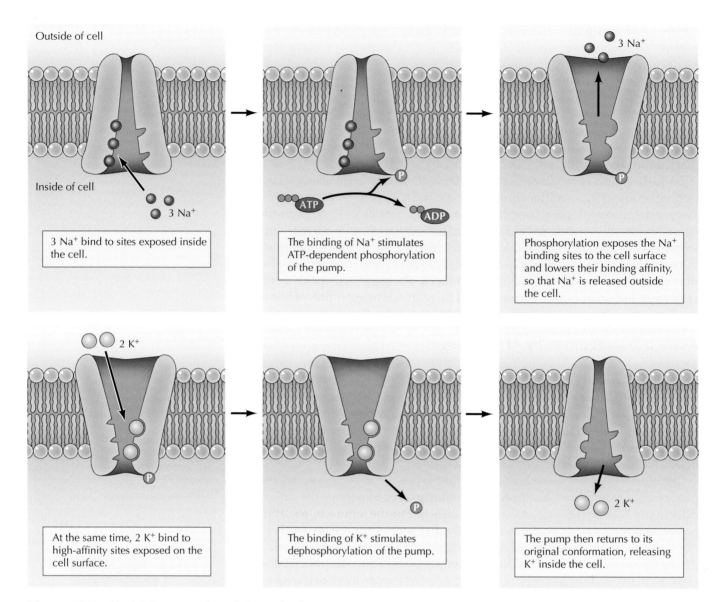

Outside of cell

Inside of cell

3 Na⁺ bind to sites exposed inside the cell.

The binding of Na⁺ stimulates ATP-dependent phosphorylation of the pump.

ATP

ADP

Phosphorylation exposes the Na⁺ binding sites to the cell surface and lowers their binding affinity, so that Na⁺ is released outside the cell.

3 Na⁺

2 K⁺

At the same time, 2 K⁺ bind to high-affinity sites exposed on the cell surface.

The binding of K⁺ stimulates dephosphorylation of the pump.

The pump then returns to its original conformation, releasing K⁺ inside the cell.

2 K⁺

Figure 14.27 Model for operation of the Na⁺-K⁺ pump

cycle transports three Na⁺ and two K⁺ ions across the plasma membrane at the expense of one molecule of ATP.

The importance of the Na⁺-K⁺ pump is indicated by the fact that it is estimated to consume more than 25% of the ATP utilized by many animal cells. One critical role of the Na⁺ and K⁺ gradients established by the pump is the propagation of electric signals in nerve and muscle. As will be discussed shortly, the Na⁺ gradient established by the pump is also utilized to drive the active transport of a variety of other molecules. Yet another important role of the Na⁺-K⁺ pump in most animal cells is to maintain osmotic balance and cell volume. The cytoplasm contains a high concentration of organic molecules, including macromolecules, amino acids, sugars, and nucleotides. In the absence of a counterbalance, this would drive the inward flow of water by osmosis, which, if unchecked would result in swelling and eventual bursting of the cell.

Figure 14.28 Ion gradients across the plasma membrane of a typical mammalian cell   The concentrations of Na^+ and Cl^- are higher outside than inside the cell, whereas the concentration of K^+ is higher inside than out. The low concentrations of Na^+ and Cl^- balance the high intracellular concentration of organic compounds, equalizing the osmotic pressure and preventing the net influx of water.

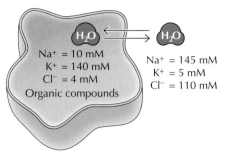

$Na^+ = 10\ mM$
$K^+ = 140\ mM$
$Cl^- = 4\ mM$
Organic compounds

$Na^+ = 145\ mM$
$K^+ = 5\ mM$
$Cl^- = 110\ mM$

The required counterbalance is provided by the ion gradients established by the Na^+-K^+ pump (Figure 14.28). In particular, the pump establishes a higher concentration of Na^+ outside than inside the cell. As already discussed, the flow of K^+ through open channels further establishes an electric potential across the plasma membrane. This membrane potential in turn drives Cl^- out of the cell, so the concentration of Cl^- (like that of Na^+) is about ten times higher in extracellular fluids than in the cytoplasm. These differences in ion concentrations balance the high concentrations of organic molecules inside cells, equalizing the osmotic pressure and preventing the net influx of water.

The active transport of Ca^{2+} across the plasma membrane is driven by a Ca^{2+} pump that is structurally related to the Na^+-K^+ pump (Figure 14.29) and is similarly powered by ATP hydrolysis. Calcium pumps transport Ca^{2+} from the cytosol out of the cell or into the ER lumen, so intracellular Ca^{2+} concentrations are extremely low: approximately 0.1 mM in comparison with extracellular concentrations of about 1 mM. This low intracellular concentration of Ca^{2+} makes the cell sensitive to small increases in intracellular Ca^{2+} levels. Such transient and highly localized increases in intracellular Ca^{2+} play important roles in cell signaling as noted already with respect to muscle contraction (see Figure 13.28) and discussed further in Chapter 16.

Similar ion pumps in the plasma membranes of bacteria, yeasts, and plant cells are responsible for the active transport of H^+ out of the cell. In addition, H^+ is actively pumped out of cells lining the stomach, resulting in the acidity of gastric fluids. Structurally distinct pumps are responsible for the active transport of H^+ into lysosomes and endosomes (see Figure 11.39). Yet a third type of H^+ pump is exemplified by the ATP synthases of mitochondria and chloroplasts: In these cases, the pumps can be viewed as operating in reverse, with the movement of ions down the electrochemical gradient being used to drive ATP synthesis.

The largest family of membrane transporters consists of the **ABC transporters**, so called because they are characterized by highly conserved ATP-binding domains or *ATP-binding cassettes* (ABC) (Figure 14.30). More than 100 family members have been identified in both prokaryotic and eukaryotic cells; *E. coli* has 79 and humans have 48 known ABC transporter genes. Different members of the family function either as importers that bring nutrients into cells (in prokaryotes) or as exporters that pump toxic substances out of cells (in both prokaryotes and eukaryotes). The ABC transporters are composed of two ATP-binding domains and two transmembrane domains. Their substrate binding site alternates between outward facing and inward facing, depending on ATP binding and hydrolysis. Thus, for an importer, the

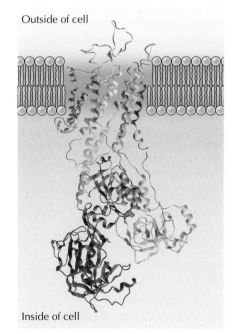

Outside of cell

Inside of cell

Figure 14.29 Structure of the Ca^{2+} pump   The cytosolic subunits of the Ca^{2+} pump are shown in red, blue, and yellow and the transmembrane subunits in tan, green, and purple. (After C. Olesen et al., 2007. *Nature* 450: 1037.)

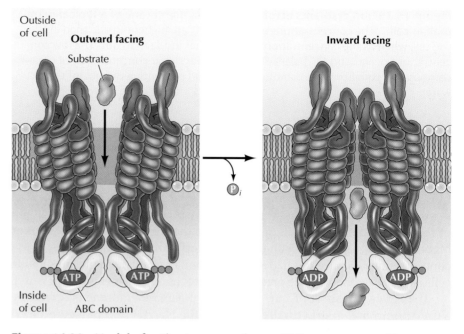

Figure 14.30 Model of active transport by an ABC transporter The transporter consists of two transmembrane domains (each containing six membrane-spanning α helices) and two cytosolic ATP-binding domains (ABC domains). The transporter illustrated is an importer, with its substrate-binding site facing outward with ATP bound to the ABC domain. Hydrolysis of ATP to ADP changes the conformation of the transporter, releasing its substrate inside the cell.

ATP-bound transporter faces outward and binds its substrate outside of the cell. Hydrolysis of ATP to ADP changes the conformation of the transporter, so that it now faces inward and releases its substrate into the cytoplasm. The transporter thus uses the free energy stored as ATP to control the internal composition of the cell.

The first eukaryotic ABC transporter was discovered as the product of a gene (called the multidrug resistance, or *mdr*, gene) that makes cancer cells resistant to a variety of drugs used in chemotherapy. Two MDR transporters have now been identified. They are normally expressed in a variety of cells, where they function to remove potentially toxic foreign compounds. For example, expression of an MDR transporter in capillary endothelial cells of the brain appears to play an important role in protecting the brain from toxic chemicals. Unfortunately, the MDR transporters are frequently expressed at high levels in cancer cells where they recognize a variety of drugs and pump them out of cells, thereby making the cancer cells resistant to a broad spectrum of chemotherapeutic agents and posing a major obstacle to successful cancer treatment.

Another medically important member of the ABC transporter family is the gene responsible for cystic fibrosis. Although it is a member of the ABC family, the product of this gene (called the cystic fibrosis transmembrane conductance regulator, or CFTR) functions as a bicarbonate and Cl⁻ channel in epithelial cells, and defective Cl⁻ transport is characteristic of the disease. The CFTR Cl⁻ channel is also unusual in that it appears to require both ATP hydrolysis and cAMP-dependent phosphorylation in order to open. Most

Molecular Medicine

Cystic Fibrosis

The Disease

Cystic fibrosis is a recessive genetic disease affecting children and young adults. It is the most common lethal inherited disease of Caucasians, with approximately one in 2500 newborns affected, although it is rare in other races. The characteristic dysfunction in cystic fibrosis is the production of abnormally thick sticky mucus by several types of epithelial cells, including the cells lining the respiratory and gastrointestinal tracts. The primary clinical manifestation is respiratory disease resulting from obstruction of the pulmonary airways by thick plugs of mucus, followed by the development of recurrent bacterial infections. In most patients, the pancreas is also involved because the pancreatic ducts are blocked by mucus. Sweat glands also function abnormally, and the presence of excessive salt in sweat is diagnostic of cystic fibrosis.

Management of the disease includes physical therapy to promote bronchial drainage, antibiotic administration, and pancreatic enzyme replacement. Although such treatment has extended the survival of affected individuals to about 30 years of age, cystic fibrosis is ultimately fatal, with lung disease being responsible for 95% of mortality.

Molecular and Cellular Basis

The hallmark of cystic fibrosis is defective chloride (Cl^-) transport in affected epithelia, including sweat ducts and the cells lining the respiratory tract. In 1984, it was demonstrated that Cl^- channels fail to function normally in epithelial cells from cystic fibrosis patients. The molecular basis of the disease was then elucidated in 1989 with the isolation of the cystic fibrosis gene as a molecular clone. The sequence of the gene revealed that it encodes a protein (called CFTR for cystic fibrosis transmembrane conductance regulator) belonging to the ABC transporter family. A variety of subsequent studies then demonstrated that CFTR functions as a bicarbonate and Cl^- channel and that the inherited mutations responsible for cystic fibrosis result directly in defective

Cl^- transport. About 70% of cystic fibrosis cases are the result of a single mutation (deletion of phenylalanine 508) that disrupts folding of the CFTR protein.

Prevention and Treatment

As with other inherited diseases, isolation of the cystic fibrosis gene opens the possibility of genetic screening to identify individuals carrying mutant alleles. In some populations, the frequency of heterozygote carriers of mutant genes is as high as one in 25 individuals, suggesting the possibility of general population screening to identify couples at risk and provide genetic counseling. In addition, understanding the function of CFTR as a Cl^- channel has suggested new approaches to treatment, including gene therapy and the use of drugs that modify the processing of the mutant CFTR proteins or that increase Cl^- transport by mutant receptors.

The potential of gene therapy has been supported by experiments demonstrating that introduction of a normal CFTR gene into cultured cells of cystic fibrosis patients is sufficient to restore Cl^- channel function. Studies with experimental animals further demonstrated that viral vectors can transmit CFTR cDNA to the respiratory epithelium, and the first human trial of gene therapy for cystic fibrosis was initiated in 1993. Trials to date have demonstrated that the CFTR cDNA can be safely delivered to and expressed in the bronchial epithelial cells of cystic fibrosis patients. However, the

efficiency of gene transfer has been too low to provide any clinical benefit.

In contrast, exciting progress has been made in the development of small molecules for cystic fibrosis treatment. The first effective drug against cystic fibrosis (ivacaftor) was identified in 2009 as a result of screening for small molecules that targeted specific CFTR mutants. Ivacaftor is active against a mutant CFTR with aspartic acid rather than glycine at position 551 (the G551D variant). The G551D CFTR, which accounts for about 5% of cystic fibrosis cases, is transported normally to the plasma membrane but is defective in Cl^- transport. Ivacaftor increases Cl- transport by the mutant G551D CFTR to a level that significantly benefits patients. It is also active against several other mutant CFTRs, and was approved for clinical use by the FDA in 2012. Further work has identified a drug (lumacaftor) that facilitates folding of the common phenylalanine 508 deletion CFTR. Lumacaftor received FDA approval in July, 2015 for treatment of patients with this common defect in CFTR.

References

Armstrong, D. K., S. Cunningham, J. C. Davies and E. W. F. W. Alton. 2014. Gene therapy in cystic fibrosis. *Arch. Dis. Child.* 99: 465–468.

Cutting, G. R. 2015. Cystic fibrosis genetics: from molecular understanding to clinical application. *Nature Rev. Genet.* 16: 45–56.

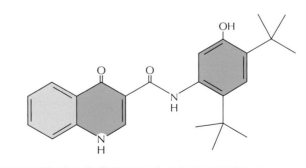

Structure of ivacaftor, the first small molecule drug for cystic fibrosis.

cases of cystic fibrosis are the result of a mutation in CFTR that interferes with proper folding of the protein.

Active transport driven by ion gradients

The ion pumps and ABC transporters discussed in the previous section utilize energy derived directly from ATP hydrolysis to transport molecules against their electrochemical gradients. Other molecules are transported against their concentration gradients using energy derived not from ATP hydrolysis but from the coupled transport of a second molecule in the energetically favorable direction. The Na^+ gradient established by the Na^+-K^+ pump provides a source of energy that is frequently used to power the active transport of sugars, amino acids, and ions in mammalian cells. The H^+ gradients established by the H^+ pumps of bacteria, yeast, and plant cells play similar roles.

The epithelial cells lining the intestine provide a good example of active transport driven by the Na^+ gradient. These cells use active-transport systems in the apical domains of their plasma membranes to take up dietary sugars and amino acids from the lumen of the intestine. The uptake of glucose, for example, is carried out by a transporter that coordinately transports two Na^+ ions and one glucose molecule into the cell (**Figure 14.31**). The flow of Na^+ down its electrochemical gradient provides the energy required to take up dietary glucose and to accumulate high intracellular glucose concentrations. Glucose is then released into the underlying connective tissue (which contains

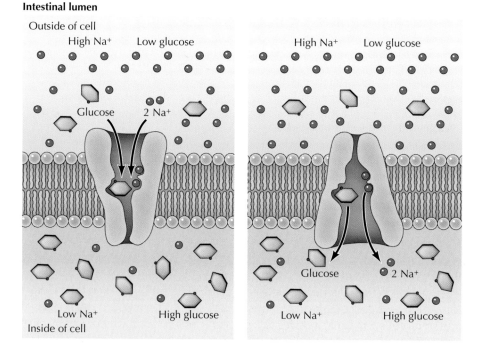

Figure 14.31 Active transport of glucose Active transport driven by the Na^+ gradient is responsible for the uptake of glucose from the intestinal lumen. The transporter coordinately binds and transports one glucose molecule and two Na^+ ions into the cell. The transport of Na^+ in the energetically favorable direction drives the uptake of glucose against its concentration gradient.

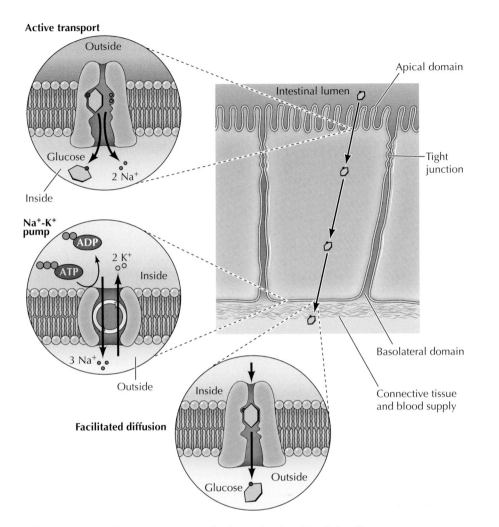

Figure 14.32 Glucose transport by intestinal epithelial cells A transporter in the apical domain of the plasma membrane is responsible for the active uptake of glucose (by cotransport with Na^+, as in Figure 14.31) from the intestinal lumen. As a result, dietary glucose is absorbed and concentrated within intestinal epithelial cells. Glucose is then transferred from these cells to the underlying connective tissue and blood supply by facilitated diffusion, mediated by a transporter in the basolateral domain of the plasma membrane. The system is driven by the Na^+-K^+ pump, which is also found in the basolateral domain.

blood capillaries) at the basolateral surface of the intestinal epithelium, where it is transported down its concentration gradient by facilitated diffusion (**Figure 14.32**). The uptake of glucose from the intestinal lumen and its release into the circulation provides a good example of the polarized function of epithelial cells, which results from the specific localization of active transport and facilitated diffusion carriers to the apical and basolateral domains of the plasma membrane, respectively.

The coordinated uptake of glucose and Na^+ is an example of **symport**, the transport of two molecules in the same direction. In contrast, the facilitated diffusion of glucose is an example of **uniport**, the transport of only a single molecule. Active transport can also take place by **antiport**, in which two

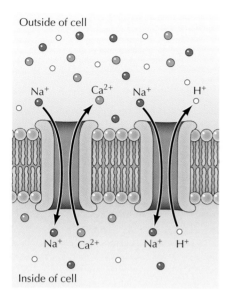

Outside of cell

Na+   Ca²+   Na+   H+

Na+   Ca²+   Na+   H+

Inside of cell

Animation 14.3

sites.sinauer.com/cooper7e/a14.3

Endocytosis In endocytosis, the plasma membrane pinches off small vesicles or large vesicles (phagocytosis) to take in extracellular materials.

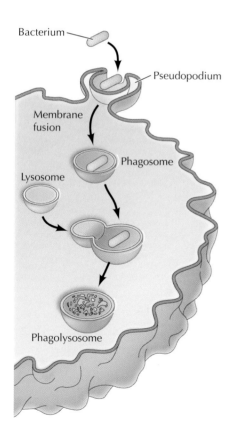

Bacterium

Pseudopodium

Membrane fusion

Phagosome

Lysosome

Phagolysosome

Figure 14.33 Examples of antiport Ca²+ and H+ are exported from cells by antiporters, which couple their export to the energetically favorable import of Na+.

molecules are transported in opposite directions (**Figure 14.33**). For example, Ca²+ is exported from cells not only by the Ca²+ pump but also by a Na+-Ca²+ antiporter that transports Na+ into the cell and Ca²+ out. Another example is provided by the Na+-H+ exchange protein, which functions in the regulation of intracellular pH. The Na+-H+ antiporter couples the transport of Na+ into the cell with the export of H+, thereby removing excess H+ produced by metabolic reactions and preventing acidification of the cytoplasm.

Endocytosis

The carrier and channel proteins discussed in the preceding section transport small molecules through the phospholipid bilayer. Eukaryotic cells are also able to take up macromolecules and particles from the surrounding medium by a distinct process called **endocytosis**. In endocytosis the material to be internalized is surrounded by an area of plasma membrane, which then buds off inside the cell to form a vesicle containing the ingested material. The term "endocytosis" was coined by Christian de Duve in 1963 to include both the ingestion of large particles (such as bacteria) and the uptake of fluids or macromolecules in small vesicles. The former of these activities is known as **phagocytosis** (cell eating) and occurs largely in specialized types of cells. Other forms of endocytosis take place by several different mechanisms in all eukaryotic cells.

Phagocytosis

During phagocytosis cells engulf large particles such as bacteria, cell debris, or even intact cells (**Figure 14.34**). Binding of the particle to receptors on the surface of the phagocytic cell triggers the extension of pseudopodia—an actin-based movement of the cell surface, as discussed in Chapter 13. The pseudopodia eventually surround the particle and their membranes fuse to form a large intracellular vesicle (>0.5 μm in diameter) called a **phagosome**. The phagosomes then fuse with lysosomes, producing **phagolysosomes**, in which the ingested material is digested by the action of lysosomal acid hydrolases (see Chapter 11).

The ingestion of large particles by phagocytosis plays distinct roles in different kinds of cells (**Figure 14.35**). Many amoebas use phagocytosis to capture food particles, such as bacteria or other protozoans. In multicellular animals, the major roles of phagocytosis are to provide a defense against invading microorganisms and to eliminate aged or damaged cells from the body. In mammals, phagocytosis of microorganisms is the function of three types of white blood cells—macrophages, neutrophils, and dendritic cells—which are frequently referred to as "professional phagocytes." These cells play critical roles in the body's defense systems by eliminating microorganisms

Figure 14.34 Phagocytosis Binding of a bacterium to the cell surface stimulates the extension of a pseudopodium, which eventually engulfs the bacterium. Fusion of the pseudopodium membranes then results in formation of a large intracellular vesicle (a phagosome). The phagosome fuses with lysosomes to form a phagolysosome within which the ingested bacterium is digested.

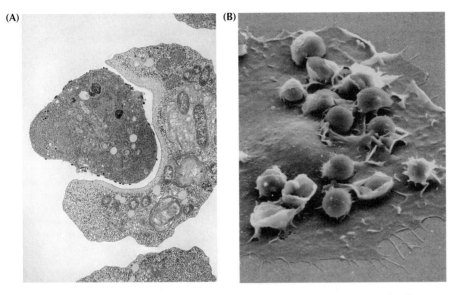

Figure 14.35 Examples of phagocytic cells (A) Electron micrograph of an amoeba engulfing another protist. (B) Scanning electron micrograph of a macrophage ingesting red blood cells. False color has been added to the micrograph. (B, courtesy of Joel Swanson, University of Michigan.)

from infected tissues. In addition, macrophages as well as other cell types, including fibroblasts and epithelial cells, eliminate aged or dead cells from tissues throughout the body, as discussed in Chapter 18. A striking example of the scope of this activity is provided by the macrophages of the human spleen and liver, which are responsible for the disposal of more than 10^{11} aged blood cells on a daily basis.

Macropinocytosis is the uptake of extracellular fluids, rather than particles, in large vesicles (0.2–10 μm in diameter). Like phagocytosis, macropinocytosis results from actin-based movements of the plasma membrane, in this case leading to the formation of lamellipodia (sheet-like projections of the plasma membrane; see Figure 13.19). Lamellipodia usually retract back into the cytoplasm. However, they sometimes curve into open cups of the plasma membrane, which can be followed by membrane fusion to form a large intracellular vesicle.

Clathrin-mediated endocytosis

The most-studied form of endocytosis is **clathrin-mediated endocytosis**, which provides a mechanism for the selective uptake of specific macromolecules. Extracellular material is taken up into transport vesicles, which then carry their cargo to endosomes and lysosomes. The mechanisms of cargo selection, vesicle budding, and vesicle fusion involved in endocytosis are similar to those involved in vesicular transport in the secretory pathway, as discussed in Chapter 11. The macromolecules to be internalized first bind to specific cell surface receptors that are concentrated in specialized regions of the plasma membrane called **clathrin-coated pits** (Figure 14.36). Internalization signals on these receptors bind to cytosolic adaptor proteins, which in turn bind clathrin on the cytosolic side of the membrane, similar to the formation of vesicles budding from the *trans*-Golgi network (see Figure 11.36). Clathrin assembles into a basketlike structure that distorts the membrane, forming

Video 14.3

sites.sinauer.com/cooper7e/v14.3
Clathrin-Mediated Endocytosis A clathrin-coated pit first forms on the surface of the membrane, then buds and pinches off inside the cell, forming the vesicle.

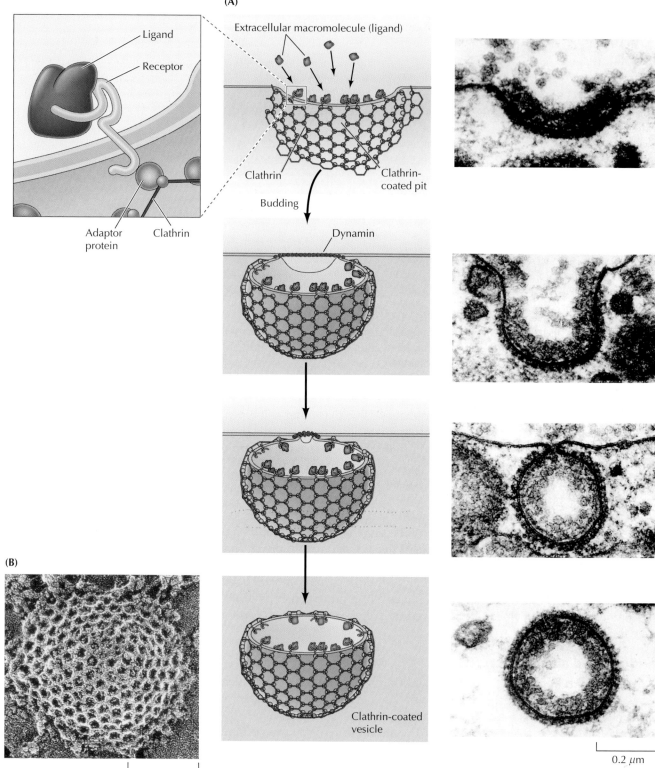

(A)

Extracellular macromolecule (ligand)

Ligand

Receptor

Clathrin

Clathrin-coated pit

Adaptor protein

Clathrin

Budding

Dynamin

Clathrin-coated vesicle

(B)

0.1 μm

0.2 μm

Figure 14.36 Clathrin-coated vesicle formation
(A) Extracellular macromolecules (ligands) bind to cell surface receptors that are concentrated in clathrin-coated pits. Adaptor proteins bind both to clathrin and to internalization signals in the cytoplasmic tails of receptors. With the assistance of the GTP-binding protein dynamin, these pits bud from the plasma membrane to form intracellular clathrin-coated vesicles. Electron micrographs (far right) show the formation of a clathrin-coated vesicle from a clathrin-coated pit. (B) Electron micrograph of a clathrin-coated pit showing the basketlike structure of the clathrin network. (A, from M. M. Perry, 1979. *J. Cell Science* 34: 266; B, courtesy of John E. Heuser, Washington University School of Medicine.)

invaginated pits, which bud from the membrane to form **clathrin-coated vesicles** containing the receptors and their bound macromolecules. The GTP-binding protein **dynamin** plays a key role in vesicle budding. Dynamin assembles into rings around the necks of invaginated pits and undergoes a conformational change, driven by the hydrolysis of GTP, that promotes fission of the underlying membrane to produce free clathrin-coated vesicles inside the cell. Extracellular fluids are also incorporated into vesicles as they bud from the plasma membrane, so clathrin-mediated endocytosis results in the nonselective uptake of extracellular fluids and their contents in addition to the internalization of specific macromolecules.

The uptake of cholesterol by mammalian cells provided a key model for understanding clathrin-mediated endocytosis at the molecular level. Cholesterol is transported through the bloodstream in the form of lipoprotein particles, the most common of which is called **low-density lipoprotein**, or **LDL**. Each LDL particle has a core of cholesterol esters surrounded by a coat of cholesterol and phospholipids and a single molecule of apolipoprotein B100 (**Figure 14.37**). Studies in the laboratories of Michael Brown and Joseph Goldstein demonstrated that the uptake of LDL by mammalian cells requires the binding of LDL to a specific cell surface receptor that is concentrated in clathrin-coated pits and internalized by endocytosis. The receptor is then recycled to the plasma membrane while LDL is transported to lysosomes where cholesterol is released for use by the cell.

The key insights into this process came from studies of patients with the inherited disease known as familial hypercholesterolemia. Patients with this disease have very high levels of serum cholesterol and suffer heart attacks early in life. Brown and Goldstein found that cells of these patients are unable to internalize LDL from extracellular fluids, resulting in the accumulation of high levels of cholesterol in the circulation. Further experiments demonstrated that cells of normal individuals possess a receptor for LDL, which is concentrated in clathrin-coated pits, and that familial hypercholesterolemia results from inherited mutations in the LDL receptor. These mutations are of two types. Cells from most patients with familial hypercholesterolemia simply fail to bind LDL, demonstrating that a specific cell surface receptor was required for LDL uptake. In addition, a few patients were identified whose cells bound LDL but were unable to internalize it. The LDL receptors

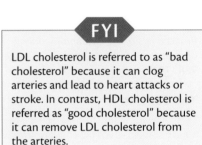

Animation 14.4

sites.sinauer.com/cooper7e/a14.4
Clathrin-Coated Pits and Vesicles
Receptor molecules in the plasma membrane cluster in clathrin-coated pits, become bound to specific macromolecules from outside the cell, and then pinch off in the form of clathrin-coated vesicles.

FYI

LDL cholesterol is referred to as "bad cholesterol" because it can clog arteries and lead to heart attacks or stroke. In contrast, HDL cholesterol is referred as "good cholesterol" because it can remove LDL cholesterol from the arteries.

FYI

Statins are an important class of drugs, used in the treatment of hypercholesterolemia. They work by inhibiting the enzyme HMG-CoA reductase and blocking cholesterol biosynthesis.

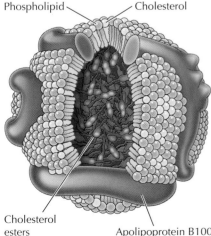

Phospholipid   Cholesterol

Cholesterol esters

Apolipoprotein B100

Figure 14.37 Structure of LDL Each particle of LDL contains approximately 1500 molecules of cholesterol esters in an oily core. The core is surrounded by a coat containing 500 molecules of cholesterol, 800 molecules of phospholipid, and a single molecule of apolipoprotein B100 that winds around the particle.

Michael S. Brown   Joseph L. Goldstein

Key Experiment

The LDL Receptor

Familial Hypercholesterolemia: Defective Binding of Lipoproteins to Cultured Fibroblasts Associated with Impaired Regulation of 3-Hydroxy-3-Methylglutaryl Coenzyme A Reductase Activity

Michael S. Brown and Joseph L. Goldstein

University of Texas Southwestern Medical School, Dallas

Proceedings of the National Academy of Science USA, 1974, Volume 71, pages 788–792

The Context

Familial hypercholesterolemia (FH) is a genetic disease in which patients have greatly elevated levels of serum cholesterol and suffer from heart attacks early in life. Michael Brown and Joseph Goldstein began their efforts to understand this disease in 1972 with the idea that cholesterol overproduction results from a defect in the control mechanisms that normally regulate cholesterol biosynthesis. Consistent with this hypothesis, they found that addition of low-density lipoprotein (LDL) to the culture medium of normal human fibroblasts inhibits the activity of 3-hydroxy-3-methylglutaryl coenzyme A reductase (HMG-CoA reductase), the rate-limiting enzyme in the cholesterol biosynthetic pathway. In contrast, HMG-CoA reductase activity is unaffected by addition of LDL to the cells of FH patients, resulting in overproduction of cholesterol by FH cells.

Perhaps surprisingly, subsequent experiments indicated that this abnormality in HMG-CoA reductase regulation is not the result of a mutation in the HMG-CoA reductase gene. Instead, the abnormal regulation of HMG-CoA reductase appeared to be due to the inability of FH cells to extract cholesterol from LDL. In 1974 Brown and Goldstein demonstrated that the lesion in FH cells is a defect in LDL binding to a receptor on the cell surface. This identification of the LDL receptor led to a series of groundbreaking experiments in which Brown, Goldstein, and their colleagues delineated the pathway of clathrin-mediated endocytosis.

The Experiments

In their 1974 paper, Brown and Goldstein reported the results of experiments in which they investigated the binding of radiolabeled LDL to fibroblasts from either normal individuals or FH patients. Small amounts of radioactive LDL were added to the culture media, and the amount of radioactivity bound to the cells was determined after varying times of incubation (see figure). Increasing amounts of radioactive LDL bound to normal cells as a function of incubation time. Importantly, addition of excess unlabeled LDL reduced the binding of radioactive LDL, indicating that binding was due to a specific interaction of LDL with a limited number of sites on the cell surface. The specificity of the interaction was further demonstrated by the failure of excess amounts of other lipoproteins to interfere with LDL binding.

In contrast to these results with normal fibroblasts, the cells of FH patients failed to specifically bind radioactive LDL. It therefore appeared that normal fibroblasts possessed a specific LDL receptor that was absent or defective in FH cells. Brown and Goldstein concluded that the defect in LDL binding observed in FH cells "may represent the primary genetic lesion in this disorder," accounting for the inability of LDL to inhibit HMG-CoA reductase and the resultant overproduction of cholesterol. Additional experiments showed that the LDL bound to normal fibroblasts is associated with the membrane fraction of the cell, suggesting that the LDL receptor is a cell surface protein.

The Impact

Following this identification of the LDL receptor, Brown and Goldstein demonstrated that LDL bound to the cell surface is rapidly internalized and degraded to free cholesterol in lysosomes. In collaboration with Richard Anderson, they further established that the LDL receptor is internalized by endocytosis from coated pits. In addition, their early studies demonstrated that the

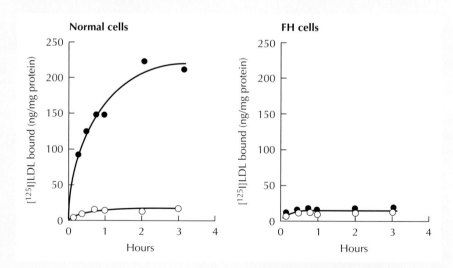

Normal cells

FH cells

Time course of the binding of radioactive LDL to normal and FH cells Cells were incubated with radioactive [125I]LDL in the presence (open circles) or absence (filled circles) of excess unlabeled LDL. Cells were then harvested, and the amount of bound radioactive LDL was determined. Data are presented as nanograms of LDL bound per milligram of cell protein.

of these patients failed to concentrate in coated pits, providing direct evidence for the central role of coated pits in clathrin-mediated endocytosis.

The mutations that prevent the LDL receptor from concentrating in coated pits lie within the cytoplasmic tail of the receptor and can be as subtle as the change of a single tyrosine to cysteine (**Figure 14.38**). Further studies have defined the internalization signal of the LDL receptor as a sequence of six amino acids, including the essential tyrosine. Similar internalization signals, frequently including tyrosine residues, are found in the cytoplasmic tails of other receptors taken up via clathrin-coated pits.

Following their internalization, clathrin-coated vesicles rapidly shed their coats and fuse with early endosomes, in which their contents are sorted for recycling to the plasma membrane or for transport to lysosomes (see Figure 11.40). The molecules to be recycled (for example, cell surface receptors) are then passed to recycling endosomes and back to the plasma membrane. In contrast, molecules destined for degradation are transported to multivesicular bodies and then to late endosomes and lysosomes.

Figure 14.38 The LDL receptor
(A) The LDL receptor includes 700 extracellular amino acids, a transmembrane α helix of 22 amino acids, and a cytoplasmic tail of 50 amino acids. The N-terminal 292 amino acids constitute the LDL-binding domain. (B) Six amino acids within the cytoplasmic tail define the internalization signal, first recognized because the mutation of tyrosine (Tyr) to cysteine (Cys) in a case of familial hypercholesterolemia prevented concentration of the receptor in coated pits.

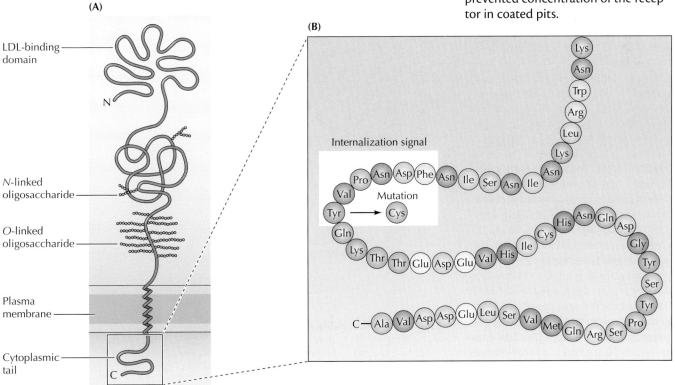

An important feature of early endosomes is that they maintain an acidic internal pH (about 6.0 to 6.2) as the result of the action of a membrane H$^+$ pump. This acidic pH leads to the dissociation of many ligands from their receptors within the early endosome compartment. Following this uncoupling, the receptors and their ligands can be transported to different intracellular destinations. A classic example is provided by LDL, which dissociates from its receptor within early endosomes (**Figure 14.39**). The receptor is then returned to the plasma membrane. In contrast, LDL remains with other soluble contents of the endosome to be targeted to lysosomes, where its degradation releases cholesterol.

Recycling to the plasma membrane is the major fate of membrane proteins taken up by clathrin-mediated endocytosis, with many receptors (like the LDL receptor) being returned to the plasma membrane following dissociation of their bound ligands in early endosomes. The recycling of these receptors results in the continuous internalization of their ligands. Each LDL receptor,

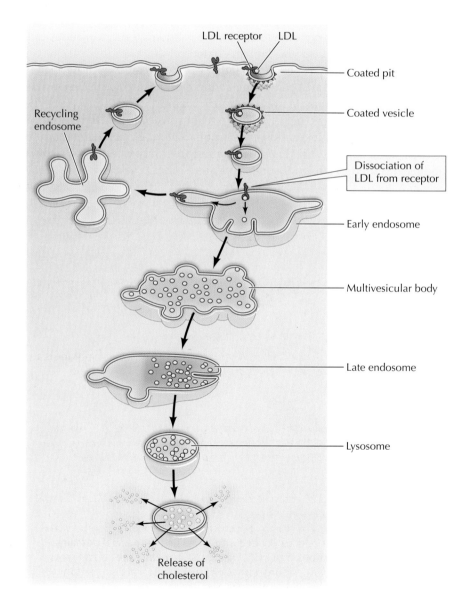

Figure 14.39 Sorting in early endosomes LDL bound to its receptor is internalized in clathrin-coated vesicles, which shed their coats and fuse with early endosomes. At the acidic pH of early endosomes, LDL dissociates from its receptor and the endocytosed materials are sorted for degradation in lysosomes or recycling to the plasma membrane. The receptor is transported to recycling endosomes and from there to the plasma membrane. LDL is carried to late endosomes and lysosomes, where LDL is degraded and cholesterol is released.

for example, makes a round-trip from the plasma membrane to endosomes and back approximately every ten minutes. The importance of the recycling pathway is further emphasized by the magnitude of membrane traffic resulting from endocytosis. Clathrin-coated pits typically occupy 1 to 2% of the surface area of the plasma membrane and are estimated to have a lifetime of 1 to 2 minutes. From these figures, one can calculate that clathrin-mediated endocytosis alone results in the internalization of an area of cell surface equivalent to the entire plasma membrane approximately every 2 hours. Most of this internalized membrane is replaced as a result of recycling; only about 5% of the cell surface is newly synthesized per hour.

FYI

Certain bacterial protein toxins, like cholera toxin, bind to a glycolipid found in lipid rafts and are taken up by endocytosis. Once inside endocytic vesicles the toxin commandeers the secretory pathway to gain entry into the cytoplasm, where it exerts its toxic effect.

Clathrin-independent endocytosis

Cells also possess several pathways of **clathrin-independent endocytosis**. In contrast with clathrin-mediated endocytosis, clathrin-independent endocytosis does not appear to involve selection of specific plasma membrane receptors or the formation of coated vesicles. One pathway of clathrin-independent endocytosis is macropinocytosis, as discussed earlier. Internalization of caveolae (see Figure 14.15) provides another mechanism of clathrin-independent endocytosis, although it appears that most caveolae are stable plasma membrane domains rather than being subject to rapid internalization. Several other pathways of clathrin-independent endocytosis have also been defined based on the cargo being transported and the cellular molecules involved in vesicle formation. For example, one pathway of clathrin-independent endocytosis mediates the uptake of GPI-anchored plasma membrane proteins. Endocytosis of these proteins is dependent upon plasma membrane cholesterol and may be mediated by the clustering of these proteins in lipid rafts. One potential role of clathrin-independent endocytosis is the gradual turnover of plasma membrane proteins via uptake and degradation in lysosomes. However, both the mechanisms and the biological roles of clathrin-independent endocytosis remain active areas of investigation.

SUMMARY

KEY TERMS

Structure of the Plasma Membrane

- **The lipid bilayer:** The fundamental structure of the plasma membrane is a phospholipid bilayer, which also contains glycolipids and cholesterol.

phosphatidylcholine, phosphatidyletha-nolamine, phosphatidylserine, sphingo-myelin, phosphatidylinositol, glycolipid, cholesterol, lipid raft

- **Plasma membrane proteins:** Associated proteins are responsible for carrying out specific membrane functions. Membranes are viewed as fluid mosaics in which proteins are inserted into phospholipid bilayers. See Video 14.1.

fluid mosaic model, peripheral membrane protein, integral membrane protein, transmembrane protein, glycosylphospha-tidylinositol (GPI) anchor, glycocalyx

- **Plasma membrane domains:** Although proteins are free to diffuse laterally through the phospholipid bilayer, the mobility of many proteins is restricted and plasma membranes are composed of distinct domains. Plasma membranes of epithelial cells are divided into apical and basolateral domains with different functions. The mobility of plasma membrane proteins is also restricted by their associations with the cytoskeleton and discrete domains consisting of lipids and other proteins. See Video 14.2.

apical domain, basolateral domain, lipid raft, caveolae

SUMMARY	KEY TERMS

Transport of Small Molecules

- **Facilitated diffusion and carrier proteins:** The passage of most biological molecules is mediated by carrier or channel proteins that allow polar and charged molecules to cross the plasma membrane without interacting with its hydrophobic interior.

- **Ion channels:** Ion channels mediate the rapid passage of selected ions across the plasma membrane. They are particularly well characterized in nerve and muscle cells, where they are responsible for the transmission of electric signals. See Animation 14.1.

- **Active transport driven by ATP hydrolysis:** Energy derived from ATP hydrolysis can drive the transport of molecules against their electrochemical gradients. See Animation 14.2.

- **Active transport driven by ion gradients:** Ion gradients are frequently used as a source of energy to drive the active transport of other molecules.

Endocytosis

- **Phagocytosis:** Cells ingest large particles, such as bacteria and cell debris, by phagocytosis. See Animation 14.3.

- **Clathrin-mediated endocytosis:** The best-characterized form of endocytosis is clathrin-mediated endocytosis, which provides a mechanism for the selective uptake of specific macromolecules into clathrin-coated vesicles. See Animation 14.4 and Video 14.3.

- **Clathrin-independent endocytosis:** Cells also possess several pathways of endocytosis that do not depend on clathrin.

KEY TERMS

facilitated diffusion, carrier protein, channel protein

aquaporin, ion channel, ligand-gated channel, voltage-gated channel, Nernst equation, action potential

active transport, ion pump, Na^+-K^+ pump, Na^+-K^+ ATPase, ABC transporter

symport, uniport, antiport

endocytosis, phagocytosis, phagosome, phagolysosome, macropinocytosis

clathrin-mediated endocytosis, clathrin-coated pit, clathrin-coated vesicle, dynamin, low-density lipoprotein (LDL)

clathrin-independent endocytosis

Refer To
The Cell
Companion Website
sites.sinauer.com/cooper7e
for quizzes, animations, videos, flashcards, and other study resources.

Questions

1. How are peripheral membrane proteins distinguished from integral membrane proteins?

2. How did the cell fusion experiments of Frye and Edidin support the fluid mosaic model of membrane structure? What results would they have obtained if they had incubated fused cells at 2°C?

3. What are lipid rafts and what are the cellular processes they are involved in?

4. How does association with the cytoskeleton contribute to establishing domains of the plasma membrane?

5. The concentration of K^+ is about 20 times higher inside squid axons than in extracellular fluids, generating an equilibrium membrane potential of –75 mV. What would be the expected equilibrium membrane potential if the K^+ concentration were only 10 times higher inside than outside the cell? Why does the actual resting membrane potential (–60 mV) differ from the K^+ equilibrium potential of –75 mV? (Given: R = 1.98 × 10^{-3} kcal /mol/ deg, T = 298 K, ln(x) = 2.3 $\log_{10}(x)$, F = 23 kcal /V/ mol)

6. Curare binds to nicotinic acetylcholine receptors and prevents them from opening. How would it affect the contraction of muscles?

7. How does the selectivity filter of K$^+$ channels differentiate between K$^+$ and Na$^+$ ions?

8. How can glucose be transported against its concentration gradient without the direct expenditure of ATP in intestinal epithelial cells?

9. How does the *mdr* gene confer drug resistance upon cancer cells?

10. How did Brown and Goldstein determine that LDL binds to a limited number of specific binding sites on the surface of normal cells?

11. What have studies on cells from children with familial hypercholesterolemia (FH) told us about the mechanisms of clathrin-mediated endocytosis?

12. How would an inhibitor of actin polymerization affect phagocytosis? Would it affect clathrin-mediated endocytosis?

References and Further Reading   (Key review articles for each major section are highlighted in **bold**.)

Structure of the Plasma Membrane

Cao, X., M. A. Surma and K. Simons. 2012. Polarized sorting and trafficking in epithelial cells. *Cell Res.* 22: 793–805. [R]

Diesenhofer, J., O. Epp, K. Miki, R. Huber and H. Michel. 1985. The structure of the protein subunits in the photosynthetic reaction centre of *Rhodopseudomonas viridis* at 3-Å resolution. *Nature* 318: 618–624. [P]

Engelman, D. M. 2005. Membranes are more mosaic than fluid. *Nature* 438: 578–580. [R]

Hansen, C. G. and B. J. Nichols. 2010. Exploring the caves: cavins, caveolins and caveolae. *Trends Cell Biol.* 20: 177–186.

Head, B. P., H. H. Patel and P. A. Insel. 2014. Interactions of membrane/lipid rafts with the cytoskeleton: impact on signaling and function. *Biochim. Biophys. Acta* 1838: 532–545. [R]

Li, L., X. Shi, X. Guo, H. Li and C. Xu. 2014. Ionic protein-lipid interaction at the plasma membrane: what can the charge do? *Trends Biochem. Sci.* 39: 130–140. [R]

Lingwood, D. and K. Simons. 2010. Lipid rafts as a membrane-organizing principle. *Science* 327: 46–50. [R]

Moraes, I., G. Evans, J. Sanchez-Weatherby, S. Newstead and P. D. S. Stewart. 2014. Membrane protein structure determination--the next generation. *Biochim. Biophys. Acta* 1838: 78–87. [R]

Nassoy, P. and C. Lamaze. 2012. Stressing caveolae: new role in cell mechanics. *Trends Cell Biol.* 22: 381–389. [R]

Nicolson, G. L. 2014. The fluid-mosaic model of membrane structure: still relevant to understanding the structure, function and dynamics of biological membranes after more than 40 years. *Biochim. Biophys. Acta* 1838: 1451–1466. [R]

Owen, D. M., A. Magenau, D. Williamson and K. Gaus. 2012. The lipid raft hypothesis revisited - New insights on raft composition and function from super-resolution fluorescence microscopy. *Bioessays* 34: 739–747. [R]

Parton, R. G. and M. A. del Pozo. 2013. Caveolae as plasma membrane sensors, protectors and organizers. *Nature Rev. Mol. Cell Biol.* 14: 98–112. [R]

Rees, D. C., H. Komiya, T. O. Yeates, J. P. Allen and G. Feher. 1989. The bacterial photosynthetic reaction center as a model for membrane proteins. *Ann. Rev. Biochem.* 58: 607–633. [R]

Rodriguez-Boulon, E. and I. G. Macara. 2014. Organization and execution of the epithelial polarity programme. *Nature Rev. Mol. Cell Biol.* 15: 225–242. [R]

Simons, K. and M. J. Gerl. 2010. Revitalizing membrane rafts: new tools and insights. *Nature Rev. Mol. Cell Biol.* 11: 688–699. [R]

Singer, S. J. 1990. The structure and insertion of integral proteins in membranes. *Ann. Rev. Cell Biol.* 6: 247–296. [R]

Singer, S. J. and G. L. Nicolson. 1972. The fluid mosaic model of the structure of cell membranes. *Science* 175: 720–731. [P]

van Meer, G., D. R. Voelker and G. W. Feigenson. 2008. Membrane lipids: where they are and how they behave. *Nature Rev. Mol. Cell Biol.* 9: 112–124. [R]

Transport of Small Molecules

Bell, G. I., C. F. Burant, J. Takeda and G. W. Gould. 1993. Structure and function of mammalian facilitative sugar transporters. *J. Biol. Chem.* 268: 19161–19164. [R]

Borgnia, M., S. Nielsen, A. Engel and P. Agre. 1999. Cellular and molecular biology of the aquaporin water channels. *Ann. Rev. Biochem.* 68: 425–458. [R]

Catterall, W. A. 2010. Ion channel voltage sensors: structure, function, and pathophysiology. *Neuron* 67: 915–928. [R]

Changeux, J.-P. and A. Taly. 2008. Nicotinic receptors, allosteric proteins and medicine. *Trends Mol. Med.* 14: 93–102. [R]

Gadsby, D. C. 2007. Structural biology: ion pumps made crystal clear. *Nature* 450: 957–959. [R]

Hodgkin, A. L. and A. F. Huxley. 1952. A quantitative description of membrane current and its application to conduction and excitation in nerve. *J. Physiol.* 117: 500–544. [P]

MacLennan, D. H., W. J. Rice and N. M. Green. 1997. The mechanism of Ca^{2+} transport by sarco(endo)plasmic reticulum Ca^{2+}-ATPases. *J. Biol. Chem.* 272: 28815–28818. [R]

Maurel, C. 2007. Plant aquaporins: novel functions and regulation properties. *FEBS Lett.* 581: 2227–2236. [R]

Morth, J. P., B. P. Pedersen, M. S. Toustrup-Jensen, T. L. Sorensen, J. Petersen, J. P. Andersen, B. Vilsen and P. Nissen. 2007. Crystal structure of the sodium-potassium pump. *Nature* 450: 1043–1049. [P]

Olesen, C., M. Picard, A. M. Winther, C. Gyrup, J. P. Morth, C. Oxvig, J. V. Moller and P. Nissen. 2007. The structural basis of calcium transport by the calcium pump. *Nature* 450: 1036–1042. [P]

Pedersen, B. P., M. J. Buch-Pedersen, J. P. Morth, M. G. Palmgren and P. Nissen. 2007. Crystal structure of the plasma membrane proton pump. *Nature* 450: 1111–1114. [P]

Rees, D. C., E. Johnson and O. Lewinson. 2009. ABC transporters: the power to change. *Nature Rev. Mol. Cell Biol.* 10: 218–227. [R]

Shinoda, T., H. Ogawa, F. Cornelius and C. Toyoshima. 2009. Crystal structure of the sodium-potassium pump at 2.4 A resolution *Nature* 459: 446–450. [P]

Sun, L., X. Zeng, C. Yan, X. Sun, X. Gong, Y. Rao and N. Yan. 2012. Crystal structure of a bacterial homologue of glucose transporters GLUT1-4. *Nature* 490: 361–366. [P]

Thorens, B. and M. Mueckler. 2010. Glucose transporters in the 21st Century. *Am. J. Physiol. Endocrinol. Metab.* 298: E141–E145. [R]

Toyoshima, C. and G. Inesi. 2004. Structural basis of ion pumping by Ca^{2+}-ATPase of the sarcoplasmic reticulum. *Ann. Rev. Biochem.* 73: 269–292. [R]

Ward, J. M., P. Maser and J. I. Schroeder. 2009. Plant ion channels: gene families, physiology, and functional genomics analyses. *Ann. Rev. Physiol.* 71: 59–82. [R]

Endocytosis

Brown, M. S. and J. L. Goldstein. 1986. A receptor-mediated pathway for cholesterol homeostasis. *Science* 232: 34–47. [R]

Campelo, F. and V. Malhotra. 2012. Membrane fission: the biogenesis of transport carriers. *Ann. Rev. Biochem.* 81: 407–427. [R]

Doherty, G. J. and H. T. McMahon. 2009. Mechanisms of endocytosis. *Ann. Rev. Biochem.* 78: 857–902. [R]

Ferguson, S. M. and P. De Camilli. 2012. Dynamin, a membrane-remodelling GTPase. *Nature Rev. Mol. Cell Biol.* 13: 75–88. [R]

Flannagan, R. S., V. Jaumouillé and S. Grinstein. 2012. The cell biology of phagocytosis. *Ann. Rev. Pathol. Mech. Dis.* 7: 61–98. [R]

Kirchhausen, T., D. Owen and S. C. Harrison. 2014. Molecular structure, function, and dynamics of clathrin-mediated membrane traffic. *Cold Spring Harbor Perspect. Biol.* 6:a016725. [R]

Kumari, S., M. G. Swetha and S. Mayor. 2010. Endocytosis unplugged: multiple ways to enter the cell. *Cell Res.* 20: 256–275. [R]

Maldonado-Báez, L., C. Williamson and J. G. Donaldson. 2013. Clathrin-independent endocytosis: a cargo-centric view. *Exp. Cell Res.* 319: 2759–2769. [R]

Mayor, S. and R. E. Pagano. 2007. Pathways of clathrin-independent endocytosis. *Nature Rev. Mol. Cell Biol.* 8: 603–612. [R]

McMahon, H. T. and E. Boucrot. 2011. Molecular mechanism and physiological functions of clathrin-mediated endocytosis. *Nature Rev. Mol. Cell Biol.* 125: 517–533. [R]

Schmid, S. L. and V. A. Frolov. 2011. Dynamin: functional design of a membrane fission catalyst. *Ann. Rev. Cell Dev. Biol.* 27: 79–105. [R]

Swanson, J. A. 2008. Shaping cups into phagosomes and macropinosomes. *Nature Rev. Mol. Cell Biol.* 9: 639–649. [R]

Traub, L. M. and J. S. Bonifacino. 2013. Cargo recognition in clathrin-mediated endocytosis. *Cold Spring Harbor Perspect. Biol.* 5:a016790. [R]

CHAPTER 15

Cell Walls, the Extracellular Matrix, and Cell Interactions

Although cell boundaries are defined by the plasma membrane, many cells are surrounded by an insoluble array of secreted macromolecules. Cells of bacteria, fungi, algae, and higher plants are surrounded by rigid cell walls, which are an integral part of the cell. Although animal cells are not surrounded by cell walls, most of the cells in animal tissues are embedded in an extracellular matrix consisting of secreted proteins and polysaccharides. The extracellular matrix not only provides structural support to cells and tissues, but also plays important roles in regulating cell behavior. Interactions of cells with the extracellular matrix anchor the cytoskeleton and regulate cell shape and movement. Likewise, direct interactions between cells are key to the organization of cells in the tissues of both plants and animals, as well as providing channels through which cells can communicate with their neighbors.

Cell Walls

The rigid cell walls that surround bacteria and many types of eukaryotic cells (fungi, algae, and higher plants) determine cell shape and prevent cells from swelling and bursting as a result of osmotic pressure. Despite their common functions, the cell walls of bacteria and eukaryotes are structurally very different. Bacterial cell walls consist of polysaccharides cross-linked by short peptides, which form a covalent shell around the entire cell. In contrast, the cell walls of eukaryotes are composed principally of polysaccharides embedded in a gel-like matrix. Rather than being fixed structures, plant cell walls can be modified both during development of the plant and in response to signals from the environment, so the cell walls of plants play critical roles in determining the organization of plant tissues and the structure of entire plants.

Bacterial cell walls

The rigid cell walls of bacteria provide protection against osmotic pressure and determine the characteristic shapes of different kinds of bacterial cells. For example, some bacteria (such as *E. coli*) are rod-shaped, whereas others are spherical (e.g., *Pneumococcus* and *Staphylococcus*) or spiral-shaped (e.g., the spirochete *Treponema pallidum*, which causes syphilis). In addition, the structure of their cell walls divides bacteria into two broad classes that can be distinguished by a staining procedure known as the Gram stain, developed by Hans Christian Gram in 1884 (**Figure 15.1**). Gram-negative bacteria (such as *E. coli*) have a dual-membrane system in which the plasma membrane is surrounded by a permeable outer membrane. These bacteria have thin cell walls located between their inner and outer membranes. In contrast,

Gram-negative

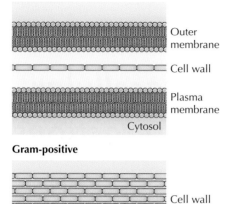

Outer membrane

Cell wall

Plasma membrane

Cytosol

Gram-positive

Cell wall

Plasma membrane

Cytosol

Figure 15.1 Bacterial cell walls The plasma membrane of Gram-negative bacteria is surrounded by a thin cell wall beneath the outer membrane. Gram-positive bacteria lack outer membranes and have thick cell walls.

Gram-positive bacteria (such as the common human pathogen *Staphylococcus aureus*) have only a single plasma membrane, which is surrounded by a much thicker cell wall.

Despite these structural differences, the principal component of the cell walls of both Gram-positive and Gram-negative bacteria is a **peptidoglycan** (**Figure 15.2**), which consists of linear polysaccharide chains cross-linked by short peptides. Because of this cross-linked structure, the peptidoglycan forms a strong covalent shell around the entire bacterial cell. Interestingly, the unique structure of their cell walls also makes bacteria vulnerable to some antibiotics. Penicillin, for example, inhibits the enzyme responsible for forming cross-links between different strands of the peptidoglycan, thereby interfering with cell wall synthesis and blocking bacterial growth.

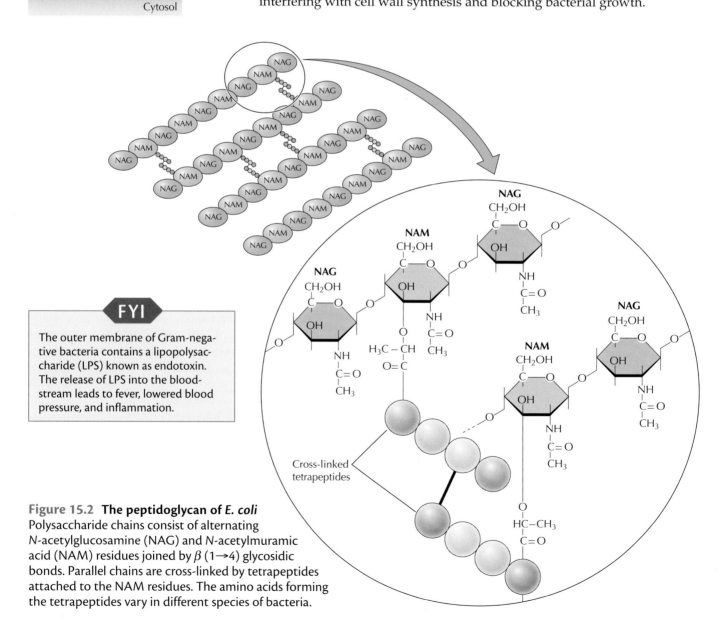

FYI

The outer membrane of Gram-negative bacteria contains a lipopolysaccharide (LPS) known as endotoxin. The release of LPS into the bloodstream leads to fever, lowered blood pressure, and inflammation.

Figure 15.2 The peptidoglycan of *E. coli* Polysaccharide chains consist of alternating *N*-acetylglucosamine (NAG) and *N*-acetylmuramic acid (NAM) residues joined by β (1→4) glycosidic bonds. Parallel chains are cross-linked by tetrapeptides attached to the NAM residues. The amino acids forming the tetrapeptides vary in different species of bacteria.

Figure 15.3 Bacterial cytoskeletal proteins regulate cell wall synthesis
FtsZ, a homolog of eukaryotic tubulin, forms a ring structure at the site of cell division and directs synthesis of a new cell wall between daughter cells. It is found in spherical, rod-shaped, and spiral-shaped bacteria. Rod-shaped bacteria also contain MreB, a homolog of actin that directs cell wall synthesis during cell elongation. Intermediate-filament related proteins (e.g., crescentin) are responsible for curvature of the cell wall in curved or spiral-shaped bacteria.

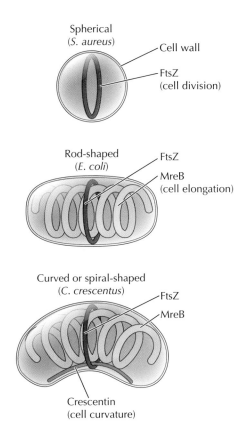

The cell wall is continuously enlarged during bacterial growth and division, so the maintenance of the characteristic shape of bacterial cells is governed by the mechanisms that control cell wall synthesis. Interestingly, this is determined by bacterial cytoskeletal proteins (**Figure 15.3**). The first cytoskeletal protein to be discovered in bacteria, FtsZ, is a homolog of eukaryotic tubulin, which is found in nearly all bacteria (spherical, rod-shaped and spiral-shaped). It forms a ring structure at the site where cell division occurs, directing synthesis of the new cell wall that separates the newly formed daughter cells. Rod-shaped bacteria, such as *E. coli*, also contain the cytoskeletal protein MreB, a homolog of actin, which directs cell wall synthesis during cell elongation. Finally, curved or spiral-shaped bacteria contain intermediate-filament related proteins (such as crescentin) that are responsible for curvature of the cell wall.

Eukaryotic cell walls

In contrast with bacteria, the cell walls of eukaryotes (including fungi, algae, and higher plants) are composed principally of polysaccharides (**Figure 15.4**). The basic structural polysaccharide of fungal cell walls is **chitin**, which also forms the shells of crabs and the exoskeletons of insects and other arthropods. Chitin is a linear polymer of *N*-acetylglucosamine residues. The cell walls of most algae and higher plants are composed principally of **cellulose**, which is the most abundant polymer on Earth. Cellulose is a linear polymer of glucose residues, often containing more than 10,000 glucose monomers. The sugar

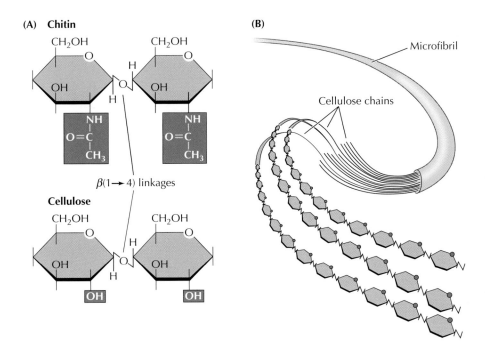

Figure 15.4 Polysaccharides of cell walls   (A) Chitin, the principal polysaccharide of fungal cell walls and the exoskeletons of crabs and insects, is a linear polymer of *N*-acetylglucosamine residues, whereas cellulose is a linear polymer of glucose. As in peptidoglycan, the carbohydrate monomers are joined by β (1→4) linkages, allowing the polysaccharides to form long, straight chains. (B) Parallel chains of cellulose associate to form microfibrils.

Hemicellulose   **Pectin**

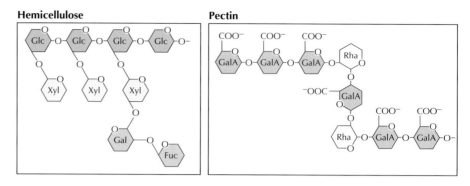

Figure 15.5 Structures of hemicellulose and pectin   A representative hemicellulose (xyloglucan) consists of a backbone of glucose (Glc) residues with side chains of xylose (Xyl), galactose (Gal), and fucose (Fuc). A representative pectin (rhamnogalacturonan) consists of a backbone of galacturonic acid (GalA) and rhamnose (Rha) residues, to which numerous side chains are also attached.

residues in both chitin and cellulose are joined by β (1→4) linkages, which allow the polysaccharides to form long straight chains. Following transport across the plasma membrane to the extracellular space, 36 single cellulose chains associate in parallel with one another to form 3-nm microfibrils. Cellulose microfibrils can extend for many micrometers in length.

Within the plant cell wall, cellulose microfibrils are embedded in a matrix consisting of proteins and two other types of polysaccharides: **hemicelluloses** and **pectins** (Figure 15.5). Hemicelluloses are highly branched polysaccharides that are hydrogen-bonded to cellulose microfibrils (Figure 15.6). This stabilizes the cellulose microfibrils into a tough fiber, which is responsible for the mechanical strength of plant cell walls. The cellulose microfibrils and hemicelluloses are embedded in a gel-like matrix formed by pectins, which are branched polysaccharides containing a large number of negatively-charged galacturonic acid residues. Because of these multiple negative charges, pectins bind positively-charged ions (such as Ca^{2+}) and trap water molecules to form gels. An illustration of their gel-forming properties is seen in jams and jellies that are produced by the addition of pectins to fruit juice. Finally, cell walls

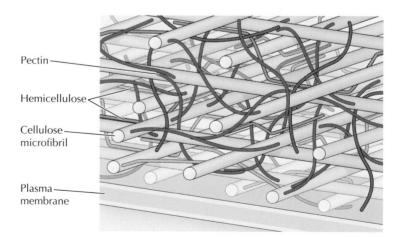

Pectin

Hemicellulose

Cellulose microfibril

Plasma membrane

Figure 15.7 Primary and secondary cell walls Secondary cell walls are laid down between the primary cell wall and the plasma membrane. Secondary walls frequently consist of three layers, which differ in the orientation of their cellulose microfibrils. Electron micrographs show cellulose microfibrils in primary and secondary cell walls. (Primary wall, courtesy of F. C. Steward.)

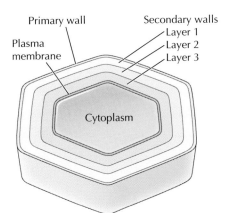

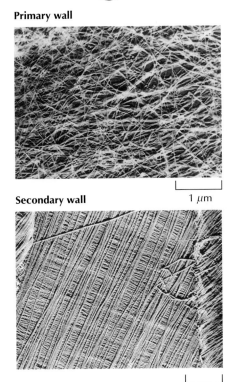

Primary wall

Secondary wall

contain a variety of glycoproteins that are incorporated into the matrix and are thought to provide further structural support.

Both the structure and function of cell walls change as plant cells develop. The walls of growing plant cells (called **primary cell walls**) are relatively thin, allowing the cell to expand in size. Once cells have ceased growth, they frequently lay down **secondary cell walls** between the primary cell wall and the plasma membrane (**Figure 15.7**). Primary and secondary cell walls differ in composition as well as in thickness. Primary cell walls contain approximately equal amounts of cellulose, hemicelluloses, and pectins. In contrast, the more rigid secondary walls generally lack pectins. Many secondary walls are further strengthened by **lignin**, a complex polymer of hydrophobic phenolic residues, which inserts into the spaces between the other polymers. Lignin is responsible for much of the strength and density of wood. Rather than forming a gel, lignin dehydrates the cell wall, making it water-tight and resistant to degradation by pathogens—allowing trees to survive for hundreds of years. The orientation of cellulose microfibrils also differs in primary and secondary cell walls. The cellulose fibers of primary walls appear to be randomly arranged, whereas those of secondary walls are highly ordered (see Figure 15.7). Secondary walls are frequently laid down in layers in which the cellulose fibers differ in orientation, forming a laminated structure that greatly increases cell wall strength. Such secondary cell walls, which are both thicker and more rigid than primary walls, are particularly important in cell types responsible for conducting water and providing mechanical strength to the plant.

One of the critical functions of plant cell walls is to prevent cell swelling as a result of osmotic pressure. In contrast to animal cells, plant cells (like bacteria) do not maintain an osmotic balance between their cytosol and extracellular fluids. Consequently, osmotic pressure continually drives the flow of water into the cell. This water influx is tolerated by plant cells because their rigid cell walls prevent swelling and bursting. Instead, an internal hydrostatic pressure (called **turgor pressure**) builds up within the cell, eventually equalizing the osmotic pressure and preventing the further influx of water.

Turgor pressure is responsible for much of the rigidity of plant tissues, as is readily apparent from examination of a dehydrated, wilted plant. In addition, turgor pressure provides the basis for a form of cell growth that is unique to plants. In particular, plant cells frequently expand by taking up water without synthesizing new cytoplasmic components (**Figure 15.8**). Cell expansion by this mechanism is signaled by plant hormones (**auxins**) that activate proteins called expansins. The expansins act to weaken a region of the cell wall, allowing turgor pressure to drive the expansion of the cell in that direction. As this occurs, the water that flows into the cell accumulates within a large central vacuole, so the cell expands without increasing the volume of its cytosol. Such expansion can result in a tenfold to hundredfold increase in the size of plant cells during development.

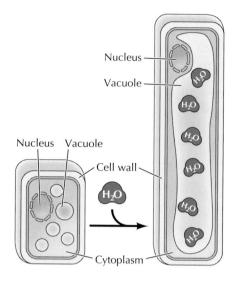

Nucleus
Vacuole

Nucleus   Vacuole

Cell wall

H_2O

Cytoplasm

Figure 15.8 Expansion of plant cells   Turgor pressure drives the expansion of plant cells during growth by the uptake of water, which is accumulated in a large central vacuole.

As cells expand, new components of the cell wall are deposited outside the plasma membrane. While matrix components, including hemicelluloses and pectins, are synthesized in the Golgi apparatus and secreted, cellulose is synthesized by the plasma membrane enzyme complex **cellulose synthase** (**Figure 15.9**). Cellulose synthase is a transmembrane enzyme that synthesizes cellulose from UDP-glucose in the cytosol. The growing cellulose chain remains bound to the enzyme as it is synthesized and translocated across the plasma membrane to the exterior of the cell through a pore created by multiple enzyme subunits. A similar mechanism is used to synthesize chitin and hyaluronan, a component of the extracellular matrix discussed later in this chapter.

In expanding cells, the newly synthesized cellulose microfibrils are typically deposited perpendicular to the direction of cell elongation—an orientation that determines the direction of further cell expansion. Interestingly, the cellulose microfibrils in elongating cell walls are laid down in parallel to cortical microtubules underlying the plasma membrane. These microtubules define the orientation of newly synthesized cellulose microfibrils by guiding the

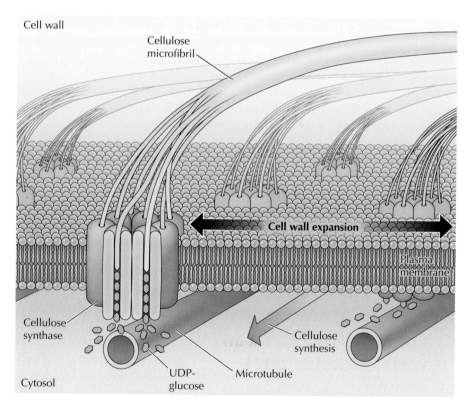

Cell wall

Cellulose microfibril

Cell wall expansion

Plasma membrane

Cellulose synthase

Cellulose synthesis

Cytosol

UDP-glucose

Microtubule

Figure 15.9 Cellulose synthesis during cell elongation   Cellulose synthases are transmembrane enzymes that synthesize cellulose from UDP-glucose. UDP-glucose binds to cellulose synthase in the cytosol and the growing cellulose chain is translocated to the outside of the cell. Complexes of cellulose synthase track microtubules beneath the plasma membrane such that new cellulose microfibrils are laid down at right angles to the direction of cell elongation.

Animation 15.1

sites.sinauer.com/cooper7e/a15.1
Cellulose Synthesis during Elongation   Cell wall formation in plants requires the synthesis of cellulose chains by enzyme complexes in the plasma membrane.

(A) Cellulose synthase   (B) Microtubules   (C) Merged

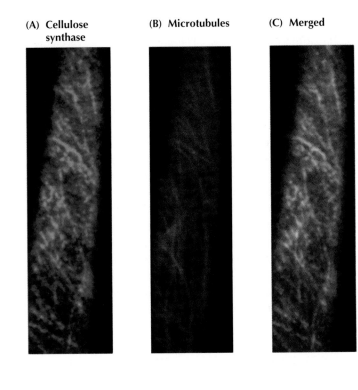

Figure 15.10 Association of cellulose synthase with microtubules   The association of cellulose synthase (green fluorescence) with microtubules (red fluorescence) is visualized in *Arabidopsis* shoot cells. The images of cellulose synthase and microtubules are combined in the merged image to illustrate their co-localization. (From Paredez, A., A. Wright and D. W. Ehrhardt. 2006. *Curr. Opin. Plant Biol.* 9: 574.)

movement of cellulose synthase complexes in the membrane (**Figure 15.10**). The cortical microtubules thus define the direction of cell wall growth, which in turn determines the direction of cell expansion and ultimately the shape of the entire plant. This is illustrated by the effect of a mutation in a microtubule-associated protein, CLASP, on the growth of *Arabidopsis*. CLASP binds to the plus ends of microtubules to regulate their stability and to track the plasma membrane (discussed in Chapter 13). Inactivation of CLASP interferes with cell expansion, resulting in stunted plants (**Figure 15.11**).

Cell walls from different plant tissues, such as leaves, stems, roots, and flowers, consist primarily of cellulose but differ in their matrix components and in the organization of their cellulose fibrils. Plant species express from nine to 18 different cellulose synthase enzymes, and each plasma membrane enzyme complex contains at least three different forms. This allows the variable organization of cell walls in different tissues.

The Extracellular Matrix and Cell–Matrix Interactions

Although animal cells are not surrounded by cell walls, most animal cells in tissues are embedded in an **extracellular matrix** that fills the spaces between cells and binds cells and tissues together. There are several types of extracellular matrices, which consist of a variety of secreted proteins and polysaccharides. One type of extracellular matrix is exemplified by the thin, sheetlike **basal laminae**, previously called "basement

Figure 15.11 Effect of mutation in a microtubule-associated protein on *Arabidopsis* growth   The plant on the left is normal. The plant on the right carries a mutation in the microtubule-associated protein CLASP, which interferes with normal cell expansion and growth. (Courtesy of Chris Ambrose, University of Saskatchewan.)

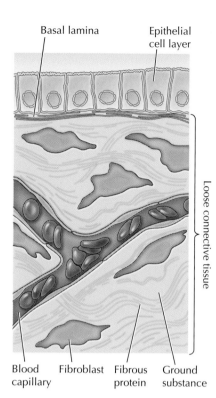

Basal lamina   Epithelial cell layer

Loose connective tissue

Blood capillary   Fibroblast   Fibrous protein   Ground substance

Figure 15.12 Examples of extracellular matrix Sheets of epithelial cells rest on a thin layer of extracellular matrix called a basal lamina. Beneath the basal lamina is loose connective tissue, which consists largely of extracellular matrix secreted by fibroblasts. The extracellular matrix contains fibrous structural proteins embedded in a gel-like polysaccharide ground substance.

membranes," upon which layers of epithelial cells rest (**Figure 15.12**). In addition to supporting sheets of epithelial cells, basal laminae surround muscle cells, adipose (fat) cells, and peripheral nerves. Extracellular matrix, however, is most abundant in connective tissues. For example, the loose connective tissue beneath epithelial cell layers consists predominantly of an extracellular matrix in which fibroblasts are distributed. Other types of connective tissue, such as bone, tendon, and cartilage, similarly consist largely of extracellular matrix, which is principally responsible for their structure and function. Several of the proteins and polysaccharides found in the extracellular matrix are also found closely associated with the plasma membrane, where they form a polysaccharide-rich coat termed the glycocalyx (see Figure 14.11).

Matrix structural proteins

Extracellular matrices are composed of tough fibrous proteins embedded in a gel-like polysaccharide ground substance—a design basically similar to that of plant cell walls. In addition to fibrous structural proteins and polysaccharides, the extracellular matrix contains adhesion proteins that link components of the matrix both to one another and to attached cells. The differences between various types of extracellular matrices result from both quantitative variations in the types or amounts of these different constituents and from modifications in their organization. For example, tendons contain a high proportion of fibrous proteins, whereas cartilage contains a high concentration of polysaccharides that form a firm compression-resistant gel. In bone, the extracellular matrix is hardened by deposition of calcium

(A) Collagen triple helix

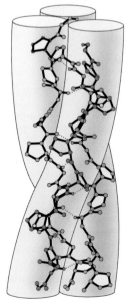

(B) Amino acid sequence

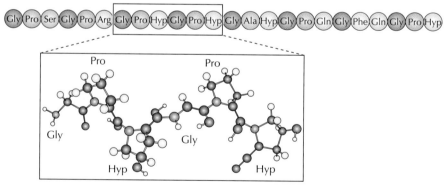

Gly Pro Ser Gly Pro Arg Gly Pro Hyp Gly Pro Hyp Gly Ala Hyp Gly Pro Gln Gly Phe Gln Gly Pro Hyp

Pro   Pro

Gly   Gly

Hyp   Hyp

Figure 15.13 Structure of collagen (A) Three polypeptide chains coil around one another in a characteristic triple helix structure. (B) The amino acid sequence of a collagen triple helix domain consists of Gly-X-Y repeats, in which X is frequently proline and Y is frequently hydroxyproline (Hyp).

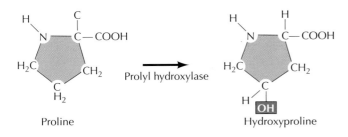

Figure 15.14 Formation of hydroxyproline
Prolyl hydroxylase converts proline residues in collagen to hydroxyproline.

phosphate crystals. The sheetlike structure of basal laminae results from a matrix composition that differs from that found in connective tissues.

The major structural protein of the extracellular matrix is **collagen**, which is the single most abundant protein in animal tissues (**Figure 15.13**). The collagens are a large family of proteins containing more than 40 different members. They are characterized by the formation of triple helices in which three polypeptide chains are wound tightly around one another in a ropelike structure. The triple helix domain of collagen consists of repeats of the amino acid sequence Gly-X-Y. A glycine (the smallest amino acid, with a side chain consisting only of a hydrogen) is required in every third position in order for the polypeptide chains to pack together close enough to form the collagen triple helix. Proline is frequently found in the X position and hydroxyproline in the Y position; because of their ring structure these amino acids stabilize the helical conformations of the polypeptide chains.

The unusual amino acid hydroxyproline is formed within the endoplasmic reticulum by modification of proline residues that have already been incorporated into collagen polypeptide chains (**Figure 15.14**). Lysine residues in collagen are also frequently converted to hydroxylysines. The hydroxyl groups of these modified amino acids are thought to stabilize the collagen triple helix by forming hydrogen bonds between polypeptide chains. These amino acids are rarely found in other proteins, although hydroxyproline is also common in some of the glycoproteins of plant cell walls. There are 28 different types of collagen, with different roles in various kinds of extracellular matrices (**Table 15.1**). The most abundant type of collagen (type I collagen) is one of the fibril-forming collagens that are the basic structural components of connective tissues. The polypeptide chains of these collagens consist of approximately 1000 amino acids or 330 Gly-X-Y repeats. After being secreted from the cell these collagens

Prolyl hydroxylase, the enzyme that catalyzes the modification of proline to hydroxyproline residues in collagen, requires vitamin C for its activity. Vitamin C deficiency causes scurvy, a disorder characterized by skin lesions and blood vessel hemorrhages as a result of weakened connective tissue.

Table 15.1   Representative Members of the Collagen Family

Collagen class	Types	Role
Fibril-forming	I, II, III, V, XI, XXIV, XXVII	Formation of fibers in most connective tissues, including skin, bone and cartilage
Network-forming	IV	Formation of basal laminae
Fibril-associated	IX, XII, XIV, XVI, XIX, XX, XXI, XXII, XXVI	Association of collagen fibrils with other extracellular matrix components
Anchoring	VII	Attachment of basal laminae to underlying connective tissue
Transmembrane	XIII, XVII, XXIII	Cell surface molecules with transmembrane domains

Figure 15.15 Collagen fibrils (A) Collagen molecules assemble in a regular staggered array to form fibrils. The molecules overlap by one-fourth of their length, and there is a short gap between the N terminus of one molecule and the C terminus of the next. The assembly is strengthened by covalent cross-links between side chains of lysine or hydroxylysine residues, primarily at the ends of the molecules. (B) Electron micrograph of collagen fibrils. The staggered arrangement of collagen molecules and the gaps between them are responsible for the characteristic cross-striations in the fibrils.

(A)

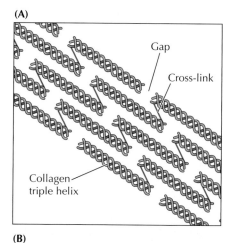

(B)

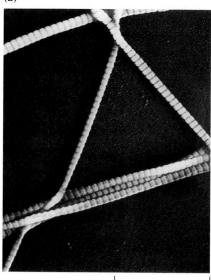

 1 μm

(A)

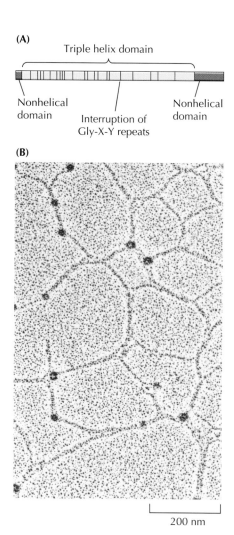

200 nm

assemble into **collagen fibrils** in which the triple helical molecules are associated in regular staggered arrays (**Figure 15.15**). These fibrils do not form within the cell because the fibril-forming collagens are synthesized as soluble precursors (**procollagens**) that contain nonhelical segments at both ends of the polypeptide chain. Procollagen is cleaved to collagen after its secretion, so the assembly of collagen into fibrils takes place only outside the cell. The association of collagen molecules in fibrils is further strengthened by the formation of covalent cross-links between the side chains of lysine and hydroxylysine residues. Frequently, the fibrils further associate with one another to form collagen fibers, which can be several microns in diameter.

In addition to the fibril-forming collagens, connective tissues contain fibril-associated collagens, which bind to the surface of collagen fibrils and link them both to one another and to other matrix components (see Table 15.1). Basal laminae form from type IV collagen, which is a network-forming collagen (**Figure 15.16**). The Gly-X-Y repeats of type IV collagen are frequently interrupted by short nonhelical sequences. Because of these interruptions, type IV collagen is more flexible than the fibril-forming collagens. Consequently, it assembles into two-dimensional cross-linked networks instead of fibrils. Other types of collagen form anchoring fibrils, which link basal laminae to underlying connective tissues, and some collagens are transmembrane proteins that participate in cell–matrix interactions.

Connective tissues also contain **elastic fibers**, which are particularly abundant in organs that regularly stretch and then return to their original shape. The

Figure 15.16 Type IV collagen (A) The Gly-X-Y repeat structure of type IV collagen (yellow) is interrupted by multiple nonhelical sequences (bars). (B) Electron micrograph of a type IV collagen network. (B, P. D. Yurchenco and J. C. Schittny, 1990. *FASEB J.* 4: 1577.)

Triple helix domain

Nonhelical domain

Interruption of Gly-X-Y repeats

Nonhelical domain

lungs, for example, stretch each time a breath is inhaled and return to their original shape with each exhalation. Elastic fibers are composed principally of a protein called **elastin**, which is cross-linked into a network by covalent bonds formed between the side chains of lysine residues (similar to those found in collagen). This network of cross-linked elastin chains behaves like a rubber band, stretching under tension and then snapping back when the tension is released.

Matrix polysaccharides

The fibrous structural proteins of the extracellular matrix are embedded in gels formed from polysaccharides called **glycosaminoglycans** (**GAGs**), which consist of repeating units of disaccharides (**Figure 15.17**). One sugar of the disaccharide is either *N*-acetylglucosamine or *N*-acetylgalactosamine and the second is usually acidic (either glucuronic acid or iduronic acid). With the exception of hyaluronan, these sugars are modified by the addition of sulfate groups. Consequently, GAGs are highly negatively charged. Like the pectins of plant cell walls, they bind positively charged ions and trap water molecules to form hydrated gels, thereby providing mechanical support to the extracellular matrix. The common sulfated GAGs are dermatan sulfate, chondroitin sulfate, keratan sulfate, and heparan sulfate.

Hyaluronan is the only GAG that occurs as a single long polysaccharide chain. Like cellulose and chitin, it is synthesized at the plasma membrane by a transmembrane hyaluronan synthase. All of the other GAGs are linked to proteins to form **proteoglycans**, which can consist of up to 95% polysaccharide by weight. Proteoglycans can contain as few as one or as many as more than 100 GAG chains attached to serine residues of a core protein. A variety of core proteins (ranging from 10 to >500 kd) have been identified, so the proteoglycans are a diverse group of macromolecules. In addition to being components of the extracellular matrix, some proteoglycans such as

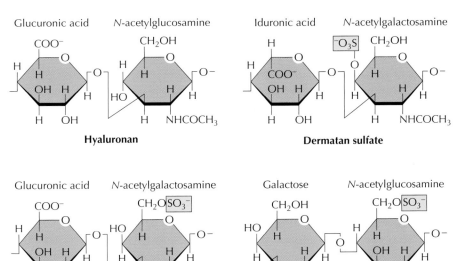
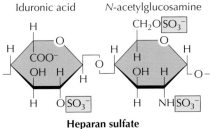

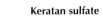
Figure 15.17 Major types of glycosaminoglycans   Glycosaminoglycans (GAGs) consist of repeating disaccharide units. With the exception of hyaluronan, the sugars frequently contain sulfate.

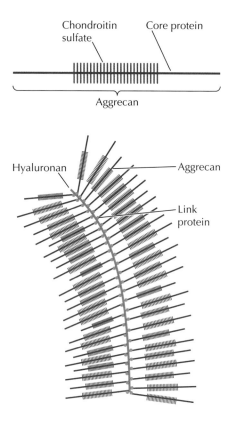

Figure 15.18 Complexes of aggrecan and hyaluronan Aggrecan is a large proteoglycan consisting of more than 100 chondroitin sulfate chains joined to a core protein. Multiple aggrecan molecules bind to long chains of hyaluronan (green), forming large complexes in the extracellular matrix of cartilage. This association is stabilized by link proteins (orange).

syndecans (with transmembrane α helices) and glypicans (with GPI anchors) are cell surface proteins that function along with integrins in cell–cell and cell–matrix adhesion.

A number of proteoglycans interact with hyaluronan to form large complexes in the extracellular matrix. A well-characterized example is aggrecan, the major proteoglycan of cartilage (**Figure 15.18**). More than 100 chains of chondroitin sulfate are attached to a core protein of about 250 kd, forming a proteoglycan of about 3000 kd. Multiple aggrecan molecules then associate with chains of hyaluronan, forming large aggregates (>100,000 kd) that become trapped in the collagen network. Proteoglycans also interact with both collagen and other matrix proteins to form gel-like networks in which the fibrous structural proteins of the extracellular matrix are embedded. For example, perlecan (the major heparan sulfate proteoglycan of basal laminae) binds to both type IV collagen and to the adhesion protein laminin, which is discussed shortly.

Adhesion proteins

Adhesion proteins, the third class of extracellular matrix constituents, are responsible for linking the components of the matrix to one another and to the surfaces of cells. They interact with collagen and proteoglycans to specify matrix organization and are the major binding sites for cell surface receptors, such as integrins (discussed in the next section).

The prototype of these molecules is **fibronectin**, which is the principal adhesion protein of connective tissues. Fibronectin is a dimeric glycoprotein consisting of two polypeptide chains, each containing nearly 2500 amino acids (**Figure 15.19**). Within the extracellular matrix, fibronectin is often cross-linked into fibrils. Fibronectin has binding sites for both collagen and proteoglycans so it cross-links these matrix components. A distinct site on the fibronectin molecule is recognized by integrins. Fibronectin proteins vary greatly from tissue to tissue but all are derived by alternative splicing of the mRNA of a single gene.

The principal components of basal laminae are distinct adhesion proteins of the **laminin** family (**Figure 15.20**), which are the major organizers of basal lamina assembly. Laminins are cross- or T-shaped heterotrimers of α, β, and γ subunits, which are the products of five α genes, four β genes, and three γ genes. Like type IV collagen, laminins can self-assemble into meshlike networks. The different laminin subunits also have binding sites for cell surface receptors (e.g., integrins) and for proteoglycans. In addition, laminins are tightly associated with another adhesion protein, called **nidogen**, which also

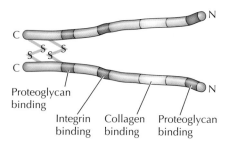

Figure 15.19 Structure of fibronectin Fibronectin is a dimer of similar polypeptide chains joined by disulfide bonds near the C terminus. Sites for binding to proteoglycans, integrins, and collagen are indicated. The molecule also contains additional binding sites that are not shown.

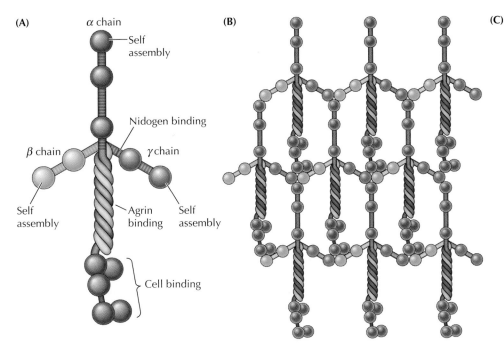

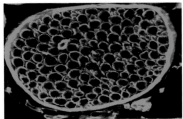

Figure 15.20 Laminins (A) Laminins consist of three polypeptide chains designated α, β, and γ. Each polypeptide consists of rod-like domains with interspersed globular domains. The polypeptide chains are linked by a triple helical region formed by the rod-like domains. Some of the binding sites for self-assembly, nidogen, agrin (a proteoglycan), and cell surface receptors are indicated. (B) Assembly of laminins into a network. (C) A laminin network in the basal lamina surrounding a nerve, stained with fluorescent antibody against the laminin β chain. (From M. Durbeej, 2010. *Cell Tissue Res.* 339: 262)

binds to type IV collagen. As a result of these multiple interactions, laminin, nidogen, collagen, and the proteoglycans form cross-linked networks within basal laminae. Since the different laminin trimers exhibit tissue-specific distributions, it is not surprising that mutations in the laminin genes underlie many human congenital diseases, including a form of muscular dystrophy.

Cell–matrix interactions

The major cell surface receptors responsible for the attachment of cells to the extracellular matrix are the **integrins**. The integrins are a family of transmembrane proteins consisting of two subunits, designated α and β (**Figure 15.21**), with combinations of 18 α subunits and eight β subunits forming 24 different integrins. The integrins bind to short amino acid sequences present in multiple components of the extracellular matrix, including collagen, fibronectin, and laminin. The first such integrin-binding site to be characterized was the sequence Arg-Gly-Asp in fibronectin, which is recognized

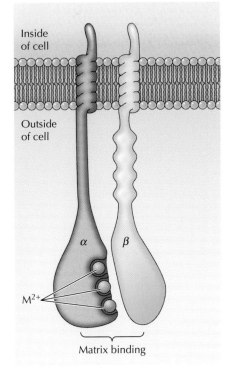

Figure 15.21 Structure of integrins The integrins are heterodimers of two transmembrane subunits, designated α and β. The α subunit binds divalent cations, designated M^{2+}. The matrix-binding region is composed of portions of both subunits.

The Characterization of Integrin

Structure of Integrin, a Glycoprotein Involved in the Transmembrane Linkage between Fibronectin and Actin

John W. Tamkun, Douglas W. DeSimone, Deborah Fonda, Ramila S. Patel, Clayton Buck, Alan F. Horwitz and Richard O. Hynes

Massachusetts Institute of Technology, Cambridge, MA (JWT, DWD, DF, RSP, and ROH), The Wistar Institute, Philadelphia (CB), and The University of Pennsylvania School of Medicine, Philadelphia (AFH)

Cell, Volume 46, 1986, pages 271–282

The Context

The molecular basis of cell adhesion to the extracellular matrix has been of interest to cell biologists since it was observed that adhesion to the matrix was reduced in cancer cells, potentially related to their abnormal growth and ability to metastasize or spread throughout the body. In the late 1970s and early 1980s work from several laboratories, including that of Richard Hynes, established that there was a physical link between the stress fibers of the actin cytoskeleton and fibronectin in the extracellular matrix. Several of the cytoskeletal proteins involved in this linkage, including vinculin and talin, had been identified, but the critical transmembrane proteins that linked cells to the extracellular matrix remained unknown.

Candidates for transmembrane proteins involved in cell adhesion to the extracellular matrix had been identified using antibodies prepared by several groups of scientists, including Alan Horwitz and Clayton Buck. These antibodies recognized a complex of 140-kd glycoproteins that appeared to be transmembrane proteins. Immunofluorescence and immunoelectron microscopy localized these glycoproteins to points of cell adhesion to the matrix. Additional studies showed that the 140-kd glycoproteins bound to fibronectin and were involved in cell adhesion. It thus appeared that these 140-kd glycoproteins were likely candidates for the transmembrane proteins

responsible for cell–matrix adhesion. In the experiments described by Tamkun et al., these antibodies were used to isolate a molecular clone encoding one of these glycoproteins, thereby providing the first molecular characterization of integrin.

The Experiments

To isolate a molecular clone encoding one of the 140-kd glycoproteins, Tamkun et al. prepared a cDNA library from mRNA of chick embryo fibroblasts. This cDNA library was prepared in a bacteriophage λ expression vector that directed high-level transcription and translation of the eukaryotic cDNA inserts in *E. coli* (see Figure 4.21). The cDNA library contained approximately 100,000 independent cDNA inserts, sufficient to represent clones of all the mRNAs expressed in the chick embryo fibroblasts from which it was prepared. The library was then screened with antibodies against the 140-kd glycoproteins to identify those recombinant phages carrying the cDNAs of interest. Plaques produced by individual recombinant phages were transferred to nitrocellulose filters as would be done to screen a recombinant library by nucleic acid hybridization (see Figure 4.25). However, to screen the expression library, the filters were probed with an antibody to identify clones expressing the desired protein. By such antibody screening, Tamkun et al. successfully identified several recombinant clones expressing proteins that were recognized by antibodies against the 140-kd glycoproteins.

Their next challenge was to determine whether the cloned cDNAs actually encoded one or more of the 140-kd

Richard O. Hynes

glycoproteins. To do this they used proteins expressed from the clones to purify antibodies that recognized the cloned proteins. In addition, they injected rabbits with the cloned proteins to raise new antibodies specifically directed against the proteins encoded by the cloned cDNAs. The resulting antibodies recognized one of the proteins of the 140-kd complex from chick embryo fibroblasts in immunoblot assays, establishing the relationship of the cloned cDNAs to this protein. In addition, antibodies purified against the cloned proteins stained cells similarly to the original antibody in immunofluorescence assays, yielding a staining pattern corresponding to the sites of stress fiber attachment to the extracellular matrix (see figures). The results of both immunoblotting and immunofluorescence thus indicated that the cloned cDNAs encoded one of the proteins of the 140-kd glycoprotein complex.

The cDNA was then sequenced and found to encode a protein consisting of 803 amino acids. The protein included an amino terminal signal sequence and an apparent transmembrane α helix of 23 hydrophobic amino acids near the carboxy

(A)

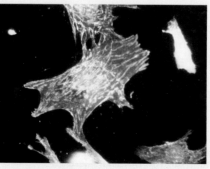

(B)

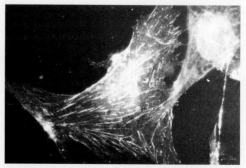

Immunofluorescence of chick embryo fibroblasts stained with (A) the original antiserum to the 140-kd glycoprotein complex or (B) with antibodies purified against the protein expressed from the cDNA clones.

terminus. The sequence predicted a short cytosolic domain and a large extracellular domain containing multiple glycosylation sites consistent with the expected structure of a transmembrane glycoprotein.

The Impact

Hynes and his colleagues concluded that they had cloned a cDNA encoding a transmembrane protein that functioned to link fibronectin in the extracellular matrix to the cytoskeleton. They named the protein integrin "to denote its role as an integral membrane complex involved in the transmembrane association between the extracellular matrix and the cytoskeleton." The initial cloning of integrin led to our current understanding of the molecular basis of stable cell junctions. At focal adhesions, integrins link the extracellular matrix to actin filaments. In addition, integrins mediate attachment of epithelial cells to the extracellular matrix at hemidesmosomes, where they link the extracellular matrix to intermediate filaments. Thus, as suggested by Hynes and his colleagues, integrins play a general role in cell–matrix adhesion. Subsequent studies have shown that integrins also play an important role as signaling complexes in cells, relaying signals from outside of the cell to control multiple aspects of cell movement, proliferation, and survival (discussed in Chapter 16). The characterization of integrins thus opened the door to understanding not only the nature of cell attachment to the matrix but also to elucidating novel signaling mechanisms that regulate cell behavior.

---

by several members of the integrin family. Other integrins, however, bind to distinct peptide sequences, including recognition sequences in collagens, laminin, and many different proteoglycans. Transmembrane proteoglycans on the surface of a variety of cells also bind to components of the extracellular matrix and modulate cell–matrix interactions.

In addition to attaching cells to the extracellular matrix the integrins serve as anchors for the cytoskeleton (**Figure 15.22**). The resulting linkage of the cytoskeleton to the extracellular matrix is responsible for the stability

Snake venoms contain proteins called disintegrins, that bind integrins and interfere with their ability to bind proteins of the extracellular matrix. Disrupting these connections between cells and the extracellular matrix helps snakes liquefy their prey.

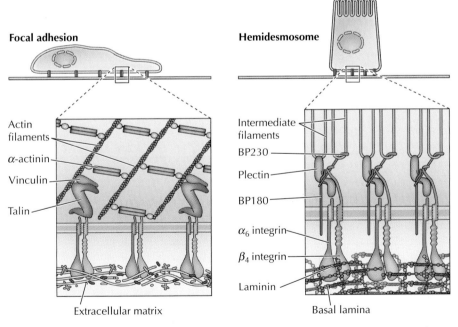

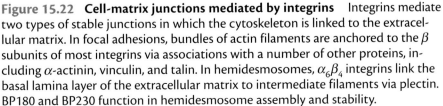

Figure 15.22 Cell-matrix junctions mediated by integrins Integrins mediate two types of stable junctions in which the cytoskeleton is linked to the extracellular matrix. In focal adhesions, bundles of actin filaments are anchored to the β subunits of most integrins via associations with a number of other proteins, including α-actinin, vinculin, and talin. In hemidesmosomes, $\alpha_6\beta_4$ integrins link the basal lamina layer of the extracellular matrix to intermediate filaments via plectin. BP180 and BP230 function in hemidesmosome assembly and stability.

of cell–matrix junctions. Distinct interactions between integrins and the cytoskeleton are found at two types of cell–matrix junctions: **focal adhesions** and **hemidesmosomes**, both of which were introduced in Chapter 13. Focal adhesions attach a variety of cells, including fibroblasts, to the extracellular matrix. At focal adhesions, the cytoplasmic domains of the β subunits of integrins anchor the actin cytoskeleton by associating with bundles of actin filaments through actin-binding proteins (such as α-actinin, vinculin, and talin). Hemidesmosomes mediate epithelial cell attachments at which a specific integrin (designated $\alpha_6\beta_4$) is linked through plectin to intermediate filaments instead of to actin. This is mediated by the long cytoplasmic tail of the β_4 integrin subunit. The $\alpha_6\beta_4$ integrin also binds to laminin, so hemidesmosomes anchor epithelial cells to the basal lamina.

Regulated changes in integrin activity underlie the rapid assembly and disassembly of focal adhesions during cell movement (see Figure 13.20). The ability of integrins to reversibly bind matrix components is dependent on their ability to change conformation between inactive and active states, which can be triggered by signals from either inside or outside the cell. In the inactive state, integrins are unable to bind to the matrix because the head groups containing the ligand-binding site are held close to the cell surface (**Figure 15.23**). When activated, the integrins undergo a conformational change that extends head groups into the matrix and allows ligand binding. This initial interaction of integrins with the extracellular matrix recruits additional integrins to the site of adhesion, leading to the development of small clusters of integrins called **focal complexes**. The focal complexes then develop into focal adhesions by the recruitment of α-actinin, vinculin, talin, and formin (which initiates actin bundle formation; see Figure 13.4). Focal adhesions can be very stable interactions involved in tissue structure, or they can turn over rapidly as cells move. During cell migration, the formation of new focal complexes at the leading edge of the cell results in the loss of tension at the

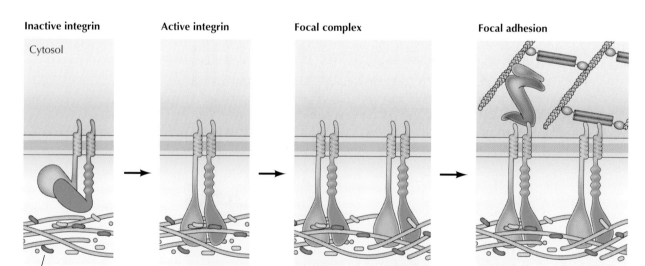

Figure 15.23 Integrin activation and formation of focal adhesions In the inactive state, integrin head groups are held close to the cell surface. Signals from the cytosol activate integrins, extending the head groups and enabling them to bind to the extracellular matrix. This leads to recruitment of additional integrins to form a focal complex, which develops into a focal adhesion (see Figure 15.22).

old focal adhesions, leading to the inactivation of integrin binding to the extracellular matrix.

Just as plant cells modify their cell walls to grow, animal cells modify the extracellular matrix as they grow and migrate. Several types of enzymes that digest glycosaminoglycans as well as a large family of proteases are secreted by cells. The first of these proteases to be identified was an enzyme that digests collagen (a collagenase), discovered in 1962 by Jerome Gross and Charles Lapiere in frog tadpole tails during metamorphosis. This enzyme was the founding member of what became the matrix metalloprotease family, which consists of 23 proteases in humans. These enzymes digest a variety of matrix proteins, including collagens and laminin, as well as cell surface receptors and adhesion molecules. The metalloproteases not only play important roles in the normal movements of cells during development, but also in the growth and metastasis of cancers.

FYI

Cancer cells secrete proteases that digest proteins of the extracellular matrix, allowing the cancer cells to invade surrounding tissue and metastasize to other parts of the body.

Cell–Cell Interactions

Direct interactions between cells, as well as between cells and the extracellular matrix, are critical to the development and function of multicellular organisms. Some cell–cell interactions are transient, such as the interactions between cells of the immune system and the interactions that direct white blood cells to sites of tissue inflammation. In other cases, stable cell–cell junctions play a key role in the organization of cells in tissues. For example, several different types of stable cell–cell junctions are critical to the maintenance and function of epithelial cell sheets. In addition to mediating cell adhesion, specialized types of junctions provide mechanisms for rapid communication between cells. Plant cells also associate with their neighbors not only by interactions between their cell walls but also by specialized junctions between their plasma membranes.

Adhesion junctions

Cell–cell adhesion is a selective process such that cells adhere only to other cells of specific types. This selectivity was first demonstrated in classical studies of embryo development, which showed that cells from one tissue (e.g., liver) specifically adhere to cells of the same tissue rather than to cells of a different tissue (e.g., brain). Such selective cell–cell adhesion is mediated by transmembrane proteins called **cell adhesion molecules**, which can be divided into four major groups: the **selectins**, the integrins, the **immunoglobulin (Ig) superfamily** (so named because family members contain structural domains similar to immunoglobulins), and the **cadherins** (Table 15.2). Cell

Table 15.2   Cell Adhesion Molecules

Family	Ligands recognized	Stable cell junctions
Selectins	Carbohydrates	No
Integrins	Extracellular matrix	Focal adhesions and hemidesmosomes
	Members of Ig superfamily	Not usually
Ig superfamily	Integrins	Not usually
	Other Ig superfamily proteins	No
Cadherins	Other cadherins	Adherens junctions and desmosomes

adhesion mediated by the selectins, integrins, and most cadherins requires Ca^{2+}, Mg^{2+}, or Mn^{2+}, so many adhesive interactions between cells are divalent cation-dependent.

The selectins mediate transient interactions between leukocytes and endothelial cells or blood platelets. There are three members of the selectin family: L-selectin, which is expressed on leukocytes; E-selectin, which is expressed on endothelial cells; and P-selectin, which is expressed on platelets. The selectins recognize cell surface carbohydrates (**Figure 15.24**). One of their critical roles is to initiate the interactions between leukocytes and endothelial cells during the migration of leukocytes from the circulation to sites of tissue inflammation. The selectins mediate the initial adhesion of leukocytes to endothelial cells. This is followed by the formation of more stable adhesions in which integrins on the surface of leukocytes bind to intercellular adhesion molecules (ICAMs), which are members of the Ig superfamily expressed on the surface of endothelial cells. The firmly attached leukocytes are then able to penetrate the walls of capillaries and enter the underlying tissue by migrating between endothelial cells.

The binding of ICAMs to integrins is an example of a **heterophilic interaction** in which an adhesion molecule on the surface of one cell (e.g., an ICAM) recognizes a different molecule on the surface of another cell (e.g., an integrin). Other members of the Ig superfamily mediate **homophilic interactions** in which an adhesion molecule on the surface of one cell binds to the same molecule on the surface of another cell. Such homophilic binding can lead to selective adhesion between cells of the same type. For example, neural cell adhesion molecules (N-CAMs) are members of the Ig superfamily expressed

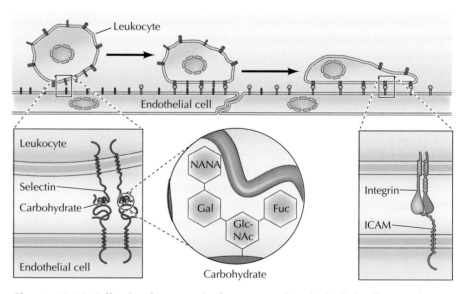

Figure 15.24 Adhesion between leukocytes and endothelial cells Leukocytes leave the circulation at sites of tissue inflammation by interacting with the endothelial cells of capillary walls. The first step in this interaction is the binding of leukocyte selectins to carbohydrates (oligosaccharide ligands) on the endothelial cell surface. The oligosaccharide contains N-acetylglucosamine (GlcNAc), fucose (Fuc), galactose (Gal), and sialic acid (N-acetylneuraminic acid, NANA). This step is followed by more stable interactions between leukocyte integrins and intercellular adhesion molecules (ICAMs)—members of the Ig superfamily—on endothelial cells.

on nerve cells, and homophilic binding between N-CAMs contributes to the formation of selective associations between nerve cells during development. There are more than 100 members of the Ig superfamily, which mediate a variety of cell–cell interactions.

The fourth class of cell adhesion molecules is the cadherins. Cadherins are involved in selective adhesion between embryonic cells, formation of specific synapses in the nervous system, and are the proteins primarily responsible for the maintenance of stable junctions between cells in tissues. Cadherins are a large family of proteins (more than 100 members) that share a highly conserved extracellular domain that mediates largely homophilic interactions. For example, E-cadherin is expressed on epithelial cells so homophilic interactions between E-cadherins lead to the selective adhesion of epithelial cells to one another. It is noteworthy that loss of E-cadherin can contribute to the development of cancers arising from epithelial tissues, illustrating the importance of cell–cell interactions in controlling cell behavior. Different members of the cadherin family, such as N-cadherin (neural cadherin) and P-cadherin (placental cadherin), mediate selective adhesion of other cell types. There are several subfamilies of cadherins (classic cadherins, desmosomal cadherins, and protocadherins) which differ primarily in their transmembrane and cytosolic domains.

Cell–cell interactions mediated by the selectins, integrins, and most members of the Ig superfamily are generally transient, although Ig superfamily proteins (N-CAM, for example) participate in forming stable junctions between neurons at synapses. However, stable adhesion junctions involving the cytoskeletons of adjacent cells are usually based on cadherins. As discussed in Chapter 13, these cell–cell junctions are of two types: **adherens junctions** and **desmosomes**. At these junctions, classic and desmosomal cadherins are linked to actin bundles and intermediate filaments, respectively, by the interaction of their cytosolic tails with β-catenin or desmoplakin. The role of the cadherins in linking the cytoskeletons of adjacent cells is thus analogous to that of the integrins in forming stable junctions between cells and the extracellular matrix.

The basic structural unit of an adherens junction includes β-catenin, p120, and α-catenin, in addition to the classic transmembrane cadherins (**Figure 15.25**). β-catenin and p120 are related members of the **armadillo protein family** (named after the *Drosophila* β-catenin protein). They bind to the cytosolic tail of cadherins and act to maintain stability of the junction. In addition, β-catenin binds to α-catenin (which despite its name is not part of the armadillo family). α-catenin binds to vinculin, which binds to actin, thereby linking adherens junctions to the actin cytoskeleton. The catenins thus serve to link the actin cytoskeleton of one cell, through the transmembrane cadherins, to the actin cytoskeleton of an adjacent cell.

Figure 15.25 Adherens junctions In adherens junctions, cadherins link the actin filaments (cytoskeleton) of one cell to the actin filaments of another. The classic transmembrane cadherins bind β-catenin and p120, which regulates the stability of the junction. β-catenin also binds α-catenin, which mediates the association of actin filaments with adherens junctions by interacting with vinculin.

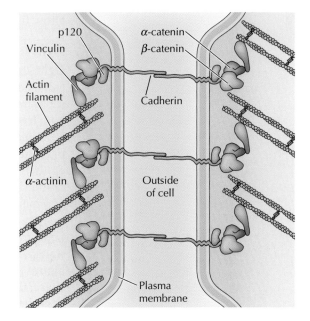

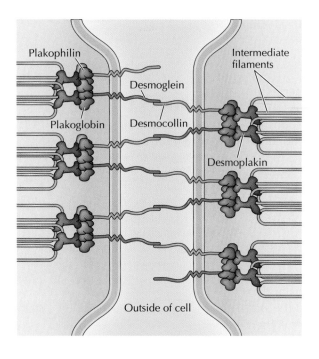

Plakophilin

Desmoglein

Plakoglobin Desmocollin

Desmoplakin

Intermediate filaments

Outside of cell

Figure 15.26 Desmosomes In desmosomes, the desmosomal cadherins (desmoglein and desmocollin) link to intermediate filaments of adjacent cells. Plakoglobin and plakophilin bind to the cytosolic tails of the cadherins, and to the plakin family protein desmoplakin, which binds to intermediate filaments.

In addition to the cadherins, a second type of cell adhesion molecule, **nectin**, is also present at adherens junctions. There are four known nectins and a related family of nectin-like proteins, which are members of the Ig superfamily. Like cadherins, nectins interact in a homophilic manner. However, nectins can also bind heterophilically to different nectins in the membrane of the apposing cell. Such heterophilic binding of nectins can mediate associations between different types of cells, for example, between neurons and glial cells in the brain. Like cadherins, nectins form links to the actin cytoskeleton. In addition to forming heterotypic cell junctions, homotypic cell junctions are often intiated by nectins, which then recruit cadherins to the adhesion sites.

In contrast to adherens junctions, desmosomes directly link the intermediate filament (rather than the actin) cytoskeletons of adjacent cells (**Figure 15.26**). The desmosomal cadherins, **desmoglein** and **desmocollin**, bind across the junction. The armadillo family proteins plakoglobin and plakophilin bind to the cytosolic tails of the desmosomal cadherins and provide a direct link to the intermediate filament-binding protein, desmoplakin. Desmoplakin is related to plectin, which functions analogously to bind intermediate filaments at hemidesmosomes (see Figure 15.22). The strength of desmosomal links between cells is a property of both the intermediate filaments to which they are attached and the multiple interactions between plakoglobin, plakophilin, and desmoplakin that link the intermediate filaments to the cadherins. Multiple genes for the desmosomal cadherins and the armadillo family proteins allow tissue-specific variation in desmosome properties.

Tight junctions

Tight junctions are critically important to the function of epithelial cell sheets as barriers between fluid compartments. For example, the intestinal epithelium separates the lumen of the intestine from the underlying connective tissue, which contains blood capillaries. Tight junctions play two roles in allowing epithelia to fulfill such barrier functions. First, tight junctions form seals that prevent the free passage of molecules (including ions) between the cells of epithelial sheets. Second, tight junctions separate the apical and basolateral domains of the plasma membrane by preventing the free diffusion of lipids and membrane proteins between them. Consequently, specialized transport systems in the apical and basolateral domains are able to control the traffic of molecules between distinct extracellular compartments, such as the transport of glucose between the intestinal lumen and the blood supply (see Figure 14.32). While tight junctions are very effective seals of the extracellular space, they provide minimal adhesive strength between the apposing cells,

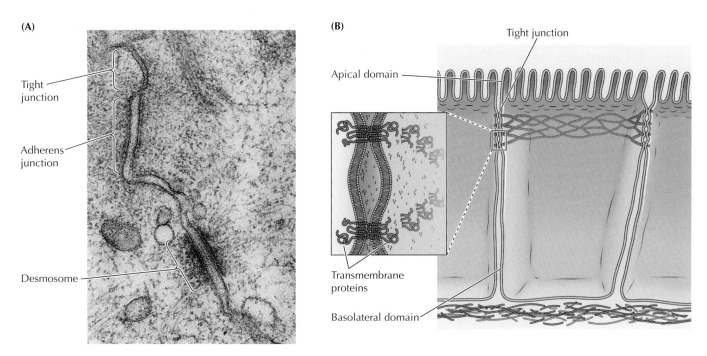

Figure 15.27 A junctional complex   (A) Electron micrograph of epithelial cells joined by a junctional complex, including a tight junction, an adherens junction, and a desmosome. (B) Tight junctions are formed by interactions between strands of transmembrane proteins on adjacent cells.

so they are usually associated with adherens junctions and desmosomes in a **junctional complex** (**Figure 15.27**).

Tight junctions are the closest known contacts between adjacent cells. They were originally described as sites of apparent fusion between the outer leaflets of the plasma membranes, although it is now clear that the membranes do not fuse. Instead, tight junctions are formed by a network of protein strands that continues around the entire circumference of the cell (see Figure 15.27B). Each strand in these networks is composed of transmembrane proteins of the occludin, claudin, and junctional adhesion molecule (JAM) families (**Figure 15.28**). All three of these proteins bind to similar proteins on adjacent cells, thereby sealing the space between their plasma membranes. The cytosolic tails of occludins, claudins, and JAMs are also associated with proteins of the zonula occludens family, which link the tight junction complex to the actin cytoskeleton and hold the tight junction in place on the plasma membrane.

Gap junctions

The activities of individual cells in multicellular organisms need to be closely coordinated. This can be accomplished by signaling molecules that are released from one cell and act on another, as discussed in Chapter 16. However, within an individual tissue, such as the liver, cells are often linked by **gap junctions**, which provide direct connections between the cytoplasms of adjacent cells. Gap junctions are regulated channels through the plasma membrane that, when open, allow ions and small molecules (less than approximately 1000 daltons) to diffuse between neighboring cells. Consequently, gap junctions couple both the metabolic activities and the electric responses of the cells

FYI

The epithelial cell layers constitute significant barriers against invading microorganisms. Pathogenic bacteria have evolved strategies that disrupt junctional complexes, allowing them to penetrate between the cells of epithelial sheets.

Figure 15.28 Tight junction proteins There are three major transmembrane proteins in a tight junction: occludin, claudin, and the junctional adhesion molecule (JAM). All three transmembrane proteins interact with similar molecules on the apposing cell and with zonula occludens proteins that link to actin filaments.

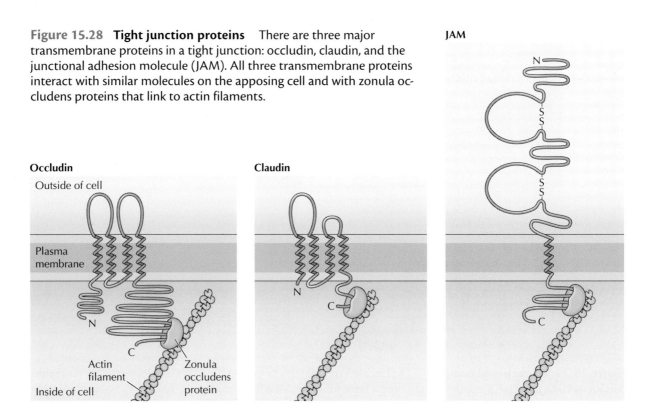

they connect. Most cells in animal tissues—including epithelial cells, endothelial cells, and the cells of cardiac and smooth muscle—communicate by gap junctions. In electrically excitable cells, such as heart muscle cells, the direct passage of ions through gap junctions couples and synchronizes the contractions of neighboring cells. Gap junctions also allow the passage of some intracellular signaling molecules, such as cAMP and Ca^{2+}, between adjacent cells, potentially coordinating the responses of cells in tissues.

Gap junctions are constructed of transmembrane proteins of the **connexin** family, which consists of at least 21 different human proteins. Six connexins assemble to form a cylinder with an aqueous pore in its center (**Figure 15.29**). Such an assembly of connexins, called a **connexon**, in the plasma membrane of one cell then aligns with a connexon of an adjacent cell, forming a channel between the two cytoplasms. The plasma membranes of the two cells are separated by a gap corresponding to the space occupied by the connexin extracellular domains—hence the term "gap junction," which was coined by electron microscopists. Many cells express more than one member of the connexin family, and combinations of different connexin proteins may give rise to gap junctions with varying properties.

Specialized assemblies of gap junctions occur on specific nerve cells and form an **electrical synapse**. The individual connexons within the electrical synapse can be opened or closed in response to several types of signals but, when open, allow the rapid passage of ions between the two nerve cells. The

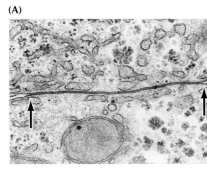

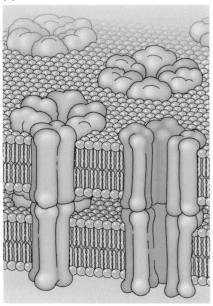

Figure 15.29 Gap junctions (A) Electron micrograph of a gap junction (arrows) between two liver cells. (B) Gap junctions consist of assemblies of six connexins (called a connexon), which form open channels through the plasma membranes of adjacent cells.

◄ **Molecular Medicine** ►

Gap Junction Diseases

The Diseases

Several unrelated human diseases have been found to result from mutations in genes that encode the connexin proteins of gap junctions. The first of these diseases to be described is the X-linked form of Charcot-Marie-Tooth disease (CMT), which was mapped to mutations in the gene encoding connexin 32 in 1993. CMT is an inherited disease leading to progressive degeneration of peripheral nerves, with slow loss of muscle control and eventual muscle degeneration. There are now more than 200 different mutations in connexin 32 known to cause CMT. CMT can also be caused by mutations in several different genes encoding myelin proteins, defects which lead directly to degeneration of myelinated nerves. In addition, one form of CMT has been mapped to a mutation in the human gene for nuclear lamin A (see Chapter 10 Molecular Medicine, Nuclear Lamina Diseases).

The finding that CMT could be caused by mutations in a connexin implied that gap junctions played a critical role in myelinated nerves. Moreover, CMT was only the first of several inherited diseases that now have been traced back to mutations in genes encoding human connexins. The most common pathological consequences of connexin mutations are cataracts, skin disorders, and deafness.

Molecular and Cellular Basis

The human genome contains 21 genes encoding different connexins (Cx genes), which are divided into three subfamilies (α, β, and γ). Mutations in eight of these genes have been identified as causes of human disease (see table). For example, mutations in Cx32/β1 are associated with both Charcot-Marie-Tooth disease and deafness. The connexin proteins are widely expressed; Cx32/β1 is expressed in peripheral nerve, liver, and brain tissues, and Cx43/α1 (which is also associated with deafness) is expressed in over 35 different human tissues. Thus it is somewhat surprising that there are fewer than ten different types of human diseases known to be caused by connexin

mutations. Since most tissues express several different connexin proteins, the simple explanation for the lack of a defect in many tissues resulting from the loss of any single connexin is that other connexins compensate for the one that is missing. The question then is why any mutation in a single connexin gene causes a human disease.

The initial discovery of Cx32/β1 mutations in CMT was based on eight different mutations in various families: six single base changes that resulted in nonconservative amino acid substitutions, one frameshift that yielded a shorter protein, and one mutation in the promoter region of the gene. These eight mutations in Cx32/β1, as well as 262 others that have been subsequently discovered, cause clinically identical CMT, suggesting that myelinated nerve might be a tissue particularly sensitive to gap junction defects. Three other tissues also appear sensitive to connexin mutations: the lens of the eye, the sensory epithelium of the inner ear, and the skin. The sensitivity of the lens is easiest to understand, because the lens fiber cells lose most of their organelles during development and fill up with crystalline protein to allow the passage of light. They must obtain nutrients and ions from the lens epithelial cells through gap junctions, and loss of this input leads to cataract formation. The sensitivity of the sensory epithelium of the inner ear is related to the need for the epithelial cells to rapidly exchange K^+ via gap junctional communication. Less clear is the basis for overgrowth of the outer layers of the skin, but it is thought that gap junctions play a role in the balance between proliferation and differentiation,

and gap junctional mutations perturb the balance.

Each of these tissues expresses additional connexins, and their failure to compensate for loss of the mutant connexin appears to be based on two phenomena. Although gap junctions can be formed from combinations of different connexins, not all connexins are able to cooperate to form functional connexons. Thus the connexins in sensitive tissues may not be able to compensate effectively for one that is mutated. In addition, gap junction connexons are assembled not at the cell surface but in the Golgi apparatus or earlier in the secretory process. Thus a single mutant connexin that cannot be properly processed and exported might act as a dominant negative and interfere with the processing of the normal connexins expressed in that tissue. This has been reported for a mutant of Cx46/α3 that causes congenital cataracts: the mutant protein fails to exit the ERGIC (ER–Golgi intermediate compartment) or Golgi complex and traps normal connexins along with it. It thus appears that the relationship of particular disorders to connexin mutations may result from the gap junctional requirements of a given tissue, as well as the nature of interactions between the connexins that are expressed in that tissue.

Prevention and Treatment

The identification of several human diseases caused by mutations in connexin genes illustrates the importance of gap junctions to normal tissue function. Like other junctions critical to tissue structure or function, gap junctions would seem to

Gap Junction Mutations in Human Disease

Disease	Connexin protein[a]
Charcot-Marie-Tooth disease	Cx32/β1
Deafness	Cx26/β2, Cx30/β6, Cx31/β3, Cx32/β1, Cx43/α1
Skin diseases	
Erythrokeratoderma variabilis	Cx31/β3, Cx30.3/β4
Vohwinkel syndrome	Cx26/β2
Cataracts	Cx46/α3, Cx50/α8

[a]Cx proteins are named either by molecular weight or as a member of one subfamily

(Continued on next page)

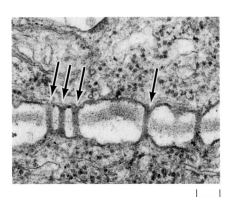

be obvious locations for hereditary human diseases, yet such diseases are uncommon. Recent discoveries on the processing and assembly of gap junction connexons have started to explain the bases of these rare diseases. While it is thought that all the human connexin genes have been described, the tissue distribution of the proteins—especially during embryonic development—remains poorly understood. New knowledge about the distribution and interactions of the connexin proteins will provide a basis for understanding each of the pathologies resulting from mutations in connexin genes. Only then can possible compensatory therapies—for example, regulating cell proliferation in the skin or modulating the exchange of K⁺ in the inner ear—be developed.

Reference
Pfenniger, A., A. Wohlwend and B. R. Kwak. 2011. Mutations in connexin genes and disease. *Eur. J. Clin. Invest* 41: 103–116.

Molecular Medicine

Figure 15.30 Plasmodesmata
Electron micrograph of plasmodesmata (arrows).

0.1 μm

importance of gap junctions—especially in the nervous system—is illustrated by the number of human diseases associated with connexin mutations.

Plasmodesmata

Adhesion between plant cells is mediated by their cell walls rather than by transmembrane proteins. In particular, a specialized pectin-rich region of the cell wall called the **middle lamella** acts as a glue to hold adjacent cells together. Because of the rigidity of plant cell walls, stable associations between plant cells do not require the formation of cytoskeletal links such as those provided by the desmosomes and adherens junctions of animal cells. However, adjacent plant cells communicate with each other through cytoplasmic connections called **plasmodesmata** (singular, plasmodesma) (**Figure 15.30**). Although distinct in structure, plasmodesmata function analogously to gap junctions as a means of direct communication between adjacent cells in tissues. In addition, plasmodesmata play an important role in signaling in plants by allowing regulatory molecules, such as transcription factors and miRNAs, to travel directly between cells.

Plasmodesmata can form from incomplete separation of daughter cells following plant cell mitosis. In addition, new plasmodesmata can form between adjacent cells. At each plasmodesma, the plasma membrane of one cell is continuous with that of its neighbor, creating a channel between the two cytosols (**Figure 15.31**). An extension of the smooth endoplasmic reticulum passes through the pore, leaving a ring of surrounding cytoplasm through

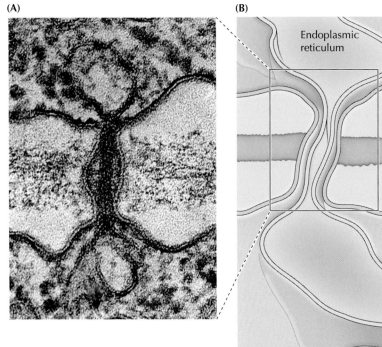

(A)

(B)

Endoplasmic reticulum

Cytosol 1

Cell wall 1

Middle lamella

Cell wall 2

Plasma membrane

Cytosol 2

Figure 15.31 Structure of a plasmodesma In a plasmodesma, the plasma membranes of neighboring cells are continuous, forming cytoplasmic channels through the adjacent cell walls. An extension of the endoplasmic reticulum usually passes through the channel. (A, From L. Tilney et al., 1991. *J. Cell Biol.* 112: 739–748.)

which ions and small molecules are able to pass freely between the cells. Plasmodesmata are dynamic structures that can open or close in response to appropriate stimuli, permitting the regulated passage of macromolecules and larger structures between adjacent cells. In addition, proteins and lipids can be targeted to plasmodesmata in response to specific signals. Plasmodesmata thus play a key role in cell signaling and plant development by controlling the trafficking of regulatory molecules, including transcription factors and regulatory RNAs, between cells.

SUMMARY	KEY TERMS

Cell Walls

- *Bacterial cell walls:* The principal component of bacterial cell walls is a peptidoglycan consisting of polysaccharide chains cross-linked by short peptides.

  peptidoglycan

- *Eukaryotic cell walls:* The cell walls of fungi, algae, and higher plants are composed of fibrous polysaccharides (e.g., cellulose) embedded in a gel-like matrix of polysaccharides and proteins. Their rigid cell walls allow plant cells to expand rapidly by the uptake of water. See Animation 15.1.

  chitin, cellulose, hemicellulose, pectin, primary cell wall, secondary cell wall, lignin, turgor pressure, auxin, cellulose synthase

The Extracellular Matrix and Cell–Matrix Interactions

- *Matrix structural proteins:* The major structural proteins of the extracellular matrix are members of the large collagen protein family. Collagens form the fibrils that characterize the extracellular matrix of connective tissues, as well as forming networks in basal laminae.

  extracellular matrix, basal lamina, collagen, collagen fibril, procollagen, elastic fiber, elastin

- *Matrix polysaccharides:* Polysaccharides in the form of glycosaminoglycans and proteoglycans make up the bulk of the extracellular matrix. They bind to and modify the collagen fibrils and interact with all other matrix molecules.

  glycosaminoglycan (GAG), proteoglycan

- *Adhesion proteins:* Adhesion proteins link the components of the extracellular matrix to one another and are the major binding sites for integrins, which mediate most cell–matrix adhesions.

  fibronectin, laminin, nidogen

- *Cell–matrix interactions:* Integrins are the major cell surface receptors that attach cells to the extracellular matrix. At focal adhesions and hemidesmosomes, integrins provide stable links between the extracellular matrix and the actin and intermediate filament cytoskeletons, respectively. See Videos 15.1 and 15.2.

  integrin, focal adhesion, hemidesmosome

Cell–Cell Interactions

- *Adhesion junctions:* Selective cell–cell interactions are mediated by four major groups of cell adhesion proteins: selectins, integrins, immunoglobulin (Ig) superfamily members, and cadherins. The cadherins link the cytoskeletons of adjacent cells at stable cell–cell junctions. See Video 15.3.

  cell adhesion molecule, selectin, immunoglobulin (Ig) superfamily, cadherin, heterophilic interaction, homophilic interaction, adherens junction, desmosome, armadillo protein family, nectin, desmoglein, desmocollin

- *Tight junctions:* Tight junctions prevent the free passage of molecules between epithelial cells and separate the apical and basolateral domains of epithelial cell plasma membranes.

  tight junction, junctional complex

SUMMARY

- *Gap junctions:* Gap junctions are regulated channels connecting the cytosols of adjacent animal cells. Electrical synapses are gap junctions that mediate signaling between cells of the nervous system.

- *Plasmodesmata:* Adjacent plant cells are linked by cytoplasmic connections called plasmodesmata.

KEY TERMS

gap junction, connexin, connexon, electrical synapse

middle lamella, plasmodesma

Refer To

The Cell

Companion Website

sites.sinauer.com/cooper7e

for quizzes, animations, videos, flashcards, and other study resources.

Questions

1. How do the cell walls and adjacent membranes differ between Gram-positive and Gram-negative bacteria?

2. How do bacterial cytoskeletal proteins determine cell shape?

3. An important function of the Na^+-K^+ pump in animal cells is the maintenance of osmotic equilibrium. Why is this unnecessary for plant cells?

4. How do hemicelluloses impart structural strength to plant cell walls?

5. What is the importance of selectively targeting different glucose transporters to the apical and basolateral domains of the plasma membrane of intestinal epithelial cells? What is the role of tight junctions in this process?

6. How would an inhibitor of the enzyme lysyl hydroxylase (the enzyme responsible for hydroxylating lysine residues in collagen) affect the stability of collagen synthesized by a cell?

7. Why do fibril-forming collagens not assemble into triple helices inside cells?

8. What property of glycosaminoglycans (GAGs) allows them to form hydrated gels?

9. You are studying a transporter found only in the apical plasma membrane of epithelial cells. You treat the epithelial cells with a synthetic peptide that is similar to the extracellular domain of a claudin family protein. The transporter is now found in both the apical and basolateral domains of the plasma membrane. What is the most likely mechanism by which the peptide disturbs the localization of the transporter?

10. You have overexpressed a dominant negative E-cadherin lacking an extracellular domain in epithelial cells. How will this mutant affect cell–cell adhesion?

11. A mutation in an epithelial cell leads to the expression of $\alpha_6\beta_4$ integrin with a deleted cytoplasmic domain. How will this mutation affect adhesion of the epithelial cell to the basal lamina?

12. What is an electrical synapse? What function does it serve?

13. How are gap junctions and plasmodesmata similar? Are they likely to be analogous or homologous structures in animals and plants?

References and Further Reading   (Key review articles for each major section are highlighted in **bold**.)

Cell Walls

Cabeen, M. T. and C. Jacobs-Wagner. 2007. Skin and bones: the bacterial cytoskeleton, cell wall, and cell morphogenesis. *J. Cell Biol. 179*: 381–387. [R]

Celler, K., R. L. Koning, A. J. Koster and G. P. van Wezel. 2013. Multidimensional view of the bacterial cytoskeleton. *J. Bacteriol. 195*: 1627–1636. [R]

Cosgrove, D. J. 2005. Growth of the plant cell wall. *Nature Rev. Mol. Cell Biol. 6*: 850–861. [R]

Lerouxel, O., D. M. Cavalier, A. H. Liepman and K. Keegstra. 2006. Biosynthesis of plant cell wall polysaccharides—a complex process. *Curr. Opin. Plant Biol. 9*: 621–630. [R]

Merzendorfer, H. 2011. The cellular basis of chitin synthesis in fungi and insects: common principles and differences. *Eur. J. Cell Biol. 90*: 759–769. [R]

Oda, Y. and H. Fukuda. 2012. Secondary cell wall patterning during xylem differentiation. *Curr. Opin. Plant Biol. 15*: 38–44. [R]

Paredez, A., A. Wright and D. W. Ehrhardt. 2006. Microtubule cortical array organization and plant cell morphogenesis. *Curr. Opin. Plant Biol. 9*: 571–578. [R]

Pauly, M. and K. Keegstra. 2008. Cell-wall carbohydrates and their modification as a resource for biofuels. *Plant J. 54*: 559–568. [R]

Somerville, C. 2006. Cellulose synthesis in higher plants. *Ann. Rev. Cell Dev. Biol. 22*: 53–78. [R]

Wasteneys, G. O. and J. C. Ambrose. 2009. Spatial organization of plant cortical microtubules: close encounters of the 2D kind. *Trends Cell Biol. 19*: 62–71. [R]

The Extracellular Matrix and Cell–Matrix Interactions

Barczyk, M., S. Carracedo and D. Gullberg. 2010. Integrins. *Cell Tissue Res. 339*: 269–280. [R]

Berrier, A. L. and K. M. Yamada. 2007. Cell–matrix adhesion. *J. Cell Physiol. 213*: 565–573. [R]

Bishop, J. R., M. Schuksz and J. D. Esko. 2007. Heparan sulphate proteoglycans fine-tune mammalian physiology. *Nature 446*: 1030–1037. [R]

Bouvard, D., J. Pouwels, N. De Franceschi and J. Ivaska. 2013. Integrin inactivators: balancing cellular functions *in vitro* and *in vivo*. *Nature Rev. Mol. Cell Biol. 14*: 430–442. [R]

Calderwood, D. A., I. D. Campbell and D. R. Critchley. 2013. Talins and kindlins: partners in integrin-mediated adhesion. *Nature Rev. Mol. Cell Biol. 14*: 503–517. [R]

Durbeej, M. 2010. Laminins. *Cell Tissue Res. 339*: 259–268. [R]

Hynes, R. O. 2009. The extracellular matrix: not just pretty fibrils. *Science 326*: 1216–1219. [R]

Lecuit, T. 2005. Adhesion remodeling underlying tissue morphogenesis. *Trends Cell Biol. 15*: 34–42. [R]

Leitinger, B. 2011. Transmembrane collagen receptors. *Ann. Rev. Cell Dev. Biol. 27*: 25–56. [R]

Moser, M., K. R. Legate, R. Zent and R. Fassler. 2009. The tail of integrins, talin, and kindlins. *Science 324*: 895–899. [R]

Mouw, J. K., G. Ou and V. M. Weaver. 2014. Extracellular matrix assembly: a multiscale deconstruction. *Nature Rev. Mol. Cell Biol. 15*: 771–785. [R]

Page-McCaw, A., A. J. Ewald and Z. Werb. 2007. Matrix metalloproteinases and the regulation of tissue remodelling. *Nature Rev. Mol. Cell Biol. 8*: 221–233. [R]

Rozario, T. and D. W. DeSimone. 2010. The extracellular matrix in development and morphogenesis: a dynamic view. *Dev. Biol. 341*, 126–140. [R]

Shattil, S. J., C. Kim and M. H. Ginsberg. 2010. The final steps of integrin activation: the end game. *Nature Rev. Mol. Cell Biol. 11*: 288–300. [R]

Tamkun, J. W., D. W. DeSimone, D. Fonda, R. S. Patel, C. Buck, A. F. Horwitz, and R. O. Hynes. 1986. Structure of integrin, a glycoprotein involved in the transmembrane linkage between fibronectin and actin. *Cell 46*: 271–282. [P]

Yurchenco, P. D. 2011. Basement membranes: cell scaffoldings and signaling platforms. *Cold Spring Harbor Perspect. Biol. 3*:a004911. [R]

Cell–Cell Interactions

Bergoffen, J., S. S. Scherer, S. Wang, M. O. Scott, L. J. Bone, D. L. Paul, K. Chen, M. W. Lensch, P. F. Chance, and K. H. Fischbeck. 1993. Connexin mutations in X-linked Charcot-Marie-Tooth disease. *Science 262*: 2039–2042. [P]

Elias, L. A., and A. R. Kriegstein. 2008. Gap junctions: multifaceted regulators of embryonic cortical development. *Trends Neurosci. 31*: 243–250. [R]

Green, K. J., and C. L. Simpson. 2007. Desmosomes: new perspectives on a classic. *J. Invest Dermatol. 127*: 2499–2515. [R]

Harris, T. J. C. and U. Tepass. 2010. Adherens junctions: from molecules to morphogenesis. *Nature Rev. Cell Mol. Biol. 11*: 502–514. [R]

Kragler, F. 2013. Plasmodesmata: intercellular tunnels facilitating transport of macromolecules in plants. *Cell Tissue Res. 352*: 49–58. [R]

Laird, D. W. 2006. Life cycle of connexins in health and disease. *Biochem. J. 394*: 527–543. [R]

Lee, J. Y., and H. Lu. 2011. Plasmodesmata: the battleground against intruders. *Trends Plant Sci. 16*: 201–210. [R]

Lucas, W. J., B. K. Ham and J. Y. Kim. 2009. Plasmodesmata - bridging the gap between neighboring plant cells. *Trends Cell Biol. 19*: 495–503. [R]

Nekrasova, O. and K. J. Green. 2013. Desmosome assembly and dynamics. *Trends Cell Biol. 23*: 537–546. [R]

Niessen, C. M. 2007. Tight junctions/adherens junctions: basic structure and function. *J. Invest Dermatol. 127*: 2525–2532. [R]

Pokutta, S., and W. I. Weis. 2007. Structure and mechanism of cadherins and catenins in cell–cell contacts. *Ann. Rev. Cell Dev. Biol. 23*: 237–261. [R]

Rikitake, Y., K. Mandai and Y. Takai. 2012. The role of nectins in different types of cell–cell adhesion. *J. Cell Sci. 125*: 3713–3722. [R]

Sotomayor, M., R. Gaudet and D. P. Corey. 2014. Sorting out a promiscuous superfamily: towards cadherin connectomics. *Trends Cell Biol. 24*: 524–536. [R]

Takeichi, M. 2014. Dynamic contacts: rearranging adherens junctions to drive epithelial remodelling. *Nature Rev. Mol. Cell Biol. 15*: 397–410. [R]

van Steensel, M. A. 2004. Gap junction diseases of the skin. *Am. J. Med. Genet. C. Semin. Med. Genet. 131C*: 12–19. [R]

Wei, C. J., X. Xu and C. W. Lo. 2004. Connexins and cell signaling in development and disease. *Ann. Rev. Cell Dev. Biol. 20*: 811–838. [R]

PART IV

Cell Regulation

Cell Signaling

All cells receive and respond to signals from their environment. Even the simplest bacteria sense and swim toward high concentrations of nutrients, such as glucose or amino acids. Many bacteria and unicellular eukaryotes also respond to signaling molecules secreted by other cells, allowing for cell–cell communication. Mating between yeast cells, for example, is signaled by peptides that are secreted by one cell and bind to receptors on the surface of another. It is in multicellular organisms, however, that cell–cell communication reaches its highest level of sophistication. Whereas the cells of prokaryotes and unicellular eukaryotes are largely autonomous, the behavior of each individual cell in multicellular plants and animals must be carefully regulated to meet the needs of the organism as a whole. This is accomplished by a variety of signaling molecules that are secreted or expressed on the surface of one cell and bind to receptors expressed by other cells, thereby integrating and coordinating the functions of the many individual cells that make up organisms as complex as human beings.

The binding of most signaling molecules to their receptors initiates a series of intracellular reactions that regulate virtually all aspects of cell behavior, including metabolism, movement, proliferation, survival, and differentiation. Understanding the molecular components of these pathways and how they are regulated has thus become a major area of research in contemporary cell biology. Interest in this area is further heightened by the fact that many cancers arise as a result of a breakdown in the signaling pathways that control normal cell proliferation and survival. In fact, many of our current insights into cell signaling mechanisms have come from the study of cancer cells—a striking example of the fruitful interplay between medicine and basic research in cell and molecular biology.

Signaling Molecules and Their Receptors

Many different kinds of molecules transmit information between the cells of multicellular organisms. Although all these molecules act as ligands that bind to receptors expressed by their target cells, there is considerable variation in the structure and function of the different types of molecules that serve as signal transmitters. Structurally, the signaling molecules used by plants and animals range in complexity from simple gases to proteins. Some of these molecules carry signals over long distances, whereas others act locally to convey information between neighboring cells. In addition, signaling molecules differ in their mode of action on their target cells. Some signaling molecules are able to cross the plasma membrane and bind to intracellular receptors

(A) Direct cell–cell signaling

(B) Signaling by secreted molecules

Endocrine signaling

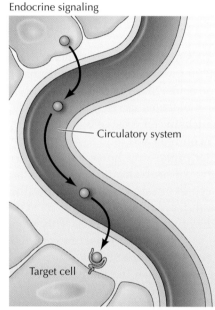

Circulatory system

Target cell

Paracrine signaling

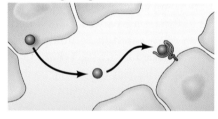

Autocrine signaling

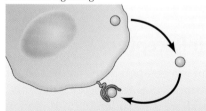

Figure 16.1 Modes of cell–cell signaling Cell signaling can take place either through (A) direct cell–cell contacts or (B) through the action of secreted signaling molecules. In endocrine signaling, hormones are carried through the circulatory system to act on distant target cells. In paracrine signaling, a molecule released from one cell acts locally to affect nearby target cells. In autocrine signaling, a cell produces a signaling molecule to which it also responds.

in the cytoplasm or nucleus, whereas most bind to receptors expressed on the target cell surface. This section discusses the major types of signaling molecules and some of the intracellular receptors with which they interact. Subsequent discussion in this chapter focuses on cell surface receptors and the mechanisms by which they function to regulate cell behavior.

Modes of cell–cell signaling

Cell signaling can result either from the direct interaction of a cell with its neighbor or from the action of secreted signaling molecules (Figure 16.1). Signaling by direct cell–cell (or cell–matrix) interactions plays a critical role in regulating the behavior of cells in animal tissues. For example, the integrins and cadherins (discussed in the previous chapter) function not only as cell adhesion molecules but also as signaling molecules that bind to adjacent cells or the extracellular matrix and regulate cell proliferation and survival in response to cell–cell and cell–matrix contacts. In addition, cells express a variety of cell surface receptors that interact with signaling molecules on the surface of neighboring cells. Signaling via such direct cell–cell interactions plays a critical role in regulating the many interactions between different types of cells that take place during embryonic development as well as in the maintenance of adult tissues.

The multiple varieties of signaling by secreted molecules are frequently divided into three general categories based on the distance over which signals are transmitted. In **endocrine signaling**, the signaling molecules (**hormones**) are secreted by specialized endocrine cells and carried through the circulation to act on target cells at distant body sites. A classic example is provided by the steroid hormone estrogen, which is produced by the ovaries and stimulates development and maintenance of the female reproductive system and secondary sex characteristics. In animals, more than 50 different hormones are produced by endocrine glands, including the pituitary, thyroid, parathyroid, pancreas, adrenal glands, and gonads.

In contrast with hormones, some signaling molecules act locally to affect the behavior of nearby cells. In **paracrine signaling**, a molecule released by one cell acts on neighboring target cells. An example is provided by the action of neurotransmitters in carrying signals between nerve cells at a synapse. Finally, some cells respond to signaling molecules that they themselves produce. One important example of such **autocrine signaling** is the response of cells of the vertebrate immune system to foreign antigens. Certain types of T lymphocytes respond to antigenic stimulation by synthesizing a growth factor that drives their own proliferation, thereby increasing the number of responsive T lymphocytes and amplifying the immune response. It is also noteworthy that abnormal autocrine signaling frequently contributes to the uncontrolled growth of cancer cells (see Chapter 19). In this situation, a cancer cell produces a growth factor to which it also responds, thereby continuously driving its own unregulated proliferation.

Steroid hormones and the nuclear receptor superfamily

As already noted, all signaling molecules act by binding to receptors expressed by their target cells. In many cases these receptors are expressed on the target cell surface, but some receptors are intracellular proteins located in the cytosol or in the nucleus. These intracellular receptors respond to small hydrophobic signaling molecules that are able to diffuse across the plasma membrane. The **steroid hormones** are the classic examples of this group of signaling molecules, which also includes thyroid hormone, vitamin D_3, and retinoic acid (**Figure 16.2**).

The steroid hormones (including testosterone, estrogen, progesterone, the corticosteroids, and ecdysone) are all synthesized from cholesterol. **Testosterone**, **estrogen**, and **progesterone** are the sex steroids, which are produced by the gonads. The **corticosteroids** are produced by the adrenal gland. They include the **glucocorticoids**, which act on a variety of cells to stimulate production of glucose, and the **mineralocorticoids**, which act on the kidney to regulate salt and water balance. **Ecdysone** is an insect hormone that plays a key role in development by triggering the metamorphosis of larvae to adults. The **brassinosteroids** are plant-specific steroid hormones that control a number of developmental processes, including cell growth and differentiation.

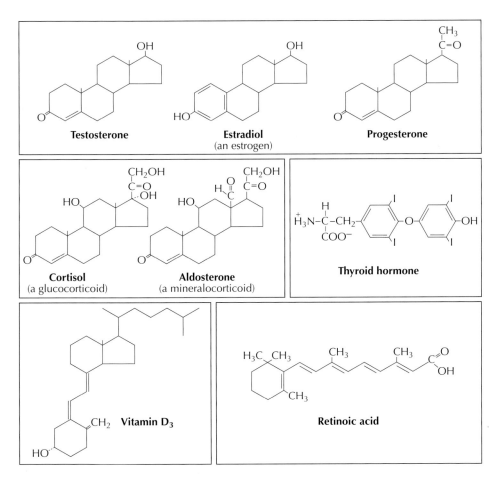

Figure 16.2 Structure of steroid hormones, thyroid hormone, vitamin D_3, and retinoic acid   The steroids include the sex hormones (testosterone, estrogen, and progesterone), glucocorticoids, and mineralocorticoids.

Although thyroid hormone, vitamin D_3, and retinoic acid are structurally and functionally distinct from the steroids, they share a common mechanism of action in their target cells. **Thyroid hormone** is synthesized from tyrosine in the thyroid gland; it plays important roles in development and regulation of metabolism. **Vitamin D_3** regulates Ca^{2+} metabolism and bone growth. **Retinoic acid** and related compounds (**retinoids**) synthesized from vitamin A play important roles in vertebrate development.

Because of their hydrophobic character, the steroid hormones are able to enter cells by diffusing across the plasma membrane. The passage of thyroid hormones across the plasma membrane is facilitated by carrier proteins. Once inside the cell, the steroids, thyroid hormone, vitamin D_3, and retinoic acid all bind to intracellular receptors that are expressed by the hormonally responsive target cells. These receptors, which are members of a family of proteins known as the **nuclear receptor superfamily**, are transcription factors that contain related domains for ligand binding, DNA binding, and transcriptional activation. Ligand binding regulates their function as activators or repressors of their target genes, so the steroid hormones and related molecules directly regulate gene expression.

Ligand binding has distinct effects on different receptors. Some members of the nuclear receptor superfamily are inactive in the absence of hormone. Glucocorticoid receptor, for example, is bound to Hsp90 chaperones in the absence of hormone (**Figure 16.3**). Glucocorticoid binding induces a conformational change in the receptor, displacing Hsp90 and leading to the

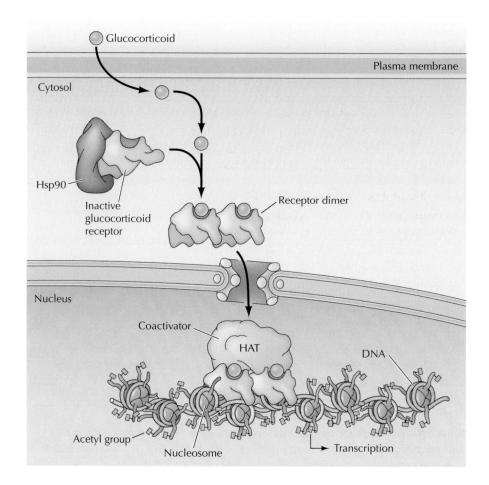

Figure 16.3 Glucocorticoid action
Glucocorticoids diffuse across the plasma membrane and bind to the glucocorticoid receptor. In the absence of ligand, the receptor is bound to Hsp90 in the cytoplasm. Glucocorticoid binding displaces the receptor from Hsp90 and allows the formation of receptor dimers. The activated receptors translocate to the nucleus, bind DNA, and associate with coactivators with histone acetyltransferase (HAT) activity to stimulate transcription of their target genes.

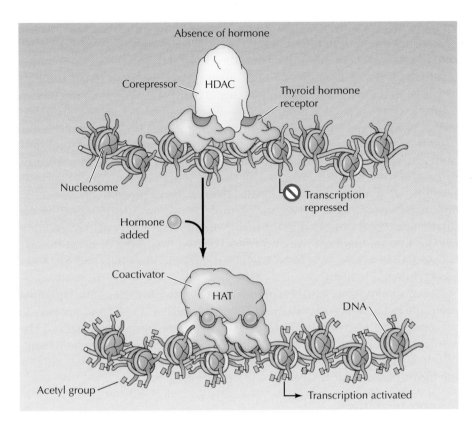

Absence of hormone

Corepressor — HDAC

Thyroid hormone receptor

Nucleosome

Transcription repressed

Hormone added

Coactivator

HAT

DNA

Acetyl group

Transcription activated

Figure 16.4 Gene regulation by the thyroid hormone receptor Thyroid hormone receptor binds DNA in either the presence or absence of hormone. However, hormone binding changes the function of the receptor from a repressor to an activator of target gene transcription. In the absence of hormone, the receptor associates with corepressors with histone deacetylase (HDAC) activity. In the presence of hormone, the receptor associates with coactivators with histone acetyltransferase (HAT) activity.

formation of receptor dimers that bind to regulatory DNA sequences and activate transcription of target genes. In other cases, the receptor binds DNA in either the presence or absence of hormone, but hormone binding alters the activity of the receptor as a transcriptional regulatory molecule. For example, in the absence of hormone, thyroid hormone receptor is associated with a corepressor complex and represses transcription of its target genes (**Figure 16.4**). Hormone binding induces a conformational change that results in the interaction of the receptor with coactivators rather than corepressors, leading to transcriptional activation of thyroid hormone-inducible genes.

Nitric oxide and carbon monoxide

The simple gas nitric oxide (NO) is a major paracrine signaling molecule in the nervous, immune, and circulatory systems. Like the steroid hormones, NO is able to diffuse directly across the plasma membrane of its target cells. The molecular basis of NO action, however, is distinct from that of steroid action; rather than binding to a receptor that regulates transcription, NO alters the activity of intracellular target enzymes.

Nitric oxide is synthesized from the amino acid arginine by the enzyme nitric oxide synthase (**Figure 16.5**). Once synthesized, NO diffuses out of the cell and can act locally to affect nearby cells. Its action is restricted to such local effects because NO is extremely unstable, with a half-life of only a few seconds. The major intracellular

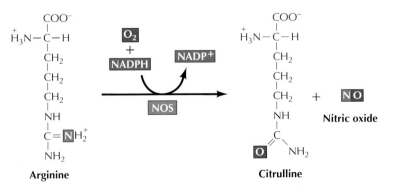

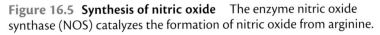

Figure 16.5 Synthesis of nitric oxide The enzyme nitric oxide synthase (NOS) catalyzes the formation of nitric oxide from arginine.

target of NO is **guanylyl cyclase**. NO binds to a heme group at the active site of this enzyme, stimulating synthesis of the second messenger **cyclic GMP** (a **second messenger** is a molecule that relays a signal from a receptor to a target inside the cell). In addition, NO can directly modify some target proteins by nitrosylation of cysteine residues (see Figure 9.43). A well-characterized example of NO action is signaling the dilation of blood vessels. The first step in this process is the release of neurotransmitters, such as acetylcholine, from the termini of nerve cells in the blood vessel wall. These neurotransmitters act on endothelial cells to stimulate NO synthesis. NO then diffuses to neighboring smooth muscle cells where it activates guanylyl cyclase, resulting in synthesis of cyclic GMP, which induces muscle cell relaxation and blood vessel dilation. NO, for example, is responsible for signaling the dilation of blood vessels that leads to penile erection. It is also interesting to note that the medical use of nitroglycerin in treatment of heart disease is based on its conversion to NO, which dilates blood vessels and increases blood flow to the heart.

Another simple gas, carbon monoxide (CO), also functions as a signaling molecule in the nervous system. CO is closely related to NO and appears to act similarly as a neurotransmitter and mediator of blood vessel dilation. The synthesis of CO in brain cells, like that of NO, is stimulated by neurotransmitters. In addition, CO can stimulate guanylate cyclase, which also represents the major physiological target of CO signaling.

Neurotransmitters

The **neurotransmitters** carry signals between neurons or from neurons to other types of target cells (such as muscle cells). They are a diverse group of small hydrophilic molecules including acetylcholine, glycine, glutamate, dopamine, epinephrine (adrenaline), norepinephrine, serotonin, histamine, and γ-aminobutyric acid (GABA) (**Figure 16.6**). The release of neurotransmitters is signaled by the arrival of an action potential at the terminus of a neuron (see Figure 14.21). The neurotransmitters then diffuse across the synaptic cleft and bind to receptors on the target cell surface. Note that some neurotransmitters can also act as hormones—for example, epinephrine functions both as a neurotransmitter and as a hormone produced by the adrenal gland to signal glycogen breakdown in muscle cells.

Because the neurotransmitters are hydrophilic, they are unable to cross the plasma membrane of their target cells. Therefore, in contrast with steroid hormones and NO or CO, the neurotransmitters act by binding to cell surface receptors. Many neurotransmitter receptors are ligand-gated ion channels, such as the acetylcholine receptor discussed in Chapter 14 (see Figure 14.22). Neurotransmitter binding to these receptors induces a conformational change that opens ion channels, directly resulting in changes in ion flux in the target cell. Other neurotransmitter receptors are coupled to G proteins—a major group of signaling molecules (discussed later in the chapter) that link cell surface receptors to a variety of intracellular responses. In the case of neurotransmitter receptors, the associated G proteins frequently act to regulate ion channels.

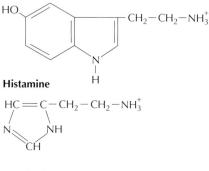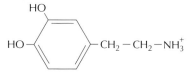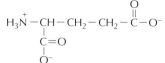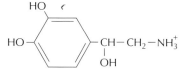

Figure 16.6 Structure of representative neurotransmitters The neurotransmitters are hydrophilic molecules that bind to cell surface receptors.

Table 16.1 Representative Peptide Hormones, Neuropeptides, and Polypeptide Growth Factors

Signaling molecule	Size^a	Activities^b
Peptide hormones		
Insulin	A = 21, B = 30	Regulation of glucose uptake; stimulation of cell proliferation
Glucagon	29	Stimulation of glucose synthesis
Growth hormone	191	General stimulation of growth
Follicle-stimulating hormone	$\alpha = 92, \beta = 118$	Stimulation of the growth of oocytes hormone (FSH) and ovarian follicles
Prolactin	198	Stimulation of milk production
Neuropeptides and neurohormones		
Substance P	11	Sensory synaptic transmission
Oxytocin	9	Stimulation of smooth muscle contraction
Vasopressin	9	Stimulation of water reabsorption in the kidney
Enkephalin	5	Analgesic
β-Endorphin	31	Analgesic
Growth factors		
Nerve growth factor (NGF)	118	Differentiation and survival of neurons
Epidermal growth factor (EGF)	53	Proliferation of many types of cells
Platelet-derived growth factor (PDGF)	A = 125, B = 109	Proliferation of fibroblasts and other cell types
Interleukin-2	133	Proliferation of T lymphocytes
Erythropoietin	166	Development of red blood cells

^{a}Size is indicated in number of amino acids. Some hormones and growth factors consist of two different polypeptide chains, which are designated either A and B or α and β.

^{b}Most of these hormones and growth factors have other activities in addition to those indicated.

Peptide hormones and growth factors

The widest variety of signaling molecules in animals are peptides, ranging in size from only a few to more than 100 amino acids. This group of signaling molecules includes peptide hormones, neuropeptides, and a diverse array of polypeptide growth factors (Table 16.1). Well-known examples of **peptide hormones** include insulin, glucagon, and the hormones produced by the pituitary gland (growth hormone, follicle-stimulating hormone, prolactin, and others).

Neuropeptides are secreted by some neurons instead of the small molecule neurotransmitters discussed in the previous section. Some of these peptides, such as the **enkephalins** and **endorphins**, function not only as neurotransmitters at synapses but also as **neurohormones** that act on distant cells. The enkephalins and endorphins have been widely studied because of their activity as natural analgesics that decrease pain responses in the central nervous system. Discovered during studies of drug addiction, they are naturally occurring compounds that bind to the same receptors on the surface of brain cells as morphine does.

The polypeptide **growth factors** include a wide variety of signaling molecules that control animal cell growth and differentiation. The first of these factors (**nerve growth factor**, or **NGF**) was discovered by Rita Levi-Montalcini in the 1950s. NGF is a member of a family of polypeptides (called **neurotrophins**) that regulate the development and survival of neurons. During the course of experiments on NGF, Stanley Cohen serendipitously discovered an unrelated factor (called **epidermal growth factor**, or **EGF**) that stimulates

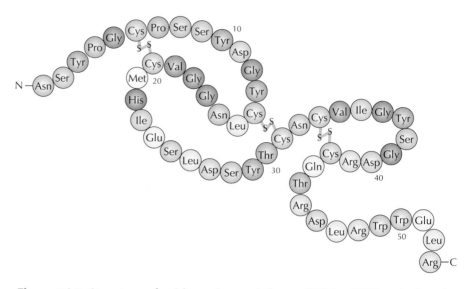

Figure 16.7 Structure of epidermal growth factor (EGF) EGF is a single polypeptide chain of 53 amino acids. Disulfide bonds between cysteine residues are indicated. (After G. Carpenter and S. Cohen, 1979. *Ann. Rev. Biochem.* 48: 193.)

cell proliferation. EGF, a 53-amino-acid polypeptide (**Figure 16.7**), has served as the prototype of a large array of growth factors that play critical roles in controlling animal cell proliferation, both during embryonic development and in adult organisms.

A good example of growth factor action is provided by the activity of **platelet-derived growth factor (PDGF)** in wound healing. PDGF is stored in blood platelets and released during blood clotting at the site of a wound. It then stimulates the proliferation and movement of fibroblasts in the vicinity of the clot, thereby contributing to regrowth of the damaged tissue. In addition, PDGF plays critical roles in the development of a variety of embryonic tissues. Members of another large group of polypeptide growth factors (called **cytokines**, such as erythropoietin and interleukin-2) regulate the development and differentiation of blood cells and control the activities of lymphocytes during the immune response. Other polypeptide growth factors (**membrane-anchored growth factors**) remain associated with the plasma membrane rather than being secreted into extracellular fluids, therefore functioning specifically as signaling molecules during direct cell–cell interactions.

Peptide hormones, neuropeptides, and growth factors are unable to cross the plasma membrane of their target cells, so they act by binding to cell surface receptors, as discussed later in this chapter. As might be expected from the critical roles of polypeptide growth factors in controlling cell proliferation, abnormalities in growth factor signaling are the basis for a variety of diseases, including many kinds of cancer. For example, abnormal expression of the EGF receptor is an important factor in the development of many human cancers, and inhibitors of the EGF receptor appear to be promising agents for cancer treatment (see Chapter 19).

Eicosanoids

Several types of lipids serve as signaling molecules that, in contrast with the steroid hormones, act by binding to cell surface receptors. The most important

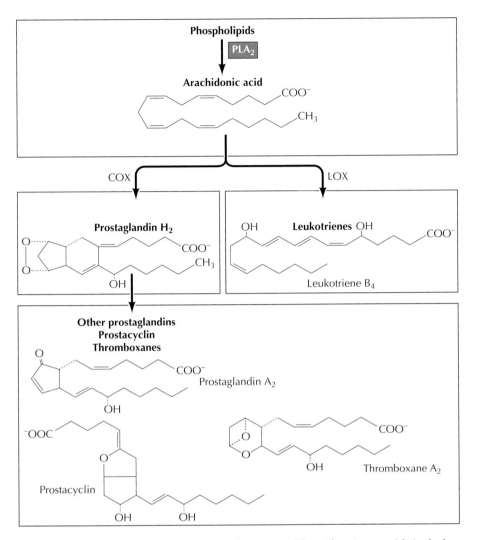

Figure 16.8 Synthesis and structure of eicosanoids   The eicosanoids include the prostaglandins, prostacyclin, thromboxanes, and leukotrienes. They are synthesized from arachidonic acid, which is formed by the hydrolysis of phospholipids catalyzed by phospholipase A_2 (PLA_2). Arachidonic acid can then be metabolized via two alternative pathways—one pathway (initiated by cyclooxygenase [COX]) leads to synthesis of prostaglandins, prostacyclin, and thromboxanes, while the other pathway (initiated by lipoxygenase [LOX]) leads to synthesis of leukotrienes.

of these molecules are members of a class of lipids called the **eicosanoids**, which includes **prostaglandins**, **prostacyclin**, **thromboxanes**, and **leukotrienes** (**Figure 16.8**). The eicosanoids are rapidly broken down and therefore act locally in autocrine or paracrine signaling pathways. They stimulate a variety of responses in their target cells, including blood platelet aggregation, inflammation, and smooth-muscle contraction.

All eicosanoids are synthesized from arachidonic acid, which is formed from phospholipids. The first step in the pathway leading to synthesis of either prostaglandins, prostacyclin, or thromboxanes is the conversion of arachidonic acid to prostaglandin H_2. Interestingly, the enzyme that catalyzes this reaction (cyclooxygenase or COX) is the target of aspirin and other nonsteroidal anti-inflammatory drugs (NSAIDs). By inhibiting synthesis of the prostaglandins, aspirin reduces inflammation and pain. By inhibiting synthesis of thromboxane, aspirin also reduces platelet aggregation and blood clotting. Because of this activity, small daily doses of aspirin are frequently prescribed for prevention of strokes. In addition, aspirin and other NSAIDs have been found to reduce the frequency of colon cancer in both animal models and humans, possibly by inhibiting the synthesis of prostaglandins that act to stimulate cell proliferation and promote cancer development. It is noteworthy that there are two forms of cyclooxygenase: COX-1 and COX-2. COX-1 is thought to be principally responsible for the normal physiological production of prostaglandins and COX-2 for the increased prostaglandin production associated with inflammation and disease. In addition, many cancers express elevated levels of COX-2, which may contribute to tumor proliferation. Consequently, selective inhibitors of COX-2 have been developed based on the rationale that such drugs would be more effective and have fewer side effects than aspirin or conventional NSAIDs, which inhibit both COX-1 and COX-2. However, COX-2 inhibitors are still associated with serious side effects, including an increased risk of cardiovascular disease, which limit their use.

Plant hormones

Plant growth and development are regulated by steroids, peptide hormones, and a group of small molecules called **plant hormones**. The levels of these molecules within the plant are typically modified by environmental factors, such as light or infection, so they coordinate the responses of tissues in different parts of the plant to environmental signals.

The plant hormones are classically divided into five major groups: **gibberellins, auxins, ethylene, cytokinins,** and **abscisic acid** (Figure 16.9). The first plant hormone to be identified was auxin, with the early experiments leading to its discovery having been performed by Charles Darwin in the 1880s. One of the effects of auxins is to induce plant cell elongation by weakening the cell wall (see Figure 15.8). In addition, auxins regulate many other aspects of plant development, including cell division and differentiation. The other plant hormones likewise have multiple effects in their target tissues, including stem elongation (gibberellins), fruit ripening (ethylene), cell division (cytokinins), and the onset of dormancy (abscisic acid). In addition to the five classical plant hormones, recent studies have identified several other compounds that function as hormones in plants, including nitric oxide and the plant steroid hormones (brassinosteroids) discussed in earlier sections.

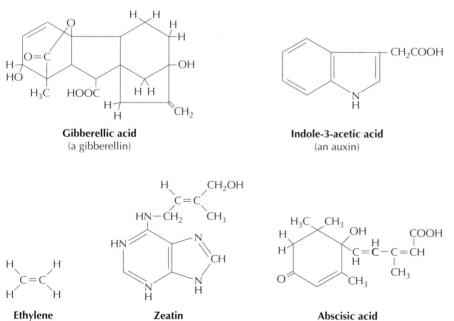

Gibberellic acid
(a gibberellin)

Indole-3-acetic acid
(an auxin)

Ethylene

Zeatin
(a cytokinin)

Abscisic acid

Figure 16.9 Plant hormones
Structures of the five classical plant hormones are illustrated.

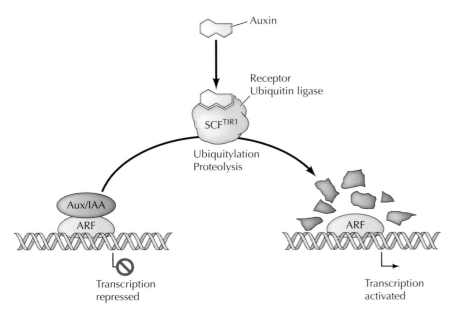

Figure 16.10 Auxin signaling   In the absence of auxin (left), the promoters of auxin-regulated genes are bound by a transcription factor (ARF) in association with a corepressor (Aux/IAA). When present, auxin binds to a receptor with ubiquitin ligase activity (SCF^{TIR1}), stimulating ubiquitylation and proteolysis of Aux/IAA. ARF then activates transcription of auxin-induced genes.

The signaling pathways triggered by some plant hormones, including ethylene, cytokinins, and the brassinosteroids, use mechanisms that are conserved in animal cells, such as activation of protein kinases (see Figure 9.41). Other plant hormones, such as auxin and giberellin, stimulate signaling pathways involving activation of protein degradation that are unique to plants (**Figure 16.10**). Auxin, for example, controls gene expression by binding to and activating a receptor associated with a ubiquitin ligase (see Figure 9.46). In the absence of auxin, auxin-regulated genes are bound by a transcription factor (auxin response factor, or ARF), which is associated with a corepressor (Aux/IAA). Stimulation of the SCF^{TIR1} ubiquitin ligase by auxin leads to degradation of the Aux/IAA corepressor, allowing ARF to activate transcription of its target genes.

G Proteins and Cyclic AMP Signaling

Most ligands responsible for cell–cell signaling (including neurotransmitters, peptide hormones, and growth factors) bind to receptors on the surface of their target cells. Consequently, a major challenge in understanding cell–cell signaling is unraveling the mechanisms by which cell surface receptors transmit the signals initiated by ligand binding. As discussed in Chapter 14, some neurotransmitter receptors are ligand-gated ion channels that directly control ion flux across the plasma membrane. Other cell surface receptors, including the receptors for peptide hormones and growth factors, act instead by regulating the activity of intracellular target enzymes. These enzymes then transmit signals from the receptor to a series of additional intracellular targets—a process called **intracellular signal transduction**. The targets of such signaling pathways frequently include transcription factors that function to regulate gene expression. Ligand binding to a receptor on the surface of the

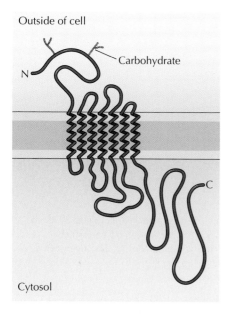

Outside of cell

Carbohydrate

N

Cytosol

Animation 16.2

sites.sinauer.com/cooper7e/a16.2

Signal Transduction The largest family of cell surface receptors transmits signals inside the cell by activating G proteins, which bind GTP and then activate effector proteins.

FYI

Dogs have approximately three times more genes encoding functional odorant receptors than humans.

Figure 16.11 Structure of a G protein-coupled receptor The G protein-coupled receptor is characterized by seven transmembrane α helices.

cell thus initiates a chain of intracellular reactions, ultimately reaching the target cell nucleus and altering programs of gene expression. The functions of some of the major types of cell surface receptors and their associated signaling pathways are discussed in this and the following sections of this chapter.

G proteins and G protein-coupled receptors

The largest family of cell surface receptors transmits signals to intracellular targets via the intermediary action of guanine nucleotide-binding proteins called **G proteins**. More than 1000 such **G protein-coupled receptors** have been identified, including the receptors for many hormones and neurotransmitters. In addition, the G protein-coupled receptor family includes a large number of receptors that are responsible for smell, sight, and taste.

The G protein-coupled receptors are structurally and functionally related proteins characterized by seven membrane-spanning α helices (**Figure 16.11**). The binding of ligands to the extracellular domain of these receptors induces a conformational change that allows the cytosolic domain of the receptor to activate a G protein associated with the inner face of the plasma membrane. The activated G protein then dissociates from the receptor and carries the signal to an intracellular target, which may be either an enzyme or an ion channel.

The discovery of G proteins came from studies of hormones (such as epinephrine) that regulate the synthesis of **cyclic AMP (cAMP)** in their target cells. As discussed below, cAMP is an important second messenger that mediates cellular responses to a variety of hormones. In the 1970s, Martin Rodbell and his colleagues made the key observation that GTP is required for hormonal stimulation of **adenylyl cyclase** (the enzyme responsible for cAMP formation). This finding led to the discovery that a guanine nucleotide-binding protein (the aforementioned G protein) is an intermediary in adenylyl cyclase activation (**Figure 16.12**). Since then, an array of G proteins have been found to act as physiological switches that regulate the activities of a variety of intracellular targets in response to extracellular signals.

Figure 16.12 Hormonal activation of adenylyl cyclase Binding of hormone promotes the interaction of the receptor with a G protein. The activated G protein α subunit then dissociates from the receptor and stimulates adenylyl cyclase, which catalyzes the conversion of ATP to cAMP.

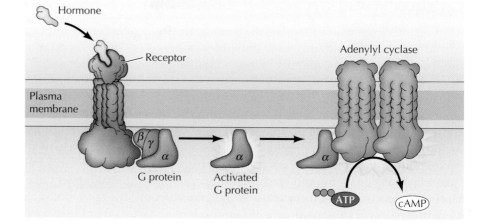

G Protein-Coupled Receptors and Odor Detection

Linda Buck and Richard Axel
Columbia University, New York
Cell, 1991, Volume 65, pages 175–187

The Context

The sense of smell is one of the key systems through which animals perceive their environment. Peripheral neurons in the nose can recognize thousands of different odor molecules, and then transmit signals to the brain where this information is processed. In order to understand the sense of smell at the molecular level, it was essential to identify the receptors on olfactory neurons that were responsible for detection of odorant molecules. In 1991, Linda Buck and Richard Axel identified a large family of G protein-coupled receptors that were responsible for odor detection and recognition.

Previous studies had established that odor molecules were recognized by receptors on the cilia of olfactory neurons. Importantly, it had also been shown that exposure of isolated cilia from rat olfactory neurons to many different odorants led to the stimulation of adenylyl cyclase and elevation of cyclic AMP (cAMP). In addition, it was found that these odorant-stimulated increases in cAMP were dependent on the presence of GTP, suggesting that odorant molecules activated G protein-coupled receptors that stimulated adenylyl cyclase. The increased cAMP then opened Na^+ channels in the plasma membrane of the olfactory neurons, initiating a nerve impulse.

Buck and Axel sought to isolate molecular clones of the genes encoding odorant receptors, based on the hypothesis that these receptors were members of the G protein-coupled receptor superfamily. Their success in these experiments not only validated this hypothesis, but also provided the foundation for an understanding of the molecular basis of odor discrimination.

The Experiments

The approach that Buck and Axel took to clone the genes for odorant receptors was based on three assumptions. First, that the odorant receptors were members of the G protein-coupled receptor superfamily and shared conserved sequences with other family members. Second, that the odorant receptors were themselves a large gene family, required to detect a large number of odorants. And third, that these receptors were expressed only in olfactory neurons.

Oligonucleotide primers corresponding to conserved sequences of known G protein-coupled receptors were therefore used to amplify mRNAs of olfactory neurons by reverse transcription, followed by PCR (polymerase chain reaction) (see Figure 4.22). A very large number of PCR fragments were obtained in these experiments, consistent with the expression of a large family of G protein-coupled receptor genes in olfactory neurons. The amplified PCR fragments were then cloned in plasmid vectors, and 18 distinct cDNA clones were isolated for further characterization. Use of these clones as probes in Northern blots demonstrated

Linda Buck   Richard Axel

Photo credits: Linda Buck: Roland Morgan/Fred Hutchinson Cancer Research Center; Richard Axel: Don Hamerman/Columbia University Medical Center.

that homologous mRNAs were expressed in olfactory neurons but not in several other types of cells, including brain, heart, kidney, liver, lung, ovary, retina, and spleen. Importantly, nucleotide sequencing of 10 complete clones showed that they encoded a distinct family of proteins with the seven transmembrane α helices characteristic of G protein-coupled receptors (see figure). Finally, hybridization of the cDNA clones to genomic DNA indicated that the odorant receptors were encoded by a large multigene family consisting of hundreds of genes.

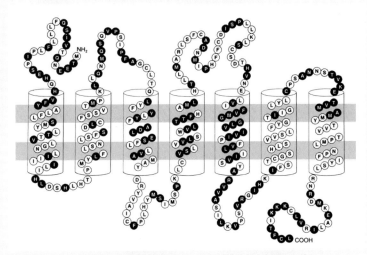

The odorant receptor protein family   The protein encoded by one of the odorant receptor cDNA clones is shown spanning the plasma membrane with seven transmembrane α helices. The amino acids that are conserved in other odorant receptor clones are shown in white and the more variable amino acids in black.

(Continued on next page)

The Impact

The odorant receptors identified by Buck and Axel are the largest family of G protein-coupled receptors. Genome sequencing has now shown that mice and dogs have over 1000 odorant receptor genes. Interestingly, the odorant receptor gene family is smaller in humans, consisting of less than 400 functional genes, consistent with the reduced sense of smell in humans compared with dogs and rodents.

The identification of odorant receptors has served as a critical first step in understanding how the sense of smell works. Importantly, individual olfactory neurons express only a single odorant receptor. Each receptor binds to several different odorant molecules, so each odorant is detected by a specific combination of receptors on a distinct set of olfactory neurons, which then transmit information to the brain. However, we have yet to

understand the network of nerve connections through which the information from olfactory neurons is carried to the brain, enabling the brain to integrate these nerve impulses and interpret specific smells. Deciphering such networks of neuronal connections remains a fundamental area of investigation in understanding information processing in the nervous system.

G proteins consist of three subunits designated α, β, and γ (**Figure 16.13**). They are frequently called **heterotrimeric G proteins** to distinguish them from other guanine nucleotide-binding proteins, such as Ras (see Figure 9.40) and the Ras-related GTP binding proteins that are involved in nuclear import (Ran, see Figure 10.10), vesicular transport (Arf and Rab, see Figures 11.36 and 11.37) and cell movement (Rho, see Figure 13.21). The α subunit of the heterotrimeric G proteins binds guanine nucleotides, which regulate G protein activity. In the resting state, α is bound to GDP in a complex with β and γ. Hormone binding induces a conformational change in the receptor, such that the cytosolic domain of the receptor interacts with the G protein. The activated receptor then acts as a guanine nucleotide exchange factor to stimulate the release of bound GDP and its exchange for GTP. The activated GTP-bound α subunit then dissociates from β and γ, which remain together and function as a $\beta\gamma$ complex. Both the

Figure 16.13 Regulation of G proteins

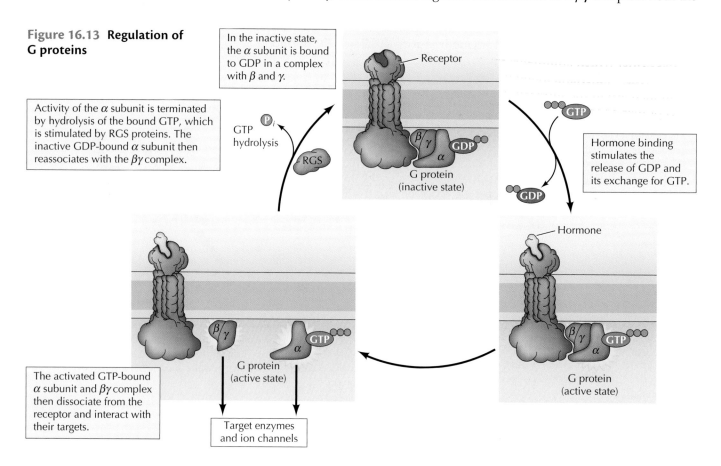

In the inactive state, the α subunit is bound to GDP in a complex with β and γ.

Activity of the α subunit is terminated by hydrolysis of the bound GTP, which is stimulated by RGS proteins. The inactive GDP-bound α subunit then reassociates with the $\beta\gamma$ complex.

Hormone binding stimulates the release of GDP and its exchange for GTP.

The activated GTP-bound α subunit and $\beta\gamma$ complex then dissociate from the receptor and interact with their targets.

GTP hydrolysis

RGS

Receptor

G protein (inactive state)

Hormone

G protein (active state)

Target enzymes and ion channels

active GTP-bound α subunit and the $\beta\gamma$ complex then interact with their targets to elicit an intracellular response. The activity of the α subunit is terminated by hydrolysis of the bound GTP, which is stimulated by RGS (*regulator of G protein signaling*) proteins, which act as GTPase activating proteins (GAPs) for the heterotrimeric G proteins. The inactive α subunit (now bound to GDP) then reassociates with the $\beta\gamma$ complex, ready for the cycle to start anew.

The human genome encodes 21 different α subunits, 6 β subunits, and 12 γ subunits. Different G proteins associate with different receptors, so this array of G proteins couples receptors to distinct intracellular targets. For example, the G protein associated with the epinephrine receptor is called G_s because its α subunit *stimulates* adenylyl cyclase (see Figure 16.12). Other G protein α and $\beta\gamma$ subunits act instead to inhibit adenylyl cyclase or to regulate the activities of other target enzymes.

In addition to regulating target enzymes, both the α and $\beta\gamma$ subunits of some G proteins directly regulate ion channels. A good example is provided by the action of the neurotransmitter acetylcholine on heart muscle, which is distinct from its effects on nerve and skeletal muscle. The nicotinic acetylcholine receptor on nerve and skeletal muscle cells is a ligand-gated ion channel (see Figure 14.22). Heart muscle cells have a different acetylcholine receptor, which is G protein-coupled. This G protein is designated G_i because its α subunit *inhibits* adenylyl cyclase. In addition, the G_i $\beta\gamma$ subunits act directly to open K+ channels in the plasma membrane, which has the effect of slowing heart muscle contraction.

The cAMP pathway: Second messengers and protein phosphorylation

The role of cAMP, a major target of G protein signaling in mammalian cells, was discovered in 1958 by Earl Sutherland Jr. during studies of the hormone epinephrine, which stimulates the breakdown of glycogen to glucose in muscle cells. Sutherland found that the action of epinephrine was mediated by an increase in the intracellular concentration of cAMP, leading to the concept that cAMP is a second messenger in hormonal signaling (the first messenger being the hormone itself). cAMP is formed from ATP by the action of adenylyl cyclase (which is stimulated by G_s) and degraded to AMP by **cAMP phosphodiesterase** (Figure 16.14).

Most effects of cAMP in animal cells are mediated by the action of **cAMP-dependent protein kinase**, or **protein kinase A**, an enzyme discovered by Donal Walsh and Ed Krebs in 1968. The inactive form of protein kinase A is a tetramer consisting of two regulatory and two catalytic subunits (Figure 16.15). Cyclic

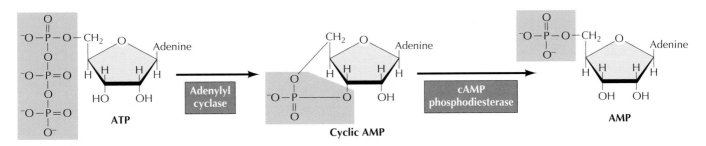

Figure 16.14 Synthesis and degradation of cAMP Cyclic AMP (cAMP) is synthesized from ATP by adenylyl cyclase and degraded to AMP by cAMP phosphodiesterase.

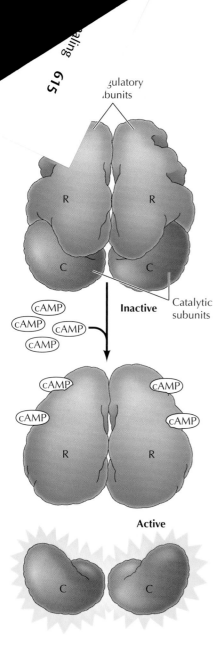

Figure 16.15 Regulation of protein kinase A The inactive form of protein kinase A consists of two regulatory (R) and two catalytic (C) subunits. Binding of cAMP to the regulatory subunits induces a conformational change that leads to dissociation of the catalytic subunits, which are then enzymatically active.

AMP binds to the regulatory subunits, leading to their dissociation from the catalytic subunits. The free catalytic subunits are then enzymatically active and able to phosphorylate serine residues on their target proteins.

In the regulation of glycogen breakdown by epinephrine, the key target of protein kinase A is another protein kinase, phosphorylase kinase, which is phosphorylated and activated by protein kinase A (**Figure 16.16**). Phosphorylase

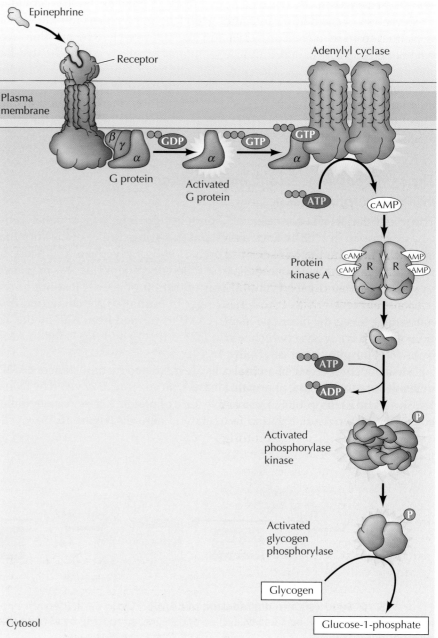

Figure 16.16 Regulation of glycogen metabolism by epinephrine Receptor stimulation by epinephrine leads to G protein-mediated activation of adenylyl cyclase. cAMP activates protein kinase A, which phosphorylates and activates phosphorylase kinase. Phosphorylase kinase then phosphorylates and activates glycogen phosphorylase, which catalyzes the breakdown of glycogen to glucose-1-phosphate.

kinase in turn phosphorylates and activates glycogen phosphorylase, which catalyzes the breakdown of glycogen to glucose-1-phosphate. In addition, protein kinase A phosphorylates and inhibits the enzyme glycogen synthase, which catalyzes glycogen synthesis. Elevation of cAMP and activation of protein kinase A thus blocks further glycogen synthesis at the same time as it stimulates glycogen breakdown.

The chain of reactions leading from the epinephrine receptor to glycogen phosphorylase provides a good illustration of signal amplification during intracellular signal transduction. Each molecule of epinephrine activates only a single receptor. Each receptor, however, may activate up to 100 molecules of G_s. Each molecule of G_s then stimulates the enzymatic activity of adenylyl cyclase, which can catalyze the synthesis of many molecules of cAMP. Signal amplification continues as each molecule of protein kinase A phosphorylates many molecules of phosphorylase kinase, which in turn phosphorylate many molecules of glycogen phosphorylase. Hormone binding to a small number of receptors thus leads to activation of a much larger number of intracellular target enzymes.

In many animal cells, increases in cAMP activate the transcription of specific target genes that contain a regulatory sequence called the **cAMP response element**, or **CRE**. In this case, the signal is carried from the cytoplasm to the nucleus by the catalytic subunit of protein kinase A, which is able to enter the nucleus following its release from the regulatory subunit. Within the nucleus, protein kinase A phosphorylates a transcription factor called **CREB** (for CRE-binding protein), leading to the recruitment of coactivators and transcription of cAMP-inducible genes (**Figure 16.17**). Such regulation of gene expression by cAMP plays important roles in controlling the proliferation, survival, and differentiation of a wide variety of animal cells, as well as being implicated in learning and memory.

It is important to recognize that protein kinases, such as protein kinase A, do not function in isolation

Animation 16.3

sites.sinauer.com/cooper7e/a16.3

Signal Amplification In a signal transduction cascade, each enzyme activated at a stage in the cascade may activate numerous molecules of the next enzyme, quickly amplifying the response to the receptor-bound ligand.

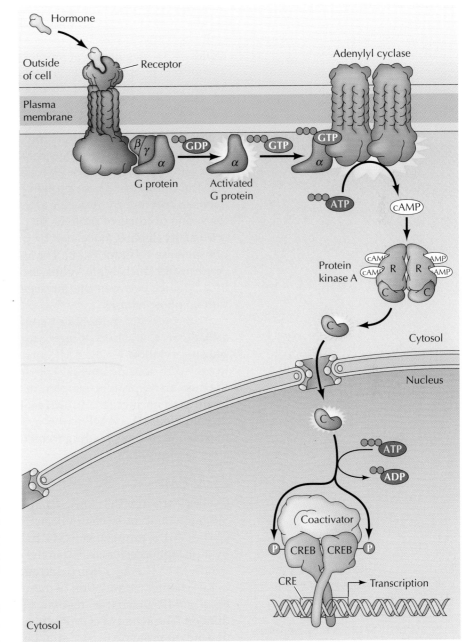

Figure 16.17 Cyclic AMP-inducible gene expression Receptor stimulation leads to G protein-mediated activation of adenylyl cyclase, synthesis of cAMP, and activation of protein kinase A. The free catalytic subunit of protein kinase A translocates to the nucleus and phosphorylates the transcription factor CREB (CRE-binding protein), leading to the recruitment of coactivators and expression of cAMP-inducible genes.

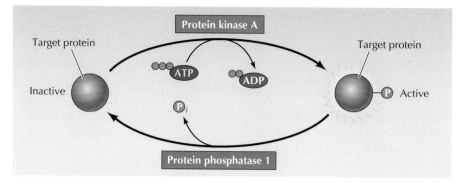

Figure 16.18 Regulation of protein phosphorylation by protein kinase A and protein phosphatase 1 The phosphorylation of target proteins by protein kinase A is reversed by the action of protein phosphatase 1.

within the cell. To the contrary, protein phosphorylation is rapidly reversed by the action of protein phosphatases, which remove phosphate groups from either phosphorylated tyrosine or serine/threonine residues in their substrate proteins. These protein phosphatases serve to terminate the responses initiated by receptor activation of protein kinases. For example, the serine residues of proteins that are phosphorylated by protein kinase A are usually dephosphorylated by the action of a phosphatase called protein phosphatase 1 (**Figure 16.18**). The levels of phosphorylation of protein kinase A substrates (such as phosphorylase kinase and CREB) are thus determined by a balance between the intracellular activities of protein kinase A and protein phosphatases.

Although most effects of cAMP are mediated by protein kinase A, cAMP can also directly regulate ion channels, independent of protein phosphorylation. Cyclic AMP functions in this way as a second messenger involved in sensing smells. The odorant receptors in sensory neurons in the nose are G-protein-coupled receptors that stimulate adenylyl cyclase, leading to an increase in intracellular cAMP. Rather than stimulating protein kinase A, cAMP in this system directly opens Na^+ channels in the plasma membrane, leading to membrane depolarization and initiation of a nerve impulse.

Cyclic GMP

Cyclic GMP (cGMP) is also an important second messenger in animal cells, although its roles have not been as extensively studied as those of cAMP. Cyclic GMP is formed from GTP by guanylyl cyclases and degraded to GMP by a phosphodiesterase. Guanylyl cyclases can be activated by peptide ligands, as well as by nitric oxide and carbon monoxide, as discussed earlier in this chapter. Stimulation of these guanylyl cyclases leads to elevated levels of cGMP, which then mediate biological responses, such as blood vessel dilation. The action of cGMP is frequently mediated by activation of cGMP-dependent protein kinases, although cGMP also regulates ion channels and phosphodiesterases.

One well-characterized role of cGMP is in the vertebrate eye, where it serves as the second messenger responsible for converting the visual signals received as light to nerve impulses. The photoreceptor in rod cells of the retina

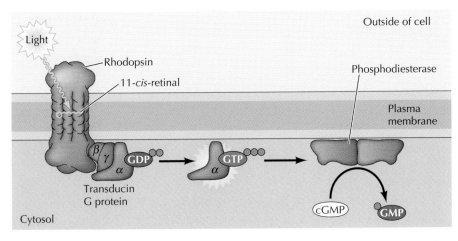

Figure 16.19 Role of cGMP in photoreception   Absorption of light by 11-*cis*-retinal activates the G protein-coupled receptor rhodopsin. The α subunit of transducin then dissociates from the βγ subunit and stimulates cGMP phosphodiesterase, leading to a decrease in intracellular levels of cGMP.

is a G protein-coupled receptor called **rhodopsin** (Figure 16.19). Rhodopsin is activated as a result of the absorption of light by the associated small molecule 11-*cis*-retinal, which then isomerizes to all-*trans*-retinal, inducing a conformational change in the rhodopsin protein. Rhodopsin then activates the G protein **transducin**, and the α subunit of transducin stimulates the activity of **cGMP phosphodiesterase**, leading to a decrease in the intracellular level of cGMP. This change in cGMP level in retinal rod cells is translated to a nerve impulse by a direct effect of cGMP on ion channels in the plasma membrane, similar to the action of cAMP in sensing smells.

Tyrosine Kinases and Signaling by MAP Kinase, PI 3-Kinase, and Phospholipase C/Calcium Pathways

In contrast with the G protein-coupled receptors, other cell surface receptors are directly linked to intracellular enzymes. The largest family of such enzyme-linked receptors are **tyrosine kinases**, which phosphorylate their substrate proteins on tyrosine residues (receptor tyrosine kinases). Many other receptors are not themselves tyrosine kinases, but act by stimulating intracellular tyrosine kinases with which they are noncovalently associated (non-receptor tyrosine kinases). Tyrosine kinases have been particularly well studied as key elements of signaling pathways involved in the control of animal cell growth and differentiation. Indeed, the first tyrosine kinase was discovered during studies of the oncogenic protein of Rous sarcoma virus (see Chapter 9 Key Experiment), directly linking tyrosine phosphorylation to the abnormal growth of cancer cells. As will be discussed in Chapter 19, many of the drugs currently being developed as new cancer treatments are inhibitors of tyrosine kinases or their downstream signaling pathways.

Receptor tyrosine kinases

The **receptor tyrosine kinases** include the receptors for most polypeptide growth factors, clearly establishing tyrosine phosphorylation as a key signaling

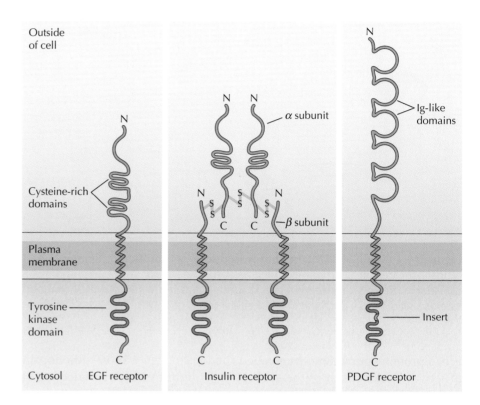

Figure 16.20 Organization of receptor tyrosine kinases Each receptor consists of an N-terminal extracellular ligand-binding domain, a single transmembrane α helix, and a cytosolic C-terminal domain with tyrosine kinase activity. The structures of three distinct subfamilies of receptor tyrosine kinases are shown. The EGF receptor and insulin receptor both have cysteine-rich extracellular domains, whereas the PDGF receptor has immunoglobulin (Ig)-like domains. The PDGF receptor is also noteworthy in that its kinase domain is interrupted by an insert of approximately 100 amino acids unrelated to those found in most other tyrosine kinase catalytic domains. The insulin receptor is unusual in being a dimer of two pairs of polypeptide chains (designated α and β).

Video 16.1

sites.sinauer.com/cooper7e/v16.1

Receptor Tyrosine Kinases Receptor tyrosine kinases bind many polypeptide growth factors, cytokines, and hormones to initiate signal transduction.

mechanism in the response of cells to growth factor stimulation. The human genome encodes 58 receptor tyrosine kinases, including the receptors for EGF, NGF, PDGF, insulin, and many other growth factors. All these receptors share a common structural organization: an N-terminal extracellular ligand-binding domain, a single transmembrane α helix, and a cytosolic C-terminal domain with tyrosine kinase activity (**Figure 16.20**). Most of the receptor tyrosine kinases consist of single polypeptides, although the insulin receptor and some related receptors are dimers consisting of two polypeptide chains. The binding of ligands (e.g., growth factors) to the extracellular domains of these receptors activates their cytosolic kinase domains, resulting in phosphorylation of both the receptors themselves and intracellular target proteins that propagate the signal initiated by growth factor binding.

The first step in signaling from most receptor tyrosine kinases is ligand-induced receptor dimerization (**Figure 16.21**). Some growth factors, such as PDGF and NGF, are themselves dimers consisting of two identical polypeptide chains; these growth factors directly induce dimerization by simultaneously

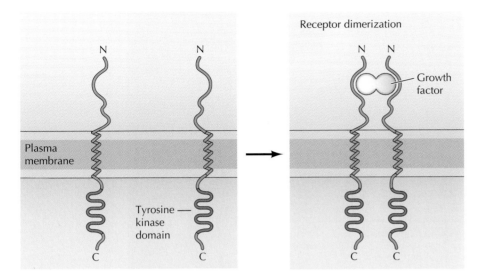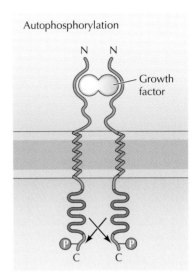

Figure 16.21 Dimerization and autophosphorylation of receptor tyrosine kinases   Growth factor binding induces receptor dimerization, which results in receptor autophosphorylation as the two polypeptide chains cross-phosphorylate one another.

binding to two different receptor molecules. Other growth factors (such as EGF) are monomers but lead to receptor dimerization as a result of inducing conformational changes that promote protein–protein interactions between different receptor polypeptides.

Ligand-induced dimerization then leads to **autophosphorylation** of the receptor as the dimerized polypeptide chains cross-phosphorylate one another (see Figure 16.21). Such autophosphorylation plays two key roles in signaling from these receptors. First, phosphorylation of tyrosine residues within the catalytic domain increases protein kinase activity. Second, phosphorylation of tyrosine residues outside of the catalytic domain creates specific binding sites for additional proteins that transmit intracellular signals downstream of the activated receptors.

The association of these downstream signaling molecules with receptor tyrosine kinases is mediated by protein domains that bind to specific phosphotyrosine-containing peptides (**Figure 16.22**). The first of these domains to be characterized are called **SH2 domains** (for *Src* *h*omology 2) because they were initially recognized in tyrosine kinases related to Src, the oncogenic protein of Rous sarcoma virus. SH2 domains consist

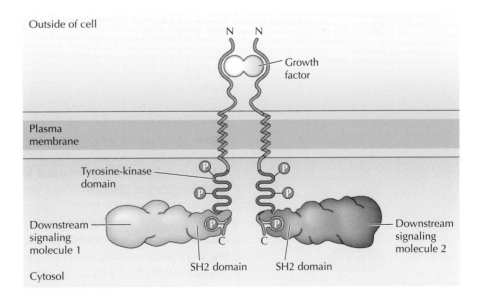

Figure 16.22 Association of downstream signaling molecules with receptor tyrosine kinases   SH2 domains bind to specific phosphotyrosine-containing peptides of the activated receptors.

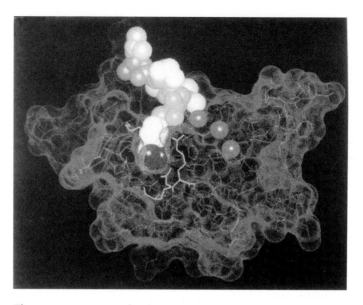

Figure 16.23 Complex between an SH2 domain and a phosphotyrosine peptide The polypeptide chain of the Src SH2 domain is shown in red with its surface indicated by green dots. Purple spheres indicate a groove on the surface. The amino acid residues that interact with the phosphotyrosine are shown in blue. The phosphotyrosine-containing peptide is shown as a space-filling model, including yellow and white spheres that indicate the backbone and sidechain atoms, respectively, and red spheres that indicate the phosphate group. (From G. Waksman and 13 others, 1992. *Nature* 358: 646.)

of approximately 100 amino acids and bind to specific short peptide sequences containing phosphotyrosine residues (**Figure 16.23**). The resulting association of SH2-containing proteins with activated receptor tyrosine kinases can have several effects: It localizes these proteins to the plasma membrane, leads to their association with other proteins, promotes their phosphorylation, and stimulates their enzymatic activities. The association of these proteins with autophosphorylated receptors thus represents the first step in the intracellular transmission of signals initiated by the binding of growth factors to the cell surface.

Nonreceptor tyrosine kinases

Rather than possessing intrinsic enzymatic activity, many receptors act by stimulating intracellular tyrosine kinases with which they are noncovalently associated. These **nonreceptor tyrosine kinases** include members of the **cytokine receptor superfamily**, which encompass the receptors for most cytokines (e.g., erythropoietin and interleukin-2) and for some polypeptide hormones (e.g., growth hormone). Like receptor tyrosine kinases, the cytokine receptors contain N-terminal extracellular ligand-binding domains, single transmembrane α helices, and C-terminal cytosolic domains. However, the cytosolic domains of the cytokine receptors are devoid of any known catalytic activity. Instead, the cytokine receptors function in association with nonreceptor tyrosine kinases, which are activated as a result of ligand binding.

The first step in signaling from cytokine receptors is thought to be ligand-induced receptor dimerization and cross-phosphorylation of the associated

FYI

Cytokine receptors are used by human immunodeficiency virus (HIV) as cell surface receptors for infection of immune cells.

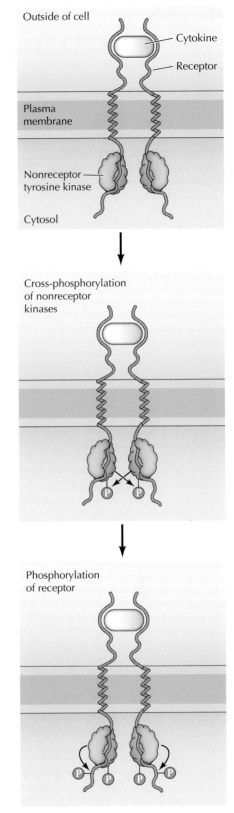

Figure 16.24 Activation of nonreceptor tyrosine kinases Ligand binding induces receptor dimerization and leads to the activation of associated nonreceptor tyrosine kinases as a result of cross-phosphorylation. The activated kinases then phosphorylate tyrosine residues of the receptor, creating phosphotyrosine-binding sites for downstream signaling molecules.

nonreceptor tyrosine kinases (**Figure 16.24**). These activated kinases then phosphorylate the receptor, providing phosphotyrosine-binding sites for the recruitment of downstream signaling molecules that contain SH2 domains. Combinations of cytokine receptors plus associated nonreceptor tyrosine kinases thus function analogously to the receptor tyrosine kinases discussed in the previous section.

The kinases associated with cytokine receptors belong to the **Janus kinase** (or **JAK**) family, which consists of four related nonreceptor tyrosine kinases. The key targets of the JAK kinases are the **STAT proteins** (signal *t*ransducers and *a*ctivators of *t*ranscription), which were originally identified in studies of cytokine receptor signaling (**Figure 16.25**). The STAT proteins are a family of seven transcription factors that contain SH2 domains. They are inactive in unstimulated cells where they are localized to the cytoplasm. Stimulation of cytokine receptors leads to recruitment of STAT proteins, which bind via their SH2 domains to phosphotyrosine-containing sequences in the cytoplasmic domains of the receptors. Following their association with activated receptors,

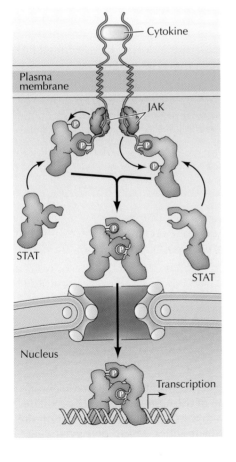

Figure 16.25 The JAK/STAT pathway The STAT proteins are transcription factors with SH2 domains that mediate their binding to phosphotyrosine-containing sequences. In unstimulated cells, STAT proteins are inactive in the cytosol. Stimulation of cytokine receptors leads to the binding of STAT proteins to phosphotyrosine-binding sites on the receptor, where they are phosphorylated by the receptor-associated JAK tyrosine kinases. The phosphorylated STAT proteins then dimerize and translocate to the nucleus where they activate the transcription of target genes.

the STAT proteins are phosphorylated by JAK family kinases. Tyrosine phosphorylation promotes the dimerization of STAT proteins, which then translocate to the nucleus where they stimulate transcription of their target genes. Further studies have shown that STAT proteins are also activated downstream of receptor tyrosine kinases, where their phosphorylation may be catalyzed either by the receptors themselves or by associated nonreceptor kinases. The STAT transcription factors thus serve as direct links between both cytokine and growth factor receptors on the cell surface and regulation of gene expression in the nucleus.

Additional nonreceptor tyrosine kinases belong to the **Src** family, which consists of Src and eight closely related proteins. As already noted, Src was initially identified as the oncogenic protein of Rous sarcoma virus and was the first protein shown to possess tyrosine kinase activity, so it has played a pivotal role in experiments leading to our current understanding of cell signaling. Members of the Src family play key roles in signaling downstream of cytokine receptors, receptor tyrosine kinases, antigen receptors on B and T lymphocytes, and receptors involved in cell–cell and cell–matrix interactions.

As discussed in Chapter 15, the integrins are the major receptors responsible for the attachment of cells to the extracellular matrix (see Figure 15.22). In addition to this structural role, the integrins serve as receptors that activate intracellular signaling pathways, thereby controlling cell movement and other aspects of cell behavior (including cell proliferation and survival) in response to cell–matrix interactions. Like cytokine receptors, the integrins have short cytoplasmic tails that lack any intrinsic enzymatic activity. However, tyrosine phosphorylation is an early response to the interaction of integrins with extracellular matrix components, suggesting that the integrins are linked to nonreceptor tyrosine kinases. One mode of signaling downstream of integrins involves activation of a nonreceptor tyrosine kinase called **FAK** (*f*ocal *a*dhesion *k*inase) (**Figure 16.26**). As its name implies, FAK is localized to

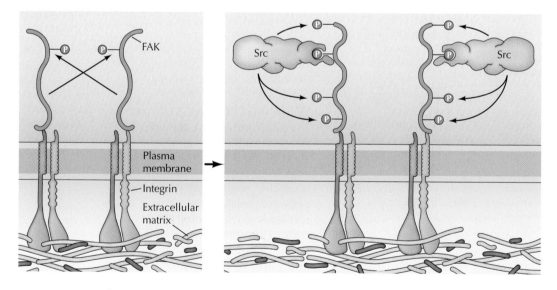

Figure 16.26 Integrin signaling Binding of integrins to the extracellular matrix leads to integrin clustering and activation of the nonreceptor tyrosine kinase FAK by autophosphorylation. Src then binds to the FAK autophosphorylation site and phosphorylates FAK on additional tyrosine residues, which serve as binding sites for downstream signaling molecules.

Molecular Medicine

Cancer: Signal Transduction and the *ras* Oncogenes

The Disease

Cancer claims the lives of approximately one out of every four Americans, accounting for approximately 590,000 deaths each year in the United States. There are more than 100 different kinds of cancer, but some are more common than others. In this country the most common lethal cancers are those of the lung and colon/rectum, which together account for approximately 35% of all cancer deaths. Other major contributors to cancer mortality include cancers of the breast, pancreas, and prostate, which are responsible for approximately 6.9%, 6.9%, and 4.7% of U.S. cancer deaths, respectively.

The common feature of all cancers is the unrestrained proliferation of cancer cells, which eventually spread throughout the body, invading normal tissues and organs and leading to death of the patient. Surgery and radiotherapy are effective treatments for localized cancers but are unable to reach cancer cells that have spread to distant body sites. Treatment of these cancers therefore requires the use of chemotherapeutic drugs. Unfortunately, the commonly available chemotherapeutic agents are not specific for cancer cells. Most act by either damaging DNA or interfering with DNA synthesis, so they also kill rapidly dividing normal cells, such as the epithelial cells that line the digestive tract and the blood-forming cells of the bone marrow. The resulting toxicity of these drugs limits their effectiveness, and many cancers are not eliminated by doses of chemotherapy that can be tolerated by the patient. Consequently, although major progress has been made in cancer treatment, nearly half of all patients diagnosed with cancer ultimately die of their disease.

Molecular and Cellular Basis

The identification of viral genes that can convert normal cells to cancer cells, such as the *src* gene of Rous sarcoma virus provided the first demonstration that cancers can result from the action of specific genes (oncogenes). The subsequent discovery that viral oncogenes are related to genes of normal cells then engendered the hypothesis that non-virus-induced cancers (including most human cancers) might arise as a result of mutations in normal cell genes, giving rise to oncogenes of cellular rather than viral origin. Such cellular oncogenes were first identified in human cancers in 1981. The following year, human oncogenes of bladder, lung, and colon cancers were found to be related to the *ras* genes previously identified in rat sarcoma viruses.

Although many different genes are now known to play critical roles in cancer development, mutations of the *ras* genes remain one of the most common genetic abnormalities in human tumors. Mutated *ras* oncogenes are found in about 25% of all human cancers, including approximately 25% of lung cancers, 50% of colon cancers, and more than 90% of pancreatic cancers. Moreover, the action of *ras* oncogenes has clearly linked the development of human cancer to abnormalities in the signaling pathways that regulate cell proliferation. The mutations that convert normal *ras* genes to oncogenes substantially decrease GTP hydrolysis by the Ras proteins. Consequently, the mutated oncogenic Ras proteins remain locked in the active GTP-bound form, rather than alternating normally between inactive and active states in response to extracellular signals. The oncogenic Ras proteins thus continuously stimulate the ERK signaling pathway and drive cell proliferation, even in the absence of the growth factors that would be required to activate Ras and signal proliferation of normal cells.

Prevention and Treatment

The discovery of mutated oncogenes in human cancers raised the possibility of

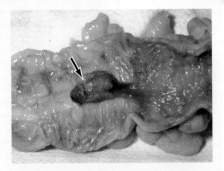

A human colon polyp (an early stage of colon cancer) The *ras* oncogenes contribute to the development of about half of all colon cancers. (E. P. Ewing Jr., Centers for Disease Control.)

developing drugs specifically targeted against the oncogene proteins. In principle, such drugs might act selectively against cancer cells with less toxicity toward normal cells than that of conventional chemotherapeutic agents. Because *ras* is frequently mutated in human cancers, the Ras proteins and other elements of Ras signaling pathways have attracted considerable interest as potential drug targets.

As discussed in Chapter 19, effective drugs that selectively target cancer cells have been recently developed against some tyrosine kinase oncogenes that act upstream of Ras. A variety of additional drugs are targeted against protein kinases activated downstream of Ras, such as Raf. The identification of oncogenes in human tumors has thus opened new strategies to rational development of drugs that act effectively and selectively against human cancer cells by targeting the signaling pathways that are responsible for cancer development.

References

Der, C. J., T. G. Krontiris and G. M. Cooper. 1982. Transforming genes of human bladder and lung carcinoma cell lines are homologous to the *ras* genes of Harvey and Kirsten sarcoma viruses. *Proc. Natl. Acad. Sci. USA* 79: 3637–3640.

Parada, L. F., C. J. Tabin, C. Shih and R. A. Weinberg. 1982. Human EJ bladder carcinoma oncogene is homologue of Harvey sarcoma virus *ras* gene. *Nature* 297: 474–478.

focal adhesions and rapidly becomes tyrosine-phosphorylated following the binding of integrins to extracellular matrix components. Like other tyrosine kinases, the activation of FAK involves autophosphorylation induced by the clustering of integrins bound to the extracellular matrix. Autophosphorylation of FAK creates docking sites for signaling molecules containing SH2 domains, including members of the Src family as well as other downstream signaling molecules.

Interactions between cells, as well as between cells and the extracellular matrix, play critical roles in multicellular organisms. For example, such interactions control the organization of cells in tissues, including the formation of epithelial cell sheets. As discussed in Chapter 15, cell–cell interactions are mediated by four major families of cell adhesion molecules: integrins, selectins, Ig superfamily members, and cadherins (see Table 15.2). Like integrins (discussed above), members of the Ig superfamily and cadherins are linked to signaling pathways that control multiple aspects of cell behavior, at least in part via stimulation of nonreceptor tyrosine kinases, including members of the Src family.

MAP kinase pathways

MAP kinase signaling is one of the major pathways of intracellular signal transduction activated downstream of both receptor and nonreceptor tyrosine kinases. The MAP kinase pathway refers to a cascade of protein kinases that are highly conserved in evolution and play central roles in signal transduction in all eukaryotic cells ranging from yeasts to humans. The central elements in the pathway are a family of serine/threonine kinases called the **MAP kinases** (for *m*itogen-*a*ctivated *p*rotein *k*inases) that are activated in response to a variety of growth factors and other signaling molecules. In yeasts, MAP kinase pathways control a variety of cellular responses, including mating, cell shape, and sporulation. In higher eukaryotes (including *C. elegans*, *Drosophila*, frogs, and mammals), MAP kinases are ubiquitous regulators of cell growth and differentiation.

The MAP kinases that were initially characterized in mammalian cells belong to the **ERK** (*e*xtracellular signal-*r*egulated *k*inase) family. The central role of ERK signaling in mammalian cells emerged from studies of the **Ras** proteins, which were first identified as the oncogenic proteins of tumor viruses that cause sarcomas in rats (hence the name Ras, from *rat* *s*arcoma virus). Interest in Ras intensified considerably in 1982 when mutations in *ras* genes were first implicated in the development of human cancers (discussed in Chapter 19). The importance of Ras in intracellular signaling was then indicated by experiments showing that microinjection of active Ras protein directly induces proliferation of normal mammalian cells. Conversely, interference with Ras function by either microinjection of anti-Ras antibody or expression of a dominant negative Ras mutant blocks growth-factor-induced cell proliferation. Thus Ras is not only capable of inducing the abnormal growth characteristic of cancer cells but is also required for the response of normal cells to growth factor stimulation.

The Ras proteins are guanine nucleotide-binding proteins that function analogously to the α subunits of G proteins, alternating between inactive GDP-bound and active GTP-bound forms (Figure 16.27). In contrast with the G protein α subunits, Ras functions as a monomer rather than in association with $\beta\gamma$ subunits. Ras activation is mediated by **guanine nucleotide exchange**

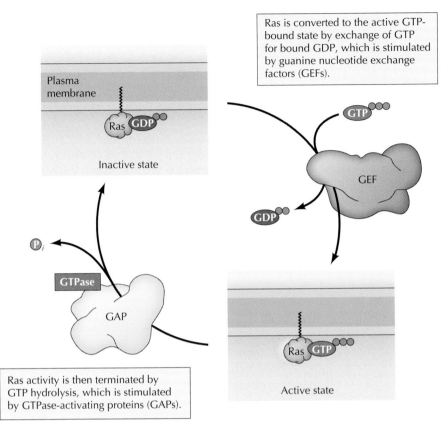

Ras is converted to the active GTP-bound state by exchange of GTP for bound GDP, which is stimulated by guanine nucleotide exchange factors (GEFs).

Plasma membrane

Ras GDP

Inactive state

GTP

GEF

GDP

P_i

GTPase

GAP

Ras GTP

Active state

Ras activity is then terminated by GTP hydrolysis, which is stimulated by GTPase-activating proteins (GAPs).

Figure 16.27 Regulation of Ras proteins   Ras proteins alternate between inactive GDP-bound and active GTP-bound states.

factors (GEFs) that stimulate the release of bound GDP and its exchange for GTP. Activity of the Ras-GTP complex is then terminated by GTP hydrolysis, which is stimulated by the interaction of Ras-GTP with **GTPase-activating proteins (GAPs)**. It is interesting to note that the mutations of *ras* genes in human cancers have the effect of inhibiting GTP hydrolysis by the Ras proteins. These mutated Ras proteins therefore remain continuously in the active GTP-bound form, driving the unregulated proliferation of cancer cells even in the absence of growth factor stimulation.

The best understood mode of Ras activation is that mediated by receptor tyrosine kinases (**Figure 16.28**). Autophosphorylation of these receptors leads to the binding of Ras GEFs via their SH2 domains. This localizes the GEFs to the plasma membrane where they are able to interact with Ras proteins, which are anchored to the inner leaflet of the plasma membrane by lipids attached to the Ras C terminus (see Figure 14.10). The GEFs then stimulate the exchange of GDP for GTP, resulting in formation of the active Ras-GTP complex. In its active GTP-bound form, Ras interacts with a number of effector proteins, including the **Raf** serine/threonine kinase. Interaction with Ras recruits Raf from the cytosol to the plasma membrane, where Raf is activated by dimerization and phosphorylation. Raf then phosphorylates and activates a second protein kinase called **MEK (MAP kinase/ERK kinase)**. MEK is a dual-specificity protein kinase that activates members of the ERK

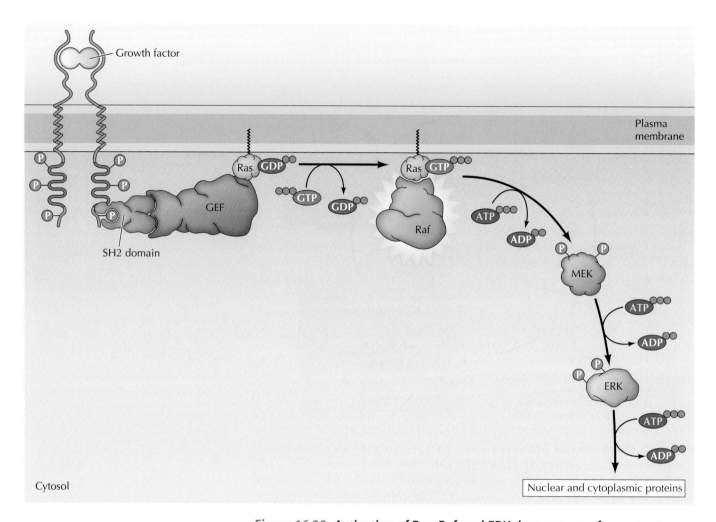

Figure 16.28 Activation of Ras, Raf, and ERK downstream of receptor tyrosine kinases Growth factor binding to a receptor tyrosine kinase leads to autophosphorylation and formation of binding sites for the SH2 domain of a guanine nucleotide exchange factor (GEF). This interaction recruits the GEF to the plasma membrane where it can stimulate Ras GDP/GTP exchange. The activated Ras-GTP complex then binds to the Raf protein kinase, leading to Raf activation. Raf phosphorylates and activates MEK, a dual-specificity protein kinase that activates ERK by phosphorylation on both threonine and tyrosine residues (Thr-183 and Tyr-185). ERK then phosphorylates a variety of nuclear and cytoplasmic target proteins.

family by phosphorylation of both threonine and tyrosine residues separated by one amino acid (e.g., threonine-183 and tyrosine-185 of ERK2). Once activated, ERK phosphorylates a variety of target proteins, including other protein kinases and transcription factors.

A fraction of activated ERK translocates to the nucleus where it regulates transcription factors by phosphorylation (**Figure 16.29**). In this regard, it is notable that a primary response to growth factor stimulation is the rapid transcriptional induction of a family of approximately 100 genes called **immediate early genes**. The induction of many immediate early genes is mediated by a regulatory sequence called the **serum response element (SRE)**, which is recognized by a complex of transcription factors including the

Figure 16.29 Induction of immediate-early genes by ERK Activated ERK translocates to the nucleus where it phosphorylates the transcription factor Elk-1. Elk-1 binds to the serum response element (SRE) in a complex with serum response factor (SRF). Phosphorylation stimulates the activity of Elk-1 as a transcriptional activator, leading to immediate-early gene induction.

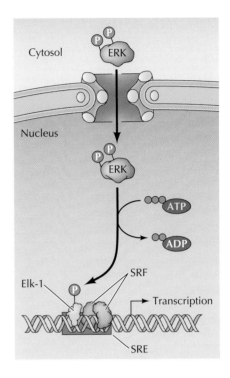

serum response factor (SRF) and **Elk-1**. ERK phosphorylates and activates Elk-1, providing a direct link between the ERK family of MAP kinases and immediate early gene induction. Many immediate early genes themselves encode transcription factors, so their induction in response to growth factor stimulation leads to altered expression of a battery of other downstream genes called **secondary response genes**. As discussed in Chapter 17, these alterations in gene expression directly link ERK signaling to the stimulation of cell proliferation induced by growth factors.

Both yeasts and mammalian cells have multiple MAP kinase pathways that control distinct cellular responses. Each cascade consists of three protein kinases: a terminal MAP kinase and two upstream kinases (analogous to Raf and MEK) that regulate its activity. In the yeast *S. cerevisiae*, five different MAP kinase cascades regulate mating, sporulation, filamentation, cell wall remodeling, and response to high osmolarity. In mammalian cells, multiple groups of MAP kinases have been identified. In addition to members of the ERK family, these include the JNK and p38 MAP kinases, which are preferentially activated in response to inflammatory cytokines and cellular stress (e.g., ultraviolet irradiation) (Figure 16.30). The JNK and p38 MAP kinase cascades are activated by members of the Rho subfamily of small GTP-binding proteins (including Rac, Rho, and Cdc42) rather than by Ras. Whereas ERK signaling principally leads to cell proliferation, differentiation, and survival, the JNK and p38 MAP kinase

Figure 16.30 Pathways of MAP kinase activation in mammalian cells In addition to ERK, mammalian cells contain JNK and p38 MAP kinases. Activation of JNK and p38 is mediated by members of the Rho subfamily of small GTP-binding proteins (Rac, Rho, and Cdc42), which stimulate protein kinase cascades parallel to that responsible for ERK activation. The protein kinase cascades leading to JNK and p38 activation appear to be preferentially activated by inflammatory cytokines or cellular stress and generally lead to inflammation and cell death.

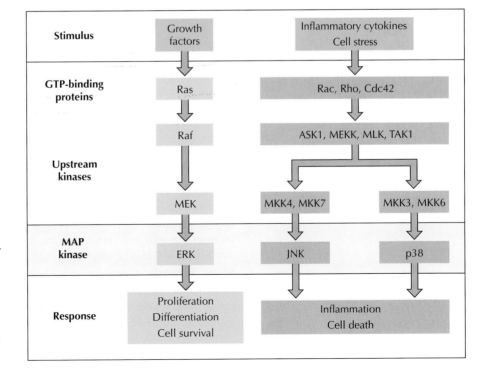

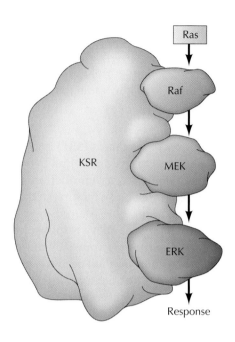

Figure 16.31 A scaffold protein for the ERK MAP kinase cascade The KSR scaffold protein binds Raf, MEK, and ERK, organizing these components of the ERK MAP kinase pathway into a signaling complex.

pathways often lead to inflammation and cell death. Like ERK, the JNK and p38 MAP kinases can translocate to the nucleus and phosphorylate transcription factors that regulate gene expression. Multiple MAP kinase pathways thus function in all types of eukaryotic cells to control cellular responses to diverse environmental signals.

The specificity of MAP kinase signaling is maintained at least in part by the organization of the components of each MAP kinase cascade as complexes that are associated with **scaffold proteins**, which organize complexes of signaling molecules. For example, the KSR scaffold protein organizes ERK and its upstream activators Raf and MEK into a signaling cassette (**Figure 16.31**). As a result of the specific association of these protein kinases on the KSR scaffold, activation of Raf by Ras leads to specific and efficient activation of MEK, which in turn activates ERK. Following phosphorylation by MEK, ERK dissociates from KSR and can translocate to the nucleus. Distinct scaffold proteins are involved not only in the organization of other MAP kinase signaling complexes but also in the association of other downstream signaling molecules with their receptors and targets. The physical association of signaling pathway components as a result of their interaction with scaffold proteins plays an important role in determining the specificity of signaling pathways within the cell.

The PI 3-kinase/Akt and mTOR pathways

A second major pathway of intracellular signaling downstream of tyrosine kinases (and also stimulated by some G proteins) is based on the use of a second messenger derived from the membrane phospholipid **phosphatidylinositol 4,5-bisphosphate (PIP$_2$)**. PIP$_2$ is a component of the plasma membrane, localized to the inner leaflet of the phospholipid bilayer (see Figure 14.2). It can be phosphorylated on the 3 position of inositol by the enzyme **phosphatidylinositide (PI) 3-kinase**.

The form of PI 3-kinase activated by tyrosine kinases has SH2 domains and is recruited and activated by association with receptor tyrosine kinases

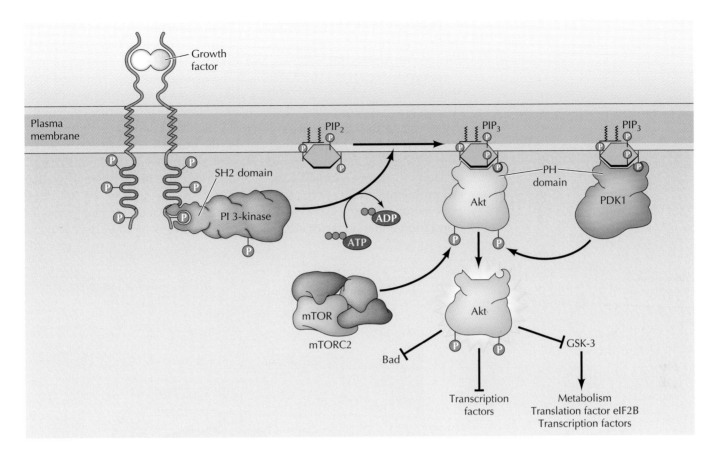

Figure 16.32 The PI 3-kinase/Akt pathway   PI 3-kinase is recruited to activated receptor tyrosine kinases via its SH2 domain. PI 3-kinase phosphorylates the 3 position of inositol, converting PIP_2 to PIP_3. Akt is recruited to the plasma membrane by binding to PIP_3 via its pleckstrin homology (PH) domain. It is then activated as a result of phosphorylation by another protein kinase (PDK1) that also binds PIP_3, as well as by mTOR in the mTORC2 complex. Akt then phosphorylates a number of target proteins, including direct regulators of cell survival (such as Bad, see Chapter 18), several transcription factors, and other protein kinases, including GSK-3 (which is inhibited by Akt phosphorylation). GSK-3 phosphorylates metabolic enzymes, the translation initiation factor eIF2B, and transcription factors.

(Figure 16.32). An alternative isoform of PI 3-kinase is activated by G proteins. Phosphorylation of PIP_2 yields the second messenger **phosphatidylinositol 3,4,5-trisphosphate (PIP_3)**, which is critical for signaling cell proliferation and survival. A key target of PIP_3 is a serine/threonine kinase called **Akt**. PIP_3 binds to a domain of Akt known as the pleckstrin homology (PH) domain. This interaction recruits Akt to the inner face of the plasma membrane where it is phosphorylated and activated by another protein kinase (called PDK1) that also contains a PH domain and binds PIP_3. The formation of PIP_3 thus results in the association of both Akt and PDK1 with the plasma membrane, leading to phosphorylation and activation of Akt. Activation of Akt also requires phosphorylation at a second site by a distinct protein kinase (mTORC2), which is also stimulated by growth factors.

Once activated, Akt phosphorylates a number of target proteins, including proteins that are direct regulators of cell survival (such as Bad, discussed in Chapter 18), transcription factors, and other protein kinases. The critical transcription factors targeted by Akt include members of the FOXO family (**Figure 16.33**). Phosphorylation of FOXO by Akt creates a binding site for cytosolic chaperones (14-3-3 proteins) that sequester FOXO in an inactive form in the cytoplasm. In the absence of growth factor signaling and Akt activity, FOXO is released from 14-3-3 and translocates to the nucleus, stimulating transcription of genes that inhibit cell proliferation or induce cell death. Another target of Akt is the protein kinase GSK-3, which regulates metabolism as well as cell proliferation and survival. Like FOXO, GSK-3 is inhibited by Akt phosphorylation. The targets of GSK-3 include several transcription factors and the translation initiation factor eIF2B. Phosphorylation of eIF2B leads to a global

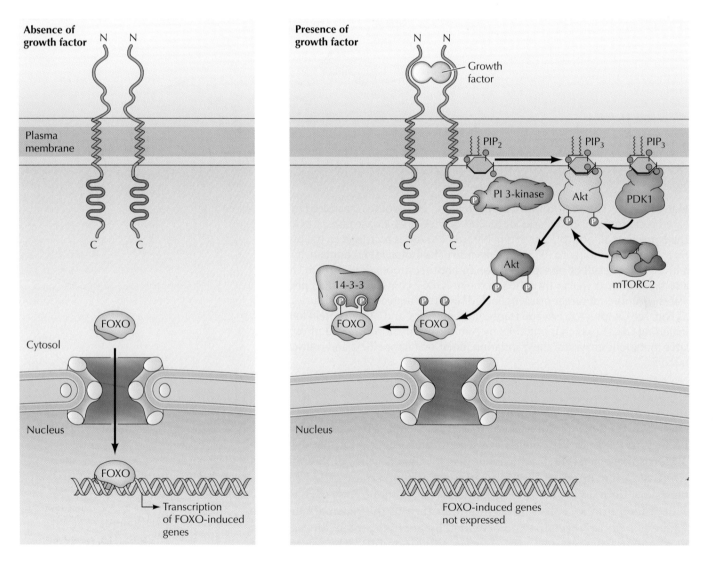

Figure 16.33 Regulation of FOXO In the absence of growth factor stimulation, the FOXO transcription factor translocates to the nucleus and induces target gene expression. Growth factor stimulation leads to activation of Akt, which phosphorylates FOXO. This creates binding sites for the cytosolic chaperone 14-3-3, which sequesters FOXO in an inactive form in the cytoplasm.

downregulation of translation initiation (see Figure 9.21), so GSK-3 provides a link between growth factor signaling and control of cellular protein synthesis.

The **mTOR** pathway is a central regulator of cell growth that couples the control of protein synthesis to the availability of growth factors, nutrients, and energy (**Figure 16.34**). This is accomplished via the regulation of mTOR by multiple signals, including the PI 3-kinase/Akt pathway. The mTOR protein kinase exists in two distinct complexes in cells. As discussed above, the mTORC2 complex is one of the protein kinases that phosphorylates and

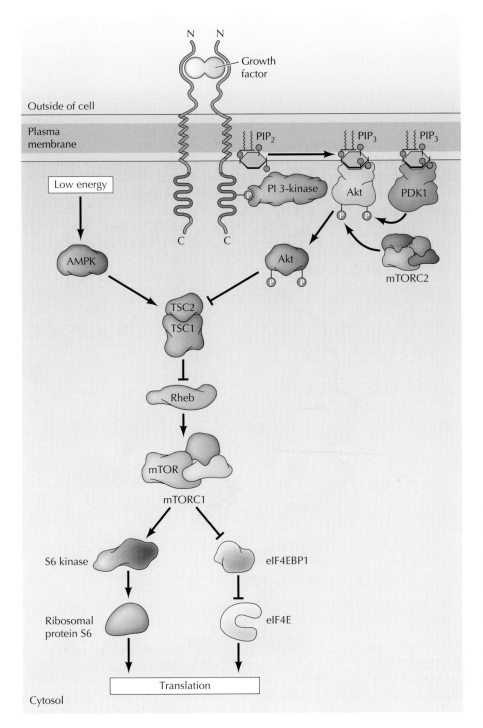

Figure 16.34 The mTOR pathway
Akt inhibits the TSC1/2 complex, leading to activation of Rheb and mTORC1 in response to growth factor stimulation. In contrast, AMPK activates the TSC1/2 complex, leading to inhibition of Rheb and mTORC1 if cellular energy stores are depleted. mTORC1 stimulates translation by phosphorylating S6 kinase (which phosphorylates ribosomal protein S6) and by phosphorylating eIF4E binding protein-1 (4E-BP1), relieving inhibition of translation initiation factor eIF4E.

activates Akt (see Figure 16.32). In contrast, the mTORC1 complex is activated downstream of Akt and functions to regulate protein synthesis. mTORC1 is regulated by the Ras-related GTP-binding protein Rheb, which is in turn regulated by the GTPase-activating protein complex TSC1/2 (comprised of TSC1 and TSC2 subunits). Akt phosphorylates and inhibits the activity of TSC1/2, leading to activation of mTORC1 in response to growth factor stimulation. In addition, TSC1/2 is regulated by another protein kinase called the AMP-activated kinase (AMPK). AMPK senses the energy state of the cell and is activated by a high ratio of AMP to ATP. Under these conditions, AMPK phosphorylates and activates TSC1/2, leading to inhibition of mTORC1 when cellular energy stores are depleted. mTORC1 is also regulated by the availability of amino acids, which are required for mTORC1 activity.

mTORC1 phosphorylates at least two well-characterized targets that function to regulate protein synthesis: S6 kinase and eIF4E binding protein-1 (4E-BP1). S6 kinase controls translation by phosphorylating the ribosomal protein S6 as well as other proteins involved in translational regulation. The eIF4E binding protein controls translation by interacting with initiation factor eIF4E, which binds to the 5' cap of mRNAs (see Figure 9.22). In the absence of mTORC1 signaling, nonphosphorylated 4E-BP1 binds to eIF4E and inhibits translation by interfering with the interaction of eIF4E with eIF4G. Phosphorylation of 4E-BP1 by mTORC1 prevents its interaction with eIF4E, leading to increased rates of translation initiation.

In addition to stimulating protein synthesis, mTORC1 inhibits the degradation of cellular proteins by regulating autophagy (**Figure 16.35**). As discussed

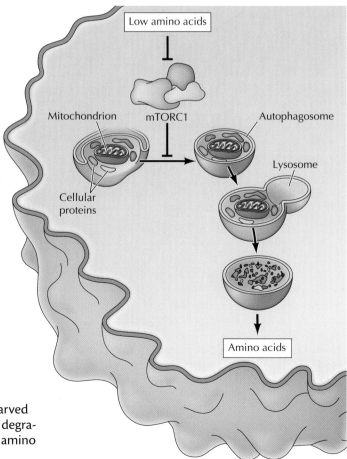

Figure 16.35 Regulation of autophagy by mTOR The mTORC1 protein kinase inhibits autophagy. When a cell is starved of nutrients (e.g., amino acids), mTORC1 is inhibited and the degradation of cellular proteins by autophagy provides a source of amino acids.

in Chapter 9, autophagy results in the degradation of cellular proteins by uptake into lysosomes (see Figure 9.48). When cells are starved of nutrients (e.g., amino acids), mTORC1 activity is decreased, thereby stimulating autophagy and allowing cells to degrade nonessential proteins so their amino acids can be reutilized.

Phospholipase C and Ca²⁺

Another major pathway of intracellular signaling, activated downstream of both tyrosine kinases and G-protein coupled receptors, is based on the hydrolysis of PIP$_2$ by **phospholipase C (PLC)**—a reaction that produces two distinct second messengers, **diacylglycerol (DAG)** and **inositol 1,4,5-trisphosphate (IP$_3$)** (**Figure 16.36**). The form of phospholipase C (PLC-γ) activated downstream of tyrosine kinases contains SH2 domains that mediate its association with activated receptor tyrosine kinases. This interaction localizes PLC-γ to the plasma membrane as well as leading to its tyrosine phosphorylation, which increases its catalytic activity. DAG and IP$_3$ then stimulate distinct downstream

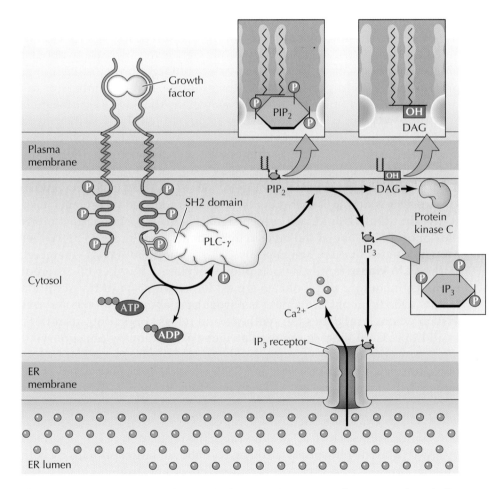

Figure 16.36 Activation of phospholipase C by tyrosine kinases Phospholipase C-γ (PLC-γ) binds to activated receptor tyrosine kinases via its SH2 domains. Tyrosine phosphorylation increases PLC-γ activity, stimulating the hydrolysis of phosphatidylinositol 4,5-bisphosphate (PIP$_2$) to diacylglycerol (DAG) and inositol trisphosphate (IP$_3$). DAG activates members of the protein kinase C family. IP$_3$ binds to receptors that are ligand-gated Ca²⁺ channels in the endoplasmic reticulum membrane. Opening these channels allows the efflux of Ca²⁺ from the endoplasmic reticulum to the cytosol.

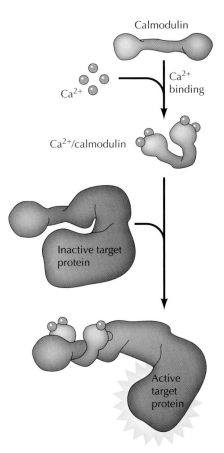

Calmodulin

Ca²⁺

Ca²⁺ binding

Ca²⁺/calmodulin

Inactive target protein

Active target protein

Ca²⁺/calmodulin-dependent protein kinase

Figure 16.37 Function of calmodulin Calmodulin is a dumbbell-shaped protein with four Ca²⁺-binding sites. The active Ca²⁺/calmodulin complex binds to a variety of target proteins, including Ca²⁺/ calmodulin-dependent protein kinases.

signaling pathways (protein kinase C and Ca²⁺ mobilization, respectively), so PIP₂ hydrolysis triggers a two-armed cascade of intracellular signaling.

The diacylglycerol produced by hydrolysis of PIP₂ remains associated with the plasma membrane and activates serine/threonine kinases belonging to the **protein kinase C** family, many of which play important roles in the control of cell growth and differentiation. The other second messenger produced by PIP₂ cleavage, IP₃, is a small polar molecule that is released into the cytosol, where it acts to signal the release of Ca²⁺ from intracellular stores. As noted in Chapter 14, the cytosolic concentration of Ca²⁺ is maintained at an extremely low level (about 0.1 μM) as a result of Ca²⁺ pumps that actively export Ca²⁺ from the cell. Ca²⁺ is pumped not only across the plasma membrane but also into the endoplasmic reticulum, which therefore serves as an intracellular Ca²⁺ store. IP₃ acts to release Ca²⁺ from the endoplasmic reticulum by binding to receptors that are ligand-gated Ca²⁺ channels. As a result, cytosolic Ca²⁺ levels increase to about 1 μM, which affects the activities of a variety of target proteins, including protein kinases and phosphatases. For example, some members of the protein kinase C family require Ca²⁺ as well as diacylglycerol for their activation, so these protein kinases are regulated jointly by both arms of the PIP₂ signaling pathway. One of the major Ca²⁺-binding proteins that mediates the effects of Ca²⁺ is **calmodulin**, which is activated when the concentration of cytosolic Ca²⁺ increases to about 0.5 μM (**Figure 16.37**). Ca²⁺/calmodulin then binds to a variety of target proteins, including protein kinases. One example of such a Ca²⁺/calmodulin-dependent protein kinase is myosin light-chain kinase, which signals actin-myosin contraction by phosphorylating one of the myosin light chains (see Figure 13.31). Other protein kinases that are activated by Ca²⁺/calmodulin include members of the **CaM kinase** family, which phosphorylate a number of different proteins, including metabolic enzymes, ion channels, and transcription factors. One form of CaM kinase is particularly abundant in the nervous system where it regulates the synthesis and release of neurotransmitters. In addition, CaM kinases can regulate gene expression by phosphorylating transcription factors. Interestingly, one of the transcription factors phosphorylated by CaM kinase is CREB, which is phosphorylated at the same site by protein kinase A. This phosphorylation of CREB illustrates one of many intersections between the Ca²⁺ and cAMP signaling pathways. Other examples include the regulation of adenylyl cyclases and phosphodiesterases by Ca²⁺/calmodulin, the regulation of Ca²⁺ channels by cAMP, and the phosphorylation of a number of target proteins by both protein kinase A and Ca²⁺/calmodulin-dependent protein kinases. The cAMP and Ca²⁺ signaling pathways thus function coordinately to regulate many cellular responses.

Increases in intracellular Ca²⁺ can result not only from the release of Ca²⁺ from intracellular stores, but also from the uptake of extracellular Ca²⁺ via regulated channels in the plasma membrane. The entry of extracellular Ca²⁺ is particularly important in the electrically excitable cells of nerve and muscle

FYI

Toxic snake venoms contain phospholipases. Hydrolysis of phospholipids by venom from rattlesnakes and cobras leads to the rupture of red blood cell membranes.

Figure 16.38 Regulation of intracellular Ca²⁺ in electrically excitable cells Membrane depolarization leads to the opening of voltage-gated Ca²⁺ channels in the plasma membrane causing the influx of Ca²⁺ from extracellular fluids. The resulting increase in intracellular Ca²⁺ then signals the further release of Ca²⁺ from intracellular stores by opening distinct Ca²⁺ channels (ryanodine receptors) in the endoplasmic reticulum membrane. In muscle cells, ryanodine receptors in the sarcoplasmic reticulum may also be opened directly in response to membrane depolarization.

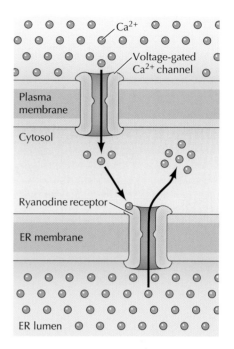

in which voltage-gated Ca²⁺ channels in the plasma membrane are opened by membrane depolarization (Figure 16.38). The resulting increases in intracellular Ca²⁺ then trigger the further release of Ca²⁺ from intracellular stores by activating distinct Ca²⁺ channels known as **ryanodine receptors**. One effect of increases in intracellular Ca²⁺ in neurons is to trigger the release of neurotransmitters, so Ca²⁺ plays a critical role in converting electric to chemical signals in the nervous system. In muscle cells Ca²⁺ is stored in the sarcoplasmic reticulum from which it is released by the opening of ryanodine receptors in response to changes in membrane potential. This release of stored Ca²⁺ leads to large increases in cytosolic Ca²⁺, which trigger muscle contraction (see Figure 13.28). Cells thus utilize a variety of mechanisms to regulate intracellular Ca²⁺ levels, making Ca²⁺ a versatile second messenger that controls a wide range of cellular processes.

Receptors Coupled to Transcription Factors

The cAMP, MAP kinase, and PI 3-kinase pathways are examples of indirect connections between the cell surface and the nucleus in which a cascade of second messengers and protein kinases ultimately leads to transcription factor phosphorylation. Several other types of growth factor receptors (discussed in this section) are more directly coupled to transcription factors that play key roles in cell proliferation, differentiation, and survival.

The TGF-β/Smad pathway

The TGF-β/Smad pathway is similar to JAK/STAT signaling (see Figure 16.25) in that a protein kinase associated with a receptor directly phosphorylates and activates a transcription factor. However, the receptors for **transforming growth factor β (TGF-β)** and related polypeptides are protein kinases that phosphorylate serine or threonine, rather than tyrosine, residues on their substrate proteins. TGF-β is the prototype of a family of polypeptide growth factors that control proliferation and differentiation of a variety of cell types. The cloning of the first receptor for a member of the TGF-β family in 1991 revealed that it is the prototype of a unique receptor family with a cytosolic serine/threonine kinase domain. Since then, receptors for additional TGF-β family members have similarly been found to be serine/threonine kinases.

The TGF-β receptors are composed of two distinct polypeptides (designated type I and type II) that become associated following ligand binding (Figure 16.39). The type II receptor then phosphorylates the type I receptor,

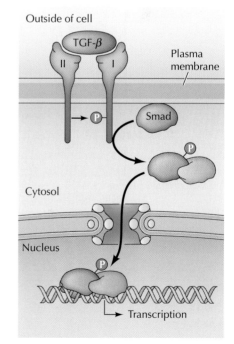

Figure 16.39 Signaling from TGF-β receptors TGF-β receptors are dimers of type I and II polypeptides. The type II receptor phosphorylates and activates type I, which then phosphorylates a Smad protein. Phosphorylated Smads form complexes and translocate to the nucleus to activate transcription of target genes.

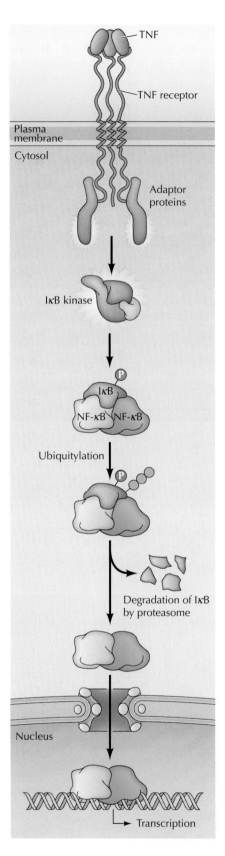

Figure 16.40   NF-κB signaling from the TNF receptor   In the inactive state, homo- or heterodimers of NF-κB are bound to IκB in the cytoplasm. Activation of the tumor necrosis factor (TNF) receptor (consisting of three polypeptide chains) leads to the recruitment of adaptor proteins that activate IκB kinase. Phosphorylation marks IκB for ubiquitylation and degradation by the proteasome, allowing NFκB to translocate to the nucleus and activate transcription of its target genes.

which in turn phosphorylates transcription factors of the **Smad** family. The phosphorylated Smads form complexes that translocate to the nucleus and stimulate expression of target genes. It should be noted that there are at least 30 different members of the TGF-β family in humans, which elicit different responses in their target cells. This is accomplished by combinatorial interactions of seven different type I receptors and five type II receptors, which lead to activation of different members of the Smad family (a total of eight family members).

NF-κB signaling

NF-κB signaling is another example of a signaling pathway that targets a specific family of transcription factors, in this case by stimulating ubiquitylation and degradation of an inhibitory subunit. The **NF-κB** family consists of five transcription factors that play key roles in the immune system and in inflammation as well as in regulation of proliferation and survival of many types of animal cells. Members of this transcription factor family are activated in response to a variety of stimuli, including cytokines, growth factors, bacterial and viral infections, and DNA damage.

The signaling pathways that lead to NF-κB activation have been particularly well-studied downstream of the receptor for **tumor necrosis factor (TNF)**, a cytokine that induces inflammation and cell death, and of the **Toll-like receptors**, which recognize a variety of molecules associated with pathogenic bacteria and viruses. In unstimulated cells, NF-κB proteins are bound to inhibitory **IκB** proteins that maintain NF-κB in an inactive state in the cytosol (**Figure 16.40**). Activation of the TNF and Toll-like receptors results in the recruitment of adaptor proteins that activate the IκB kinase, which phosphorylates IκB. This phosphorylation targets IκB for ubiquitylation and degradation by the proteasome, freeing NF-κB to translocate to the nucleus and induce expression of its target genes.

The Hedgehog, Wnt, and Notch pathways

The **Hedgehog** and **Wnt** pathways are closely connected signaling systems that activate transcription factors by inhibiting their ubiquitylation and degradation. Both Hedgehog and Wnt pathways were first described in *Drosophila*, but members of the Hedgehog and Wnt families have been found to control a wide range of events during the development of both vertebrate and invertebrate embryos. Examples of the processes regulated by these signaling pathways include the determination of cell types and establishment of cell patterning during the development of limbs, the nervous system, the skeleton, lungs, hair, teeth, and gonads. In addition, both the Hedgehog and Wnt pathways play key roles in regulating the proliferation of stem cells in adult tissues (see Chapter 18).

The *hedgehog* genes (one in *Drosophila* and three in mammals) encode secreted proteins that are modified by the addition of lipids and act on adjacent

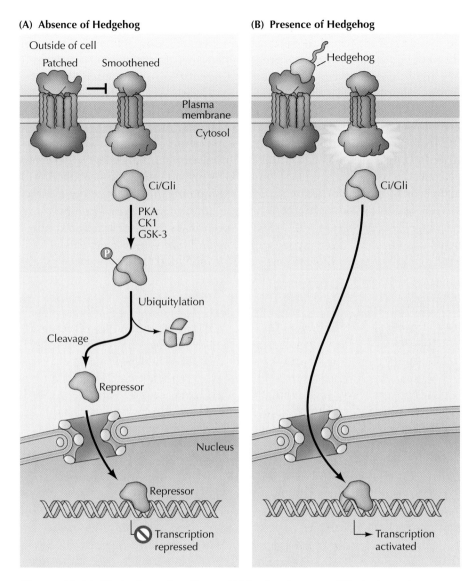

(A) Absence of Hedgehog

Outside of cell

Patched   Smoothened

Plasma membrane

Cytosol

Ci/Gli

PKA
CK1
GSK-3

P

Ubiquitylation

Cleavage

Repressor

Nucleus

Repressor

⊘ Transcription repressed

(B) Presence of Hedgehog

Hedgehog

Ci/Gli

→ Transcription activated

Figure 16.41 Hedgehog signaling (A) In the absence of Hedgehog, the Hedgehog receptor (Patched) acts to inhibit Smoothened and the Ci/Gli transcription factors are phosphorylated by protein kinase A (PKA), casein kinase-1 (CK1), and glycogen synthase kinase-3 (GSK-3). The phosphorylated Ci/Gli proteins are then ubiquitylated and cleaved to fragments that act as repressors. (B) Binding of Hedgehog inhibits Patched, resulting in activation of Smoothened. This inhibits the phosphorylation and cleavage of the Ci/Gli proteins, which translocate to the nucleus and activate transcription of their target genes.

cells. The receptor for Hedgehog is a transmembrane protein (Patched) that acts to inhibit a second transmembrane protein (Smoothened) (Figure 16.41). The binding of Hedgehog inhibits Patched, leading to activation of Smoothened, which then initiates a signaling pathway that activates a transcription factor called Ci in *Drosophila* or Gli in mammals. In the absence of Smoothened signaling, the Ci/Gli proteins are phosphorylated by a series of protein kinases, including protein kinase A, casein kinase-1, and GSK-3. Phosphorylation leads to ubiquitylation of the Ci/Gli proteins, which are then cleaved to fragments that act as repressors. Signaling from Smoothened

FYI

Hedgehog was first identified in *Drosophila* and received its name because mutant embryos have hair-like bristles that resemble hedgehog spines.

inhibits phosphorylation and subsequent cleavage of the Ci/Gli proteins, allowing them to function as transcriptional activators. However, the molecular events downstream of Smoothened are incompletely characterized and may differ between *Drosophila* and mammals, so much remains to be learned about this important signaling pathway.

The Wnt proteins are a family of secreted growth factors that bind to a complex of receptors of the Frizzled and LRP families (**Figure 16.42**). Signaling from Frizzled and LRP leads to stabilization of β-catenin, which was discussed in Chapter 15 as a transmembrane protein that links cadherins to actin at adherens junctions (see Figure 15.25). Importantly, linking cadherins to actin is only one role of β-catenin. In Wnt signaling, β-catenin acts as a direct

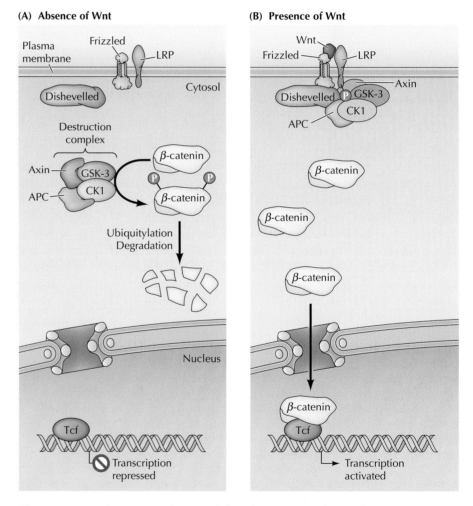

Figure 16.42 The Wnt pathway (A) In the absence of Wnt, β-catenin is phosphorylated by GSK-3 in a complex with casein kinase-1, axin, and APC (the destruction complex), leading to β-catenin ubiquitylation and degradation. (B) Wnt polypeptides bind to Frizzled and LRP receptors, leading to recruitment of Dishevelled, inactivation of the destruction complex, and stabilization of β-catenin. β-catenin then translocates to the nucleus and forms a complex with Tcf transcription factors, converting them from repressors to activators of their target genes.

Figure 16.43 Notch signaling Notch serves as a receptor for direct cell–cell signaling by transmembrane proteins (e.g., Delta) on neighboring cells. The binding of Delta leads to proteolytic cleavage of Notch by γ-secretase. This releases the Notch intracellular domain, which translocates to the nucleus and interacts with the CSL transcription factor to induce gene expression.

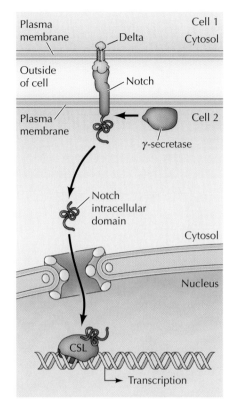

regulator of gene expression by functioning as a transcriptional activator. In the absence of Wnt signaling, β-catenin is phosphorylated by GSK-3 in a complex with casein kinase-1 and the proteins axin and APC (called the destruction complex). Like phosphorylation of the Ci/Gli proteins, phosphorylation of β-catenin results in its ubiquitylation and degradation. The binding of Wnt leads to the association of Dishevelled (a cytoplasmic protein) with Frizzled, which then disrupts the destruction complex and inhibits GSK-3, preventing the degradation of β-catenin. β-catenin then translocates to the nucleus and forms a complex with members of the Tcf family of transcription factors. In the absence of β-catenin, Tcf factors act as repressors. The association of β-catenin converts Tcf family members to activators, leading to the expression of target genes encoding other cell signaling molecules and a variety of transcription factors that control cell fate.

The **Notch** pathway is another highly conserved signaling pathway that controls cell fate during animal development. Notch signaling is an example of direct cell–cell interactions during development. Notch is a large protein with a single transmembrane domain that serves as a receptor for signaling by transmembrane proteins (e.g., Delta) on the surface of adjacent cells (**Figure 16.43**). Stimulation of Notch initiates a novel and direct pathway of transcriptional activation. In particular, ligand binding leads to proteolytic cleavage of Notch by γ-secretase, and the intracellular domain of Notch is then translocated into the nucleus. The Notch intracellular domain then interacts with a transcription factor (called CSL in mammals) and converts it from a repressor to an activator of its target genes. As in the Wnt signaling pathway, the Notch target genes include genes encoding other transcriptional regulatory proteins, which act to determine cell fate.

Signaling Dynamics and Networks

We have so far discussed signaling in terms of linear pathways that transmit information from the environment to intracellular targets. However, signaling within the cell is far more complicated. First, the activities of individual pathways are regulated by feedback loops that control the extent and duration of signaling activity. In addition, signaling pathways do not operate in isolation; rather, there is frequent crosstalk between different pathways, so that intracellular signal transduction ultimately needs to be understood as an integrated network of connected pathways. As discussed in Chapter 5, computational modeling of the dynamic behavior of such signaling networks (see Figure 5.20) is a focus of current research in systems biology.

Feedback loops and signaling dynamics

The activity of signaling pathways is controlled by **feedback loops**, which are similar in principle to feedback regulation of metabolic pathways (see

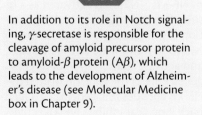

FYI

In addition to its role in Notch signaling, γ-secretase is responsible for the cleavage of amyloid precursor protein to amyloid-β protein (Aβ), which leads to the development of Alzheimer's disease (see Molecular Medicine box in Chapter 9).

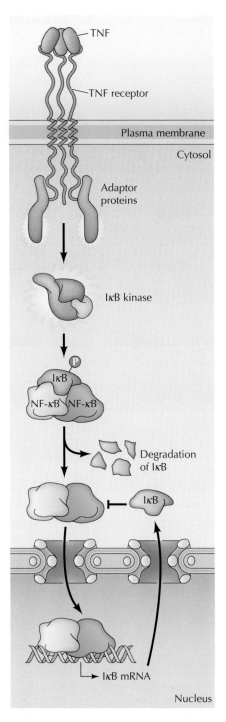

Figure 16.44 Feedback inhibition of NF-κB NF-κB is activated as a result of phosphorylation and degradation of IκB (see Figure 16.40), allowing NF-κB to translocate to the nucleus and activate transcription of target genes. One of the genes activated by NF-κB encodes IκB, generating a feedback loop that inhibits NF-κB activity.

Figure 9.39). A good example of a negative feedback loop is provided by the NF-κB pathway (**Figure 16.44**). NF-κB is activated by signals that lead to proteolysis of the inhibitor IκB, allowing NF-κB to translocate to the nucleus and induce expression of its target genes. One of the target genes induced by NF-κB encodes IκB, so NF-κB signaling leads to the synthesis of new IκB, which inhibits continued NF-κB activity. This regulation is critical because the extent and duration of NF-κB activity can determine the transcriptional response of the cell. For example, some target genes are induced by transient NF-κB activity, persisting for only 30 to 60 minutes, whereas the induction of other genes requires several hours of sustained NF-κB signaling.

Signaling by the ERK MAP kinase provides another example of the importance of the duration of signaling. In a well-studied model of cell differentiation in response to nerve growth factor (NGF), ERK signaling can lead either to cell proliferation or to neuronal differentiation depending on the duration of ERK activity. In particular, transient activation of ERK (for 30 to 60 minutes) stimulates cell proliferation, but sustained activation of ERK for 2 to 3 hours induces differentiation of the NGF-treated cells into neurons. Although the mechanism by which these differences in the duration of ERK activity lead to such distinct biological outcomes remains to be understood, it is clear that quantitative considerations of signaling activity are critical to cell response.

Networks and crosstalk

Crosstalk refers to the interaction of one signaling pathway with another, which integrates the activities of different pathways within the cell. For example, there is extensive crosstalk between the PI 3-kinase/Akt/mTORC1 and Ras/Raf/MEK/ERK pathways (**Figure 16.45**). These two signaling pathways are activated downstream of tyrosine kinase receptors and play critical roles in the control of cell proliferation and survival. The crosstalk between them includes both positive and negative points of regulation, which serve to coordinate their activities within the cell.

An interesting example in which regulation of the duration of signaling is combined with crosstalk has come from studies of G protein-coupled receptors. These receptors are linked to MAP kinase signaling by β-arrestins, which were first identified as regulatory proteins that turn off signaling from G protein-coupled receptors to G proteins (**Figure 16.46**). The signaling activity of these receptors is turned off as a result of phosphorylation by a family of protein kinases called GRKs (G protein-coupled receptor kinases), followed by association of β-arrestin with the phosphorylated receptor. However, the β-arrestins do more than turn off signaling to G proteins: They also act as signaling molecules themselves to stimulate additional downstream pathways, including nonreceptor tyrosine kinases (e.g., Src family members) and MAP kinase pathways. In particular, β-arrestin serves as a scaffold protein for Raf-MEK-ERK signaling, directly linking this MAP kinase pathway to G protein-coupled receptors.

The extensive crosstalk between individual signal transduction pathways means that multiple pathways interact with one another to form **signaling networks** within the cell. A full understanding of cell signaling will therefore need to go beyond the analysis of individual pathways to the development of network models that predict the dynamic behavior of the interconnected signaling pathways that ultimately result in a biological response.

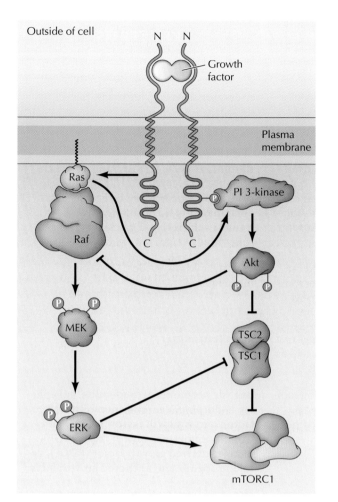

Figure 16.45 Crosstalk between the ERK and PI 3-kinase signaling pathways   The Ras/Raf/MEK/ERK and PI 3-kinase/Akt/mTORC1 pathways are connected by both positive and negative crosstalk, including activation of PI 3-kinase by Ras, inhibition of Raf by Akt, inhibition of TSC1/2 by ERK, and activation of mTORC1 by ERK.

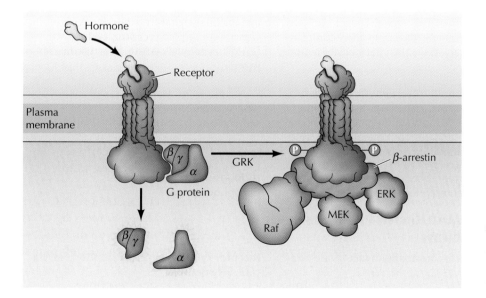

Figure 16.46 Crosstalk between G protein-coupled receptors and ERK signaling by β-arrestin   Ligand binding stimulates G protein-coupled receptors, leading to G protein activation. The activity of the receptors is turned off as a result of phosphorylation by GRKs and association of β-arrestin with the phosphorylated receptor. β-arrestin also acts as a scaffold protein for Raf, MEK, and ERK, linking G protein-coupled receptors to the ERK signaling pathway.

SUMMARY	KEY TERMS

Signaling Molecules and Their Receptors

- *Modes of cell–cell signaling:* Most signaling molecules are secreted by one cell and bind to receptors expressed by a target cell. Cell–cell signaling is divided into three general categories (endocrine, paracrine, and autocrine signaling) based on the distance over which signals are transmitted.

  endocrine signaling, hormone, paracrine signaling, autocrine signaling

- *Steroid hormones and the nuclear receptor superfamily:* The steroid hormones, thyroid hormone, vitamin D$_3$, and retinoic acid are small hydrophobic molecules that diffuse across the plasma membrane of their target cells and bind to intracellular receptors. Members of the nuclear receptor superfamily function as transcription factors to directly regulate gene expression in response to ligand binding. See Animation 16.1.

  steroid hormone, testosterone, estrogen, progesterone, corticosteroid, glucocorticoid, mineralocorticoid, ecdysone, brassinosteroid, thyroid hormone, vitamin D$_3$, retinoic acid, retinoid, nuclear receptor superfamily

- *Nitric oxide and carbon monoxide:* The simple gases nitric oxide and carbon monoxide are important paracrine signaling molecules in the nervous system and other cell types. They activate guanylyl cyclase, leading to synthesis of cyclic GMP.

  guanylyl cyclase, cyclic GMP, second messenger

- *Neurotransmitters:* Neurotransmitters are small hydrophilic molecules that carry signals between neurons or between neurons and other target cells at a synapse. Many neurotransmitters bind to ligand-gated ion channels.

  neurotransmitter

- *Peptide hormones and growth factors:* The widest variety of signaling molecules in animals are peptides, ranging from only a few to more than 100 amino acids. This group of molecules includes peptide hormones, neuropeptides, and growth factors.

  peptide hormone, neuropeptide, enkephalin, endorphin, neurohormone, growth factor, nerve growth factor (NGF), neurotrophin, epidermal growth factor (EGF), platelet-derived growth factor (PDGF), cytokine, membrane-anchored growth factor

- *Eicosanoids:* The eicosanoids are a class of lipids that function in paracrine and autocrine signaling.

  eicosanoid, prostaglandin, prostacyclin, thromboxane, leukotriene

- *Plant hormones:* Small molecules known as plant hormones regulate plant growth and development.

  plant hormone, gibberellin, auxin, ethylene, cytokinin, abscisic acid

G Proteins and Cyclic AMP Signaling

- *G proteins and G protein-coupled receptors:* The largest family of cell surface receptors, including the receptors for many hormones and neurotransmitters, transmit signals to intracellular targets via the intermediary action of G proteins. See Animation 16.2.

  intracellular signal transduction, G protein, G protein-coupled receptor, cyclic AMP (cAMP), adenylyl cyclase, heterotrimeric G protein

- *The cAMP pathway: Second messengers and protein phosphorylation:* Cyclic AMP is an important second messenger in the response of animal cells to a variety of hormones and odorants. Most actions of cAMP are mediated by protein kinase A, which phosphorylates both metabolic enzymes and the transcription factor CREB. See Animation 16.3.

  cAMP phosphodiesterase, cAMP-dependent protein kinase, protein kinase A, cAMP response element (CRE), CREB

- *Cyclic GMP:* Cyclic GMP is also an important second messenger in animal cells. Its best-characterized role is in visual reception in the vertebrate eye.

  rhodopsin, transducin, cGMP phosphodiesterase

Tyrosine Kinases and Signaling by MAP Kinase, PI 3-Kinase, and Phospholipase C/Calcium Pathways

- *Receptor tyrosine kinases:* The receptors for most growth factors are tyrosine kinases. See Video 16.1.

  receptor tyrosine kinase, autophosphorylation, SH2 domain

SUMMARY

KEY TERMS

- **Nonreceptor tyrosine kinases:** Many receptors act in association with non-receptor tyrosine kinases. Cytokine receptors are associated with JAK family tyrosine kinases, which phosphorylate and activate STAT transcription factors. Nonreceptor tyrosine kinases of the Src family function downstream of a variety of cytokine and growth factor receptors as well as integrins and other cell adhesion molecules.

nonreceptor tyrosine kinase, cytokine receptor superfamily, Janus kinase (JAK), STAT protein, Src, FAK

- **MAP kinase pathways:** The MAP kinase pathways are conserved chains of protein kinases activated downstream of a variety of extracellular signals. In animal cells, the best-characterized forms of MAP kinase are coupled to tyrosine kinase receptors by the small GTP-binding protein Ras, which initiates a protein kinase cascade leading to MAP kinase (ERK) activation. ERK then phosphorylates a variety of cytosolic and nuclear proteins, including transcription factors that mediate immediate early gene induction. Other MAP kinase pathways mediate responses of mammalian cells to inflammation and stress. Components of MAP kinase pathways are organized by scaffold proteins, which play an important role in maintaining the specificity of MAP kinase signaling.

MAP kinase, ERK, Ras, guanine nucleotide exchange factor (GEF), GTPase-activating protein (GAP), Raf, MEK (MAP kinase/ERK kinase), immediate early gene, serum response element (SRE), serum response factor (SRF), Elk-1, secondary response gene, scaffold protein

- **The PI 3-kinase/Akt and mTOR pathways:** The plasma membrane phospholipid PIP$_2$ can be phosphorylated to the distinct second messenger PIP$_3$ by PI 3-kinase. This leads to activation of the serine/threonine kinase Akt, which plays a key role in cell survival. One of the targets of Akt signaling is the protein kinase mTOR, which is a central regulator of cell growth and couples protein synthesis and autophagy to the availability of growth factors, nutrients, and cellular energy.

phosphatidylinositol 4,5-bisphosphate (PIP$_2$), phosphatidylinositide (PI) 3-kinase, phosphatidylinositol 3,4,5-trisphosphate (PIP$_3$), Akt, mTOR

- **Phospholipase C and Ca^{2+}:** Phospholipids and Ca^{2+} are common second messengers activated downstream of both tyrosine kinases and G proteins. Hydrolysis of phosphatidylinositol 4,5-bisphosphate (PIP$_2$) yields diacylglycerol (DAG) and inositol 1,4,5-trisphosphate (IP$_3$), which activate protein kinase C and mobilize Ca^{2+} from intracellular stores, respectively. Increased levels of cytosolic Ca^{2+} then activate a variety of target proteins, including Ca^{2+}/calmodulin-dependent protein kinases. In electrically excitable cells of nerve and muscle, levels of cytosolic Ca^{2+} are increased by the opening of voltage-gated Ca^{2+} channels in the plasma membrane and ryanodine receptors in the endoplasmic and sarcoplasmic reticula.

phospholipase C (PLC), diacylglycerol (DAG), inositol 1,4,5-trisphosphate (IP$_3$), protein kinase C, calmodulin, CaM kinase, ryanodine receptor

Receptors Coupled to Transcription Factors

- **The TGF-β/Smad pathway:** Members of the TGF-β receptor family are serine/threonine kinases that directly phosphorylate and activate Smad transcription factors.

transforming growth factor β (TGF-β), Smad

- **NF-κB signaling:** Members of the NF-κB family of transcription factors are activated in response to cytokines, growth factors, and a variety of other stimuli. Their activation is mediated by phosphorylation and degradation of inhibitory IκB subunits.

NF-κB, tumor necrosis factor, Toll-like receptor, IκB

- **The Hedgehog, Wnt, and Notch pathways:** The Hedgehog, Wnt, and Notch pathways play key roles in determination of cell fate and patterning during animal development. The Hedgehog and Wnt signaling pathways both act by preventing proteolytic cleavage of transcription factors in complexes in the cytoplasm. Notch signaling is mediated by direct cell–cell interactions, which induce proteolytic cleavage of Notch, followed by translocation of the Notch intracellular domain to the nucleus where it interacts with a transcription factor to affect expression of target genes.

Hedgehog, Wnt, Notch

SUMMARY	KEY TERMS

Signaling Dynamics and Networks

- *Feedback loops and signaling dynamics:* The activity of signaling pathways within the cell is regulated by feedback loops that control the extent and duration of signaling. Quantitative differences in activity can be critical to the biological outcome of signaling pathways.

- *Networks and crosstalk:* Different signaling pathways interact by crosstalk to regulate each other's activity. The extensive crosstalk between individual pathways leads to the formation of complex signaling networks. A full understanding of signaling within the cell will require the development of quantitative network models.

feedback loop

crosstalk, signaling network

Refer To

The Cell

Companion Website

sites.sinauer.com/cooper7e

for quizzes, animations, videos, flashcards, and other study resources.

Questions

1. What is the difference between paracrine and endocrine signaling?

2. How does signaling by hydrophobic molecules like steroid hormones differ from signaling by peptide hormones?

3. How does aspirin reduce inflammation and blood clotting?

4. Hormones that activate a receptor coupled to G_s stimulate the proliferation of thyroid cells. How would inhibitors of cAMP phosphodiesterase affect the proliferation of these cells?

5. The epinephrine receptor is coupled to G_s, whereas the acetylcholine receptor (on heart muscle cells) is coupled to G_i. Suppose you were to construct a recombinant molecule containing the extracellular sequences of the epinephrine receptor joined to the cytosolic sequences of the acetylcholine receptor. What effect would epinephrine have on cAMP levels in cells expressing such a recombinant receptor? How would acetylcholine affect these cells?

6. How would overexpression of protein phosphatase 1 affect the induction of cAMP-inducible genes in response to hormone stimulation of target cells? Would protein phosphatase 1 affect the function of cAMP-gated ion channels?

7. Platelet-derived growth factor (PDGF) is a dimer of two polypeptide chains. How would PDGF monomers affect signaling by the PDGF receptor?

8. You have generated a truncated version of the EGF receptor that lacks the tyrosine kinase domain. Expression of this truncated receptor inhibits the response of cells to EGF. Why?

9. A specific member of the STAT family induces certain liver genes in response to stimulation by a cytokine. How would the induction of these genes be affected if you overexpressed a dominant-negative mutant of JAK?

10. You are studying an immediate early gene that is induced via the Ras/Raf/ MEK/ERK pathway in response to growth factor treatment of fibroblasts. How would expression of siRNA against the Ras guanine nucleotide exchange factor affect the induction of this gene?

11. How does the PI 3-kinase/Akt pathway regulate cellular protein synthesis in response to growth factor stimulation?

12. You are studying a gene that is induced by the Wnt signaling pathway. To determine the role played by β-catenin, you make different site-specific mutants of β-catenin. You find that changing a specific lysine (Lys) residue to arginine (Arg) leads to nuclear accumulation of β-catenin and constitutive expression of the gene (even in the absence of Wnt stimulation). What is the most likely mechanism by which this mutation alters β-catenin activity?

References and Further Reading   (Key review articles for each major section are highlighted in **bold**.)

Signaling Molecules and Their Receptors

Andrea, J., R. Gallini and C. Betsholtz. 2008. Role of platelet-derived growth factors in physiology and medicine. *Genes Dev.* 22: 1276–1312. [R]

Arai, K., F. Lee, A. Miyajima, S. Miyatake, N. Arai and T. Yokota. 1990. Cytokines: Coordinators of immune and inflammatory responses. *Ann. Rev. Biochem.* 59: 783–836. [R]

Calabrese, V., C. Mancuso, M. Calvani, E. Rizzarelli, D. A. Butterfield and A. M. G. Stella. 2007. Nitric oxide in the central nervous system: neuroprotection versus neurotoxicity. *Nature Rev. Neurosci.* 8: 766–775. [R]

Cao, Y. 2013. Multifarious function of PDGFs and PDGFRs in tumor growth and metastasis. *Trends Mol. Med.* 19: 460–473. [R]

Cohen, P. 2006. The twentieth century struggle to decipher insulin signaling. *Nature Rev. Mol. Cell Biol.* 7: 867–873. [R]

Cohen, S. 2008. Origins of growth factors: NGF and EGF. *J. Biol. Chem.* 283: 33793–33797. [R]

Cunningham, T. J. and G. Duester. 2015. Mechanisms of retinoic acid signalling and its roles in organ and limb development. *Nature Rev. Mol. Cell Biol.* 16: 110–123. [R]

Hess, D. T., A. Matsumoto, S.-O. Kim, H. E. Marshall and J. S. Stamler. 2005. Protein *S*-nitrosylation: Purview and parameters. *Nature Rev. Mol. Cell Biol.* 6: 150–166. [R]

Hetherington, A. M. and L. Bardwell. 2011. Plant signaling pathways: a comparative evolutionary overview. *Curr. Biol.* 21: R317–R319. [R]

Hill, B. G., B. P. Dranka, S. M. Bailey, J. R. Lancaster Jr. and V. M. Darley-Usmar. 2010. What part of NO don't you understand? Some answers to the cardinal questions in nitric oxide biology. *J. Biol. Chem.* 285: 19699–19704. [R]

Levi-Montalcini, R. 1987. The nerve growth factor 35 years later. *Science* 237: 1154–1162. [R]

Li, L. and P. K. Moore. 2007. An overview of the biological significance of endogenous gases: new roles for old molecules. *Biochem. Soc. Tran.* 35: 1138–1141. [R]

McDonnell, D. P. and J. D. Norris. 2002. Connections and regulation of the human estrogen receptor. *Science* 296: 1642–1644. [R]

Nagy, L. and J. W. R. Schwabe. 2004. Mechanism of the nuclear receptor molecular switch. *Trends Biochem. Sci.* 29: 317–324. [R]

Ornitz, D. M. and P. J. Marie. 2015. Fibroblast growth factor signaling in skeletal development and disease. *Genes Dev.* 29: 1463–1486. [R]

Santner, A. and M. Estelle. 2009. Recent advances and emerging trends in plant hormone signaling. *Nature* 459: 1071–1078. [R]

Sonoda, J., L. Pei and R. M. Evans. 2008. Nuclear receptors: decoding metabolic disease. *FEBS Let.* 582: 2–9. [R]

Turner, N., and R. Grose. 2010. Fibroblast growth factor signaling: from development to cancer. *Nature Rev. Cancer* 10: 116–129. [R]

Wang, D. and R. N. DuBois. 2010. Eicosanoids and cancer. *Nature Rev. Cancer* 10: 181–193. [R]

G Proteins and Cyclic AMP Signaling

Audet, M. and M. Bouvier. 2012. Restructuring G-protein-coupled receptor activation. *Cell* 151: 14–23. [R]

Conti, M. and J. Beavo. 2007. Biochemistry and physiology of cyclic nucleotide phosphodiesterases: essential components in cyclic nucleotide signaling. *Ann. Rev. Biochem.* 76: 481–511. [R]

Hofmann, F. 2005. The biology of cyclic GMP-dependent protein kinases. *J. Biol. Chem.* 280: 1–4. [R]

Hofmann, K. P., P. Scheerer, P. W. Hildebrand, H.-W. Choe, J. H. Park, M. Heck and O. P. Ernst. 2009. A G protein-coupled receptor at work: the rhodopsin model. *Trends Biochem. Sci.* 34: 540–552. [R]

Lefkowitz, R. J. 2004. Historical review: A brief history and personal retrospective of seven-transmembrane receptors. *Trends Pharmacol. Sci.* 25: 413–422. [R]

Malbon, C. C. 2005. G proteins in development. *Nature Rev. Mol. Cell Biol.* 6: 689–701. [R]

Mayr, B. and M. Montminy. 2001. Transcriptional regulation by the phosphorylation-dependent factor CREB. *Nature Rev. Mol. Cell Biol.* 2: 599–609. [R]

Neves, S. R., P. T. Ram and R. Iyengar. 2002. G protein pathways. *Science* 296: 1636–1639. [R]

Oldham, W. M. and H. E. Hamm. 2008. Heterotrimeric G protein activation by G-protein-coupled receptors. *Nature Rev. Mol. Cell Biol.* 9: 60–71. [R]

Siderovski, D. P. and F. S. Willard. 2005. The GAPs, GEFs, and GDIs of heterotrimeric G-protein alpha subunits. *Intl. J. Biol. Sci.* 1: 51–66. [R]

Taylor, S. S., C. Kim, C. Y. Cheng, S. H. J. Brown, J. Wu and N. Kannan. 2008. Signaling through cAMP and cAMP-dependent protein kinase: diverse strategies for drug design. *Biochim. Biophys. Acta* 1784: 16–26. [R]

Tyrosine Kinases and Signaling by MAP Kinase, PI 3-Kinase, and Phospholipase C/Calcium Pathways

Baker, S. J., S. G. Rane and E. P. Reddy. 2007. Hematopoietic cytokine receptor signaling. *Oncogene* 26: 6724–6737. [R]

Brazil, D. P., Z.-Z. Yang, and B. A. Hemmings. 2004. Advances in protein kinase B signaling: *AKT*ion on multiple fronts. *Trends Biochem. Sci.* 29: 233–242. [R]

Cavallaro, U. and E. Dejana. 2011. Adhesion molecule signaling: not always a sticky business. *Nature Rev. Mol. Cell Biol.* 12: 189–197. [R]

Clapham, D. E. 2007. Calcium signaling. *Cell* 131: 1047–1058. [R]

Corcoran, E. E. and A. R. Means. 2001. Defining Ca²⁺/calmodulin-dependent protein kinase cascades in transcriptional regulation. *J. Biol. Chem.* 276: 2975–2978. [R]

Efeyan, A., W. C. Comb and D. M. Sabatini. 2015. Nutrient-sensing mechanisms and pathways. *Nature* 517: 302–310. [R]

Eijkelenboom, A. and B. M. T. Burgering. 2013. FOXOs: signalling integrators for homeostasis maintenance. *Nature Rev. Mol. Cell Biol.* 14: 83–97. [R]

Good, M. C., J. G. Zalatan and W. A. Lim. 2011. Scaffold proteins: hubs for controlling the flow of cellular information. *Science* 332: 680–686. [R]

Hunter, T. and B. M. Sefton. 1980. Transforming gene product of Rous sarcoma virus phosphorylates tyrosine. *Proc. Natl. Acad. Sci. U.S.A.* 77: 1311–1315. [P]

Irvine, R. F. 2003. 20 years of Ins(1,4,5)P₃ and 40 years before. *Nature Rev. Mol. Cell Biol.* 4: 586–590. [R]

Johnson, G. L. and R. Lapadat. 2002. Mitogen-activated protein kinase pathways mediated by ERK, JNK, and p38 protein kinases. *Science* 298: 1911–1912. [R]

Jope, R. S. and G. V. W. Johnson. 2004. The glamour and gloom of glycogen synthase kinase-3. *Trends Biochem. Sci.* 29: 95–102. [R]

Lemmon, M. A. and J. Schlessinger. 2010. Cell signaling by receptor tyrosine kinases. *Cell* 141: 1117–1134. [R]

Manning, G., D. B. Whyte, R. Martinez, T. Hunter and S. Sudarsanam. 2002. The protein kinase complement of the human genome. *Science* 298: 1912–1934. [R]

Matallanas, D., M. Birtwistle, D. Romano, A. Zebisch, J. Rauch, A. von Kriegsheim and W. Kolch. 2011. Raf family kinases: old dogs have learned new tricks. *Genes Cancer* 2: 232–260. [R]

McKay, M. M. and D. K. Morrison. 2007. Integrating signals from RTKs to ERK/MAPK. *Oncogene* 26: 3113–3121. [R]

Mihaylova, M. M. and R. J. Shaw. 2011. The AMPK signaling pathway coordinates cell growth, autophagy and metabolism. *Nature Cell Biol.* 13: 1016–1023. [R]

Miranti, C. K. and J. S. Brugge. 2002. Sensing the environment: A historical perspective on integrin signal transduction. *Nature Cell Biol.* 4: E83–E90. [R]

Mitra, S. K. and D. D. Schlaepfer. 2006. Integrin-regulated FAK-Src signaling in normal and cancer cells. *Curr. Opin. Cell Biol.* 18: 516–523. [R]

Parsons, S. J. and J. T. Parsons. 2004. Src family kinases, key regulators of signal transduction. *Oncogene* 23: 7906–7909. [R]

Raman, M., W. Chen and M. H. Cobb. 2007. Differential regulation and properties of MAPKs. *Oncogene* 26: 3100–3112. [R]

Schindler, C., D. E. Levy and T. Decker. 2007. JAK-STAT signaling: from interferons to cytokines. *J. Biol. Chem.* 282: 20059–20063. [R]

Stepniak, E., G. L. Radice and V. Vasioukhin. 2009. Adhesive and signaling functions of cadherins and catenins in vertebrate development. *Cold Spring Harb. Perspect. Biol.* 1:a002949. [R]

Turjanski, A. G., J. P. Vaque and J. S. Gutkind. 2007. MAP kinases and the control of nuclear events. *Oncogene* 26: 3240–3253. [R]

Vanhaesebroeck, B., L. Stephens and P. Hawkins. 2012. PI3K signaling: the path to discovery and understanding. *Nature Rev. Mol. Cell Biol.* 13: 195–203. [R]

Venkatachalam, K., D. B. van Rossum, R. L. Patterson, H.-T. Ma and D. L. Gill. 2002. The cellular and molecular basis of store-operated calcium entry. *Nature Cell. Biol.* 4: E263–272. [R]

Zalk, R., S. E. Lahnart and A. R. Marks. 2007. Modulation of the ryanodine receptor and intracellular calcium. *Ann. Rev. Biochem.* 76: 367–385. [R]

Zoncu, R., A. Efeyan and D. M. Sabatini. 2011. mTOR: from growth signal integration to cancer, diabetes and ageing. *Nature Mol. Cell. Biol.* 12: 21–35. [R]

Receptors Coupled to Transcription Factors

Briscoe, J. and P. P. Thérond. 2013. The mechanisms of Hedgehog signalling and its roles in development and disease. *Nature Rev. Mol. Cell Biol.* 14: 416–429. [R]

Clevers, H., K. M. Loh and R. Nusse. 2014. An integral program for tissue renewal and regeneration: Wnt signaling and stem cell control. *Science* 346: 1248012. [R]

Gay, N. J. and M. Gangloff. 2007. Structure and function of Toll receptors and their ligands. *Ann. Rev. Biochem.* 76: 141–165. [R]

Hayden, M. S. and S. Ghosh. 2012. NF-κB, the first quarter-century: remarkable progress and outstanding questions. *Genes Dev.* 26:203–234. [R]

Ng, J. M. Y. and T. Curran. 2011. The Hedgehog's tale: developing strategies for targeting cancer. *Nature Rev. Cancer* 11: 493–501. [R]

Ingham, P. W., Y. Nakano and C. Seger. 2011. Mechanisms and functions of Hedgehog signaling across the metazoa. *Nature Rev. Genet.* 12: 393–406. [R]

Kitisin, K., T. Saha, T. Blake, N. Golestaneh, M. Deng, C. Kim, Y. Tang, K. Shetty, B. Mishra and L. Mishra. 2007. TGF-β signaling in development. *Science STKE* 2007: cm1. [R]

Kopan, R. and M. X. G. Ilagan. 2009. The canonical Notch signaling pathway: unfolding the activation mechanism. *Cell* 137: 216–233. [R]

Massagué, J., J. Seoane and D. Wotton. 2005. Smad transcription factors. *Genes Dev.* 19: 2783–2810. [R]

Niehrs, C. 2012. The complex world of WNT receptor signalling. *Nature Rev. Mol. Cell Biol.* 13: 767–779. [R]

Schmierer, B. and C. S. Hill. 2007. TGFβ-SMAD signal transduction: molecular specificity and functional flexibility. *Nature Rev. Mol. Cell Biol.* 8: 970–982. [R]

Signaling Dynamics and Networks

Aldridge, B. B., J. M. Burke, D. A. Lauffenburger and P. K. Sorger. 2006. Physicochemical modelling of cell signalling pathways. *Nature Cell Biol.* 8: 1195–1203. [R]

Hoffmann, A., A. Levchenko, M. L. Scott and D. Baltimore. 2002. The IκB—NF-κB signaling module: Temporal control and selective gene activation. *Science* 298: 1241–1245. [P]

Lefkowitz, R. J. and S. K. Shenoy. 2005. Transduction of receptor signals by β-arrestins. *Science* 308: 512–517. [R]

Ma'ayan, A., S. L. Jenkins, S. Neves, A. Hasseldine, E. Grace, B. Dubin-Thaler, N. J. Eungdamrong, G. Weng, P. T. Ram, J. J. Rice, A. Kershenbaum, G. A. Stolovitzky, R. D. Blitzer, and R. Iyengar. 2005. Formation of regulatory patterns during signal propagation in a mammalian cellular network. *Science* 309: 1078–1083. [P]

Marshall, C. J. 1995. Specificity of receptor tyrosine kinase signaling: Transient versus sustained extracellular signal-regulated kinase activation. *Cell* 80: 179–185. [R]

Mendoza, M. C., E. E. Er and J. Blenis. 2011. The Ras-ERK and PI3K-mTOR pathways: cross-talk and compensation. *Trends Biochem. Sci.* 36: 320–328. [R]

Papin, J. A., T. Hunter, B. O. Palsson and S. Subramaniam. 2005. Reconstruction of cellular signaling networks and analysis of their properties. *Nature Rev. Mol. Cell Biol.* 6: 99–111. [R]

Purvis, J. E. and G. Lahav. 2013. Encoding and decoding cellular information through signaling dynamics. *Cell* 152: 945–956. [R]

Zhu, X., M. Gerstein and M. Snyder. 2007. Getting connected: analysis and principles of biological networks. *Genes Dev.* 21: 1010–1024. [R]

CHAPTER 17

The Cell Cycle

Self-reproduction is perhaps the most fundamental characteristic of cells—as may be said for all living organisms. All cells reproduce by dividing in two, with each parental cell giving rise to two daughter cells on completion of each cycle of cell division. These newly formed daughter cells can themselves grow and divide, giving rise to a new cell population formed by the growth and division of a single parental cell and its progeny. In the simplest case, such cycles of growth and division allow a single bacterium to form a colony consisting of millions of progeny cells during overnight incubation on a plate of nutrient agar medium. In a more complex case, repeated cycles of cell growth and division result in the development of a single fertilized egg into the approximately 10^{14} cells that make up the human body.

The division of all cells must be carefully regulated and coordinated with both cell growth and DNA replication in order to ensure the formation of progeny cells containing intact genomes. In eukaryotic cells, progression through the cell cycle is controlled by a series of protein kinases that have been conserved from yeasts to mammals. In higher eukaryotes, this cell cycle machinery is itself regulated by the growth factors that control cell proliferation, allowing the division of individual cells to be coordinated with the needs of the organism as a whole. Not surprisingly, defects in cell cycle regulation are a common cause of the abnormal proliferation of cancer cells, so studies of the cell cycle and cancer have become closely interconnected, similar to the relationship between studies of cancer and the cell signaling pathways discussed in Chapter 16.

The Eukaryotic Cell Cycle

The division cycle of most cells consists of four coordinated processes: cell growth, DNA replication, distribution of the duplicated chromosomes to daughter cells, and cell division. In bacteria, cell growth and DNA replication take place throughout most of the cell cycle, and duplicated chromosomes are distributed to daughter cells in association with the plasma membrane. In eukaryotes, however, the cell cycle is more complex and consists of four discrete phases. Although cell growth is usually a continuous process, DNA is synthesized during only one phase of the cell cycle, and the replicated chromosomes are then distributed to daughter nuclei by a complex series of events preceding cell division. Progression between these stages of the cell cycle is controlled by a conserved regulatory apparatus, which not only coordinates the different events of the cell cycle but also links the cell cycle with extracellular signals that control cell proliferation.

Video 17.1

sites.sinauer.com/cooper7e/v17.1

Mitosis in Real Time Mitosis and cytokinesis together define the mitotic phase of an animal cell cycle.

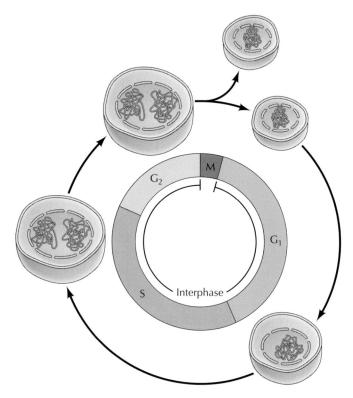

Figure 17.1 Phases of the cell cycle The division cycle of most eukaryotic cells is divided into four discrete phases: M, G₁, S, and G₂. M phase (mitosis) is usually followed by cytokinesis. S phase is the period during which DNA replication occurs. The cell grows throughout interphase, which includes G₁, S, and G₂. The relative lengths of the cell cycle phases shown here are typical of rapidly replicating mammalian cells.

Phases of the cell cycle

A typical eukaryotic cell cycle is illustrated by human cells in culture, which divide approximately every 24 hours. As viewed in the microscope, the cell cycle is divided into two basic parts: **mitosis** and **interphase**. Mitosis (nuclear division) is the most dramatic stage of the cell cycle, corresponding to the separation of daughter chromosomes and usually ending with cell division (**cytokinesis**). However, mitosis and cytokinesis last only about an hour, so approximately 95% of the cell cycle is spent in interphase—the period between mitoses. During interphase, the chromosomes are decondensed and distributed throughout the nucleus, so the nucleus appears morphologically uniform. At the molecular level, however, interphase is the time during which both cell growth and DNA replication occur in an orderly manner in preparation for cell division.

The cell grows at a steady rate throughout interphase, with most dividing cells doubling in size between one mitosis and the next. In contrast, DNA is synthesized during only a portion of interphase. The timing of DNA synthesis thus divides the cycle of eukaryotic cells into four discrete phases (**Figure 17.1**). The **M phase** of the cycle corresponds to mitosis, which is usually followed by cytokinesis. This phase is followed by the **G₁ phase** (gap 1), which corresponds to the interval (gap) between mitosis and initiation of DNA replication. During G₁, the cell is metabolically active and continuously grows but does not replicate its DNA. G₁ is followed by **S phase** (synthesis), during which DNA replication takes place. The completion of DNA synthesis is followed by the **G₂ phase** (gap 2), during which cell growth continues and proteins are synthesized in preparation for mitosis.

The duration of these cell cycle phases varies considerably in different kinds of cells. For a typical proliferating human cell with a total cycle time of 24 hours, the G₁ phase might last about 11 hours, S phase about 8 hours, G₂ about 4 hours, and M about 1 hour. Other types of cells, however, can divide much more rapidly. Budding yeasts, for example, can progress through all four stages of the cell cycle in only about 90 minutes. Even shorter cell cycles (30 minutes or less) occur in early embryo cells shortly after fertilization of the egg (**Figure 17.2**). In this case, however, cell growth does not take place. Instead, these early embryonic cell cycles rapidly divide the egg cytoplasm

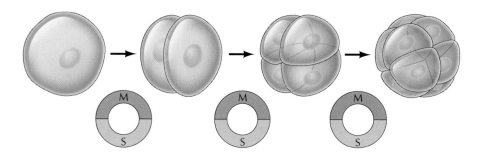

Figure 17.2 Embryonic cell cycles During early embryonic cell cycles, the egg cytoplasm is rapidly divided into smaller cells. The cells do not grow during these cycles, which lack G₁ and G₂ phases, and consist simply of short S phases alternating with M phases.

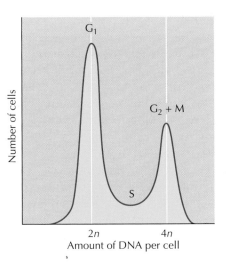

Figure 17.3 Determination of cellular DNA content A population of cells is labeled with a fluorescent dye that binds DNA. The cells are then passed through a flow cytometer, which measures the fluorescence intensity of individual cells. The data are plotted as cell number versus fluorescence intensity, which is proportional to DNA content. The distribution shows two peaks, corresponding to cells with DNA contents of $2n$ and $4n$; these cells are in the G_1 and G_2/M phases of the cycle, respectively. Cells in S phase have DNA contents between $2n$ and $4n$ and are distributed between these two peaks.

into smaller cells. There is no G_1 or G_2 phase, and DNA replication occurs very rapidly in these early embryonic cell cycles, which therefore consist of very short S phases alternating with M phases.

In contrast to the rapid proliferation of embryonic cells, some cells in adult animals cease division altogether (e.g., nerve cells) and many other cells divide only occasionally, as needed to replace cells that have been lost because of injury or cell death. Cells of the latter type include skin fibroblasts, as well as the cells of some internal organs, such as the liver. As discussed further in the next section, these cells exit G_1 to enter a quiescent stage of the cycle called G_0, where they remain metabolically active but no longer proliferate. Some cells that are arrested in G_0 can re-enter the cell cycle when called on to do so by appropriate extracellular signals. However, other types of cells, including most differentiated cells in adult animals, permanently withdraw from the cell cycle and are incapable of resuming proliferation.

Analysis of the cell cycle requires identification of cells at the different stages discussed above. Although mitotic cells can be distinguished microscopically, cells in other phases of the cycle (G_1, S, and G_2) must be identified by biochemical criteria. In most cases, cells at different stages of the cell cycle are distinguished by their DNA content (**Figure 17.3**). For example, animal cells in G_1 are diploid (containing two copies of each chromosome), so their DNA content is referred to as $2n$ (n designates the haploid DNA content of the genome). During S phase, replication increases the DNA content of the cell from $2n$ to $4n$, so cells in S have DNA contents ranging from $2n$ to $4n$. DNA content then remains at $4n$ for cells in G_2 and M, decreasing to $2n$ after cytokinesis. Experimentally, cellular DNA content can be determined by incubation of cells with a fluorescent dye that binds to DNA, followed by analysis of the fluorescence intensity of individual cells in a **flow cytometer** or **fluorescence-activated cell sorter**, thereby distinguishing and allowing isolation of cells in the G_1, S, and G_2/M phases of the cell cycle.

Regulation of the cell cycle by cell growth and extracellular signals

The progression of cells through the division cycle is regulated by extracellular signals from the environment, as well as by internal signals that monitor and coordinate the various processes that take place during different cell cycle phases. An example of cell cycle regulation by extracellular signals is provided by the effect of growth factors on animal cell proliferation. In

Animation 17.2

sites.sinauer.com/cooper7e/a17.2
Embryonic Cell Cycles During early embryonic cell cycles, the cells do not grow, but instead divide into progressively smaller cells.

(A)

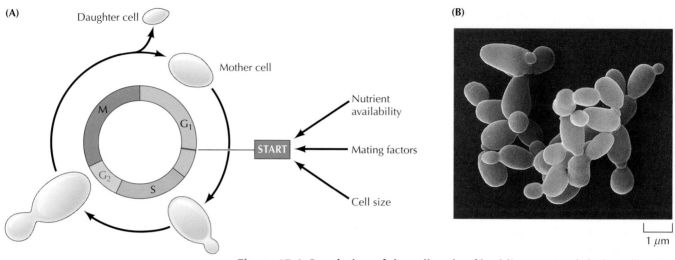

(B)

1 μm

Figure 17.4 Regulation of the cell cycle of budding yeast (A) The cell cycle of *Saccharomyces cerevisiae* is regulated primarily at a point in late G$_1$ called START. Passage through START is controlled by the availability of nutrients, mating factors, and cell size. Note that these yeasts divide by budding. Buds form just after START and continue growing until they separate from the mother cell after mitosis. The daughter cell formed from the bud is smaller than the mother cell and therefore requires more time to grow during the G$_1$ phase of the next cell cycle. Although G$_1$ and S phases occur normally, the mitotic spindle begins to form during S phase, so the cell cycle of budding yeast lacks a distinct G$_2$ phase. (B) Scanning electron micrograph of *S. cerevisiae*. The size of the bud reflects the position of the cell in the cycle.

addition, different cellular processes, such as cell growth, DNA replication, and mitosis, all must be coordinated during cell cycle progression. This is accomplished by a series of control points that regulate progression through various phases of the cell cycle.

A major cell cycle regulatory point in many types of cells occurs late in G$_1$ and controls progression from G$_1$ to S. This regulatory point was first defined by studies of budding yeast (*Saccharomyces cerevisiae*), where it is known as **START** (Figure 17.4). Once cells have passed START, they are committed to entering S phase and undergoing one cell division cycle. However, passage through START is a highly regulated event in the yeast cell cycle, controlled by external signals such as the availability of nutrients, as well as by cell size. For example, if yeasts are faced with a shortage of nutrients, they arrest their cell cycle at START and enter a resting state rather than proceeding to S phase. Thus START represents a decision point at which the cell determines whether sufficient nutrients are available to support progression through the rest of the division cycle. Polypeptide factors that signal yeast mating also arrest the cell cycle at START, allowing haploid yeast cells to fuse with one another instead of progressing to S phase.

In addition to serving as a decision point for monitoring extracellular signals, START is the point at which cell growth is coordinated with DNA replication and cell division. The importance of this regulation is particularly evident in budding yeasts in which cell division produces progeny cells of very different sizes: a large mother cell and a small daughter cell. In order for yeast cells to maintain a constant size, the small daughter cell must grow more than the large mother cell does before they divide again. Thus cell size must be monitored in order to coordinate cell growth with other cell cycle events. This regulation is accomplished by a control mechanism that requires each cell to reach a minimum size before it can pass START. Consequently, the small daughter cell spends a longer time in G_1 and grows more than the mother cell.

The proliferation of most animal cells is similarly regulated in the G_1 phase of the cell cycle. In particular, a decision point in late G_1, called the **restriction point** in animal cells, functions analogously to START in yeasts (**Figure 17.5**). In contrast with yeasts, however, the passage of animal cells through the cell cycle is regulated primarily by the extracellular growth factors that signal cell proliferation, rather than by the availability of nutrients. In the presence of the appropriate growth factors, cells pass the restriction point and enter S phase. Once it has passed through the restriction point, the cell is committed to proceed through S phase and the rest of the cell cycle, even in the absence of further growth factor stimulation. On the other hand, if appropriate growth factors are not available in G_1, progression through the cell cycle stops at the restriction point. Such arrested cells then enter a quiescent stage of the cell cycle called **G_0** in which they can remain for long periods of time without proliferating. G_0 cells are metabolically active, although they cease growth and have reduced rates of protein synthesis. As already noted, many cells in animals remain permanently in G_0, whereas others can resume proliferation if stimulated by appropriate growth factors or other extracellular signals. For example, skin fibroblasts are arrested in G_0 until they are stimulated to divide as required to repair damage resulting from a wound. The proliferation of these cells is triggered by platelet-derived growth factor, which is released from blood platelets during clotting and signals the proliferation of fibroblasts in the vicinity of the injured tissue.

Although the proliferation of most cells is regulated primarily in G_1, some cell cycles are instead controlled principally in G_2. One example is the cell cycle of the fission yeast *Schizosaccharomyces pombe* (**Figure 17.6**). In contrast with *Saccharomyces cerevisiae*, the cell cycle of *S. pombe* is regulated

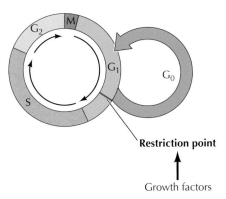

Figure 17.5 Regulation of animal cell cycles by growth factors   The availability of growth factors controls the animal cell cycle at a point in late G_1 called the restriction point. If growth factors are not available during G_1, the cells enter a quiescent stage of the cycle called G_0.

Video 17.2

sites.sinauer.com/cooper7e/v17.2

Fission Yeast Division Fission yeast divide by medial fission to produce two daughter cells of equal sizes, which makes them a powerful tool in cell cycle research.

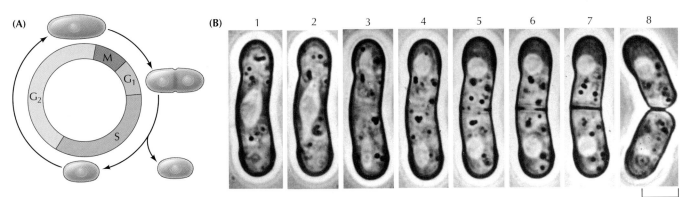

Figure 17.6 Cell cycle of fission yeast   (A) Fission yeasts grow by elongating at both ends and divide by forming a wall through the middle of the cell. In contrast with the cycle of budding yeasts, the cell cycle of fission yeasts has normal G_1, S, G_2, and M phases. Note that cytokinesis occurs in G_1. The length of the cell indicates its position in the cycle. (B) Light micrographs showing successive stages of mitosis and cytokinesis of *Schizosaccharomyces pombe*. (B, courtesy of C. F. Robinow, University of Western Ontario.)

primarily by control of the transition from G_2 to M, which is the principal point at which cell size and nutrient availability are monitored. In animals, the primary example of cell cycle control in G_2 is provided by oocytes. Vertebrate oocytes can remain arrested in G_2 for long periods of time (several decades in humans) until their progression to M phase is triggered by hormonal stimulation. Extracellular signals can thus control cell proliferation by regulating progression from the G_2 to M as well as the G_1 to S phases of the cell cycle.

Cell cycle checkpoints

The controls discussed in the previous section regulate cell cycle progression in response to cell size and extracellular signals, such as nutrients and growth factors. In addition, the events that take place during different stages of the cell cycle must be coordinated with one another so that they occur in the appropriate order. For example, it is critically important that the cell not begin mitosis until replication of the genome has been completed. The alternative would be a catastrophic cell division in which the daughter cells failed to inherit complete copies of the genetic material. In most cells, this coordination between different phases of the cell cycle is dependent on a series of **cell cycle checkpoints** that prevent entry into the next phase of the cell cycle until the events of the preceding phase have been completed.

Several cell cycle checkpoints, called **DNA damage checkpoints**, function to ensure that damaged DNA is not replicated and passed on to daughter cells (**Figure 17.7**). These checkpoints sense damaged or incompletely replicated DNA and coordinate further cell cycle progression with the completion of DNA replication or repair. DNA damage checkpoints function in G_1, S, and G_2 phases of the cell cycle. For example, the checkpoint in G_2 prevents the initiation of mitosis if the cell contains DNA that has not been completely replicated or contains unrepaired lesions. Such damaged DNA activates a signaling pathway that leads to cell cycle arrest. Operation of the G_2 checkpoint therefore prevents the initiation of M phase until the genome has been completely replicated and any damage repaired. Only then is the inhibition of G_2 progression relieved, allowing the cell to initiate mitosis and distribute the completely replicated chromosomes to daughter cells. Likewise, arrest at the G_1 checkpoint allows repair of any DNA damage to take place before the cell enters S phase, where the damaged DNA would be replicated. The S-phase checkpoint not only provides continual monitoring of the integrity of DNA to ensure that damaged DNA is repaired before it is replicated, but also provides a quality control monitor to promote the repair of any errors that occur during DNA replication, such as the incorporation of incorrect bases or incomplete replication of DNA segments.

Another important cell cycle checkpoint that maintains the integrity of the genome occurs toward the end of mitosis. This checkpoint, called the **spindle assembly checkpoint**, monitors the alignment of chromosomes on the mitotic

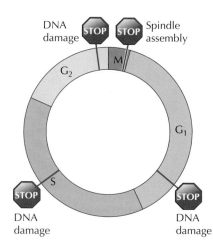

Figure 17.7 Cell cycle checkpoints Several checkpoints function to ensure that complete genomes are transmitted to daughter cells. DNA damage checkpoints in G_1, S, and G_2 lead to cell cycle arrest in response to damaged or unreplicated DNA. The spindle assembly checkpoint arrests mitosis if the chromosomes are not properly aligned on the mitotic spindle.

spindle, thus ensuring that a complete set of chromosomes is distributed accurately to the daughter cells. The failure of one or more chromosomes to align properly on the spindle causes mitosis to arrest at metaphase, prior to the segregation of the newly replicated chromosomes to daughter nuclei. As a result of the spindle assembly checkpoint, the chromosomes do not separate until a complete complement of chromosomes has been organized for distribution to each daughter cell.

Regulators of Cell Cycle Progression

One of the most exciting developments in contemporary cell biology has been the elucidation of the molecular mechanisms that control the progression of eukaryotic cells through the division cycle. Our current understanding of cell cycle regulation has emerged from a convergence of results obtained through experiments on organisms as diverse as yeasts, sea urchins, frogs, and mammals. These studies have revealed that the cell cycle of all eukaryotes is controlled by a conserved set of protein kinases, which are responsible for triggering the major cell cycle transitions.

Protein kinases and cell cycle regulation

Three initially distinct experimental approaches contributed to identification of the key molecules responsible for cell cycle regulation. The first of these avenues of investigation originated with studies of frog oocytes (**Figure 17.8**). These oocytes are arrested in the G_2 phase of the cell cycle until hormonal stimulation (progesterone) triggers their entry into the M phase of meiosis (discussed later in this chapter). In 1971, two independent teams of researchers (Yoshio Masui and Clement Markert, as well as Dennis Smith and Robert Ecker) found that oocytes arrested in G_2 could be induced to enter M phase by microinjection of cytoplasm from oocytes that had been hormonally stimulated. It thus appeared that a cytoplasmic factor present in hormone-treated oocytes was sufficient to trigger the transition from G_2 to M in oocytes that had not been exposed to hormone. Because the entry of oocytes into meiosis is frequently referred to as oocyte maturation, this cytoplasmic factor was called **maturation promoting factor (MPF)**. Further studies showed, however, that the activity of MPF is not restricted to the entry of oocytes into meiosis. To the contrary, MPF is also present in somatic cells, where it induces entry into M phase of the mitotic cycle. Rather than being specific to oocytes, MPF thus appeared to act as a general regulator of the transition from G_2 to M.

The second approach to understanding cell cycle regulation was the genetic analysis of yeasts, pioneered by Lee Hartwell and his colleagues in

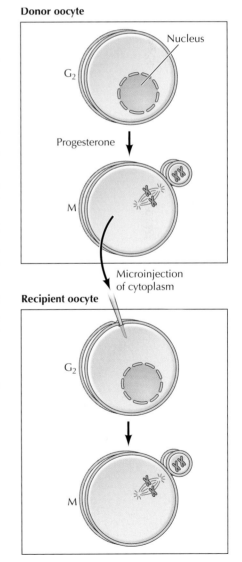

Figure 17.8 Identification of MPF Frog oocytes are arrested in the G_2 phase of the cell cycle, and entry into the M phase of meiosis is triggered by the hormone progesterone. In the experiment diagrammed here, G_2-arrested oocytes (recipient oocytes) were microinjected with cytoplasm extracted from oocytes that had undergone the transition from G_2 to M (donor oocytes). Such cytoplasmic transfers induced the G_2 to M transition in the absence of hormonal stimulation, demonstrating that a cytoplasmic factor, called maturation promoting factor (MPF), is sufficient to induce entry into the M phase of meiosis.

The Discovery of MPF

Cytoplasmic Control of Nuclear Behavior during Meiotic Maturation of Frog Oocytes

Yoshio Masui and Clement L. Markert

Yale University, New Haven, CT

Journal of Experimental Zoology, 1971, Volume 177, pages 129–146

The Context

Nuclear transplantation and cell fusion experiments performed in the 1960s indicated that nuclei transferred into cells at different stages of the mitotic cell cycle generally adopt the behavior of the host cell. Thus it appeared that mitotic activity of the nucleus was regulated by the cytoplasm. However, the existence of postulated cytoplasmic factors that control the mitotic activity of nuclei remained to be demonstrated by a direct experimental approach. This demonstration was provided by the studies of Masui and Markert, who investigated the role of cytoplasmic factors in regulating nuclear behavior during meiosis of frog oocytes.

Several features of frog oocyte meiosis suggested that it is controlled by cytoplasmic factors. In particular, meiosis of frog oocytes is arrested at the end of the prophase of meiosis I. Treatment with the hormone progesterone triggers the resumption of meiosis, which is equivalent to the G_2 to M transition of somatic cells. The oocytes then undergo a second arrest at the metaphase of meiosis II, where they remain until fertilization. Masui and Markert hypothesized that the effects of both hormone treatment and fertilization on meiosis were due to changes in the cytoplasm that secondarily controlled behavior of the nucleus. They tested this hypothesis directly by transferring cytoplasm from hormone-stimulated oocytes into unstimulated oocytes. These experiments demonstrated that a cytoplasmic factor is responsible for the induction of meiosis following hormone treatment.

The Experiments

Because of both their large size and their ability to survive injection with glass micropipettes, frog oocytes appeared to provide a particularly suitable experimental system for testing the activity of cytoplasmic factors. The basic design of Masui and Markert's experiments was to remove cytoplasm from donor oocytes that had been treated with progesterone to induce the resumption of meiosis. Varying amounts of this cytoplasm were then injected into untreated recipient oocytes. The key result was that cytoplasm removed six or more hours after hormone treatment of donor oocytes induced the resumption of meiosis in injected recipients (see figure). In contrast, injection of cytoplasm from control oocytes that had not been exposed to progesterone had no effect on the recipients. It thus appeared that hormone-treated oocytes contained a cytoplasmic factor that could induce the resumption of meiosis in recipients that had never been exposed to progesterone.

Control experiments ruled out the possibility that progesterone itself is the meiosis-inducing factor in donor cytoplasm. In particular, it was demonstrated that the injection of progesterone into recipient oocytes fails to induce meiosis. Only external application of the hormone is effective, indicating that progesterone acts on a cell surface receptor to activate a distinct cytoplasmic factor. Similar experiments performed independently by Dennis Smith and Robert Ecker in 1971 led to the same conclusion.

Interestingly, the action of progesterone in this system is distinct from its action in most cells, where it diffuses across the plasma membrane and binds to an intracellular receptor (see Chapter 16). In oocytes, however, progesterone clearly acts on the cell surface to activate a distinct factor in the oocyte cytoplasm. Since the resumption of oocyte meiosis is commonly referred to as oocyte maturation, Masui and Markert coined the term "maturation promoting factor" for their newly discovered regulator of meiosis.

Yoshio Masui

Clement Markert

The Impact

Following its discovery in frog oocytes, MPF was also found to be present in somatic cells, where it induces the G_2 to M transition of mitosis. Thus MPF appeared to be a general regulator of entry into the M phase of both mitotic and meiotic cell cycles. The eventual purification of MPF from frog oocytes in 1988 then converged with yeast genetics and studies of sea urchin embryos to reveal the identity of this critical cell cycle regulator. Namely, MPF

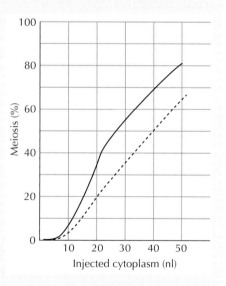

Recipient oocytes were injected with the indicated amounts of cytoplasm from oocytes that had been treated with progesterone. Donor cytoplasm was either withdrawn from the central region of oocytes with a micropipette (solid line) or prepared by homogenization of whole oocytes (dashed line). Results are presented as the percentage of injected oocytes that were induced to resume meiosis.

was found to be a dimer of cyclin B and the Cdk1 protein kinase. Further studies have established that both cyclin B and Cdk1 are members of large families of proteins, with different cyclins and Cdk1-related protein kinases functioning analogously to MPF in the regulation of other cell cycle transitions. The discovery of MPF in frog oocytes thus paved the way to understanding a cell cycle regulatory apparatus that is conserved throughout all eukaryotes.

the early 1970s. Studying the budding yeast *Saccharomyces cerevisiae*, these investigators identified temperature-sensitive mutants that were defective in cell cycle progression. The key characteristic of these mutants (called *cdc* for *cell division cycle* mutants) was that they underwent growth arrest at specific points in the cell cycle. For example, a particularly important mutant designated *cdc28* caused the cell cycle to arrest at START, indicating that the Cdc28 protein is required for passage through this critical regulatory point in G_1 (**Figure 17.9**). A similar collection of cell cycle mutants was isolated in the fission yeast *Schizosaccharomyces pombe* by Paul Nurse and his collaborators. These mutants included *cdc2*, which arrests the *S. pombe* cell cycle both in G_1 and at the G_2 to M transition (the major regulatory point in fission yeast). Comparative analysis showed that *S. cerevisiae cdc28* and *S. pombe cdc2* are functionally homologous genes, which are required for passage through START as well as for entry into mitosis in both species of yeasts. Further studies demonstrated that *cdc2* and *cdc28* encode a protein kinase—the first indication of the prominent role of protein phosphorylation in regulating the cell cycle. In addition, related genes were identified in other eukaryotes, including humans. The protein kinase encoded by the yeast *cdc2* and *cdc28*

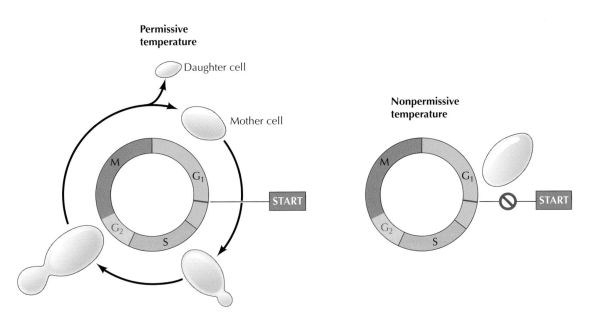

Figure 17.9 Properties of S. cerevisiae cdc28 mutants The temperature-sensitive *cdc28* mutant replicates normally at the permissive temperature. At the nonpermissive temperature, however, progression through the cell cycle is blocked at START.

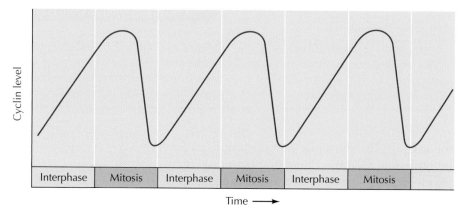

Figure 17.10   Accumulation and degradation of cyclins in sea urchin embryos   The cyclins were identified as proteins that accumulate throughout interphase and are rapidly degraded toward the end of mitosis.

genes has since been shown to be a conserved cell cycle regulator in all eukaryotes, which is known as **Cdk1**.

The third line of investigation that eventually converged with the identification of MPF and yeast genetics emanated from studies of protein synthesis in early sea urchin embryos. Following fertilization, these embryos go through a series of rapid cell divisions. Intriguingly, studies with protein synthesis inhibitors had revealed that entry into M phase of these embryonic cell cycles requires new protein synthesis. In 1983, Tim Hunt and his colleagues identified two proteins that display a periodic pattern of accumulation and degradation in sea urchin and clam embryos. These proteins accumulate throughout interphase and are then rapidly degraded toward the end of each mitosis (**Figure 17.10**). Hunt called these proteins **cyclins** (the two proteins were designated cyclin A and cyclin B) and suggested that they might function to induce mitosis, with their periodic accumulation and destruction controlling entry and exit from M phase. Direct support for such a role of cyclins was provided in 1986, when Joan Ruderman and her colleagues showed that microinjection of cyclin A into frog oocytes is sufficient to trigger the G_2 to M transition.

These initially independent approaches converged dramatically in 1988 when MPF was purified from frog eggs in the laboratory of James Maller. Molecular characterization of MPF in several laboratories then showed that this conserved regulator of the cell cycle is composed of two key subunits: Cdk1 and cyclin B (**Figure 17.11**). Cyclin B is a regulatory subunit required for catalytic activity of the Cdk1 protein kinase, consistent with the notion that MPF activity is controlled by the periodic accumulation and degradation of cyclin B during cell cycle progression.

A variety of further studies have confirmed this role of cyclin B, as well as demonstrating the regulation of MPF by phosphorylation and

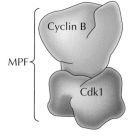

Figure 17.11   Structure of MPF   MPF is a dimer consisting of cyclin B and the Cdk1 protein kinase. Cyclin B is required for catalytic activity of Cdk1.

Key Experiment

The Identification of Cyclin

Cyclin: A Protein Specified by Maternal mRNA in Sea Urchin Eggs That Is Destroyed at Each Cleavage Division

Tom Evans, Eric T. Rosenthal, Jim Youngblom, Dan Distel, and Tim Hunt

Marine Biological Laboratory, Woods Hole, MA

Cell, 1983, Volume 33, pages 389–396

The Context

Tim Hunt and his colleagues undertook an analysis of protein synthesis in sea urchin eggs following fertilization, with the goal of identifying newly synthesized proteins that might be required for cell division during embryo development. A variety of earlier experiments had shown that translation of maternal mRNAs, stored in unfertilized eggs, was activated at fertilization, leading to striking changes in the profile of protein synthesis following fertilization. In addition, studies with inhibitors of translation had established that protein synthesis was required for cell division during early embryonic development. For example, inhibition of protein synthesis following fertilization of sea urchin eggs blocked nuclear envelope breakdown, chromosome condensation, and spindle formation. These results suggested that embryonic cell divisions required proteins that were synthesized from maternal mRNAs in fertilized eggs.

By studies of protein synthesis in fertilized eggs, Hunt and colleagues therefore hoped to identify newly synthesized proteins that were involved in cell division. Their experiments led to the identification of cyclins and elucidation of the fundamental mechanism that drives the division cycle of eukaryotic cells.

The Experiments

Unfertilized and fertilized sea urchin eggs were incubated in the presence of radioactive methionine, and cell extracts were analyzed by gel electrophoresis to identify newly synthesized proteins. Several prominent proteins were found to be

synthesized in fertilized eggs, as had previously been seen in eggs of other organisms. Moreover, careful examination of the profile of radiolabeled proteins at different times after fertilization revealed a surprising result. By analyzing the proteins present at 10-minute intervals after fertilization, Hunt and his colleagues identified a protein that was prominent in newly fertilized eggs, but then almost disappeared 85 minutes after fertilization. This protein again became abundant 95 minutes after fertilization, and declined again 30 minutes later. Because of these cyclic oscillations in its abundance, the protein was named "cyclin."

A further striking result was obtained when the oscillations in levels of cyclin were compared with the timing of cell divisions in the developing sea urchin embryos (see figure). The first division of the fertilized egg to a two-cell embryo took place 85 minutes after fertilization, coinciding with the decline of cyclin levels. The levels of cyclin then increased following this first cell division, and declined again 125 minutes after fertilization, coincident with the second cycle of embryonic cell division. Additional experiments demonstrated that cyclin was continuously synthesized after fertilization, but was rapidly degraded prior to each cell division.

The Impact

Based on its periodic degradation, coincident with the onset of embryonic cell

Tim Hunt

division, Hunt and colleagues concluded that it was "difficult to believe that the behavior of the cyclins is not connected with the processes involved in cell division." However, in the absence of experiments demonstrating the biological activity of cyclin, they had no direct evidence for this conclusion. Nonetheless, the activity of MPF was known to oscillate similarly during the cell cycle, and additional experiments had demonstrated that the formation of active MPF required protein synthesis. These similarities between MPF and cyclin led Hunt and his colleagues to suggest a direct relationship between MPF, defined as a biological activity, and cyclin, defined as a protein whose levels oscillated during cell cycle progression.

Further experiments of Joan Ruderman and her colleagues demonstrated that injection of cyclin was sufficient to induce the entry of frog oocytes into meiosis, showing that cyclin had the biological

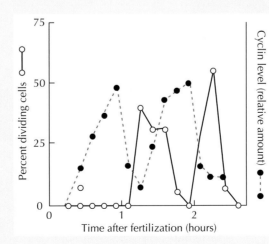

Sea urchin eggs were fertilized and radioactive methionine was added 5 minutes after fertilization. Samples were then harvested at 10-minute intervals and analyzed by gel electrophoresis to determine the levels of cyclin. In parallel, samples were examined microscopically to quantitate the percentage of cells undergoing cell division.

(Continued on next page)

activity of MPF. The biochemical relationship between cyclin and MPF was then established by the purification of MPF from frog eggs in the laboratory of James Maller and the subsequent characterization of

MPF as a dimer of cyclin B and Cdk1. The periodic degradation of cyclin B, discovered by Tim Hunt and colleagues, controls the activity of Cdk1 during entry and exit from mitosis. Moreover, other cyclins play similar

roles in regulating transitions between other phases of the cell cycle. The discovery of cyclin as an oscillating protein in sea urchin embryos was thus key to elucidating the molecular basis of cell cycle regulation.

dephosphorylation of Cdk1 (**Figure 17.12**). In mammalian cells, cyclin B is synthesized and forms complexes with Cdk1 during G_2. As these complexes form, Cdk1 is phosphorylated at two critical regulatory positions. One of these phosphorylations occurs on threonine-161 and is required for Cdk1 kinase activity. The second is a phosphorylation of tyrosine-15 and of the adjacent threonine-14 in vertebrates. Phosphorylation of tyrosine-15, catalyzed by a protein kinase called Wee1, inhibits Cdk1 activity and leads to the accumulation of inactive Cdk1/cyclin B complexes throughout G_2. The transition from G_2 to M is then brought about by activation of the Cdk1/cyclin B complex as

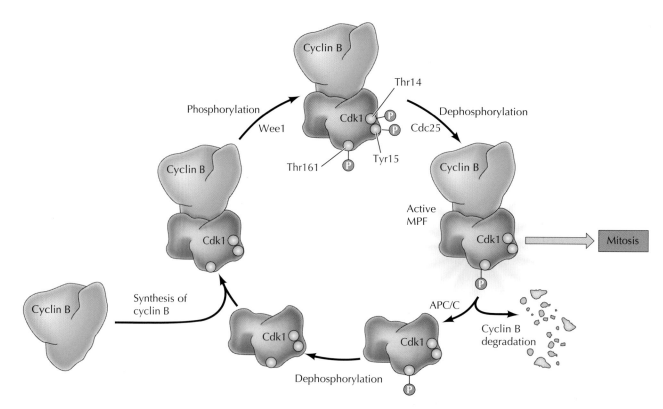

Figure 17.12 MPF regulation Cdk1 forms complexes with cyclin B during G_2. Cdk1 is then phosphorylated on threonine-161 (Thr161), which is required for Cdk1 activity, as well as on tyrosine-15 (Tyr15)—and threonine-14 (Thr14) in vertebrate cells—which inhibits Cdk1 activity. Dephosphorylation of Tyr15 and Thr14 activates MPF at the G_2 to M transition. MPF activity is then terminated toward the end of mitosis by proteolytic degradation of cyclin B, which is followed by dephosphorylation of Cdk1.

a result of dephosphorylation of threonine-14 and tyrosine-15 by a protein phosphatase called Cdc25.

Once activated, the Cdk1 protein kinase phosphorylates a variety of target proteins that initiate the events of M phase, which are discussed later in this chapter. In addition, Cdk1 activity triggers the degradation of cyclin B, which occurs as a result of ubiquitin-mediated proteolysis. Ubiquitylation of cyclin B is mediated by a ubiquitin ligase called the **anaphase-promoting complex/cyclosome (APC/C)**, which is activated as a result of phosphorylation by Cdk1/cyclin B. The resulting proteolytic destruction of cyclin B then inactivates Cdk1, leading the cell to exit mitosis, undergo cytokinesis, and return to interphase. As discussed further in the following sections, the APC/C ubiquitin ligase plays a central role in cell cycle progression by degrading cyclins and other key regulatory proteins not only in M phase, but also during other phases of the cell cycle.

Families of cyclins and cyclin-dependent kinases

The structure and function of MPF (Cdk1/cyclin B) provide not only a molecular basis for understanding entry and exit from M phase but also the foundation for elucidating the regulation of other cell cycle transitions. The insights provided by characterization of the Cdk1/cyclin B complex have thus had a sweeping impact on understanding cell cycle regulation. In particular, further research has established that both Cdk1 and cyclin B are members of families of related proteins, with different members of these families controlling progression through distinct phases of the cell cycle.

As discussed earlier, Cdk1 controls passage through START as well as entry into mitosis in yeasts. It does so, however, in association with distinct cyclins (**Figure 17.13**). In particular, the G$_2$ to M transition is driven by Cdk1 in association with the mitotic B-type cyclins (Clb1, Clb2, Clb3, and Clb4). Passage through START, on the other hand, is controlled by Cdk1 in association with a distinct class of cyclins called **G$_1$ cyclins** (or **Cln's**). Cdk1 then associates with different B-type cyclins (Clb5 and Clb6), which are required for progression through S phase. These associations of Cdk1 with distinct B-type and G$_1$ cyclins direct Cdk1 to phosphorylate different substrate proteins, as required for progression through specific phases of the cell cycle.

The cell cycles of higher eukaryotes are controlled not only by multiple cyclins but also by multiple Cdk1-related protein kinases, which are known as **Cdk's** for *cyclin-dependent kinases*. These multiple members of the Cdk family associate with specific cyclins to drive progression through the different stages of the cell cycle (see Figure 17.13). For example, progression from G$_1$ to S is regulated principally by Cdk4, Cdk6, and Cdk2, in association with cyclins D and E. Complexes of Cdk4 and Cdk6 with the D-type cyclins (D1, D2, and D3) play a critical role in progression through the restriction

Figure 17.13 Complexes of cyclins and cyclin-dependent kinases In yeast, passage through START is controlled by Cdk1 in association with G$_1$ cyclins (Cln1, Cln2, and Cln3). Complexes of Cdk1 with distinct B-type cyclins (Clb's) then regulate progression through S phase and entry into mitosis. In animal cells, progression through the G$_1$ restriction point is controlled by complexes of Cdk4 and Cdk6 with D-type cyclins. Cdk2/cyclin E complexes are required for the G$_1$ to S transition. Cdk2/cyclin A complexes are then required for progression through S phase and G2. Cdk1/cyclin A and Cdk1/cyclin B complexes drive the G$_2$ to M transition and progression through mitosis.

Animation 17.4
sites.sinauer.com/cooper7e/a17.4
Cyclins, Cdk's, and the Cell Cycle
Maturation promoting factor (MPF)—consisting of cyclin B and Cdk1—regulates the transition from G$_2$ to M phase of the cell cycle.

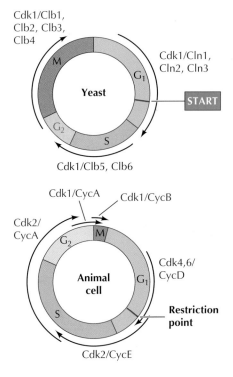

point in G_1. The E-type cyclins (E1 and E2) are then expressed, and Cdk2/cyclin E complexes are required for the G_1 to S transition and initiation of DNA synthesis. Complexes of Cdk2 with A-type cyclins (A1 and A2) are activated later in S phase and function through G_2. The initiation of mitosis and passage through M phase are then driven by Cdk1 in complexes with A- and B-type cyclins (B1, B2, and B3).

Although different Cdk/cyclin complexes normally function at different stages of the cell cycle, studies of Cdk's and cyclins in genetically modified mice have revealed a surprisingly high level of plasticity, allowing different cyclins and Cdk's to compensate for the loss of one another in mutant mice. Most strikingly, it has been shown that cells lacking all of the Cdk's that normally function during interphase (Cdk4, Cdk6, and Cdk2) are still capable of proliferation. Analysis of these cells indicates that, in the absence of other Cdk's, Cdk1 binds to all of the cyclins and is able to drive progression through all stages of the cell cycle. In contrast, mice lacking Cdk1 are unable to develop. Thus, Cdk1 is the only essential Cdk in mammalian cells (as in yeast) and is capable of substituting for the other Cdk's in their absence.

The activity of Cdk's during cell cycle progression is regulated by four molecular mechanisms (Figure 17.14). As already discussed for Cdk1, the first level of regulation involves the association of Cdk's with their cyclin partners. Thus the formation of specific Cdk/cyclin complexes is controlled by cyclin synthesis and degradation. Second, activation of Cdk/cyclin complexes requires phosphorylation of a conserved Cdk threonine residue around position 160. This activating phosphorylation of the Cdk's is catalyzed by an enzyme called CAK (for Cdk-activating kinase), which is itself composed of a Cdk (Cdk7) complexed with cyclin H. Complexes of Cdk7 and cyclin H are also associated with the transcription factor TFIIH, which is required for initiation of transcription by RNA polymerase II (see Chapter 8), so this member of the Cdk family participates in transcription as well as cell cycle regulation.

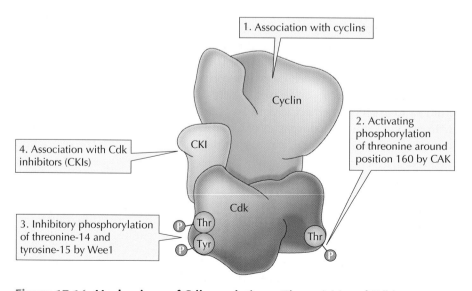

Figure 17.14 Mechanisms of Cdk regulation The activities of Cdk's are regulated by four molecular mechanisms.

Table 17.1   Cdk Inhibitors

Inhibitor	Cdk or Cdk/ cyclin complex	Cell cycle phase inhibited
Ink4 family (p15, p16, p18, p19)	Cdk4 and Cdk6	G_1
Cip/Kip family (p21, p27, p57)	Cdk2/cyclin E	G_1
	Cdk2/cyclin A	S, G_2

In contrast with the activating phosphorylation by CAK, the third mechanism of Cdk regulation (which affects only Cdk1 and Cdk2) involves inhibitory phosphorylation of tyrosine residues near the Cdk amino terminus, catalyzed by the Wee1 protein kinase. Both Cdk1 and Cdk2 are inhibited by phosphorylation of tyrosine-15, and the adjacent threonine-14 in vertebrates. These Cdk's are then activated by dephosphorylation of these residues by members of the Cdc25 family of protein phosphatases.

In addition to regulation of the Cdk's by phosphorylation, their activities are also controlled by the binding of inhibitory proteins (called **Cdk inhibitors** or **CKIs**). In mammalian cells, two families of Cdk inhibitors are responsible for regulating different Cdk's (Table 17.1). Members of the Ink4 family specifically bind to and inhibit Cdk4 and Cdk6, so the Ink4 CKIs act to inhibit progression through the restriction point in G_1. In contrast, members of the Cip/Kip family bind jointly to both cyclins and Cdk's and inhibit complexes of Cdk2 with cyclin A or cyclin E, thereby inhibiting entry into S phase and progression through the S and G_2 phases of the cell cycle. Control of the Ink4 and Cip/Kip proteins thus provides an additional mechanism for regulating Cdk activity. The combined effects of these multiple modes of Cdk regulation are responsible for controlling cell cycle progression in response both to checkpoint controls and to the variety of extracellular stimuli that regulate cell proliferation.

Growth factors and the regulation of G_1 Cdk's

As discussed earlier, the proliferation of animal cells is regulated largely by a variety of extracellular growth factors that control the progression of cells through the restriction point in late G_1. In the absence of growth factors, cells are unable to pass the restriction point and instead become quiescent, frequently entering the resting state known as G_0, from which they can reenter the cell cycle in response to growth factor stimulation. This control of cell cycle progression by extracellular growth factors implies that the intracellular signaling pathways stimulated downstream of growth factor receptors (discussed in the preceding chapter) ultimately act to regulate components of the cell cycle machinery.

One critical link between growth factor signaling and cell cycle progression is provided by the D-type cyclins (Figure 17.15). Cyclin D synthesis is induced in response to growth factor stimulation, in part by signaling through the Ras/Raf/MEK/ERK pathway, and cyclin D continues to be synthesized as long as growth factors are present. However, cyclin D is also rapidly degraded during G_1 as a result of ubiquitylation by the APC/C ubiquitin ligase, so its intracellular concentration rapidly falls if growth factors are removed. Thus, as long as growth factors are present through G_1, complexes of Cdk4, 6/cyclin D drive cells through the restriction point. On the other hand, if growth factors are removed prior to this key regulatory point in the cell cycle, the

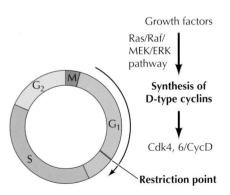

Figure 17.15 Induction of D-type cyclins   Growth factors regulate cell cycle progression through the G_1 restriction point by inducing synthesis of D-type cyclins via the Ras/Raf/MEK/ERK signaling pathway (see Figures 16.28 and 16.29).

levels of cyclin D rapidly fall and cells are unable to progress through G_1 to S, instead becoming quiescent and entering G_0. The inducibility and rapid turnover of cyclin D1 thus integrates growth factor signaling with the cell cycle machinery, allowing the availability of extracellular growth factors to control the progression of cells through G_1.

Since cyclin D is a critical target of growth factor signaling, it might be expected that defects in cyclin D regulation would contribute to the loss of growth regulation characteristic of cancer cells. Consistent with this expectation, many human cancers have been found to arise as a result of defects in cell cycle regulation, just as many others result from abnormalities in the intracellular signaling pathways activated by growth factor receptors (see Chapter 16). For example, mutations resulting in continual unregulated expression of cyclin D contribute to the development of a variety of human cancers, including lymphomas and breast cancers. Similarly, mutations that inactivate the Ink4 Cdk inhibitors that bind to Cdk4 and Cdk6 are commonly found in human cancer cells.

The connection between cyclin D, growth control, and cancer is further fortified by the fact that a key substrate protein of Cdk4, 6/cyclin D complexes is itself frequently mutated in a wide array of human tumors. This protein, designated **Rb**, was first identified as the product of a gene responsible for retinoblastoma, a rare inherited childhood eye tumor (see Chapter 19). Further studies then showed that mutations resulting in the absence of functional Rb protein are not restricted to retinoblastoma but also contribute to a variety of common human cancers. *Rb* is the prototype of a **tumor suppressor gene**—a gene whose inactivation leads to tumor development. Whereas oncogene proteins such as Ras (see Chapter 16) and cyclin D drive cell proliferation, the proteins encoded by many tumor suppressor genes (including Rb and the Ink4 Cdk inhibitors) act as brakes that slow down cell cycle progression.

Further studies have revealed that Rb and related members of the Rb family play a key role in coupling the cell cycle machinery to the expression of genes required for cell cycle progression and DNA synthesis (Figure 17.16). The activity of Rb proteins is regulated by changes in phosphorylation as cells progress through the cycle. In particular, Rb becomes phosphorylated by Cdk4, 6/cyclin D complexes as cells pass through the restriction point in G_1. In its underphosphorylated form (present in G_0 or early G_1), Rb binds to

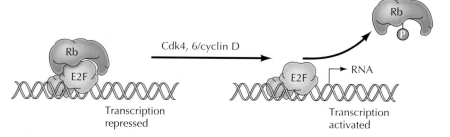

Figure 17.16 Cell cycle regulation of Rb and E2F   In its underphosphorylated form, Rb binds to members of the E2F family, repressing transcription of E2F-regulated genes. Phosphorylation of Rb by Cdk4, 6/cyclin D complexes results in its dissociation from E2F in late G_1. E2F then stimulates expression of its target genes, which encode proteins required for cell cycle progression.

members of the **E2F** family of transcription factors, which regulate expression of several genes involved in cell cycle progression, including the gene encoding cyclin E. E2F binds to its target sequences in either the presence or absence of Rb. However, Rb acts as a repressor, so the Rb/E2F complex suppresses transcription of E2F-regulated genes. Phosphorylation of Rb by Cdk4, 6/cyclin D complexes results in its dissociation from E2F, which then activates transcription of its target genes. Rb thus acts as a molecular switch that converts E2F from a repressor to an activator of genes required for cell cycle progression. The control of Rb by Cdk4, 6/cyclin D phosphorylation in turn couples this critical regulation of gene expression to the availability of growth factors in G_1.

Progression through the restriction point and entry into S phase is mediated by the activation of Cdk2/cyclin E complexes (**Figure 17.17**). This results in part from the synthesis of cyclin E, which is stimulated by E2F following phosphorylation of Rb (see Figure 17.16). In addition, the activity of Cdk2/cyclin E is inhibited in G_0 or early G_1 by the Cdk inhibitor p27 (see Table

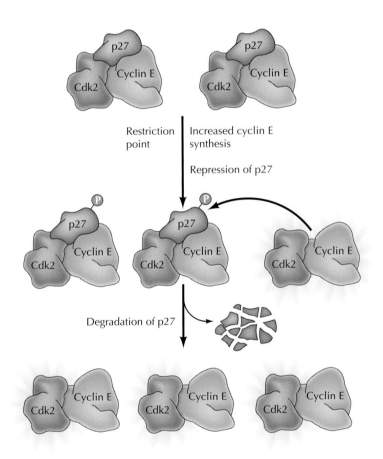

Figure 17.17 Activation of Cdk2/cyclin E   In G_1, Cdk2/cyclin E complexes are inhibited by the Cdk inhibitor p27. Passage through the restriction point induces the synthesis of cyclin E via activation of E2F. In addition, growth factor signaling inhibits synthesis of p27. As Cdk2 becomes activated, it phosphorylates and targets p27 for degradation, resulting in full activation of Cdk2/cyclin E complexes and entry into S phase.

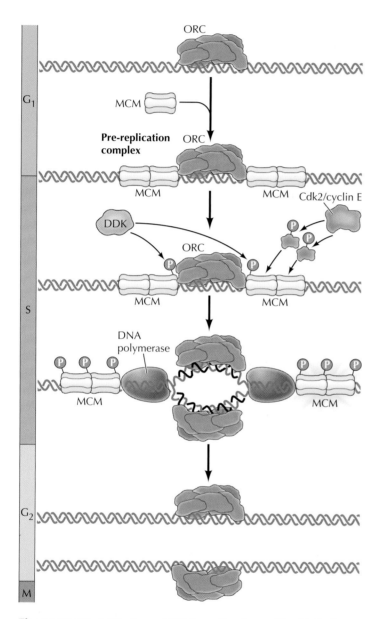

Figure 17.18 Initiation of DNA replication The MCM helicase proteins bind to origins of replication together with ORC (origin recognition complex) proteins in G₁ to form a pre-replication complex. DNA replication is initiated in S phase by Cdk2/cyclin E and the DDK protein kinase. DDK phosphorylates MCM proteins and Cdk2 phosphorylates additional proteins that join the complex and activate MCM. Activation of MCM initiates DNA replication and the MCM proteins move away from the origin with the replication fork. The high activity of Cdk's prevents the MCM proteins from reassociating with origins during S, G₂ and M, so pre-replication complexes can only re-form during G₁.

17.1). This inhibition of Cdk2 by p27 is relieved by multiple mechanisms as cells progress through G₁. First, growth factor signaling via both the Ras/Raf/MEK/ERK and PI 3-kinase/Akt pathways reduces both the transcription and translation of p27, lowering the levels of p27 within the cell. In addition, once Cdk2 becomes activated, it brings about the complete degradation of p27 by phosphorylating it and targeting it for ubiquitylation. This positive autoregulation then results in full Cdk2/cyclin E activation and progression to S phase. The APC/C ubiquitin ligase is also inhibited by Cdk2, so high levels of cyclins are maintained through S and G₂.

S phase and regulation of DNA replication

Cdk2/cyclin E complexes initiate S phase by activating DNA synthesis at replication origins. As discussed in Chapter 7, mammalian cells use thousands of origins to replicate their DNA (see Figure 7.13). The initiation of replication at each of these origins must be carefully controlled so that each segment of the genome is replicated once and only once during the S phase of each cell cycle. Thus, once a segment of DNA has been replicated in S phase, control mechanisms exist to prevent reinitiation of DNA replication until the cell cycle has been completed and the cell has passed through mitosis.

DNA replication is initiated by the activity of MCM helicase proteins, which is regulated by cyclin/Cdk complexes at different stages of the cell cycle (**Figure 17.18**). The MCM helicase proteins bind to replication origins, together with the origin recognition complex (ORC) proteins (see Figure 7.15), during G₁. They remain inactive as a pre-replication complex throughout G₁ and become activated when the cell enters S phase. This activation of MCM in S phase results from the action of Cdk2/cyclin E, which phosphorylates several activating proteins that are recruited to the pre-replication complex. As noted earlier, activation of Cdk2/cyclin E also leads to inhibition of the APC/C ubiquitin ligase. This inhibition of APC/C leads to activation of a second protein kinase, DDK, which phosphorylates MCM proteins directly. Activation of the MCM helicase initiates DNA replication and the MCM proteins move away from the origin with the replication fork. The high activity of Cdk's during S, G₂ and M phases prevents the MCM proteins from reassociating with replication origins, so pre-replication complexes can only re-form during G₁, when Cdk activity is low. Thus, once an origin becomes active during S phase, replication at that origin cannot initiate again until the cell passes through mitosis and enters the G_1 phase of the next cell cycle.

DNA damage checkpoints

Cell proliferation is regulated not only by growth factors but also by a variety of signals that act to inhibit cell cycle progression. DNA damage checkpoints play a critical role in maintaining the integrity of the genome by arresting cell cycle progression in response to damaged or incompletely replicated DNA. These checkpoints, which are operative in G_1, S, and G_2 phases of the cell cycle, allow time for the damage to be repaired before DNA replication or cell division proceeds (see Figure 17.7).

Cell cycle arrest is mediated by two related protein kinases, designated **ATR** and **ATM**, that are activated in response to DNA damage. ATR and ATM then activate a signaling pathway that leads not only to cell cycle arrest, but also to the activation of DNA repair and, in some cases, programmed cell death. The importance of this DNA damage response is emphasized by the fact that these proteins were initially identified because mutations in the gene encoding ATM are responsible for the disease ataxia telangiectasia, which results in defects in the nervous and immune systems as well as a high frequency of cancer in affected individuals.

Both ATR and ATM are components of protein complexes that recognize damaged or unreplicated DNA (**Figure 17.19**). ATR is activated by single-stranded or unreplicated DNA, while ATM is activated by double-strand breaks. Once activated by DNA damage, ATR and ATM phosphorylate and activate the **checkpoint kinases** Chk1 and Chk2, respectively. Chk1 and Chk2 induce cell cycle arrest by phosphorylating and inhibiting or inducing the degradation of Cdc25 phosphatases. The Cdc25 phosphatases are required to activate Cdk2 and Cdk1 by removing inhibitory phosphorylations (see Figure 17.14) during cell cycle progression. DNA damage thus leads to

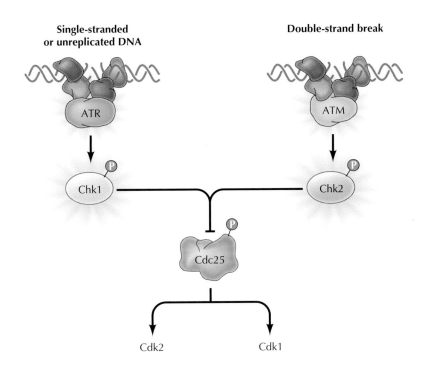

Single-stranded or unreplicated DNA

Double-strand break

ATR

ATM

Chk1

Chk2

Cdc25

Cdk2   Cdk1

Figure 17.19 Cell cycle arrest at the DNA damage checkpoints   The ATR and ATM protein kinases are activated in complexes of proteins that recognize damaged DNA. ATR is activated by single-stranded or unreplicated DNA, and ATM principally by double-strand breaks. ATR and ATM then phosphorylate and activate the Chk1 and Chk2 protein kinases, respectively. Chk1 and Chk2 phosphorylate and inhibit the Cdc25 protein phosphatases. Cdc25 phosphatases are required to activate both Cdk2 and Cdk1, so their inhibition leads to arrest at the DNA damage checkpoints in G_1, S, and G_2.

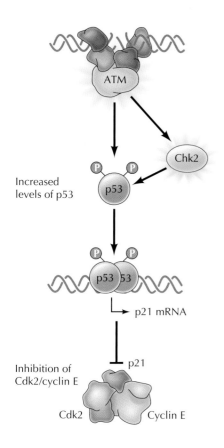

Increased levels of p53

Inhibition of Cdk2/cyclin E

Cdk2   Cyclin E

Figure 17.20 Role of p53 in cell cycle arrest The protein p53 plays a key role in cell cycle arrest at DNA damage checkpoints in mammalian cells. Phosphorylation by ATM and Chk2 stabilize p53, resulting in rapid increases in p53 levels in response to DNA damage. The protein p53 then activates transcription of the gene encoding the Cdk inhibitor p21, leading to inhibition of Cdk2/cyclin E or cyclin A complexes and cell cycle arrest.

inhibition of both Cdk2, resulting in cell cycle arrest in G_1 and S, and Cdk1, resulting in arrest in G_2.

In mammalian cells, arrest at DNA damage checkpoints is also mediated by the action of an additional protein known as **p53**, which is phosphorylated by both ATM and Chk2 (**Figure 17.20**). Phosphorylation stabilizes p53, which is otherwise rapidly degraded, resulting in a rapid increase in p53 levels in response to damaged DNA. The p53 protein is a transcription factor, and increased p53 levels lead to the induction of the Cdk inhibitor p21, which inhibits complexes of Cdk2 with cyclin E or cyclin A (see Table 17.1).

Interestingly, the gene encoding p53 is frequently mutated in human cancers. Loss of p53 function as a result of these mutations prevents cell cycle arrest in response to DNA damage, so the damaged DNA is replicated and passed on to daughter cells instead of being repaired. This inheritance of damaged DNA results in an increased frequency of mutations and general instability of the cellular genome, which contributes to cancer development. Mutations in the *p53* gene are among the most common genetic alterations in human cancers (see Chapter 19), illustrating the critical importance of cell cycle regulation in the life of multicellular organisms.

The Events of M Phase

M phase is the most dramatic period of the cell cycle, involving a major reorganization of virtually all cell components. During mitosis (nuclear division), the chromosomes condense, the nuclear envelope of most cells breaks down, the cytoskeleton reorganizes to form the mitotic spindle, and the chromosomes move to opposite poles. Chromosome segregation is then usually followed by cell division (cytokinesis). Although some of these events have been discussed in previous chapters, they are reviewed here in the context of a coordinated view of M phase and the action of MPF (Cdk1/cyclin B).

Stages of mitosis

Although many of the details of mitosis vary among different organisms, the fundamental processes that ensure the faithful segregation of sister chromatids are conserved in all eukaryotes. These basic events of mitosis include chromosome condensation, formation of the mitotic spindle, and attachment of chromosomes to the spindle microtubules. Sister chromatids then separate from each other and move to opposite poles of the spindle, followed by the formation of daughter nuclei.

Mitosis is conventionally divided into four primary stages—**prophase**, **metaphase**, **anaphase**, and **telophase**—which are illustrated for an animal cell in **Figures 17.21** and **17.22**. The beginning of prophase is marked by the appearance of condensed chromosomes, each of which consists of two sister chromatids (the daughter DNA molecules produced in S phase). These

Animation 17.5

sites.sinauer.com/cooper7e/a17.5

Mitosis in an Animal Cell Mitosis is the division of the nucleus. It consists of four phases—prophase, metaphase, anaphase, and telophase—and is followed by the division of the cytoplasm, called cytokinesis.

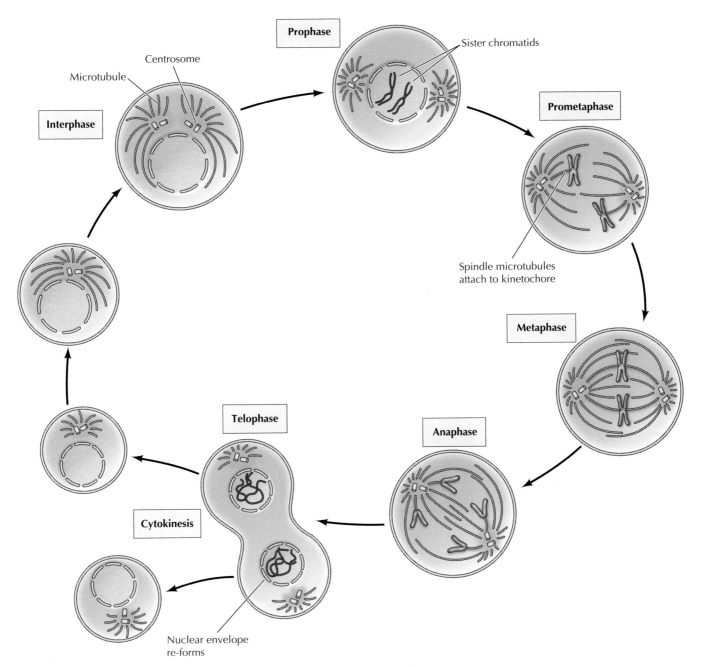

Figure 17.21 Stages of mitosis in an animal cell During prophase, the chromosomes condense and centrosomes move to opposite sides of the nucleus, initiating formation of the mitotic spindle. Breakdown of the nuclear envelope then allows spindle microtubules to attach to the kinetochores of chromosomes. During prometaphase the chromosomes shuffle back and forth between the centrosomes and the center of the cell, eventually aligning in the center of the spindle (metaphase). At anaphase, the sister chromatids separate and move to opposite poles of the spindle. Mitosis then ends with re-formation of nuclear envelopes and chromosome decondensation during telophase, and cytokinesis yields two interphase daughter cells. Note that each daughter cell receives one centrosome, which duplicates prior to the next mitosis.

Video 17.3

sites.sinauer.com/cooper7e/v17.3

Mitosis Mitosis is the most dramatic stage of the cell cycle, corresponding to the separation of daughter chromosomes and usually ending with cell division.

Mitosis

Interphase

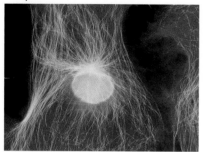

Early prophase

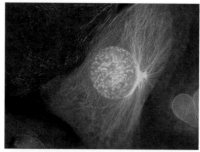

Late prophase

Prometaphase

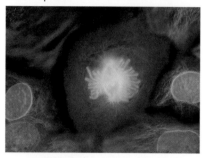

Metaphase

Early anaphase

Late anaphase

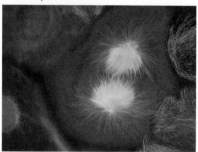

Telophase

Figure 17.22 Fluorescence micrographs of chromatin, keratin, and micro-tubules during mitosis in newt lung cells Chromatin is stained blue, keratin is stained red, and microtubules are stained green/yellow.

Video 17.4

sites.sinauer.com/cooper7e/v17.4
Mitosis with Fluorescent Proteins
During mitosis, the chromosomes condense, the nuclear envelope breaks down, the cytoskeleton forms the mitotic spindle, and the chromosomes move to opposite poles.

newly replicated DNA molecules remain intertwined throughout S and G_2, becoming untangled during the process of chromatin condensation. The condensed sister chromatids are then held together at the **centromere**, which (as discussed in Chapter 6) is a chromosomal region to which proteins bind to form the **kinetochore**—the site of eventual attachment of the spindle microtubules. In addition to chromosome condensation, cytoplasmic changes leading to the development of the mitotic spindle initiate during prophase. The **centrosomes** (which had duplicated during interphase) separate and move to opposite sides of the nucleus. There they serve as the two poles of the **mitotic spindle**, which begins to form during late prophase.

In higher eukaryotes, the end of prophase corresponds to the breakdown of the nuclear envelope. However, this disassembly of the nucleus is not a universal feature of mitosis and does not occur in all cells. Some unicellular

Closed mitosis

Spindle

Open mitosis

Spindle

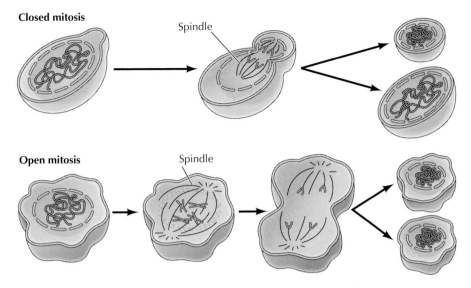

Figure 17.23 Closed and open mitosis In closed mitosis, the nuclear envelope remains intact as chromosomes migrate to opposite poles of a spindle within the nucleus, which then divides in two. In open mitosis, the nuclear envelope breaks down and then re-forms around the two sets of separated chromosomes.

eukaryotes (e.g., yeasts) undergo so-called closed mitosis, in which the nuclear envelope remains intact (**Figure 17.23**). In closed mitosis, the daughter chromosomes migrate to opposite poles of the nucleus, which then divides in two. In these cells, the spindle pole bodies are embedded within the nuclear envelope, and the nucleus divides in two following migration of daughter chromosomes to opposite poles of the spindle.

Following completion of prophase, the cell enters **prometaphase**—a transition period between prophase and metaphase. During prometaphase the microtubules of the mitotic spindle attach to the kinetochores of condensed chromosomes. The kinetochores of sister chromatids are oriented on opposite sides of the chromosome, so they attach to microtubules emanating from opposite poles of the spindle. The chromosomes shuffle back and forth until they eventually align on the metaphase plate in the center of the spindle. At this stage, the cell has reached metaphase (see Figure 17.21).

Most cells remain only briefly at metaphase before proceeding to anaphase. The transition from metaphase to anaphase is triggered by breakage of the link between sister chromatids, which then separate and move to opposite poles of the spindle. Mitosis ends with telophase, during which nuclei re-form and the chromosomes decondense. Cytokinesis usually begins during late anaphase and is almost complete by the end of telophase, resulting in the formation of two interphase daughter cells.

Entry into mitosis

Mitosis involves dramatic changes in multiple cellular components, leading to a major reorganization of the entire structure of the cell. As discussed earlier in this chapter, these events are triggered by activation of the Cdk1/cyclin B protein kinase (MPF). Cdk1/cyclin B acts as a master regulator of the M phase transition, both by activating other mitotic protein kinases and by directly

Figure 17.24 Mitotic protein kinases The mitotic protein kinases Cdk1, Aurora, and Polo-like kinases are activated in a positive feedback loop at the onset of M phase. They induce multiple nuclear and cytoplasmic changes during mitosis by phosphorylating proteins such as condensins, cohesins, components of the nuclear envelope, Golgi matrix proteins, and proteins associated with centrosomes, kinetochores, and microtubules.

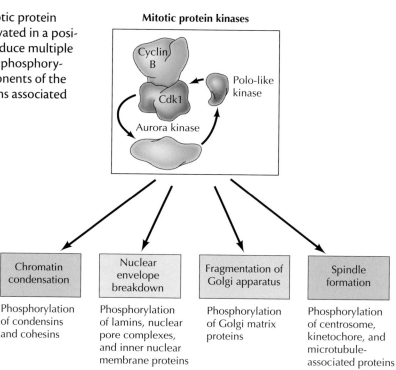

phosphorylating some of the structural proteins involved in this cellular reorganization (**Figure 17.24**). In particular, mitotic protein kinases of the **Aurora kinase** (Aurora A and Aurora B) and **Polo-like kinase** families are activated coordinately with Cdk1 to signal entry into M phase. The Aurora and Polo-like kinases function with Cdk1 in a positive feedback loop, with Cdk1 activating Aurora kinases, which activate Polo-like kinases, which in turn activate Cdk1. All of these protein kinases then play multiple roles in mitosis.

The condensation of interphase chromatin to form the compact chromosomes of mitotic cells is a key event in mitosis, critical in enabling the chromosomes to move along the mitotic spindle without becoming broken or tangled. As discussed in Chapter 6, the chromatin in interphase nuclei condenses nearly a thousandfold during the formation of metaphase chromosomes. Such highly condensed chromatin cannot be transcribed, so transcription ceases as chromatin condensation takes place. Despite the fundamental importance of this event, we do not fully understand the structure of metaphase chromosomes nor the molecular mechanism of chromatin condensation. However, it has been established that chromatin condensation is driven by protein complexes called **condensins**, which are members of a class of "structural maintenance of chromatin" (SMC) proteins that play key roles in the organization of eukaryotic chromosomes.

Both condensins and another family of SMC proteins, called **cohesins**, contribute to chromosome segregation during mitosis (**Figure 17.25**). Cohesins bind to DNA in S phase and maintain the linkage between sister chromatids following DNA replication.

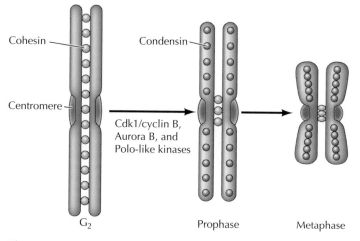

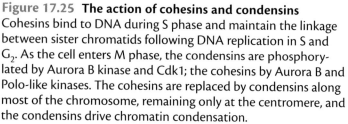

Figure 17.25 The action of cohesins and condensins Cohesins bind to DNA during S phase and maintain the linkage between sister chromatids following DNA replication in S and G₂. As the cell enters M phase, the condensins are phosphorylated by Aurora B kinase and Cdk1; the cohesins by Aurora B and Polo-like kinases. The cohesins are replaced by condensins along most of the chromosome, remaining only at the centromere, and the condensins drive chromatin condensation.

As the cell enters M phase, the condensins are activated as a result of phosphorylation by both Cdk1 and Aurora B kinase. The condensins then replace the cohesins (which are phosphorylated by Aurora B and Polo-like kinases) along most of the length of the chromosome, so that the sister chromatids remain linked only at the centromere. The condensins also induce chromatin condensation, leading to the formation of metaphase chromosomes. Phosphorylation of histone H3 serine-10 by Aurora B kinase also plays an important role in mitotic chromosome condensation.

Breakdown of the nuclear envelope, which is one of the most dramatic events of mitosis, involves changes in all of its components: the nuclear membranes fragment, the nuclear pore complexes dissociate, and the nuclear lamina depolymerizes. Depolymerization of the nuclear lamina (the meshwork of filaments underlying the nuclear membrane) results from phosphorylation of the lamins by Cdk1/cyclin B (**Figure 17.26**). Phosphorylation causes the lamin filaments to break down into individual lamin dimers, leading directly to depolymerization of the nuclear lamina. Cdk1 also phosphorylates several proteins in the inner nuclear membrane and the nuclear pore complex, leading to disassembly of nuclear pore complexes and detachment of the inner

Figure 17.26 Breakdown of the nuclear envelope Cdk1/cyclin B phosphorylates the nuclear lamins as well as proteins of the nuclear pore complex and inner nuclear membrane. Phosphorylation of the lamins causes the filaments that form the nuclear lamina to dissociate into free lamin dimers.

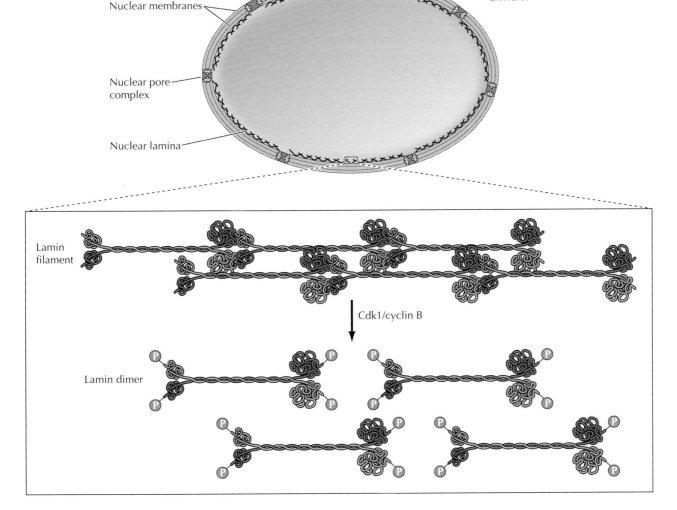

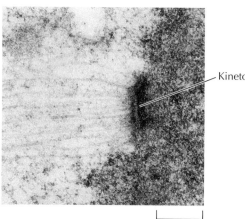

Kinetochore

0.5 μm

Figure 17.27 Electron micrograph of microtubules attached to the kinetochore of a chromosome

nuclear membrane from lamins and chromatin. In animal cells, integral proteins of the nuclear membrane are then absorbed into the endoplasmic reticulum, which remains as an intact network and is distributed to daughter cells at mitosis.

The Golgi apparatus fragments into small vesicles at mitosis, which may either be absorbed into the endoplasmic reticulum or distributed directly to daughter cells at cytokinesis. Breakdown of the Golgi apparatus into vesicles is mediated by phosphorylation of multiple Golgi matrix proteins by Cdk1 and Polo-like kinases.

The reorganization of the cytoskeleton that culminates in formation of the mitotic spindle results from the dynamic instability of microtubules (see Chapter 13). At the beginning of prophase, the centrosomes, which were duplicated during S phase, separate and move to opposite sides of the nucleus. Prior to separation, they undergo a process of maturation during which they enlarge and recruit γ-tubulin and other proteins needed for spindle assembly. Centrosome maturation, separation, and spindle assembly are driven by Aurora A and Polo-like kinases, which are located at centrosomes. The rate of microtubule turnover increases five- to tenfold during mitosis, resulting in depolymerization and shrinkage of the interphase microtubules. This increased turnover is thought to result from phosphorylation of microtubule-associated proteins by mitotic protein kinases, including Cdk1, Aurora A, and Polo-like kinases. The number of microtubules emanating from the centrosomes also increases, so the interphase microtubules are replaced by large numbers of short microtubules radiating from the centrosomes.

The breakdown of the nuclear envelope then allows some of the spindle microtubules to attach to chromosomes at their kinetochores (**Figure 17.27**), initiating the process of chromosome movement that characterizes prometaphase. The proteins assembled at the kinetochore include Aurora B kinase, which is involved in the interactions of chromosomes with microtubules. Once microtubules attach to the chromosomes, microtubule motors direct the movement of chromosomes toward the minus ends of the spindle microtubules, which are anchored in the centrosome. The action of these proteins, which draw chromosomes toward the centrosome, is opposed by plus-end-directed motor proteins and by the growth of the spindle microtubules, which pushes the chromosomes away from the spindle poles. Consequently, the chromosomes in prometaphase shuffle back and forth between the centrosomes and the center of the spindle.

Microtubules from opposite poles of the spindle eventually attach to the two kinetochores of sister chromatids (which are located on opposite sides of the chromosome), and the balance of forces acting on the chromosomes leads to their alignment on the metaphase plate in the center of the spindle (**Figure 17.28**). As discussed in Chapter 13, the spindle consists of kinetochore and chromosomal microtubules, which are attached to the chromosomes, as well as interpolar microtubules, which overlap with one another in the center of the cell. In addition, short astral microtubules radiate outward from the centrosomes toward the cell periphery.

(A)

Interpolar microtubules

Chromosomal microtubules

Astral microtubules

Kinetochore microtubules

(B)

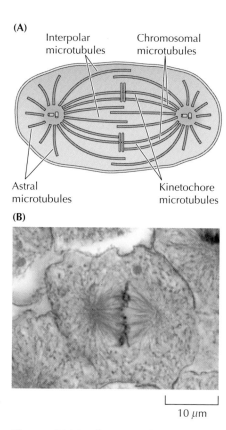

10 μm

Figure 17.28 The metaphase spindle (A) The spindle consists of four kinds of microtubules. Kinetochore and chromosomal microtubules are attached to chromosomes; interpolar microtubules overlap in the center of the cell; and astral microtubules radiate from the centrosome to the cell periphery. (B) A whitefish cell at metaphase.

The spindle assembly checkpoint and progression to anaphase

As discussed earlier in this chapter, the spindle assembly checkpoint monitors the alignment of chromosomes on the metaphase spindle. Once this has been accomplished, the cell proceeds to initiate anaphase and complete mitosis. The progression from metaphase to anaphase results from ubiquitin-mediated proteolysis of key regulatory proteins, triggered by activation of the APC/C ubiquitin ligase. The APC/C is inhibited by Cdk2/cyclin E and A complexes during the S and G_2 phases of the cell cycle. At the beginning of mitosis, however, activation of the APC/C is induced as a result of phosphorylation by Cdk1/cyclin B. The APC/C remains inhibited, however, until the cell passes the spindle assembly checkpoint, after which activation of the ubiquitin degradation system brings about the transition from metaphase to anaphase and progression through the rest of mitosis.

The spindle assembly checkpoint is remarkable in that the presence of even a single unaligned chromosome is sufficient to prevent activation of the APC/C. The checkpoint is mediated by a complex of proteins (called the **mitotic checkpoint complex** or **MCC**) that is formed at unattached kinetochores and inhibits the APC/C (**Figure 17.29**). The mitotic checkpoint complex

Video 17.5

sites.sinauer.com/cooper7e/v17.5

Spindle Assembly Checkpoint The spindle assembly checkpoint monitors the alignment of chromosomes on the metaphase spindle.

FYI

Abnormalities in chromosome segregation resulting from failures of the spindle assembly checkpoint are common in cancer cells and are thought to play an important role in the development of many tumors (see Chapter 19).

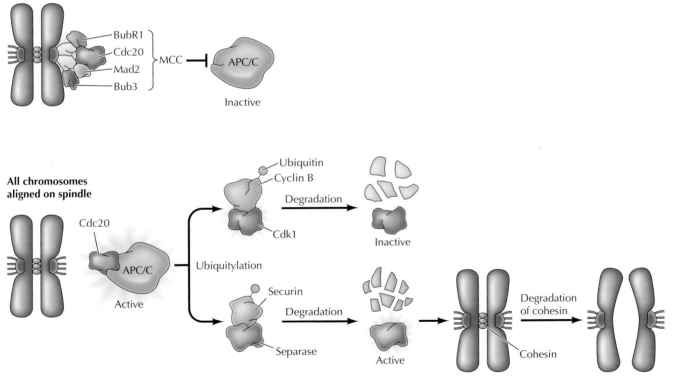

Figure 17.29 The spindle assembly checkpoint   Progression to anaphase is mediated by activation of the APC/C ubiquitin ligase. Unattached kinetochores lead to the assembly of a protein complex (the mitotic checkpoint complex, MCC) that inhibits APC/C. Once all chromosomes are aligned on the spindle, the inhibitory complex is no longer formed and APC/C is activated. APC/C ubiquitylates cyclin B, leading to its degradation and inactivation of Cdk1. In addition, APC/C ubiquitylates securin (an inhibitory subunit of a protease called separase), leading to activation of separase. Separase degrades cohesin, breaking the link between sister chromatids and initiating anaphase.

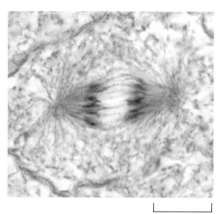

Figure 17.30 A whitefish cell at anaphase

consists of four proteins: BubR1, Bub3, Mad2, and Cdc20. The protein Cdc20 is a required activator of the APC/C, but its activity is blocked when bound by Mad and Bub proteins in the MCC. Once microtubules have attached to the kinetochores, MCC complexes are no longer formed and Cdc20 is able to activate rather than inhibit the APC/C.

Activation of the APC/C results in ubiquitylation and degradation of two key target proteins that trigger the metaphase to anaphase transition. The onset of anaphase results both from degradation of cyclin B, leading to inactivation of Cdk1, and from degradation of a component of the cohesins, which maintain the connection between sister chromatids while they are aligned on the metaphase plate (see Figure 17.25). Cohesin degradation is not catalyzed directly by the APC/C, which instead degrades a protein called securin that is an inhibitory subunit of a protease called separase. Degradation of securin results in the activation of separase, which in turn degrades cohesin. Cleavage of cohesin breaks the linkage between sister chromatids, allowing them to segregate by moving to opposite poles of the spindle (**Figure 17.30**). The separation of chromosomes during anaphase then proceeds as a result of the action of several types of motor proteins associated with the spindle microtubules (see Figures 13.51 and 13.52).

Once the cell has progressed to anaphase, the APC/C also triggers the degradation of Aurora and Polo-like kinases, allowing the cell to exit mitosis and return to interphase. Many of the cellular changes involved in these transitions are simply the reversal of the events induced by Cdk1, Aurora, and Polo-like kinases during mitosis. For example, reassembly of the nuclear envelope, chromatin decondensation, and the return of microtubules to an interphase state result directly from loss of activity of mitotic kinases and dephosphorylation of proteins that had been phosphorylated at the beginning of mitosis. As discussed next, inactivation of Cdk1 also triggers cytokinesis.

Cytokinesis

The completion of mitosis is usually accompanied by cytokinesis, giving rise to two daughter cells. In most cases, cytokinesis initiates shortly after the onset of anaphase and is triggered by the inactivation of Cdk1. In contrast, Aurora and Polo-like kinases remain active in anaphase and play important roles in coordinating nuclear and cytoplasmic division of the cell.

As discussed in Chapter 13, cytokinesis of yeast and animal cells is mediated by a **contractile ring** of actin and myosin II filaments that forms

(A)

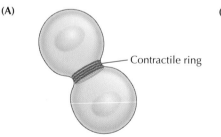

Contractile ring

(B)

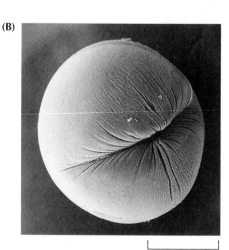

Figure 17.31 Cytokinesis of animal cells (A) Cytokinesis results from contraction of a ring of actin and myosin filaments, which pinches the cell in two. (B) Scanning electron micrograph of a frog egg undergoing cytokinesis.

Figure 17.32 Cytokinesis in higher plants   Golgi vesicles carrying cell wall precursors associate with interpolar microtubules at the former site of the metaphase plate. Fusion of these vesicles yields a membrane-enclosed, disk-like structure (the early cell plate) that expands outward and fuses with the parental plasma membrane. The daughter cells remain connected at plasmodesmata.

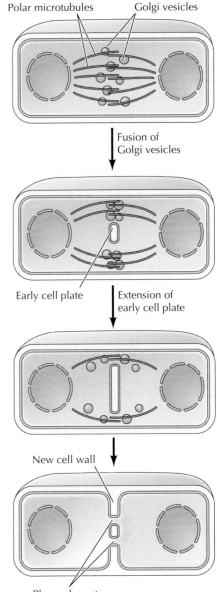

Polar microtubules   Golgi vesicles

Fusion of Golgi vesicles

Early cell plate   Extension of early cell plate

New cell wall

Plasmodesmata

beneath the plasma membrane (**Figure 17.31**). The formation of this ring is activated by Aurora and Polo-like kinases at a location determined by the position of the mitotic spindle, so the cell is eventually cleaved in a plane that passes through the metaphase plate perpendicular to the spindle. Cleavage proceeds as contraction of the actin-myosin filaments pulls the plasma membrane inward, eventually pinching the cell in half. The bridge between the two daughter cells is then broken, and the plasma membrane is resealed.

The mechanism of cytokinesis is different for higher plant cells. Rather than being pinched in half by a contractile ring, these cells divide by forming new cell walls and plasma membranes inside the cell (**Figure 17.32**). In early telophase, vesicles carrying cell wall precursors from the Golgi apparatus associate with remnants of the spindle microtubules and accumulate at the former site of the metaphase plate. These vesicles then fuse to form a large, membrane-enclosed, disk-like structure, and their polysaccharide contents assemble to form the matrix of a new cell wall (called a cell plate). The cell plate expands outward, perpendicular to the spindle, until it reaches the plasma membrane. The membrane surrounding the cell plate then fuses with the parental plasma membrane, dividing the cell in two. Connections between the daughter cells (plasmodesmata, see Figure 15.31) are formed as a result of incomplete vesicle fusion during cytokinesis.

Meiosis and Fertilization

The somatic cell cycles discussed so far in this chapter result in diploid daughter cells with identical genetic complements. Meiosis, in contrast, is a specialized kind of cell cycle that reduces the chromosome number by half, resulting in the production of haploid daughter cells. Unicellular eukaryotes, such as yeasts, can undergo meiosis as well as reproduce by mitosis. Diploid *Saccharomyces cerevisiae*, for example, undergo meiosis and produce spores when faced with unfavorable environmental conditions. In multicellular plants and animals, however, meiosis is restricted to the germ cells, where it is key to sexual reproduction. Whereas somatic cells undergo mitosis to proliferate, the germ cells undergo meiosis to produce haploid gametes (the sperm and the egg). The development of a new progeny organism is then initiated by the fusion of these gametes at fertilization.

The process of meiosis

In contrast with mitosis, **meiosis** results in the division of a diploid parental cell into haploid progeny, each containing only one member of the pair of homologous chromosomes that were present in the diploid parental cell (**Figure 17.33**). This reduction in chromosome number is accomplished by two sequential rounds of nuclear and cell division (called meiosis I and meiosis II), which follow a single round of DNA replication. Like mitosis, meiosis I initiates after S phase has been completed and the parental chromosomes have replicated to produce identical sister chromatids. The pattern of chromosome segregation in meiosis I, however, is dramatically

Animation 17.7

sites.sinauer.com/cooper7e/a17.7

Meiosis In meiosis, a cell divides to produce daughter cells with half the number of chromosomes as the parent cell.

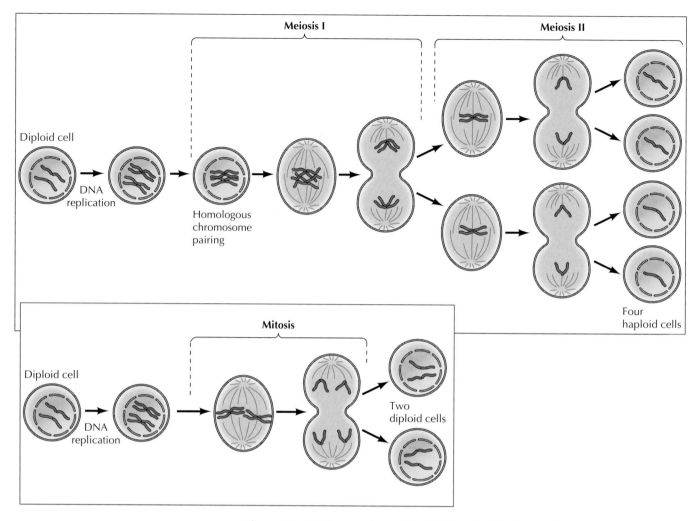

Figure 17.33 Comparison of meiosis and mitosis Both meiosis and mitosis initiate after DNA replication, so each chromosome consists of two sister chromatids. In meiosis I, homologous chromosomes pair with each other and then segregate to different cells. Sister chromatids then separate during meiosis II, which resembles a normal mitosis. Meiosis thus gives rise to four haploid daughter cells.

Animation 17.8

sites.sinauer.com/cooper7e/a17.8
Meiosis I and Mitosis Compared
One difference between mitosis and meiosis can be seen at metaphase—in mitosis, homologous chromosomes line up separately on the metaphase plate, whereas in metaphase of meiosis I, homologous chromosomes line up in pairs on the metaphase plate.

different from that of mitosis. During meiosis I, homologous chromosomes first pair with one another and then segregate to different daughter cells. Sister chromatids remain together, so completion of meiosis I results in the formation of daughter cells containing a single member of each chromosome pair (consisting of two sister chromatids). Meiosis I is followed by meiosis II, which resembles mitosis in that the sister chromatids separate and segregate to different daughter cells. Completion of meiosis II thus results in the production of four haploid daughter cells, each of which contains only one copy of each chromosome.

The pairing of homologous chromosomes after DNA replication results from recombination between chromosomes of paternal and maternal origin, so genetic recombination is critically linked to chromosome segregation during meiosis. Recombination between homologous chromosomes takes place during an extended prophase of meiosis I, which is divided into five stages

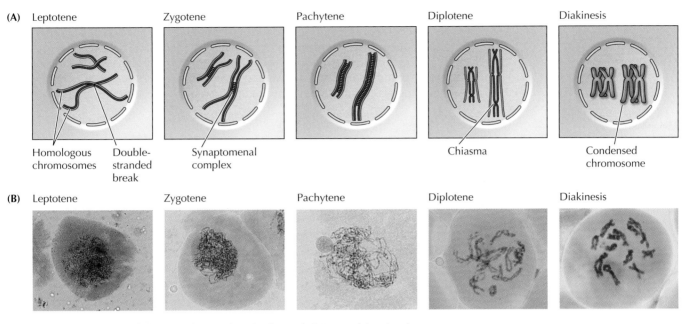

Figure 17.34 Stages of the prophase of meiosis I (A) Recombination between homologous chromosomes is initiated by the formation of double-strand breaks in leptotene. A zipperlike structure, the synaptomental complex, then forms between homologous chromosomes during zygotene and pachytene. The completion of recombination leads to the formation of chiasmata at the sites of crossing over and the synaptomenal complex disappears at the diplotene stage. The chromosomes then become condensed at diakinesis in transition to metaphase. (B) Micrographs illustrating the morphology of chromosomes of the lily.

(**leptotene**, **zygotene**, **pachytene**, **diplotene**, and **diakinesis**) on the basis of chromosome morphology (**Figure 17.34**). Recombination occurs at a high frequency during meiosis, and is initiated by double-strand breaks that are induced early in meiotic prophase (leptotene) by a highly conserved endonuclease called Spo11. As discussed in Chapter 7, the formation of double-strand breaks leads to the formation of single-strand regions that invade a homologous chromosome by complementary base pairing (see Figure 7.29). The close association of homologous chromosomes (**synapsis**) begins during the zygotene stage. During this stage, a zipperlike protein structure, called the **synaptomenal complex**, forms along the length of the paired chromosomes. This complex keeps the homologous chromosomes closely associated and aligned with one another through the pachytene stage, which can persist for several days. Recombination between homologous chromosomes is completed by the end of pachytene, leaving the chromosomes linked at the sites of crossing over (**chiasmata**; singular, chiasma). The synaptonemal complex disappears at the diplotene stage and the homologous chromosomes separate along their length. Importantly, however, they remain associated at the chiasmata, which is critical for their correct alignment at metaphase. At this stage, each chromosome pair (called a bivalent) consists of four chromatids with clearly evident chiasmata. Diakinesis, the final stage of prophase I, represents the transition to metaphase during which the chromosomes become fully condensed.

Animation 17.9

sites.sinauer.com/cooper7e/a17.9

Prophase I of Meiosis Prophase I of meiosis consists of five stages, during which chromosomes condense and homologous chromosomes pair with each other and recombine.

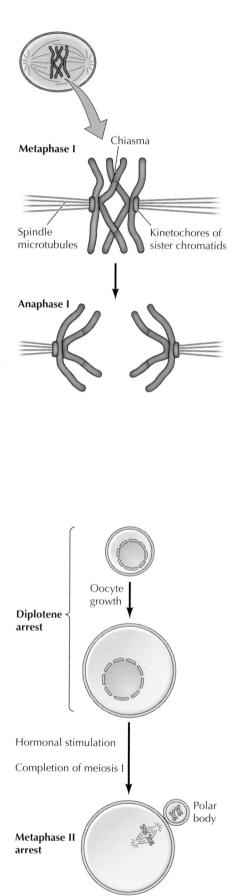

Metaphase I

Chiasma

Spindle microtubules

Kinetochores of sister chromatids

Anaphase I

Diplotene arrest

Oocyte growth

Hormonal stimulation

Completion of meiosis I

Metaphase II arrest

Polar body

Figure 17.35 Chromosome segregation in meiosis I At metaphase I, the kinetochores of sister chromatids are either fused or adjacent to one another. Microtubules from the same pole of the spindle therefore attach to the kinetochores of sister chromatids, while microtubules from opposite poles attach to the kinetochores of homologous chromosomes. Chiasmata are disrupted at anaphase I, and homologous chromosomes move to opposite poles of the spindle.

At metaphase I, the bivalent chromosomes align on the spindle. In contrast to mitosis (see Figure 17.28), the kinetochores of sister chromatids are adjacent to each other and oriented in the same direction, while the kinetochores of homologous chromosomes are pointed toward opposite spindle poles (**Figure 17.35**). Consequently, microtubules from the same pole of the spindle attach to sister chromatids, while microtubules from opposite poles attach to homologous chromosomes. Anaphase I is initiated by disruption of the chiasmata at which homologous chromosomes are joined. The homologous chromosomes then separate, while sister chromatids remain associated at their centromeres. At completion of meiosis I each daughter cell has therefore acquired one member of each homologous pair, consisting of two sister chromatids.

Meiosis II initiates immediately after cytokinesis, usually before the chromosomes have fully decondensed. In contrast to meiosis I, meiosis II resembles a normal mitosis. At metaphase II, the chromosomes align on the spindle with microtubules from opposite poles of the spindle attached to the kinetochores of sister chromatids. The link between the centromeres of sister chromatids is broken at anaphase II, and sister chromatids segregate to opposite poles. Cytokinesis then follows, giving rise to haploid daughter cells.

Regulation of oocyte meiosis

Vertebrate oocytes (developing eggs) have been particularly useful models for research on the cell cycle in part because of their large size and ease of manipulation in the laboratory. A notable example, discussed earlier in this chapter, is provided by the discovery and subsequent purification of MPF (Cdk1/cyclin B) from frog oocytes. Meiosis of these oocytes, like those of other species, is regulated at two unique points in the cell cycle, and studies of oocyte meiosis have illuminated novel mechanisms of cell cycle control.

The first regulatory point in oocyte meiosis is in the diplotene stage of the first meiotic division (**Figure 17.36**). Oocytes can remain arrested at this stage for long periods of time—up to 50 years in humans. During this diplotene arrest, the oocyte chromosomes decondense and are actively transcribed. This transcriptional activity is reflected in the tremendous growth of oocytes during this period. Human oocytes, for example, are about 100 μm in diameter (more than 100 times the volume of a typical somatic cell). Frog oocytes are even larger, with diameters of approximately 1 mm. During this period of cell growth, the oocytes accumulate stockpiles of materials, including RNAs

Figure 17.36 Meiosis of vertebrate oocytes Meiosis is arrested at the diplotene stage, during which oocytes grow to a large size. Oocytes then resume meiosis in response to hormonal stimulation and complete the first meiotic division, with asymmetric cytokinesis giving rise to a small polar body. Most vertebrate oocytes are then arrested again at metaphase II.

and proteins that are needed to support early development of the embryo. As noted earlier in this chapter, early embryonic cell cycles then occur in the absence of cell growth, rapidly dividing the fertilized egg into smaller cells (see Figure 17.2).

Oocytes of different species vary as to when meiosis resumes and fertilization takes place. In some animals, oocytes remain arrested at the diplotene stage until they are fertilized, only then proceeding to complete meiosis. However, the oocytes of most vertebrates (including frogs, mice, and humans) resume meiosis in response to hormonal stimulation and proceed through meiosis I prior to fertilization. Cell division following meiosis I is asymmetric, resulting in the production of a small **polar body** and an oocyte that retains its large size. The oocyte then proceeds to enter meiosis II without having re-formed a nucleus or decondensed its chromosomes. Most vertebrate oocytes are then arrested again at metaphase II, where they remain until they are fertilized or degenerate.

Like the M phase of somatic cells, the meiosis of oocytes is controlled by the activity of Cdk1/cyclin B complexes. The regulation of Cdk1 during oocyte meiosis, however, displays unique features that are responsible for the progression from meiosis I to meiosis II and for metaphase II arrest (Figure 17.37). Hormonal stimulation of diplotene-arrested oocytes initially triggers the resumption of meiosis by activating Cdk1, as at the G_2 to M transition of somatic cells. As in mitosis, Cdk1 then induces chromosome condensation, nuclear envelope breakdown, and formation of the spindle. Activation of the APC/C ubiquitin ligase then leads to the metaphase to anaphase transition

FYI

Animals can be cloned by the procedure of somatic cell nuclear transfer in which the nucleus of a somatic cell is transferred to a metaphase II oocyte from which the normal chromosomes have been removed. The oocyte is then stimulated to divide and, upon implantation into a surrogate mother, can give rise to an animal genetically identical to the donor of the somatic cell nucleus. Since the cloning of Dolly the sheep in 1997, this technology has been used to create cloned offspring of several mammalian species. (See Chapter 18.)

Animation 17.10

sites.sinauer.com/cooper7e/a17.10

Polar Body Formation During meiosis in female vertebrates, the meiotic divisions are often unequal, resulting in a single large egg and much smaller polar bodies.

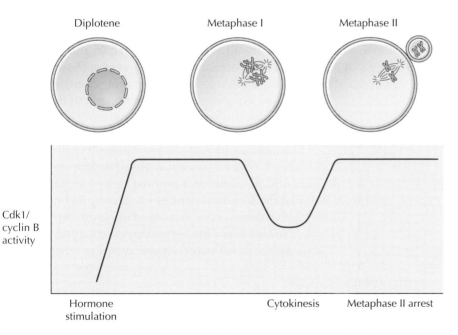
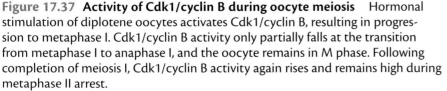

Figure 17.37 Activity of Cdk1/cyclin B during oocyte meiosis Hormonal stimulation of diplotene oocytes activates Cdk1/cyclin B, resulting in progression to metaphase I. Cdk1/cyclin B activity only partially falls at the transition from metaphase I to anaphase I, and the oocyte remains in M phase. Following completion of meiosis I, Cdk1/cyclin B activity again rises and remains high during metaphase II arrest.

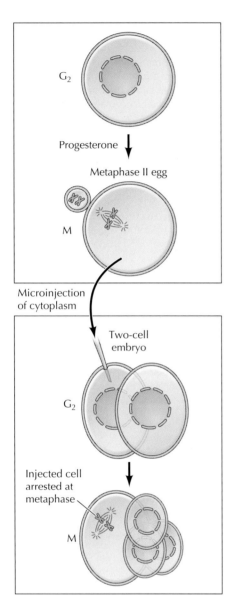

Microinjection of cytoplasm

Figure 17.38 Identification of cytostatic factor Cytoplasm from a metaphase II egg is microinjected into one cell of a two-cell embryo. The injected embryo cell arrests at metaphase, while the uninjected cell continues to divide. A factor in metaphase II egg cytoplasm (cytostatic factor) therefore has induced metaphase arrest of the injected embryo cell.

of meiosis I, accompanied by a decrease in the activity of Cdk1. However, in contrast to mitosis, Cdk1 activity is only partially decreased, so the oocyte remains in M phase, chromatin remains condensed, and nuclear envelopes do not re-form. Following cytokinesis, Cdk1 activity again rises and remains high while the egg is arrested at metaphase II. A regulatory mechanism unique to oocytes thus acts to maintain Cdk1 activity during the metaphase to anaphase transition of meiosis I and subsequent metaphase II arrest, preventing the inactivation of Cdk1 that would result from cyclin B proteolysis during a normal M phase.

The factor responsible for metaphase II arrest was first identified by Yoshio Masui and Clement Markert in 1971, in the same series of experiments that led to the discovery of MPF. In this case, however, cytoplasm from an egg arrested at metaphase II was injected into an early embryo cell that was undergoing mitotic cell cycles (**Figure 17.38**). This injection of egg cytoplasm caused the embryonic cell to arrest at metaphase, indicating that the arrest was induced by a cytoplasmic factor present in the egg. Because this factor acted to arrest mitosis, it was called **cytostatic factor (CSF)**.

Subsequent experiments identified a serine/threonine kinase known as **Mos** as an essential component of CSF. Mos is specifically synthesized in oocytes around the time of completion of meiosis I and is required for the maintenance of Cdk1/cyclin B activity during the metaphase to anaphase transition of meiosis I as well as during metaphase II arrest. The action of Mos results from activation of the ERK MAP kinase, which plays a central role in the cell signaling pathways discussed in the previous chapter. In oocytes, however, ERK plays a different role: It activates another protein kinase called Rsk, which maintains the activity of MPF by inhibiting cyclin B degradation (**Figure 17.39**). Inhibition of cyclin B degradation is mediated by inhibition of APC/C by a protein called Emi2/Erp1, which is phosphorylated by Rsk and inhibits APC/C via interaction with Cdc20—similar to arrest at the spindle assembly checkpoint by the mitotic checkpoint complex (see Figure 17.29). Oocytes can remain arrested at this point in the meiotic cell cycle for several days, awaiting fertilization.

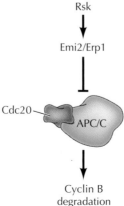

Figure 17.39 Maintenance of Cdk1/cyclin B activity by the Mos protein kinase The Mos protein kinase maintains Cdk1/cyclin B activity by inhibiting the degradation of cyclin B by the anaphase-promoting complex/cyclosome (APC/C). The action of Mos is mediated by MEK, ERK, and Rsk protein kinases. Rsk phosphorylates the protein Emi2/Erp1, which inhibits the APC/C by binding to Cdc20.

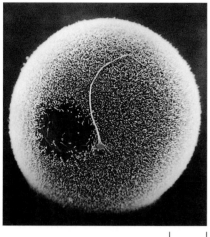

Figure 17.40 Fertilization Scanning electron micrograph of a human sperm fertilizing an egg.

10 μm

In vitro fertilization (IVF) is widely used to help infertile couples. A variety of reproductive disorders can be dealt with by this procedure in which metaphase II eggs are recovered from the ovary, fertilized *in vitro*, and then returned to the Fallopian tube or uterus of the mother. The first baby resulting from *in vitro* fertilization was born in 1978, and tens of thousands of infertile couples have since been treated by IVF.

Fertilization

At **fertilization**, the sperm binds to a receptor on the surface of the egg and fuses with the egg plasma membrane, initiating the development of a new diploid organism containing genetic information derived from both parents (**Figure 17.40**). Not only does fertilization lead to the mixing of paternal and maternal chromosomes, but it also induces a number of changes in the egg cytoplasm that are critical for further development. These alterations activate the egg, leading to the completion of oocyte meiosis and initiation of the mitotic cell cycles of the early embryo.

A key signal resulting from the binding of a sperm to its receptor on the plasma membrane of the egg is an increase in the level of Ca^{2+} in the egg cytoplasm, probably as a consequence of stimulation of the hydrolysis of phosphatidylinositol 4,5-bisphosphate (PIP_2) (see Figure 16.36). One effect of this elevation in intracellular Ca^{2+} is the induction of surface alterations that prevent additional sperm from entering the egg. Because eggs are usually exposed to large numbers of sperm at one time, this is a critical event in ensuring the formation of a normal diploid embryo. These surface alterations result from the Ca^{2+}-induced exocytosis of secretory vesicles that are present in large numbers beneath the egg plasma membrane. Release of the contents of these vesicles alters the extracellular coat of the egg so as to block the entry of additional sperm.

The increase in cytosolic Ca^{2+} following fertilization also signals the completion of meiosis (**Figure 17.41**). In eggs arrested at metaphase II, the metaphase to anaphase transition is triggered by activation of the APC/C, resulting from Ca^{2+}-dependent phosphorylation and degradation of the

Figure 17.41 Fertilization and completion of meiosis Fertilization induces the transition from metaphase II to anaphase II, leading to completion of oocyte meiosis and emission of a second polar body (which usually degenerates). The sperm nucleus decondenses, so the fertilized egg (zygote) contains two haploid nuclei (male and female pronuclei). In mammals, the pronuclei replicate DNA as they migrate toward each other. They then initiate mitosis, with male and female chromosomes aligning on a common spindle. Completion of mitosis and cytokinesis thus gives rise to a two-cell embryo, with each cell containing a diploid genome.

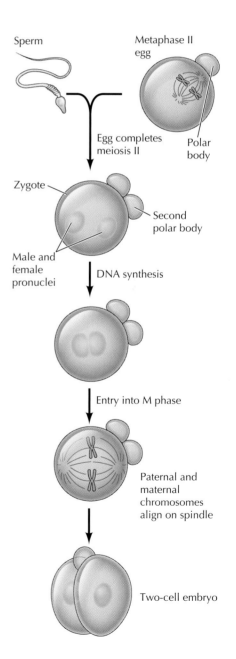

APC/C inhibitor Emi2/Erp1 responsible for maintaining metaphase II arrest (see Figure 17.39). The resultant degradation of cyclin B and cohesin leads to completion of the second meiotic division, with asymmetric cytokinesis (as in meiosis I) giving rise to a second small polar body.

Following completion of oocyte meiosis, the fertilized egg (now called a **zygote**) contains two haploid nuclei (called **pronuclei**), one derived from each parent. In mammals, the two pronuclei then enter S phase and replicate their DNA as they migrate toward each other. As they meet, the zygote enters M phase of its first mitotic division. The two nuclear envelopes break down, and the condensed chromosomes of both paternal and maternal origin align on a common spindle. Completion of mitosis then gives rise to two embryonic cells, each containing a new diploid genome. These cells then commence the series of embryonic cell divisions that eventually lead to the development of a new organism.

SUMMARY	KEY TERMS

The Eukaryotic Cell Cycle

- **Phases of the cell cycle:** Eukaryotic cell cycles are divided into four discrete phases: M, G_1, S, and G_2. M phase consists of mitosis, which is usually followed by cytokinesis. S phase is the period of DNA replication. See Animations 17.1, 17.2, and Video 17.1.

  mitosis, interphase, cytokinesis, M phase, G_1 phase, S phase, G_2 phase, flow cytometer, fluorescence-activated cell sorter

- **Regulation of the cell cycle by cell growth and extracellular signals:** Extracellular signals and cell size regulate progression through specific control points in the cell cycle. See Video 17.2.

  START, restriction point, G_0

- **Cell cycle checkpoints:** Checkpoints and feedback controls coordinate the events that take place during different phases of the cell cycle and arrest cell cycle progression if DNA is damaged. See Animation 17.3.

  cell cycle checkpoint, DNA damage checkpoint, spindle assembly checkpoint

Regulators of Cell Cycle Progression

- **Protein kinases and cell cycle regulation:** MPF is the key molecule responsible for regulating the G_2 to M transition in all eukaryotes. MPF is a dimer of the Cdk1 protein kinase and cyclin B. See Animation 17.4.

  maturation promoting factor (MPF), Cdk1, cyclin, anaphase-promoting complex/cyclosome (APC/C)

- **Families of cyclins and cyclin-dependent kinases:** Distinct pairs of cyclins and Cdk1-related protein kinases regulate progression through different stages of the cell cycle. The activity of Cdk's is regulated by association with cyclins, activating and inhibitory phosphorylations, and the binding of Cdk inhibitors.

  G_1 cyclin, Cln, Cdk, Cdk inhibitor (CKI)

- **Growth factors and the regulation of G_1 Cdk's:** Growth factors stimulate animal cell proliferation by inducing synthesis of the D-type cyclins. Cdk4, 6/cyclin D complexes then act to drive cells through the restriction point in G_1. A key substrate of Cdk4, 6/cyclin D complexes is the tumor suppressor protein Rb, which regulates transcription of genes required for cell cycle progression, including cyclin E. Activation of Cdk2/cyclin E complexes is then responsible for entry into S phase.

  Rb, tumor suppressor gene, E2F

- **S phase and regulation of DNA replication:** Cdk2/cyclin E, together with the DDK protein kinase, initiates DNA replication by activating the MCM helicase at origins of replication. Once MCM has been activated, reinitiation of replication is prevented by high Cdk activity until the cell has passed through mitosis.

SUMMARY	**KEY TERMS**

- **DNA *damage checkpoints:*** DNA damage or incompletely replicated DNA arrest cell cycle progression in G_1, S, and G_2. Cell cycle arrest is mediated by protein kinases that are activated by DNA damage and inhibit Cdc25 phosphatases, which are required for Cdk activation. In mammalian cells, arrest at the G_1 checkpoint is also mediated by p53, which induces synthesis of the Cdk inhibitor p21.

ATM, ATR, checkpoint kinase, p53

The Events of M Phase

- ***Stages of mitosis:*** Mitosis is conventionally divided into four stages: prophase, metaphase, anaphase, and telophase. The basic events of mitosis include chromosome condensation, formation of the mitotic spindle, nuclear envelope breakdown, and attachment of spindle microtubules to chromosomes at the kinetochore. Sister chromatids then separate and move to opposite poles of the spindle. Finally, nuclei re-form, the chromosomes decondense, and cytokinesis divides the cell in half. See Animation 17.5, and Videos 17.3,17.4.

prophase, metaphase, anaphase, telophase, centromere, kinetochore, centrosome, mitotic spindle, prometaphase

- ***Entry into mitosis:*** M phase is initiated by activation of Cdk1/cyclin B, Aurora, and Polo-like kinases, which are responsible for chromatin condensation, nuclear envelope breakdown, fragmentation of the Golgi apparatus, and reorganization of microtubules to form the mitotic spindle. The attachment of spindle microtubules to the kinetochores of sister chromatids then leads to their alignment on the metaphase plate.

Aurora kinase, Polo-like kinase, condensin, cohesin

- ***The spindle assembly checkpoint and progression to anaphase:*** Activation of the APC/C ubiquitin ligase leads to degradation of key regulatory proteins at the metaphase to anaphase transition. The activity of APC/C is inhibited until the cell passes the spindle assembly checkpoint and all chromosomes are properly aligned on the spindle. Ubiquitin-mediated proteolysis initiated by the APC/C then leads to the degradation of cohesin, breaking the link between sister chromatids at the onset of anaphase. The APC/C also ubiquitylates cyclin B, leading to inactivation of Cdk1 and exit from mitosis. See Videos 17.5 and 17.6.

mitotic checkpoint complex (MCC)

- ***Cytokinesis:*** Inactivation of Cdk1/cyclin B also triggers cytokinesis. In yeast and animal cells, cytokinesis results from contraction of a ring of actin and myosin filaments. In higher plant cells, cytokinesis results from the formation of a new cell wall and plasma membrane inside the cell. See Animation 17.6.

contractile ring

Meiosis and Fertilization

- ***The process of meiosis:*** Meiosis is a specialized cell cycle that gives rise to haploid daughter cells. A single round of DNA synthesis is followed by two sequential cell divisions. During meiosis I, homologous chromosomes first form pairs and then segregate to different daughter cells. Meiosis II then resembles a normal mitosis in which sister chromatids separate. See Animations 17.7–17.9.

meiosis, leptotene, zygotene, pachytene, diplotene, diakinesis, synapsis, synaptomenal complex, chiasmata

- ***Regulation of oocyte meiosis:*** Meiosis of vertebrate oocytes is regulated at two unique points in the cell cycle: the diplotene stage of meiosis I and metaphase of meiosis II. Metaphase II arrest results from inhibition of the APC/C by a protein kinase expressed in oocytes. See Animation 17.10.

polar body, cytostatic factor (CSF), Mos

- ***Fertilization:*** Fertilization triggers the resumption of oocyte meiosis by Ca^{2+}-dependent activation of the APC/C. The fertilized egg then contains two haploid nuclei, which form a new diploid genome and initiate embryonic cell divisions.

fertilization, zygote, pronucleus

Questions

1. In what ways are cells in G_0 and G_1 similar? How do they differ?

2. Radiation damages DNA and arrests cell cycle progression at checkpoints in G_1, S, and G_2. Why is this advantageous for the cell?

3. What are the mechanisms that regulate the activity of cyclin-dependent kinases (Cdk's)?

4. The spindle assembly checkpoint delays the onset of anaphase until all chromosomes are properly aligned on the spindle. What would be the result if a failure of this checkpoint allowed anaphase to initiate while one chromosome was attached to microtubules from only a single centrosome?

5. What cellular processes would be affected by expression of siRNA targeted against Cdk7?

6. The Cdk inhibitor p16 binds specifically to Cdk4, 6/cyclin D complexes. What would be the predicted effect of overexpression of p16 on cell cycle progression? Would overexpression of p16 affect a tumor cell lacking functional Rb protein?

7. *In vitro* mutagenesis of cloned lamin cDNAs has been used to generate mutants that cannot be phosphorylated by Cdk1. How would expression of these mutant lamins affect nuclear envelope breakdown at the end of prophase?

8. What substrates are phosphorylated by Cdk1/cyclin B to initiate mitosis?

9. A mutant of cyclin B resistant to ubiquitylation by the anaphase-promoting complex/cyclosome (APC/C) has been generated. How would expression of this mutant cyclin B affect the events at the metaphase to anaphase transition?

10. How does the activity of anaphase-promoting complex/cyclosome (APC/C) lead to the separation of sister chromatids?

11. Homologous recombination has been used to inactivate the *mos* gene in mice. What effect would you expect this to have on oocyte meiosis?

References and Further Reading   (Key review articles for each major section are highlighted in **bold**.)

The Eukaryotic Cell Cycle

Cimprich, K. A. and D. Cortez. 2008. ATR: an essential regulator of genome integrity. *Nature Rev. Mol. Cell Biol.* 9: 616–627. [R]

Forsburg, S. L. and P. Nurse. 1991. Cell cycle regulation in the yeasts *Saccharomyces cerevisiae* and *Schizosaccharomyces pombe*. *Ann. Rev. Cell Biol.* 7: 227–256. [R]

Harper, J. W. and S. J. Elledge. 2007. The DNA damage response: ten years after. *Mol. Cell* 28: 739–745. [R]

Hartwell, L. H. and T. A. Weinert. 1989. Checkpoints: Controls that ensure the order of cell cycle events. *Science* 246: 629–634. [R]

Jackson, S. P. and J. Bartek. 2009. The DNA-damage response in human biology and disease. *Nature* 461: 1071–1078. [R]

Morgan, D. O. 2007. *The Cell Cycle: Principles of Control.* New Science Press: London.

Norbury, C. and P. Nurse. 1992. Animal cell cycles and their control. *Ann. Rev. Biochem.* 61: 441–470. [R]

Pardee, A. B. 1989. G_1 events and the regulation of cell proliferation. *Science* 246: 603–608. [R]

Russell, P. 1998. Checkpoints on the road to mitosis. *Trends Biochem. Sci.* 23: 399–402. [R]

Regulators of Cell Cycle Progression

Arias, E. E. and J. C. Walter. 2007. Strength in numbers: preventing rereplication via multiple mechanisms in eukaryotic cells. *Genes Dev.* 21: 497–518. [R]

Blow, J. J. and A. Dutta. 2005. Preventing re-replication of chromosomal DNA. *Nature Rev. Mol. Cell Biol.* 6: 476–486. [R]

Besson, A., S. F. Dowdy and J. M. Roberts. 2008. CDK inhibitors: cell cycle regulators and beyond. *Dev. Cell* 14: 159–169. [R]

Bloom, J. and F. R. Cross. 2007. Multiple levels of cyclin specificity in cell-cycle control. *Nature Rev. Mol. Cell Biol.* 8: 149–159. [R]

Boutros, R., V. Lobjois and B. Ducommun. 2007. CDC25 phosphatases in cancer cells: key players? Good targets? *Nature Rev. Cancer* 7: 495–507. [R]

Bracken, A. P., M. Ciro, A. Cocito and K. Helin. 2004. E2F target genes: Unraveling the biology. *Trends Biochem. Sci.* 29: 409–417. [R]

Chen, H.-Z., Tsai, S.-Y. and G. Leone. 2009. Emerging roles of E2Fs in cancer: an exit from cell cycle control. *Nature Rev. Cancer* 9: 785–797. [R]

Dick, F. A. and S. M. Rubin. 2013. Molecular mechanisms underlying RB protein function. *Nature Rev. Mol. Cell Biol.* 14: 297–306. [R]

Evans, T., E. T. Rosenthal, J. Youngblom, D. Distel and T. Hunt. 1983. Cyclin: A protein specified by maternal mRNA in sea urchin eggs that is destroyed at each cleavage division. *Cell* 33: 389–396. [P]

Harper, J. W. and S. J. Elledge. 2007. The DNA damage response: ten years after. *Mol. Cell* 28: 739–745. [R]

Hartwell, L. H., R. K. Mortimer, J. Culotti and M. Culotti. 1973. Genetic control of the cell division cycle in yeast: V. Genetic analysis of *cdc* mutants. *Genetics* 74: 267–287. [P]

Hills, S. A. and J. F. X. Diffley. 2014. DNA replication and oncogene-induced replicative stess. *Current Biol.* 24: R435–R444. [R]

Lohka, M. J., M. K. Hayes and J. L. Maller. 1988. Purification of maturation-promoting factor, an intracellular regulator of early mitotic events. *Proc. Natl. Acad. Sci. U.S.A.* 85: 3009–3013. [P]

Malumbres, M. and M. Barbacid. 2009. Cell cycle, CDKs and cancer: a changing paradigm. *Nature Rev. Cancer* 9: 153–167. [R]

Masui, Y. and C. L. Markert. 1971. Cytoplasmic control of nuclear behavior during meiotic maturation of frog oocytes. *J. Exp. Zool.* 177: 129–146. [P]

Murray, A. W. 2004. Recycling the cell cycle: Cyclins revisited. *Cell* 116: 221–234. [R]

Pines, J. 2011. Cubism and the cell cycle: the many faces of the APC/C. *Nature Rev. Mol. Cell Biol.* 12: 427–438. [R]

Reinhardt, H. C. and M. B. Yaffe. 2009. Kinases that control the cell cycle in response to DNA damage: Chk1, Chk2, and MK2. *Current Opin. Cell Biol.* 21: 245–255. [R]

Sclafani, R. A. and T. M. Holzen. 2007. Cell cycle regulation of DNA replication. *Ann. Rev. Genet.* 41: 237–280. [R]

Smith, L. D. and R. E. Ecker. 1971. The interaction of steroids with *Rana pipiens* oocytes in the induction of maturation. *Dev. Biol.* 25: 232–247. [P]

Swenson, K. I., K. M. Farrell and J. V. Ruderman. 1986. The clam embryo protein cyclin A induces entry into M phase and the resumption of meiosis in *Xenopus* oocytes. *Cell* 47: 861–870. [P]

Vousden, K. H. and D. P. Lane. 2007. p53 in health and disease. *Nature Rev. Mol. Cell Biol.* 8: 275–283. [R]

The Events of M Phase

Archambault, V. and D. M. Glover. 2009. Polo-like kinases: conservation and divergence in their functions and regulation. *Nature Rev. Mol. Cell Biol.* 10: 265–275. [R]

Barr, F. A., H. H. W. Sillje and E. A. Nigg. 2004. Polo-like kinases and the orchestration of cell division. *Nature Rev. Mol. Cell Biol.* 5: 429–440. [R]

Carmena, M. and W. C. Earnshaw. 2003. The cellular geography of Aurora kinases. *Nature Rev. Mol. Cell Biol.* 4: 842–854. [R]

Cheeseman, I. M. and A. Desai. 2008. Molecular architecture of the kinetochore-microtubule interface. *Nature Rev. Mol. Cell Biol.* 9: 33–45. [R]

De Gramont, A. and O. Cohen-Fix. 2005. The many phases of anaphase. *Trends Biochem. Sci.* 30: 559–568. [R]

Foley, E. A. and T. M. Kapoor. 2013. Microtubule attachment and spindle assembly checkpoint signaling at the kinetochore. *Nature Rev. Mol. Cell Biol.* 14: 25–37. [R]

Glotzer, M. 2005. The molecular requirements for cytokinesis. *Science* 307: 1735–1739. [R]

Guttinger, S., E. Laurell and U. Kutay. 2009. Orchestrating nuclear envelope disassembly and reassembly during mitosis. *Nature Rev. Mol. Cell Biol.* 10: 178–191. [R]

Jongsma, M. L. M., I. Berlin and J. Neefjes. 2015. On the move: organelle dynamics during mitosis. *Trends Cell Biol.* 25: 112–124. [R]

Jurgens, G. 2005. Plant cytokinesis: Fission by fusion. *Trends Cell Biol.* 15: 277–283. [R]

Kline-Smith, S. L. and C. E. Walczak. 2004. Mitotic spindle assembly and chromosome segregation: Refocusing on microtubule dynamics. *Mol. Cell* 15: 317–327. [R]

Lara-Gonzalez, P., F. G. Westhorpe and S. S. Taylor. 2012. The spindle assembly checkpoint. *Current Biol.* 22: R966–R980. [R]

Lens, S. M. A., E. E. Voest and R. H. Medema. 2010. Shared and separate functions of polo-like kinases and aurora kinases in cancer. *Nature Rev. Cancer* 10: 825–841. [R]

Losada, A. and T. Hirano. 2005. Dynamic molecular linkers of the genome: The first decade of SMC proteins. *Genes Dev.* 19: 1269–1287. [R]

Lowe, M. and F. A. Barr. 2007. Inheritance and biogenesis of organelles in the secretory pathway. *Nature Rev. Mol. Cell Biol.* 8: 429–439. [R]

Ma, H. T. and R. Y. C. Poon. 2011. How protein kinases coordinate mitosis in animal cells. *Biochem. J.* 435: 17–31. [R]

Musacchio, A. and E. D. Salmon. 2007. The spindle-assembly checkpoint in space and time. *Nature Rev. Mol. Cell Biol.* 8: 379–393. [R]

Peters, J.-M., A. Tedeschi and J. Schmitz. 2008. The cohesin complex and its roles in chromosome biology. *Genes Dev.* 22: 3089–3114. [R]

Petronczki, M., P. Lenart and J.-M. Peters. 2008. Polo on the rise—from mitotic entry to cytokinesis with Plk1. *Dev. Cell* 14:646–659. [R]

Rieder, C. L. and A. Khodjakov. 2003. Mitosis through the microscope: Advances in seeing inside live dividing cells. *Science* 300: 91–96. [R]

Sivakumar, S. and G. J. Gorbsky. 2015. Spatiotemporal regulation of the anaphase-promoting complex in mitosis. *Nature Rev. Mol. Cell Biol.* 16: 82–94. [R]

Sullivan, M. and D. O. Morgan. 2007. Finishing mitosis, one step at a time. *Nature Rev. Mol. Cell Biol.* 8: 894–903. [R]

Walczak, C. E., S. Cai and A. Khodjakov. 2010. Mechanisms of chromosome behaviour during mitosis. *Nature Mol. Cell Biol.* 11: 91–102. [R]

Wood, A. J., A. F. Severson and B. J. Meyer. 2010. Condensin and cohesin complexity: the expanding repertoire of functions. *Nature Rev. Genet.* 11: 391–404. [R]

Zitouni, S., C. Nabais, S. C. Jana, A. Guerrero and M. Bettencourt-Dias. 2014. Polo-like kinases: structural variations lead to multiple functions. *Nature Rev. Mol. Cell Biol.* 15: 433–452. [R]

Meiosis and Fertilization

Baudat, F., Y. Imai and B. de Massy. 2013. Meiotic recombination in mammals: localization and regulation. *Nature Rev. Genet.* 14: 794–806. [R]

Bhalla, N. and A. F. Dernburg. 2008. Prelude to a division. *Ann. Rev. Cell Biol.* 24: 397–424. [R]

Clift, D. and M. Schuh. 2013. Restarting life: fertilization and the transition from meiosis to mitosis. *Nature Rev. Mol. Cell Biol.* 14: 549–562. [R]

Hawley, R. S. and W. D. Gilliland. 2009. Homologue pairing: getting it right. *Nature Cell Biol.* 11: 917–918. [R]

Page, S. L. and R. S. Hawley. 2003. Chromosome choreography: The meiotic ballet. *Science* 301: 785–789. [R]

Pawlowski, W. P. and W. Z. Cande. 2005. Coordinating the events of the meiotic prophase. *Trends Cell Biol.* 15: 674–681. [R]

Pesin, J. A. and T. L. Orr-Weaver. 2008. Regulation of APC/C activators in mitosis and meiosis. *Ann. Rev. Cell Biol.* 24: 475–499. [R]

Petronczki, M., M. F. Siomos and K. Nasmyth. 2003. Un ménage à quatre: The molecular biology of chromosome segregation in meiosis. *Cell* 112: 423–440. [R]

Cell Death and Cell Renewal

C ell death and cell proliferation are balanced throughout the life of multicellular organisms. Animal development begins with the rapid proliferation of embryonic cells, which then differentiate to produce the many specialized types of cells that make up adult tissues and organs. Whereas the nematode *C. elegans* consists of only 959 somatic cells, humans possess a total of approximately 10^{14} cells, consisting of more than 200 differentiated cell types. Starting from only a single cell—the fertilized egg—all the diverse cell types of the body are produced and organized into tissues and organs. This complex process of development involves not only cell proliferation and differentiation but also cell death. Although cells can die as a result of unpredictable traumatic events, such as exposure to toxic chemicals, most cell deaths in multicellular organisms occur by a normal physiological process of programmed cell death, which plays a key role both in embryonic development and in adult tissues.

In adult organisms, cell death must be balanced by cell renewal, and most tissues contain stem cells that are able to replace cells that have been lost. Abnormalities of cell death are associated with a wide variety of illnesses, including cancer, autoimmune disease, and neurodegenerative disorders, such as Parkinson's and Alzheimer's disease. Conversely, the ability of stem cells to proliferate and differentiate into a wide variety of cell types has generated enormous interest in the possible use of these cells to replace damaged tissues. The mechanisms and regulation of cell death and cell renewal have therefore become areas of research at the forefront of biology and medicine.

Programmed Cell Death

Programmed cell death is carefully regulated so that the fate of individual cells meets the needs of the organism as a whole. In adults, programmed cell death is responsible for balancing cell proliferation and maintaining constant cell numbers in tissues undergoing cell turnover. For example, about 5×10^{11} blood cells are eliminated daily in humans by programmed cell death, balancing their continual production in the bone marrow. In addition, programmed cell death provides a defense mechanism by which damaged and potentially dangerous cells can be eliminated for the good of the organism as a whole. Virus-infected cells frequently undergo programmed cell death, thereby preventing the production of new virus particles and limiting spread of the virus through the host organism. Other types of cellular insults, such as DNA damage, also induce programmed cell death. In the case of DNA damage, programmed cell death may eliminate cells carrying potentially

Video 18.1

sites.sinauer.com/cooper7e/v18.1

Apoptosis Programmed cell death is an active process which usually proceeds by a distinct series of cellular changes known as apoptosis.

harmful mutations, including cells with mutations that might lead to the development of cancer.

During development, programmed cell death plays a key role by eliminating unwanted cells from a variety of tissues. For example, programmed cell death is responsible for the elimination of larval tissues during amphibian and insect metamorphosis, as well as for the elimination of tissue between the digits during the formation of fingers and toes. Another well-characterized example of programmed cell death is provided by development of the mammalian nervous system. Neurons are produced in excess, and up to 50% of developing neurons are eliminated by programmed cell death. Those that survive are selected for having made the correct connections with their target cells, which secrete growth factors that signal cell survival by blocking the neuronal cell death program. The survival of many other types of cells in animals is similarly dependent on growth factors or contacts with neighboring cells or the extracellular matrix, so programmed cell death is thought to play an important role in regulating the associations between cells in tissues.

The events of apoptosis

In contrast with the accidental death of cells that results from an acute injury (**necrosis**), programmed cell death is an active process, which usually proceeds by a distinct series of cellular changes known as **apoptosis**, first described in 1972 (**Figure 18.1**). During apoptosis, chromosomal DNA is usually fragmented as a result of cleavage between nucleosomes. The chromatin condenses and the nucleus then breaks up into small pieces. Finally, the cell itself shrinks and breaks up into membrane-enclosed fragments called apoptotic bodies.

Apoptotic cells and cell fragments are efficiently recognized and phagocytosed by both macrophages and neighboring cells, so cells that die by apoptosis are rapidly removed from tissues. In contrast, cells that die by necrosis swell and lyse, releasing their contents into the extracellular space and causing inflammation. The removal of apoptotic cells is mediated by the expression of so-called "eat me" signals on the cell surface. These signals include phosphatidylserine, which is normally restricted to the inner leaflet of the plasma membrane (see Figure 14.2). During apoptosis, phosphatidylserine becomes expressed on the cell surface, where it is recognized by receptors expressed by phagocytic cells (**Figure 18.2**).

Pioneering studies of programmed cell death during the development of *C. elegans* provided the critical initial insights that led to understanding the molecular mechanism of apoptosis. These studies in the laboratory of Robert Horvitz initially identified three genes that play key roles in regulating and executing apoptosis. During normal nematode development, 131 somatic cells out of a total of 1090 are eliminated by programmed cell death, yielding the 959 somatic cells in the adult worm. The death of these cells is highly specific, such that the same cells always die in developing embryos. Based on this developmental specificity, Horvitz undertook a genetic analysis of cell death in *C. elegans* with the goal of identifying the genes responsible for these developmental cell deaths. In 1986 mutagenesis of *C. elegans* identified two genes that were required for developmental *cell* death (*ced-3* and *ced-4*). If either *ced-3* or *ced-4* was inactivated by mutation, the normal programmed cell deaths did not take place. A third gene, *ced-9*, functioned as a negative regulator of apoptosis. If *ced-9* was inactivated by mutation, the cells that would normally survive failed to do so. Instead, they also underwent apoptosis, leading to death of the developing animal. Conversely, if *ced-9* was expressed at an abnormally high level, the normal programmed cell deaths

(A)

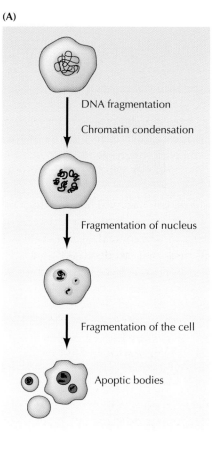

DNA fragmentation

Chromatin condensation

Fragmentation of nucleus

Fragmentation of the cell

Apoptic bodies

(B)

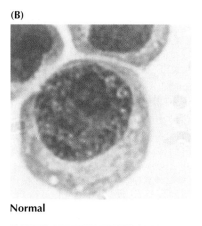

Normal

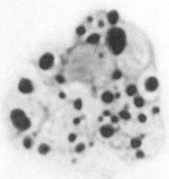

Apoptotic

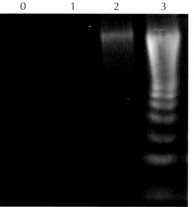

0   1   2   3

Figure 18.1 Apoptosis (A) Diagrammatic representation of the events of apoptosis. (B) Light micrographs of normal and apoptotic human cells illustrating chromatin condensation and nuclear fragmentation. (C) Gel electrophoresis of DNA from apoptotic cells, showing its degradation to fragments corresponding to multiples of 200 base pairs (the size of nucleosomes) at 0–3 hours following induction of apoptosis. (B, courtesy of D. R. Green/La Jolla Institute for Allergy & Immunology; C, courtesy of Ken Adams, Boston University.)

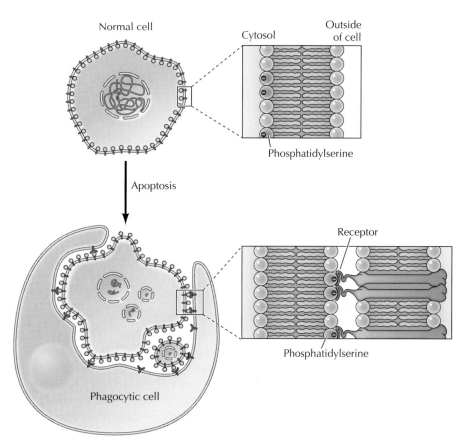

Normal cell

Cytosol   Outside of cell

Phosphatidylserine

Apoptosis

Receptor

Phosphatidylserine

Phagocytic cell

Figure 18.2 Phagocytosis of apoptotic cells Apoptotic cells and cell fragments are recognized and engulfed by phagocytic cells. One of the signals recognized by phagocytes is phosphatidylserine on the cell surface. In normal cells, phosphatidylserine is restricted to the inner leaflet of the plasma membrane, but it becomes expressed on the cell surface during apoptosis.

Identification of Genes Required for Programmed Cell Death

Genetic Control of Programmed Cell Death in the Nematode *C. elegans*

Hilary M. Ellis and H. Robert Horvitz

Massachusetts Institute of Technology, Cambridge, MA

Cell, 1986, Volume 44, pages 817–829

The Context

By the 1960s cell death was recognized as a normal event during animal development, implying that it was a carefully regulated process with specific cells destined to die. The simple nematode *C. elegans*, which has been a critically important model system in developmental biology, proved to be key to understanding both the regulation and the mechanism of such programmed cell deaths. Microscopic analysis in the 1970s established a complete map of *C. elegans* development so that the embryonic origin and fate of each cell was known. Importantly, *C. elegans* development included a very specific pattern of programmed cell deaths. In particular, John Sulston and H. Robert Horvitz reported in 1977 that the development of adult worms (consisting of 959 somatic cells) involved the programmed death of 131 cells out of 1090 that were initially produced. The same cells died in all embryos, indicating that the death of these cells was a normal event during development, with cell death being a specific developmental fate. It was also notable that all of these dying cells underwent a similar series of morphological changes, suggesting that these programmed cell deaths occurred by a common mechanism.

Based on these considerations, Horvitz undertook a genetic analysis with the goal of characterizing the mechanism and regulation of programmed cell death during *C. elegans* development. In the experiments reported in this 1986 paper, Hilary Ellis and Horvitz identified two genes that were required for all of the programmed cell deaths that took place during development of the nematode. The identification and characterization of these genes was a critical first step

leading to our current understanding of the molecular biology of apoptosis.

The Experiments

Cells undergoing programmed cell death in *C. elegans* can readily be identified as highly refractile cells by microscopic examination, so Ellis and Horvitz were able to use this as an assay to screen for mutant animals in which the normal cell deaths did not occur. To isolate mutants that displayed abnormalities in cell death, they treated nematodes with the chemical mutagen ethyl methanesulfonate, which reacts with DNA. The progeny of approximately 4000 worms were examined to identify dying cells, and two mutant strains were found in which the expected cell deaths did not take place (see figure). Both of these mutant strains harbored recessive mutations of the same gene, which was called *ced-3*. Further studies indicated that the mutations in *ced-3* blocked all of the 131 programmed cell deaths that would normally occur during development.

Continuing studies identified an additional mutation that was similar to *ced-3* in preventing programmed cell death. However, this mutation was in a different gene, which was located on a different chromosome than *ced-3*. This second gene was called *ced-4*. Similar to mutations in *ced-3*, recessive mutations in *ced-4* were found to block all programmed cell deaths in the worm.

The Impact

The isolation of *C. elegans* mutants by Ellis and Horvitz provided the first identification of genes that were involved in the process of programmed cell death. The proteins encoded by the *ced-3* and *ced-4* genes, as well as by the *ced-9* gene (which was subsequently identified by Horvitz and colleagues), proved to be prototypes of the central regulators and effectors of apoptosis that are highly conserved in evolution. The cloning and sequencing of *ced-3* revealed that it was related to a protease that had been previously identified in mammalian cells, and which became the first member of the caspase family. The *C. elegans ced-9* gene was related to the *bcl-2* oncogene, first isolated from a human B cell lymphoma, which had the unusual property of inhibiting apoptosis rather than

H. Robert Horvitz

stimulating cell proliferation. And *ced-4* was found to encode an adaptor protein related to mammalian Apaf-1, which is required for caspase activation. The identification of these genes in *C. elegans* thus led the way to understanding the molecular basis of apoptosis, with broad implications both for development and for the maintenance of normal adult tissues. Since abnormalities of apoptosis contribute to a wide variety of diseases, including cancer, autoimmune disease, and neurodegenerative disorders, the seminal findings of Ellis and Horvitz have impacted a wide range of areas in biology and medicine.

(A)

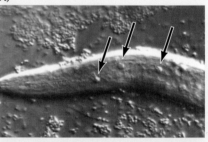

(B)

Photomicrographs of a normal worm (A) and a *ced-3* mutant (B) Dying cells are highly refractile and are indicated by arrows in (A). These cells are not present in the mutant animal.

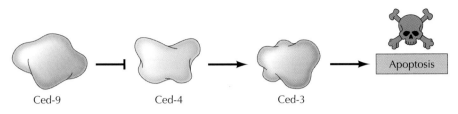

Ced-9   Ced-4   Ced-3

Figure 18.3 Programmed cell death in C. elegans   Genetic analysis identified three genes that play key roles in programmed cell death during development of *C. elegans*. Two genes, *ced-3* and *ced-4*, are required for cell death, whereas *ced-9* inhibits cell death. The Ced-9 protein inhibits Ced-4, which activates Ced-3.

failed to occur. Further studies indicated that the proteins encoded by these genes acted in a pathway with Ced-4 acting to stimulate Ced-3, and Ced-9 inhibiting Ced-4 (Figure 18.3). Genes related to *ced-3*, *ced-4*, and *ced-9* have also been identified in *Drosophila* and mammals and found to encode proteins that represent conserved effectors and regulators of apoptosis induced by a variety of stimuli.

Caspases: The executioners of apoptosis

The molecular cloning and nucleotide sequencing of the *ced-3* gene indicated that it encoded a protease, providing the first insight into the molecular mechanism of apoptosis. We now know that Ced-3 is the prototype of a family of more than a dozen proteases, known as **caspases** because they have cysteine (C) residues at their active sites and cleave after aspartic acid (Asp) residues in their substrate proteins. The caspases are the ultimate effectors or executioners of programmed cell death, bringing about the events of apoptosis by cleaving more than 100 different target proteins (Figure 18.4). One key target of the caspases is an inhibitor of a DNase, which when activated is responsible for fragmentation of nuclear DNA. In addition, caspases cleave nuclear lamins, leading to fragmentation of the nucleus; cytoskeletal proteins, leading to disruption of the cytoskeleton, blebbing (irregular bulging) of the plasma membrane, and cell fragmentation; and Golgi matrix proteins,

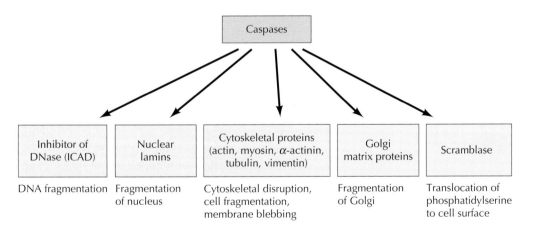

| Inhibitor of DNase (ICAD) | Nuclear lamins | Cytoskeletal proteins (actin, myosin, α-actinin, tubulin, vimentin) | Golgi matrix proteins | Scramblase |

DNA fragmentation | Fragmentation of nucleus | Cytoskeletal disruption, cell fragmentation, membrane blebbing | Fragmentation of Golgi | Translocation of phosphatidylserine to cell surface

Figure 18.4 Caspase targets   Caspases cleave over 100 cellular proteins to induce the morphological alterations characteristic of apoptosis. Caspase targets include an inhibitor of DNase (ICAD), nuclear lamins, cytoskeletal proteins, Golgi matrix proteins, and a scramblase that promotes translocation of phosphatidyl-serine to the outer leaflet of the plasma membrane.

leading to fragmentation of the Golgi apparatus. Caspases also cleave and activate a "scramblase" that translocates phosphatidylserine from the inner to the outer leaflet of the plasma membrane.

Ced-3 is the only caspase in *C. elegans*. However, *Drosophila* and mammals contain families of at least seven caspases—classified as either *initiator* or *effector* caspases—that function in a cascade to bring about the events of apoptosis. All caspases are synthesized as inactive precursors (called procaspases) that can be converted to the active form by proteolytic cleavage, catalyzed by other caspases. Initiator caspases are activated directly in response to the various signals that induce apoptosis, as discussed later in this chapter. The initiator caspases then cleave and activate the effector caspases, which are responsible for digesting the cellular target proteins that mediate the events of apoptosis (see Figure 18.4). The activation of an initiator caspase therefore starts off a chain reaction of caspase activation leading to death of the cell.

Genetic analysis in *C. elegans* initially suggested that Ced-4 functioned as an activator of the caspase Ced-3. Subsequent studies have shown that Ced-4 and its mammalian homolog (Apaf-1) bind to caspases and promote their activation. Ced-4 and Apaf-1 are in turn controlled by proteins encoded by members of a large gene family related to *ced-9*, as discussed below.

Central regulators of apoptosis: the Bcl-2 family

The third gene identified as a key regulator of programmed cell death in *C. elegans, ced-9*, was found to be closely related to a mammalian gene called *bcl-2*, which was first identified in 1985 as an oncogene that contributed to the development of human B cell lymphomas (cancers of B lymphocytes). In contrast with other oncogene proteins, such as Ras, that stimulate cell proliferation (see Molecular Medicine, Chapter 16), **Bcl-2** was found to inhibit apoptosis. Ced-9 and Bcl-2 were thus similar in function, and the role of Bcl-2 as a regulator of apoptosis first focused attention on the importance of cell survival in cancer development. As discussed further in the next chapter, we now recognize that cancer cells are generally defective in the normal process of programmed cell death and that their inability to undergo apoptosis is as important as their uncontrolled proliferation in the development of malignant tumors.

Mammals encode a family of approximately 20 proteins related to Bcl-2, which are divided into three functional groups (Figure 18.5). Some members of the Bcl-2 family (antiapoptotic family members)—like Bcl-2 itself—function as inhibitors of apoptosis and programmed cell death. Other family members, however, are proapoptotic proteins that act to induce caspase activation and promote cell death. There are two groups of these proapoptotic proteins, which differ in function as well as in their extent of homology to Bcl-2. Bcl-2 and the other antiapoptotic family members share four conserved regions called Bcl-2 homology (BH) domains. One group of proapoptotic family members, the proapoptotic effector proteins also have four BH domains, whereas the second group, the "BH3-only" proteins, have only one (the BH3 domain).

The fate of the cell—life or death—is determined by the balance of activity of these three groups of

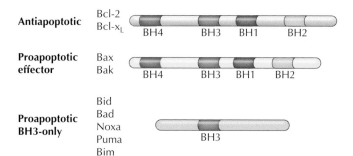

Figure 18.5 The Bcl-2 family The Bcl-2 family of proteins is divided into three functional groups. Antiapoptotic proteins (e.g., Bcl-2 and Bcl-x$_L$) and proapoptotic effector proteins (Bax and Bak) have four Bcl-2 homology domains (BH1–BH4). The proapoptotic BH3-only proteins (e.g., Bid, Bad, Noxa, Puma, and Bim) have only one homology domain (BH3).

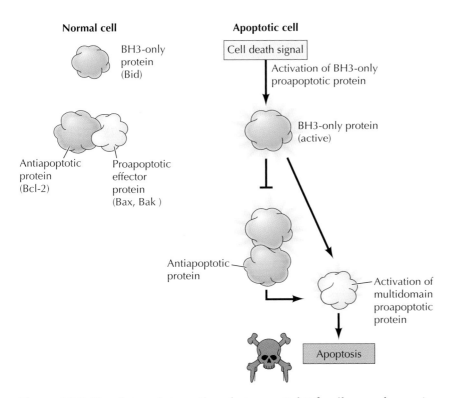

Figure 18.6 Regulatory interactions between Bcl-2 family members   In normal cells, the BH3-only proapoptotic proteins (e.g., Bid) are inactive, and the proapoptotic effector proteins (Bax and Bak) are inhibited by interaction with antiapoptotic proteins (e.g., Bcl-2). Cell death signals activate the BH3-only proteins, which then antagonize the antiapoptotic proteins as well as activating Bax and Bak directly, leading to cell death.

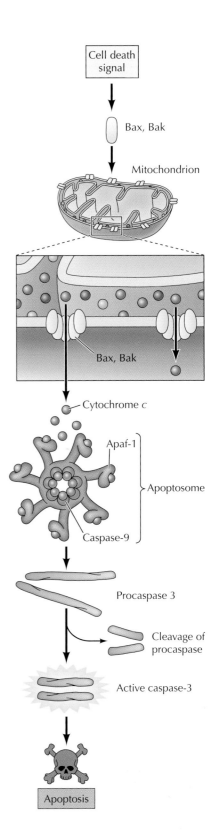

proapoptotic and antiapoptotic Bcl-2 family members, which act to regulate one another (**Figure 18.6**). The proapoptotic effector proteins, Bax and Bak, are the downstream effectors that directly induce apoptosis. They are inhibited by interactions with the *antiapoptotic* family members, such as Bcl-2. The proapoptotic *BH3-only* family members are upstream members of the cascade, regulated by the signals that induce cell death (e.g., DNA damage) or cell survival (e.g., growth factors). When activated, the BH3-only family members initiate apoptosis by two mechanisms: (1) they antagonize the antiapoptotic Bcl-2 family members and (2) they activate the proapoptotic effector proteins, Bax and Bak.

In mammalian cells, members of the Bcl-2 family act at mitochondria, which play a central role in controlling programmed cell death (**Figure 18.7**).

Figure 18.7 The mitochondrial pathway of apoptosis   In mammalian cells, many cell death signals induce apoptosis as a result of damage to mitochondria. When active, the proapoptotic effector proteins Bax and Bak form oligomers in the outer membrane of mitochondria, resulting in the release of cytochrome *c* from the intermembrane space. Release of cytochrome *c* leads to the formation of apoptosomes containing Apaf-1 and caspase-9 in which caspase-9 is activated. Caspase-9 then activates downstream caspases, such as caspase-3, by proteolytic cleavage.

Animation 18.2

sites.sinauer.com/cooper7e/a18.2

The Mitochondrial Pathway of Apoptosis Many forms of cell stress activate the intrinsic pathway of apoptosis—a pathway that leads to the release of cytochrome *c* from mitochondria, the activation of caspase-9, and the subsequent death of the cell.

When activated, Bax and Bak oligomerize to form pores in the mitochondrial outer membrane, leading to the release of cytochrome *c* from the mitochondrial intermembrane space. The release of cytochrome *c* then triggers caspase activation. In particular, the key initiator caspase in mammalian cells (caspase-9) is activated by forming a complex with Apaf-1 in a multisubunit complex called the **apoptosome**. In mammals, formation of this complex also requires cytochrome *c*. Under normal conditions of cell survival, cytochrome *c* is localized to the mitochondrial intermembrane space (see Figure 12.1) while Apaf-1 and caspase-9 are found in the cytosol, so caspase-9 remains inactive. Activation of Bax or Bak results in the release of cytochrome *c* to the cytosol, where it binds to Apaf-1 and triggers apoptosome formation and caspase-9 activation. Once activated, caspase-9 cleaves and activates downstream effector caspases, such as caspase-3 and caspase-7, eventually resulting in cell death.

Caspases are also regulated by a group of proteins called the **IAP** (for *i*nhibitor of *ap*optosis) family. Members of the IAP family directly interact with caspases and suppress apoptosis by either inhibiting caspase activity or by targeting caspases for ubiquitylation and degradation in the proteasome. IAPs are present in both *Drosophila* and mammals (but not *C. elegans*), and regulation of their activity or expression provides another mechanism for controlling apoptosis. Regulation of IAPs is particularly important in *Drosophila*, where initiator caspases are chronically activated but held in check by IAPs (Figure 18.8). Many signals that induce apoptosis in *Drosophila* function by activating proteins that inhibit the IAPs, thus leading to caspase activation. In mammalian cells, the permeabilization of mitochondria by Bax or Bak results not only in the release of cytochrome *c* but also of IAP inhibitors that help to stimulate caspase activity.

Signaling pathways that regulate apoptosis

Programmed cell death is regulated by the integrated activity of a variety of signaling pathways, some acting to induce cell death and others acting to promote cell survival. These signals control the fate of individual cells, so that cell survival or elimination is determined by the needs of the organism as a whole. The pathways that induce apoptosis in mammalian cells are grouped as *intrinisic* or *extrinsic* pathways, which differ in their involvement of Bcl-2 family proteins and in the identity of the caspase that initiates cell death. The intrinsic pathway is activated by DNA damage and other forms of cell stress, whereas the extrinsic pathway is activated by signals from other cells.

One important role of apoptosis is the elimination of damaged cells, so apoptosis is stimulated by many forms of cell stress, including DNA damage, viral infection, and growth factor deprivation. These stimuli activate the intrinsic pathway of apoptosis, which leads to release of cytochrome *c* from mitochondria and activation of caspase-9 (see Figure 18.7). As illustrated in the following examples, the multiple signals that activate this pathway converge on regulation of the BH3-only members of the Bcl-2 family.

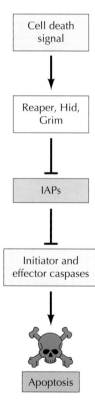

Figure 18.8 Regulation of caspases by IAPs in *Drosophila* Many signals that induce apoptosis in *Drosophila* function by activating members of a family of proteins (Reaper, Hid, and Grim) that inhibit the IAPs, resulting in caspase activation. Both initiator and effector caspases are inhibited by IAPs

Figure 18.9 Role of p53 in DNA damage-induced apoptosis   DNA damage leads to activation of the ATM and Chk2 protein kinases, which phosphorylate and stabilize p53, resulting in rapid increases in p53 levels. The p53 protein then activates transcription of genes encoding the proapoptotic BH3-only proteins PUMA and Noxa, leading to cell death.

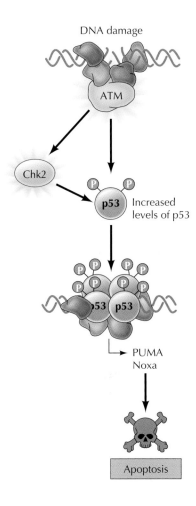

DNA damage is a particularly dangerous form of cell stress, because cells with damaged genomes may have suffered mutations that can lead to the development of cancer. DNA damage is thus one of the principal triggers of programmed cell death, leading to the elimination of cells carrying potentially harmful mutations. As discussed in Chapter 17, several cell cycle checkpoints halt cell cycle progression in response to damaged DNA, allowing time for the damage to be repaired. In mammalian cells, a major pathway leading to cell cycle arrest in response to DNA damage is mediated by the transcription factor **p53**. The ATM and Chk2 protein kinases, which are activated by DNA damage, phosphorylate and stabilize p53. The resulting increase in p53 leads to transcriptional activation of p53 target genes. These include the Cdk inhibitor p21, which inhibits Cdk2/cyclin E complexes, halting cell cycle progression in G_1 (see Figure 17.20). However, activation of p53 by DNA damage can also lead to apoptosis (**Figure 18.9**). The induction of apoptosis by p53 results, at least in part, from transcriptional activation of genes encoding the proapoptotic BH3-only Bcl-2 family members PUMA and Noxa. Increased expression of these BH3-only proteins leads to activation of Bax and Bak, release of cytochrome *c* from mitochondria, and activation of caspase-9. Thus p53 mediates both cell cycle arrest and apoptosis in response to DNA damage. Whether DNA damage in a given cell leads to apoptosis or reversible cell cycle arrest depends on the extent of damage and the resulting level of p53 induction, as well as on the influence of other life/death signals being received by the cell.

Growth factor deprivation is another form of cell stress that activates the intrinsic pathway of apoptosis. In this case, apoptosis is controlled by signaling pathways that promote cell survival by inhibiting apoptosis in response to growth factor stimulation. These signaling pathways control the fate of a wide variety of cells whose survival is dependent on extracellular growth factors or cell–cell interactions. As already noted, a well-characterized example of programmed cell death in development is provided by the vertebrate nervous system. About 50% of neurons die by apoptosis, with the survivors having received sufficient amounts of survival signals from their target cells. These survival signals are polypeptide growth factors related to nerve growth factor (NGF), which induces both neuronal survival and differentiation by activating a receptor tyrosine kinase. Other types of cells are similarly dependent upon growth factors or cell contacts that activate nonreceptor tyrosine kinases associated with integrins. Indeed, most cells in higher animals are programmed to undergo apoptosis unless cell death is actively suppressed by survival signals from other cells.

One of the major intracellular signaling pathways responsible for promoting cell survival is initiated by the enzyme **PI 3-kinase**, which is activated by either tyrosine kinases or G protein-coupled receptors. PI 3-kinase phosphorylates the membrane phospholipid PIP_2 to form PIP_3, which activates the serine/threonine kinase **Akt** (see Figure 16.32). Akt then phosphorylates a number of proteins that regulate apoptosis (Figure 18.10). One key substrate for Akt is the proapoptotic BH3-only Bcl-2 family member called Bad. Phosphorylation of Bad by Akt creates a binding site for 14-3-3 chaperone proteins that sequester Bad in an inactive form, so phosphorylation of Bad by Akt inhibits apoptosis and promotes cell survival. Bad is similarly phosphorylated by protein kinases of other growth factor-induced signaling pathways, including the Ras/Raf/MEK/ERK pathway, so it serves as a convergent regulator of growth factor signaling in mediating cell survival.

Other targets of Akt, including the FOXO transcription factors, also play key roles in cell survival. Phosphorylation of FOXO by Akt creates a binding site for 14-3-3 proteins, which sequester FOXO in an inactive form in the cytoplasm (see Figure 16.33). In the absence of growth factor signaling and Akt activity, FOXO is released from 14-3-3 and translocates to the nucleus, stimulating transcription of proapoptotic genes, including the gene

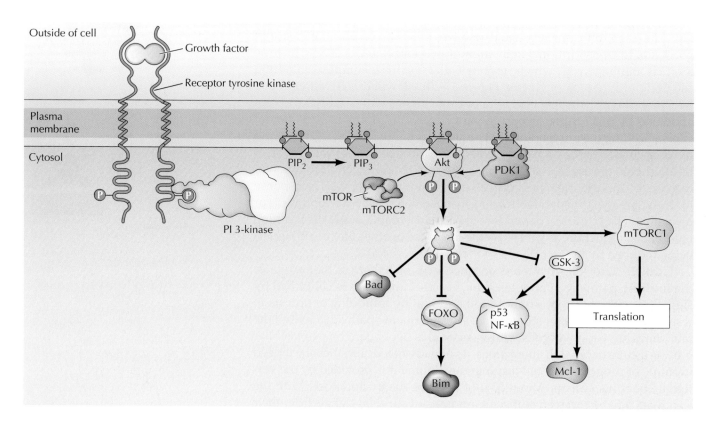

Figure 18.10 The PI 3-kinase pathway and cell survival Many growth factors that signal cell survival activate receptor tyrosine kinases, leading to activation of PI 3-kinase, formation of PIP$_3$, and activation of the protein kinase Akt. Akt then phosphorylates a number of proteins that contribute to cell survival. Phosphorylation of the BH3-only protein Bad maintains it in an inactive state, as does phosphorylation of the FOXO transcription factors. In the absence of Akt signaling, Bad promotes apoptosis and FOXO stimulates transcription of another proapoptotic BH3-only protein, Bim. Additional targets of Akt include the proapoptotic protein kinase GSK-3 and additional transcription factors, such as p53 and NF-κB. Phosphorylation by GSK-3 also targets the antiapoptotic protein Mcl-1 for degradation, and translational regulation by both GSK-3 and mTORC1 (see Figure 16.34) affect cell survival.

encoding the BH3-only protein, Bim. Akt and its downstream target GSK-3 also regulate other transcription factors with roles in cell survival, including p53 and NF-κB, which control the expression of additional Bcl-2 family members. In addition, the antiapoptotic Bcl-2 family member Mcl-1 is targeted for ubiquitylation and degradation by GSK-3, and its level is also modulated via translational regulation by both GSK-3 and the mTOR pathway (see Figure 16.34). These multiple effects on members of the Bcl-2 family converge to regulate the intrinsic pathway of apoptosis, controlling the activation of caspase-9 and cell survival in response to growth factor stimulation.

In contrast with the cell stress and growth factor signaling pathways that regulate the intrinsic pathway of apoptosis, some secreted polypeptides activate receptors that induce cell death via the extrinsic pathway of apoptosis. These receptors directly activate a distinct initiator caspase—caspase-8 (Figure 18.11). The polypeptides that signal cell death by this pathway belong to the **tumor necrosis factor (TNF)** family. They bind to members of the TNF receptor family, which can induce apoptosis in a variety of cell types. One of the best characterized members of this family is the cell surface receptor Fas, which plays important roles in controlling cell death in the immune system. For example, apoptosis induced by activation of Fas is responsible for killing target cells of the immune system, such as cancer cells or virus-infected cells, as well as for eliminating excess lymphocytes at the end of an immune response.

TNF and related family members consist of three identical polypeptide chains, and their binding induces receptor trimerization. The cytoplasmic portions of the receptors bind adaptor molecules that in turn bind caspase-8. This leads

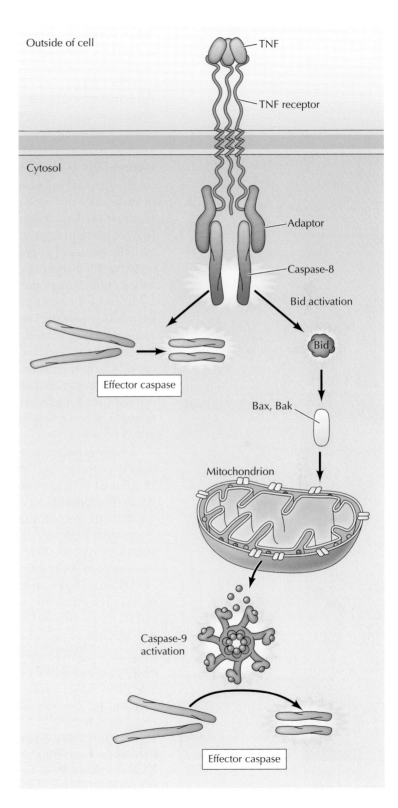

Figure 18.11 Cell death receptors TNF and other cell death receptor ligands consist of three polypeptide chains, so their binding to cell death receptors induces receptor trimerization. Caspase-8 is recruited to the receptor and activated by adaptor molecules (the extrinsic pathway of apoptosis). Once activated, caspase-8 can directly cleave and activate effector caspases. In addition, caspase-8 cleaves the BH3-only protein Bid, which activates the mitochondrial pathway of apoptosis, leading to caspase-9 activation (see Figure 18.7).

to activation of caspase-8, which can then cleave and activate downstream effector caspases. In some cells, caspase-8 activation and subsequent activation of caspases-3 and -7 is sufficient to induce apoptosis directly. In other cells, however, amplification of the signal is needed. This results from caspase-8 cleavage of the proapoptotic BH3-only protein Bid, leading to Bid activation, permeabilization of mitochondria, and activation of caspase-9, thus amplifying the caspase cascade initiated by direct activation of caspase-8 at cell death receptors.

Alternative pathways of programmed cell death

Although apoptosis is the most common form of regulated or programmed cell death, programmed cell death can also occur by alternative, non-apoptotic mechanisms. One of these alternative pathways of regulated cell death is **autophagy**. As discussed in Chapter 9, autophagy provides a mechanism for the gradual turnover of the cell's components by the uptake of proteins or organelles into vesicles (autophagosomes) that fuse with lysosomes (see Figure 9.48). One role of autophagy is to promote cell survival under conditions of nutrient deprivation. Activation of autophagy under these conditions leads to increased degradation of cellular proteins and other macromolecules, allowing their components to be degraded as a source of energy or reutilized for essential functions.

In some circumstances, however, autophagy can lead to cell death by degrading essential cell components. Autophagy as well as apoptosis is responsible for programmed cell death leading to the elimination of larval tissues during metamorphosis in *Drosophila*. In mammals, autophagy may contribute to cell death induced by a variety of pathological treatments, such as drug treatments or infection with some viruses. Autophagic cell death does not require caspases and, rather than possessing the distinct morphological features of apoptosis, the dying cells are characterized by an accumulation of lysosomes.

Death by necrosis can also be a programmed cellular response, rather than simply representing uncontrolled cell lysis as the result of an acute injury. In contrast with unregulated necrosis, these forms of regulated necrotic cell death (**necroptosis**) are induced as a programmed response to stimuli such as infection or DNA damage. Under these conditions, necroptosis provides an alternative pathway of cell death if apoptosis does not occur. For example, stimulation of the TNF receptor leads to cell death by necroptosis as well as by apoptosis. In addition to simulating apoptosis via activation of caspase-8 (see Figure 18.11), stimulation of the TNF receptor also leads to activation of receptor interacting protein kinase-3 (RIPK3) and phosphorylation of its substrate protein MLKL (**Figure 18.12**). Phosphorylated MLKL acts as an executioner of necroptosis by forming oligomers that associate with and disrupt the plasma membrane. Necroptosis appears to play an important role in inflammation and may be involved in a variety of diseases that involve inflammation and tissue damage.

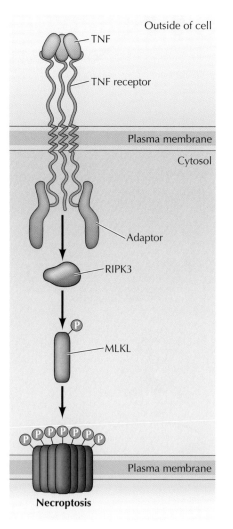

Figure 18.12 Necroptosis Activation of the TNF receptor leads to stimulation of receptor interacting protein kinase-3 (RIPK3) and phosphorylation of MLKL. Phosphorylated MLKL forms oligomers that disrupt the plasma membrane, causing cell death by necroptosis.

Stem Cells and the Maintenance of Adult Tissues

Early development is characterized by the rapid proliferation of embryonic cells, which then differentiate to form the specialized cells of adult tissues and organs. As cells differentiate, their rate of proliferation usually decreases, and most cells in adult animals are arrested in the G_0 stage of the cell cycle. However, cells are lost throughout life, either due to injury or programmed cell death. In order to maintain a constant number of cells in adult tissues and organs, cell death must be balanced by cell proliferation. In order to maintain this balance, most tissues contain cells that are able to proliferate as required to replace cells that have died. Moreover, in some tissues a subpopulation of cells divide continuously throughout life to replace cells that have a high rate of turnover in adult animals. Cell death and cell renewal are thus carefully balanced to maintain properly sized and functioning adult tissues and organs.

Proliferation of differentiated cells

Most types of differentiated cells in adult animals are no longer capable of proliferation. If these cells are lost, they are replaced by the proliferation of less differentiated cells derived from self-renewing stem cells, as discussed in the following section. Other types of differentiated cells, however, retain the ability to proliferate as needed to repair damaged tissue throughout the life of the organism. These cells enter the G_0 stage of the cell cycle but resume proliferation as needed to replace cells that have been injured or died.

Cells of this type include fibroblasts, which are dispersed in connective tissues where they secrete collagen (Figure 18.13). Skin fibroblasts are normally arrested in G_0 but rapidly proliferate if needed to repair damage resulting from a cut or wound. Blood clotting at the site of a wound leads to the release of platelet-derived growth factor (PDGF) from blood platelets. As discussed

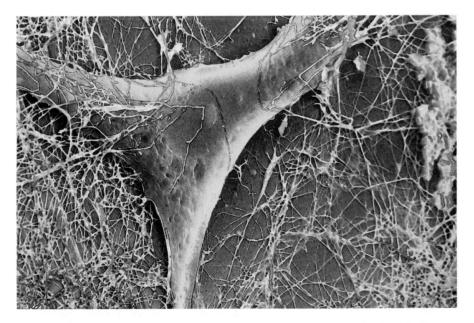

Figure 18.13 A skin fibroblast   Scanning electron micrograph of a skin fibroblast surrounded by collagen fibrils.

Figure 18.14 Endothelial cells
Electron micrograph of a capillary. The capillary is lined by a single endothelial cell surrounded by a thin basal lamina.

Capillary lumen

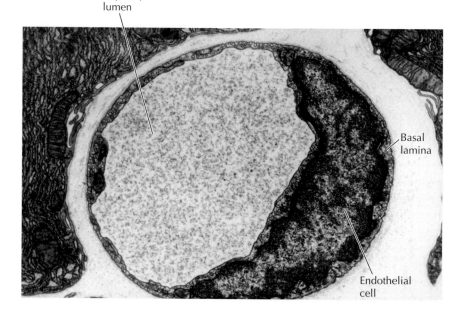
Basal lamina

Endothelial cell

in Chapter 16, PDGF activates a receptor tyrosine kinase, stimulating both the proliferation of fibroblasts and their migration into the wound where their proliferation and secretion of collagen contributes to repair and regrowth of the damaged tissue.

The endothelial cells that line blood vessels (**Figure 18.14**) are another type of fully differentiated cell that remains capable of proliferation. Proliferation of endothelial cells allows them to form new blood vessels as needed for repair and regrowth of damaged tissue. Endothelial cell proliferation and the resulting formation of blood capillaries is triggered by a growth factor (vascular endothelial growth factor, or VEGF) produced by cells of the tissue that the new capillaries will invade. The production of VEGF is in turn triggered by a lack of oxygen, so the result is a regulatory system in which tissues that have a low oxygen supply resulting from insufficient circulation stimulate endothelial cell proliferation and recruit new capillaries (**Figure 18.15**). Smooth muscle cells, which form the walls of larger blood vessels (e.g., arteries) as well as the contractile portions of the digestive and respiratory tracts and other internal

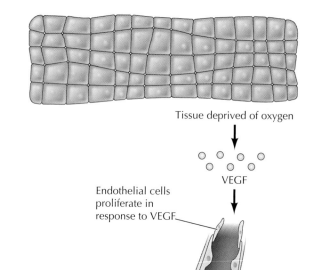

Tissue deprived of oxygen

VEGF

Endothelial cells proliferate in response to VEGF

Capillary

Figure 18.15 Proliferation of endothelial cells
Endothelial cells are stimulated to proliferate by vascular endothelial growth factor (VEGF). VEGF is secreted by cells deprived of oxygen, leading to the outgrowth of new capillaries into tissues lacking an adequate blood supply.

Figure 18.16 Liver regeneration   Liver cells are normally arrested in G_0 but resume proliferation to replace damaged tissue. If two-thirds of the liver of a rat is surgically removed, the remaining cells proliferate to regenerate the entire liver in a few days.

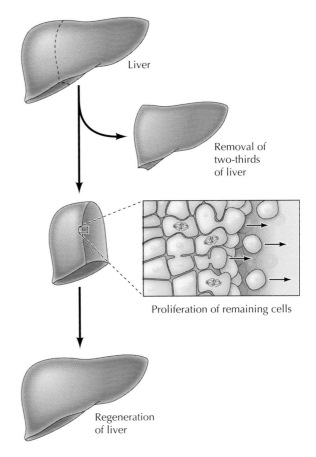

Liver

Removal of two-thirds of liver

Proliferation of remaining cells

Regeneration of liver

organs, are also capable of resuming proliferation in response to growth factor stimulation.

The epithelial cells of some internal organs are also able to proliferate to replace damaged tissue. A striking example is provided by liver cells, which are normally arrested in the G_0 phase of the cell cycle. However, if large numbers of liver cells are lost (e.g., by surgical removal of part of the liver), the remaining cells are stimulated to proliferate to replace the missing tissue (**Figure 18.16**). For example, surgical removal of two-thirds of the liver of a rat is followed by rapid proliferation of the remaining cells, leading to regeneration of the entire liver within a few days.

Stem cells

Most fully differentiated cells in adult animals, however, are no longer capable of cell division. Nonetheless, they can be replaced by the proliferation of a subpopulation of less differentiated self-renewing cells called **stem cells** that are present in most adult tissues. Because they retain the capacity to proliferate and replace differentiated cells throughout the lifetime of an animal, stem cells play a critical role in the maintenance of most tissues and organs.

The key property of stem cells is that they divide to produce one daughter cell that remains a stem cell and one that divides and differentiates (**Figure 18.17**). Stem cells typically give rise to rapidly proliferating cells (called transit-amplifying cells) that undergo several divisions before they differentiate, so division of a stem cell leads to the production of multiple differentiated cells. Because the division of stem cells produces new stem cells as well as differentiated daughter cells, stem cells are self-renewing populations that can serve as a source for the production of differentiated cells throughout life. The role of stem cells is particularly evident in the case of several types of differentiated cells, including blood cells, sperm, epithelial cells of the skin, and epithelial cells lining the digestive tract—all of which have short life spans and must be replaced by continual cell proliferation in adult animals. In all of these cases, the fully differentiated cells do not themselves proliferate; instead, they are continually renewed by the proliferation of stem cells that then differentiate to maintain a stable number of differentiated cells. Stem cells have also been identified in a variety of other adult tissues, including skeletal muscle and the nervous system, where they function to replace damaged tissue.

Stem cells were first identified in the hematopoietic (blood-forming) system by Ernest McCulloch and James Till in 1961 in experiments showing that single cells derived from mouse bone marrow could proliferate and give rise to multiple differentiated types of blood cells. Hematopoietic stem cells are well-characterized and the production of blood cells provides a good

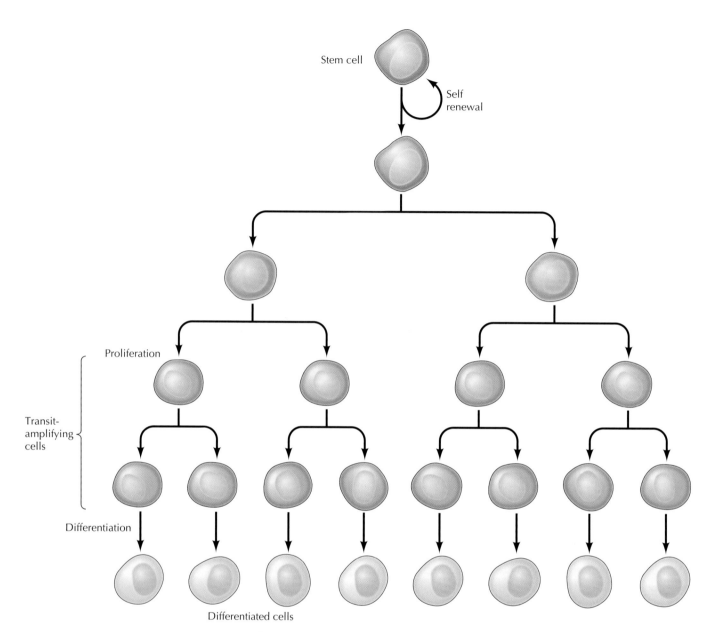

Figure 18.17 Stem cell proliferation Stem cells divide to form one daughter cell that remains a stem cell and a second that gives rise to transit-amplifying cells, which go through several cell divisions and then differentiate.

example of the role of stem cells in maintaining differentiated cell populations. There are several distinct types of blood cells with specialized functions: erythrocytes (red blood cells) that transport O_2 and CO_2; granulocytes and macrophages, which are phagocytic cells; platelets (which are fragments of megakaryocytes) that function in blood coagulation; and lymphocytes that are responsible for the immune response. All these cells have limited life spans ranging from less than a day to a few months, and all are derived from the same population of hematopoietic stem cells. More than 100 billion blood cells are lost every day in humans, and must be continually produced from

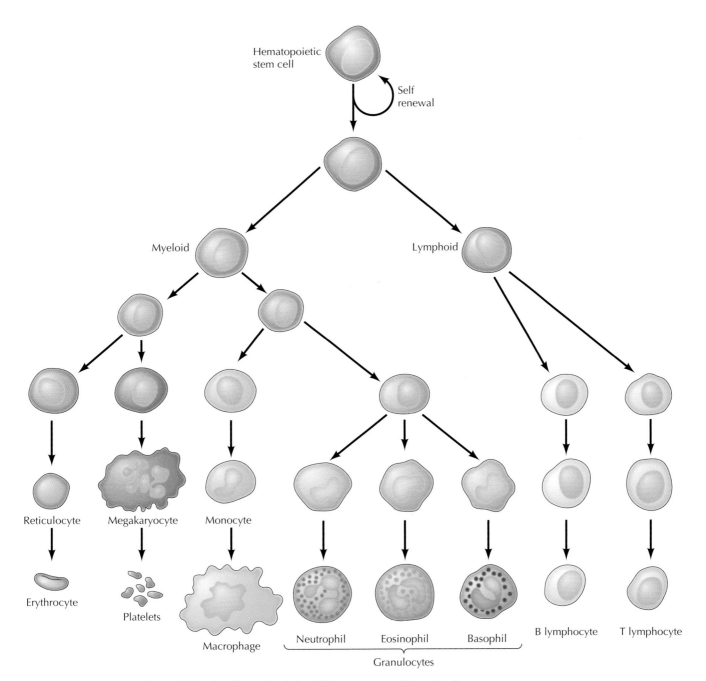

Figure 18.18 Formation of blood cells   All of the different types of blood cells develop from hematopoietic stem cells in the bone marrow. The precursors of differentiated cells undergo several rounds of cell division (not shown) before they differentiate.

hematopoietic stem cells in the bone marrow (**Figure 18.18**). Descendants of the hematopoietic stem cell continue to proliferate and undergo several rounds of division as they become committed to specific differentiation pathways, as determined by growth factors that direct cell differentiation. Once they become fully differentiated, blood cells cease proliferation, so the maintenance of differentiated blood cell populations is dependent on continual division of the self-renewing hematopoietic stem cell.

The intestine provides an excellent example of stem cells in the self-renewal of an epithelial tissue. The intestine is lined by a single layer of epithelial cells that are responsible for the digestion of food and absorption of nutrients. These intestinal epithelial cells are exposed to an extraordinarily harsh environment and have a lifetime of only a few days before they die by apoptosis and are shed into the digestive tract. Renewal of the intestinal epithelium is therefore a continual process throughout life. New cells are derived from the continuous but slow division of stem cells at the bottom of intestinal crypts (**Figure 18.19**). The stem cells give rise to a population of

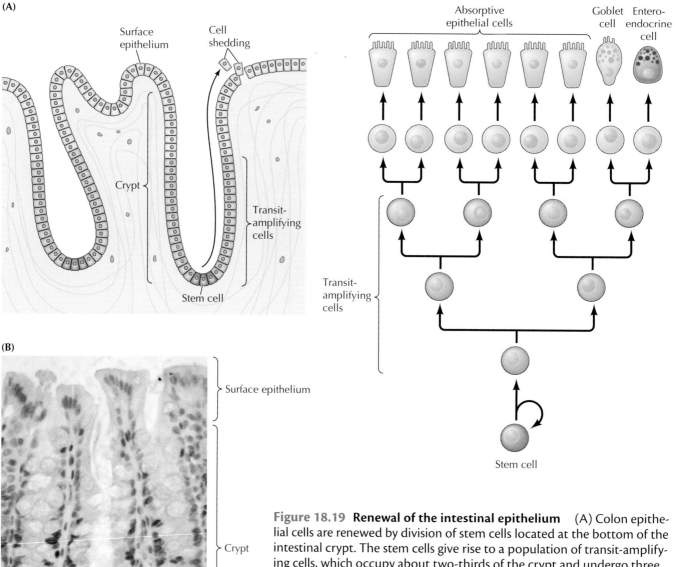

Figure 18.19 Renewal of the intestinal epithelium (A) Colon epithelial cells are renewed by division of stem cells located at the bottom of the intestinal crypt. The stem cells give rise to a population of transit-amplifying cells, which occupy about two-thirds of the crypt and undergo three to four divisions before differentiating into the three cell types of the surface epithelium (absorptive epithelial cells, goblet cells, and enteroendocrine cells). The surface epithelial cells continually undergo apoptosis and are shed into the intestinal lumen. (B) Micrograph of a colon crypt and surface epithelium. Proliferating cells are stained with antibody against a cell cycle protein (brown nuclei). (B, from F. Radtke and H. Clevers, 2005. *Science* 307: 1904.)

transit-amplifying cells, which divide rapidly and occupy about two-thirds of the crypt. The transit-amplifying cells proliferate for three to four cell divisions and then differentiate into the three cell types of the colon surface epithelium: absorptive epithelial cells and two types of secretory cells, called goblet cells and enteroendocrine cells. The small intestine also contains a fourth cell type, Paneth cells, which secrete antibacterial agents. Each crypt contains self-renewing stem cells, which can give rise to all of the different types of cells in the intestinal epithelium.

Stem cells are also responsible for continuous renewal of the skin and hair. Like the lining of the intestine, the skin and hair are exposed to a harsh external environment—including ultraviolet radiation from sunlight—and are continuously renewed throughout life. The skin consists of three major cell lineages: the epidermis, hair follicles, and sebaceous glands, which release oils that lubricate the skin surface. Each of these three cell populations is maintained by their own stem cells (**Figure 18.20**). The epidermis is a multilayered epithelium, which is undergoing continual cell renewal. In humans, the epidermis turns over every two weeks, with cells being sloughed from the surface. These cells are replaced by epidermal stem cells, which reside in a single basal layer. The epidermal stem cells give rise to transit-amplifying cells, which undergo three to six divisions before differentiating and moving outward to the surface of the skin. The stem cells responsible for producing hair reside in a region of the hair follicle called the bulge. The bulge stem cells give rise to transit-amplifying matrix cells, which proliferate and differentiate to form the hair shaft. Finally, a distinct population of stem cells resides at the base of the sebaceous gland. It is notable that, if the skin is injured, stem cells of the bulge can also give rise to epidermis and sebaceous glands, demonstrating their activity as multipotent stem cells from which both skin and hair can be derived.

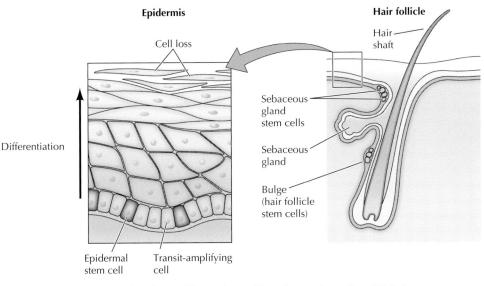

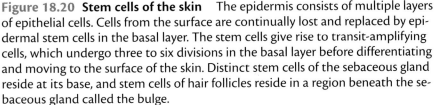

Figure 18.20 Stem cells of the skin   The epidermis consists of multiple layers of epithelial cells. Cells from the surface are continually lost and replaced by epidermal stem cells in the basal layer. The stem cells give rise to transit-amplifying cells, which undergo three to six divisions in the basal layer before differentiating and moving to the surface of the skin. Distinct stem cells of the sebaceous gland reside at its base, and stem cells of hair follicles reside in a region beneath the sebaceous gland called the bulge.

Skeletal muscle provides an example of the role of stem cells in the repair of damaged tissue, in contrast with the continual cell renewal just described in the hematopoietic system, intestinal epithelium, and skin. Skeletal muscle is composed of large multinucleated cells (muscle fibers) formed by cell fusion during development (see Figure 13.22). Although skeletal muscle is normally a stable tissue with little cell turnover, it is able to regenerate rapidly in response to injury or exercise. This regeneration is mediated by the proliferation of satellite cells, which are the stem cells of adult muscle. Satellite cells are located beneath the basal lamina of muscle fibers (Figure 18.21). They are normally quiescent, arrested in the G_0 phase of the cell cycle, but are activated to proliferate in response to injury or exercise. Once activated, the satellite cells give rise to progeny that undergo several divisions and then differentiate and fuse to form new muscle fibers. The continuing capacity of skeletal muscle to regenerate throughout life is due to self-renewal of the satellite stem cell population.

Stem cells have also been found in many other adult tissues, including the brain and heart, and it is possible that most—if not all—tissues contain stem cells with the potential of replacing cells that are lost during the lifetime of the organism. It appears that stem cells reside within distinct microenvironments, called **niches**, which provide the environmental signals that maintain stem cells throughout life and control the balance between their self-renewal and differentiation. Stem cells are rare in adult mammalian tissues, however, so the precise identification of stem cells and their niches represents a major challenge in the field of stem cell biology. For example, although the role of stem cells in

(A)

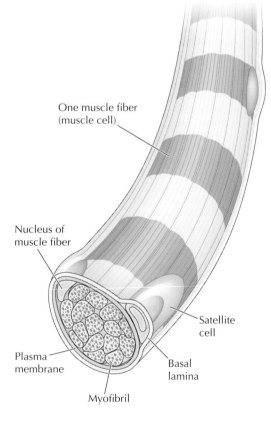

One muscle fiber
(muscle cell)

Nucleus of
muscle fiber

Plasma
membrane

Satellite
cell

Basal
lamina

Myofibril

(B)

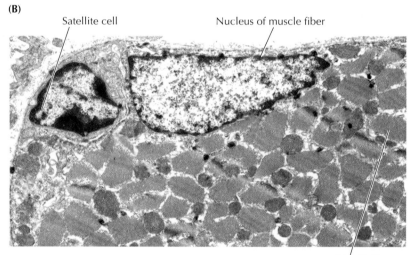

Satellite cell

Nucleus of muscle fiber

Myofibril

Figure 18.21 Muscle satellite cells (A) The stem cells of skeletal muscle are the satellite cells, located beneath the basal lamina of muscle fibers. (B) Electron micrograph showing a satellite cell and the nucleus of a muscle fiber. (B, from S. Chargé and M. Rudnicki, 2003. *Physiol. Rev.* 84: 209; courtesy of Sophie Chargé and Michael Rudnicki.)

maintenance of the intestinal epithelium has long been recognized, the intestinal stem cells at the base of the crypt (see Figure 18.19) were first identified by studies of Hans Clevers and his colleagues in 2007. Signaling by the Wnt pathway (see Figure 16.42) plays a major role in controlling the proliferation of these stem cells, and the Wnt polypeptides secreted by adjacent epithelial cells and fibroblasts of the underlying connective tissue are responsible for intestinal stem cell maintenance (**Figure 18.22**). Wnt signaling is also involved in regulation of several other types of stem cells, including stem cells of the skin and hematopoietic system. In addition, signaling by the TGF-β, Hedgehog, and Notch pathways (see Figures 16.39, 16.41, and 16.43), as well as interactions with the extracellular matrix, play important roles in stem cell regulation, although the precise roles of these factors in regulating different types of stem cells within their distinct niches remains to be understood.

Medical applications of adult stem cells

The ability of adult stem cells to repair damaged tissue clearly suggests their potential utility in clinical medicine. If these stem cells could be isolated and propagated in culture, they could in principal be used to replace damaged tissue and treat a variety of disorders, such as diabetes or degenerative diseases like muscular dystrophy, Parkinson's or Alzheimer's. In some cases, the use of stem cells derived from adult tissues may be the optimal approach for such stem cell therapies, although the use of embryonic or induced pluripotent stem cells (discussed in the next section of this chapter) is likely to provide a more versatile approach to treatment of a wider variety of disorders.

A well-established clinical application of adult stem cells is **hematopoietic stem cell transplantation** (or **bone marrow transplantation**), which plays an important role in the treatment of a variety of cancers. As discussed in Chapter 19, most cancers are treated by chemotherapy with drugs that kill rapidly dividing cells by damaging DNA or inhibiting DNA replication. These drugs do not act selectively against cancer cells and so are also toxic to those normal tissues that are dependent on continual renewal by stem cells, such as blood, skin, hair, and the intestinal epithelium. The hematopoietic stem cells are among the most rapidly dividing cells of the body, so the toxic effects of anticancer drugs on these cells frequently limit the effectiveness of chemotherapy in cancer treatment. Hematopoietic stem cell transplantation provides an approach to bypassing this toxicity, thereby allowing the use of higher drug doses to treat the patient's cancer more effectively. In this procedure, the patient is treated with high doses of chemotherapy that would normally not be tolerated because of toxic effects on the hematopoietic system (**Figure 18.23**). The potentially lethal damage is repaired, however, by transferring new hematopoietic stem cells (obtained either from bone marrow or peripheral blood) to the patient following completion of chemotherapy, so that the normal hematopoietic system is restored. In some cases, the stem cells are obtained from the patient prior to chemotherapy, stored, and

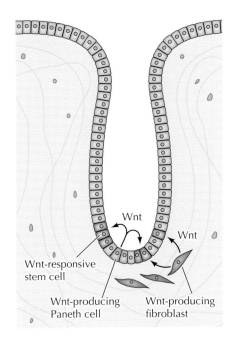

Figure 18.22 The intestinal stem cell niche   Stem cells of the small intestine are maintained by Wnt polypeptides secreted by adjacent Paneth cells and fibroblasts in the underlying connective tissue.

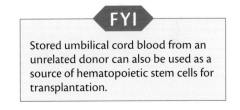

Stored umbilical cord blood from an unrelated donor can also be used as a source of hematopoietic stem cells for transplantation.

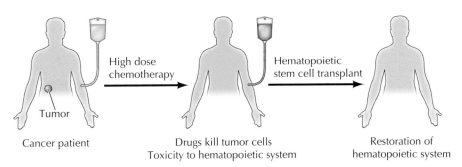

Figure 18.23 Hematopoietic stem cell transplantation   A cancer patient is treated with high doses of chemotherapy, which effectively kill tumor cells but normally would not be tolerated because of potentially lethal damage to the hematopoietic system. This damage is then repaired by transplantation of new hematopoietic stem cells.

then returned to the patient once chemotherapy is completed. However, it is important to ensure that these cells are not contaminated with cancer cells. Alternatively, the stem cells to be transplanted can be obtained from a healthy donor (usually a close relative) whose tissue type closely matches the patient. In addition to their use in cancer treatment, hematopoietic stem cell transfers are used to treat patients with diseases of the hematopoietic system, such as aplastic anemia, hemoglobin disorders, and immune deficiencies.

Epithelial stem cells have also found clinical application in the form of skin grafts that are used to treat patients with burns, wounds, and ulcers. One approach to these procedures is to culture epidermal skin cells to form an epithelial sheet, which can then be transferred to the patient. Because the patient's own skin can be used for this procedure, it eliminates the potential complication of graft rejection by the immune system. The possibilities of using adult stem cells for similar replacement therapies of other diseases, including diabetes, Parkinson's, and muscular dystrophies, are also being pursued. However, these clinical applications of adult stem cells are limited by the difficulties in isolating and culturing the appropriate stem cell populations.

Pluripotent Stem Cells, Cellular Reprogramming, and Regenerative Medicine

While adult stem cells are difficult to isolate and culture, it is relatively straightforward to isolate and propagate the stem cells of early embryos (**embryonic stem cells**). These cells can be grown indefinitely as pure stem cell populations while maintaining the ability to give rise to all of the differentiated cell types of adult organisms (**pluripotency**). Consequently, there has been an enormous interest in embryonic stem cells not only from the standpoint of basic science, but also because of their potential use as a source of cells for transplantation therapy. Continuing studies of embryonic stem cells have revealed that adult cells can also be reprogrammed either to display the pluripotency characteristic of embryonic stem cells or to behave as a different type of differentiated cell. These studies have illuminated novel aspects of the control of cell fate, and are likely to find major applications in regenerative medicine.

Embryonic stem cells

Embryonic stem cells were first cultured from mouse embryos in 1981 (**Figure 18.24**). They can be propagated indefinitely in culture and, if reintroduced

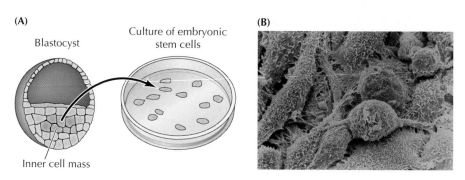

Figure 18.24 Culture of mammalian embryonic stem cells   (A) Embryonic stem cells are cultured from the inner cell mass of an early embryo (blastocyst). (B) Scanning electron micrograph of cultured embryonic stem cells.

Culture of Embryonic Stem Cells

Isolation of a Pluripotent Cell Line from Early Mouse Embryos Cultured in Medium Conditioned by Teratocarcinoma Stem Cells

Gail R. Martin

University of California at San Francisco

Proceedings of the National Academy of Sciences, USA, 1981, Volume 78, pages 7634–7638

The Context

The cells of early embryos are unique in their ability to proliferate and differentiate into all of the types of cells that make up the tissues and organs of adult animals. In 1970 it was found that early mouse embryos frequently developed into tumors if they were removed from the uterus and transplanted to an abnormal site. These tumors, called teratocarcinomas, contained cells that were capable of forming an array of different tissues as they grew within the animal. In addition, cells from teratocarcinomas (called embryonal carcinoma cells) could be isolated and grown in tissue culture. These cells resembled normal embryo cells and could be induced to differentiate into a variety of cell types in culture. Some embryonal carcinoma cells could also participate in normal development of a mouse if they were injected into early mouse embryos (blastocysts) that were then implanted into a foster mother.

The ability of embryonal carcinoma cells to differentiate into a variety of cell types and to participate in normal mouse development suggested that these tumor-derived cells might be closely related to normal embryonic stem cells. However, the events that occurred during the establishment of teratocarcinomas in mice were unknown. Gail Martin hypothesized that the embryonal carcinoma cells found in teratocarcinomas were essentially normal embryo cells that proliferated abnormally simply because, when they were removed from the uterus and transplanted to an abnormal site, they did not receive the appropriate signals to induce normal differentiation. Based on this hypothesis, she attempted to culture cells from mouse embryos with the goal of isolating normal embryonic stem cell lines. Her experiments, together with similar work by Martin Evans and Matthew Kaufman (Establishment in culture of pluripotential cells from mouse embryos, *Nature*, 1981, 292: 154–156), demonstrated that pluripotent stem cells could be cultured directly from normal mouse embryos. The isolation of these embryonic stem cell lines paved the way to genetic manipulation and analysis of mouse development, as well as to the possible use of human embryonic stem cells in transplantation therapy.

Gail R. Martin

The Experiments

Based on the premise that embryonal carcinoma cells were derived from normal embryonic stem cells, Martin attempted to culture cells from normal mouse blastocysts. Starting with cells from approximately 30 embryos, she initially isolated four colonies of growing cells after a week of culture. These cells could be repeatedly passaged into mass cultures, and new cell lines could be reproducibly derived when the experiment was repeated with additional mouse embryos.

The cell lines derived from normal embryos (embryonic stem cells) closely resembled the embryonal carcinoma cells derived from tumors. Most importantly, the embryonic stem cells could be induced to differentiate in culture into a variety of cell types, including endodermal cells, cartilage, and neuron-like cells (see figure). Moreover, if the embryonic stem cells were injected into a mouse, they formed

(A)   **(B)**   **(C)**

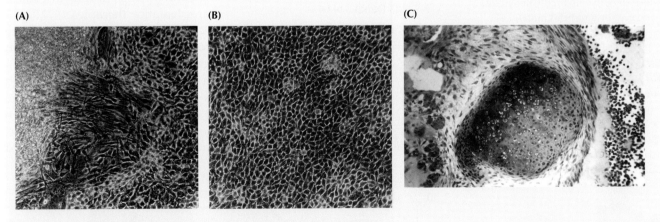

Embryonic stem cells differentiate in culture to a variety of cell types, including neuron-like cells (A), endodermal cells (B), and cartilage (C).

(Continued on next page)

Key Experiment

tumors containing multiple differentiated cell types. It thus appeared that embryonic stem cell lines, which retained the ability to differentiate into a wide array of cell types, could be established in culture from normal mouse embryos.

The Impact

The establishment of embryonic stem cell lines has had a major impact on studies of mouse genetics and development as well as opening new possibilities for the treatment of a variety of human diseases. Subsequent experiments demonstrated that embryonic stem cells were pluripotent and could

participate in normal mouse development following their injection into mouse embryos. Since gene transfer techniques could be used to introduce or mutate genes in cultured embryonic stem cells, these cells have been used to investigate the role of a variety of genes in mouse development. As discussed in Chapter 4, any gene of interest can be inactivated in embryonic stem cells by homologous recombination with a cloned DNA, and the role of that gene in mouse development can then be determined by introducing the altered embryonic stem cells into mouse embryos.

In 1998 two groups of researchers developed the first lines of human embryonic stem cells. More recently, in 2006, the induction of pluripotent stem cells from adult fibroblasts was described. Because of the proliferative and differentiative capacity of these cells, they offer the hope of providing new therapies for the treatment of a variety of diseases. Although a number of technical problems remain to be addressed, transplantation therapies based on the use of pluripotent stem cells may provide the best hope for eventual treatment of diseases such as Parkinson's, Alzheimer's, diabetes, and spinal cord injuries.

---

into early embryos, are able to give rise to cells in all tissues of the mouse. Thus they are pluripotent and retain the capacity to develop into all of the different types of cells in adult tissues and organs. In addition, they can be induced to differentiate into a variety of different types of cells in culture.

As discussed in Chapter 4, mouse embryonic stem cells have been an important experimental tool in cell biology because they can be used to introduce altered genes into mice (see Figure 4.35). Moreover, they provide an outstanding model system for studying the molecular and cellular events associated with embryonic cell differentiation, so embryonic stem cells have long been of considerable interest to cell and developmental biologists. Interest in these cells reached a new peak of intensity, however, in 1998 when two groups of researchers reported the isolation of stem cells from human embryos, raising the possibility of using embryonic stem cells in clinical transplantation therapies.

Mouse embryonic stem cells are grown in the presence of a growth factor called LIF (for leukemia inhibitory factor), which signals through the JAK/STAT pathway (see Figure 16.25) and is required to maintain these cells in their undifferentiated state (**Figure 18.25**). If LIF is removed

Video 18.3

sites.sinauer.com/cooper7e/v18.3

Differentiated Cardiac Beating Cells Cells differentiate into a wide range of cell types, including blood cells, epithelial cells, adipocytes, vascular smooth muscle cells, nerve cells, and even beating heart muscle cells.

Figure 18.25 Differentiation of embryonic stem cells Mouse embryonic stem cells are maintained in the undifferentiated state in the presence of LIF. If LIF is removed from the culture medium, the cells aggregate to form embryoid bodies and then differentiate into a variety of cell types.

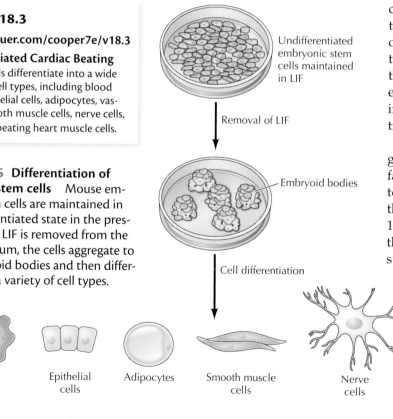

Undifferentiated embryonic stem cells maintained in LIF

Removal of LIF

Embryoid bodies

Cell differentiation

Blood cells

Epithelial cells

Adipocytes

Smooth muscle cells

Nerve cells

Heart muscle cells

from the medium, the cells aggregate into structures that resemble embryos (embryoid bodies) and then differentiate into a wide range of cell types, including blood cells, epithelial cells, adipocytes, vascular smooth muscle cells, nerve cells, and even beating heart muscle cells. Human embryonic stem cells do not require LIF but are similarly maintained in the undifferentiated state by other growth factors.

Importantly, embryonic stem cells can be directed to differentiate along specific pathways by the addition of appropriate growth factors or small molecules (e.g., retinoic acid) to the culture medium. It is thus possible to derive populations of specific types of cells, such as heart cells or nerve cells, for transplantation therapy. For example, methods have been developed to direct the differentiation of both mouse and human embryonic stem cells into cardiomyocytes, various types of neurons, and insulin-producing pancreatic β cells. These differentiated cells have been used for treatment of mouse models of cardiac arrhythmia, Parkinson's disease, spinal cord injury, blindness due to macular degeneration, and diabetes. Continuing research is focused on the development of protocols to increase the efficiency of differentiation of embryonic stem cells along specific pathways, thereby producing populations of differentiated cells that can be used for transplantation therapy for a variety of diseases.

> **Video 18.4**
>
> sites.sinauer.com/cooper7e/v18.4
>
> **Controlling Stem Cells with Light** Methods have been developed to direct the differentiation of both mouse and human embryonic stem cells into various types of neurons.

> **Video 18.5**
>
> sites.sinauer.com/cooper7e/v18.5
>
> **Somatic Cell Nuclear Transfer** Somatic cell nuclear transfer is used to produce clones of cells with the same genetic background as the donor nucleus.

Somatic cell nuclear transfer

The isolation of human embryonic stem cells in 1998 followed the first demonstration that the nucleus of an adult mammalian cell could give rise to a viable cloned animal. In 1997 Ian Wilmut and his colleagues initiated a new era of regenerative medicine with the cloning of Dolly the sheep (**Figure 18.26**). Dolly arose from the nucleus of a mammary epithelial cell that was transplanted into an unfertilized egg in place of the normal egg nucleus—a process called **somatic cell nuclear transfer**. It is interesting to note that this type of experiment was first carried out in frogs by John Gurdon in 1962. The fact that it took almost 40 years before it was successfully performed in mammals attests to the technical difficulty of the procedure. Since the initial success of Wilmut and his colleagues, transfer of nuclei from adult somatic cells into enucleated eggs has been used to create cloned offspring of a variety of mammalian species, including sheep, mice, pigs,

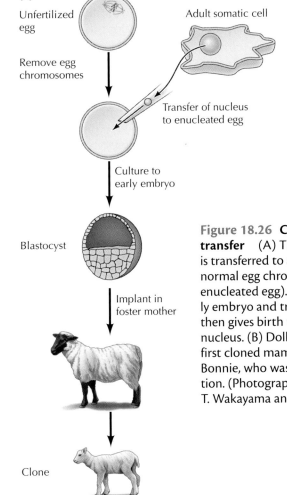

Figure 18.26 Cloning by somatic cell nuclear transfer (A) The nucleus of an adult somatic cell is transferred to an unfertilized egg from which the normal egg chromosomes have been removed (an enucleated egg). The egg is then cultured to an early embryo and transferred to a foster mother, who then gives birth to a clone of the donor of the adult nucleus. (B) Dolly (the adult sheep, left) was the first cloned mammal. She is shown with her lamb, Bonnie, who was produced by normal reproduction. (Photograph by Roddy Field; courtesy of T. Wakayama and R. Yanagimachi.)

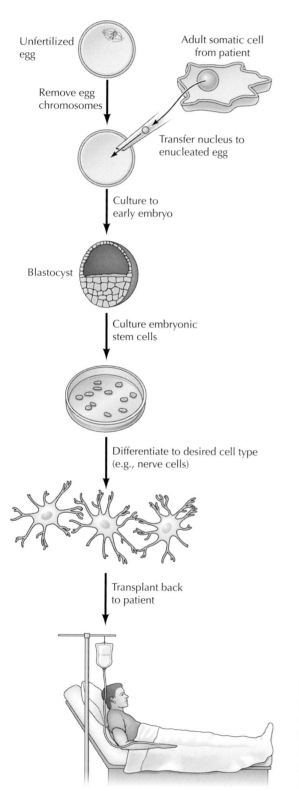

Figure 18.27 **Therapeutic cloning**   In therapeutic cloning, the nucleus of a patient's cell would be transferred to an enucleated egg, which would be cultured to an early embryo. Embryonic stem cells would then be derived, differentiated into the desired cell type, and transplanted back into the patient. The transplanted cells would be genetically identical to the recipient (who was the donor of the adult nucleus), so complications of immune rejection would be avoided.

cattle, goats, rabbits, and cats. However, cloning by somatic cell nuclear transfer in mammals remains an extremely inefficient procedure, such that only 1 or 2% of embryos generally give rise to live offspring. Moreover, the small number of animals that survive beyond birth often suffer a variety of abnormalities, resulting in a limited life span. Dolly, for example, lived only 6 years—about half the lifetime of a normal sheep. Both the inefficiency of cloning by somatic cell nuclear transfer and the abnormalities prevalent in cloned animals are thought to reflect the difficulties in completely reprogramming the epigenetic state of an adult nucleus, including reversal of DNA methylation.

Animal cloning by somatic cell nuclear transfer, together with the properties of embryonic stem cells, opens the possibility of **therapeutic cloning** (Figure 18.27). In therapeutic cloning, a nucleus from an adult human cell would be transferred to an enucleated egg, which would then be used to produce an early embryo in culture. Embryonic stem cells could then be cultured from the cloned embryo and used to generate appropriate types of differentiated cells for transplantation therapy. Human embryonic stem cells were first produced by somatic cell nuclear transfer in 2013, supporting the possibility of this approach. The major advantage provided by therapeutic cloning is that the embryonic stem cells derived by this procedure would be genetically identical to the recipient of the transplant, who was the donor of the adult somatic cell nucleus. This bypasses the barrier of the immune system in rejecting the transplanted tissue.

The possibility of therapeutic cloning provides a general approach to treatment of the wide variety of devastating disorders for which stem cell transplantation therapy could be applied. However, although some success has been achieved in animal models, there remain major obstacles that would need to be overcome before therapeutic cloning could be applied to humans. Substantial improvements would be needed to overcome the low efficiency with which embryos are generated by somatic cell nuclear transfer. In addition, therapeutic cloning by somatic cell nuclear transfer raises ethical concerns, not only with respect to the possibility of cloning human beings (**reproductive cloning**), but also with respect to the destruction of embryos that serve as the

Figure 18.28 Induced pluripotent stem cells   Adult mouse fibroblasts in culture are converted to pluripotent stem cells by infection with retroviral vectors (see Figure 4.30) carrying genes for four transcription factors: Oct4, Sox2, Klf4, and c-Myc.

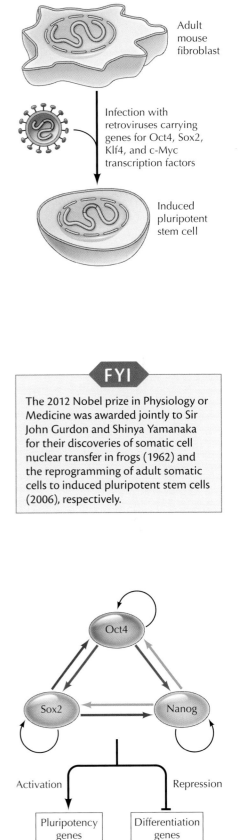

Adult mouse fibroblast

Infection with retroviruses carrying genes for Oct4, Sox2, Klf4, and c-Myc transcription factors

Induced pluripotent stem cell

source of embryonic stem cells. These concerns may be alleviated by recent advances in reprogramming somatic cells to a pluripotent state resembling embryonic stem cells.

Induced pluripotent stem cells

Given both the technical and ethical difficulties in the derivation of embryonic stem cells by somatic cell nuclear transfer, a major advance in the field has come from studies demonstrating that adult somatic cells can be directly converted to pluripotent stem cells in culture. This circumvents the need for derivation of embryos and provides a direct mechanism for converting somatic cells to stem cells, which, like embryonic stem cells, have the potential to develop into all tissues of an organism.

The conversion (or reprogramming) of somatic cells to pluripotent stem cells was first reported by Kazutoshi Takahashi and Shinya Yamanaka in 2006. They found that mouse fibroblasts could be reprogrammed to cells resembling embryonic stem cells (called **induced pluripotent stem cells**) by the action of only four transcription factors introduced by retroviral gene transfer (**Figure 18.28**). Subsequent studies have shown that induced pluripotent stem cells, like embryonic stem cells, are capable of differentiating into all cell types when introduced into early mouse embryos. Importantly, further research has also demonstrated that adult human fibroblasts can be reprogrammed to pluripotency by a similar procedure. Thus, it is now possible to convert skin cells from a patient directly to induced pluripotent stem cells in culture, providing a new route to the derivation of pluripotent stem cells for use in transplantation therapy.

The finding that only four key transcription factors were sufficient to reprogram adult somatic cells to pluripotent stem cells was remarkable, and stimulated a great deal of investigation into the transcriptional programs that control cell fate. Reprogramming of fibroblasts to induced pluripotent stem cells can be induced not only by the four transcription factors initially used by Takahashi and Yamanaka (see Figure 18.28), but also by several alternative combinations of transcription factors. All of these combinations of transcription factors that induce pluripotency activate a transcriptional program that is also expressed in embryonic cells and maintains the pluripotent state. Three core transcription factors (Oct4, Sox2, and Nanog) play central roles in this program (**Figure 18.29**). These factors form an autoregulatory loop, in which they act together to positively regulate each other's expression. In addition, they activate expression of other genes that maintain the pluripotent state,

FYI

The 2012 Nobel prize in Physiology or Medicine was awarded jointly to Sir John Gurdon and Shinya Yamanaka for their discoveries of somatic cell nuclear transfer in frogs (1962) and the reprogramming of adult somatic cells to induced pluripotent stem cells (2006), respectively.

Oct4

Sox2   Nanog

Activation   Repression

Pluripotency genes   Differentiation genes

Figure 18.29 Pluripotency transcriptional program   The three core transcription factors Oct4, Sox2, and Nanog maintain the pluripotent state. These factors form positive autoregulatory loops as well as activate genes that maintain pluripotency and repress genes that induce differentiation.

while repressing genes that enable differentiation along specific cell lineages. The positive autoregulation at the core of this transcriptional program maintains pluripotency, while allowing the cells to undergo differentiation in response to appropriate signals.

It is noteworthy that the autoregulatory maintenance of pluripotency has allowed investigators to overcome a major obstacle to the potential therapeutic use of induced pluripotent stem cells. Namely, some of the transcription factors (e.g., c-Myc) that were initially used to reprogram fibroblasts (see Figure 18.28) can act as oncogenes, resulting in a high risk of cancer following transplantation of induced pluripotent stem cells produced in this manner. Additionally, the retroviral vectors that were used to introduce genes into fibroblasts can themselves cause harmful mutations leading to cancer development. However, methods have now been developed to establish pluripotent stem cells by transient expression of the reprogramming factors, without the need for stable integration of viral genomes or foreign genes. Once the recipient somatic cell is reprogrammed to the pluripotent state, autoregulation maintains pluripotency without the need for continued expression of the factors that initially induced the reprogramming. Induced pluripotent stem cells generated by transient expression of reprogramming factors thus provide a ready source of pluripotent stem cells for transplantation therapy.

Transdifferentiation of somatic cells

The success in reprogramming somatic cells to induced pluripotent stem cells raised the possibility that somatic cells might alternatively be reprogrammed to other types of differentiated cells—a phenomenon termed **transdifferentiation**. Transdifferentiation was initially demonstrated in 1987, when Harold Weintraub and his colleagues showed that the transcription factor MyoD was sufficient to induce the differentiation of fibroblasts into muscle cells. Based on the success in reprogramming fibroblasts to pluripotent stem cells with combinations of three to four transcription factors, scientists have more recently investigated the possibility that combinations of transcription factors might induce the differentiation of fibroblasts into other cell types with potential utility in transplantation therapy. Notably, success has been obtained in inducing the differentiation of mouse fibroblasts into both heart muscle cells and nerve cells with appropriate combinations of only three transcription factors in each case (**Figure 18.30**). Differentiated blood cells can also be reprogrammed either to other blood cell types or to hematopoietic stem cells, potentially providing a new source of patient-specific cells for hematopoietic stem cell transplantation (see Figure 18.23). Transdifferentiation thus provides a direct route to generating desired types of differentiated cell populations, bypassing the need for pluripotent stem cells and allowing the direct conversion of adult cells from a patient into differentiated cells that could be used for transplantation therapy.

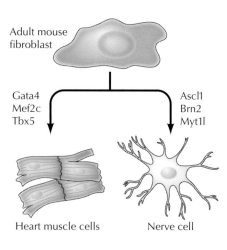

Figure 18.30 Transdifferentiation Combinations of three transcription factors convert fibroblasts into other differentiated cell types, including heart muscle and nerve cells.

SUMMARY	KEY TERMS

Programmed Cell Death

- *The events of apoptosis:* Programmed cell death plays a key role in both the maintenance of adult tissues and embryonic development. In contrast with the accidental death of cells from an acute injury, most programmed cell death takes place by the active process of apoptosis. Apoptotic cells and cell fragments are then efficiently removed by phagocytosis. Genes responsible for the regulation and execution of apoptosis were initially identified by genetic analysis of *C. elegans*. See Animation 18.1 and Video 18.1.

programmed cell death, necrosis, apoptosis

- *Caspases: The executioners of apoptosis:* The caspases are a family of proteases that are the effectors of apoptosis. Caspases are classified as either initiator or effector caspases, and both function in a cascade leading to cell death. The effector caspases cleave more than 100 cellular proteins to bring about the events of apoptosis.

caspase

- *Central regulators of apoptosis: The Bcl-2 family:* Members of the Bcl-2 family are central regulators of caspase activation and apoptosis. Some members of the Bcl-2 family function to inhibit apoptosis (antiapoptotic), whereas others act to promote it (proapoptotic). Signals that control programmed cell death alter the balance between proapoptotic and antiapoptotic Bcl-2 family members, which regulate one another. In mammalian cells, proapoptotic Bcl-2 family members act at mitochondria, where they promote the release of cytochrome *c*, leading to caspase activation. Caspases are also regulated directly by inhibitory IAP proteins. See Animation 18.2.

Bcl-2, apoptosome, IAP

- *Signaling pathways that regulate apoptosis:* A variety of signaling pathways regulate apoptosis by controlling the expression or activity of proapoptotic members of the Bcl-2 family. These pathways include DNA damage-induced activation of the tumor suppressor p53, growth factor-stimulated activation of PI 3-kinase/Akt signaling, and activation of death receptors by polypeptides that induce programmed cell death. See Video 18.2.

p53, PI 3-kinase, Akt, tumor necrosis factor (TNF)

- *Alternative pathways of programmed cell death:* Autophagy and regulated necrosis provide alternatives to apoptosis for induction of programmed cell death.

autophagy, necroptosis

Stem Cells and the Maintenance of Adult Tissues

- *Proliferation of differentiated cells:* Most cells in adult animals are arrested in the G_0 stage of the cell cycle. A few types of differentiated cells, including skin fibroblasts, endothelial cells, smooth muscle cells, and liver cells are able to resume proliferation as required to replace cells that have been lost because of injury or cell death.

- *Stem cells:* Most differentiated cells do not themselves proliferate but can be replaced via the proliferation of stem cells. Stem cells divide to produce one daughter cell that remains a stem cell and another that divides and differentiates. Stem cells have been identified in a wide variety of adult tissues, including the hematopoietic system, skin, intestine, skeletal muscle, brain, and heart.

stem cell, niche

- *Medical applications of adult stem cells:* The ability of stem cells to repair damaged tissue suggests their potential use in clinical medicine. Adult stem cells are used to repair damage to the hematopoietic system in hematopoietic stem cell transplantation. However, clinical applications of adult stem cells are limited by difficulties in isolating and culturing these cells.

hematopoietic stem cell transplantation, bone marrow transplantation

SUMMARY	KEY TERMS

Pluripotent Stem Cells, Cellular Reprogramming, and Regenerative Medicine

- **Embryonic stem cells:** Embryonic stem cells are cultured from early embryos. They can be readily grown in the undifferentiated state in culture while retaining the ability to differentiate into all of the different types of cells in an organism, so they offer considerable advantages over adult stem cells for many clinical applications. See Videos 18.3 and 18.4.

embryonic stem cell, pluripotency

- **Somatic cell nuclear transfer:** Mammals have been cloned by somatic cell nuclear transfer in which the nucleus of an adult somatic cell is transplanted into an enucleated egg. This opens the possibility of therapeutic cloning in which embryonic stem cells would be derived from a cloned embryo and used for transplantation therapy of the donor of the adult nucleus. Although many obstacles need to be overcome, the possibility of therapeutic cloning holds promise for the development of new treatments for a variety of devastating diseases. See Video 18.5.

somatic cell nuclear transfer, therapeutic cloning, reproductive cloning

- **Induced pluripotent stem cells:** Adult somatic cells can be converted to pluripotent stem cells in culture by four key transcription factors, providing an alternative to the use of embryonic stem cells for transplantation therapy.

induced pluripotent stem cell

- **Transdifferentiation of somatic cells:** Somatic cells can also be converted directly into other differentiated cell types, including heart muscle and nerve cells, by appropriate combinations of transcription factors.

transdifferentiation

Refer To

The Cell

Companion Website

sites.sinauer.com/cooper7e

for quizzes, animations, videos, flashcards, and other study resources.

Questions

1. Why is cell death via apoptosis more advantageous to multicellular organisms than cell death via necrosis?

2. What molecular mechanisms regulate caspase activity?

3. You have expressed mutants of nuclear lamins in human fibroblasts. The Asp residue in the caspase cleavage site has been mutated to Glu in these lamins. How would these mutant lamins affect the progression of apoptosis?

4. How do Bcl-2 family proteins regulate apoptosis in mammalian cells?

5. How does p53 activation in response to DNA damage affect cell cycle progression and cell survival?

6. You have constructed a Bad mutant in which the Akt phosphorylation site has been mutated such that Akt no longer phosphorylates it. How would expression of this mutant affect cell survival?

7. How would expression of siRNA targeted against 14-3-3 proteins affect apoptosis?

8. How would siRNA against Ced-3 affect the development of *C. elegans*?

9. Many adult tissues contain terminally differentiated cells that are incapable of proliferation. However, these tissues can regenerate following damage. What gives these tissues their regenerative capabilities?

10. What is the critical property of stem cells?

11. What are the potential advantages of embryonic stem cells as compared with adult stem cells for therapeutic applications?

12. What are the potential advantages of induced pluripotent stem cells compared with embryonic stem cells?

References and Further Reading (Key review articles for each major section are highlighted in **bold**.)

Programmed Cell Death

Czabotar, P. E., G. Lessene, A. Strasser and J. M. Adams. 2014. Control of apoptosis by the BCL-2 protein family: implications for physiology and therapy. *Nature Rev. Mol. Cell Biol.* 15: 49–63. [R]

Danial, N. K. and S. J. Korsmeyer. 2004. Cell death: Critical control points. *Cell* 116: 205–219. [R]

Ellis, H. M. and H. R. Horvitz. 1986. Genetic control of programmed cell death in the nematode *C. elegans*. *Cell* 44: 817–829. [P]

Fuchs, Y. and H. Steller. 2011. Programmed cell death in animal development and disease. *Cell* 147. 742–758. [R]

Gyrd-Hansen, M. and P. Meier. 2010. IAPs: from caspase inhibitors to modulators of NF-κB, inflammation and cancer. *Nature Rev. Cancer* 10: 561–574. [R]

Hay, B. A. and M. Guo. 2006. Caspase-dependent cell death in *Drosophila*. *Ann. Rev. Cell Dev. Biol.* 22: 623–650. [R]

Hengartner, M. O. 2000. The biochemistry of apoptosis. *Nature* 407: 770–776. [R]

Holcik, M. and N. Sonenberg. 2005. Translational control in stress and apoptosis. *Nature Rev. Mol. Cell Biol.* 6: 318–327. [R]

Jope, R. S. and G. V. W. Johnson. 2004. The glamour and gloom of glycogen synthase kinase-3. *Trends Biochem. Sci.* 29: 95–102. [R]

Lauber, K., S. G. Blumenthal, M. Waibel and S. Wesselborg. 2004. Clearance of apoptotic cells: Getting rid of the corpses. *Mol. Cell* 14: 277–287. [R]

Levine, B. and G. Kroemer. 2008. Autophagy in the pathogenesis of disease. *Cell* 132: 27–42. [R]

Moldoveanu, T., A. V. Follis, R. W. Kriwacki and D. R. Green. 2014. Many players in BCL-2 family affairs. *Trends Biochem. Sci.* 39: 101–111. [R]

Riedl, S. J. and G. S. Salvesen. 2007. The apoptosome: signaling platform of cell death. *Nature Rev. Mol. Cell Biol.* 8: 405–413. [R]

Samejima, K. and W. C. Earnshaw. 2005. Trashing the genome: The role of nucleases during apoptosis. *Nature Rev. Mol. Cell Biol.* 6: 677–688. [R]

Suzuki, J., D. P. Denning, E. Imanishi, H. R. Horvitz and S. Nagata. 2013. Xk-related protein 8 and CED-8 promote phosphatidylserine exposure in apoptotic cells. *Science* 341: 403–406. [P]

Taylor, R. C., S. P. Cullen and S. J. Martin. 2008. Apoptosis: controlled demolition at the cellular level. *Nature Rev. Mol. Cell Biol.* 9: 231–241. [R]

Vandanabeele, P., L. Galluzzi, T. V. Berghe and G. Kroemer. 2010. Molecular mechanisms of necroptosis: an ordered cellular explosion. *Nature Rev. Mol. Cell Biol.* 11: 700–714. [R]

Vaux, D. L. and J. Silke. 2005. IAPs, RINGs and ubiquitylation. *Nature Rev. Mol. Cell Biol.* 6: 287–297. [R]

Vousden, K. H. and X. Lu. 2002. Live or let die: The cells response to p53. *Nature Rev. Cancer* 2: 594–604. [R]

Wajant, H. 2002. The Fas signaling pathway: More than a paradigm. *Science* 296: 1635–1636. [R]

Stem Cells and the Maintenance of Adult Tissues

Blanpain, C. and E. Fuchs. 2009. Epidermal homeostasis: a balancing act of stem cells in the skin. *Nature Rev. Mol. Cell Biol.* 10: 207–217. [R]

Carmeliet, P. 2003. Angiogenesis in health and disease. *Nature Med.* 9: 653–660. [R]

Clevers, H. 2013. The intestinal crypt, a prototype stem cell compartment. *Cell* 154: 274–284. [R]

Clevers, H., K. M. Loh and R. Nusse. 2014. An integral program for tissue renewal and regeneration: Wnt signaling and stem cell control. *Science* 346: 1248012. [R]

Daley, G. Q. and D. T. Scadden. 2008. Prospects for stem cell-based therapy. *Cell* 132: 544–548. [R]

Dhawan, J. and T. A. Rando. 2005. Stem cells in postnatal myogenesis: Molecular mechanisms of satellite cell quiescence, activation and replenishment. *Trends Cell Biol.* 15: 666–672. [R]

Fuchs, E. and V. Horsley. 2008. More than one way to skin… *Genes Dev.* 22: 976–985. [R]

Jones, D. L. and A. J. Wagers. 2008. No place like home: anatomy and function of the stem cell niche. *Nature Rev. Mol. Cell Biol.* 9: 11–21. [R]

Kuang, S. and M. A. Rudnicki. 2008. The emerging biology of satellite cells and their therapeutic potential. *Trends Mol. Med.* 14: 82–91. [R]

McCulloch, E. A. and J. E. Till. 2005. Perspectives on the properties of stem cells. *Nature Med.* 11: 1026–1028. [R]

Morrison, S. J. and A. C. Spradling. 2008. Stem cells and niches: mechanisms that promote stem cell maintenance throughout life. *Cell* 132: 598–611. [R]

Orkin, S. H. and L. I. Zon. 2008. Hematopoiesis: an evolving paradigm for stem cell biology. *Cell* 132: 631–644. [R]

Shi, X. and D. J. Garry. 2006. Muscle stem cells in development, regeneration, and disease. *Genes Dev.* 20: 1692–1708. [R]

Simons, B. D. and H. Clevers. 2011. Strategies for homeostatic stem cell self-renewal in adult tissues. *Cell* 145: 851–862. [R]

Taub, R. 2004. Liver regeneration: From myth to mechanism. *Nature Rev. Mol. Cell Biol.* 5: 836–847. [R]

Van Berlo, J. H. and J. D. Molkentin. 2014. An emerging consensus on cardiac regeneration. *Nature Medicine* 20: 1386–1393. [R]

Wu, S. M., K. R. Chien and C. Mummery. 2008. Origins and fates of cardiovascular progenitor cells. *Cell* 132: 537–543. [R]

Zhao, C., W. Deng and F. H. Gage. 2008. Mechanisms and functional implications of adult neurogenesis. *Cell* 132: 645–660. [R]

Pluripotent Stem Cells, Cellular Reprogramming, and Regenerative Medicine

Cohen, D. E. and D. Melton. 2011. Turning straw into gold: directing cell fate for regenerative medicine. *Nature Rev. Genet.* 12: 243–252. [R]

Davis, R. L., H. Weintraub and A. B. Lassar. 1987. Expression of a single transfected cDNA converts fibroblasts to myoblasts. *Cell* 51: 987–1000. [P]

Fox, I. J., G. Q. Daley, S. A. Goldman, J. Huard, T. J. Kamp and M. Trucco. 2014. Use of differentiated pluripotent stem cells in replacement therapy for treating disease. *Science* 345: 1247391. [R]

Hanna, J. H., K. Saha and R. Jaenisch. 2010. Pluripotency and cellular reprogramming: facts, hypotheses, unresolved issues. *Cell* 143: 508–525. [R]

Ieda, M., J.-D. Fu, P. Delgado-Olguin, V. Vedantham, Y. Hayashi, B. G. Bruneau and D. Srivastava. 2010. Direct reprogramming of fibroblasts into functional cardiomyocytes by defined factors. *Cell* 142: 375–386. [P]

Ladewig, J., P. Koch and O. Brüstle. 2013. Leveling Waddington: the emergence of direct programming and the loss of cell fate hierarchies. *Nature Rev. Mol. Cell Biol.* 14: 225–236. [R]

Pagliuca, F. W., J. R. Milman, M. Gürtier, M. Segel, A. Van Devort, J. H. Ryu, Q. P. Peterson, D. Greiner and D. A. Melton. 2014. Generation of functional human pancreatic β cells *in vitro*. *Cell* 159: 428–439. [P]

Park, I.-H., R. Zhao, J. A. West, A. Yabuuchi, H. Huo, T. A. Ince, P. H. Lerou, M. W. Lensch and G. Q. Daley. 2007. Reprogramming of human somatic cells to pluripotency with defined factors. *Nature* 451: 141–147. [P]

Shamblott, M. J., J. Axelman, S. Wang, E. M. Bugg, J. W. Littlefield, P. J. Donovan, P. D. Blumenthal, G. R. Huggins and J. D. Gearhart. 1998. Derivation of pluripotent stem cells from cultured human primordial germ cells. *Proc. Natl. Acad. Sci. USA* 95: 13726–13731. [P]

Tabar, V. and L. Studer. 2014. Pluripotent stem cells in regenerative medicine: challenges and recent progress. *Nature Rev. Genet.* 15: 82–92. [R]

Takahashi, K., K. Tanabe, M. Ohnuki, M. Narita, T. Ichisaka, K. Tomoda and S. Yamanaka. 2007. Induction of pluripotent stem cells from adult human fibroblasts by defined factors. *Cell* 131: 861–872. [P]

Takahashi, K. and S. Yamanaka. 2006. Induction of pluripotent stem cells from mouse embryonic and adult fibroblast cultures by defined factors. *Cell* 126: 663–676. [P]

Thompson, J. A., J. Itskovitz-Eldor, S. S. Shapiro, M. A. Waknotz, J. J. Swiergiel, V. S. Marshall and J. M. Jones. 1998. Embryonic stem cell lines derived from human blastocysts. *Science* 282: 1145–1147. [P]

Vierbuchen, T., A. Ostermeier, Z. P. Pang, Y. Kokubu, T. C. Sudhof and M. Wernig. 2010. Direct conversion of fibroblasts to functional neurons by defined factors. *Nature* 463: 1035–1042. [P]

Wilmut, I., N. Beaujean, P. A. de Sousa, A. Dinnyes, T. J. King, L. A. Paterson, D. N. Wells and L. E. Young. 2002. Somatic cell nuclear transfer. *Nature* 419: 583–586. [R]

Wilmut, I., A. E. Schnieke, J. McWhir, A. J. Kind and K. H. S. Campbell. 1997. Viable offspring derived from fetal and adult mammalian cells. *Nature* 385: 810–813. [P]

Wu, S. M. and K. Hechedlinger. 2011. Harnessing the potential of induced pluripotent stem cells for regenerative medicine. *Nature Cell Biol.* 13: 497–505. [R]

Young, R. A. 2011. Control of the embryonic stem cell state. *Cell* 144: 940–954. [R]

Yu, J., M. A. Vodyanik, K. Smuga-Otto, J. Antosiewicz-Bourget, J. L. Frane, S. Tian, J. Nie, G. A. Jonsdottir, V. Ruotti, R. Stewart, I. I. Slukvin and J. A. Thomson. 2007. Induced pluripotent stem cell lines derived from human somatic cells. *Science* 318: 1917–1920. [P]

Cancer

Cancer is a particularly appropriate topic for the concluding chapter of this book because it results from a breakdown of the regulatory mechanisms that govern normal cell behavior. As discussed in preceding chapters, the proliferation, differentiation, and survival of individual cells in multicellular organisms are carefully regulated to meet the needs of the organism as a whole. This regulation is lost in cancer cells, which grow and divide in an uncontrolled manner, ultimately spreading throughout the body and interfering with the function of normal tissues and organs.

Because cancer results from defects in fundamental cell regulatory mechanisms, it is a disease that ultimately has to be understood at the molecular and cellular levels. Indeed, understanding cancer has been an objective of molecular and cellular biologists for many years. Importantly, studies of cancer cells have also illuminated the mechanisms that regulate normal cell behavior. In fact, many of the proteins that play key roles in cell signaling, regulation of the cell cycle, and control of programmed cell death were first identified because abnormalities in their activities led to the uncontrolled proliferation of cancer cells. The study of cancer has thus contributed significantly to our understanding of normal cell regulation, as well as vice versa.

The Development and Causes of Cancer

The fundamental abnormality resulting in the development of cancer is the continual unregulated proliferation of cancer cells. Rather than responding appropriately to the signals that control normal cell behavior, cancer cells grow and divide in an uncontrolled manner, invading normal tissues and organs and eventually spreading throughout the body. The generalized loss of growth control exhibited by cancer cells is the net result of accumulated abnormalities in multiple cell regulatory systems and is reflected in several aspects of cell behavior that distinguish cancer cells from their normal counterparts.

Types of cancer

Cancer can result from abnormal proliferation of any of the different kinds of cells in the body, so there are more than 100 distinct types of cancer, which can vary substantially in their behavior and response to treatment. The most important issue in cancer pathology is the distinction between benign and malignant tumors. A **tumor** is any abnormal proliferation of cells, which may be either benign or malignant. A **benign tumor**, such as a common skin

Video 19.1

sites.sinauer.com/cooper7e/v19.1
Normal vs. Cancer Cells Rather than responding appropriately to the signals that control normal cell behavior, cancer cells grow and divide in an uncontrolled manner.

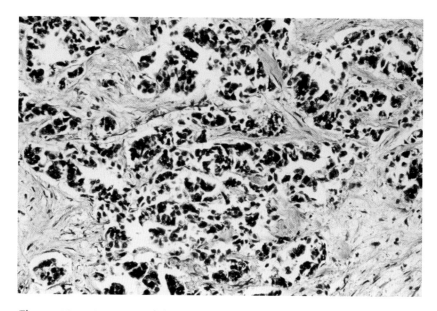

Figure 19.1 A cancer of the pancreas   Light micrograph of a section through a pancreas showing pancreatic cancer. The cancer cells have dark purple nuclei and are invading the normal tissue (pink).

Animation 19.1

sites.sinauer.com/cooper7e/a19.1

Metastasis of a Cancer A cancer begins when one altered cell gives rise to a proliferative cell population, forming a primary tumor from which cells can break away and form secondary tumors elsewhere in the body.

wart, remains confined to its original location, neither invading surrounding normal tissue nor spreading to distant body sites. A **malignant tumor**, however, is capable of both invading surrounding normal tissue (**Figure 19.1**) and spreading throughout the body via the circulatory or lymphatic systems (**metastasis**). Only malignant tumors are properly referred to as cancers, and it is their ability to invade and metastasize that makes cancer so dangerous. Whereas benign tumors can usually be removed surgically, the spread of malignant tumors to distant body sites frequently makes them resistant to such localized treatment.

Both benign and malignant tumors are classified according to the type of cell from which they arise. Most cancers fall into one of three main groups: carcinomas, sarcomas, and leukemias or lymphomas. **Carcinomas**, which include approximately 90% of human cancers, are malignancies of epithelial cells. **Sarcomas**, which are rare in humans, are solid tumors of connective tissues, such as muscle, bone, cartilage, and fibrous tissue. **Leukemias** and **lymphomas**, which account for approximately 8% of human malignancies, arise from the blood-forming cells and from cells of the immune system, respectively. Tumors are further classified according to tissue of origin (e.g., lung or breast carcinomas) and the type of cell involved. For example, fibrosarcomas arise from fibroblasts, and erythroid leukemias from precursors of erythrocytes (red blood cells).

Although there are many kinds of cancer, only a few occur frequently (**Table 19.1**). More than 1.5 million cases of cancer are diagnosed annually in the United States, and nearly 600,000 Americans die of cancer each year. Cancers of 12 different body sites account for almost 80% of this total cancer incidence. The four most common cancers—accounting for almost half of all cancer cases—are those of the breast, lung, prostate, and colon/rectum. Lung cancer, by far the most lethal, is responsible for more than 25% of all cancer deaths.

Table 19.1 Most Frequent Cancers in the United States

Cancer site	Cases per year		Deaths per year	
Breast	234,200	(14.1%)	40,700	(6.9%)
Lung	221,200	(13.3%)	158,000	(26.8%)
Prostate	220,800	(13.3%)	27,500	(4.7%)
Colon/rectum	102,500	(6.2%)	49,700	(8.4%)
Lymphomas	80,900	(4.9%)	20,900	(3.5%)
Bladder	74,000	(4.5%)	16,000	(2.7%)
Skin (melanoma)	73,900	(4.5%)	9,900	(1.7%)
Uterus	67,700	(4.1%)	14,300	(2.4%)
Thyroid	62,500	(3.8%)	1,900	(0.3%)
Kidney	61,600	(3.7%)	14,100	(2.4%)
Leukemias	54,300	(3.3%)	24,500	(4.2%)
Pancreas	49,000	(2.9%)	40,600	(6.9%)
Subtotal	1,303,000	(78.6%)	418,100	(70.9%)
All sites	1,658,000	(100%)	589,400	(100%)

Source: American Cancer Society, *Cancer Facts & Figures 2015.*

The development of cancer

One of the fundamental features of cancer is tumor clonality—the development of tumors from single cells that begin to proliferate abnormally. The single-cell origin of many tumors was first demonstrated by analysis of X chromosome inactivation (**Figure 19.2**). As discussed in Chapter 6, one member of the X chromosome pair is inactivated by being converted to heterochromatin in female cells. X inactivation occurs randomly during embryonic development, so one X chromosome is inactivated in some cells, while the other X chromosome is inactivated in other cells. Thus, if a female is heterozygous for an X chromosome gene, different alleles will be expressed in different cells. Normal tissues are composed of mixtures of cells with different inactive X chromosomes, so expression of both alleles is detected in normal tissues of heterozygous females. In contrast, tumor tissues generally express only one allele of a heterozygous X chromosome gene. The implication is that all of the cells constituting such a tumor were derived from a single cell of origin in which the pattern of X inactivation was fixed before the tumor began to develop.

The clonal origin of tumors does not, however, imply that the original progenitor cell that gives rise to a tumor has initially acquired all of the characteristics of a cancer cell. On the contrary, the development of cancer is a multistep process in which cells gradually become malignant through a progressive series of alterations. One indication of the multistep development of cancer is that most cancers develop late in life (**Figure 19.3**). This large increase in cancer incidence with age is due to the fact that most cancers develop as a consequence of multiple abnormalities, which accumulate over periods of many years.

Figure 19.2 Tumor clonality Normal tissue is a mosaic of cells in which different X chromosomes (X_1 and X_2) have been inactivated. Tumors develop from a single initially altered cell, so each tumor cell displays the same pattern of X inactivation (X_1 inactive, X_2 active).

Video 19.2

sites.sinauer.com/cooper7e/v19.2
Breast Cancer Cells Dividing Tumor cells migrate over one another and grow in a disordered, multilayered pattern.

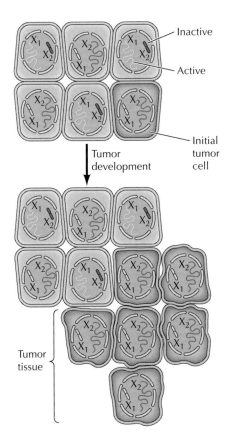

Figure 19.3 Increased rate of cancer with age The probability of developing cancer for men and women at selected age intervals. (Data from American Cancer Society, *Cancer Facts and Figures 2015.*)

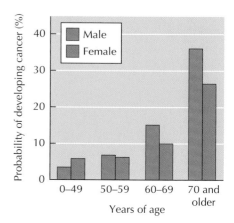

Initiation

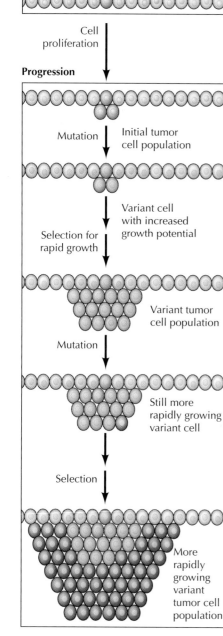

At the cellular level, the development of cancer is viewed as a multistep process involving mutation and selection for cells with progressively increasing capacity for proliferation, survival, invasion, and metastasis (**Figure 19.4**). The first step in the process, **tumor initiation**, is thought to be the result of a genetic alteration leading to abnormal proliferation of a single cell. Cell proliferation then leads to the outgrowth of a population of clonally derived tumor cells. **Tumor progression** continues as additional mutations occur within cells of the tumor population. Tumor cells typically exhibit a high level of genetic instability, resulting in a high frequency of mutations and chromosome abnormalities. Some of these mutations confer a selective advantage to the cell, such as more rapid growth, and the descendants of a cell bearing such a mutation will consequently become dominant within the tumor population. The process is called **clonal selection**, since new clones of tumor cells evolve on the basis of increased growth rate or other properties (such as survival, invasion, or metastasis) that confer a selective advantage. Clonal selection continues throughout tumor development, so tumors continuously become more rapid-growing and increasingly malignant.

Studies of colon carcinomas have provided a clear example of tumor progression during the development of a common human malignancy (**Figure 19.5**). The earliest stage in tumor development is increased proliferation of colon epithelial cells. One of the cells within this proliferative cell population is then thought to give rise to a small benign neoplasm (an **adenoma** or **polyp**). Further rounds of clonal selection lead to the growth of adenomas of increasing size and proliferative potential. Malignant carcinomas then arise from the benign adenomas, indicated by invasion of the tumor cells through the basal lamina into underlying connective tissue. The cancer cells then continue to proliferate and spread through the connective tissues of the colon wall. Eventually the cancer cells penetrate the wall of the colon and invade other abdominal organs, such as the bladder or small intestine. In addition, the cancer cells invade blood and lymphatic vessels, allowing them to metastasize throughout the body.

Figure 19.4 Stages of tumor development The development of cancer initiates when a single mutated cell begins to proliferate abnormally. Additional mutations followed by selection for more rapidly growing cells within the population then result in progression of the tumor to increasingly rapid growth and malignancy.

Figure 19.5 Development of colon carcinomas A single initially altered cell gives rise to a proliferative cell population, which progresses first to benign adenomas of increasing size and then to a malignant carcinoma. The cancer cells then invade the underlying connective tissue and penetrate blood and lymphatic vessels, thereby spreading throughout the body.

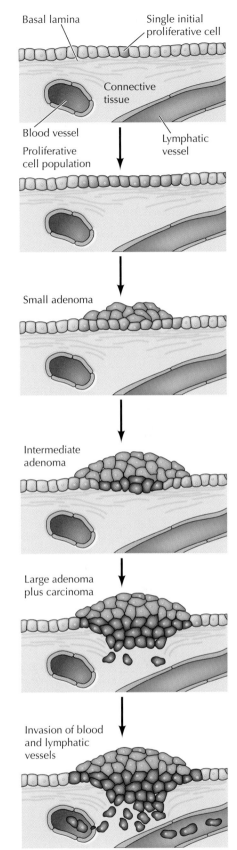

Causes of cancer

Substances that cause cancer, called **carcinogens**, have been identified both by studies in experimental animals and by epidemiological analysis of cancer frequencies in human populations (e.g., the high incidence of lung cancer among cigarette smokers). Since the development of malignancy is a complex multistep process, many factors may affect the likelihood that cancer will develop, and it is overly simplistic to speak of single causes of most cancers. Nonetheless, many agents, including radiation, chemicals, and viruses, have been found to induce cancer in both experimental animals and humans.

Radiation and most chemical carcinogens (**Figure 19.6**) act by damaging DNA and inducing mutations. Some of the carcinogens that contribute to human cancers include solar ultraviolet radiation (the major cause of skin cancer), carcinogenic chemicals in tobacco smoke, and aflatoxin (a potent liver carcinogen produced by some molds that contaminate improperly stored supplies of peanuts and other grains). The carcinogens in tobacco smoke (including benzo(α)pyrene, dimethylnitrosamine, and nickel compounds) are the major identified causes of human cancer. Smoking is the undisputed cause of nearly 90% of lung cancers, as well as being implicated in cancers of the oral cavity, pharynx, larynx, esophagus, and other sites. In total, it is estimated that smoking is responsible for nearly one-third of all cancer deaths—an impressive toll for a single carcinogenic agent.

Other carcinogens contribute to cancer development by stimulating cell proliferation, rather than by inducing mutations. Such compounds are referred to as **tumor promoters**, since the increased cell division they induce facilitates the outgrowth of a proliferative cell population during early stages of tumor development. Moreover, the cell proliferation induced by tumor promoters

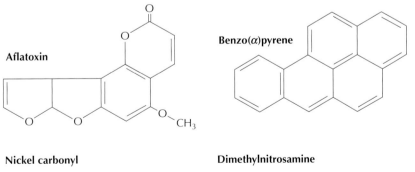

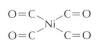

Figure 19.6 Structure of representative chemical carcinogens Aflatoxin is a carcinogen produced by molds found in peanuts and grains; benzo(α)pyrene, dimethylnitrosamine, and nickel carbonyl are carcinogens found in tobacco smoke.

leads to mutations that unavoidably occur during DNA replication. Hormones, particularly estrogens, are important as tumor promoters in the development of some human cancers. The proliferation of cells of the uterine endometrium, for example, is stimulated by estrogen, and exposure to excess estrogen significantly increases the likelihood that a woman will develop endometrial cancer. The risk of endometrial cancer is therefore substantially increased by long-term postmenopausal estrogen replacement therapy with high doses of estrogen alone. Fortunately, this risk is minimized by administration of progesterone to counteract the stimulatory effect of estrogen on endometrial cell proliferation. However, oral contraceptives and hormone replacement therapy with combinations of estrogen and progesterone still lead to an increased risk of breast cancer. Additional carcinogens that act as tumor promoters in humans include asbestos, some viruses that cause chronic inflammation and the bacterium *Helicobacter pylori*, which causes stomach cancer.

Several kinds of viruses cause cancer, both in experimental animals and in humans. The common human cancers caused by viruses include liver cancer and cervical carcinoma, which together account for 10 to 20% of worldwide cancer incidence. Studies of these viruses are important not only for identifying causes of human cancer but, as discussed later in this chapter, for playing a key role in elucidating the molecular events responsible for the development of cancers induced by both viral *and* nonviral carcinogens.

Properties of cancer cells

The uncontrolled growth of cancer cells results from accumulated abnormalities affecting many of the cell regulatory mechanisms that have been discussed in preceding chapters. This relationship is reflected in several aspects of cell behavior that distinguish cancer cells from their normal counterparts. Cancer cells typically display abnormalities in the mechanisms that regulate normal cell proliferation, differentiation, and survival. Taken together, these characteristic properties of cancer cells provide a description of malignancy at the cellular level.

The uncontrolled proliferation of cancer cells *in vivo* is mimicked by their behavior in cell culture. A primary distinction between cancer cells and normal cells in culture is that normal cells display **density-dependent inhibition** and **contact inhibition** of their growth and movement (**Figure 19.7**). Normal cells proliferate until they reach a finite cell density, which is determined both by the availability of growth factors in the culture medium and by inhibition resulting from contacts with neighboring cells. They then cease proliferating and become quiescent, arrested in the G_0 stage of the cell cycle (see Figure 17.5). The proliferation of most cancer cells, however, is not normally restricted by either growth factor availability or cell–cell contact. Rather than responding to the signals that cause normal cells to cease

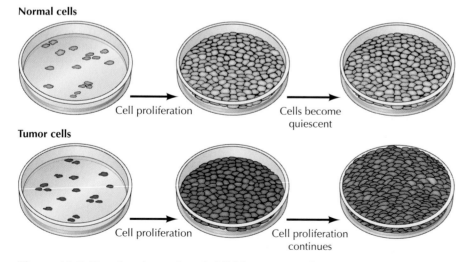

Normal cells

Cell proliferation → Cells become quiescent

Tumor cells

Cell proliferation → Cell proliferation continues

Figure 19.7 Density-dependent inhibition Normal cells proliferate in culture until they reach a finite cell density at which point they become quiescent. Tumor cells, however, continue to proliferate independent of cell density.

Normal cells

Tumor cells

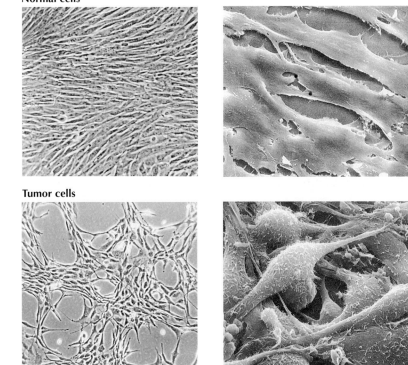

Figure 19.8 Contact inhibition Light micrographs (left) and scanning electron micrographs (right) of normal fibroblasts and tumor cells. The migration of normal fibroblasts is inhibited by cell contact, so they form an orderly side-by-side array on the surface of a culture dish. Tumor cells, however, are not inhibited by cell contact, so they migrate over one another and grow in a disordered, multilayered pattern. (Courtesy of Lan Bo Chen, Dana-Farber Cancer Institute.)

proliferation and enter G_0, cancer cells generally continue growing to high cell densities in culture, mimicking their uncontrolled proliferation *in vivo*. Similarly, normal fibroblasts migrate across the surface of a culture dish until they make contact with a neighboring cell. Further cell migration is then inhibited, and normal cells adhere to each other, forming an orderly array of cells on the culture dish surface (**Figure 19.8**). Tumor cells, in contrast, continue moving after contact with their neighbors, migrating over adjacent cells, and growing in disordered, multilayered patterns.

A related difference between normal cells and cancer cells is that many cancer cells can grow in the absence of growth factors required by their normal counterparts, contributing to the unregulated proliferation of tumor cells both *in vitro* and *in vivo*. In some cases, cancer cells produce growth factors that stimulate their own proliferation (**Figure 19.9**). Such abnormal production of a growth factor by a responsive cell leads to continuous autostimulation of cell division (**autocrine growth stimulation**), and the cancer cells are therefore less dependent on growth factors from other, physiologically normal sources. In other cases, the reduced growth factor dependence of cancer cells results from abnormalities in intracellular signaling systems, such as unregulated activity of growth factor receptors or other proteins (e.g., Ras proteins or protein kinases) that were discussed in Chapter 16 as elements of signal transduction pathways leading to cell proliferation.

Animation 19.2

sites.sinauer.com/cooper7e/a19.2
Density-Dependent Inhibition In culture, normal cells proliferate until they reach a certain cell density, but tumor cells continue to proliferate independent of cell density.

Figure 19.9 Autocrine growth stimulation A cell produces a growth factor to which it and neighboring cells also respond, resulting in continuous stimulation of cell proliferation.

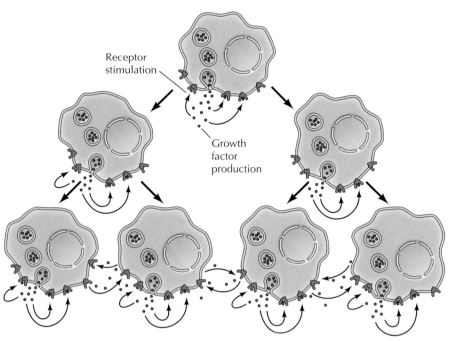

Cancer cells are also less stringently regulated than normal cells by cell–cell and cell–matrix interactions. Most cancer cells are less adhesive than normal cells, often as a result of reduced expression of cell surface adhesion molecules. For example, loss of E-cadherin, the principal adhesion molecule of epithelial cells (see Figure 15.25), is important in the development of carcinomas (epithelial cancers). As a result of reduced expression of cell adhesion molecules, cancer cells are comparatively unrestrained by interactions with other cells and tissue components, contributing to the ability of malignant cells to invade and metastasize. The reduced adhesiveness of cancer cells also results in morphological and cytoskeletal alterations: Many tumor cells are rounder than normal, in part because they are less firmly attached to either the extracellular matrix or neighboring cells.

Two additional properties of cancer cells affect their interactions with other tissue components, thereby playing important roles in invasion and metastasis. First, malignant cells generally secrete proteases that digest extracellular matrix components, allowing the cancer cells to invade adjacent normal tissues. Secretion of proteases that digest collagen, for example, appears to be an important determinant of the ability of carcinomas to penetrate through basal laminae and invade underlying connective tissue (see Figure 19.5). Second, cancer cells secrete growth factors that promote the formation of new blood vessels (**angiogenesis**). Angiogenesis is needed to support the growth of a tumor beyond the size of about a million cells, at which point new blood vessels are required to supply oxygen and nutrients to the proliferating tumor cells. Such blood vessels are formed in response to growth factors, secreted by the tumor cells, that stimulate proliferation of endothelial cells in the walls of capillaries in surrounding tissue, resulting in the outgrowth of new capillaries into the tumor (see Figure 18.15). The formation of such new blood vessels is important not only in supporting tumor growth but also in metastasis. The actively growing new capillaries formed in response to angiogenic stimulation are easily penetrated by the tumor cells, providing a ready opportunity for cancer cells to enter the circulatory system and begin the metastatic process.

Another general characteristic of most cancer cells is that they fail to differentiate normally. Such defective differentiation is closely coupled to abnormal proliferation since, as discussed in Chapter 18, most fully differentiated cells cease cell division. Rather than carrying out their normal differentiation program, cancer cells are usually blocked at an early stage of differentiation, consistent with their continued active proliferation.

The leukemias provide a particularly good example of the relationship between defective differentiation and malignancy. All of the different types of blood cells are derived from a common hematopoietic stem cell in the bone marrow (see Figure 18.18). Descendants of these cells then become committed to specific differentiation pathways. Some cells, for example, differentiate to form erythrocytes, whereas others differentiate to form lymphocytes, granulocytes, or macrophages. Progenitor cells of each of these types undergo several rounds of division as they differentiate, but once they become fully differentiated, cell division ceases. Leukemic cells, in contrast, fail to undergo terminal differentiation (**Figure 19.10**). Instead, they become arrested at early stages of maturation, at which point they retain their capacity for proliferation and continue to reproduce.

The growth of leukemias, and possibly some solid tumors, may be driven by the proliferation of a subpopulation of cancer stem cells, rather than by continued proliferation of all cells in the tumor. In this model, the cancer stem cells (like the stem cells of normal adult tissues; see Figure 18.17) divide to produce more stem cells as well as differentiated tumor cells that lack the potential for continual self-renewal. A clear example is provided by chronic myeloid leukemia, which arises from oncogenic transformation of the hematopoietic stem cells, which then give rise to differentiated leukemic cells of both myeloid and lymphoid lineages.

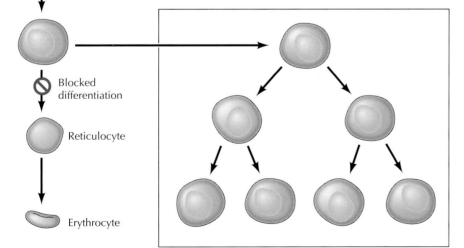

Leukemic cells fail to differentiate and continue to divide.

Figure 19.10 Defective differentiation and leukemia Different types of blood cells develop from the hematopoietic stem cell in the bone marrow. The precursors of differentiated cells undergo several rounds of cell division as they mature, but cell division ceases at the terminal stages of differentiation. The differentiation of leukemic cells is blocked at early stages of maturation, consistent with their continued proliferation.

Evidence for cancer stem cells has also been obtained in other leukemias and some solid tumors, although the extent to which cancer stem cells contribute to the growth of most tumors remains to be established.

As discussed in Chapter 18, **programmed cell death** or **apoptosis** is an integral part of the differentiation program of many cell types, including blood cells. Many cancer cells fail to undergo apoptosis and therefore exhibit increased life spans compared with their normal counterparts. This failure of cancer cells to undergo programmed cell death contributes substantially to tumor development. For example, the survival of many normal cells is dependent on signals from either growth factors or from the extracellular matrix that prevent apoptosis. In contrast, tumor cells are often able to survive in the absence of growth factors required by their normal counterparts. Such a failure of tumor cells to undergo apoptosis when deprived of normal environmental signals may be important not only in primary tumor development but also in the survival and growth of metastatic cells in abnormal tissue sites. Normal cells also undergo apoptosis following DNA damage, while many cancer cells fail to do so. In this case, the failure to undergo apoptosis contributes to the resistance of cancer cells to irradiation and many chemotherapeutic drugs, which act by damaging DNA. Abnormal cell survival, as well as cell proliferation, thus plays a major role in the unrelenting growth of cancer cells in an animal.

The continued proliferation of cancer cells also requires that they effectively maintain the ends of their chromosomes, which are lost as normal cells divide. The ends of eukaryotic chromosomes (telomeres) consist of repeated sequences that are replicated by a special enzyme called telomerase (see Figure 7.16). Normal cells only contain sufficient telomerase to maintain the ends of their chromosomes for a limited number of divisions. As a result, normal cells gradually lose their telomeres as they divide, leading to cessation of replication (senescence) or death. In contrast, cancer cells generally acquire the capacity for unlimited replication as a result of expression of high levels of telomerase, allowing them to maintain the ends of their chromosomes for an indefinite number of divisions.

Transformation of cells in culture

The study of tumor induction by radiation, chemicals, or viruses requires experimental systems in which the effects of a carcinogenic agent can be reproducibly observed and quantitated. Although the activity of carcinogens can be assayed in intact animals, such experiments are difficult to quantitate and control. The development of *in vitro* assays to detect the conversion of normal cells to tumor cells in culture, a process called **cell transformation**, therefore represented a major advance in cancer research. Such assays are designed to detect transformed cells, which display the *in vitro* growth properties of tumor cells, following exposure of a culture of normal cells to a carcinogenic agent. Their application has allowed experimental analysis of cell transformation to reach a level of sophistication that could not have been attained by studies in whole animals alone.

The first and most widely used assay of cell transformation is the focus assay, which was developed by Howard Temin and Harry Rubin in 1958. The focus assay is based on the ability to recognize a group of transformed cells as a morphologically distinct "focus" against a background of normal cells on the surface of a culture dish (**Figure 19.11**). The focus assay takes advantage of the altered morphology of cancer cells, as well as loss of contact and density-dependent inhibition. The result is the formation of a colony of morphologically-altered transformed cells that overgrow the background of normal cells in the culture. Such foci of transformed cells can usually be detected and quantified within a week or two after exposure to a carcinogenic agent. In general, cells transformed *in vitro* are able to form tumors following inoculation into susceptible animals, supporting *in vitro* transformation as a valid indicator of the formation of cancer cells.

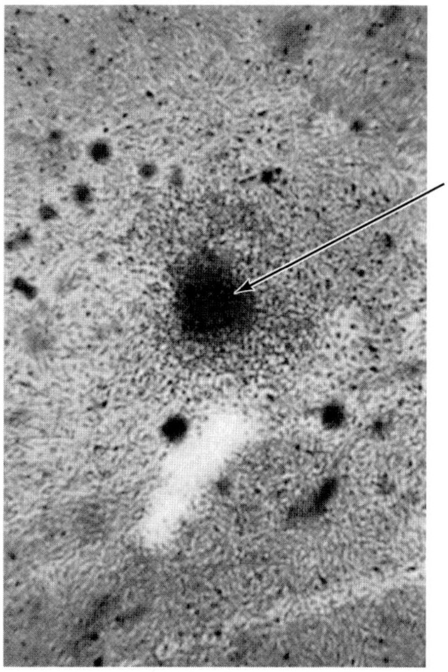

Tumor Viruses

Members of several families of viruses, called **tumor viruses**, are capable of directly causing cancer in either experimental animals or humans (Table 19.2). The viruses that cause human cancer include hepatitis B and C viruses (liver cancer), papillomaviruses (cervical and other anogenital cancers), Merkel cell polyomavirus (Merkel cell carcinoma), herpesviruses such as Epstein-Barr virus (Burkitt's lymphoma and nasopharyngeal carcinoma) and Kaposi's sarcoma-associated herpesvirus (Kaposi's sarcoma), as well as retroviruses such as human T-cell lymphotropic virus (adult T-cell leukemia). In addition, HIV is indirectly responsible for the cancers that develop in AIDS patients as a result of immunodeficiency.

As already noted, tumor viruses not only are important as causes of human disease but have also played a critical role in cancer research by serving as models for cellular and molecular studies of cell transformation. The small size of their genomes has made tumor viruses readily amenable to molecular analysis, leading to the identification of viral genes responsible for cancer induction and paving the way to our current understanding of cancer at the molecular level.

Figure 19.11 The focus assay
A focus of chicken embryo fibroblasts (arrow) induced by Rous sarcoma virus. (From H. M. Temin and H. Rubin, 1958. *Virology* 6: 669.)

Hepatitis B and C viruses

The **hepatitis B** and **C viruses** are the principal causes of liver cancer, which is the third leading cause of cancer deaths worldwide. Both hepatitis B and C viruses specifically infect liver cells and can lead to long-term chronic infections of the liver. Such chronic infections are associated with a high risk of liver cancer, which eventually develops in 10–20% of people chronically infected with hepatitis B and in about 5% of people chronically infected with hepatitis C.

Table 19.2 Tumor Viruses

Virus family	Human tumors	Genome size (kb)
DNA Genomes		
Hepatitis B viruses	Liver cancer	3
Polyomaviruses and SV40	Merkel cell carcinoma	5
Papillomaviruses	Cervical carcinoma	8
Adenoviruses	None	35
Herpesviruses	Burkitt's lymphoma, nasopharyngeal carcinoma, Kaposi's sarcoma	100–200
RNA Genomes		
Hepatitis C virus	Liver cancer	10
Retroviruses	Adult T-cell leukemia	9–10

FYI

The Pap smear was developed by George Papanicolaou in the 1930s. Uterine cells are collected and smeared on a microscope slide. Since cancer cells have altered morphology they can be readily detected upon microscopic examination.

The molecular mechanisms by which the hepatitis viruses cause liver cancer are not fully understood. Hepatitis B is a DNA virus with a genome of only 3 kb. Cell transformation by hepatitis B virus may be mediated by a viral protein (called HBx) that inhibits the p53 tumor suppressor protein and affects expression of several cellular genes that drive cell proliferation. In addition, the development of cancers induced by hepatitis B virus is driven by the continual proliferation of liver cells that results from chronic tissue damage and inflammation. Hepatitis C is an RNA virus with a genome of approximately 10 kb. Cell proliferation in response to chronic inflammation is a major contributor to cancer development in people chronically infected with hepatitis C virus, although it is also possible that some hepatitis C virus proteins directly stimulate the proliferation of infected liver cells.

Small DNA tumor viruses

The small DNA tumor viruses include several families of viruses (the **polyomaviruses**, **papillomaviruses,** and **adenoviruses**) with genomes ranging from 5 to 35 kb. The papillomaviruses are important causes of human cancer. Approximately 150 different types of human papillomaviruses, which infect epithelial cells of several tissues, have been identified. Some of these viruses cause only benign tumors (such as warts), whereas others are causative agents of malignant carcinomas, particularly cervical and other anogenital cancers. The mortality from cervical cancer is relatively low in the United States, in large part as a result of early detection and curative treatment made possible by the Pap smear. Vaccines against the most common types of human papillomaviruses are also effective in preventing cervical cancer. In other parts of the world, however, cervical cancer remains common; it is the second most common cancer among women and is responsible for 5 to 10% of worldwide cancer incidence.

The most recently described human tumor virus is **Merkel cell polyomavirus**, which was identified in 2008 as the cause of a rare skin cancer, Merkel cell carcinoma. Other polyomaviruses, SV40, and the adenoviruses are not associated with human cancer, but they have been critically important as models for understanding the molecular basis of cell transformation. The utility of these viruses in cancer research has stemmed from the availability of good cell culture assays for both virus replication and transformation, as well as from the small size of their genomes.

Most polyomaviruses, SV40, and adenoviruses do not induce tumors or transform cells of their natural host species in which they replicate. Instead, virus replication leads to cell lysis and release of progeny virus particles (Figure 19.12). Since the cell is killed as a consequence of virus replication, it cannot become transformed. The transforming potential of these viruses is revealed, however, by infection of a cell in which virus replication is blocked. This might occur if the cell is lacking host factors that are required for virus replication, or if the viral genome integrates into cellular DNA rather than replicating. Under these circumstances of nonproductive infection, expression of specific viral genes results in transformation of the infected cell.

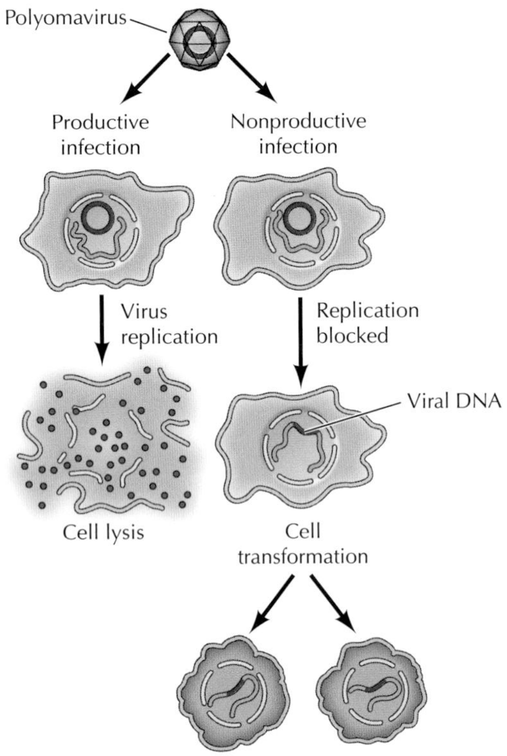

Figure 19.12 Polyomavirus replication and transformation Productive infection results in virus replication, cell lysis, and release of progeny virus particles. In a nonproductive infection, virus replication is blocked and portions of the viral DNA can become integrated into the host genome, leading to cell transformation.

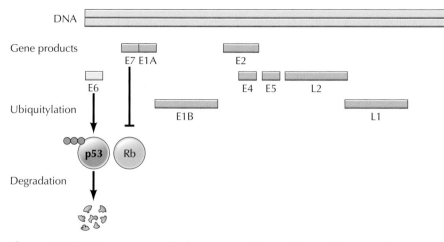

Figure 19.13 The genome of a human papillomavirus   Gene products are designated E (early) or L (late). Transformation results from the action of E6 and E7. E6 stimulates ubiquitylation and degradation of p53, and E7 inhibits Rb.

The viral genes that lead to cell transformation are the same as those that function in early stages of lytic infection (leading to cell lysis). The genomes of the small DNA tumor viruses are divided into early and late regions. The early region is expressed immediately after infection and is required for synthesis of viral DNA. The late region is not expressed until after viral DNA replication has begun and includes genes encoding structural components of the virus particle. During lytic infection these early-region proteins fulfill multiple functions required for virus replication, including stimulating host cell gene expression and DNA synthesis. Since virus replication is dependent on host cell enzymes (e.g., DNA polymerase), such stimulation of the host cell is a critical event in the viral life cycle. As discussed in Chapter 18, most cells in an animal are nonproliferating and therefore must be stimulated to divide in order to induce the enzymes needed for viral DNA replication. This stimulation of cell proliferation by the early gene products can lead to transformation if the viral DNA becomes stably integrated and expressed in a cell in which virus replication is blocked.

The early region proteins of small DNA viruses induce transformation by interacting with host proteins that regulate cell proliferation. For example, cell transformation by human papillomaviruses results from expression of two early-region genes, *E6* and *E7* (**Figure 19.13**). The E6 and E7 proteins bind to and inactivate the host cell tumor suppressor proteins p53 and Rb, which are key regulators of cell proliferation and survival. In particular, E6 stimulates the degradation of p53 by ubiquitin-mediated proteolysis, and E7 inhibits Rb. The transforming proteins of SV40 and adenoviruses similarly target p53 and Rb, indicating a key role for inactivation of these tumor suppressors in cell transformation.

Herpesviruses

The **herpesviruses** are among the most complex animal viruses, with genomes of 100 to 200 kb. Several herpesviruses induce tumors in animal species, including frogs, chickens, and monkeys. In addition, two members of the herpesvirus family, **Kaposi's sarcoma-associated herpesvirus** and **Epstein-Barr**

virus, cause human cancers. Kaposi's sarcoma-associated herpesvirus plays a critical role in the development of Kaposi's sarcomas, and Epstein-Barr virus has been implicated in several human malignancies, including Burkitt's lymphoma in some regions of Africa, B-cell lymphomas in AIDS patients and other immunosuppressed individuals, and nasopharyngeal carcinoma in China.

In addition to its association with these human malignancies, Epstein-Barr virus is able to transform human B lymphocytes in culture. Partly because of the complexity of the genome, however, the molecular biology of Epstein-Barr virus replication and transformation remains to be fully understood. The main Epstein-Barr virus transforming protein (LMP1) mimics a cell surface receptor on B lymphocytes and functions by activating signaling pathways that stimulate cell proliferation and inhibit apoptosis. Several other viral genes may also contribute to transformation of lymphocytes, but their functions in the transformation process have not been established.

Kaposi's sarcoma-associated herpesvirus DNA is regularly found in Kaposi's sarcoma cells, and several viral genes that affect cell proliferation and survival are expressed in these tumors. A noteworthy feature of Kaposi's sarcoma cells is that they secrete a variety of cytokines and growth factors that drive tumor development. Interestingly, the transforming proteins of Kaposi's sarcoma-associated herpesvirus appear to act at least in part by stimulating growth factor secretion.

Retroviruses

Retroviruses cause cancer in a variety of animal species, including humans. One human retrovirus, human T-cell lymphotropic virus type I (HTLV-I), is the causative agent of adult T-cell leukemia, which is common in parts of Japan, the Caribbean, and Africa. Transformation of T lymphocytes by HTLV-I results from expression of the viral gene *tax*, which encodes a regulatory protein affecting expression of several cellular growth control genes. AIDS is caused by another retrovirus, HIV. In contrast with HTLV-I, HIV does not cause cancer by directly converting a normal cell into a tumor cell. However, AIDS patients suffer a high incidence of some malignancies, particularly lymphomas and Kaposi's sarcoma. These cancers, which are also common among other immunosuppressed individuals, are associated with infection by other viruses (e.g., Epstein-Barr virus and Kaposi's sarcoma-associated herpesvirus) and apparently develop as a secondary consequence of immunosuppression in AIDS patients.

Different retroviruses differ substantially in their oncogenic potential. Most retroviruses contain only three genes (*gag*, *pol*, and *env*) that are required for virus replication but play no role in cell transformation (Figure 19.14). Retroviruses of this type induce tumors only rarely, if at all, as a consequence of mutations resulting from the integration of proviral DNA within or adjacent to cellular genes.

Other retroviruses, however, contain specific genes responsible for induction of cell transformation and are potent carcinogens. The prototype of these highly oncogenic retroviruses is **Rous sarcoma virus (RSV)**, first isolated from a chicken sarcoma by Peyton Rous in 1911. More than 50 years later, studies of RSV led to identification of the first viral oncogene, which has provided a model for understanding many aspects of tumor development at the molecular level.

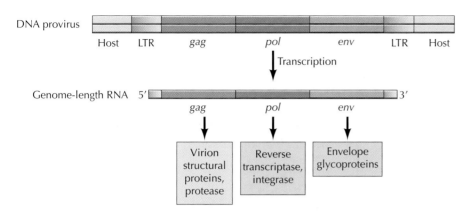

Figure 19.14 A typical retrovirus genome   The DNA provirus, integrated into cellular DNA, is transcribed to yield genome-length RNA. This primary transcript serves as the genomic RNA for progeny virus particles and as mRNA for the *gag* and *pol* genes. In addition, the full-length RNA is spliced to yield mRNA for *env*. The *gag* gene encodes the structural proteins of the virus particles and viral protease, *pol* encodes reverse transcriptase and integrase, and *env* encodes envelope glycoproteins.

Oncogenes

Cancer results from alterations in critical regulatory genes that control cell proliferation, differentiation, and survival. Studies of tumor viruses revealed that specific genes (called **oncogenes**) are capable of inducing cell transformation, thereby providing the first insights into the molecular basis of cancer. However, the majority (approximately 80%) of human cancers are not induced by viruses and apparently arise from other causes, such as radiation and chemical carcinogens. Therefore, in terms of our overall understanding of cancer, it has been critically important that studies of viral oncogenes also led to the identification of cellular oncogenes, which are involved in the development of non-virus-induced cancers. The key link between viral and cellular oncogenes was provided by studies of the highly oncogenic retroviruses.

Retroviral oncogenes

Viral oncogenes were first defined in RSV, which transforms chicken embryo fibroblasts in culture and induces large sarcomas within 1 to 2 weeks after inoculation into chickens (**Figure 19.15**). In contrast, the closely related avian leukosis virus (ALV) replicates in the same cells as RSV without inducing transformation. This difference in transforming potential suggested the possibility that RSV contains specific genetic information responsible for transformation of infected cells. A direct comparison of the genomes of RSV and ALV was consistent with this hypothesis: The genomic RNA of RSV is about 10 kb, whereas that of ALV is smaller, about 8.5 kb.

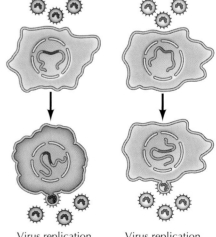

Figure 19.15 Cell transformation by RSV and ALV   Both RSV and ALV infect and replicate in chicken embryo fibroblasts, but only RSV induces cell transformation.

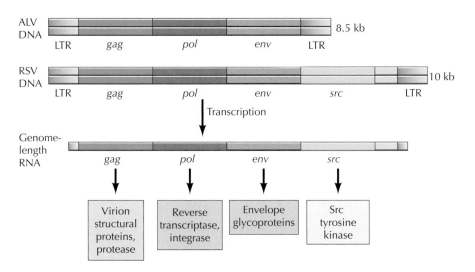

Figure 19.16 The RSV and ALV genomes RSV contains an additional gene, *src* (not present in ALV), which encodes the Src tyrosine kinase. The other three retrovirus genes (*gag*, *pol*, and *env*, see Figure 19.14) are required for replication, but play no role in cell transformation.

In the early 1970s, Peter Vogt and Steven Martin isolated both deletion mutants and temperature-sensitive mutants of RSV that were unable to induce transformation. Importantly, these mutants still replicated normally in infected cells, indicating that RSV contains genetic information that is required for transformation but not for virus replication. Further analysis demonstrated that both the deletion and the temperature-sensitive RSV mutants define a single gene responsible for the ability of RSV to induce tumors in birds and transform fibroblasts in culture. Because RSV causes sarcomas, its oncogene was called **src**. The *src* gene is an addition to the genome of RSV; it is not present in ALV (**Figure 19.16**). It encodes a 60-kd protein that was the first tyrosine kinase to be identified (see the Key Experiment in Chapter 9).

More than 40 different highly oncogenic retroviruses have been isolated from a variety of animals, including chickens, turkeys, mice, rats, cats, and monkeys. All of these viruses, like RSV, contain at least one oncogene (in some cases two) that is not required for virus replication but is responsible for cell transformation. In some cases, different viruses contain the same oncogenes, but more than two dozen distinct oncogenes have been identified among this group of viruses (**Table 19.3**). Like *src*, many of these genes (such as **ras** and **raf**) encode proteins that are now recognized as key components of signaling pathways that stimulate cell proliferation (see Figure 16.28).

Table 19.3 Representative Retroviral Oncogenes

Oncogene	Virus	Species
abl	Abelson leukemia	Mouse
akt	AKT8	Mouse
erbA	Avian erythroblastosis-ES4	Chicken
erbB	Avian erythroblastosis-ES4	Chicken
fos	FBJ murine osteogenic sarcoma	Mouse
jun	Avian sarcoma-17	Chicken
mos	Moloney sarcoma	Mouse
myc	Avian myelocytomatosis	Chicken
p3k	Avian sarcoma-16	Chicken
raf	3611 murine sarcoma	Mouse
rasH	Harvey sarcoma	Rat
rasK	Kirsten sarcoma	Rat
rel	Reticuloendotheliosis	Turkey
sis	Simian sarcoma	Monkey
src	Rous sarcoma	Chicken

Proto-oncogenes

An unexpected feature of retroviral oncogenes is their lack of involvement in virus replication. Since most viruses are streamlined to replicate as efficiently as possible, the existence of viral oncogenes that are not an integral part of the virus life cycle seems paradoxical. Scientists were thus led to question where the retroviral oncogenes had originated and how they had become incorporated into viral genomes—a line of investigation that ultimately led to the identification of cellular oncogenes in human cancers.

The first clue to the origin of oncogenes came from the way in which the highly oncogenic retroviruses were isolated. The isolation of Abelson leukemia virus is a typical example (**Figure 19.17**). More than 150 mice were inoculated with a nontransforming virus containing only the *gag*, *pol*, and *env* genes required for virus replication. One of these mice developed a lymphoma from which a new, highly oncogenic virus (Abelson leukemia virus), which now contained an oncogene (*abl*), was isolated. The scenario suggested the hypothesis that the retroviral oncogenes are derived from genes of the host cell. A host cell gene that can drive cell proliferation occasionally becomes incorporated into a viral genome, yielding a new, highly oncogenic virus with an oncogene derived from the host cell.

The critical prediction of this hypothesis was that normal cells contain genes that are closely related to the retroviral oncogenes. This was definitively demonstrated in 1976 by Harold Varmus, J. Michael Bishop, and their colleagues, who showed that a cDNA probe for the *src* oncogene of RSV hybridized to closely-related sequences in the DNA of normal chicken cells. Moreover, *src*-related sequences were also found in normal DNAs of a wide range of other vertebrates (including humans) and thus appeared to be highly conserved in evolution. Similar experiments with probes for the oncogenes of other highly oncogenic retroviruses have yielded comparable results, and it is now firmly established that the retroviral oncogenes were derived from closely related genes of normal cells.

The normal-cell genes from which the retroviral oncogenes originated are called **proto-oncogenes**. They are important cell regulatory genes, in many cases encoding proteins that function in the signal transduction pathways controlling normal cell proliferation (e.g., *src*, *ras*, and *raf*). The oncogenes are abnormally expressed or mutated forms of the corresponding proto-oncogenes. As a consequence of such alterations, the oncogenes induce abnormal cell proliferation and tumor development.

An oncogene incorporated into a retroviral genome differs in several respects from the corresponding proto-oncogene. First, the viral oncogene is transcribed under the control of viral promoter and enhancer sequences, rather than being controlled by the normal transcriptional regulatory sequences of the proto-oncogene. Consequently, oncogenes are usually expressed at much

Figure 19.17 Isolation of Abelson leukemia virus   The highly oncogenic virus Ab-MuLV was isolated from a rare tumor that developed in a mouse that had been inoculated with a nontransforming virus (Moloney murine leukemia virus, or Mo-MuLV). Mo-MuLV contains only the *gag*, *pol*, and *env* genes required for virus replication. In contrast, Ab-MuLV has acquired a new oncogene (*abl*), which is responsible for its transforming activity. The *abl* oncogene replaced some of the viral replicative genes (*pol* and *env*) and is fused with a partially deleted *gag* gene (designated Δ *gag*) in the Ab-MuLV genome.

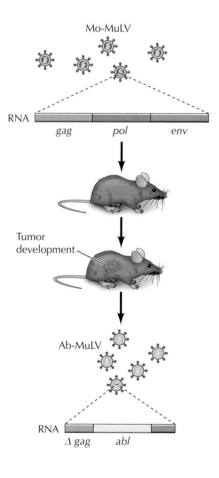

Key Experiment

The Discovery of Proto-Oncogenes

DNA Related to the Transforming Gene(s) of Avian Sarcoma Viruses Is Present in Normal Avian DNA

Dominique Stehelin, Harold E. Varmus, J. Michael Bishop and Peter K. Vogt

Department of Microbiology, University of California at San Francisco (DS, HEV, and JMB) and Department of Microbiology, University of California at Los Angeles (PKV)

Nature, Volume 260, 1976, pages 170–173

The Context

Genetic analysis of RSV (Rous sarcoma virus) defined the first viral oncogene (*src*) as a gene that was specifically responsible for cell transformation but was not required for virus replication. The origin of highly oncogenic retroviruses from tumors of infected animals then suggested the hypothesis that retroviral oncogenes are derived from related genes of host cells. Consistent with this suggestion, normal cells of several species were found to contain retrovirus-related DNA sequences that could be detected by nucleic acid hybridization. However, it was unclear whether these sequences were related to the retroviral oncogenes or to the genes required for virus replication.

Harold Varmus, J. Michael Bishop, and their colleagues resolved this critical issue by exploiting the genetic characterization of the *src* oncogene. In particular, Peter Vogt had previously isolated transformation-defective mutants of RSV that had sustained deletions of approximately 1.5 kb, corresponding to most or all of the *src* gene. Stehelin and collaborators used these mutants to prepare a cDNA probe that specifically represented *src* sequences. The use of this defined probe in nucleic acid hybridization experiments allowed them to definitively demonstrate that normal cells contain *src*-related DNA sequences.

The Experiments

First, reverse transcriptase was used to synthesize a radioactive cDNA probe composed of short single-strand DNA fragments complementary to the entire genomic RNA of RSV. This probe was then hybridized to an excess of RNA isolated from a transformation-defective deletion mutant. Fragments of cDNA that were complementary to the viral replication genes hybridized to the transformation-defective RSV RNA, forming RNA–DNA duplexes. In contrast, cDNA fragments that were complementary to *src* were unable to hybridize and remained single-strand. This single-strand DNA was then isolated to provide a specific probe for *src* oncogene sequences. As predicted from the size of deletions in the transformation-defective RSV mutants, the *src*-specific probe was homologous to about 1.5 kb of RSV RNA.

The radioactive *src* cDNA was then used as a hybridization probe to attempt to detect related DNA sequences in normal avian cells. Strikingly, the *src* cDNA hybridized extensively to normal chicken DNA, as well as to DNA of other avian species (see figure). These experiments thus demonstrated that normal cells contain DNA sequences that are closely related to the *src* oncogene, supporting the hypothesis that retroviral oncogenes originated from cellular genes that became incorporated into viral genomes.

The Impact

Stehelin and colleagues concluded their 1976 paper by raising the possibility that the cellular *src* sequences are "involved in the normal regulation of cell growth and development or the transformation of cell behavior by physical, chemical or viral agents." This prediction has been strikingly substantiated, and the discovery of cellular *src* sequences has opened new doors to understanding both the regulation of normal cell proliferation and the molecular basis of human cancer. Studies of oncogene and proto-oncogene proteins,

J. Michael Bishop   Harold Varmus

including Src itself, have proven critical in unraveling the signaling pathways that control the proliferation and differentiation of normal cells. The discovery of the *src* proto-oncogene further suggested that non-virus-induced tumors could arise as a result of mutations in related cellular genes, leading directly to the discovery of oncogenes in human tumors. By unifying studies of tumor viruses, normal cells, and non-virus-induced tumors, the results of Varmus, Bishop, and their colleagues have had an impact on virtually all aspects of studies of cell regulation and cancer research.

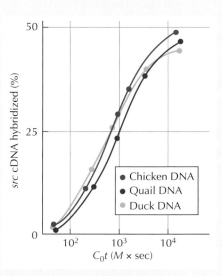

Hybridization of *src*-specific cDNA to normal chicken, quail, and duck DNA.

Figure 19.18 The Raf oncogene protein The Raf proto-oncogene protein consists of an amino-terminal regulatory domain and a carboxy-terminal protein kinase domain. In the viral Raf oncogene protein, the regulatory domain has been deleted and replaced by partially deleted viral Gag sequences (Δ Gag). As a result, the Raf kinase domain becomes constitutively active, causing cell transformation.

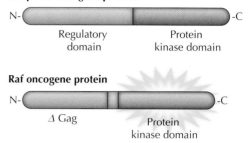

Raf proto-oncogene protein

Raf oncogene protein

higher levels than the proto-oncogenes and are sometimes transcribed in inappropriate cell types. In some cases, such abnormalities of gene expression are sufficient to convert a normally functioning proto-oncogene into an oncogene that drives cell transformation.

In addition to such alterations in gene expression, oncogenes frequently encode proteins that differ in structure and function from those encoded by their normal homologs. Many oncogenes, such as *raf*, are expressed as fusion proteins with viral sequences at the amino terminus (**Figure 19.18**). Recombination events leading to the generation of such fusion proteins often occur during the capture of proto-oncogenes by retroviruses, and sequences from both the amino and carboxy termini of proto-oncogenes are frequently deleted during the process. Such deletions may result in the loss of regulatory domains that control the activity of the proto-oncogene proteins, thereby generating oncogene proteins that function in an unregulated, hyperactive manner. For example, the viral *raf* oncogene encodes a fusion protein in which amino-terminal sequences of the normal Raf protein have been deleted. These amino-terminal sequences are critical to the normal regulation of Raf protein kinase activity, and their deletion results in unregulated (constitutive) activity of the oncogene-encoded Raf protein. This hyperactivity of Raf drives abnormal cell proliferation, resulting in transformation.

Many other oncogenes differ from their corresponding proto-oncogenes by point mutations, resulting in single amino acid substitutions in the oncogene products. In some cases, such amino acid substitutions (like the deletions already discussed) lead to unregulated activity of the oncogene proteins. An important example of such point mutations is provided by the *ras* oncogenes, which are discussed in the next section in terms of their role in human cancers.

Oncogenes in human cancer

Understanding the origin of retroviral oncogenes raised the question of whether non-virus-induced tumors contain cellular oncogenes that were generated from proto-oncogenes by mutations or by DNA rearrangements during tumor development. Direct evidence for the involvement of cellular oncogenes in human tumors was first obtained by gene transfer experiments in 1981 in the laboratories of Robert Weinberg and Geoffrey Cooper (one of the authors of this book). DNA of a human bladder carcinoma was found to efficiently induce transformation of recipient mouse cells in culture, indicating that the human tumor contained a biologically active cellular oncogene (**Figure 19.19**). Both gene

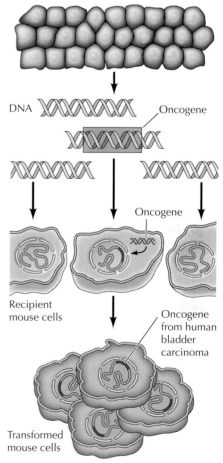

Figure 19.19 Detection of a human tumor oncogene by gene transfer DNA extracted from a human bladder carcinoma induced transformation of recipient mouse cells in culture. Transformation resulted from integration and expression of an oncogene derived from the human tumor.

transfer assays and alternative experimental approaches have since led to the detection of active cellular oncogenes in human tumors of many different types (Table 19.4).

Some of the oncogenes identified in human tumors are cellular homologs of oncogenes that were previously characterized in retroviruses, whereas others are new oncogenes first discovered in human cancers. The first human oncogene identified in gene transfer assays was subsequently identified as the human homolog of the *ras*H oncogene of Harvey sarcoma virus (see Table 19.3). Three closely related members of the *ras* gene family (*ras*H, *ras*K, and *ras*N) are the oncogenes most frequently encountered in human tumors. These genes are involved in approximately 25% of all human malignancies, including about 50% of colon and 25% of lung carcinomas.

The *ras* oncogenes arise from the proto-oncogenes present in normal cells as a result of mutations that occur during tumor development. The *ras* oncogenes differ from their proto-oncogenes by point mutations resulting in single amino acid substitutions at critical positions. The first such mutation discovered was the substitution of valine for glycine at position 12 (**Figure 19.20**). Other amino acid substitutions at position 12 (as well as at positions 13 and 61) are also frequently encountered in *ras* oncogenes in human tumors. The mutations that convert *ras* proto-oncogenes to oncogenes are caused by

---

Table 19.4 Representative Oncogenes of Human Tumors

Oncogene	Type of cancer	Activation mechanism
abl	Chronic myeloid leukemia, acute lymphocytic leukemia	Translocation
bcl-2	Follicular B-cell lymphoma	Translocation
CCND1	Parathyroid adenoma, B-cell lymphoma	Translocation
CCND1	Squamous cell, bladder, breast, esophageal, liver, and lung carcinomas	Amplification
cdk4	Melanomas	Point mutation
CTNNB1	Colon carcinoma	Point mutation
erbB	Gliomas, many carcinomas	Amplification
erbB	Lung carcinomas	Point mutation
erbB-2	Breast and ovarian carcinomas	Amplification
c-*myc*	Burkitt's lymphoma	Translocation
c-*myc*	Breast and lung carcinomas	Amplification
L-*myc*	Lung carcinoma	Amplification
N-*myc*	Neuroblastoma, lung carcinoma	Amplification
PDGFR	Chronic myelomonocytic leukemia	Translocation
PDGFR	Gastrointestinal stromal tumors	Point mutation
PI3K	Breast carcinoma	Point mutation
	Ovarian, gastric, lung carcinoma	Amplification
PML/RARα	Acute promyelocytic leukemia	Translocation
B-*raf*	Melanoma, colon carcinoma	Point mutation
*ras*H	Thyroid carcinoma	Point mutation
*ras*K	Colon, lung, pancreatic, and thyroid carcinomas	Point mutation
*ras*N	Acute myeloid and lymphocytic leukemias, thyroid carcinoma	Point mutation

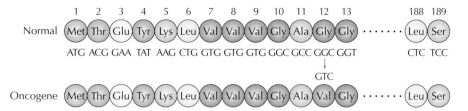

Figure 19.20 Point mutations in *ras* oncogenes A single nucleotide change, which alters codon 12 from GGC (Gly) to GTC (Val), is responsible for the transforming activity of the *ras*H oncogene detected in bladder carcinoma DNA.

chemical carcinogens, providing a direct link between the mutagenic action of carcinogens and cell transformation.

As discussed in Chapter 16, the *ras* genes encode guanine nucleotide-binding proteins that function in transduction of mitogenic signals from a variety of growth factor receptors. The activity of the Ras proteins is controlled by GTP or GDP binding, such that they alternate between active (GTP-bound) and inactive (GDP-bound) states (see Figure 16.27). The mutations characteristic of *ras* oncogenes have the effect of maintaining the Ras proteins constitutively in the active GTP-bound conformation. In large part, this effect is a result of nullifying the response of oncogenic Ras proteins to GAP (GTPase-activating protein), which stimulates hydrolysis of bound GTP by normal Ras. Because of the resulting decrease in their intracellular GTPase activity, the oncogenic Ras proteins remain in the active GTP-bound state and drive unregulated cell proliferation.

Point mutations are only one of the ways in which proto-oncogenes are converted to oncogenes in human tumors. Many cancer cells display abnormalities in chromosome structure, including translocations, duplications, and deletions. The gene rearrangements resulting from chromosome translocations frequently lead to the generation of oncogenes. In some cases, analysis of these rearrangements has implicated already known oncogenes in tumor development. In other cases, novel oncogenes have been discovered by molecular cloning and analysis of rearranged DNA sequences.

The first characterized example of oncogene activation by chromosome translocation was the involvement of the **c-*myc*** oncogene in human Burkitt's lymphomas and mouse plasmacytomas, which are malignancies of antibody-producing B lymphocytes (**Figure 19.21**). Both of these tumors are characterized by

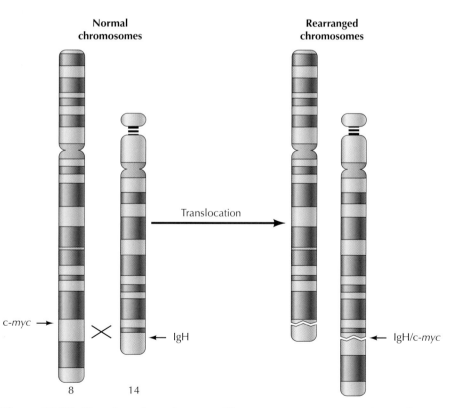

Figure 19.21 Translocation of c-*myc* The c-*myc* proto-oncogene is translocated from chromosome 8 to the immunoglobulin heavy-chain locus (IgH) on chromosome 14 in Burkitt's lymphomas, resulting in abnormal c-*myc* expression.

chromosome translocations involving the genes that encode immunoglobulins. For example, virtually all Burkitt's lymphomas have translocations of a fragment of chromosome 8 to one of the immunoglobulin gene loci, which reside on chromosomes 2 (κ light chain), 14 (heavy chain), and 22 (λ light chain). The fact that the immunoglobulin genes are actively expressed in these tumors suggested that the translocations activate a proto-oncogene from chromosome 8 by inserting it into the immunoglobulin loci. This possibility was investigated by analysis of tumor DNAs with probes for known oncogenes, leading to the finding that the c-*myc* proto-oncogene was located at the chromosome 8 translocation breakpoint in Burkitt's lymphomas. These translocations inserted c-*myc* into an immunoglobulin locus, where it was expressed in an unregulated manner. Such uncontrolled expression of the c-*myc* gene, which encodes a transcription factor normally induced in response to growth factor stimulation, is sufficient to drive cell proliferation and contribute to tumor development.

Translocations of other proto-oncogenes frequently result in rearrangements of coding sequences, leading to the formation of abnormal gene products. The prototype is translocation of the **abl** proto-oncogene from chromosome 9 to chromosome 22 in chronic myeloid leukemia (**Figure 19.22**). This translocation leads to fusion of *abl* with its translocation partner, a gene called *bcr*, on chromosome 22. The result is production of a Bcr/Abl fusion protein in which the normal amino terminus of the Abl proto-oncogene protein has been

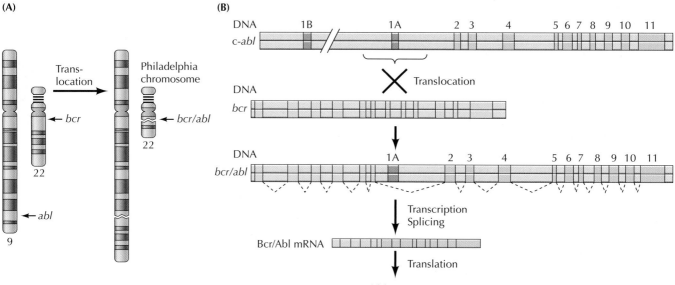

Figure 19.22 Translocation of abl   (A) The *abl* oncogene is translocated from chromosome 9 to chromosome 22, forming the Philadelphia chromosome in chronic myeloid leukemia (CML). (B) The *abl* proto-oncogene, which contains two alternative first exons (1A and 1B), is joined to the middle of the *bcr* gene on chromosome 22. Exon 1B is deleted as a result of the translocation. Transcription of the fused gene initiates at the *bcr* promoter and continues through *abl*. Splicing then generates a fused Bcr/Abl mRNA in which *abl* exon 1A sequences are also deleted and *bcr* sequences are joined to *abl* exon 2. The Bcr/Abl mRNA is translated to yield a recombinant Bcr/Abl fusion protein.

replaced by Bcr amino acid sequences. The fusion of Bcr sequences results in unregulated activity of the Abl tyrosine kinase, leading to cell transformation.

A distinct mechanism by which oncogenes are activated in human tumors is gene amplification, which results in elevated gene expression. DNA amplification (see Figure 7.40) is common in tumor cells, occurring more than 1000 times more frequently than in normal cells, and amplification of oncogenes may play a role in the progression of many tumors to more rapid growth and increasing malignancy. For example, c-*myc* and two other members of the *myc* family (N-*myc* and L-*myc*) are frequently amplified in a variety of human cancers, including breast and lung carcinomas. N-*myc* is also amplified in some neuroblastomas, where its amplification is associated with rapidly growing aggressive tumors. Amplification of another oncogene, **erbB-2**, which encodes a receptor tyrosine kinase, is similarly related to rapid growth of breast and ovarian carcinomas.

Functions of oncogene products

The viral and cellular oncogenes have defined a large group of genes (about 150 in total) that can contribute to the abnormal behavior of malignant cells. As already noted, many of the proteins encoded by proto-oncogenes regulate normal cell proliferation; in these cases, the elevated expression or hyperactivity of the corresponding oncogene proteins drives the uncontrolled proliferation of cancer cells. Other oncogene products contribute to other aspects of the behavior of cancer cells, such as failure to undergo programmed cell death or defective differentiation.

The function of oncogene proteins in regulation of cell proliferation is illustrated by their activities in growth factor-stimulated pathways of signal transduction, such as the activation of ERK signaling downstream of receptor tyrosine kinases (**Figure 19.23**). The oncogene proteins within this pathway include polypeptide growth factors, growth factor receptors, intracellular signaling proteins, and transcription factors.

The action of growth factors as oncogene proteins results from their abnormal expression, leading to a situation in which a tumor cell produces a growth factor to which it also responds. The result is autocrine stimulation of the growth factor-producing cell (see Figure 19.9), which drives abnormal cell proliferation and contributes to the development of a wide variety of human tumors.

A large group of oncogenes encode growth factor receptors, most of which are tyrosine kinases. These receptors can be converted to oncogene proteins by alterations of their amino-terminal domains, which would normally bind extracellular growth factors. For example, the receptor for platelet-derived growth factor (PDGF) is converted to an oncogene in some human leukemias by a chromosome translocation in which the normal amino terminus of the PDGF receptor (PDGFR) is replaced by the amino terminal sequences of a

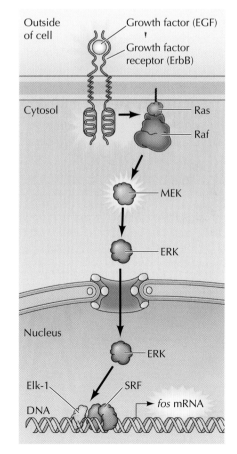

Figure 19.23 Oncogenes and the ERK signaling pathway Oncogene proteins act as growth factors (e.g., EGF), growth factor receptors (e.g., ErbB), and intracellular signaling molecules (Ras, Raf, and MEK). Ras, Raf, and MEK activate the ERK MAP kinase (see Figures 16.28 and 16.29), leading to the induction of additional genes (e.g., *fos*) that encode potentially oncogenic transcriptional regulatory proteins.

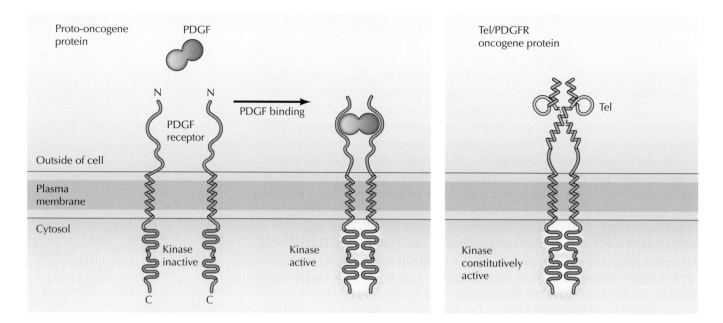

Figure 19.24 Mechanism of Tel/PDGFR oncogene activation The normal PDGF receptor (PDGFR) is activated by dimerization induced by PDGF binding. The Tel/PDGFR oncogene encodes a fusion protein in which the normal extracellular domain of PDGFR is replaced by the amino terminal sequences of the Tel transcription factor, which include its helix-loop-helix dimerization domain (see Figure 8.26D). These sequences dimerize in the absence of PDGF, leading to constitutive activation of the oncogene protein kinase.

transcription factor called Tel (**Figure 19.24**). The Tel sequences of the resulting Tel/PDGFR fusion protein dimerize in the absence of growth factor binding, resulting in constitutive activity of the intracellular kinase domain and unregulated production of a proliferative signal from the oncogene protein. Alternatively, genes that encode receptor tyrosine kinases can be activated by gene amplification or by point mutations that result in unregulated kinase activity. Other oncogenes (including *src* and *abl*) encode nonreceptor tyrosine kinases that are constitutively activated by deletions or mutations of regulatory sequences.

The Ras proteins play a key role in mitogenic signaling by coupling growth factor receptors to activation of the Raf serine/threonine kinase, which initiates a protein kinase cascade leading to activation of the ERK MAP kinase (see Figure 16.28). As discussed earlier, the mutations that convert *ras* proto-oncogenes to oncogenes result in constitutive Ras activity, which leads to activation of the ERK pathway. Genes encoding Raf and MEK can also be converted to oncogenes by mutations that result in unregulated protein kinase activity, which similarly leads to constitutive ERK activation.

The ERK pathway ultimately leads to the phosphorylation of transcription factors and alterations in gene expression. As might therefore be expected, many oncogenes encode transcriptional regulatory proteins that are normally induced in response to growth factor stimulation. For example, transcription of the *fos* proto-oncogene is induced as a result of phosphorylation of Elk-1 by ERK (see Figure 19.23). **Fos** and the product of another proto-oncogene, **Jun**, are components of the AP-1 transcription factor, which activates transcription of a number of target genes, including cyclin D, in growth factor-stimulated cells (**Figure 19.25**). Constitutive activity of AP-1, resulting from unregulated expression of either the Fos or Jun oncogene proteins, is sufficient to drive abnormal cell proliferation, leading to cell transformation. The Myc proteins similarly function as transcription factors regulated by mitogenic stimuli, and abnormal expression of *myc* oncogenes contributes to the development of a variety of human tumors. Other transcription factors are frequently activated as oncogenes by chromosome translocations in human leukemias and lymphomas.

The signaling pathways activated by growth factor stimulation ultimately regulate components of the cell cycle machinery that promote progression through the restriction point in G_1. The D-type cyclins are induced in response

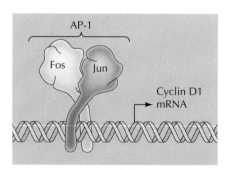

Figure 19.25 The AP-1 transcription factor Fos and Jun dimerize to form AP-1, which activates transcription of cyclin D1 and other growth factor-inducible genes.

to growth factor stimulation (at least in part via activation of the AP-1 transcription factor) and play a key role in coupling growth factor signaling to cell cycle progression. Perhaps not surprisingly, the gene encoding cyclin D1 is a proto-oncogene, which can be activated as an oncogene (called **CCND1**) by chromosome translocation or gene amplification. These alterations lead to constitutive expression of cyclin D1, which then drives cell proliferation in the absence of normal growth factor stimulation. The catalytic partner of cyclin D1, Cdk4, is also activated as an oncogene by point mutations in melanomas.

Components of other signaling pathways discussed in Chapter 16, including G protein-coupled signaling pathways, the NF-κB pathway, and the Hedgehog, Wnt, and Notch pathways, can also act as oncogenes. For example, Wnt proteins were identified as oncogenes in mouse breast cancers, and activating mutations frequently convert the downstream target of Wnt signaling, β-catenin, to an oncogene (*CTNNB1*) in human colon cancers (**Figure 19.26**). These activating mutations stabilize β-catenin, which then forms a complex

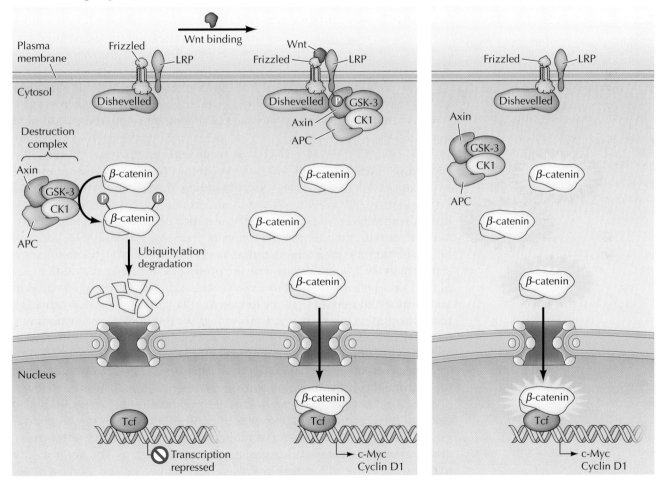

Figure 19.26 Oncogenic activity of β-catenin   (A) In a normal cell in the absence of Wnt stimulation, β-catenin is phosphorylated by GSK-3 in a complex with axin, APC and casein kinase-1 (the destruction complex), leading to β-catenin ubiquitylation and degradation. Wnt polypeptides bind to Frizzled and LRP receptors, leading to recruitment of Dishevelled, inactivation of the destruction complex, and stabilization of β-catenin. β-catenin then translocates to the nucleus and forms a complex with Tcf transcription factors, converting them from repressors to activators of their target genes, including genes encoding c-Myc and cyclin D1. (B) Mutations that prevent degradation of β-catenin in the absence of Wnt signaling convert it to an oncogene.

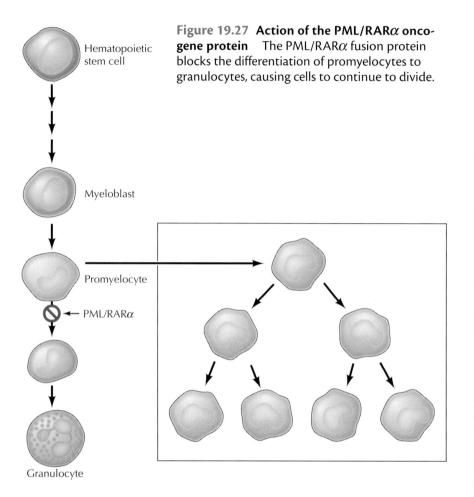

Figure 19.27 Action of the PML/RARα onco-gene protein The PML/RARα fusion protein blocks the differentiation of promyelocytes to granulocytes, causing cells to continue to divide.

Hematopoietic stem cell

Myeloblast

Promyelocyte

PML/RARα

Granulocyte

with Tcf and stimulates transcription of target genes. The targets of β-catenin/Tcf include the genes encoding c-Myc and cyclin D1, leading to unregulated cell proliferation. It is noteworthy that Wnt signaling normally promotes the proliferation of stem cells and their progeny during continuing cell re-newal in the colon (see Figure 18.22), indicating that colon cancer results from abnormal activity of the same pathway that signals physiologically normal proliferation of colon epithelial cells.

Although many oncogenes stimu-late cell proliferation, the oncogenic activity of some transcription factors instead results from inhibition of cell differentiation. As noted in Chapter 16, thyroid hormone and retinoic acid induce differentiation of a variety of cell types. These hormones cross the plasma membrane and bind to intra-cellular receptors that act as transcrip-tional regulatory molecules. Mutated forms of both the thyroid hormone receptor **ErbA** and the retinoic acid receptor **PML/RARα** act as oncogene proteins in chicken erythroleukemia and human acute promyelocytic leu-kemia, respectively. In both cases, the mutated oncogene receptors appear to interfere with the action of their normal homologs, thereby blocking cell differentiation and maintaining the leukemic cells in an actively proliferating state (**Figure 19.27**). In the case of acute promyelocytic leukemia, high doses of retinoic acid can overcome the effect of the PML/RARα oncogene protein and induce differentiation of the leukemic cells. This biological observation has a direct clinical correlate: Patients with acute promyelocytic leukemia can be treated by administration of retinoic acid, which induces differentiation and blocks continued cell proliferation.

As discussed earlier in this chapter, the failure of cancer cells to undergo programmed cell death (or apoptosis) is a critical factor in tumor devel-opment, and several oncogenes encode proteins that act to promote cell survival (**Figure 19.28**). The survival of most animal cells is dependent on growth factor stimulation, so those oncogenes that encode growth factors, growth factor receptors, and signaling proteins such as Ras act not only to stimulate cell proliferation but also to prevent cell death. As discussed in Chapter 18, the PI 3-kinase/Akt signaling pathway plays a key role in preventing apoptosis of many growth factor-dependent cells, and the genes encoding **PI 3-kinase** and **Akt** act as oncogenes in both retroviruses and hu-man tumors. The downstream targets of PI 3-kinase/Akt signaling include a proapoptotic member of the Bcl-2 family, Bad (which is inactivated as a result of phosphorylation by Akt), as well as the FOXO transcription factor

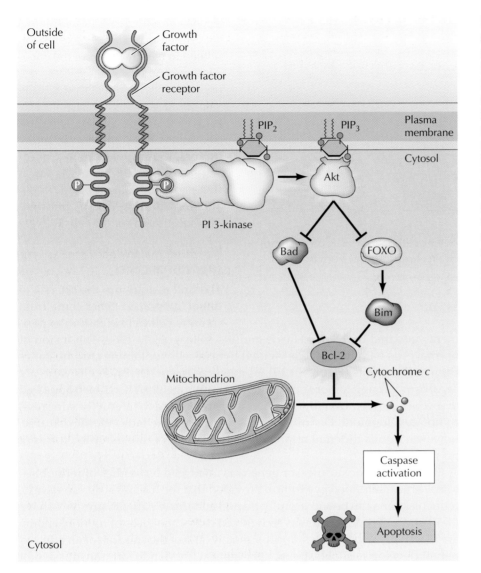

Figure 19.28 Oncogenes and cell survival   The oncogene proteins that signal cell survival include growth factors, growth factor receptors, PI 3-kinase, Akt, and Bcl-2. The targets of Akt include the proapoptotic Bcl-2 family member Bad and the FOXO transcription factor, which stimulates transcription of another proapoptotic Bcl-2 family member, Bim. Phosphorylation by Akt inhibits both Bad and FOXO, promoting cell survival. The antiapoptotic protein Bcl-2 also functions as an oncogene by promoting cell survival and inhibiting the release of cytochrome *c* from mitochondria. Proteins with oncogenic potential are highlighted with a yellow glow.

(which regulates expression of the proapoptotic Bcl-2 family member, Bim). In addition, it is notable that **Bcl-2** itself was first discovered as the product of an oncogene in human lymphomas. The *bcl*-2 oncogene is generated by a chromosome translocation that results in elevated expression of Bcl-2, which blocks apoptosis and maintains cell survival under conditions that normally induce cell death. The identification of *bcl*-2 as an oncogene not only provided the first demonstration of the significance of programmed cell death in the development of cancer but also led to the discovery of the role of *bcl*-2 and related genes as central regulators of apoptosis in organisms ranging from *C. elegans* to humans.

Tumor Suppressor Genes

The activation of cellular oncogenes represents only one of two distinct types of genetic alterations involved in tumor development; the other is inactivation of tumor suppressor genes. Oncogenes drive abnormal cell proliferation as

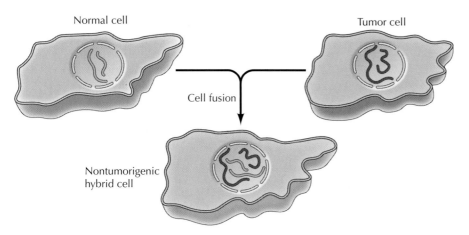

Figure 19.29 Suppression of tumorigenicity by cell fusion   Fusion of tumor cells with normal cells yields hybrids that contain chromosomes from both parents. Such hybrids are usually nontumorigenic.

Video 19.3

sites.sinauer.com/cooper7e/v19.3
Tumor Suppressor Gene Regulation The disruption of tumor suppressor gene expression causes the cancerous cell to divide when it should not.

---

a consequence of genetic alterations that either increase gene expression or lead to hyperactivity of the oncogene-encoded proteins. Tumor suppressor genes represent the opposite side of cell growth control, normally acting to inhibit cell proliferation and tumor development. In many tumors, these genes are lost or inactivated, thereby removing negative regulators of cell proliferation and contributing to the abnormal proliferation of tumor cells.

Identification of tumor suppressor genes

The first insight into the activity of **tumor suppressor genes** came from somatic cell hybridization experiments initiated by Henry Harris and his colleagues in 1969. The fusion of normal cells with tumor cells yielded hybrid cells containing chromosomes from both parents (**Figure 19.29**). In most cases, such hybrid cells were not capable of forming tumors in animals (nontumorigenic). Therefore it appeared that genes derived from the normal cell parent acted to inhibit (or suppress) tumor development. Definition of these genes at the molecular level came, however, from a different approach—the analysis of rare inherited forms of human cancer.

The first tumor suppressor gene was identified by studies of retinoblastoma, a rare childhood eye tumor. Provided that the disease is detected early, retinoblastoma can be successfully treated and many patients survive to have families. Consequently, it was recognized that some cases of retinoblastoma are inherited. In these cases, approximately 50% of the children of an affected parent develop retinoblastoma, consistent with Mendelian transmission of a single dominant gene that confers susceptibility to tumor development (**Figure 19.30**).

Although susceptibility to retinoblastoma is transmitted as a dominant trait, inheritance of the susceptibility gene is not sufficient to transform a normal retinal cell into a tumor cell. All retinal cells in a patient inherit the susceptibility gene, but only a small fraction of these cells give rise to tumors. Thus tumor development requires additional events beyond inheritance of tumor susceptibility. In 1971 Alfred Knudson proposed that the development of retinoblastoma requires two mutations, which are now known to correspond to the loss of both of the functional copies of the tumor susceptibility gene (the **Rb** tumor suppressor gene) that would be present on homologous chromosomes of a normal diploid cell (**Figure 19.31**). In inherited retinoblastoma, one defective copy of *Rb* is genetically transmitted. The loss of this single *Rb* copy is not by itself sufficient to trigger tumor development, but retinoblastoma almost always develops in these individuals as a result

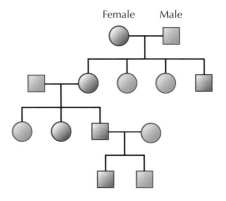

Figure 19.30 Inheritance of retinoblastoma   Susceptibility to retinoblastoma is transmitted to approximately 50% of offspring. Affected and normal individuals are indicated by purple and green symbols, respectively.

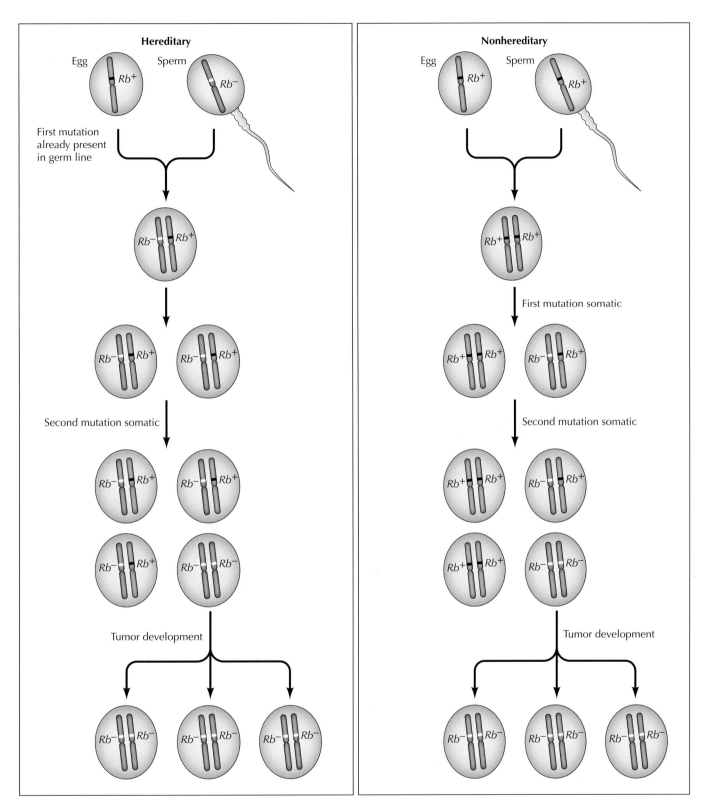

Figure 19.31 Mutations of *Rb* during retinoblastoma development   In hereditary retinoblastoma, a defective copy of the *Rb* gene (*Rb⁻*) is inherited from the affected parent. A second somatic mutation, which inactivates the single normal *Rb⁺* copy in a retinal cell, then leads to the development of retinoblastoma. In nonhereditary cases, which are rare, two normal *Rb⁺* genes are inherited, and retinoblastoma develops only if two somatic mutations inactivate both copies of *Rb* in the same cell.

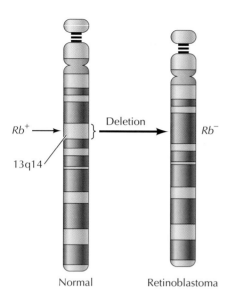

Figure 19.32 *Rb* deletions in retinoblastoma Many retinoblastomas have deletions of the chromosomal locus (13q14) that contains the *Rb* gene.

of a second somatic mutation leading to the loss of the remaining normal *Rb* allele. Affected individuals often develop multiple tumors in both eyes, with each tumor resulting from an independent second somatic mutation. Noninherited retinoblastoma, in contrast, is rare, since its development requires two independent somatic mutations to inactivate both normal copies of *Rb* in the same cell.

The functional nature of the *Rb* gene as a negative regulator of tumorigenesis was initially indicated by observations of chromosome morphology. Visible deletions of chromosome 13q14 were found in some retinoblastomas, suggesting that loss (rather than activation) of the *Rb* gene led to tumor development (**Figure 19.32**). Gene-mapping studies further indicated that tumor development resulted from loss of normal *Rb* alleles in the tumor cells, consistent with the function of *Rb* as a tumor suppressor gene. Isolation of the *Rb* gene as a molecular clone in 1986 then firmly established that *Rb* is consistently lost or mutated in retinoblastomas. Gene transfer experiments also demonstrated that introduction of a normal *Rb* gene into retinoblastoma cells reverses their tumorigenicity, providing direct evidence for the activity of *Rb* as a tumor suppressor.

Although *Rb* was identified in a rare childhood cancer, it is also involved in more common tumors of adults. In particular, studies of the cloned gene have established that *Rb* is lost or inactivated in many bladder, breast, and lung carcinomas. The significance of the *Rb* tumor suppressor gene thus extends beyond retinoblastoma, and mutations of the *Rb* gene contribute to development of a substantial fraction of human cancers. In addition, as noted earlier in this chapter, the Rb protein is a key target for the oncogene proteins of several DNA tumor viruses, including SV40, adenoviruses, and human papillomaviruses, which bind to Rb and inhibit its activity (**Figure 19.33**). Transformation by these viruses thus results, at least in part, from inactivation of Rb at the protein level rather than from mutational inactivation of the *Rb* gene.

Characterization of *Rb* as a tumor suppressor gene served as the prototype for the identification of additional tumor suppressor genes that contribute to the development of many different human cancers (**Table 19.5**). Some of these genes were identified as the causes of rare inherited cancers, playing a role similar to that of *Rb* in hereditary retinoblastoma. Other tumor suppressor genes have been identified as genes that are frequently deleted or mutated in common noninherited cancers of adults, such as colon carcinoma. In either case, it appears that most tumor suppressor genes are involved in the development of both inherited and noninherited forms of cancer. Indeed, mutations of some tumor suppressor genes appear to be the most common molecular alterations leading to human tumor development.

The second tumor suppressor gene to have been identified is *p53*, which is frequently inactivated in a wide variety of human cancers, including

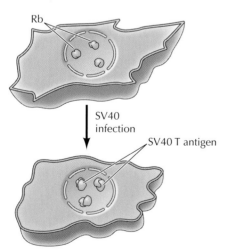

Rb

SV40
infection

SV40 T antigen

Cell transformation

Figure 19.33 Interaction of Rb proteins with oncogene proteins of DNA tumor viruses The oncogene proteins of several DNA tumor viruses (e.g., SV40 T antigen) induce transformation by binding to and inactivating Rb proteins.

Table 19.5   Representative Tumor Suppressor Genes

Gene	Type of cancer
APC	Colon/rectum carcinoma
BRCA1	Breast and ovarian carcinomas
BRCA2	Breast carcinoma
INK4	Melanoma, lung carcinoma, brain tumors, leukemias, lymphomas
p53	Brain tumors; breast, colon/rectum, esophageal, liver, and lung carcinomas; sarcomas; leukemias and lymphomas
PTCH	Basal cell carcinoma
PTEN	Brain tumors; melanoma; prostate, endometrial, kidney, and lung carcinomas
Rb	Retinoblastoma; sarcomas; bladder, breast, and lung carcinomas
Smad2	Colon/rectum carcinoma
Smad4	Colon/rectum carcinoma, pancreatic carcinoma
TβRII	Colon/rectum carcinoma, gastric carcinoma

FYI

p53 was initially thought to be an oncogene because mutated p53 genes found in many cancer cells induced transformation in gene transfer assays. Subsequent studies revealed that p53 was actually a tumor suppressor and that the mutant p53 genes found in many tumors acted as dominant negatives that induced transformation by interfering with normal p53 function.

leukemias, lymphomas, sarcomas, brain tumors, and carcinomas of many tissues, including breast, colon, and lung. In total, mutations of *p53* play a role in about 50% of all cancers, making it the most common target of genetic alterations in human malignancies. It is also of interest that inherited mutations of *p53* are responsible for genetic transmission of a rare hereditary cancer syndrome in which affected individuals develop any of several different types of cancer. In addition, the p53 protein (like Rb) is a target for the oncogene proteins of SV40, adenoviruses, and human papillomaviruses.

Like *p53*, the *INK4* and *PTEN* tumor suppressor genes are very frequently mutated in several common cancers, including lung cancer, prostate cancer, and melanoma. Other tumor suppressor genes that are involved in the Wnt (*APC*) and TGF-β pathways (*TβRII, Smad2,* and *Smad4*) are frequently inactivated in colon cancers. In addition to being involved in noninherited cases of this common adult cancer, inherited mutations of the *APC* gene are responsible for a rare hereditary form of colon cancer, called familial adenomatous polyposis. Individuals with this condition develop hundreds of benign colon adenomas (polyps), some of which inevitably progress to malignancy. Inherited mutations of two other tumor suppressor genes, *BRCA1* and *BRCA2*, are responsible for hereditary cases of breast cancer, which account for about 5% of total breast cancer incidence. Mutations in additional tumor suppressor genes are responsible for other rare inherited cancer syndromes and more than 50 tumor suppressor genes have now been implicated in the development of common types of human cancer.

Functions of tumor suppressor gene products

In contrast to proto-oncogene and oncogene proteins, the proteins encoded by most tumor suppressor genes inhibit cell proliferation or survival. Inactivation of tumor suppressor genes therefore leads to tumor development by eliminating negative regulatory proteins. In many cases, tumor suppressor proteins inhibit the same cell regulatory pathways that are stimulated by the products of oncogenes.

The protein encoded by the **PTEN** tumor suppressor gene is an interesting example of antagonism between oncogene and tumor suppressor gene products

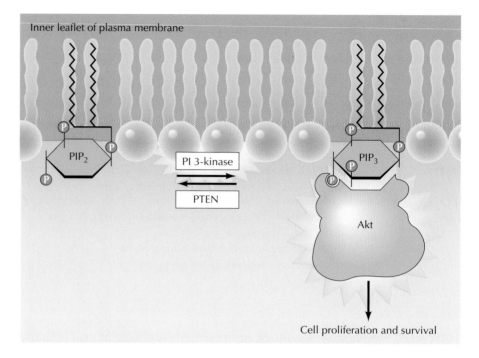

Figure 19.34 Suppression of cell proliferation and survival by PTEN The tumor suppressor protein PTEN is a lipid phosphatase that dephosphorylates PIP_3 at the 3 position of inositol, yielding PIP_2. PTEN thus counters the action of the oncogene proteins PI 3-kinase and Akt, which promote cell proliferation and survival.

(Figure 19.34). The PTEN protein is a lipid phosphatase that dephosphorylates the 3 position of phosphatidylinositides, such as phosphatidylinositol 3,4,5-bisphosphate (PIP_3). By dephosphorylating PIP_3, PTEN antagonizes the activities of PI 3-kinase and Akt, both of which can act as oncogenes by promoting cell survival, as well as stimulating cell proliferation. Conversely, inactivation or loss of the PTEN tumor suppressor protein can contribute to tumor development as a result of increased levels of PIP_3, activation of Akt, and inhibition of programmed cell death.

Proteins encoded by both oncogenes and tumor suppressor genes also function in the Hedgehog signaling pathway (see Figure 16.41). *Patched* (*PTCH*, which encodes the receptor for Hedgehog) is a tumor suppressor gene in basal cell carcinomas, whereas the gene encoding Smoothened (which is inhibited by Patched) is an oncogene. In addition, the Gli transcription factors, which are activated by Smoothened, were first identified as the proteins encoded by an oncogene (*gli*) that was amplified in glioblastomas (malignant brain tumors).

Several tumor suppressor genes encode transcriptional regulatory proteins. A good example is provided by the *Smad2* and *Smad4* tumor suppressor genes, which encode transcription factors that are activated by TGF-β signaling and lead to inhibition of cell proliferation (see Figure 16.39). Target genes that are induced by the Smad proteins include the Cdk inhibitors p15, p21, p27, and p57 (see Table 17.1), arresting cell proliferation in the G_1 phase of the

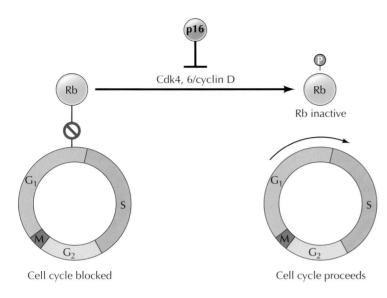

p16

Cdk4, 6/cyclin D

Rb

Rb

Rb inactive

G₁

S

M

G₂

Cell cycle blocked

G₁

S

M

G₂

Cell cycle proceeds

Figure 19.35 Inhibition of cell cycle progression by Rb and p16 Rb inhibits progression past the restriction point in G₁. Cdk4, 6/cyclin D complexes promote passage through the restriction point by phosphorylating and inactivating Rb. The activity of Cdk4, 6/cyclin D is inhibited by p16. Rb and p16 are tumor suppressors, whereas Cdk4 and cyclin D1 are oncogenes.

cell cycle. Consistent with the activity of the TGF-β pathway in inhibiting cell proliferation, the TGF-β receptor is also encoded by a tumor suppressor gene (*TβRII*).

The products of the *Rb* and *INK4* tumor suppressor genes regulate cell cycle progression at the same point as that affected by Cdk4 and cyclin D, both of which can act as oncogenes (**Figure 19.35**). Rb inhibits passage through the restriction point in G₁ by repressing transcription of a number of genes involved in cell cycle progression and DNA synthesis (see Figure 17.16). In normal cells, passage through the restriction point is regulated by Cdk4, 6/cyclin D complexes, which phosphorylate and inactivate Rb. Mutational inactivation of *Rb* in tumors thus removes a key negative regulator of cell cycle progression. The *INK4* tumor suppressor gene, which encodes the Cdk inhibitor p16, also regulates passage through the restriction point. As noted in Chapter 17, p16 inhibits Cdk4, 6/cyclin D activity (see Table 17.1). Inactivation of *INK4* therefore leads to elevated activity of Cdk4, 6/cyclin D complexes, resulting in uncontrolled phosphorylation of Rb.

The *p53* gene product regulates both cell cycle progression and apoptosis. DNA damage leads to rapid induction of p53, which activates transcription of both cell cycle inhibitory and proapoptotic genes (**Figure 19.36**). p53 blocks cell cycle progression in response to DNA damage by inducing the Cdk inhibitor p21 (see Figure 17.20). The p21 protein blocks cell cycle progression in G₁ by inhibiting Cdk2/cyclin E complexes, and the resulting cell cycle arrest presumably allows time for damaged DNA to be repaired before it is replicated. Loss of p53 prevents this damage-induced cell cycle arrest, leading to increased mutation frequencies and a general instability of the cell genome. As noted earlier, genetic instability is a common property of

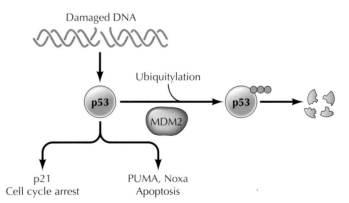

Damaged DNA

Ubiquitylation

p53

MDM2

p53

p21
Cell cycle arrest

PUMA, Noxa
Apoptosis

Figure 19.36 Action of p53 p53 is required for both cell cycle arrest and apoptosis induced by DNA damage. Cell cycle arrest is mediated by induction of the Cdk inhibitor p21 and apoptosis by induction of the proapoptotic Bcl-2 family members PUMA and Noxa. The MDM2 oncogene protein is a ubiquitin ligase that targets p53 for degradation.

cancer cells, which contributes to further alterations in oncogenes and tumor suppressor genes during tumor progression.

More extensive DNA damage can result in a higher level of p53 induction, leading to apoptosis. Induction of apoptosis by p53 is mediated in part by activating the transcription of proapoptotic members of the Bcl-2 family (PUMA and Noxa) that induce programmed cell death. Unrepaired DNA damage normally induces apoptosis of mammalian cells, a response that is presumably advantageous to the organism because it eliminates cells carrying potentially deleterious mutations (e.g., cells that might develop into cancer cells). Cells lacking p53 fail to undergo apoptosis in response to agents that damage DNA, including radiation and many of the drugs used in cancer chemotherapy. This failure to undergo apoptosis in response to DNA damage contributes to the resistance of many tumors to chemotherapy. In addition, loss of p53 appears to interfere with apoptosis induced by other stimuli, such as growth factor deprivation and oxygen deprivation. These effects of p53 on cell survival are thought to account for the high frequency of *p53* mutations in human cancers. Consistent with a central role for *p53* as a tumor suppressor, the ubiquitin ligase that targets p53 for degradation is encoded by an oncogene (called *MDM2*) whose over-expression enhances cell proliferation and survival by lowering p53 levels (see Figure 19.36).

The products of the *BRCA1* and *BRCA2* genes, which are responsible for some inherited breast and ovarian cancers, appear to be involved in checkpoint control of cell cycle progression and repair of double-strand breaks in DNA. *BRCA1* and *BRCA2* thus function as **stability genes**, which act to maintain the integrity of the genome. Mutations in genes of this type lead to the development of cancer not as a result of direct effects on cell proliferation or survival but because their inactivation leads to a high frequency of mutations in oncogenes or tumor suppressor genes. Other stability genes whose loss contributes to the development of human cancers include the *ATM* gene, which acts at the DNA damage checkpoint (see Figure 17.19), the mismatch repair genes that are defective in some inherited colorectal cancers (see Molecular Medicine in Chapter 7), and the nucleotide excision repair genes that are mutated in xeroderma pigmentosum (see Figure 7.23).

As discussed in previous chapters, microRNAs (miRNAs) have become recognized as major regulators of gene expression in eukaryotic cells. In animal cells, miRNAs act post-transcriptionally, targeting mRNAs to inhibit translation and induce mRNA degradation (see Figure 9.20). It is thought that miRNAs contribute to the regulation of approximately half of all protein-coding genes, clearly suggesting the potential involvement of miRNAs in cancer. Consistent with this possibility, many tumors have characteristic alterations in miRNA expression compared with normal cells. In general, the expression of miRNAs is significantly lower in tumors, suggesting that many miRNAs might act as tumor suppressors. One example of such a tumor suppressor miRNA is *let-7*, which targets oncogenes including c-*myc* and *ras*K (**Figure 19.37**). Another miRNA that acts as a tumor suppressor (*miR-34*) is induced by p53 and targets mRNAs encoding several proteins that stimulate cell cycle progression (e.g., Cdk4, Cdk6, and cyclin E) and promote cell survival (e.g., Bcl-2). However, not all miRNAs act as tumor suppressors—some can act as oncogenes instead. For example, the miRNAs designated *miR-17-92* are

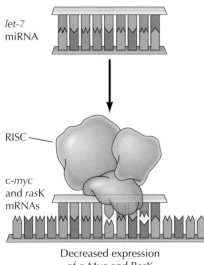

let-7
miRNA

RISC

c-myc
and rasK
mRNAs

Decreased expression
of c-Myc and RasK
oncogene proteins

Figure 19.37 Action of *let-7* miRNA as a tumor suppressor   The *let-7* miRNA targets mRNAs encoding the c-Myc and RasK oncogene proteins.

amplified in several types of tumors and target mRNAs encoding proteins that inhibit cell cycle progression (e.g., the Cdk inhibitor p21) or promote apoptosis (e.g., the proapoptotic Bcl-2 family member Bim).

Cancer genomics

As discussed earlier, the development of cancer is a multistep process in which normal cells gradually progress to malignancy. Mutations resulting in both the activation of oncogenes and the inactivation of tumor suppressor genes are critical steps in tumor initiation and progression. Accumulated damage to multiple genes is needed to eventually result in the increased proliferation, survival, invasiveness, and metastatic potential of a malignant cell.

Recent studies have used large-scale genome sequencing to systematically analyze the spectrum of mutations in oncogenes and tumor suppressor genes that are found in thousands of individual cancers. These large-scale sequencing studies have identified approximately 150 genes that are mutated in human cancers and contribute to tumor development. About half of these genes are dominantly-acting oncogenes and half are tumor suppressor genes. Consistent with the multistep nature of tumor development, many cancers contain mutations in two to six different genes that contribute to cancer development.

Although a large number of oncogenes and tumor suppressor genes have been identified in human cancers, only a subset of these are involved in a high fraction of individual cancers of any particular type. For example, the results of sequence analysis of multiple different colon and breast cancers are summarized in Figure 19.38. In analysis of more than 100 colon cancers, only a few genes were consistently mutated in a high fraction of individual tumors, including the *PI3K* and *ras*K oncogenes, the *APC* and *p53* tumor suppressor genes, and the gene encoding a ubiquitin ligase that targets cyclin E for degradation (*FBXW7*) (**Figure 19.38A**). Mutations in many other genes were found less frequently, typically in less than 5% of individual tumors. A similar pattern of mutations was found in breast cancers, which contained

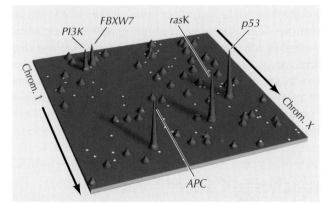
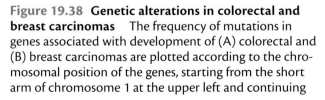
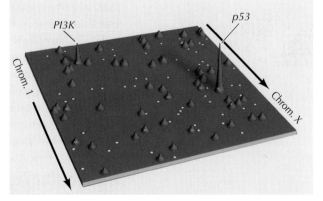

Figure 19.38 Genetic alterations in colorectal and breast carcinomas The frequency of mutations in genes associated with development of (A) colorectal and (B) breast carcinomas are plotted according to the chromosomal position of the genes, starting from the short arm of chromosome 1 at the upper left and continuing in the direction of the arrow. Positions that follow the front edge of the plot are continued in the next row at the back edge, and chromosomes are aligned end to end. The height of the peak at each position corresponds to the frequency at which the gene is mutated. (From L. D. Wood et al., 2007, *Science* 318: 1108.)

Figure 19.39 Pathways affected by human oncogenes and tumor suppressor genes The oncogenes and tumor suppressor genes mutated in human cancers can be organized into 12 pathways that confer a selective growth advantage by regulating cell survival, genome maintenance, and cell fate. (From B. Vogelstein et al., 2013, *Science* 339: 1546.)

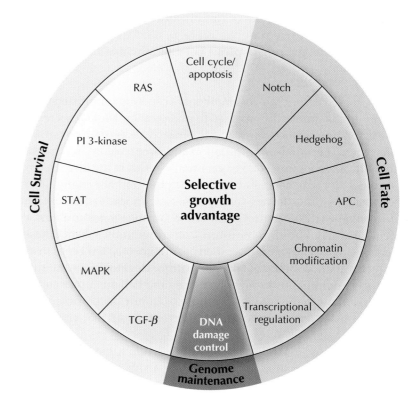

frequent mutations in *PI3K* and *p53*, combined with less frequent mutations in multiple other genes (**Figure 19.38B**).

Although individual cancers contain mutations in different genes, the same signaling pathway is frequently affected by independent mutations in different tumors. Thus, mutations in several distinct oncogenes and tumor suppressor genes can have similar effects on cell behavior and tumor development. For example, some colon cancers have mutations leading to activation of a member of the *raf* oncogene family (B-*raf*), rather than mutations of *ras*K. Since Raf is immediately downstream of Ras, activation of either *ras*K or B-*raf* leads to stimulation of ERK signaling in the tumor cells. Likewise, some tumors have activating mutations of the *PI3K* oncogene, whereas others have inactivating mutations of the *PTEN* tumor suppressor gene—both of which affect cell survival by leading to activation of the Akt protein kinase. The large number of different mutations in tumors thus affect a much smaller number of complementary pathways that regulate cell proliferation, survival, and genome stability (**Figure 19.39**). Accumulated damage to multiple oncogenes and tumor suppressor genes in these distinct regulatory pathways appears to be responsible for the progressive loss of growth control that leads to the development of cancer.

Molecular Approaches to Cancer Treatment

A great deal has been learned about the molecular defects responsible for the development of human cancers. However, cancer is more than a topic of scientific interest. It is a dreaded disease that claims the lives of nearly one out of every four Americans. Translating our understanding of cancer into practical improvements in cancer prevention and treatment therefore represents a major challenge for current and future research. Fortunately, advances in elucidating the molecular biology of cancer have begun to contribute to the development of new approaches to its treatment, which promise to ultimately yield major advances in dealing with this disease.

Prevention and early detection

The most effective way to deal with cancer would be to prevent development of the disease. A second-best, but still effective, alternative would be to reliably detect early premalignant stages of tumor development that could be easily treated. Many cancers can be cured by localized treatments, such as surgery or radiation, if they are detected before they spread throughout the body by metastasis. For example, early premalignant stages of colon cancer (adenomas) are usually completely curable by relatively minor surgical procedures (Figure 19.40). The cure rate for early carcinomas that remain localized to their site of origin is also high, about 90%. However, survival rates drop to about 70% for patients whose cancers have spread to adjacent tissues and lymph nodes, and to 13% for patients with metastatic colon cancer. Early detection of cancer can thus be a critical determinant of the outcome of the disease.

The major application of molecular biology to prevention and early detection may lie in the identification of individuals with inherited susceptibilities to cancer development. Such inherited cancer susceptibilities can result from mutations in tumor suppressor genes, in at least two oncogenes (*ret* and *cdk4*), and in stability genes, such as *BRCA1* and *BRCA2* or the mismatch repair genes responsible for development of hereditary nonpolyposis colon cancer (see Molecular Medicine in Chapter 7). Mutations in these genes can be detected by genetic testing, allowing the identification of high-risk individuals before disease develops.

Some patients at high cancer risk, for example with inherited mutations of *BRCA1* or *BRCA2*, may choose prophylactic surgery to prevent cancer from developing. In addition, careful monitoring of high-risk individuals may allow early detection and more effective treatment of some types of cancer. For example, colon adenomas can be detected by colonoscopy and removed prior to the development of malignancy. Patients with familial adenomatous polyposis (resulting from inherited mutations of the *APC* tumor suppressor gene) typically develop hundreds of adenomas within the first 20 years of life, so the colons of these patients are usually removed before the inevitable progression of some of these polyps to malignancy. However, patients with hereditary nonpolyposis colon cancer develop a smaller number of polyps later in life and may therefore benefit from routine colonoscopy and drugs, such as nonsteroidal anti-inflammatory drugs (NSAIDs), which inhibit colon cancer development.

The direct inheritance of cancers resulting from defects in known genes is a rare event, constituting about 5% of total cancer incidence. The most common inherited cancer susceptibility is hereditary nonpolyposis colon cancer, which accounts for about 15% of colon cancers and 1 to 2% of all cancers in the United States. Mutations in the *BRCA1* and *BRCA2* tumor suppressor genes are also relatively common, accounting for approximately 5% of all breast cancers. However, additional genes that confer a more moderate increase in cancer susceptibility are thought to contribute to the development of a larger fraction of common adult malignancies. The continuing identification of such cancer susceptibility genes is an important undertaking with clear practical implications. The reliable identification of susceptible individuals,

FYI

Prophylactic surgery for women who inherit mutated *BRCA* genes was spotlighted by actress Angelina Jolie's decision to have a preventive double mastectomy in 2013.

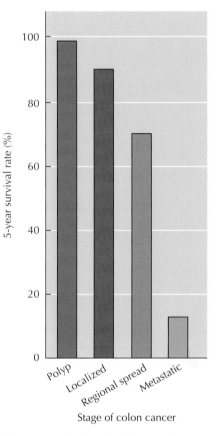

Figure 19.40 Survival rates of patients with colon carcinoma Five-year survival rates are shown for patients diagnosed with adenomas (polyps), with carcinoma still localized to its site of origin, with carcinoma that has spread regionally to adjacent tissues and lymph nodes, and with metastatic carcinoma. (Data from American Cancer Society, *Cancer Facts & Figures 2015*.)

if followed by appropriate early detection measures, might ultimately make a significant impact on cancer mortality.

Treatment

The most critical question, however, is whether the discovery of oncogenes and tumor suppressor genes will allow the development of new drugs that act selectively against cancer cells. Most of the drugs traditionally used in cancer treatment either damage DNA or inhibit DNA replication. Consequently, these drugs are toxic not only to cancer cells but also to normal cells, particularly those normal cells that are continually replaced by the division of stem cells (e.g., hematopoietic cells, epithelial cells of the gastrointestinal tract, and hair follicle cells). The action of anticancer drugs against these normal cell populations accounts for most of the toxicity associated with these drugs and limits their effective use in cancer treatment.

One alternative approach to cancer therapy is the use of drugs that inhibit tumor growth by interfering with angiogenesis (blood vessel formation), rather than acting directly against cancer cells. As noted earlier in this chapter, the formation of new blood vessels is needed to supply the oxygen and nutrients required for tumor growth. Promoting angiogenesis is thus critical to tumor development, and tumor cells secrete a number of growth factors, including VEGF, that stimulate the proliferation of capillary endothelial cells, resulting in the outgrowth of new capillaries into the tumor (see Figure 18.15). The importance of angiogenesis was first recognized by Judah Folkman in 1971, and continuing research by Folkman and his colleagues led to the development of drugs that inhibit angiogenesis by blocking the proliferation of endothelial cells. Because these drugs act specifically to inhibit the formation of new blood vessels, they are less toxic to normal cells than standard anticancer agents. Angiogenesis inhibitors showed promising results in both animal tests and clinical trials that evaluated their effectiveness against human cancers. In 2004, positive clinical results led the U.S. Food and Drug Administration (FDA) to approve the use of the first angiogenesis inhibitor, a monoclonal antibody against VEGF, for treatment of colon cancer. This monoclonal antibody is now also used against glioblastomas, kidney cancers, and lung cancers. In addition, two small molecule inhibitors of the VEGF receptor (sunitinib and sorafenib) have been approved for cancer treatment.

The most promising strategy for achieving more selective cancer treatment is the development of drugs targeted specifically against the oncogenes that drive tumor growth. Unfortunately, from the standpoint of cancer treatment, oncogenes are not unique to tumor cells. Since proto-oncogenes play important roles in normal cells, general inhibitors of oncogene expression or function are likely to act against normal cells as well as tumor cells. The exploitation of oncogenes as targets for anticancer drugs is therefore not a straightforward proposition, but several promising advances indicate that it is possible to develop selective oncogene-targeted therapies (Table 19.6).

The first therapeutic regimen targeted against a specific oncogene is used for the treatment of acute promyelocytic leukemia. This leukemia is characterized by a chromosome translocation in which the gene that encodes the retinoic acid receptor (*RARα*) is fused to another gene (*PML*) to form the *PML/RARα* oncogene. The PML/RARα protein is thought to function as a transcriptional repressor that blocks cell differentiation. These leukemic cells, however, differentiate in response to treatment with high doses of retinoic acid, which binds to and inactivates the PML/RARα oncogene protein. Such

Table 19.6   Representative Oncogene-Targeted Therapies

Drug	Target protein	Tumor types
Retinoic acid	PML/RARα	Acute promyelocytic leukemia
Herceptin	ErbB-2	Breast cancer
Erbitux	ErbB	Colorectal cancer
Imatinib	Abl	Chronic myeloid leukemia
	Kit	Gastrointestinal stromal tumors
	PDGFR	Gastrointestinal stromal tumors, chronic myelomonocytic leukemia, hypereosinophilic syndrome, dermatofibrosarcoma protuberans
Gefitinib	ErbB	Lung cancer
Erlotinib	ErbB	Lung cancer
Vemurafenib	B-Raf	Melanoma
Crizotinib	Alk	Lung cancer
Vismodegib	SMO	Skin cancer (basal cell carcinoma)
Trametinib	MEK	Melanoma
Idelalisib	PI3K	Chronic lymphocytic leukemia
Everolimus	mTOR	Renal cell carcinoma, breast cancer

treatment with retinoic acid results in remission of the leukemia in most patients, although this favorable response is temporary and patients eventually relapse. However, combined treatment with retinoic acid and standard chemotherapeutic agents significantly reduces the incidence of relapse, so the use of retinoic acid is of substantial benefit in the treatment of acute promyelocytic leukemia. The therapeutic activity of retinoic acid was observed prior to identification of the *PML/RARα* oncogene, so its effectiveness against leukemic cells expressing this oncogene protein was discovered by chance rather than by rational drug design. Nonetheless, the use of retinoic acid for treatment of acute promyelocytic leukemia provides the first example of a clinically useful drug targeted against an oncogene protein.

Herceptin, a monoclonal antibody against the ErbB-2 oncogene protein, was the first drug developed against a specific oncogene to achieve FDA approval for clinical use in cancer treatment. The ErbB-2 protein is overexpressed in 25–30% of breast cancers as a result of amplification of the *erb*B-2 gene. It was first found that an antibody against the extracellular domain of ErbB-2 (a receptor tyrosine kinase) inhibited the proliferation of tumor cells in which ErbB-2 was overexpressed. These results led to the development and clinical testing of Herceptin, which was found to significantly reduce tumor growth and prolong patient survival in clinical trials involving over 600 women with metastatic breast cancers that overexpressed the ErbB-2 protein. Based on these results, Herceptin was approved by the FDA in 1998 for treatment of metastatic breast cancers that express elevated levels of ErbB-2. Erbitux, a monoclonal antibody against the EGF receptor (the ErbB oncogene protein), was also approved by the FDA in 2004 for use in treatment of advanced colorectal cancer and, more recently, for head and neck tumors.

The therapeutic use of monoclonal antibodies is limited to extracellular targets, such as growth factors or cell surface receptors. A more widely applicable area of drug development is the identification of small molecule inhibitors of oncogene proteins, including the protein kinases that play key roles in signaling the proliferation and survival of cancer cells. The pioneering advance in this area was the development of a selective inhibitor of the Bcr/Abl tyrosine kinase, which is generated by the Philadelphia chromosome translocation in chronic myeloid leukemia (CML) (see Figure 19.22). Brian Druker and his colleagues developed a potent and specific inhibitor of the Bcr/Abl protein kinase and showed that this compound (called imatinib or Gleevec) effectively blocks proliferation of chronic myeloid leukemia cells. Based on these results, a clinical trial of imatinib was initiated in 1998. The responses to imatinib were remarkable, and the drug had minimal side effects. The striking success of imatinib in these clinical studies served as the basis for its rapid approval by the FDA in 2001 for use in the treatment of chronic myeloid leukemia. Since that time, the use of imatinib has decreased mortality from chronic myeloid leukemia by approximately 80% (Figure 19.41). Although some patients relapse and develop resistance to the drug, imatinib is unquestionably a highly effective therapy for this leukemia. Interestingly, resistance to imatinib most often results from mutations of the Bcr/Abl protein kinase domain that prevent imatinib binding. By analysis of these resistant mutants, it has been possible to design new inhibitors, which are currently being tested in clinical trials to determine their effectiveness against leukemias that have become resistant to imatinib.

Imatinib is also a potent inhibitor of the PDGF receptor and Kit tyrosine kinases, and it has proven to be an effective therapy for tumors in which the genes encoding these protein kinases are mutationally activated as oncogenes. *Kit* is activated as an oncogene by point mutations resulting in constitutive protein kinase activity in approximately 90% of gastrointestinal stromal tumors, which are tumors of stromal connective tissue of the stomach and small intestine. Many of the gastrointestinal stromal tumors that do not have activating mutations of Kit instead have activating mutations of the PDGF receptor. Consequently, gastrointestinal stromal tumors are highly responsive to imatinib. In addition, imatinib is active against three other types of tumors in which the PDGF receptor is activated as an oncogene, including chronic

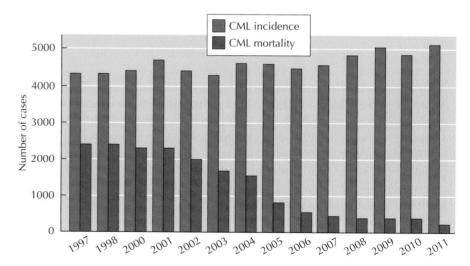

Figure 19.41   Effect of imatinib on mortality from chronic myeloid leukemia The annual number of cases and deaths from chronic myeloid leukemia (CML) in the United States are plotted from 1997 to 2011. Incidence of the disease has not changed significantly, but mortality has declined by about 80% since imatinib was approved for treatment in 2001. (From W. R. Sellers, 2011, *Cell* 147: 26.)

Molecular Medicine

Imatinib: Cancer Treatment Targeted against the *bcr/abl* Oncogene

The Disease

Chronic myeloid leukemia (CML) accounts for approximately 12% of leukemias in adults. In 2015, it is estimated that about 6700 cases of CML will be diagnosed in the United States, and that there will be about 1100 deaths from this disease.

CML originates from the hematopoietic stem cell of the bone marrow. It is a slowly progressing disease, which is clinically divided into two stages: chronic phase and blast crisis. The chronic phase of CML can persist for years and is associated with minimal symptoms. Eventually, however, patients progress to an acute life-threatening stage of the disease known as the blast crisis. Blast crisis is characterized by the accumulation of large numbers of rapidly proliferating leukemic cells, called blasts. Prior to the development of imatinib, patients in blast crisis were treated with standard chemotherapeutic drugs, which could induce remission to the chronic phase of the disease. Standard chemotherapy was also used during the chronic phase of CML, but did not usually succeed in eliminating the leukemic cells. CML can also be treated by transplantation of bone marrow stem cells, which can be curative for about half of patients in the chronic phase of the disease.

Molecular and Cellular Basis

Activation of the *abl* oncogene by translocation from its normal locus on chromosome 9 to chromosome 22 is a highly reproducible event in CML, occurring in about 95% of these leukemias. This translocation results in formation of a fusion between *abl* and the *bcr* gene on chromosome 22, yielding the *bcr/abl*

oncogene. The oncogene expresses a Bcr/Abl fusion protein in which Bcr amino acid sequences replace the first exon of Abl (see Figure 19.22). The Bcr/Abl protein is a constitutively active tyrosine kinase, which leads to leukemia by activating a variety of downstream signaling pathways.

Because activation of the *bcr/abl* oncogene is such a reproducible event in the development of CML, it appeared to be a good candidate against which to develop a selective tyrosine kinase inhibitor that might be of clinical use. These studies led to the development of imatinib (also called Gleevec) as the first therapeutic drug successfully designed as a selective inhibitor of an oncogene protein.

Prevention and Treatment

The development of imatinib started with the identification of 2-phenyl-aminopyrimidine as a nonspecific inhibitor of protein kinases. A series of related compounds were then synthesized and optimized for activity against different targets, including the Abl tyrosine kinase. Among many compounds screened in these investigations, imatinib was found to be a potent and specific inhibitor of Abl and two other tyrosine kinases: the platelet-derived growth factor receptor and Kit. Further studies demonstrated that imatinib specifically inhibited the proliferation of cells transformed by *bcr/abl* oncogenes, including cells from CML patients in culture. In addition, imatinib prevented tumor formation by *bcr/abl*-transformed cells in mice. In contrast, normal cells or cells transformed by other oncogenes were not affected by imatinib, demonstrating its specificity against the Bcr/Abl tyrosine kinase.

Based on these results, an initial phase I clinical study of imatinib was initiated in June 1998. The results were strikingly positive. Of 54 chronic-phase patients treated with imatinib, 53 responded to

the drug. Moreover, responses to imatinib were seen in more than half of treated blast crisis patients. The success of these initial studies prompted expanded Phase II studies involving over 1000 patients. These studies confirmed the promising results of the phase I study, with 95% of chronic phase patients and approximately 50% of blast crisis patients responding to imatinib. Moreover, in contrast with conventional chemotherapeutic drugs, imatinib had minimal side effects. These clinical studies clearly demonstrated that imatinib is a highly effective therapy for CML, and led to accelerated FDA approval of imatinib in May 2001—a milestone in the translation of basic science to clinical practice. Since its approval in 2001, the use of imatinib has decreased mortality from CML by approximately 80%.

Reference

Druker, B. J. 2002. Inhibition of the Bcr-Abl tyrosine kinase as a therapeutic strategy for CML. *Oncogene* 21: 8541–8546.

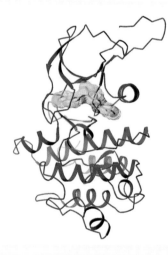

Crystal structure of the catalytic domain of Abl complexed with a derivative of imatinib. (From T. Schindler et al. 2000. *Science* 289: 1938.)

myelomonocytic leukemia in which it is activated by fusion with the Tel transcription factor (see Figure 19.24).

Two small molecule inhibitors of the EGF receptor (gefitinib and erlotinib) have shown striking activity against a subset of lung cancers in which the EGF receptor is activated by point mutations. It is noteworthy that the rationale for treatment of lung cancers with these inhibitors was the fact that EGF receptors are overexpressed in most lung cancers, rather than the fact that they can be mutationally activated as oncogenes. In contrast with the general effectiveness of imatinib against chronic myeloid leukemia, clinical studies indicated that gefitinib or erlotinib were effective in only about 10% of lung cancer patients, although the responses in these patients were quite dramatic. A major advance was thus made in 2004 when two groups of researchers found that the subset of lung cancers that responded to these inhibitors were those in which mutations resulted in constitutive activation of the EGF receptor tyrosine kinase. These results indicated that inhibition of the EGF receptor was an effective treatment for tumors in which it had been mutated to act as an oncogene—but not for tumors expressing the normal protein. More recently, an inhibitor of another tyrosine kinase (Alk) has been found to be effective against a distinct subset of lung cancers in which Alk has been activated as an oncogene by translocation. Additional tyrosine kinase inhibitors are effective against tumors in which the Met and Ret tyrosine kinases are activated oncogenes.

An inhibitor of B-Raf (vemurafenib) has shown striking results in the treatment of melanoma. *B-Raf* is activated as an oncogene in about 60% of human melanomas by a mutation that changes valine at position 600 to glutamic acid. Vemurafenib has a higher affinity for B-Raf proteins with glutamic acid at this position, so it selectively inhibits mutated B-Raf oncogene proteins and is highly effective against melanomas with this mutation. It received FDA approval for treatment of advanced melanomas in 2011. Other oncogene-targeted drugs that have been approved for cancer treatment include inhibitors of MEK, PI 3-kinase, and mTOR (see Figure 16.32), as well as an inhibitor of Hedgehog signaling, used for treatment of basal cell carcinomas with oncogenic mutations of the Hedgehog receptor Smoothened (see Figure 16.41).

The effectiveness of oncogene inhibitors in cancer treatment supports the hypothesis that tumors with activated oncogenes are particularly susceptible to inhibitors of those oncogenes. This is also consistent with a variety of experiments in animal model systems. The sensitivity of tumors to inhibition of activated oncogenes has been referred to as **oncogene addiction**. It is thought that an activated oncogene becomes a major driving force in the tumor cell, such that other signaling pathways in the tumor cell become secondary in importance. Consequently, the proliferation and survival of a tumor cell may become dependent on continuing activity of the oncogene, whereas normal cells have alternative signaling pathways that can compensate if any one pathway is blocked. A cancer cell with an activated oncogene may therefore be selectively sensitive to inhibition of that oncogene-encoded protein.

Despite the initially striking responses of tumors to oncogene-targeted therapies, the effectiveness of many of these drugs is limited by the development of resistance. Whereas imatinib is effective for long-term treatment of chronic myeloid leukemia (as illustrated in Figure 19.41), most patients with lung cancer or melanoma only respond to inhibitors of the EGF receptor or B-Raf for about a year before resistance develops. Resistance can result from several mechanisms, including mutations in the targeted kinase, the activation of other tyrosine kinases, and the activation of downstream oncogenic pathways. For example, oncogenic activation of Ras or PI 3-kinase can result in resistance to EGF receptor inhibitors. Attempts to prevent the development of drug resistance, possibly by combining multiple targeted therapies, is therefore a major focus of ongoing research.

Although much remains to be done, the effectiveness of the drugs discussed above suggests that the continuing exploitation of oncogenes as targets for drug development has the potential to lead to a new generation of drugs that act selectively against cancer cells. The apparent dependence of cancer cells on mutationally-activated oncogenes offers the promise that the use of oncogene-targeted drugs combined with genomic sequencing of the tumors of individual patients may lead to major advances in personal cancer treatment. Although the eventual impact of molecular biology on the treatment of cancer remains to be seen, it is clear that the rational design of drugs targeted against specific oncogene proteins will play an important role.

SUMMARY	KEY TERMS

The Development and Causes of Cancer

- **Types of cancer:** Cancer can result from the abnormal proliferation of any type of cell. The most important distinction for the patient is between benign tumors, which remain confined to their site of origin, and malignant tumors, which can invade normal tissues and spread throughout the body. See Animation 19.1 and Video 19.1.

  cancer, tumor, benign tumor, malignant tumor, metastasis, carcinoma, sarcoma, leukemia, lymphoma

- **The development of cancer:** Tumors develop from single cells that begin to proliferate abnormally. Additional mutations lead to the selection of cells with progressively increasing capacities for proliferation, survival, invasion, and metastasis. See Video 19.2.

  tumor initiation, tumor progression, clonal selection, adenoma, polyp

- **Causes of cancer:** Radiation and many chemical carcinogens act by damaging DNA and inducing mutations. Other carcinogens contribute to the development of cancer by stimulating cell proliferation. Viruses also cause cancer in both humans and other species.

  carcinogen, tumor promoter

- **Properties of cancer cells:** The uncontrolled proliferation of cancer cells is reflected in reduced requirements for extracellular growth factors and lack of inhibition by cell–cell contact. Many cancer cells are also defective in differentiation, consistent with their continued proliferation *in vivo*. The characteristic failure of cancer cells to undergo apoptosis also contributes substantially to tumor development. See Animation 19.2.

  density-dependent inhibition, contact inhibition, autocrine growth stimulation, angiogenesis, programmed cell death, apoptosis

- **Transformation of cells in culture:** The development of *in vitro* assays for cell transformation has allowed the conversion of normal cells into tumor cells to be studied in cell culture.

  cell transformation

Tumor Viruses

- **Hepatitis B and C viruses:** The hepatitis B and C viruses cause liver cancer in humans.

  tumor virus, hepatitis B virus, hepatitis C virus

- **Small DNA tumor viruses:** Papillomaviruses cause a variety of tumors, including cervical carcinoma in humans, and Merkel cell polyomavirus causes a rare human skin cancer. Although other small DNA tumor viruses do not cause human cancer, they are important models for studying the molecular biology of cell transformation. Transforming proteins of these viruses frequently interact with the cellular Rb and p53 tumor suppressor proteins.

  polyomavirus, papillomavirus, adenovirus, Merkel cell polyomavirus

- **Herpesviruses:** The herpesviruses, which are among the most complex animal viruses, cause cancer in several species, including humans.

  herpesvirus, Kaposi's sarcoma-associated herpesvirus, Epstein-Barr virus

SUMMARY	KEY TERMS

- **Retroviruses:** Retroviruses cause cancer in humans and a variety of other animals. Some retroviruses contain specific genes responsible for inducing cell transformation, and studies of these highly oncogenic retroviruses have led to the characterization of both viral and cellular oncogenes.

retrovirus, Rous sarcoma virus (RSV)

Oncogenes

- **Retroviral oncogenes:** The first oncogene to be identified was the *src* gene of RSV. Subsequent studies have identified more than two dozen distinct oncogenes in different retroviruses.

oncogene, *src, ras, raf*

- **Proto-oncogenes:** Retroviral oncogenes originated from closely related genes of normal cells, called proto-oncogenes.

proto-oncogene

- **Oncogenes in human cancer:** A variety of oncogenes are activated by point mutations, DNA rearrangements, and gene amplification in human cancers. Some of these human tumor oncogenes, such as the *ras* genes, are cellular homologs of oncogenes that were first described in retroviruses.

*c-myc, abl, erb*B-2

- **Functions of oncogene products:** Many oncogene proteins function as elements of signaling pathways that stimulate cell proliferation. The genes that encode cyclin D1 and Cdk4 can also act as oncogenes by stimulating cell cycle progression. Other oncogene proteins interfere with cell differentiation, and oncogenes encoding PI 3-kinase, Akt, and Bcl-2 inhibit apoptosis.

Fos, Jun, CCND1, ErbA, PML/RARα, PI 3-kinase, Akt, Bcl-2

Tumor Suppressor Genes

- **Identification of tumor suppressor genes:** In contrast with oncogenes, tumor suppressor genes inhibit tumor development. The prototype tumor suppressor gene, *Rb*, was identified by studies of inheritance of retinoblastoma. Loss or mutational inactivation of *Rb* and other tumor suppressor genes, including *p53*, contributes to the development of a wide variety of human cancers. See Video 19.2.

tumor suppressor gene, *Rb, p53*

- **Functions of tumor suppressor gene products:** The proteins encoded by most tumor suppressor genes act as inhibitors of cell proliferation or survival. The Rb, INK4, and p53 proteins are negative regulators of cell cycle progression. In addition, p53 is required for apoptosis induced by DNA damage and other stimuli, so its inactivation contributes to enhanced tumor cell survival. Some genes, such as *BRCA1* and *BRCA2*, act to maintain genomic stability rather than directly influencing cell proliferation.

PTEN, stability gene

- **Cancer genomics:** Mutations in both oncogenes and tumor suppressor genes contribute to the progressive development of human cancers. Accumulated damage to multiple such genes results in the abnormalities of cell proliferation, differentiation, and survival that characterize the cancer cell. The many different genes that can contribute to tumor development affect the activities of a relatively small number of cell regulatory pathways.

Molecular Approaches to Cancer Treatment

- **Prevention and early detection:** Many cancers can be cured if they are detected at early stages of tumor development. Genetic testing to identify individuals with inherited cancer susceptibilities may allow early detection and more effective treatment of high-risk patients.

- **Treatment:** The development of drugs targeted against specific oncogenes is beginning to lead to the discovery of new therapeutic agents that act selectively against cancer cells.

oncogene addiction

Questions

1. How does a benign tumor differ from a malignant tumor?

2. What is the role of clonal selection in the development of cancer?

3. How do estrogens increase the risk of cancer?

4. How does autocrine growth stimulation contribute to the progression of tumors?

5. What properties of cancer cells give them the ability to metastasize?

6. Why do AIDS patients have a high incidence of some types of cancer?

7. How can a proto-oncogene be converted to an oncogene without a change or mutation in its coding sequence? Explain two ways by which this can occur.

8. What effect would overexpression of the *INK4* tumor suppressor gene have on tumor cells that express inactive Rb?

9. Which would you expect to be more sensitive to treatment with radiation—tumors with wild-type *p53* genes or tumors with mutated *p53* genes?

10. What is the mode of action of imatinib? How do some tumors develop resistance to this drug?

11. What is "oncogene addiction," and why is this concept important for selecting molecular targets for cancer therapy?

12. Would an activating mutation of Raf confer resistance to an inhibitor of the MEK protein kinase?

Refer To

The Cell
Companion Website

sites.sinauer.com/cooper7e

for quizzes, animations, videos, flashcards, and other study resources.

References and Further Reading
(Key review articles for each major section are highlighted in **bold**.)

The Development and Causes of Cancer

Adams, J. M. and A. Strasser. 2008. Is tumor growth sustained by rare cancer stem cells or dominant clones? *Cancer Res.* 68: 4018–4021. [R]

Colditz, G. A., T. A. Sellers and E. Trapido. 2006. Epidemiology—Identifying the causes and preventability of cancer? *Nature Rev. Cancer* 6: 75–83. [R]

Elinav, E., R. Nowarski, C. A. Thaiss, B. Hu, C. Jin and R. A. Flavell. 2013. Inflammation-induced cancer: crosstalk between tumours, immune cells and microorganisms. *Nature Rev. Cancer.* 13: 759–771. [R]

Hanahan, D. and R. A. Weinberg. 2011. Hallmarks of cancer: the next generation. *Cell* 144: 646–674. [R]

Kuilman, T., C. Michaloglou, W. J. Mooi and D. S. Peeper. 2010. The essence of senescence. *Genes Dev.* 24: 2463–2479. [R]

Nowell, P. C. 1986. Mechanisms of tumor progression. *Cancer Res.* 46: 2203–2207. [R]

Peto, J. 2001. Cancer epidemiology in the last century and the next decade. *Nature* 411: 390–395. [R]

Raff, M. C. 1992. Social controls on cell survival and cell death. *Nature* 356: 397–400. [R]

Rosen, J. M. and C. T. Jordan. 2009. The increasing complexity of the cancer stem cell paradigm. *Science* 324: 1670–1673. [R]

Tenen, D. G. 2003. Disruption of differentiation in human cancer: AML shows the way. *Nature Rev. Cancer* 3: 89–101. [R]

Vanharanta, S. and J. Massagué. 2013. Origins of metastatic traits. *Cancer Cell* 24: 410–421. [R]

Wan, L., K. Pantel and Y. Kang. 2013. Tumor metastasis: moving new biological insights into the clinic. *Nature Med.* 19: 1450–1464. [R]

Tumor Viruses

Boshoff, C. and R. Weiss. 2002. AIDS-related malignancies. *Nature Rev. Cancer* 2: 373–382. [R]

Bouchard, M. J. and R. J. Schneider. 2004. The enigmatic X gene of hepatitis B virus. *J. Virol.* 78: 12725–12734. [R]

Coffin, J. M., S. H. Hughes and H. E. Varmus, eds. 1997. *Retroviruses.* Woodbury, N.Y.: Cold Spring Harbor Laboratory Press.

Damania, B. 2004. Oncogenic γ-herpesviruses: Comparison of viral proteins involved in tumorigenesis. *Nature Rev. Microbiol.* 2: 656–668. [R]

Ganem, D. 2006. KSHV infection and the pathogenesis of Kaposi's sarcoma. *Ann. Rev. Pathol. Mech. Dis.* 1: 273–296. [R]

Grassmann, R., M. Aboud and K.-T. Jeang. 2005. Molecular mechanisms of cellular transformation by HTLV-1 Tax. *Oncogene* 24: 5976–5985. [R]

Helt, A.-M. and D. A. Galloway. 2003. Mechanisms by which DNA tumor virus oncoproteins target the Rb family of pocket proteins. *Carcinogenesis* 24: 159–169. [R]

Matsuoka, M. and K.-T. Jeang. 2007. Human T-cell leukemia virus type 1 (HTLV-1) infectivity and cellular transformation. *Nature Rev. Cancer* 7: 270–280.[R]

Mesri, E. A., E. Cesarman and C. Boshoff. 2010. Kaposi's sarcoma and its associated herpesvirus. *Nature Rev. Cancer* 10: 707–719. [R]

Moody, C. A. and L. A. Laimins. 2010. Human papillomavirus oncoproteins: pathways to transformation. *Nature Rev. Cancer* 10: 550–560. [R]

Moore, P. S. and Y. Chang. 2010. Why do viruses cause cancer? Highlights of the first century of human tumour virology. *Nature Rev. Cancer* 10: 878–889. [R]

Polk, D. B. and R. M. Peek Jr. 2010. *Helicobacter pylori*: gastric cancer and beyond. *Nature Rev. Cancer* 10: 403–414. [R]

Saenz-Robles, M. T., C. S. Sullivan and J. M. Pipas. 2001. Transforming functions of simian virus 40. *Oncogene* 20: 7899–7907. [R]

Young, L. S. and A. B. Rickinson. 2004. Epstein-Barr virus: 40 years on. *Nature Rev. Cancer*. 4: 757–768. [R]

zur Hausen, H. 2002. Papillomaviruses and cancer: From basic studies to clinical application. *Nature Rev. Cancer* 2: 342–350. [R]

Oncogenes

Albertson, D. G. 2006. Gene amplification in cancer. *Trends Genet.* 22: 447–455. [R]

Blume-Jensen, P. and T. Hunter. 2001. Oncogenic kinase signaling. *Nature* 411: 355–365. [R]

Clevers, H. 2006. Wnt/β-catenin signaling in development and disease. *Cell* 127: 469–479. [R]

Dang, C. V. 2012. *MYC* on the path to cancer. *Cell* 149: 22–35. [R]

Danial, N. K. and S. J. Korsmeyer. 2004. Cell death: Critical control points. *Cell* 116: 205–219. [R]

Der, C. J., T. G. Krontiris and G. M. Cooper. 1982. Transforming genes of human bladder and lung carcinoma cell lines are homologous to the *ras* genes of Harvey and Kirsten sarcoma viruses. *Proc. Natl. Acad. Sci. USA* 79: 3637–3640. [P]

Dorsam, R. T. and J. S. Gutkind. 2007. G-protein-coupled receptors and cancer. *Nature Rev. Cancer* 7:79–94. [R]

Downward, J. 2006. Prelude to an anniversary for the *RAS* oncogene. *Science* 314: 433–434. [R]

Krontiris, T. G. and G. M. Cooper. 1981. Transforming activity of human tumor DNAs. *Proc. Natl. Acad. Sci. USA* 78: 1181–1184. [P]

Leder, P., J. Battey, G. Lenoir, C. Moulding, W. Murphy, H. Potter, T. Stewart and R. Taub. 1983. Translocations among antibody genes in human cancer. *Science* 222: 765–771. [R]

Martin, G. S. 1970. Rous sarcoma virus: A function required for the maintenance of the transformed state. *Nature* 227: 1021–1023. [P]

Mitelman, F., B. Johansson and F. Mertens. 2007. The impact of translocations and gene fusions on cancer causation. *Nature Rev. Cancer* 7: 233–245. [R]

Nussenzweig, A. and M. C. Nussenzweig. 2010. Origin of chromosomal translocations in lymphoid cancer. *Cell* 141: 27–38. [R]

Pylayeva-Gupta, Y., E. Grabocka and D. Bar-Sagi. 2011. RAS oncogenes: weaving a tumorigenic web. *Nature Rev. Cancer* 11: 761–774. [R]

Shih, C., L. C. Padhy, M. Murray and R. A. Weinberg. 1981. Transforming genes of carcinomas and neuroblastomas introduced into mouse fibroblasts. *Nature* 300: 539–542. [P]

Stehelin, D., H. E. Varmus, J. M. Bishop and P. K. Vogt. 1976. DNA related to the transforming gene(s) of avian sarcoma viruses is present in normal avian DNA. *Nature* 260: 170–173. [P]

Stephen, A. G., D. Esposito, R. K. Bagni and F. McCormick. 2014. Dragging Ras back in the ring. *Cancer Cell* 25: 272–281. [R]

Tabin, C. J., S. M. Bradley, C. I. Bargmann, R. A. Weinberg, A. G. Papageorge, E. M. Scolnick, R. Dhar, D. R. Lowy and E. H. Chang. 1982. Mechanism of activation of a human oncogene. *Nature* 300: 143–149. [P]

Thorpe, L. M., H. Yuzugullu and J. J. Zhao. 2015. PI3K in cancer: divergent roles of isoforms, modes of activation and therapeutic targeting. *Nature Rev. Cancer* 15: 7–24. [R]

Vanhasesebroeck, B., L. Stephens and P. Hawkins. 2012. PI3K signaling: the path to discovery and understanding. *Nature Rev. Mol. Cell Biol.* 13: 195203. [R]

Vogelstein, B. and K. W. Kinzler. 2004. Cancer genes and the pathways they control. *Nature Med.* 10: 789–799. [R]

Vogt, P. K. 1971. Spontaneous segregation of nontransforming viruses from cloned sarcoma viruses. *Virology* 46: 939–946. [P]

Vogt, P. K. 2012.Retroviral oncogenes: a historical primer. *Nature Rev. Cancer* 12: 639–648. [R]

Tumor Suppressor Genes

Adams, B. D., A. L. Kasinski and F. J. Slack. 2014. Aberrant regulation and function of microRNAs in cancer. *Curr. Biol.* 24: R762–R776. [R]

Chan, E., D. E. Prado and J. B. Weidhaas. 2011. Cancer microRNAs: from subtype profiling to predictors of response to therapy. *Trends Mol. Med.* 17:235–243. [R]

Friend, S. H., R. Bernards, S. Rogelj, R. A. Weinberg, J. M. Rapaport, D. M. Albert and T. P. Dryja. 1986. A human DNA segment with properties of the gene that predisposes to retinoblastoma and osteosarcoma. *Nature* 323: 643–646. [P]

Hammond, S. M. 2007. MicroRNAs as tumor suppressors. *Nature Genet.* 39: 582–583. [R]

Harris, H., O. J. Miller, G. Klein, P. Worst and T. Tachibana. 1969. Suppression of malignancy by cell fusion. *Nature* 223: 363–368. [P]

Kastan, M. B. and J. Bartek. 2004. Cell-cycle checkpoints and cancer. *Nature* 432: 316–323. [R]

Knudson, A. G. 1976. Mutation and cancer: A statistical study of retinoblastoma. *Proc. Natl. Acad. Sci. USA* 68: 820–823. [P]

Lee, W.-H., R. Bookstein, F. Hong, L.-J. Young, J.-Y. Shew and E. Y.-H. P. Lee. 1987. Human retinoblastoma susceptibility gene: Cloning, identification, and sequence. *Science* 235: 1394–1399. [P]

Levine, A. J. and M. Oren. 2009. The first 30 years of p53: growing ever more complex. *Nature Rev. Cancer* 9: 749–758. [R]

Manning, A. L. and N. J. Dyson. 2011. pRB, a tumor suppressor with a stabilizing presence. *Trends Cell Biol.* 21: 433–441. [R]

Muller, P. A. J. and K. H. Vousden. 2014. Mutant p53 in cancer: new functions and therapeutic opportunities. *Cancer Cell* 25: 304–317. [R]

Narod, S. A. and W. D. Foulkes. 2004. *BRCA1* and *BRCA2*: 1994 and beyond. *Nature Rev. Cancer* 4: 665–676. [R]

Negrini, S., V. S. Gorgoulis and T. D. Halazonetis. 2010. Genomic instability—an evolving hallmark of cancer. *Nature Rev. Mol. Cell Biol.* 11: 220–228. [R]

Pardali, K. and A. Moustakas. 2007. Actions of TGF-β as a tumor suppressor and pro-metastatic factor in human cancer. *Biochim. Biophys. Acta* 1775: 21–62. [R]

Sherr, C. J. 2004. Principles of tumor suppression. *Cell* 116: 235–246. [R]

Song, M. S., L. Salmena and P. P. Pandolfi. 2012. The functions and regulation of the PTEN tumour suppressor. *Nature Rev. Mol. Cell Biol.* 13: 283–296. [R]

Stratton, M. R. 2011. Exploring the genomes of cancer cells: progress and promise. *Science* 331: 1553–1558. [R]

Ventura, A. and T. Jacks. 2009. MicroRNAs and cancer: short RNAs go a long way. *Cell* 136: 586–591. [R]

Vogelstein, B., N. Papadopoulos, V. E. Velculescu, S. Zhou, L. A. Diaz Jr. and K. W. Kinzler. 2013. Cancer genome landscapes. *Science* 339: 1546–1558. [R]

Vousden, K. H. and C. Prives. 2009. Blinded by the light: the growing complexity of p53. *Cell* 137: 413–431. [R]

Wood, L. D. and 41 others. 2007. The genomic landscapes of human breast and colorectal cancers. *Science* 318: 1108–1113. [P]

Molecular Approaches to Cancer Treatment

Amakye, D., Z. Jagani and M. Dorsch. 2013. Unraveling the therapeutic potential of the Hedgehog pathway in cancer. *Nature Med.* 19: 1410–1422 [R]

Arteaga, C. L. and J. A. Engelman. 2014. ERBB receptors: from oncogene discovery to basic science to mechanism-based cancer therapeutics. *Cancer Cell* 25: 282–303. [R]

Chong, C. R. and P. A. Jänne. 2013. The quest to overcome resistance to EGFR-targeted therapies in cancer. *Nature Med.* 19: 1389–1400. [R]

de The, H. and Z. Chen. 2010. Acute promyelocytic leukemia: novel insights into the mechanisms of cure. *Nature Rev. Cancer* 10: 775–783. [R]

Druker, B. J. 2002. Inhibition of the Bcr-Abl tyrosine kinase as a therapeutic strategy for CML. *Oncogene* 21: 8541–8546. [R]

Engelman, J. A. 2009. Targeting PI3K signaling in cancer: opportunities, challenges and limitations. *Nature Rev. Cancer* 9: 550–562. [R].

Flaherty, K. T., F. S. Hodi and D. E. Fisher. 2012. From genes to drugs: targeted strategies for melanoma. *Nature Rev. Cancer* 12: 349–361. [R]

Fruman, D. A. and C. Rommel. 2014. PI3K and cancer: lessons, challenges and opportunities. *Nature Rev. Drug Disc.* 13: 140–156. [R]

Gschwind, A., O. M. Fischer and A. Ullrich. 2004. The discovery of receptor tyrosine kinases: Targets for cancer therapy. *Nature Rev. Cancer* 4: 361–370. [R]

Haber, D. A., N. S. Gray and J. Baselga. 2011. The evolving war on cancer. *Cell* 145: 19–24. [R]

Herbst, R. S., M. Fukuoka and J. Baselga. 2004. Gefitinib—A novel targeted approach to treating cancer. *Nature Rev. Cancer* 4: 956–965. [R]

Holderfield, M., M. M. Deuker, F. McCormick and M. McMahon. 2014. Targeting RAF kinases for cancer therapy: BRAF-mutated melanoma and beyond. *Nature Rev. Cancer* 14: 455-467. [R]

Imai, K. and A. Takaoka. 2006. Comparing antibody and small-molecule therapies for cancer. *Nature Rev. Cancer* 6: 714–727. [R]

Janne, P. A., N. Gray and J. Settleman. 2009. Factors underlying sensitivity of cancers to small-molecule kinase inhibitors. *Nature Rev. Drug Disc.* 8: 709–723. [R]

Kerbel, R. S. 2006. Antiangiogenic therapy: a universal chemosensitization strategy for cancer? *Science* 312: 1171–1175. [R]

Lo, J. A. and D. E. Fisher. 2014. The melanoma revolution: from UV carcinogenesis to a new era in therapeutics. *Science* 346: 945–949. [R]

Luo, J., N. L. Solimini and S. J. Elledge. 2009. Principles of cancer therapy: oncogene and non-oncogene addiction. *Cell* 136: 823–837. [R]

O'Hare, T., A. S. Corbin and B. J. Druker. 2006. Targeted CML therapy: Controlling drug resistance, seeking cure. *Curr. Opin. Gen. Dev.* 16: 92–99. [R]

Rahman, N. 2014. Realizing the promise of cancer predisposition genes. *Nature* 505: 302–308. [R]

Samatar, A. A. and P. I. Poulikakos. 2014. Targeting RAS-ERK signalling in cancer: promises and challenges. *Nature Rev. Drug Disc.* 13: 928–942. [R]

Sawyers, C. L. 2004. Targeted cancer therapy. *Nature* 432: 294–297. [R]

Sellers, W. R. 2011. A blueprint for advancing genetics-based cancer therapy. *Cell* 147: 26–31. [R]

Sharma, S. V. and J. Settleman. 2007. Oncogene addiction: setting the stage for molecularly targeted cancer therapy. *Genes Dev.* 21: 3214–3231. [R]

Varmus, H. 2006. The new era in cancer research. *Science* 312: 1162–1165. [R]

Weinstein, I. B. 2002. Addiction to oncogenes—The Achilles heal of cancer. *Science* 297: 63–64. [R]

Weisberg, E., P. W. Manley, S. W. Cowan-Jacob, A. Hochhaus and J. D. Griffin. 2007. Second generation inhibitors of BCR-ABL for the treatment of imatinib-resistant chronic myeloid leukemia. *Nature Rev. Cancer* 7: 345–356. [R]

Answers to Questions

Chapter 1

1. Stanley Miller's experiments showed that organic molecules, including several amino acids, can be formed spontaneously from a mixture of reducing gases.

2. RNA is uniquely capable of both serving as a template and catalyzing the chemical reactions required for its own replication.

3. Mitochondria and chloroplasts contain their own DNA and ribosomes, are similar in size to bacteria, and divide like bacteria. The ribosomal proteins and RNAs of these organelles are also more closely related to those of bacteria than to those encoded by eukaryotic nuclear genomes.

4. Because O_2 became abundant in Earth's atmosphere as a result of photosynthesis.

5. The volume of a sphere is given by the formula $4/3\pi r^3$, where r = the radius. Since the volume of the cells is proportional to the cube of their radii, the relative volume of a macrophage compared to *S. aureus* is the ratio of the cubes of their radii: $25^3/0.5^3 = 125,000$.

6. Yeast provides the simplest system for studying eukaryotic DNA replication.

7. Mice.

8. The refractive index of air is 1.0; the refractive index of oil is approximately 1.4. Since resolution $= 0.61/\eta \sin \alpha$ (η is the refractive index), viewing a specimen through air rather than through oil changes the limit of resolution from about 0.2 μm to about 0.3 μm.

9. Resolution $= 0.61\lambda/\text{NA}$. Taking $\lambda = 0.5$ μm for visible light, the objective with an NA $=1.3$ will resolve objects about 0.23 μm apart, while the objective of NA $= 1.1$ will resolve objects about 0.27 μm apart. Magnification of the images can be obtained by other means (such as projecting on a screen or using an ocular lens with higher magnification). Thus the better choice would be the objective with 60× magnification and the numerical aperture of 1.3.

10. GFP-tagged proteins can be expressed in live cells, so they can be seen without the need to fix and kill the cells for staining with a fluorescent antibody. Movement of GFP-tagged proteins can therefore be followed in living cells.

11. Velocity centrifugation separates organelles on the basis of size and shape, whereas equilibrium centrifugation separates them on the basis of density (independent of size and shape).

12. Serum contains growth factors that are required to stimulate division of most animal cells in culture.

13. Primary cell cultures are cells grown directly from an organism or tissue. Immortal cell lines have the ability to proliferate indefinitely in culture.

14. These cells retain the ability to differentiate into all of the different cell types of adult organisms, so they offer the potential of being used in transplantation therapies for treatment of a variety of diseases.

Chapter 2

1. Water is a polar molecule whose hydrogens and oxygen can form hydrogen bonds with each other (to make water a liquid at most of Earth's temperatures), as well as with polar organic molecules and inorganic ions. In contrast, nonpolar molecules or parts of molecules cannot interact with water, forcing them to associate with each other to form important structures like cell and organelle membranes.

2. Glycogen is a polymer of glucose residues connected primarily by $\alpha(1 \rightarrow 4)$ glycosidic bonds, and it has occasional $\alpha(1 \rightarrow 6)$ bonds that produce branches. Cellulose is a straight, unbranched polymer with $\alpha(1 \rightarrow 4)$ glycosidic bonds between the glucose residues.

3. Fats accumulate in fat droplets of cells and are an efficient form of energy storage. Phospholipids function as the major components of cell membranes.

4. Nucleotides function as carriers of chemical energy (e.g., ATP) and as intracellular signaling molecules (e.g., cyclic AMP).

5. No. The information in nucleic acids is conveyed by the sequence of its bases. If the bases were removed, we would be left with a linear sugar phosphate chain with little informational content.

6. Christian Anfinsen and his colleagues showed that denatured ribonuclease can spontaneously refold into the active enzyme in the absence of other cellular constituents. This showed that the information required for folding was contained in the primary sequence of ribonuclease.

7. The side chain of cysteine contains a sulfhydryl group that can form covalent disulfide bonds with another cysteine residue, stabilizing the structure of cell surface and secreted proteins.

8. Cholesterol can be incorporated into membranes, where it regulates membrane fluidity. Additionally, cholesterol is used in the synthesis of steroid hormones.

9. The α-helical structure allows the CO and NH groups of peptide bonds to form hydrogen bonds with each other, thereby neutralizing their polar character. β barrels are also capable of traversing lipid bilayers.

10. d. Alanine is the only amino acid on the list with a hydrophobic side chain capable of interacting with the fatty acids of membrane lipids.

11. Aspartate is an acidic amino acid that interacts with basic amino acids in the substrates of trypsin. Substitution of aspartate with lysine (a basic amino acid) would interfere with substrate binding and catalysis.

12. The side chain of histidine can be either uncharged or positively charged at physiological pH, thus allowing it to be used for exchange of hydrogen ions.

13. Channel proteins form open pores through the membrane, whereas carrier proteins selectively bind and transport small molecules.

Chapter 3

1. An energetically unfavorable reaction can be coupled to an energetically favorable reaction with a high negative free energy change (often the hydrolysis of ATP), such that the combined reaction is energetically favorable.

2. The reaction catalyzed by phosphofructokinase is fructose-6-phosphate + ATP → fructose-1,

6-bisphosphate + ADP. The standard free energy change can be calculated as the sum of the standard free energy change of fructose-1,6-bisphosphate formation from fructose-6-phosphate and phosphate ($\Delta G^{o\prime}$ = +4 kcal/mol) and the standard free energy change of ATP hydrolysis (–7.3 kcal/mol), giving a standard free energy change of –3.3 kcal/mol for the coupled reaction.

3. Substituting the given values into the equation:

$$\Delta G = \Delta G^{o} + RT \ln \frac{[B][C]}{[A]}$$

gives the following result: ΔG = –1.93 kcal/mol. The reaction will therefore proceed from left to right, with A being converted to B plus C within the cell.

4. Phosphofructokinase is inhibited by high concentrations of ATP. Thus, glycolysis will be inhibited in response to an increased cellular concentration of ATP.

5. Under anaerobic conditions, glucose is metabolized only through glycolysis, with a net production of 2 ATP molecules per glucose molecule. Under aerobic conditions, glucose is completely oxidized to produce 36–38 molecules of ATP.

6. Under anaerobic conditions the NADH produced during glycolysis is used to reduce pyruvate to ethanol or lactate, thereby regenerating NAD^{+}.

7. Since lipids are more reduced than fatty acids, their oxidation yields much more energy per molecule.

8. The proton gradient of 1 pH unit corresponds to a tenfold higher $[H^{+}]_{o}$. By substituting the above values in the equation

$$\Delta G = RT \ln [H^{+}]_{i}/[H^{+}]_{o}$$

we get an answer of approximately –1.4 kcal/mol.

9. Coenzyme Q accepts a pair of electrons from either complex I or complex II and passes them to complex III. Coenzyme Q also binds two protons from the mitochondrial matrix, carries them across the inner membrane, and releases them to the intermembrane space, contributing to the generation of a proton gradient. Cytochrome c is a small protein in the intermembrane space that picks up a pair of electrons from complex III and transfers them to complex IV.

10. Two high-energy electrons are required to split each molecule of H_2O, so 24 high-energy electrons are required for the synthesis of each molecule of glucose. The passage of these electrons through the two photosystems generates 12 molecules of NADPH and between 12 and 18 molecules of ATP,

depending on proton pumping at the cytochrome *bf* complex. Since 18 molecules of ATP are required for the Calvin cycle, the synthesis of glucose may require additional ATPs produced by cyclic electron flow.

11. In the light reactions, energy derived from the absorption of light by chlorophylls is used to split water to $2H^+ + \frac{1}{2} O_2 + 2e^-$. The electrons enter an electron transport chain that results in synthesis of ATP and NADPH. In the dark reactions, ATP and NADPH drive the synthesis of glucose from CO_2 and H_2O.

12. Some reactions in glycolysis involve a large decrease in free energy and are not easily reversible. Gluconeogenesis bypasses these reactions by other reactions that are driven by the expenditure of ATP and NADH.

Chapter 4

1. You would breed flies with mutations in the two different genes. If mutations in the two genes are frequently inherited together, the genes would be linked and reside on the same chromosome.

2. Half the DNA will be intermediate density and half will be light.

3. Addition or deletion of one or two nucleotides would alter the reading frame of the gene, resulting in a completely different amino acid sequence of the encoded protein.

4. To clone this piece of human DNA, a yeast artificial chromosome would need telomeres and a centromere in order to replicate as a linear chromosome-like molecule, as well as an origin of replication and a cleavage site for *Eco*RI.

5. UGC also encodes cysteine, so this mutation would have no effect on enzyme function. UGA, however, is a stop codon, so this mutation would lead to a truncated protein that would be inactive.

6. The restriction map is

```
                              EcoRI
         2.5 kb          0.5 kb |  1 kb
  _____|_____|_____
                            |
                          HindIII
```

7. The haploid sperm contains a single starting copy of the DNA sequence. Each cycle of PCR amplifies the starting material twofold, so 10 cycles yields 2^{10}, or approximately a thousand, copies. Amplification for 30 cycles yields 2^{30}, or more than a billion, copies.

8. From Table 4.3, cosmids carry inserts of 30–45 kb. From Table 4.2, the recognition sequence of *Bam*HI is 6 base pairs long, which occurs with a random frequency of once every 4000 base pairs of DNA ($4^6 = 4096$). Dividing 30,000–45,000 by 4000, we expect about 10 *Bam*HI fragments in a cosmid insert.

9. The human genome is approximately 3×10^6 kb, and the size of an insert for a BAC vector is 120–300 kb (see Table 4.3). Dividing the size of the genome by the size of an insert indicates that a minimum of 10,000 to 25,000 BAC clones would be needed to cover the genome.

10. Influenza virus replicates via RNA-directed RNA synthesis, so actinomycin D (which inhibits DNA-directed RNA synthesis) will not affect influenza virus replication.

11. The cloning vector's critical feature would be a selectable marker, such as drug resistance, that allows you to select stably transfected cells.

12. Proteins are dissolved in a solution containing the negatively charged detergent SDS. Each protein binds many molecules of SDS, which imparts a net negative charge to the protein. The proteins can then be separated by gel electrophoresis according to their size.

13. The CRISPR/Cas system targets specific DNA sequences by homologous pairing with an RNA molecule. Since RNAs homologous to any desired target sequence can easily be synthesized, this is a highly effective method for introducing targeted gene mutations in mammalian cells.

Chapter 5

1. Protein-coding sequences can be identified as long stretches of nucleotide sequence that can encode polypeptides because they do not contain chain-terminating codons.

2. The fraction of protein-coding sequence is much lower in the human genome: approximately 1% as compared with 70% in *S. cerevisiae*.

3. The number of protein-coding genes in apples is more than twice the number of protein-coding genes in humans.

4. A protein is digested to yield small peptides, whose mass-to-charge ratio is determined by mass spectrometry. The experimentally determined mass spectrum can then be compared to a database of all known proteins to identify the protein of interest.

5. Mitochondria from both cancer cells and normal cells can be isolated by subcellular fractionation. Their protein compositions can then be determined by shotgun mass spectrometry.

6. A protein can be immunoprecipitated from a cell extract under gentle conditions, so that the immunoprecipitated target protein remains associated with the proteins it normally interacts with inside the cell. The immunoprecipitated protein complexes can then be analyzed, for example by mass spectrometry, to identify the interacting proteins that were immunoprecipitated together with the target protein.

7. Both breast cancer and normal cells could be screened to identify siRNAs that interfere with the growth of breast cancer but not normal cells. Such siRNAs would identify genes specifically required for breast cancer cell growth.

8. Next-generation sequencing can be used to sequence all of the RNAs expressed in a cell, thereby identifying all transcribed sequences.

9. Unlike protein-coding sequences, regulatory sequences are short, poorly defined sequences that occur frequently by chance in large genomes, making it difficult to identify functional regulatory sequences. Approaches used to identify functional regulatory sequences include looking for clusters of regulatory elements, for sequences that are conserved in evolution, and for sequences that are present in coordinately regulated genes.

10. By designing synthetic (unnatural) biological systems, synthetic biologists hope to gain a better understanding of the principles that govern the behavior of existing (natural) cells.

Chapter 6

1. Organisms with larger genomes than their complexity have larger amounts of noncoding DNA.

2. Phillip Sharp and coworkers made hybrids between adenovirus hexon mRNA and a single strand of viral DNA. Upon observation under an electron microscope, they saw that the hexon mRNA hybridized with separate regions of viral DNA and the intervening unhybridized DNA formed loops. The loops of single-stranded DNA corresponded to introns that were spliced out of the long primary transcript.

3. Through alternative splicing, more than one protein can be synthesized from the same gene, thereby increasing the diversity of proteins expressed from a limited number of genes.

4. Because of their distinct base composition, simple-sequence repeats (e.g., ACAAACT) differ in density from the average AT/GC ratio of bulk nuclear DNA. Sequences containing tandem copies of these repeats will separate upon CsCl

density-gradient centrifugation into satellite bands that are distinct from the main band of genomic DNA.

5. Noncoding RNAs can regulate gene expression at either the level of transcription (long noncoding RNAs) or by affecting mRNA translation and stability (miRNAs).

6. A processed pseudogene would not contain introns.

7. A centromere of *S. cerevisiae* is small (125 base pairs) and presumably binds a limited number of kinetochore proteins to form a single microtubule-binding site. Centromeres of animal chromosomes are much larger, with repeated sequences that form more extensive kinetochores and provide attachment sites for multiple microtubules.

8. Cutting the plasmid with a restriction endonuclease creates a linear chromosome with two free ends but no telomeres. The plasmid genes are quickly lost because of the instability of the chromosome ends. To test this, you could attach telomere sequences to the ends of the linearized plasmid and see if this allows the plasmid to be stably inherited.

9. One molecule of histone H1 binds genomic DNA approximately every 200 bp, so the number of histone H1 molecules bound to the yeast genome can be determined by dividing the size of the yeast genome (12 Mb) by 200 bp to yield a value of 60,000.

10. From Table 6.1, we can calculate the average length of a human intron by dividing the total length of intron sequence by the number of introns in an average human gene: approximately 5800 bp.

11. From Table 6.1, the exon sequence of an average human gene totals 4300 bp. This would be the expected average length of complete cDNA inserts in your library.

12. Plants contain many duplicated genes as a result of duplications of their entire genomes.

Chapter 7

1. At high temperature the mutant would have defects in replacing RNA primers with DNA in Okazaki fragments and in filling gaps in DNA following excision repair.

2. Topoisomerase I cuts one strand of the DNA double helix and allows it to swivel around the other strand to relieve twist tension. Topoisomerase II cuts both strands of the DNA double helix and can pass another DNA molecule through the cut to untangle intertwined molecules (e.g., during mitosis).

3. From Table 5.1, the size of the yeast genome is 12 Mb (12 million base pairs), while Okazaki fragments are approximately 1–3 kb (1,000–3,000 base pairs) in length. We can calculate the number of Okazaki fragments synthesized during replication by dividing the size of the genome by the length of Okazaki fragments, giving an answer of 4,000–12,000 fragments.

4. DNA polymerases are unable to initiate DNA synthesis *de novo*: they can only add nucleotides to a preexisting primer strand. Primase initiates *de novo* and synthesizes short RNA primers, which can then be extended by a DNA polymerase.

5. 3′ to 5′ exonuclease activity is required for the excision of mismatched bases in newly synthesized DNA during proofreading. A mutant *E. coli* with a DNA polymerase III lacking this activity would have a high frequency of mutations each time the DNA is replicated.

6. You would insert the DNA sequence into a plasmid that lacks an origin of replication and determine whether the recombinant plasmid is able to transform mutant yeast that require a plasmid gene for their growth. Only plasmids with a functional origin will yield a high frequency of transformed yeast colonies.

7. DNA polymerases are unable to replicate the ends of linear DNA molecules; yeast cells have evolved telomerase to maintain the ends of their linear chromosomes. Since the *E. coli* genome is a circular DNA molecule and has no ends, special mechanisms to replicate the ends of linear DNA are not required.

8. Double-strand breaks can be repaired by recombinational repair, in which the missing portion of one chromosome can be recovered from a homologous chromosome. Single-strand breaks can be repaired by excision repair, since an undamaged complementary strand is available for use as a template to direct synthesis of the excised portion of the damaged strand.

9. The high frequency of skin cancer results from DNA damage induced by solar UV irradiation, which is subject to repair by the nucleotide-excision repair system. The lack of elevated incidence of other cancers may suggest that similar types of damage are not frequent in internal organs, and that most cancers of these organs result from other types of mutations (e.g., the incorporation of mismatched bases during DNA replication).

10. The cellular processes that might be affected by these drugs include maintenance of

telomeres by telomerase and the transposition of retrotransposons.

11. Instead of using methylation of parental strands, the mismatch repair system in humans uses single-strand breaks to identify newly replicated DNA. Thus, a homolog of *MutH* is not required in human cells.

12. The mouse would be immunodeficient, lacking both B and T lymphocytes, as a result of being unable to rearrange its immunoglobulin and T cell receptor genes.

Chapter 8

1. Sigma factors bind to sequences upstream of the transcription start site, bringing RNA polymerase specifically to the promoter region to initiate transcription.

2. The most common type of termination involves transcription of a GC-rich inverted repeat that forms a stable stem-loop structure by complementary base pairing. Formation of this structure disrupts association of the mRNA with the DNA and terminates transcription.

3. Binding of the repressor to the operator interferes with the binding of RNA polymerase to the promoter, thereby blocking transcription.

4. The promoter containing the TATA box can be transcribed *in vitro* in the presence of either TBP or TFIID. However, the Inr promoter requires TFIID, since the Inr sequence is recognized by TAFs rather than by TBP.

5. The activity of enhancers depends neither on their distance nor their orientation with respect to the transcription start site. Promoters are defined as being near the transcription start site.

6. The sequence element would be a potential binding site for a tissue-specific repressor.

7. Insulators divide chromosomes into individual domains of chromatin structure that can be euchromatin or heterochromatin but cannot spread beyond an insulator. Also, they prevent an enhancer in one domain from acting on a promoter in the next domain.

8. Many lncRNAs form complexes with proteins that modify chromatin and recruit these chromatin-modifying complexes to their target genes. In addition, some lncRNAs interact with transcription factors, Mediator, or components of the basal transcription machinery, and other lncRNAs interact with proteins that regulate mRNA processing or translation.

9. The specific high-affinity binding of Sp1 to the GC box DNA sequence was exploited for its biochemical purification by DNA-affinity chromatography. To demonstrate that the purified protein is Sp1, you could carry out *in vitro* transcription assays.

10. The anti-Sm antiserum would bind to the snRNPs and prevent them from binding splice sites, thereby inhibiting splicing.

11. The nucleosomes flanking promoters and enhancers have different histone modifications that can be distinguished by chromation immunoprecipitation. Promoters are marked by the trimethylated form of H3 lysine-4, whereas enhancers are characterized by the monomethylated form of this lysine.

12. Splicing factors direct snRNPs to the correct splice sites by binding to specific sequences in the pre-mRNA.

13. Apo-B100 is synthesized in the liver by translation of the unedited mRNA. The shorter Apo-B48 is synthesized in the intestine by translation of an edited mRNA in which the editing reaction has generated a stop codon.

14. Nonsense-mediated mRNA decay is a quality control process by which mRNAs that lack complete open-reading frames are degraded. This process eliminates defective mRNAs that would have given rise to abnormal truncated proteins.

Chapter 9

1. Many tRNAs are able to recognize more than one codon because the third base of their anticodon can "wobble" or form hydrogen bonds in nonstandard ways with bases other than the usual complementary pairs.

2. A Shine-Dalgarno sequence is needed to bind to bacterial ribosomes.

3. Ribosomes depleted of most of their proteins can still synthesize polypeptides, whereas treatment with RNase completely abolishes translational activity. In addition, high-resolution structural analysis of the 50S ribosomal subunit showed that the site at which the peptidyl transferase reaction occurs is composed of rRNA, not ribosomal protein.

4. Polyadenylation is an important translational regulatory mechanism in early development. Its inhibition would block the translation of many oocyte mRNAs following fertilization.

5. Chaperones are proteins that aid the proper folding of other proteins. Elevated temperatures can denature proteins. Heat-shock proteins are chaperones that aid in the refolding of these denatured proteins, thereby restoring protein function.

6. Phospholipase treatment would release a GPI-anchored protein, but not a transmembrane protein, from the cell surface.

7. The degradation of cyclin B, which allows cells to exit mitosis, requires a specific 9-amino-acid sequence called the destruction box. Mutations in this sequence prevent ubiquitylation and degradation of cyclin B and block exit from mitosis.

8. No. In addition to targeting proteins for degradation, ubiquitylation of a protein can have other functions, including targeting proteins for endocytosis and serving as a modification of histones that regulates transcription.

9. miRNAs can target specific mRNAs via the RISC complex, leading to either mRNA degradation or inhibition of translation.

10. Sequences in the 3' UTRs can regulate mRNA stability, localization, and translation by serving as binding sites for regulatory factors and miRNAs.

11. Signal sequences that target proteins to specific cellular compartments are often removed by the action of signal peptidases. Additionally, many proteins are synthesized as larger precursor proteins and are proteolytically processed to yield the mature protein.

12. A "decoding center" in the small ribosomal subunit recognizes correct codon-anticodon base pairs and discriminates against mismatches. Insertion of a correct aminoacyl tRNA into the A site induces a conformational change that causes the hydrolysis of GTP bound to eEF1α and the release of the elongation factor bound to GDP.

13. PDI catalyzes the formation of disulfide bonds. Inhibiting PDI with the siRNA will interfere with the correct folding of RNase, so you will detect much lower RNase activity in the medium of cells expressing the siRNA.

Chapter 10

1. Prokaryotic mRNAs are translated as they are transcribed. Separation of the site of transcription from the site of translation in eukaryotes allows regulation of mRNAs by posttranscriptional processes, such as alternative splicing, polyadenylation, and regulated transport to the cytoplasm. In addition, transcription can be regulated by modulating the nuclear localization of transcription factors.

2. Lamins form a filamentous network that supports and stabilizes the nuclear envelope. Lamins also

provide binding sites for chromatin to attach to the inside of the nuclear envelope. In addition, many proteins that participate in transcription, DNA replication, and chromatin modification interact with lamins.

3. The 15 kd but not the 100 kd protein will be able to enter the nucleus, since proteins smaller than approximately 40 kd can pass freely through the nuclear pore complex.

4. The distribution of Ran/GTP across the nuclear envelope determines the directionality of nuclear transport.

5. One example of this regulation by nuclear import is NF-κB. In unstimulated cells, IκB binds to NF-κB, and the complex cannot be imported to the nucleus. Upon stimulation, IκB is phosphorylated and degraded by ubiquitin-mediated proteolysis. This exposes a nuclear localization signal on NF-κB, allowing it to enter the nucleus and activate transcription of target genes.

6. The transcription factor could no longer be phosphorylated at these sites, so it would be constitutively imported to the nucleus and activate target gene expression.

7. Inactivating the nuclear export signal would result in retention of the protein in the nucleus.

8. You would label newly replicated DNA with bromodeoxyuridine, which is incorporated into DNA in place of thymidine. The bromodeoxyuridine-labeled DNA can be localized with antibodies against bromodeoxyuridine and fluorescence microscopy.

9. Smith and colleagues fused the T antigen amino acid sequence 126 to 132 to normally cytoplasmic proteins, β-galactosidase, and pyruvate kinase. These fusion proteins accumulated in the nucleus.

10. Nuclear speckles contain concentrated populations of snRNPs and are thought to be storage sites for splicing components.

11. Small nucleolar RNAs (snoRNAs) localize to the nucleolus where they participate in the cleavage and modification of rRNAs.

12. Exportin-t is required for the export of tRNAs from the nucleus, so inhibiting exportin-t function would prevent tRNA export and lead to inhibition of translation.

Chapter 11

1. George Palade and coworkers labeled pancreatic acinar cells with a pulse of radioactive amino acids, which were incorporated into proteins. Autoradiography showed that the labeled proteins were first detected in the rough ER. After a short "chase" with nonradioactive amino acids, the labeled proteins had moved to the Golgi apparatus and after longer periods, the labeled proteins were found in secretory vesicles and then outside of cells.

2. When an mRNA encoding a secreted protein is translated *in vitro* on free ribosomes, a larger protein results than when the same mRNA is translated in the presence of microsomes from the rough ER. In the latter case, the signal sequence is cleaved off by a signal peptidase in the rough ER vesicles.

3. Cotranslational translocation involves binding of the nascent polypeptide to the signal recognition particle and translocation through a translocon driven by the process of protein synthesis. Post-translational translocation targets a polypeptide to the ER after synthesis is complete and does not require SRP. A Sec62/63 complex recognizes a polypeptide to be incorporated, and inserts it into a translocon. The polypeptide is pulled through into the ER lumen by the chaperone BiP.

4. Proteins bound for the Golgi apparatus are unable to enter the ER in a Sec61 mutant and therefore remain in the cytosol.

5. Carbohydrate groups are added within the lumen of the ER and Golgi apparatus, both of which are topologically equivalent to the exterior of the cell.

6. Mutation of the KDEL sequence would inhibit the return of this ER-resident protein from the Golgi to the ER, so it would be secreted from the cell. Inactivation of the KDEL receptor protein would result in secretion of all KDEL-containing ER proteins.

7. Initially, a signal sequence targets the nascent lysosomal polypeptide to the rough ER. As it enters the ER, an *N*-linked oligosaccharide is added to the protein. After it moves to the Golgi, the three mannoses that are normally removed are not, but are instead converted to mannose-6-phosphate. These mannose-6-phosphates are recognized by a receptor in the *trans* Golgi network, which directs the transport of these proteins to lysosomes. The normally cytosolic protein lacks a signal sequence and does not enter the ER. Therefore addition of a lysosome-targeting signal would have no effect. In contrast, such an addition would direct a normally secreted protein to lysosomes from the Golgi apparatus.

8. In the absence of mannose-6-phosphate formation in the Golgi apparatus, lysosomal proteins would be secreted.

9. Glycolipids and sphingomyelin are produced by addition of sugars or phosphorylcholine to ceramide on the cytosolic and lumenal surfaces, respectively, of the Golgi apparatus. Glucosylceramide is then flipped to the lumenal surface. After vesicular transport and fusion with the plasma membrane, these lipids are located on the outer half of the plasma membrane.

10. The diagnosis is Gaucher disease. Since this disease involves a mutation in the lysosomal hydrolase glucocerebrosidase, you might suggest enzyme replacement therapy.

11. The enzymes destined for lysosomes are acid hydrolases that are not active at the neutral pH of the cytoplasm, ER, or Golgi apparatus. The acid hydrolases are activated by the acid pH of lysosomes, which is maintained by lysosomal proton pumps.

12. The strong interaction between the coiled-coil domains of vesicle and target SNAREs places the two membranes nearly in direct contact. This produces membrane instability and causes the membranes to fuse.

Chapter 12

1. The surface area of the mitochondrial inner membrane is high due to the formation of cristae. Also, the protein content of the membrane is unusually high (>70%) and includes many enzymes and electron carriers.

2. Mitochondrial tRNAs are able to do an extreme form of wobble in which U in the anticodon can pair with any of the four bases in the third codon position of mRNA, allowing four codons to be recognized by a single tRNA.

3. Hsp70 proteins in the cytosol maintain newly synthesized mitochondrial proteins in an unfolded state so they can be inserted into the Tom complex and be imported as an unfolded chain. Mitochondrial matrix Hsp70 chaperones bind to the polypeptide chain as it emerges from the Tim complex and use ATP hydrolysis to pull the polypeptide into the matrix.

4. The energy for metabolite transport is provided by the electrochemical gradient across the mitochondrial inner membrane.

5. The inner membrane of mitochondria and the thylakoid membrane of chloroplasts are the sites of electron transport and oxidative phosphorylation.

6. In contrast to mitochondria, there is no electric potential across the chloroplast membrane. Therefore the charge of transit peptides does not contribute to protein translocation.

7. It will be translocated across the chloroplast membranes but remain in the chloroplast stroma instead of being translocated into the thylakoid lumen.

8. Chromoplasts contain carotenoids instead of chlorophyll and do not contain a thylakoid membrane.

9. Most peroxisome transmembrane proteins are synthesized in the ER and transported in vesicles that fuse to form peroxisomes. Other transmembrane proteins are synthesized on free ribosomes in the cytosol and targeted to the peroxisome membrane by a membrane-targeting signal.

10. Pex5 is the major cytosolic receptor for import of peroxisome matrix proteins, so depletion of Pex5 with siRNA would prevent the import of most matrix proteins.

Chapter 13

1. The asymmetrical actin monomers associate in a head-to-tail fashion to form actin filaments. Since all the monomers are oriented in the same direction, the filament has a distinct polarity. The polarity of actin filaments defines the direction of myosin movement. If actin filaments were not polar, the unidirectional movement of myosin that results in the sliding of actin and myosin filaments could not take place.

2. Cytochalasin binds to the barbed ends of actin filaments and blocks their elongation, so it would lead to depolymerization of treadmilling filaments. Phalloidin binds to actin filaments and inhibits their depolymerization, so it would cause filaments to stop treadmilling but remain present and grow longer.

3. Cofilin severs actin filaments, generating new filament ends which can either grow or depolymerize. Profilin binds ADP-actin and stimulates the exchange of ADP for ATP, forming ATP-actin monomers that can join growing filaments. The Arp2/3 complex initiates the formation of branches.

4. The I band and the H zone shorten during contraction. The A band doesn't shorten because it is occupied by thick myosin filaments.

5. The contraction of smooth muscle cells is regulated by the phosphorylation of the myosin regulatory light chain by myosin light-chain kinase, which in turn is regulated by association with the

calcium-binding protein calmodulin. An increase in the cytosolic concentration of calcium leads to the binding of calmodulin to myosin light-chain kinase.

6. The *in vitro* movement of microtubules required ATP and was inhibited by the nonhydrolyzable ATP analog AMP-PNP. Importantly, organelles remained attached to microtubules in the presence of AMP-PNP, suggesting that the motor proteins responsible for organelle movement might also remain attached.

7. Colcemid inhibits the polymerization of microtubules and would inhibit the transport of secretory vesicles along microtubules.

8. γ-tubulin plays a key role in the formation of microtubule organizing centers. γ-tubulin associates with other proteins to form the γ-tubulin ring complex, which functions as a seed for the nucleation of new microtubules.

9. Intermediate filaments are not required for the growth of cells in culture, so siRNA against vimentin would have no effect.

10. Dimers of cytoskeletal intermediate filaments assemble in a staggered antiparallel manner to form tetramers, which can then assemble end-to-end to form protofilaments. Since they are assembled from antiparallel tetramers, intermediate filaments do not have distinct ends and are apolar.

11. By linking the microtubule doublets in cilia together, nexin converts the sliding of individual microtubules to a bending motion that leads to the beating of cilia. If it were eliminated, the microtubules would simply slide past one another.

Chapter 14

1. Peripheral membrane proteins can be removed from a membrane by a high-salt wash or by solutions of extreme pH that do not disrupt the phospholipid bilayer. Integral membrane proteins can only be extracted from membranes by detergents that disrupt the phospholipid bilayer.

2. Frye and Edidin fused mouse and human cells and examined the distribution of membrane proteins after staining with anti-mouse and anti-human antibodies labeled with different fluorescent dyes. After a brief incubation at 37°C the proteins were intermixed, demonstrating that the proteins could diffuse laterally in a fluid membrane. If incubated at 2°C, the proteins would remain separated because the membrane is not fluid at this temperature.

3. Lipid rafts are discrete membrane domains that are enriched in cholesterol and sphingolipids. Lipid rafts are believed to play important roles in cell movement, endocytosis, and cell signaling.

4. Transmembrane proteins that are associated with the cytoskeleton are not free to move in the lipid bilayer. In addition to being restricted in their own mobility, transmembrane proteins that are anchored to the cytoskeleton may act as barriers that limit the mobility of other membrane proteins

5. Given $C_o/C_i = 10$, the K^+ equilibrium potential calculated from the Nernst equation is about –59 mV. The actual resting membrane potential differs from the K^+ equilibrium potential because resting squid axons are more permeable to K^+ than to other ions.

6. The opening of nicotinic acetylcholine receptors is required for membrane depolarization in muscle cells. Thus curare blocks the contraction of muscle cells in response to acetylcholine.

7. The K^+ channel contains a selectivity filter lined with carbonyl oxygen atoms. The pore is just wide enough to allow passage of dehydrated K^+ ions from which all water molecules have been displaced as a result of association with the carbonyl oxygen atoms. Hydrated Na^+ ions are too small to interact with the carbonyl oxygen atoms and remain associated with water molecules. This complex is too large to pass through the channel pore.

8. The uptake of glucose against its concentration gradient is coupled to the transport of Na^+ ions in the energetically favorable direction.

9. The *mdr* gene encodes an ABC transporter that is frequently overexpressed in cancer cells. The transporter can recognize a variety of drugs and pump them out of the cell, conferring resistance to chemotherapeutic drugs.

10. The addition of excess unlabeled LDL reduced the binding of labeled LDL to the surface of normal cells. This indicated that the labeled and unlabeled LDL were competing for a limited number of specific binding sites on the surface of normal cells.

11. Two types of mutations in the LDL receptor resulting in the inability to take up LDL were identified in FH patients. Cells from most FH patients failed to bind LDL, demonstrating that a specific receptor was required for LDL uptake. Other mutant receptors bound LDL, but failed to cluster in coated pits, demonstrating the role of coated pits in receptor-mediated endocytosis.

12. Phagocytosis requires actin-based extension of the plasma membrane, so it would be blocked by inhibition of actin polymerization. Clathrin-mediated endocytosis is independent of actin and would not be affected.

Chapter 15

1. Gram-positive bacteria have a single membrane—the plasma membrane—surrounded by a thick cell wall. Gram-negative bacteria have both a plasma membrane and an outer membrane, separated by a thin cell wall.

2. Bacterial cytoskeletal proteins determine cell shape by controlling cell wall synthesis.

3. The rigid plant cell wall prevents cell swelling and allows the buildup of turgor pressure.

4. Hemicelluloses are branched polysaccharides that hydrogen bond to the surface of cellulose microfibrils. This interaction stabilizes the cellulose microfibrils into tough fibers.

5. The correct localization of these glucose transporters is necessary for the polarized function of intestinal epithelial cells in transferring glucose from the intestinal lumen to the blood supply. Tight junctions prevent the diffusion of these transporters between domains of the plasma membrane, as well as sealing the spaces between cells of the epithelium.

6. Hydroxylysine stabilizes the collagen triple helix by forming hydrogen bonds between the polypeptide chains. Inhibition of lysyl hydroxylase would therefore decrease the stability of collagen fibrils.

7. Fibril-forming collagens are synthesized as soluble precursors known as procollagens. Procollagens have nonhelical segments on their ends, which prevent the assembly of fibrils. Only after procollagen is secreted outside the cell are the non-helical segments removed and the fibrils assembled.

8. GAGs contain acidic sugar residues, which are modified by the addition of sulfate groups. This imparts a high negative charge to GAGs, so they bind positively charged ions and trap water molecules to form hydrated gels.

9. The peptide most likely disrupts tight junctions, thereby allowing free diffusion of the transporter between the apical and basolateral domains of the plasma membrane.

10. Since E-cadherins mediate selective adhesion of epithelial cells, overexpression of a dominant negative version would disrupt interactions between neighboring cells.

11. The cytoplasmic domain of $\alpha_6\beta_4$ integrin is required for its interaction with the intermediate filament cytoskeleton through plectin. The mutation would therefore disrupt hemidesmosome formation and lead to decreased cell–matrix interaction.

12. Electrical synapses are specialized gap junctions found in nerve cells. They allow the rapid passage of ions between cells, thereby conducting nerve impulses.

13. Gap junctions and plasmodesmata are similar in that they both provide channels between the cytoplasms of adjacent cells. They are likely to be analogous rather than homologous structures in animals and plants, since their structures are extremely different.

Chapter 16

1. In paracrine signaling, molecules released by one or a few cells affect nearby cells. In endocrine signaling, hormones are carried throughout the body to act on any target cell that has a receptor for that hormone.

2. Hydrophobic molecules like steroid hormones can diffuse through the plasma membrane and bind to cytosolic or nuclear receptors. Hydrophilic molecules (like peptide hormones) cannot cross the plasma membrane, so they act by binding to receptors on the cell surface.

3. Aspirin inhibits the enzyme cyclooxygenase, which catalyzes the first step in the synthesis of prostaglandin and thromboxanes from arachidonic acid.

4. Inhibition of cAMP phosphodiesterase would result in elevated levels of cAMP, which would stimulate cell proliferation.

5. The recombinant molecule would function as an epinephrine receptor coupled to G_i. Epinephrine would therefore inhibit adenylyl cyclase, lowering intracellular cAMP levels. Acetylcholine would have no effect, since it would not bind to the recombinant receptor.

6. Protein phosphatase 1 dephosphorylates serine residues that are phosphorylated by protein kinase A. Cyclic AMP-inducible genes are activated by CREB, which is phosphorylated by protein kinase A, so overexpression of protein phosphatase 1 would inhibit their induction. However, protein phosphatase 1 would not affect the activity of cAMP-gated ligand channels, since these channels are opened directly by cAMP binding rather than by protein phosphorylation.

7. PDGF monomers would not induce receptor dimerization. Since this is the first critical step in signaling from receptor tyrosine kinases, they would be unable to stimulate the PDGF receptor.

8. The truncated receptor would act as a dominant-negative mutant because it would dimerize with the normal receptor. The dimers would be inactive because they would be unable to cross-phosphorylate.

9. JAKs phosphorylate and activate STATs in response to cytokine signaling. Expression of a dominant-negative JAK would inhibit STAT activation and block gene induction.

10. A GEF is required for the activation of Ras, so siRNA against GEF would inhibit induction of the immediate early gene.

11. Akt phosphorylates the protein kinase GSK-3, which regulates the translation factor eIF2B. In addition, Akt regulates the mTOR protein kinase, which regulates translation by phosphorylating S6 kinase and the eIF4E binding protein 4E-BP1.

12. The mutation most likely alters a ubiquitylation site on β-catenin, leading to its stabilization and nuclear accumulation. In the nucleus, β-catenin forms a complex with Tcf transcription factors and induces gene expression.

Chapter 17

1. G_0 and G_1 cells both have the same amount of DNA ($2n$) and both are metabolically active. G_0 cells differ from G_1 cells in that they are in a quiescent state and do not proliferate unless stimulated to re-enter the cell cycle.

2. Cell cycle arrest at the G_1 and S phase checkpoints allows time for damaged DNA to be repaired before it is replicated. Arrest at the G_2 checkpoint allows time for DNA breaks or other damage to be repaired before mitosis occurs, preventing the damaged DNA from being passed on to daughter cells.

3. Cdk's are regulated by four different mechanisms: association with cyclins, activating phosphorylations, inhibitory phosphorylations, and association with Cdk inhibitors.

4. One daughter cell would receive two copies of the misaligned chromosome; the other daughter cell would receive none.

5. The Cdk7/cyclin H complex is a Cdk-activating kinase and is also required for initiation of transcription by RNA polymerase II. Thus inhibiting Cdk7 would lead to cell cycle arrest and the inhibition of transcription.

6. In a normal cell, overexpression of p16 would inhibit cell cycle progression at the restriction point in G_1. Because Rb is the principal target of Cdk4, 6/cyclin D complexes, a tumor cell lacking functional Rb would be unaffected by p16 overexpression.

7. The phosphorylation of nuclear lamins is required for nuclear lamina breakdown. Expression of mutant lamins lacking a phosphorylation site for Cdk1 would prevent breakdown of the nuclear envelope.

8. Cdk1/cyclin B phosphorylates several structural proteins directly to alter their properties and initiate mitosis. Among them are condensins, nuclear lamins, Golgi matrix proteins, and microtubule-associated proteins. In addition, Cdk1/cyclin B activates Aurora and Polo-like protein kinases.

9. Anaphase would initiate normally. However, Cdk1/cyclin B would remain active, so re-formation of nuclei, chromosome decondensation, and cytokinesis would not occur.

10. The anaphase-promoting complex/cyclosome (APC/C) degrades securin, leading to activation of the protease separase. Separase then degrades cohesins, breaking the link between sister chromatids.

11. Oocytes of these mice would fail to arrest at metaphase II.

Chapter 18

1. Apoptotic cells are efficiently removed from tissues by phagocytosis, whereas cells that die by necrosis release their contents into the extracellular space and cause inflammation.

2. Caspases are synthesized as long inactive precursors that are activated in complexes (e.g., the apoptosome) or converted to active enzymes by proteolytic cleavage. In addition, cells contain IAPs (*inhibitors of apoptosis*) that associate with caspases and inhibit their activity.

3. The cleavage of lamins by caspases is required for nuclear fragmentation during apoptosis. The mutated lamins will not be cleaved by caspases, so their expression will block nuclear fragmentation.

4. Proapoptotic effector proteins of the Bcl-2 family induce apoptosis by promoting the release of cytochrome *c* from mitochondria, which leads to caspase activation. The activity of the proapoptotic effector proteins is regulated by antiapoptotic and BH3-only members of the Bcl-2 family.

5. Activation of p53 in response to DNA damage leads to the expression of its target genes, which include the Cdk inhibitor p21 and the BH3-only Bcl-2 family members PUMA and Noxa. p21

induces cell cycle arrest and the BH3-only Bcl-2 family members induce apoptosis.

6. The mutant Bad would no longer be maintained in an inactive state by 14-3-3 proteins, so it will act to induce apoptosis.

7. 14-3-3 proteins sequester proapoptotic proteins, such as Bad and FOXO transcription factors, in an inactive state. Cells expressing siRNA against 14-3-3 proteins will therefore have an increased rate of apoptosis.

8. Ced-3 is the only caspase in *C. elegans*. Mutating it leads to the survival of all the cells that would normally die by apoptosis during development, and siRNA against Ced-3 would have the same effect.

9. These tissues also contain stem cells that retain the ability to proliferate and replace differentiated cells.

10. The critical characteristic of stem cells is their capacity for self-renewal. They divide to produce one daughter cell that remains a stem cell and one that divides and differentiates.

11. Embryonic stem cells are easier to isolate and culture and are capable of giving rise to all of the differentiated cell types in an adult organism.

12. Induced pluripotent stem cells can be derived directly from adult somatic cells without the need for derivation of embryos.

Chapter 19

1. A benign tumor remains confined to its original location, whereas a malignant tumor can invade surrounding normal tissue and metastasize to other parts of the body.

2. As a tumor progresses, mutations occur within cells of the tumor population. Some of these mutations confer a selective advantage to the cells in which they occur and allow them to outgrow other cells in the tumor. This process is called clonal selection, and it leads to the continuing development of more rapidly growing and increasingly malignant tumors.

3. Estrogens act as tumor promoters by stimulating the proliferation of estrogen responsive cells, such as breast and endometrial cells.

4. Autocrine growth stimulation is a positive feedback system in which tumor cells produce growth factors that stimulate their own proliferation.

5. Cancer cells are less adhesive than normal cells and are not as strictly regulated by cell–cell and cell–matrix interactions. In addition, cancer cells secrete proteases that degrade components of the extracellular matrix, facilitating invasion of adjacent tissues. Finally, cancer cells secrete angiogenic factors that promote the formation of new blood vessels that supply tumors with oxygen and nutrients and facilitate metastasis by allowing the cancer cells easy access to the circulatory system.

6. AIDS patients are immunosuppressed and are therefore susceptible to infection by oncogenic viruses.

7. A proto-oncogene may be expressed at abnormal levels or in abnormal cell types. These changes in expression can convert a proto-oncogene to an oncogene even though a structurally normal protein is produced. A proto-oncogene can be activated in this manner either by a translocation that puts it under the control of an active promoter or by gene amplification.

8. *INK4* encodes the Cdk inhibitor p16, which inhibits Cdk4, 6/cyclin D complexes. Since Rb is the critical target of Cdk4, 6/cyclin D, overexpression of p16 would not affect the proliferation of cells with inactivated Rb.

9. p53 is required for cell cycle arrest and apoptosis in response to DNA damage induced by ionizing radiation. Thus tumors with wild-type *p53* genes will be more sensitive to radiation.

10. Imatinib is a specific inhibitor of the Bcr/Abl tyrosine kinase expressed in chronic myeloid leukemia cells. Inhibition of Bcr/Abl blocks proliferation of these tumor cells. Resistance to imatinib is most often caused by mutations in Bcr/Abl that prevent binding of the drug.

11. It is believed that the proliferation and survival of tumor cells become dependent on activated oncogenes, with other signaling pathways becoming secondary in importance. This dependence of tumor cells on activated oncogenes has been termed oncogene addiction. It suggests that drugs against an activated oncogene would selectively target tumor cells, while normal cells would be able to compensate by using alternative signaling pathways.

12. No. MEK is downstream of Raf, so oncogenic activation of Raf would have no effect if MEK is inhibited.

Glossary

A

α-actinin An actin-binding protein that cross-links actin filaments into contractile bundles.

α-helix A coiled secondary structure of a polypeptide chain formed by hydrogen bonding between amino acids separated by four residues.

ABC transporter One of a large family of membrane transport proteins characterized by a highly conserved ATP binding domain.

abl A proto-oncogene that encodes a tyrosine kinase and is activated by chromosome translocation in chronic myeloid leukemia.

abscisic acid A plant hormone.

actin An abundant 43-kd protein that polymerizes to form cytoskeletal filaments.

actin-binding protein A protein that binds actin and regulates the assembly, disassembly, and organization of actin filaments.

actin bundle Actin filaments that are cross-linked into closely packed arrays.

actin-bundling protein A protein that cross-links actin filaments into bundles.

actin network Actin filaments that are cross-linked into loose three-dimensional meshworks.

action potential Nerve impulses that travel along axons.

activation energy The energy required to raise a molecule to its transition state to undergo a chemical reaction.

activation-induced deaminase (AID) An enzyme expressed in B lymphocytes that deaminates cytosine in DNA to form uracil in the variable regions of immunoglobulin genes. AID is required for both class switch recombination and somatic hypermutation.

active site The region of an enzyme that binds substrates and catalyzes an enzymatic reaction.

active transport The transport of molecules in an energetically unfavorable direction across a membrane coupled to the hydrolysis of ATP or other source of energy.

adenine A purine that base-pairs with either thymine or uracil.

adenoma A benign tumor arising from glandular epithelium.

adenosine 5′-triphosphate (ATP) See ATP.

adenovirus A widely-studied DNA tumor virus.

adenylyl cyclase An enzyme that catalyzes the formation of cyclic AMP from ATP.

adherens junction A region of cell–cell adhesion at which the actin cytoskeleton is anchored to the plasma membrane.

Akt A serine/threonine kinase that is activated by PIP_3 and plays a key role in signaling cell survival.

allele One copy of a gene.

allosteric regulation The regulation of enzymes by small molecules that bind to a site distinct from the active site, changing the conformation and catalytic activity of the enzyme.

alternative splicing The generation of different mRNAs by varying the pattern of pre-mRNA splicing.

amino acid Monomeric building blocks of proteins, consisting of a carbon atom bound to a carboxyl group, an amino group, a hydrogen atom, and a distinctive side chain.

aminoacyl tRNA synthetase An enzyme that joins a specific amino acid to a tRNA molecule carrying the correct anticodon sequence.

amphipathic A molecule that has both hydrophobic and hydrophilic regions.

amyloid A fibrous aggregate of misfolded protein.

amyloplast A plastid that stores starch.

anaphase The phase of mitosis during which sister chromatids separate and move to opposite poles of the spindle.

anaphase A The movement of daughter chromosomes toward the spindle poles during mitosis.

anaphase B The separation of the spindle poles during mitosis.

anaphase-promoting complex/cyclosome (APC/C) A ubiquitin ligase that triggers progression from metaphase to anaphase by signaling the degradation of cyclin B and cohesins.

angiogenesis The formation of new blood vessels.

ankyrin A protein that binds spectrin and links the actin cytoskeleton to the plasma membrane.

antibody A protein (also called immunoglobulin) produced by B lymphocytes that binds to a foreign molecule.

anticodon The nucleotide sequence of transfer RNA that forms complementary base pairs with a codon sequence on messenger RNA.

antigen A molecule against which an antibody is directed.

antiport The transport of two molecules in opposite directions across a membrane.

antisense nucleic acid A nucleic acid (either RNA or DNA) that is complementary to an mRNA of interest and is used to block gene expression.

AP endonuclease A DNA repair enzyme that cleaves next to apyrimidinic or apurinic sites in DNA.

apical domain The exposed free surface of a polarized epithelial cell.

apoptosis An active process of programmed cell death, characterized by cleavage of chromosomal DNA, chromatin condensation, and fragmentation of both the nucleus and the cell.

apoptosome A protein complex in which caspase-9 is activated to initiate apoptosis following the release of cytochrome *c* from mitochondria.

aquaporin A channel protein through which water is able to rapidly cross the plasma membrane.

Arabidopsis thaliana A small flowering plant used as a model for plant molecular biology and development.

archaebacteria One of two major groups of prokaryotes; many species of archaebacteria live in extreme conditions similar to those prevalent on primitive Earth.

armadillo protein family A family of proteins, including β-catenin, that links cadherins to the cytoskeleton at stable cell–cell junctions.

Arp2/3 complex A protein complex that binds actin filaments and initiates the formation of branches.

astral microtubule A microtubule of the mitotic spindle that extends to the cell periphery.

ATM A protein kinase that recognizes damaged DNA and leads to cell cycle arrest.

ATP (adenosine 5′-triphosphate) An adenine-containing nucleoside triphosphate that serves as a store of free energy in the cell.

ATP synthase A membrane-spanning protein complex that couples the energetically favorable transport of protons across a membrane to the synthesis of ATP.

ATR A protein kinase related to ATM that leads to cell cycle arrest in response to DNA damage.

Aurora kinase A protein kinase family involved in mitotic spindle formation, kinetochore function, and cytokinesis.

autocrine growth stimulation Stimulation of cell proliferation as a result of growth factor production by a responsive cell.

autocrine signaling A type of cell signaling in which a cell produces a growth factor to which it also responds.

autonomously replicating sequence (ARS) An origin of DNA replication in yeast.

autophagosome A vesicle containing internal organelles enclosed by fragments of cytoplasmic membranes that fuses with lysosomes.

autophagy The degradation of cytoplasmic proteins and organelles by their enclosure in cytoplasmic vesicles that fuse with lysosomes.

autophosphorylation A reaction in which a protein kinase catalyzes its own phosphorylation.

auxin A plant hormone that controls many aspects of plant development.

axonemal dynein The type of dynein found in cilia and flagella.

axoneme The fundamental structure of cilia and flagella composed of a central pair of microtubules surrounded by nine microtubule doublets.

B

β barrel A transmembrane domain formed by the folding of β sheets into a barrel-like structure.

β sheet A sheetlike secondary structure of a polypeptide chain, formed by hydrogen bonding between amino acids located in different regions of the polypeptide.

bacteria One of two major groups of prokaryotes, including most common species.

bacterial artificial chromosome (BAC) A type of vector used for cloning large fragments of DNA in bacteria.

bacteriophage A bacterial virus.

basal body A structure similar to a centriole that initiates the growth of axonemal microtubules and anchors cilia and flagella to the surface of the cell.

basal lamina A sheetlike extracellular matrix that supports epithelial cells and surrounds muscle cells, adipose cells, and peripheral nerves.

base-excision repair A mechanism of DNA repair in which single damaged bases are removed from a DNA molecule.

basolateral domain The surface region of a polarized epithelial cell that is in contact with adjacent cells or the extracellular matrix.

Bcl-2 A member of a family of proteins that regulate programmed cell death.

benign tumor A tumor that remains confined to its site of origin.

bioinformatics The use of computational methods to analyze large amounts of biological data, such as genome sequences.

bone marrow transplantation A clinical procedure in which transplantation of bone marrow stem cells is used in the treatment of cancer and diseases of the hematopoietic system.

brassinosteroid A plant steroid hormone.

bright-field microscopy The simplest form of light microscopy in which light passes directly through a cell.

brush border The surface of a cell (e.g., an intestinal epithelial cell) containing a layer of microvilli.

C

cadherin A cell adhesion molecule that forms stable cell–cell junctions at adherens junctions and desmosomes.

Caenorhabditis elegans A nematode used as a simple multicellular model for development.

Cajal body A nuclear body involved in assembly of snRNPs and other RNA-protein complexes.

callus An undifferentiated mass of plant cells in culture.

calmodulin A calcium-binding protein.

calnexin A chaperone involved in glycoprotein folding in the ER lumen.

calreticulin A chaperone involved in glycoprotein folding in the ER membrane.

Calvin cycle A series of reactions by which six molecules of CO_2 are converted into glucose.

CaM kinase A member of a family of protein kinases that are activated by the binding of Ca^{2+}/calmodulin.

cAMP-dependent protein kinase See protein kinase A.

cAMP phosphodiesterase An enzyme that degrades cyclic AMP.

cAMP response element (CRE) A regulatory sequence that mediates the transcriptional response of target genes to cAMP.

cancer A malignant tumor.

carbohydrate A molecule with the formula $(CH_2O)_n$. Carbohydrates include both simple sugars and polysaccharides.

carcinogen A cancer-inducing agent.

carcinoma A cancer of epithelial cells.

cardiolipin A phospholipid containing four hydrocarbon chains.

carrier protein A protein that selectively binds and transports small molecules across a membrane.

caspase A member of a family of proteases that bring about programmed cell death.

catalase An enzyme that decomposes hydrogen peroxide.

catenin A group of cytoplasmic proteins (including α-catenin and β-catenin) that link actin filaments to cadherins at adherens junctions.

caveolae Small invaginations of the plasma membrane that may be involved in endocytosis.

CCND1 The gene encoding cyclin D1, which is an oncogene in a variety of human cancers.

Cdk A member of a family of cyclin-dependent protein kinases that control the cell cycle of eukaryotes.

Cdk1 A serine/threonine kinase that is a key regulator of mitosis in eukaryotic cells.

Cdk inhibitor (CKI) Member of a family of proteins that bind Cdk's and inhibit their activity.

cDNA A DNA molecule that is complementary to an mRNA molecule, synthesized *in vitro* by reverse transcriptase.

cell adhesion molecule A transmembrane protein that mediates cell–cell interactions.

cell cortex The actin network underlying the plasma membrane.

cell cycle checkpoint A regulatory point that prevents entry into the next phase of the cell cycle until the events of the preceding phase have been completed.

cell line Cells that can proliferate indefinitely in culture.

cell transformation The conversion of normal cells to tumor cells in culture.

cell wall A rigid, porous structure forming an external layer that provides structural support to bacteria, fungi, and plant cells.

cellulose The principal structural component of the plant cell wall, a linear polymer of glucose residues linked by β $(1{\rightarrow}4)$ glycosidic bonds.

cellulose synthase An enzyme that catalyzes the synthesis of cellulose.

CENP-A A variant of histone H3 that is present at centromeres.

central dogma The concept that genetic information flows from DNA to RNA to proteins.

centriole A cylindrical structure consisting of nine triplets of microtubules in the centrosomes of most animal cells.

centromere A specialized chromosomal region that connects sister chromatids and attaches them to the mitotic spindle.

centrosome The microtubule-organizing center in animal cells.

cGMP phosphodiesterase An enzyme that degrades cGMP.

channel protein A protein that forms pores through a membrane.

chaperone A protein that facilitates the correct folding or assembly of other proteins.

checkpoint kinase A protein kinase (Chk1 or Chk2) that brings about cell cycle arrest in response to damaged DNA.

chemiosmotic coupling The generation of ATP from energy stored in a proton gradient across a membrane.

chiasmata Sites of recombination that link homologous chromosomes during meiosis.

chitin A polymer of *N*-acetylglucosamine residues that is the principal component of fungal cell walls.

chlorophyll The major photosynthetic pigment of plant cells.

chloroplast The organelle responsible for photosynthesis in the cells of plants and green algae.

cholesterol A lipid consisting of four hydrocarbon rings. Cholesterol is a major constituent of animal cell plasma membranes and the precursor of steroid hormones.

chromatin The fibrous complex of eukaryotic DNA and histone proteins.

chromatin immunoprecipitation A method for determining regions of DNA that bind transcription factors within a cell.

chromatin remodeling factor A protein that disrupts chromatin structure by altering the contacts between DNA and histones.

chromatosome A chromatin subunit consisting of 166 base pairs of DNA wrapped around a histone core and held in place by a linker histone.

chromoplast A plastid that contains carotenoids.

chromosomal microtubule Microtubule of the mitotic spindle that attaches to the ends of condensed chromosomes.

chromosome A carrier of genes, consisting of long DNA molecules and associated proteins.

cilium A microtubule-based projection of the plasma membrane that moves a cell through fluid or fluid over a cell.

circumferential belt A beltlike structure around epithelial cells in which a contractile bundle of actin filaments is linked to the plasma membrane.

***cis*-acting control element** A regulatory DNA sequence that serves as a protein binding site and controls the transcription of adjacent genes.

citric acid cycle A series of reactions (also called the Krebs cycle) in which acetyl CoA is oxidized to CO_2. The central pathway of oxidative metabolism.

class switch recombination A type of region-specific recombination responsible for the association of rearranged immunoglobulin V(D)J regions with different heavy-chain constant regions.

clathrin A protein that coats the cytoplasmic surface of cell membranes and assembles into basketlike lattices that drive vesicle budding.

clathrin-coated pit A specialized region of the plasma membrane that contains receptors for macromolecules to be taken up by endocytosis.

clathrin-coated vesicle A transport vesicle coated with clathrin.

clathrin-independent endocytosis One of several pathways of endocytosis not mediated by specific plasma membrane receptors or the formation of coated vesicles.

clathrin-mediated endocytosis The selective uptake of specific macromolecules by receptors that concentrate in clathrin-coated pits.

Cln A yeast cyclin (also called G_1 cyclin) that controls passage through START.

clonal selection The process by which new clones of tumor cells evolve on the basis of increased growth rate or other properties (such as survival, invasion, or metastasis) that confer a selective advantage.

c-myc A proto-oncogene that encodes a transcription factor and is frequently activated by chromosome translocation or gene amplification in human tumors.

coactivator A protein that interacts with a transcription factor to stimulate transcription.

codon The basic unit of the genetic code; one of the 64 nucleotide triplets that code for an amino acid or stop sequence.

coenzyme Low-molecular-weight organic molecules that work together with enzymes to catalyze biological reactions.

coenzyme A (CoA-SH) A coenzyme that functions as a carrier of acyl groups in metabolic reactions.

coenzyme Q A small lipid-soluble molecule (also called ubiquinone) that carries electrons between protein complexes in the mitochondrial electron transport chain.

cofilin An actin-binding protein that severs actin filaments.

cohesin A protein that maintains the connection between sister chromatids.

colcemid A drug that inhibits the polymerization of microtubules.

colchicine A drug that inhibits the polymerization of microtubules.

collagen The major structural protein of the extracellular matrix.

collagen fibril A fibril formed by the assembly of collagen molecules in a regularly staggered array.

condensin A protein complex that drives metaphase chromosome condensation.

confocal microscopy A form of microscopy in which fluorescence microscopy is combined with electronic image analysis to obtain images with increased contrast and detail.

connexin A member of a family of transmembrane proteins that form gap junctions.

connexon A cylinder formed by six connexins in the plasma membrane.

contact inhibition The inhibition of movement or proliferation of normal cells that results from cell–cell contact.

contractile bundle A bundle of actin filaments that interacts with myosin II and is capable of contraction.

contractile ring A structure of actin and myosin II that forms beneath the plasma membrane during mitosis and mediates cytokinesis.

COPI and **COPII** The two proteins other than clathrin that coat transport vesicles (COP indicates coat protein).

COPI-coated vesicle A transport vesicle coated with COPI that mediates retrieval from the Golgi apparatus to the endoplasmic reticulum.

COPII-coated vesicle A transport vesicle coated with COPII that mediates transport from the endoplasmic reticulum to the Golgi apparatus.

corepressor A protein that associates with repressors to inhibit gene expression, often by modifying chromatin structure.

corticosteroid Steroid hormone produced by the adrenal gland.

cosmid A vector that contains bacteriophage λ sequences, antibiotic resistance sequences, and an origin of replication. It can accommodate large DNA inserts of up to 45 kb.

CREB Cyclic AMP response element-binding protein. A transcription factor that is activated by cAMP-dependent protein kinase.

CRISPR/Cas system A system for introducing targeted mutations into mammalian genes. It consists of CRISPR RNAs that recognize specific target sequences and Cas proteins that cleave the targeted DNA.

cristae Folds in the inner mitochondrial membrane extending into the matrix.

crosstalk A regulatory mechanism in which one signaling pathway controls the activity of another.

CTCF An architectural protein involved in regulating the organization of chromatin.

cyanobacteria The largest and most complex prokaryotes in which photosynthesis is believed to have evolved.

cyclic AMP (cAMP) Adenosine monophosphate in which the phosphate group is covalently bound to both the 3′ and 5′ carbon atoms, forming a cyclic structure and an important second messenger in the response of cells to a variety of hormones.

cyclic electron flow An electron transport pathway associated with photosystem I that produces ATP without the synthesis of NADPH.

cyclic GMP (cGMP) Guanosine monophosphate in which the phosphate group is covalently bound to both the 3′ and 5′ carbon atoms, forming a cyclic structure and an important second messenger in the response of cells to a variety of hormones, and in vision.

cyclin Member of a family of proteins that regulate the activity of Cdk's and control progression through the cell cycle.

cytochalasin A drug that blocks the elongation of actin filaments.

cytochrome _bf_ complex A protein complex in the thylakoid membrane that carries electrons during photosynthesis.

cytochrome _c_ A mitochondrial peripheral membrane protein that carries electrons during oxidative phosphorylation.

cytochrome oxidase A protein complex in the electron transport chain that accepts electrons from cytochrome _c_ and transfers them to O_2.

cytokine A growth factor that regulates blood cells and lymphocytes.

cytokine receptor superfamily A family of cell surface receptors that act by stimulating the activity of intracellular tyrosine kinases.

cytokinesis Division of a cell following mitosis or meiosis.

cytokinin A plant hormone that regulates cell division.

cytoplasmic dynein The form of dynein associated with microtubules in the cytoplasm.

cytosine A pyrimidine that base-pairs with guanine.

cytoskeleton A network of protein filaments that extends throughout the cytoplasm of eukaryotic cells. It provides the structural framework of the cell and is responsible for cell movements.

cytostatic factor (CSF) A cytoplasmic factor that arrests oocyte meiosis at metaphase II.

D

2′–deoxyribose The five-carbon sugar found in DNA.

dark reactions The series of reactions that convert carbon dioxide and water to carbohydrates during photosynthesis. See Calvin cycle.

density-dependent inhibition The cessation of the proliferation of normal cells in culture at a finite cell density.

density-gradient centrifugation A method of separating particles by centrifugation through a gradient of a dense substance, such as sucrose or cesium chloride.

deoxyribonucleic acid (DNA) The genetic material of the cell.

desmin An intermediate filament protein expressed in muscle cells.

desmocollin A type of transmembrane cadherin that links intermediate filament cytoskeletons of adjacent cells at desmosomes.

desmoglein A type of transmembrane cadherin that links intermediate filament cytoskeletons of adjacent cells at desmosomes.

desmosome A region of contact between epithelial cells at which keratin filaments are anchored to the plasma membrane. See also hemidesmosome.

diacylglycerol (DAG) A second messenger formed from the hydrolysis of PIP_2 that activates protein kinase C.

diakinesis The final stage of the prophase of meiosis I during which the chromosomes fully condense and the cell progresses to metaphase.

dideoxynucleotide A nucleotide that lacks the normal 3′ hydroxyl group of deoxyribose and is used as a chain-terminating nucleotide in DNA sequencing.

differential centrifugation A method used to separate the components of cells on the basis of their size and density.

differential interference-contrast microscopy A type of microscopy in which variations in density or thickness between parts of the cell are converted to differences in contrast in the final image.

diploid An organism or cell that carries two copies of each chromosome.

diplotene The stage of meiosis I during which homologous chromosomes separate along their length but remain associated at chiasmata.

DNA-affinity chromatography A method used to isolate DNA-binding proteins based on their binding to specific DNA sequences.

DNA damage checkpoint A cell cycle checkpoint that ensures that damaged DNA is not replicated and passed on to daughter cells.

DNA glycosylase A DNA repair enzyme that cleaves the bond linking a purine or pyrimidine to the deoxyribose of the backbone of a DNA molecule.

DNA ligase An enzyme that seals breaks in DNA strands.

DNA microarray A glass slide or membrane filter onto which oligonucleotides or fragments of cDNAs are printed at a high density, allowing simultaneous analysis of thousands of genes by hybridization of the microarray with fluorescent probes.

DNA polymerase An enzyme catalyzing the synthesis of DNA.

DNase hypersensitive site A nucleosome-free region of chromatin, characteristic of promoters and enhancers, that is sensitive to digestion by DNase.

DNA transposon A transposable element that moves via a DNA intermediate.

dolichol phosphate A lipid molecule in the endoplasmic reticulum upon which oligosaccharides are assembled for the glycosylation of proteins.

domain A folded three-dimensional region of a protein that forms the basic unit of tertiary structure.

dominant The allele that determines the phenotype of an organism when more than one allele is present.

dominant inhibitory mutant A mutant (also called dominant negative mutant) that interferes with the function of the normal allele of the gene.

double-strand break Damage that results in breaks in both complementary strands of DNA.

Drosophila melanogaster A species of fruit fly commonly used for studies of animal genetics and development.

dynamic instability The alternation of microtubules between cycles of growth and shrinkage.

dynamin A membrane-associated GTPase involved in vesicle budding.

dynein A motor protein that moves along microtubules towards the minus end.

dystrophin A cytoskeletal protein of muscle cells.

E

E2F A family of transcription factors that regulate the expression of genes involved in cell cycle progression and DNA replication.

ecdysone An insect steroid hormone that triggers metamorphosis.

eicosanoid A class of lipids, including prostaglandins, prostacyclins, thromboxanes, and leukotrienes, that act in autocrine and paracrine signaling.

elaioplast A plastid that stores lipids.

elastic fiber A protein fiber present in the extracellular matrix of connective tissues in organs that stretch and then return to their original shape.

elastin The principal component of elastic fibers.

electrical synapse A specialized assembly of gap junctions that allows the rapid passage of ions between nerve cells.

electrochemical gradient A difference in chemical concentration and electric potential across a membrane.

electron microscopy A type of microscopy that uses an electron beam to form an image. In transmission electron microscopy, a beam of electrons is passed through a specimen stained with heavy metals. In scanning electron microscopy, electrons scattered from the surface of a specimen are analyzed to generate a three-dimensional image.

electron tomography A method used to generate three-dimensional images by computer analysis of multiple two-dimensional images obtained by electron microscopy.

electron transport chain A series of carriers through which electrons are transported from a higher to a lower energy state.

electrophoretic-mobility shift assay An assay for the binding of a protein to a specific DNA sequence.

electroporation The introduction of DNA into cells by exposure to a brief electric pulse.

Elk-1 A transcription factor that is activated by ERK phosphorylation and induces expression of immediate early genes.

elongation factor A protein involved in the elongation phase of transcription or translation.

embryonic stem (ES) cell A stem cell cultured from an early embryo.

endocrine signaling A type of cell–cell signaling in which endocrine cells secrete hormones that are carried by the circulation to distant target cells.

endocytosis The uptake of extracellular material in vesicles formed from the plasma membrane.

endoplasmic reticulum (ER) An extensive network of membrane-enclosed tubules and sacs involved in protein sorting and processing as well as in lipid synthesis.

endorphin A neuropeptide that acts as a natural analgesic.

endosome A vesicular compartment involved in the sorting and transport to lysosomes of material taken up by endocytosis.

endosymbiosis A symbiotic relationship in which one cell resides within a larger cell.

enhancer A transcriptional regulatory sequence that can be located at a site distant from the promoter.

enkephalin A neuropeptide that acts as a natural analgesic.

enzyme A protein or RNA that catalyzes a biological reaction.

epidermal growth factor (EGF) A growth factor that stimulates cell proliferation.

epigenetic inheritance The transmission of information from parent to progeny that is not contained within the sequence of DNA.

epithelial cell A type of cell that forms sheets (epithelial tissue) that cover the surface of the body and line internal organs.

Epstein-Barr virus A human herpesvirus that causes B-cell lymphomas.

equilibrium centrifugation The separation of particles on the basis of density by centrifugation to equilibrium in a gradient of a dense substance.

ER-associated degradation (ERAD) A process in which misfolded proteins in the ER are identified and returned to the cytosol for degradation by the ubiquitin-proteasome system.

*erb*B-2 A proto-oncogene encoding a receptor tyrosine kinase that is frequently amplified in breast and ovarian carcinomas.

ErbA A proto-oncogene protein corresponding to the thyroid hormone receptor.

ERK A member of the MAP kinase family that plays a central role in growth factor-induced cell proliferation.

erythrocyte A red blood cell.

***Escherichia coli* (*E. coli*)** A species of bacteria that has been extensively used as a model system for molecular biology.

estrogen A steroid hormone produced by the ovaries.

ethylene A plant hormone responsible for fruit ripening.

etioplast An intermediate stage of chloroplast development in which chlorophyll has not been synthesized.

euchromatin Decondensed, transcriptionally active interphase chromatin.

eukaryotic cell A cell that has a nuclear envelope, cytoplasmic organelles, and linear chromosomes.

excinuclease The protein complex that excises damaged DNA during nucleotide-excision repair in bacteria.

exon A segment of a gene that is included in a spliced mRNA.

exonuclease An enzyme that hydrolyzes DNA molecules in either the 5′ to 3′ or 3′ to 5′ direction.

exportin A karyopherin that recognizes nuclear export signals and directs transport from the nucleus to the cytosol.

expression vector A vector used to direct expression of a cloned DNA fragment in a host cell.

extracellular matrix Secreted proteins and polysaccharides that fill spaces between cells and bind cells and tissues together.

F

facilitated diffusion The transport of molecules across a membrane by carrier or channel proteins.

FADH2 See flavin adenine dinucleotide.

FAK (focal adhesion kinase) A nonreceptor tyrosine kinase that plays a key role in integrin signaling.

fats See triacylglycerols.

fatty acid A long hydrocarbon chain usually linked to a carboxyl group (COO^-).

feedback inhibition A type of allosteric regulation in which the product of a metabolic pathway inhibits the activity of an enzyme involved in its synthesis.

feedback loop A regulatory mechanism in which a downstream element of a signaling pathway controls the activity of an upstream component of the pathway.

feedforward relay A regulatory mechanism in which one element of a signaling pathway stimulates a downstream component.

fertilization The union of a sperm and an egg.

fibroblast A cell type found in connective tissue.

fibronectin The principal adhesion protein of the extracellular matrix.

filamentous [F] actin Actin monomers polymerized into filaments.

filamin An actin-binding protein that cross-links actin filaments into networks.

filopodium A thin projection of the plasma membrane supported by actin bundles.

fimbrin An actin-bundling protein involved in formation of cell surface projections.

flagellum A microtubule-based projection of the plasma membrane that is responsible for cell movement.

flavin adenine dinucleotide (FADH2) A coenzyme that functions as an electron carrier in oxidation/reduction reactions.

flippase A protein that catalyzes the translocation of lipids across the membrane of the endoplasmic reticulum.

flow cytometer An instrument that measures the fluoresence intensity of individual cells.

fluid mosaic model A model of membrane structure in which proteins are inserted in a fluid phospholipid bilayer.

fluorescence-activated cell sorter An instrument that sorts individual cells on the basis of their fluorescence intensity.

fluorescence microscopy Type of microscopy in which molecules are detected based on the emission of fluorescent light.

fluorescence recovery after photobleaching (FRAP) A method used to study the movement of proteins within living cells.

fluorescence resonance energy transfer (FRET) A method used to study protein interactions within living cells.

focal adhesion A site of attachment of cells to the extracellular matrix at which integrins are linked to bundles of actin filaments.

formin An actin-binding protein that nucleates and polymerizes actin filaments.

Fos A transcription factor, encoded by a proto-oncogene, that is induced in response to growth factor stimulation.

freeze fracture Method of electron microscopy in which specimens are frozen in liquid nitrogen and then fractured to split the lipid bilayer, revealing the interior faces of cell membranes.

G

γ-tubulin ring complex A protein complex that nucleates the formation of microtubules.

G protein A family of cell signaling proteins regulated by guanine nucleotide binding.

G protein-coupled receptor A receptor characterized by seven membrane-spanning α helices. Ligand binding causes a conformational change that activates a G protein.

G_0 A quiescent state in which cells remain metabolically active but do not proliferate.

G_1 cyclin A yeast cyclin (also called Cln) that controls passage through START.

G_1 phase The phase of the cell cycle between the end of mitosis and the beginning of DNA synthesis.

G_2 phase The phase of the cell cycle between the end of S phase and the beginning of mitosis.

gap junction A plasma membrane channel forming a direct cytoplasmic connection between adjacent cells.

gel electrophoresis A method in which molecules are separated based on their migration in an electric field.

gene A functional unit of inheritance, corresponding to a segment of DNA that encodes a polpeptide or RNA molecule.

gene amplification An increase in the number of copies of a gene resulting from the repeated replication of a region of DNA.

gene family A group of related genes that have arisen by duplication of a common ancestor.

gene transfer The introduction of foreign DNA into a cell.

general transcription factor A transcription factor that is part of the general transcription machinery.

genetic code The correspondence between nucleotide triplets and amino acids in proteins.

genomic imprinting The regulation of genes whose expression depends on whether they are maternally or paternally inherited, controlled by DNA methylation.

genotype The genetic composition of an organism.

gibberellin A plant hormone.

Gibbs free energy (G) The thermodynamic function that combines the effects of enthalpy and entropy to predict the energetically favorable direction of a chemical reaction.

globular [G] actin Monomers of actin that have not been assembled into filaments.

glucocorticoid A steroid produced by the adrenal gland that acts to stimulate production of glucose.

gluconeogenesis The synthesis of glucose.

glycerol phospholipid A phospholipid consisting of two fatty acids bound to a glycerol molecule.

glycocalyx A carbohydrate coat covering the cell surface.

glycogen A polymer of glucose residues that is the principal storage form of carbohydrates in animals.

glycolipid A lipid consisting of two hydrocarbon chains linked to a polar head group containing carbohydrates.

glycolysis The anaerobic breakdown of glucose.

glycoprotein A protein linked to oligosaccharides.

glycosaminoglycan (GAG) A gel-forming polysaccharide of the extracellular matrix.

glycosidic bond The bond formed between sugar residues in oligosaccharides or polysaccharides.

glycosylation The addition of carbohydrates to proteins.

glycosylphosphatidylinositol (GPI) anchor A glycolipid containing phosphatidylinositol that anchors proteins to the external face of the plasma membrane.

glyoxylate cycle The conversion of fatty acids to carbohydrates in plants.

glyoxysome A peroxisome in which the reactions of the glyoxylate cycle take place.

Golgi apparatus A cytoplasmic organelle involved in the processing and sorting of proteins and lipids. In plant cells, it is also the site of the synthesis of cell wall polysaccharides.

Golgi complex See Golgi apparatus.

granulocyte A type of blood cell involved in inflammatory reactions.

green fluorescent protein (GFP) A protein from jellyfish that is commonly used as a marker for fluorescence microscopy.

growth factor A polypeptide that controls animal cell growth and differentiation.

GTPase-activating protein (GAP) A protein that stimulates GTP hydrolysis by the small GTP-binding proteins.

guanine A purine that base-pairs with cytosine.

guanine nucleotide exchange factor (GEF) A protein that acts on small GTP-binding proteins to stimulate the exchange of bound GDP for GTP.

guanylyl cyclase An enzyme that catalyzes the formation of cyclic GMP from GTP.

H

haploid An organism or cell that has one copy of each chromosome.

Hedgehog A secreted signaling molecule that stimulates a pathway regulating cell fate during embryonic development.

helicase An enzyme that catalyzes the unwinding of DNA.

helix-loop-helix A transcription factor DNA-binding domain formed by the dimerization of two polypeptide chains. The dimerization domains of these proteins consist of two helical regions separated by a loop.

helix-turn-helix A transcription factor DNA-binding domain in which three or four helical regions contact DNA.

hematopoietic stem cell transplantation A clinical procedure in which transplantation of hematopoietic stem cells is used in the treatment of cancer and diseases of the hematopoietic system.

hemicellulose A polysaccharide that cross-links cellulose microfibrils in plant cell walls.

hemidesmosome A region of contact between cells and the extracellular matrix at which keratin filaments are attached to integrin.

hepatitis B virus Member of a family of DNA viruses that infect liver cells and can lead to the development of liver cancer.

hepatitis C virus Member of a family of RNA viruses that infect liver cells and can lead to the development of liver cancer.

herpesvirus Member of a family of DNA viruses, some members of which induce cancer.

heterochromatin Condensed, transcriptionally inactive chromatin.

heterophilic interaction An interaction between two different types of cell adhesion molecules.

heterotrimeric G protein A guanine nucleotide-binding protein consisting of three subunits.

high-energy bond A chemical bond that releases a large amount of free energy when it is hydrolyzed.

histone acetylation The modification of histones by the addition of acetyl groups to specific lysine residues.

histone Member of a family of proteins that package DNA in eukaryotic chromosomes.

homeodomain A type of DNA binding domain found in transcription factors that regulates gene expression during embryonic development.

homologous recombination Recombination between segments of DNA with homologous nucleotide sequences.

homophilic interaction An interaction between cell adhesion molecules of the same type.

hormone A signaling molecule produced by an endocrine gland that acts on cells at distant body sites.

hydrophilic Soluble in water.

hydrophobic Not soluble in water.

I

IAP Inhibitor of *apoptosis*. A member of a family of proteins that inhibit apoptosis by interacting with caspases.

IκB An inhibitory subunit of NF-κB transcription factors.

immediate early gene A gene whose transcription is rapidly induced in response to growth factor stimulation.

immunoblotting A method that uses antibodies to detect proteins separated by SDS-polyacrylamide gel electrophoresis. Also called Western blotting, it is a variation of Southern blotting.

immunoglobulin See antibody.

immunoglobulin (Ig) superfamily A family of cell adhesion molecules containing structural domains similar to immunoglobulins.

immunoprecipitation The use of antibodies to isolate proteins.

importin A karyopherin that recognizes nuclear localization signals and directs nuclear import.

induced fit A model of enzyme action in which the configurations of both the enzyme and the substrate are altered by substrate binding.

induced pluripotent stem cell An adult somatic cell that has been reprogrammed in culture to resemble an embryonic stem cell.

initiation factor A protein that functions in the initiation stage of translation.

***in situ* hybridization** The use of radioactive or fluorescent probes to detect RNA or DNA sequences in chromosomes or intact cells.

***in vitro* mutagenesis** The introduction of mutations into cloned DNA *in vitro*.

***in vitro* translation** Protein synthesis in a cell-free extract.

inositol 1,4,5-trisphosphate (IP$_3$) A second messenger, formed from the hydrolysis of PIP$_2$, that signals the release of calcium ions from the endoplasmic reticulum.

insulator A sequence that divides chromatin into independent domains and prevents an enhancer from acting on a promoter in a separate domain.

integral membrane protein A protein embedded within the lipid bilayer of cell membranes.

integrin A transmembrane protein that mediates the adhesion of cells to the extracellular matrix.

intermediate filament A cytoskeletal filament about 10 nm in diameter that provides mechanical strength to cells in tissues. See also keratin and neurofilament.

interphase The period of the cell cycle between mitoses that includes G$_1$, S, and G$_2$ phases.

interpolar microtubule One of a class of microtubules of the mitotic spindle that overlap in the center of the cell and push the spindle poles apart.

intracellular signal transduction A chain of reactions that transmits chemical signals from the cell surface to their intracellular targets.

intron A noncoding sequence that interrupts exons in a gene.

ion channel A protein that mediates the rapid passage of ions across a membrane by forming open pores through the phospholipid bilayer.

ion pump A protein that couples ATP hydrolysis to the transport of ions across a membrane.

J

Janus kinase (JAK) A family of nonreceptor tyrosine kinases associated with cytokine receptors.

Jun A transcription factor, encoded by a proto-oncogene, that is activated in response to growth factor stimulation.

junctional complex A region of cell–cell contact containing a tight junction, an adherens junction, and a desmosome.

K

Kaposi's sarcoma-associated herpesvirus A human herpesvirus that causes Kaposi's sarcoma.

karyopherin A nuclear transport receptor.

keratin A type of intermediate filament protein of epithelial cells.

kilobase (kb) One thousand nucleotides or nucleotide base pairs.

kinesin A motor protein that moves along microtubules, usually toward the plus end.

kinetochore A specialized structure consisting of proteins attached to a centromere that mediates the attachment and movement of chromosomes along the mitotic spindle.

kinetochore microtubule A microtubule of the mitotic spindle that attaches to condensed chromosomes at their centromeres.

knockout Inactivation of a chromosomal gene by homologous recombination with a cloned mutant allele.

Krebs cycle See citric acid cycle.

L

lagging strand The strand of DNA synthesized opposite to the direction of movement of the replication fork by ligation of Okazaki fragments.

lamellipodium A broad, actin-based extension of the plasma membrane involved in the movement of fibroblasts.

lamin An intermediate filament protein that forms the nuclear lamina.

lamina-associated domain (LAD) A region of heterochromatin associated with the nuclear lamina.

laminin The principal adhesion protein of basal laminae.

leading strand The strand of DNA synthesized continuously in the direction of movement of the replication fork.

leptotene The initial stage of the extended prophase of meiosis I during which homologous chromosomes pair before condensation.

leucine zipper A protein dimerization domain containing repeated leucine residues that is found in many transcription factors.

leucoplast A plastid that stores energy sources in nonphotosynthetic plant tissues.

leukemia Cancer arising from the precursors of circulating blood cells.

leukotriene An eicosanoid synthesized from arachidonic acid.

ligand-gated channel An ion channel that opens in response to the binding of signaling molecules.

light reactions The reactions of photosynthesis in which solar energy drives the synthesis of ATP and NADPH.

lignin A polymer of phenolic residues that strengthens secondary cell walls.

LINE (long interspersed element) Member of a family of highly repeated retrotransposons in mammalian genomes.

lipid Member of a group of hydrophobic molecules that function as energy storage molecules, signaling molecules, and the major components of cell membranes.

lipid raft A discrete plasma membrane domain formed as a cluster of cholesterol and sphingolipids.

liposome A lipid vesicle used to introduce DNA into mammalian cells.

lock-and-key model A model of enzyme action in which the substrate fits precisely into the enzyme active site.

long noncoding RNA (lncRNA) An RNA longer than 200 nucleotides that regulates gene expression.

low-density lipoprotein (LDL) A lipoprotein particle that transports cholesterol in the circulation.

lymphocyte A blood cell that functions in the immune response. B lymphocytes produce antibodies and T lymphocytes are responsible for cell mediated immunity.

lymphoma A cancer of lymphoid cells.

lysosomal storage disease A family of diseases characterized by the accumulation of undegraded material in the lysosomes of affected individuals.

lysosome A cytoplasmic organelle containing enzymes that break down biological polymers.

M

7-methylguanosine cap A structure consisting of GTP and methylated sugars that is added to the 5′ ends of eukaryotic mRNAs.

M phase The mitotic phase of the cell cycle.

macrophage A type of white blood cell specialized for phagocytosis.

macropinocytosis The uptake of fluids in large vesicles.

malignant tumor A tumor that invades normal tissue and spreads throughout the body.

mannose-6-phosphate A modified mannose residue that targets proteins to lysosomes.

MAP kinase A family of mitogen-activated protein-serine/threonine kinases that are ubiquitous regulators of cell growth and differentiation.

mass spectrometry A method for identifying compounds based on accurate determination of their mass. Mass spectrometry is commonly used for protein identification.

matrix The inner mitochondrial space.

matrix processing peptidase (MPP) The protease that cleaves presequences from proteins imported to the matrix of mitochondria.

maturation promoting factor (MPF) A complex of Cdk1 and cyclin B that promotes entry into the M phase of either mitosis or meiosis.

Mediator A complex of proteins that stimulates transcription of eukaryotic protein-coding genes and allows them to respond to gene-specific regulatory factors.

meiosis The division of diploid cells to haploid progeny, consisting of two sequential rounds of nuclear and cellular division.

MEK (MAP kinase/ERK kinase) A dual-specificity protein kinase that phosphorylates and activates members of the ERK family of MAP kinases.

membrane-anchored growth factor A growth factor associated with the plasma membrane that functions as a signaling molecule during cell–cell contact.

Merkel cell polyomavirus A papillomavirus that causes a rare skin cancer (Merkel cell carcinoma) in humans.

messenger RNA (mRNA) See mRNA.

messenger RNP (mRNP) See mRNP.

metal shadowing An electron microscopic technique in which the surface of a specimen is coated with a thin layer of evaporated metal.

metaphase The phase of mitosis during which the chromosomes are aligned on a metaphase plate in the center of the cell.

metastasis Spread of cancer cells through the blood or lymphatic system to other organ sites.

microfilament A cytoskeleton filament composed of actin.

microRNA (miRNA) A naturally-occurring short noncoding RNA that acts to regulate gene expression.

microsome A small vesicle formed from the endoplasmic reticulum when cells are disrupted.

microtubule A cytoskeletal component formed by the polymerization of tubulin into rigid, hollow rods about 25 nm in diameter.

microtubule-associated protein (MAP) A protein that binds to microtubules and modifies their stability.

microtubule-organizing center An anchoring point near the center of the cell from which most microtubules extend outward.

microvillus An actin-based protrusion of the plasma membrane, abundant on the surfaces of cells involved in absorption.

middle lamella A region of the plant cell wall that acts as a glue to hold adjacent cells together.

mineralocorticoid A steroid hormone produced by the adrenal gland that acts on the kidney to regulate salt and water balance.

mismatch repair A repair system that removes mismatched bases from newly synthesized DNA strands.

mitochondria Cytoplasmic organelles responsible for synthesis of most of the ATP in eukaryotic cells by oxidative phosphorylation.

mitosis Nuclear division. The most dramatic stage of the cell cycle, corresponding to the separation of daughter chromosomes and usually ending with cell division (cytokinesis).

mitotic checkpoint complex (MCC) A complex of proteins that is formed at unattached kinetochores and inhibits progression to anaphase until all chromosomes are aligned on the metaphase spindle.

mitotic spindle An array of microtubules extending from the spindle poles that is responsible for separating daughter chromosomes during mitosis. See also kinetochore microtubule, interpolar microtubule, chromosomal microtubule, and astral microtubule.

molecular clone See recombinant molecule.

molecular cloning The insertion of a DNA fragment of interest into a DNA molecule (vector) that is capable of independent replication in a host cell.

molecular motor A protein that generates force and movement by converting chemical energy to mechanical energy.

monocistronic Messenger RNAs that encode a single polypeptide chain.

monoclonal antibody An antibody produced by a clonal line of B lymphocytes.

monocyte A type of blood cell involved in inflammatory reactions.

monosaccharide A simple sugar with the basic formula of $(CH_2O)_n$.

Mos A protein kinase that is required for progression from meiosis I to meiosis II and for maintenance of metaphase II arrest in vertebrate oocytes.

mRNA An RNA molecule that serves as a template for protein synthesis.

mRNP Messenger ribonucleoprotein particles, consisting of mRNA associated with proteins.

mTOR A protein kinase involved in regulation of protein synthesis and autophagy in response to growth factors, nutrients, and energy availability.

multi-photon excitation microscopy A form of fluorescence microscopy in which the specimen is illuminated with a wavelength of light such that excitation of the fluorescent dye requires the simultaneous absorption of two or more photons.

muscle fiber A large cell of skeletal muscle, formed by the fusion of many individual cells during development.

mutation A genetic alteration.

myofibril A bundle of actin and myosin filaments in muscle cells.

myosin A protein that interacts with actin as a molecular motor.

myosin I A type of myosin that acts to transport cargo along actin filaments.

myosin II A type of myosin that produces contraction by sliding actin filaments.

myosin light-chain kinase A protein kinase that activates myosin II by phosphorylating its regulatory light chain.

N

Na⁺-K⁺ ATPase See Na⁺-K⁺ pump.

Na⁺-K⁺ pump An ion pump that transports Na⁺ out of the cell and K⁺ into the cell.

NAD⁺ See nicotinamide adenine dinucleotide.

NADP reductase An enzyme that transfers electrons from ferrodoxin to NADP⁺, yielding NADPH.

necroptosis A form of necrotic cell death that is induced as a programmed response to stimuli such as infection or DNA damage.

necrosis Accidental death of cells resulting from an acute injury.

nectin A cell adhesion molecule involved in the formation of adherens junctions.

Nernst equation The relationship between ion concentration and membrane potential.

nerve growth factor (NGF) A polypeptide growth factor that regulates the development and survival of neurons.

nested gene A gene contained within an intron of a larger gene.

network An array of metabolic or regulatory pathways that interact within a cell.

neurofilament (NF) protein A member of a family of intermediate filament proteins of many types of mature nerve cells.

neurohormone Peptides that are secreted by neurons and act on distant cells.

neuron A nerve cell specialized to receive and transmit signals throughout the body.

neuropeptide A peptide signaling molecule secreted by neurons.

neurotransmitter A small, hydrophilic molecule that carries a signal from a stimulated neuron to a target cell at a synapse.

neurotrophin A member of a family of polypeptides that regulates neuron development and survival.

nexin A protein that links microtubule doublets to each other in the axoneme.

next-generation sequencing New methods that allow rapid sequencing of billions of bases of DNA.

NF-κB A family of transcription factors that are activated in response to a variety of stimuli.

niche A microenvironment that maintains stem cells in tissues.

nicotinamide adenine dinucleotide (NAD$^+$) A coenzyme that functions as an electron carrier in oxidation/reduction reactions.

nidogen An extracellular matrix protein that interacts with laminins and type IV collagen in basal laminae.

nitrogen fixation The reduction of atmospheric nitrogen (N_2) to NH_3.

N-myristoylation The addition of myristic acid (a 14-carbon fatty acid) to the N-terminal glycine residue of a polypeptide chain.

nonreceptor tyrosine kinase An intracellular tyrosine kinase.

nonsense-mediated mRNA decay Degradation of mRNAs that lack complete open-reading frames.

Northern blotting A method in which mRNAs are separated by gel electrophoresis and detected by hybridization with specific probes.

Notch A transmembrane receptor in a signaling pathway that regulates cell fate as a result of cell–cell interactions during development.

nuclear body A discrete organelle within the nucleus.

nuclear envelope The barrier separating the nucleus from the cytoplasm, composed of an inner and outer membrane, a nuclear lamina, and nuclear pore complexes.

nuclear export signal An amino acid sequence that targets proteins for transport from the nucleus to the cytosol.

nuclear lamina A meshwork of lamin filaments providing structural support to the nucleus.

nuclear localization signal An amino acid sequence that targets proteins for transportation from the cytoplasm to the nucleus.

nuclear membrane One of the membranes forming the nuclear envelope; the outer nuclear membrane is continuous with the endoplasmic reticulum and the inner nuclear membrane is adjacent to the nuclear lamina.

nuclear pore complex A large structure forming a transport channel through the nuclear envelope.

nuclear receptor superfamily A family of transcription factors that includes the receptors for steroid hormones, thyroid hormone, retinoic acid, and vitamin D_3.

nuclear transport receptor A protein that recognizes nuclear localization signals and mediates transport across the nuclear envelope.

nucleic acid hybridization The formation of double-stranded DNA and/or RNA molecules by complementary base pairing.

nucleolar organizing region A chromosomal region containing the genes for ribosomal RNAs.

nucleolus The nuclear site of rRNA transcription, processing, and ribosome assembly.

nucleolus-associated domain (NAD) A region of heterochromatin associated with the nucleolus.

nucleoside A purine or pyrimidine base linked to a sugar (ribose or deoxyribose).

nucleosome The basic structural unit of chromatin consisting of DNA wrapped around a histone core.

nucleosome core particle Particle containing 146 base pairs of DNA wrapped around an octamer consisting of two molecules each of histones H2A, H2B, H3, and H4.

nucleotide A phosphorylated nucleoside.

nucleotide-excision repair A mechanism of DNA repair in which oligonucleotides containing damaged bases are removed from a DNA molecule.

nucleus The most prominent organelle of eukaryotic cells; contains the genetic material.

O

Okazaki fragment A short DNA fragment synthesized to form the lagging strand of DNA.

oligonucleotide A short polymer of only a few nucleotides.

oligosaccharide A short polymer of only a few sugars.

oncogene A gene capable of inducing one or more characteristics of cancer cells.

oncogene addiction The dependence of cancer cells on the continuing activity of oncogenes.

open-reading frame A stretch of nucleotide sequence that does not contain stop codons and can encode a polypeptide.

operator A regulatory sequence of DNA that controls transcription of an operon.

operon A group of adjacent genes transcribed as a single mRNA.

origin of replication A specific DNA sequence that serves as a binding site for proteins that initiate replication.

origin recognition complex (ORC) A protein complex that initiates DNA replication at eukaryotic origins.

oxidative metabolism The use of molecular oxygen as an electron acceptor in the breakdown of organic molecules.

oxidative phosphorylation The synthesis of ATP from ADP, coupled to the energetically favorable transfer of electrons to molecular oxygen as the final acceptor in an electron transport chain.

P

P1 artificial chromosome (PAC) A vector used for cloning large fragments of DNA in *E. coli.*

p53 A transcription factor (encoded by the *p53* tumor suppressor gene) that arrests the cell cycle in G_1 in response to damaged DNA and is required for apoptosis induced by a variety of stimuli.

pachytene The stage of meiosis I during which recombination takes place between homologous chromosomes.

palmitoylation The addition of palmitic acid (a 16-carbon fatty acid) to cysteine residues of a polypeptide chain.

papillomavirus A member of a family of DNA viruses, some of which cause cervical and other anogenital cancers in humans.

paracrine signaling Local cell–cell signaling in which a molecule released by one cell acts on a neighboring target cell.

parallel bundle A type of actin bundle, containing closely spaced actin filaments aligned in parallel.

passive transport The transport of molecules across a membrane in the energetically favorable direction.

pectin A gel-forming polysaccharide in plant cell walls.

peptide bond The bond joining amino acids in polypeptide chains.

peptide hormone A signaling molecule composed of amino acids.

peptidoglycan The principal component of bacterial cell walls consisting of linear polysaccharide chains cross-linked by short peptides.

peptidyl prolyl isomerase An enzyme that facilitates protein folding by catalyzing the *cis-trans* isomerization of prolyl peptide bonds.

pericentriolar material The material in the centrosome that initiates microtubule assembly.

peripheral membrane protein A protein that is indirectly associated with cell membranes by protein-protein interactions.

peroxisome A cytoplasmic organelle specialized for carrying out oxidative reactions.

peroxisome biogenesis disorder A disease resulting from a mutation in one of the Pex proteins responsible for peroxisome assembly.

Pex protein A peroxisome transmembrane protein that is involved in peroxisome assembly.

phagocytosis The uptake of large particles, such as bacteria, by a cell, sometimes called "cell eating."

phagolysosome A lysosome that has fused with a phagosome or autophagosome.

phagosome A vacuole containing a particle taken up by phagocytosis.

phalloidin A drug that binds to actin filaments and prevents their disassembly.

phase-contrast microscopy A type of microscopy in which variations in density or thickness between parts of the cell are converted to differences in contrast in the final image.

phenotype The physical appearance of an organism.

phosphatidylcholine A glycerol phospholipid with a head group formed from choline.

phosphatidylethanolamine A glycerol phospholipid with a head group formed from ethanolamine.

phosphatidylinositide (PI) 3-kinase An enzyme that phosphorylates PIP_2, yielding the second messenger phosphatidylinositol 3,4,5-trisphosphate (PIP_3).

phosphatidylinositol A glycerol phospholipid with a head group formed from inositol.

phosphatidylinositol 3,4,5-triphosphate (PIP_3) A second messenger formed by phosphorylation of PIP_2.

phosphatidylinositol 4,5-bisphosphate (PIP_2) A phospholipid component of the inner leaflet of the plasma membrane. Hormones and growth factors stimulate its hydrolysis by phospholipase C, yielding the second messengers diacylglycerol and inositol trisphosphate, and its phosphorylation by PI 3-kinase, yielding the second messenger PIP_3.

phosphatidylserine A glycerol phospholipid with a head group formed from serine.

phosphodiester bond A bond between the 5'-phosphate of one nucleotide and the 3'-hydroxyl of another.

phospholipase C An enzyme that hydrolyzes PIP_2 to form the second messengers diacylglycerol and inositol trisphosphate.

phospholipid One of a family of molecules that are the principal components of cell membranes, consisting of two hydrocarbon chains (usually fatty acids) joined to a polar head group containing phosphate.

phospholipid bilayer The basic structure of biological membranes in which the hydrophobic tails of phospholipids are buried in the interior of the membrane and their polar head groups are exposed to the aqueous solution on either side.

phospholipid transfer protein A protein that transports phospholipid molecules between cell membranes.

phosphorylation The addition of a phosphate group to a molecule.

photocenter An assembly of photosynthetic pigments in the thylakoid membrane of chloroplasts.

photoreactivation A mechanism of DNA repair in which solar energy is used to split pyrimidine dimers.

photorespiration A process that metabolizes a byproduct of photosynthesis.

photosynthesis The process by which cells harness energy from sunlight and synthesize glucose from CO_2 and water.

photosystem I A protein complex in the thylakoid membrane that uses energy absorbed from sunlight to synthesize NADPH.

photosystem II A protein complex in the thylakoid membrane that uses energy absorbed from sunlight to synthesize ATP.

PI 3-kinase See phosphatidylinositide (PI) 3-kinase.

plakin A member of a family of proteins that link intermediate filaments to other cellular structures.

plant hormone Member of a group of small molecules that coordinate the responses of plant tissues to environmental signals.

plasma membrane A phospholipid bilayer with associated proteins that surrounds the cell.

plasmalogen A phospholipid that has an ether bond and an ester bond.

plasmid A small, circular DNA molecule capable of independent replication in a host cell.

plasmodesma A cytoplasmic connection between adjacent plant cells formed by a continuous region of the plasma membrane.

plastid Member of a family of plant organelles including chloroplasts, chromoplasts, leucoplasts, amyloplasts, and elaioplasts.

platelet-derived growth factor (PDGF) A growth factor released by platelets during blood clotting to stimulate the proliferation of fibroblasts.

pluripotency The capacity to develop into all of the different types of cells in adult tissues and organs.

PML/RARα An oncogene protein formed by translocation of the retinoic acid receptor in acute promyelocytic leukemia.

polar body A small cell formed by asymmetric cell division following meiosis of oocytes.

Polo-like kinase A protein kinase involved in mitotic spindle formation, kinetochore function, and cytokinesis.

poly-A tail A tract of about 200 adenine nucleotides added to the 3′ ends of eukaryotic mRNAs.

polyadenylation The process of adding a poly-A tail to a pre-mRNA.

polycistronic Messenger RNAs that encode multiple polypeptide chains.

Polycomb body A nuclear body enriched in Polycomb proteins.

Polycomb proteins A complex of repressor proteins that methylate histone H3 lysine 27.

polymerase chain reaction (PCR) A method for amplifying a region of DNA by repeated cycles of DNA synthesis *in vitro*.

polynucleotide A polymer containing up to millions of nucleotides.

polyomavirus A widely-studied DNA tumor virus.

polyp A benign tumor projecting from an epithelial surface.

polypeptide A polymer of amino acids.

polysaccharide A polymer containing hundreds or thousands of sugars.

polysome A series of ribosomes translating a messenger RNA.

porin A member of a class of proteins that cross membranes as β barrels and form channels in the outer membranes of some bacteria, mitochondria, and chloroplasts.

pre-mRNA The primary transcript, which is processed to form messenger RNAs in eukaryotic cells.

pre-rRNA The primary transcript, which is cleaved to form individual ribosomal RNAs (the 28S, 18S, and 5.8S rRNAs of eukaryotic cells).

pre-tRNA The primary transcript, which is cleaved to form transfer RNAs.

prenylation The addition of specific types of lipids (prenyl groups) to C-terminal cysteine residues of a polypeptide chain.

presequence An amino-terminal sequence that targets proteins to mitochondria.

primary cell wall The wall of growing plant cells.

primary culture The initial cell culture established from a tissue.

primary structure The sequence of amino acids in a polypeptide chain.

primase An RNA polymerase used to initiate DNA synthesis.

prion A misfolded protein capable of self-replication.

processed pseudogene A pseudogene that has arisen by reverse transcription of mRNA.

procollagen Soluble precursor to the fibril-forming collagens.

product A compound formed as a result of an enzymatic reaction.

profilin An actin-binding protein that stimulates the assembly of actin monomers into filaments.

progesterone A steroid hormone produced by the ovaries.

programmed cell death A normal physiological form of cell death characterized by apoptosis.

prokaryotic cell A cell lacking a nuclear envelope and cytoplasmic organelles.

prometaphase A transition period between prophase and metaphase during which the microtubules of the mitotic spindle attach to the kinetochores and the chromosomes shuffle until they align in the center of the cell.

promoter A DNA sequence at which RNA polymerase binds to initiate transcription.

pronucleus One of the two haploid nuclei in a newly fertilized egg.

proofreading The selective removal of mismatched bases by DNA polymerase.

prophase The beginning phase of mitosis, marked by the appearance of condensed chromosomes and the development of the mitotic spindle.

proplastid A small undifferentiated organelle that can develop into different types of mature plastids.

prostacyclin An eicosanoid formed from prostaglandin H_2.

prostaglandin A family of eicosanoid lipids involved in signaling inflammation.

prosthetic group A small molecule bound to a protein.

proteasome A large protease complex that degrades proteins tagged by ubiquitin.

protein A polypeptide with a unique amino acid sequence.

protein disulfide isomerase (PDI) An enzyme that catalyzes the formation and breakage of disulfide (S–S) linkages.

protein kinase An enzyme that phosphorylates proteins by transferring a phosphate group from ATP.

protein kinase A A protein kinase (also called cAMP-dependent protein kinase) regulated by cyclic AMP.

protein kinase C Member of a family of serine/threonine kinases that are activated by diacylglycerol and Ca^{2+} and function in intracellular signal transduction.

protein misfolding disease A disease resulting from defective protein folding.

protein phosphatase An enzyme that reverses the action of protein kinases by removing phosphate groups from phosphorylated amino acid residues.

proteoglycan A protein linked to glycosaminoglycans.

proteolysis Degradation of polypeptide chains.

proteome All of the proteins expressed in a given cell.

proteomics Large-scale analysis of cell proteins.

proto-oncogene A normal cell gene that can be converted into an oncogene.

pseudogene A nonfunctional gene copy.

pseudopodium An actin-based extension of the plasma membrane responsible for phagocytosis and amoeboid movement.

PTEN A lipid phosphatase that dephosphorylates PIP_3 and acts as a tumor suppressor.

purine One of the types of bases present in nucleic acids. The purines are adenine and guanine.

pyrimidine One of the types of bases present in nucleic acids. The pyrimidines are cytosine, thymine, and uracil.

pyrimidine dimer A common form of DNA damage caused by UV light in which adjacent pyrimidines are joined to form a dimer.

Q

quaternary structure The interactions between polypeptide chains in proteins consisting of more than one polypeptide.

R

Rab A family of small GTP-binding proteins that play key roles in vesicular transport.

Rad51 The eukaryotic homolog of RecA.

Raf A serine/threonine kinase (encoded by the *raf* oncogene) that is activated by Ras and leads to activation of the ERK MAP kinase.

raf Gene encoding Raf proteins.

Ran A small GTP-binding protein involved in nuclear import and export.

Ras A family of small GTP binding proteins (encoded by the *ras* oncogenes) that couple growth factor receptors to intracellular targets, including the Raf serine/threonine kinase and the ERK MAP kinase pathway.

ras Gene encoding Ras proteins.

Rb Tumor suppressor gene that encodes the Rb protein.

Rb A transcriptional regulatory protein that controls cell cycle progression and is encoded by a tumor suppressor gene that was identified by the genetic analysis of retinoblastoma.

real-time PCR A method in which the polymerase chain reaction (PCR) is used to quantitate the amount of target DNA or RNA present in a sample.

RecA A protein that promotes the exchange of strands between homologous *E. coli* DNA molecules during recombination.

receptor tyrosine kinase A membrane-spanning tyrosine kinase that is a receptor for extracellular ligands.

recessive An allele that is masked by a dominant allele.

recombinant DNA library A collection of genomic or cDNA clones.

recombinant molecule A DNA insert joined to a vector.

recombinational repair The repair of damaged DNA by recombination with an undamaged homologous DNA molecule.

release factor A protein that recognizes stop codons and terminates translation of mRNA.

replication factory A large complex containing clustered sites of DNA replication in eukaryotic cells.

replication fork The region of DNA synthesis where the parental strands separate and two new daughter strands elongate.

repressor A regulatory molecule that blocks transcription.

reproductive cloning The use of nuclear transfer to create a cloned organism.

resolution The ability of a microscope to distinguish objects separated by small distances.

restriction endonuclease An enzyme that cleaves DNA at a specific sequence.

restriction map The locations of restriction endonuclease cleavage sites on a DNA molecule.

restriction point A regulatory point in animal cell cycles that occurs late in G_1. After this point, a cell is committed to entering S and undergoing one cell division cycle.

retinoic acid A signaling molecule synthesized from vitamin A.

retinoid A molecule related to retinoic acid.

retrotransposition Movement of a transposable element via reverse transcription of an RNA intermediate.

retrotransposon A transposable element that moves via reverse transcription of an RNA intermediate.

retrovirus A virus that replicates by making a DNA copy of its RNA genome by reverse transcription.

retrovirus-like element A retrotransposon that is structurally similar to a retrovirus.

reverse transcriptase A DNA polymerase that uses an RNA template.

reverse transcription Synthesis of DNA from an RNA template.

Rho A family of small GTP-binding proteins involved in regulation of the cytoskeleton.

rhodopsin A G protein-coupled photoreceptor in retinal rod cells that activates transducin in response to light absorption.

ribonucleic acid (RNA) A polymer of ribonucleotides.

ribose The five-carbon sugar found in RNA.

ribosomal RNA (rRNA) The RNA component of ribosomes.

ribosome A particle composed of RNA and proteins that is the site of protein synthesis.

ribozyme An RNA enzyme.

RNA editing RNA processing events other than splicing that alter the protein coding sequences of mRNAs.

RNA interference (RNAi) The degradation of mRNAs by short complementary double-stranded RNA molecules.

RNA polymerase An enzyme that catalyzes the synthesis of RNA.

RNA-seq Global analysis of RNAs by next-generation sequencing.

RNA splicing The joining of exons in a precursor RNA molecule.

RNase H An enzyme that degrades the RNA strand of RNA-DNA hybrid molecules.

RNase P A ribozyme that cleaves the 5′ end of pre-tRNAs.

RNA world An early stage of evolution based on self-replicating RNA molecules.

rough ER The region of the endoplasmic reticulum covered with ribosomes and involved in protein metabolism.

Rous sarcoma virus (RSV) An acutely transforming retrovirus in which the first oncogene was identified.

rRNA See ribosomal RNA.

ryanodine receptor A type of calcium channel in muscle and nerve cells that opens in response to changes in membrane potential.

S

S phase The phase of the cell cycle during which DNA replication occurs.

Saccharomyces cerevisiae A frequently studied budding yeast.

sarcoma A cancer of cells of connective tissue.

sarcomere The contractile unit of muscle cells composed of interacting myosin and actin filaments.

sarcoplasmic reticulum A specialized network of membranes in muscle cells that stores a high concentration of Ca^{2+}.

satellite DNA Simple-sequence repetitive DNA with a buoyant density differing from the bulk of genomic DNA.

scaffold protein A protein that binds to components of signaling pathways, leading to their organization in specific signaling cassettes.

scanning electron microscopy See electron microscopy.

SDS-polyacrylamide gel electrophoresis (SDS-PAGE) A commonly used method to separate proteins by gel electrophoresis on the basis of size.

second messenger A compound whose metabolism is modified as a result of a ligand-receptor interaction; it functions as a signal transducer by regulating other intracellular processes.

secondary cell wall A thick cell wall laid down between the plasma membrane and the primary cell wall of plant cells that have ceased growth.

secondary response gene A gene whose induction following growth factor stimulation of a cell requires protein synthesis.

secondary structure The regular arrangement of amino acids within localized regions of a polypeptide chain. See α helix and β sheet.

secretory pathway The movement of secreted proteins from the endoplasmic reticulum to the Golgi apparatus and then, within secretory vesicles, to the cell surface.

secretory vesicle A membrane-enclosed sac that transports proteins from the Golgi apparatus to the cell surface.

selectin A type of cell adhesion molecule that recognizes oligosaccharides exposed on the cell surface.

self-splicing The ability of some RNAs to catalyze the removal of their own introns.

semiconservative replication The process of DNA replication in which the two parental strands separate and serve as templates for the synthesis of new progeny strands.

serine/threonine kinase A protein kinase that phosphorylates serine and threonine residues.

serum response element (SRE) A regulatory sequence that is recognized by the serum response factor and mediates the transcriptional induction of many immediate early genes in response to growth factor stimulation.

serum response factor (SRF) A transcription factor that binds to the serum response element.

SH2 domain A protein domain of approximately 100 amino acids that binds phosphotyrosine-containing peptides.

Shine-Dalgarno sequence The sequence prior to the initiation site that correctly aligns bacterial mRNAs on ribosomes.

short interfering RNA (siRNA) A short noncoding double-strand RNA that acts in the RNA interference pathway.

signaling network The interconnected network formed by the interactions of multiple signaling pathways within a cell.

signal patch A recognition determinant formed by the three-dimensional folding of a polypeptide chain.

signal peptidase An enzyme that removes the signal sequence of a polypeptide chain by proteolysis.

signal recognition particle (SRP) A particle composed of proteins and RNA that binds to signal sequences and targets polypeptide chains to the endoplasmic reticulum.

signal sequence A hydrophobic sequence at the amino terminus of a polypeptide chain that targets it for secretion in bacteria or incorporation into the endoplasmic reticulum in eukaryotic cells.

simian virus 40 (SV40) A widely-studied DNA tumor virus.

simple-sequence repeat Member of a class of repeated DNA sequences consisting of tandem arrays of thousands of copies of short sequences.

SINE (short interspersed element) Member of a family of highly repeated retrotransposons in mammalian genomes.

single-stranded DNA-binding protein A protein that stabilizes unwound DNA by binding to single-stranded regions.

siRNA See short interfering RNA.

site-specific recombination Recombination mediated by proteins that recognize specific DNA sequences.

sliding filament model The model of muscle contraction in which contraction results from the sliding of actin and myosin filaments relative to each other.

Smad A family of transcription factors activated by TGF-β receptors.

small nuclear ribonucleoprotein particle (snRNP) Complex of an snRNA with proteins.

small nuclear RNA (snRNA) A nuclear RNA ranging in size from 50 to 200 bases.

small nucleolar RNA (snoRNA) A small RNA present in the nucleolus that functions in pre-rRNA processing.

smooth ER The region of the endoplasmic reticulum that is the major site of lipid synthesis in eukaryotic cells.

SNARE A transmembrane protein that mediates fusion of vesicle and target membranes.

somatic cell nuclear transfer The basic procedure of animal cloning in which the nucleus of an adult somatic cell is transferred to an enucleated egg.

somatic hypermutation The introduction of multiple mutations within rearranged immunoglobulin variable regions to increase antibody diversity.

Southern blotting A method in which radioactive or fluorescent probes are used to detect specific DNA fragments that have been separated by gel electrophoresis.

speckle A nuclear body that contains components of the mRNA splicing machinery.

spectrin A major actin-binding protein of the cell cortex.

sphingomyelin A phospholipid consisting of two hydrocarbon chains bound to a polar head group containing serine.

spindle assembly checkpoint A cell cycle checkpoint that monitors the alignment of chromosomes on the metaphase spindle.

spliceosome A large complex of snRNAs and proteins that catalyzes the splicing of pre-mRNAs.

src 1. Gene encoding Src protein. 2. The oncogene of Rous sarcoma virus.

Src A nonreceptor tyrosine kinase encoded by *src*.

SRP receptor A protein on the membrane of the endoplasmic reticulum that binds the signal recognition particle (SRP).

SRP RNA The small cytoplasmic RNA component of SRP.

stability gene A gene that acts to maintain the integrity of the genome and whose loss can lead to the development of cancer.

starch A polymer of glucose residues that is the principal storage form of carbohydrates in plants.

START A regulatory point in the yeast cell cycle that occurs late in G_1. After this point a cell is committed to entering S and undergoing one cell division cycle.

STAT protein Member of a family of transcription factors that have an SH2 domain and are activated by tyrosine phosphorylation, which promotes their translocation from the cytoplasm to the nucleus.

stem cell A cell that divides to produce daughter cells that can either differentiate or remain as stem cells.

stereocilium A specialized microvillus of auditory hair cells.

steroid hormone A member of a group of hydrophobic hormones, including estrogen and testosterone, that are derivatives of cholesterol.

steroid hormone receptor A transcription factor that regulates gene expression in response to steroid hormones.

stress fiber A bundle of actin filaments anchored at sites of cell adhesion to the extracellular matrix.

stroma The compartment of chloroplasts that lies between the envelope and the thylakoid membrane.

stromal processing peptidase (SPP) The protease that cleaves transit peptides from proteins imported to the chloroplast stroma.

substrate A molecule acted upon by an enzyme.

SUMO *S*mall *u*biquitin-related *mo*difier. A ubiquitin-like protein that serves as a marker for protein localization and regulates protein activity.

super-resolution light microscopy New techniques that break the diffraction barrier of the light microscope and increase the resolution of fluorescence microscopy to the range of 10-100 nm.

symport The transport of two molecules in the same direction across a membrane.

synapse The junction between a neuron and another cell, across which information is carried by neurotransmitters.

synapsis The association of homologous chromosomes during meiosis.

synaptic vesicle A secretory vesicle that releases neurotransmitters at a synapse.

synaptomenal complex A zipperlike protein structure that forms along the length of paired homologous chromosomes during meiosis.

synthetic biology An engineering approach to design and create new biological systems.

systems biology A new field of biology in which large-scale experimental approaches are combined with quantitative analysis and modeling to study complex biological systems.

T

T cell receptor A T lymphocyte surface protein that recognizes antigens expressed on the surface of other cells.

talin A protein that mediates the association of actin filaments with integrins at focal adhesions.

tandem mass spectrometry A method in which sequential analyses by mass spectrometry is used to determine the amino acid sequence of a peptide.

TATA box A regulatory DNA sequence found in the promoters of many eukaryotic genes transcribed by RNA polymerase II.

TATA-binding protein (TBP) A basal transcription factor that binds directly to the TATA box.

taxol A drug that binds to and stabilizes microtubules.

TBP-associated factor (TAF) A polypeptide associated with TBP in the general transcription factor TFIID.

telomerase A reverse transcriptase that synthesizes telomeric repeat sequences at the ends of chromosomes from its own RNA template.

telomere A repeat of simple-sequence DNA that maintains the ends of linear chromosomes.

telophase The final phase of mitosis, during which the nuclei re-form and chromosomes decondense.

tertiary structure The three-dimensional folding of a polypeptide chain that gives the protein its functional form.

testosterone A steroid hormone produced by the testes.

tethering factor A protein that mediates the initial interactions between transport vesicles and their target membranes.

therapeutic cloning A procedure in which nuclear transfer into oocytes could be used to produce embryonic stem cells for use in transplantation therapy.

thromboxane An eicosanoid involved in blood clotting.

thylakoid membrane The innermost membrane of chloroplasts that is the site of electron transport and ATP synthesis.

thymine A pyrimidine found in DNA that base-pairs with adenine.

thyroid hormone A hormone synthesized from tyrosine in the thyroid gland.

Tic complex The protein translocation complex of the chloroplast inner membrane.

tight junction A continuous network of protein strands around the circumference of epithelial cells, sealing the space between cells and forming a barrier between the apical and basolateral domains.

Tim complex The protein translocation complex of the mitochondrial inner membrane.

Toc complex The protein translocation complex of the chloroplast outer membrane.

Toll-like receptor Member of a family of receptors that recognize a variety of molecules associated with pathogenic bacteria and viruses.

Tom complex The protein translocation complex of the mitochondrial outer membrane.

topoisomerase An enzyme that catalyzes the reversible breakage and rejoining of DNA strands.

***trans* Golgi network** The Golgi compartment within which proteins are sorted and packaged to exit the Golgi apparatus.

transcription The synthesis of an RNA molecule from a DNA template.

transcription-coupled repair The preferential repair of damage to transcribed strands of DNA.

transcription factor A protein that regulates the activity of RNA polymerase.

transcription factory A clustered site in the nucleus where transcription of multiple genes occurs.

transcriptome All of the RNAs that are transcribed in a cell.

transcriptional activator A transcription factor that stimulates transcription.

transdifferentiation Differentiation of one type of somatic cell into another differentiated cell type.

transducin A G protein that stimulates cGMP phosphodiesterase when it is activated by rhodopsin.

transfection The introduction of a foreign gene into eukaryotic cells.

transfer RNA (tRNA) An RNA molecule that functions as an adaptor between amino acids and mRNA during protein synthesis.

transformation The transfer of DNA between genetically distinct bacteria. See also cell transformation.

transforming growth factor β (TGF-β) A polypeptide growth factor that generally inhibits animal cell proliferation.

transgenic mouse A mouse that carries foreign genes that are incorporated into the germ line.

transient expression The expression of unintegrated plasmid DNAs that have been introduced into cultured cells.

transition state A high energy state through which substrates must pass during the course of an enzymatic reaction.

transit peptide An N-terminal sequence that targets proteins for import into chloroplasts.

translation The synthesis of a polypeptide chain from an mRNA template.

translesion DNA synthesis A form of repair in which specialized DNA polymerases replicate across a site of DNA damage.

translocon The membrane channel through which polypeptide chains are transported into the endoplasmic reticulum.

transmembrane protein An integral membrane protein that spans the lipid bilayer and has portions exposed on both sides of the membrane.

transmission electron microscopy See electron microscopy.

transposable element A DNA sequence that can move to different positions in the genome.

treadmilling A dynamic behavior of actin filaments and microtubules in which the loss of subunits from one end of the filament is balanced by their addition to the other end.

triacylglycerol Three fatty acids linked to a glycerol molecule.

tRNA See transfer RNA.

tropomyosin A fibrous protein that binds actin filaments and regulates contraction by blocking the interaction of actin and myosin.

troponin A complex of proteins that binds to actin filaments and regulates skeletal muscle contraction.

tubulin A cytoskeletal protein that polymerizes to form microtubules.

tumor Any abnormal proliferation of cells.

tumor initiation The first step in tumor development, resulting from abnormal proliferation of a single cell.

tumor necrosis factor (TNF) A polypeptide growth factor that induces inflammation and programmed cell death.

tumor progression The accumulation of mutations within cells of a tumor population, resulting in increasingly rapid growth and malignancy.

tumor promoter A compound that leads to tumor development by stimulating cell proliferation.

tumor suppressor gene A gene whose inactivation leads to tumor development.

tumor virus A virus capable of causing cancer in animals or humans.

turgor pressure The internal hydrostatic pressure within plant cells.

two-dimensional gel electrophoresis A method for separating cell proteins based on both charge and size.

tyrosine kinase A protein kinase that phosphorylates tyrosine residues.

U

3′ untranslated region (UTR) A noncoding region at the 3′ end of mRNA.

5′ untranslated region (UTR) A noncoding region at the 5′ end of mRNA.

ubiquinone See coenzyme Q.

ubiquitin A highly conserved protein that acts as a marker to target other cellular proteins for rapid degradation.

ultracentrifuge A centrifuge that rotates samples at high speeds

unconventional myosin A type of myosin other than that found in muscle cells (myosin II).

unfolded protein response (UPR) A cellular stress response in which an excess of unfolded proteins in the endoplasmic reticulum leads to general inhibition of protein synthesis, increased expression of chaperones, and increased proteasome activity.

uniport The transport of a single molecule across a membrane.

uracil A pyrimidine found in RNA that base-pairs with adenine.

V

vacuole A large membrane-enclosed sac in the cytoplasm of eukaryotic cells. In plant cells, vacuoles function to store nutrients and waste products, to degrade macromolecules, and to maintain turgor pressure.

vector A DNA molecule used to direct the replication of a cloned DNA fragment in a host cell.

velocity centrifugation The separation of particles based on their rates of sedimentation.

video-enhanced microscopy The combined use of video cameras with the light microscope to allow the visualization of small objects.

villin The major actin-bundling protein of intestinal microvilli.

vimentin An intermediate filament protein found in a variety of different kinds of cells.

vinblastine A drug that inhibits microtubule polymerization.

vincristine A drug that inhibits microtubule polymerization.

vinculin A protein that mediates the association of actin filaments with integrins at focal adhesions.

vitamin D3 A vitamin that regulates calcium metabolism and bone growth by stimulating the activity of a member of the nuclear receptor superfamily.

voltage-gated channel An ion channel that opens in response to changes in electric potential.

W

WASP A protein that stimulates actin filament branching.

Western blotting See immunoblotting.

Wnt A secreted signaling molecule that stimulates a pathway regulating cell fate during embryonic development.

X

X chromosome inactivation A dosage compensation mechanism in which most of the genes on one X chromosome are inactivated in female cells.

X-ray crystallography A method in which the diffraction pattern of X-rays is used to determine the arrangement of individual atoms within a molecule.

Xenopus laevis An African clawed frog used as a model system for developmental biology.

Y

yeast The simplest unicellular eukaryotes. Yeasts are important models for studies of eukaryotic cells.

yeast artificial chromosome (YAC) A vector that can replicate as a chromosome in yeast cells and can accommodate very large DNA inserts (hundreds of kb).

yeast two-hybrid system A method in which yeast genetics is used to identify proteins that interact with one another.

Z

zebrafish A species of small fish used for genetic studies of vertebrate development.

zinc finger domain A type of DNA binding domain consisting of loops containing cysteine and histidine residues that bind zinc ions.

zygote A fertilized egg.

zygotene The stage of meiosis I during which homologous chromosomes become closely associated.

Illustration Credits

Chapter 1

1.5: © Biophoto Associates/Science Source. **1.8:** © Medical-on-Line/Alamy. **1.9A:** © M. I. Walker/Science Source. **1.9B:** © Aaron J. Bell/Science Source. **1.11B:** © David Scharf/Science Source. **1.12A:** © Dr. Ken Wagner/Visuals Unlimited, Inc. **1.12B:** © Phil Gates/Biological Photo Service. **1.12C:** © Alfred Owczarzak/Biological Photo Service. **1.12D:** © J. Robert Waaland/Biological Photo Service. **1.13A(i):** © G. W. Willis/Visuals Unlimited, Inc. **1.13A(ii):** © G. W. Willis /Visuals Unlimited, Inc. **1.13A(iii):** © Carolina Biological Supply Company/Visuals Unlimited, Inc. **1.13B:** © Biophoto Associates/Science Source. **1.13C:** © G. W. Willis/Visuals Unlimited, Inc. **1.14:** © A. M. Siegelman/Visuals Unlimited, Inc. **1.15:** © Biophoto Associates/Science Source. **1.17, 1.18, 1.20B:** David McIntyre. **1.23:** © destillat/istock. **1.24:** © G. W. Willis/Visuals Unlimited, Inc. **1.27B:** © Conly L. Rieder/Biological Photo Service. **1.32:** © Dr. Torsten Wittmann/Science Source. **1.34:** © Don W. Fawcett/Visuals Unlimited, Inc. **1.36:** © Don W. Fawcett, J. Heuser/Science Source. **1.37B:** © Don W. Fawcett/Science Source. **1.38:** © David Phillips/Visuals Unlimited, Inc. **1.42:** © CNRI/SPL/Science Source. **1.44:** © Dr. Brad Mogen/Visuals Unlimited, Inc. **1.45B:** © Linda Stannard/Science Source. **1.46:** © E. C. S. Chen/Visuals Unlimited, Inc. **Page 39, Gey:** Courtesy of The Alan Mason Chesney Medical Archives of The Johns Hopkins Medical Institutions.

Chapter 2

Page 58, Anfinsen: Courtesy of the National Library of Medicine. **Page 73, Singer:** Courtesy of S. J. Singer. **Page 73, Nicolson:** Courtesy of The Institute of Molecular Medicine, Huntington Beach, CA.

Chapter 3

Page 93, Mitchell: © UPI/Corbis-Bettmann.

Chapter 4

4.28: © Dr. Torsten Wittmann/Science Source. **Page 125, Temin:** Courtesy of the National Cancer Institute. **Page 151, Fire:** © Linda A. Cicero/Stanford News Service. **Page 151, Mello:** Courtesy of the University of Massachusetts.

Chapter 5

Page 161, Venter: William Geiger Photography. **Page 161, Lander:** © Rick Friedman/Corbis. **Page 180:** Courtesy of Dr. Mae Melvin/CDC.

Chapter 6

6.20: © Andrew Syred/Science Source. **6.22:** © James Cavallini/Science Source. **Page 190, Sharp:** © Donna Coveney, MIT.

Chapter 7

7.30: From S. C. West, 2003. Nature Rev. Mol. Cell Biol. 4: 1. Courtesy of A. Stasiak and S. West. **Page 241:** © David M. Martin, M.D./SPL/Science Source. **Page 246, Tonegawa:** © Donna Coveney/MIT.

Chapter 8

Page 303, Steitz: Courtesy of the Yale University Public Affairs Office.

Chapter 9

9.26B: © Scientifica/ADEAR/Visuals Unlimited, Inc. **Page 342:** © Dr. Thomas Deerinck/Visuals Unlimited, Inc.

Chapter 10

10.1A: © David M. Phillips/Science Source. **10.2:** © Don Fawcett/Science Source. **10.19:** © Dr. Donald Fawcett/Visuals Unlimited, Inc. **10.30:** Courtesy of Zheng'an Wu and Joseph Gall, Carnegie Institution.

Chapter 11

11.1A: © Richard Rodewald, University of Virginia/Biological Photo Service. **11.1B:** © Don Fawcett/Science Source. **11.32:** © Dr. Thomas Deerinck/Visuals Unlimited, Inc. **11.38:** © K. G. Murti/Visuals Unlimited, Inc.

Chapter 12

12.1: © K. R. Porter/Science Source. **12.11:** © E. H. Newcombe/Biological Photo Service. **12.16A:** © Biophoto Associates/Science Source. **12.16B:** © Dr. Jeremy Burgess/Science Source. **12.19:** © John N. A. Lott/Biological Photo Service. **12.20:** © Don Fawcett/Science Source.

Chapter 13

13.11: © Susumu Nishinaga/Science Source. **13.14:** © Science Source. **13.19A:** © K. Wassermann/Visuals Unlimited, Inc. **13.19B:** © Stanley Flegler/Visuals Unlimited, Inc. **13.19C:** © Don Fawcett/Science Source. **13.23A:** © Frank A. Pepe/Biological Photo Service. **13.40A:** © Cytographics. **13.40B:** © Don Fawcett/Science Source. **13.45A:** © SPL/Science Source.

13.45B: © Charles Daghlian/Science Source. **13.46A:** © K. G. Murti/Visuals Unlimited, Inc. **13.47A:** © Conly L. Rieder/Biological Photo Service. **13.47B:** © W. L. Dentler/Biological Photo Service. **13.49:** From C. L. Rieder and S. S. Bowser, 1985, *J. Histochem. Cytochem.* 33: 165/Biological Photo Service. **13.54:** © Don W. Fawcett/Science Source. **13.55A:** © Don W. Fawcett/Science Source.

Chapter 14

14.7: © Harold H. Edwards/Visuals Unlimited, Inc. **14.11:** © Don Fawcett/Visuals Unlimited, Inc. **14.35A:** © R. N. Band and H. S. Pankratz/Biological Photo Service.

Chapter 15

15.7 Secondary: © Biophoto Associates/Science Source. **15.15B:** © Don Fawcett/Visuals Unlimited, Inc. **15.27A:** © Don Fawcett/Science Source. **15.29A:** © Don Fawcett and R. Wood/Science Source. **15.30:** © E. H. Newcomb, Univ. of Wisconsin/Biological Photo Service. **Page 613, Buck:** Courtesy of Roland Morgan/Fred Hutchinson Cancer Research Center. **Page 613, Axel:** Courtesy of Don Hamerman/Columbia University Medical Center.

Chapter 17

17.4B: © David M. Phillips/Visuals Unlimited, Inc. **17.22, 17.27:** © Conly L. Rieder/Biological Photo Service. **17.28B:** © Michael Abbey/Science Source. **17.30:** © Michael Abbey/Science Source. **17.31B:** © David M. Phillips/Visuals Unlimited, Inc. **17.34:** © C. Hasenkampf/Biological Photo Service. **17.40:** © David M. Philips/Visuals Unlimited, Inc.

Chapter 18

18.13: © CMEABG-UCBL/Science Source. **18.14:** © Dr. Don W. Fawcett/Visuals Unlimited, Inc. **18.24B:** © Yorgos Nikas/Science Source. **Page 694, Horvitz:** © Donna Coveney, MIT.

Chapter 19

19.1: © Astrid and Hanns-Frieder Michler/SPL/Science Source. **Page 740, Bishop:** © Reuters/Corbis-Bettman. **Page 740, Varmus:** © Reuters/Corbis-Bettman.

Index

J

Jacob, François, 121, 263
JAK/STAT pathway, *623*, 714
Janus kinase (JAK), 623, *623*
Jun proto-oncogene, 746
Junctional adhesion molecules (JAM), 591
Junctional complexes, 591, *591*
Junk DNA, 195

K

Kabsch, Wolfgang, 480
Kadonaga, James, 279
Kaposi's sarcoma, 733, 736
Kaposi's sarcoma-associated herpesvirus, 733, 735–736
Karyopherins, 377, *377*
KASH proteins, *370*
Kaufman, Matthew, 713
Keasling, Jay, 179
Kemphues, Kenneth, 151
Kendrew, John, 60
Keratan sulfate, *425*, *581*
Keratins
    filaments, *522*, 522–524
    function of, *524*
    metaphase, *672*
    mutant, 525–526
    types of, 520–521
α-Ketoglutarate, 265
Kilobases, definition of, 191
Kinesin II, 512
Kinesins, *508*
    function of, 508
    isolation of, 509–510
    plus-end, 519
    structure of, 511
    translocation, 510
Kinetochores, 209, 672
    microtubule attachment to, *676*
    microtubules, 518
    proteins of, 212
Kirschner, Marc, 502, 518
Kit tyrosine kinase, 762, 763
Knockout mice, 147
Kornberg, Arthur, 217
Krebs, Ed, 352, 615
Krebs cycle, glycolysis and, 84–88, 107
Kuru, 343

L

L-myc oncogene, *742*
lac operon
    control of, 264, *264*

E. coli, 325
    negative control of, *263*
Lacks, Henrietta, 39
Lactobacillus casei, 105
Lactose metabolism, *263*
Lagging strands
    description of, 220
    synthesis of, 224
Lamellipodia, 490, 561
Lamin A protein, *371*
Lamin B receptor (LBR), *370*, 372
Lamina, diseases related to, 371
Lamina-associated domains (LADs), 384
Laminins
    basal lamina, 582
    composition of, *583*
    self-assembly of, 582–583
Lamins
    assembly of, *370*
    cleavage of, 695
    description of, 369
Lander, Eric, 161, *161*
Lapiere, Charles, 587
Lasek, Ray, 510
Laskey, Ron, 338
Leading strands, *219*, 219–220, 224
Leber's hereditary optic neuropathy (LHON), 451, 452–453
    basis of, 452, *452*
    description of, 452
    prevention of, 452
    treatment of, 452–453
Leibler, Stanislas, 178
Leptotene, 681, *681*
Lerner, Michael, 303
let-7 miRNA, 756
Leucine
    hydrocarbon side chains, 56
    structure of, *57*
Leucine zippers, 311
    composition of, *281*
    DNA-binding domains, 281–282
Leucoplasts, 465
Leukemia, differentiation in, *731*
Leukemia inhibitory factor (LIF), 714–715
Leukemias
    definition of, 724
    differentiation in, 731
Leukocytes, adhesion of, *588*
Leukotrienes, 609
Levi-Montalcini, Rita, 607
Ligand-gated channels, 547
Light, diffraction of, 23

Light-chain genes, Ig, 248
Light reactions, 96, *97*
Lignin, 575
Lily genome, *204*
LINC complexes, 370
Lipid bilayers, 6–7
    freeze fracture of, *32*
    plasma membrane, 531–534, *532*
Lipid rafts
    caveolae, *543*, 543–544
    components of, 542
    formation of, 534
    micrograph of, *543*
    structure of, *534*
Lipids
    attachment of, 347–350
    biosynthesis of, 108
    cellular, 77
    energy from, 89, 107
    export from ERs, 420–421
    metabolism of, 426–427
    plasma membrane, *533*
    roles of, 50–54
    secretory pathway transport of, 427
    synthesis of, 68–69
Lipopolysaccharide (LPS), 572
Liposomes, 142
Lipoxygenase (LOX), 609
Liver cancer, 733–734
Liver regeneration, *705*
LMNA gene, *371*
Lock-and-key model, 64
Long interspersed elements (LINEs), 198, 200
Long noncoding RNAs (lncRNAs), 197, 293, 294, *294*, 312
Low-density lipoprotein (LDL) receptors, 564–565, *565*
Low-density lipoproteins (LDLs), 563, *563*
LRP receptors, 640
Lumacaftor, 557
Lung cancer
    incidence of, 724–725
    treatment of, 763
Lungfish genome, *204*
Lymphocytes
    function of, 17
    micrographs of, *16*
Lymphomas, definition of, 724
Lysine
    charge of, 56
    structure of, *57*
Lysosomal proteolysis, 358–359
Lysosomal storage diseases, 436–437

About the Book

Editor: Andrew D. Sinauer
Project Editor: Chelsea D. Holabird
Copy Editor: Kathie Gow
Production Manager: Christopher Small
Photo Researcher: David McIntyre
Book Design and Production: Jen Basil-Whittaker
Illustration Program: Dragonfly Media Group
Indexer: Sharon Hughes
Cover and Book Manufacturer: R. R. Donnelley

The Genetic Code

First position	Second position				Third position
	U	C	A	G	
U	Phe	Ser	Tyr	Cys	U
	Phe	Ser	Tyr	Cys	C
	Leu	Ser	Stop	Stop	A
	Leu	Ser	Stop	Trp	G
C	Leu	Pro	His	Arg	U
	Leu	Pro	His	Arg	C
	Leu	Pro·	Gln	Arg	A
	Leu	Pro	Gln	Arg	G
A	Ile	Thr	Asn	Ser	U
	Ile	Thr	Asn	Ser	C
	Ile	Thr	Lys	Arg	A
	Met	Thr	Lys	Arg	G
G	Val	Ala	Asp	Gly	U
	Val	Ala	Asp	Gly	C
	Val	Ala	Glu	Gly	A
	Val	Ala	Glu	Gly	G

The Amino Acids

Amino acid	Three-letter abbreviation	One-letter abbreviation
Alanine	Ala	A
Arginine	Arg	R
Asparagine	Asn	N
Aspartic acid	Asp	D
Cysteine	Cys	C
Glutamic acid	Glu	E
Glutamine	Gln	Q
Glycine	Gly	G
Histidine	His	H
Isoleucine	Ile	I
Leucine	Leu	L
Lysine	Lys	K
Methionine	Met	M
Phenylalanine	Phe	F
Proline	Pro	P
Serine	Ser	S
Threonine	Thr	T
Tryptophan	Trp	W
Tyrosine	Tyr	Y
Valine	Val	V